Polymer Data Handbook

Second Edition

Polymer Data Handbook

Second Edition

EDITED BY James E. Mark

OXFORD
UNIVERSITY PRESS
2009

OXFORD
UNIVERSITY PRESS

Oxford University Press, Inc., publishes works that further
Oxford University's objective of excellence
in research, scholarship, and education.

Oxford   New York
Auckland   Cape Town   Dar es Salaam   Hong Kong   Karachi
Kuala Lumpur   Madrid   Melbourne   Mexico City   Nairobi
New Delhi   Shanghai   Taipei   Toronto

With offices in
Argentina   Austria   Brazil   Chile   Czech Republic   France   Greece
Guatemala   Hungary   Italy   Japan   Poland   Portugal   Singapore
South Korea   Switzerland   Thailand   Turkey   Ukraine   Vietnam

Published by Oxford University Press, Inc.
198 Madison Avenue, New York, New York 10016
www.oup.com

Oxford is a registered trademark of Oxford University Press

Library of Congress Cataloging-in-Publication Data
Polymer data handbook / edited by James E. Mark.—2nd ed.
    p. cm.
Includes bibliographical references and index.
ISBN 978-0-19-518101-2 (cloth : alk. paper)
    1. Polymers—Handbooks, manuals, etc. I. Mark, James E., 1934–
    TA455.P58P675 2009
    668.9—dc22   2008053004

1   3   5   7   9   8   6   4   2

Printed in the United States of America
on acid-free paper

Preface to the First Edition

The *Polymer Data Handbook* offers, in a standardized and readily accessible tabular format, concise information on the syntheses, structures, properties, and applications of the most important polymeric materials. Those included are currently in industrial use or they are under study for potential new applications in industry and in academic laboratories. Considerable thought was given to the criteria for selecting the polymers included in this volume. The first criterion was current commercial importance—the use of the polymer in commercial materials—for example, as a thermoplastic, a thermoset, or an elastomer. The second criterion was novel applications—a polymer that is promising for one or more purposes but not yet of commercial importance—for example, because of its electrical conductivities, its nonlinear optical properties, or its suitability as a preceramic polymer. The hope is that some readers will become interested enough in these newer materials to contribute to their further development and characterization. Finally, the handbook includes some polymers simply because they are unusually interesting—for example, those utilized in fundamental studies of the effects of chain stiffness, self-assembly, or biochemical processes.

Based on these three criteria, more than two hundred polymers were chosen for inclusion in this work. The properties presented for each polymer include some of great current interest, such as surface and interfacial properties, pyrolyzability, electrical conductivity, nonlinear optical properties, and electroluminescence. Not all the properties are available for all the polymers included, and some properties may not even be relevant for certain polymer classes. Some polymers exhibit properties shown by few others—such as electroluminescence—and those have been presented as 'Properties of Special Interest.'

The handbook entries were written by authors carefully chosen for their recognized expertise in their specific polymers. The authors were asked to be highly selective, to choose and document those results that they considered to have the highest relevance and reliability. All the entries were then reviewed carefully by one or more referees, to ensure the highest quality and significance. Care was taken to achieve maximum consistency between entries, especially with regard to terminology, notations, and units. The goal was to facilitate searches in the printed version of the handbook and electronically on the online site.

Grateful acknowledgment is made here to the important contributions of the anonymous referees. It is also my real pleasure to thank a number of people at Oxford University Press for their help: specifically, Robert L. Rogers and Sean Pidgeon contributed greatly to the initiation and formulation of the basic structure of the handbook, and Matthew Giarratano carried out its implementation. It is appropriate here to thank my wife Helen for the kind of support, tangible and intangible, that makes an intimidating project, like this one, doable and sometimes even a pleasant experience.

James E. Mark
University of Cincinnati
October 1998

Preface to the Second Edition

This second edition of the Polymer Data Handbook again offers information on the syntheses, structures, properties, and applications of a wide variety of important polymers.

The updating for this edition involved the significant expansions of almost all the original articles and the incorporation of new articles on other polymers, particularly those that became important or interesting in the years since the appearance of the first edition.

It is a real pleasure to thank all the authors for their contributions, and Jeremy Lewis at Oxford University Press for his help in preparing this volume. I also wish to thank my wife Helen for her encouragement, understanding, and support throughout the project.

James E. Mark
The University of Cincinnati
December 2007

Contents

Polymer Data Handbook

Second Edition

Acrylonitrile–butadiene Elastomers

Shuhong Wang

Acronym NBR

Class Chemical copolymers

Trade Names Chemigum® (Eliokem) Hycar® (Emerald Performance Materials) JSR (Japan Synthetic Rubber Co.) Krynac® (Lanxess) NIPOL (Nippon Zeon Co.) Nysyn® (DSM Copolymer Rubber and Chemical Co.) Paracril® (Uniroyal Chemical Co.)

Structure

$$-[CH_2-CH=CH-CH_2]_m\text{-}[CH_2-CH]_n\text{-}$$
$$\underset{C\equiv N}{|}$$

Major Applications Hoses for transportation of oil, fuel, chemicals, and solutions. Oil-drilling industry. Powder and particulate forms in cements and adhesives. Modification of PVC and ABS to improve impact resistance.

Properties of Special Interest Special-purpose, oil-resistant rubbers. Balance of low-temperature, oil, fuel, and solvent resistance. Good abrasion resistance, gas permeability, and thermal stability. Good strength.

Property	Units	Conditions	Value	Reference
Density	g/cm^3	26–27% ACN	0.92	(1)
Glass transition temperature, T_g	°C	~20% ACN	−60	(2)
		~30% ACN	−42	(2)
		~34% ACN	−35	(2)
		~40% ACN	−18	(2)
		~48% ACN	−10	(2)
Service temperature (max)	°C	9% N	100	(3)
Solubility parameter	$MPa^{1/2}$	25% ACN, 25°C, Calc.	18.93	(4)
Theta temperature, θ	°C	26% ACN, cyclohexane/MEK (64/36)	20.0	(5)
		40% ACN, cyclohexane/MEK (52.5/47.5)	22.0	

Property	Units	Conditions	Value	Reference
Volume swell	%	Black loaded vulcanizate, 72 h at room temperature, or 100°C with *		(2)

Solvent	17% ACN	34% ACN	37% ACN
Lard*	18	−2	−3
Butter fat*	29	−3	−3
Lanolin*	20	0	−1.5
Margarine*	24	−5	−5
Stearic acid*	26	23	−2
Oleic acid	20	3	0
Cod liver oil	5	0	0
Dehydrogenated corn oil	3	0	0
Automobile lube oil (SAE-20)	0	0	0
Automobile hydraulic fluid	8	8	6
Jet aircraft fuel			
18% aromatic, 28% olefin	60	14	11
21% aromatic, 0.1% olefin	38	9	5
Ethylene glycol	0	0	0
Automobile gasoline	39	8	6
Skydrol hydraulic fluid	112	59	41
Dioctyl phthalate	52	6	2
Dibutyl phthalate	119	76	52
Tricresyl phosphate	50	21	16
Butyl carbitol formal (polyether)	92	32	21
Bis(dimethyl benzyl)ether	147	45	29
Liquid polyester	−2	0	−3
Triglycol dioctylate	83	12	5
Tributoxy ethyl phosphate	67	29	17

Property	Units	Conditions	Value	Reference
Tensile strength	MPa	Unfilled, vulc. (26~27% ACN)	4~7	(1)
Ultimate elongation	%	—	350~800	(1)

** NBR Black formulation							
ACN %	40	33	33	33	33	27	20
Polymer Mooney	60	30	50	70	85	50	40

Property	Units								Reference
Tensile strength	MPa	17.9	15.8	16.0	17.6	19.5	14.2	13.4	(6)
Ultimate elongation	%	466	478	433	357	439	334	387	(6)
Modulus, 100%	MPa	3.6	3.1	3.2	3.9	3.5	3.7	2.9	(6)

Property	Units	Conditions					Value		Reference
	ACN %	40	33	33	33	33	27	20	
	Polymer Mooney	60	30	50	70	85	50	40	
Modulus, 200%	MPa	8.6	7.0	7.7	9.5	8.9	8.5	7.0	(6)
Modulus, 300%	MPa	13.0	10.5	11.7	14.8	14.1	12.8	10.5	(6)
Hardness	Shore A	68	67	66	67	66	67	64	(6)
		Oven aging at 100°C 70 h							
% Tensile change	%	3	5	5	1	−9	8	−1	(6)
% Elongation change	%	−12	−17	−15	−10	−25	−10	−21	(6)
% Hardness change	%	4	4	4	4	4	4	3	(6)
		Oven aging at 121°C 70 h							
% Tensile change	%	3	9	6	8	1	16	4	(6)
% Elongation change	%	−24	−21	−21	−8	−21	−10	−24	(6)
% Hardness change	%	6	6	6	5	5	5	5	(6)
		Fluid aging at 121°C in ASTM oil No.1							
% Tensile change	%	6	12	15	9	8	6	13	(6)
% Elongation change	%	−24	−26	−11	−13	−18	−18	−17	(6)
% Hardness change	%	9	9	9	7	8	5	−2	(6)
Volume swell	%	−6.5	−5.9	−5.2	−5.2	−4.6	−2.6	0.9	(6)
		Fluid aging at 121°C in ASTM oil No.3							
% Tensile change	%	1	11	8	8	−1	0	−27	(6)
% Elongation change	%	−20	−11	−4	1	−16	−11	−35	(6)
% Hardness change	%	3	0	0	0	1	−6	0	(6)
Volume swell	%	1.8	5.6	7.8	8.2	6.6	18	35	(6)
		Fluid aging at 23°C in ASTM Fuel B							
% Tensile change	%	−43	−43	−42	−43	−46	−43	−54	(6)
% Elongation change	%	−42	−40	−40	−41	−45	−44	−59	(6)
% Hardness change	%	−9	−12	−10	−9	−9	−13	−14	(6)
Volume swell	%	18	26	28	28	28	38	53	(6)
		Fluid aging at 23°C in ASTM Fuel C							
% Tensile change	%	−54	−51	−57	−55	−58	−58	−66	(6)
% Elongation change	%	−58	−52	−58	−54	−59	−61	−72	(6)
% Hardness change	%	−11	−15	−12	−10	−10	−13	−13	(6)
Volume swell	%	37	45	50	48	46	68	94	(6)
		Fluid aging at 100°C in distilled water							
% Tensile change	%	−5	−8	−2	8	−8	−3	5	(6)
% Elongation change	%	−18	−26	−18	−1	−23	−16	−8	(6)
% Hardness change	%	0	−1	0	0	0	0	0	(6)
Volume swell	%	3.6	3.6	4.4	3.2	3.9	2.4	2.1	(6)
Compression set	%	100°C, 70 h, (ASTM D395, method B)							
		10.1	12.5	10.8	8.4	13.2	10.1	11.2	(6)
		121°C, 70 h, (ASTM D395, method B)							
		24.0	26.0	23.0	20.1	23.9	24.0	25.3	(6)

5

Property	Units	Conditions					Value		Reference
	ACN %	40	33	33	33	33	27	20	
	Polymer Mooney	60	30	50	70	85	50	40	
Rebound	%	Goodyear–Healey method, 23°C							
		42	57	58	59	57	61	64	(6)
		Goodyear–Healey method, 100°C							
		60	74	76	77	76	78	79	(6)
Brittle temperature	°C	−27.5	−36.5	−38.3	−38.9	−38.9	−50.9	−54.5	(6)
Gehman temperature		Torsion							
T (2)	°C	−4	−15	−16	−17	−16	−21	−27	(6)
T (5)	°C	−11	−20	−22	−22	−21	−25	−32	(6)
T (10)	°C	−14	−22	−24	−24	−23	−28	−34	(6)
T (100)	°C	−18	−28	−31	−29	−29	−33	−41	(6)
Low temperature retraction, TR-10	°C	50% elongation							
		−21	−27	−29	−29	−27	−32	−42	(6)

** NBR compound formulation—Polymer: 100 phr, N774: 60 phr, ZnO: 4 phr, Wingstay 100: 2 phr, Paraplex G-25: 5 phr, TP 95 Plasticizer: 7 phr, METHYL TUADS: 2 phr, AMAX: 2 phr, Stearic acid: 0.5 phr, Sulfur: 0.4 phr.

References

1. In: Mark JE, ed. *Physical Properties of Polymers Handbook.* 2nd ed. New York, NY: Springer Science Business Media, LLC; 2007.
2. *Bayer/Lanxess Nitrile Handbook.*
3. Ohm RF. 'Introduction to the Structure and Properties of Rubber'. In: *The Vanderbilt Rubber Handbook.* 13th ed. 1990.
4. Small PA. *J. Appl. Chem.* 1953;3:71.
5. Poddubnyi IYa, Grechanovskii VA, Podalinskii AV. *J. Polym. Sci., Part C.* 1968;16:3109.
6. Purdon JR. 'Nitrile Elastomers'. In: *The Vanderbilt Rubber Handbook.* 13th ed.

Alkyd Resins

Mee Y. Shelley and Jennifer L. Braun

Trade Names Plaskon, Durez, Glaskyd

Class Thermoset polymers (polyesters modified with monobasic fatty acids)

Principal Components Fatty acids and oils (e.g., lauric, palmitic, stearic, oleic, linoleic, linolenic, eleostearic, and licanic acids). Polyhydric alcohols (e.g., glycerol, pentaerythritol, ethylene glycol). Polybasic acids (e.g., phthalic acid/anhydride, maleic acid/anhydride, fumaric acid/anhydride).

Major Applications Paints, brushing enamels, and clear varnish. Industrial coatings (spraying, dipping, flow coating, roller coating). Industrial baking finishes.

Properties of Special Interest Rapid drying. Good adhesion. Flexibility. Mar resistance and durability. Ester groups can be hydrolyzed under alkaline conditions.

Property	Units	Conditions	Value	Reference
Processing temperature	K	Molding, mineral filled (granular and putty)		(1)
		Compression	405–450	
		Injection	410–470	
		Transfer	430–460	
		Molding, glass fiber-reinforced		(1)
		Compression	420–450	
		Injection	410–470	
		Unspecified	425–440	(2)
Molding pressure	MPa	Molding, mineral filled (granular and putty)	14–140	(1)
		Molding, glass fiber-reinforced	14–170	
Compression ratio		Molding, mineral filled (granular and putty)	1.8–2.5	(1)
		Molding, glass fiber-reinforced	1–11	
Linear mold shrinkage	ratio	Mineral-filled	0.005	(8)
		Short glass fiber reinforced	0.002	(8)
		Mineral and glass fiber reinforced	0.004	(8)
		Molding, mineral filled (granular and putty)	0.003–0.010	(1)
		Molding, glass fiber-reinforced	0.001–0.010	(1)
		Unspecified	0.002–0.007	(2)
Density	g cm^{-3}	General range	0.835–0.837	(7)
		Mineral-filled	1.9	(8)
		Short glass fiber reinforced	2.06	(8)

Property	Units	Conditions	Value	Reference
		Mineral and glass fiber-reinforced	2.2	(8)
		Molding, mineral filled (granular and putty)	1.6–2.3	(1)
		Molding, glass fiber-reinforced	2.0–2.3	(1, 3)
		Unspecified	2.05–2.16	(2)
		Coating	1.2	(3)
Water absorption	%	Mineral-filled	0.2	(8)
		Short glass fiber reinforced	0.15	(8)
		Mineral and glass fiber reinforced	0.19	(8)
		Molding, mineral filled (granular and putty), 1/8 in. thick specimen, 24 h	0.05–0.5	(1)
		Molding, glass fiber-reinforced, 1/8 in. thick specimen, 24 h	0.03–0.5	(1)
		Coating	2	(3)
Tensile strength at break	MPa	General range	20.0–26.0	(7)
		Mineral-filled	45	(8)
		Short glass fiber reinforced	60	(8)
		Mineral and glass fiber reinforced	72	(8)
		Molding, mineral filled (granular and putty)	20–60	(1)
		Molding, glass fiber-reinforced	30–66	(1)
		Molding, glass fiber-filled	41	(3)
		Unspecified	40–60	(2)
		Coating	35	(3)
Elongation	%	General range	32–60	(7)
		Mineral-filled	0.6	(8)
		Short glass fiber reinforced	1	(8)
		Mineral and glass fiber reinforced	0.8	(8)
		Coating	65	(3)
		Molding, glass fiber-filled	2	

Solubility parameters[4,5]

Conditions	Hansen parameters $(MPa)^{1/2}$			
	δ_d	δ_p	δ_h	δ_t
Long oil (66% oil length, Plexal P65, Polyplex)	20.42	3.44	4.56	21.20
Short oil (coconut oil 34% phthalic anhydride; Plexal C34)	18.50	9.21	4.91	21.24

Property	Units	Conditions	Value	Reference
Tensile yield strength	MPa	Unspecified	45–48	(2)
Compressive strength (rupture or yield)	MPa	Molding, mineral filled (granular and putty)	83–260	(1)
		Molding, glass fiber-reinforced	100–250	(1)
		Unspecified	150–190	(2)

Property	Units	Conditions	Value	Reference
Flexural strength (rupture or yield)	MPa	Molding, mineral filled (granular and putty)	40–120	(1)
		Molding, glass fiber-reinforced	60–180	(1)
		Unspecified	60–160	(2)
		Molding, glass fiber-filled	103	(3)
Tensile modulus	MPa	Molding, mineral filled (granular and putty)	3,000–20,000	(1)
		Molding, glass fiber-reinforced	14,000–19,000	
Compressive modulus	MPa	Molding, mineral filled (granular and putty)	14,000–20,000	(1)
		Molding, glass fiber-filled	140	(3)
Flexural modulus	MPa	General range	641–1770	(7)
		Mineral-filled	7000	(8)
		Short glass fiber-reinforced	8000	(8)
		Mineral and glass fiber-reinforced	8600	(8)
		Molding, mineral filled (granular and putty), 296 K	14,000	(1)
		Molding, glass fiber-reinforced, 296 K	14,000	(1)
		Unspecified	14,000–20,000	(2)
Impact strength, Izod	$J\,m^{-1}$	Molding, mineral filled (granular and putty)	16–27	(1)
		Molding, glass fiber-reinforced	27–850	(1)
		Unspecified	17–400	(2)
		Mineral-filled	20	(8)
		Short glass fiber-reinforced	265	(8)
		Mineral and glass fiber-reinforced	240	(8)
Hardness	Rockwell	General range	62–90	(7)
	Surface	Mineral-filled	RE98	(8)
	Surface	Short glass fiber reinforced	RM110	(8)
	Surface	Mineral and glass fiber reinforced	RM125	(8)
	Rockwell	Molding, mineral filled (granular and putty)	E98	(1)
	Rockwell	Molding, glass fiber-reinforced	E95	(1)
	Rockwell	Molding, glass fiber-filled	E80	(3)
	Shore	Coating	D80	(3)
Deflection temperature	K	Under flexural load, 0.45 MPa		
		General range	348–363	(7)
		Mineral-filled	533+	(8)
		Short glass fiber reinforced	533+	(8)
		Mineral and glass fiber reinforced	533+	(8)
		Under flexural load, 1.80 MPa		
		Mineral-filled	483	(8)
		Short glass fiber reinforced	513	(8)
		Mineral and glass fiber reinforced	493	(8)
		Molding, mineral filled (granular and putty) under flexural load, 1.82 MPa	450–530	(1)
		Molding, glass fiber-reinforced under flexural load, 1.82 MPa	480–530	(1)
		Molding, glass fiber-filled, 1.82 MPa	470	(3)

Property	Units	Conditions	Value	Reference
Maximum resistance to continuous heat	K	Coating	360	(3)
		Molding, glass fiber-filled	470	
Thermal conductivity	W m^{-1} K^{-1}	Granular and putty, mineral filled	0.5–1.0	(1)
		Glass fiber-reinforced	0.6–1.0	
Dielectric strength	V mil^{-1}	Molding, mineral filled (granular and putty)	350–450	(1)
		Molding, glass fiber-reinforced	259–530	(1)
		Glass-filled	375	(6)
		Mineral-filled	400	(6)
	MV/m	Mineral-filled	13	(8)
		Short glass fiber reinforced	13	(8)
		Mineral and glass fiber reinforced	13	(8)
Volume resistivity	ohm cm	Glass-filled	10^{15}	(6)
		Mineral-filled	10^{14}	
		Mineral-filled	10^{14}	(8)
		Short glass fiber reinforced	10^{15}	(8)
		Mineral and glass fiber reinforced	10^{15}	(8)
Dielectric constant	–	At 1 kHz		
		Mineral-filled	6.5	(8)
		Short glass fiber reinforced	6.5	(8)
		Mineral and glass fiber reinforced	6.1	(8)
		At 1 MHz		
	–	Glass-filled, 1 MHz	4.6	(6)
		Mineral-filled, 1 MHz	4.7	(6)
		Unspecified, 1 MHz	4.7–6.7	(2)
		Coating	4	(3)
Dissipation factor	–	At 1 kHz		
		Mineral-filled	0.02	(8)
		Short glass fiber reinforced	0.017	(8)
		Mineral and glass fiber reinforced	0.012	(8)
		At 1 MHz		
		Glass-filled	0.02	(6)
		Mineral-filled	0.02	(6)
		Unspecified	0.009–0.02	(2)
Coefficient of thermal expansion	°C^{-1}	General range	0.00038	(7)
		Mineral-filled	4×10^{-5}	(8)
		Short glass fiber reinforced	4.5×10^{-5}	(8)
		Mineral and glass fiber reinforced	3.75×10^{-5}	(8)
Oxygen index	%	Mineral-filled	29	(8)
		Short glass fiber reinforced	30	(8)
		Mineral and glass fiber reinforced	27	(8)
Flammability	UL94	Mineral-filled	V1	(8)
		Short glass fiber reinforced	V1	(8)
		Mineral and glass fiber reinforced	V2	(8)

Property	Units	Conditions	Value	Reference
Melting temperature	K	Mineral-filled	313–353	(8)
		Short glass fiber reinforced	313–343	(8)
		Mineral and glass fiber reinforced	313–353	(8)

References

1. Kaplan WA, et al. eds. *Modern Plastics Encyclopedia '97*. New York: McGraw-Hill; *Modern Plastics*, Mid-November 1996.
2. *Plastics Digest, Thermoplastics and Thermosets*, 15th ed., vol. 1. D.A.T.A. Englewood: Business Publishing; 1994.
3. Seymour RB. *Polymers for Engineering Applications*. Washington, D.C.: ASM International; 1987.
4. Hansen CM. *Skand. Tidskr, Färg Lack*. 1971;17:69.
5. Du Y, Xue Y, Frisch HL. In: Mark JE, ed. *Physical Properties of Polymers Handbook*. New York: Wiley-Interscience; 1996:227–239.
6. Harper CA, ed. *Handbook of Plastics, Elastomer, and Composites*. 3d ed. New York: McGraw-Hill; 1996.
7. ©1986–2007 IDES Inc., www.ides.com
8. The Rubber and Plastics Research Association, www.rapra.co.uk

Amino Resins

N. Venkatasubramanian, Zongwu Bai and Milind Sohoni

Alternative Names Melamines, urea resins

Trade Names Resimene (Solutia, Inc.), Cymel (Cytek Industries, Inc.)

Class Thermoset polymers; chemical copolymers

Typical Comonomers Melamines, urea, formaldehyde, ethylene urea, benzoguanamine, thiourea, acetoguanamine

Polymerizations Condensation

Major Applications Molding resins, adhesives, coatings, treatment of paper, wood and textiles, automobile tires

Properties of Special Interest Hardness, nonflammability, arc resistance, thermal properties, lightfastness

Properties of amino–formaldehyde molding compounds[1]

Property	Units	Resin and Filler	
		Urea–Formaldehyde, Alpha–Cellulose	Melamine–Formaldehyde, Alpha–Cellulose
Pigmentation and coloring possibilities	–	Unlimited	Unlimited
Appearance	–	Translucent to opaque	Translucent to opaque
Molding qualities	–	Excellent	Excellent
Type of resin	–	Thermosetting	Thermosetting
Molding temperature	°F (°C)	275–300 (135–177)	280–370 (138–188)
Molding pressure	psi	2000–8000	1500–8000
Mold shrinkage	in in^{-1}	0.006–0.014	0.005–0.015
Specific gravity	–	1.47–1.52	1.47–1.52
Tensile strength	psi	$6–13 \times 10^3$	$7–13 \times 10^3$
Flexural strength	psi	$10–16 \times 10^3$	$10–16 \times 10^3$
Notched Izod impact strength	ft-lb in^{-1}	0.25–0.4	0.24–0.35
Rockwell hardness	–	M 110–M 120	M 110–M 125
Thermal expansion	°C^{-1}	$2.2–3.6 \times 10^{-6}$	4.0×10^{-6}
Deflection temperature under load	°F	260–290	410
Dielectric strength, short time, 0.125 in thickness	V mil^{-1}	300–400	300–400
Dielectric constant	–	6–8	7.2–8.4
Dissipation factor	–	0.025–0.035	0.027–0.045
Arc resistance	s	80–150	110–180

Property	Units	Resin and Filler	
		Urea–Formaldehyde, Alpha–Cellulose	Melamine–Formaldehyde, Alpha–Cellulose
Cold-water absorption, room temp.			
24 h, 0.125 in thickness	%	0.4–0.8	0.1–0.6
7 days	mg $(100 \text{ cm}^2)^{-1}$	800	270
Boiling water test, 10 min, 100°C	%	3.4	0.4
Burning rate	–	Self-extinguishing	Self-extinguishing
Effect of sunlight	–	Pastels turn gray	Slight color change

Curing range of urea– and melamine–formaldehyde molding compounds[1]

Cure time (min)	Cure Temperature (°C)							
	0.5	1	1.5	2	3	4	6	8
Urea–formaldehyde base								
Upper limit	–	170	167	163	158	154	148	145
Optimum temperature	–	169	164	160	155	151	145	140
Lower limit	–	167	160	156	150	145	139	135*
Melamine–formaldehyde								
Upper limit	187	182	179	177	172	169	165	161
Optimum temperature	175	167	159	154	146	140	130*	120*
Lower limit	172	155	145	138	125*	120*	115*	110*

* Value extrapolated.

Rate constants for urea–formaldehyde reactions at 35°C and pH 4.0[3]

Reaction*	Rate Constant K, L (s mol)$^{-1}$
U + F → UF	4.4×10^{-4}
UF + U → U—CH$_2$—U	3.3×10^{-4}
UF + UF → U—CH$_2$—UF	0.85×10^{-4}
UF$_2$ + UF → FU—CH$_2$—UF	0.5×10^{-4}
UF$_2$ + UF$_2$ → FU—CH$_2$—UF$_2$	$<3 \times 10^{-6}$

* U = urea, F = formaldehyde.

^{1}H NMR chemical shifts† for melamine resins[2]

Proton	Chemical Shift	Structure
—NH$_2^*$	5.8–6.2	Broad singlet
—NH*	7.2–7.4	Broad singlet
—N—CH$_2$OH*	5.4–5.6	Broad triplet
—N—CH$_2^*$OH	–	–
—N—CH$_2^*$OR	4.6–5.1	Broad peak
—N—CH$_2^*$—N—	–	–
—O—CH$_3^*$	3.2	Singlet

Proton	Chemical Shift	Structure
$-O-CH_2^*CH_3$	3.0–3.2	Quadruplet
$-O-CH_2CH_3^*$	1.2	Triplet
$-O-CH_2^*CH_2CH_3$	3.5	Triplet
$-O-CH_2CH_2^*CH_3$	1.5	Multiplet
$-O-CH_2CH_2CH_3^*$	1.1	Triplet
$-O-CH^*(CH_3)_2$	3.8–4.0	Multiplet
$-O-CH(CH_3^*)_2$	1.0–1.1	Doublet
$-O-CH_2^*CH_2CH_2CH_3$	3.0–3.3	Triplet
$-O-CH_2CH_2^*CH_2^*CH_3$	1.0–1.5	Multiplet
$-O-CH_2CH_2CH_2CH_3^*$	0.7–1.0	Triplet
$-O-CH_2^*CH(CH_3)_2$	3.5	Multiplet
$-O-CH_2CH^*(CH_3)_2$	1.5	Multiplet
$-O-CH_2CH(CH_3^*)_2$	0.8	Doublet

† Chemical shifts in ppm relative to TMS.

^{13}C NMR chemical shifts† for melamine resins[5]

Carbon Atom	Chemical Shift
$\overset{N}{\underset{N}{}}C^*-NH_2$	167.4
$\overset{N}{\underset{N}{}}C^*-NH(CH_2O-)$	166.0–166.6
$\overset{N}{\underset{N}{}}C^*-N(CH_2O-)_2$	165.4–165.8
$-NHCH_2OC^*H_2O-$	93.0
$-N(C^*H_2OCH_3)_2$	76.8
$-NHC^*H_2OCH_3$	72.6
$-N(C^*H_2OC_4H_9)_2$	74.4
$-NHC^*H_2OC_4H_9$	71.0
$-NHC^*H_2OC^*H_2NH-$	68.0–69.0
$-NHC^*H_2OH$	64.5
$-N(CH_2OC^*H_3)_2$	55.4
$-NHCH_2OC^*H_3$	54.5
$-NCH_2OC^*H_2CH_2CH_2CH_3$	66.9
$-NCH_2OCH_2^*CH_2CH_2CH_3$	31.4
$-NCH_2OCH_2CH_2C^*H_2CH_3$	18.9
$-NCH_2OCH_2CH_2CH_2C^*H3$	13.7
$-NCH_2OCH_2C^*H(CH_3)_2$	28.1
$-NCH_2OCH_2CH(C^*H_3)_2$	18.9

† Chemical shifts in ppm relative to TMS.

Melamine/formaldehyde reactions[2]

1.	$-NCH_2OCH_3 + ROH \rightarrow -NCH_2OR + CH_3OH$
2.	$2-NCH_2OCH_3 + H_2O \rightarrow -NCH_2N- + H_2C=O + 2CH_3OH$
3.	$-NCH_2OCH_3 + -NH \rightarrow -NCH_2N- + CH_3OH$
4.	$2-NCH_2OCH_3 \rightarrow -NCH_2N- + CH_3OCH_2OCH_3$
5.	$-NCH_2OCH_3 + -NCH_2OH \rightarrow -NCH_2OCH_2N- + CH_3OH$
6.	$-NCH_2OCH_3 + H_2O \rightarrow -NCH_2OH + CH_3OH$
7.	$-NCH_2OH \rightarrow -NH + H_2C=O$
8.	$-NCH_2OH + -NH \rightarrow -NCH_2N- + H_2O$
9.	$2-NCH_2OH \rightarrow -NCH_2N- + H_2C=O + H_2O$
10.	$-NCH_2OH + ROH \rightarrow -NCH_2OR + H_2O$
11.	$2-NCH_2OH \rightarrow -NCH_2OCH_2N- + H_2O$

Reaction constants for urea–formaldehyde at 35°C and pH 7.0[4]

Reaction	Second Order Reaction Velocity, Constant, k_1 ($l\,mol^{-1}\,s^{-1}$)	Equilibrium Constant $K(k_2k_1^{-1})$ ($mol\,l^{-1}$)
1. $U + F \underset{R_2}{\overset{R_1}{\rightleftharpoons}} UF$	0.9×10^{-4}	0.036
2. $UF + F \underset{R_2}{\overset{R_1}{\rightleftharpoons}} UF_2$	0.38×10^{-4}	0.22
3. $UF_2 + F \underset{R_2}{\overset{R_1}{\rightleftharpoons}} UF_3$	0.1×10^{-4}	1.2

Properties of melamine–formaldehyde laminates[1]

Property	Units	Melamine–Formaldehyde Laminate	
		Cellulose Paper Base	Glass Fabric Base
Coloring possibilities	–	Unlimited	Unlimited
Appearance	–	Translucent/opaque	Translucent/opaque
Laminating temperature	°F	270–320	270–300
Laminating pressure	psi	500–1800	1000–1800
Specific gravity		1.4–1.5	1.82–1.98
Tensile strength	psi	$10–25 \times 10^3$	$25–40 \times 10^3$
Flexural strength	psi	$14–20 \times 10^3$	$40–65 \times 10^3$
Notched Izod impact Strength	ft-lb in^{-1}	0.3–1.5	5–15
Rockwell hardness	–	M 110–M 125	M 115–M 125
Water absorption, 24 h, room temp., 0.125 in thickness	%	1.0–2.0	1.0–2.5
Effect of sunlight	–	Slight color change	Slight color change
Machining qualities	–	Fair	Fair
Thermal expansion	°C^{-1}	$0.7–2.5 \times 10^{-5}$*	$0.7–1.2 \times 10^{-5}$
Resistance to heat (continuous)	°F	210–260	300
Burning rate	–	~Nil	Nil
Dielectric strength, short time	V mil^{-1}	400–700	200–500
Dielectric constant, at 10^6 cps	–	6.4–8.5	6.0–9.0
Dissipation factor, at 10^6 cps	–	0.035–0.05	0.011–0.025
Arc resistance	s	100	175–200

* Cotton fabric filler.

Rate constants for melamine–formaldehyde resins at pH 7.7 [6]

Reaction	Temp. (°C)	Second Order Rate Constant of Forward Reaction, k_1	First Order Rate Constant of Reverse Reaction, k_2
1. $M + F \rightleftharpoons MF$	50	1.4×10^{-3}	0.3×10^{-4}
	70	6.1×10^{-3}	3.5×10^{-4}
2. $MF + F \rightleftharpoons MF_2$	50	1.0×10^{-3}	1.4×10^{-4}
	70	5.4×10^{-3}	6.6×10^{-4}
3. $MF_2 + F \rightleftharpoons MF_3$	50	1.8×10^{-3}	–
	70	7.4×10^{-3}	–

Typical properties of filled amino resin molding compounds [3]

Property	Units	Urea Alpha–Cellulose	Melamine Alpha–Cellulose	Macerated Fabric	Asbestos	Glass Fiber
Physical						
Specific gravity	–	1.47–1.52	1.47–1.52	1.5	1.7–2.0	1.8–2.0
Water absorption, 24 h, 3.2 mm thick	%	0.48	0.1–0.6	0.3–0.6	0.08–0.14	0.09–0.21
Mechanical						
Tensile strength	MPa (10^3 psi)	38–48 (5.5–7)	48–90 (7–13)	55–69 (8–10)	38–45 (5.5–6.5)	35–70 (5–10)
Elongation	%	0.5–1.0	0.6–0.9	0.6–0.8	0.3–0.45	
Tensile modulus	GPa (10^5 psi)	9–9.7 (13–14)	9.3 (13.5)	9.7–11 (14–16)	13.5 (19.5)	16.5 (24)
Hardness, Rockwell M	–	110–120	120	120	110	115
Flexural strength	MPa (10^3 psi)	70–124 (11–18)	83–104 (12–15)	83–104 (12–15)	52–69 (7.4–10)	90–165 (13–24)
Flexural modulus	GPa (10^5 psi)	9.7–10.3 (14–15)	7.6 (11)	9.7 (14)	12.4 (18)	16.5 (24)
Notch Impact strength	Jm^{-1} (ft-lb in^{-1})	14–18 (0.27–0.34)	13–19 (0.24–0.35)	32–53 (0.6–1.0)	16–21 (0.3–0.4)	32–1000 (0.6–18)
Thermal						
Thermal conductivity	10^{-4} J-cm s^{-1} cm^{-2} °C^{-1*}	42.3	29.3–42.3	44.3	54.4–71	48.1
Coefficient of thermal expansion	10^{-5} cm cm^{-1} °C^{-1*}	2.2–3.6	2.0–5.7	2.5–2.8	2.0–4.5	1.5–1.7
Deflection temperature at 1.8 MPa (264 psi)	°C	130	182	154	129	204
Flammability class	–	VO†	VO†	–	–	VO
Continuous no-load service temperature	°C	77‡	99‡	121	149	149–204

Property	Units	Urea	Melamine			
		Alpha–Cellulose	Alpha–Cellulose	Macerated Fabric	Asbestos	Glass Fiber
Electrical						
Dielectric strength	V/0.00254 cm					
Short time, 3.2 mm thick		330–370	270–300	250–350	410–430	170–300
Step by step		220–250	240–270	200–300	280–320	170–240
Dielectric constant, 22.8°C	–					
60 Hz		7.7–7.9	8.4–9.4	7.6–12.6	6.4–10.2	9.7–11.1
10^3 HZ		–	7.8–9.2	7.1–7.8	9.0	–
Dissipation factor, 22.8°C	–					
60 Hz		0.034–0.043	0.030–0.083	0.07–0.34	0.07–0.17	0.14–0.23
10^3 Hz		–	0.015–0.036	0.03–0.05	0.07	–
Volume resistivity, 22.8°C, 50% rh	ohm cm	0.5–5.0×10^{11}	0.8–2.0×10^{12}	1.0–3.0×10^{11}	1.2×10^{12}	0.9–2.0×10^{11}
Arc resistance	s	80–100	125–136	122–128	20–180	180–186

* To convert J to cal divide by 4.184.

† Applies to specimens thicker than 1.6 mm.

‡ Based on no color change.

^{13}C NMR spectra of Urea Formaldehyde (UF), Melamine Formaldehyde (MF) and Melamine Urea Formaldehyde (MUF) resins as well as ^{15}N NMR spectra of MUF resin[7]

Property	Units	Structure	Resin	Value
^{13}C NMR (solid-state CP-MAS) chemical shift*	ppm	—NH—CH$_2$NH—	UF	48
			MF	49
			MUF	48
		—N(CH$_2$—)CH$_2$—NH— (branched methylene)	MF	55
			MUF	55
		—N(CH$_2$—)CH$_2$N(CH$_2$—)— (branched methylene)	MF	62
		—NHCH$_2$OH	MF	66
		—N(CH$_2$—)CH$_2$OH (branched methylol)	MF	72
		—N—CO—N— (urea carbonyl)	UF	160
			MUF	161
		—N=C(NR$_2$)—N= (triazine carbon)	MF	167
			MUF	166
^{13}C NMR (liquid-state) chemical shift**	ppm	—NH—CH$_2$—NH—	UF	49.1
			MF	49.8

Property	Units	Structure	Resin	Value
			MUF	49.1
		$-N(CH_2-)CH_2-NH-$	UF	55.6
		(branched methylene)	MUF	55.7
		$-NHCH_2OH$	UF	67
			MF	67
			MUF	66.8
		$-N(CH_2-)CH_2OH$	UF	73.8
		(branched methylol)	MF	73.2
			MUF	73.8
		$-NH-CH_2OCH_2NH-$	UF	71.1
		(linear dimethylene ether)	MF	71.0
			MUF	71.1
		$-N(CH_2-)CH_2OCH-NH-$	UF	77.5
		(branched dimethylene ether)		
		$-NH-CH_2OCH_2OH$	UF	71.1
		(linear hemiformal of methylol)	MF	71.0
			MUF	71.1
		$-N(CH_2-)CH_2OCH_2OH$	UF	77.5
		(branched hemiformal of methylol)		
		$HOCH_2NHCONHCH_2OH$	UF	162
		(N, N'-dimethylolurea carbonyl)	MUF	162
		$H_2NCONHCH_2OH$	UF	163.5
		(monomethylol urea carbonyl)	MUF	163.5
		$-N=C(NR_2)-N=$	MF	168.2
		(triazine carbon)	MUF	168.3
^{15}N NMR (liquid-state) chemical shift[***]	ppm	$-NHCONH_2$	MUF	55.1
		(methylenediurea and chain ends)		
		$H_2NCONHCH_2OH$	MUF	55.8
		(monomethyloyurea)		
		$-NH-CH_2OCH_2NH-$	MUF	73.4
		(linear dimethylene ether linkage within urea dimers)		
		$-NHCH_2NH-$	MUF	74.3
		(linear methylene linkage within urea dimers)		
		$HOCH_2NHCONHCH_2OH$	MUF	80.0, 80.2
		(N,N'-dimethylolurea)		

* Solids obtained by gelling of liquid resins by curing at 100°C; external chemical shift reference tetramethylsilane (TMS, 0 ppm).

** Neat liquid resin samples; chemical shift reference 3-(trimethylsilyl)propionic acid (TMSPA, 0 ppm).

*** Neat liquid resin samples; external chemical shift reference NH_4^+ (NH_4NO_3, 0 ppm).

Melamine–urea–formaldehyde resin (MUF) preparation-FTIR data[8]

Property	Units	Structure/Description	Value
IR bands	Wavenumbers (cm^{-1})	—OH stretch	3578
		NH stretch (urea)	3200–3440
		Monosubstituted urea	3320–3360
		N—H stretch	
		Asymmetric monosubstituted urea—NH$_2$ stretch	3400–3440
		C=O stretch, primary amide —NH$_2$ (urea)	1656–1660
		C=O stretch, secondary amide —NH— (methylol and methylene urea)	1644
		Primary amine —NH$_2$ (urea NH bending)	1629
		Secondary amine —NH— (methylol and methylene melamine)	1625
		—C—NH— of melamine	1556
		Secondary amide —NH— (methylol and methylene urea)	1544
		N—C—N of a methylene bridge	1513
		—CH$_2$— deformation of methylol groups on urea	1459
		—N—C— of substituted melamine	1374
		Asymmetric N—C—N stretching of substituted melamine	1363
		—CH$_2$— (wagging) of methylol groups on urea	1336–1339
		—CH$_2$— of —CH$_2$—O—CH$_2$—	1282–1285
			1258–1262
		N—C—N symmetric stretch (urea, monosubstituted urea)	1140–1190
		C—O—C of —CH$_2$—O—CH$_2$	1150, 1135, 1100

Thermal properties of wood treated with melamine formaldehyde (MF) resin and melamine formaldehyde resin modified with phosphoric acid (MFP) as flame retardants[9]

Property	Units	Conditions	Value	Peak Nature
Peak temperature (maxima, DTA, air)	°C	Wood, no flame retardant	350, 456	Exo
		Wood, MF-treated	337, 524	Exo
		Wood, treated with	187	Endo
		MFP	330, 511	Exo

Thermal properties of wood treated with melamine formaldehyde (MF) resin and melamine formaldehyde resin modified with phosphoric acid (MFP) as flame retardants[9]

Property	Units	Conditions	Value
Activation energy E_a (Thermal degradation, TGA in air)	KJ (mol^{-1})	Wood, no flame retardant	
		(I stage)	67
		(II stage)	122
		(III stage)	19
		Wood, treated with MF	
		(I stage)	27
		(II stage)	96
		(III stage)	18
		Wood, treated with MFP	
		(I stage)	52
		(II stage)	72
		(III stage)	15
Frequency factor Z	s^{-1}	Wood, no flame retardant	
		(I stage)	3.6×10^2
		(II stage)	8.1×10^7
		(III stage)	2.0×10^{-2}
		Wood, MF treated	
		(I stage)	3.4×10^{-1}
		(II stage)	3.6×10^5
		(III stage)	2.0×10^{-2}
		Wood, treated with MFP	
		(I stage)	1.1×10^2
		(II stage)	1.3×10^4
		(III stage)	1.2×10^{-2}
Limiting-Oxygen-Index (LOI)	%	Wood, no flame retardant	18
		Wood, MF treated	27.5
		Wood, MFP treated	38
Char yield	Wt %	Wood, no flame retardant	6
		Wood, MF treated	20
		Wood, MFP treated	27.6

References

1. Widmer G. In: *Encyclopedia of Polymer Science and Technology*, vol. 2. New York: John Wiley and Sons; 1965:54.
2. Bauer DR. *Progress in Organic Coatings.* 1986;14:193.
3. Updegraff IH, Moore ST, Herbes WF, Roth PB. In: Kroschwitz JI, ed. *Kirk-Othmer Encyclopedia of Chemical Technology*, 3rd ed. vol. 2, New York: John Wiley and Sons; 1978:440.
4. Vale CP, Taylor WGK. *Aminoplastics.* London: Iliffe Books; 1964:24.

5. Christensen G. *Progress in Organic Coatings.* 1980;8:211.
6. Vale CP, Taylor WGK. *Aminoplastics.* London: Iliffe Books; 1964:47.
7. Angelatos AS, Burgar MI, Dunlop N, Separovic F. *Journal of Applied Polymer Science.* 2004;91:3504–3512.
8. Kandelbauer A, Despres A, Pizzi A, Taudes I. *Journal of Applied Polymer Science.* 2007;106:2192–2197.
9. Gao M, Pan DX, Sun CY. *Journal of Fire Sciences.* 2003;21:189–201.

Amylopectin

W. Brooke Zhao, John F. Kadla and Jennifer L. Braun

Class Carbohydrate polymers

Structure

Major Applications Thickeners, stabilizers, and adhesives.

Properties of Special Interests The highly branched nature of amylopectin accounts for the extreme brittleness of its films and extrudates. The extensive branching reduces chain entanglements usually required in high polymers to achieve satisfactory film properties.

Preparative Techniques Fractionation of starches. Native starches usually contain about 70–80% amylopectin. Genetic modification can result in starches having virtually no amylose content, such as waxy maize.

Property	Units	Conditions	Value	Reference
Molecular weight	$g\ mol^{-1}$	Ranges	4.5×10^4–4.2×10^8	(1)
(of repeat unit)		Method: DMSO, light scattering		(2)
		Barley	4.0×10^7	
		Pea		
		Smooth	5.0×10^7	
		Wrinkled	5.0×10^7	
		Potato I	4.4×10^7	
		Potato II	6.5×10^6	
		Tapioca	4.5×10^7	
		Waxy maize	4.0×10^7	
		Waxy maize, sheared	1.0×10^6	
		Wheat	4.0×10^7	
Polydispersity index	–	Range, depending on source	300–500	(1)
(M_w/M_n)		In DMSO, GPC ($M_w = 15.96 \times 10^6$; $M_n = 8.5 \times 10^6$)	1.88	(3)

Property	Units	Conditions		Value	Reference
Degree of polymerization	–	Depending on plant source and methods of extraction		$280–(1.45 \times 10^6)$	(1)
NMR	ppm	^{13}C chemical shift Solid state CP/MAS 25.18 MHz		101.9–100.3 (C-1) 63.1 (C-6)	(4)
		See also data for "a" and "4e" units of glycogen, p. 193			
Surface tension	mN m^{-1}	–		35	(5)
Specific rotation [α]	Degrees	Solvent	λ (nm)		
		Water	135	+200	(6)
				+192	(7)
		1M NaOH	134	+163	(6)
		1N KOH	–	+160	(7)
		Ethylenediamine	–	+173	(7)
		Ethylenediamine hydrate	–	+182	(7)
		Formamide	–	+167	(7)
				+192	(7)
Refractive index increment dn/dc	ml g^{-1}	0.5 N KCl		0.156	(8)
		1N KOH		0.142	(7)
		Ethylenediamine		0.098 ± 0.001	(7)
		Ethylenediamine hydrate		0.092 ± 0.003	(7)
		Formamide		0.069 ± 0.002	(7)
		Water		0.151, 0.155	(9)
Common solvents		Dimethyl sulfoxide, ethylene-diamine (hydrate and anhydrate), chloral hydrate, and hydrazine hydrate			(2)
Dilute-limit self diffusion coefficient	m^2 s^{-1}	D_0			
		In DMSO		8×10^{-13}	(10)
		In d-DMSO		$(3.2 \pm 0.7) \times 10^{-11}$	(11)
		In H$_2$O		$(1.0 \pm 0.2) \times 10^{-11}$	(11)
		Mass-weighted average molar mass ($D_0(M_w)$)			
		In d-DMSO		$(9 \pm 2) \times 10^{-12}$	(11)
		In D$_2$O		$(2.8 \pm 0.6) \times 10^{-12}$	(11)
		Mass-z average molar mass ($D_0(M_z)$)			
		In DMSO, 24°C		$(13 \pm 3) \times 10^{-13}$	(11)
				$(9 \pm 2) \times 10^{-13}$	(12)
Diffusion coefficient	–	–		9×10^{-12}	(10)
$J = a/b$ (ratio of semi-axes of the particles)				38	(10)
				28	(3)
Sedimentation coefficient (s_0)	Svedberg	Plant source Rongotes		88 ± 3 115 ± 4 103 ± 1	(13)

Property	Units	Conditions		Value	Reference
		Crossbow		65 ± 7	
				67 ± 1	
				73 ± 1	
		Aotea		105 ± 2	
		Karamu		87 ± 8	
		Hilgendorf		98 ± 4	
		Waxy maize		50	(15)
		Pea		51	(15)
		Maize		79	(15)
Hydrodynamic volume $(a^2/b)^{1/3}$	–	In DMSO		18(7)	(10)
Hydrodynamic radius (R_D)	nm	In DMSO		22(3)	(10)
		In d-DMSO		14	(11)
		In D_2O		82	(11)
		In 0.5M NaOH (10°C)		172	(18)
		In 0.5M NaOH (17°C)		175	(18)
		In 0.5M NaOH (25°C)		282	(18)
		In 90% w/w DMSO/H_2O (25°C)		190	(18)
Solvation coefficient (h)	gg^{-1}	Amylopectin/H_2O		0.25 ± 0.04	(11)
		Amylopectin/DMSO		0.6 ± 0.2	
Radius of gyration	Å	Solvent for light scattering	M_w		(7)
		1N KOH	8.0×10^7	2050	
			1.0×10^8	2060, 2120	
		Ethylenediamine	7.5×10^7	2150, 2120	
		Ethylenediamine hydrate	9.5×10^8	2050, 2090	
		Formamide	1.66×10^8	2960, 2920	
		Water	4.3×10^7	1540, 1630	
		0.5M NaOH (10°C)	1.29×10^8	2230	(18)
		0.5M NaOH (17°C)	1.42×10^8	2280	(18)
		0.5M NaOH (25°C)	5.30×10^8	2760	(18)
		90% w/w DMSO/H_2O (25°C)	1.50×10^8	2380	(18)
Second virial coefficient A_2	mol cm^3 g^{-2}	Solvent for light scattering	M_w		(7)
		1N KOH	8.0×10^7	9.6×10^7	
			1.0×10^8	7.6×10^7	
		Ethylenediamine	7.5×10^7	2.4×10^7	
		Ethylenediamine hydrate	9.5×10^8	2.9×10^7	
		Formamide	1.66×10^8	8.0×10^7	
		Water	4.3×10^7	0	
		0.5M NaOH (10°C)	1.29×10^8	9.4×10^{-8}	(18)
		0.5M NaOH (17°C)	1.42×10^8	8.8×10^{-8}	(18)
		0.5M NaOH (25°C)	5.30×10^8	2.2×10^{-8}	(18)
		90% w/w DMSO/H_2O (25°C)	1.50×10^8	5.5×10^{-8}	(18)

Amylopectin

Property	Units	Conditions	Value	Reference
Thermal conductivity	$W\,m\,K^{-1}$		1.37×10^{-4}	(14)
Thermal diffusivity	m^2/sec		$(1.53 \pm 0.04) \times 10^{-7}$	(14)
O_2 permeability	$cm^2\,Pa^{-1}\,sec^{-1}$	$20°C$, 50% rh	10^{-16}	(16)
Tan δ	–	6% halberd in water at $25°C$	1.8	(17)
	–	6% halberd in 0.05M urea at $25°C$	1.7	(17)
	–	6% halberd in 0.05M NaOH at $25°C$	1.6	(17)
Glass transition	K	DSC, potato ($M_n = 6 \times 10^6$)	378	(19)
Melting Point	K	DSC, potato ($M_n = 6 \times 10^6$)	433–439	(19)

Mark-Houwink parameters

Solvent	Temperature (°C)	$K_m \times 10^6$ (dL/g)	a	Molecular Weight Range	Method of Calibration	Reference
0.5N KOH	25	8.2	0.69	170,000–267,000	Viscometry	(20)
DMSO	23	1.118	0.281	152,000–3,760,000	Viscometry	(21)
25% (v/v) DMSO/ 2M aq KSCN	23	0.292	0.297	152,000–1,210,000	Viscometry	(21)

References

1. Powell EL. In: Whistler RL and Bemiller JN, eds. *Industrial Gums: Polysaccharides and Their Derivatives*, 2d ed. New York: Academic Press; 1973.
2. Young Austin H. In: Whistle RL, Bemiller JN, Paschall EF, eds. *Starch: Chemistry and Technology.* 2d ed. Orlando, Fla.: Academic Press; 1984 (and references therein).
3. Salemis P, Rinaudo M. *Polym. Bull.* 1984;11:397.
4. Hewitt JM, Linder M, Perez S, Buleen A. *Carbohydr. Res.* 1986;154:1.
5. Ray BR, Anderson JR, Scholtz JJ. *J. Phys. Chem.* 1958;62:1220.
6. Neely WB. *J. Org. Chem.* 1961;26:3015.
7. Stacy CJ, Foster JF. *J. Polym. Sci.* 20 (1956): 57.
8. Brice BA, Halwer M. *J. Opt. Soc. Amer.* 1951;41:1033.
9. Debye P. *J. Phys. Coll. Chem.* 1947;51:18.
10. Callaghan PT, Lelievre J, Lewis JA. *Carbohydr. Res.* 1987;162:83.
11. Collaghan PT, Lelievre J. *Biopolymers.* 1985;24:441.
12. Dickenson E, Lelievre J, Stainsby G, Waight S. In: *Progress in Food and Nutrition Science: Gums and Stabilizers for the Food Industry. Part II. Applications of Hydrocolloids.* Oxford: Pergamon Press; 1984.
13. Lelievre J, Lewis JA, Marsden K. *Carbohydr. Res.* 1986;153:195.
14. Rodríguez P, de la Cruz GG. *J. Food Eng.* 2003;58:205.
15. Tester RF, Patel T, Harding SE. *Carbohydrate Research.* 2006;341:130.

16. Forssell P, Lahtinen R, Lahelin M, Myllärinen P. *Carbohydrate Polymers.* 2002;47:125.
17. Tako M, Hizukuri S. *Stach/Stärke.* 2003;55:345.
18. Yang C, Meng B, Chen M, Liu X, Hua Y, Ni Z. *Carbohydrate Polymers.* 2006;64:190.
19. Al-Gham di A, Melibari M, Al-Saigh ZY. *J. Polym. & Env.* 2005;13(4):319.
20. Liu ZQ, Yi XS, Yi F. *Stach/Stärke.* 1999;51:406.
21. Cornell HJ, Rix CJ, McGrane SJ. *Stach/Stärke.* 2002;54:517.

Amylose

W. Brooke Zhao, John F. Kadla and Jennifer L. Braun

Class Carbohydrate polymers

Structure

Major Applications Adhesives, food, pharmaceutical, gels and foams, coating, and biodegradable packaging films.

Properties of Special Interests The linear glucan chains in amylose are responsible for its film-forming ability.

Preparative Techniques Fractionation of starches. Native starches contain about 20–30% amylose. Genetic modification can result in high amylose content (up to 80%).

Property	Units	Conditions		Value	Reference
Molecular weight	g mol^{-1}	Range		3.2×10^4–3.6×10^6	(1)
		Source	Methods		(2)
		Apple	Anaerobic, viscosity	2.4×10^5	
		Banana	Anaerobic, viscosity	2.7×10^5	
		Broad bean	Anaerobic, viscosity	2.9×10^5	
		Barley	Anaerobic, viscosity	3.0×10^5	
			DMSO, light scattering	2.11×10^6	
		Iris (rhizome)	Anaerobic, viscosity	2.9×10^5	
		Mango seed	Anaerobic, viscosity	2.9×10^5	
		Oat	Anaerobic, viscosity	2.1×10^5	
			DMSO, light scattering	2.19×10^6	
		Parsnip	Anaerobic, viscosity	7.1×10^5	
		Pea			
		Smooth	Anaerobic, viscosity	2.1×10^5	
		Wrinkled	Anaerobic, viscosity	1.6×10^5	
		Potato	Anaerobic, light scattering	4.9×10^5	
		Potato	DMSO, light scattering	1.9×10^6	

Property	Units	Conditions		Value	Reference
		Rubber seed	Anaerobic, light scattering	2.4×10^5	
		Rye	DMSO, light scattering	2.5×10^6	
		Sweet corn	Anaerobic, light scattering	1.8×10^5	
		Wheat	Anaerobic, light scattering	3.4×10^5	
		Wheat I	DMSO, light scattering	1.33×10^6	
		Wheat II	DMSO, light scattering	2.65×10^6	
Polydispersity index (M_w/M_n)	–	In DMSO, GPC, $M_w = 2.83 \times 10^5$, $M_n = 1.53 \times 10^5$		1.85	(3)
Degree of polymerization (DP)	–	Depending on plant source and extracting methods		200–22,000	(1)
Polymorphs	–	Alkali amylose after kept at 80% or higher relative humidity at 85–90°C		A-amylose	(2)
		Alkali amylose after kept at 80% or higher relative humidity at room temperature		B-amylose	
		V_h form after extensive drying		V_a-amylose	
		Crystallized from n-butanol		V_h-amylose	

Polymorphs	Lattice	Cell Dimension (Å)			Cell Angle	Helix Symmetry	Interchain and Intersheet Spacings (Å)			
		a	b	c	γ		$d_{\uparrow\downarrow}$	d_{110}	h	Ref.
A	Orthorhombic	11.90	17.70	10.52	90	$2 \times 6/1$ in 21.04 A repeat	10.66	9.87	3.51	(4)
B	Orthorhombic	18.50	18.50	10.40	90	$2 \times 6/1$ in 20.8 A repeat	10.68	9.25	3.47	(4)
B	Hexagonal	18.52	18.52	10.57	120	2 parallel 6/1 helices				(32)
V_a	Orthorhombic	12.97	22.46	7.91	90	21 (~6/5)	12.97	11.23	1.32	(4)
V_h	Orthorhombic	13.7	23.7	8.05	90	6/5	13.69	11.86	1.34	(4)

Property	Units	Conditions	Value	Reference
Infrared absorption	cm^{-1}	OH stretching		(5)
		V-amylose (crystalline)	3,500–3,300 (broad)	
		B-amylose (crystalline)	3,500–3,300 (broad), 1,122	
		Amorphous	3,500–3,300 (sharper)	
		CH_2OH bending, V → B	1,263 → 1,254	
		CH_2 skeletal, V → B	946 → 936	
NMR	ppm	^1H chemical shift		
		DMSO-d_6 (100°C)	5.07 (H-1), 3.30 (H-2), 3.64 (H-3), 3.32 (H-4), 3.4 (H-5) 3.7 (H-6)	(6)
		D_2O, 500 MHz (75°C)	5.896 (d) (H-1), 4.162 (dd) (H-2), 4.478 (dd) (H-3), 4.162 (t) (H-4), 4.350 (H-5), 4.406 (dd) (H-6$_a$), 4.328 (dd) (H-6$_b$)	(7)

Property	Units	Conditions	Value		Reference
		See also data for "a" units of glycogen, p. 193			
		^{13}C chemical shift	100.4 (C-1), 72.6 (C-2), 73.7 (C-3), 79.4 (C-4), 72.1 (C-5), 61.2 (C-6)		(6)
		See also data for "a" units of glycogen, p. 193			
		Solid state CP/MAS			(8)
		A-amylose	102.30, 101.32, 100.05 (t) (C-1), 63.67, 62.73 (shoulder) (C-6)		
		B-amylose	101.71, 100.74 (d) (C-1) 62.69 (C-6)		
		V_h-amylose	103.85 (C-1), 62.21 (C-6)		
		V_a-amylose	103.76 (C-1), 61.79 (C-6)		
Spin-spin coupling constant $^3J_{HH}$	Hz	D$_2$O, 500 MHz (75°C)	4.0 (H-1), 9.9 (H-2), 9.1 (H-3), 9.3 (H-4), 2.0 4.7 (H-5), 12 (H-6)		(7)
Dissociation constant pK_a	–	pH = 11.2 ± 0.1	12.5 ± 0.2		(9)
		pH = 12.5 ± 1	13.0 ± 0.1		
Degree of dissociation α	–	pH = 11.2 ± 0.1	0.05		(9)
		pH = 12.5 ± 1	0.26		
Electrophoretic mobility U	cm^{2}V^{-1}s^{-1}	pH = 11.2 ± 0.1	3.5		(9)
		pH = 12.5 ± 1	18.4		
Common solvents		Alkaline solutions, aqueous chloral hydrate, formamide, dichloroacetic acid, pyrrolidine, dimethyl sulfoxide, acetamide, ethylenediamine, piperazine, formic acid, and urea			(2)
Theta temperature Θ	K	0.33 M KCl	293–296		(10)
		DMSO/acetone 43.5%	298		(11)

Mark-Houwink parameters: K and a	–		$K \times 10^5$ (ml g^{-1})	a	(2)
		Water	13.2	0.68	
		0.5 N NaOH	1.44	0.93	
			3.64	0.85	
		0.15 N NaOH	8.36	0.77	
		0.2 N KOH	6.92	0.78	
		0.5 N KOH	8.5	0.76	
		1.0 N KOH	1.18	0.89	
		0.33 N KCl	113	0.50	
			112	0.50	
			115	0.50	
		0.5 N KCl	55	0.53	
			55	0.53	
		Aqueous KCl (acetate buffer)	59	0.53	
		Dimethyl sulfoxide	1.25	0.87	
			30.6	0.64	
			15.1	0.70	
			3.95	0.82	
		Ethylenediamine	15.5	0.70	
		Formamide	22.6	0.67	
			30.5	0.62	

Property	Units	Conditions	Value	Reference
Flexibility parameter	–	KCl · KOH, 25°C		(12)
		Hydrodynamic data in Θ condition	$\lambda_\theta = 2.70$	
		Stockmayer-Fixman plot	$\lambda_\theta = 2.58$	
		Extrapolation of λ tending to zero	$\lambda_{\theta,1/2} = 1.34$	
			$\lambda_{\theta,1/3} = 2.18$	
			$\lambda_{\theta,1/4} = 2.69$	
Unperturbed chain dimension $\langle S \rangle^2/P$	Å^2	Solvent		(13)
		$Me_2SO\text{-}H_2O$	12.2	
		Five solvents	12.2	
		$Me_2SO\text{-}KCl$	12.2	
		$Me_2SO\text{-}MeOH$	14.6	
		$H_2O\text{-}KCl$	14.6	
		0.5N NaOH	14.9	
		H_2O	14.7	
		Formamide	14.7	
		Me_2SO	14.7	
Second virial coefficient A_2	$mol\ cm^3g^{-2}$	DP = 3,650, 20°C		
		DMSO/44% acetone	-2.3×10^{-6}	(14)
		DMSO/42% acetone	3.45×10^{-5}	(14)
		DMSO/39% acetone	1.322×10^{-4}	(14)
		Formamide	2.19×10^{-4}	(14)
		DMSO/30% acetone	2.59×10^{-4}	(14)
		DMSO/20% acetone	3.92×10^{-4}	(14)
		DMSO/10% acetone	4.78×10^{-4}	(14)
		DMSO	5.35×10^{-4}	(14)
		Ethylenediamine	5.64×10^{-4}	(14)
		0.5N NaOH	4.88×10^{-4}	(14)
		Water	1.10×10^{-4}	(14)
		1N KOH	8.9×10^{-5}	(11)
		0.5N KCl, 31°C	2.89×10^{-5}	(11)
		0.5N KCl, 28°C	1.41×10^{-5}	(11)
Expansion coefficient α	–	DP = 3,650, 20°C		(14)
		DMSO/44% acetone	0.96	
		MSO/42% acetone	1.1	
		DMSO/39% acetone	1.25	
		Formamide	1.47	
		DMSO/30% acetone	1.53	
		DMSO/20% acetone	1.73	
		DMSO/10% acetone	1.88	
		DMSO	2.0	
		Ethylenediamine	2.08	
		0.5N NaOH	1.86	
		Water	1.59	

Amylose

Property	Units	Conditions		Value	Reference
Radius of gyration	Å	Solvent for L.S.	M_w		(11)
		DMSO	2.22×10^6	935	
		1 N KOH	2.23×10^6	912	
		0.5 N KCl,31°C	2.44×10^6	763	
		0.5 N KCl, 28°C	–	745	
		DMSO	1.35×10^6	724	
		DMSO	1.05×10^6	656	
		DMSO	8.47×10^5	610	
		DMSO	5.52×10^5	543	
		DMSO	2.70×10^5	425	
		DMSO	1.46×10^5	334	
Universal constant Θ	–	–		2.1×10^{21}	(15)
				3.6×10^{21}	(16)
Length of Kuhn statistical segment	Å	Aqueous, viscosity		21.1	(10)
		Aqueous, sedimentation		17.3	(10)
		DMSO		95	(10)
		Helical region, efficient bond length $b_0 = 1.33$ Å			(17)
		0.33 M KCl, viscosity		24	
		0.33 M KCl, sedimentation		26	
		0.2 M KOH, viscosity		74	
		0.2 M KOH, sedimentation		70	
		Nonhelical region, efficient bond length $b_0 = 4.41$ Å			(17)
		0.33 M KCl, viscosity		90	
		0.33 M KCl, sedimentation		86	
		0.2 M KOH, viscosity		240	
		0.2 M KOH, sedimentation		230	
Glass transition temperature T_g	K	Extrapolation data from substituted amylose		317	(18)
Melting temperature T_m	K	Extrapolation data from substituted amylose		527	(18)
Pyrolysis		Acidic catalyst, 79–120°C, 3–8 h		White dextrins	(19)
		Acidic catalyst, 150–270°C, 6–18 h		Yellow or canary dextrins	
		220°C, 10–20 h		British gums	
Pyrolysis weight loss	%	Amount (mg)	Temp, range		(20)
		80	240–265	10	
			265–300	59	
		100	300–350	11	
			240–265	10	
			265–300	58	
			300–350	12	

Property	Units	Conditions		Value	Reference
Enthalpy of hydration	kJ mol^{-1}	50–95°C, V$_h$ (helix diameter, 13.7 Å) $\leftrightarrow$ H$_2$O + V$_a$ (helix diameter, 13.0 Å)		43.5	(21)
Thermal conductivity	Wm K^{-1}			1.16×10^{-4}	(31)
Thermal diffusivity	m^2/sec			$(1.32 \pm 0.06) \times 10^{-7}$	(31)
Specific rotation [α]	Degree	Solvent	λ (nm)		
		Water	135	+ 200	(22)
			546	+ 232, 236	(23)
		0.5 M KCl	134	+ 201	(22)
			546	+ 200	(11)
		8 M urea	132	+ 200	(22)
		1 M NaOH	132	+ 162	(22)
		DMSO	210	+ 175	(22)
			546	+ 225, 226	(23)
			546	+ 171	(11)
		Formamide	546	+ 238, 239	(23)
		Ethylenediamine	546	+ 191, 195	(23)
		Hexamethylphosphoramide	546	+ 210, 212	(23)
		1 N KOH	546	+ 156	(11)
		0.5 N KOH	546	+ 174	(11)
Refractive index n_2^{46}	–	–		1.5198	(24)
Refractive index increment dn/dc	ml g^{-1}	DMSO, $\lambda = 436$ nm		$0.0676 \pm 3\%$	(11)
		DMSO, $\lambda = 546$ nm		$0.0659 \pm 3\%$	(11)
		1 N KOH		0.146	(25)
Sedimentation coefficient	Svedberg	Ultracentrifugation		10.2	(26)
		Source			(27)
		Rongotea		5.0 ± 0.2	
				5.2 ± 0.5	
				4.5 ± 0.1	
		Crossbow		2.6 ± 0.4	
				3.2 ± 0.1	
		Aotea		3.3 ± 0.1	
		Karamu		4.0 ± 0.7	
		Hilgendorf		5.7 ± 0.9	
				2.9 ± 0.6	
		Pea		14	(33)
		Maize		12	(33)
Segment mobility ms	–	Ultracentrifugation		0 (rigid)	(26)
Segment size l_m	Å	Ultracentrifugation		1.3	(26)
Surface tension	mNm^{-1}	–		37	(28)

Property	Units	Conditions		Value		Reference
Tensile strength	MPa	50% relative humidity, 72°F				(29)
		Source	DP	Dry ($\times 10^{-2}$)	Wet ($\times 10^{-3}$)	
		Tapioca	2,110	6.08	1.3	
		White potato	1,610	5.79	2.1	
		Wheat	1,230	6.47	1.5	
		Sweet potato	1,215	6.27	2.2	
		Tapioca	1,205	6.86	1.8	
		Tapioca	915	7.06	1.9	
		Corn	820	7.15	2.0	
		Corn	505	6.66	1.0	
		Corn	435	6.96	0.2	
		Corn	420	7.25	0.6	
		Corn	400	7.45	1.0	
		Corn	310	5.19	0.5	
		Corn	265	6.47	–	
		Corn	230	1.86	–	
Elongation at break	%	50% relative humidity, 72°F				(29)
		Source	DP	Dry	Wet	
		Tapioca	2,110	13	39	
		White potato	1,610	9	57	
		Wheat	1,230	13	19	
		Sweet potato	1,215	14	42	
		Tapioca	1,205	18	38	
		Tapioca	915	14	18	
		Corn	820	13	15	
		Corn	505	6	6	
		Corn	435	7	5	
		Corn	420	8	5	
		Corn	400	10	6	
		Corn	310	6	4	
		Corn	265	9	–	
		Corn	230	1	–	
Tear strength	g	50% relative humidity, 72°F				(29)
		Source	DP			
		Tapioca	2,110	8		
		White potato	1,610	10		
		Wheat	1,230	10		
		Sweet potato	1,215	9		
		Tapioca	1,205	8		
		Tapioca	915	–		
		Corn	820	8		
		Corn	505	6		
		Corn	435	7		
		Corn	420	4		
		Corn	400	7		

Property	Units	Conditions		Value	Reference
		Corn	310	5	
		Corn	265	3	
		Corn	230	–	
Permeability constant	$mol\,cm^{-1}s^{-1}$ $mm\,Hg^{-1}$	Water at 25°C			(30)
		Relative humidity			
		1–53%		3.1 ± 10^{-11}	
		29–1%		1.5 ± 10^{-10}	
		1–100%		2.7 ± 10^{-10}	
			Vapor pressure		(30)
		Organic vapor	(cm Hg) at 35°C		
		Methanol	20.4	2.5 ± 10^{-12}	
		Ethanol	10.4	5.8 ± 10^{-14}	
		1-Propanol	3.74	8.6 ± 10^{-14}	
		1-Butanol	1.31	2.5 ± 10^{-13}	
		Acedic acid	2.67	5.6 ± 10^{-14}	
		Ethyl acetate	16.5	4.4 ± 10^{-14}	
		Acetone	34.6	3.4 ± 10^{-14}	
		Carbon tetrachloride	17.6	1.6 ± 10^{-14}	
		Benzene	14.8	6.3 ± 10^{-14}	
		Benzaldehyde	0.16	8.0 ± 10^{-13}	
		Gas at 25°C			(30)
		Air		0	
		Oxygen		0	
		Nitrogen		0	
		Carbon dioxide		2.6 ± 10^{-16}	
		Ammonia		1.1 ± 10^{-12}	
		Sulfur dioxide		7.8 ± 10^{-14}	

References

1. Howe-Grant M. In: Kroschwitz JI, ed. *Kirk-Othmer Encyclopedia of Chemical Technology.* 4th ed. Vol. 4, New York: John Wiley and Sons; 1992.
2. Young AH. In: Whistler RL, Bemiller JN, Paschall EF, eds. *Starch: Chemistry and Technology.* 2d ed. Orlando, Fla.: Academic Press; 1984, (and references therein).
3. Salemis P, Rinaudo M. *Polym. Bull.* 11(1984):397.
4. Sarko A, Zugenmaier P. In: French AD, Gardner KK, eds. *Fiber Diffraction Methods,* (ACS Symposium Series 141), Washington, D.C.: American Chemical Society; 1980.
5. Casu B, Reggiani M. *J. Polym. Sci., Part C.* 1964;7:171.
6. Gagnaire D, Mancier D, Vincendon M. *Org. Mag. Res.* 1978;11:1978.
7. Neszmelyi A, Hollo J. *Starch/Starke.* 1990;5:167.
8. Horii F, Yamamoto H, Hirai A, Kitamaru R. *Carbohydr. Res.* 1987;160:29.
9. Doppert HL, Staverman AJ. *J. Polym. Sci., Part A-1.* 1966;4:2367, 2373.
10. Banks W, Greenwood CT. *Macromol. Chem.* 1963;67:49.
11. Everett WW, Foster JF. *J. Am. Chem. Soc.* 1959;81:3459.
12. Gonzalez C, Zamora F, Guzman GM, Leon LM. *J. Macromol. Sci. Phys.* 1987;B26(3):257.

13. Burchard W. In: *Solution Properties of Natural Polymers* (Special Publication Number 23). London: The Chemical Society; 1967.
14. Burchard W. *Makromol. Chem.* 1963;59:16.
15. Mandelkern L, Flory PJ. *J. Chem. Phys.* 1952;20:212.
16. Flory PJ, Fox TG. *J. Am. Chem. Soc.* 1951;73:1904.
17. Banks W, Greenwood CT. *Eur. Polym. J.* 1969;5:649.
18. Cowie JMG, Toporowski PM, Costaschuk F. *Makromol. Chem.* 1969;121:51.
19. Greenwood CT. In: *Advances in Carbohydrate Chemistry*. Vol. 22. New York: Academic Press; 1967.
20. Desai DH, Patel KC, Patel RD, Vidyanagar V. *Die Starke.* 1976;11:377.
21. Nicolson PC, Yuen GU, Zaslow B. *Biopolymers.* 1966;4:677.
22. Neely WB. *J. Org. Chem.* 1961;26:3015.
23. Dintzis FR, Tobin R, Babcock GE. *Biopolymers.* 1971;10:379.
24. Van Wijk R, Staverman AJ. *J. Polym. Sci., Part A-2.* 1966;4:1012.
25. Foster JF, Sterman MD. *J. Polym. Sci.* 1956;21:91.
26. Elmgren, H. *Carbohydr. Res.* 160 (1987): 227.
27. Lelievre J, Lewis JA, Marson K. *Carbohydr. Res.* 1986;153:195.
28. Ray BR, Anderson JR, Scholtz JJ. *J. Phys. Chem.* 1958;62:1220.
29. Wolff IA, et al. *Ind. Eng. Chem.* 1951;43:915.
30. Rankin JC, Wolff IA, Davis HA, Rist CE. *Ind. Eng. Chem.* 1958;3:120.
31. Rodríguez P, de la Cruz GG. *J. Food Eng.* 2003;58:205.
32. Takahashi Y, Kumano T, Nishikawa S. *Macromolecules.* 2004;37:6827.
33. Tester RF, Patel T, Harding SE. *Carbohydrate Research.* 2006;341:130.

Aromatic Polyamides

Shaw Ling Hsu

Acronyms and Trade Names Ultramid T (BASF), nylon 6 copolymer, nylon 6/6T, polyphthalamide

Class Polyamides, aromatic nylon

Structure These aromatic nylons consist of aliphatic and aromatic building blocks and incorporate the repeat units of nylon 6 [poly(caprolactam)] and nylon 6T [poly(hexamethylene terephthalamide)]. Because of this composition, Ultramid T resins are often designated as nylon 6/6T materials. The basic structure is nylon 6/6T, with the majority component being nylon 6.

$$-\text{N}\!\!-\!\!(\text{CH}_2)_{\overline{6}}\; \overset{\displaystyle O}{\overset{\|}{\text{C}}}-$$
$$\underset{\text{H}}{|}$$

Nylon 6

$$-\text{N}\!\!-\!\!(\text{CH}_2)_{\overline{6}}\; \text{N}-\overset{\displaystyle O}{\overset{\|}{\text{C}}}-\!\!\left\langle \text{C}_6\text{H}_4 \right\rangle\!\!-\overset{\displaystyle O}{\overset{\|}{\text{C}}}-$$
$$\underset{\text{H}}{|}\underset{\text{H}}{|}$$

Nylon 6T

Major Properties of Special Interest Extremely high melting temperatures can be achieved by adjusting the relative amount of the aromatic component. Good mechanical properties are exhibited at elevated temperatures. Good resistance to chemicals. Good dielectric properties. Dimensional stability in the presence of moisture. Tough and strong material.

Major Applications High temperature applications, low moisture absorption, good impact resistance, good dielectric properties, good resistance to chemicals, easy to process.

Automobile parts (e.g. radiator and ventilation systems, fuel supply systems), electronics housing, plug and socket connectors, printed circuit boards, tennis rackets, golf clubs.

Infrared Absorptions Overall, the infrared spectrum greatly resembles those found for other polyamides. The principal spectroscopic features can be definitively assigned.[1–9] The N–H stretching, amide I, II bands are found at 3305, 1627, and 1545 cm^{-1}, respectively. Bands characteristic of Nylon 6 copolymer are symmetric and asymmetric methylene stretching vibrations are found in the 3000 cm^{-1} range.[2,5,6,8–14] Methylene bending vibrations in the 1400 cm^{-1} region can also be definitively assigned. Other spectroscopic features can be linked to the presence of the nylon 6T component.[15,16] Vibrations at 862 cm^{-1}, 1019 cm^{-1}, and 1498 cm^{-1}, along with broad features at $\sim$1300 cm^{-1}, are assignable to para-disubstituted aromatic units.[2,7,17]

NMR The amide proton peaks are found in the range of 6–7 ppm and the methylene proton peaks in the range of 1–4 ppm. Both assignments fall in the range of peak positions listed for those groups in standard NMR tables.[18]

Thermal Transitions A broad melting behavior starts at approximately 210°C and ends at approximately 290°C. These trends are consistent with the melting behaviors reported for nylon 6/6T copolymers.[19]

T_m is 310–340°C.[20,21]

Glass Transition Temperature 100°C[21]

Degradation Of Aromatic Polyamides Plasma treatment can modify aromatic nylons reducing the relative concentration of amide units relative to that in untreated Nylon 66 copolymer.[22]

When these polyamides are dissolved in acidic solution, the polyamide is gradually degraded.

Density 1.18 g/cm^3 [20]

Tensile Strength 100 MPa (dry); 90 MPa (moist)[20]

Tensile Modulus 3.2 GPa (dry and moist)[20]

Heat Deflection Temperature@*264 psi* 100°C[20]

Dielectric Constant (1.0 MHz) 4.0[20]

Moisture Absorption 1.8% (23°C/50% RH)[20]; 7.0% (23°C, saturation)[21]

Linear Thermal Expansion Coefficient 0.2% at 60°C[20]

Solvent Hexafluoroisopropanol (HFIP)

References

1. Wobkemeier M, Hinrichsen G. *Polymer Bulletin.* 1989;21:607.
2. Kohan MI, ed. *Nylon Plastics.* New York: Wiley-Interscience; 1973.
3. Miyazawa T, Blout ER. *J. Am. Chem. Soc.* 1961;83:712.
4. Miyazawa T. *J. Chem. Phys.* 1960;32:1647.
5. Bradbury EM, Elliot A. *Polymer.* 1963;4:47.
6. Jakes J, Krimm S. *Spectrochim Acta.* 1971;27A:19–34.
7. Sadtler Research Laboratories, D. D7529K. pp. D7527K.
8. Chen C-C. Ph D Thesis, University of Massachusetts, 1996.
9. Arimoto H. *J. Polym. Sci. Part A.* 1964;2:2283.
10. Snyder RG, Sachtschneider JH. *Spectrochim. Acta.* 1964;20:853.
11. Snyder RG. *J. Chem. Phys.* 1965;42:1744.
12. Snyder RG. *J. Chem. Phys.* 1967;47:1316.
13. Snyder RG. *Macromolecules* 1990;23:2081.
14. Miyake A. *J. Polym. Sci.* 1960;54:223.
15. Keske RG. In: Salamone JC, ed. *Polymeric Materials Encyclopedia.* Baco Raton, Florida: CRC Press; 1996.
16. Blinne G, Baierweck P, Gotz W, Kopietz M. *Kunststoffe.* 1989;79:814.

17. Colthup NB, Daly LH, Wiberley SE. *Introduction to Infrared and Raman spectroscopy.* New York: Academic Press; 1990.
18. Gordon AJ. *The Chemist's Companion: a Handbook of Practical Data, Techniques, and References.* New York: John Wiley & Sons; 1972.
19. Ajroldi G, Stea G, Mattiussi A, Fumagalli M. *J. Appl. Polym. Sci.* 1973;17:3187–3197.
20. 'BASF Product literature'.
21. Kohan MI, ed. *Nylon Plastics Handbook.* Munich: Hanser; 1995.
22. Inagaki N, Tasaka S, Kawai H. *J. Polym. Sci Part A: Polymer Chem.* 1995;33:2001–2011.

Barex

Zhengcai Pu

Acronyms, Alternative Name Nitrile barrier resin

Class Acrylonitrile–methyl acrylate copolymer

Major Applications Extruded and calendered sheet, blown film, injection and extrusion blown bottles, and injection molded and engineered components; used in packaging industry for bags, bottles, and containers to retain aromas and flavors to enhance product integrity and extend shelf life.

Properties of Special Interest High permeation barrier to oxygen and other gases, high chemical resistance to withstand many of most aggressive chemicals.

Producers and/or Suppliers INEOS USA LLC

Property	Units	Conditions	Value	Reference
Bulk density	g cm^{-3}	ASTM D1895	0.66	(1–3, 5)
			0.4	(4)
Density	g cm^{-3}	ASTM D792	1.15	(1–3)
			1.13	(4)
			1.11	(5)
Elongation, yield	%	STM D638	3	(1, 3)
			4	(2, 4, 5)
Flexural modulus	GPa	ASTM D790	3.38	(1, 3)
			3.31	(2)
			2.96	(4)
			2.69	(5)
Flexural strength, yield	MPa	ASTM D790	96.5	(1–3)
			95.8	(4)
			94.5	(5)
Gas permeability				
Oxygen	cm^3-mm/ m^2-24 h-bar	296 K, 100% RH, ASTM D3985	0.3, 0.45, 0.6	(1–5)
		264–322 K, ASTM D3985	0.15–1.2	(6)
Nitrogen		296 K, 100% RH, ASTM D3985	0.08, 0.12, 0.16	(1–5)
Carbon dioxide		296 K, 100% RH, ASTM D3985	0.45, 0.52, 0.6	(1–5)
Water vapor		311 K, 90% RH, ASTM F1249–90	2.0, 2.5, 3.0	(1–5)

Property	Units	Conditions	Value	Reference
Hardness (Rockwell)		ASTM D785, M scale	60	(1–3)
			50	(4)
			45	(5)
Heat deflection temperature	K	ASTM D648		(1–5)
		455 KPa	344–350	
		1820 KPa	339–342	
Impact strength (Izod)	$J\,m^{-1}$	Notched, ASTM D256	267	(1, 3)
			133.5	(2)
			320	(4)
			481	(5)
Linear thermal expansion	K^{-1}	ASTM D696, 293–353 K	6.65×10^{-5}	(1–5)
Melt index	$g\,min^{-1}$	200°C , 27.5 lbs., 0.0824" D×0.3145" L, ASTM D1238	0.3	(1, 3–5)
			1.2	(2)
Mold shrinkage		ASTM D955	$2-5 \times 10^{-3}$	(1–5)
Specific heat	$J\,g^{-1}\,K^{-1}$	ASTM C351, 293 K	0.41	(1–5)
Tensile strength, yield	MPa	ASTM D638	65.5	(1–3)
			55.2	(4)
			51.7	(5)
Thermal conductivity	$W\,m^{-1}\,K^{-1}$	ASTM C177	0.25	(1–5)
Transmittance	%	Innovene Test Method, 0.010" thick sheet	92.5	(1–5)
Yield	m^2-25μm/kg		34.2	(1–3)
			34.8	(4)
			35.0	(5)

Chemical resistance[7,8]

*Exposure time in all testing was 1 year.

Solvent	Temperature (°C)	Appearance Change	% Weight Change
Alcohols			
Ethyl alcohol	23, 38	None	−0.16, +3.07
Ethylene glycol	23, 38	None	0.90, 3.67
Isopropyl alcohol	23, 38	None	0.03, 3.16
Hydrocarbons			
Benzene	23, 38	None	−1.06, −6.39
Toluene	23, 38	None	−1.91, −3.24
Xylene	23, 38	None	−1.19, −0.99
D-limonene	23, 38	None	
Halogenated hydrocarbons			
Carbon tetrachloride	23, 38	None	−2.80, −4.66
Methylene chloride	23, 38	None	
Trichloroethylene	23, 38	None	−3.63, −7.82
1,1,1-trichloroethane	23, 38	None	−1.01, −4.48

Solvent	Temperature (°C)	Appearance Change	% Weight Change
Ketones			
Acetone	23, 38	Frosted, softened	
Methyl ethyl ketone	23, 38	Frosted, softened	
Methyl isobutyl ketone	23, 38	None	$-0.81, -12.0$
Esters			
Butyl acetate	23, 38	None	$-0.42, -0.91$
Ethyl acetate	23	None	-1.31
	38	Frosted, softened	
Acids			
Acetic acid (100%)	23	None	-0.38
	38	Frosted, softened	
Hydrochloric acid (10%)	23, 38	None	$-3.03, -6.10$
Nitric acid (10%)	23	None	-3.28
	38	Yellowed	
Phosphoric acid (30%)	23, 38	None	$-2.92, -8.13$
Sulfuric acid (30%)	23, 38	None	$-1.94, -3.54$
Bases			
Ammonium hydroxide (10%)	23	None	-3.75
	38	Softened	
Barium hydroxide (saturated)	23, 38	None	$-4.12, -8.50$
Calcium hydroxide (saturated)	23, 38	None	$-4.19, -9.13$
Potassium hydroxide (10%)	23	Slight frost	-3.86
	38	Frosted, softened	
Sodium hydroxide (10%)	23	Slight frost	-3.00
	38	Frosted, Softened	

References

1. INEOS Barex. http://ineosbarex.avenit.de/files/upload/Barex210E_06_06.pdf. Retrieved July, 2007.
2. INEOS Barex. http://ineosbarex.avenit.de/files/upload/Barex210Injection_06_06.pdf. Retrieved July, 2007.
3. INEOS Barex. http://ineosbarex.avenit.de/files/upload/Barex210Film_06_06.pdf. Retrieved July, 2007.
4. INEOS Barex. http://ineosbarex.avenit.de/files/upload/Barex214Calender_06_06.pdf. Retrieved July, 2007.
5. INEOS Barex. http://ineosbarex.avenit.de/files/upload/Barex218E_06_06.pdf. Retrieved July, 2007.
6. INEOS Barex. http://ineosbarex.avenit.de/files/upload/BarexBarrierProps_06_06.pdf. Retrieved July, 2007.
7. INEOS Barex. http://ineosbarex.avenit.de/files/upload/BarexChemResist_06_06.pdf. Retrieved July, 2007.
8. INEOS Barex. http://ineosbarex.avenit.de/files/upload/BarexChemResistBook_06_06.pdf. Retrieved July, 2007.

Benzimidazobenzophenanthroline-type Ladder Polymer (BBL) and Semi-ladder Polymer (BBB)

N. Venkatasubramanian, Thuy D. Dang, Vladyslav Kholodovych and William J. Welsh

Acronyms, Alternative Names BBL: Poly(benzimidazobenzophenanthroline), Oxobenz(de)imidazobenzimidazoisoquinoline ladder polymer, Poly[(7-oxo-7H, 10H-benz(de)-imidazo(4′,5′:5,6)-benzimidazo(2,1-a)isoquinoline-3,4:10,11-tetrayl)-10-carbonyl], *ladder*-poly[(1,4,5,8-naphthalenetetracarboxylic acid)-*alt*-(1,2,4,5-tetraaminobenzene)]

BBB: Poly[6,9-dihydro-6,9-dioxobisbenzimidazo[2,1-*b*:1′,2′-*j*] benzo [1mn] phenanthroline-2,12-diyl]

Class Heteroaromatic ladder and semi-ladder polymers, double chain conjugated polymers (BBL), double-stranded conjugated polymers (BBL)

Structure

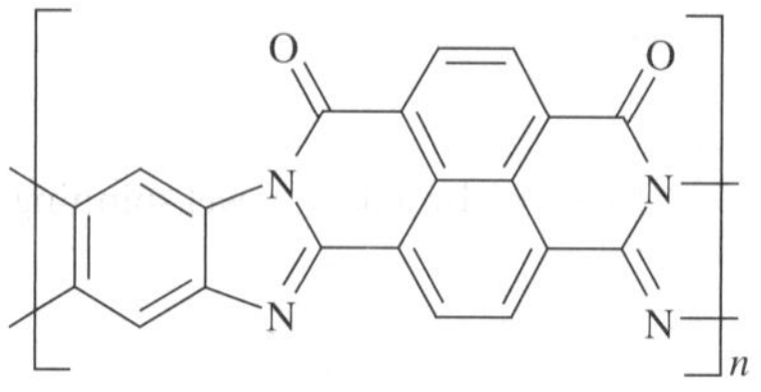

BBL

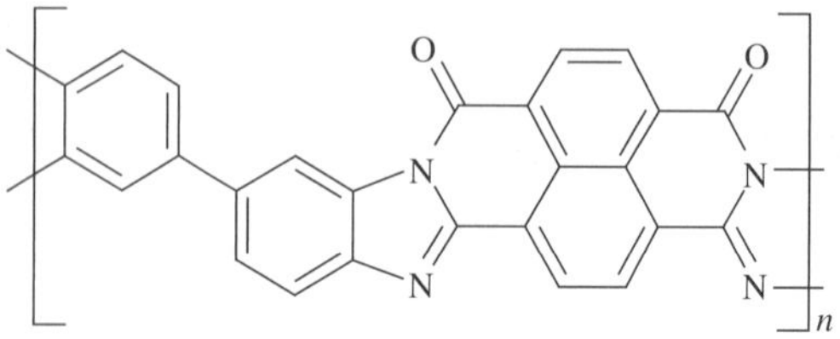

BBB

Major Applications of Interest High-performance fibers, films, and coatings for structural and opto-electronic applications; fire-resistant materials and humidity sensors/detectors.

Benzimidazobenzophenanthroline-type Ladder Polymer

Synthesis

BBL Polycondensation of 1,2,4,5-tetraaminobenzene (free tetraamine) with 1,4,5,8-naphthalenetetracarboxylic acid in polyphosphoric acid (PPA) medium at high temperatures (180–190°C) in dilute (1–2 wt % polymer) solutions[1] or that of 1,2,4,5-tetraaminobenzene tetrahydrochloride (TABH) with 1,4,5,8-naphthalenetetracarboxylic acid in PPA medium generated with 'P$_2$O$_5$' adjustment[2] for the polymerization conducted at high concentrations (~10 wt % polymer) to provide a liquid crystalline polymer solution in PPA.[3]

BBB Polycondensation of 1,4,5,8-naphthalenetetracarboxylic acid (NTCA) with 3,3′-diaminobenzidine (DABD).[4,5] Also reported for BBB synthesis is a polycondensation procedure involving 1,4,5,8-naphthalenetetracarboxylic dianhydride (NTDA) with 3,3′-diaminobenzidine in N,N-dimethylacetamide (DMAc) at room temperature to produce polybenzimidazobenzophenanthroline (PBIPA) precursor which was thermally cured to obtain the semi-ladder polymer network.[6]

Structural Features BBL forms liquid crystalline solutions (lyotropic mesophase) owing to its rod-like extended chain structure. BBB, in contrast to BBL, is a semi-ladder/linear polymer that does not form nematic mesophase in solution[7]. A scanning tunneling microscopy (STM) study of BBB films demonstrated structural features that include helical fibers coiled into strands and strands of coiled strings.[8]

Processing BBL can be processed into films, coatings, and fibers from both isotropic and anisotropic solutions. Thin films were cast from 0.5 wt % solution of BBL in methanesulfonic acid (MSA).[9] Films were also extruded from PPA solution (10 wt % polymer) by a dry-jet wet spin technique at 90°C and at 300 psi through a slit die 0.45 mm wide.[3,10] Films and coatings were also fabricated via the formation of soluble polymer–Lewis acid complexes using AlCl$_3$, GaCl$_3$, and FeCl$_3$ as Lewis acids and niroalkanes such as nitromethane, nitroethane, and 1-nitro- and 2-nitropropanes as solvents.[11,12] BBL fibers were fabricated from nematic liquid crystalline PPA solution (10 wt %) by a dry-jet wet spinning process.[13]

Properties of Special Interest Though both BBL and BBB exhibit similar thermal/thermo-oxidative stabilities, solubilities, as well as many opto-electronic attributes, BBL, being the ladder polymer, has received much more attention and has been the material of choice for extensive research related to many device applications.

BBL High thermal and thermo-oxidative stabilities, exceptional resistance to organic solvents, good mechanical properties as fibers and films; films are purple in color with metallic luster or golden yellow and metallic in appearance; conjugated ladder polymer with interesting opto-electronic properties due to its structural and molecular order; displays dramatically enhanced thermally induced electronic conductivity [14–16]; n-type (electron transport) organic semi-conductor based on electrochemical doping experiments,[17–20] photo-induced electron transfer and photoconductivity studies [21–24]; NLO material with large third-order optical non-linearities[25]; has been evaluated for potential utilization in many opto-electronic devices that include p–n junctions, organic thin film transistors, light emitters (OLEDs), tunable electrochromic and electroluminescent devices, polymer-based light sensors, xerographic imaging systems, and photovoltaic (solar) cells.[26–32]

Solubility Both BBL and BBB are soluble in strong, corrosive protonic acid media such as RSO$_3$H where R = —CH$_3$, —CF$_3$, or —Cl as well as PPA; soluble in polar aprotic organic solvents (such as nitroalkanes containing metal halide Lewis acids (e.g. AlCl$_3$, GaCl$_3$, and FeCl$_3$)).[11,12]

Benzimidazobenzophenanthroline-type Ladder Polymer

Polymerization Conditions (BBL)	Property	Units	Measurement	Value	Reference
Free tetramine monomer, dilute solution process	Intrinsic viscosity	dl g^{-1}	100 % MSA	5.0	(1)
Tetraamine tetrahydro-chloride monomer, 10 wt % final polymer concentration	Intrinsic viscosity	dl g^{-1}	MSA, 30°C	18.0, 23.5	(3)
Tetraamine tetrahydro-chloride monomer, 12 wt % final polymer concentration*	Intrinsic viscosity	dl g^{-1}	MSA, 30°C	20.0	(33)

* In this case, polymerization was conducted in the presence of 5 wt % multi-walled carbon nanotubes (MWNT).

Property	Units	Conditions	Value	Reference
Intrinsic viscosity	dl g^{-1}	BBL, in MSA	8.2	(34)
		BBL in MSA	1.34–6.12	(1)
		BBL in MSA, 30°C	4.78	(11)
		BBL in MSA, 30°C	7.91	(11)
		BBL in MSA, 30°C	32	(20)
		BBL in MSA, 25°C	2.87	(35)
		BBL in MSA, 25°C	4.10	(35)
		BBB in 96 % H_2SO_4 30°C	2.66	(11)
		BBB in 96 % H_2SO_4 30°C	2.72	(11)
		BBB in MSA	2.66	(1)
		BBB, in 100 % MSA 30°C	5.24	(11)
		BBB, in 30 wt % AlCl$_3$/ nitromethane, 30°C	4.21	(11)
		BBB, in 18.2 wt % AlCl$_3$/ nitromethane, 30°C	3.53	(11)
		BBB in 9.0 wt % AlCl$_3$/ nitromethane, 30°C	3.54	(11)
		BBB in 30.0 wt % GaCl$_3$/ nitromethane, 30°C	5.51	(11)
		BBB in 18.2 wt % GaCl$_3$/ nitromethane, 30°C	4.62	(11)
		BBB in 9.0 wt % GaCl$_3$/ nitromethane, 30°C	5.23	(11)
Glass transition temperature (T_g)	°C	BBL (estimated)	≥ 500	(36)
		BBL (estimated)	≥ 550	(37)
		BBB (estimated)	≥ 500	(36)
		BBB (estimated)	≥ 550	(38)
Thermal stability	°C	BBL, in nitrogen (TGA)	700	(1, 10)
		BBB, in nitrogen	690–710	(39)
		BBB, in nitrogen	700	(1)
Thermo-oxidative stability	°C	BBL, in air (TGA)	600	(1, 10)
		BBB, in air (TGA)	570	(1)
Density	g cm^{-3}	BBL, cast film	1.31	(39, 10)
		deposited film*	0.94	(39, 10)

Property	Units	Conditions	Value	Reference
Tensile modulus	GPa	BBL, cast film	7.6	(39, 10)
		deposited film*	3.7	(39 ,10)
		fiber, heat treated**	120.0	(10, 14)
Tensile strength	MPa	BBL, cast film	114.0	(39, 10)
		deposited film*	66.0	(39, 10)
		fiber, heat treated**	830.0	(10, 14)
Compressive strength	MPa	BBL fiber, heat treated**	410.0	(10, 14)
Rupture elongation	%	BBL	2.9	(1)

* Polymer precipitate from dilute MSA (methanesulfonic acid) solutions deposited on to a sintered glass filter to aggregate into a thin film under vacuum.
** Annealed at 300°C.

Unit cell dimensions

Lattice	Cell dimensions (')	Cell angles (°)	Method	Reference
Orthorhombic	$a = 7.87$; $b = 3.37$; $c = 11.97$	$\alpha = \beta = \gamma = 90$	X-ray diffraction, fiber	(9)

Electrical conductivity of uniaxial BBL films as a function of incorporated MSA (methanesulfonic acid) content (Reference 3)

MSA content (%)	MSA/BBL molar ratio	$\sigma \parallel$ (longitudinal) (S/cm)	$\sigma_\perp$ (transverse) (S/cm)	$\sigma \parallel / \sigma_\perp$
0.11	0.004	1.8×10^{-8}	6.8×10^{-10}	26
3.09	0.11	5.7×10^{-8}	1.3×10^{-9}	44
6.15	0.23	1.5×10^{-6}	4.4×10^{-7}	3.4
11.0	0.43	3.0×10^{-4}	6.3×10^{-5}	4.8
19.4	0.84	2.0×10^{-2}	1.4×10^{-3}	16
23.9	1.10	1.1×10^{-2}	6.8×10^{-4}	16
33.0	1.71	3.4×10^{-2}	1.1×10^{-2}	2.7
51.0	3.62	7.6×10^{-2}	1.6×10^{-2}	4.8

Room temperature electrical conductivity of chemically[40] and electrochemically[41] doped BBL ladder polymer (compilation from Reference 10)

Dopant	σ_{RT} (S/cm)
None	10^{-9}–10^{-12}
I_2, Br_2	10^{-10}
BF_3	10^{-7}
AsF_5	10^{-3}
SO_3	10^{-2}
H_2SO_4	2.1
K (metal)	2×10^{-1}
K^+	3.2×10^{-2}–1.1

Benzimidazobenzophenanthroline-type Ladder Polymer

Dopant	σ_{RT} (S/cm)
ClO_4^- ($AgClO_4$)	3.9×10^{-5}
(Li^+, Na^+, K^+, Et_4N^+, $n\text{-}Bu_4N^+$)*	10^{-2}–10^{-1}
(Li^+, Na^+, K^+, Et_4N^+, $n\text{-}Bu_4N^+$)*	5–20**
(BF_4^-, ClO_4^-, HSO_4^-)*	10^{-2}–10^{-1}

* Electrochemically doped, ** Oriented BBL.

Property	Units	Conditions	Value	Reference
IR (wavenumbers) (intensity)	cm^{-1}	Cast BBL film 1800–600 cm^{-1} region	1745 (vw); 1703 (vs); 1647, 1637 (vw); 1627, 1609 (vw); 1580 (w); 1561 (m); 1533 (vw); 1500 (s); 1458 (m); 1413 (s); 1370 (vs); 1320 (vs); 1305 (vs); 1237 (vs); 1223 (s); 1189 (vw); 1171 (s); 1152 (w); 1127 (m); 1104 (w); 1080 (m); 1025 (w); 995 (vs); 919 (m); 884 (m): 866 (vs); 843, 831 (m); 774 (m); 759 (s); 733 (m); 719 (w); 668 (w)	(19)
Electronic absorption	wavelength (nm)	BBL solution (MSA) BBL	$\lambda_{max} = 544$	(11)
		nitromethane/$FeCl_3$	546	(11)
		nitromethane/$AlCl_3$	546	(11)
		nitromethane/$SbCl_3$	550	(11)
		nitrobenzene/$AlCl_3$	557	(11)
		nitrobenzene/$SbCl_5$	557	(11)
		thin film (cast from nitromethane/$AlCl_3$)	560	(11)
		BBB film (cast from nitromethane/$AlCl_3$)	540	(11)
Transition energy	eV	BBL, in methanesulfonic acid	2.28	(11, 42)
		in organic solvent/Lewis acid	2.27–2.23	(11, 42)
		BBL molecular analog, ab initio Hartree–Fock calculation	2.40	(42)
Energy band gap	eV	BBL, optical absorption edge at $\sim$ 700 nm (thin film cast from nitromethane/$AlCl_3$)	1.77	(11)
		BBB, optical absorption edge at $\sim$ 700 nm (thin film cast from nitromethane/$AlCl_3$)	1.77	(11)

Property	Units	Conditions	Value	Reference
		BBB/BBL	~ 2.5	(8)
		BBL	2–2.2	(41)
		AsF_5-doped BBL	1.5	(41)
Spin density	spins/mol	BBB films: heated below 980 K (EPR data)	$\sim 10^{22}$	(8)
		BBB films: heated above 980 K	$<10^{20}$	(8)
Mean distance between conductive structures	Å	BBB film, X-ray diffraction	8.2	(8)
Next-neighbor distance within the polymeric features	Å	BBB film, X-ray diffraction	3.2	(8)
Binding energies (XPS Mg Kα spectrum of the nitrogen emission)	eV	BBB film annealed to 980 K	398.7 400.8	(8)
Diameter of the helical chain	nm	BBB molecular modeling	4–10 nm	(8)
d-spacing of the first equatorial Bragg reflection (side-by-side packing)	Å	BBL film	7.5	(41)
d-spacing of the second equatorial Bragg reflection (face-to-face packing)	Å	BBL film	3.3	(41)
Volume occupied by a single monomer unit	nm^3	BBL (calculated)	0.144 ± 0.003	(20)
		BBB (calculated)	0.165 ± 0.003	(20)
Intermolecular interaction energy two chains packed face-to-face	kJ/mol	BBL (calculated)	156.5	(36)
		BBB (calculated)	188.7	(36)
Electron affinity (EA)	eV	BBL, solid-state, estimated from measured redox potentials of BBL	~ 4.0	(24)
		BBL film, cyclic voltammetry	4.2	(43)
		BBB film, cyclic voltammetry	4.5	(43)
Ionization potential (IP)	eV	BBL, solid-state, estimated from measured redox potentials of BBL	~ 5.9	(24)
Third-order nonlinear optical susceptibility (χ^3)	esu	Off-resonant ($\lambda = 1.064$ µm) (degenerate four-wave mixing technique), BBL film	1.5×10^{-11}	(25)
		BBL film at 1.695 µm	6.4×10^{-11}	(44)
		BBB film at 1.064 µm	5.5×10^{-12}	(25)
		BBB film at 1.695 µm	3.15×10^{-11}	(44)
		BBL film (electrochemically doped) measured at 1.064 µm	2.0×10^{-11}	(25)
Electrical conductivity	$ohm^{-1}cm^{-1}$	2–20 µm film BBL	10^{-12}	(25)
		4 µm BBL film	10^{-14}	(34)

Benzimidazobenzophenanthroline-type Ladder Polymer

Property	Units	Conditions	Value	Reference
Conductivity	S/cm	Pristine BBL	10^{-14}	(15, 8)
		and BBB	10^{-12}	(41)
		BBL, at room temperature	3×10^{-10}	(42)
		BBB film, Ar bombarded	$\geq 10^2$	(8)
		BBB film, heated to 500 K	10^{-9}	(8)
		BBB film, heated to 800 K	10^{-6}	(8)
		BBB film, heated to > 950 K	50	(8)
Conductivity (DC, at RT)	S/cm	BBL fibers (heat-treated, 350°C)	3.0×10^{-4}	(14)
		Heat-treated, 600°C	6×10^{-7}	(14)
Conductivity (DC, 600 K)	S/cm	Cast BBL film ($\sigma_{\parallel}$)	$\sim 10^{-3}$	(16)
		Cast BBL film ($\sigma_{\perp}$)	$\sim 10^{-5}$	(16)
		Oriented BBL fiber	~ 0.1	(16)
Conductivity	S/cm	Deposited BBL film (Krypton ion-implantation, 190 keV ion energy, beam current density 0.12 μAmp/cm^2, dose 4×10^{16} ions/cm^2)	$\sim 10^{-3}$	(45)
		Deposited BBL film (Krypton ion-implantation, 200 keV ion energy, beam current density 2.0 μAmp/cm^2, dose 4×10^{16} ions/cm^2)	$\sim 10^2$	(46)
Conductivity	S/cm	BBL film (Argon ion-implantation, 200 keV ion energy, beam current density 2.0 μAmp/cm^2, dose 4×10^{16} ions/cm^2)	224	(47)
		BBL film (Krypton ion-implantation, same bombardment conditions)	136	(47)
		BBL film (Boron ion-implantation, same bombardment conditions)	56	(47)
Conductivity (300 K)	S/cm	Deposited BBL film (Krypton ion-implantation, bombardment conditions same as above)	~ 185	(48)
Conductivity (dark, DC)	S/cm	BBL thin film, cast from AlCl$_3$/nitromethane, (under vacuum, 0 % relative humidity)	4×10^{-14}	(49)
		(room ambient conditions, 50 % relative humidity)	5×10^{-11}	(49)
Conductivity (DC, RT)	S/cm	BBL films, complexed with poly(styrenesulfonic acid), $\geq$ 70 mol % singly protonated	2.0	(43)
Dielectric constant at 10^5 Hz	—	Sandwich device Au/BBL/Au (frequency range 10 Hz–1 MHz), 0 % relative humidity	3.0	(49)
		10 % relative humidity	3.75	(49)
		50 % relative humidity	5.2	(49)
Electrochemical Volts reduction potential (n-doping) (cyclic voltammetry)		BBL film on Ge electrode, (SCE reference electrode) First reduction wave	≈ -0.6	(19)

Property	Units	Conditions	Value	Reference
		Second reduction wave	≈ -0.9	(18, 19)
		Third reduction wave	≈ -1.1	(18, 19)
		Fourth reduction wave	≈ -1.2	(18, 19)
		BBL film on Pt electrode (SCE reference electrode) First reduction wave	≈ -0.52	(18, 19)
		Second reduction wave	≈ -0.82	(18, 19)
		Third reduction wave	≈ -1.0	(18, 19)
		Fourth reduction wave	≈ -1.1	(18, 19)
IR bands (spectroelectro-chemistry)	cm^{-1}	BBL film, first reduction in situ ATR-FTIR difference spectra	> 2000 (br); 1649, 1522 (s); 1349 (m); 1255 (s); 1150, 1066 (s)	(19)
		Second reduction	1614, 1390 (m); 1278, 1219 (s); 1099, 1028 (s)	(19)
		Third reduction	> 2000 (br); 1509, 1466 (m); 1369 (vs); 1312 (w); 1240, 1168 (m); 1109, 1070 (s); 1002, 968 (s); 896 (s); 801 (m); 771 (w); 732 (m); 654 (m)	(19)
		Fourth reduction	1593, 1379 (m); 1363 (s); 1345 (m); 1261 (w)	(19)
Raman characteristic frequencies	cm^{-1}	Neutral BBL film ($\lambda_{excitation} = 514$ nm)	1597, 1539, 1389, 1282	(19)

References

1. Arnold FE, Van Deusen RL. *Macromolecules.* 1969;2(5):497.
2. Wolfe JF. *Encyclopedia of Polymer Science and Technology.* Vol 11. New York: Wiley Interscience. 1985; 601.
3. Wang CS, Burkett J, Arnold FE. *Polymer Preprints* (ACS). 1994;35(1):321.
4. Van Deusen RL. *J. Polym. Sci., Polym. Lett.* 1966;4:211.
5. Van Deusen RL, Goins OK, Sicree AJ. *J. Polym. Sci. A-1.* 1968;6:1777.
6. Zhou W, Lu F. *J. Appl. Polym. Sci.* 1995;58(9):1561.

7. Berry GC. *Discussions of the Faraday Society.* 1970;49:121.
8. Munz AW, Schmeisser D, Gopel W. *Chem. Mater.* 1994;6;2288.
9. Song HH, Fratini AV, Chabynic M, Price GE, Agrawal AK, Wang CS, Burkett J, Dudis DS, Arnold FE. *Synthetic Metals.* 1995;69:533.
10. Wang CS. *Trends in Polymer Science.* 1993;1(7):199.
11. Jenekhe SA, Johnson PO. *Macromolecules.* 1990;23:4419.
12. Jenekhe SA, Peterson JR. *United States Patents.* 5,114,610 (1992); 4,945,156 (1990).
13. Wang CS. *Trends in Polymer Science.* 1997;5(5):138.
14. Wang CS, Lee CY-C, Arnold FE. *Mater. Res. Soc. Symp. Proc.* 1992;247:747.
15. Agrawal AK, Wang CS, Song HH. *Mater. Res. Soc. Symp. Proc.* 1994;328:279.
16. Narayan KS, Alagiriswamy AA, Spry RJ. *Physical Review B.* 1999;59(15):10 054.
17. Wilbourn K, Murray RW. *Macromolecules.* 1988;21:89.
18. Yohannes T, Neugebauer H, Jenekhe SA, Sariciftci NS. *Synthetic Metals.* 2001;116:241.
19. Yohannes T, Neugebauer H, Luzzati S, Catellani M, Jenekhe SA, Sariciftci NS. *J. Phys. Chem. B.* 2000;104:9430.
20. Quinto M, Jenekhe SA, Bard AJ. *Chem. Mater.* 2001;13:2824.
21. Narayan KS, Taylor BE, Spry RJ, Ferguson JB. *Journal of Luminescence.* 1994;60&61:482.
22. Narayan KS, Taylor BE, Spry RJ, Ferguson JB. *J. Appl. Phys.* 1995;77(8):3938.
23. Antoniadis H, Abkowitz MA, Osaheni JA, Jenekhe SA, Stolka M. *Synthetic Metals.* 1993;60:149.
24. Jenekhe SA, De Paor LR, Chen XL, Tarkka RM. *Chem. Mater.* 1996;8(10):2401.
25. Lindle JR, Bartoli FJ, Hoffman CA, Kim O-K, Lee YS, Shirk JS, Kafafi ZH. *Appl. Phys. Lett.* 1990;56(8):712.
26. Stolka M, Abkowitz MA, Ong BS, Jenekhe SA. *US Patent.* 1993;5,248,580.
27. Osaheni JA, Jenekhe SA, Peristein J. *J. Phys. Chem.* 1994;98:12727.
28. Manoj AG, Alagiriswamy AA, Narayan KS. *J. Appl. Phys.* 2003;94(6):4088.
29. Babel A, Jenekhe SA. *Adv. Mater.* 2002;14(5):371.
30. Narayan KS, Manoj AG, Singh ThB, Alagiriswamy AA. *Thin Solid Films.* 2002;417:75.
31. Manoj AG, Narayan KS. *Optical Materials.* 2002;21:417.
32. Gowrishankar V, Luscombe CK, McGehee MD, Frechet JMJ. *Solar Energy Materials & Solar Cells.* 2007;91:807.
33. Kumar S, Arnold FE, Dang TD. *United States Patent.* 2005:6,900,264.
34. Abkowitz MA, Antoniadis H, Jenekhe SA, Stolka M. *United States Patent.* 1994;5,373,738.
35. Janietz S, Sainova D. *Macromol. Rapid Commun.* 2006;27:943.
36. Jenekhe SA, Roberts MF. *Macromolecules.* 1993;26:4981.
37. Cassidy PE. *Thermally Stable Polymers.* New York: Marcel Dekker; 1980.
38. Berry GC. *J. Polym. Sci., Polym. Phys. Ed.* 1976;14:451.
39. Arnold FE, Van Deusen RL. *J. Appl. Poly. Sci.* 1971;15:2035.
40. Kim O-K. *United States Patent.* 1986;4,620,942.
41. Hong SY, Kertesz M, Lee YS, Kim O-K. *Macromolecules.* 1992;25:5424.
42. Yeates AT, Dudis DS, Connolly JW. *Synthetic Metals.* 2001; 116:289.
43. Alam MM, Jenekhe SA. *J. Phys. Chem. B.* 2002;106:11172.
44. Jenekhe SA, Roberts M, Agrawal AK, Meth JS, Vanherzeele H. *Mater. Res. Soc. Symp. Proc., (Opt. Electr. Prop. Polym.).* NY, USA: Rochester; 1991; Vol 214. 55.
45. Wang CS, Burkett J, Lee CY-C, Arnold FE. *Polym. Mater. Sci. Eng.* 1991;64:171.

46. Wang CS, Lee CY-C, Arnold FE. *Polym. Mater. Sci. Eng.* 1992;66:291.

47. Jenekhe, S. A. and Tibbetts, S. J. *J. Poly. Sci., Part B: Poly. Phys.,* 1988;26:201.

48. Du G, Prigodin VN, Burns A, Joo J, Wang CS, Epstein AJ. *Physical Review B.* 1998;58(8):4485.

49. Antoniadis H, Abkowitz MA, Osaheni JA, Jenekhe SA, Stolka M. *Chem. Mater.* 1994;6:63.

Bisphenol-A Polysulfone

Sizhu Wu and Tarek M. Madkour

Acronym, Trade Names PSF, Udel P1700, and P3500 (Amoco)

Class Poly(ether sulfones)

Synthesis Polycondensation

Structure

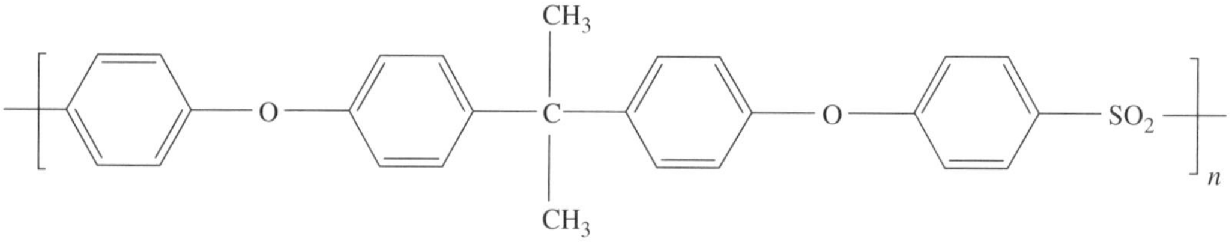

Major Applications Medical and household appliances that are sterilizable by hot air and steam such as corrosion-resistant piping and injection molded engineering parts. Also used in electric and electronic applications and as membranes for reverse gas streams and gas separation.

Properties of Special Interest High-performance thermoplastic of relatively low flammability. Amorphous, high-creep resistance, and stable electrical properties over wide temperature and frequency ranges. Transparent with good thermal and hydrolytic resistance. High alkaline stability.

Property	Units	Conditions	Value	Reference
Number average Molecular weights	–		4000~26 000	(1, 2)
Molecular weight distribution	–	–	$M_w/M_n = 3.15$	(3)
Infrared bands (frequency)	cm^{-1}	Group assignments		(4, 5)
		SO$_2$ scissors deformation	560	
		Aromatic ring bend	690	
		Para out-of-plane aromatic CH wag	834	
		Para in-plane aromatic CH bend	1014, 1105	
		SO$_2$ symmetric stretch	1151, 1175	
		Aryl—O-aryl C—O stretch	1244	
		SO$_2$ asymmetric stretch	1294, 1325	

Bisphenol-A Polysulfone

Property	Units	Conditions	Value	Reference
Infrared bands (frequency)	cm^{-1}	Group assignments		(4, 5)
		CH_3 symmetric (umbrella) deformation	1365	
		Para aromatic ring semicircle stretch	1410, 1490, 1505	
		Para aromatic ring quadrant stretch	1585	
		CH_3 symmetric stretch	2875	
		CH_3 asymmetric stretch	2970	
		Aromatic CH stretches	3000–3200	
Thermal expansion coefficient	K^{-1}	1 atm and 20°C	2.1×10^4	(6)
Isothermal compressibility	bar^{-1}	20°C	2.2×10^5	(6)
Density	$g\,cm^{-3}$	–	1.24	(7)
Solubility parameter	$(MPa)^{1/2}$	–	20.26	(8)
Glass transition temperature	K	Forced oscillation dynamic – mechanical analysis	459	(9)
Sub-Tg transition temperature	K	β-relaxation temperature	358	(9)
		γ-relaxation temperature	193	
Heat deflection temperature	K	(1.82 MPa)	447	(10)

Mechanical properties[7,10,11]

Property	Units	Resin		
		Neat Resin	30% Glass Fiber Reinforced	30% Carbon Fiber Reinforced
Tensile modulus	MPa	2482	–	–
Tensile strength	MPa	69.0	120	190
Maximum extensibility $(L/L_0)_r$	%	3.0	2.31	1.02
Flexural modulus	MPa	2758	6747	13 069
Flexural strength	MPa	103	208	244
Notched Izod impact strength	$J\,m^{-1}$	80.4	400	118
Unnotched Izod impact strength	$J\,m^{-1}$	–	1049	456
Hardness	Shore D	69	85	87

Property	Units	Conditions	Value	Reference
WLF parameters: C_1 and C_2	–	–	$C_1 = 15.1$	(12)
	–	–	$C_2 = 49.0$	
Dielectric strength	$MV\,m^{-2}$	–	14.6	(7)

Bisphenol-A Polysulfone

Property	Units	Conditions	Value			Reference
Resistivity	Ohm cm	–	580			(13)
Thermal conductivity	$W\,m^{-1}K^{-1}$	–	0.26			(14)
Melt index	$g\,(10\,min)^{-1}$	–	6.5			(7)
Water absorption	%	24 h	0.3			(4)
Intrinsic viscosity	$cm^3\,g^{-1}$	25°C in chloroform	End group			(15)
Viscosity	Pa s	25°C in N,N-dimethylacetamide/ acetone	0.02~0.73			(2)
			NH_2	Cl	t-Butyl	
		Mol. Wt. = 5720	0.16	0.16	–	
		Mol. Wt. = 9934	0.23	–	0.24	
		Mol. Wt. = 17 500	0.29	–	0.30	
		Mol. Wt. = 21 230	0.34	–	0.33	
Gas permeability		P_{CO_2} (barrier)	5.6			(16)
		P_{He} (barrier)	1.3~3.1			(16,17)

References

1. Ibieta JB, Kalika DS, Penn LS. *J. Polym. Sci.* 1998;[A1] 36:1309–1316.
2. Yuan XY, Zhang YY, et al. *Polym. Int.* 2004;53:1704–1710.
3. Ohtani HJ, Ishida YY J. *Anal. Appl. Pyrolysis.* 2001;61:35–44.
4. Colthup N, Daly L, Wiberley S. *Introduction to Infrared and Raman Spectroscopy.* 2nd ed. New York: Academic Press; 1975.
5. Pouchert C. *The Aldrich Library of FT-IR Spectra.* Milwaukee: Aldrich Chemical; 1985.
6. Zoller PJ. *Polym. Sci., Polym. Phys. Ed.* 1978;16:1261.
7. Elias H, Vohwinkel F. *New Commercial Polymers 2.* New York: Gordon and Breach Science Publishers; 1986, Chapter 8.
8. Matsuura T, Blais P, Sourirajan SJ. *Appl. Polym. Sci.* 1976;20:1515.
9. Aitken C, McHattie J, Paul D. *Macromolecules.* 1992;10:2910.
10. Ma C. *In Proc. of the Natl. SAMPE Symp. Exhib.* 30 (Adv. Technol. Mater. Processes), 1985, p. 543.
11. Hisue E, Miller R. In Proc. of the Natl. SAMPE Symp. Exhib., 30 (Adv. Technol. Mater. Processes), 1985, p. 1035.
12. Hwang E, Inoue T, Osaki K. *Polym. Eng. Sci.* 1994;34:135.
13. Brandrup J, Immergut EH, eds. *Polymer Handbook.* 3rd ed. New York: John Wiley and Sons; 1989.
14. Mark H, et al. eds. *Kirk-Othmer: Encyclopedia of Chemical Technology.* 3rd ed. New York: Wiley-Interscience; 1984.
15. Yoon T, et al. *Macromol. Symp.* 98 (35th IUPAC International Symposium on Macromolecules), 1995, p. 673.
16. Wang XY, Veld PJ, et al. *Polymer.* 2005;46:9155.
17. Hu CC, Chang CS, et al. *J. Memb. Sci.* 2003;226:51–61.
18. Robertson JE, Ward TC. *Polymer.* 2000;41:6251–6262.

Carbon Nanotube-containing Polymers

Warren T. Ford and Maxim N. Tchoul

Trade Names NanoSolve (Zyvex), Nano In (Nanoledge), RTP compounds (RTP Co.), Stat-Kon (General Electric)

Class Polymer composites

Structures Single-walled carbon nanotubes (SWNT), mixture of 0.7–2 nm diameters and of chiralities of tubular graphene prepared catalytically by electric arc deposition (arc), pulsed laser vaporization (PLV), high pressure carbon monoxide chemical vapor deposition (HiPco), supported Co–Mo catalyst chemical vapor deposition (CoMoCat), multi-walled carbon nanotubes (MWNT), concentric graphene tubes 10–50 nm diameter prepared by chemical vapor deposition

Major Applications Composite reinforcement, electronics, electrostatic dissipation, photovoltaic materials, photoemissive materials

Properties of Special Interest Electrical conductivity, thermal conductivity, mechanical strength and flexibility, electronic absorption and emission

CNT Type	CNT Wt %	Fabrication Method	Analyses	Tensile Modulus, Strength (GPa)	Other Properties	Reference
Polyethylene SWNT (PLV)	1–30	Hot coagulation, compression molding, drawn fiber	Crystallinity	—	Electrical conductivity, thermal conductivity	1,2
Polypropylene MWNT	0.25–15	Solution, extrusion	Optical microscopy	—	Shear flow, electrical conductivity	3,4
Polyisobutylene MWNT	0.025–10	Solution	—	—	Shear flow, dielectric constant, electrical conductivity	4
Polystyrene MWNT	1	Solution	TEM	1.6, 0.016	—	5
Polyacrylonitrile MWNT	3–20	Electrospinning	TEM, AFM, WAXD, TGA, TMA	6.4–14.5, 0.285–0.37	Electrical conductivity, thermal expansion coefficient	6
Polycyanoacrylate resin MWNT on silica sheet		Bulk polymerization	SEM	—	Complex modulus, fatigue	7,8

CNT Type	CNT Wt %	Fabrication Method	Analyses	Tensile Modulus, Strength (GPa)	Other Properties	Reference
Poly(methyl methacrylate)						
SWNT (HiPco)	0.1–7	Melt spinning compression molding	TGA, SEM, TEM Raman, SAXS, WAXS, UV-vis-NIR, AFM, optical microscopy	5–8, –	Electrical conductivity, thermal conductivity, flammability, melt rheology	9–12
Poly(vinyl alcohol)						
SWNT (HiPco)	60	Solution drawn fiber	—	–, 1.8	—	13
SWNT MWNT	50	Hot drawn fiber	XRD	35–45, 1.4–1.8	Electrical conductivity	14,15
SWNT (HiPco)	1–5	Solution cast film	Vis-NIR, Raman	4.0, 0.148	—	16
Poly (N-vinylcarbazole)						
SWNT (HiPco)	5	Grafting to	Vis-NIR, Raman, TGA	—	Optical limiting	17
Polycarbonate						
SWNT (HiPco)	1–2	Solution	SEM	—	Loss modulus	18
Nylon-6						
SWNT (arc, oxidized)	0.1–1.5	in situ polymerization, fiber	IR, Raman, DTA, TGA, AFM, SEM, TEM	0.58–1.5, 0.07–0.11		19,20
Nylon-6,6						
SWNT (HiPco, oxidized)	1.6–5.5	Interfacial polymerization, films, fibers	DSC, optical microscopy	1.5–3, –	Intrinsic viscosity electrical conductivity	21
Nylon-6,10						
SWNT (HiPco functionalized)	0.05–1.0	in situ interfacial polymerization, fibers, films	IR, Raman, XPS, SEM, WAXS, SAXS	1.2–1.3, 0.08–0.18	Intrinsic viscosity	22
Polyetherimide						
MWNT (oxidized)	0.14–0.38	Solution	UV-vis-NIR, IR, TEM, TGA	3.7–4.4, 0.13–0.19	—	23
Poly(ethylene glycol)						
SWNT (arc, oxidized)	71	Grafted	UV-vis-NIR, IR, TGA, AFM, SEM	—	Zeta potential	24
SWNT (HiPco)	0.05–0.5	Surfactant solutions	Vis-NIR, Raman, DSC, DMA	—	Viscoelastic properties, electrical conductivity, T_m	25
Polyimide						
SWNT (arc)	—	Solution	Vis-NIR, Raman, SEM, TEM	—	Luminescence	26
SWNT (oxidized PLV)	0.02–1.0	in situ polymerization	Vis-NIR, Raman, SEM, TEM, TGA, optical microscopy	4.5–7.2, 0.12	Electrical conductivity	27
Poly(p-phenylene ethynylene)						
SWNT (HiPco)	60	Solution	UV-vis, fluorescence, ^{1}H NMR	—	—	28

CNT Type	CNT Wt %	Fabrication Method	Analyses	Tensile Modulus, Strength (GPa)	Other Properties	Reference
Polyurethanes						
Morthane® elastomer						
MWNT	1–25	Solution	SEM, WAXS, DSC	0.02–0.16, –	Electrical conductivity	29
Poly[4,4′-methylenebis(cyclohexyl isocyanate)-*alt*-tetramethylene oxide]						
SWNT (CoMoCat)	0.1–2.5	Solution	Raman, UV-vis, DMA, DSC, WAXS, SAXS	0.15–0.3, –	Electrical conductivity dynamic modulus over −100 to + 150°C, T_g	30
Epoxy resins						
SWNT (HiPco)	0.01–0.05	High shear	IR, SEM, TGA, optical microscopy	—	Flexural modulus 1.15–1.25 GPa	31
SWNT (HiPco)	20	Infiltration into SWNT paper	SEM, optical microscopy	—	Thermal conductivity	11
MWNT	5	Solution	SEM	3.7, –	Compression modulus 4.5 GPa	32
CNT yarns						
MWNT	100	Drawn fiber	SEM	1.2, 0.5–0.7	—	33
SWNT (HiPco)	100	Drawn from fuming H_2SO_4	SEM, WAXS	120, 0.116	Electrical conductivity, thermal conductivity	34
CNT sheets						
MWNT	100	Drawn aerogel, densified	SEM	—	Electrical conductivity, optical transmittance	35

References

1. Haggenmueller R, Fischer JE, Winey KI. *Macromolecules*. 2006;39:2964–2971.
2. Haggenmueller R, Guthy C, Lukes JR, Fischer JE, Winey KI. *Macromolecules*. 2007;40:2417–2421.
3. Kharchenko SB, Douglas JF, Obrzut J, Grulke EA, Migler KB. *Nature Mater*. 2004;3:564–568.
4. Hobbie EK, Obrzut J, Kharchenko SB, Grulke EA. *J. Chem. Phys*. 2006;125:044712/044711-044712/044713.
5. Qian D, Dickey EC, Andrews R, Rantell T. *Appl. Phys. Lett*. 2000;76:2868–2870.
6. Ge JJ, Hou H, Li Q, et al. *J. Am. Chem. Soc*. 2004;126:15754–15761.
7. Suhr J. Koratkar N, Keblinski P, Ajayan P. *Nature Mater*. 2005;4:134–137.
8. Suhr J, Victor P, Ci L, et al. *Nature Nanotech*. 2007;2:417–421.
9. Kashiwagi T, Fagan J, Douglas JF, et al. *Polymer*. 2007;48:4855–4866.
10. Du F, Fischer JE, Winey KI. *J. Polym. Sci. Part B: Polym. Phys*. 2003;41:3333–3338.
11. Du F, Guthy C, Kashiwagi T, Fischer JE, Winey KI. *J. Polym. Sci. Part B: Polym. Phys*. 2006;44:1513–1519.
12. Du F, Scogna RC, Zhou W, Brand S, Fischer JE, Winey KI. *Macromolecules*. 2004;37:9048–9055.
13. Dalton AB, Collins S, Munoz E, et al. *Nature*. 2003;423:703.
14. Miaudet P, Badaire S, Maugey M, et al. *Nano Letters*. 2005;5:2212–2215.
15. Miaudet P, Bartholome C, Derre A, et al. *Polymer*. 2007;48:4068–4074.
16. Zhang X, Liu T, Sreekumar TV, et al. *Nano Letters*. 2003;3:1285–1288.

17. Wu W, Zhang S, Li Y, et al. *Macromolecules*. 2003;36:6286–6288.
18. Koratkar NA, Suhr J, Joshi A, et al. *Appl. Phys. Lett.* 2005;87:063102/063101-063102/063103.
19. Gao J, Itkis ME, Yu A, Bekyarova E, Zhao B, Haddon RC. *J. Am. Chem. Soc.* 2005;127:3847–3854.
20. Gao J, Zhao B, Itkis ME, et al. *J. Am. Chem. Soc.* 2006;128:7492–7496.
21. Haggenmueller R, Du F, Fischer JE, Winey KI. *Polymer*. 2006;47:2381–2388.
22. Moniruzzaman M, Chattopadhay J, Billups WE, Winey KI. *Nano Letters*. 2007;7:1178–1185.
23. Ge JJ, Zhang D, Li Q, et al. *J. Am. Chem. Soc.* 2005;127:9984–9985.
24. Zhao B, Hu H, Yu A, Perea D, Haddon RC. *J. Am. Chem. Soc.* 2005;127:8197–8203.
25. Chatterjee T, Yurekli K, Hadjiev VG, Krishnamoorti R. *Adv. Funct. Mater.* 2005;15:1832–1838.
26. Zhou B, Lin Y, Hill DE, et al. *Polymer*. 2006;47:5323–5329.
27. Yu A, Hu H, Bekyarova E, et al. *Composites Sci. Tech.* 2006;66:1190–1197.
28. Chen J, Liu H, Weimer WA, Halls MD, Waldeck DH, Walker GC. *J. Am. Chem. Soc.* 2002;124:9034–9035.
29. Koerner H, Liu W, Alexander M, Mirau P, Dowty H, Vaia RA. *Polymer*. 2005;46:4405–4420.
30. Buffa F, Abraham GA, Grady BP, Resasco D. *Journal of Polymer Science, Part B: Polymer Physics*. 2007;45:490–501.
31. Moniruzzaman M, Du F, Romero N, Winey KI. *Polymer*. 2006;47:293–298.
32. Schadler LS, Giannaris SC, Ajayan PM. *Appl. Phys. Lett.* 1998;73:3842–3844.
33. Atkinson KR, Hawkins SC, Huynh, C, et al. *Physica B: Condens. Mat.* 2007;394:339–343.
34. Ericson LM, Fan H, Peng H, et al. *Science*. 2004;305:1447–1450.
35. Zhang M, Fang S, Zakhidov AA, et al. *Science*. 2005;309:1215–1219.

Carborane-containing Polymers

Edward N. Peters

General Comments The term 'carborane' is commonly used in a generic sense to describe cage compounds composed of boron, hydrogen, and carbon whose molecular geometries are polyhedra or polyhedral fragments. The polyhedral $C_2B_nH_{n+2}$ species (where $n = 5$ and 10) have been used in the preparation of carborane–siloxane polymers. The incorporation of the carborane moiety into the siloxane backbone significantly enhances the overall thermal stability. Properties of polymers prepared from dodecahedral and heptahedral will be summarized in this section.

DODECACARBORANE–SILOXANE

Acronyms, Trade Names 10-SiB, Dexsil (Olin Corp.)

Class Cage structure polymers, D_n-Dodecacarborane–siloxane

Structure $[-Si(CH_3)_2CB_{10}H_{10}CSi(CH_3)_2-O-\{Si(CH_3)_2-O-\}_{n-1}-]$
where $CB_{10}H_{10}C =$

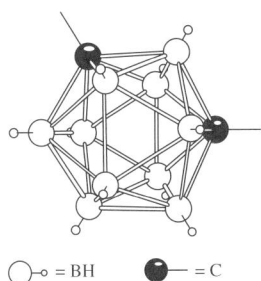

$\bigcirc\!\!-\!\circ = BH$ $\bullet\!\!-\!\!- = C$

Major Applications Liquid phase in gas chromatography. High temperature elastomer used to formulate gaskets, O-rings, and wire coatings. Can be fabricated like conventional silicones.

Properties of Special Interest Elastomeric. Very high thermal stability. Outstanding flame resistance.

Carborane-containing Polymers

IR (characteristic frequencies)*[1]

Assignment n	C—H	B—H	CH$_3$	Si—O	Si—C
1	2963, 2904	2600	1410, 1262	1090	800
3	2960, 2900	2595	1411, 1263	1090, 1048	801
4	2963, 2904	2600	1411, 1260	1095, 1047	800
5	2963, 2904	2594	1410, 1263	1065, 1030	800

* Wave numbers (cm^{-1}) for [$-$Si(CH$_3$)$_2$CB$_{10}$H$_{10}$CSi(CH$_3$)$_2$–O–{Si(CH$_3$)$_2$–O–}$_{n-1}$–]

Property	Units	Conditions	Value	Reference
NMR	^{1}H-NMR in CDCl$_3$ at 60 MHz	–		(2)
	^{1}H-NMR in CDCl$_3$ at 100 MHz	–		(1)
Solvents	–	25°C	Diethyl ether	(2)
	–	25°C	Chlorobenzene	(3)
Nonsolvents	–	25°C	Methanol	(3)
Mark Houwink constants: K and a	K = ml/g a = None	–	$K = 1.02 \times 10^{-4}$ $a = 0.72$	

Property	Units	Conditions*				Value	Reference
Density	g cm^{-1}	R$_1$	R$_2$	R$_3$	n		
		CH$_3$	CH$_3$	CH$_3$ (67) Phenyl (33)	2	1.074	(4)
		CH$_3$	Phenyl (33) CH$_3$ (67)	Phenyl (33) CH$_3$ (67)	2	1.123	(4)
Glass transition temperature	K	R$_1$	R$_2$	R$_3$	n		
		CH$_3$	CH$_3$	–	1	298	(5)
		CH$_3$	CH$_3$	CH$_3$	2	243	(5)
						223	(5)
		CH$_3$	CH$_3$	CH$_3$	3	205	(6)
		CH$_3$	CH$_3$	CH$_3$	4	203	(6)
		CH$_3$	CH$_3$	CH$_3$	5	185	(6)
		CH$_3$	CH$_3$	Phenyl	2	261	(7)
		CH$_3$	CH$_3$	CH$_3$ (33) Phenyl (67)	2	251	(7)
		CH$_3$	CH$_3$	CH$_3$ (67) Phenyl (33)	2	236	(7)
		CH$_3$	Phenyl	Phenyl	2	295	(7)

Property	Units	Conditions[*]				Value	Reference
Glass transition temperature	K	R_1	R_2	R_3	n		
		CH_3	Phenyl (33) CH_3 (67)	Phenyl (33) CH_3 (67)	2	248	(7)
		CH_3	Phenyl (24) CH_3 (76)	Phenyl (24) CH_3 (76)	2	240	(7)
		$CH_2CH_2CF_3$	$CH_2CH_2CF_3$	–	1	301	(8)
		CH_3	CH_3	$CH_2CH_2CF_3$	2	244	(9)
		$CH_2CH_2CF_3$	$CH_2CH_2CF_3$	CH_3	2	261	(9)
		$CH_2CH_2CF_3$	$CH_2CH_2CF_3$	$CH_2CH_2CF_3$	2	270	(9)
		$CH_2CH_2CF_3$	$CH_2CH_2CF_3$	$CH_2CH_2CF_3$	3	270	(8)
Melting temperature	K	R_1	R_2	R_3	n		
		CH_3	CH_3	–	1	513	(5)
		CH_3	CH_3	CH_3	2	339	(5)
						341, 363	(5)
		CH_3	CH_3	CH_3	3	313	(6)
Tensile modulus[†]	MPa	R_1	R_2	R_3	n		
		CH_3	CH_3	CH_3 (33) Phenyl (67)	2	3.45	(10)
		CH_3	Phenyl (33)	Phenyl (33)	2	2.97	(11)
Tensile strength[†]	MPa	CH_3	CH_3	CH_3 (33) Phenyl (67)	2	3.58	(10)
		CH_3	Phenyl (33) CH_3 (67)	Phenyl (33) CH_3 (67)	2	5.10	(11)
Maximum extensibility[†]	%	CH_3	CH_3	CH_3 (33) Phenyl (67)	2	130	(10)
		CH_3	Phenyl (33) CH_3 (67)	Phenyl (33) CH_3 (67)	2	220	(11)
Dielectric constant ε[†]		R_1	R_2	R_3	n		
		CH_3	CH_3	CH_3	2	2.27	(12)
		CH_3	CH_3	CH_3	4	5.92	(12)
Loss factor tan δ[†]		R_1	R_2	R_3	n		
		CH_3	CH_3	CH_3	2	0.0053	(12)
		CH_3	CH_3	CH_3	4	0.52	(12)
Pyrolyzability	%	800°C in argon					
		R_1	R_2	R_3	n		
		CH_3	CH_3	–	1	20	(6)
		CH_3	CH_3	CH_3	2	29	(6)
		CH_3	CH_3	CH_3	3	36	(6)
		CH_3	CH_3	CH_3	4	47	(6)
		CH_3	CH_3	CH_3	5	48	(6)

Property	Units	Conditions*				Value	Reference
	%	800°C in air					
		R_1	R_2	R_3	n		
		CH_3	CH_3	Phenyl	2	4	(7)
		CH_3	CH_3	CH_3 (33) Phenyl (67)	2	5	(7)
		CH_3	CH_3	CH_3 (67) Phenyl (33)	2	6	(7)
Flammability[†] Oxygen index	%	R_1	R_2	R_3	n		
		CH_3	CH_3	Phenyl	2	62	(10)
TGA: 5% weight loss temperature in air	K	R_1	R_2	R_3	n		
		CH_3	CH_3		-1	>973	(13)
		CH_3	CH_3	CH_3 (67) Phenyl (33)	2	1023	(7)
		CH_3	CH_3	CH_3 (33) Phenyl (67)	2	>1073	(7)
		CH_3	Phenyl	Phenyl	2	>1073	(7)
		CH_3	CH_3	CH_3	3	793	(2)

* For the

$$-\!\!\left[\!SiCB_{10}H_{10}CSi\!-\!O\!-\!\{Si\!-\!O\!-\!\}_{n-1}\right]\!\!-$$

with substituents R_1, CH_3 (on first Si); R_1, CH_3 (on second Si); R_2, R_3 — polymer series.

[†] Mechanical properties: for resins with 30 phr trimethylsilated amorphous silica, 2.5 phr ferric oxide, and cured with 2.5 phr dicumyl peroxide.

Synthesis

n	Solvent	Catalyst	Temp. (°C)	Monomers	Ref.
1	–	$FeCl_3$	175–225	1,7-bis-(methoxydimethylsilyl)-m-carborane 1,7-bis-(chlorodimethylsilyl)-m-carborane	(14)
2	Chlorobenzene	–	-10	1,7-bis-(hydroxyldimethyl)-m-carborane bis(N-phenyl-N′tetramethyleneureido)silane	(3)
3	Diethyl ether/THF/water	–	25	1,7-bis-(chloro-1,1,3,3-tetramethyldisilyl)-m-carborane	(2)

HEPTACARBORANE–SILOXANE

Acronyms, Trade Names 5-SiB, Pentasil (Chemical Systems Inc.)

Class D_n-heptacarborane=siloxane

Structure $[-Si(CH_3)_2CB_5H_5CSi(CH_3)_2-O-\{Si(CH_3)_2-O-\}_{n-1}-]$
where $CB_5H_5C =$

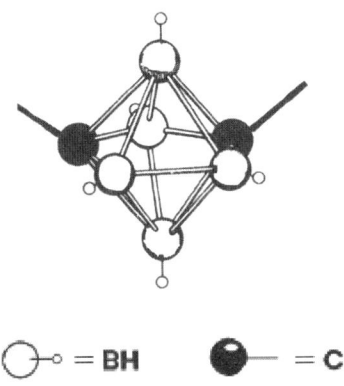

$\bigcirc\!\!-\!\circ = \mathbf{BH}$ $\quad\bullet\!\!-\! = \mathbf{C}$

Properties of Special Interest Elastomeric. High thermal stability.

Property	Units	Conditions*	Value	Reference
Glass transition temperature	K	n		(15)
		1	213	
		2	185	
		3	177	
		4	171	
		5	165	
Melting temperature	K	n		(15)
		1	343	

References

1. Mohadger Y, Roller MB, Gillham JK. *J. Applied Polymer Sci.* 1973;17:2635.
2. Knollmueller KO, Scott RN, Kwasnik H, Sieckhaus JF. *J. Polym. Sci.: Part A-1.* 1971;9:1071.
3. Hedaya E, Kawakami JH, Kopf PW, et al. *J. Polym. Sci., Polym. Chem. Ed.* 1977;15:2229.
4. Peters EN. *Ind. Eng. Chem. Prod. Res. Dev.* 1984;23:28.
5. Zaganiaris EJ, Sperling LH, Tobolsky AV. *J. Macromol. Sci. Chem.* 1967;A-1(6):1111.
6. Peters EN, Kawakami JH, Kwaitkowski GT, McNeil DW, Hedaya E. *J. Polymer Sci., Polym. Phys. Ed.* 1977;15:723.
7. Scott RN, Knollmueller KO, Hooks H, Siekhaus JF. *J. Polym. Sci., A-1.* 1972;10:2303.
8. Roller MB, Gillham JK. *J. Appl. Poly. Sci.* 1973;17:2141.
9. Peters EN, Stewart DD, Bohan JJ, Moffitt R, Beard CD. *J. Polym. Sci., Polym. Chem. Ed.* 1977;15:973.

10. Peters EN, Hedaya E, Kawakami JH, Kwaitkowski GT, McNeil DW, Tulis R. *Rubber Chem. Technol.* 1975;48:14.
11. Peters EN, Stewart DD, Bohan JJ, McNeil DW. *J. Elastomers Plast.* 1978;10:29.
12. Schroeder H, Schaffling OG, Larchar TB, Frulla FF, Heying TL. *Rubber Chem. Technol.* 1966;39:1184.
13. Roller MB, Gillham JK. *J. Appl. Poly. Sci.* 1973;17:2623.
14. Papetti S, Schaeffer BB, Gray AP, Heying TL. *J. Polym. Sci.: Part A-1.* 1966;4:1623.
15. Roller MB, Gillham JK. *J. Appl. Poly. Sci.* 1972;16:3095.

Carbosilane Dendrimers

Aziz M. Muzafarov

Acronym PCSD

Class Dendritic polymers

Structure

$$Si - \left\{ \left[CH_2CH_2CH_2\underset{|}{\overset{\underset{|}{CH_3}}{Si}} - \rule{1cm}{0.4pt} \right]_{2^{(G-1)}-1} \left[CH_2CH_2CH_2\underset{|}{\overset{\underset{|}{CH_3}}{Si}} \underset{CH_2CH=CH_2}{\overset{CH_2CH=CH_2}{<}} \right]_{2^{(G-1)}} \right\}_4$$

G – generation number

Introduction Carbosilane dendrimers[1−3] have appeared as a logical evolution of the synthesis of organosiloxane dendrimers. High reactivity of the functional groups used in the previous case was even improved by switching to organometallic reagents instead of silanolates, while a molecular skeleton became less polar. The controllable synthesis and a variety of suitable reagents attract much attention to this class of silicon-containing dendrimers and extend flexibility of synthetic approach in the area. A variety of combinations of branching unit parameters used resulted in the synthesis of a long list of similar dendritic structures with small but important differences[4−9]. Recently the general approach to the carbosilane dendrimer preparation was significantly enriched by introducing heterocyclic units between the silicon branching centers[10−12].

Related Polymers Due to quite inert molecular skeleton carbosilane dendrimers, widely used as multifunctional cores, resulted in majority of core-shell structures targeting different applications. Syntheses of multiarm stars using 'grafting to'[13−15] or 'polymerization from'[16−19] approaches were the initial motivations in the synthesis of carbosilane dendrimers resulting in the series of star polymers and copolymers of this type. The examples of LC[19−22], fluorohydrocarbon[23−29], polyhydroxy[9,30−33], as well as organometallic[6,34−42] outer shells using almost the same carbosilane cores are widely presented in the literature.

Preparative Techniques Hydrosilylation as well as Grignard reactions, both well-known and widely studied, were used in the dendrimer synthesis in different modifications. In some cases hydrosilylation was alternated with other organometallic species treatment, i.e. with lithium phenylacetylide[7,44]. The usual α - substitution was observed in most cases as summarized in ref. 44. Preparative GPC technique was shown to be an effective method for the isolation of pure dendrimers[22,45].

Major Applications A number of practical applications is proposed for carbosilane dendrimers. Multifunctional scaffold for catalysts or initiators of different kinds of chemical processes[2,13,14,17,34,35,46–50], molecular precursors for the ceramics of Si-C[4] or hybrid types[51–54], materials for organic electronics[10–12,55] are highlighted as the most promising.

Properties of Special Interest Intrinsic viscosity and bulk density values of the carbosilane dendrimers demonstrate independent character against the generation number. Direct measurements of these parameters were done based on a representative homologous series of nonfunctional dendrimers [45].

Intrinsic viscosity and density values of polybutylcarbosilane dendrimers[45]

Dendrimer	M (calculation)	$[\eta]$(dl g^{-1})	d_4^{25}(g cm^{-3})
G-3(Bu)$_{32}$	4241.4	0.036	0.8767
G-4(Bu)$_{64}$	8795.6	0.040	–
G-5(Bu)$_{128}$	17 903.9	0.042	0.8877
G-6(Bu)$_{256}$	36 120.7	0.036	–
G-7(Bu)$_{512}$	72 554.1	0.046	0.8734
G-8(Bu)$_{1024}$	145 420.9	0.038	–
G-9(Bu)$_{2048}$	291 154.5	0.043	0.8838
G-10(Bu)$_{4096}$	582 621.8	–	–

The consecutive study of the homologous series of functional polyallyl- and nonfunctional polybutyl-carbosilane dendrimers allowed discovering the peculiar changes in neat dendrimer state. The dendrimer consistency changes from viscous liquid to wax-like after the fifth generation, while T_g values remain constant. Good solubility of the dendrimers in most of the organic solvents was also not affected by the generation. The changes in the aggregate state coincide with appearance of the second transition on the temperature dependence of heat capacity of carbosilane dendrimers, which was clearly detected for the dendrimer of sixth and higher generations with terminal butyl groups[56,57].

Glass transition thermodynamic parameters of the G-1–G-5 carbosilane dendrimers with allyl terminal groups ($p = 0.1$ MPa)[58]

Dendrimer	ΔT_{glass}	T_{glass}°	$\Delta C_p^\circ (T_{glass}^\circ)$	S_{conf}°
	K		kJ K^{-1} mol^{-1}	
G-1(All)$_8$	150–160	154 ± 1	0.405	0.044
G-2(All)$_{16}$	170–180	172 ± 1	0.810	0.206
G-3(All)$_{32}$	170–180	173 ± 1	1.640	0.417
G-4(All)$_{64}$	170–180	172 ± 1	3.660	0.931
G-5(All)$_{128}$	160–170	162 ± 1	7.922	2.016

Carbosilane Dendrimers

Glass transition and glassy state characteristics of the G-3(Bu)$_{32}$–G-6(Bu)$_{256}$ carbosilane dendrimers with butyl terminal groups ($p = 0.1$ MPa)[56]

Dendrimer	Gross-formulae	M (g/mol)	$T_g^\circ \pm 0.5$(K)	ΔC_p° (kJK^{-1}mol^{-1})	S_{conf}° (kJK^{-1} mol^{-1})
G-3(Bu)$_{32}$	C$_{240}$H$_{540}$Si$_{29}$	4241.4	179.8	1.904	0.485
G-4(Bu)$_{64}$	C$_{496}$H$_{1116}$Si$_{61}$	8795.5	186.0	3.82	0.973
G-5(Bu)$_{128}$	C$_{1008}$H$_{2268}$Si$_{125}$	17 903.8	186.5	9.29	2.366
G-6(Bu)$_{256}$	C$_{2032}$H$_{4572}$Si$_{253}$	36 120.3	186.2	17.26	4.395

References

1. van der Made, et al. *J. Chem. Soc. Chem. Commun.* 1992:1400.
2. Zhou LL, et al. *Macromolecules.* 1993;26:963.
3. Muzafarov AM, et al. *Polym. Sci.* 1993;35(11):1575.
4. Seyferth D, et al. *Organometallics.* 1994;13:2682.
5. Gossage RA, et al. *Tetrahedron Lett.* 1998;39:2397.
6. Alonso B, et al. *J. Chem. Soc. Chem. Commun.* 1994:2575.
7. Kim C, et al. *Bull. Korean Chem. Soc.* 1996;17:592.
8. Kim C, et al. *J. Organomet. Chem.* 1998;563:43.
9. Lorenz K. *Macromolecules.* 1995;28:6657.
10. Nakayama J. *Tetrahedron Lett.* 1997;38:6043.
11. Ponomarenko SA, et al. *Russian Chem. Bull.* 2005;54(3):684.
12. Ponomarenko SA, et al. *Polym. Mater. Sci. Eng.* 2007;96:298.
13. Polyakov DK, et al. *Polym. Sci.* 1998;A40(9):876.
14. Meijboom R, et al. *Organometallics.* 2003;22:1811.
15. Hovestad NJ, et al. *Macromolecules.* 2000;33:4048.
16. Vasilenko NG, et al. *Polym. Sci.* 1997;A39(9):977.
17. Vasilenko NG, et al. *Macromol. Chem. Phys.* 1998;199:889.
18. Comanita B, et al. *Macromolecules.* 1999;32:1069.
19. Frey H, et al. *Polym. Mater Sci. Eng.* 1995;73:127.
20. Ponomarenko SA, et al. *Liq Cryst.* 1996;21:1.
21. Collaud Coen M, et al. *Macromolecules.* 1996;29:8069.
22. Ponomarenko SA, et al. *Polym. Sci.* 1998;A40:763.
23. Lorenz K, et al. *Macromolecules.* 1997;30:6860.
24. Omotowa BA, et al. *J. Am. Chem. Soc.* 1999;121:11 130.
25. Omotowa BA, et al. *Macromolecules.* 2003;36:8336.
26. Casado MA, et al. *Chem. Commun.* 2001:313.
27. Kuklin AI, et al. *Polym. Sci.* 2002;A44(12):1273.
28. Shumilkina NA, et al. *Doklady Chemistry.* 2005;403(2):155.
29. Shumilkina NA, et al. *Polym. Sci.* 2006;A48(12):1240.
30. Getmanova EV, et al. *Reactive & Functional Polymers.* 1997;33:289.
31. Krska SW, et al. *J. Am. Chem. Soc.* 1998;120:3604.
32. Comanita B, et al. *Des Monomers Polym.* 1999;2:111.
33. Getmanova EV, et al. *Russian Chem. Bull.* 2004;53(1):137.
34. Knapen JWJ, et al. *Nature.* 1994;72:659.
35. van Koten G, et al. *Polym. Mater. Sci. Eng.* 1995;73:228.
36. Lobete F, et al. *J. Organomet. Chem.* 509 (1996):109.
37. Cuadrado I, et al. *Inorg. Chim. Acta.* 1996;251:5.
38. Jutzi P, et al. *Angew. Chem. Int. Ed.* 1996;35:2118.

39. Cuadrado I, et al. *J. Am. Chem. Soc.* 1997;119:7613.

40. Losada J, et al. *Anal. Chim. Acta.* 1997;338:191.

41. Hovestad NJ, et al. *Organometallics.* 1999;18:2970.

42. Hovestad NJ, et al. *Angew. Chem. Int. Ed.* 1999;38:1655.

43. Vodopryanov EA, et al. *Russian Chem. Bull.* 2004;53(2):358.

44. Frey H, et al. *Top. Curr. Chem.* 2000;210:69.

45. Tatarinova EA, et al. *Russian Chem. Bull.* 2004;53(11):2591.

46. Dani P, et al. *J. Am. Chem. Soc.* 1997;119(11):317.

47. Hoare JL, et al. *Organometallics.* 1997;16:4167.

48. Gossage RA, et al. *Acc. Chem. Res.* 1998;31:423.

49. Kleij AW, et al. *Organometallics.* 1999;18:277.

50. Hovestad NJ, et al. *Polym. Mater Sci. Eng.* 1999;80:53.

51. Boury B, et al. *Chem. Mater.* 1998;10:1795.

52. Kriesel JW, et al. *Chem. Mater.* 1999;11:1190.

53. Kriesel JW, et al. *Chem. Mater.* 2000;12:1171.

54. Kriesel JW, et al. *Adv. Mater.* 2001;13:1645.

55. Borshchev OV, et al. *PMSE Preprints.* 2007;96:720.

56. Smirnova NN, et al. *Thermochimica Acta.* 2006;440:188.

57. Smirnova NN, et al. *Russian Chem. Bull.* 2007;56(10):1991.

58. Lebedev BV, et al. *Russian Chem. Bull.* 2003;52(3):545.

Cellulose

Rachel Mansencal, John F. Kadla and Jennifer L. Braun

Alternative Names Rayon, cellophane, regenerated cellulose[1]

Class Carbohydrate polymers; polysaccharides

Structure

Functions It is the basic structural material of the cell walls of all higher land plants and of some seaweeds.[2−8]

Natural Sources Wood (coniferous, deciduous), bamboo, cotton, hemp, straw, jute, flax, reed, sisal. Cellulose is isolated from the plant cell walls and is never in a pure form in nature. Always associated with lignin and hemicellulose.[2−4]

Source[4]	Cellulose (%)
Cotton	94
Hemp	77
Flax	75
Kapok	75
Sisal	75
Ramie	73
Jute	63
Wood (coniferous or deciduous)	50
Bamboo	40–50
Straw	40–50

Biosynthesis Depends on the system.[6−8]

Commercial Uses Natural cellulose is used as fuel and lumber. Purified cellulose is employed for production of paper and textiles. Derivatives of cellulose are used in plastics, films, foils, glues, and varnishes. Most of the cellulose is used in paper and paperboard manufacture.[4]

Extraction The separation process of cellulose from hemicellulose and lignine is by pulping. The two different kinds of pulping are mechanical and chemical.[2−4,6]

Cellulose

Properties of Special Interest Cellulose is the most abundant macromolecular material naturally occurring in plant cell walls. Semicrystalline natural polymer. Very difficult to dissolve.[2–7]

Property	Units	Conditions	Value	Reference
Average molecular weight	$g\ mol^{-1}$	–	$\approx 10_6$	(6)
Density	$g\ cm^{-3}$	Crystalline portion (by X-ray)	1.590–1.630	(4)
		Amorphous portion (by X-ray)	1.482–1.489	
		Cellulose I	1.582–1.630	(5, 39–41)
		Cellulose II	1.583–1.62	(5, 40)
		Cellulose IV	1.61	(5)
		Cotton	1.545–1.585	(5, 42–44)
		Ramie	1.55	(40)
		Flax	1.541	(5)
		Hemp	1.541	(5)
		Jute	1.532	(5)
		Wood pulps	1.535–1.547	(5, 40, 45)
Crystallinity Fraction (by X-ray)	–	Cellulose (*valonia ventricosa*)	0.68	(5, 10)
		Different wood pulps	0.62–0.70	(5, 10, 11)
		Ramie	0.60–0.71	(5, 10, 11)
		Native (average)	0.70	(4)
		Cotton	0.70	(70)
		Fiber G (high wet modulus rayon)	0.53	(70)
		Fortisan	0.50	(70)
		Mercerized ramie	0.49	(70)
		Bacterial Cellulose	0.40	(70)
		Viscose rayons	0.40	(70)
Cellulose fibril size	nm	Subelementary	1.5	(4)
		Elementary	3.5	
Optical refractive index	–	$n_D^{\parallel}$	1.618	(4)
			1.599	
			1.600	
			1.595	
		$n_D^{\perp}$	1.543	
			1.532	
			1.531	
			1.534	
Solubility	–	Water, organic solvent, dilute acid, alkalies	Insoluble	(5)
		Cuprammonium hydroxide	Soluble (complex formation)	
		Cupriethylenediamine hydroxide		
		Cadmium ethylene diamine hydroxide		
		Iron sodium tartrate complex		
Solubility parameters	$(MPa)^{1/2}$		32.02	(1)

Cellulose

Unit cell dimensions

Polymorph	Space Group	Chains per Unit Cell	Unit Cell Dimensions						Reference
			a (Å)	b (Å)	c (Å)	α (°)	β (°)	γ (°)	
I_α	P1	1	6.717	5.962	10.400	118.08	114.80	80.37	(64)
I_β	P2$_1$	2 parallel	7.784	8.201	10.38	90.0	90.0	96.5	(65)
II	P2$_1$	2 antiparallel	8.01	9.04	10.36	90.0	90.0	117.1	(65)
II	P2$_1$	2 antiparallel	8.10	9.03	10.31	90.0	90.0	117.10	(66)
III$_1$	P2$_1$	1	4.450	7.850	10.31	90.0	90.0	105.10	(65)

Property	Units	Conditions	Value	Reference
Polymorphs		Cellulose I, II, III, IV, III-1, III-2, IV-1, IV-2		(2)
Thermal conductivity λ_c	Wm^{-1} K^{-1}	Cotton, 293 K	0.071	(1, 5, 13)
		Rayon	0.054–0.07	(1, 5, 14)
		Sulfite pulp, wet	0.8	(1, 5, 15)
		Sulfite pulp, dry	0.067	(1, 5, 15)
		Laminated kraft paper	0.13	(1, 5, 16)
		Different papers, 303–333 K	0.029–0.17	(1, 5, 17)
		Ramie, 293 K	0.8	(71)

Thermal Expansion Coefficient of Cellulose Polymorphs

Polymorph	Axis				c		Reference
	a		b				
	p	q	p	q	p	q	
I_β	7.26×10^{-2}	2.44	0	0.5			(68)
II	7.32×10^{-2}	−6.57	13.5×10^{-2}	−5.02	0	−0.31	(69)
III$_1$	0	7.6	0	0.8			(68, 69)

$$\alpha = (pt + q) \times 10^{-5} \,°C^{-1}, \quad 25°C \leq t \leq 200°C$$

Property	Units	Conditions	Value	Reference
Thermal expansion coefficient (linear expansion) for different papers	K^{-1} ($\times 10^{-6}$)	Machine direction	2–7.5	(5, 18)
		Cross-machine direction	7.9–16.2	
Specific heat	Jg^{-1}K^{-1}	–	1.22	(4)
Heat of combustion	kJ g^{-1}	–	17.43	(4)
Dielectric constant	–	Crystalline portion	5.7	(4)
Isolation resistance	ohm cm	–	2×10^4	(4)

Property	Units	Conditions	Value	Reference
Insulating value	kV cm^{-1}	–	500	(4)
Thermal decomposition	K	–	523	(4)
Start of thermal degradation	K	Linters	498	(19)
		Bleached sulfite pulp	498	(19)
		Kraft pulp	513	(19)
		Filter paper (under nitrogen)	493	(20)
Fast endothermal degradation	K	Linters	≈573	(19)
		Bleached sulfite pulp	≈603	(19)
		Cotton (under nitrogen)	563	(4)
		Cellulose powder (thermogravimetry)	563	(21, 22)
Ignition temperature	K	Cotton	663, 673	(14, 23)
		Viscose rayon	693	(23)
Self ignition temperature	K	Cotton	673	(4)
External ignition temperature	K	Cotton	623	(4)
Maximum flame temperature	K	Cotton		
		19% O_2	1,123	(4, 5)
		25% O_2	1,323	(5, 24)
Heat capacity	kJ kg^{-1} K^{-1}	Cellulose	1.34	(5, 25)
		Cotton	1.22	(5, 26)
		Mercerized cotton	1.235	(5, 26)
		Ramie	1.775	(5, 27)
		Flax	1.344–1.348	(5, 28)
		Hemp	1.327–1.352	(5, 28)
		Jute	1.357	(5, 28)
		Viscose rayon	1.357	(5, 28)
		Paper	1.17–1.32	(5)
Heat of crystallization	kJ kg^{-1}	Cellulose I	121.8	(5)
		Cellulose II	134.8	(5)
Heat of recrystallization	kJ kg^{-1}	Amorphous cellulose → Cellulose I	41.9	(5, 29)
Heat of transition	kJ kg^{-1}	Cellulose I → Cellulose II	38.1	(5, 30)
Heat of formation	kJ kg^{-1}	–	5949.7	(5, 31)
Heat of solution of dry material	kJ kg^{-1}	Cotton in cupriethylendiamine	108.0	(5, 32)
		Cotton in Et$_3$PhNOH	142.5	(5, 33)
		Rayon in Et$_3$PhNOH	95.5	(5, 34)
		Cellulose II in Et$_3$PhNOH	182.7	(5, 33)
Yields of scission $G(S)$	µmol J^{-1}	Electron beam or γ-irradiation	11	(5, 35)
Glass transition temperature	K	–	503	(5)
			493–518	
Secondary transition	K	–	292–296	(5)
			298	

Property	Units	Conditions	Value		Reference
Tensile strength	MPa		Dry	Wet	(4)
		Ramie	900	1,060	
		Cotton	200–800	200–800	
		Flax	824	863	
		Viscose rayon	200–400	100–200	
		Viscose rayon, highly oriented	610	520	
		Cellulose acetate	150–200	100–120	
		Wood fiber	164		(72)
Relative wet/dry strength	%	Ramie	117		(4)
		Cotton	105		
		Flax	105		
		Viscose rayon	50		
		Viscose rayon, highly oriented	86		
		Cellulose acetate	65		
Extension at break	%		Dry	Wet	(4)
		Ramie	2.3	2.4	
		Cotton	16–12	6–13	
		Flax	1.8	2.2	
		Viscose rayon	8–26	13–43	
		Viscose rayon, highly oriented	9	9	
		Cellulose acetate	21–30	29–30	
		Tencel	9.8		(73)
		Lyocell	7.5		(74)
Elastic modulus	MPa	Native flax	78,000–108,000		(4)
		Native hemp	59,000–78,000		
		Native ramie	48,000–69,000		
		Mercerized ramie	80,000		
		Oriented rayon	33,000		
		Cellulose acetate film	4,000		
		Tunicate cellulose whiskers	143,000		(75)
		Bacterial cellulose	78,000		(74)
		Cellulose II	98,000		(76)
		Amorphous cellulose	10,000		(76)
		Cotton	4,000		(72)
		Wood fiber	2,000		(72)

Void system determination by X ray small angle scattering

Cellulose	Relative Internal Surface (Å^2 Å^{-3})	Specific Internal Surface (m^2 g^{-1})	Conditions	Reference
Microcrystalline	0.09273	2.93	Average values	(5, 36, 37)
	0.0714	1.74		
Microfine	0.07232	1.10	–	(5)
	0.12800	2.08		

Property	Units	Conditions	Value	Reference
Permeability to gases	–	Cellulose, 25°C, pressure not specified	$H_2, N_2, O_2, CO_2,$ SO_2, H_2S, NH_3	(5, 38)
Heat of adsorption of water, ΔH_{ads}	Jg^{-1}	Cotton, 25°C	384	(4)
		Holocellulose, 25°C	344	
		Bleached sulfite pulp, 20°C	348	
		Cellophane, 25°C	358	
		Viscose rayon, 25°C	397	

Characteristic IR Absorption Frequencies

Type of Motion	Wavenumber (cm^{-1})[72]	
	Cellulose I	Cellulose II
γ_{OH} (hydrogen bonded)	3352	3447
γ_{CH}	2901	2892
δ_{CH2} (sym) at C-6	1431	1419
δ_{CH}	1373	1376
δ_{CH2} (wagging) at C-6	1319	1317
δ_{CH}	1282	1278
δ_{COH} in plane at C-6	1236	1228
δ_{COH} in plane at C-6	1202	1200
γ_{COC} at β-glucosidic linkage	1165	1162
γ_{CO} at C-6	1032	1019
γ_{CO} at C-6	983	993
γ_{COC} at β-glucosidic linkage; γ_{COC}, γ_{CCO}, and γ_{CCH} at C-5 and C-6	897	894

Ring Carbon	^{18}C-NMR Chemical Shift (ppm)[78]		
	Cellulose I$_{\alpha}$	Cellulose I$_{\beta}$	Cellulose II
C1	101.41	100.45	102.21
C1$'$	100.94	99.59	101.47
C2	70.40	71.88	72.75
C2$'$	68.09	73.85	73.86
C3	70.30	70.74	72.57
C3$'$	71.98	70.59	69.67
C4	75.52	75.79	78.01
C4$'$	76.11	75.43	77.39
C5	72.44	74.45	75.44
C5$'$	71.65	73.03	73.01
C6	59.28	61.17	59.73
C6$'$	59.89	55.89	63.14

Mark-Houwink parameter*: K and $a^{(70)}$

Solvent	Temperature (°C)	Shear Rate (s^{-1})	$K_m \times 10^6$ (dL/g)	a	Molecular Weight Range	Method of Calibration
Cuoxam	25	500	33.0	0.76	24,000–1,100,000	Viscometry
Cuen	25	vs. 0	47.3	0.76	49,000–970,000	Viscometry
	25	vs. 0	10.1	0.9	40,000–1,300,000	UC on nitrate
	25	vs. 0	29	0.8	52,000–150,000	K, OS
Cadoxen	25		38.5	0.76	9700–970,000	LS
FeTNa(EWNN)	25	400	51.8	0.78	37,000–65,000	LS
	25	400	76.5	0.74	>160,000	LS

Property	Units	Conditions		Value	Reference
Martin coefficient K'	–	Cellulose			(5)
		Solvent		0.13–0.15	
			Cuene*	0.1303	
			Cuoxam†	0.112	
Huggins coefficient K''	–	Cellulose			(5)
		Solvent			
			Cuoxam†	0.37	
			Cadoxene‡	0.26–0.39	
		Phosphoric acid, 278 K			(77)
			73% P_2O_5 by mass	2.26	
			74.3% P_2O_5 by mass	1.53	
			75% P_2O_5 by mass	1.29	
			76% P_2O_5 by mass	1.00	
		Phosphoric acid, 293 K			(77)
			73% P_2O_5 by mass	2.78	
			74.3% P_2O_5 by mass	2.4	
			75% P_2O_5 by mass	1.97	
			76% P_2O_5 by mass	1.65	
Schulz-Blaschke coefficient K'''	–	Cellulose			
		Solvent		0.33	(5)
			Cuene*	0.29	(5, 52)
			Cuoxam†	0.1552	(5, 53)
				0.287	(5, 52)
			Cadoxene‡	0.280	(5)
Second virial coefficient A_2	mol cm^3 g^{-2} ($\times 10^4$)	Cellulose		16.1	(5)
		Hydrolyzed linters; cadmium ethylene diamine solvent; 25°C; $M = (225 - 945) \times 10^{-3}$ g mol^{-1}; light scattering			
		Sulfite pulp; $M = 215 \times 10^{-3}$ g mol^{-1}; light scattering		12.1	
Sedimentation coefficients s_O	s $\times 10^{13}$	Cellulose in solution Cuene*; 25°C			(1, 5)
			$M = 175 \times 10^{-3}$ g mol^{-1}	5.5	
			$M = 9.5 \times 10^{-3}$ g mol^{-1}	8.3	

Property	Units	Conditions	Value	Reference
		Cadoxene[‡]; 12°C		(1, 5, 54)
		$M = 33.6 \times 10^{-3}$ g mol^{-1}	1.25	
		$M = 24.5 \times 10^{-3}$ g mol^{-1}	1.13	
		$M = 18.8 \times 10^{-3}$ g mol^{-1}	1.04	
		$M = 10.1 \times 10^{-3}$ g mol^{-1}	0.74	
Diffusion coefficients D_0	cm^{3}s ($\times 10^7$)	Cellulose in solution		(1,5)
		Cuene*; 25°C		
		$M = 175 \times 10^{-3}$ g mol^{-1}	1.2	
		$M = 9.5 \times 10^{-3}$ g mol^{-1}	0.95	
Frictional ratios υ_2	cm^3 g^{-1}	Cellulose in solution; cuene*; 25°C;	0.65	(1, 5)
		$M = 175 \times 10^{-3}$ g mol^{-1}		
Specific resistance ρ	ohm cm	–	10^{18}	(5, 55)
Dielectric constant ε	–	106 kHz	5.5–8.1	(5, 56)
Dielectric loss factor tan δ	–	20°C, 0.1 kHz	0.015	(5)
		20°C, 1 kHz	0.02	
		20°C, 10 kHz	0.03	
		20°C, 10^2 kHz	0.045	
		20°C, 10^3 kHz	0.065	
		20°C, 10^4 kHz	0.08	
		20°C, 10^5 kHz	0.07	
Dielectric strength	kV mm^{-1}	Dry (native cellulose fiber)	50	(5, 57)
Zeta-potential	mV	Fines from filter paper, Whatman No. 1	21.0	(5, 58)
Surface tension	mN m^{-1}	Contact angle method, at 20°C		(5, 59)
		Cellulose regenerated from cotton	42	
		Cellulose regenerated from wood pulp	36–42	

* Cuene: cupriethylenediamine.
[†] Cuoxam: cuprammonium hydroxide.
[‡] Cadoxene: cadmiumethylenediamine.

Specific refractive index increment in dilute solution, dn/dc (mlg^{-1})

Solvent	$\lambda_0 = 436$ nm	$\lambda_0 = 546$ nm	Temp. (°C)	Reference
Acetone	0.111	–	25	(1, 60)
Cadoxene*	0.186	0.183	25	(1, 12, 54)
Cadoxene*, (5% Cd)/water (1:1 vol)	0.190	0.189	25	(1, 61)
0.237 M Cd	0.1317	0.1927	25	(1, 62)
Cuoxam[†] 0.205 M Cu	0.117	0.233	25	(1, 5)
Cuoxam[†] 0.0518 M Cu	0.1352	0.2574	25	(1, 62)
FeTNa	0.110	0.245	25	(1, 63)

* Cadoxene: cadmiumethylenediamine.
[†] Cuoxam: cuprammonium hydroxide.

Polymorph	Poisson's Ratio (200/400)	Reference
I$_\beta$	0.38	(67)
II	0.30	(67)

References

1. Zhao W, Mark JE. In: Mark JE, ed. *Physical Properties of Polymers Handbook*. Woodbury, N.Y.: AIP Press; 1996.

2. Huang Y, Chen J. In: Salamone JC, ed. *Polymeric Materials Encyclopedia*. Vol 2. Boca Raton, Fla.: CRC Press; 1996.

3. James DW Jr, Preiss J, Elbein AD. In: Aspinall GO, ed. *The Polysaccharides*. Vol 3. New York: Academic Press; 1985.

4. Dane JR. In: Mark, HF, et al. eds. *Encyclopedia of Polymer Science and Engineering*. 2d ed. Vol 3. New York: John Wiley and Sons; 1989.

5. Gröbe A. In: Branrup J, Immergut EH, eds. *Polymer Handbook*. 3d ed. New York: John Wiley and Sons; 1989.

6. Tarchevsky IA, Marchenko GN, eds. *Cellulose: Biosynthesis and Structure*. New York: Springer-Verlag; 1991.

7. Brown RM Jr, ed. *Cellulose and Other Natural Polymer Systems*. New York: Plenum Press; 1982.

8. Kennedy JF, Phillips GO, Williams PA, eds. *Cellulose, Structural and Functional Aspects*. Chichester: Ellis Horwood Ltd.; 1989.

9. Kolpak FJ, Blackwell J. *Macromolecules*. 1976;273:1.

10. Hermans PH, Weidinger A. *J. Polym. Sci.* 1950;5:565.

11. Hermans PH. *Makromol. Chem.* 1951;6:25.

12. Henley D. *Swensk Papperstidn.* 1960;63:143.

13. Hammons MA, Reeves WA. *Textiles Chem. Colourists.* 1982;14:26/210.

14. Goerlach H. *Chemiefasern.* 1972;22(6):524.

15. Guthrie JC. *J. Textile Inst.* 1949;40:T489.

16. Terada T, Ito N, Goto Y. *Kami Pa Gikyoshi.* 1969;23:191.

17. Terasaki K, Matsuura K. *Kami Pa Gikyoshi.* 1972;26(4):173.

18. Kubat J, Martin-Loef S, Soeremark Ch. *Swensk Paperstidn.* 1969;72:763.

19. Otmar T, Dreilheller H, Grossberger G. *Ger. Offen.* 1971;1:964.

20. Broido A, Martin SB. *U.S. Dept. Com., Office Tech. Serv.* 1961;AD 268:729.

21. Fu YL, Shafizadeh F. *Carbohydr. Res.* 1973;29(1):113.

22. Shafizadek F, Sekiguchi Y. *Carbon.* 1983;21:511.

23. *The Flammability of Textile Fibers*. Bull. X-45. E. I. DuPont de Nemours, Wilmington, 1955.

24. Miller B, et al. *Textile Res. J.* 1976;46:531.

25. National Research Council (U.S.). *International Critical Tables*. Vol II. New York: McGraw-Hill; 1926–1930:237.

26. Magne FC, Portas HJ, Wakeham H. *J. Am. Chem. Soc.* 1947;69:1896.

27. Mikhailov NV, Fainberg EZ. *Vysokomol. Soedin.* 1962;4:230.

28. Goetze W, Winkler F. *Faserforsch. Textiltechn.* 1967;18:119.

29. Hermans PH, Weidinger A. *J. Am. Chem. Soc.* 1946;68:2547.

30. Lauer K. *Kolloid-Z.* 1951;121:139.

31. Jessup RS, Proser EI. *J. Res. Natl. Bur. Std.* 1950:44.

32. Calvet E, Hermans PH. *J. Polym. Sci.* 1951;6:33.

33. Lipatov SM, Zharkovskii DV, Zagraevskaya IM. *Kolloidn. Zh.* 1959;21:526.

34. Mikhailov NV, Fainberg EZ. *J. Polym. Sci.* 1958;30:259.

35. Charlesby A. *J. Polym. Sci.* 1955;15:263.

36. Schurz J, Janosi A. *Das Papier.* 1982;36:584.

37. Schurz J, Janosi A. *Holzforschung.* 1982;36:307.

38. Simril VL, Hershberger A. *Modern Plastics.* 1950;27:95.

39. Kast W, Schwarz R. *Z. Electrochem.* 1952;56:228.
40. Hermans PH. *Contribution to the Physics of Cellulose Fibers.* New York: Elsevier; 1946.
41. Lyons WJ. *J. Chem. Phys.* 1941;9:377.
42. Stamm AJ, Hansen LA. *J. Phys. Chem.* 1937;41:1007.
43. Wakeham H. *Textile Res. J.* 1949;19:595.
44. Hermans PH, Hermans JJ, Vermas D. *J. Polymer Sci.* 1946;1:149, 156, 162.
45. Brenner FC, Frilette V, Mark H. *J. Am. Chem. Soc.* 1948;70:877.
46. Tsvetkov VS. *Rigid-chain Polymer Molecules.* Moscow: Nauka; 1985.
47. Staudinger H, Daumiller G. *Ann. Chem.* 1937;529:219.
48. Cumberbirch RJE, Harland WG. *J. Textile Inst.* 1958;49:T679.
49. Immergut EH, Schurz J. Mark HF. *Monatsh. Chem.* 1953;84:219.
50. Immergut EH, Ranby BG, Mark HF. *Ind. Eng. Chem.* 1953;45:2483.
51. Prati G, Errani L. *Tincoria.* 1962;59:233, 279.
52. Marx M, Schulz GV. *Makromol. Chem.* 1959;31:140.
53. Schulz GV, Blaschke F. *J. Prakt. Chem.* 1941;158:130.
54. Brown W, Wirkstroem R. *Eur. Polym. J.* 1965;1:1.
55. Murphy EJ. *Can. J. Phys.* 1963;141:1022.
56. Claussnitzer W. In: *Landolt-Boerstein, Zalhenwerte und Funktionen.* 6th ed. Vol IV. Part 3. Berlin: Springer-Verlag; 1957.
57. Meyer K, Mark H. *Makromoleculare Chemie*, 2d ed. Leipzig: Akad. Verlag; 1950.
58. McKenzie AW. *APPITA.* 1968;21(4):104.
59. Luner P, Sandell M. *J. Polym. Sci.* 1969;c28:115.
60. Marx-Figini M, Penzel E. *Makromol. Chem.* 1965;87:307.
61. Huglin MB, O'Donohue SJ, Sasia PM. *J. Polym. Sci., Polym., Phys. Ed.* 1988;26:1067.
62. Vink H, Dahlström G. *Makromol. Chem.* 1967;109:249.
63. Valtasaari L. *Tappi.* 1965;48:627.
64. Nishiyama Y, Sugiyama J, Chanzy H, Langan P. *J. Am. Chem. Soc.* 2003;125(47):14300.
65. Wada M, Nishiyama Y Langan L. *Macromolecules.* 2006;39:2947.
66. Langan P, Nishiyama Y, Chanzy H. *J. Am. Chem. Soc.* 1999;121:9940.
67. Nakamura K, Wada M, Kuga S, Okano T. *J. Polym. Sci: Part B: Polym. Phys.* 2004;42:1206.
68. Wada M. *J. Polym. Sci. Part B: Polym. Phys.* 2002;40:1095.
69. Hori R, Wada M. *Cellulose.* 2006;13:281.
70. Krässig HA. *Cellulose.* Switzerland: Gordon & Breach Science Publishers; 1993:123.
71. Yamanaka A, et al. *J. Appl. Polym. Sci.* 2006;100:5007.
72. Kompella MK, Lambros J. *Polymer Testing.* 2002;21:523
73. Abdullah I, Blackburn RS, Russell SJ, Taylor J. *J. Appl. Polym. Sci.* 2006;99:1496.
74. Guhados G, Wan W, Hutter JL. *Langmuir.* 2005;21:6642.
75. Šturcová A, Davies GR, Eichhom SJ. *Biomacromolecules.* 2005; 6:1055.
76. Gindl W, Keckes J. *Composites Sci. & Technol.* 2006;66:2049.
77. Ramzi M, Vroege GJ, Lekkerkerker HNW. *Macromol. Symp.* 2001;166:209.
78. Oh SY, et al. *Carbohydrate Research.* 2005;340:2376.
79. Hesse S, Jäger C. *Cellulose.* 2005;12:5.

Cellulose Acetate

Yong Yang

Acronym CA

Class Carbohydrate polymers

Structure

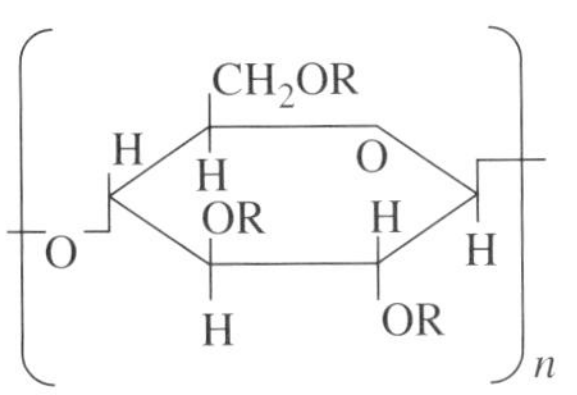

R is –COCH$_3$ or H

Major Applications Textile fibers, cigarette filters, plastics for molding and extrusion, films for photography and recording tape, LCD display, drug release modifier, sheeting, lacquers, protective coatings for paper, metal, and glass, adhesive for photographic film, ink reservoirs for fiber tip pens, absorbent cloths, and wipes.

Preparative Techniques Cellulose acetate is made from processed wood pulp (cellulose). The pulp is processed using acetic anhydride to form acetate flake from which products are made. Another technique for producing cellulose acetate involves treating cotton with acetic acid, using sulfuric acid as a catalyst.

Properties of Special Interest White, ordorless, nontoxic, wide range of solvent tolerance, biodegradable.

Property	Units	Conditions	Value	Reference
Molecular weight of repeat unit	g mol^{-1}	Degree of substitution DS : 3.0	288.25	(1–3)

Infrared characteristic absorption frequencies

Frequency (cm^{-1})	Assignment
~3400	(OH) stretching
2950	(CH$_3$) asymmetric stretching
2860	(CH$_3$) symmetric stretching
1750	(C=O) stretching
1432	(CH$_3$) asymmetric deformation
1370	(CH$_3$) symmetric deformation
1235	Acetate C—C—O stretching
1050	(C—O) stretching
603	Structural factor

Property	Units	Conditions	Value	Reference
NMR		^{13}C and ^{1}H		(4)
Thermal expansion coefficient	K^{-1}	Sheet	$(10 - 15) \times 10^{-5}$	(5)
		<190 K	8.21×10^{-5}	(6)
		>190 K	26.6×10^{-5}	(6)
Density	$g\,cm^{-3}$		$1.29 - 1.30$	(7)

Solvent and Nonsolvent

DS	Solvent	Nonsolvent	Reference
0.6–0.8	Water		(8)
1.3–1.7	2-Methoxyethanol	Acetone, water	(8–10)
2–2.5	Acetic acid*, acetone*, acrylic acid*, aniline, benzyl alcohol, cyclohexanone, *p*-chlorophenol*, *m*-cresol*, dichloroacetic acid*, diethanolamine, difluoroacetic acid*, *N,N*-dimethylacetamide*, dimethylformamide*, 1,5-dimethyl-2-pyrrolidone*, dimethylsulfoxide*, 1,4-dioxane*, ethylene glycol ether, ethyl acetate, formic acid*, glycol sulfite*, hexafluoroisopropanol*, methyl acetate, *n*-methylpyrrolidone-2*, naphthol*, nitrobenzene/ ethyl acetate, nitromethane*, phenol*, phosphoric acid*, pyridine*, tetrafluoro-*n*-propanol*, tetrafluoroisopropanol*, trifluoroacetic acid*, trifluoroethanol*	Hydrocarbons, aliphatic ethers, weak mineral acids	
3.0	Acetic acid* acetone*, acetone/water (8:2), aniline*, chloroform, *m*-cresol*, dichloroacetic acid*, dichloromethane*, *N,N*-dimethylacetamide*, dimethylformamide*, dimethylsulfoxide*, 1,4-dioxane*, ethyl acetate, ethylene carbonate, ethylene glycol ether acetates, methyl acetate*, methylene chloride, methylene chloride/ethanol (8:2), nitromethane*, 3-picoline*, 4-picoline*, *n*-propyl acetate*, pyridine*, tetrachloroethane*, tetrahydrofuran, trifluoroacetic acid*, trifluoroethane, trifluoroethanol*	Aliphatic hydrocarbons, benzene, dichloroethane, chlorobenzene, *o*-chlorotoluene, ethanol, aliphatic ethers, weak mineral acids	

* Forms liquid crystalline mesophase.

Solubility Parameters δ

DS	$\delta\,[(MPa)^{1/2}]$	Solvent	Method	Reference
1.9	27.2		Heat of solution/solvation	(11)
2.3	23.0	Acetone	Osmotic pressure	(12)
	21.2	*m*-Cresol	Osmotic pressure	

Cellulose Acetate

DS	$\delta \ [(MPa)^{1/2}]$	Solvent	Method	Reference
	22.5	Dioxane	Osmotic pressure	
	22.6	Methyl acetate	Osmotic pressure	
	21.9	α-Picoline	Osmotic pressure	
	22.4	β-Picoline	Osmotic pressure	
	22.0	γ-Picoline	Osmotic pressure	
	22.5	Pyridine	Osmotic pressure	
2.3	24.7		Gel swelling	(13)
2.4	21.7		Intrinsic viscosity max.	(14)
2.5	27.8		Heat of solution/solvation	(11)
2.8	27.8		Gel swelling	(13)

Polymer–liquid Interaction Parameters $\chi\,(\phi_2)$

Solvent	DS	Temperature (K)	$\chi\,(0)$	$\chi\,(0.2)$	$\chi\,(0.4)$	$\chi\,(0.6)$	Reference
Acetone	2.3	298–318	0.44				(12, 15–19)
	2.5	303		0.30	0.51		
Acetice acid	2.3	298–318		0.40			
Aniline	2.3	298–318		0.34–0.375			
Chloroform	3.0	298	0.34				
	3.0	303		0.36	0.45	0.51	
Dichloromethane	3.0	298	0.3			0.49	
1, 4-Dioxane	2.3	298–318	0.38				
	2.5	303	0.31	0.51			
Methyl acetate	2.3	298–318	0.45				
	2.5	303		0.43	0.59		
Nitromethane	2.3	298–318	0.43				
2-Picoline	2.3	298	0.36				
3-Picoline	2.3	298	0.285				
4-Picoline	2.3	298	0.26				
Pyridine	2.3	298–318	0.28				
	2.5	303		0.07	0.09		
Tetrahydrofuran	2.5	286	0.442				

Second Virial Coefficients A_2

	Solvent	Temperature (K)	MW10^{-3} (g mol^{-1})	$A_2\,10^4$ (mol cm^3 g^{-2})	Reference
Cellulose acetate	Acetone	RT	60–173	5.8–9.4	(20)
Cellulose diacetate	Acetone	285.3	94	4.1	(21)

Solvent	Temperature (K)	$MW10^{-3}$ $(g\,mol^{-1})$	$A_2\,10^4$ $(mol\,cm^3\,g^{-2})$	Reference
(DS = 2.46)	298.6		3.8	
	311.0		3.6	
	363.2		3.5	
	323.5		3.4	
Butanone	303	71	−0.5	(22)
	313		−0.25	
	323		0	
	333		0.25	
	313	92	−0.25	
	323	92	0	
	323	141	−0.21	

Mark–Houwink Parameters, K and a

Solvent	DS	Temperature (K)	$MW\,10^{-4}$ $(g\,mol^{-1})$	$K10^3$ $(ml\,g^{-1})$	a	Reference
Acetone	2.0	298	27	133	0.616	(23, 24)
	2.25–2.38	303	2.6–15	16	0.82	
	3.0	293	14	2.38	1.0	
		298	18	8.97	0.90	
		298	30	3.30	0.760	
		298	39	14.9	0.82	
		298	69	28.9	0.725	
Acetone/methylene chloride	3.0	298	1.4–13	2.2	0.95	
Acetone/water (80/20)	3.0	293	11	2.65	1.0	
Chloroform	3.0	293	13	2.2	0.95	
		298	69	45.4	0.649	
		303	18	14.4	0.800	
		303	18	4.5	0.9	
o-Cresol	3.0	303	18	6.15	0.9	
Dichloromethane	3.0	293	69	24.7	0.704	
Dimethylacetamide	0.49	298	15	191	0.6	
	1.75	298	14	95.8	0.65	
	2.0	298	19	39.5	0.738	
	3.0	298	69	26.4	0.750	
Dimethyl sulfoxide	0.49	298	15	171	0.61	
Ethanol/methylene chloride (20/80 by vol)	3.0	298	30	13.9	0.834	
Formaldehyde	0.49	298	15	20.9	0.60	
Methylene chloride	3.0	298	DP: 150–560	1.45*	0.83	

Cellulose Acetate

Solvent	DS	Temperature (K)	MW 10^{-4} (g mol^{-1})	$K10^3$ (ml g^{-1})	a	Reference
Tetrachloroethane	2.86	298		5.8	0.90	
Tetrahydrofuran	2.0	298	30	51.3	0.688	
Trifluoroacetic acid	2.0	298	19	52.7	0.696	
	3.0	298	69	39.6	0.706	
Water	0.49	298	15	20.9	0.60	

* From $\eta = K(DP)^a$, DP: degree of polymerization.

Unit cell Dimension of Cellulose triacetate (CTA)

	Lattice	Monomers per Unit Cell	Cell Dimension (Å) a	b	c	Space Group	Density (gcm^{-3})	Reference
CTA I	Orthogonal	4	23.63	6.27	10.43	P2$_1$	1.239	(25)
	Orthorhombic	16	44.3	13.45	10.47	P2$_1$	1.228	(26)
CTA II	Orthorhombic	8	24.68	11.52	10.54	P2$_1$2$_1$2$_1$	1.278	(27)

Property	Units	Conditions	Value	Reference
Theta temperature Θ	K	DS = 2.46		
		Acetone	428	(28)
		Butanone	310	(28)
			323	(22)
		Cellulose triacetate, acetone	300	(29)
Characteristic ratio $<r^2>_0/nl^2$		Cellulose diacetate, 298 K, light scattering , acetone	26.3	(23)
		THF	13.2	(23)
Persistence length	Å	Acetone	55.6	(30)
		Trifluoroethanol	59.7	(30)
Chain conformation		CTA I & II	2$_1$ helix	(26)
Glass transition temperature	K	Conflicting data	243– 473	(31)
		DS 1.8–2.9	453– 462	(32)
Melting point	K	CTA I , annealed at 250°C for 15–30 min, DSC, 20 deg/min	580	(26)
		CTA II annealed at 250°C for 15–30 min, DSC, 20 deg/min	582	(26)
		DS = 2.3–2.5	508–528	(28)
		DS = 2.9	553–573	(32)
Heat capacity (of repeat unit)	kJ K^{-1}mol^{-1}	Sheet	0.36–0.60	(5)
		Molding	0.36–0.51	(5)

Property	Units	Conditions	Value	Reference
Deflection temperature	K	1.82 MPa	321–364	(5)
		0.455 MPa	326–371	(5)
Tensile modulus	MPa	Sheet	$(2.1–4.1) \times 10^3$	(5)
		Molding, lightly cross-linked, $\overline{Mc} = 12\ 300$ g mol^{-1}	$(0.45–2.8) \times 10^3$ 2300	(5) (33)
Tensile strength	MPa	Molding, lightly cross-linked, $\overline{Mc} = 12\ 300$ g mol^{-1}	14–248 10	(5) (33)
Maximum extensibility	%	Sheet	20–50	(5)
		Molding	60–70	(5)
Compressive strength	MPa	Molding, ASTM D695	14–248	(5)
Flexural yield strength	MPa	Sheet	41–69	(5)
		Molding	14–110	(5)
Impact strength	J m^{-1}	Molding ,1/2 by 1/2 in. notched bar , Izod test, ASTM D256	21–278	(5)
Hardness		Rockwell, R scale, sheet	85–120	(5)
		molding	34–125	(5)
Index of refraction n			1.47–1.48	(34)

Refractive Index Increment dn/dc (in (ml g^{-1}))

DS	Solvent	Temperature (K)	dn/dc (λ_0 nm)	
0.49	DMA		0.068 (436)	(35)
0.49	Formamide		0.069 (436)	
0.49	Water		0.131 (436)	
1.75	DMA	298	0.046 (436)	
2.45	THF	298	0.071 (436)	
2.45	Trifluoroethanol	298	0.157 (436)	
2.46	Acetone	298	0.122 (436)	(21, 22)
2.46	Acetone	298	0.109 (546)	(21)
3	DMA	298	0.040 (436)	(34)

Resistivity (Ohm cm) of Cellulose acetate Fiber (35)

% Relative Humidity	Commercial	Purified
45	967 000	81 500 500
55	424 000	6 040 000
65	150 000	448 000
75	28 900	33 200
85	1610	2460
95	11	39

Property	Units	Conditions	Value	Reference

Permeability Coefficient P (in $m^3(STP)\ m\ s^{-1}\ m^{-2}\ Pa^{-1}$)

Permeant	Temperature (K)	$P \times 10^{17}$	Reference
H_2	293	2.63	(36)
		22.1–9.5	(37)
He	293	10.2	(36)
N_2^*	303	0.21	(36)
O_2^*	303	0.585	(36)
CO_2^*	303	17.3	(36)
		63.4–73.7	(37)
H_2O	298	4130	(36)
H_2O^*	298	5500	(36)
H_2S	303	2.63	(36)
H_2S^*	303	4.58	(36)
$C_2H_4O^*$	303	30.0	(36)
CH_3Br^*	303	4.2	(36)

* Film with plasticizer.

Property	Units	Conditions	Value	Reference
Surface tension	mN m^{-1}	Contact angle	45.9	(38)
Dielectric constant @1MHz			0.5	(39)
Dielectric strength	kV mm^{-1}		11	(39)
Dissipation factor		@ 1kHz	0.06	(39)
Volume resistivity	Ohm cm		5×10^{12}	(39)
Thermal conductivity	W m K^{-1}	293K	0.20	(40)
Water absorption	%	25% relative humidity	0.6	(7)
		50% relative humidity	2.0	
		75% relative humidity	3.8	
		95% relative humidity	7.8	
Flammability	cm/min		1.27– 5.08	(37)

References

1. Noda I, Dowrey AE, Marcott C. Group Frequency Assignments for Major Infrared Bands Observed in Common Synthetic Polymers. In: Mark JE, ed. *Physical Properties of Polymer Handbook*. New York: American Institute of Physics Press, Woodbury; 1996.

2. Zhbankov RG. In: Stepanov ABI, ed. *Infrared Spectra of Cellulose and Its Derivatives*. New York: Consultants Bureau Pub.; 1966

3. Blackwell J, Marchessault RH. Structure of Cellulose and Its Derivatives. Infrared Spectroscopy Structure Studies. *High Polym.* 1971;5:1.

4. Doyle S, Pethrick RA. ^{13}C Nuclear Magnetic Resonance Studies of Cellulose Acetate in the Solution and Solid States. *Polymer.* 1986;27:19.

5. Rudd GE, Sampson RN. In: Harper CA, ed. *Handbook of Plastics, Elastomers, and Composites*. New York: McGraw-Hill; 1992.

6. Biermann CJ, Narayan R. Crosslinking of Cellulose Acetate with Phosphorus Pentoxide. *J. Polym. Sci.: Pt. C: Polym. Let.* 1987;25:89.

7. Bogan RT, Kuo CM, Brewer RJ. Cellulose Derivatives, Ether. In: Mark H, et al. eds. *Encyclopedia of Chemical Technology.* Vol 5. New York: John Wiley and Sons (1979).

8. Fuchs O. Solvents and Non-Solvents for Polymers. In: Brandrup J, Immergut EH, eds. *Polymer Handbook.* 3rd ed. New York: John Wiley & Sons; 1989:VII/379.

9. Aharoni SM. Rigid Backbone Polymers XIII: Effects of the Nature of the Solvent on the Lytropic Mesomorphicity of Cellulose Acetate. *Mol. Cryst. Liq. Crysl. Lett.* 1980;56:237.

10. Gray DG. Liquid Crystalline Cellulose Derivatives. *J. Appl. Polym. Sci. Appl. Polym. Symp.* 1983;37:179.

11. Shvarts AG. The Compatibility of High Polymers. *Kolloidn. Zh.* 1956;18:755.

12. Moore WR, Epstein JA, Brown AM, Tidswell BM. Cellulose Derivatives - Solvent Interaction. *J. Polym. Sci.* 1957;23(103):23.

13. Golender BA, Larin PP, Tashmukhamedov SA. Study of the Physicochemical Properties of Multicomponent Polymer Systems Based on Cellulose Acetates. *Polym. Sci. U.S.S.R.* 1976;18:1522.

14. Barton AFM. Cellulose Acetate. *CRC handbook of Polymer-Liquid Interaction and Solubility Parameters.* Boca Raton, Florida: CRC Press; 1990.

15. Orwoll RA, Arnold PA. Solubility Parameters. In: Mark JE, ed. *Physical Properties of Polymer Handbook.* Woodbury, New York: American Institute of Physics Press; 1996.

16. Gundert F, Wolf BA. (1989). Polymer-Solvent Interaction Parameters. In: Brandrup J, Immergut EH, eds. *Polymer Handbook.* 3rd ed. New York: John Wiley & Sons; VII/173.

17. Moore WR, Tidswell BM. Thermodynamic Properties of Solutions of Cellulose Derivatives. I. Dilute Solutions of Secondary Cellulose Acetate. *J. Polym. Sci.* 1958;27:459.

18. Moore WR, Shuttleworth R. Thermodynamic Properties of Solutions of Cellulose Acetate and Cellulose Nitrate. *J. Polym. Sci., Polym. Chem. Ed.* 1963;1:733.

19. Moore WR, Tidswell BM. Thermodynamics Properties of Dilute Solutions of Cellulose Derivatives. *J. Polym. Sci.* 1958;29:37.

20. Lechner MD, Steinmeier DG. Sedimentation Coefficients, Diffusion Coefficients, Partial Specific Volumes, Frictional Ratios, and Second Virial Coefficients of Polymers in Solution. In: Brandrup J, Immergut EH, eds. *Polymer Handbook..* 3rd ed. New York: John Wiley & Sons; 1989:VII/61.

21. Suzuki H, Miyazaki Y, Kamide K. Temperature Dependence of Limiting Viscosity Number and Radias of Gyration of Cellulose Diacetate in Acetone. *Euro. Polym. J.* 1980;16:703.

22. Suzuki H, Muraoka YK, Saitoh M. Light Scattering Study on Cellulose Diacetate in 2-Butanone. *Euro. Polym. J.* 1982;18:831.

23. Kurata M, Tsunashima Y. Viscosity-Molecular Weight Relationships and Unperturbed Dimension of Linear Chain Molecules. In: Brandrup J, Immergut EH, eds. *Polymer Handbook.* 3rd ed. New York: John Wiley & Sons; 1989:VII/46.

24. Gröbe A. Properties of Cellulose Materials. In: Brandrup J, Immergut EH, eds. *Polymer Handbook.* 3rd ed. New York: John Wiley & Sons; 1989:V/117.

25. Spanovic AT, Sarka A. Molecular and Crystal Structure of Cellulose Triacetate I: A Parallel Chain Structure. *Polymer.* 1978;19:3.

26. Roche E, Chanzy H, Bouldenlle M, Marchessault RH. Three Dimensional Crystalline Structure of Cellulose Acetate II. *Macromolecules.* 1978;11:86.

27. Zugenmaier P. Structural Investigation on Cellulose Derivatives. *J. Appl. Polym. Sci. Polym. Symp.* 1983;37:223.

28. Suzuki H, Kamide K, Saitoh M. Lower Critical Solution Temperature Study on Cellulose Diacetate – Acetone Solution. *Euro. Polym. J.* 1982;18:123.

29. Elias H-G. Theta-Solvents. In: Brandrup J, Immergut EH, eds. *Polymer Handbook*. 3rd ed. New York: John Wiley & Sons; 1989:VII/205.

30. Gilbert RD, Patton PA. Liquid Crystal Formation in Cellulose and Cellulose Derivatives. *Prog. Polym. Sci.* 1983;9:115.

31. Peyser P. Glass Transition Temperatures of Polymers. In: Brandrup J, Immergut EH, eds. *Polymer Handbook*. 3rd ed. New York: John Wiley & Sons; 1989:VI/258.

32. 'Eastman Cellulose Esters for Pharmaceutical Drug Delivery'. Eastman Publication PCI-105C, October, 2005, *Eastman Chemical Company*.

33. Yang Yong. *Ph. D. Thesis*. Cincinnati, Ohio: University of Cincinnati; 1993.

34. Seard GA, Sanders JR. Cellulose Acetate and Triacetate Fibers. In: Mark H, et al. eds. *Encyclopedia of Chemical Technology*. New York: John Wiley & Sons; 1979:V5.

35. Huglin MB. Specific Refractive Index Increments of Polymers in Dilute Solution. In: Brandrup J, Immergut EH, eds. *Polymer Handbook*. 3rd ed. New York: John Wiley & Sons; 1989:VII/466.

36. Pauly S. Permeability and Diffusion Data. In: Brandrup J, Immergut EH, eds. *Polymer Handbook*. 3rd ed. New York: John Wiley & Sons; 1989:VI/451.

37. Seard GA. Cellulose Esters, Organic. In: Mark HF, Bikales NM, Overberger CG, Menges G, Kroschwitz JI, eds. *Encyclopedia of Polymer Science and Engineering*. New York: Wiley-Interscience; 1985:V3.

38. Wu S. Surface and Interfacial Tensions of Polymers, Oligomers, Plasticizers, and Organic Pigments. In: Brandrup J, Immergut EH, eds. *Polymer Handbook*. 3rd ed. New York: John Wiley & Sons; 1989:VI/411.

39. 'Cellulose Acetate', Technical Data Sheet. *Goodfellow Corporation*. Pennsylvania, USA: Oakdale.

40. Yang Yong. Thermal Conductivity. In: Mark JE, ed. *Physical Properties of Polymer Handbook*. Woodbury, New York: American Institute of Physics Press (1996).

Cellulose Butyrate

Yong Yang

Acronym EC

Class Carbohydrate polymers

Structure

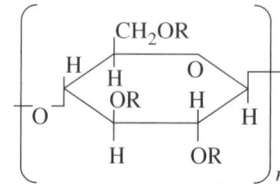

R is $-COC_3H_7$ or H

Major Application Used as cellulose acetate butyrate in lacquers, coatings, hot-melt adhesives, and plastics.

Properties of Special Interest Good tolerance for inexpensive lacquer solvents and common diluents.

Property	Units	Conditions	Value	Reference
Molecular weight of repeat unit	$g\ mol^{-1}$	Degree of substitution DS : 3.0	372.41	
Density	$g\ cm^{-3}$		1.17	(1)

Infrared characteristic absorption frequencies[2]

Frequency (cm^{-1})	Assignment
2960	(C_3H_7) stretching
2940	(C_3H_7) stretching
2870	(C_3H_7) stretching
1750	(C=O) stretching
1460	(C_3H_7) stretching
1420	(C_3H_7) deformation
1380	(C_3H_7) deformation
1370	(C_3H_7) deformation
1310	(C_3H_7) deformation
1250	(C_3H_7) deformation
1170	Structural factors
1080	Structural factors

Property	Units	Conditions	Value	Reference
Solubility parameter δ	$(MPa)^{1/2}$		17–24	(3)
Theta temperature Θ	K	Dodecane/tetralin (75:25 vol)	395	(3)
		Tetrachloroethane	329.7	(4)

Solvent and nonsolvent for cellulose tributyrate[4,5]

Solvent	Nonsolvent
Benzene, chloroform, cyclohexanone , dodecane/tetralin (3:1, $>130°C$), tetrachloroethane, xylene (hot)	Cyclohexane, diethyl ether, 2-ethylhexanol, hexane, methanol

Mark–Houwink parameters for cellulose tributyrate, K and a[6]

Solvent	Method	Temp. (K)	MW10^{-4} (g mol^{-1})	$K 10^3$ (ml g^{-1})	a
Butanone	Light scattering	303	6–32	4.3	0.87
	Osmometry	303	8–22	18.2	0.80
Tributyrin	Light scattering	273	6–32	5.3	0.87
	Light scattering	298	6–32	5.6	0.85
	Light scattering	323	6–32	6.1	0.82
	Light scattering	343	6–32	6.2	0.80
Dodecane/tetralin (75/25 by vol.)	Osmometry	403	11–21	82	0.50

Unit cell dimension of cellulose tributyrate[6,7]

Lattice	Monomers per Unit Cell	Cell Dimension (Å)		
		a	b	c
Orthorhombic	16	31.3	25.6	10.36

Degree of crystallinity of cellulose tributyrate[8]

Annealing Temperature (K)	Annealing Hours	Crystallinity (%)
298	18	36
363	136	40
373	72	39
383	18	41
393	18	43
403	18	43
413	18	45

Property	Units	Conditions	Value	Reference
Chain conformation			2_1	(9)
Heat of fusion	KJ mol^{-1}		12.6	(9)
	(of repeat unit)		12.8	(8)
Density (crystalline)	g cm^{-3}		1.192	(9)
Glass transition temperature	K	DS =3.0	388	(10)
		DS = 3.0, 100% amorphous, DSC	354	(8)
Melting point	K		206–207	(9)
			354	(8)
Heat capacity	KJ mol^{-1}		0.108	(8)
	(of repeat unit)			
Tensile strength	(MPa)$^{1/2}$		30.4	(1)
Water absorption		25% relative humidity	0.1	(1)
		50%	0.2	
		75%	0.7	
		95%	1.0	

Refractive index increment dn/dc (in (ml g^{-1}))[11]

Solvent	DS	Temperature (K)	dn/dc (λ_0 nm)
Bromoform	3.0	294	−0.11 (546)
Dimethylformamide	3.0	314	0.0442 (436)
			0.0478 (546)
Dioxane/water (93.5/6.5 vol.)	3.0	336	0.104 (546)
Methyl ethyl ketone	3.0	294	0.078 (546)

References

1. Bogan RT, Kuo CM, Brewer RJ. Cellulose Derivatives, Ether. In: Mark H, et al. eds. *Encyclopedia of Chemical Technology*. Vol 5. New York: John Wiley & Sons; 1979.
2. Zhbankov RG. In: Stepanov ABI, ed. *Infrared Spectra of Cellulose and Its Derivatives*. New York: Consultants Bureau Pub.; 1966.
3. Barton AFM. Cellulose Butyrate. *CRC handbook of Polymer-Liquid Interaction and Solubility Parameters*, Boca Raton, Florida: CRC Press; 1990.
4. Elias HG. Theta-Solvents. In: Brandrup J, Immergut EH, eds. *Polymer Handbook*. 3rd ed. New York: John Wiley & Sons; 1989:VII/205.
5. Fuchs O. Solvents and Non-Solvents for Polymers. In: Brandrup J, Immergut EH, eds. *Polymer Handbook*. 3rd ed. New York: John Wiley & Sons; 1989:VII/379.
6. Kurata M, Tsunashima Y. Viscosity-Molecular Weight Relationships and Unperturbed Dimension of Linear Chain Molecules. In: Brandrup J, Immergut EH, eds. *Polymer Handbook*. 3rd ed. New York: John Wiley & Sons; 1989:VII/31.
7. Zugenmaier P. Structural Investigation on Cellulose Derivatives. *J. Appl. Polym. Sci. Polym. Symp.* 1983;37:223.

8. Piana U, Pizzoli M, Buchanan CM. Thermodynamic Parameters Relative to the Melting of Cellulose Tributyrate. *Polymer.* 1995;36(2):373.

9. Miller RL. Crystallographic Data for Various Polymers. In: Brandrup J, Immergut EH, eds. *Polymer Handbook.* 3rd ed. New York: John Wiley & Sons; 1991:VI/88.

10. Peyser P. 'Glass Transition Temperatures of Polymers. In: Brandrup J, Immergut EH, eds. *Polymer Handbook.* 3rd ed. New York: John Wiley & Sons; 1989:VI/258.

11. Huglin MB. Specific refractive Index Increments of Polymers in Dilute Solution. In: Brandrup J, Immergut EH, eds. *Polymer Handbook.* 3rd ed. New York: John Wiley & Sons; 1989:VII/409.

Cellulose Nitrate

Yong Yang

Acronym CN

Class Carbohydrate polymers

Structure

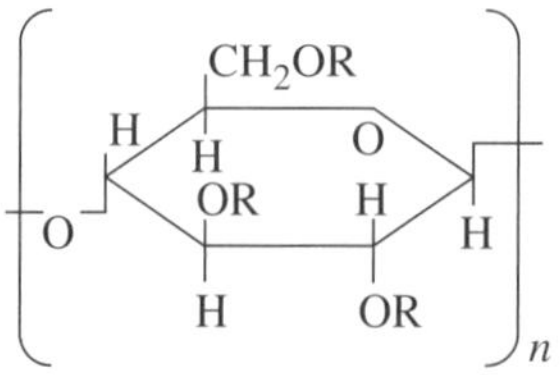

R is $-NO_2$ or H

Major Applications Photographic films, guncotton, membranes, protective and decorative lacquer coatings, rotogravure and flexographic inks, leather finishes, fabric and household adhesives, explosives, propellants, plastics.

Preparation Techniques Cellulose nitrate is prepared by reacting cellulose with nitric acid to convert the cellulose into cellulose nitrate.

$$\text{Cellulose} + HNO_3 \xrightarrow{H_2SO_4} \text{Cellulose Nitrate} + H_2O$$

Sulfuric acid is present to prevent the water produced in the reaction from diluting the concentrated nitric acid.

Properties of Special Interest Soluble in a wide variety of organic solvents, fast solvent release under ambient drying conditions, durability, toughness, clarity.

Property	Units	Conditions	Value	Reference
Molecular weight (of repeat unit)	$g\ mol^{-1}$	Degree of substitution DS : 3.0	297.13	
IR (Characteristic Absorption Frequencies)	(cm^{-1})	Assignment		(1, 2)
		(OH) stretching	3450	
		(CH$_2$) stretching	2970	
		(CH$_2$) stretching	2940	
		(ONO$_2$) stretching	1650	
		(ONO$_2$) stretching	1280	
		(ONO$_2$) stretching	840	
		(C–C–O) stretching	1070	
Thermal expansion coefficient	K^{-1}		$(8\text{--}12) \times 10^{-5}$	(3)

Cellulose Nitrate

Solvents and nonsolvents

DS	Solvent	Nonsolvent	Reference
1.00	Water		(4–6)
1.83–2.32	Acetone*, acetic acid (glacial), lower alcohols, alcohol/diethyl ether, amyl acetate, n-butyl acetate*, butyl lactate, γ-butyrolactin*, cyclopentanone*, diethyl acetate*, diethyl ketone*, N, N-dimethylacetamide*, dimethyl carbonate*, dimethyl cyanamide*, dimethylformamide*, dimethyl maleate*, dimethylsulfoxide*, 2-ethoxyethyl acetate, ethyl acetate*, ethyl amyl ketone, ethylene glycol ethers, ethyl lactate, 2-hexanone*, methyl acetate*, methyl ethyl ketone*, methyl propyl ketone*, n-methylpyrrolidone-2*, 2-octanone*, 1-pentanone*, n-pentyl acetate*, pyridine*	Higher alcohols, higher carboxylic acids, higher ketones, tricresyl phosphate	
2.48	Acetone*, cyclohexanone, ethanol/diethyl ether, ethyl butyrate, ethylene carbonate, ethylene glycol ether acetates, ethyl lactate, halogenated hydrocarbons, methyl acetate*, methyl amyl ketone*, furan derivatives, nitrobenzene	Alcohols, aliphatic hydrocarbons, aromatic hydrocarbons, carboxylic acids, dil, ethylene glycol, diethyl ether, water	

* Forms liquid crystalline mesophase.

Solubility parameters δ

DS	$\delta[(MPa)^{1/2}]$	Reference
2.21	21.7	(7)
	30.39	(7)
	23.5	(7)
2.08	21.93	(7)
2.21	21.44	(8)

Polymer–liquid interaction parameters χ (ϕ_2) at various volume fractions of polymer ϕ_2

Solvent	DS	Temperature (K)	$\chi(0)$	$\chi(0.2)$	$\chi(0.4)$	$\chi(0.6)$	$\chi(0.8)$	$\chi(1.0)$	Reference
Acetone	2.4	298	0.27						(8–10)
		303	0.24	0.05					
	2.6	293		0.14	0.06	−0.37	−1.24		
Acetonitrile	2.6	293			0.59	0.42	0.12	−0.1	
Amyl acetate	2.4	298	0.02						

Solvent	DS	Temperature (K)	$\chi(0)$	$\chi(0.2)$	$\chi(0.4)$	$\chi(0.6)$	$\chi(0.8)$	$\chi(1.0)$	Reference
2-Butanone	2.4	298	0.21						
Butyl acetate	2.4	298	0.015						
Cyclopentanone	2.6	293		0.42	0.07	−0.71	−2.4		
2,4-Dimethyl-3-pentanone	2.6	293	−0.89	−1.8	−1.7				
1,4-Dioxane	2.4	293			1.2	−0.25	−1.7		
Ethyl acetate	2.4	298	0.02						
	2.6	293		0.04	−0.43	−1.35			
Ethyl formate	2.6	293		−0.08	−0.14	−0.42	−3.2		
Ethyl n-propyl ether	2.6	293				1.20			
2-Heptanone	2.4	298	0.02						
2-Hexanone	2.4	298	0.15						
Isopentyl acetate	2.6	293			−0.89	−1.8	−3.3		
Isoproyl ketone	2.6	293			0.62	−0.08	−1.7		
Methyl acetate	2.4	298	0.30						
		303	0.17	−0.06					
Methyl t-butyl ketone	2.6	293			.016	−1.5	−2.8	−3.7	
Methyl isopropyl ketone	2.6	293			−0.5	−0.52	−1.6		
Nitromethane	2.6	293			0.66	0.64	0.60	0.45	
2-Octanone	2.4	298	0.16						
Propyl acetate	2.4	298	0.13						
	2.6	293			−0.38	−0.83	−2.0	−4.1	

Second virial coefficients A_2

	Solvent	Temperature (K)	MW 10^{-3} (g mol^{-1})	Method	$A_2 10^4$ (mol cm^3 g^{-2})	Reference
DS = 2.91	Acetone	298	81–3850	Light scattering	8.2–10.8	(11)
DS = 2.55	Acetone	298	141–1700	Light scattering	12.5–13.3	
DS = 2.78	Acetone	RT	61.6–2482	Osmometry	0.24	
		298	77–2640	Light scattering	6.10	
		–	780	Light scattering	11.2	
From raw cotton						
DS = 2.82	Acetone	288	22.8–417	Osmometry	0.24	
DS = 2.87	Ethyl acetate		~1000	Light scattering	6.2–7.0	(12)
From cotton	Acetone	293	31–661	Osmometry	0.28	(11)
	Butyl acetate	293	150–400	Light scattering	0.5–1.0	
		298	30–360	Osmometry	0.3–3.5	
From viscose rayon	Ethyl acetate	303	71.5	Osmometry	44.1	
From chemical cotton			295–450	Osmometry	25.7–28.5	
DS = 2.39	Butanone	298	130	Osmometry	10.8	

Cellulose Nitrate

Mark–Houwink parameters, K and a

Polymer	Solvent	Temperature (K)	MW 10^{-4} g mol^{-1}	K 10^3 ml g^{-1}	a	Method	Reference
Cellulose trinitrate	Acetone	293	250	2.80	1.00	Sedimentation	(13)
		298	265	1.69	1.00	Light scattering	
		298	250	1.66	0.86	Light scattering	
		298	32	10.8	0.89	Light scattering	
(DS=2.55)		298	200	5.70	0.90	Light scattering	
(DS=2.91)		298	400	6.93	0.91	Light scattering	
		298	50	7.00	0.933	Osmometry	
		298	100	11.0	0.91	Osmometry	
		298	26	23.5	0.78	Osmometry	
	Butyl acetate	298	50	5.68	0.969	Osmometry	
	Butyl formate	298	26	23	0.81	Osmometry	
	Cyclohexanone	298	22	2.24	0.810	Osmometry	
	Ethyl acetate	298	100	3.8	1.03	Osmometry	
		298	26	8.3	0.90	Osmometry	
		298	250	1.66	0.86	Light scattering	
		303	57	2.50	1.01	Light scattering	
	Ethyl butyrate	298	50	3.64	1.0	Osmometry	
	Ethyl formate	298	26	30	0.79	Osmometry	
	Ethyl lactate	298	65	12.2	0.92	Osmometry	
	2-Heptanone	298	26	5.0	0.93	Osmometry	
	Methyl acetate	298	22	18.3	0.835	Osmometry	
	Nitrobenzene	298	22	6.1	0.945	Osmometry	
	Pentyl acetate	298	26	1.1	1.04	Osmometry	

Persistence length

Solvent		Temperature (K)	Persistence length (nm)	Method	Reference
DS = 2.91	Acetone	298	970		(13)
DS = 2.55	Acetone	298	530		
DS = 2.75	Ethyl acetate	303	700		
DS = 2.26	Acetone	293	0.48		
Cellulose trinitrate	Acetone	298	360		
		295	0.26 ± 0.01		
		293	0.40–0.70		
Cellulose trinitrate	Acetone		13.2		(14)
	Ethyl acetate		11.8		

Unit cell dimension of cellulose trinitrate

Lattice	Monomers per Unit Cell	Cell Dimension (Å)			γ	Tm (K)	Heat of Fusion (KJ mol^{-1})	Chain Conformation	Reference
		a	b	c					
Orthorhombic	10	12.25	25.4	9.0		697	3.8	5_1	(15,16)
Orthorhombic	10	9.0	14.6	25.4		700	6.3	5_2	(16)
Monoclinic (CTNII)	10	12.3	8.55	25.4	91°				(17)

Cellulose Nitrate

Property	Units	Conditions	Value	Reference
Specific gravity	g cm^{-1}	DS = 2.20–2.32	1.58–1.65	(4)
Huggins constants, k' and k''				(11)
Glass transition temperature	K		326, 329	(18)
Melting point	K	Dinitrate	~473	(19)
		Trinitrate	~553	(19)
Heat capacity	KJ K^{-1} mol^{-1}		0.37–0.50	(3)
Deflection temperature	K	at 1820 KPa	60–71	(3)
Tensile modulus	MPa		1310–1520	(3, 20)
Tensile strength	MPa	RS, 296 K, 50% relative humidity	62–110	(4)
			48.3–55.2	(20)
Maximum extensibility	%	RS, 296 K, 50% relative humidity	13–14	(4)
			40–45	(3)
Compressive strength	MPa	ASTM D695	152–241	(3)
Flexural yield strength	MPa		62–75.9	(3)
Impact strength	J M^{-1}	1/2 by 1/2 in. notched bar, Izod test, ASTM D256	267–374	(3)
Hardness		RS, Sward, % on glass	90	(4)
		Rockwell, R scale	95–115	(3)
Index of refraction n			1.51	(4)

Refractive index increment dn/dc (in (ml g^{-1}))[12,21]

Solvent	DS	Temperature (K)	dn/dc (λ_0 nm)
Acetone	1.96	298	0.1022 (436), 0.0998 (546)
	2.23		0.1010 (436), 0.0985 (546)
	2.26–2.35	293	0.107 (436), 0.0950 (546)
	2.43	298	0.0968 (436)
	2.55		0.1151(436)
	3.0	298	0.0930 (436), 0.0903 (546), 0.098 (1086)
Ethyl acetate	2.05	293	0.103 (546)
	2.77	298	0.102 (436)
		293	0.107 (436)
	2.87		0.105 (436,546)
	3.0	303	0.102 (436)
		298	0.107 (436)
		293	0.105 (436), 0.103–0.105 (546)

Property	Units	Conditions	Value	Reference
Dielectric constant ε''		293–298 K		
		60 Hz	7–7.5	(4)
		1000 Hz	7	
		1 000 000 Hz	6	
Power factor	%	293–298 K,		
		60 Hz	3–5	(4)
		1000 Hz	3–6	(4)
Surface tension	mNm^{-1}		38	(22)
Thermal conductivity	$W\ mK^{-1}$		0.23	(23)
Water absorption	%	294 K, 24 h, 80% relative humidity	1.0	(4)
Compatible polymers		Cellulose acetate, ethyl cellulose, ethylhydroxyethylcellulose, poly(carprolacton), poly(vinyl acetate)		(20, 24)
Cost	$ kg^{-1}$	in 30% isopropanol	3.7–5.5	
Supplier		Hercules Inc., 1313 North Market Street, Wilmington, DE 19894		

Permeability coefficient P (in $m^3(STP)\ m\ s^{-1}\ m^{-2}\ Pa^{-1}$)[25]

Permeant	Temperature (K)	$P \times 10^{17}$
H_2	293	1.5
He	298	5.18
N_2	298	0.087
O_2	298	1.46
Ar	298	0.0825
CO_2	298	1.59
NH_3	298	42.8
H_2O	298	4720
SO_2	298	1.32
C_2H_6	298	0.0473
CH_3H_8	298	0.0063

References

1. Zhbankov RG. In: Stepanov ABI, ed. *Infrared Spectra of Cellulose and Its Derivatives.* New York: Consultants Bureau Pub.; 1966.
2. Julian JM, Anderson DG, Brandau AH, McGinn JR, Millon AM. In: Brezinski DR, ed. *An Infrared Spectroscopy for the Coatings Industry.* 4th ed. Blue Bell, Pennsylvania: Federation of Societies for Coatings Technology; 1991.
3. Rudd GE, Sampson RN. In: Harper CA, ed. *Handbook of Plastics, Elastomers, and Composites.* New York: McGraw-Hill; 1992.
4. Hercules. *Nitrocellulose. Chemical and Physical Properties.* Wilmington, Delaware: Hercules, Inc.; 1996.
5. Fuchs O. Solvents and Non-Solvents for Polymers. In: Brandrup J, Immergut EH, eds. *Polymer Handbook.* 3rd ed. New York: John Wiley & Sons; 1989:VII/400.

6. Gray DG. Liquid Crystalline Cellulose Derivatives. *J. Appl. Polym. Sci. Appl. Polym. Symp.* 1983;37:179.

7. Grulke EA. Solubility Parameter Values. In: Brandrup J, Immergut EH, eds. *Polymer Handbook*. 3rd ed. New York: John Wiley & Sons; 1989:VII/555.

8. Du Y, Xue Y, Frish HL. Solubility Parameters. In: Mark JE, ed. *Physical Properties of Polymer Handbook*. Woodbury, NY: American Institute of Physics Press; 1996.

9. Gundert F, Wolf BA. Polymer-Solvent Interaction Parameters. In: Brandrup J, Immergut EH, eds. *Polymer Handbook*. 3rd ed. New York: John Wiley & Sons; 1989:VII/173.

10. Orwoll RA. The Polymer-Solvent Interaction parameter chi. *Rubber Chem. Technol.* 1977;50:451.

11. Lechner MD, Steinmeier DG. Sedimentation Coefficients, Diffusion Coefficients, Partial Specific Volumes, Frictional Ratios, and Second Virial Coefficients of Polymers in Solution. In: Brandrup J. and Immergut EH eds. *Polymer Handbook*. 3rd ed. New York: John Wiley & Sons; 1989:VII/61.

12. Holt C, Mackie W, Sellen DB. Configuration of Cellulose Trinitrate in Solution. *Polymer.* 1976;17:1027.

13. Kurata M, Tsunashima Y. Viscosity-Molecular Weight Relationships and Unperturbed Dimension. In: Brandrup J, Immergut EH, eds. *Polymer Handbook*. 3rd ed. New York: John Wiley & Sons; 1989:VII/46.

14. Gilbert RD, Patton PA. Liquid Crystal Formation in Cellulose and Cellulose Derivatives. *Prog. Polym. Sci.* 1983;9:115.

15. Miller RL. Crystallographic Data for Various Polymers. In: Brandrup J, Immergut EH, eds. *Polymer Handbook*, 3rd ed. New York: John Wiley & Sons; 1991:VI/1.

16. Meader D, Atkins EDT, Happey. Cellulose Trinitrate, Molecular Conformation and Packing Consideration. *Polymer.* 1978;19:1371.

17. Marchessault RH, Sundarajan PR. 1983. Cellulose. *The Polysaccharides*. Orlando, Florida: Academic Press.

18. Peyser P. Glass Transition Temperatures of Polymers. In: Brandrup J, Immergut EH, eds. *Polymer Handbook*. 3rd ed. New York: John Wiley & Sons; 1989:VT/209.

19. Colborne RS. Melting Points of Cellulose Nitrates. *J. Appl. Polym. Sci.* 2003;12(4):761.

20. Bogan RT, Kuo CM, Brewer RJ. Cellulose Derivatives, Ether. In: Mark H, et al eds. *Encyclopedia of Chemical Technology*. Vol 5. New York: John Wiley & Sons; 1979.

21. Huglin MB. Specific Refractive Index Increments of Polymers in Dilute Solution. In: Brandrup J, Immergut EH, eds. *Polymer Handbook*. 3rd ed. New York: John Wiley & Sons; 1989: VII/409.

22. Wu S. Surface and Interfacial Tensions of Polymers, Oligomers, Plasticizers, and Organic Pigments. In: Brandrup J, Immergut EH eds. *Polymer Handbook*. 3rd ed. New York: John Wiley & Sons. 1989:VI/411.

23. Yang, Yong. Thermal Conductivity. In: Mark JE ed. *Physical Properties of Polymer Handbook*. Woodbury, New York: American Institute of Physics Press; 1996.

24. Krause S. Compatible Polymers. In: Brandrup J, Immergut EH, eds. *Polymer Handbook*. 3rd ed. New York: John Wiley & Sons; 1989:VI/352.

25. Pauly S. Permeability and Diffusion Data. In: Brandrup J, Immergut EH eds. *Polymer Handbook*. 3rd ed. New York: John Wiley & Sons; 1989:VI/451.

Chitin

Rachel Mansencal, John F. Kadla and
Jennifer L. Braun

Class Carbohydrate polymers; polysaccharides

Structure

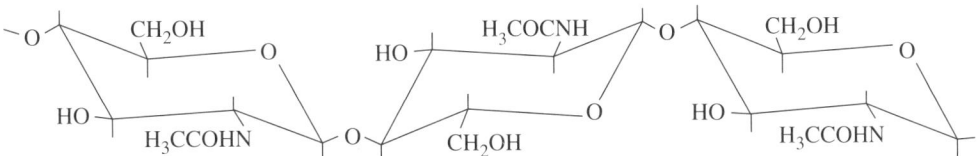

Natural Resources Chitin is a biopolymer found in crustaceans shells (crab, shrimp, prawn, lobster) in some mollusks (krill, oyster, clam shells, squid skeleton). It is also found in fungi (mushrooms, yeast) and in various insects (cockroaches, silkworms, spiders, beetles).[1−2]

Biosynthesis[1−2]

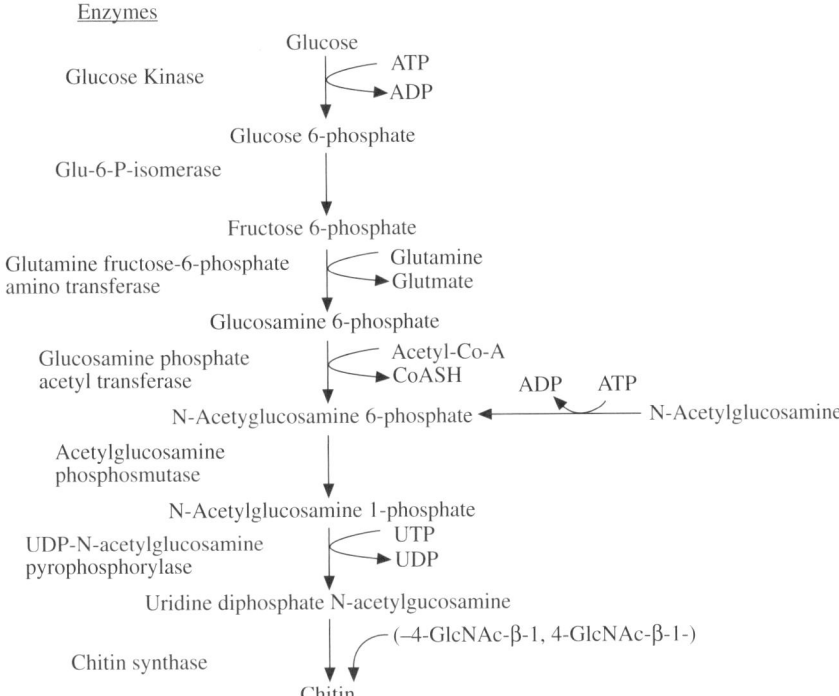

Chitin

Extraction Chitin is produced by removing calcium carbonate and proteins from the shells.[1]

Major Applications Biomedical (wound and burn healing, treatment of fungal infections, antitumor agent, hemostatic agent, etc.); cosmetics (additives); biotechnology (enzyme and cell immobilization); industry (paper industry, food industry, etc.); agriculture and environmental protection.[1−3]

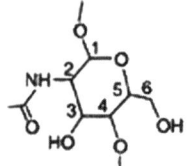

Properties of Special Interest Natural resources; basic polysaccharides; nontoxic; biodegradability; bioactivity; biosynthesis; interesting derivatives (chitosan); toughness; graft copolymerization; chelating ability for transition metal cations; immobilizes enzymes by chemical linking or adsorption; chiral polymer.[1−4]

Property	Units	Conditions	Value	Reference
Infrared absorption				

Type of Motion	Wavenumber (cm^{-1})[14]	
	α-chitin	β-chitin
ν_{OH}	3479, 3448	3479, 3426
ν^{AS}_{NH}	3268	3290
ν^{S}_{NH}	3106	3102
ν^{AS}_{CH3}	2965	2962
ν^{S}_{CH2}	2927	2929
ν^{AS}_{CH3}	2883	2880
$\nu_{C=O}$ (amide I)	1660, 1627	1656
ν_{C-N} (C−N−H) + δ_{NH} (amide II)	1558	1556
δ_{CH2}	1422	1424
δ_{CH} + δ_{C-CH3}	1376	1376
ν_{C-N} + δ_{NH} (amide III)	1312	1314
δ_{NH}	1255	1262
ν^{AS}_{C-O-C} (ring)	1157	1155
ν_{C-O}	1072	1069
ν_{C-O}	1113	1111
ν_{C-O}	1021	1032
γ_{CH3}	957	948
γ_{CH} (C1 axial)(β bond)	896	902
ρ_{CH2}	746	
ν_{NH} (amide V)	698	692
ν_{C-O}	610	616
ν_{C-C}	566	

Property	Units															Conditions	Value	Reference
NMR Chemical Shift (ppm)[15–16]																		
	C-1		C-2		C-3		C-4		C-5		C-6		CH$_2$		C=O			
	^{12}C	^{1}H	^{12}C	^{1}H	^{12}C	^{1}H	^{12}C	^{1}H	^{12}C	^{1}H	^{12}C	^{1}H	^{12}C	^{1}H	^{12}C			
α	104.0	4.6	54.8	3.3	73.4	3.1	82.9	3.0	75.6	3.0	60.6	3.1, 3.8	22.6	1.1	173.0			
α	104.0		54.6		73.0		82.8		75.4		60.5		22.5		172.6			
β	104.1	3.7	55.2	3.3	74.2	3.1	83.4	3.1	73.6	3.1	60.8	3.0	22.8	1.5	173.6			
β	104.2		55.2		74.8		84.1		74.8		60.9		22.5		173.1			
γ	103.7		54.8		73.1		82.9		75.4		61.1		22.6		173.4			

Source	Polymorph	Nitrogen (%)	Acetylation (%)	Moisture (%)	Ash (%)	Reference
King crab	α	6.8	79.5	3.8	0.59	(14)
Shrimp	α	6.8	78.7	5.9	0.49	(14)
Other crab	α	6.6	83.9	5.2	0.32	(14)
Squid	β	6.7	95.1	13.6	0.27	(14)
Lobster	α	6.8	84.6	6.9	0.50	(14)

Property	Units	Conditions	Value	Reference
Molecular weight	g mol^{-1}	Native chitin	>10^6	(1–4)
		Commercial chitin	(1–5) × 10^5	
Transition metals	μg g^{-1}	—	<5.0	(1–4)
Protein content	%	Amino-acid catalyst	<0.5	(4)
Solubility	—	Aq. NaOH	—	(20)
		Formic acid	—	(16)
		5 wt % LiCl/DMAc	—	(21, 22)
		Dichloroacetic acid	—	(23)
		Trichloroacetic acid	—	(23)
		Lithium thiocyanate	—	(24)
		LiCl/N-methylpyrrolidinone	—	(24)
Dissociation constant K_a	—	—	6.0–7.0	(19)
Solubility parameter	(J$^{1/2}$cm$^{-3/2}$)	—	41.15	(25)
Onset of thermal decomposition	K	—	523	(16)
Elastic modulus	MPa	Crab shell	41,000	(28)
Biodegradability (effective microorganisms)	Chitinoclastic bacteria	Normal	Very slow	(2)
	Chitinase associated with chitobiases, β-N, acetylhexosaminidases, and lysozymes	pH = 4.0–7.0	Most active	(2)
Toxicity	g kg^{-1} body weight	LD 50	16	(1)

Unit cell dimensions

Polymorph	Space Group	Chains Per Unit Cell	Unit Cell Dimensions						Reference
			a (Å)	b (Å)	c (Å)	α (°)	β (°)	γ (°)	
α	$P2_12_12_1$	2 parallel, 2 antiparallel	4.74	18.86	10.32	90.0	90.0	90.0	(14)
β	$P2_1$	2 Parallel	Vary with degree of hydration						(18)

Major X-ray reflections

α-chitin[18]		β-chitin (anhydrous)[18]	
2θ	hkl	2θ	hkl
9.39	0 2 0	9.7	0 0 1
12.72	0 2 1	13.0	0 1 1
19.30	1 1 0	18.5	1 0 0
26.37	0 1 3	20.2	0 2 1

Mark-Houwink constants

Solvent	Temperature (°C)	$K_m \times 10^6$ (dL/g)	a	Molecular Weight Range	Method of Calibration	Reference
2.77M NaOH	20	100.0	0.68	100,000–1,200,000	LS & viscometry	(20)
Formic acid	25	89.3	0.71	52,000–70,000	Viscometry	(16)
5 wt % LiCl/DM Ac	30	7.6	0.95	80,000–710,000	LS & viscometry	(21)
		21.0	0.88	59,000–1,400,000	Viscometry	(22)

Thermal Expansion Coefficient of Chitin Polymorphs
$\alpha = (pt + q) \times 10^{-5}$°$C^{-1}$, 25°C $\leq$ t $\leq$ 250 °C

Polymorph	Axis				Reference
	a		b		
	p	q	p	q	
α	0.0	6.0	0.0	5.7	(26)
β	0.0	4.0	3.0	−14.6	(26)

Complex Dielectric Constant

Frequency (kHz)	Complex Dielectric Constant	Reference
200	5.6	(27)
2000	5.4	(27)

References

1. Mark HF, et al. *Encyclopedia of Polymer Science and Engineering.* 2d ed. Vol 3. New York: John Wiley and Sons; 1989.

2. Salamone JC, ed. *Polymeric Materials Encyclopedia.* Vol 2. Boca Raton, Fla.: CRC Press; 1996.

3. Zikakis John, ed. *Chitin, Chitosan and Related Enzymes.* Orlando: Academic Press; 1984.

4. Muzzarelli RAA. *Natural Chelating Polymers.* Oxford: Pergamon Press; 1973.

5. Gow NAR, et al. *Carbohydr. Res.* 1987;165:105.

6. Huong DM, Dung NX, Luyen DV. *Journal of Chemistry.* 1989;27(3):20.

7. Saito H, Tabeta R, Hirano S. *Chem. Lett.* 1981:1479.

8. Saito H, Tabeta R, Ogawa R. *Macromolecules* 1987;20:2424.

9. Hirai A, Odani H, Nakajima A. *Polym. Bull.* 1991;26(1):87.

10. Persson JE, Domard A, Chanzy H. *Int. J. Biol. Macromol.* 1992;14(2):221.

11. Minke R, Blackwell J. *J. Mol. Biol.* 1978;120:167.

12. Gardner KH, Blackwell J. *Biopolymers.* 1975;14(8):1581.

13. Muzzarelli RAA. In: Aspinall GO, ed. *The Polysaccharides.* Vol 3. New York: Academic Press; 1982.

14. Càrdenas G, Cabrera G, Taboada E, Miranda SP. *J. Appl. Polym. Sci.* 2004;93:1876.

15. Kono H. *Biopolymers.* 2004;75(3):255.

16. Jang MK, Kong BG, Jeong JI, Lee CH, Nah JW. *J. Polym. Sci.: Part A: Polym. Chem.* 2004;42:3423.

17. Goodrich JD, Winter WT. *Biomacromolecules.* 2007;8:252.

18. Saito Y, Kumagai H, Wada M, Kuga S. *Biomacromolecules* 2002;3(3):407.

19. Elvers B, Hankins S, Russey W, eds. *Ullmanns Encykl. Ind. Chem.* VCH, 1994.

20. Einbu A, Naess SN, Elgsaeter A, Vårum KM. *Biomacromolecules.* 2004;5:2048.

21. Poirier M, Charlet G. *Carbohydrate Polymers.* 2002;50:363.

22. Noishiki Y, Takami H, Nishiyama Y, Wada M, Okada S, Kuga S. *Biomacromolecules.* 2003;4:896.

23. Roberts GAF. *Chitin Chemistry.* McMillan; 1992.

24. Rutherford FA, Austin PR. In: Muzzarelli RAA, Panser ER, eds. *Proc. Int. Conf. Chitin/Chitosan.* 1st ed. 1978:182.

25. Ravindra R, Krovvidi KR, Khan AA. *Carbohydrate Polymers.* 1998;36:121.

26. Wada M, Saito Y. *J. Polym. Sci.: Part B: Polym. Phys.* 2001;39:168.

27. Seoudi RA, Nada MA, Elmongy SA, Hamed SS. *J. Appl. Polym. Sci.* 2005;98:936.

28. Nishino T, Matsui R, Nakamae K. *J. Polym. Sci.: Part B: Polym. Phys.* 1999;37:1191.

Collagen

Jagath K. Premachandra and Chandima Kumudinie Jayasuriya

Class Polypeptides and proteins

Structure The most common type of collagen, Collagen I, is composed of two kinds of polypeptide helices, $\alpha 1$ and $\alpha 2$, in a 2:1 ratio respectively, to form a triple helix. The $\alpha 1$ and $\alpha 2$ chains of tropocollagen have a regularly repeating sequence of amino acid residues in which glycine is found at every third residue. This sequence can be written $(GLY–X–Y)_n$, where X and Y are often proline and hydroxyproline respectively.[1]

Functions An extracellular protein, which is responsible for the strength and flexibility of connective tissue. Accounts for 25–30% of the protein in an animal.[2] Major component in all mammalian tissues including skin, bone cartilage, tendons, and ligaments.[3]

Major Applications Biomaterial applications such as dermal implant, carrier of drugs, cell culture matrix, wound dressing, material for hybrid organ, drug delivery system, soft contact lens, tissue implants, cardiovascular graft, artificial heart, etc. Synthetic sausage casings in food industry.[3,4]

Major types of collagen[1,5]

Molecular Formula	Composition	Tissue Distribution
$(\alpha 1)_2\alpha 2$ Type I chains	Low hydroxylysine, low carbohydrate, broad fibrils	Tendons, bone, skin, ligaments cornea, internal organs (comprises 90% of body collagen)
$(\alpha 1)_3$ Type II chains	High carbohydrate, high hydroxylysine, thin fibrils	Cartilage, vitreous body of eye, intervertebral disc
$(\alpha 1)_3$ Type III chains	Low carbohydrate, high hydroxylysine, high hydroxyproline	Blood vessels, skin, internal organs
$(\alpha 1)_3$ Type IV chains	high hydroxylysine, high carbohydrate	Basal laminae
$(\alpha 1)_2\alpha 2$ Type V chains	High carbohydrate, high hydroxylysine	Many tissues in small amounts

Collagen

Biosynthesis of Collagen[6]

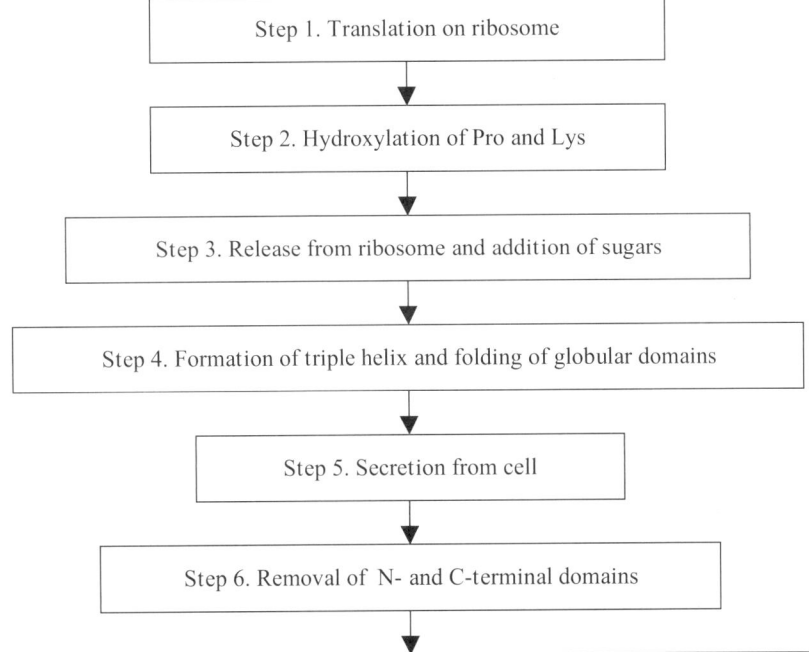

Step 1. Translation on ribosome

Step 2. Hydroxylation of Pro and Lys

Step 3. Release from ribosome and addition of sugars

Step 4. Formation of triple helix and folding of globular domains

Step 5. Secretion from cell

Step 6. Removal of N- and C-terminal domains

Step 7. Deamination of lysine residues to form aldehyde and formation of cross-links

Selected amino acid analysis of collagens[7]

Amino Acid*	Collagen					
	α (I)[a]	α2[a]	α2[b]	α (II)[c]	α (III)[a]	α (IV)[d]
3-Hydroxyproline	1	0	1	2	0	11
4-Hydroxyproline	96	82	86	99	125	130
Aspartic acid	41	47	49	42	42	51
Threonine	16	19	19	20	13	23
Serine	37	35	37	26	39	37
Glutamic acid	71	68	73	90	71	84
Proline	133	120	107	121	107	61
Glycine	336	337	324	332	350	310
Alanine	115	105	102	100	96	33
Cysteine	0	0	0	0	2	8.0
Valine	20	33	37	18	14	29
Methionine	7	5	4.6	9	8	10
Isoleucine	7	15	17	9	13	30
Leucine	20	30	34	25	22	54
Tyrosine	1.9	4.6	3.0	1	3	6
Phenylalanine	12	12	12	13	8	27
Hydroxylysine	5.4	7.6	11.5	14	5	44.6

Collagen

Amino Acid*	Collagen					
	α (I)[a]	$\alpha 2$[a]	$\alpha 2$[b]	α (II)[c]	α (III)[a]	α (IV)[d]
Lysine	30	22	21	22	30	10
Histidine	2	10	7.9	2	6	10
Arginine	49	51	53	51	46	33
Galactose	–	–	1.0	–	–	34.0
Glucose	–	–	0.65	–	–	2.0

* Values expressed as residues per 1000 amino acids.
[a] Human skin; [b] Human cornea; [c] Human cartilage; and [d] Human glomerular basement membrane.

Property	Units	Conditions	Value	Reference
Typical molecular weight range of polymer M_w	g mol^{-1} ($\times 10^5$)	Calf skin, solvents: citrate pH 3.7, phosphate pH 7.4, sedimentation equilibrium	2.5–3.1	(8)
		Calf skin, solvents: citrate pH 3.7, viscosity	3.5	(8)
		Dogfish shark skin, solvent: formate pH 3.8, sedimentation equilibrium	<3.5	(8)
		Chick cartilage and skin, sedimentation equilibrium.	2.62–3.07	(9)
		Type 1 collagen, rat tail tendon, aggregates in 0.01 M HCl	8.05	(10)
		Type 1 collagen, single molecules	2.82	(10)
IR (characteristic absorption frequencies)	cm^{-1}	N—H stretch	3330	(11)

Structural Information	Reference
Collagen I is composed of two kinds of polypeptide helices, $\alpha 1$ and $\alpha 2$, in a 2:1 ratio respectively to form a triple helix. The $\alpha 1$ and $\alpha 2$ chains of tropocollagen have a regularly repeating sequence of amino acid residues in which glycine is found at every third residue. This sequence can be written $(GLY–X–Y)_n$, where X and Y are often proline and hydroxyproline.	(1)
Arranged in fibrils, composed of microfibrils, characteristic striation with a repeat distance of about 670 Å.	(2)
Consists of macrofibrils, fibrils, and subfibrils of diameters $\sim 10^4$, 10^3, and 10^2 Å respectively, spun collagen fiber after thermal treatment at 170°C, by scanning electron microscopy.	(12)
Helical structure, when heated above 40°C, the helices loosen and form thread-like chains and the collagen becomes gelatin.	(3)
Triple helix, by optical rotatory dispersion.	(4)
Rodlike with a length ~ 2800 Å, by light scattering.	(11)
Helical rod, length ~ 3000 Å and width ~ 15 Å.	(13)
Native collagen type I fibrils, using AFM revealed that banding of the shell also occur in the core	(14)
Single pulse of UV radiation can affect the conformation and photostability of collagen peptide. UV laser light is capable of inducing conformational changes in the irradiated collagen films.	(15)
Supramolecular structure of adsorbed collagen is highly dependent on the underlying substrate surface chemistry.	(16)

Modeling Studies	Reference
Multi-scale modeling of mechanical behavior of collagen fiber networks: Volume-averaging theory for studying the mechanics of collagen networks	(17)
Molecular multiscale modeling of nanomechanical properties of collagen fibrils with varying cross-link densities. Confirmed the significance of cross-links in collagen fibrils in improving its mechanical strength.	(18)

Property	Units	Conditions	Value	Reference
Linear thermal expansion coefficient	K^{-1}	Whale ligament	~ 0	(19)
		Rat tail tendons	3.24×10^{-4}	(19)
Density	g cm^{-3}	At 25°C, in 8 M LiBr–diethyleneglycol monobutyl ether		
		Rat tail tendon from a 2-month-old rat	1.30	
		Rat tail tendon from a 10-month-old rat	1.30	
		Whale ligament	1.32	
		Rat tail tendon cross-linked with 1,3-bis(vinylsulfonyl)-2 propanol (BVSP)	1.33	
		At 25°C, in 8 M LiBr–diethyleneglycol nonomethyl ether, cross-linked with BVSP, rat tail tendon	1.34	(19)

Property	Units	Conditions	Value	Reference
Solubility		Collagen in its mature form is insoluble under physiological conditions, can be denatured by heat, mild acid, or alkaline treatment.		(2)
Second virial coefficient	Mol cm^3 g^{-2}	Solvent: 2 M KCNS at 25°C	3.0×10^{-4}	(11)
Degree of crystallinity	%	X-ray diffraction from collagen fibers	20–40	(11)
Denaturation temperature	K	Noncross-linked, spun collagen fibers before thermal treatment	316	(12)
		100°C for 30 min	314	
		170°C for 30 min	312	
		3 wt% glutaraldehyde cross-linked before thermal treatment	337	(12)
		100°C for 30 min	337	
		170°C for 30 min	318	
		7 wt% Cr-tanned before thermal treatment	365	(12)
		100°C for 30 min	360	
		170°C for 30 min	353	

Property	Units	Conditions	Value	Reference
		Type I collagen	312.6	(20)
		Type II collagen	314.0	(20)
		Type III collagen	312.2	(20)
		In 0.1 M acetic acid by optical rotation		
		Chick cartilage	313.7	
		Chick skin	315.5	
		Lamb anterior lens capsule by circular dichorism spectroscopy	313	(21)
		Pig kidney collagen	310	(22)
		Muscle layer collagen in Ascaris	313	(22)
		Intact collagen		(23)
		Bovine semimembranosus	343.6	
		Bovine longissimus dorsi	344.3	
		Bovine longissimus dorsi, after 1-week storage	340.2	
		Bovine longissimus dorsi	344.5	
		Rat skin	341.3	
		Bovine tendon	340.7	
		Cod skin	323.6	
		Tropocollagen		(23)
		Calf skin	323.6	
		Rat skin	329.2	
		Type IV procollagen, optical rotatory dispersion, at neutral pH		(24)
		Heating rate $\sim$0.17°C min^{-1}	308–315, 321	
		Heating rate $\sim$0.027°C min^{-1}	307–314, 320	
		At neutral pH, DSC	309.0, 315.1, 321.0	
		In 10 mM acetic acid	308.6, 311.9, 314.7, 323.0	
		Soluble collagen, conc = 0.86 mg ml^{-1}, in 0.15 M potassium acetate buffer, pH 4.7, solvent:		(25)
		Ethylene glycol (1 M)	312.3	
		2-Methoxyethanol (1M)	312.0	
		Control	311.9	
		2-Ethoxyethanol (1 M)	311.1	
		2-Butoxyethanol	308.3	
		Type IV collagen, bovine anterior lens capsules	327.4, 361.8	(26)
Transition enthalpy	kJ mol^{-1}	In 50 mM sodium citrate buffer, pH 3.9, in kJmol^{-1} in tripeptide units		(20)
		Type I collagen	17.0	
		Type II collagen	17.5	
		Type III collagen	16.5	
Heat capacity	kJ K^{-1}mol^{-1}	Native hydrated	1.60×10^{-3}	(27)
		Native anhydrous	1.22×10^{-3}	

Collagen

Property	Units	Conditions	Value	Reference
Tensile modulus	MPa	Gauge length 2.0 cm, strain rate 50% min^{-1} Uncross-linked, unstretched fiber, wet diameter $d = 327 \mu$m	1.8 ± 0.3	(28)
		10% stretched fiber, uncross-linked, $d = 253~\mu$m	5.7 ± 2.5	
		30% stretched fiber, uncross-linked, $d = 173~\mu$m	20.8 ± 4.34	
		50% stretched fiber, uncross-linked, $d = 147~\mu$m	46.0 ± 19.9	
		Unstretched fiber, cross-linked, $d = 94~\mu$m	383 ± 112	
		10% stretched fiber, cross-linked, $d = 95.6~\mu$m	429 ± 111	
		30% stretched fiber, cross-linked, $d = 86.1~\mu$m	726 ± 120	
		50% stretched fiber, cross-linked, $d = 80.3~\mu$m	766 ± 111	
		Gauge length $= 1$ cm, elongation rate $= 100$ mm min^{-1}		(29)
		Collagen–poly(lactic acid) (PLA) composites, 50% collagen fiber and 50% PLA matrix (w/w)	~37	
		Collagen–collagen composites, 50% collagen fiber and 50% collagen matrix (w/w)	~18	(29)
		Uncross-linked collagen matrix	10	(29, s30)
		Native collagen type I fibrils	Same tensile modulus on the shell and in the core	(14)
Shear modulus	MPa	At 25°C, in 8 M LiBr-diethyleneglycol monobutyl ether		(19)
		Rat tail tendon from a 2-month-old rat	2.91×10^{-3}	
		Rat tail tendon from a 10-month-old rat	0.129	
		Whale ligament	0.172	
		Rat tail tendon cross-linked with 1,3-bis(vinylsulfonyl)-2-propanol (BVSP)	0.622	
		At 25°C in 8 M LiBr-diethyleneglycol nonomethyl ether, rat tail tendon cross-linked with BVSP	0.602	(19)
Tensile strength	MPa	Gauge length 2.0 cm, strain rate 50% min^{-1} Uncross-linked, unstretched fiber, wet diameter $d = 327~\mu$m	0.91 ± 0.21	(28)
		10% stretched fiber, uncross-linked, $d = 253~\mu$m	2.0 ± 1.2	
		30% stretched fiber, uncross-linked, $d = 173~\mu$m	5.9 ± 1.3	
		50% stretched fiber, uncross-linked, $d = 147~\mu$m	7.2 ± 1.3	
		Unstretched fiber, cross-linked, $d = 94~\mu$m	46.8 ± 17.1	
		10% stretched fiber, cross-linked, $d = 95.6~\mu$m	51.6 ± 17.0	
		30% stretched fiber, cross-linked, $d = 86.1~\mu$m	71.5 ± 18.3	
		50% stretched fiber, cross-linked, $d = 80.3~\mu$m	68.8 ± 15.8	
		Gauge length $= 1$cm, elongation rate $= 100$ mm min^{-1}		(29)
		Collagen–poly(lactic acid) (PLA) composites, 50% collagen fiber and 50% PLA matrix (w/w)	~13	

Property	Units	Conditions	Value	Reference
		Collagen–collagen composites, 50% collagen fiber and 50% collagen matrix (w/w)	~7	(29)
		Uncross-linked, collagen matrix	5	(29, 30)
		Gauge length = 2 cm, strain rate 1 mm min^{-1}		(12, 31)
		Uncross-linked fiber	~350	
		0.1 wt% glutaraldehyde (GA) cross-linked fiber	~300	
		1 wt% GA cross-linked fiber	~320	
		0.7 wt% Cr-tanned fiber	~425	
		7 wt% Cr-tanned fiber	~400	
Ultimate elongation	%	Gauge length = 2.0 cm, strain rate 50% min^{-1}		(28)
		Uncross-linked, unstretched fiber, wet diameter $d = 327\ \mu$m	68.0 ± 6.87	
		10% stretched fiber, uncross-linked, $d = 253\ \mu$m	45.1 ± 8.97	
		30% stretched fiber, uncross-linked, $d = 173\ \mu$m	32.5 ± 4.76	
		50% stretched fiber, uncross-linked, $d = 147\ \mu$m	24.1 ± 5.67	
		Unstretched fiber, cross-linked, $d = 94\ \mu$m	15.6 ± 2.66	
		10% stretched fiber, cross-linked, $d = 95.6\ \mu$m	15.5 ± 2.61	
		30% stretched fiber, cross-linked, $d = 86.1\ \mu$m	12.3 ± 1.75	
		50% stretched fiber, cross-linked, $d = 80.3\ \mu$m	11.6 ± 2.49	
		Gauge length = 1 cm, elongation rate= 100 mm min^{-1}		(29)
		Collagen–poly(lactic acid) (PLA) composites, 50% collagen fiber and 50% PLA matrix (w/w)	~20	
		Collagen–collagen composites, 50% collagen fiber and 50% collagen matrix (w/w)	~24	
		Gauge length = 2 cm, strain rate 1 mm min^{-1}		(12, 31)
		Uncross-linked fiber	~20	
		0.1 wt% glutaraldehyde (GA) cross-linked fiber	~30	
		1 wt% GA cross-linked fiber	~24	
		0.7 wt% Cr-tanned fiber	~22	
		7 wt% Cr-tanned fiber	~18	
Optical rotation	Degrees	At 589 μm		(22)
		Muscle layer collagen in Ascaris	−400	
		Pig kidney collagen	−380	
		Ascaris muscle layer, denatured	−150	
		Pig kidney, denatured	−130	
			−380 to −420	(13)
Electrical conductivity σ	S cm^{-1}	From bovine corium, dissolved in 1mM HCl, conc. 0.19% at 5°C, temp. range ~20–50°C, heating rate = 0.3°Cmin^{-1}	~(1.5–2.25) ×10^{-4}*	(32)
		Pepsin-solubilized prepared collagen, conc. 0.20% in 1mM HCl, temp. range ~20–50°C, heating rate = 0.3°Cmin^{-1}	~(1.75–2.5) ×10^{-4}	(32)
Permeability	cm S^{-1}	Collagen/ poly(vinyl alcohol) (PVA) cross-linked films to NaCl at 37°C		(33)

Property	Units	Conditions	Value	Reference
Pyroelectric coefficient	cm^{-2} K^{-1}	Native collagen	0.37×10^{-4}	(34, 35)
		Anionic collagen	1.16×10^{-4}	
		Anionic collagen: vinyledene fluoride-trofluoroethylene composites (1:1)	1.89×10^{-4}	
		At 0 wt% PVA content	~5.5	
		At 50 wt% PVA content	~15	
		At 80 wt% PVA content	~17.5	
Transitional diffusion coefficient	$cm^2\ S^{-1}$	Type I collagen rat tail tendon, aggregates in 0.01 M HCl	4.5×10^{-8}	(10)
		Type I collagen, single molecules	7.8×10^{-8}	
Intrinsic viscosity	$dl\ g^{-1}$	Ascaris muscle layer, in 0.5 M NaCl at 25°C	16	(22)
		Pig kidney, in 0.15 M sodium acetate, pH = 4.0, at ~25°C	12	
Average hydrophobicity	–	In chicken tendon	880	(34)
		$\alpha-$fraction in calf skin	880	
		Spongin B in sponge	760	
		Strugeon swim bladder	770	
Adhesion		Native collagen type I fibrils	Higher adhesion in the core compared to the shell	(14)
		Mineral deposition to anchor collagen fibrils to a solid surface: Attachment of collagen gel to polystyrene and poly(ether ether ketone) by enzyme-induced deposition		(36)

* σ increased with temperature then decreased stepwise at ~40°C, and then increased again.

References

1. Rawn JD. *Biochemistry*. Burlington, N.C.: Neil Patteerson Publishers; 1989.
2. Scott T, trans. *Concise Encyclopedia of Biochemistry*. Berlin: Walter de Gruter and Co.; 1983:101.
3. Itoh H, Miyata T. In: Salamone JC, ed. *Polymeric Materials Encyclopedia*. Vol 2. Boca Raton, Fla.: CRC Press; 1996:1287–1290.
4. Piez KA. In: Mark HF, et al. *Encyclopedia of polymer Science and Engineering*. 2nd ed. Vol 3. New York: John Wiley and Sons; 1985:699–727.
5. Stryer L. *Biochemistry*. 2nd ed. San Fransisco: W. H. Freeman and Co.; 1975.
6. Mathews CK, Van Holde KE. *Biochemistry*. 2nd ed. Menlo Park, Calif.: Benjamin/Cummings Publishing; 1996:178–180.
7. Fasman GD, ed. *Handbook of Biochemistry and Molecular Biology, Proteins*. 3rd ed. Vol 3. Cleveland: CRC Press; 1976:520–521.
8. Von Hippel PH. In: Ramachandran GN, ed. *Treatise on Collagen*. London: Academic Press; 1967.
9. Igarashi S, Trelstad RL, Kang AJ. *Biochim. Biophys. Acta*. 1973;295:514.
10. Silver FH, Trelstad RL. *J. Biol. Chem*. 1980;255: 9427.
11. Walton AG, Blackwell J. *Biopolymers*. New York: Academic Press; 1973.
12. Takaku K. et al. *J. Appl. Polym. Sci*.59 (1996): 887.
13. Hashemeyer RH, Haschemyer AEV. *Proteins: A Guide to study by Physical and Chemical Methods*. New York: John Wiley and Sons; 1973:410–419.

14. Strasser S, Zink A, Janko M, Heckl WM, Thalhammer S. *Biochem. Biophys. Res. Commun.* 2007;354:27.
15. Wisniewski M, Sionkowska A, Kaczamarek H, Lazare S, Tokarev V, Belin CJ. *Photochem. and Photobio. A: Chemistry.* 2007;188:192.
16. Elliot JT, Woodward JT, Umarji A, Mei Y, Tona A. *Biomaterials.* 2007;28:576.
17. Stylianopoulos T, Barocas VH. *Computer Methods in Applied Mechanics and Engineering.* 2007;196:2981.
18. Buehler MJ. *J. Mech. Behaviour of Biomed. Mat.* 2007:doi: 10.1016/j.jmbbm.2007.04.001.
19. Honda I, Arai K. *J. Appl. Polym. Sci.* 1996;62:1577.
20. Davis JM, Bachinger HP. *J. Biol. Chem.* 1993;268:25965.
21. Gelman RA, et al. *Biochim. Biophys. Acta.* 1976;427:492.
22. Fujimoto D. *Biochim. Biophys Acta.* 1968;168:537.
23. McClain PE, Wiley ER. *J. Biol. Chem.* 1972;247:692.
24. Davis JM, Boswell BA, Bachinger HPJ. *Biol. Chem.* 1989;264:8956.
25. Hart GJ, Russell AE, Coope DR. *Biochem. J.* 1971;125:599.
26. Bailey AJ, et al. *Biochem. J.* 1993;296:489.
27. Fasman GD, ed. *Handbook of Biochemistry and Molecular Biology, Proteins.* 3rd ed. Vol 1. Cleveland: CRC Press; 1976:109–110.
28. Pins GD, et al. *J. Appl. Polym. Sci.* 1997;63:1429.
29. Dunn MG, et al. *J. Appl. Polym. Sci.* 1997;62:1423.
30. Dunn MG. Avasarala PN, Zawadski JP. *J. Biomed. Mater. Res.* 1993;27:1545.
31. Takaku K, Kuriyama T, Narisawa I. *J. Appl. Polym. Sci.* 1996;61:2437.
32. Matsushita S, et al. *J. Appl.Polym. Sci.* 1996;50:1969.
33. Giusti P, Lazzeri L, Cascone MG. In: Salamone JC, ed. *Polymeric Materials Encyclopedia.* Vol 1. Boca Raton, Fla.: CRC Press; 1996:538–549.
34. Goissis G, Piccirili L, Goes JC, De G, Plepis AM, Das-Gupta DK. *Artificial Organs.* 1998;22(3):203.
35. Plepis AM, Goissis GG, Das-Gupta DK. *Polym. Eng. Sci.* 1996;36:293.
36. Fasman GD, ed. *Handbook of Biochemistry and Molecular Biology, Proteins.* 3rd ed. Vol 1. Cleveland: CRC Press; 1976:217.
37. Berendsen AD, Smit TH, Hoeben KA, Walboomers XF, Bronckers ALJJ, Everts V. *Biomaterials.* 2007;28:3530.

Cyclic Poly(dimethylsiloxane)

Stephen J. Clarson

Acronym Cyclic PDMS

Class Cyclic polymers, Inorganic polymers

Structure $-[(CH_3)_2SiO]_x-$

Cyclic poly(dimethylsiloxanes) and their Properties Polymer molecules may have a variety of architectural structures such as linear, ring, star, branched and ladder chains, and also three-dimensional network structures. The first synthetic cyclic polymers to be prepared and characterized were the poly(dimethysiloxanes) PDMS which were reported in 1977. Since then, a number of other cyclic polymers have been synthesized which include cyclic polystyrene, cyclic poly(phenylmethylsiloxane), cyclic poly(2-vinylpyridine), cyclic polybutadiene, and cyclic poly(vinylmethylsiloxane). The preparation of cyclic poly(dimethysiloxanes) is achieved by isolating the distribution of cyclic PDMS $-[(CH_3)_2SiO]_x-$ from PDMS ring–chain equilibration reactions carried out either in the bulk state or in solution. The successful utilization of such reactions for preparing large ring molecules is largely due to the extensive experiments performed on characterizing this system. Also, there is a good theoretical understanding of the reactions through the Jacobson–Stockmayer cyclization theory when used in conjunction with the rotational isomeric state model for PDMS. After attaining an equilibrium distribution of rings, vacuum fractional distillation and preparative gel permeation chromatography (GPC) may be used to prepare sharp fractions of the cyclic siloxanes having narrow molar mass distributions. Such methods allow the preparation of cyclic PDMS samples containing up to 1000 skeletal bonds, on average, on a gram scale. The molar mass for each polymer and the polydispersity may then be characterized using techniques such as gas chromatography (GC), high performance liquid chromatography (HPLC), analytical gel permeation chromatography (GPC), and other methods.

Some selected properties of the cyclic poly(dimethylsiloxanes) are given below, which include their solution, bulk, and surface properties. It is also highlighted where significant differences are seen when compared to their linear polymeric PDMS analogs. Detailed calculations for the molar cyclization constants for ring–chain equilibration reactions and their dependence on the conformations of poly(dimethylsiloxane) chains and on their distributions have been described by Semlyen, and this approach also enables a number of properties of the rings to be theoretically calculated. The area of topological entrapment of ring polymers into network structures has also been described in the literature, which is an area that is not accessible to linear polymers unless they undergo end-cyclizing chemistry. This concept of topological threading is somewhat general for ring molecules as it may also be utilized in the preparation of novel catenanes and rotaxanes.

Cyclic Poly(dimethylsiloxane)

Major Applications Ring-opening polymerization of small rings to give linear PDMS high polymers. Copolymerization with other siloxane small rings to give copolymers of controlled composition. Both the homopolymers and copolymers are widely used as silicone fluids, elastomers, and resins.

Selected Properties of the Cyclic poly(dimethylsiloxanes) (r) and their Properties Compared to the Linear poly(dimethylsiloxanes) (l)

Properties	Values
Characteristic ratio $\langle r^2 \rangle / nl^2$ derived from molar cyclization equilibrium constants in the bulk state at 383 K	6.8
Density at 298 K ($x = 95$)	971.67 kg m^{-3}
Glass transition temperature $T_g(\infty)$	149.8 K
Melting point T_{m1} ($M_n = 24370$ g mol^{-1})	227.0 K
Melting point T_{m2} ($M_n = 24370$ g mol^{-1})	237.8 K
Raman absorption ν_s(Si–O) Crystalline region	466 cm^{-1}
Raman absorption ν_s(Si–O) Amorphous region	486 cm^{-1}
Activation energy for viscous flow E_{visc} (∞)	15.5 kJ
Static dielectric permittivity ε_o at 298 K ($x = 95$)	2.757
Root mean square dipole moment $10^{30} \langle \mu^2 \rangle^{1/2}$ at 298 K ($x = 95$)	14.3 C m
Refractive index (632.8 nm) at 298 ($x = 95$)	1.4025
Refractive index (436.0 nm) at 298 ($x = 95$)	1.4140
Onset temperature for thermal depolymerization under N$_2$	623 K
Intrinsic viscosities $[\eta]_r/[\eta]_l$ in butanone (θ-solvent) at 293 K	0.67
Intrinsic Viscosities $[\eta]_r/[\eta]_l$ in cyclohexane at 298 K	0.58
Intrinsic viscosities in $[\eta]_r/[\eta]_l$ in bromocyclohexane (θ-solvent) at 301 K	0.66
Diffusion coefficients D_r/D_l in PDMS networks at 296 K	1.18 ± 0.03
Diffusion coefficients D_l/D_r in toluene at 298 K	0.84 ± 0.01
Mean square radius of gyration $\langle s^2 \rangle_{z,l}/\langle s^2 \rangle_{z,r}$ in benzene d$_6$ at 292 K	1.90
Translational friction coefficients f_r/f_l in toluene at 298 K	0.83 ± 0.01
Number-average molar masses of PDMS rings and chains with the same GPC retention values M_r/M_l	1.24 ± 0.04
Melt viscosities at η_r/η_l for $M_w = 24000$ g mol^{-1}	0.45 ± 0.02
Molar mass dependence on bulk dimensions (small angle neutron scattering)	$R_g \propto M_\omega^{-0.4}$

Selected Bibliography

Arrighi V, Gagliardi S, Dagger AC, Semylen JA, Higgins JS, Shenton MJ. *Macromolecules.* 2004;37:8057.

Bannister DJ, Semlyen JA. *Polymer.* 1981;22:377–381.

Barbarin-Castillo J-M, McLure IA, Clarson SJ, Semlyen JA. *Polym. Communications.* 1987;28:212.

Beevers MS, Mumby SJ, Clarson SJ, Semlyen JA. *Polymer.* 1983;24:1565.

Brown JF, Slusarczuk GMJ. *J. Am. Chem. Soc.* 1965;87:931.

Clarson SJ. *New Journal Chem.* 1993;17:711.

Clarson SJ, Dodgson K, Semlyen JA. *Polymer.* 1985;26:930.

Clarson SJ, Mark JE, Semlyen JA. *Polym. Communications.* 1986;27:244.

Clarson SJ, Mark JE, Semlyen JA. *Polym. Communications.* 1987;28:151.

Clarson SJ, Rabolt JF. *Macromolecules.* 1993;26:2621.

Clarson SJ, Semlyen JA, Horska J, Stepto RFT. *Polym. Communications.*1986; 27:31.

Clarson SJ, Semlyen JA. *Polymer.* 1986;27:91.

Clarson SJ, Semlyen JA, eds. *Siloxane Polymers.* Prentice Hall; 1993.

Dagger AC, Arrighi V, Gagliardi S, Shenton MJ, Clarson SJ, Semlyen JA. In: Clarson SJ, Fitzgerald JJ, Owen MJ, Smith SD, Van Dyke MA, eds. *Synthesis and Properties of Silicones and Silicone-Modified Materials.* ACS Symposium Series Volume 838, Washington, DC, p. 96.

Di Marzio EA, Guttman CM. *Macromolecules* 1987;20:1403.

Dodgson K, Semlyen JA. *Polymer.* 1977;18:1265.

Edwards CJC, Bantle S, Burchard W, Stepto RFT, Semlyen JA. *Polymer.* 1982;23:873–876.

Edwards CJC, Stepto RFT, Semlyen JA. *Polymer.* 1980;21:781–786.

Edwards CJC, Stepto RFT, Semlyen JA. *Polymer.* 1982;23:865–868.

Edwards CJC, Stepto RFT, Semlyen JA. *Polymer.* 1982;23:869–872.

Edwards CJC, Stepto RFT. In: Semlyen JA, ed. *Cyclic Polymers.* Barking, UK: Elsevier; 1986:135–165.

Flory PJ, Semlyen JA. *J. Am. Chem. Soc.* 1965;88:3209.

Garrido L, Mark JE, Clarson SJ, Semlyen JA. *Polym. Communications.* 1984;;25:218.

Garrido L, Mark JE, Clarson SJ, Semlyen JA. *Polym. Communications.* 1985;26:53.

Garrido L, Mark JE, Clarson SJ, Semlyen JA. *Polym. Communications.* 1985;26:55.

Goodwin AA, Beevers MS, Clarson SJ, Semlyen JA. *Polymer.* 1996;37(13):2603–2607.

Granick S, Clarson SJ, Formoy TR, Semlyen JA. *Polymer.* 1985;26:925.

Kuo CM, Clarson SJ, Semlyen JA. *Polymer.* 1994;35:4623.

Orrah DJ, Semlyen JA, Ross-Murphy SB. *Polymer.* 1988;29:1455–1458.

Pham-Van-Cang C, Bokobza L, Monnerie L. et al. *Polymer.* 1987;28:1561.

Semlyen JA, Wright PV. *Polymer.* 1969;10:543.

Wright PV. In: Ivin KJ, Saegusa T. eds. *Ring Opening Polymerization.* Vol 2. New York: Elsevier; 1984:324.

Cyclic Poly(phenylmethylsiloxanes)

Stephen J. Clarson

Acronym Cyclic PPMS, Cyclic PMPS

Class Cyclic polymers, Inorganic polymers

Structure $-[(C_6H_5)(CH_3)SiO]_x-$

Properties and Applications of Cyclic poly(phenylmethylsiloxanes) The molar
cyclization constants from ring–chain equilibration reactions of poly(phenylmethylsiloxane)
PPMS in both the bulk state and in solution were investigated in detail by Beevers and
Semlyen. Based upon these studies, Clarson and Semlyen have described the scaling
up of such reactions to successfully isolate cyclic poly(phenylmethylsiloxanes)
$-[(C_6H_5)(CH_3)SiO]_x-$ from ring–chain equilibration reactions carried out in toluene
solution at 383 K. Following fractionation, a variety of investigations on the physical
properties of these cyclic polymers have been carried out. The properties have also been
compared with their linear polymer analogs. It should be noted that the large rings are atactic
due to the equilibration used in their preparation. It is possible to obtain the stereoisomers of
the small rings for this system. Polymerization of isolated cyclic stereoisomers can lead to
stereoregular PMPS. Although a rotational isomeric state model has been developed for the
PPMS system by Mark and Ko, no detailed calculations of the properties of the rings using
this model have been described so far.

Anionic polymerization of cyclic PMPS allows for the preparation of organic-siloxane
block copolymers. Well-defined styrene–PMPS diblocks have been prepared and investigated.

Major Applications Ring-opening polymerization of small rings to give linear PPMS high
polymers. Copolymerization with other siloxane small rings to give copolymers of controlled
composition.

Properties of Special Interest Viscous fluids having good thermal stability. Certain
stereoisomers when highly pure (ref. 1, 2, 9) are solids at room temperature. Elastomers
have been studied for various applications including membranes for organic / water
separations.

Preparative Techniques Ring–chain equilibration reactions (see ref. 1, 2, 4, 5).

Selected Properties of the Cyclic poly(phenylmethylsiloxanes) (r) and their Properties Compared to
the Linear Poly(phenylmethylsiloxanes) (l)

Property	Units	Conditions	Value	Reference
Characteristic ratio $\langle r^2 \rangle / n l^2$ derived from molar cyclization equilibrium constants		Bulk state at 383 K	10.7	(2,3)

Cyclic Poly(phenylmethylsiloxanes)

Property	Units	Conditions	Value	Reference
Characteristic ratio $\langle r^2 \rangle / nl^2$ derived from molar cyclization equilibrium constants		Toluene at 383 K	10.4	(2,3)
Critical dilution point	% Volume polymer	Toluene at 383 K	52%	(2,3)
Glass transition temperature $T_g(\infty)$	K	DSC	244.9	(5,9)
Characteristic ratio $\langle r^2 \rangle / nl^2$ derived from GPC		Toluene at 292 K	8.8	(5,6)
Mean square radius of gyration $\langle s^2 \rangle_{z,l} / \langle s^2 \rangle_{z,r} /$ in		Benzene d$_6$ at 292 K	2.0	(5,8)
Dipole moment $\mu^2 (x = 5)$	10^{-31} C m		5.01	(11)
Number-average molar masses of PDMS rings and chains with the same GPC retention values M_r / M_l		Toluene at 292 K	1.25 ± 0.05	(5,6)
Rate of ring opening polymerization	min^{-1}	P$_4$/phosphazine base/bulk at 293K	1.6	(12)
Rate of ring opening polymerization	min^{-1}	KOH/phosphazine base/toluene at 293K	11	(12)
^{1}H NMR peak assignments	ppm	Cyclic trimer and tetramer isomers		(17)
Enthalpy change for the formation of the *cis*-trimer	kJ mol^{-1}		27	(2,3)
Enthalpy change for the formation of the *trans*-trimer	kJ mol^{-1}		22	(2,3)
Enthalpy change for the formation of the *cis*-tetramer	kJ mol^{-1}		8	(2,3)

References

1. Hickton HJ, Holt A., Homer J, Jarvie AWJ. *Chem. Soc.* 1966;(C):149.
2. Beevers MS and Semlyen JA *Polymer.* 1971;12:373–382.
3. Beevers MS. *D. Phil. Thesis.* University of York; 1972.
4. Mark JE, Ko JH. *J. Polym. Sci. Polym. Phys. Ed.* 1975;13:2221.
5. Clarson SJ. *D. Phil. Thesis.* University of York:1985.
6. Clarson SJ, Semlyen JA. *Polymer.* 1986;27:1633–1636.
7. Semlyen JA. *Makromol. Chem., Macromol. Symp.* 1986;6:155–163.
8. Clarson SJ, Dodgson K, Semlyen JA. *Polymer.* 1987;28:189–192.
9. Clarson SJ, Semlyen JA, Dodgson K. *Polymer.* 1991;32:2823.
10. Clarson SJ, Semlyen JA. *Siloxane Polymers.* Prentice Hall 1993.
11. Goodwin AA, Beevers MS, Clarson SJ, Semlyen JA. *Polymer.* 1996;37(13)2597–2602.
12. Van Dyke MA, Clarson SJJ. *Inorg. Organometallic Polym.* 1998;8(2):111.
13. Gerharz B, Fischer EW, Fytas G. *Polymer.* 1991;32:469.
14. Clarson SJ, Stuart JO, Selby CE, Sabata A, Smith SD, Ashraf A. *Macromolecules.* 1995;28:674.
15. Jadhav AV, Patwardhan SV, Ahn HW, et al. In: Clarson SJ, Fitzgerald JJ, Owen MJ, Smith SD, Van Dyke MA, eds. *Science and Technology of Silicones and Silicone-modified Materials.* Vol 964. Washington, DC: ACS Symposium Series; 116.

16. Zhang XK, Poojari Y, Clarson SJ. In: Clarson SJ, Fitzgerald JJ, Owen MJ, Smith, SD, Van Dyke MA eds. *Science and Technology of Silicones and Silicone-modified Materials.* Vol 964. Washington, DC: ACS Symposium Series; 165.
17. Ahn HW, Clarson SJ. *J. Inorg. Organometallic Polym.* 200111(4):203.
18. Ahn HW, Clarson SJ. In: Clarson SJ, Fitzgerald JJ, Owen MJ, Smith SD, Van Dyke MA eds. *Synthesis and Properties of Silicones and Silicone-Modified Materials.* Vol 838. Washington, DC: ACS Symposium Series; 40.

Cyclic Poly(vinylmethylsiloxanes)

Stephen J. Clarson

Acronym Cyclic PVMS

Class Cyclic polymers, Inorganic polymers

Structure $-[(CH_2{=}CH)(CH_3)SiO]_x-$

Applications and Properties of Cyclic poly(vinylmethylsiloxanes) The cyclic poly(vinylmethylsiloxanes) are an interesting cyclic polymer system in that they contain a reactive pendent group. Thus, possible chemistries include hydrogenation which yields cyclic poly(ethylmethylsiloxane) $-[(CH_3CH_2)(CH_3)SiO]_x-$ (this route has also been used to deuterate the rings for neutron scattering investigations $-[(CH_3CHD)(CH_2D)SiO]_x-$). Other useful reactions are with molecules containing terminal or pendent Si–H groups which can readily be attached by hydrosilation chemistry. This functional ring system also shows that one can directly prepare elastomeric network structures having none of the usual network defects (dangling chain ends, etc.). Although there are a limited number of studies of these novel functional rings to date, the large rings have been successfully isolated from ring–chain equilibration reactions carried out in solution. Following fractionation, some investigations of the physical properties of these cyclic polymers have been carried out and have also been compared with their linear polymer analogs.

Major Applications Ring-opening polymerization of small rings to give linear PVMS high polymers. Cyclic vinylmethylsiloxane rings are miscible with many other small siloxane rings, thus the preparation of a wide variety of silicone copolymers is possible in the bulk or solution state. Both the homopolymers and copolymers are widely used for preparing silicone elastomers.

Properties of Special Interest Viscous fluids having good thermal stability.

Preparative Techniques Ring–chain equilibration reactions (see ref. 3 and 5).

Selected properties of the cyclic polymers (r) and their properties compared to the linear poly(vinylmethylsiloxanes) (l)

Properties	Values
Characteristic ratio $\langle r^2 \rangle / nl^2$ derived from molar cyclization equilibrium constants in 50% toluene solution at 383 K	7.8
Intrinsic viscosities $[\eta]_r/[\eta]_l$ in toluene at 298 K	0.69
Density at 298 K ($M_n = 5430$ g mol^{-1} and $M_n/M_n = 1.06$)	1006.0 kg m^{-3}
Refractive index (589.3 nm) at 298 K ($M_n = 5430$ g mol^{-1} and $M_n/M_n = 1.06$)	1.4458

Cyclic Poly(vinylmethylsiloxanes)

Properties	Values
Refractive index (589.3 nm) at 303 K ($M_n = 5430$ g mol^{-1} and $M_n/M_n = 1.06$)	1.4421
Refractive index (589.3 nm) at 313 K ($M_n = 5430$ g mol^{-1} and $M_n/M_n = 1.06$)	1.4380
Refractive index (589.3 nm) at 298 K ($M_n = 11\,440$ g mol^{-1} and $M_n/M_n = 1.14$)	1.4465
Refractive index (589.3 nm) at 303 K ($M_n = 11\,440$ g mol^{-1} and $M_n/M_n = 1.14$)	1.4427
Refractive index (589.3 nm) at 313 K ($M_n = 11\,440$ g mol^{-1} and $M_n/M_n = 1.14$)	1.4385
Glass transition temperature T_g ($M_n = 5430$ g mol^{-1} and $M_n/M_n = 1.06$)	144.5 K
Glass transition temperature T_g ($M_n = 11\,440$ g mol^{-1} and $M_n/M_n = 1.14$)	144.7 K
Melt viscosity $10^3 \eta_r$ at 298 K ($M_n = 5430$ g mol^{-1} and $M_n/M_n = 1.06$)	75.9 kg m^{-1} s^{-1}
Activation energy for viscous flow ($M_n = 5430$ g mol^{-1} and $M_n/M_n = 1.06$)	16.75 kJ

Selected Bibliography

1. Kantor SW, Osthoff RC, Hurd DTJ. *Am. Chem. Soc.* 1995;77:1685.
2. Hampton JF. US Patent 3.465.016
3. Formoy, T. R. D. Phil. Thesis, University of York (1985).
4. Semlyen JA. *Makromol. Chem., Macromol. Symp.* 1986;6:155–163.
5. Formoy TR, Semlyen JA. *Polymer Comm.* 1989;30:86–89.
6. Kendrick TC, Parbhoo B, White JW. In: Patai S, Rappoport Z, eds. *The Chemistry of Organic Silicon Compounds.* Vol 2. New York: John Wiley & Sons; 1989:1289.
7. Semlyen JA. In: Clarson SJ, Semlyen JA, eds. *Siloxane Polymers.* Englewood Cliffs, New Jersey: Prentice Hall; 1993:135.
8. Kennan JJ. In: Clarson SJ, Semlyen JA. eds. *Siloxane Polymers.* Englewood Cliffs, New Jersey: Prentice Hall; 1993:72.

Elastic, Plastic, and Hydrogel-forming Protein-based Polymers

Dan W. Urry and Chi-Hao Luan

Acronyms, Trade names	Class	Representative Structure
	Elastomer	poly($G\beta G\alpha P$)
	Plastic	poly(AVGVP)
	Hydrogel	poly(GGAP)

Major Applications[1]	Medical:	Nonmedical:
	■ Soft tissue augmentation	■ Controlled release of herbicides, pesticides, fertilizers, and growth factors
	■ Cell attachment to elastic matrices	
	■ Prevention of postsurgical adhesions	■ Food product additives
		■ Material coating
	■ Tissue reconstruction	■ Transducers (sensors/actuators)
	■ Coatings on catheters, leads, and tubings	■ Molecular machines
		■ Biodegradable plastics
	■ Drug delivery	■ Controllable super absorbents
	■ Biosensors	
Properties of Special Interest[2,3]	■ Water soluble below a critical temperature	
	■ Hydrophobic folding and assembly (inverse temperature transition)	
	■ Biocompatible	
	■ Biodegradable (chemical clocks enabling proteolytic degradation)	
	■ Relatively low cost when microbially produced	
	■ To perform free energy transduction involving the intensive variables of mechanical force, temperature, pressure, chemical potential, electrochemical potential, and light	
	■ Thermoplastics (regular and inverse)	
Synthesis	■ Chemical synthesis using solution and solid phase methods	
	■ Microbial synthesis using gene construction and expression in the cells of animals and plants	
Supplier	Bioelastics Inc, 2423 Vestavia Drive, Vestavia Hills, Alabama 35216–1333	

Table 1A. Hydrophobicity scale for protein-based polymers using the properties of the inverse temperature transition of elastic protein-based polymers, poly[f_v (GVGVP), f_x(GXGVP)][a]

Residue X Three-letter Abbr.	One-letter Symbol	T_t,°C (in PBS)[2]	T_b,°C (in H$_2$0)[4,5]	ΔH[4,5], kcal/mol[e] ±0.05	ΔS[4,5], cal/mol[e]-K ±0.05
Lys (dihydro NMeN)[b]		−130	–	–	–
Trp	(W)	−90	−105	2.10	7.37
Tyr	(Y)	−55	−75	1.87	6.32
Phe	(F)	−30	−45	1.93	6.61
His (imidazole)	(H°)	−10	–	–	–
Leu	(L)	5	5	1.51	5.03
Ile	(I)	10	10	1.43	4.60
Lys (6-OH tetrahydro NMeN)[b]	–	15	–	–	–
Met	(M)	20	15	1.00	3.29
Val	(V)	24	26	1.20	3.90
Glu(COOCH$_3$)	(E^m)	25	–	–	–
Glu(COOH)	(E°)	30	20	0.96	3.14
Cys	(C)	30	–	–	–
His (imidazolium)	(H$^+$)	−30	–	–	–
Lys(NH$_2$)	(K°)	35	40	0.71	2.26
Pro	(P)	−8^c,40^d	40^d	0.92	2.98
Asp(COOH)	(D°)	45	40	0.78	2.57
Ala	(A)	45	50	0.85	2.64
HyP	–	50	–	–	–
Thr	(T)	50	60	0.82	2.60
Asn	(N)	50	50	0.71	2.29
Ser	(S)	50	60	0.59	1.86
Gly	(G)	55	55	0.70	2.25
Arg	(R)	60	–	–	–
Gln	(Q)	60	70	0.55	1.76
Tyr(ϕ-O$^-$)	(Y$^-$)	120	140	0.31	0.94
Lys(NH$_3^+$)	(K$^+$)	120	–	–	–
Lys(NMeN,oxidized)[b]	–	120	–	–	–
Asp(COO$^-$)	(D$^-$)	170	–	–	–
Glu(COO$^-$)	(E$^-$)	250	–	–	–
Ser(PO$_4^=$)	–	1000	–	–	–

[a] T_t and T_b are the on-set temperature for the hydrophobic folding and assembly by inverse temperature transition, in PBS (0.15 N NaCl, 0.01 M phosphate) as determined by light scattering and in water as determined by DSC, respectively. Both values are linearly extrapolated to $f_x = 1$ and rounded to a number divisible by 5. ΔH and ΔS are the values at $f_x = 0.2$ on the curve for a linear fit of the DSC-derived endothermic heats and entropies of the transitions for aqueous polymers.

[b] NMeN is for N-methyl nicotinamide pendant on a lysyl side chain, i.e. N-methyl-nicotinate attached by amide linkage to the ε-NH$_2$ of Lys. N-methyl-1,6-dihydronicotinamide (dihydro NMeN) is the most hydrophobic reduced state, and the second reduced state is N-methyl-6-OH, 1,4,5,6-tetrahydronicotinamide (6-OH tetrahydro NMeN).

[c] The calculated T_t value for Pro comes from poly(GVGVP) when the experimental values of Val and Gly are used. This hydrophobicity value of −8°C is unique to the β-spiral structure where there is hydrophobic contact between the Val$_i^1\gamma$CH$_3$ and the adjacent Pro$_i^2\delta$CH$_2$ and the interturn Pro$_{i+3}^2\beta$CH$_2$ moieties, where the subscripts i and $i+3$ identify the pentamer number.

[d] The experimental value determined from poly[f_v(GVGVP), f_p(GVGPP)].

[e] Per mole of pentamer.

Table 1B. Hydrophobicity scale for amino acid residues in terms of $\Delta G°_{HA}$, the change in Gibbs free energy of hydrophobic association with the corresponding $T_{b(t)}$ [6]

Residue X	T_b °C	$\Delta G°_{HA}$ (GXGVP) kcal/mol-pentamer
W: Trp	−105	−7.00
F: Phe	−45	−6.15
Y: Tyr	−75	−5.85
H°: His°	−10 (T_t)	−4.80 (from graph)
L: Leu	5	−4.05
I: Ile	10	−3.65
V: Val	26	−2.50
M: Met	15	−1.50
H+: His+	30 (T_t)	−1.90 (from graph)
C: Cys	30 (T_t)	−1.90 (from graph)
E°: Glu(COOH)	20 (2)	−1.30 (−1.50)
P: Pro	40	−1.10
A: Ala	50	−0.75
T: Thr	60	−0.60
D°: Asp(COOH)	40	−0.40
K°: Lys(NH$_2$)	40 (38)	−0.05 (−0.60)
N: Asn	50	−0.05
G: Gly	55	0.00
S: Ser	60	+0.55
R: Arg	60 (T_t)	+0.80 (from graph)
Q: Gln	70	+0.75
Y−: Tyr(ϕ-O$^-$)	140	+1.95
D−: Asp(COO$^-$)	170 (T_t)	≈ +3.4 (from graph)
K+: Lys(NH$_3^+$)	(104)	(+2.94)
E−: Glu(COO$^-$)	(218)	(+3.72)
Ser (PO$_4^=$)	860 (T_t)	≈ +8.0 (from graph)

Data within parentheses utilized microbial preparations of poly(30 mers), e.g. (GVGVP GVGVP GXGVP GVGVP GVGVP GVGVP)$_n$, with n of the order of 30. The notation (from graph) indicates that the value of T_t was used to obtain $\Delta G°_{HA}$(GXGVP) from the experimental sigmoid curve of T_t versus $\Delta G°_{HA}$ from reference 6, Figure 5.10, and T_t-values are adapted from reference 6, Table 5.3.

Table 1C. Hydrophobicity scale (approximate T_t and $\Delta G°_{HA}$ values) for chemical modifications and prosthetic groups of proteins.[a] T_t = Temperature of inverse temperature transition for poly[f_v(VPGVG), f_x(VPGXG)] [6]

Residue X	$\Delta G°_{HA}$ (kcal/mol)[g]	T_t, linearly extrapolated to $f_x = 1$
Lys (dihydro NMeN)[b,d]	−7.0	−130°C
Glu(NADH)[c]	−5.5	−30°C
Lys (6–OH tetrahydro NMeN)[b,d]	−3.0	15°C
Glu(FADH$_2$)	−2.5	25°C
Glu(AMP)	+1.0	70°C
Ser(−O−SO$_3$H)	+1.5	80°C

Table 1C. *(Continued)*

Residue X	$\Delta G°_{HA}$ (kcal/mol)[g]	T_t, linearly extrapolated to $f_x = 1$
Thr(–O–SO$_3$H)	+2.0	100°C
Glu(NAD)[c]	+2.0	120°C
Lys(NMeN,oxidized)[b,d]	+2.0	120°C
Glu(FAD)	+2.0	120°C
Tyr(–O–SO$_3$H)[e]	+2.5	140°C
Tyr(–O–NO$_2^-$)[f]	+3.5	220°C
Ser(PO$_4^=$)	+8.0	860°C

[a] Usual conditions are 40 mg/ml polymer, 0.15 N NaCl, and 0.01 M phosphate at pH 7.4.

[b] NMeN is for N-methyl nicotinamide at a lysyl side chain, i.e. N-methyl-nicotinate attached by amide linkage to the e-NH$_2$ of Lys and the most hydrophobic reduced state is N-methyl-1,6-dihydronicotinamide (dihydro NMeN), and the second reduced state is N-methyl-6-OH 1,4,5,6-tetrahydronicotinamide or (6-OH tetrahydro NMeN).

[c] For the oxidized and reduced nicotinamide adenine dinucleotides, the conditions were 2.5 mg/ml polymer, 0.2 M sodium bicarbonate buffer at pH 9.2.

[d] For the oxidized and reduced N-methyl nicotinamide, the conditions were 5.0 mg/ml polymer, 0.1 M potassium bicarbonate buffer at pH 9.5, 0.1 M potassium chloride.

[e] The pK$_a$ of polymer bound –O–SO$_3$H is 8.2.

[f] The pK$_a$ of Tyr(–O–NO$_2$) is 7.2.

[g] Approximate values of $\Delta G°_{HA}$ using the T_t-values in the right column in combination with the T_b versus $\Delta G°_{HA}$ values from reference 6, Table 5.2 and Figure 5.10.

Table 2A. Hydrophobic-induced pK shifts on polytricosamers and polymers of random mixtures of composite pentamers[a,[7]]

	pKa
Glu-containing Polymer	
poly[0.8(GVGVP), 0.2(GEGVP)]	4.3
poly[GEGFP GVGVP GVGVP GVGVP GFGFP GFGFP]	7.7
poly[(GEGFP), 3(GVGVP), 2(GFGFP)][b]	4.7
poly[GEGVP GFGFP GFGVP GVGVP GFGFP GVGVP]	7.8
poly[GEGFP GVGVP GVGFP GFGFP GVGVP GVGFP]	8.1
poly[(GEGFP), 2(GVGVP), 2(GVGFP), (GFGFP)][b]	5.2
Asp-containing Polymer	
poly[GDGFP GVGVP GVGFP GFGFP GVGVP GVGFP]	10.1
poly[(GDGFP), 2(GVGVP), 2(GVGFP), (GFGFP)][b]	4.6
poly[GDGFP GVGVP GVGVP GVGVP GFGFP GFGFP]	9.5
poly[(GDGFP), 3(GVGVP), 2(GFGFP)][b]	5.2
poly[GDGVP GFGFP GFGVP GVGVP GFGFP GVGVP]	6.7

[a] Experimental conditions: 40 mg/ml at 20°C, M.W. of the polypeptides >50 kDa.

[b] Random mixture of pentamers comprising associated polytricosapeptide.

Table 2B. Hydrophobic-induced pK shift for poly$[f_V(GVGIP), f_X(GXGIP)]$ where X = E, D, K, and f_X varies from 0.06 to 1.0[8−10]

f_E	0.06	0.15	0.31	0.37	0.42	0.49	0.70	0.77	0.90	1.00
pK*	6.08	5.70	4.90	4.70	4.55	4.50	4.40	4.35	4.35	4.35
pK**	6.61	5.92	5.03	4.80	4.48	4.40	4.35	4.40	4.55	4.70

f_D	0.06	0.08	0.19	0.28	0.35	0.51	0.73	0.84	0.89	1.00
pK*	5.4	5.2	4.7	4.3	4.1	4.0	3.9	3.9	3.9	3.8
pK**	6.0	5.0	4.5	4.2	3.9	3.8	3.9	4.1	4.2	4.6

f_K	0.06	0.09	0.14	0.22	0.41	0.59	0.76	0.88	0.91	1.00
pK*	8.60	8.90	9.13	9.38	9.59	9.68	9.70	9.65	9.60	9.40
pK**	8.18	8.65	8.85	9.11	9.43	9.55	9.60	9.62	9.58	9.20

* in 0.15 N NaCl; ** in H_2O; M.W. of the polypeptides >50 kDa; at 37°C.

Table 2C. Stretch-induced pK shifts for cross-linked Glu-containing polymers

Polymer	ΔpK	Force		
		dynes/cm^2	gram	pK
X^{20}-poly[0.8(GVGVP), 0.2(GEGVP)][11]	0.0	0.0	0.0	3.99
	0.85	–	1.0	4.84
X^{20}-poly[0.82(GVGIP), 0.18(GEGIP)] [12]	0.0	0.0	0.0	6.2
	0.37	3.6×10^5	1.0	6.57
	0.65	5.4×10^5	1.5	6.85
	1.25	6.4×10^5	1.75	7.45
	1.95	7.3×10^5	2.0	8.15
	2.8	8.0×10^5	2.2	9.0

X^{20} indicates 20 Mrad γ-irradiation cross-linked polymer matrix.

Table 2D. Additivity of hydrophobic-induced and stretch-induced[12] pKa shifts for the pentamer composition set of (GVGIP), (GEGIP), and (GVGVP) polymers, which use replacement of Val by Ile for six GVP to GIP positions in the repeating polytricosapeptide, culminating in (GVGIP GVGIP GEGIP GVGIP GVGIP GVGIP)$_{23}$, for the hydrophobic-induced pKa shifts, and the mean of poly[0.83(GVGIP),0.17(GEGIP)], once cross-linked to form the elastomer of the same ratio of (GVGIP) to (GEGIP) for the stretch-induced pKa shifts. The result is (with a step due to the constraints of cross-linking) that stretching simply adds on to the hydrophobic-induced pKa shift.

Protein-based Polymer Composition		f	pKa	ΔpKa	ΔG_{ap}
I : (GVGVP GVGVP GEGVP GVGVP GVGVP GVGVP)$_{36}$:	E/0I	0	4.5	0.5	0.7
II : (GVGIP GVGIP GEGIP GVGIP GVGIP GVGIP)$_{23}$:	E/6I	0	5.4	1.4	1.9
V′ : X^{20}-poly[0.83(GVGIP), 0.17(GEGIP)]:	E/6I	0	6.3	2.3	3.1
V′ : X^{20}-poly[0.83(GVGIP), 0.17(GEGIP)]:	E/6I	3.6	6.6	2.6	3.5

Table 2D. (Continued)

Protein-based Polymer Composition		f	pKa	ΔpKa	ΔG_{ap}
$\mathbf{V'}$: X^{20}-poly[0.83(GVGIP), 0.17(G$\underline{E}$GIP)]:	E/6I	5.4	6.9	2.9	4.0
$\mathbf{V'}$: X^{20}-poly[0.83(GVGIP), 0.17(G$\underline{E}$GIP)]:	E/6I	6.4	7.4	3.4	4.6
$\mathbf{V'}$: X^{20}-poly[0.83(GVGIP), 0.17(G$\underline{E}$GIP)]:	E/6I	7.3	8.2	4.2	5.7
$\mathbf{V'}$: X^{20}-poly[0.83(GVGIP), 0.17(G$\underline{E}$GIP)]:	E/6I	8.0	9.0	5.0	6.8

$\Delta G_{ap}(=2.3 RT \Delta pKa)$ is in kcal/mol-carboxylate. f is force in units of 10^5 dynes/cm^2.
X^{20} is a dose for forming cross-linked elastic matrix.

Table 2E. Additivity of hydrophobic-induced[13] and stretch-induced[14] pKa shifts using the E/nF set of polymers with $n = 0, 2, 3, 4,$ and 5 for the hydrophobic-induced set and 20 Mrad γ-cross-linked E/4F, i.e. X^{20}–(GVGVP GVGFP G$\underline{E}$GFP GVGVP GVGFP GVGFP)$_{41}$, for the stretch-induced pKa shifts. The result is (with a step due to the constraints of cross-linking while in the protonated, phase-separated state) that stretching simply adds on to the hydrophobic-induced pKa shift. This demonstrates the ΔG_{ap}-Additivity Principle for hydrophobic- and stretch-induced pKa shifts.

Protein-based Polymer Composition		f	pKa	ΔpKa	ΔG_{ap}
$\mathbf{I}$: (GVGVP GVGVP G$\underline{E}$GVP GVGVP GVGVP GVGVP)$_{36}$	E/0F	0	4.5	0.5	0.7
$\mathbf{II}$: (GVGVP GVGFP G$\underline{E}$GFP GVGVP GVGVP GVGVP)$_{40}$	E/2F	0	4.8	0.8	1.1
$\mathbf{III}$: (GVGVP GVGVP G$\underline{E}$GVP GVGVP GVGFP GVGFP)$_{39}$	E/3F	0	5.2	1.2	1.6
$\mathbf{IV}$: (GVGVP GVGFP G$\underline{E}$GFP GVGVP GVGFP GVGFP)$_{41}$	E/4F	0	5.6	1.6	2.2
$\mathbf{V}$: (GVGVP GVGFP G$\underline{E}$GFP GVGVP GVGFP GFGFP)$_{42}$	E/5F	0	6.4	2.4	3.3
$\mathbf{IV'}$: X^{20}–(GVGVP GVGFP G$\underline{E}$GFP GVGVP GVGFP GVGFP)$_{41}$	E/4F	0	7.8	3.8	5.2
$\mathbf{IV'}$: X^{20}–(GVGVP GVGFP G$\underline{E}$GFP GVGVP GVGFP GVGFP)$_{41}$	E/4F	3.6	8.1	4.1	5.6
$\mathbf{IV'}$: X^{20}–(GVGVP GVGFP G$\underline{E}$GFP GVGVP GVGFP GVGFP)$_{41}$	E/4F	5.4	8.4	4.4	6.0
$\mathbf{IV'}$: X^{20}–(GVGVP GVGFP G$\underline{E}$GFP GVGVP GVGFP GVGFP)$_{41}$	E/4F	6.4	8.9	4.9	6.7
$\mathbf{IV'}$: X^{20}–(GVGVP GVGFP G$\underline{E}$GFP GVGVP GVGFP GVGFP)$_{41}$	E/4F	7.3	9.3	5.3	7.2
$\mathbf{IV'}$: X^{20}–(GVGVP GVGFP G$\underline{E}$GFP GVGVP GVGFP GVGFP)$_{41}$	E/4F	8.0	9.8	5.8	7.9

ΔG_{ap} $(=2.3 RT \Delta pKa)$ is in kcal/mol-carboxylate. f is force in units of 10^5 dynes/cm^2.
X^{20} is a 20 Mrad γ-irradiation dose for forming cross-linked elastic matrix.

Table 2F. Comparison of increases in hydrophobicity on pKa and reduction potential shifts and on positive cooperativity for different functional groups in the basis set:[15, 16]

Model protein I : [GVGVP GVGVP GΦGVP GVGVP GVGVP GVGVP]$_n$(GVGVP); Φ/0F
Model protein II : [GVGVP GVGFP GΦGFP GVGVP GVGVP GVGVP]$_n$(GVGVP); Φ/2F
Model protein III : [GVGVP GVGVP GΦGVP GVGVP GVGFP GFGFP]$_n$(GVGVP); Φ/3F
Model protein IV : [GVGVP GVGFP GΦGFP GVGVP GVGFP GVGFP]$_n$(GVGVP); Φ/4F
Model protein V : [GVGVP GVGFP GΦGFP GVGVP GVGFP GFGFP]$_n$(GVGVP); Φ/5F

Φ(function)	K{NMeN}—(redox)[15]		K(lysine Amino, —NH$_2$)[16]		E(glutamic Acid, COOH)[15]	
Energy Term kcal/mol	$-nF\Delta E$ Redox Shift	$(\partial \Delta G/\partial \alpha')$ Slope; (n)*	$-2.3RT\Delta pK$ pKa Shift	$(\partial \Delta G/\partial \alpha)$ Slope; (n)†	$-2.3RT\Delta pK$ pKa Shift	$(\partial \Delta G/\partial \alpha)$ Slope; (n)§
Φ/0F	0	0 (1.0)	0	0 (0.9)	-0.8	-1.0 (1.5)
Φ/2F	-0.8	-0.5 (1.2)	-0.3	-0.3 (1.1)	-1.1	-1.3 (1.6)

Table 2F. *(Continued)*

Φ(function)	K{NMeN}—(redox)[15]		K(lysine Amino, —NH$_2$)[16]		E(glutamic Acid, COOH)[15]	
Energy Term kcal/mol	$-nF\Delta E$ Redox Shift	$(\partial\Delta G/\partial\alpha')$ Slope; (n)*	$-2.3RT\Delta$pK pKa Shift	$(\partial\Delta G/\partial\alpha)$ Slope; (n)†	$-2.3RT\Delta$pK pKa Shift	$(\partial\Delta G/\partial\alpha)$ Slope; (n)§
$\Phi/3F$	-1.7	-1.1 (1.5)	-0.6	-0.7 (1.2)	-1.5	-1.7 (1.9)
$\Phi/4F$	-2.5	-2.7 (4.8)	-1.4	-1.2 (2.1)	-2.2	-2.2 (2.7)
$\Phi/5F$	—	—	-2.0	-2.1 (2.7)	-3.0	-3.0 (8.0)

For the analysis of the titration data, we begin with the Henderson–Hasselbalch/Hill Equation, that is, pH = pK$_o$ + $(1/n)\log[\alpha/(1-\alpha)]$ where the Hill coefficient term, $1/n$, is rewritten, following Harris and Rice and others, as $(\partial\Delta G/\partial\alpha)_T/2.3RT$, i.e. pH = pK$_o$ + $\log[\alpha/(1-\alpha)] + (\partial\Delta G/\partial\alpha)_T/2.3RT$. Now to include charge–charge (c–c) and apolar–polar (a–p) repulsion effects of protein-based polymers, we write pH = pK$_o$ + ΔpK$_{c-c}$ + ΔpK$_{a-p}$ + $\log[\alpha/(1-\alpha)] + \{[(\partial\Delta G/\partial\alpha)_T]_{c-c} + (\partial\Delta G/\partial\alpha)_T]_{a-p}\}/2.3RT$. For the compositions noted above charge–charge repulsion is negligible to give, pH = pK$_o$ + ΔpK$_{a-p}$ + $\log[\alpha/(1-\alpha) + \{[(\partial\Delta G/\partial\alpha)_T]_{a-p}/2.3RT$. For redox titrations, $E = E_o + \Delta E_{a-p} + (2.3RT/nF)\log[\alpha'/(1-\alpha')] + (1/nF)[(\partial\Delta G/\partial\alpha')_T]_{a-p}$ provides the analogous equation.

The numbers in parentheses in the table are the Hill coefficients, n, and Hill plots of $\log[\alpha/(1-\alpha)]$ versus pH for acid–base titrations and $\log[\alpha'/(1-\alpha')]$ versus ΔE for the redox titrations are used, respectively, to calculate $(\partial\Delta G/\partial\alpha)$ and $(\partial\Delta G/\partial\alpha')$. $\partial\alpha$ and $\partial\alpha'$ are determined by graphical means using the y-axis intercept values of $\log[\alpha/(1-\alpha)] = 1$ ($\alpha = 0.91$) and -1 ($\alpha = 0.9$), giving the graphically derived values of the divisors of $(0.91 - 0.09) = 0.82$ for both $\Delta\alpha$ and $\Delta\alpha'$. The corresponding x-axis reference values are a ΔpH of 2 with $\Delta G = -2.3RT\Delta$pH for the acid–base titrations and a ΔE of 59 mV with $\Delta G = -nF\Delta E$ for the redox titrations. The expressions become $(\partial\Delta G/\partial\alpha) = -2.3RT(2 - \Delta$pH$^{expt})/0.82$ with ΔpH^{expt} being the run corresponding to the $\Delta\alpha = 0.82$ rise for a specific experimental slope and $(\partial\Delta G/\partial\alpha') = -nF(59\text{ mV} - \Delta E^{expt})/0.82$ with ΔE^{expt} being the run corresponding to the rise of $\Delta\alpha' = 0.82$ for a specific experimental slope. As ΔpH^{expt} approaches 2 and as ΔE^{expt} approaches 59, the slope approaches one and the terms $(\partial\Delta G/\partial\alpha)$ and $(\partial\Delta G/\partial\alpha')$ approach zero. The sign on the free energies are written for the effect in which an increase in hydrophobicity lowers the free energy on going from the more polar to the less polar state, i.e. to the uncharged or the reduced state. This results from a decrease in ΔG due to the relaxation (decrease) of the apolar–polar repulsion.

* The Hill coefficient, n, for the reference state composition, K{NMeN}/0F, i.e. N-methyl nicotinamide (NMeN) attached by amide linkage to lysine (K) of Model protein I, is taken as 1.0 after including the term, $1/n$ of $2.3RT/nF$, for the number of electrons transferred in the overall reaction.

† The Hill coefficient, n, for K/0F was experimentally found to be 0.9, possibly due to the effect of Cl$^-$–NH$_3^+$ ion pair formation during the titration. Since the interest is the change due to addition of more hydrophobic Phe (F) residues, K/0F is taken as the reference state.

§ The pKa for the reference state of an unperturbed glutamic acid is known to be in the range of 3.8–4.0. If the pKa of 3.9 is used with the more refined value of 4.55 obtained from the Hill plot of E/0F, then $-2.3RT(4.55 - 3.9) = -0.886$ kcal/mol-E. Thus, as included in the above table the value of -0.9 becomes a reasonable estimation of the value of $-2.3RT\Delta$pK for E/0F.

Table 3A. Composition effect on inverse temperature transition in water [a,b,3,5]

Polymer	T_b (°C)	T_m (°C)	ΔQ (cal/g)	ΔH (kcal/mol)	ΔS (cal/mol-K)
poly(AVGIP)	18.4	19.7	10.39	4.54	15.46
poly(AVGVP)	33.6	34.7	6.20	2.62	8.47
poly(GVGIP)	11.2	13.3	6.31	2.67	9.32
poly(GVGVP)	27.8	30.0	2.69	1.10	3.59
(GVGVP)$_{251}$	26.9	29.0	3.21	1.32	4.31
poly[(GVGVP), (GVGIP)]	18.9	20.3	4.24	1.77	5.98
poly[0.8(GVGVP), 0.2(GFGVP)]	12.9	16.0	4.58	1.92	6.54

Table 3A. *(Continued)*

Polymer	T_b (°C)	T_m (°C)	ΔQ (cal/g)	ΔH (kcal/mol)	ΔS (cal/mol-K)
poly[0.8(GVGVP), 0.2(GAGVP)]	31.8	34.9	1.99	0.80	2.57
poly(GVGVAP) (irreversible)	29.0	35.3	1.61	0.77	2.50

[a] Examples with different repeat compositions from some one thousand polymer preparations.

[b] T_b, T_m, ΔQ, ΔH, and ΔS are on-set temperature, maximum heat absorption temperature, heat, enthalpy, and entropy of the inverse temperature transition as determined by DSC, respectively. ΔH and ΔS are values per mole of repeating peptide. This is true for the DSC data reported in all tables in this chapter.

Table 3B. Poly[0.82(GVGIP),0.18(GEGIP)]: pH effect on inverse temperature transition[3]

pH	α	T_b, °C	T_m, °C	ΔQ, cal/g	ΔH, kcal/mol	ΔS, cal/mol-K
2.3	0.00	10.3	12.9	4.00	1.71	5.92
3.5	0.04	15.2	18.2	2.80	1.20	4.10
4.2	0.11	20.7	24.9	1.51	0.64	2.14
4.6	0.13	24.0	32.0	1.19	0.51	1.66
5.1	0.20	26.0	37.0	0.60	0.26	0.83

α is the degree of ionization of the Glu side chain.

Table 3C. Solute effect on inverse temperature transition[2,5,17]

Polymer	Solute	δT_t °C/[M]	$\delta \Delta Q$ cal/g-[M]	Linearity
poly(GVGVP)	Na_3PO_4 (pH > 8)	−140.0		Yes
	$(NH4)_2SO_4$	−69.0		Yes
	Na_2CO_3 (pH > 8)	−28.0		Yes
	NaCl	−13.9	1.25	Yes
	$CaCl_2$	−6.6		Yes
	NaBr	−3.5		Yes
	NaI, NaSCN	3.5		No
	Sodium dodecyl sulfate	∼600.0		
poly(GGIP)	NaCl	−13.4	1.35	Yes
poly(GGVP)	NaCl	−15.6	0.52	Yes
poly(GVGLP)	NaCl	−12.6	2.99	Yes
poly(VPGVGVPGG)	NaCl	−15.9	0.96	Yes
poly(AVGVP)	NaCl	−14.7	1.80	Yes

Table 3D. Lowering transition temperature by charge neutralization[5]

Polymer	[NaCl]/[N]	T_t, °C	T_t, °C
Poly[0.8(GVGVP),0.2(GEGVP)]		pH < 3	pH > 7
	0.0	26.4	>100
	0.15	24.5	91.5

Table 3D. (*Continued*)

Polymer	[NaCl]/[N]	T_t, °C	T_t, °C
	0.20	−28.0	73.7
	0.25	–	52.0
	0.50	19.9	47.0[a]
	1.0	13.6	35.8[a]
	1.5	–	26.0[a]
		$\delta T_t/[N] = -12.8$	$\delta T_t/[N]^a = -21.0$
Poly[0.85(GVGVP),0.15(GDGVP)]		pH < 3	pH > 7
	0.0	28.3	>100
	0.15	24.5	75.0
	0.25	–	55.0
	0.50	21.5	50.3[a]
	0.75	–	44.0[a]
	1.0	15.5	40.0[a]
	1.5		31.0[a]
		$\delta T_t/[N] = -12.3$	$\delta T_t/[N]^a = -19.0$
Poly[0.76(GVGVP),0.24(GKGVP)]		pH < 12	pH > 6
	0.0	30.4	>100
	0.05	–	70.0
	0.10	–	58.4
	0.125	–	53.0
	0.20	28.0	43.5
	0.25	–	35.2
	0.50	–	32.0[a]
	1.0	17.8	23.3[a]
	1.5	–	16.0[a]
		$\delta T_t/[N] = -12.6$	$\delta T_t/[N]^a = -16.0$

[a] Value for [NaCl] <0.25, at lower salt concentration, the slope is much steeper.

Table 3E. Enhanced charge neutralization effect by $CaCl_2$ on lowering T_t [5]

Polymer	[CaCl$_2$]/[N]	T_t, °C	T_t, °C
Poly[0.8(GVGVP),0.2(GEGVP)]		pH = 2	pH = 7
	0.0	26.4	> 100.0
	0.0125		78.6
	0.025		70.0
	0.05		55.4
	0.1	25.5	51.5
	0.2		47.4[a]
	0.3	24.0	45.1[a]
	0.4		43.5[a]
	0.7	21.7	39.6[a]
		$\delta T_t/[N] = -6.8$	$\delta T_t/[N]^a = -13.6$

[a] Value for [CaCl$_2$] <0.1, at lower salt concentration, the slope is much steeper.

Table 3F. Urea and guanidine salt effect on inverse temperature transition[5]

Polymer	Solute	Range [M]	$<\delta T_t>$ °C/[M]	$<\delta \Delta Q>$ cal/g-[M]	Linearity
poly(GVGVP)	Guanidine$_2$.H$_2$SO$_4$	[0, 1]	−11.4	1.06	Yes*
	Guanidine.HCl	[0, 2]	10.3	−1.01	Yes*
	Urea	[0, 3]	5.3	−0.57	Yes*
	Dimethyl Urea	[0, 1]	12.2	−2.28	Yes
poly(GVGIP)	Guanidine$_2$.H$_2$SO$_4$	[0, 1]	−8.9	0.76	Yes
	Guanidine.HCl	[0, 3]	6.8	−1.52	Yes*
	Urea	[0, 3]	4.0	−0.95	Yes

$<\delta T_t>$, average slope over the listed range.   Yes*with small nonlinearity.
pH (Guanidine$_2$.H$_2$SO$_4$) = 7.6.

Table 3G. Cosolvent effect on inverse temperature transition of Poly(GVGVP) in water [5]

Cosolvent	Volume %	T_b, °C	ΔQ, cal/g	ΔH, kcal/mol	ΔS, cal/mol-K
DMSO					
	35	16.7	0.38	0.15	0.53
	30	21.1	0.51	0.21	0.70
	20	29.5	0.91	0.37	1.21
	10	30.0	1.73	0.71	2.30
Acetone					
	20	31.8	0.99	0.40	1.30
	10	29.0	2.10	0.86	2.79
Dioxane					
	20	41.1	0.44	0.18	0.56
	10	31.6	1.42	0.58	1.87
H$_2$O		26.7	2.90	1.20	3.90

Table 3H. Cosolvent effect on inverse temperature transition of Poly(AVGVP) in water [5]

Cosolvent	Volume %	T_b, °C	ΔQ, cal/g	ΔH, kcal/mol	ΔS, cal/mol-K
EtOH					
	15	27.2	4.51	1.91	6.31
	10	30.9	5.37	2.27	7.42
	5	32.8	6.65	2.81	9.14
Ethylene glycol					
	20	23.2	3.89	1.65	5.52
	10	29.6	5.54	2.35	7.70
Acetone					
	30	31.2	1.54	0.65	2.13
	20	32.7	3.31	1.40	4.55
H$_2$O		33.1	7.20	3.05	9.89

Table 3I. Cosolvent effect on inverse temperature transition of Poly(GVGIP) in water [5]

Acetone (volume %)	T_b, °C	ΔQ, cal/g	ΔH, kcal/mol	ΔS, cal/mol-K
20	5.0	2.71	1.15	4.06
10	10.2	3.91	1.66	5.77
5	11.5	4.53	1.92	6.67
0	11.8	6.40	2.71	9.41

Table 3J. Solvent deuteration effect on inverse temperature transition[18]

Polymer	Solvent	T_b, °C	ΔQ, cal/g	ΔH, kcal/mol
Poly(GVGVP)	D_2O	27.2	3.25	1.33
	H_2O	28.8	2.62	1.07
Poly(GVGLP)	D_2O	13.7	6.70	2.83
	H_2O	15.5	6.02	2.55
Poly(GVGIP)	D_2O	9.0	6.76	2.86
	H_2O	10.6	6.17	2.61
Poly(AVGVP)	D_2O	31.7	7.17	3.04
	H_2O	34.2	6.46	2.73
	D_2O, 1.0 M Urea	36.5	6.07	2.57
	H_2O, 1.0 M Urea	38.5	5.27	2.23
	D_2O, 2.0 M Urea	40.8	5.09	2.16
	H_2O, 2.0 M Urea	43.0	4.37	1.85
	D_2O, 3.0 M Urea	45.8	3.94	1.67
	H_2O, 3.0 M Urea	48.5	3.55	1.51
	D_2O, 0.5 N NaCl	24.5	7.50	3.17
	H_2O, 0.5 N NaCl	26.1	6.93	2.93
	D_2O, 1.0 N NaCl	17.6	9.24	3.91
	H_2O, 1.0 N NaCl	19.1	8.79	3.72

Table 3K. Alcohol effect on inverse temperature transition of Poly(GVGVP) in water [5]

| Alcohol | | T_b | ΔQ | ΔH | ΔS |
Mole %	Volume %	(°C)	(J/g)	(kJ/mol)	(J/mol-K)
Methanol					
16.01	30	13.6	1.21	0.50	1.67
10.0	20	23.5	3.47	1.42	4.73
4.71	10	27.8	7.07	2.89	9.46
2.29	5	28.2	9.87	4.06	13.18
Ethanol					
8.00	22	18.3	1.76	0.71	2.47
7.16	20	20.5	2.01	0.84	2.76
5.16	15	26.4	4.44	1.80	5.98
3.31	10	28.8	7.28	2.97	9.71
1.60	5	29.1	10.04	4.10	13.35

Table 3K. *(Continued)*

Alcohol		T_b	ΔQ	ΔH	ΔS
Mole %	Volume %	(°C)	(J/g)	(kJ/mol)	(J/mol-K)
iso-Propanol					
5.55	20	18.4	3.01	1.21	4.14
3.26	15	26.0	4.56	1.84	6.19
2.08	10	28.1	7.70	3.14	10.33
n-Propanol					
5.69	20	11.9	3.51	1.42	4.94
4.08	15	19.8	7.03	2.89	9.71
2.61	10	25.5	9.50	3.89	12.80
tert-Butanol					
4.56	20	14.0	3.14	1.30	4.39
3.26	15	21.2	6.15	2.51	8.45
2.08	10	26.6	8.95	3.68	12.01
1.0	5	28.7	11.46	4.69	15.23
Ethylene glycol					
12.16	30	14.9	2.09	0.84	2.93
7.47	20	22.2	4.56	1.84	6.19
3.46	10	26.1	7.70	3.14	10.33
1.67	5	27.0	10.08	4.14	13.51
Glycerol					
9.57	30	9.4	6.53	2.68	9.25
5.82	20	20.1	7.28	2.97	10.00
2.67	10	24.2	9.04	3.68	12.22
1.28	5	26.4	10.50	4.31	14.06
H_2O		27.5	12.34	5.06	16.44

Table 4. Physical properties of elastic and plastic protein-based polymers

Polymer	Property	Conditions	Value
Elastic polymer			
X^{20}-poly (GVGVP)[1]	Young's modulus (dynes/cm^2)	50% extension, 37°C	1.0×10^6
X^{20}-(GVGVP)$_{251}^{[1]}$	Young's modulus (dynes/cm^2)	50% extension, 37°C	1.6×10^6
	Max. extensibility(L/L_0)	–	>2.5
	Entropic elasticity (f_S)	$f_S = (1 - f_E/f)$	<0.1
	Tensile strength		
X^{20}-poly (GVGIP)[1]	Young's modulus (dynes/cm^2)	90% extension, 37°C	3.9×10^6
	Max. extensibility (L/L_0)		>2.6
	Entropic elasticity (f_S)	$f_S = (1 - f_E/f)$	<0.1
	Tensile strength		
X^{20}-poly[3(GVGVP),(GFGVP)][1]	Young's modulus (dynes/cm^2)	80% extension, 37°C	4.9×10^6
	Transition temperature	40 mg/ml H_2O	13°C

Table 4. *(Continued)*

Polymer	Property	Conditions	Value
	Max. extensibility (L/L_0)		>2
	Entropic elasticity (f_S)	$f_S = (1 - f_E/f)$	<0.05
	Tensile strength		
X^{20}-poly (GGVP)	Young's modulus (dynes/cm^2)	20% extension, 64°C	8.2×10^5
	Transition temperature	40 mg/ml H_2O	48°C
X^{20}-poly(VPGFGVGAG)[19]	Young's modulus (dynes/cm^2)	5% extension, 37°C	6.8×10^7
	Transition temperature	40 mg/ml H_2O	8°C
Plastic polymer			
X^{20}-poly (AVGVP)[3]	Young's modulus (dynes/cm^2)	7% extension, 37°C	2.0×10^8
	Tensile strength		
Hydrogel polymer		Density	
X^{20}-poly (GVGVP)[1]	<25°C	~40 mg/ml at 4°C	
X^{20}-poly (GVGIP)[1]	<10°C	~80 mg/ml at 4°C	
X^{20}-poly (AVGVP)[1]	<30°C	~40 mg/ml at 4°C	
X^{20}-poly (GGXP), X = A, V, I	<50°C	~20 mg/ml at 4°C	
poly[0.8(AVGVP),0.2(AFGVP)]	Forms gel at concentrations >20 mg/ml and >18°C		
poly (AVGIP)	Forms gel at concentrations >20 mg/ml and >18°C		

Table 5. Reversible contraction–relaxation of cross-linked polymer matrices.

Elastic Polymer	Variable	Length
Thermally driven		
X^{20}-poly(GVGVP)[20]	40°C	4.4 cm
(3 g load)	20°C	5.4 cm
pH driven		
X^{20}-poly[0.8(GVGVP), 0.2(GEGVP)][2]	pH = 2	4.2 mm
(3 g load)	pH = 7	8.2 mm
Salt driven		
X^{20}-poly(GVGVP)[20]	1 N NaCl	5.15 cm
(at 20°C, 3 g load)	H_2O	5.50 cm
Organic solvent driven		
X^{20}–(VPGVG)$_{251}$[5]	H_2O	23.5 mm
(at 23°C, no load)	20 volume% EtOH	21.0 mm
Pressure driven (as in Table 6)		
Redox driven (as in Table 6)		

Table 6. Examples of free energy transduction effected by elastin protein-based polymers.

Transduction / Elastic Polymer	Intensive Variable	Property Measured
Thermo-mechanical	Temperature	Force, dynes/cm^2
X^{20}-poly(GVGVP)[21]	36°C	1.0×10^6

Table 6. *(Continued)*

Transduction / Elastic Polymer	Intensive Variable	Property Measured
	Temperature	Force, dynes/cm^2
(at constant length)	5°C	4.0×10^5
X^{20}-poly[0.9(GVGVP),0.1(GEGVP)][21]	37°C	2.3×10^6
(in pbs at 37°C, constant length.)	3°C	4.0×10^5
Chemo-mechanical	Chemical potential	Force, dynes/cm^2
X^{20}-poly[0.80(GVGVP),0.20(GEGVP)][22]	pH = 2.1	5.1×10^5
(in pbs at 37°C, constant length.)	pH = 7.4	$<1.0 \times 10^4$
X^{20}-poly(GVGVP)[23]	[.15 N NaCl, .01 M	2.1 g
(at 25°C, constant length)	phosphate] in H$_2$O	0
Baro-mechanical	Pressure	Length, mm
poly[0.79(GVGVP), 0.21(GVGFP)][24,25]	68 atm	(16.2)[24], (38.5)[25]
(at 12.6°C, 1 gram constant force)	1 atm	(15.2)[24], (37.0)[25]
Electro-mechanical	Redox	Length, cm
X^{20}-poly[0.70(GVGVP),0.30(GVGK{NMeN}P)][25]	Reduced	(3.2)[25], (1.92)[26]
X^{20}-poly[0.73(GVGVP),0.27(GVGK{NMeN}P)][26]	Oxidized	(4.0)[25], (2.16)[26]
Photo-mechanical	Photon	T_t
poly[0.8(GVGVP),0.2(GVGE{AzB}P)][27]	Dark	32°C
	350 nm light	42°C
Mechano-chemical	Mechanical force	pK of Glu(E)
X^{20}-poly[0.82(GVGIP),0.18(GEGIP)][12]	0	6.2
	8.0×10^5 dynes/cm^2	9.0
Electro-chemical	Redox	pK of Asp(D)
poly(GDGFP GVGVP GVGVP GFGVP GVGVP GVG	Reduced	11.0
K{NMeN}P)[28]	Oxidized	8.5

K{NMeN} is *N*-methyl nicotinamide attached as amide to Lys; E{AzB} is azobenzene attached as ester to Glu.

Table 7. Physical properties of the synthetic poly(W4) from human elastin[29]

Polymer	Condition	T_b°C	ΔQ cal/g	ΔH kcal/mole[a]	ΔH kcal/mole[b]	ΔS kcal/mole[b]-K
poly(W4)	pH = 2.0	25.5	1.92	0.86	8.79	28.63
	pH = 4.3	27.8	1.45	0.65	6.63	21.43
	pH = 5.8	28.0	0.32	0.14	1.48	4.62
	pK of the Glu = 4.84					
X^{20}-poly(W4)	Young's modulus = 3.4×10^5 dynes/cm^2 at 37°C in pbs					
	Entropic elasticity indicated by a f_E/f value of <0.1 from 35°C to 50°C.					

W4(6-56): GLVPGGPGFPGGVVGVPGAGVPGVGVPGAGIPVVPGAGIPGAAVPGVVSPE
[a] Per mole of pentapeptide.   [b] Per mole of W4.

Table 8. Dielectric relaxation of poly(GVGIP) in water as a function of temperature[30]

Temp. °C	$\Delta \varepsilon$	τ, ns	α	σ, mS/m	ε^∞ at 100 MHz ± 0.5
0[a]	26 ± 3	139 ± 6	0.36 ± 0.08	1.2 ± 0.2	32.3
6[a]	26 ± 4	87 ± 10	0.3 ± 0.1	1.8 ± 0.6	30.8

Table 8. *(Continued)*

Temp. °C	$\Delta\varepsilon$	τ, ns	α	σ, mS/m	ε^{∞} at 100 MHz $\pm$ 0.5
12	37 ± 2	49 ± 3	0.11 ± 0.03	0.5 ± 0.1	26.3
18	34 ± 2	44 ± 3	0.08 ± 0.02	0.6 ± 0.1	24.7
24	37 ± 3	44 ± 3	0.08 ± 0.07	0.6 ± 0.1	23.7
30	38 ± 2	38 ± 2	0.08 ± 0.02	0.6 ± 0.1	23.2
36	42 ± 3	36 ± 2	0.09 ± 0.02	0.6 ± 0.1	23
42	43 ± 3	35 ± 2	0.09 ± 0.02	0.7 ± 0.1	22.7
48	48 ± 3	39 ± 2	0.10 ± 0.02	0.7 ± 0.2	22.6
54	50 ± 3	35 ± 3	0.10 ± 0.02	0.9 ± 0.2	22.6
60	55 ± 4	31 ± 2	0.11 ± 0.03	1.0 ± 0.2	22.4

[a] The dielectric parameters, obtained using the Cole–Cole function (see reference 30), of the sample at 0 and 6°C are only suggestive since the relevant dielectric relaxation is not yet very well resolved, i.e. the transition has not sufficiently progressed at such a low temperature.

Note: The minimal dielectric constant at 36°C is 65, that is, 23 for the high frequency (100 MHz) limit plus 42 for the relaxation in the low MHz range.

Table 9. Comparison of the volume fraction of the pentameric protein-based polymer analogs in water (60°C) computed by means of the Maxwell–Fricke equation using the dielectric relaxation data[30]

Protein-based Polymer	poly(GVGVP)	poly(GVGLP)	poly(GVGIP)
Volume fraction of analog	0.31 ± 0.02	0.50 ± 0.02	0.54 ± 0.03
Mass fraction of analog	0.37 ± 0.02[a]	0.57 ± 0.02	0.61 ± 0.03

[a] By direct analytical measurement the weight percent of phase separated polymer in water at 31°C was found to be 37% (See reference 31).

References

1. Urry DW, Nicol A, McPherson, DT, et al. Properties, Preparations and Applications of Bioelastic Materials. In: *Handbook of Biomaterials and Bioengineering – Part A: Materials*. New York: Marcel Dekker, Inc.; 1995:1619–1673.

2. Urry DW. Molecular Machines: How Motion and Other Functions of Living Organisms Can Result from Reversible Chemical Changes. *Angew. Chem.* (German) 1993;105:859–883; *Angew. Chem. Int. Ed. Engl.* 1993;32:819–841.

3. Urry DW, Luan C-H, Harris CM, Parker TM. Protein-based Materials with a Profound Range of Properties and Applications: The Elastin ΔT_t Hydrophobic Paradigm. In: Kevin McGrath and David Kaplan, eds. *Protein-based Materials*. Boston, Massachusetts: Birkhauser Press; 1997.

4. Urry DW, Luan C-H. Proteins: Structure, Folding and Function. In: Giorgio Lenaz, ed. *Bioelectrochemistry: Principles and Practice*. Basel, Switzerland: Birkhäuser Verlag AG; 1995:105–182.

5. Luan C-H, Urry DW. Elastic, Plastic, and Hydrogel Protein-based Polymers. In: James E. Mark, ed. *Polymer Data Handbook*. New York, Oxford: Oxford University Press; 1999:78–89.

6. Urry DW. What Sustains Life? Consilient mechanisms for protein-based machines and materials. Springer (Birkhauser), LLC, New York, (2006). ISBN-10: 0-8176-4346-X, ISBN-13: 978-08176-4346-1, Figure 5.10 and Tables 5.2 and 5.3.

7. Urry DW, McPherson DT, Xu J, Daniell H, Guda C, Gowda DC, Jing N, Parker, TM. Protein-Based Polymeric Materials: Syntheses and Properties. In: *Polymeric Materials Encyclopedia: Synthesis, Properties and Applications.* Boca Raton, Florida: CRC Press; 1996:7263–7279.

8. Urry DW, Peng SQ, Parker TM. Delineation of Electrostatic-and Hydrophobic-induced pKa Shifts in Polypentapeptides: The Glutamic Acid Residue. *J. Am. Chem. Soc.* 1993;115:7509–7510.

9. Urry DW, Peng SQ, Parker TM, Gowda DC, Harris RD. Relative Significance of Electrostatic- and Hydrophobic-Induced pK Shifts in a Model Protein: The Aspartic Acid Residue. *Angew. Chem.* (German) 1993;105:1523–1525; *Angew. Chem. Int. Ed. Engl.* 1993;32:1440–1442.

10. Urry DW, Peng SQ, Gowda DC, Parker TM, Harris RD. Comparison of Electrostatic- and Hydrophobic-induced pKa Shifts in Polypentapeptides: The Lysine Residue. *Chemical Physics Letters.* 1994;225:97–103.

11. Urry DW, Peng SQ, Hayes L, Jaggard J, Harris RD. A New Mechanism of Mechanochemical Coupling: Stretch-Induced Increase in Carboxyl pK_a as Diagnostic. *Biopolymers.* 1990;30:215–218.

12. Urry DW, Peng SQ. Non-linear Mechanical Force-induced pKa Shifts: Implications for Efficiency of Conversion to Chemical Energy. *J. Am. Chem. Soc.* 1995;117:8478–8479.

13. Urry DW. What Sustains Life? Consilient mechanisms for protein-based machines and materials. New York: Springer (Birkhauser), LLC; 2006: ISBN-10: 0-8176-4346-X, ISBN-13: 978-08176-4346-1, Figure 1.2.

14. Hayes LC, Urry DW (2007) unpublished data.

15. Hayes LC, Woods TC, Gowda DC, Urry DW. Effect of the Hydrophobicity of Elastic Model Proteins on Redox Potential and Hill coefficients. (in preparation).

16. Woods TC, Hayes LC, Xu J, McPherson DT, Urry DW. Lys-containing Elastic Protein-based Polymers: Biosynthesis and Supra-linear Increases in ΔpKa and in Positive Cooperativity with Linear Increases in Hydrophobicity. (in preparation).

17. Luan C-H, Parker TM, Prasad KU, Urry DW. DSC Studies of NaCl Effect on the Inverse Temperature Transition of Some Elastin-based Polytetra, Polypenta-, and Polynonapeptides. *Biopolymers.* 1991;31:465–475.

18. Luan C-H, Urry DW. Solvent Deuteration Enhancement of Hydrophobicity: DSC Study of the Inverse Temperature Transition of Elastin-based Polypeptides. *J. Phys. Chem.* 1991;95:7896–7900.

19. Urry DW, Jaggard J, Harris RD, Chang DK, Prasad KU. The Poly (nonapeptide) of Elastin: A New Elastomeric Polypeptide Biomaterial. In: Charles G Gebelein, Richard L Dunn, eds. *Progress in Biomed. Polym.* New York: Plenum Publishing Corp.;1990:171–178.

20. Urry DW. Free Energy Transduction in Polypeptides and Proteins Based on Inverse Temperature Transitions. *Prog. Biophys. Molec. Biol.* 1992;57:23–57.

21. Peng SQ, Gowda DC, Parker TM, Urry DW. 2007; unpublished data.

22. Urry DW, Haynes B, Zhang H, Harris RD, Prasad KU. Mechanochemical Coupling in Synthetic Polypeptides by Modulation of an Inverse Temperature Transition. *Proc. Natl. Acad. Sci. USA.* 1988;85:3407–3411.

23. Urry DW, Harris RD, Prasad KU. Chemical Potential Driven Contraction and Relaxation by Ionic Strength Modulation of an Inverse Temperature Transition. *J. Am. Chem. Soc.* 1988;110:3303–3305.

24. Urry DW, Hayes LC, Parker TM, Harris RD. Baromechanical Transduction in a Model Protein by the ΔT_t Mechanism. *Chem. Phys. Letters.* 1993;201:336–340.

25. Hayes LC, Gowda DC, Parker TM, Urry DW. 2007: unpublished data.

26. Urry DW, Hayes LC, Channe Gowda DC. Electromechanical Transduction: Reduction-driven Hydrophobic Folding Demonstrated in a Model Protein to Perform Mechanical Work. *Biochem. Biophys. Res. Comm.* 1994;204:230–237.

27. Strzegowski LA, Martinez MB, Gowda DC, Urry DW, Tirrell DA. Photomodulation of the Inverse Temperature Transition of a Modified Elastin Poly(Pentapeptide). *J. Am. Chem. Soc.* 1994;116:813–814.

28. Urry DW, Hayes LC, Gowda DC, Peng SQ, Jing N. Electro-chemical Transduction in Elastic Protein-based Polymers. *Biochem. Biophys. Res. Commun.* 1995;210:1031–1039.

29. Gowda DC, Luan C-H, Furner RL, et al. Synthesis and Characterization of Human Elastin W_4 Sequence, *Int. J. Pept. Protein Res.* 1995;46:453–463.

30. Buchet R, Luan C-H, Prasad KU, Harris RD, Urry DW. Dielectric Relaxation Studies on Analogs of the Polypentapeptide of Elastin. *J. Phys. Chem.* 1988;92:511–517.

31. Urry DW, Trapane TL, Prasad KU. Phase-Structure Transitions of the Elastin Polypentapeptide-Water System Within the Framework of Composition- Temperature Studies. *Biopolymers.* 1985;24:2345–2356.

Epoxy Resins

Mee Y. Shelley and Jennifer L. Braun

Trade Names Araldite, DER, Epi-Cure, Epi-Res, Epikote, Epon, Epotuf, etc.

Class Thermoset polymers, after curing (the uncured base resins are thermoplastic)

Major Resin Types DGEBA (diglycidyl ether of bisphenol A), novolacs, peracid resins, hydantoin resins, etc.

Other Ingredients in Epoxy Formulation Diluents, resinuous modifiers (to affect flexibility, toughness, peel strength, adhesion, etc.), fillers, colorants and dyes, other additives (e.g., rheological additives, flame retardants).

Major Applications Protective coatings (for appliance, automotive primers, pipes, etc.). Encapsulation of electrical and electronic devices. Adhesives. Bonding materials for dental uses. Replacement of welding and riveting in aircraft and automobiles. In composites for materials in space industry, printed circuitry, pressure vessels and pipes. Construction uses such as flooring, paving, and airport runway repair.

Properties of Special Interest Wide range of properties depending on the formulation and processing. Chemical and weathering resistance, toughness, durability. Excellent adhesion to a variety of surfaces. Good electrical and thermal insulation. Better mechanical properties than most other castable plastics. Discolor when exposed to UV. Some are skin sensitizers.

Property	Units	Conditions	Value	Reference
Specific gravity	–	General range	1.01–2.00	(19)
		Unfilled	1.2–1.3	(1)
		Bisphenol molding compounds (glass fiber-reinforced/mineral-filled)	1.6–2.1	(2)
		Bisphenol molding compounds (low density glass sphere-filled)	0.75–1.0	(2)
		Sheet molding compounds	0.1	(2)
		Novolac molding compounds	1.6–2.05	(2)
		Surface coats	1.20–1.76	(21)
		Laminates	0.47–1.40	(21)
		Casting resins, unfilled	1.13	(21)
		Casting resins, glass-filled	2.0	(20)
		Casting resins, aluminum-filled	1.44–1.62	(21)
		Casting resins, mineral-filled	1.92	(20)
		Casting resins, silica-filled	1.6–2.0	(2)
		Casting resins, aluminum-filled	1.4–1.8	(2)

Property	Units	Conditions	Value	Reference
		Casting resins, flexibilized	0.96–1.35	(2)
		Casting resins, cycloaliphatic	1.16–1.21	(2)
Water absorption	%	1/8 in. thick specimen, 24 h		
		Bisphenol molding compounds	0.04–1.0	(2)
		Sheet molding compounds	1.4–1.6	(2)
		Novolac molding compounds	0.04–0.29	(2)
		Casting resins, glass-filled	0.1	(20)
		Casting resins, mineral-filled	0.1	(20)
		Casting resins, silica-filled	0.04–0.1	(2)
		Casting resins, aluminum-filled	0.1–4.0	(2)
		Casting resins, flexibilized	0.27–0.5	(2)
		Filament wound (80 wt% glass fiber-reinforced)	0.50	(3, 4)
Impact strength, Izod	$J\,m^{-1}$	General range	0.345–53.7	(19)
		Unfilled	10–50	(1)
		Silica-filled	16–24	(2)
		Aluminum-filled	21–85	(2)
		Flexibilized	120–270	(2)
		Bisphenol molding compounds (glass fiber-reinforced)	16–530	(2)
		Bisphenol molding compounds (mineral-filled)	16–27	(2)
		Bisphenol molding compounds (low density glass sphere-filled)	8–13	(2)
		Sheet molding compounds (glass fiber-reinforced)	1,600–2,100	(2)
		Sheet molding compounds (carbon fiber-reinforced)	800–1,100	(2)
		Novolac molding compounds	16–27	(2)
		Filament wound (80 wt% glass fiber-reinforced)	2,400	(3, 4)
		Casting resins, glass-filled	20	(20)
		Casting resins, mineral-filled	20	(20)
Hardness	Rockwell	General range	109–118	(19)
	Shore	Casting resin, unfilled	92D	(21)
	Shore	Casting resins, aluminum-filled	87D–92D	(21)
	Rockwell	Casting resins, mineral-filled	106	(20)
	Rockwell	Casting resins, glass-filled	115	(20)
	Shore	Casting resins, flexibilized	D65–89	(2)
	Barcol	Novolac molding compounds	70–78	(2)
	Rockwell	Filament wound (80 wt% glass fiber-reinforced)	M98	(3, 4)
	Shore	Surface coats	80D–92D	(21)
	Shore	Laminates	60D–94D	(21)
Fracture toughness	$J\,cm^{-3}m^{1/2}$	Unspecified	0.6	(1)

Property	Units	Conditions	Value	Reference
Tensile modulus	MPa	Unfilled	3,000–5,000	(1)
		Casting, unfilled	2,400	(2)
		Bisphenol molding compounds (glass fiber-reinforced)	21,000	(2)
		Bisphenol molding compounds (mineral-filled)	2,400	(2)
		Sheet molding compounds (glass fiber-reinforced)	14,000–28,000	(2)
		Sheet molding compounds carbon fiber-reinforced)	70,000	(2)
		Novolac molding compounds	14,500–16,600	(2)
		Filament wound (80 wt% glass fiber-reinforced)	27,600	(3, 4)
Compressive modulus	MPa	Casting, flexibilized	7–2,400	(2)
		Casting, cycloaliphatic	3,400	
		Bisphenol molding compounds (mineral-filled)	4,500	
		Novolac molding compounds (mineral- and glass-filled, high temperature)	4,550	
Stress at break	MPa	Unfilled	30–90	(1)
Tensile strength at break	MPa	General range	0.00621–69.8	(19)
		Casting resins, unfilled	28–90	(2)
		Casting resins, silica-filled	48–90	(2)
		Casting resins, aluminum-filled	48–83	(2)
		Casting resins, flexibilized	14–70	(2)
		Casting resins, cycloaliphatic	55–83	(2)
		Casting resin, unfilled	68	(21)
		Casting resins, glass-filled	75	(20)
		Casting resins, aluminum-filled	34–55	(21)
		Casting resins, mineral-filled	50	(20)
		Surface coats	28–49	(21)
		Laminates	6–309	(21)
		Bisphenol molding compounds (glass fiber-reinforced/mineral-filled)	28–140	(2)
		Bisphenol molding compounds (low density glass sphere-filled)	17–28	(2)
		Sheet molding compounds (glass fiber-reinforced)	140–240	(2)
		Sheet molding compounds (carbon fiber-reinforced)	280–340	(2)
		Novolac molding compounds	34–110	(2)
		Filament wound (80 wt% glass fiber-reinforced)	552	(3, 4)
Elongation at break	%	General range	0.40–7.9	(19)
		Unfilled	1–2	(1)

Property	Units	Conditions	Value	Reference
		Bisphenol molding compounds (filled with glass fiber)	4	(2)
		Sheet molding compounds	0.5–2	(2)
		Casting resins, unfilled	3–6	(2)
		Casing resins, aluminum-filled	0.5–3	(2)
		Casting resins, glass-filled	0.8	(20)
		Casting resins, mineral-filled	1	(20)
		Casing resins, silica-filled	1–3	(2)
		Casing resins, flexibilized	20–85	(2)
		Casing resins, cycloaliphatic	2–10	(2)
		Filament wound (80 wt% glass fiber-reinforced)	1.6	(3, 4)
Flexural strength	MPa	Casting resin, unfilled	134	(21)
		Casting resins, aluminum-filled	52–80	(21)
		Surface coats	51–241	(21)
		Laminates	11–365	(21)
		Bisphenol molding compounds	34–200	(2)
		Sheet molding compounds	340–660	
		Novolac molding compounds	70–150	
		Casting resins and compounds	55–170	
Flexural modulus	MPa	Casting resins, glass-filled	14,000	(20)
		Casting resins, mineral-filled	10,000	(20)
		Bisphenol molding compounds	3,400–31,000	(2)
		Sheet molding compounds	14,000–34,000	(2)
		Novolac molding compounds	9,700–17,000	(2)
		Filament wound (80 wt% glass fiber-reinforced)	34,500	(3, 4)
Compressive strength	MPa	Casting, unfilled	100–170	(2)
		Casting, silica or alumina-filled	100–240	(2)
		Casting, flexibilized	7–97	(2)
		Casting, cycloaliphatic	100–140	(2)
		Casting resin, unfilled	109	(21)
		Casting resins, aluminum-filled	79–129	(21)
		Surface coats	94–190	(21)
		Laminates	34–200	(21)
		Bisphenol molding compounds (glass fiber-reinforced/mineral-filled)	120–280	(2)
Compressive strength	MPa	Bisphenol molding compounds (low density glass sphere-filled)	70–100	(2)
		Sheet molding compounds (glass fiber-reinforced)	140–210	(2)
		Sheet molding compounds (carbon fiber-reinforced)	210–280	(2)
		Novolac molding compounds	170–330	(2)
		Filament wound (80 wt% glass fiber-reinforced)	310	(3, 4)

Surface tension

Polymer	Temp. (K)	Value (mN m^{-1})	Reference
DGEBA with 6 wt% N-N-diethylaminopropylamine, cured	293	46.8	(5)
DGEBA with stoichiometric amount of triethylenetetramine, cured	293	39.1	(5)
DGEBA, 2,3-(diglycidoxy-1,4-phenylene)propane, chain extended with bisphenol A	296	51.2	(5, 6)

Interfacial tension[5]

Polymer	Temp. (K)	Value (mN m^{-1})
Poly(butadiene) vs. epoxy resin*	296	1.77
	328	1.40
Poly(butadiene-stat-acrylonitrile) 18 wt% AN vs. epoxy resin*	293	1.23
	328	0.57
Poly(butadiene-stat-acrylonitrile) 26 wt% AN vs. epoxy resin*	328	0.58

* Epoxy resin: DGEBA, chain extended with bisphenol A.

Property	Units	Conditions	Value	Reference
Volume resistivity	ohm cm	Filled with glass	10^{16}	(7)
		Filled with mineral	10^{16}	
		Casting resins, glass-filled	10^{15}	(20)
		Casting resins, mineral-filled	10^{15}	(20)
Dielectric strength	V mil^{-1}	Filled with glass	360	(7)
		Filled with mineral	400	
	MV/m	Casting resins, glass-filled	16	(20)
		Casting resins, mineral-filled	16	(20)
Dielectric constant		At 1 MHz		(7)
		Filled with glass	4.6	
		Filled with mineral	5.0	
		At 1 kHz		
		Casting resins, glass-filled	4.5	(20)
		Casting resins, mineral-filled	4	(20)
Thermal conductivity	Wm^{-1} K^{-1}	Casting grade, 293 K	0.19	(8, 9)
		Casting grade, 300–500 K	0.19–0.34	(9, 10)
		Foam, $d = 0.032$–0.048 g cm^{-3}	0.016–0.022	(9, 11)
		Foam, $d = 0.080$–0.128 g cm^{-3}	0.035–0.040	(9, 11)
		Filled with 50% aluminum	1.7–3.4	(9, 12)
		Filled with 25% Al_2O_3	0.35–0.52	(9, 12)
		Filled with 50% Al_2O_3	0.52–0.69	(9, 12)
		Filled with 75% Al_2O_3	1.4–1.7	(9, 12)
		Filled with 30% mica	0.24	(9, 8)
		Filled with 50% mica	0.39	(9, 8)

Epoxy Resins

Property	Units	Conditions	Value	Reference
		Filled with silica	0.42–0.84	(9, 8, 12)
		Filament wound (80 wt% glass fiber-reinforced)	1.77	(3, 4)
Deflection temperature	K	Under flexural load, 0.45 MPa		
		General range	321.3–516.2	(19)
		Casting resins, glass-filled	533+	(20)
		Casting resins, mineral-filled	533+	(20)
		Under flexural load, 1.82 MPa		(2)
		General range	320.1–522.2	(19)
		Bisphenol molding compounds, glass fiber-reinforced/mineral-filled	380–530	
		Bisphenol molding compounds, low density glass sphere-filled	370–390	
		Sheet molding compounds	560	
		Novolac molding compounds	420–530	
		Casting resins, glass-filled	427	(20)
		Casting resins, mineral-filled	411	(20)
		Casting resins and compounds (unfilled)	320–560	
		Casting resins and compounds (silica-filled)	340–560	
		Casting resins and compounds (aluminum-filled)	360–590	
		Casting resins and compounds (flexibilized)	296–390	
		Casting resins and compounds (cycloaliphatic)	370–510	

Radiation resistance, half-value dose in air*

Conditions	Dose rate (Gy h^{-1})	Value (MGy)	Reference
Filled with glass fiber	$\geq 10^5$	25–100+	(13, 14)
Filled with graphite	$\geq 10^5$	50	(13, 14)
Filled with mineral flour	$\geq 10^5$	10–30	(13, 14)
Filled with mineral flour (85% quartz sand)	500	7	(13, 15)
Filled with cotton	$\geq 10^5$	1	(13, 14)

* Defined as the absorbed dose that reduces a property (flexural strength) to 50% of the initial value.

Property	Units	Conditions	Value	Reference
Solubility parameters	(MPa)$^{1/2}$			
Hildebrand parameter δ		–	22.3	(16, 17)
Hansen parameters		Epikote 1001 (Shell)		(17, 18)
δ_d			20.36	
δ_p			12.03	
δ_h			11.48	
δ_t			26.29	

Property	Units	Conditions	Value	Reference
Processing temperature	K	Bisphenol molding compounds		(2)
		Compression	390–440	
		Transfer	390–470	
		Sheet molding compounds		
		Compression	390–440	
		Transfer	405–440	
		Novolac molding compounds		
		Compression	410–460	
		Injection	420–450	
		Transfer	390–480	
Molding pressure	MPa	Bisphenol molding compounds	0.7–34	(2)
		Sheet molding compounds	3.4–14	
		Novolac molding compounds	1.7–21	
Compression ratio	–	Bisphenol molding compounds	2.0–7.0	(2)
		Sheet molding compounds	2.0	
		Novolac molding compounds	1.5–2.5	
Mold shrinkage (linear)	–	Bisphenol molding compounds	0.001–0.01	(2)
		Sheet molding compounds	0.001	
		Novolac molding compounds	0.004–0.008	
		Casting resins	0.0005–0.01	
		Casting resins, glass-filled	0.5	(20)
		Casting resins, mineral-filled	0.6	(20)
		Casting resins, aluminum-filled	0.06–0.5	(21)
		Surface coats	0.00–0.04	(21)
		Laminates	0.0–0.5	(21)
Coeff. of Thermal Expansion	$°C^{-1}$	General range	5.8×10^{-6}–0.00012	(19)
		Casting resin, unfilled	2.2×10^{-7}	(21)
		Casting resins, glass-filled	2.3×10^{-5}	(20)
		Casting resins, mineral-filled	5.7×10^{-5}	(20)
		Casting resins, aluminum-filled	5.1×10^{-5}–9.4×10^{-5}	(21)
		Surface coats	1.3×10^{-5}–1.3×10^{-4}	(21)
		Laminates	5.1×10^{-6}–5.0×10^{-5}	(21)
Oxygen index	%	Casting resins, glass-filled	25	(20)
		Casting resins, mineral-filled	25	(20)
Flammability	UL 94	Casting resins, glass-filled	V0	(20)
		Casting resins, mineral-filled	V0	(20)
Dissipation factor	–	At 1 kHz		
		Casting resins, glass-filled	0.01	(20)
		Casting resins, mineral-filled	0.01	(20)

References

1. Brostow W, Kubát J, Kubát MM. In: Mark JE, ed. *Physical Properties of Polymers Handbook*. New York: Wiley-Interscience; 1996:313–334.

2. Kaplan WA, et al., eds. *Modern Plastics Encyclopedia '97.* New York: McGraw-Hill; 1996: *Modern Plastics*, Mid-November.

3. Rosato D. In: Mark HF, et al. eds. *Encyclopedia of Polymer Science and Engineering.* Vol 14. New York: John Wiley and Sons; 1988:350–391.

4. *Fiberglas Plus Design: A Comparison of Materials and Processes for Fiber Glass Composites.* Owens-Corning Fiberglas Corp., July 1985.

5. Wu S. In: Brandrup J, Immergut EH, eds. *Polymer Handbook.* 3d ed. New York: Wiley-Interscience; 1989:VI 411–434.

6. Sohn JE, et al. *Polym. Mater. Sci. Eng.* 1983;49:449.

7. Harper CA, ed. *Handbook of Plastics, Elastomer, and Composites.* 3d ed. New York: McGraw-Hill; 1996.

8. Thompson EV. In: Mark HF, et al., eds. *Encyclopedia of Polymer Science and Engineering.* Vol 16. New York: John Wiley and Sons; 1989:711–747.

9. Yang Y. In: Mark JE, ed. *Physical Properties of Polymers Handbook.* New York: Wiley-Interscience; 1996:111–117.

10. Chern BC, et al. In: Klemens PG, Chu TK, eds. *Thermal Conductivity 14, Proceedings of the 14th International Thermal Conference.* New York: Plenum; 1975.

11. Mark HF, et al., eds. *Encyclopedia of Chemical Technology.* 3d ed. New York: Wiley-Interscience; 1978.

12. Goodman I, Sidney H, eds. *Handbook of Thermoset Plastics.* Park Ridge, N.J.: Noyes; 1986.

13. Wündrich K. In: Brandrup J, Immergut EH, eds. *Polymer Handbook.* 3d ed. New York: Wiley-Interscience; 1989:VI 463–474.

14. Schönbacher H, Stolarz-Izycka A. *CERN.* 1979;79–08.

15. Rauhut K, Rösinger S, Wilski H. *Kunststoffe.* 1980;70:89.

16. Tobolsky AV. *Properties and Structure of Polymers.* New York: John Wiley and Sons; 1960:64–66.

17. Du Y, Xue Y, Frisch HL. In: Mark JE, ed. *Physical Properties of Polymers Handbook.* New York: Wiley-Interscience; 1996:227–239.

18. Hansen CM. *Skand. Tidskr, Färg Lack.* 1971;17:69.

19. ©1986–2007 IDES Inc., www.ides.com

20. The Rubber and Plastics Research Association. www.rapra.co.uk

21. Ad-Tech product data sheet, www.ad-tech.com

Ethylcellulose

Yong Yang

Acronym EC

Class Carbohydrate polymers

Structure

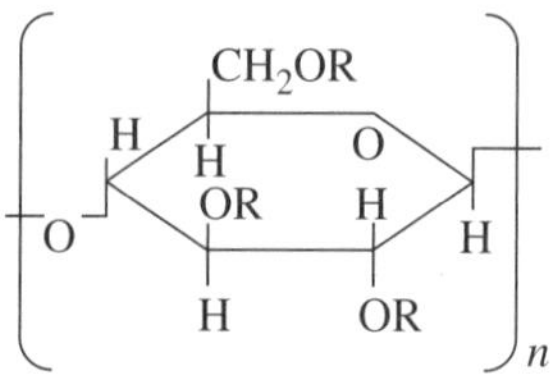

R is $-CH_2CH_3$ or H

Major Applications Lacquers for wood, plastic, and paper, varnishes, hot melts, adhesives, thickener for coatings and inks, tablet coatings and binding, encapsulation.

Preparative Techniques Ethylcellulose is prepared by reacting ethyl chloride with alkali cellulose, as expressed by the following reaction:

$$RONa + C_2H_5CI \rightarrow ROC_2H_5 + NaCI$$

where R represents the cellulose radical.

Properties of Special Interest Low temperature flexibility, soluble in a variety of organic solvents, clarity.

Property	Units	Conditions	Value	Reference
Molecular weight of repeat unit		Degree of substitution DS : 3.0	246.30	
Infrared characteristic absorption frequencies				(1,2)

Frequency (cm^{-1})	Assignment
2970	(C_2H_5) stretching
2870	(C_2H_5) stretching
2900	(C_2H_5) stretching
2870	(C_2H_5) stretching
1490	(C_2H_5) deformation
1450	(C_2H_5) deformation
1410	(C_2H_5) deformation
1380	(C_2H_5) deformation

Ethylcellulose

Property	Units	Conditions	Value	Reference
Frequency (cm^{-1})	Assignment			
1320	(C$_2$H$_5$) deformation			
1280	(C$_2$H$_5$) deformation			
1109	(C—O) stretching			

Types of ethylcellulose   (3)

Type	Ethoxyl Content, %	Ethoxyl Groups per Anhydroglucose Unit
K	45.0–47.2	2.22–2.41
N	48.0–49.5	2.46–2.58
T	49.6–51.5	2.58–2.73
X	50.5–52.5	2.65–2.81

Property	Units	Conditions	Value	Reference
Thermal expansion coefficient	K^{-1}	Sheet	$(5–15) \times 10^{-5}$	(3)
Heat capacity	kJ K^{-1} mol^{-1} (of repeat unit)		0.31–0.77	(4)
Specific gravity	g cm^{-3}		1.14	(3)

Solvents and nonsolvents

DS	Solvent	Nonsolvent	Reference
0.5–0.7	Aqueous alkali	Water	(3, 5, 6)
1.0–1.5	Acetic acid, formic acid, pyridine, water (cold)	Ethanol	
2	Chloroform, chlorohydrins, dichloroethylene, ethanol, methylene chloride, tetrahydrofuran	Alcohols, carbon tetrachloroethylene, diethyl ether, esters, ketones, hydrocarbons, water	
2.3	Acetic esters, alcohols, alkyl halogenids, benzene, carbon disulfide, furan derivatives, ketones, nitromethane	Acetone (cold), ethylene glycol	
2.4–2.6	Acetic acid*, acetone, amyl acetate, amyl alcohol, hexane, nitromethane, petroleum ether, benzene, benzyl acetate, benzyl alcohol, butanol, butyl acetate, butyl cellosolve, butyl lactate, carbon tetrachloride*, cellosolve, cellosolve acetate, chloroform*, m-cresol*, cyclohexanol, cyclohexanone, dichloroacetic acid*, dichloromethane*, 1,5-dimethyl-2-pyrrolidone*, dioxane, ethanol, ethylene chloride, ethyl acetate, ethyl ether, ethyl formate, ethyl lactate, formic acid*, hexone, isopropyl acetate, methanol, methyl ethyl ketone, methyl acetate, methyl cellosolve acetate, methylene dichloride, methyl formate, 1-nitropropane, 2-nitropropane, pentachloroethane, phenol*, propanol, propyl acetate, tetralin, toluene, trichloroethylene*, trifluoroacetic acid*, trifluoroethanol*, xylene	Hexane, petroleum ether, VM&P naphtha, vasol	

DS	Solvent	Nonsolvent	Reference
3	Alcohols, ester, benzene, methylene Chloride, tetrahydrofurfuryl alcohol	Carbon tetrachloride, diols, hydrocarbons, n-propyl ether	

* Forms liquid crystal mesophase.

Solubility parameters δ

Solvent	DS	δ (MPa)$^{1/2}$	Temperature (K)	Reference
	2.4–2.6	21.1		(7)
Poor H-bonding solvent	2.4–2.6	20		(8)
		16.6–22.7		(9)
	2.6–2.8	17.4–19.4		(9)
Moderate H-bonding solvent	2.4–2.6	21		(8)
		15.1–22.2		(9)
Strong H-bonding solvent	2.4–2.6	19.4–29.7		(9)
	2.6–2.8	19.4–23.3		(9)
Acetone	2.5	19.4	298	(10)
n-Amyl acetate	2.5	18.7	298	(10)
Benzene	2.5	20.6	298	(10)
n-Butyl acetate	2.5	18.7	298	(10)
Ethyl acetate	2.5	19.1	298	(10)
Methyl acetate	2.5	19.3	298	(10)
Methyl n-amyl ketone	2.5	18.8	298	(10)
Methyl ethyl ketone	2.5	19.1	298	(10)
Methyl n-propyl ketone	2.5	19.0	298	(10)
n-Propyl acetate	2.5	18.8	298	(10)
Tetrachloromethane	2.5	20.6	298	(10)
Toluene	2.5	20.4	298	(10)
Trichloromethane	2.5	18.6	298	(10)

Polymer–liquid interaction parameter χ

Solvent	DS	Temperature (K)	χ	Reference
Acetone	2.45	298	0.46	(10, 11)
n-Amyl acetate	2.45	298	0.28	
Benzene	2.45	298	0.48	
n-Butyl acetate	2.45	298	0.24	
Ethyl acetate	2.45	298	0.40	
Methyl acetate	2.45	298	0.41	
Methyl n-amyl ketone	2.45	298	0.38	
Methyl ethyl ketone	2.45	298	0.42	
Methyl n-propyl ketone	2.45	298	0.37	
n-Propyl acetate	2.45	298	0.33	
Tetrachloromethane	2.45	298	0.46	
Toluene	2.45	298	0.47	
Trichloromethane	2.45	298	0.34	
Water	1.4	328	1.1	

Ethylcellulose

Mark–Houwink parameters, K and a

Solvent	Temperature (K)	MW × 10^{-4} g mol^{-1}	Method	K × 10^3 ml g^{-1}	a	Reference
Acetone	293	8	Osmometry	1.51	1.05	(12)
Benzene	293	8	Osmometry	1.34	1.07	
	298	14	Osmometry	29.2	0.81	
	333	14	Osmometry	35.8	0.78	
Butanone	298	14	Osmometry	18.2	0.84	
	333	14	Osmometry	26.7	0.79	
Butyl acetate	298	14	Osmometry	14.0	0.87	
	333	14	Osmometry	18.1	0.83	
Chloroform	298	14	Osmometry	11.8	0.89	
	319	14	Osmometry	9.3	0.90	
Ethyl acetate	298	14	Osmometry	10.7	0.89	
	333	14	Osmometry	14.0	0.85	
Methanol	298	14	Light scattering	52.3	0.65	
Nitroethane	298	14	Osmometry	4.2	0.96	
	333	14	Osmometry	22.6	0.79	

Unit cell dimension of triethylcellulose

Lattice	Monomers per Unit Cell	Cell Dimension (Å)			Density (gcm^{-3})	Chain Conformation	Reference
		a	b	c			
Orthorhombic	12	15.64	27.09	15.0	1.158	3_2	(13)

Property	Units	Conditions	Value	Reference
Martin coefficient k'		Toluene/ethanol: 80/20	0.111	(14)
Glass transition temperature	K		316	(15)
		DS = 2.45–2.60	393–397	(3)
Deflection temperature	K	1.82 MPa	319–362	(4)
Tensile modulus	MPa		897–2069	(4)
Tensile strength	MPa		14–55	(4)
			47–72	(3)
Maximum extensibility	%		15–100	(4)
		Conditioned at 298 K, 50% relative humidity	7–30	(3)
Compressive strength	MPa	ASTM D695	69–241	(4)
Flexural yield strength	MPa		28–83	(4)
Flexural strength	MPa		62–69	(16)
Impact strength	J m^{-1}	1/2 by 1/2 in. notched bar , Izod test, ASTM D256, molding sheet	107–455	(4)
			16–91	(4)

Ethylcellulose

Property	Units	Conditions	Value	Reference
Hardness		Rockwell, R scale	50–115	(4)
		Sward, 3-mil film	52–61	(3)
Index of refraction n			1.47	(4)
Refractive index increment dn/dc	ml g^{-1}	MeOH, $\lambda = 436$ nm, 298 K	0.130	(17)
Dielectric constant ε''		298 K, 1 MHz	2.8–3.9	(3)
		298 K, 1000 Hz	3.0–4.1	(3)
		298 K, 60 Hz	2.5–4.0	(3)
Dielectric strength		V/mil, 10-mil film, ASTM D 149-64, step by step	1500	(3)
Power factor		298 K, 1 Hz	0.002–0.02	(3)
		298 K, 60 Hz	0.005–0.02	(3)
Resistivity	ohm cm^{-1}		10^{12}–10^{14}	(3)
Surface tension	mN m^{-1}	Contact angle	32	(18)
Thermal conductivity	W m^{-1}K^{-1}		0.21	(19)
Water absorption	%	24 h at 50% relative humidity	2	(3)

Permeability coefficient P (in m^3(STP) m s^{-1} m^{-2} Pa^{-1})

Permeant	Temperature (K)	$P \times 10^{17}$	Reference
H_2	293	65.3	(20)
He	298	40.1	(20)
N_2	298	3.32	(20)
O_2	298	19.0	(20)
Ar	298	7.65	(20)
CO_2	298	84.8	(20)
SO_2	298	198	(20)
NH_3	298	529	(20)
H_2O	298	6700	(20)
C_2H_6	298	6.9	(20)
C_3H_8	298	2.78	(20)
n-C_4H_{10}	298	2.9	(20)
n-C_5H_{12}	298	2.78	(20)
n-C_6H_{14}	298	5.75	(20)

References

1. Zhbankov RG. In: Stepanov ABI, ed. *Infrared Spectra of Cellulose and Its Derivatives.* New York: Consultants Bureau Pub.; 1966.
2. Prouchert CJ. *The Aldrich Library of FT-IR Spectra.* 1st ed. Wisconsin, USA: Aldrich Chem. Co., Inc.; 1985.
3. Auqalon. *Ethylcellulose, Chemical and Physical Properties.* Pub. 250-42A. Wilmington, Delaware: Hercules, Inc.; 2002.

4. Rudd GE, Sampson RN. In: Harper CA, ed. *Handbook of Plastics, Elastomers, and Composites.* New York: McGraw-Hill; 1992.

5. Fuchs O. Solvents and Non-Solvents for Polymers. In: Brandrup J, Immergut EH, eds. *Polymer Handbook.* 3rd ed. New York: John Wiley & Sons; 1989:VII/399.

6. Gray DG. Liquid Crystalline Cellulose Derivatives. *J. Appl. Polym. Sci. Appl. Polym. Symp.* 1983;37:179.

7. Burrell H. Solubility Parameters. *Interchem. Rew.* 1955;14:3.

8. Kent DJ, Rowe RC. Solubility Studies on Ethyl Cellulose Used in Film Coating. *J. Pharm. Pharmocol.*1978;30:808.

9. Grulke EA. Solubility Parameter Values. In: Brandrup J, Immergut EH, eds. *Polymer Handbook.* 3rd ed. New York: John Wiley & Sons; 1989:VII/555.

10. Barton AFM. Ethyl Cellulose. *CRC handbook of Polymer-Liquid Interaction and Solubility Parameters.* Boca Raton, Florida: CRC Press; 1990.

11. Moore WR, Epstein JA, Brown AM, Tidswell BM. Cellulose Derivatives-Solvent Interaction. *J. Polym. Sci.* 1957;23:23.

12. Kurata M, Tsunashima Y. Viscosity-Molecular Weight Relationships and Unperturbed Dimension. In: Brandrup J, Immergut EH, eds. *Polymer Handbook.* 3rd ed. New York: John Wiley & Sons; 1989:VII/31.

13. Zugenmaier P. Structural Investigation on Cellulose Derivatives. *J. Appl. Polym. Sci. Polym. Symp.* 1983;37:223.

14. Gröbe A. Properties of Cellulose Materials. In: Brandrup J, Immergut EH, eds. *Polymer Handbook.* 3rd ed. New York: John Wiley & Sons; 1989:V/117.

15. Peyser P. Glass Transition Temperatures of Polymers. In: Brandrup J, Immergut EH, eds. *Polymer Handbook.* 3rd ed. New York: John Wiley & Sons; 1989:VT/209.

16. Haynes W. *Cellulose, the Chemical that Grows.* Garden City, New York: Doubleday & Co., Inc.; 1953.

17. Huglin MB. Specific Refractive Index Increments of Polymers in Dilute Solution. In: Brandrup J, Immergut EH, eds. *Polymer Handbook.* 3rd ed. New York: John Wiley & Sons; 1989:VII/409.

18. Wu S. Surface and Interfacial Tensions of Polymers, Oligomers, Plasticizers, and Organic Pigments. In: Brandrup J, Immergut EH, eds. *Polymer Handbook.* 3rd ed. New York: John Wiley & Sons; 1989:VI/411.

19. Yang Yong. Thermal Conductivity. In: Mark JE, ed. *Physical Properties of Polymer Handbook.* Woodbury, New York: American Institute of Physics Press; 1996.

20. Pauly S. Permeability and Diffusion Data. In: Brandrup J, Immergut EH, eds. *Polymer Handbook.* 3rd ed. New York: John Wiley & Sons; 1989:VI/451.

Ethylene Acid Copolymer Metal Salts (Ionomers)

Barry Morris, John Pennias and David Walsh

Trade Name Surlyn® (DuPont)

Class Ethylene copolymers

Structure $(CH_2-CH_2)_n-(CH_2-C(CH_3)COO^-M^+)_n$ (random, partially neutralized)

General Information Surlyn® (Surlyn® is a registered trademark of the DuPont Company and Surlyn® ionomers are only available from DuPont.) ionomers consist of copolymers of ethylene and methacrylic acid, partially neutralized by a variety of metals such as sodium, zinc, and less commonly, lithium and magnesium. Other manufacturers make ionomers based on acrylic acid copolymers. The neutralization leads to a complex morphology where the ionic groups form clusters. The clusters disrupt the natural spherulitic structure of the polyethylene segments and result in high clarity and improved toughness[1].

The base resin of the ionomer, the acid copolymer, has many properties similar to low-density polyethylene, which is described in the relevant chapter of this book. It too is prepared at high pressure and high temperature in solution in excess ethylene. It has short- and long-chain branching, and has a similar range of molecular weight and polydispersity. Since ionomers are insoluble in nearly all common solvents, it is not possible to use the normal molecular weight characterization methods. One can however react it with acid and recover the original base resin. Before analyzing this by GPC, it is necessary to convert it to the (usually methyl) ester. There is however some indication that this series of reactions can alter the original structure.

In this chapter, a range of values for each material property is given, which may be quite large, but is known to vary in a generally predictable way with the structure of the ionomer.

The major variables in the ionomers are acid content, degree of neutralization, and choice of metal. Higher acid content leads to lower melting point, increased hardness, more oil and grease resistance, better optical properties, higher tensile strength, lower extension, and higher impact resistance. Higher percent neutralization leads to lower melt index, increased hardness, more oil resistance, better optical properties, higher tensile strength, improved abrasion resistance, and more moisture uptake. Sodium ionomers have more oil resistance, lower haze, and higher stress crack resistance than zinc ionomers. Zinc ionomers have less water sensitivity and higher impact strength.

Ionomers are often used in packaging film and coating applications for their heat sealability, hot tack, seal through contamination, oil resistance, adhesion, and toughness. These properties of ionomers are also controlled by the above variables. Increasing the acid level generally increases hot tack and decreases heat seal initiation temperature and adhesion to polyethylene. Increasing neutralization generally decreases film blocking, tear strength, and adhesion to polyethylene. A higher un-neutralized acid level generally provides better adhesion to aluminium foil and nylon, and higher flex durability, tear resistance, and notch sensitivity. The two primary cations used for packaging applications are sodium and zinc. Sodium ionomers generally have greater film gloss and drawability (extrusion coating) and lower film haze than zinc ionomers. Zinc ionomers have better foil and nylon adhesion and less neck-in (extrusion coating).

Sometimes acrylate comonomers are added to the acid copolymer. The acrylate lowers the melting point and heat seal initiation temperature, improves flex crack resistance, increases tear resistance, and lowers modulus.

Major Applications Major applications are in moldings, film and laminates, and as modifiers for other polymers. Molding applications are in golf balls and other sporting goods, bottles and caps, and in general consumer goods. Film and laminate applications make use of the melt strength, sealing capability, and adhesion to various substrates for the packaging industry. Ionomers are also used as wear layers in flooring constructions, and in automotive exterior paintless systems. They are used as modifiers for a wide range of polymers including polyamides, polyurethanes, and polyesters.

Property	Units	Conditions	Value	Reference
Melt flow index	g/10 min	D-1238	0.6–14	(2)
Density	g/cc	D-792	0.94–0.95	(2)
Linear coefficient of expansion	10^{-5} cm/cm.°C	D-696 (−20–32°C)	10–17	(2)
Melt point	°F (°C)	D-3418	190–212 (88–100)	(2)
Freeze point	°F (°C)	D-3418	129–174 (54–81)	(2)
Vicat softening point	°F (°C)	D-1525	133–174 (56–78)	(2)
Heat deflection temperature	°F (°C)	D-648	104–118 (40–48)	(2)
Specific heat	cal/g.°C	20°C	0.36–0.43	(2)
		150°C	0.44–0.55	(2)
Thermal conductivity	10^{-4} cal/cm.s.K		5.6–6.6	(2)
Flammability	mm/min	D-635	18–28	(2)
Flammability	Pass or fail	Motor vehicle standard 302	Pass	(2)
Notched Izod	J/cm	D-256	365–no break	(2)
Hardness, Shore D	–	D-2240	54–68	(2)
Tensile strength	MPa	D-638, IV bars, 5 cm/min,	21–38	(2)
Yield strength	MPa	D-638, IV bars, 5 cm/min,	8–19	(2)
Elongation to break	%	D-638, IV bars, 5 cm/min,	285–530	(2)
Tensile impact	kJ/m^2	D-1822S at 23°C	335–1670	(2)
		D-1822S at −40°C	295–1040	(2)
Flexural modulus	MPa	D-790 at 23°C	100–460	(2)
Abrasion resistance	NBS index	D-1630	120–640	(2)
Blown film**				(2)
Ultimate tensile strength, MD	psi (MPa)	D-882	3500–5900 (24–41)	(2)
TD	psi (MPa)	D-882	3300–5500 (23–38)	(2)
Ultimate elongation, MD	%	D-882	250–500	(2)
TD	%	D-882	250–450	(2)
Secant modulus, MD	psi (MPa)	D-882	6600–70 000 (46–480)	(2)
TD	psi (MPa)	D-882	6500–59 000 (45–410)	(2)
Graves tear, MD	g	D-1004	420–700	(2)
TD	g	D-1004	440–760	(2)
Elmendorf tear strength, MD	g/mil (N/mm)	D-1922	7–97 (2.7–37)	(2)
TD	g/mil (N/mm)	D-1922	7–97 (2.7–37)	(2)

Ethylene Acid Copolymer Metal Salts (Ionomers)

Property	Units	Conditions	Value	Reference
Spencer impact	in-lb/mil (J/mm)	D-3410	4.0–10 (18–44)	(2)
Dart drop strength	g/mil (g/μm)	D-1709(B)	150–350 (5.9–14)	(2)
Gloss, 20°		D-2547	25–105	(2)
Haze	%	D-1003	2.0–7.0	(2)
Clarity	%	D-1746	20–80	(2)
Permeability - WVTR (2-mil film)	g/100 in^2-day-atm (g/m^2-day-atm)	38°C, 100% RH	0.6–2.0 (9–31)	(2)(3)
Permeability - OTR (2-mil film)	cc/100 in^2-day-atm (cc/m^2-day-atm)	23°C, 50% RH	165–295 (2600–4600)	(2)(3)

** 2-mil (50-µm) film, 3:1 BUR.

References

1. Longworth R. In: Holliday L, ed. *Ionic Polymers.* Barking, U.K.: Applied Science Publishers; 1975, Chapter 2.
2. DuPont product literature.
3. Groenewege MP, et al. In: Raff RA, Doak KW, eds. *Crystalline Olefin Polymers.* New York: IntersciencePublishers; 1965.

Ethylene–Propylene–Diene Monomer Elastomers

Witold Brostow, Tea Datashvili, Gary W. Ver Strate
and David J. Lohse

Acronyms and Alternative Name EP, EPM, EPR (as copolymer), EPDM (contains a third monomer, which provides up to 10 weight % of an olefin site for cross-linking), ethene–propene–diene elastomers.[1–7]

Class Chemical copolymers, polyolefin copolymer; ter-polymer elastomer.

Structure Ethylene–propylene rubbers use the same chemical building blocks or monomers as polyethylene (PE) and polypropylene (PP) thermoplastic polymers. These ethylene (C_2) and propylene (C_3) monomers are combined in a random manner to produce rubbery and stable polymers. A wide family of ethylene–propylene elastomers can be produced—ranging from amorphous to semi-crystalline structures—depending on polymer composition and how the monomers are combined. These polymers are also produced in unusually wide range of Mooney viscosities (or molecular weights).

The ethylene and propylene monomers combine to form a chemically saturated, stable polymer backbone providing very good heat, oxidation, ozone, and weather aging. A third, nonconjugated diene monomer can be terpolymerized in a controlled manner to maintain a saturated backbone and place the reactive unsaturation in a side chain available for vulcanization or polymer modification chemistry. The terpolymers are referred to as EPDM (or ethylene–propylene–diene with 'M' referring to the saturated backbone structure). An EPDM polymer structure is illustrated in Fig. 26.1. The ethylene–propylene copolymers are called EPM.

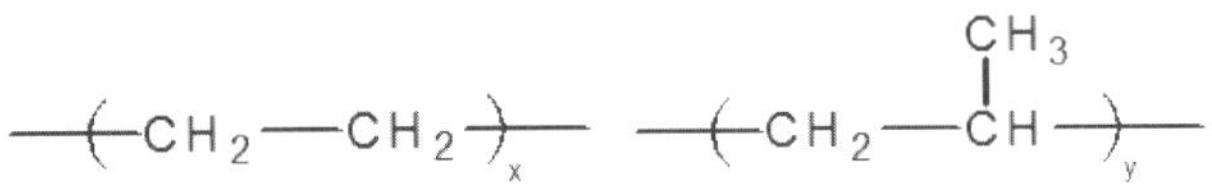

ENB is shown as the third monomer[2–6]

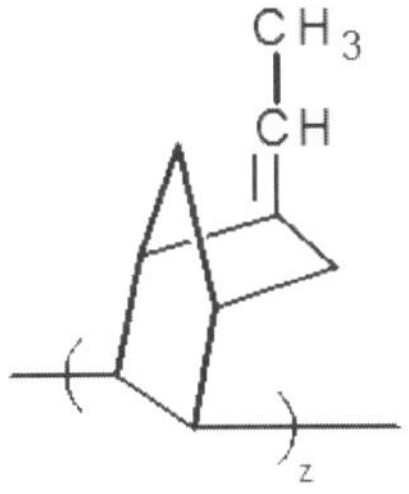

Figure 1. *EPDM Polymer Structure*

Typical Comonomers The most widely used diene termonomers are primarily ethylidene norbornene (ENB) followed by dicyclopentadiene (DCPD), 1,4 hexadiene (1,4 HD), vinyl norbornene (VNB), or norbornadiene (NBD).[2-9] Each diene incorporates into the system with a different tendency for introducing long-chain branching (LCB) or polymer side chains that influence processing and rates of vulcanization by sulfur or peroxide cures.

Properties of Special Interest Ethylene–propylene rubbers and elastomers (also called EPDMs and EPMs) continue to be one of the most widely used and fastest growing synthetic rubbers having both specialty and general-purpose applications. Sales had reached to 870 metric tons (or 1.9 billion pounds) already in 2000 since commercial introduction in the early 1960s. Polymerization and catalyst technologies in use today provide the ability to design polymers to meet specific and demanding application and processing needs.

As noted above, ethylene–propylene rubbers are valuable for their resistance to heat, oxidation, ozone, and weather aging due to their stable saturated polymer backbone structure. Properly pigmented black and nonblack compounds are color stable. As nonpolar elastomers, they have good electrical resistivity as well as resistance to polar solvents such as water, acids, alkalis, phosphate esters, and many ketones and alcohols. Amorphous or low crystalline grades have good low temperature flexibility with glass transition points of $\approx -60°C$.

Heat aging resistance up to $130°C$ can be achieved with properly selected sulfur acceleration systems while heat resistance at $160°C$ can be obtained with peroxide cured compounds. Compression set resistance is good, particularly at high temperatures, if sulfur donor or peroxide cure systems are used.

These materials can have high tensile and tear properties, good abrasion resistance, as well as improved oil swell resistance and flame retardance.

High plateau modulus values permit high filler loadings and cost-effective compounds, which are chemically inert, semi-crystalline grades have high green strength.[2-5,8]

A general summary of properties is presented in the table below.

Chemistry and Manufacturing Processes Specialized catalysts are used to polymerize the monomers into controlled polymer structures. Since their introduction, ethylene–propylene elastomers have used a family of catalysts referred to as Ziegler–Natta—named after their original creators. Improvements in catalysts and processes have provided increased productivity while maintaining control of polymer structure.

New families of catalysts, referred to as metallocene catalysts (e.g. bis(η–cyclopentadienyl)zirconium Cp_2Zr-derived metallocene/alumoxane, hydrocarbon solution, 80–120°C) have been developed and are in commercial use.[2-8]

There are three major commercial processes for manufacturing ethylene–propylene rubbers: solution, slurry (suspension), and gas-phase. The manufacturing systems vary with each of the several producers. There are differences in the product grade slates made by each producer and process, but generally speaking all are capable of making a variety of EPDM and EPM polymers. The physical forms range from solid to friable bales, pellets and granular forms and oil blends.

The solution polymerization process is the most widely used since it is versatile in making a wide range of polymers. Ethylene, propylene, and catalyst systems are polymerized in an excess of hydrocarbon solvent. Stabilizers and oils, if used, can be added directly after polymerization. The solvent and unreacted monomers are then flashed off with hot water or steam, or with mechanical devolatilization. The polymer, which is in crumb form, is dried with dewatering on screens, in mechanical presses, or drying ovens. The crumbs are formed into wrapped bales or extruded into pellets. The high viscosity, crystalline polymers are sold in loosely compacted, friable bales or as pellets. The amorphous polymer grades are typically sold in solid bales.

The slurry (or suspension) process is a modification of bulk polymerization. The monomers and catalyst system are injected into a reactor filled with propylene. The polymerization takes place immediately, forming crumbs of polymer that are not soluble in propylene. Slurry polymerization reduces the need for solvent and solvent handling equipment, and the low viscosity of the slurry helps to control temperature and handle the product. The process is not limited by solution viscosity, so that high molecular weight polymers can be produced. Flashing off the propylene and termonomer completes the process before forming and packaging.

Gas-phase polymerization technology was also developed for the manufacture of ethylene–propylene rubbers. The reactor consists of a vertical fluidized bed. Monomers and nitrogen in gas form along with catalyst are fed to the reactor and solid product is removed periodically. Heat of reaction is removed through the use of the circulating gas that also serves to fluidize the polymer bed. Solvents are not used, eliminating the need for solvent stripping, washing, and drying. The process is not limited by solution viscosity, so here also high molecular weight polymers can be produced. Continuous injection of a substantial amount of carbon black used as a partitioning aid is necessary to prevent the polymer granules from sticking to each other and to the reactor walls. Products are made in a granular form to enable rapid mixing.

Typical Areas of Application Versatility in polymer design and performance has resulted in wide usage in automotive weather-stripping and seals, glass-run channel, radiator, garden and appliance hose, tubing, belts, electrical insulation, roofing membrane, rubber mechanical goods, plastic impact modification, thermoplastic vulcanizates, and motor oil additive applications. Moreover, EPDMs are constituents of elastomers for medical applications such as non-reusable syringes.[40]

Suppliers Major producers and suppliers of EPDM and EPM are Lanxess Polymers, Crompton Corp., ExxonMobil Chemical Co., DSM Elastomers, Dupont Dow Elastomers, Herdillia, JSR, Kumho Polychem, Mitsui Chemicals, Polimeri Europa, and Sumitomo Chemical Co. Wide ranges of grades are available worldwide to provide solutions to many product requirements.

Property	Units	Conditions	Value	Reference
Ceiling temperature	K	Polymerization at <1MPa up to this temperature is possible	>440	(8)
Molecular weight (of repeat unit)	g mol^{-1}	50 mol% ethene, 1 mol% diene	~35 (average)	(4–7)
Tacticity (stereoregularity)	% stereo-regular propene in elastomer grades	Vanadium or metallocene catalysis	0	(4–7, 10)
Head-to-head contents	%	Vanadium or metallocene catalysis	<3	(4–7, 10)
Degree of branching	% of molecules having long branches	Vanadium catalysis mostly by cationic diene coupling; metallocene by end group copolymerization; can be controlled (information available from manufacturers)	0–100	(4–7)

Property	Units	Conditions	Value	Reference
Typical molecular weight range of polymer $\_M_n$	g mol^{-1}	500–5000 as dispersants 2000–300 000 as elastomers 5000–5 000 000 in blends	–	(4–7, 9)
Typical polydispersity index ($M_w = M_n$)	Dimensionless	Controlled by catalyst choice, reactor type	1.5–50	(4–7)
Morphology in multiphase systems	As shown	Semi-crystalline copolymers, blends, block copolymers		(4–6, 11, 12)
		Lamellae, width	50–150 Å	
		Lamellae, length	0.01–2 μm	
		Elastomer blends, major dimension	0.5–10 μm	
IR (characteristic absorption frequencies)	cm^{-1}	20°C, films		(4–6, 13)
		–CH_2	720	
		Isopropyl –CH_3	1145	
		–CH_3	1370	
UV (characteristic absorption frequencies)	nm	20°C in solution (broad, maximum, depends on diene type)	<200	(4–6)
NMR	ppm (chemical shift)	^{13}C, ^{1}H (see references for extensive detailed work)	Specific carbons or protons have specific shifts	(5, 10, 38)
Thermal expansion coefficients	K^{-1}	1 atmosphere, no crystallinity $(1/V)\,(dV/dT)_P$	7×10^{-4}	(4, 14, 15)
Compressibility coefficients	bar^{-1}	20°C, $(1/V)\,(dV/dP)_T$	5.8×10^{-8}	(4, 14, 15)
Reducing temperature T^*	K	150–250°C, 10–200 MPa	6800	(14)
Reducing pressure P^*	MPa	150–250°C, 10–200 MPa	444–465	(14)
Reducing volume V^*	cm^{3}g^{-1}	150–250°C, 10–200 MPa	1.000	(14)
Density (amorphous)	gcm^{-3}	20°C, no diene present, < 55 wt. % ethene, the dienes ENB, DCPD, and VNB raise density as does crystallinity	0.854	(4, 14)
Solvents	–	Ambient	Aliphatic, aromatic, halogenated hydrocarbons	(4, 5, 16)
Nonsolvents	–	Ambient	Water, organic acids, ketones, polar halogenated hydrocarbons	(4, 5, 16)
Solubility parameter	(MPa)$^{1/2}$	By SANS (depending on ethene contents)	16–17	(13)
		By GLC	16	(17)

Property	Units	Conditions	Value	Reference
Theta temperature θ	K	Phenyl ethyl ether, depends on exact composition	353	(4, 5, 16, 18–20)
		n-Octyl acetate	300	
		n-Decyl acetate	278	
		n-Hexyl acetate	334	
Interaction parameter χ	Dimensionless	n-Heptane, 300 K	$(0.35 + 0.08 \times V_{polymer})$	(4, 5, 16)
Second virial coefficients	mol cm^{3}g^{-2}	Trichlorobenzene, 135°C	(9.9×10^{-3}) $M^{-0.18}$	(5, 21)
Mark–Houwink parameters: K and a	$K = $ ml g^{-1} $a = $ None	Trichlorobenzene, 135°C	$K = 2.9 \times 10^{-2}$ $a = 0.726$	(5, 21)
Huggins constants k'	–	Theta solvents, depends on polymer molecular weight	0.4–0.8	(20)
Characteristic ratio	–	Phenyl ethyl ether, 80°C	6.9	(18)
		SANS, 20°C	6.9	(19)
Lattice	-	Methylene units crystallize into a polyethene lattice, methyl groups can incorporate somewhat	Orthorhombic	(4, 16)
Unit cell dimensions	′	1 atm, 20°C, –CH2– sequences only, methyl group causes expansion	$a = 7.418$ $b = 4.946$ $c = 2.546$	(4, 5, 16, 22)
Unit cell contents	Number of mers	Ethene crystallinity	2	(4, 5, 16, 22)
Degree of crystallinity	%	Depends on ethene content	0–50	(4, 6, 36, 37)
Heat of fusion (of repeat units)	kJ mol$^-$ cal g^{-1}	DSC on samples annealed at 20°C > 48 h, varies with ethene content	0–4.4 0–35	(4, 5)
Density (crystalline)	g cm^{-1}	1 atm, 20°C	0.997 (can be reduced by defects to 0.99 and below)	(4, 5, 22)
Glass transition temperature	K	1 atm, DSC, dynamic mechanical, depends on ethene content, lowest at about 50% ethene; crystallinity confuses the issue at high ethene contents	213–240	(4, 5, 23, 39)
Melting point	K	1 atm, DSC, depends on ethene content; often melts just above last annealing	218–373	(4, 5)

Property	Units	Conditions	Value	Reference
		temperature; will crystallize down to Tg		
Heat capacity (of repeat units)	kJK^{-1} mol^{-1}	DSC, 1 atm	0.078	(4, 5, 14)
Polymers with which miscible	–	MW < 150 000	Head-to-head polypropylene	(24)
		MW < 100 000	Ethylene–butene copolymers of similar comonomer content	
Tensile modulus	MPa	20°C, low strain rate, filled compound 25% rubber, 50% carbon black, 25% oil tested at $\sim 1s^{-1}$	3–7	(4, 5, 8, 25)
Shear modulus	MPa	20°C, low strain rate, filled compound 25% rubber, 50% carbon black, 25% oil	1–4	(4, 5, 8, 25)
Storage modulus	MPa	High molecular weight gum rubber, 20°C, 1 Hz	0.16	(8, 17)
Loss modulus, tan δ	–	High molecular weight gum rubber, 20°C, 1 Hz	0.2	(8, 17)
Tensile strength	MPa	Dependent on compounding and test conditions, typical compounds at 20°C, 1 s $^{-1}$	0.5–50 5–25	(4, 5, 8, 25, 26)
Maximum extensibility $(L/L_0)_r$	%	Dependent on compounding and test conditions	200–800	(4, 5, 8, 25)
Compression Set B	%	Dependent on compounding and test conditions	20–60	(2, 3, 5)
Hardness	Shore A values	Dependent on compounding and test conditions	10–100	(4, 5, 25)
Poisson's ratio	–	0–50°C, strained at 100 s^{-1} or less	0.49	(8, 13)
Plateau modulus	MPa	20–150°C	1.6	(8, 17)
Entanglement molecular weight	g mol^{-1}	20°C, Me = $\rho R T G_N^0$	1,300	(4, 5, 8, 23)
Index of refraction n$_D$	–	1 atm, no diene, 23°C No crystallinity, 90°C 125°C	1.4740 1.4524 1.4423	(4, 5)
Refractive index increment dn/dc	ml g^{-1}	Trichlorobenzene, 135°C	−0.104	(5)
Dielectric constant ε'	–	20°C, 10^3 Hz, 1 atm, depends on the compound, a good insulator	2.8	(4, 5, 16, 25)
Dielectric loss ε''	–	20°C, 10^3 Hz, 1 atm, depends on the compound	0.2 0.25	(4, 5, 16)

Property	Units	Conditions	Value	Reference
Resistivity	log R, ohm cm	1 atm, 20°C, depends on the compound, generally a good insulator	3–14	(4, 16, 28–30)
Molar polarizability α	cm^3	Frequency	4.4×10^{-26}	(4, 5)
Surface tension	mNm^{-1}	20°C, 1 atm, increases with ethene content	29.4–36.8	(30, 31)
Interfacial tension	mNm^{-1}	With PDMS, 20°C	3.2–5.3	(30, 31)
		With PS, 140°C	5.1–5.9	
Permeability coefficient	([m^3] [m])/([S] [m^2] [^{T}Pa])	He 25°C	16.0–24.0 ($\times 10^{-17}$)	(32)
		N$_2$ 25°C	3.7–4.1 ($\times 10^{-17}$)	
Thermal conductivity	W m^{-1} K^{-1}	20°C, 1 atm	0.355	(4, 5, 33)
Melt viscosity	Pa s	Newtonian at 100°C	$(4 \times 10^{-5})M^{3.6}$	(5, 23)
Melt index	g	2.2 kg, 190°C, depends completely on polymer molecular weight	0.001–50	(4, 5, 25)
Mooney viscosity	Mooney units	125°C	5–100	(4, 5, 25)
Pyrolyzability, amount of product remaining	%	>500°C	<.3%	(4, 5, 34)
Severe decomposition	K	N$_2$ blanket	580	(4, 34)
Maximum use temperature	K	Open atmosphere	450	(4, 5, 25)
Minimum use temperature	K	Open atmosphere	223	(4, 5, 25)
Decomposition temperature	K	Nitrogen atmosphere, 1 min	570	(4, 5, 34)
Scission, G factor, G(s)	mol J^{-1}	γ irradiation, depends on ethene and diene content	1.1–5.9 ($\times 10^{-8}$)	(4, 26, 35)
Cross-linking, G factor, G(x)	mol J^{-1}	γ irradiation, depends on ethene and diene content	2.7–22.6 ($\times 10^{-8}$)	(4, 26, 35)
Gas evolution, G factor, G(gas)	mol J^{-1}	γ irradiation, depends on ethene and diene content	3.4×10^{-8}	(26)
Water absorption	% volume increase	168 h, 55% ethylene glycol in water, boiling	+1	(25)
	% tensile change		−1	

References

1. D1418 Rubber and Rubber Latices – Nomenclature. American Society for Testing and Materials, Philadelphia, (1981).
2. Karpeles R, Grossi AV. In: Bhowmick AK, Stephens HL, eds. *Handbook of Elastomers.* 2nd ed. New York: Marcel Decker; 2001.
3. Riedel JA, Laan RV. *The Vanderbilt Rubber Handbook.* 13th ed. Norwalk, Connecticut: R.T. Vanderbilt Co.; 1990.

4. Baldwin FP, Strate GV. *Rubber Reviews.* 1972;44:709.
5. Ver Strate G. In: Mark HF, et al., eds. *Encyclopedia of Polymer Science and Engineering,* 2nd ed. New York: John Wiley and Sons; 6(1987):522.
6. Cesca SJ. *Macromol. Sci.* 1972;10:1.
7. Banzi V, et al. *Angew Makromol. Chem.* 1995;229:113.
8. Mark J, Erman B, Eirich F. *Science and Technology of Rubber.* 2nd ed. New York: Academic Press. (1994):70, 157, 211, 495.
9. See the proceedings of Flexpo Conferences, Chemical Market Resources, Houston, (281) 333–3313.
10. Trillo I, et al. *Macromolecules.* 1995;28:342, (and references therein).
11. Strate GV, et al. *Flexpo '97.* Houston: Chemical Market Resources; 1997.
12. Hess WM, Herd CR, Vegvari PC. *Rubber Reviews.*1993; 66:330.
13. D3900-94 Determination of Ethylene Units in EPM and EPDM. American Society for Testing and Materials.
14. Walsh DJ et al. *Macromolecules.* 1993;25:5236.
15. Krishnamoorti R. *Ph.D. Thesis.* Princeton; 1994.
16. Brandrup J, Immergut E, eds. *Polymer Handbook.* 3rd ed. New York: John Wiley and Sons; 1989.
17. Schuster RH, Issel HM, Peterseim V. *Rubber Chem. and Tech.* 1996;69:769.
18. Bruckner S. et al. *Eur. Polymer J.* 1974;10:347.
19. Zirkel A, et al. *Macromolecules.* 1992;25:954.
20. Mays JW, Fetters LJ. *Macromolecules.* 1989;22:921.
21. Scholte ThG, et al. *J Appl. Polymer Sci.*1984;29:3763.
22. Wunderlich B. *Crystals of Linear Macromolecules.* Washington D.C.: American Chemical Society; 1973.
23. Gotro JE, Graessley WW. *Macromolecules.* 1984;17:2767.
24. Krishnamoorti R, et al. *Macromolecules.* 1994;27:3073.
25. *Vistalon Users Guide.* Houston: Exxon Chemical Co.
26. Bohm GA, Tveekrem JO. *Rubber Rev.* 1982;55:575.
27. Ferry JD. *Viscoelastic Properties of Polymers.* 3d ed. New York: John Wiley and Sons; 1980.
28. Aminabhavi TM, Cassidy PE, Thompson CM. *Rubber Reviews.* 1990;63:451.
29. Thompson CM, Allen JS. *Rubber Chem. & Tech.* 1994;67:107.
30. Wu S. *Polymer Interface and Adhesion.* New York: Marcel Dekker; 1982.
31. Roe RJ. *J. Colloid Interface Sci.* 1969;31:228.
32. Paul DR, Dibenedetto ATJ. *Polymer Sci., C.* 1965;10:17.
33. Mark HF, et al. eds. *Encyclopedia of Chemical Technology.* New York: Wiley-Interscience; 1978.
34. Sircar A. *Rubber Rev.* 1992;65:503.
35. Odian G, Lamparella D, Canamare J. *J. Polymer Sci., C.* 1968;16:3619.
36. Brostow W, D'Souza NA, Ramamurthy AC, Wang Y, Tomas RL, Favro LD. *Polymer Eng. & Sci.* 1996;36:2.
37. Brostow W, D'Souza NA, Galina H, Ramamurthy AC. *Polymer Eng. & Sci.* 1996;36:8.
38. Mark JE, ed. *Physical Properties of Polymers Handbook.* 2nd ed. New York: Springer Science; 2007.
39. Brostow W, ed. *Performance of Plastics.* Munich-Cincinnati: Hanser; 2000.
40. Brostow W, Pietkiewicz D, Wisner SR. *Adv. Polymer Tech.* 2007;26:56.

Ethylene-Vinyl Acetate Copolymer

Ping Xu

Acronyms, Trade Names EVA; Elvax® (DuPont); Levapren® (Lanxess); Microthene®, Ultrathene®, Plexar (Equistar Chemicals); Airflex® (Air Products); Novatec® (Japan Polychem)

Class Chemical Copolymers

Structure $[-CH_2-CH_2-]_m-[-CH_2-CH-]_n$
$$\begin{array}{c} | \\ O \\ | \\ C=O \\ | \\ CH_3 \end{array}$$

Major Applications Film extrusion, packaging, wire and cable insulation, adhesives, coatings, and compounding.

Properties of Special Interest Flexibility and toughness, good adhesion, and stress crack resistance.

Property	Units	Conditions*	Value	Reference
Linear thermal expansion coefficient	K^{-1}	ASTM D696, no composition given	$16–25 \times 10^{-5}$	(1)
Density	$g\,cm^{-3}$	ASTM D792, 9–28% vinyl acetate	0.93–0.95	(2)
Solubility parameter	$(MPa)^{1/2}$	Halogenated aliphatic and aromatic liquids, 20°C		(3)
		30% vinyl acetate	19.0	
		40% vinyl acetate	19.2	
		67% vinyl acetate	19.0	
		Halogenated aliphatic and aromatic liquids, 30°C		
		30% vinyl acetate	18.8	
		40% vinyl acetate	18.9	
		67% vinyl acetate	18.9	
Interaction parameter χ	–	29% vinyl acetate, 150°C, inverse GC, infinite solution		(4, 5)
		Acetaldehyde	0.16	

Property	Units	Conditions*	Value	Reference
		Acetic acid	1.12	
		Benzene	−0.02	
		1-Butanol	0.65	
		2-Butanol	0.51	
		Cyclohexane	0.07	
Interaction parameter χ	–	Dioxane	0.45	
		Ethanol	1.28	
		Hexane	0.25	
		Methanol	1.69	
		Octane	0.23	
		2-Propanol	0.93	
		Tetrahydrofuran	0.25	
		m-Xylene	−0.02	
Glass transition temperature	K	30% vinyl acetate, $M_n = 27{,}000$ g mol^{-1}, $M_w = 110{,}000$ g mol^{-1}	231	(3)
		40% vinyl acetate, $M_n = 25{,}000$ g mol^{-1}, $M_w = 130{,}000$ g mol^{-1}	235	
Melting point	K	30% vinyl acetate, $M_n = 27{,}000$ g mol^{-1}, $M_w = 110{,}000$ g mol^{-1}	345	(3)
		40% vinyl acetate, $M_n = 25{,}000$ g mol^{-1}, $M_w = 130{,}000$ g mol^{-1}	318	
Brittleness temperature	K	ASTM D746		(2)
		9% vinyl acetate, melt index = 2.2 g/10 min	<197	
		9% vinyl acetate, melt index = 9.8 g/10 min	<197	
		15% vinyl acetate, melt index = 8.2 g/10 min	<197	
		15% vinyl acetate, melt index = 30 g/10 min	<197	
		18% vinyl acetate, melt index = 1.5 g/10 min	<197	
		18% vinyl acetate, melt index = 30 g/10 min	<197	
		19% vinyl acetate, melt index = 0.45 g/10 min	<197	
		19% vinyl acetate, melt index = 30 g/10 min	<197	
		28% vinyl acetate, melt index = 3.1 g/10 min	<197	
Vicat softening temperature	K	ASTM D1525, ring and ball method		(2)
		9% vinyl acetate, melt index = 2.2 g/10 min	356	
		9% vinyl acetate, melt index = 9.8 g/10 min	348	
		15% vinyl acetate, melt index = 8.2 g/10 min	339	
		15% vinyl acetate, melt index = 30 g/10 min	334	
		18% vinyl acetate, melt index = 1.5 g/10 min	334	
		18% vinyl acetate, melt index = 30 g/10 min	327	
		19% vinyl acetate, melt index = 0.45 g/10 min	335	
		19% vinyl acetate, melt index = 30 g/10 min	331	
		28% vinyl acetate, melt index = 3.1 g/10 min	322	
Tensile strength at break	MPa	ASTM D638		(2)
		9% vinyl acetate, melt index = 2.2 g/10 min	13.9	
		9% vinyl acetate, melt index = 9.8 g/10 min	11.7	

Property	Units	Conditions*	Value	Reference
		15% vinyl acetate, melt index = 8.2 g/10 min	12.8	
		15% vinyl acetate, melt index = 30 g/10 min	10.4	
		18% vinyl acetate, melt index = 1.5 g/10 min	13.5	
		18% vinyl acetate, melt index = 30 g/10 min	9.0	
		19% vinyl acetate, melt index = 0.45 g/10 min	19.3	
		19% vinyl acetate, melt index = 30 g/10 min	8.1	
		28% vinyl acetate, melt index = 3.1 g/10 min	15.2	
Elongation at break	%	ASTM D638		(2)
		9% vinyl acetate, melt index = 2.2 g/10 min	740	
		9% vinyl acetate, melt index = 9.8 g/10 min	675	
		15% vinyl acetate, melt index = 8.2 g/10 min	730	
		15% vinyl acetate, melt index = 30 g/10 min	750	
		18% vinyl acetate, melt index = 1.5 g/10 min	850	
		18% vinyl acetate, melt index = 30 g/10 min	700	
		19% vinyl acetate, melt index = 0.45 g/10 min	740	
		19% vinyl acetate, melt index = 30 g/10 min	680	
		28% vinyl acetate, melt index = 3.1 g/10 min	750	
1% Secant modulus	MPa	ASTM D638		(2)
		9% vinyl acetate, melt index = 2.2 g/10 min	75.9	
		9% vinyl acetate, melt index = 9.8 g/10 min	93.1	
		19% vinyl acetate, melt index = 0.45 g/10 min	33.1	
		19% vinyl acetate, melt index = 30 g/10 min	29.7	
		28% vinyl acetate, melt index = 3.1 g/10 min	18.6	
Dart drop impact	$F_{50}g^{-3}$	ASTM D1709		(2)
		9% vinyl acetate, melt index = 2.2 g/10 min	300	
		9% vinyl acetate, melt index = 9.8 g/10 min	305	
		15% vinyl acetate, melt index = 8.2 g/10 min	310	
		18% vinyl acetate, melt index = 1.5 g/10 min	>600	
Flexural modulus	MPa	ASTM D790, no composition given	53.1	(1)
Hardness	Shore D/A values	ASTM D2240		(2)
		9% vinyl acetate, , melt index = 2.2 g/10 min	93 (A)	
		9% vinyl acetate, melt index = 9.8 g/10 min	34 (D)	
		15% vinyl acetate, melt index = 30 g/10 min	30 (D)	
		18% vinyl acetate, melt index = 1.5 g/10 min	42 (D)	
		18% vinyl acetate, melt index = 30 g/10 min	30 (D)	
		19% vinyl acetate, melt index = 30 g/10 min	88 (A)	
		28% vinyl acetate, melt index = 3.1 g/10 min	78 (A)	
Dielectric strength	$V\ mil^{-1}$	ASTM D149, no composition given, 0.31-cm thick specimen	620–760	(1)
Water absorption	%	ASTM D570, no composition given, 24 h	0.005–0.13	(1)

* The melt index values were obtained with ASTM D1238.

References

1. *Modern Plastics Encyclopedia '96.* New York: McGraw-Hill.
2. *Ultrathene® High Ethylene Vinyl Acetate Copolymers, Resins, Key Properties and Applications.* Quantum Chemical Corporation, USI Division product bulletin, Cincinnati, 1992.
3. Lath D, Lathova E, Cowie JMG. In: *Prepr. Short Contrib. Bratislava IUPAC 5th Int. Conf. Modif. Polym.* 1979;2:225.
4. Dincer S, Bonner DC. *Macromolecules.* 1978;11:107.
5. Aspler JS. *Chromatogr. Sci.* 1985;29:399.

Ethylene-Vinyl Alcohol Copolymer

Ping Xu

Acronyms, Trade Names EVOH; Eval® (Eval); Soarnol® (Nippon Gohei); Clarene® (Mitsui)

Class Chemical copolymers

Structure $[-CH_2-CH_2-]_m-[-CH_2-CH-]_n$
                                                                          |
                                                                        OH

Major Applications Coextrusion, film lamination, coatings, and food packaging.

Properties of Special Interest Superior barrier properties to gases, fragrances, solvents, etc.

Property	Units	Conditions*	Value	Reference
Linear thermal expansion coefficient	K^{-1}	32 mol% vinylalcohol, melt index = 3.8 g/10 min	11×10^{-5}	(1)
		38 mol% vinyl alcohol, melt index = 3.8 g/10 min	12×10^{-5}	
		44 mol% vinyl alcohol, melt index = 13.0 g/10 min	13×10^{-5}	
Density	$g\,cm^{-3}$	ASTM D1505		(1)
		27 mol% vinyl alcohol, melt index = 3.0 g/10 min	1.20	
		32 mol% vinyl alcohol, melt index = 3.8 g/10 min	1.19	
		38 mol% vinyl alcohol, melt index = 3.8 g/10 min	1.17	
		44 mol% vinyl alcohol, melt index = 13.0 g/10 min	1.14	
Interaction parameter χ	–	No composition given, 20°C, water	1.2–1.8	(2)
Heat of fusion	Jg^{-1}	32 mol% vinyl alcohol, melt index = 3.8 g/10 min	81.9	(1)
		38 mol% vinyl alcohol, melt index = 3.8 g/10 min	81.1	
		44 mol% vinyl alcohol, melt index = 13.0 g/10 min	79.8	

Property	Units	Conditions*	Value	Reference
Heat of combustion	Jg^{-1}	32 mol% vinyl alcohol, melt index = 3.8 g/10 min	30,037	(1)
		38 mol% vinyl alcohol, melt index = 3.8 g/10 min	31,200	
		44 mol% vinyl alcohol, melt index = 13.0 g/10 min	32,366	
Glass transition temperature	K	Dynamic viscoelasticity		(1)
		27 mol% vinyl alcohol, melt index = 3.0 g/10 min	345	
		32 mol% vinyl alcohol, melt index = 3.8 g/10 min	342	
		38 mol% vinyl alcohol, melt index = 3.8 g/10 min	335	
		44 mol% vinyl alcohol, melt index = 13.0 g/10 min	328	
Melting point	K	DSC		(1)
		27 mol% vinyl alcohol, melt index = 3.0 g/10 min	464	
		32 mol% vinyl alcohol, melt index = 3.8 g/10 min	454	
		38 mol% vinyl alcohol, melt index = 3.8 g/10 min	448	
		44 mol% vinyl alcohol, melt index = 13.0 g/10 min	437	
Tensile modulus	MPa	ASTM D638		(1)
		27 mol% vinyl alcohol, melt index = 3.0 g/10 min	3,138	
		32 mol% vinyl alcohol, melt index = 3.8 g/10 min	2,648	
		38 mol% vinyl alcohol, melt index = 3.8 g/10 min	2,352	
		44 mol% vinyl alcohol, melt index = 13.0 g/10 min	2,062	
Tensile strength at break	MPa	ASTM D638		(1)
		27 mol% vinyl alcohol, melt index = 3.0 g/10 min	71.6	
		32 mol% vinyl alcohol, melt index = 3.8 g/10 min	71.6	
		38 mol% vinyl alcohol, melt index = 3.8 g/10 min	46.1	
		44 mol% vinyl alcohol, melt index = 13.0 g/10 min	51.0	

Property	Units	Conditions*	Value	Reference
Elongation at break	%	ASTM D638		(1)
		27 mol% vinyl alcohol, melt index = 3.0 g/10 min	200	
		32 mol% vinyl alcohol, melt index = 3.8 g/10 min	230	
		38 mol% vinyl alcohol, melt index = 3.8 g/10 min	280	
		44 mol% vinyl alcohol, melt index = 13.0 g/10 min	380	
Izod impact strength	J m^{-1}	ASTM D255, notched		(1)
		27 mol% vinyl alcohol, melt index = 3.0 g/10 min	58.7	
		32 mol% vinyl alcohol, melt index = 3.8 g/10 min	90.7	
		38 mol% vinyl alcohol, melt index = 3.8 g/10 min	64.1	
		44 mol% vinyl alcohol, melt index = 13.0 g/10 min	53.4	
Rockwell hardness	–	ASTM D785		(1)
		27 mol% vinyl alcohol, melt index = 3.0 g/10 min	104	
		32 mol% vinyl alcohol, melt index = 3.8 g/10 min	100	
		38 mol% vinyl alcohol, melt index = 3.8 g/10 min	93	
		44 mol% vinyl alcohol, melt index = 13.0 g/10 min	88	
Taber abrasion	mg	ASTM D1175, 1,000 times		(1)
		32 mol% vinyl alcohol, melt index = 3.8 g/10 min	1.2	
		38 mol% vinyl alcohol, melt index = 3.8 g/10 min	2.0	
		44 mol% vinyl alcohol, melt index = 13.0 g/10 min	2.2	
Bending strength	MPa	ASTM D790		(1)
		27 mol% vinyl alcohol, melt index = 3.0 g/10 min	149	
		32 mol% vinyl alcohol, melt index = 3.8 g/10 min	128	
		38 mol% vinyl alcohol, melt index = 3.8 g/10 min	108	
Surface resistivity	ohm	Various films	$1.9–2.7 \times 10^{15}$	(1)

Property	Units	Conditions*	Value	Reference
Volume resistivity	ohm cm	Various films	$0.47–1.2 \times 10^{13}$	(1)
Thermal conductivity	$Wm^{-1} K^{-1}$	32 mol% vinyl alcohol, melt index = 3.8 g/10 min	0.35	(1)
		44 mol% vinyl alcohol, melt index = 13.0 g/10 min	0.36	
Water permeability	cm^3 25 μm $(m^2 \text{ day atm})^{-1}$	Eval films, 40°C	21.7–124	(3)
Oxygen permeability	cm^3 25 μm $(m^2 \text{ day atm})^{-1}$	Eval films, 23°C	0.095–1.8	(3)
Water absorption	%	24 h, Eval® F resins	0.19–7.7	(4)

* The melt index values were obtained with ASTM D1238.

References

1. *Eval® Ethylene Vinyl Alcohol Copolymers Resins: Resins, Key Properties and Applications.* Eval Company of America product bulletin, Lisle, Ill., 2000.
2. Barton AFM. *CRC Handbook of Polymer-Liquid Interaction Parameters and Solubility Parameters.* Boca Raton, Fla.: CRC Press; 1990.
3. *Permeability and Other Film Properties of Plastics and Elastomers.* New York: Plastics Design Library; 1995.
4. Elias HG, Vohwinkel F. *New Commercial Polymers 2.* New York: Gordon and Breach Science Publishers; 1986.

Fluorosiloxane Polymers

Michael J. Owen

Acronyms, Trade Names LS 'Low swell', FS 'Fluorosilicone', FluoroPOSS [polyhedral oligomeric silsesquioxanes]

Class Polysiloxanes

Structure Poly(methyltrifluoropropylsiloxane) [PMTFPS] $[CH_3\{CF_3(CH_2)_2\}SiO]$ Polymer **I**,

Poly(methylnonafluorohexylsiloxane) [PMNFHS] $[CH_3\{CF_3(CF_2)_3(CH_2)_2\}SiO]$ Polymer **II**,

$[CF_3(CF_2)_7(CH_2)_2]_8Si_8O_{12}$ [FD$_8$T$_8$] Polymer **III**.

Major Applications The commercial applications of fluorosilicones are dominated by PMTFPS (Polymer **I**). Its major use is as elastomers and sealants in applications exposed to hydrocarbon fuels and oils, and organic solvents; predominantly in the automotive and aerospace industries. Other uses of PMTFPS include antifoam fluids, lubricants, protective gels, and as a cosmetic barrier cream ingredient. Longer fluorocarbon side-chain fluorosiloxanes such as PMNFHS (Polymer **II**) are used in release coatings for polydimethylsiloxane-based silicone pressure-sensitive adhesives. Emerging fluorosiloxanes include fluoroPOSS materials, such as Polymer **III**, with potential as nanoscopic fillers and polymer surface modifying additives.

Properties of Special Interest Excellent solvent resistance combined with good thermal stability. Elastomers based on PMTFPS have the widest hardness range and broadest operating service temperature range of any fuel-resistant elastomer. PMTFPS has a surface energy comparable to methylsiloxanes (higher liquid values, lower solid values). More highly fluorinated fluorosiloxanes such as PMNFHS have significantly lower surface energy. The 154 deg water contact angle for FD$_8$T$_8$ fluoroPOSS (Polymer **III**) shows it to be an exceptionally hydrophobic crystalline substance. However, when micron-scale roughness from processing technique is avoided leaving only nano-scale roughness from molecular structure, the lower 124 deg value results.

Property	Units	Conditions	Value	Reference
Density	g cm^{-3}	Polymer **I**, MW = 14 000	1.30	(1)
		25°C	1.292	(2)
	g ml^{-1}	Polymer **III**	2.067	(3)
Solubility parameter	(MPa)$^{1/2}$	Polymer **I**, Conditions not given	17.88	(2)
Theta temperatures	K	Polymer **I**, Cyclohexyl acetate	298	(4)
		Methyl hexanoate	345.8	

Property	Units	Conditions	Value	Reference
Mark–Houwink parameters: K and α	$K = $ ml g^{-1} $\alpha = $ None	Polymer **I**, Methyl hexanoate, 72.8°C Cyclohexyl acetate, 25°C Ethyl acetate, 25°C	$K = 4.45 \times 10^{-4}, \alpha = 0.50$ $K = 4.10 \times 10^{-4}, \alpha = 0.50$ $K = 5.92 \times 10^{-5}, \alpha = 0.70$	(4)
Glass transition temperature	K	Polymer **I**, Atactic, DSC Made from *trans* trimer isomer (100%), DSC Made from *cis* trimer isomer, (96%), DSC Polymer **II**, DSC	203 204.2 207.2 198	(5) (6) (7)
Melting temperature	K	Polymer **I**, made from *trans* trimer isomer (100%), DSC Polymer **I**, made from *cis* trimer isomer (96%), DSC	268.6 321	(6)
Tensile strength	MPa	Polymer **I**, range for typical filled commercial elastomers 22°C 204°C	5.5–11.7 2.4–4.1	(8)
Maximum extensibility	%	Polymer **I**, range for typical filled commercial elastomers 22°C 204°C	100–600 90–300	(8)
Index of refraction	–	Polymer **I**, MW = 14 000 Polymer **II**, MW = 30 200	1.383 1.350	(1) (10)
Dielectric constant	–	Polymer **I**, 100 Hz	6.85	(9)
Loss factor	–	Polymer **I**, 100 Hz	0.109	(9)
Liquid surface tension	mN m^{-1}	RT, 'infinite' MW Polymer **I** Polymer **II**	24.4 19.2	(10) (7)
Solid surface tension	mN m^{-1}	Owens and Wendt method Polymer **I** Polymer **II** Polymer **III**	13.6 9.5 8.7	(11) (7) (12)
Contact angle	deg	Water, advancing Polymer **I** Polymer **II** Polymer **III**	104 115 154/124	(11) (7) (3, 12)

Property	Units	Conditions	Value	Reference
Permeability Coefficient	$m^3(STP)$ m $s^{-1}m^{-2}$ Pa^{-1}	Polymer **I**, He, 100 psi, 35°C	1.85×10^{-15}	(5)
		O_2, 100 psi, 35°C	1.63×10^{-15}	
		CO_2, 100 psi, 35°C	1.04×10^{-14}	
		CH_4, 100 psi, 35°C	1.51×10^{-15}	

References

1. Larsen GL, Smith C. *Silicon Compounds: Register and Review*. 5th ed. Piscataway, N.J.: Huls America Inc.; 1987:275.
2. Stern SA, Bhide BD. *J. Appl. Polym. Sci.* 1989;38:2131.
3. Mabry JM, Vij A, Viers BD, et al. In: Clarson SJ, Fitzgerald JJ, Owen MJ, Smith SD, Van Dyke ME, eds. *Science and Technology of Silicones and Silicone-Modified Materials*. ACS Symp. Ser. 964, Oxford University Press; 2007:290.
4. Buch RR, Klimisch HM, Johansson OK. *J. Polym. Sci., Part A-2*. 1969;7:563.
5. Stern SA, Shah VM, Hardy BJ. *J. Polym. Sci., Part B*. 1987;25:1263.
6. Kuo C-M, Saam JC, Taylor RB. *Polymer International*. 1994;33:187.
7. Kobayashi H, Owen MJ. *Macromolecules*. 1990;23:4929.
8. Maxson MT. *Gummi Fasern Kunststoffe*. 1995;12:873.
9. Ku CC, Liepens R. *Electrical Properties of Polymers*. Munich: Hanser Publishers; 1987:326.
10. Kobayashi H, Owen MJ. *Makromol. Chem*. 1993;194:1785.
11. Owen MJ. *J. Appl. Polym. Sci.* 1988;35:895.
12. Mabry JM. Oral Presentation at *Silicones and Silicone-Modified Materials Symposium IV*, *ACS 232nd* Fall National Meeting, San Francisco, CA, Sept. 2006. See Mabry, J. M., A. Vij, and B. D. Viers. *Polym. Preprints*. 2006;47(2):1216.

Fullerene-containing Polymers

Warren T. Ford

Class Cage structure polymers

Structures

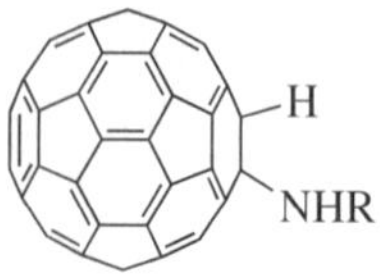

1

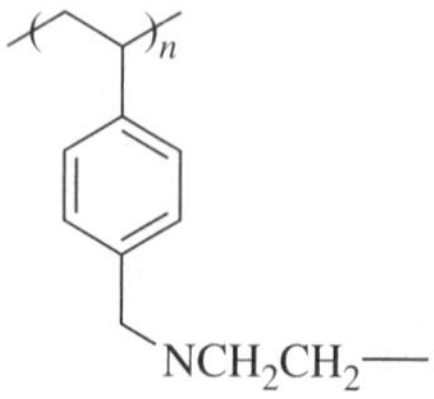

R_1

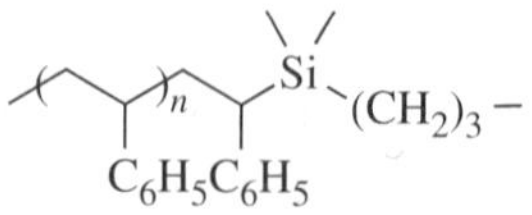

R_2

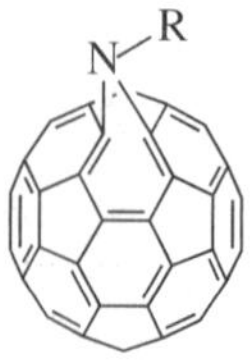

2

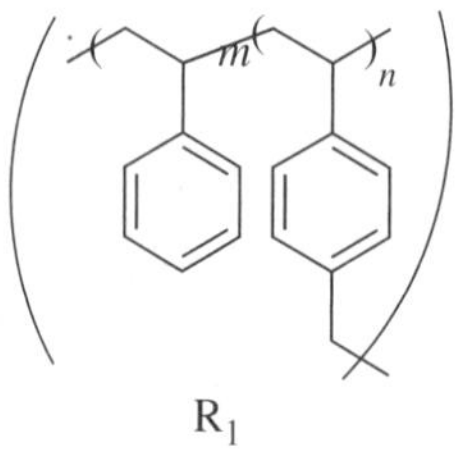

R_1

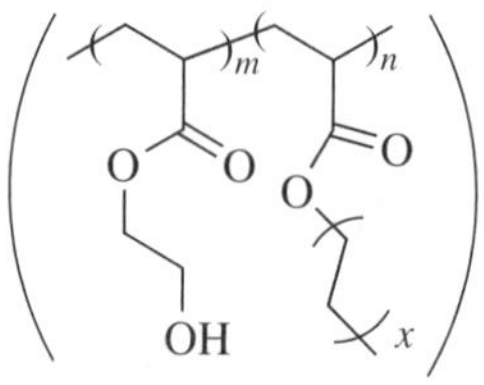

R_2, $x = 1, 3$

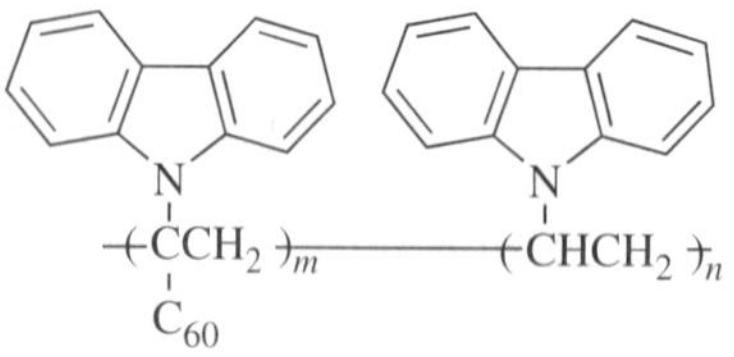

3

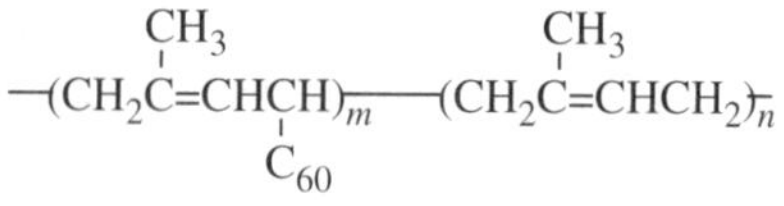

4

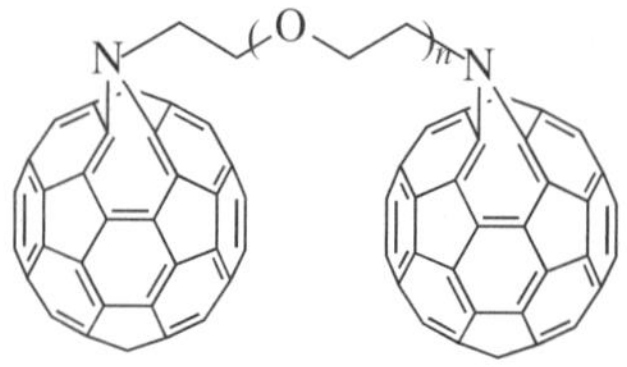

5

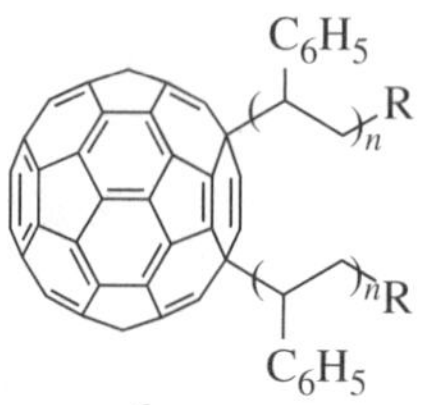

6

$C_{60}(PS)_6$

PS = polystyrene

7

$C_{60}(PS)_2$, $C_{60}(PS)_4$

8

$C_{60}(PS)_6PI$

PI = polyisoprene

9

$C_{60}(PS_a)_n(PS_b)_{6-n}$

$C_{60}(PS)_n(PI)_{6-n}$

$C_{60}(PS)_n(PMMA)_{6-n}$

PMMA = poly(methyl methacrylate)

10

$C_{60}(PS\text{-}b\text{-}PI)_6$

11

$(C_{60})_y(PMMA)_x$

12

$(C_{60})_x(PC)$, PC = bisphenol-A polycarbonate, $x = 0.1\text{--}2.3$

13

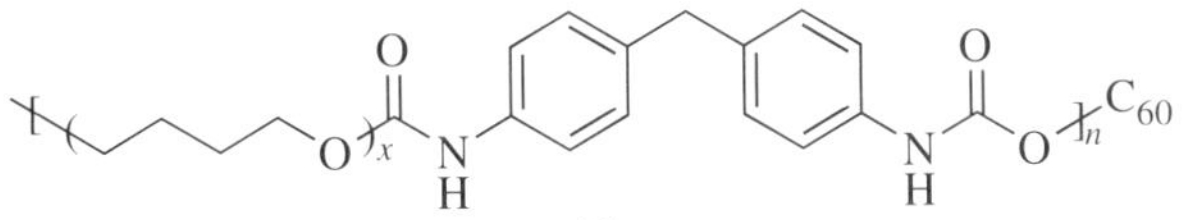

14

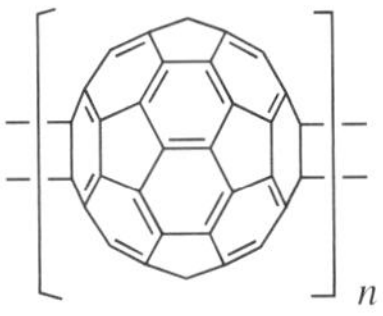

15

Major Applications Electrical and optical materials, cross-linking of elastomers, low-friction films, photovoltaic materials; none are commercial.

Properties of Special Interest Electrical conductivity, photoconductivity, nonlinear optical activity.

Structure	C_{60} (wt %)	M_n (g mol^{-1})	M_w/M_n	Solvents	Analyses	Ref.
1(R1)	18	20 000	NR	Toluene, CS$_2$	^{13}C MAS NMR, IR, UV-vis, TGA	1
1(R2)	4.3	12 100	1.24	Toluene	UV, ^{13}C NMR, DSC	2
2(R1)	29	12 300	1.24	Toluene	UV-vis, IR, ^{1}H NMR, DSC, TGA	3
2(R2)	0.6–3.0	88 000–142 000	2.0–3.0	DMF, methanol	UV-vis, IR, DSC, TGA	4
3	2–15	NR	NR	DMF > CS$_2$, CHCl$_3$ > THF	UV-vis, IR, DSC, TGA, ESR, CV, ^{13}C NMR, SEM, XRD	5
4	10	53 000	1.8	THF, cyclohexane	UV-vis, ^{1}H NMR, DSC, TGA	6
5	36	3900		THF, H$_2$O	UV-vis, XPS, DSC, optical limiting	7
6	3.5–26		≤1.2	THF, o-C$_6$H$_4$Cl$_2$, cyclohexane	UV-vis, CV	8
7	0.036–2.4	30 000 2 000 000	≤1.1	THF, CS$_2$, toluene	UV-vis, LS, SANS, TGA	9–12
8	0.5–4.2	17 000–137 000	1.2–1.4	THF, toluene, C$_6$H$_5$Cl,	UV-vis, LS, TGA, optical limiting	9,11–13

Structure	C_{60} (wt %)	M_n (g mol^{-1})	M_w/M_n	Solvents	Analyses	Ref.
9	0.5–1.2	58 000–136 000	1.3	Toluene	UV-vis, LS	14
10	0.5–1.7	43 000–127 000	1.1	THF, toluene	UV-vis, LS	15
11	0.15–0.85	85 000–492 000	1.07–1.28	THF, toluene	LS, DSC, SAXS, TEM	16
12	0.5–12	4800–34,000	1.5–2.0	THF, o-$C_6H_4Cl_2$, toluene	UV-vis, LS, ^{1}H NMR, ESR	17,18
13	0.46–5.76	16 400–28 800	1.8–2.4	$CHCl_2CHCl_2$, THF	UV-vis, IR, DSC	19
14	NR	18 000	1.45	THF:DMF (3:1)	IR, DSC, TMA	20
15 photo polymer of C_{60}	100	$(720)_n$ $n = 1$–21	–	Insoluble	MS, SEM, IR, Raman, XRD, LDMS	21–23
15 pressure polymer of C_{60}	100	$(720)_n$	–	Insoluble	XRD, Raman, IR, ^{13}C MAS NMR, LDMS	23,24

References

1. Geckeler KE, Hirsch A. *J. Am. Chem. Soc.* 1993;115:3850–3851.
2. Weis C, Friedrich C, Muelhaupt R, Frey H. *Macromolecules.* 1995;28:403–405.
3. Hawker CJ. *Macromolecules.* 1994;27:4836–4837.
4. Goh HW, Goh SH, Xu GQ. *J. Polym. Sci. Part A: Polym. Chem.* 2002;40:1157–1166.
5. Chen Y, Huang Z-E, Cai R-F. *J. Polym. Sci. Part B: Polym. Phys.* 1996;34:631–640.
6. Dai L, Mau AWH, Griesser HJ, Spurling TH, and White JW *J. Phys. Chem.* 1995;99:17302–17304.
7. Huang XD, Goh SH. *Macromolecules.* 2000;33:8894–8897.
8. Okamura H, Terauchi T, Minoda M, Fukuda T, Komatsu K. *Macromolecules.* 1997;30:5279–5284.
9. Mathis C, Schmaltz B, Brinkmann M. *C. R. Chim.* 2006;9:1075–1084.
10. Ederle Y, Mathis C. *Macromolecules.* 1997;30:4262-4267.
11. Pozdnyakov OF, Pozdnyakov AO, Schmaltz B, Mathis C. *Polymer.* 2006;47:1028–1035.
12. Venturini J, Koudoumas E, Couris S, Janot JM, Seta P, Mathis C, Leach S. *J. Mater. Chem.* 2002;12:2071–2076.
13. Audouin F, Nuffer R, Mathis C. *J. Polym. Sci. Part A: Polym. Chem.* 2004;42:4820–4829.
14. Ederle Y, Mathis C. *Macromolecules.* 1999;32:554–558.
15. Audouin F, Renouard T, Schmaltz B, Nuffer R, Mathis C. *Polymer.* 2005;46:8519–8527.
16. Schmaltz B, Brinkmann M, Mathis C. *Macromolecules.* 2004;37:9056–9063.
17. Ford WT, Graham TD, Mourey TH. *Macromolecules.* 1997;30:6422–6429.
18. Ford WT, Nishioka T, McCleskey SC, Mourey TH, Kahol P. *Macromolecules.* 2000;33:2413–2423.
19. Tang BZ, Peng H., Leung SM, et al. *Macromolecules.* 1998;31:103–108.
20. Chiang LY, Wang LY, Kuo C-S. *Macromolecules.* 1995;28:7574–7576.

21. Cornett DS, Amster IJ, Duncan MA, Rao AM, Eklund PC. *J. Phys. Chem.* 1993;97:5036–5039.
22. Rao AM, Zhou P, Wang KA, et al. *Science.* 1993;259:955–957.
23. Rao AM, Eklund PC. *Springer Ser. Mater. Sci.* 2000;38:145–184.
24. Nunez-Regueiro M, Marques L, Hodeau JL, Bethoux O, Perroux M. *Phys. Rev. Lett.* 1995;74:278–281.

Gelatin

W. Brooke Zhao and Zongming Gao

Gelatin (also gelatine, from French gélatine) is a translucent brittle solid substance, colorless or slightly yellow, nearly tasteless and odorless, which is created by prolonged boiling of connective tissue such as skin, cartilage, and bones obtained from the animal processing industry.

Property	Units	Conditions	Value		Reference
Amino acid composition	Numbers per 1,000 residues		Type A	Type B	(1)
		Alaine	112	117	
		Arginine	49	48	
		Asparagine	16	0	
		Aspartic acid	29	46	
		Cysteine	–	0	
		Glutamic acid	48	72	
		Glutamine	48	0	
		Glycine	330	335	
		Histidine	4	4.2	
		Hydroxyproline	91	93	
		Hydroxylysine	6.4	4.3	
		Isoleucine	10	11	
		Leucine	24	24.3	
		Lysine	27	28	
		Methionine	3.6	3.9	
		Phenylalanine	14	14	
		Proline	132	124	
		Serine	35	33	
		Threonine	18	18	
		Tryptophan	–	–	
		Tyrosine	2.6	1.2	
		Valine	26	22	
Molecular weight (of repeat unit)	$g\,mol^{-1}$	Alpha	9.65×10^4		(1)
		Beta	1.93×10^5		
		Gamma	2.89×10^5		
		Limed-ossein	2.2×10^5		
		Weight-average	$(1->5) \times 10^5$		
		Number-average	$(5-15) \times 10^4$		

Property	Units	Conditions	Value		Reference
Gel rigidity	Bloom	Depending on source and extracting conditions	75–330		(1)
Density	g cm^{-3}	–	1.35		(2)
IR (characteristic absorption frequencies)	cm^{-1}	–C=O stretching	1,650		(3)
		–NH stretching	3,300		
Common solvents		Water (warm), acetic acid, trifluro-ethanol, formamide, ethylene glycol, glycerol, dimethyl sulfoxide			(1)
Common nonsolvents		Ethanol, acetone, tetrahydrafuran			(1)
Isoionic pH	–	Type A		4.8–5.2	(1)
		Type B		7–9	
pK$_a$ of the ionizable side groups of gelatin	–	Anionic amino acid	Conc. (mmol g^{-1})		(4)
		Aspartic acid	0.50, 0.32	4–4.5	
		Glutamic acid	0.78, 0.52	4.5	
		Tyrosine	0.011–0.044	10	
		α–COOH	0.011	3.6	
		Cationic amino acid			
		Lysine	0.30	10–10.4	
		Hydroxylysine	0.054	9.5	
		Arginine	0.53	>12	
		Histidine	0.044	6.5–7	
		α-NH$_2$	0.011	7.8	
Mark–Houwink parameters: K and a	K = ml g^{-1} a = None	Calf skin gelatin		$K = 1.66 \times 10^{-5}$ $a = 0.885$	(5)
		Pig skin		$K = 1.10 \times 10^{-4}$ $a = 0.74$	(6)
Radius of gyration	Å	Alpha		138	(1)
		Beta		215	
		Gamma		257	

Property	Units	Conditions		Value	Reference
Radius of gyration	Å	Solvent	M_w (ossein gelatin)		(7)
		1.0 M KCNS, 25°C, pH = 5.1	90,000	175	
		0.15 M NaCl, 40°C			
		pH = 5.1	89,000	165	
		pH = 3.1	88,000	175	

Property	Units	Conditions		Value	Reference
			M_w (Type B)		(8)
		0.05 M phosphate	2.7 × 105	302	
		10 M KCNS	3.02 × 105	302	
		10 M KCNS	3.83 × 105	242	
		10 M KCNS	5.96 × 105	280	
		Alcohol-water ratio	M_w (bovine corium extract)		(9)
		2:1 ~ 2.5:1	8.33 × 10^6	2,410	
		2.5:1 ~ 3.0:1	7.45 × 10^5	444	
		2.5:1 ~ 3.0:1	3.45 × 10^5	314	
		2.5:1 ~ 3.0:1	2.32 × 10^5	371	
		3.0:1 ~ 3.5:1	2.02 × 10^5	345	
			$M_w × 10^{-5}$		
		0.2 M KCl	3.3 (acid-precursor)	452	
		0.2 M KCl	3.3 (alkali-precursor)	447	
		Rousselot gelatin, photographic grade, $M_w = 1.9 × 10^5$, $M_w/M_n = 2.3$		350 ± 40	(11)
Radius of gyration of the cross-section R_c	Å	Rousselot gelatin, photographic grade, $M_w = 1.9 × 10^5$, $M_w/M_n = 2.3$		3.2 ± 1	(11)
Root-mean-square end-to-end distance $(\overline{r^2})^{1/2}$	Å	Solvent	M_w (Type B gelatin)		(8)
		0.05 M phosphate	2.7 × 10^5	740	
		1.0 M KCNS	3.02 × 10^5	740	
		1.0 M KCNS	3.83 × 10^5	590	
		1.0 M KCNS	5.96 × 10^5	685	
Expansion coefficient α	–	Solvent	M_w (Type B gelatin)		(8)
		1.0 M KCNS	3.83 × 10^5	1.25	
		1.0 M KCNS	5.96 × 10^5	1.25	
Universal constant Φ	–	Solvent	M_w (Type B gelatin)		(8)
		1.0 M KCNS	3.83 × 10^5	1.29 × 10^{21}	
		1.0 M KCNS	5.86 × 10^5	1.63 × 10^{21}	
Second virial coefficient A_2	mol cm^3 g^{-2}	Solvent	M_w (ossein gelatin)		(7)
		1.0 M KCNS, 25°C pH = 5.1 0.15 M Nalco, 40°C	9 × 10^4	2.6 × 10^{-4}	
		pH = 5.1	8.9 × 10^4	2.9 × 10^{-4}	
		pH = 3.1	8.8 × 10^4	6.0 × 10^{-4}	
			M_w (Type B)		
		0.05 M phosphate	2.7 × 10^5	2.4 × 10^{-4}	
		1.0 M KCNS	3.02 × 10^5	2.6 × 10^{-4}	
		1.0 M KCNS	3.83 × 10^5	3.3 × 10^{-4}	
		1.0 M KCNS	5.96 × 10^5	2.4 × 10^{-4}	
		Rousselot gelatin, photographic grade, $M_w = 1.9 × 10^5$, $M_w/M_n = 2.3$		(3 ± 1) × 10^{-4}	(11)

Property	Units	Conditions		Value	Reference
Persistence length l	Å	Rousselot gelatin, photographic grade, $M_w = 1.9 \times 10^5$, $M_w/M_n = 2.3$		20 ± 3	(11)
Mass per unit length g mol^{-1}	Å^{-1}	Rousselot gelatin, photographic grade, $M_w = 1.9 \times 10^5$, $M_w/M_n = 2.3$		28 ± 8	(11)
Screen length ξ	Å	Rousselot gelatin, photographic grade, $M_w = 1.9 \times 10^5$, $M_w/M_n = 2.3$			(11)
		1%		70 ± 10	
		2%		51 ± 5	
		5%		35 ± 3	
Hydrodynamic screen length ξ_h	Å	Quasi-elastic light scattering dilute solution, Rousselot gelatin, photographic grade, $M_w = 1.9 \times 10^5$, $M_w/M_n = 2.3$, concentration ranges from 5 to15%		$25–100$	(12)
Sizes of inhomogeneities a	Å	Rousselot gelatin, photographic grade, $M_w = 1.9 \times 10^5$, $M_w/M_n = 2.3$			(11)
		1%		–	
		2%		220	
		5%		135	
z-average self-diffusion coefficient $(D_0)_z$	cm^{2}s^{-1}	Quasi-elastic light scattering, dilute solution grade, Rousselot gelatin, photographic $M_w = 1.9 \times 10^5$, $M_w/M_n = 2.3$		2×10^7 (fast-mode) 3.5×10^8 (slow-mode)	(12)
Hydrodynamic radius R_h	Å	Quasi-elastic light scattering, Rousselot gelatin, photographic grade, $M_w = 1.9 \times 10^5$, $M_w/M_n = 2.3$			(12)
		In dilute solution		220	
		In semi-dilute solution		210	
Flory–Huggins inter action parameter χ	–	Type A gelatin			(13)
		283.15 K		0.497–0.498	
		293.15 K		0.497	
Self-diffusion coefficient D	cm^{2}s^{-1}	Gelatin, quasi-elastic neutron scattering	Volume fraction		(14)
			0.031	1.36×10^7	
			0.063	1.01×10^7	
			0.129	7.1×10^6	
			0.199	6.1×10^6	
			0.274	5.3×10^6	
			0.353	4.9×10^6	
Collective self-diffusion coefficient D_{coll}	cm^{2}s^{-1}	Gelatin, quasi-elastic neutron scattering	Volume fraction		(14)
			0.031	4.8×10^6	
			0.063	5.2×10^6	
			0.129	5.2×10^6	
			0.199	5.4×10^6	
			0.274	5.1×10^6	
			0.353	4.8×10^6	

Property	Units	Conditions		Value	Reference
Single-particle self-diffusion coefficient D_{s-p}	$cm^2 s^{-1}$	Gelatin, quasi-elastic neutron scattering	Volume fraction 0.031	8.8×10^6	
			0.063	4.9×10^6	
			0.129	1.9×10^6	
			0.199	7.0×10^5	
			0.274	2.0×10^5	
			0.353	1.0×10^5	
Residence time τ_0	s	Gelatin, quasi-elastic neutron scattering	Volume fraction		(14)
			0.031	2.0	
			0.063	2.7	
			0.129	9.8	
			0.199	22	
			0.274	66	
			0.353	165	
Specific optical rotation $[\alpha]$	Degree	Alpha		-137	(1)
		Beta		-137	
		Gamma		-137	
		Limed-ossein		-137	
Refractive index	–	Dry, $\lambda = 546.1$ mm		1.54	(15)

Property	Units	Conditions	Value	Reference
Refractive increment dn/dc	–	$\lambda = 300$ nm	0.187	(16)
		$\lambda = 436$ nm		
		Ossein gelatin		(7)
		H_2O, 25°C	0.194	
		0.1 M NaCl, 25°C	0.1925	
		1.0 M NaCl, 25°C	0.186	
		1.0 M KCNS, 25°C	0.185	
		2.0 M KCNS, 25°C	0.173	
		Bovine corium extract		
		H_2O, 40°C	0.192	(9)
		0.25 M NaCl, 40°C	0.192	(9)
		0.10 M KCl, 40°C	0.192	(9)
		0.25 M KCl, 40°C	0.192	(9)
		1.0 M KCl, 40°C	0.176	(17)
		Type B gelatin		(8)
		1.0 M KCNS, 30°C	0.172	
		0.1 M KH_2PO_4, 30°C	0.172	
		0.1 M K_2HPO_4, 30°C	0.172	
		$\lambda = 632.8$ nm, Rousselot gelatin, photographic grade, $M_w = 1.92 \times 10^5$, $M_w/M_n = 2.3$	0.18	(11)

Gelatin

Property	Units	Conditions	Value	Reference
Glass transition temperature T_g	K	Dilatometry	368	(18)
		Viscoelastic	393	
		Viscoelastic	463 (calculated)	
		Viscoelastic	453	
		Thermomechanical	473	
		DTA, viscoelastic	448 ± 10 (uncross-linked)	
		DTA, viscoelastic	469 ± 3 (cross-linked)	
Melting temperature T_m	K	Dilatometry	418	(18)
		DTA, X-Ray	491	
		Viscoelastic	493	
		DTA, TGA	503 (calculated)	
		DSC	503	
		DTA, X-ray	503	
Activation energy of hydrolysis	kJ mol^{-1}	pH		(19)
		3.05	107.2	
		3.60	107.2	
		4.75	72.2	
		7.10	72.2	
		8.50	72.2	
		9.35	72.2	
		9.85	108.8	

Property	Units	Conditions	Value	Reference
Contact angle of water	Degrees	15% aqueous gelatin gel, 20°C	75	(20)
		Gelatin hydrogel		
		Air-equilibrated surface	110	
		Fresh-cut wet surface	36	
		Gelatin film, Langmuir–Bloggett process		(22)
		Surface in contact with air	110	
		Surface in contact with benzene	90	
Tensile strength	MPa	Type B, uniaxially oriented in water, v_2 at stretching $= 0.2$–0.25	Stretching ratio (%)	(23)
			0	25.61
			65	41.57
			120	63.39
			145	84.84
			180	27.87($\perp$)*
			87	74.94
		Biaxial Type A, uniaxially oriented in water, v_2 at stretching $= 0.2$–0.25	0	28.36
			75	54.48
			125	80.88
			155	109.19

Property	Units	Conditions		Value	Reference
			190	139.80	
			110	60.46($\perp$)*	
			160	44.31($\perp$)*	
		Biaxial	50	50.68	
			90	104.47	
			110	128.37	
Tensile modulus	MPa	Type B, uniaxially oriented in water, ν_2 at stretching = 0.2–0.25	Stretching ratio (%)		(23)
			0	473	
			65	690	
			120	790	
			145	1300	
			180	930($\perp$)*	
			87	1345	
		Biaxial Type A, uniaxially oriented in water, ν_2 at stretching = 0.2–0.25	0	631	
			75	890	
			125	1090	
			155	1240	
			190	1690	
			110	1100($\perp$)*	
			160	1040($\perp$)*	
		Biaxial	50	1090	
			90	1200	
			110	1570	
Elongation at break	%	Type B, uniaxially oriented in water, ν_2 at stretching = 0.2–0.25	Stretching ratio (%)		(23)
			0	7.7	
			65	8.3	
			120	12.3	
			145	8.3	
			180	3.6($\perp$)*	
			87	15.5	
		Biaxial Type A, uniaxially oriented in water ν_2 at stretching = 0.2–0.25	0	6.7	
			75	11.4	
			125	14.9	
			155	21.6	
			190	22.5	
			110	7.7($\perp$)*	
			160	5.5($\perp$)*	
		Biaxial	50	5.86	
			90	14.5	
			110	18.4	
Toughness	MPa	Type B, uniaxially oriented in water, ν_2 at stretching = 0.2–0.25	Stretching ratio (%)		(23)
			0	1.24	
			65	2.31	
			120	5.00	

Property	Units	Conditions		Value	Reference
			145	4.05	
			180	0.60(⊥)*	
			87	8.93	
		Biaxial Type A, uniaxially oriented in water, v_2 at stretching = 0.2–0.25	0	1.26	
			75	4.15	
			125	8.42	
			155	17.13	
			190	23.85	
			110	3.00(⊥)*	
			160	1.45(⊥)*	
		Biaxial	50	1.86	
			90	9.51	
			110	17.56	

* Property measured at the direction perpendicular to the orientation direction.

Class Polypeptides and proteins[24]

Structure[25] Gelatin contains a large number of glycine (almost 1 in 3 residues, arranged every third residue), proline, and 4-hydroxyproline residues. A typical structure is
—Ala—Gly—Pro—Arg—Gly—Glu—4Hyp—Gly—Pro—.

Physical Properties Gelatin is a protein produced by partial hydrolysis of collagen extracted from connective tissues of animals such as porcine and bovines. The natural molecular bonds between individual collagen strands are broken down into a form that rearranges more easily. Gelatin melts when heated and solidifies when cooled again. Together with water it forms a semi-solid colloidal gel. Gelatin forms a solution of high viscosity in water, which sets to a gel on cooling and its chemical composition is, in many respects, closely similar to that of its parent collagen. Gelatin solutions show viscoelastic flow and streaming birefringence. If gelatin is put into contact with cold water, some of the material dissolves. The solubility of the gelatin is determined by the method of manufacture. Typically, gelatin can be dispersed in a relatively concentrated acid. Such dispersions are stable for 10–15 days, with little or no chemical changes, and are suitable for coating purposes or for extrusion into a precipitating bath. Gelatin is also soluble in most polar solvents. Gelatin gels exist over only a small temperature range, the upper limit being the melting point of the gel which depends on gelatin grade and concentration and the lower limit the ice point at which ice crystallizes out. The mechanical properties are very sensitive to temperature variations, previous thermal history of the gel, and time. Gelatin concentration and the temperature have important effects on viscosity. The higher they are, the higher viscosity is obtained.[26]

Gelatin Testing Methods[27]

Identification

Gelatin gives the normal positive trichloroacetic acid, biuret, and ninhydrin tests for protein. The precipitate with 5% tannic acid is a particularly sensitive test for very dilute solutions of gelatin. In addition the thermally reversible gelation of a 6% solution in water between 10 and 60°C is unique for this protein.

Gelatin

Gel Strength

The most important attribute of gelatin is its gel strength and when determined by the standard method[28], is called the Bloom Strength. This is the force in grams required to press a 12.5 mm diameter plunger 4 mm into 112 g of a standard $6^{2/3}$% w/v gelatin gel at 10°C. Several penetrometer-type instruments have been adapted to determine Bloom Strength.

A frequent question is how to substitute gelatin of one Bloom Strength for a gelatin of another. As a guide one can say:

$$C \times B^{1/2} = k \ (24)$$
$$\text{or } C_1 (B_1)^{1/2} \div (B_2)^{1/2} = C_2$$

Where C = concentration, B = Bloom strength, and k = constant. However, there are other considerations besides gel strength which can invalidate such a substitution calculation. For example, in a gummy formulation, the texture using 250 Bloom gelatin is far shorter than when 180 Bloom gelatin is used.

Viscosity

From the point of view of functionality, the solution viscosity of gelatin is probably the second most important parameter. The standard method calls for the viscosity of a $6^{2/3}$% solution at 60°C. Low viscosity (and a high gel strength) is required for poured confectionery, and high viscosity for film-forming applications.

In viscosity calculations, usually $C \log V = k$ but the model is not as good as is the mathematical model for Bloom calculations.

Color and Clarity

Solution color and turbidity or clarity are attributes which may or may not be important depending on the application. Poor clarity markedly affects the ability to measure color [29] and at this stage there are no internationally accepted methods for determining these attributes. However, if clarity is good, then gelatin color obeys Beer's Law.

pH

Solution pH (1%) is usually about pH 5 but can vary considerably. At this pH the viscosity of Type B gelatin is minimal and the gel strength is maximal, hence from the manufacturers' point of view it is advantageous to manufacture gelatin at this pH. However, due to the strong buffering capacity of gelatin this pH may not be the most advantageous for the customer.

Moisture

The moisture content of gelatin may be as high as 16%. However, more normally it is about 10–13% because at 13.0% moisture content the glass transition temperature[30] of gelatin is about 64°C which allows particle size reduction to be a simple operation. In addition, at 13% moisture content and 25°C gelatin is close to equilibrium with ambient air moisture contents of ca. 46% RH. At 6–8% moisture content gelatin is very hygroscopic and it becomes difficult to determine the physical attributes with accuracy.

Due to the variable granule size of gelatin, the rate of moisture loss at 105°C can be variable. Hence it is normal to add water to the gelatin powder before placing the sample in the drying oven. This means that the gelatin melts and water is lost from a uniform thin film of protein. It should be noted that metal dishes have to be used because, on drying, the film of gelatin shrinks and breaks containers of glass or ceramic.

Finally, the drying of gelatin to very low moisture contents results in cross-linking and loss of solubility. It is thus difficult to distinguish between free and bound water in gelatin.

Ash

The gelatin ash content is determined by pyrolysis at 550°C. Usually ash contents up to 2.5% can be accepted in food applications. However the nature of the ash can be important. For example, 2% $CaSO_4$ in gelatin can have excellent clarity in spite of the solubility product of the ash being exceeded (due to the crystal-habit modifying effect of gelatin); however on dilution of the gelatin in a confectionery formulation, the ash can precipitate. Furthermore, ammonia is often used as a pH modifier in gelatin preparation and salts like NH_4Cl are not determinable by pyrolysis.

Isoionic Point

The isoionic point of gelatin[31] is best determined by passing a 1% solution of the gelatin at 40°C through a mixed bed column of ion exchange resin (Rohm & Haas MB3) at a flow rate of not more than 10 bed volumes per hour and by measuring the pH of the eluate. It should be noted that on cooling, isoionic gelatin has poor clarity and the conductivity should be between 1 and 5 s/cm for Type B gelatin.

Sulphur dioxide Content

Sulphur dioxide is used as a biocide and bleach in gelatin manufacture. The nationally permitted level of residual SO_2 in gelatin is variable and the methods for its determination can give a great variation in results. It is known that gelatin promotes oscillating redox reactions [32,33] and the control of this contaminant is not easy. Hydrogen peroxide is often used to control the SO_2 content of gelatin and sometimes the permitted level of this contaminant is also specified. It is interesting to note that both H_2O_2 and SO_2 can be shown to coexist in gelatin.

Heavy Metal Content

Once again the determination of heavy metals in gelatin can be a problem because of the difficulty of completely degrading gelatin and also because the main component of the ash in gelatin can be of low solubility, like calcium sulfate, hence with a variable ability to absorb traces of heavy metals. It must be recommended that internal standards be used wherever possible.

Microbiological Properties

Gelatin is an excellent nutrient for most bacteria, hence the manufacturing processes have to carefully avoid contamination. Most countries have microbiological specifications for gelatin, but generally they are not very onerous. Total mesophyllic plate counts of 1000 are generally accepted with various countries limiting the presence of Coliforms, E. Coli, Salmonella, Clostridial spores, Staphylococci, and sometimes even Pseudomonades.

Applications[24] Probably best known as a gelling agent in cooking, different types and grades of gelatin are used in a wide range of food and nonfood products:

Food Uses

Common examples of foods that contain gelatin are gelatin desserts or jelly, trifles, aspic, marshmallows, and confectioneries such as Peeps and gummy bears. Gelatin may be used as a stabilizer, thickener, or texturizer in foods such as ice cream, jams, yogurt, cream cheese, and margarine; it is used, as well, in fat-reduced foods to simulate the mouth feel of fat and to create volume without adding calories.

Gelatin is used for the clarification of juices, such as apple juice, and of vinegar. Isinglass, from the swim bladders of fish, is still in use as a fining agent for wine and beer. Beside hartshorn jelly, from deer antlers (hence the name 'hartshorn'), isinglass was one of the oldest sources of gelatin.

Technical Uses

- Gelatin typically constitutes the shells of pharmaceutical capsules in order to make them easier to swallow. Hypromellose is the vegetarian counterpart to gelatin, but is more expensive to produce.
- Animal glues such as hide glue are essentially unrefined gelatin.
- It is used to hold silver halide crystals in an emulsion in virtually all photographic films and photographic papers. Despite some efforts, no suitable substitutes with the stability and low cost of gelatin have been found.
- Used as a carrier, coating, or separating agent for other substances, it, for example, makes beta-carotene water-soluble, thus imparting a yellow color to any soft drinks containing beta-carotene.
- Gelatin is closely related to bone glue and is used as a binder in match heads and sandpaper.
- Cosmetics may contain a nongelling variant of gelatin under the name 'hydrolysed collagen'.
- As a surface sizing it smoothes glossy printing papers or playing cards and maintains the wrinkles in crepe paper.

Other Uses

- Blocks of ballistic gelatin simulate muscle tissue as a standardized medium for testing firearms ammunition.
- Gelatin is used by synchronized swimmers to hold their hair in place during their routines as it will not dissolve in the cold water of the pool. It is frequently referred to as 'knoxing', a reference to Knox brand gelatin. Though commonly used, the owners of the trademark object to the genericized use of the term.
- When added to boiling water and cooled, unflavored gelatin can make a home-made hair styling gel that is cheaper than many commercial hair styling products, but by comparison has a shorter shelf life (about a week) when stored in this form (usually in a refrigerator). After being applied to scalp hair, it can be removed with rinsing and some shampoo.
- It is commonly used as a biological substrate to culture adherent cells.

References

1. Rose PI. In: Mark HF, et al. eds. *Encyclopedia of Polymer Science and Engineering.* Vol 7. John New York: Wiley and Sons; 1987.

2. Chien JCW. *J. Macromol. Sci. Rev. Macromol. Chem.* 1975;12:1.

3. Veis A. *Macromolecular Chemistry of Gelatin.* New York: Academic Press; 1964:49.

4. Kenchington AW, Ward A. *G. Biochem. J.* 1954;58:202.

5. Pouradier J, Venet AM. *J. Chem. Phys.* 1950;47:391.

6. Pouradier J, Venet AM. *J. Chem. Phys.* 1950;49:85.

7. Boedtker H, Doty P. *J. Phys. Chem.* 1954;58:968.

8. Gouinlock EV Jr, Flory PJ, Scheraga HA. *J. Polym. Sci.* 1955;16:383.

9. Veis A, Eggenberger DN, Cohen J. *J. Am. Chem. Soc.* 1955;77:2368.

10. Veis A, Cohen J. *J. Polym. Sci.* 1957;26:113.

11. Pezron I, Djabourov M, Leblond J. *Polymer.* 1991;32:17.

12. Herning T, Djabourov M, Leblond J, Takerkart G. *Polymer.* 1991;32:3211.

13. Holtus G, CoÈlfen H, Borchard W. *Progr. Colloid. Polym. Sci.* 1991;86:92.

14. Mel'nichenko YB, Bulavin LA. *Polymer.* 1991;32:3295.

15. Sklar E. *Photogr. Sci. Eng.* 1969;13:29.

16. Lewis MS, Piez KA. *Biochemistry.* 1964;3:1126.

17. Veis A, Cohen J. *J. Am. Chem. Soc.* 1956;78:6238.

18. Yannas IV. *J. Macromol. Sci. Revs. Macromol. Chem.* 1972;C7(1):49.

19. Marshall AS, Petrie SEB. *J. Photogr. Sci.* 1980;28:128.

20. Yasuda T, Okuno T, Yasuda H. *Langmuir.* 1994;10:2435.

21. Wolfram E, Stergiopulos C. *Acta Chim. Acad. Sci. Hung.* 1977;92:157.

22. Mironjuk NV, Summ BD. *Vysokomol Soedin., Ser.* 1982;24:391.

23. (a) Zhao W PhD. *Thesis.* University of Cincinnati; 1995. (b) Zhao W, et al. *J. Macromol. Sci. Pure Appl. Chem.* 1996;A33(5):525. (c) Zhao W, et al. *CHEMTECH.* 1996;26:32.

24. From Wikipedia, the free encyclopedia. http://en.wikipedia.org/wiki/Gelatin. Accessed March 2, 2007.

25. Water structure and behavior. http://www.lsbu.ac.uk/water/hygel.html. Accessed March 2, 2007.

26. Ward AG, Courts A. *The Science and Technology of Gelatin.* New York: Academic Press; 1977.

27. Bernard Cole, Gelatin. http://www.gelatin.co.za/gltn1.html. accessed March 2, 2007.

28. Methods for sampling and testing gelatine. BS 757 : 1975. Gr8. British Standards Institution. 2 Park St. London W1A 2BS.

29. Cole CGB, Roberts JJ. "Gelatine Colour Measurement". *Meat Science.* 1997;45(1):23–31.

30. McCormick-Goodhart M.H. Research Techniques in Photographic Conservation. *Proceedings of the Copenhagen Conference.* May 1995;65–70.

31. Veis A. *The Macromolecular Chemistry of Gelatin.* New York and London: Academic Press; 1964:107–113.

32. Chinake CR, Simoyi RHS. *Afr. J. Chem.* 1995;48:1–7.

33. Melichova Z, Olexova A, Treindel L. *Chemical Abstracts. Number 123:267635. Z. Phys. Chem. (Munich).* 1995;191(2):259–64.

Glycogen

Rachel Mansencal, John F. Kadla and
Jennifer L. Braun

Class Carbohydrate polymers; polysaccharides

Structure Branded glucan. α-D-glucopyranosyl units joined by α-D-(1 → 4) glycosidic linkages.[1–2]

Functions Biological function restricted to source of energy. Principal food-reserve materials in animals. Found in cells of vertebrates and invertebrates. No commercial use.[1–2]

Extraction Extraction with hot concentrated alkali. But extensive degradation. Milder extraction with cold trichloroacetic acid solution, dimethyl sulfoxide, or water-phenol mixtures.[1–6]

Purification After extraction, redissolution in distilled water; low speed centrifugation (100 g); precipitation with excess ethanol; high speed centrifugation (1,500 g).[2]

Properties of Special Interest Amorphous polymer; very high molecular weight; polydisperse; highly soluble; very good hydrodynamic behavior.[1–2]

Property	Units	Conditions	Value	Reference
Molecular weight	g mol^{-1}	Rabbit liver	2.7×10^8	–
Average degree of polymerization	–	–	1.7×10^6	(1–2)
Average chain length	–	Depending on the source of the glycogen and the method used	6–21	(1–2)

Glycogen

Property	Units	Conditions	Value	Reference
Average interior chain length	–	Depending on the source of the glycogen and the method used	2–5	(1–2)
Morphology	–	α-particle $\approx$ 100 β-particles β-particle diameter = 25 nm	–	(2)

Action of enzymes on glycogen[2]

Enzyme	Glucosidic Bond Attacked	Source
Phosphorylase	$(1 \rightarrow 4)$-α	Plants, microbes, mammals
Amylo-1,6-glucosidase	$(1 \rightarrow 4)$-α	Yeast, mammals
	$(1 \rightarrow 6)$-α	
Pullulanase	$(1 \rightarrow 6)$-α	*Aerobacter aerogenes*
Glucoamylase	$(1 \rightarrow 4)$-α	Microbes
Isoamylase	$(1 \rightarrow 6)$-α	*Cytophaga pseudomonas*
β-Amylase	$(1 \rightarrow 4)$-α	Sweet potato, cereals

Weight mean sedimentation coefficients of glycogen fraction[2]

Fraction Number	$s^0_{20,w}$ at Boundary (S)	Absolute Range (S)	$\bar{s}_w$ (S)	z-avg. std. diff. coeff. $\bar{D}_z(10^{-8}\text{cm}^2\text{s}^{-1})$
20	0	0–85	26	2.56 ± 0.02
19	46	0–181	85	2.7 ± 0.4
18	114	40–292	175	4.3 ± 0.3
17	225	150–419	291	5.90 ± 0.07
16	353	278–561	422	5.8 ± 0.1
15	495	420–717	570	5.1 ± 0.1
14	651	576–884	732	4.3 ± 0.1
13	818	743–1,061	904	3.6 ± 0.1
12	915	920–1,249	1,087	3.4 ± 0.1
11	1,182	1,107–1,444	1,278	3.01 ± 0.05
10	1,378	1,303–1,648	1,478	2.74 ± 0.07
9	1,581	1,507–1,858	1,658	2.73 ± 0.04
8	1,792	1,717–2,074	1,898	2.59 ± 0.04
7	2,008	1,933–2,298	2,117	2.42 ± 0.04
6	2,231	2,156–2,531	2,346	2.6 ± 0.2
5	2,464	2,389–2,781	2,588	2.27 ± 0.06
4	2,714	2,639–3,064	2,855	2.42 ± 0.06
3	2,998	2,923–3,428	3,178	2.21 ± 0.04
2	3,361	3,286–4,020	3,647	2.18 ± 0.02
1	3,953	3,878–7,077	4,335	2.5 ± 0.1
0	7,010	–	–	–

Glycogen

Hydrodynamic parameters for glycogen subfractions[2,7]

Fraction Number	Average Molecular Weight ($\times 10^{-6}$)	Scheraga-Mandelkern Function β ($\times 10^{-6}$)	Viscosity (ml g^{-1})	Frictional Ratio (f/f_{min})
20	6	0.61	14	7.1
19	20	0.71	9	5.0
18	26	1.26	6.5	2.6
17	32	1.75	6.0	1.8
1–16	48–1600	2.06 ± 0.17	6.7 ± 0.4	1.7 ± 0.1

Property	Units	Conditions	Value	Reference
Intrinsic viscosity	cm^3/g	25°C in 0.5N NaOH	6.9	(12)
2nd virial coefficient	mol mL/g^2	25°C in 0.5N NaOH	2.81×10^{-6}	(12)

Infrared Absorption

Type of Motion	Wavenumber (cm^{-1})	Reference
O—H stretch	3397	(8)
C—H stretch	2917	(8)
1st overtone of O—H bending vibration	1656	(9)
C—O stretch	1155	(8)
CH$_2$—O—CH$_2$ stretch	1091	(8)
CH$_2$—O—CH$_2$ stretch	1017	(8)

NMR chemical shifts

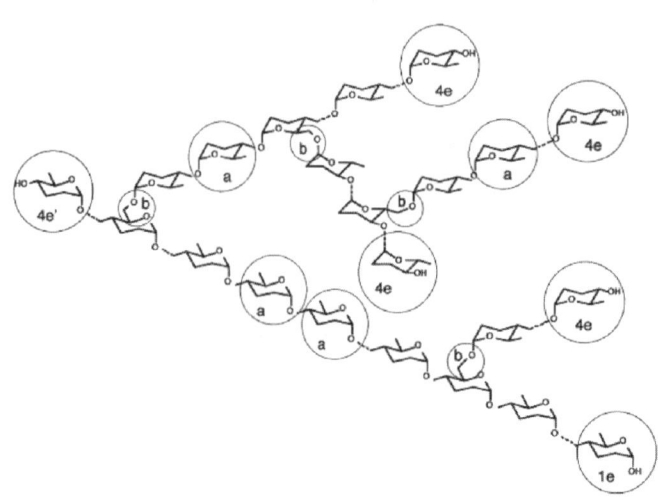

Unit	Nucleus	NMR chemical shift for DMSO -d6 at 60°C (ppm)[10–11]									
		C-1	C-2	C-3	C-4	C-5	C-6	O-2	O-3	O-4	O-6
a	^{1}H	5.11	3.33	3.68	3.36	3.62	3.64	5.25	5.23	—	4.35
	^{13}C	99.84	71.84	73.03	78.67	71.42	60.39	—	—	—	—
4e	^{1}H	5.01	3.26	3.40	3.09	3.51	3.56	5.18	4.64	4.68	4.32
	^{13}C	100.5	72.43	73.25	70.03	73.25	60.88	—	—	—	—

Radius of gyration

Solvent	M_a	Value (Å)	Reference
Water	5.741×10^6	850	(9)
0.5N NaOH	7.1×10^6	250	(12)

References

1. BeMiller JN. In: Mark HF, et al, eds. *Encyclopedia of Polymer Science and Engineering.* 2nd ed. Vol 3. John Wiley and Sons; 1989:545–551.
2. Geddes R. In: Aspinal GO, ed. *The Polysaccharides.* Vol. 3. New York: Academic Press; 1985:283–336.
3. Pfluger EFW. *Arch. Gen. Physiol.* 1909;129:362.
4. Stetten MR, Katzen HM, Setten D Jr. *J. Biol.Chem.* 1956;222:587.
5. Whistler RL, Be Miller JN. *Arch. Biochem. Biophys.* 1962;98:120.
6. Laskov R, Margoliash E. *Bull. Res. Counc. Isr.,Sect A.* 1963;11(4):351.
7. Geddes R, Harvey JD, Wills PR. *Biochem. J.* 1977;163:201.
8. Adhikary P, Tiwari KN, Singh RP. *J. Appl. Polym. Sci.* 2007;103:773.
9. Pal S, Mal D, Singh RP. *Colloids and Surfaces A: Physiochem. Eng. Aspects.* 2006;289:193.
10. Stanek M, Falk H, Huber A. *Monatshefte für Chemie.* 1998;129:355.
11. Falk H, Stanek M. *Monatshefte für Chemie.* 1997;128:777.
12. Ioan CE, Aberle T, Butchard W. *Macromolecules.* 1999;32:8655.

Hydridopolysilazane

Donna M. Narsavage-Heald

Acronym HPZ

Class Polysilazanes

Empirical Formula $(SiH)_{39.7}(Me_3Si)_{24.2}(NH)_{37.3}(N)_{22.6}$

Major Application Composites

Properties of Special Interest Preceramic polymer; melt-spinnable polymer; produces a ceramic fiber upon pyrolysis.

Preparative Technique Condensation (step) polymerization: Exothermic reaction; temperature rises to 75°C. Mixture is heated to 150°C and eventually to 200–230°C. Reaction of trichlorosilane with hexamethyldisilazane (1 : ≥3 mol ratio) in Ar purged flask at room temperature.[1]

Property	Units	Conditions	Value	Reference
Molecular weight	g mol^{-1}	GPC data	$M_n = 3,800$ $M_w = 15,100$ $M_z = 38,000$	(1)
NMR	ppm	^{1}H	0.2, broad, SiMe 1.0, broad, NH 4.8, broad, SiH	(1)
Glass transition temperature	K	TMA	368	(1)
Melt viscosity	P	Determined using a viscometer in a glove box; at 503 K	100	(1)
Pyrolyzability, nature of product	–	3°C min^{-1} to 1,200°C under high purity N$_2$	Silicon carbonitride	(1)
Pyrolyzability, amount of product	–	TGA, N$_2$ flow	74%	(1)
Pyrolyzability, impurities remaining	–	3°C min^{-1} to 1,200°C under high purity N$_2$	~5 wt% carbon, 2.2 wt% oxygen	(1)
Decomposition temperature	K	TGA	563	(1)

Hydridopolysilazane

Property	Units	Conditions	Value	Reference
Fiber spinning	–	Inert atmosphere	Fibers ≤15–20 μm obtained	–
Important patents		U.S. Patent 4,535,007		(2)
		U.S. Patent 4,543,344		(3)

References

1. Legrow GE, et al. *Am. Cer. Soc. Bull.* 1987;66(2):363–367.
2. Cannady JP. *U.S. Patent.* 4,535,007 (13 August 1985).
3. Cannady JP. *U.S. Patent.* 4,543,344 (24 September 1985).

Hydroxypropylcellulose

Yong Yang

Acronym HPC

Class Carbohydrate polymers

Structure

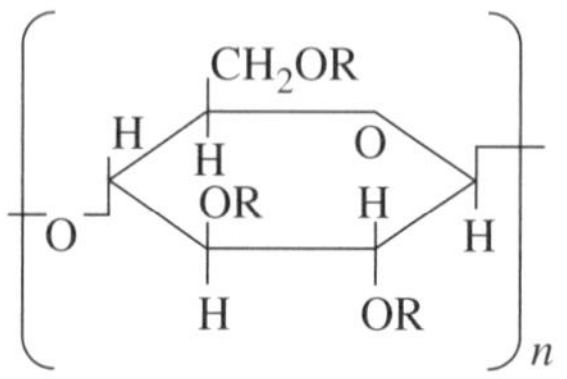

R is $CH_2CH(OR')CH_3$ or H,
$R' = R$ or H

Major Applications Lubricants for artificial eyes , pharmaceutical excipients such as tablet binder, tablet coatings, modified drug release, food additives, protective colloid for emulsion polymerization, rheology additive for coatings, inks, adhesives, cosmetics, and papers.

Properties of Special Interest Completely soluble in water and organic polar solvents such as alcohols.

Property	Units	Conditions	Value	Reference
Molecular weight of repeat unit	$g\,mol^{-1}$	Molar substitution (MS): 3.0	336	
Preparation				(1)
		Cellulose + NaOH $\rightarrow$ Na-Cellulose (Alkali cellulose)		
		Na-Cellulose + propylene oxide $\rightarrow$ Hydroxypropylcellulose		
Density	$g\,cm^{-3}$	Water cast film	1.17	(2)
NMR		^{13}C		(3)

Infrared Bands Assignment

Frequency (cm^{-1})	Assignment	
3450	(OH) side chain stretching	(2, 4)
3440	(OH) ring stretching	
2965	(CH_3) asymmetric stretching	
2930	(CH_2) asymmetric stretching	
2900	(CH_2) symmetric stretching	
2870	(CH_3) symmetric stretching and CH ring stretching	

Hydroxypropylcellulose

Property	Units	Conditions		Value	Reference
Frequency (cm^{-1})	**Assignment**				
1455	(CH$_3$) asymmetric bending deformation				
1425	(OH + CH) side chain bending deformation				
1410	(OH + CH) ring bending deformation				
1373	(CH3) symmetric bending deformation				
1324	(OH + CH) ring bending deformation				
1300	(OH + CH) side chain bending deformation				
1265	(CH) ring bending deformation				
1150	(C—O—C) ring asymmetric stretching				
1126	(C—O) side chain stretching				
1120	(C—O—C) ring asymmetric stretching				
1085	(C—O—C) side chain asymmetric stretching				
1055	(C—O) ring stretching				
1045	(C—O—C) ether bridge asymmetric stretching				

Solvents and nonsolvents

Solvents	Acetic acid*, acetone*, acetonitrile*, benzene:water (1:1), chloroform, cyclohexanone, dichloroacetice acid*, dichloromethane*, N, N-dimethylacetamide*, dimethylformamide*, dimethylsulfoxide*, dioxane*, ethanol*, ethylene glycol monomethyl ether*, formamide*, formic acid*, 2-hydroxyethyl methacrylate*, isopropanol*, methanol*, 2-methoxyethanol*, methyl ether ketone*, morphloline*, 2,2'-oxydiethanol*, 1-pentanol*, phenol*, 1,2-propanediol*, 1-propanol*, propylene glycol, pyridine*, tetrahydrofuran, triethyl phosphate*, trifluoroacetic acid*, trimethyl phosphate*, water*			(5–10)	
Nonsolvents	Aliphatic hydrocarbons, benzene, carbon tetrachloride, methyl chloroform toluene, trichloroethylene				

* Forms liquid crystalline mesophase.

Solubility parameter (δ) and interaction parameter (χ) at infinite dilution*[11,12]

Solvent	δ (MPa$^{1/2}$)	χ
Acetic acid	25.6	−2.28
Acetic anhydride	20.8	−1.65
Acetone	19.3	0.38
1-Butanol	24	0.26
n-Butyl acetate	17.2	0.14
Cyclohexane	16.3	0.96
Cyclohexanol	21.0	2.31
Cyclohexanone	20.1	0.18
n-Decane	18.8	1.83
Dichloromethane	19.8	−0.38
N, N-Dimethylformamide	29.3	−0.01
Dimethylsulfoxide	25.8	−0.19

Solvent	δ (MPa$^{1/2}$)	χ
Diethyl ether	15.1	−0.14
1,4-Dioxane	19.3	0.06
Ethanol	26	0.38
N-Heptane	14.8	0.10
Methanol	29	0.47
1-Propanol	24	0.26
2-Propanol	24	0.43
Pyridine	21.0	−0.42
Tetrachloromethane	16.8	0.45
Tetrahydrofuran	18.6	−0.12
Toluene	17.7	0.17
Trichloromethane	18.0	−0.73
Water	47	1.55

* By Inverse GC, MS: 4.0, Mw: 10^5, 323.4K.

Unit cell dimension[2]

Lattice	Monomers per unit cell	Chains per unit cell	Cell dimension (Å)		
			a	b	c
Tetragonal	6	2	11.3	11.3	15.0

Property	Units	Conditions	Value	Reference
Lower critical solution temperature (LCST)	K	Solvent: water	316–320	(22)
Mark–Houwink parameters		$[\eta] = K'_m(DP)^a$, (D P: Degree of Polymerization), ethanol, 25°C		
		K'_m	0.121	(13)
		a	1.17	(13)
Chain conformation			Irregular 3_1 helix	(2)
Degree of crystallinity	%	Water cast film	14.9	(2)
		Dioxane cast film	14.4	(21)
Heat of fusion	KJ mol^{-1} (of repeat units)	Melting point depression due to a diluent	10.6	(2)
Entropy of fusion	KJ mol^{-1} (of repeat units)	Melting point depression due to a diluent	0.021	(2)
Density (amorphous region)	g cm^{-3}	24°C	1.088	(2)
Density (crystalline region)	g cm^{-3}	MS = 4.0, calculated from crystallographic data	2.054	(2)
Glass transition temperature	K	Dynamic mechanical property measurement (DMA)	298	(14)
	K	DMA	286–288	(21)
Cross-linked with Toluene diisocyanate	K	DMA	286–306	(21)

Hydroxypropylcellulose

Property	Units	Conditions	Value	Reference
Melting point	K	MS = 4	481	(2)
Mesomeric transition temperature	K	Isotropic to cholesteric	433–473	(15)
Tensile modulus	MPa		414	(16)
		MS: 4.25, water cast	440	(17)
		MS: 4.25, MeOH cast	1240	(17)
		MS: 4.25, DMAc cast	570	(17)
			703	(8)
		MS: 4.0, lightly cross-linked, $\overline{Mc} = 1.23 \times 10^3$ g mol^{-1}	6.2×10^2	(18)
Storage modulus	MPa	288–413K, 110 Hz	$2.5 \times 10^3 - 0.3 \times 10^3$	(5)
Loss modulus	MPa	288–413K, 110 Hz	$2.6 \times 10^2 - 0.4 \times 10^2$	(5)
Tensile strength	MPa		13.8	(16)
		MS: 4.25, water cast	9.3	(17)
		MS: 4.25, MeOH cast	24	(17)
		MS: 4.25, DMAc cast	9	(17)
			16	(19)
		MS: 4.0, lightly cross-linked, $\overline{Mc} = 1.23 \times 10^3$ g mol^{-1}	16	(18)
Maximum extensibility	%		50	(20)
		Cross-head speed: 5 mm/min		
		MS: 4.25, H$_2$O cast	17.3	(17)
		MS: 4.25, MeOH cast	3.5	(17)
		MS: 4.25, DMAc cast	7.0	(17)
		Cross-head speed 2.5 mm/min		
		MS:4.0, lightly cross-linked, $\overline{Mc} = 1.23 \times 10^3$ g mol^{-1}	100	(18)
Index of refraction n			1.48	(2)
Refractive index increment dn/dc	ml g^{-1}	Ethanol, $\lambda = 546$ nm	0.120	(20)
		Water, $\lambda = 436$ nm	0.146	(20)
		$\lambda = 546$ nm	0.143	(20)
		$\lambda = 578$ nm	0.143	(20)
Dielectric constant ε''		1000 Hz, 297 K		
		101.3 kPa, 38% relative humidity	9.07	(1)
		133 Pa, 0% relative humidity	6.71	(1)
Dielectric loss ε''		1000 Hz, 297 K		
		101.3 kPa, 38% relative humidity	0.0706	(1)
		133 Pa, 0% relative humidity	0.0408	(1)
Resistivity	ohm cm^{-1}	297 K 101.3 kPa, 38% relative humidity	5×10^9	(1)
		133 Pa, 0% relative humidity	9×10^{11}	(1)
Water absorption	%	50% relative humidity, 296 K	4	(1)
		84% relative humidity, 296 K	12	(1)

References

1. Hercules. *'KLUCEL Hydroxypropylcellulose'*. Wilmington, Delaware: Hercules, Inc; 1987.

2. Samuels RJ. 'Solid-state Characterization of the Structure and Deformation Behavior of Water-soluble Hydroxypropylcellulose'. *J. Polym. Sci.* 1969;Pt. A-2, **7**: 1197.

3. Kimura K, Shigemura T, Kubo M, Maru Y. '^{13}C NMR Study of O-(2-hydroxypropyl)cellulose'. *Macromol. Chem.* 1985;**186**:61.

4. Zhbankov RG. In: Stepanov ABI, ed. *Infrared Spectra of Cellulose and its Derivatives.* New York: Consultants Bureau Pub; 1966.

5. Gray DG.'Liquid Crystalline Cellulose Derivatives'. *J. Appl. Polym. Sci. Applied. Polym. Symp.* 1983;**37**:179.

6. Fuchs O. 'Solvents and Non-Solvents for Polymers'. In: Brandrup J, Immergut EH, eds. *Polymer Handbook.* 3rd ed. New York: John Wiley & Sons; 1989:VII/379.

7. Nishio T, Yamane T, Takahashi T. Morphological Studies of Liquid-Crystalline Cellulose Derivative. I. Liquid-Crystalline Characteristics of Hydroxypropyl Cellulose in 2-Hydroxyethyl Methacrylate Solutions and in Polymer Composites Prepared by Bulk Polymerization. *J. Polym. Sci. Polym. Phys.* Ed. 1985;**23**:1043.

8. Bheda J, Fellers JF, White JL. Phase Behavior and Structure of Liquid Crystalline Solutions of Cellulose Derivatives. *Colloid & Polym Sic.* 1980;**258**:1335.

9. Werbowyj RS, Gray DG. Optical Properties of (Hydroxypropyl)cellulose Liquid Crystals. Cholesteric Pitch and Polymer Concentration. *Macromolecules.* 1984; **17**:1512.

10. Werbowyj RS, Gray DG. Ordered Phase Formation in Concentrated Hydroxypropylcellulose Solutions. *Macromolecules.* 1980;**13**:69.

11. Barton AFM. Hydroxypropyl Cellulose. *CRC handbook of Polymer-Liquid Interaction and Solubility Parameters.* Boca Raton, Florida: CRC Press; 1990.

12. Aspler JS, Gray DG. Gas Chromatographic and Static Measurements of Solute Activity for a Polymeric Liquid-Crystalline Phase. *Macromolecules.* 1979;**12**:5626; Interaction of Organic Vapors with Hydroxypropylcellulose. *Polymer.* 1982;**23**:43.

13. Gröbe A. Properties of Cellulose materials. In: Brandrup J, Immergut EH. eds. *Polymer Handbook.* 3rd ed. New York: John Wiley & Sons; 1989:V/117.

14. Suto S, Kudo M, Karasawa M. Static Tensile and Dynamic Mechanical Properties of Hydroxypropylcellulose Films Prepared under Various Conditions. *J. Appl. Polym. Sci.* 1986;**31**:1217.

15. Shimaura K, White JL, Fellers JF. Hydroxypropylcellulose, a Thermotropic Liquid Crystalline Characteristics and Structure Development in Continuous Extrusion and Melt Spinning. *J. Appl. Polym. Sci.*1981;**26**:2165.

16. Just EK, Magewicz TG. Cellulose Ethers. In: Mark HF, Bikales NM, Overberger CG, Menges G, Kroschwitz JI, eds. *Encyclopedia of Polymer Science and Engineering.* New York: Wiley-Interscience; 1985:V.3.

17. Suto S, Tashiro H, Karasawa M. Preparation of Chemically Cross-linked Hydroxypropyl Cellulose Solid Films Retaining Cholesteric Liquid Crystalline Order. *J. Appl. Polym. Sci.* 1992;**45**:1569.

18. Yang Yong Ph.D. *Thesis.* Ohio: University of Cincinnati; 1993.

19. Yanajida N, Matsuo M. Morphology and Mechanical Properties of Hydroxypropyl Cellulose Films Crosslinked in Solution. *Polymer*, 1992;**33**(5):996.

20. Huglin MB. Specific Refractive Index Increments of Polymers in Dilute Solution. In: Brandrup J, Immergut, EH, eds. *Polymer Handbook.* 3rd ed. New York: John Wiley & Sons; 1989:VII/466.

21. Rials TG, Glasser WG. Thermal and Dynamic Mechanical Properties of Hydroxypropyl Cellulose Films. *J. Appl. Polym. Sci.* 1988;**36**:749.

22. Ichikawa H, Fukumori Y. Negatively Thermosensitive Release of Drug from Microcapsules with Hydroxypropyl Cellulose Membranes Prepared by the Wurster Process. *Chem. Pharm. Bull..* 1999;**47**(8):1102.

Kevlar

Brent D. Viers

Acronym, Alternative Names PPTA, poly(*p*-phenylene terephthalamide), aramid, aramide, polyaramid, polyaramide

IUPAC Nomenclature Poly(imino-1,4-phenyleneiminocarbonyl-1,4-phenylenecarbonyl)

CAS Registry Number 24938-64-5

Class Aromatic polyamides

Structure

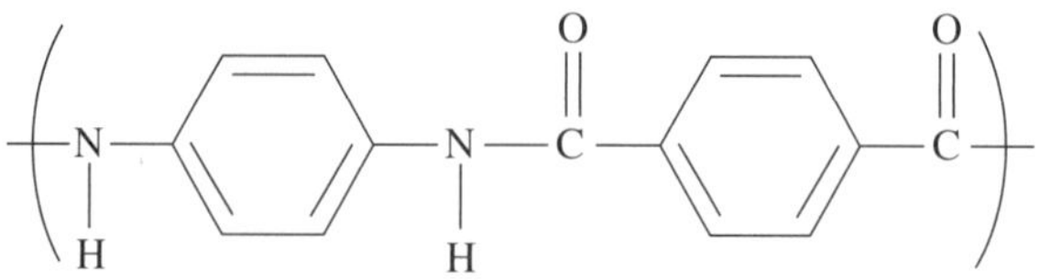

Major Applications Cut, heat, and bullet-fragment resistant apparel, brake and transmission friction parts, gaskets, ropes and cables, composites, fiber-optic cables, circuit-board reinforcement, sporting goods, tires, automotive belts and hoses.

Major Forms Continuous filament yarn, staple, wet and dry pulp floc, cord.

Properties of Special Interest High tensile strength at low weight, low elongation to break, high modulus (structural rigidity), low electrical conductivity, high chemical resistance, low thermal shrinkage, high toughness (work-to-break), excellent dimensional stability, high cut resistance, flame resistant, self-extinguishing.

Other Polymer Showing This Special Property Polybenzamide. (See also the entry on *Polybenzamide* in this handbook.)

Supplier Kevlar is a registered trademark of E. I. Dupont de Nemours.

Preparative techniques

Property	Conditions	Value	Reference
Condensation of terephthaloyl chloride and paraphenylene diamine	Interfacial polymerization Low temperature solution condensation		(1)
Direct syntheses	Yamazaki procedure (PBA): Para-aminobenzoic acid (pABA) Pyridine (as acid scavenger) (Py) N-methyl pyrrolidone (NMP)	$[pABA] = 0.75$ mol l^{-1} NMP/Py $= 3$ (v/v) 3.7% LiCl$_2$ (w/v) $[TPP]/[pABA] = 0.6$	(2)

Property	Conditions	Value	Reference
	Triphenyl phosphate activator (TPP) Lithium chloride (LiCl) Dry conditions/inert atmosphere	$T = 115°C$ (higher M_w when TPP added stepwise)	
Direct syntheses	Yamazaki procedure (PPTA) Terephalic acid (TA) p-Phenylene diamine (PPD) NMP, TPP, LiCl	$[TA] = [PPD] = 0.125 \text{ mol } l^{-1}$ NMP/Py = 1.5 (v/v) $[TPP]/[TA] = 2.0$ 2.7% LiCl, $T = 115°C$	(2)
	Higashi Procedure (PBA): As Yamazaki, with calcium chloride (CaCl$_2$)	$[pABA] = 0.27 \text{ mol } l^{-1}$ NMP/Py = 5(v/v) $[TPP]/[pABA] = 0.6$ 1.7% LiCl (w/v); 5.0% CaCl$_2$ (w/v) $T = 115°C$ (Higher M_w when TPP added stepwise)	
	Higashi procedure (PPTA): As Yamazaki, with CaCl$_2$	$[TA] = [PPD] = 0.083 \text{ mol } l^{-1}$ NMP/Py = 5 (v/v) $[TPP]/[TA] = 2.2$ 1.7% LiCl (w/v); 5.0% CaCl$_2$ (w/v) $T = 115°C$	

Effect of salt in the synthesis of PBA[2]

LiCl$_2$(% w/v)	CaCl$_2$(% w/v)	LiCl + CaCl$_2$ (% w/v)	η_{inh} (dl g^{-1})
3.7	0.0	3.7	2.00*
3.7	11.3	15.0	2.15*
0.9	2.8	3.7	0.85*
1.6	4.8	6.4	1.53†
1.4	0.0	1.4	1.82†
6.7	0.0	6.7	2.19†

* [pABA] = 0.75; NMP/Py = 3; [TPP]/[pABA] = 0.6.
† [pABA] = 0.27; NMP/Py = 5; [TPP]/[pABA] = 0.6; $T = 115°C$.

Effect of salt in the synthesis of PPTA[2]

LiCl$_2$(% w/v)	CaCl$_2$(% w/v)	LiCl + CaCl$_2$ (% w/v)	η_{inh} (dl g^{-1})
2.7	0.0	2.7	0.32*
0.7	2.0	2.7	0.21*
0.0	3.5	3.5	0.24*
5.5	0.0	5.5	1.22†
6.7	0.0	6.7	1.16†
0.0	6.7	6.7	1.26†
1.7	5.0	6.7	6.84†

* [TA] = 0.125; NMP/Py = 1.5; [TPP]/[TA] = 2.0.
† [TA] = 0.083; NMP/Py = 5; [TPP]/[TA] = 2.0.

Effect of reactant ratios in the synthesis of PBA[2]

[TPP]/[pABA]	η_{inh} (dl g^{-1})
0.4	0.31*
0.6	2.00*
0.8	71.80*
1.0	1.26*
0.4	0.10†
0.6	1.53†
0.8	1.38†
1.0	1.40†

* Yamakazi conditions: [pABA] = 0.75; NMP/Py = 3;
  3.7% LiCl.
† Higashi conditions: [pABA] = 0.27; NMP/Py = 5;
  1.7% LiCl; 5.0% CaCl$_2$; T = 115°C.

Effect of reactant ratios in the synthesis of PPTA[2]

[TPP]/[TA]	η_{inh} (dl g^{-1})
1.3	0.24*
1.7	0.36*
2.0	0.31*
2.3	70.38*
2.0	6.84†
2.2	8.15†
2.4	6.89†

* Yamakazi conditions: [TA] = 0.125; NMP/Py = 1.5;
  2.7% LiCl; T = U5°C.
† Higashi conditions: [TA] = 0.083; NMP/Py = 5;
  1.7% LiCl; 5.0% CaCl$_2$.

Effect of temperature in the synthesis of PBA[2]

Temperature (°C)	η_{inh} (dl g^{-1})*
105	1.70
110	1.66
115	2.19
120	1.53

* Yamakazi conditions: [pABA] = 0.27; NMP/Py = 5;
  [TPP]/[pABA] = 0.6; 6.7% LiCl.

Effect of temperature in the synthesis of PPTA[2]

Temperature (°C)	η_{inh} (dl g^{-1})
100	0.28*
110	0.36*
115	0.31*
120	0.37*

Temperature (°C)	η_{inh} (dl g^{-1})
107	7.71[†]
115	8.15[†]
122	6.27[†]

* Yamakazi conditions: [TA] = 0.125; NMP/Py = 1.5,
  [TPP]/[TA] = 2.0, 2.7% LiCl.
[†] Higashi conditions: [TA] = 0.083; NMP/Py = 5,
  [TPP]/[TA] = 2.2; 1.7% LiCl 5.0% CaCl$_2$.

Property	Units	Conditions	Value	Reference
Molecular weight of repeat unit	g mol^{-1}	Poly(*p*-phenylene terephthalamide)	240.2	–
Typical polydispersity index ($M_z : M_w : M_n$)	g mol^{-1} ratios	–	$>4.5 M_z : 1.6\, M_w : 1\, M_n$	(3)
			$>5.3 M_z : 1.57 M_w : 1 M_n$	(3)
			$M_w : M_n = 1.85$ ($M_w = 12,300$)	(4)
			$M_w : M_n = 1.63$ ($M_w = 6,300$)	(4)
			$M_w : M_n = 1.37$ ($M_w = 5,300$)	(4)
Morphology in multiphase systems	–	Composites	Rods (in weaves, fibers, etc.)	–
Raman (characteristic absorption frequencies)	cm^{-1}	Kevlar 29, 49 fibers		(5)
		NC torsion, CC out of plane bending	92 (m)	
		CC in plane bending; ring torsion	106 (m)	
		NH out of plane bending; NC torsion	225 (w)	
		CC in plane bending; CO out of plane bending	265 (w)	
		CC out of plane bending; ring asymmetric torsion	414 (w)	
		CC ring in plane deformation	629 (w)	
		CH out of plane deformation, CO bending	695 (w)	
		Amide V	710 (w)	
		CO in plane bending; ring asymmetric CH deformation; CN stretching	733 (w)	
Raman (characteristic absorption frequencies)	cm^{-1}	CH out of plane deformation; CCC ring puckering deformation	788 (m)	(5)

Property	Units	Conditions	Value	Reference
Raman (characteristic absorption frequencies)	cm^{-1}	1:4 substituted ring deformation	815 (vw)	
		CH out of plane deformation; ring CC stretching, ring bending, and ring torsion	843 (w)	
		Ring out of plane bending	915 (m)	
		ω_4 ring and ring CH deformation	1,182 (m)	
		ω_4 ring and ring CH in plane deformation	1,190 (sh)	
		NH bending, CH stretching	1,277 (s)	
		Ring CH bending	1,328 (s)	
		ω_3 symmetric ring puckering/aromatic CH in plane bending	1,412 (vw)	
		ω_{14} ring vibration; ring CH bending	1,517 (m)	
		Amide II vibration, $\delta(NH)$ and $\nu(CN)$	1,569 (w)	
		ω_2 (aromatic ring) CC stretching vibration	1,611 (vs)	
		Amide 1 (C=O) stretching	1,647 (m)	

Raman depolarization ratios[5]

$\Delta\nu$ (cm^{-1})	Kevlar $\rho_\perp$	Kevlar 29 $\rho_\perp$	Kevlar 49 $\rho_\perp$
1,182	0.54 ± 0.02	0.56 ± 0.02	0.30 ± 0.01
1,190	0.55 ± 0.02	0.56 ± 0.02	0.32 ± 0.01
1,277	0.57 ± 0.01	0.56 ± 0.01	0.32 ± 0.01
1,328	0.56 ± 0.01	0.54 ± 0.01	0.30 ± 0.01
1,412	0.55 ± 0.05	0.57 ± 0.05	–
1,517	0.57 ± 0.02	0.54 ± 0.02	0.32 ± 0.01
1,569	0.64 ± 0.05	0.62 ± 0.05	0.33 ± 0.04
1,611	0.53 ± 0.01	0.55 ± 0.01	0.30 ± 0.01
1,647	0.53 ± 0.02	0.56 ± 0.02	0.30 ± 0.01

Property	Units	Conditions	Value	Reference
X-ray photo-electron (XPS)	eV	Kevlar 29, 49 fibers	C 1s = 284.6 (intense)	(6)
			O 1s = 530.3 (intense)	
			N 1s = 399.7 (intense)	
		Valence band XPS is more sensitive to surface functionalized species, although the surface appears to be identical to the bulk	C(KVV) = 990 (weak Auger)	
			N(KVV) = 873 (weak Auger)	
			O(KVV) = 745 (weak Auger)	

Property	Units	Conditions		Value		Reference
Bragg spacings	–			2θ (degree) ($\lambda = 0.1542$ nm)	Intensity	(7)
		hkl	d-value (nm)			
		110	0.4327	20.53	vs	
		200	0.3935	22.60	vs	
		020	0.2590	34.63	vw	
		310	0.2340	38.46	m	
		220	0.2163	41.75	w	
		011	0.4807	18.46	vw	
		111	0.4102	21.66	ms	
		211	0.3045	29.33	s	
		021	0.2539	35.35	w	
		121	0.2417	37.20	vw	
		311	0.2302	39.12	vw	
Thermal expansion coefficients	K^{-1}	Kevlar 29 fiber ASTM D3379-75e axial thermal expansion coefficient		-3.2×10^{-6} $-2 \times 10^{-6} < \alpha < -4 \times 10^{-6}$		(8)
Solvents		Concentrated H_2SO_4 Polar aprotic solvents (NMP/DMAc) w ~5 wt% $LiCl_2$				–
Nonsolvents		Aromatics, aliphatics, water, alcohols, ethers, esters				–

Chemical resistances[9]

Chemical*	Conc. (%)	Temp. (°C)	Time (h)	Effect[†]
Acids				
Acetic	99.7	21	24	None
Acetic	40	21	1,000	Slight
Acetic	40	99	100	Appreciable
Benzoic	3	99	100	Appreciable
Chromic	10	21	1,000	Appreciable
Formic	90	21	100	None
Formic	40	21	1,000	Moderate
Formic	90	99	100	Degraded
Hydrobromic	10	21	1,000	Appreciable
Hydrochloric	37	21	24	None
Hydrochloric	10	21	100	Appreciable
Hydrochloric	10	71	10	Degraded
Hydrofluoric	10	21	100	None
Nitric	1	21	100	Slight
Nitric	10	21	100	Appreciable
Nitric	70	21	24	Appreciable
Nitric	70	99	100	Appreciable
Oxalic	10	99	100	None
Phosphoric	10	21	100	None
Phosphoric	10	21	1,000	Slight
Phosphoric	10	99	100	Appreciable
Salicylic	3	99	1,000	None

Chemical*	Conc. (%)	Temp. (°C)	Time (h)	Effect[†]
Acids				
Sulfuric	10	21	1,000	Moderate
Sulfuric	10	21	100	None
Sulfuric	10	100	10	Appreciable
Sulfuric	70	21	100	Moderate
Bases				
Ammonium hydroxide	28.5	21	24	None
Ammonium hydroxide	28	21	1,000	None
Potassium hydroxide	50	21	24	None
Sodium hydroxide	40	21	100	None
Sodium hydroxide	10	21	1,000	None
Sodium hydroxide	10	99	100	Degraded
Sodium hydroxide	10	100	10	Appreciable
Sodium hypochlorite	0.1	21	1,000	Degraded
Salt solutions				
Copper sulfate	3	21	1,000	None
Copper sulfate	3	99	100	Moderate
Ferric chloride	3	99	100	Appreciable
Sodium chloride	3	21	1,000	None
Sodium chloride	10	99	100	None
Sodium chloride	10	121	100	Appreciable
Sodium phosphate	5	99	100	Moderate
Organic solvents				
Carbon tetrachloride	100	Boiling	100	Moderate
Ethylene glycol/H_2O	50/50	99	1,000	Moderate
Brake fluid	100	113	100	Moderate

* Chemicals not listed in the table have no noticeable effect.

[†] Effect on breaking strength: none = 0–10% stress loss; slight = 11–20% stress loss; moderate = 21–40% stress loss; appreciable = 41–80% stress loss; degraded = 81–100% stress loss.

Property	Units	Conditions	Value	Reference
Phase diagrams		Solid/anisotropic solution/isotropic solution regimes		(10, 11)
Fractionation	–	Chromatography		
		90% H_2SO_4	Preparative GPC, silica gel	(12)
		96% H_2SO_4	GPC	(13)
		Tetrahydrofuran	GPC, shodex	(13)
Mark–Houwink parameters: K and a	$K = ml\,g^{-1}$ $a = $ None	(a values greater than 1.7 indicate strong aggregation effects) $1.0 < a < 1.85$		(14)
		PBA $3,100 < M_w < 13,000$	$K = 2.14$	
		Concentrated H_2SO_4, 25°C	$a = 1.203$	

Property	Units	Conditions	Value	Reference
Persistence length	Å	96% sulfuric acid, 25°		
		Electric birefringence, Kerr effect	PPTA = 300	(15)
		Light scattering, R_g	PBA = 400	(12)
			PPTA = 200	(12)
		Light scattering, depolarization ratio	PPTA = 150	(16)
		Light scattering depolarization	PPTA = 287	(17)
		Light scattering	PPTA = 450	(18)
		Light scattering	PBA = 600	(13)
			PPTA = 200	(13)
		Viscosity	PBA = 400	(13)
			PPTA = 150	(13)
		Flow birefringence	PBA = 1050	(19)
			PPTA = 650	(19)
		Flow birefringence	PBA = 435	(20)
			PPTA = 275	(20)
		Flow birefringence	PBA = 325	(21)
		Flow birefringence	PPTA = 185	(22)
		Depolarization ratio, unfractionated	PPTA = 1020	(14)
		Depolarization ratio, unfractionated oligomer $M_w < 10,900$	PPTA = 306	(14)
Viscosity vs. shear rate	poise	Kevlar-100% H_2SO_4 solutions, 25°C		(1)
		0.5 wt% Kevlar, 10^{-1} s^{-1} < γ < 10 s^{-1}	$\eta = 2.1$	
		6–8 wt% Kevlar, 10^{-2} s^{-1} < γ < 10^1 s^{-1}	$\eta = 1,100$	
		10 wt% Kevlar		
		$\gamma = 10^{-2}$ s^{-1}	$\eta = 30,000$	
		$\gamma = 10^{-1}$ s^{-1}	$\eta = 6,000$	
		$\gamma = 10^0$ s^{-1}	$\eta = 800$	
		$\gamma = 10^1$ s^{-1}	$\eta = 300$	
Viscosity vs. shear stress	poise	Kevlar-100% H_2SO_4 solutions, 25°C 6–8 wt% Kevlar		(1)
		10^1 dynes cm^{-2} < σ_{12} < 10^4 dynes cm^{-2}	$\eta = 1,100$	
		10 wt% Kevlar		
		$\sigma_{12} = 5 \times 10^2$ dynes cm^{-2}	$\eta = 40,000$	
		$\sigma_{12} = 10^3$ dynes cm^{-2}	$\eta = 4,000$	
		$\sigma_{12} = 3 \times 10^3$ dynes cm^{-2}	$\eta = 200$	
Normal stress vs. shear rate	dynes cm^{-2}	Kevlar-100% H_2SO_4 solutions, 25°C		(1)
		6 wt% Kevlar		
		$\gamma = 2$ s^{-1}	$N_1 = 3,000$	
		$\gamma = 8$ s^{-1}	$N_1 = 10^5$	
		8 wt% Kevlar		
		$\gamma = 1$ s^{-1}	$N_1 = 5,000$	
		$\gamma = 8$ s^{-1}	$N_1 = 10^5$	
		10 wt% Kevlar		
		$\gamma = 0.7$ s^{-1}	$N_1 = 2 \times 10^5$	
		$\gamma = 1$ s^{-1}	$N_1 = 5 \times 10^4$	
		$\gamma = 8$ s^{-1}	$N_1 = 8 \times 10^4$	

Property	Units	Conditions	Value	Reference
Bond lengths	Å	C(1)–C(2)	1.47	(23)
		C(2)–C(3)	1.39	
		C(3)–C(4)	1.39	
		C(1)–O(1)	1.24	
		C(1)–N(1)	1.34	
		N(1)–H(1)	1.00	
		N(1)–C(8)	1.42	
		C(8)–C(9)	1.39	
		C(9)–C(10)	1.39	
		C–H(phenyl)	1.00	
Bond angles	Degrees	C(4)–C(3)–C(2)	120	(24)
		C(3)–C(2)–C(7)	120	
		C(3)–C(2)–C(7)	120	
		C(7)–C(2)–C(1)	120	
		C(2)–C(1)–N(1)	120	
		C(2)–C(1)–O(1)	120	
		N(1)–C(1)–O(1)	115	
		C(1)–N(1)–H(1)	123	
		C(8)–N(1)–H(1)	120	
		N(1)–C(8)–C(9)	117	
		N(1)–C(8)–C(13)	120	
		C(9)–C(18)–C(13)	120	
		C(8)–C(9)–C(10)	120	
Torsional potential diagram	–	–	–	(23)
Persistence length	Å	Extended all *trans* conformation (upper bound)		(23)
		PBA (no temp, dependence)	900	
		PPTA (no temp, dependence)	410	
		Rotatable amide group (lower bound)		
		PBA at 200 K	200	
		PBA at 600 K	~0	
		PBA at 300 K	100	
		PPTA at 200 K	200	
		PPTA at 600 K	~0	
		PPTA at 300 K	100	
Maximum temperature for liquid crystallinity	K	–	600	(23)
Lattice	–	–	Monoclinic (pseudo orthorhombic)	(24)
Space group	–	–	$P2_{1/n}$–C_{2h}	(24)
Chain conformation	–	Extended *trans*	Modification1,2	(24)
Unit cell dimensions	Å	–	$a = 7.80$, $b = 5.19$, c (fiber axis) $= 12.9$	(24)

Property	Units	Conditions	Value		Reference
Unit cell angles	Degrees	–	$\gamma = 90$		(24)
Unit cell contents (number of repeat units)	–	–	2 chains/ cell		(24)
Degree of crystallinity	%	Kevlar 49 H_2SO_4 cast film Annealed 100°C, 2 h Annealed 200°C, 2 h Annealed 300°C, 2 h	 0.22 0.38 0.45		(25)
Polymorphs	–	Modification I (PBA-LiCl$_3$-DMAc) Modification II (PBA-LiCl$_2$ cocrystal) Modification III (PBA) Lyotropic nematic			(26)
Glass transition temperature	K	–	698		(27)
Melting point	K	In general beyond decomposition temperature (500°C) Modification III(PBA)	 827		(26)
Super-T_g transition temperatures	K	Modification I (PBA-LiCl$_3$-DMAc) Modification II (PBA-LiCl$_2$ cocrystal) Modification III (PBA)	1–II = 487 K I-amorphous = wash with H_2O II–III = anneal at 748 K, then cool II–III = wash with H_2O I–III = wash with H_2O and anneal >673 K		(26)
Super-T_g transition temperatures	K	Modification III (PBA)	II-nematic = 748 K III–nematic = 827 III-amorphous = heat to 873 K and cool		(26)
Sub-T_g transition temperatures	–	$f = 10{,}000$ Hz $f = 10{,}000$ Hz $f = 110$ Hz $f = 110$ Hz $f = 110$ Hz $f \sim 1$ Hz $f \sim 1$ Hz $f \sim 1$ Hz $f \sim 1$ Hz	$T_\gamma = 291$ K $T_{\gamma*} = 417$ K $T_\beta = 733$ K $T_\gamma = 333$ K $T_\delta = 243$ K $T_\beta = 816$ K $T_\gamma = 235$ K $T_{\gamma*} = 440$ K $T_\delta = 115$ K	$E_{a,\gamma} = 63$ kJ mol^{-1} $E_{a,*^*} = 92$ kJ mol^{-1} $E_{a,\beta} = 767$ kJ mol^{-1} $E_{a,\gamma} = 204$ kJ mol^{-1} $E_{a,\delta} = 52$ kJ mol^{-1} $E_{a,\beta} = 813$ kJ mol^{-1} $E_{a,\gamma} = 54$ kJ mol^{-1} $E_{a,\gamma*} = 83$ kJ mol^{-1} $E_{a,\delta} = 21$ kJ mol^{-1}	(28) (28) (29) (29) (30) (30) (30) (30) (30)

Property	Units	Conditions	Value	Reference
Polymers with which they are compatible	–	None known. Surface modifications for composites.		–
Tensile modulus	MPa	Ultimate Modulus D	$\sim\!165 \times 10^3$	(9)
		Kevlar 29 fiber	83×10^3	
		Kevlar 49 fiber	124×10^3	
		Kevlar 149 fiber	161×10^3	
		Twaron LM fiber	76×10^3	
		Twaron HM fiber	105×10^3	
Crystal modulus	MPa	Kevlar, Twaron fibers XRD	156×10^3	–
		Kevlar, Twaron fibers	220×10^3	
Shear modulus	MPa	Kevlar fibers in tension and compression	1,150	–
Storage modulus	MPa	Kevlar/100% H_2SO_2 solutions		
		8 wt% Kevlar		
		$\omega = 0.02$ Hz	$G^{|} = 1 \times 10^{-9}$	(1)
		$\omega = 0.1$ Hz	$G^{|} = 2 \times 10^{-9}$	
		$\omega = 0.5$ Hz	$G^{|} = 1 \times 10^{-8}$	
		$\omega = 1$ Hz	$G^{|} = 2 \times 10^{-8}$	
		$\omega = 5$ Hz	$G^{|} = 7 \times 10^{-8}$	
		$\omega = 10$ Hz	$G^{|} = 2 \times 10^{-7}$	
		10 wt% Kevlar		
		$\omega = 0.01$ Hz	$G^{|} = 4 \times 10^{-7}$	
		$\omega = 0.1$ Hz	$G^{|} = 5 \times 10^{-7}$	
		$\omega = 0.5$ Hz	$G^{|} = 6 \times 10^{-7}$	
		$\omega = 1$ Hz	$G^{|} = 7 \times 10^{-7}$	
		$\omega = 5$ Hz	$G^{|} = 1 \times 10^{-6}$	
		$\omega = 10$ Hz	$G^{|} = 2 \times 10^{-6}$	
Loss tangent vs. frequency	–	Kevlar/100% H_2SO_4 solutions		(1)
		8 wt% Kevlar		
		$\omega = 0.02$ Hz	$\tan \delta = 2$	
		$\omega = 0.05$ Hz	$\tan \delta = 3$	
		$\omega = 0.1$ Hz	$\tan \delta = 4$	
		$\omega = 0.5$ Hz	$\tan \delta = 5$	
		$\omega = 1$ Hz	$\tan \delta = 6$	
		$\omega = 5$ Hz	$\tan \delta = 5$	
		$\omega = 10$ Hz	$\tan \delta = 3$	
		10 wt% Kevlar		
		$\omega = 0.01$ Hz	$\tan \delta = 0.3$	
		$\omega = 0.1$ Hz	$\tan \delta = 0.4$	
		$\omega = 1$ Hz	$\tan \delta = 0.6$	
		$\omega = 10$ Hz	$\tan \delta = 0.9$	
Tensile strength	MPa	LC solution spun Kevlar fibers	2,000–3,000	–

Property	Units	Conditions	Value	Reference
Maximum extensibility $(L/L_0)_r$	%	Kevlar 29 fiber in tension	4.0	(27)
		Kevlar 49 fiber in tension	2.5	(9)
Fracture stress	MPa	Fiber in tension		–
		Kevlar 29 fiber	2,500	
		Kevlar 49 fiber	2,300	
		Kevlar 149 fiber	1,700	
		Twaron LM fiber	3,400	
		Twaron HM fiber	2,800	
Fracture strain	%	Fiber in tension		–
		Kevlar 29 fiber	2.5	
		Kevlar 49 fiber	1.8	
		Kevlar 149 fiber	1.0	
		Twaron LM fiber	2.4	
		Twaron HM fiber	2.5	
Compressive strength		Four point bend of a fiber embedded in a PMMA matrix		–
Tenacity (fiber)	MPa	Kevlar 49	2,800	(27)
		Kevlar 29	2,800	
Poisson ratio	–	–	0.36	(9)
Force-temperature relationships	–	Kevlar 49 fibers in ASTM D3379-75e; force-temperature cycling 5 gpd applied load-heat to 300°C, cool to ambient	Critical temp, for stress-drop decreases from ~493 K to 198 K as 1.0 gpd stress applied	(8)
Thermal expansivity		Axial expansivity, Kevlar fiber		(6)
		At 200 K	−0.8	
		At 450 K	−0.7	
Index of refraction n	–	$n_\perp$ = index of refraction perpendicular to fiber axis Kevlar 29	2.0499 (center of fiber) 2.0853 (fiber edge)	(31)
		$n_\parallel$ = index of refraction parallel to fiber axis	1.5886 (center of fiber) 1.6504 (edge of fiber)	
Refractive index increment dn/dc	$ml\,g^{-1}$	All values at 25°C using a 633 nm source		
		Chlorosulfonic acid	0.275	(32)
		Chlorosulfonic acid + 0.01 M $LiClSO_3$	0.287	(33)
		H_2SO_4	0.278	(32)
		96% H_2SO_4	0.309 (546 nm source)	(13)
		Methane sulfonic acid	0.254	(33)
Segmental polarizability $(\alpha_1 - \alpha_2)$, $(\alpha_\parallel - \alpha_\perp)$	cm^3	Sulfuric acid (includes form effect)	$(\alpha_1 - \alpha_2) = +5,250 \times 10^{-25}$ $(\alpha_\parallel - \alpha_\perp) = +206 \times 10^{-25}$	(34, 35)

Property	Units	Conditions	Value	Reference
		Copolymer with poly(benzamide) 1/9 PPTA/PBA ratio	$(\alpha_1 - \alpha_2)$ $= +4,380 \times 10^{-25}$	
Segmental optical anisotropy δ_0^2	–	Light depolarization, unfractionated polymer	PPTA $= 0.266$	(14)
		Light depolarization, unfractionated oligomers, $M_w < 10,900$	PPTA $= 0.357$	
Optical anisotropy Δ^2	–	Depolarization ratio, fractionated PPTA		(16)
		$M = 1,560$ g mol^{-1}	0.290	
		$M = 2,160$ g mol^{-1}	0.223	
		$M = 2,760$ g mol^{-1}	0.184	
		$M = 3,480$ g mol^{-1}	0.154	
		$M = 4,560$ g mol^{-1}	0.30	
		$M = 6,600$ g mol^{-1}	0.103	
		$M = 7,920$ g mol^{-1}	0.094	
		Unfractionated PPTA		
		$M_w = 2,160$ g mol^{-1}	0.270	(16)
		$M_w = 4,320$ g mol^{-1}	0.177	(16)
		$M_w = 1,680$ g mol^{-1}	0.294	(13)
		$M_w = 4,500$ g mol^{-1}	0.183	(13)
		$M_w = 9,350$ g mol^{-1}	0.164	(13)
		$M_w = 19,700$ g mol^{-1}	0.111	(13)
		$M_w = 35,000$ g mol^{-1}	0.105	(13)
		$M_w = 43,500$ g mol^{-1}	0.094	(13)
		$M_w = 63,000$ g mol^{-1}	0.084	(13)
Surface free energy	mJ m^{-1}	γ_s^d dispersive	40 ± 4	–
Heat of adsorption	kJ mol^{-1}	IGC adsorption		–
		Epoxystyrene on Kevlar 29	45 ± 3	
		Aniline on Kevlar 29	11	
Specific free energy of interaction	kJ mol^{-1}	ΔG_{sp}		–
		Epoxystyrene on Kevlar 29	5.6	
		Aniline on Kevlar 29	11	
Heat of hydration	kJ mol^{-1}	ΔH_H of Kevlar 29 Fiber	-60	–

Permeability coefficient (Kevlar 49 film, H_2SO_4 cast)*(25)

	Annealed 2 h at			
Gas (at 35°C)	200°C	100°C	300°C	Amorphous Kevlar/Nomex copolymer
H_2	10,000	10,000	6,000	–
He	11,500	11,000	10,100	–
CO_2	1,020	1,020	500	–

Kevlar

Gas (at 35°C)	Annealed 2 h at			Amorphous Kevlar/Nomex copolymer
	200°C	100°C	300°C	
O_2	220	220	80	2,579
N_2	20	35		–

* Values given in $cm^2(STP)\,cms^{-1}\,cm^{-2}\,cmHg^{-1}$ $(\times 10^{-15})$.

Property	Units	Conditions	Value	Reference
Temperature dependence of permeability coefficient P	$cm^2(STP)\,cms^{-1}$ $cm^{-1}\,cmHg$	Linear (Arrhenius) decay for H_2 and CO_2 Amorphous Kevlar/Nomex copolymer		(25)
		25°C	5.8×10^{-12}	
		65°C	0.9×10^{-11}	
		Kevlar 49 H_2SO_4 cast film Annealed 100°C, 2 h		
		25°C	8×10^{-13}	
		65°C	1.5×10^{-12}	
		Annealed 200°C, 2 h		
		25°C	1×10^{-12}	
		65°C	3×10^{-12}	
		Annealed 300°C, 2 h		
		25°C	3×10^{-13}	
		65°C	1×10^{-12}	
Arrhenius activation energy for permeability coefficient E_P	$kcal\,mol^{-1}$	Carbon dioxide Amorphous Kevlar/Nomex copolymer	5.5	(25)
		Kevlar 49 H_2SO_4 cast film		
		Annealed 100°C, 2 h	5.6	
		Annealed 200°C, 2 h	6.3	
		Annealed 300°C, 2 h	5.9	
Diffusion coefficient	$cm^2\,s^{-1}$	Water into Kevlar 29 fiber	0.95×10^{-12}	(36)
		Kevlar 29 H_2SO_4 film	(See table below)*	(25)

Gas	Kevlar 49 film, annealed 2 h at			Amorphous Kevlar/Nomex copolymer
	100°C	200°C	300°C	
H_2	~200	~200	~200	–
He	~800	~800	~800	–
CO_2	0.4	0.42	0.2	–
O_2	0.81	1.10	0.62	9.26
N_2	0.18	0.22	–	–

* Values given in $cm^2\,s^{-1}\,(\times 10^{10})$.

Property	Units	Conditions	Value	Reference
Temperature dependence of diffusion coefficient D	$cm^2\,s^{-1}$	Linear (Arrhenius) Relationship for H_2 and CO_2 Amorphous Kevlar/Nomex copolymer		(25)
		25°C	2×10^{-10}	
		65°C	1.62×10^{-9}	
		Kevlar 49 H_2SO_4 cast film Annealed 100°C, 2 h		
		25°C	2×10^{-11}	
		65°C	1×10^{-10}	
		Annealed 200°C, 2 h		
		25°C	2.2×10^{-11}	
		65°C	1.3×10^{-10}	
		Annealed 300°C, 2 h		
		25°C	8×10^{-12}	
		65°C	5×10^{-11}	
Arrhenius activation energy for diffusion coeffient E_D	$kcal\,mol^{-1}$	Carbon dioxide		–
		Amorphous Kevlar/Nomex copolymer	10.5	
		Kevlar 49 H_2SO_4 cast film		
		Annealed 100°C, 2 h	9.2	
		Annealed 200°C, 2 h	9.4	
		Annealed 300°C, 3 h	10.1	
Heat of sorption ΔH_s	$kcal\,mol^{-1}$	Carbon dioxide		(25)
		Amorphous Kevlar/Nomex copolymer	−5.0	
		Kevlar 49 H_2SO_4 cast film		
		Annealed 100°C, 2 h	−3.6	
		Annealed 200°C, 2 h	−3.1	
		Annealed 300°C, 2 h	−4.2	
Thermal conductivity	$W\,m^{-1}\,K^{-1}$	Phonon propagation in Kevlar 49	10	(37)
	$W\,m^{-1}\,K^{-2}$	5–250 K	$dK/dT = 1$	(37)
	$W\,m^{-1}\,K^{-1}$	Axial thermal conductivity, 125–250 K	20–30	(38)
Biodegradability, effective microorganisms	–	Degradation by A. flavus	Kevlar 29 degrades more than Kevlar 49	(39)
Degradation mechanisms		UV radiation	Critical UV window = 300–500 nm	(9)
		UV reduction of M_n	Photolytic-degradation kinetics	(40, 41)
		Hydrolytic	Concentrated H_2SO_4	(42)
			Humidity, temperature	(43)
		Atomic oxygen/UV	UV resistance mechanism	(44)
		Laser	488 nm Ar ion laser beam	(45)

Property	Units	Conditions	Value	Reference
		Photolytic	Simulated sunlight	(46)
		Thermal	Radical homolytic	(47, 48)
			High pressure	(49)
			ESR radical study	(50)
		Photochemical	Smog, ozone, temp., RH	(51)
			Oxidation in H_2SO_4	(52, 53)
Maximum use temperature	K	In air	573–623	(9)
Decomposition temperature	K	In air	700–755	(9)
Decomposition products	K	H_2, CO, CO_2, HCN, H_2O, benzene, toulene, benzonitrile	573–773	(9, 54)
		CO_2, H_2O, CO	643–723	(9, 55)
		Benzene, HCN, benzonitrile, H_2	723–823	(9, 55)
Heat of combustion	$J\,kg^{-1}$	–	35×10^6	(9)
Limiting oxygen index	–	–	29	(9)
Chemical resistance		Aqueous chlorine solutions	Kevlar degrades Nomex resists degradation	(57)
Crystallinity	%	Kevlar 29	72.2	(66)
		Kevlar 49	73.4	
		Kevlar 149	91.0	
Water sorption hysteresis	%	Kevlar 49 Polymerized PPTA	>30%sat. R.H. (water in microvoid) <30%sat. R.H (intercalated water)	(68)
Water regain	%	65% RH	3.9 (Kevlar 29) 3.7 (Kevlar 49) 0.96 (Kevlar 149)	(67)
		95% RH	6.09 (Kevlar 29) 5.79 (Kevlar 49) 2.34 (Kevlar 149)	
Surface bifurcations	nm	Inverse gas chromatography and AFM	500 nm microfibrils (50 nm height pleating)	(78)
Thermally induced structural changes		400, 440, 470 °C Kevlar 29	Shell cross-linking Core hydrogen bond disruption	(69)
Fiber morphology		Synchrotron SAXS along fiber axis	Rotational Domain Disorder between skin and core	(70)

Property	Units	Conditions	Value	Reference
Compressive strength	GPa °C^{-1}	Temperature dependence	$8.2 \times 10^{-4} - 1.2 \times 10^{-3}$	(75)
Fiber modulus	GPa	Skin/Core comparisons	Kevlar 49 Core $= 60.8$ Skin $= 13.4$	(76)
Degradation mechanisms		Simulated low earth orbit	Atomic oxygen resistance	(58)
		Simulated low earth orbit	Combined VUV and atomic oxygen degradation	(59)
	°C	$T < T_{decomposition}$	Decomposition mechanisms after thermal aging at 150–450°C	(60)
		$T \sim T_{decomposition}$	Mechanisms at 500–550°C	(77)
		Master curve for creep Time-Temperature superposition	Predictions for stress to rupture	(61)
	kJ mol^{-1}	E_a for thermal degradation	Freeman Carrol method	(62)
		100% RH (inert and oxidative atm)	110	(65)
			133 (air)	(81)
		High resolution TGA	154 (nitrogen)	
		Thermooxidative dimensional stability	Stable until 450°C. General stability trends: polyimide > heterocyclic para-aramid and polyoxadiazole > carbocyclic para-aramid > meta-aramid	(63)
	life	Torsional fatigue	UHMW-PE > Kevlar 129 > Kevlar 29 > Twaron 2000.	(64)
	life	Torsional rupture	UHMW-PE > Kevlar 129 > Kevlar 29 > Twaron 2000	(64)
		Environmental exposure	Trends for thermal, ultrasonic, and chemical exposure	(73)
		Fracture strength reduction	Vacuum vs humid environments	(74, 79)
		Hydrolysis	Tensile strength reductions in either NaOH or H$_2$SO$_4$ solutions	(80)

Kevlar

Property	Units	Conditions	Value	Reference
Tenacity	GPa	Kevlar 119	2.96 ± 0.09	(71)
		Kevlar 29	2.58 ± 0.07	
		Kevlar 49	2.40 ± 0.07	
		Kevlar 149	2.15 ± 0.06	
		Annealing effects		(72)
Modulus	GPa	Kevlar 119	61 ± 1	(71)
		Kevlar 29	71 ± 1	
		Kevlar 49	113 ± 2	
		Kevlar 149	138 ± 2	
		Annealing effects		(72)
Breakage strain	%	Kevlar 119	4.1 ± 0.1	(71)
		Kevlar 29	3.1 ± 0.1	
		Kevlar 49	2.47 ± 0.1	
		Kevlar 149	1.5 ± 0.1	
		Annealing effects		(72)
Energy to break	J	Kevlar 119	1.27 ± 0.03	(71)
		Kevlar 29	0.95 ± 0.03	
		Kevlar 49	1.23 ± 0.04	
		Kevlar 149	0.36 ± 0.01	
		Annealing effects		(72)
Lattice constants	Å	Kevlar 119	$a = 7.750 \pm 0.003$	(71)
		Kevlar 29	$a = 7.748 \pm 0.003$	
		Kevlar 49	$a = 7.784 \pm 0.003$	
		Kevlar 149	$a = 7.904 \pm 0.003$	
		Annealing effects		(72)
Lattice constants	Å	Kevlar 119	$b = 5.224 \pm 0.003$	(71)
		Kevlar 29	$b = 5.232 \pm 0.003$	
		Kevlar 49	$b = 5.232 \pm 0.003$	
		Kevlar 149	$b = 5.188 \pm 0.003$	
		Annealing effects		(72)
Lattice constants	Å	Kevlar 119	$c = 12.82 \pm 0.01$	(71)
		Kevlar 29	$c = 12.84 \pm 0.01$	
		Kevlar 49	$c = 12.88 \pm 0.01$	
		Kevlar 149	$c = 12.92 \pm 0.01$	
		Annealing effects		(72)
Paracrystalline parameter	%	Kevlar 119	$g_{II} = 1.91 \pm 0.02$	(71)
		Kevlar 29	$g_{11} = 1.92 \pm 0.02$	
		Kevlar 49	$g_{11} = 1.66 \pm 0.04$	
		Kevlar 149	$g_{11} = 1.40 \pm 0.03$	
		Annealing effects		(72)
Equatorial X-ray crystallinity	%	Kevlar 119	75.20	(71)
		Kevlar 29	75.70	
		Kevlar 49	77.00	
		Kevlar 149	77.00	
		Annealing effects		(72)

Property	Units	Conditions	Value	Reference
Apparent crystal size	nm	Kevlar 119 (002 hkl)	609.19	(71)
		Kevlar 29 (002 hkl)	656.14	
		Kevlar 49 (002 hkl)	737.15	
		Kevlar 149 (002 hkl)	1547.79	
		Annealing effects		(72)
Apparent crystal size	nm	Kevlar 119 (110 hkl)	50.00	(71)
		Kevlar 29 (110 hkl)	52.00	
		Kevlar 49 (110 hkl)	66.00	
		Kevlar 149 (110 hkl)	123.00	
		Annealing effects		(72)
Apparent crystal size	nm	Kevlar 119 (200 hkl)	45.00	(71)
		Kevlar 29 (200 hkl)	46.00	
		Kevlar 49 (200 hkl)	51.00	
		Kevlar 149 (200 hkl)	76.00	
		Annealing effects		(72)
Crystal orientation angle	°	Kevlar 119 (200 hkl)	16.2 ± 0.02	(71)
		Kevlar 29 (200 hkl)	12.2 ± 0.03	
		Kevlar 49 (200 hkl)	6.8 ± 0.01	
		Kevlar 149 (200 hkl)	6.4 ± 0.02	
		Annealing effects		(72)

References

1. Aoki H, et al. *J. Polym. Sci., Polym. Sym. (Rigid Chain Polym.: Synth. Prop.).* 1978;65:29–40.
2. Mariani A, Mazzanti SLE, Russo S. *Can. J. Chem.* 1995;73(11):1960–1965.
3. Chu B, et al. *Polym. Commun.*. 1984;25(7):211–213.
4. Ogata N, Sanui K, Kitayama S. *J. Polym. Sci., Polym. Chem. Ed.*. 1984; 22(3):863–867.
5. Edwards HGM, Hakiki S. *Br. Polym. J.* 1989;21(6):505–512.
6. Xie Y, Sherwood PMA. *Chem. Mater.* 1993;5(7):1012–1017.
7. Northolt MG. *Eur. Polym. J.* 10 (1974):799.
8. Pottick LA, Farris RJ. *Polym. Prepr. (Am. Chem. Soc., Div. Polym. Chem.).* 1984;25(2):209–210.
9. Kevlar Technical Guide. http://www.dupont.com.
10. Salaris F, et al. *Makromol. Chem.* 1976;177(10):3073–3076.
11. Lin J, Wu H, Li S. *Eur. Polym. J.* 1994;30(2):231–234.
12. Arpin M, Strazielle C. *Makromol. Chem.* 1976;177:581.
13. Arpin M, Strazielle C. *Polymer.* 1977;18:591.
14. Zero K, Aharoni SM. *Macromolecules.* 1987;20:1957–1960.
15. Tsvetkov VN. *Polym. Sci. USSR. (Engl. Trans.).* 1979;21:2879.
16. Arpin M, et al. *Polymer.* 1977;18:262.
17. Ying Q, Chu B. *Makromol. Chem. Rapid. Commun.* 1984;5:785.
18. Cotts PM, Berry GC. *J. Polym. Sci., Polym. Phys. Ed.*. 1983;21:1255.
19. Tsvetkov VN. *Polym. Sci. USSR. (Engl. Trans.).* 1977;19:2485.

20. Tsvetkov VN, Andreeva LN. *Adv. Polym. Sci.*. 1981;39:27.

21. Pogodin NV, Bogatova IN, Tsvetkov VN. *Polym. Sci. USSR (Engl. Trans.)*. 1985;27:1574.

22. Arpin M, Debeauvais F, Strazielle C. *Makromol. Chem.* 1976;177:585.

23. He C, Windle AH. *Macromol. Theory Simul.* 1995;4(2):289–304.

24. Chatzi EG, Koenig JL. *Polym.-Plast. Technol. Eng.*. 1987;26(3–4):229–270.

25. Weinkauf DH, Kim HD, Paul DR. *Macromolecules*. 1992;25(2):788–796.

26. Takase M, et al. *J. Polym. Sci., Part B: Polym. Phys.*. 1986;24(8):1675–1682.

27. Fitzgerald JA, Irwin RS. *Spec. Publ.: High Value Polym. (R. Soc. Chem.)*. 1991;87:392–419.

28. Frosini V, Butta E. *J. Polym. Sci., Polym. Lett.*. 1971;9:253.

29. Kunugi T, Watanabe H, Hashimoto M. *J. Appl. Polym. Sci.*. 1979;24:1039.

30. Badayev AS, Perepechko II, Sorokin YY. *Polym. Sci. USSR*. 1988;30:892.

31. Warner SB. *Macromolecules*. 1983;16:1546–1548.

32. Cotts PM, Berry GC. *J. Polym. Sci. Polym. Phys. Ed.*. 1983;21:189.

33. Wong CP, Ohnuma H, Berry GC. *J. Polym. Sci., Polym. Symp.*. 1978;65:173.

34. Tsvetkov VN. *Rigid Chain Polymer Molecules*. Nauka, Moscow: 1985.

35. Pogodina NV, et al. *Vysokomol. Soedin.*. 1981;23A:2185.

36. Rebouillat S, et al. *J. Appl. Polym. Sci.* 1995;58(8):1305–1315.

37. Poulaert B, et al. *Polym. Commun.*. 1985;26(5):132–133.

38. Choy CL, et al. *J. Polym. Sci., Part B: Polym. Phys.*. 1995;33(14):2055–2064.

39. Watanabe T. *Sen'i Gakkaishi*. 1991;47(8):439–441.

40. Knoff WF. *J. Appl. Polym. Sci.*. 1994;52(12):1731–1737.

41. Harris GG. *J. Ind. Fabr.*. 1982;1(1):18–28.

42. Morgan RJ, Butler NL. *Polym. Bull. (Berlin)*. 1992;27(6):689–696.

43. Morgan RJ, et al. In: *Proceedings of the 29th National SAMPE Symposium Exhib. (Technol. Vectors)*. Reno, Nev., 3–5 April 1984. Covina, Calif: Society for Advancement of Material and Process Engineering; 1984:891–900.

44. Powell SC, et al. *Polym. Prepr. (Am. Chem. Soc, Div. Polym. Chem.)*. 1991;32(1):122–123.

45. Young RJ, Lu D, Day RJ. *Polym. Intl.*. 1991;24(2):71–76.

46. Toy MS, Stringham RS. *ACS Symp. Ser. (Chem. React. Polym.)*. 1988;364:326–341.

47. Schulten HR, et al. *Angew. Makromol. Chem.* 1987;155:1–20.

48. Brown JR, Power AJ. *Polym. Degrad. Stab.* 1982;4(5):379–392.

49. Brown JR, Hodgeman DKC. *Polymer* 1982;23(3):365–368.

50. Brown JR, et al. *Text. Res. J.*. 1983;53(4):214–219.

51. Mead JW, et al. *Ind. Eng. Chem. Prod. Res. Dev.*. 1982;21(2): 158–163.

52. Toy MS. Stringham RS. *Polym. Prepr. (Am. Chem. Soc., Div. Polym. Chem.)*.1986;27(2):83–4.

53. Toy MS, Stringham RS. *Polym. Mater. Sci. Eng.*. 1986;54:312–315.

54. Krasnov Ye-P, et al. *Polym. Sci. USSR*. 1966;8:413.

55. Krasnov Ye-P, et al. *Vysokomolekul. Soedin.*. 1970;8:380.

56. Andrews MC, Lu D, Young RJ. *Polymer*. 1997;38(10):2379–2388.

57. Akdag Akin, Kocer, Hasan B, et al. *Journal of Physical Chemistry B*. 2007;111(20):5581–5586.

58. Bitetti G, Mileti S, Marchetti M, Micciche P, European Space Agency, [Special Publication] SP. SP-616(10th International Symposium on Materials in a Space Environment, 2006). 2006:44.

59. Ghosh Lipika, Fadhilah Mohammad Harris, Kinoshita Hiroshi, Ohmae Nobuo. *Polymer*. 2006;47(19):6836–6842.

60. Iyer RV, Sudhakar A,Vijayan Kalyani. *High Performance Polymers*. 2006;18(4):495–517.

61. Alwis KGNC, Burgoyne CJ,Atkins WS. *Applied Composite Materials* 2006;13(4):249–264.

62. Liu Xiaoyan, Yu Weidong. Journal of Applied Polymer Science. 2006;99(3):937–944.

63. Perepelkin KE, Pakshver EA, Andreeva IV, Malan'ina OB, Makarova RA, Oprits ZG. *Fibre Chemistry*. 2005;37(5):346–351.

64. Liu Xiaoyan, Weidong Yu, Weidong Yu. *High Performance Polymers*. 2005;17(4):593–603.

65. Bernstein Robert, Derzon Dora K, Shedd Michelle M Polymer Preprints (American Chemical Society, Division of Polymer Chemistry. 2005;46(2):676–677.

66. Yang HH. *Kevlar Aramid Fiber*. Chichester: Wiley-Interscience Publications; 1993.

67. Fukuda M, Ochi M, Miyagawa M, Kawai H. *Textile Research Journal*. 1991;61:668–680.

68. Mooney DA, McElroy JMD. *Chemical Engineering Science*. 2004;59(11):2159–2170.

69. Downing James W Jr, Newell James A. *Journal of Applied Polymer Science*. 2004;91(1):417–424.

70. Roth SM, Burghammer A, Janotta, Riekel C, *Macromolecules*. 2003;36(5):1585–1593.

71. Rao Y, Waddon AJ, Farris RJ. *Polymer*.2001;42(13):5937–5946.

72. Rao Y, Waddon AJ, Farris RJ, *Polymer*. 2001;42(13):5925–5936.

73. Vijayan Kalyani. *Metals, Materials and Processes*. 2000;12(2&3):259–268.

74. Minoshima Kohji, Maekawa Yoshihiro, Komai Kenjiro. *International Journal of Fatigue*. 2000;22(9):757–765.

75. Lacks DJ. *Materials Letters*. 2000;44(1):12–13.

76. Graham JF, McCague C, Warren OL, Norton PR, *Polymer*. 2000;41(12):4761–4764.

77. Iyer RV, Vijayan Kalyani. *Bulletin of Materials Science*. 1999;22(7):1013–1023.

78. Rebouillat S, Peng JCM, Donnet JB. *Polymer*. 1999;40(26):7341–7350.

79. Minoshima K, Tsuru K, Komai K. *Structures and Materials*. 1998;1(Damage and Fracture Mechanics):595–604

80. Lin Jeng-Shyong, Chiu Hsien-Tang. *Polymers & Polymer Composites*. 2001;9(4):239–245.

81. Li Xin-Gui, Huang Mei-Rong. *Journal of Applied Polymer Science*. 1999;71(4):565–571.

82. Rebouillat S, Donnet JB, Guo H, Wang TK. Maydown Research Centre, DuPont (UK) Ltd., Londonderry, UK. *Journal of Applied Polymer Science* 1998;67(3):487–500.

Kraton D1100 SBS Block Copolymers

C.M. Roland

Chemical Name Styrene-butadiene-styrene triblock copolymer

Structure

$$[-CH-CH_2-]_m-[-CH_2-CH=CH-CH_2-]_n-[-CH-CH_2-]_m$$
$$\qquad | \qquad\qquad\qquad\qquad\qquad\qquad\qquad | $$
$$\qquad C_6H_5 \qquad\qquad\qquad\qquad\qquad\qquad C_6H_5$$

Class Unsaturated thermoplastic elastomer

Other Manufacturers or Suppliers of SBS Block Copolymers Chevron Phillips, Dexco Polymers, Polimeri Europa, BASF, Solvent Uritem, Vita Thermoplastic Polymers, Consolidated Polymer Technologies, Denka, J-Von, Chi Mei Ind., En Chuan Chemical Ind., Firestone Polymers, Elastron Kimya, PolyOne

Major Applications Asphalt modifiers, polymer and thermoset modifiers, adhesives, sealants, coatings, molded, injected molded, and extruded items, general compounding (automotive parts, sporting goods, medical items, etc.)

Properties of Special Interest Thermoplastic elastomers (TPE) provide the mechanical properties of rubber in combination with the processing characteristics of plastics; recyclable; Kraton Ds are low-cost TPE

Supplier Manufacture of Kraton D began in 1964 by the Shell Chemical Co. In 2001 the business was sold and Kraton D is now produced by Kraton Polymers LLC. Copolymers of styrene-isoprene-styrene versions are also marketed as Kraton D(IR). The 1000 series of Kraton D are pure triblock polymers, the 1400 are diblock styrene-butadiene polymers, the 2000 series are compounded materials, and the 4000 series oil-extended.

Kraton	D1101	D1102	D1118	D1133	D1153	D1155	D1192	Ref.
S/B (wgt. %)	31/69	28/72	33/67	36/64	29/71	29/71	30/70	1,2,3
Physical form	Pellet, powder	Pellet	Pellet, powder	Pellet, powder	Pellet	Pellet	Pellet, powder	2,3
Specific gravity	0.94	0.94	0.94	0.94	0.94	0.96	0.94	3
Viscosity (25% in toluene at 25°C)	3.9 Pa s	1.2 Pa s	0.63	4.8	1.7	0.6	1.5	2,3

Kraton	D1101	D1102	D1118	D1133	D1153	D1155	D1192	Ref.
Shore A hardness	69	66	64	74	70	87	66	1,3
300% Modulus	2.8	2.8	1.2	—	2.9	2.9	—	1,4
Tensile strength	32	32	1.7	—	2.8	2.8	—	1,4
Elongation	880%	880%	600%	—	800%	800%	—	3,4,5
Permanent set at break (%)	10%	10%	40%	—	—	—	—	3
Melt index (ASTM D1238)	< 1	6–14	10	< 1	3	14	< 1	1,2,3,5

D1101 (ref. 6)	Solvent Interaction Parameter		
Solvent	308 K	328 K	348 K
Benzene	0.414	0.400	0.423
Toluene	0.363	0.355	0.348
Ethyl benzene	0.372	0.379	0.360
p-Xylene	0.433	0.429	0.409
Chloroform	0.126	0.252	0.297
Cyclohexane	0.597	0.555	0.522
Hexane	0.773	0.785	0.711
Heptane	0.812	0.777	0.762
MEK	1.16	1.17	1.17

Chemical resistance (ref. 7)

Acids	Bases	Aromatics	Aliphatics	Oil in water	Water in oil
Good	Good	None	None	Good	None

Gas permeability of D1101 (ref. 8)

Property	Units	Conditions	Value
Permeability coefficient	SI	O_2	2.0×10^{-12}
		CO_2	8.0×10^{-12}
Transmission rate	cm^2/s	O_2	2.0×10^{-7}
		CO_2	7.8×10^{-7}

Water permeability of D1101 (ref. 8)

Property	Units	Value
Permeability coefficient	SI	2.7×10^{-10}
Transmission rate	$g\ cm^{-2}\ s^{-1}$	3.0×10^{-8}

Wet chemical identification of Kraton D (ref. 9)

Step	Observed Color
Pyrolysis vapors passed into Burchfield reagent	Yellow green
Methanol added and boiled	Green

Effect of solvent on D1101 viscosity (ref. 10)

Solvent	MIBK	Toluene	Tetralin	o-Xylene	Cyclohexane
Solubility parameter (cal/ml)½	8.35	8.59	8.76	9.03	9.62
Intrinsic viscosity (dl/g)	0.31	1.04	1.17	0.91	0.44

Mechanical properties of Kraton D/Asphalt Blends (ref. 11)

Kraton D1101 (wgt. %)	67	50
Asphalt (wgt. %)	33	50
300% Modulus (MPa)	1.8	0.8
Elongation (%)	1700	1500
Tensile strength (MPa)	17.5	9.7
Shore A hardness	46	36
Permanent set (%)	50	30

Softening point of Kraton D1101 in
Asphalt/oil blends (ref. 11)

Kraton D1101 (wgt %)	Softening point (°C)
0	38
2	62
4	74
6	81
8	87
10	92
12	98

Adhesion of 10% Kraton D1101 in
Asphalt/oil blends (ref. 11)

Adherend	180 Peel strength (lbs/in)
Itself	5.5
Smooth plywood	6.8
Ground steel	5.6
Concrete	7.0
Galvanized iron	5.4

Kraton D blends with HIPS (ref. 12)

HIPS (wgt. %)	100	90	90
Kraton D1101 (wgt. %)	—	10	—
Kraton 1102 (wgt. %)	—	—	10
1/8" Notched Izod (N)	85	150	120
Flex modulus (GPa)	2100	1800	1900

Kraton D1102 (ref. 13)

M_w (kg/mol)	M_n (kg/mol)	cis-1,4 units*	trans-1,4 units*	1,2 units*	χ_{SB}[†]
68	57	47%	53%	10%	$6.59 \times 10^{-3} + 3.6/T$

* Butadiene block, [†] Styrene-butadiene interaction parameter.

Phase dissolution temperature (ref. 14)

D1101	130°C
D1102	117°C

D1102 (ref. 6)	Tear energy (KJ/m^2)			
Casting solvent	8.3 μm/s	83 μm/s	830 μm/s	8300 μm/s
Cyclohexane	21.6 ± 13.0	18.3 ± 8.7	17.3 ± 8.4	14.5 ± 4.8
Toluene	25.4 ± 7.2	21.3 ± 7.0	19.3 ± 6.7	16.1 ± 4.0
THF/MEK (90/10)	33.8 ± 11.0	32.6 ± 12.0	31.0 ± 14.0	29.0 ± 6.0

Acknowledgement

This work was supported by the Office of Naval Research.

References

1. Wilder CR. In: Bhowmick AK, Stephens HL, eds. *Handbook of Elastomers*. New York: Marcel Dekker; Chapter 9. 1988.
2. Shell Technical Bulletin #SC 1434-96. March 1996; #SC1158-93, February 1996.
3. Kraton Polymers and Compounds. www.kraton.com/content/includes/Kraton%20Typical%20Properties%20Guide.pdf. 2006.
4. Shell Technical Bulletin #SC0068-96. January, 1997.
5. Holden G. Thermoplastic Elastomers. In: Holden G, Legge NR, Quirk R, Schroeder HE, eds. 2nd ed. New York: Hanser; Chapter 16. 1996.
6. I Hadl Romdhane, Plana A, Hwang S, Danner RP, *J. Appl. Polym. Sci.* 1992;45:2049.
7. Shell Technical Bulletin #SC519-93. August, 1993.
8. Shell Technical Bulletin #SC941-87. July, 1994.
9. Braun D. Identification of Plastics. 3rd ed. New York: Hanser; 1996.
10. Paul DR, St. Lawrence JE, Troell JH. *Polym. Eng. Sci.* 1970;10:70.
11. Shell Technical Bulletin #SC0057-84. July 1984.
12. Shell Technical Bulletin #SC0165-93. July 1994.
13. Sakurai S, Mori K, Okawara A, Kimishima K, Hashimoto T. *Macromolecules.* 1992;25:2679.
14. Nakijima N. *Rubber Chem. Technol.* 1996;69:73.

Kraton G SEBS Block Copolymers

C.M. Roland

Chemical Name   Linear styrene-(ethylene-butylene)-styrene triblock copolymer

Structure $[-CH-CH_2-]_m-[[-CH_2-CH]_x-[CH-CH_2]_y]_n-[-CH-CH_2-]_m$
$$C_6H_5 \qquad\qquad CH_2\text{-}CH_3 \qquad C_6H_5$$

Class   Saturated thermoplastic elastomer

Other manufacturers/suppliers/compounders: Vita Thermoplastic Polymers, TechnoCompound, Kraiburg TPE, PolyOne, GLS, Dexco Polymers, Teknor, Kuraray

Major Applications   Asphalt modifiers, adhesives, sealants, coatings, waterproofing applications, footwear, polymer modifiers, oil gels

Properties of Special Interest: In general, thermoplastic elastomers (TPE) provide the mechanical properties of rubber in combination with the processing characteristics of plastics; recyclable; Kraton Gs are low-cost TPE with oxidative and thermal stability, good weathering, and UV and ozone resistance

Kraton G1650

Property	Units	Conditions	Value	Reference
Specific gravity	—	—	0.91	1
T_g (EB block)	°C	—	−60	2
M_w (EB block)	g/mol	—	54 000	3
M_w (S block)	g/mol	—	10 000 (×2)	3

Kraton	G1641	G1650	G1651	G1652	G1654	G1657	Ref.
Styrene (wgt %)	34	30	33	30	31	13	1,4,5
Physical form	Powder	Powder	Powder	Powder	Powder	Pellet	4,5
Specific gravity	0.92	0.91	0.91	0.91	0.91	0.89	
Viscosity (10% in toluene at 25°C)	80 mPa s	50 mPa s	1800 mPa s	30 mPa s	410 mPa s	65 mPa s	4,5
Hardness (Shore A)	52	72	61	70,77	63	47	1,5
Melt index 230°C, 5 kg (g/min)	<0.1	<0.1	<0.1	0.5	<0.1	2.2	1,4,5
300% Modulus (Mpa)	4.2	3.8, 5.5	—	4.8, 5.5		2.4	1,5
Elongation (%)	>800	500	>800	500	>800	750	1,5
Tensile strength (MPa)	>17	>27	>27	27, 31	>27	23	1,5

Kraton G SEBS Block Copolymers

Chemical resistance (ref. 6)

Acids	Bases	Aromatics	Aliphatics	Oil in water	Water in oil
Good	Good	Poor	Poor/fair	Good	Poor/fair

Gas permeability of Kraton G (ref. 7)

Property	Units	Resin	Conditions	Value
Permeability coefficient	SI	G1650	O_2	1.1×10^{-12}
			CO_2	4.4×10^{-11}
		G1651	O_2	1.0×10^{-11}
			CO_2	2.9×10^{-11}
		G1652	O_2	1.3×10^{-12}
			CO_2	4.4×10^{-11}
Transmission rate	cm^2/s	G1650	O_2	1.1×10^{-7}
			CO_2	2.7×10^{-7}
		G1651	O_2	9.8×10^{-8}
			CO_2	2.8×10^{-7}
		G1652	O_2	1.2×10^{-7}
			CO_2	3.9×10^{-7}

Water permeability of Kraton G (ref. 7)

Property	Units	Resin	Value
Permeability coefficient	SI	G1650	5.8×10^{-11}
		G1651	6.6×10^{-11}
		G1652	8.7×10^{-11}
Transmission rate	g/cm^2-s	G1650	6.4×10^{-9}
		G1651	7.3×10^{-9}
		G1652	9.7×10^{-9}

Viscosity of Kraton G1650 in toluene solutions (ref. 8)

Toluene (by wgt.)	75%	80%	85%	90%	95%
Viscosity (Pa-s)	6.8×10^{-1}	1.8×10^{-1}	3.8×10^{-2}	5.6×10^{-3}	7.5×10^{-4}

Viscosity of G1650 solutions (15% solids) (ref. 8)

Solvent	Isobutyl isobutyrate	Ethyl benzene	Cyclohexane	Methyl-N-amyl-ketone	Toluene
Viscosity (Pa-s)	1.4×10^{-1}	4.8×10^{-1}	3.6×10^{-2}	1.5×10^{-3}	4.3×10^{-2}

Kraton G SEBS Block Copolymers

Resin and oil compatibility with EB segment (ref. 9)

Polyterpenes	Incompatible
Hydrogenated resin esters	Incompatible
Saturated hydrocarbon resins	Compatible
Naphthenic oils	Incompatible
Paraffinic oils	Compatible
Low molecular weight polybutenes	Compatible
Aromatic resins	Incompatible

Kraton G/Polypropylene blends (ref. 10)

Polypropylene (wgt. %)	100	90	80	90	80
Kraton G1650 (wgt. %)	—	10	20	—	—
Kraton G1652 (wgt. %)	—	—	—	10	20
1/8" Notched Izod (N)	48	64	690	75	520
Flex Modulus (GPa)	1500	1200	940	1000	890

Effect of SEBS level on failure of PET/HDPE 50/50 blends (ref. 11)

% SEBS	Modulus (MPa)	Yield (MPa)	Elongation (%)
0	1300	26	3
5	1200	23	40
10	920	20	130
20	650	18	(no break)

Kraton G modification of mopping Asphalts (ref. 12)

	Type III Asphalt	Kraton G Modified Asphalt
Cold bond (ASTM D5147–91)	$+15°C$	$-20°C$
Elongation (%)	100	1000
Tensile strength (MPa)	0.21	0.69
Puncture sealing	Poor	very good/fast
Ring and ball softening point	$89°C$	$106°C$

Gel–Sol transition temperature—G1650/magnetite in medium viscosity paraffin oil (ref. 13)

Fe_3O_4 (wgt %)	G1650 concentration (wgt %)				
	1.7	3.8	7.0	10.6	14.5
0	$46.6°C$	$50.4°C$	$61.8°C$	$72.4°C$	$80.6°C$
22	$48.4°C$	$49.2°C$	$54.8°C$	$69.6°C$	$81.8°C$
42	$59.0°C$	$66.6°C$	$77.0°C$	$83.8°C$	$91.5°C$

Acknowledgement

The work was supported by the Office of Naval Research.

References

1. Wilder CR. In: Bhowmick AK, Stephens HL, eds. *Handbook of Elastomers*. New York: Marcel Dekker; Chapter 9. 1988.
2. Holden G. Thermoplastic Elastomers. In Holden G, Legge NR, Quirk R, Schroeder HE, eds. 2nd ed. New York: Hanser; Chapter 16. 1996.
3. Yoshimura DK, Richards WD. *Modern Plastics*. 64: March 1987.
4. Shell Technical Bulletin #SC 1434-96. March 1996; #SC1158-93. February 1996.
5. Kraton Polymers and Compounds. www.kraton.com/content/includes/Kraton%20Typical%20Properties%20Guide.pdf. 2006.
6. Shell Technical Bulletin #SC519-93. August, 1993.
7. Shell Technical Bulletin #SC941-87. July, 1994.
8. Shell Technical Bulletin #SC0072-85. July, 1994.
9. Holden G.Thermoplastic Elastomers. In Legge NR, Holden G, Schroeder HE, eds. 1st ed. New York: Hanser; Chapter 13. 1987.
10. Shell Technical Bulletin #SC0165-93. July 1994.
11. Paul DR. Thermoplastic Elastomers. In Holden G, Legge NR, Quirk R, Schroeder HE, eds. 2nd Ed. New York: Hanser; Chapter 15C. 1996.
12. Shell Technical Bulletin #SC01810-94. July 1994.
13. Lattermann G, Krekhova M. *Macro. Rapid Comm.* 2006;27:1373.

Metallophthalocyanine Polymers

Martel Zeldin and Yuli Zhang

Class Cofacial polymers

Structure

$[M(Pc)O]_n$:

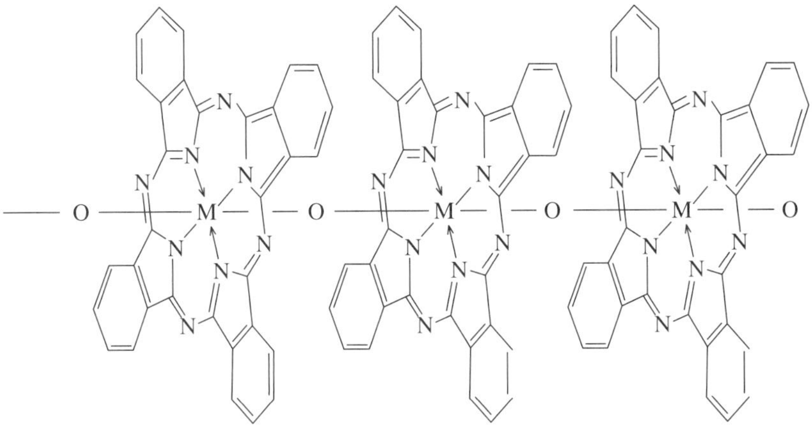

where M = Si, Ge, or Sn; Pc = phathalocyanine.

$[M'(Pc')L]_n$:

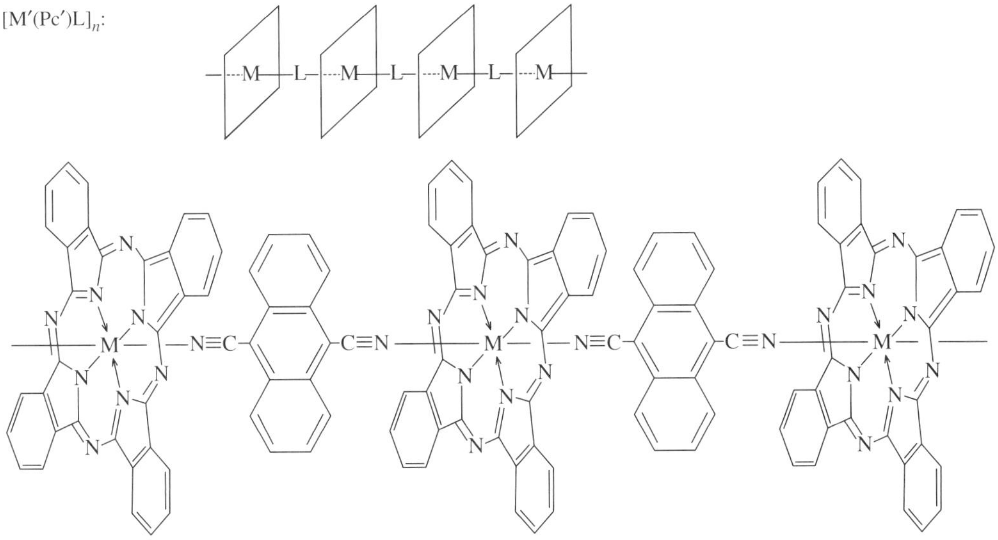

where $M' = Fe^{2+}$, Fe^{3+}, Co^{2+}, Co^{3+}, Ru^{2+}, Mn^{2+}, Mn^{3+}, or Cr^{3+}.

L =

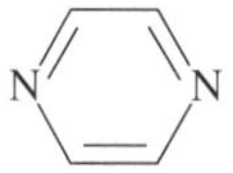

Pyrazine (pyz)

CN—⟨⟩—CN   *p*-Diisocyanobenzene (dib)

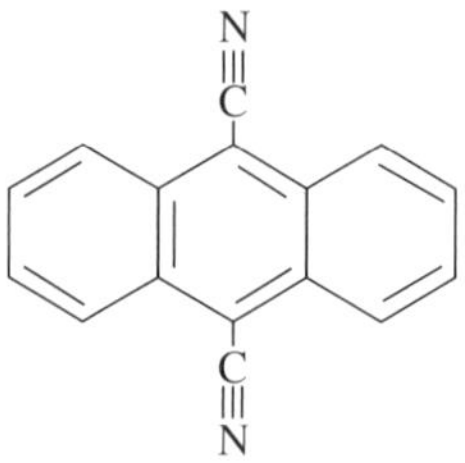

9,10-Diisocyanoanthracene (9,10-dia)

CN^{1-}, or SCN^{1-}.

$Pc' = Pc^{2-}$, R_4Pc^{2-}, R_8Pc^{2-}, $1,2\text{-}Nc^{2-}$ (1,2-naphthalocyaninato), $2,3\text{-}Nc^{2-}$ (2,3-naphthalocy-aninato), or TBP^{2-}. $R = t\text{-}Bu$, Et, OR' (R' = C_5H_{11}–$C_{12}H_{25}$) (substituted in the peripheral positions).

Synthesis Condensation of $Si(Pc)(OH)_2$, $Ge(Pc)(OH)_2$, $Sn(Pc)(OH)_2$ to form phthalocyaninato polysiloxanes, polygermyloxanes, and polystannyloxanes.[1–4]

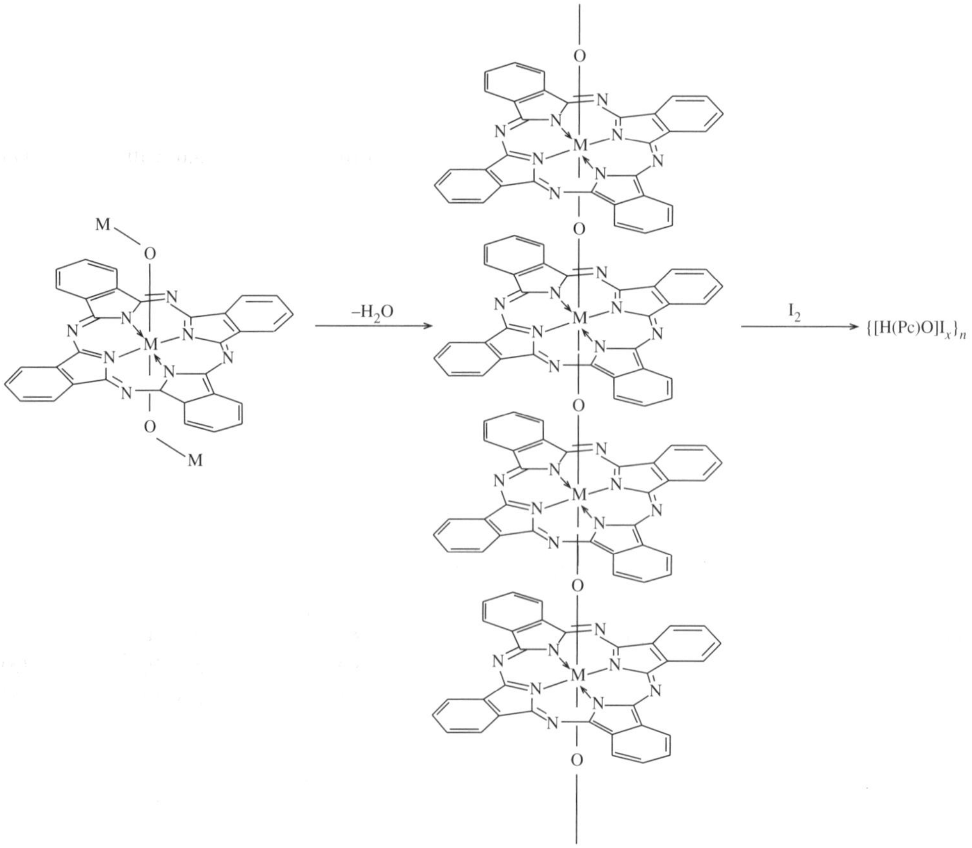

Major Applications Electrical conductors, semiconductors, and materials with photooptical properties.

Electric conductivity

Polymer	Units	y	Value	Reference
$[Si(Pc)O]_n$	σ_{RT} ohm^{-1} cm^{-1}	Nondoped	3×10^{-8}	(5)
	σ (300 K) ohm^{-1} cm^{-1}	Nondoped	5.5×10^{-6}	(6)
$\{[Si(Pc)O]I_y\}_n$	σ_{RT} ohm^{-1} cm^{-1}	0.50	2×10^{-2}	(5)
		1.40	2×10^{-1}	(5)
		4.60	1×10^{-2}	(5)
$\{[Si(Pc)O](I_3)_y\}_n$	s (300 K) ohm^{-1} cm^{-1}	0.37	5.8×10^{-1}	(6)
$\{[Si(Pc)O]Br_y\}_n$	σ_{RT} ohm^{-1} cm^{-1}	1.00	6×10^{-2}	(5)
$\{[Si(Pc)O]BF_{4y}\}_n$	σ (300 K) ohm^{-1} cm^{-1}	0.00[a]	3.0×10^{-4}	(7)
		0.00[b]	2.2×10^{-6}	(7)
		0.11	3.7×10^{-3}	(6)
		0.13	3.3×10^{-3}	(7)
		0.18	2.4×10^{-2}	(6)
		0.19	1.4×10^{-2}	(7)
		0.20	5.3×10^{-2}	(6)
		0.27	2.9×10^{-2}	(7)
		0.28	6.7×10^{-2}	(6)
		0.31	9.0×10^{-2}	(6)
		0.36	1.8×10^{-1}	(7)
		0.36	8.6×10^{-2}	(6)
		0.41	1.2×10^{-1}	(7)
		0.50	1.3×10^{-1}	(7)
$\{[Si(Pc)O]TOS_y\}_n$		0.10	5.6×10^{-4}	(7)
(TOS = p-toluenesulfonate)		0.19	1.0×10^{-2}	(7)
		0.28	2.0×10^{-2}	(7)
		0.37	3.7×10^{-2}	(7)
		0.52	4.5×10^{-2}	(7)
		0.67	4.3×10^{-2}	(7)
$\{[Si(Pc)O]SO_{4y}\}_n$		0.040	8.5×10^{-3}	(7)
		0.095	8.8×10^{-2}	(7)
$\{[Si(Pc)O]PF_{6y}\}_n$	σ (300 K) ohm^{-1} cm^{-1}	0.08	1.3×10^{-2}	(6)
		0.18	1.7×10^{-2}	(6)
		0.20	2.3×10^{-2}	(6)
		0.32	7.8×10^{-2}	(6)
$\{[Si(Pc)O]SbF_{6y}\}_n$	σ (300 K) ohm^{-1} cm^{-1}	0.39	1.5×10^{-1}	(6)
$[Ge(Pc)O]_n$	σ_{RT} ohm^{-1} cm^{-1}	Nondoped	$<10^{-8}$	(5)
$\{[Ge(Pc)O]I-y\}_n$	σ_{RT} ohm^{-1} cm^{-1}	1.80	3×10^{-2}	(5)
		1.90	5×10^{-2}	(5)
		1.94	6×10^{-2}	(5)
		2.0	1×10^{-1}	(5)
$[Sn(Pc)O]_n$	σ_{RT} ohm^{-1} cm^{-1}	Nondoped	$<10^{-8}$	(5)
$\{[Sn(Pc)O]I_y\}_n$	σ_{RT} ohm^{-1} cm^{-1}	1.2	1×10^{-6}	(5)
	σ_{RT} ohm^{-1} cm^{-1}	5.5	2×10^{-4}	(5)

Polymer	Units	y	Value	Reference
${[Fe(Pc)pyz]I_y}_n$	σ_{RT} ohm^{-1} cm^{-1}	0[c]	7.79×10^{-8}	(8)
		0.19[d]	9.31×10^{-4}	(8)
		0.38[d]	2.58×10^{-3}	(8)
		0.77[d]	8.63×10^{-3}	(8)
		2.10[d]	7.55×10^{-3}	(8)
		2.76[d]	2.33×10^{-2}	(8)
		0.38[e]	4.60×10^{-4}	(8)
		1.49[e]	5.99×10^{-3}	(8)
		2.10[e]	1.28×10^{-1}	(8)
		2.54[e]	1.90×10^{-1}	(8)
$[Fe(Pc)tz]_n$	σ_{RT} ohm^{-1} cm^{-1}	Nondoped[f]	2×10^{-2}	(9)
$[Ru(Pc)tz]_n$		Nondoped[f]	1×10^{-2}	(9)
$[Fe(Pc)Me_2tz]_n$		Nondoped[f]	4×10^{-3}	(9)
$[Ru(Pc)(NH_2)_2tz]_n$		Nondoped[f]	4×10^{-3}	(9)
$[Ru(Pc)p\text{-}(NH_2)C_6H_4]_n$		Nondoped[f]	5×10^{-9}	(9)
$[Ru(Pc)Cl_2tz]_n$		Nondoped[f]	3×10^{-3}	(9)
$[Os(Pc)pyz]_n$		Nondoped[f]	1×10^{-6}	(9)
$[Os(Pc)tz]_n$		Nondoped[f]	1×10^{-2}	(9)
$[Fe(Me_8Pc)pyz]_n$		Nondoped[f]	3×10^{-9}	(9)
$[Fe(Me_8Pc)tz]_n$		Nondoped[f]	1×10^{-2}	(9)
$[Ru(Pc)Me_2tz]_n$		Nondoped[f]	4×10^{-3}	(9)
$[Fe(CN_4Pc)pyz]_n$		Nondoped[f]	5×10^{-9}	(9)
$[Fe(CN_4Pc)tz]_n$		Nondoped[f]	1×10^{-6}	(9)
$[Fe(2,3\text{–Nc})pyz]_n$		Nondoped[f]	5×10^{-5}	(9)

[a] Orthorhombic. [b] Tetragonal. [c] Prepared by: $n Fe(Pc) + n(pyz) \xrightarrow{C_6H_6Cl \text{ or benzene}} [Fe(Pc)(pzy)]_n$.

[d] Prepared by: $[Fe(Pc)(pyz)]_n + (ny/2)I_2 \xrightarrow{benzene} {[Fe(Pc)(pyz)]I_y}_n$.

[e] Prepared by: $n[Fe(Pc)(pyz)]_n + (ny/2)I_2 \xrightarrow{CHCl_2} {[Fe(Pc)(pyz)]I_y}_n + n(pyz)$.

[f] Room temperature, pressed pellets, 1 kbar.

Thermoelectric power[7]

Polymer	y	S (300 K)[a] (μ VK^{-1})	$(\Delta S / \Delta T) \sim 300$ (mV K^{-2})	4t[b] (eV)
${[Si(Pc)O]BF_{4y}}_n$	0.13	113		
	0.19	62.9		
	0.27	43.4	0.101	2.70
	0.36	10.5	0.134	1.16
	0.41	4.6	0.135	0.81
	0.50	0.31	0.100	0.70
${[Si(Pc)O]TOS_y}_n$	0.10	284		
	0.19	114		
	0.28	82.1	0.079	4.2
	0.37	50.9	0.115	1.27
	0.52	28.2	0.109	0.63

Metallophthalocyanine Polymers

Thermoelectric power[7]

Polymer	y	S (300 K)[a] (μ VK^{-1})	$(\Delta S/\Delta T) \sim 300$ (mV K^{-2})	$4t$[b] (eV)
	0.67	26.0	0.096	0.36
$\{[Si(Pc)O]SO_{4y}\}_n$	0.095	48.6	0.17	3.8

[a] $S_{sample} = $ (Slope of voltage $\sim$ temp. data) $(S_{thermocouple}) + S_{gold}$.
[b] Tight-binding bandwidth derived from a fit to: $S = [2\pi^2 k_B^2 T \cos(\pi\rho/2)]/[3e(4t)\sin^2(\pi\rho/2)]$.

Static magnetic susceptibility

Polymer	y	χ_{Pauli}[a] (10^{-4} emu mol^{-1})	Pauli-like spins/M(Pc)[b]	A (10^{-4})	α	Curie-like spins/M(Pc)[c]	Reference
$\{[Si(Pc)O]BF_{4y}\}_n$	0.11	0.68	0.05	135	0.82	0.10	(7)
	0.19	1.40	0.11	16	0.67	0.13	(7)
	0.26	2.39	0.19	101	0.83	0.07	(7)
	0.35	2.22	0.18	92	1.00	0.024	(7)
	0.43	2.28	0.18	90	1.00	0.024	(7)
	0.50	2.32	0.18	83	1.00	0.022	(7)
$\{[Si(Pc)O]TOS_y\}_n$	0.67	3.13	0.25	124	1.00	0.032	(7)
$\{[Si(Pc)O]SO_{4y}\}_n$	0.095	1.93	0.15	116	0.82	0.09	(7)
$\{[Si(Pc)O]BF_{4y}\}_n$	0.36	2.22	0.18	–	–	–	(6)
$\{[Si(Pc)O]PF_{6y}\}_n$	0.36	2.49	0.19	–	–	–	(6)
$\{[Si(Pc)O]SbF_{6y}\}_n$	0.36	2.22	0.18	–	–	–	(6)
$\{[Si(Pc)O](I_3)_y\}_n$	0.37	2.35	0.18	–	–	–	(6)
$\{[Ge(Pc)O](I_3)_y\}_n$	0.37	2.70	0.21	–	–	–	(6)

[a] For $\chi = \chi_{Pauli} + ATs^{\alpha}$.
[b] $N_p = 3\chi_{Pauli}kT/Ng^2\mu_B^2 S(S+1)$, where $T = 298$ K.
[c] $Nc = 3AT^{1-\alpha}k/Ng^2\mu_B^2 S(S+1)$, where $T = 298$ K.

Unit cell Dimensions

Polymer	y	Space Group	z	Cell Dimensions (Å) a	b	c	Interplanar Spacing (Å)	Staggering Angle ϕ (degrees)	Reference
$[Si(Pc)O]_n$	–	Ibam	4	13.80	27.59	6.66	3.33	39	(5)
$[Ge(Pc)O]_n$	–	P4/m	1	13.27	3.53	–	3.53	0	(5)
	–	I4/m	2	18.76	3.57	–	3.57	0	(5)
$[Sn(Pc)O]_n$	–	P4/m	1	12.81	3.8	–	3.82	Probably eclipsed	(5)
$[Ga(Pc)F]_n$	–	PI		3.871	12.601	12.793	3.87	Probably eclipsed	(5)
$\{[Si(Pc)O]BF_{4y}\}_n$	0.36	P4/mcc	2	13.70	–	6.58	3.29	40	(6, 7)
	0.50	P4/mcc	2	13.96	–	6.66	–	–	(7)
$\{[Si(Pc)O]PF_{6y}\}_n$	0.36	P4/mcc	2	13.98	–	6.58	3.29	40	(6, 7)
	0.47	P4/mcc	2	14.08	–	6.63	–	–	(7)
$\{[Si(Pc)O]SbF_{6y}\}_n$	0.36	P4/mcc	2	14.31	–	6.58	3.29	40	(6, 7)
	0.41	P4/mcc	2	14.19	–	6.61	–	–	(7)

Polymer	y	Space Group	z	Cell Dimensions (Å)			Interplanar Spacing (Å)	Staggering Angle ϕ (degrees)	Reference
				a	b	c			
$\{[Si(Pc)O](I_3)_y\}_n$	0.37	$P4/mcc$	2	13.97	–	6.60	3.30	39	(6)
$\{[Si(Pc)O](Br_3)_y\}_n$	0.37	$P4/mcc$	2	13.97	–	6.60	3.30	39	(6)
$\{[Ge(Pc)O](I_3)_y\}_n$	0.36	$P4/mcc$	2	13.96	–	6.96	3.48	40	(6)
$\{[Si(Pc)O]TOS_y\}_n$	0.67	$P4/mcc$	2	14.39	–	6.64	–	–	(7)
$\{[Si(Pc)O]PYS_y\}_n$	0.22	$P4/mcc$	2	13.70	–	6.65	–	–	(7)
$\{[Si(Pc)O](CF_3SO_3)y\}_n$	0.55	$P4/mcc$	2	13.99	–	6.60	–	–	(7)
$\{[Si(Pc)O]SO_{4y}\}_n$	0.095	$P4/mcc$	2	13.86	–	6.67	–	–	(7)
$\{[Si(Pc)O]NFBS_y\}_n$	0.36	$P4/mcc$	2	14.37	–	6.63	–	–	(7)
$\{[Si(Pc)O]PFOS_y\}_n$	0.26	$P4/mcc$	2	13.91	–	6.61	–	–	(7)

Other physical properties[4]

Property	Polymer	Conditions
Color	$[Si(Pc)O]_n$	Dark purple powder
Solubility	$[Si(Pc)O]_n$	Concentrated H_2SO_4 and HSO_3CH_3:
		0.013 g in 25 ml concentrated H_2SO_4 at room temperature
		0.020 g in 25 ml concentrated H_2SO_4 at 80C

Densities

Polymer	y	Density (G CM^{-3})		Reference
		Calculated	Found	
$[Si(Pc)O]_n$	–	1.458	1.432	(4, 6)
$[Ge(Pc)O]_n$	–	1.609[a]	1.512	(4, 6)
	–	1.589[b]	–	(4, 6)
$[Sn(Pc)O]_n$	–	1.715	1.719	(4, 6)
$\{[Si(Pc)O]BF_{4y}\}_n$	0.36	1.581	1.545	(6)
$\{[Si(Pc)O]PF_{6y}\}_n$	0.36	1.573	1.563	(6)
$\{[Si(Pc)O]SbF_{6y}\}_n$	0.36	1.582	1.591	(6)
$\{[Si(Pc)O](I_3)_y\}_n$	0.37	1.802	1.744	(6)
$\{[Ge(Pc)O](I_3)_y\}_n$	0.36	1.805	1.774	(6)

[a] Space group: $P4/m$; $Z = 1$; $a = 13.27$, $c = 3.53$.
[b] Space group: $I4/m$; $Z = 2$; $a = 18.76$, $c = 3.57$.

Infrared spectroscopy[4]

Polymer	IR Spectral Data (CM^{-1})*
$[Si(Pc)O]_n$	530(m), 575(m), 646(w), 721(vs), 759(vs), 762(w), 804(w), 869(vw), 910(s), 936(vw), 1000(bd), 1043(m), 1080(vs), 1121(vs), 1164(s), 1170(sh), 1192(w), 1289(s), 1334(vs), 1351(m), 1426(vs), 1517(s), 1596(w), 1614(m)

Polymer	IR spectral data (CM^{-1})*
$[Ge(Pc)O]_n$	425(vw), 435(vw), 508(m), 572(m), 640(w), 660(vw), 725(vs), 753(m), 762(vw), 772(vw), 801(w), 865(bd), 899(s), 935(vw), 945(vw), 970(vw), 998(w), 1068(vs), 1087(vs), 1119(vs), 1162(s), 1195(w), 1284(m), 1332(vs), 1345(m), 1419(s), 1500(m), 1588(w), 1612(m)
$[Sn(Pc)O]_n$	428(w), 435(sh), 495(m), 570(m), 640(w), 660(vw), 687(vw), 716(vs), 750(s), 762(m), 769(m), 775(sh), 808(m), 825(bd), 872(w), 888(m), 950(w), 1005(vw), 1058(s), 1089(s), 1120(vs), 1168(m), 1183(w), 1263(w), 1284(m), 1293(w), 1338(vs), 1405(sh), 1580(sh), 1610(m)

* Peaks not readily assigned to M(Pc) moiety; s = strong, m = medium, w = weak, bd = broad, sh = shoulder, v = very.

Optical spectroscopy[(4)]

Compound	Absorption maximum (NM)
$[Si(Pc)O]_n$	203, 285, 335, 625
$[Ge(Pc)O]_n$	285, 350, 645
$[Sn(Pc)O]_n$	205, 290, 365, 655, 695

Class Linear polymer[10]

Synthesis

NC
NC
O
S
S
O
CN
CN
Solvent, heat, 5 h
Yield, 56-80%
R
N
N
N
N
M
N
N
N
N
O
S
S
R
n

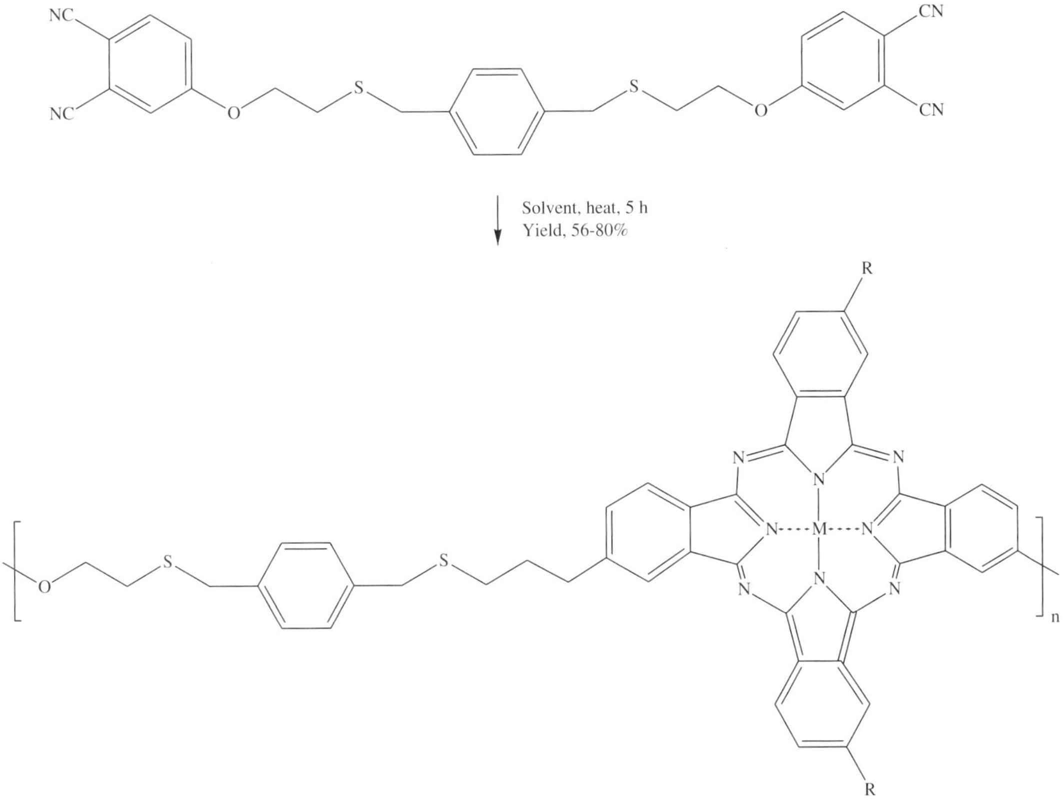

where M = Ni, Co, Fe, Pb, Cu, and Zn, and

R = [structure]

Properties: Infrared Spectra (pellet)

Metal, M	Main Frequencies, cm^{-1} (assignments)
Ni	3410 (imide N-H), 3050 (arom =CH), 2940–2830 (CH$_2$), 1770 (sym. C=O), 1710 (asym. C=O), 1604 (arom —C=C—), 1574, 1499, 1415, 1320, 1230 (Ar-O-C), 1110–1020, 720 cm^{-1}.
Co	3408 (imide N-H), 3055 (arom =CH), 2950–2845 (CH$_2$), 1775 (sym. C=O), 1715 (asym. C=O), 1608 (arom —C=C—), 1570, 1490, 1410, 1360, 1229 (Ar-O-C), 1095–1064, 754 cm^{-1}.
Fe	3417 (imide N-H), 3058 (arom =CH), 2960–2829 (CH$_2$), 1773 (sym. C=O), 1730 (asym. C=O), 1605 (arom —C=C—), 1564, 1485, 1410, 1328, 1270, 1240 (Ar-O-C), 1100–1020, 760 cm^{-1}.
Pb	3380 (imide N-H), 3056 (arom =CH), 2982–2830 (CH$_2$), 1775 (sym. C=O), 1712 (asym. C=O), 1604 (arom —C=C—), 1554, 1492, 1412, 1320, 1262, 1230 (Ar-O-C), 1105–1026, 840, 735 cm^{-1}.
Cu	3350 (imide N-H), 3057 (arom =CH), 2924–2855 (CH$_2$), 1770 (sym. C=O), 1702 (asym. C=O), 1600 (arom —C=C—), 1485, 1434, 1328, 1282, 1226 (Ar-O-C), 1188–1012, 950, 737 cm^{-1}.
Zn	3440 (imide N-H), 3062 (arom =CH), 2968–2842 (CH$_2$), 1775 (sym. C=O), 1709 (asym. C=O), 1645 (arom —C=C—), 1498, 1440, 1360, 1280, 1240 (Ar-O-C), 1190–1025, 740 cm^{-1}.

UV-Vis spectra (in H$_2$SO$_4$)

Metal	λ nm (log ε)	Ratio (UV/Vis)
Ni	796 (6.65), 706 (6.12), 616 (4.42), 496 (5.74), 306 (6.72), 252 (6.82)	1.02
Co	800 (5.89), 712 (5.39), 646 (4.75), 612 (4.88), 416 (5.45), 304 (6.19), 298 (6.08)	1.03
Fe	810 (6.18), 733 (6.24), 618 (6.25), 386 (6.60), 318 (7.01), 270 (6.79)	1.10
Pb	883 (3.93), 833 (3.85), 739 (4.74), 632 (4.89), 610 (3.83), 485 (4.57), 334 (4.95), 291 (5.26), 278 (5.24)	1.33
Cu	834 (5.37), 706 (3.90), 622 (5.85), 362 (5.91), 302 (6.24), 254 (6.23), 236 (6.85)	0.86
Zn	824 (6.40), 726 (5.73), 664 (4.81), 614 (4.97), 502 (5.52), 490 (5.50), 352 (6.12), 308 (6.47), 256 (6.49)	1.01

Electrical Conductivity (pellet)

Metal	Conductivity (S · cm^{-1})		Intrinsic Viscosity [η] (H$_2$SO$_4$)
	Argon	Vacuum	
Ni	1.03×10^{-7}	1.91×10^{-8}	1.64
Co	2.51×10^{-8}	8.90×10^{-9}	1.28
Fe	1.28×10^{-7}	8.56×10^{-8}	1.37
Pb	1.65×10^{-7}	1.26×10^{-7}	1.13
Cu	2.36×10^{-7}	5.65×10^{-8}	1.27
Zn	2.05×10^{-7}	1.54×10^{-7}	1.60

Metallophthalocyanine Polymers

Thermal Properties

Metal	Initial Decomposition Temperature (°C)	Main Decomposition Temperature (°C)
Ni	349	477
Co	334	440
Fe	321	433
Pb	325	392
Cu	305	381
Zn	415	508

Class Cyclic oligomer[11]

Synthesis

M=Cu, Co, Cr–Cl, Fe–Cl

Properties Infared Spectra (pellet)

Metal, M	Main Frequencies, cm⁻¹ (assignments)
Cu	3197 (OH), 3062 (arom =CH), 1771 (C=O), 1696 (C=N), 1465 (arom —C=C—), 1304–1355–1376 (C—N), 1060–1153 (conjugated system), 756–947 (benzene CH), 635–743 (out of plane, CH deformation in benzene).
Co	3191–3425 (OH), 3029 (arom =CH), 1770 (C=O), 1697 (C=N), 1464 (arom —C=C—), 1304–1355–1376 (C—N), 1061–1153 (conjugated system), 755–947 (benzene CH), 635–724 (out of plane, CH deformation in benzene).
Fe	3195–3448 (OH), 3029–3067 (arom =CH), 1772 (C=O), 1697 (C=N), 1466 (arom —C=C—), 1305–1376 (C—N), 1061–1154–1185 (conjugated system), 758–894–947 (benzene CH), 636–726 (out of plane, CH deformation in benzene).
Cr	3196–3455 (OH), 3031–3069 (arom =CH), 1774 (C=O), 1695 (C=N), 1467 (arom —C=C—), 1307–1357–1378 (C—N), 1061–1154 (conjugated system), 761–891–947 (benzene CH), 636–725 (out of plane, CH deformation in benzene).

Metallophthalocyanine Polymers

UV-Vis Spectra

Metal	λ_{max}(nm)
Cu	244, 322, 370, 411, 430, 712, 764
Co	233, 320, 409, 679, 709, 744, 792
Fe	222, 246, 324, 385, 413, 672, 811
Cr	230, 289, 322, 411, 490, 751

Electrical Conductivity

	ΔE (eV)			
Metal	100 Hz	1k Hz	10 kHz	100 kHz
Cu	0.2885	0.2876	0.2736	0.2316
Fe	-	-	-	0.1761

Relative Permittivity

	(Room temp. ε_r')				(High temp. ε_r')			
Metal	100 Hz	1k Hz	10 kHz	100 kHz	100 Hz	1k Hz	10 kHz	100 kHz
Cu	450.00	180.00	50.00	20.00	200.00	370.00	220.00	125.00
Fe	3.96	3.43	2.60	2.30	4.80	4.60	3.90	2.90

Class Cofacial polymer[12]

Synthesis

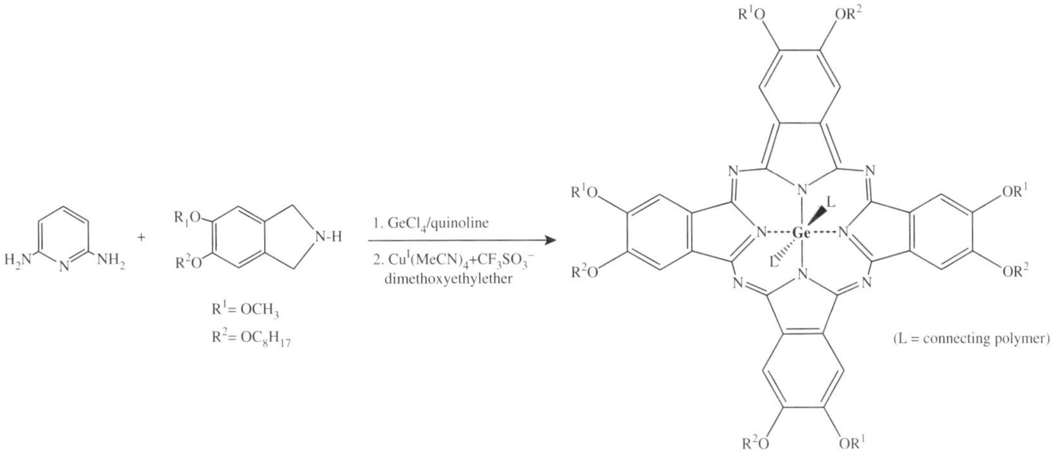

$R^1 = OCH_3$
$R^2 = OC_8H_{17}$

1. GeCl$_4$/quinoline
2. Cu^I(MeCN)$_4$+CF$_3$SO$_3^-$ dimethoxyethylether

(L = connecting polymer)

Metallophthalocyanine Polymers

Properties Infrared Spectra (pellet)

Metal	Main Frequencies, cm^{-1} (assignments)
Ge	2950 m sh, 2920 m, 2870 m sh, 2850 m, 1660 w, 1625 s, 1580 s, 1545 m, 1495 s, 1470 w sh, 1450 w, 1360 s, 1300 w, 1245 s, 1215 m, 1150 w br, 1070 s, 1030 w, 860 m br, 810 m, 770 vw, 715 m, 635 w.

UV-Vis Spectra

Metal	λ nm
Ge	276, 343, 387, 414 sh.

^{1}H-NMR Spectra (in CDCl$_3$ @ 300 MHz)

Metal	Main Shifts, δ ppm (assignments)
Ge	7.0–5.7 (m, br, ≈10H), 3.65 (s br, ≈10H), 1.78 (s br, ≈8H), 1.32 (m, ≈20H), 0.88 (s br, ≈6H).

References

1. Hanack M. In Wisian-Neilson P, Allcock HR, Wynne KJ. Eds. *Inorganic and Organometallic Polymers II: Advanced Materials and Intermediates.* Washington D.C: American Chemical Society; 1994:572.
2. Marks TJ, Schoch KF Jr, Kundalkar BR, *Synth. Metals.* 1980;1(3):337–347.
3. Joyner RJ, Kenney M. *Inorg. Chem.* 1960;82(22):5790–5791.
4. Davison JB, Wynne KJ. *Macromolec.* 1978;11(1):186–191.
5. Dirk CW, Inabe T, Schoch KF Jr. Marks TJ. *J. Amer. Chem. Soc.* 1983;105(6):1539–1550.
6. Inabe T, Gaudiello JG, Moguel MK, et al. *J. Amer. Chem. Soc.* 1986;108(24):7595–7608.
7. Almeida M, Gaudiello JG, Kellogg GE, et al. *J. Amer. Chem. Soc.* 1989;111(14):5271–5284.
8. Schoch KF Jr, Kundalkar BR, Marks TJ. *J. Amer. Chem. Soc.* 1979;101(23):7071–7073.
9. Diel BN, Inabe T, Jaggi NK. *J. Amer. Chem. Soc.* 1984;106(11):3207–3014.
10. Bilgin A, Symposium Series ACS, Yagci C, Yildiz U. *Macromolec. Chem. Phys.* 2005;206(22):2257–2268.
11. Abd El-Ghaffar MA, Youssef EAM, El-Halawany NR, Ahmed Angew MA. *Makromolekulare Chem.* 1998;254:1–9.
12. Ferencz A, Ries R, Wegner G. *Angew. Chem. Int. Ed. Engl.* 1993;32(8):1184–1187.

Methacrylate Polymers Containing Adamantane

Ruzhi Zhang

Class Vinyl polymers; acrylics

Structure

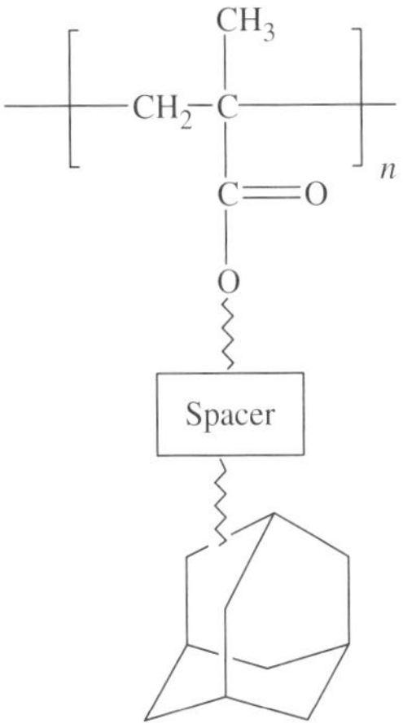

Properties of Special Interest and Major Applications Methacrylate polymers containing adamantane are of special interest due to the fact that pendant adamantanes improve physical properties such as stiffness, glass transition temperature and solubility, and reduce or eliminate crystallinity.[1] It also has high carbon to hydrogen ratio required to resist the RIE process. It is transparent at DUV (~193 nm) wavelength and copolymers containing adamantane are employed in a number of positive resists used for ArF lithography.[2,3]

Preparative Methods Methacrylate polymers containing adamantane are commonly made by free radical polymerization[3,4] or anionic polymerization.[5]

Physical constants of methacrylate polymers containing Adamantanes

Polymer of	T_g (°C)	T_d (°C)*	Comments	Reference
1-adamantyl methacrylate	—	302	CAS number [28854-38-8]; Solvents: cyclohexane, benzene, carbon tetrachloride, chloroform, dichloromethane, THF;	(5, 6)

Polymer of	T_g (°C)	T_d (°C)*	Comments	Reference
			Nonsolvents: hexane, acetone, diethyl ether, DMF, ethanol, methanol, water; ^{1}H NMR, ^{13}C NMR, FT-IR	
1-adamantylmethyl methacrylate	201	235	CAS number [258355-03-2]; ^{1}H NMR, ^{13}C NMR, FT-IR	(4)
4-(1-adamantyl)phenyl methacrylate	253	260	CAS number [258355-04-3]; ^{1}H NMR, ^{13}C NMR, FT-IR	(1, 4)
2-methyl-4-(1-adamantyl) phenyl methacrylate	250	250	^{1}H NMR, ^{13}C NMR, FT-IR	(1)
2-methyl-2-adamantyl methacrylate	160	240	—	(3)

* T_d: Decomposition temperature by TGA.

References

1. Mathias LJ, Jensen J, Thigpen K, McGowen J, McCormick D, Somlai L. *Polymer.* 2001;42:6527.
2. Willson CG, Trinque BC. *J. Photopolym. Sci. Technol.* 2003;16:621.
3. Dammel R, Cook M, Klauck-Jacobs A. *J. Photopolym. Sci. Technol.* 1999;12:433.
4. Acar HY, Jensen JJ, Thigpen K, McGowen JA, Mathias LJ. *Macromolecules.* 2000;33, 3855.
5. Ishizone T, Tajima H, Torimae H, Nakahama S. *Macromol. Chem. Phys.* 2002;203:2375.
6. Matsumoto A, Tanaka S, Otsu T. *Macromolecules.* 1991;24:4017.

Nylon 3

Junzo Masamoto

Class Aliphatic polyamides

Structure $[-CH_2CH_2CONH-]$

Major Application Thermal stabilizer for polyoxymethylene, and stabilizer for polyacetal resin. Because of high amide concentration, nylon 3 shows properties of an excellent formaldehyde scavenger.[1]

Properties of Special Interest Nylon 3 shows properties of an excellent stabilizer for polyoxymethylene. Features of nylon 3 as a stabilizer for polyoxymethylene are as follows: High thermal stability, negligible decoloration when the polymer remains for a long time in injection mold machine at its molten state, and low deposit on the mold.[1] Nylon 3 is an interesting material as an odd-numbered nylon with shortest methylene group, thus forming high glass transition temperature and high absorption of water.

Other Polymer Showing This Special Properties Formaldehyde scavenger: copolyamide composed of Nylon 6, Nylon 6,6 and Nylon 6,10.

Preparative Techniques Hydrogen transfer polymerization: Acryl amide is polymerized in the presence of a strong base catalyst (e.g., t-BuOK). The polymerization occurs with hydrogen transfer, producing nylon 3. The polymerization is conducted using inactive solvents, such as toluene, pyridine, chlorobenzene, and o-dichlorobenzene from 80 to 200°C.[1−4]

The following methods were also reported for the preparation of nylon 3: anionic ring-opening polymerization of β-lactam (β-propiolactam);[5,6] ring-opening polymerization of 8-ring dilactam (1,5-diazacyclooctane-2,6-dion);[7] thermal polymerization of ethylene cyanehydrine;[8] and alternative copolymerization of carbon monoxide and ethylene imine.[9]

The polymerization method of α-amino acid to nylon 3 poly(-β-alanine) was also studied. For example, nylon 3 was synthesized from β-alanine N-carboxyanhydride (NCA), N-dithiocarbonyl ethoxycarbonyl-β-alanine and N-carbothiophenyl-β-alanine.[10−12]

The use of various active β-alanine esters for the preparation of poly(-β-alanine) was proposed.[13] The direct condensation of β-alanine to obtain nylon 3 was also reported.[14]

Property	Units	Conditions	Value	Reference
Molecular weight (of repeat unit)	g mol^{-1}	–	71	–
Typical molecular weight range of polymer	g mol^{-1}	Light scattering	$M_w = 90,000–120,000$	(15)
		Light scattering	$M_w = 80,000$	(2)
		Viscosity	43,000	(16, 17)
		Intrinsic viscosity in 90% formic acid	54,000	(18)

Property	Units	Conditions	Value	Reference
IR (characteristic absorption frequencies)	cm^{-1}	Amide VII for the extended planar zigzag conformation of the molecular chain	240	(19)
		Skeletal vibration of delta methylene ($CH_2CONHCH_2$)	365	(19)
		Skeletal vibration of delta methylene NCH_2CH_2	465	(19)
		Amide VI	580	(19)
		Amide V	700	(19)
		NH asymmetric	3,400	(20)
		NH asymmetric	3,250	(20)
		$-NH\cdots C=O$	3,070	(20)
		$CH_2(N)$	2,927	(20)
		CH_2	2,890	(20)
		$C=O$ amide I	1,640	(20)
		Amide II	1,530	(20)
		Amide III	1,283, 1,220	(20)
		Amide IV	1,100, 1,040, 960	(20)
		Amide V	683	(20)
Raman	cm^{-1}	Amide VI	577	(21)
		Amide V	682	
		C–CO stretch	970	
		C–C stretch	1,110	
		NH wagging	1,227	
		Amide III	1,260	
		CH_2 twisting	1,293	
		CH_2 wagging	1,367	
		CH_2–CO bending	1,426	
		CH_2 bending	1,443	
		Amide I	1,630	
		Symmetric CH_2 stretching	2,853, 2,900	
		Asymmetric CH_2 stretching	2,933	
		CH_2 stretching	2,963/2,995	
		NH stretching	3,293	
NMR		60 MHz ^{1}H NMR: trifluoroacetic acid at 60°C with Varian A 60		(3)
		^{1}H NMR: D_2O (0.5% solution) at room temperature with a Bruker Model-WH 270 spectrometer		(22)
		^{13}C NMR: FSO3H, 90.5 MHz with a Bruker WH 360 FT-NMR spectrometer		(23)
Solvents		Soluble at room temperature: formic acid, dichloroacetic acid, trifluoroacetic acid		(24)
		Soluble at 60°C: chloral hydrate		
Nonsolvents		Insoluble at room temperature: water, methanol, butanol		(25)
Mark-Houwink parameters: K and a	$K = ml\,g^{-1}$ $a =$ None	0.4 mol-KCl/l-HCOOH, at 35°C	$K = 1.6 \times 10^{-4}$ $a = 0.50$	(26)

Property	Units	Conditions	Value	Reference
Huggins constants k'	–	99 wt% formic acid:	0	(27)
		90 wt% formic acid:	0.4	
		80 wt% formic acid	0.5	
		0.4 mol-KCl/99 wt%-HCOOH	0.4	
		0.4 mol-KCl/80 wt%-HCOOH	0.5	
Lattice	–	Modification I	Monoclinic	(19, 28, 29)
			Orthorhombic	(28)
Space group	–	–	P2$_1$	(19)
			C2-2	(28)
Chain conformation	–	Modification I (monoclinic), II, and III	Extended chain of planar zigzag	(19)
		Monoclinic form and orthorhombic form	Both extended	(28)
Unit cell dimensions	Å	Modification I (monoclinic)	$a = 9.33, b = 4.78$ (fiber identity period), $c = 8.73$	(19, 29)
(19, 29)		Monoclinic	$a = 9.60, b = 4.78$ (fiber identity period), $c = 8.96$	(28)
(19, 29)		Orthorhombic	$a = 9.56, b = 4.78$ (fiber identity period), $c = 7.56$	(19, 29)
Unit cell angles	degrees	Monoclinic, modification I	$\beta = 60$	(19)
		Monoclinic	$\beta = 122.5$	(28)
		Monoclinic	$\beta = 57.5$	(30)
Unit cell contents	monomeric units	Monoclinic form	4	(19, 28–30)
		Orthorhombic form	4	(28)
Degree of crystallinity	%	Dielectric relaxation of nylon 3 powder	38	(31)
Density	g cm^{-3}	Theoretical density for modification I	1.39	(19)
		Observed density for nylon 3 drawn fiber (modification I) at 25°C	1.33	(19)
		Observed density for nylon 3 undrawn fiber (modification II) at 25°C	1.32	(19)
		Observed density for nylon 3 for modification III at 25°C	1.33	(19)
		Theoretical density for monoclinic and orthorhombic	1.36	(28)
		Observed density for nylon 3		
		Form I (monoclinic)	1.30	(28)
		Form II (orthorhombic)	1.27	(28)

Property	Units	Conditions	Value	Reference
Polymorphs		Modification I (monoclinic); Modification II; Modification III; Modification IV (smectic hexagonal)		(19)
		Form I (monoclinic); Form II (orthorhombic)		(28)
Glass transition temperature	K	Tan δ maximum of drawn fiber	443–453	(32)
		DTA	384	(33)
		Dielectric relaxation of nylon 3 powder	384	(31)
		20 K min^{-1}, with a Perkin Elmer DSC 4	396	(34)
		NMR method, water insoluble nylon 3 in dry state	480	(18)
Melting point	K	With decomposition, water insoluble polymer	613	(2)
		With decomposition, water soluble polymer	598	(2)
		Hot stage microscope, water insoluble nylon 3	625	(18)
Sub T_g transition	K	NMR, local motion of the methylene groups of the polymer chain in amorphous region, molecular chain approach a rigid structure in this temperature range	77–130	(18)
Tensile modulus	MPa	Drawn and wet heat-treated fiber	8,000–12,000	(32)
Tensile strength	MPa	Drawn and wet heat-treated fiber	240–360	(32)
Yield strain	%	Drawn and wet heat-treated fiber	3	(32)
Maximum extensibility	%	Drawn and wet heat-treated fiber	10–20	(32)
Dielectric constant	–	10^{-3}–10^{-7} Hz, 30°C	4.7	(31)
Dielectric loss	–	10^{-3}–10^{-7} Hz, 30°C	10^{-2}–10^{-1}	(31)
Decomposition temperature	K	50% weight loss, 2°C min^{-1}, in air	608–613	(31)
Water absorption	%	60% RH, 25°C	7	(1)
Important Patents		U.S. Patent 4,855,365 U.S. Patent 5,015,707, assigned to Asahi Chemical		
Availability		Not commercially available (only used inside Asahi Kasei Chemicals Corporation, 1-1-2, Yuraku-cho, Chiyoda-ku, Tokyo, 100-8440 Japan)		

References

1. Masamoto, J. In: Salamone JC, ed. *Polymeric Material Encyclopedia.* Vol 6. Boca Raton, Fla: CRC Press; 1996:4672.
2. Breslow DS, Hulse GE, Matrack AS. *J. Am. Chem. Soc.* 1957; 79:3760.

3. Masamoto JK, Yamaguchi H, Kobayashi H. *Kobunshi-kagaku* (*Jpn. J. Polym. Sci. Technol.*) 1969;26:631.

4. Masamoto JC, Ohizumi, Kobayashi H. *Jpn. J. Polym.Sci. Technol.* 1969;26:638.

5. Bestian R. *Angew. Chem.* 1968;80:304.

6. Kodaira T, et al. *Bull. Chem. Soc. Jpn.* 1965;38:1788.

7. Hall HK. *J. Am. Chem. Soc.* 1958;80:604.

8. Lautenschlanger H. *U.S. Patent.* 1964;3,126,353, assigned to BASF.

9. Kagiya T, et al. *J. Polym. Sci., Part B.* 1965;3:617.

10. Birkhofer L, Modic R. *Liebigs Ann. Chem.* 1959;628:162.

11. Noguchi J, Hayakawa T. *J. Am. Chem. Soc.* 1954;76:2846.

12. Higashimura T, et al. *Makromol. Chem.* 1966;90:243.

13. Hanabusa KK, Kondo, Takemoto K. *Makromol. Chem.* 1979;180:307.

14. Sakabe HH, Nakamura, Konishi H. *Sen-i Gakkaisshi* (*J. Soc. Fiber Sci. Technol. Jpn.*) 1991;45:493.

15. Masamoto JK. Sasaguri, Kobayashi H. *J. Soc. Fiber Sci. Technol. Jpn.* 1970;26:246.

16. Munoz-Guerra S, et al. *J. Polym. Sci., Polym. Phys.Ed.* 1985;23:733.

17. Munoz-Guerra S, Prieto A. In: Dosiere M, ed. *Crystallization of Polymers*, Netherlands: Kluwer Academic Publishers; 1993:277.

18. Tsumi A, et al. *J. Polym. Sci., Part A-2.* 1996;86:493.

19. Masamoto J, et al. *J. Polym. Sci., Part A-2.* 1970;8:1703.

20. Morgenstern U, Berger W. *Makrol. Chem.* 1992;193:2561.

21. Hendra PJ, et al. *Spectrochimica Acta* 1990;46:747.

22. Verneker VRP, Shaha B. *Polym. Commun.* 1984;25:363.

23. Kricheldorf HR. *J. Polym. Sci., Polym. Chem. Ed.* 1978;16: 2253.

24. Masamoto J, et al. *J. Soc. Fiber Sci. Technol. Jpn.* 1969;25:525.

25. Masamoto JK, Yamaguchi, Kobayashi H. *J. Soc. Fiber Sci. Technol. Jpn.* 1969;25:533.

26. Masamoto JK, Sasaguri, Kobayashi H. *J. Soc. Fiber Sci. Technol. Jpn.* 1970;26 :246.

27. Masamoto J, et al. *J. Soc. Fiber Sci. Technol. Jpn.* 1970;26:239.

28. Munoz-Guerra S, et al. *J. Polym. Sci., Polym. Phys. Ed.* 1985;23:733.

29. Tadokoro E. *Structure of Crystalline Polymers.* London: John Wiley and Sons; 1979.

30. Munoz-Guerra SA, Prieto, Montserrant JM. *J. Mater. Sci.* 1992;27:89.

31. Wolfle E, Stoll B. *Colloid Polym. Sci.* 1980;258:300.

32. Masamoto J, et al. *J. Appl. Polym. Sci.* 1970;14:667.

33. Tsvetkov VN, et al. *Vysokomol. Soedin A* 1968;10:547.

34. Morgenstern U, Berger W. *Makrol. Chem.* 1992;193:2561.

35. Kricheldorf HR, Schilling G. *Makromol. Chem.* 1976;177:607.

Nylon 4,6

Dinesh V. Patwardhan

Trade Name Stanyl (DSM); TW300 (dry, unfilled Stanyl); TW241F10 (dry, 50% glass fiber-filled Stanyl). Approximately 30 other varieties of Stanyl (filled or unfilled) are available.

CA Number [1,2] 50327-22-5 [Poly(imino-1,4-butanediylimino (1,6-dioxo-1,6-hexanediyl))] 50327-77-0 [Hexanedioic acid, polymer with 1,4-butanediamine]

Class Aliphatic polyamides; Nylons

Structure $H-(HN-(CH_2)_4-HN-CO-(CH_2)_4-CO)_n-OH$

Preparative Techniques [2-5] On an industrial scale nylon 4,6 is polymerized from 1,4-diaminobutane and adipic acid; 1, 4-diaminobutane is prepared separately by reacting acrylonitrile and HCN followed by dehydrogenation. The first step in polymer synthesis involves condensing the two monomers, 1,4 diamino butane and adipic acid, to give low molecular weight pre-polymer. This reaction step is completed at a lower temperature (200°C) to avoid formation of cyclics by the amine. In a separate step, this prepolymer is molded into uniform cylindrical pellets and heated to 250°C in an atmosphere of nitrogen and steam. The use of preformed pellets is important for uniform rate of solid-state polymerization and thus the degree of polymerization. The formation of uniform pellets is especially important for nylon 4,6 because transamidation, which leads to uniform molecular weight, cannot be performed on the final melt due to sensitivity of nylon 4,6 to thermal degradation. The typical molecular weight range for nylon 4,6 thus obtained is 30 000 gmol^{-1} with polydispersity of 1.15.

Major Applications [6,7] Nylon 4,6 is often blended with glass fiber or PTFE and used in a variety of underhood automotive applications. It is also used in gears, electrical parts, and bearings.

Properties of Special Interest [2-4] Distribution of methylene moieties in nylon 4,6 exists in regular groups of four which gives it high order. As a result of this high order nylon 4,6 has higher crystallinity and a faster rate of crystallization as compared with other polyamides such as nylon 6,6 and nylon 6. Furthermore the high order manifests as higher tensile strength, heat deflection, and tenacity which makes it a better high temperature engineering plastic. On the other hand high moisture regain is a significant drawback.

Property	Units	Conditions	Value	Reference
Specific gravity/density	–	23°C dry, unfilled (TW300)	1.18	8
Melting point	K	Dry, unfilled (TW300)	563	8

Property	Units	Conditions	Value	Reference
Glass transition temperature	T_g	Dynamic	316	9
Specific heat	btu/lb-F		0.5	6
Solvent		Dry unfilled (TW300)	95% formic acid	5
IR (Characteristics absorption frequencies)	cm^{-1}	Obtained on thin films from formic acid solution N—H C—H$_2$ Amide I–VI	3300 (vs); 3070 (m); 2945 (s); 2870 (m), 1638 (vvs), 1540 (vvs); 1280 (m); 940 (w); 730 (sh); 690 (s,b); 575 (m); 520 (w)	5
NMR	ppm	Deuterated formic acid as the solvent ^{1}H ^{13}C	1.57; 1.66; 2.37; 3.27 25.19; 25.95; 35.46; 39.65	10
Crystalline state properties Crystal system	Å	– –	Monoclinic $a = 9.60$, b (chain axis) $= 14.80$, $c = 8.06$	11
Unit cell dimensions Cell angles	degrees	–	$\beta = 67$	
Brill transition temperature	degrees	in-situ X-ray diffraction	γ-phase (hexagonal structure) ($\sim$280°C) to HT α-phase ($\sim$240°C) to RT α-phase ($\sim$150°C)	12, 13
Bragg angles	degrees(Å)	Wide angle X-ray scattering, ultra electrospun nano fibers from 10 wt.% solution in formic acid, 15–20 kV	$2\theta = 20.5$ (4.33), 24(3.70)	14
Heat of fusion	kJ mol^{-1}	Sample quenched and annealed for 5 min at 279°C	15.1	5
Tensile modulus	MPa	Dry, unfilled (TW300) Dry filled (TW241F10)	3000 16 000	6, 7 7, 15
Tensile strength at break	MPa	Dry, unfilled (TW300)	99.31	6
Tensile strength at yield	MPa	Dry, unfilled (TW300) Dry filled (TW241F10)	79.31 234.5	6 15
Yield stress	MPa	Dry, unfilled (TW300)	95	8
Elongation at break	%	Dry, unfilled (TW300)	30	6
Flexural modulus	MPa	Dry, unfilled (TW300) Dry filled (TW241F10)	3100 14 000	8 7, 15

Property	Units	Conditions	Value	Reference
Flexural strength at yield	MPa	Dry, unfilled (TW300)	149.6	6
		Dry filled (TW241F10)	350	15
Impact strength	ft-lb/in	Izod, 73°F, dry unfilled	1.8	6, 7
		Dry, filled (TW241F10)	2.2	7, 15
Hardness	Shore D	Durameter	85	6
Poisson ratio	–	–	0.37	6
Dielectric constant $\acute{\varepsilon}$	–	1 kHz	3.83	6
Resistivity	Ohm cm	–	5×10^{14}	6
Surface resistivity	Ohm	–	8×10^{15}	6
Thermal Conductivity	btu-in/h-ft^2-F	–	2	6
Water absorption	%	50% relative humidity	3	8
Vicat softening temperature	K	–	560.8	6
Distortion temperature	K	–	433	16
Viscosity	MPa s	Ultra thin electrospun	11–1190	17
Conductivity	S cm^{-1}	nano fibers from	2.2–3.2 (nonlinear)	17
Surface tension	mN m^{-1}	10 wt.% solution in formic acid, trace pyridine, 6 nm to 845 nm diameter	35.5–69.5	17
Inherent viscosity	dl g^{-1}	Ubbelodhe viscometer and formic acid as solvent	1.17	17
Cost	US$, lb	Dry 30% filled, 2007 Price	~4	18

References

1. Johnson M, et al. In: Brandrup J, et al, eds. *Polymer Handbook.* 4th ed. New York: John Wiley & Sons; VIII/34.
2. Palmer RJ. In: Kroschwitz JI. *Encyclopedia of Chemical Technology.* 4th ed. Vol 19. New York: John Wiley and Sons; 1996:497.
3. Weber JN. In: Kroschwitz J I. *Encyclopedia of Chemical Technology.* 4th ed. Vol 19. New York: John Wiley and Sons; 1996:571.
4. O'Sullivan D. Chemical and Engineering News 1984;62 (21):33.
5. Gaymans RJ, et al. *J. Polym. Sci.* 1977;15:537.
6. Material Data Sheet on Stanyl TW 300 (dry). Ashland, Inc., 1996–1997.
7. DSM Engineering Plastic Web Site (ww.dsmep.com).
8. Johnson RW. et al. In: Mark H F, et al. *Encyclopedia of Polymer Science and Engineering.* Vol 11. New York: John Wiley and Sons; 1988:371.
9. Andrew R, et al. In: Brandrup J, et al, eds. *Polymer Handbook.* 4th ed. New York: John Wiley & Sons; VI/235.
10. De Vries NK. *Polymer Bull. (Berlin).* 1991;26 (4):451.
11. Franco L, et al. *Polymer* 1999;40:3255.
12. Ramesh,C, et al. *Macromolecules* 1999;32:3721.
13. Rastogi S, et al. *Macromolecules* 2004;37:8825.

14. Bergshoef M, et al. *Advanced Materials* 1999;11 (16):1362.

15. Material Data Sheet on Stanyl TW241F10 (dry). DSM Engineering Plastics, 1996.

16. Ash M, Ash I, eds. *Handbook of Plastics Compounds, Elastomers, and Resins.* New York: VCH Publishers; 1992:177.

17. Huang C, et al. *Nanotechnology* 2006;17:1558.

18. www.ides.com.

Nylon 6

Paul G. Galanty and Karl I. Jacob

Acronym, Alternative Name, Trade Names Polyamide 6 (PA-6), poly-ε-caproamide, Capron®, Ultramid®, Nylatron®

Class Aliphatic polyamides

Structure $[-NH(CH_2)_5CO-]$

Major Applications Gears, fittings, and bearings, electrical switches, bobbins, and connectors, food packaging film, monofilament for weed trimmers and fishing lines, blow and roto-molded containers, wire and cable jacketing, power tool housings, wheelchair wheels, automotive cooling fans, and other underhood parts.

Properties and Special Interest Excellent mechanical properties such as strength and stiffness especially at elevated temperature, good chemical and abrasion resistance, low coefficient of friction, good toughness and impact resistance.

Preparative Techniques (a) Hydrolytic polymerization of ε-caprolactam[1] at a temperature between 250 and 260°C, pressure cycle for hydrolysis and addition but vacuum cycle for accelerated condensation rates; (b) solid state polymerization for very high molecular weight[1] at a temperature between 140 and 170°C (lower temperature is preferred since higher temperature causes discoloration); high vacuum to reach target molecular weight; nitrogen purge at atmospheric pressure is a costly alternative.

Property	Units	Conditions	Value	Reference
Monomer			ε-Caprolactam	(2)
Molecular mass	g mol^{-1}	–	113.16	
Melting point	K	–	342.2	
Boiling point	K	–	543	
Bulk density	g cm^{-3}	–	0.6–0.7	
Polymerization heats of reaction	kcal mol^{-1}	Hydrolysis	2.1	(3)
		Addition	-4	
		Condensation	-6.1	
Repeat unit	$[-NH(CH_2)_5CO-]$			(4)
Molecular mass range	g mol^{-1}	Typical as sold	$1.8–5.2(\times 10^4)$	(3)
		Solid state	$\sim 1 \times 10^5$	
Typical polydispersity Index	–	–	1.9–2.0	(3)
IR (characteristic absorption frequencies)	cm^{-1}	Assignment		(5)
		N–H hydrogen-bonded stretch	3300	
		C=O amide I stretch	1640	
		C–N amide II stretch	1545	

Chemical resistance[6]

Chemical	Temp (°C)	Conc (%)	Rating			
			Excellent	Good	Poor	Severe Attack
Acetone	23	100	x			
Benzene	23	100	x			
Ethylene glycol/water	23	50		x		
Ethylene glycol/water	120	50				x
Formic acid	23	90				x
Gasoline	100	100	x			
Gasoline/methanol (15%)	23	15			x	
Hydrochloric acid	23	10				x
Potassium hydroxide	23	10		x		
Potassium hydroxide	60	20			x	
Trichloroethylene	23	100		x		
Water	<50	100		x		
Water (steam)	100–150	100				x

Property	Units	Conditions	Value	Reference
Lattice	–	–	Monoclinic	(7)
Unit cell dimensions	nm	α structure	$a = 0.956$, $b = 1.724$, $c = 0.801$	(7,21)
		β structure	$a = 0.960$, $b = 1.720$, $c = 0.960$	
		γ structure	$a = 0.933$, $b = 1.688$, $c = 0.920$	
Unit cell angles	Degrees	α structure	67.5	(7,21)
		β structure	60.0	
		γ structure	59.0	
Unit cell forms		Monoclinic α		(21)
		Hexagonal β		
		Monoclinic and Orthorhombic γ		
Unit cell contents	–	α structure	4	(7,21)
		γ structure	4	
Heat of fusion	Jg^{-1}	Calorimetry	188	(8)
Density	$g\,cm^{-3}$	ASTM D-792 (Dry as molded)	1.13	(9)
		35 % Mineral filler	1.42	
Amorphous density	$g\,cm^{-3}$		~1.084	
Crystalline density	$g\,cm^{-3}$	25°C	1.23	
Degree of crystallinity	%	Typically molded	~50	(10)
Isothermal crystallization parameters		n	2.8	(7,22)
		$\log K$	−2.4	
	min	$t_{(1/2)}$	6.4	
	Jg^{-1}	ΔH	26	

Property	Units	Conditions	Value	Reference
Mark–Houwink parameters (K,a)	K = ml g^{-1} a = none	In 85% formic acid 25°C	K = 0.023 a= 0.82	(11)
Polymorphs	–	–	α and γ	(7)
Refractive index	–	All thickness < 0.5 mm molded, undrawn	1.53	(12)
Glass transition temperature	K	DSC	320–330	(13)
Melting point	K	ASTM D–789 (Fisher Johns)	493	(9)
Specific heat	J g^{-1} K^{-1}	Neat 0°C Neat 120°C Neat 160°C	1.38 2.30 2.68	(14)

Thermal expansion coefficient	mm(mm K)$^{-1}$	ASTM D–696
		   Neat resin
		   33% Glass fiber
		   55% Glass fiber
		   23% Mineral filler
		   35% Mineral filler
			
			 8.30×10^{-4}
			 3.80×10^{-4}
			 1.60×10^{-5}
			 4.20×10^{-5}
			 4.50×10^{-5}
				(9,19,20)

Heat deflection
   temperature (under load)
	K	ASTM D–648
		   Neat resin
		     Load = 1.80 MPa
		     Load = 0.45 MPa
		   33% Glass fiber
		     Load = 1.80 MPa
		     Load = 0.45 Mpa
		   55% glass fiber
		     Load = 1.82 MPa
		     Load = 0.46 MPa
		   23% mineral filler
		     Load = 1.82 MPa
		     Load = 0.46 MPa
		   35% mineral filler
		     Load = 1.82 MPa
		     Load = 0.46 MPa
			
			 338
			 438
			
			 483
			 493
			
			 537
			 543
			
			 433
			 478
			
			 448
			 483
				(9,19,20)

Tensile strength (yield)	MPa	ASTM D–638
		   Neat resin
		     DAM, 23°C
		     DAM, 121°C
		     50% RH, 23°C
		   33% Glass fiber
		     DAM, 23°C
		     DAM, 121°C
		     50% RH, 23°C
		   55% Glass fiber by weight
		     DAM, −40°C
		     DAM, 23°C
		     DAM, 80°C
			
			 79
			 21
			 36
			
			 200
			 83
			 127
			
			 295
			 245
			 170
				(9,19,20)

Property	Units	Conditions	Value	Reference
		23% Mineral fiber		
		DAM, −40°C	65	
		DAM, 23°C	51	
		DAM , 80°C	28	
		35% Mineral fiber		
		DAM, −40°C	120	
		DAM, 23°C	95	
		DAM, 80°C	50	
		With water absorption		
		23% Mineral fiber		
		(1.8% water content)		
		−40°C	61	
		23°C	38	
		80°C	24	
		35% Mineral fiber		
		(2.3% water content)		
		−40°C	115	
		23°C	65	
		80°C	40	
Tensile elongation at yield	%	ASTM D–638		(9,19,20)
		Neat resin		
		DAM, 23°C	7	
		DAM, 121°C	15	
		50% RH, 23°C	7	
		33% Glass fiber		
		DAM, 23°C	3	
		DAM, 121°C	7	
		50% RH, 23°C	6	
Tensile elongation (ultimate)	%	ASTM D–638		(9,19,20)
		Neat resin		
		DAM, 23°C	70	
		DAM, 121°C	>300	
		50% RH, 23°C	260	
		33% Glass fiber		
		DAM, 23°C	3	
		DAM, 121°C	7	
		50% RH, 23°C	6	
		55% Glass fiber		
		DAM, 23°C	1.8	
		23% Mineral filler		
		DAM, 23°C	1.5	
		35% Mineral filler		
		DAM, 23°C	3.2	

Property	Units	Conditions	Value	Reference
Flexural strength	MPa	ASTM D–790		(9,19,20)
		Neat resin		
		DAM, 23°C	108	
		DAM, 121°C	17	
		50% RH, 23°C	35	
		33% Glass fiber		
		DAM, 23°C	276	
		DAM, 121°C	112	
		50% RH, 23°C	179	
		55% Glass fiber	*	
		DAM, −40°C	400	
		DAM, 23°C	330	
		DAM, 80°C	230	
		23% Mineral filler		
		DAM, −40°C	110	
		DAM, 23°C	90	
		DAM, 80°C	50	
		35% Mineral filler		
		DAM, −40°C	175	
		DAM, 23°C	145	
		DAM, 80°C	80	
Flexural modulus	MPa	ASTM D–790		(9,19,20)
		Neat resin		
		DAM, 23°C	2,829	
		DAM, 121°C	304	
		50% RH, 23°C	738	
		33% Glass fiber		
		DAM, 23°C	9384	
		DAM, 121°C	3319	
		50% RH, 23°C	5127	
		55% Glass fiber		
		DAM, −40°C	19 000	
		DAM, 23°F	17 500	
		DAM, 80°F	15 000	
		23% Mineral filler		
		DAM, −40°C	7200	
		DAM, 23°C	6100	
		DAM, 80°C	3400	
		35% Mineral filler		
		DAM, −40°C	8600	
		DAM, 23°C	7300	
		DAM, 80°C	4000	

Property	Units	Conditions	Value	Reference
Notched Izod impact strength	J m^{-1}	ASTM D–256		(9,19,20)
		Neat resin		
		DAM, 23°C	53	
		50% RH, 23°C	NB	
		33% Glass fiber		
		DAM, 23°C	117	
		50% RH, 23°C	235	
		55% Glass fiber		
		DAM, −40°C	140	
		DAM, 23°C	150	
		23% Mineral filler		
		DAM, −40°C	30	
		DAM, 23°C	40	
		35% Mineral filler		
		DAM, −40°C	58	
		DAM, 23°C	95	
Tensile modulus (as fibers)	MPa	Undrawn fiber	900	(23,24)
		Zone-drawn fiber	1400–2280	
		Zone-annealed fiber	2400–9910	
Tensile strength (break)	MPa	Undrawn fiber	118–150	(23,24)
		Zone-drawn fiber	120–310	
		Zone-annealed fiber	450–938	
Elongation to break	%	Undrawn fiber	340–620	(23,24)
		Zone-drawn fiber	86.8	
		Zone-annealed fiber	13.6	
Hardness, Rockwell	R scale	ASTM D–785, DAM, 23°C		(9,19,20)
		Neat resin	119	
		33% Glass fiber	121	
		55% Glass fiber	120	
		35% Mineral filler	120	
Compression strength	MPa	ASTM D–695, DAM, 23°C		(19,20)
		Neat resin	85	
		30% Glass fiber	155	
		23% Mineral filler	62	
		35% Mineral filler	100	
Abrasion resistance	mg(kcycles)$^{-1}$	ASTM D–1044, Taber, DAM, 23°C		(15,19,20)
		Neat resin	9	
		33% Glass fiber	30	
		23% Mineral fiber	32	
		35% Mineral filler	30	
Volume resistivity	ohm cm	ASTM D–257; DAM, 23°C		(9,19,20)
		Neat resin	1.0×10^{14}	
		33% Glass fiber	1.0×10^{15}	
		55% Glass fiber	1.0×10^{15}	
		23% Mineral filler	1.0×10^{15}	
		35% Mineral filler	1.0×10^{15}	

Property	Units	Conditions	Value	Reference
Melt flow index	$g(10\ min)^{-1}$	DAM		(19,20)
		Neat resin	44.0	
		30% Glass fiber	19.0	
		55% Glass fiber	4.0	
		23% Mineral filler	20.0	
		35% Mineral filler	3.0	
Surface resistivity	ohm	ASTM D–257, DAM, 23°C		(9,19,20)
		Neat resin	1.0×10^{15}	
		33% Glass fiber	1.0×10^{15}	
Dielectric constant	1 kHz	ASTM D–150, DAM, 23°C, 1 MHz;	3.80	
		33% Glass fiber		
		55% Glass fiber	4.3	
Dissipation factor	1 kHz	ASTM D–150; DAM, 1 MHz; 23°C,		(9,19,20)
		33% Glass fiber	0.022	
		55% Glass fiber	0.01	
Dielectric strength, Short time	$kVmm^{-1}$	ASTM D–149; DAM, 23°C, 3.2 mm		(9)
		Neat resin	460	
		33% Glass fiber	560	
Thermal conductivity	$W(mK)^{-1}$	–	0.23	(16)
Coefficient of friction	–	ASTM D–1894, Polymer to steel		(17)
		Neat resin		
		Static	0.24	
		Dynamic	0.16	
		33% Glass fiber		
		Static	0.25	
		Dynamic	0.16	
Thermal index ratings	K	UL–746B; 60 000-h		(9)
		Half-life at indicated temperature; 3.2 mm		
		Neat resin		
		Mechanical	378	
		Impact	378	
		Electrical	403	
		33% glass fiber		
		Mechanical	413	
		Impact	393	
		Electrical	413	

Property	Units	Conditions	Value	Reference
Moisture absorption	%	ASTM D–570; 23°C		(9,19,20)
		Neat resin		
		50% RH equilibrium	2.7	
		Saturation	9.5	
		33% Glass fiber		
		50% RH equilibrium	1.9	
		Saturation	6.7	
		55% Glass fiber		
		65% RH equilibrium	0.8	
		23% Mineral filler		
		65% RH equilibrium	1.8	
		35% Mineral filler		
		65% RH equilibrium	2.3	
Flammability ratings		UL–94; 3.2 mm		(9,19,20)
		Neat resin	V-2	
		33% Glass fiber	HB	
		55% Glass fiber	V-0	
Price	US $ kg^{-1}	Estimated average	1.50	
		1997 Price—Commercial quantities		(18)
		(2–5 $\times 10^5$ kg/year)		
		Neat resin	0.66–0.68	
		33% Glass fiber	0.68–0.70	
Major suppliers		Honeywell, Hopewell, VA; BASF, Freeport, TX; (Fibers) Shaw Industries, Dalton, GA; Wellman, Inc, Fort Mill, SC; (Composite resin) Toyobo & Toray, Japan		

Small uncertainties are reported with fundamental material parameters (such as crystal unit cell dimensions (21)). Large variations of properties (such as inflammability) are obtained with different additives, thus the data presented here is representative of nylon-6 systems to provide a general trend rather than a comprehensive data for all nylon-6 systems. For additional specific data for Nylon-6-based composites please see manufacturer's database.

* DAM = tested dry as molded; 50% RH = tested after equilibration in a 50% RH, 23°C room.

References

1. Kohan ML. *Nylon Plastics Handbook.* Hanser/Gardner Publications; 1995:525–526.
2. Bander J. AlliedSignal Nylon-6 Databook, Section F, 1985.
3. Kohan ML. *Nylon Plastics Handbook.* Hanser/Gardner Publications; 1995:524.
4. Kohan ML. *Nylon Plastics Handbook.* Hanser/Gardner Publications; 1995:110
5. Kohan ML. *Nylon Plastics Handbook.* Hanser/Gardner Publications; 1995:85
6. Chemical Resistance Guide. *AlliedSignal Plastics.* 1995:2–7.
7. Kohan ML. *Nylon Plastics Handbook.* Hanser/Gardner Publications; 1995:114–119.
8. Kohan ML. *Nylon Plastics Handbook.* Hanser/Gardner Publications; 1995:142.
9. Capron® Nylon Resins Product Guide. *AlliedSignal Plastics.* 1993.
10. Kohan ML. *Nylon Plastics Handbook.* Hanser/Gardner Publications; 1995:125.
11. Kohan ML. *Nylon Plastics Handbook.* Hanser/Gardner Publications; 1995:81.

12. Kohan ML. *Nylon Plastics Handbook*. Hanser/Gardner Publications; 1995:348.

13. Kohan ML. *Nylon Plastics Handbook*. Hanser/Gardner Publications; 1995:147.

14. Kohan ML. *Nylon Plastics Handbook*. Hanser/Gardner Publications; 1995:344.

15. Product Information Bulletin, 92–102, AlliedSignal Plastics, 1992.

16. Kohan ML. *Nylon Plastics Handbook*. Hanser/Gardner Publications; 1995:344.

17. Product Information Bulletin, 92–103, AlliedSignal Plastics, 1992.

18. AlliedSignal Plastics, Sales Department, 1997.

19. http://www.toyobo.co.jp/e/seihin/xj/enpla/nylon/

20. http://prospector.ides.com/

21. Parker JP, Lindenmeyer PH. On the crystal structure of Nylon – 6. *Journal of Applied Polymer Science*.1977;**21**:821–837. (online 2003).

22. Yoon K, Polk MB, Min BG, Schiraldi DA.'*Polymer International*'. 2004;**53**:2072–2078.

23. Kunugi T, Akiyama I, Hashimoto M. Preparation of high-modulus and high-strength nylon-6 fibre by the zone-annealing method. *Polymer*1982;**23**: 1193–1198.

24. Fakirov S, Evstatiev M, Schultz JM. Polyamides and polyesters with improved mechanical properties. *Journal of Applied Polymer Science*.1990;**42**:575–581.

Nylon 6,6

Brent D. Viers

Common Name Poly(hexamethylene adipamide)

IUPAC Nomenclature Poly[imino(1,6-dioxohexamethylene) imnohexamethylene], poly[iminoadipoly iminohexamethylene]

Trade Names Zytel (DuPont), Maranyl (ICI), Ultramid A (BASF)

CAS Registry Numbers 32131-17-2

Class Aliphatic polyamides

Structure

Major Applications Gear teeth, pinions, ball bearing cages, switch parts, spools, electro-insulating parts; semifinished products, pipes, profiles; machine parts, parts subject to wear such as friction bearings, roller bearing cages, engine parts, water pump impellers, and also parts of door locks; fan and blower wheels, parts of housings, fuel filters, clips, chain tension rails; sliding bearings for swivel chairs and folding tables, sliding feet and fittings, connecting parts in furniture making; patio and party furniture.

Properties of Special Interest High mechanical strength, great rigidity, good deep-drawing behavior, good dimensional stability under heat, good toughness even at low temperatures, favorable tribological properties, good resistance to chemicals, very good electro-insulating properties, good dimensional stability, rapid processing. Relatively high T_m and T_g for aliphatic polyamides (used in synthetic fibers; can be toughened as resin).

Other Polymers Showing this Special Property Poly(ε-caprolactam) (Nylon 6). (See also the entry on *Nylon 6* in this handbook.)

Property	Units	Conditions	Value	Reference
Type of polymerization	–	Interfacial, melt	Condensation between hexamethylene diamine and adipic acid	–
Enthalpy of reaction $-\Delta H_a$	kJ mol^{-1}	Decrease in temperature shifts equilibrium to higher molecular weight	25–29 (average) 42–46 (limit)	(1)
Side products		1–2% of cyclic oligomers (14 membered ring)		(1)

Property	Units	Conditions	Value	Reference
Kinetic parameters	–	Second order kinetics, not accelerated by catalysts	Conversion <90%	(2, 3)
		Third order kinetics, carboxyl catalyzed, low water	Conversion >90%	(1, 4)
Typical comonomers	–	Aliphatic amines	Heptamethylene diamine, octamethylene diamine, decamethylene diamine, cyclohexyl diamine	–
		Aliphatic diacids (sometimes diacid chlorides)	Sebacic acid, undecanedioic acid, suberic acid	
		Aromatic amines	p-Xylenediamine	
		Aromatic diacids/diacid chlorides	Terephthalic acid, phenylenedipropionic acid	
Molecular weight (of repeat unit)	g mol^{-1}	–	226.3	–
Degree of branching	Can be controlled by controlled copolymerization of polyfunctional amines/acids			–
Typical molecular weight range of polymer	g mol^{-1}	High conversion ($p > 0.99$) (lowered by monofunctional end-blockers)	12,000–20,000	(1)
Typical polydispersity index (M_w/M_n)	–	Most probable Undergoes amide interchange reactions (broadened by incorporation of multifunctional units)	2 (expected) 1.7–2.1 (by GPC)	(1)
Typical viscosity averages	–	Relative viscosity: 8.4% solution in 90% formic acid	$\eta_r = 30\text{–}70$	–
		Inherent viscosity: 0.5 g/100 cm^3 in m-cresol	$\eta_{onh} = 1$ dlg^{-1}	
IR (characteristic absorption frequencies)	cm^{-1}	NH wag (broad)	700	(5, 6)
		CH$_2$ rock	722	
		N—C=O skeletal vibration	1,170	
		N—C=O skeletal vibration	1,200	
		CH$_2$ wag	1,370	
		CH$_2$ symmetric scissors deformation (CH$_2$ next to C=O)	1,420	
		CH$_2$ symmetric scissors deformation (CH$_2$ next to N)	1,440	

Property	Units	Conditions	Value	Reference
IR (characteristic absorption frequencies)	cm^{-1}	CH_2 symmetric scissors deformation	1,460	(5, 6)
		NH bend/C—N stretch	1,540	
		Amide C=O stretch	1,640	
		CH_2 symmetric stretch	2,860	
		CH_2 asymmetric stretch	2,920	
		$2 \times$ NH bend (1,540) overtone	3,100	
		NH stretch	3,300	
NMR	s^{-1}	^{13}C NMR T_1 relaxation times		(7)
		Amorphous	1.37	
		Meso	9.2	
		Crystalline	82.5	
Thermal expansion coefficients	K^{-1} ($\times 10^{-4}$)	Zytel ASTM D 696	0.81	(8)
		Crystalline, volumetric 20°C	2.8	(9)
		Linear crystalline, 20°C	7–10	(10)
		Linear crystalline, 100°C	10–14	(10)
		Triclinic, α_a	2.1	(11)
		Triclinic, α_c	22.0	(11)
Compressibility coefficients	$(MPa)^{-1}$ ($\times 10^{-6}$)	Pressure/temperature dependence		
		50 MPa, 20°C	62	(12)
		100 MPa, 20°C	58	(12)
		150 MPa, 20°C	54	(12)
		300 MPa, 20°C	50	(12)
		50 MPa, 120°C	125	(12)
		100 MPa, 120°C	115	(12)
		150 MPa, 120°C	95	(12)
		300 MPa, 120°C	75	(12)
		>100 MPa, 200°C	>300	(13)
Molar volume	$cm^3 \, mol^{-1}$	20°C, nylon rods	193	(9)
		20°C, amorphous	207.5	(14)
		20°C, amorhous group contribution calculation	208.3	(14)
Density (amorphous)	$g \, cm^{-3}$	Zytel ASTM D 792, 23°C	1.14	–
Solvents		Room temperature: trifluoroethanol, trichloroethanol, phenols, chloral hydrate, formic acid, chloro-acetic acid, HF, HCl, methanol, H_2SO_4, phosphoric acid, benzyl alcohol, ethylene chlorohydrin, 1,3 chloropropanol, 2-butene-1,4,diol., diethylene glycol, acetic acid, formamide, DMSO		–
Nonsolvents		Hydrocarbons, aliphatic alcohols, chloroform, diethyl ether, aliphatic ketones, esters		–
Chemical resistances		Acid resistance: limited; attacked by strong acids; general order of resistance nylon 6,12 > nylon 6,6 > copolymers or nylon 6		(8)
		Base resistance: Excellent at room temperature; attacked by strong bases at elevated temperatures		

Property	Units	Conditions	Value	Reference
		Solvent resistance: generally excellent; some absorption of such polar solvents as water, alcohols, and certain halogenated hydrocarbons causing plasticization and dimension changes		
Solubility parameter	$(MPa)^{1/2}$	δ	27.8	(15)
			24.02	(16)
			22.87	(17)
		Dispersive component δ_D	18.62	(16, 17)
		Polar component δ_P	5.11	(16)
		Hydrogen bonding	14.12	(17)
		component δ_H	12.28	(16)
Theta temperature Θ	K	Carbon tetrachloride/ m-cresol/cyclohexane	293	(18)
		Formic acid/KCl/H_2O	298	(19, 20)
Second virial coefficient A_2	mol cm^3 g^{-2} ($\times 10^{-4}$)	m-Cresol, 60°C, $M_n = 18,000$	183	(21)
		Formic acid (90%), 25°C, $M_n = 18,000$	840	(22)
		Formic acid (90%)/0.2–2.5 M KCl, 25°C, $M_n = 31,000$		(20)
		At 0.2 M KCl	59.2	
		At 2.5 M KCl	7.0	
		Formic acid (90%)/2.3 M KCl, 25°C, $M_n = 31,000$	0	(20)
		Formic acid (82.5–40%), 2 M KCl, 25°C, $M_n = 31,000$		(20)
		At 82.5%	−9.4	
		At 40%	36.5	
		Formic acid (90%), 2 M KCl, $2,000 < M_n < 52,000$		(18)
		At 2,000 M_n	312	
		At 52,000 M_n	10.1	
		Formic acid (75–98%), 0.5 M NaHCOO, 2,2,3, 3-tetrafluoropropanol, 25°C, $M_n = 32,000$	1.0–4.0	(23)
		2,2,3,3-tetrafluoropropanol, 0.1 M sodium trifluroacetate, 25°C, $M_n = 62,000$	57.1	(24)
Fractionation	–	Fractional precipitation	m-Cresol/ cyclohexane	(25, 26)
			Phenol/H_2O	(27)
		Turbitimetric titration	m-Cresol/ cyclohexane	(28)
			m-Cresol/n-heptane	(29)
		Chromatography	Methylene chloride	(30)
		Gel permeation	Hexafluoroisopropanol	(31)

Property	Units	Conditions	Value	Reference
		Partition chromatography, 20°C	Formic acid/H_2O (88%)	(32)
		Sedimentation gradient: ultracentrifugation	Carbon tetrachloride/ *m*-cresol/ cyclohexane	(19)
		Continuous immiscible liquid distribution	Phenol/water	(33)
Mark-Houwink parameters: K and a	K = ml g^{-1} a = None	*o*-Chlorophenol, 25°C, $14{,}000 < M_n < 50{,}000$	$K = 168$, $a = 0.62$	(34)
		m-Cresol, 25°C, $14{,}000 < M_n < 50{,}000$	$K = 240$, $a = 0.61$	(34)
		m-Cresol, 25°C, $150 < M_n < 50{,}000$	$[\eta] = 0.5 + 0.0353 M^{0.792}$	(18)
		Dichloroactetic acid, 25°C, $150 < M_n < 50{,}000$	$[\eta] = 0.5 + 0.0352 M^{0.551}$	(18)
		2,2,3,3-Tetrafluoropropanol/ CF_3COONa(0.1 M), 25°C, $14{,}000 < M_n < 50{,}000$	$K = 114$, $a = 0.66$	(34)
		Aqueous HCOOH (90 vol%), 25°C		
		$6{,}000 < M_n < 65{,}000$	$K = 35.3$, $a = 0.786$	(34)
		$5{,}000 < M_n < 25{,}000$	$K = 110$, $a = 0.72$	(27)
		$14{,}000 < M_n < 50{,}000$	$[\eta] = 2.5 + 0.0132 M^{0.873}$	(18)
		HCOOH (90%)/HCOONa (0.1 M), 25°C		
		$10{,}000 < M_n < 50{,}000$	$K = 32.8$, $a = 0.74$	(34)
		$14{,}000 < M_n < 50{,}000$	$K = 87.7$, $a = 0.65$	(34)
		$150 < M_n < 50{,}000$	$[\eta] = 1.0 + 0.0516 M^{0.687}$	(18)
		HCOOH (90%)/KCl (2.3 M), 25°C		
		$14{,}000 < M_n < 50{,}000$	$K = 227$, $a = 0.50$ (θ)	(34)
		$150 < M_n < 50{,}000$	$K = 253$, $a = 0.50$ (θ)	(18)
		H_2SO_4 (95%), 25°C, $150 < M_n < 50{,}000$	$[\eta] = 2.5 + 0.0249 M^{0.832}$	(18)
		H_2SO_4 (96%), 25°C, $14{,}000 < M_n < 50{,}000$	$K = 115$, $a = 0.67$	(34)
		Melt polymer, high molecular weight	$a = 3.5$	(35, 36)
Huggins constants: k_H	–	Formic acid, 25°C		(37)
		$[\eta] = 83$ ml g^{-1}	0.20	
		$[\eta] = 100$ ml g^{-1}	0.22 ± 0.01	
		$[\eta] = 120$ ml g^{-1}	0.24 ± 0.02	
		$[\eta] = 140$ ml g^{-1}	0.27 ± 0.02	
		$[\eta] = 160$ ml g^{-1}	0.27 ± 0.02	
		$[\eta] = 180$ ml g^{-1}	0.28 ± 0.02	
		$[\eta] = 200$ ml g^{-1}	0.29 ± 0.01	

Property	Units	Conditions	Value			Reference
Schulz-Bakke coefficients k_{SB}	–	Formic acid, 25°C				(37)
		$[\eta] = 83$ ml g^{-1}	0.20			
		$[\eta] = 100$ ml g^{-1}	0.22 ± 0.02			
		$[\eta] = 120$ ml g^{-1}	0.24 ± 0.02			
		$[\eta] = 140$ ml g^{-1}	0.26 ± 0.02			
		$[\eta] = 200$ ml g^{-1}	0.28 ± 0.01			
Characteristic ratio $\langle r^2 \rangle_0 n l^2$	–	HCOOH (90%), 25°C	5.3			(25, 27)
		HCOOH (90%)/KCl 2.3 M, 25°C	6.85			(22)
			5.95			(38, 39)
End-to-end distance $r_0/M^{1/2}$	nm (×10^{-4})	HCOOH (90%), 25°C	890 ± 40			(25, 27)
		HCOOH (90%)/KCl 2.3 M, 25°C	1,010			(22)
			935			(38, 39)
Lattice (monoclinic, etc.)	–	–	(α) I: triclinic			–
			(α) I: monoclinic			
			(α) II: triclinic			
			(β) triclinic			
			(high temperature) triclinic (170°C)			
Space group	–	–	CI-1			–
Chain conformation (ρ_n of helix)	–	–	14*1/1			–

Property	Units	Conditions	a	b	c	Reference
Unit cell dimensions	Å					
		α I: monoclinic	15.7	10.5	17.3	(40)
		α I: triclinic	4.9	5.4	17.2	(41)
			5.00	4.17	17.3	(42)
			4.87	5.26	17.15	(43)
			4.97	5.47	17.29	(44)
		α II: triclinic	4.95	5.45	17.12	(44)
		β triclinic	4.9	8.0	17.2	(45)
		High temperature (170°C)	5	5.9	16.23	(46)

Property	Units	Conditions	α	β	γ	Reference
Unit cell angles	Degrees					
		α I: monoclinic	–	73	–	(40)
		α I: triclinic	48	77	63	(41)
			81	76	63	(42)
			50	76	64	(43)
			48	77	62	(44)
		α II: triclinic	52	80	63	(44)
		β triclinic	90	77	67	(45)
		High temperature	57	80	60	(46)

Property	Units	Conditions		Value		Reference
Unit cell contents (number of repeat units)	–	α I: monoclinic		9		(40)
		α I: triclinic		1		(42–44)
		α II: triclinic		1		(44)
		β triclinic		2		(45)
		High temperature		1		(46)
Bragg spacings	–	*hkl*	*d*-value (nm)	2θ (degrees)	Relative intensity	(45)
		002	0.641	13.83	w	
		100, 010, 110	0.390	22.96	vvs	
		015	0.335	26.65	w	
		110, 210	0.236	38.12	s	
		017, 127	0.233	38.69	w	
		117, 027	0.218	41.37	w	
		117, 227	0.194	46.71	w	
		020, 220	0.183	49.70	s	
Degree of crystallinity	As shown	General range		40–60%		(1)
		General equation based on density		$\alpha = 830-(900/\rho)\%$		(10)
		IR determination		Crystalline $= 852$ cm^{-1}		(1)
				Amorphous $= 1,140$ cm^{-1}		
Heat of fusion	kJ mol^{-1}	α I triclinic		46.5		(47)
				40		(47)
				36.8		(48)
				68		(48)
				58		(48)
				46.9		(49)
				53.2		(44)
		α II triclinic		43.4		(44)
				41.9		(50)
Heat of fusion (per repeat unit)	J g^{-1}	α II triclinic		191.9		(44)
Entropy of fusion	J K^{-1} mol^{-1}	–		83–86		(10)
				79.9		(44)
Density (crystalline)	g cm^{-3}	α I triclinic		1.220		(40)
				1.24		(41)
				1.241		(42)
				1.225		(43)
				1.204		(44)
		α II triclinic		1.152		(44)
				1.165		(45)
		β triclinic		1.25		(46)
		High temperature (170°C) triclinic		1.10		(5)
		Crystalline molded		1.09		(51)
				1.13–1.145		(51)

Property	Units	Conditions			Value	Reference
Density (amorphous)	$g cm^{-3}$	a I triclinic			1.09	(52)
					1.12	(53)
					1.069	(54)
					1.095	(50)
		a II triclinic			1.095	(44)
		b, triclinic			1.09	(53)
		Amorphous molded			0.989	(51)
		Melt, 270°C			1.248	(5)
Crystal modulus	dynes cm^{-2}	$\alpha 1$			175×10^4	(7)
Polymorphs (listing)	–	–			α I, α II, β, high temperature	–
Crystal growth activation energy	$kJ mol^{-1}$	–			64.5	(10)
Maximum crystallization rate	–	150°C			–	(7)
Growth rate (T_f = fusion temperature; T_c = crystallization temperature)	$\mu m s^{-1}$ $nm s^{-1}$	Maximum linear growth			20	(51)
		$M_n \times 10^3$	$T_f(°C)$	$T_c(°C)$		
		11.6	295	241	166.7	(55)
			295	250	13.84	(55)
			295	252	10.50	(55)
			285	247	66.08	(55)
			262 (10 min)	251	14.21 (negative spherulites)	(56)
				256	83.4	(56)
				257	13.3	(57)
				259	9.17	(57)
				261	6.67	(57)
				263	4.17	(57)
				265	2.50	(57)
		12.9	300 (30 min)	246	106.7	(58)
				248	56.34	(58)
				253	10.84	(58)
		13.7	300 (30s)	141	13,502.7	(59)
				160	13,669.4	(59)
				180	12,119.1	(59)
				199	8,901.8	(59)
				215	5,167.7	(59)
				230	2,117.1	(59)
				234	1,530.3	(59)
				237	920.18	(59)
				239.5	765.15	(59)
				241	471.76	(59)
				244	368.40	(59)

Property	Units	Conditions			Value	Reference
Growth rate	nms^{-1}	$M_n \times 10^3$	T_f (°C)	T_c (°C)		
(T_f = fusion		14.6	280	241.5	283.39	(58)
temperature;				243	230.05	(58)
T_c = crystal				245	180.86	(58)
lization				248	33.685	(58)
temperature				252	14.66	(58)
			300	241.5	204.4	(58)
				243	175.0	(58)
				245	128.3	(58)
				248	58.34	(58)
		14.6	300	252	$G = 5.501$	(60)
			315	241.5	280.0	(60)
				243	168.4	(60)
				245	113.4	(60)
				248	56.68	(60)
				252	6.335	(60)
		25.5 (M_w)		50	3,650.7 (positive spherulites)	(60)
				100	4,706.6	(60)
				142	6,751.3	(60)
				160	6,101.2	(60)
				178	5,201.0	(60)
				198	3,700.7	(60)
				200	12,900.6	(60)
				228	466.7	(60)
		25.5	300	180	11,435.6 (positive spherulites)	(60)
				200	7,951.5	(60)
				211	5,284	(60)
				220	2,733.8	(60)
				230	1,615.3	(60)
				235.5	680.13	(60)
				240	483.4	(60)
Hoffman-Lauritzen theory constants						(55)
Growth rate constant G_0	cms^{-1}	–			1.55×10^3	
Diffusion activation energy U^*	cal mol^{-1}	–			167	
Chain dimensions	Å	–			$a_0 = 4.76$, $b_0 = 3.70$	
Nucleation rate constant K_g	K^2	–			1.02×10^5	
Lateral surface free energy σ	erg cm^{-2}	–			8.0	
Fold surface free energy σ_e	erg cm^{-2}	–			40	
Melting point (equilibrium)	K	T_m (determined by $T_m - T_c$ extrapolation)			542.2	(44)

Property	Units	Conditions	Value	Reference
Deformation induced crystallization	–	Spinning effects	–	(7)
Glass transition temperature	K	Dependent on relative humidity (water plasticized)	320–330	(1)
		Oven dry	351	(7)
		50% RH	308	(7)
		100% RH	258	(7)
Melting point	K	General	537	–
		α I: monoclinic	543	
		α I: triclinic	534–574	
		α II: triclinic	542.5	
Sub-T_g transtion temperatures	K	β (plasticized glass transition) At 11 Hz	363	(61)
		–	357	(62)
		At 11 Hz γ (amide hydrogen bond motions with sorbed H_2O)	370	(63)
		At 40–600 Hz	249	(64)
		–	245	(62)
		δ (methylene group motion) At 40–600 Hz	156	(64)
		–	186	(62)
Heat capacity (of repeat units)	$kJ\,kg^{-1}K^{-1}$	DSC annealed nylon solid	1.4	(65)
Deflection temperature	K	Zytel ASTM D 648		(8)
		0.5 MPa	508	
		1.8 MPa	363	
Tensile modulus	MPa	Nylon 23°C	3,300	(10)
		Nylon 23°C moist ISO-1110	1,700	
		Nylon, 100°C	600	
Bulk modulus	MPa	Nylon dry crystalline rods	3,300	(10)
Shear modulus	MPa	23°C	1,300	(10)
		23°C (nucleated)	1,700	
		100°C	300	
		200°C	150	
Shear strength	MPa	Zytel Resins ASTM D 732, 23°C	66.8–72.4	(8)
		50% relative humidity, 23°C	63.4–68.9	
Storage modulus	MPa	0.1–110 Hz	5–100	(7)
Loss modulus	–	0.1–110 Hz, log tan δ	−1.3 to 0.9	(7)
Tensile strength	MPa	Zytel Resins ASTM D 638		(8)
		−40°C	113.8–128.9	
		23°C	82.7–90.3	
		77°C	58.6–62.1	
		121°C	42.7–47.6	

Property	Units	Conditions	Value	Reference
		50% relative humidity		
		−40°C	110.3–117.2	
		23°C	62.1–77.2	
		77°C	40.7–50.3	
		121°C	32.4–42.1	
Yield stress	MPa	Zytel Resins ASTM D 638		(8)
		−40°C	113.8–128.9	
		23°C	82.7–90.3	
		77°C	44.8–58.6	
		121°C	33.1–34.5	
		50% relative humidity		
		−40°C	110.3–117.2	
		23°C	58.6–62.1	
		77°C	39.3–40.7	
		121°C	27.6–32.4	
Yield strain $(L/L_0)_y$	%	Zytel Resins ASTM D 638		(8)
		−40°C	4–5	
		23°C	4–5	
		77°C	25–30	
		121°C	30–45	
		50% relative humidity		
		−40°C	5	
		23°C	25–30	
		121°C	30–40	
Maximum extensibility $(L/L_0)_r$	%	Zytel ASTM D 638		(8)
		−40°C	10–15	
		23°C	30–60	
		77°C	145–>300	
		121°C	200–>300	
		50% relative humidity		
		−40°C	15–35	
		23°C	200–>300	
		77°C	250–>300	
		121°C	>300	
Flexural modulus	MPa	Zytel ASTM D 790		(8)
		−40°C	3,241–3,516	
		23°C	2,827–2,964	
		77°C	689–724	
		121°C	538–552	
		50% relative humidity		
		40°C	3,447	
		23°C	1,207–1,310	
		77°C	565–586	
		121°C	414	

Property	Units	Conditions	Value	Reference
Impact strength	J m^{-1}	Zytel ASTM D 256 Izod		(8)
		−40°C	32	
		23°C	53–64	
		50% relative humidity Izod		
		−40°C	27	
		23°C	112–133	
	kJ m^{-2}	Zytel ASTM D 1822 tensile impact, 23°C		
		Long specimen	504	
		50% RH long specimen	1,470	
		Short specimen	157	
		50% RH short specimen	231	
Compressive strength	MPa	20°C nylon molded 2.5% H_2O		(10)
		1% strain	14	
		2% strain	28	
		4% strain	56	
		6% strain	70	
Hardness	–	Zytel ASTM D676 Durometer	89	(8)
		50% Relative humidity	82	
Poisson ratio	–	General extruded rod	0.41	(8)
			0.38	(10)
		100°C	0.44	(10)
		Melt	0.5	(10)
Abrasion resistance	gMHz^{-1}	Zytel Taber abrasion CS-17 wheel, 1,000 g	4–7	(8)
Refractive index increment *dn/dc*	ml g^{-1}	(All data at 25°C)	(Source wavelength noted)	
		Formic acid 90% + 0.5 M sodium formate	0.137 (436 nm)	(66)
		Trifluroethanol	0.228 (436 nm)	(66)
(Trifluoroacetylated nylon 6,6)		Acetone	0.076 (436 nm)	(67)
Refractive index increment *dn/dc*	ml g^{-1}	(All data at 25°C) Formic acid	(Source wavelength noted)	
		75%	0.144 (633 nm)	(39)
		80%	0.145 (633 nm)	(39)
		85%	0.141 (436 nm)	(68)
		90%	0.145 (633 nm)	(20, 39)
		90%	0.145 (546 nm)	(18)
		90%	0.145 (436 nm)	(68)
		95%	0.150 (633 nm)	(39)
		100%	0.157 (633 nm)	(39)
		100%	0.1525 (436 nm)	(69)
		Formic acid + KCl		(20)
		85% + 2.0 M KCl	0.124 (633 nm)	
		90% + 0.2 M KCl	0.143 (633 nm)	
		90% + 0.5 M KCl	0.140 (633 nm)	
		90% +1.0 M KCl	0.136 (633 nm)	
		90% + 1.5 M KCl	0.131 (633 nm)	

Property	Units	Conditions	Value	Reference
		90% + 2.0 M KCl	0.126 (633 nm)	
		90% + 2.5 M KCl	0.122 (633 nm)	
		95% + 2.0 M KCl	0.129 (633 nm)	
		Formic acid + sodium formate		(39)
		75% + 0.5 M NaHCOO	0.138 (633 nm)	
		80% + 0.5 M NaHCOO	0.136 (633 nm)	
		90% + 0.02 M NaHCOO	0.147 (633 nm)	
		90% + 0.05 M NaHCOO	0.146 (633 nm)	
		90% + 0.10 M NaHCOO	0.142 (633 nm)	
		90% + 0.2 M NaHCOO	0.142 (633 nm)	
		90% + 0.5 M NaHCOO	0.136 (633 nm)	
		90% + 0.75 M NaHCOO	0.130 (633 nm)	
		90% + 1.0 M NaHCOO	0.124 (633 nm)	
		95% + 0.5 M NaHCOO	0.136 (633 nm)	
		100% + 0.5 M NaHCOO	0.136 (633 nm)	
		Tetrafluropropanol	0.190 (546 nm)	(24)
		Tetrafluropropanol + 0.1 N sodium trifluroacetate buffer	0.190 (436 nm)	(24)
Birefringence	–	$n_{\parallel}$	1.582	(51)
		$n_{\perp}$	1.519	(9)
Dielectric constant ε'	–	Zytel ASTM D 150		(8)
		1×10^2 Hz	4.0	
		1×10^3 Hz	3.9	
		1×10^6 Hz	3.6	
		1×10^2 Hz	8.0	
		50% relative humidity		
		1×10^3 Hz	7.0	
		1×10^6 Hz	4.6	
		–	(See also table below)	

Dielectric constant ε'

Temp. (°C)	10^2 Hz	10^3 Hz	10^4 Hz	10^5 Hz	10^6 Hz	10^7 Hz	10^8 Hz	10^9 Hz
−30	120	105	105	130	165	160	100	49
0	110	120	135	160	200	200	160	81
30	85	125	180	215	250	255	220	135
60	810	590	460	390	370	360	320	240
90	2,000	1,450	1,300	1,450	1,600	1,300	810	440
20 (50% RH)	1,100	1,020	1,000	900	700	450	280	170

Property	Units	Conditions	Value	Reference
Dielectric tan δ	–	Nylon (coupled with above table) values given as tan $\delta \times 10^4$	See table below	(10)

Nylon 6,6

Dielectric tan δ

Temp. (°C)	10^2 Hz	10^3 Hz	10^4 Hz	10^5 Hz	10^6 Hz	10^7 Hz	10^8 Hz	10^9 Hz
−30	3.1	3.1	3.1	3.0	3.0	3.0	3.0	3.0
0	3.3	3.3	3.2	3.2	3.1	3.0	3.0	3.0
30	3.6	3.5	3.4	3.4	3.2	3.1	3.1	3.0
60	5.0	4.6	4.3	4.0	3.7	3.5	3.3	3.1
90	10	8.9	7.6	6.2	5.0	4.0	3.4	3.2
20 (50% RH)	7.5	5.9	4.8	4.1	3.7	3.4	3.3	3.2

Property	Units	Conditions	Value	Reference
Dielectric strength	V cm^{-1}	VDE 0303, part 2, IEC-243, electrode K20/P50		(10)
		Dry	120×10^{-4}	
		Dry, 100°C	40×10^{-4}	
		Moist ISO-1110	80×10^{-4}	
Dissipation factor	–	Zytel ASTM D 150		(8)
		1×10^2 Hz	0.01	
		1×10^3 Hz	0.02	
		1×10^6 Hz	0.02	
		50% relative humidity		
		1×10^2 Hz	0.2	
		1×10^3 Hz	0.2	
		1×10^6 Hz	0.1	
Resistivity	ohm cm	Zytel ASTM D 257	1×10^{15}	(8)
		Zytel ASTM D 257, 50% RH	1×10^{13}	(8)
		Nylon, 20°C, 50% RH	3×10^{11}	(10)
		Nylon, 20°C, 100% RH	1×10^9	(10)
		Nylon, 60°C	6×10^{11}	(10)
		Nylon, 100°C	3×10^9	(10)
		Nylon, 100°C, 50% RH	4×10^7	(10)
Thermally stimulated current	–	Relaxation, humidity effects	–	(7)
Surface tension	mN m^{-1}	Nylon, $M_n = 17{,}000$, $M_w = 35{,}000$		
		20°C	46.5	(70)
		150°C	38.1	
		200°C	34.8	
		280°C	29.6	
		325°C	26.7	
		$-d\gamma/dT$	0.065	(41, 70)
		γ_{LV} at 20°C	46.4	(41)
		Zisman critical wetting surface tension, γ_c	42.5	(71)

Nylon 6,6

Property	Units	Conditions	Value	Reference
Contact angle θ	Degrees	Water	72	(72)
Surface free energy	mJ m^{-2}	Dispersive, γ^D	40.8	(72)
		Polar, γ^P	6.2	(72)
		Lifschitz-van Der Waals, γ^{LW}	36.4	(73)
		Lewis Acid Base, γ^{AB}	1.3	(73)
		Electron acceptor parameter, γ^+	0.02	(73)
		Electron donor parameter, γ^-	21.6	(73)
Interfacial tension	mN m^{-1}	Polyethylene, γ_{12} at 20°C	14.9	(70)
	mN m^{-1} K^{-1}	$-d\gamma/dT$	0.018	
Adhesive bond strength	MPa	Nylon-aluminum tensile	68	(74)
		Nylon-steel tensile	70	
		Nylon-copper tensile	76	
Diffusion coefficient	cm^2 s^{-1}	H_2O, 20°C	0.02×10^{-8}	(10)
		H_2O, 60°C	3.5×10^{-8}	(10)
		H_2O, 100°C	25×10^{-8}	(10)
		CO_2, 5°C, undrawn fiber	1.8×10^{-10}	(75)
		CO_2, 25°C, undrawn fiber	8.3×10^{-10}	(75)
		CO_2, 5°C, drawn fiber	1.8×10^{-10}	(75)
		CO_2, 25°C, drawn fiber	4.8×10^{-10}	(75)
Activation energy for diffusion	kJ mol^{-1}	H_2O	58	(10)
Permeability coefficient	cm^3 (STP) cm s^{-1} cm^{-2} Pa^{-1}	CO_2, 5°C, undrawn fiber	0.018×10^{-13}	(75)
		CO_2, 25°C, undrawn fiber	0.052×10^{-13}	(75)
		CO_2, 5°C, drawn fiber	0.023×10^{-13}	(75)
		CO_2, 25°C, undrawn fiber	0.071×10^{-13}	(75)
	cm^3 (NPT) m^{-2} mil^{-1} atm^{-1}	CO_2	140	(7)
		O_2	80	(7)
		N_2	5	(7)
Activation energy for permeation	–	CO_2	–	(75)
Solubility coefficient	cm^3 (STP) cm^{-3} Pa^{-1}	CO_2, 5°C, undrawn fiber	9.97×10^{-6}	(75)
		CO_2, 25°C, undrawn fiber	6.32×10^{-6}	
		CO_2, 5°C, drawn fiber	12.8×10^{-6}	
		CO_2, 25°C, undrawn fiber	14.8×10^{-6}	
Thermal conductivity	W m^{-1} K^{-1}	Zytel resins	0.25	–
Melt viscosity	Pa s	Newtonian (shear stress <30 kPa) $[\eta] = 1.09$ dl g^{-1}, $M_n = 14{,}000$	40–1,000	(1)
		270°C	110	
		280°C	70	
		290°C	50	
Speed of sound	m s^{-1}	Longitudinal; density = 1.147 g cm^{-3}	2,710	(76)
		Shear; density = 1.147 g cm^{-3}	1,120	

Property	Units	Conditions	Value	Reference
Biodegradability, effective microorganisms	–	Wood Rotting Basidiomycetes	–	(7)
Heat of combustion	kJ kg^{-1}	–	–31.400	(51)
Decomposition products	K	H_2O, CO_2, cyclopentanone, hydrocarbons	583–653	(77)
		H_2O, CO_2, NH_3, cyclic monomomer, cyclopentanone, cyclopentylidinicyclopentanone, cyclopentylcyclopentanane, hexylamine, hexamethyleneimine, hexamethylene diamine	578	(78)
Cross-linking, G factor	mol J^{-1}	Electron beam/γ irradiation	0.50	(79)
Gas evolution, G factor	mol J^{-1}	–	0.70	(79)
Water absorption	%	Zytel ASTM D 570		(8)
		24 h immersion, 23°C	1.2	
		Saturation, 23°C	8.5	
		Annealed (Karl Fisher method)	7	(1)
Solvent absorption	%	Ethanol, 20°C, saturation	9–12	(10)
		Butanol, 20°C, saturation	4–8	
		Glycol, 20°C, saturation	2–10	
		Methanol, 20°C, saturation	9–14	
		Propanol, 20°C, saturation	9–12	
Oxygen index	%	Zytel ASTM D 2863	28–31	(8)
Hoffman-Lauritzen constants	Erg cm^{-2}	Solution crystal	73.6	81
		Melt crystallized (1 h averaged for multiple undercoolings)	23.7	
Fold surface free energy σ_e		Melt crystallized (24 h averaged for multiple undercoolings)	25.4	
Hoffman-Lauritzen constants	Erg cm^{-2}	Positively Birefringent Crystals	110.1	82
Hydrogen Bond (fold) surface free energy σ_{HB}				
Hoffman-Lauritzen constants	Erg cm^{-2}	Positively Birefringent Crystals	12	82
Lateral surface free energy σ_{VDW}				
T_m–T_c relations	°C (Tm)	Melt Quenched for 220°C<T_c<250°C	259.4°C < T_m < 279.1°C	81

Property	Units	Conditions	Value	Reference
Degree of crystallinity	%	Melt Quenched for $220°C < T_c < 250°C$	$0.26 < \chi < 0.31$	81
Crystal Interface Thickness	nm	Melt Quenched for $175°C < T_c < 250°C$	Ca. ½ repeat unit (0.6 nm) (interface assumed to chemically enrich in NH)	81
Crystal core thickness	nm	Melt Quenched for $175°C < T_c < 250°C$	$1.5 (175°C) < l_c < 2.0$ $250°C$	81
Crystal Lamella thickness	nm	Melt Quenched for $175°C < T_c < 250°C$	ca. 2 repeat units (constant long period)	81
Hoffman-Lauritzen constants		"Pseudo" Regime I/II transition for positively birefringent crystals	$T < 239°C$ (Regime 1) $T > 239°C$ (Regime 2)	82
Rigid – Amorphous Phase transitions	°K	Mobile amorphous Ordered semicrystal Rigid Amorphous	323 337 (transition 337–342) 370 (transition 340–400)	83
Brill Transition	°C	Spherulitic (processing dependent) Transcrystalline	139–230 180	88
Brill Transition Activation Energy	kJ mol^{-1}	Spherulitic Transcrystalline	28 77	88
Diffusion constant	m^2 s^{-1}	H_2SO_4 infinite dilution (concentration dependence)	2.08×10^{-13} $D = 2.08 \times 10^{-13} e^{0.41c}$	84
Diffusion Activation Energy	kJ mol^{-1}	H_2O	58	95
Infra Red Characteristic Absorption Frequencies	cm^{-1}	NH stretch (vs σ)	3307	102
		N—H stretch + amide (I + II) overtone (w σ)	3195	
		N—H stretch + amide (II) overtone (m σ)	3058	
		Amide I. CO stretch (vs σ)	1645	
		Amide II. In plane NH (m π)	1556	
		Amide II. In plane NH (m π)	1539	
		Amide III. CN stretch coupled with hydrocarbon skeleton (m π)	1370	
		Amide III. CN stretch coupled with hydrocarbon skeleton (m π)	1275	
		Amide III. CN stretch coupled with hydrocarbon skeleton crystalline (m π)	1202	
		Amide V. NH scissor (s σ)	595	
		CH$_2$ scissor next to NH (s σ)	1474	
		CH$_2$ scissor not next to NH (w σ)	1466	
		CH$_2$ scissor next to CO (m σ)	1417	

Nylon 6,6

Property	Units	Conditions	Value	Reference
Brill IR bands		"Regular Fold" (vw)	1327	
		CH$_2$ twisting (vw)	1307	
		"Regular Fold" (vw sh)	1223	
		Skeletal C—C stretch (vw)	1068, 1042, 1014	
		CH$_2$ rocking (w-vw)	987, 906	
		C—CO crystallinity stretch (m)	943	
^{13}C NMR Characteristic frequencies	Hz	C1	40.03 (trans) 43.36 (cis)	104
		C2	29.08 (trans) 29.99 (cis)	
		C3	26.49	
		C1′	36.10 (trans) 31.68 (cis)	
		C2′	25.39 (trans) 24.43 (cis)	
		C7′	176.15 (trans) 178.35 (cis)	
^{13}C NMR endgroups		αNH$_2$	40.67	
		βNH$_2$	27.67	
		ωNH$_2$	39.06	
		αCO$_2$H	34.98	
		βCO$_2$H	24.89	
		7′CO$_2$H	180.44	
Surface Free Energy	mJ m^{-2}	Dispersive, γ_d	37.4 (70°C)	86
			42.7 (80°C)	
			46.8 (90°C)	
			47.5 (110°C)	
Surface Free Energy		Lewis Acid Number, K$_a$	2.03	86
		Lewis Base Number, K$_b$	7.17	
Fiber Modulus	GPa	Single Fiber (30 μm)	4.7 ± 0.5	85
Failure Stress	GPa	Single Fiber (30 μm)	1.10–1.19 ± 0.6	85
Failure Strain	%	Single Fiber (30 μm)	21–24% (±2.1)	85
Elongation at break	%	Electrospun nanofiber (550 nm)	66	87
Young's Modulus	MPa	Electrospun nanofiber (550 nm)	453	87
Yield Strength	MPa	Electrospun nanofiber (550 nm)	110	87
Degradation Mechanisms		Sulfuric acid concentration (M_n decrease)	$-dM/dt = 5 \times 10^{-10}$ $M^{2.5}$ C$^{1.5}$	84
		Sulfuric Acid (<1.0 M)	Neutron Activation to determine S content	92
		Fatigue Failure (single fiber)	Differing crystal/amorphous microstructure at fracture	85
		Ethylene Glycol Glycolysis	250–275°C	89
		Electron Beam (ca. 20–500 kGy)	Ambient and high T (393 K) 200 kGy shows least degradation	90

Property	Units	Conditions	Value	Reference
		Intense Electric Arc	30 kA, 5–11 kJ/shot	91
		Photooxidation	Hindered Amine Stabilized	93
		Photooxidation	MALDI analysis of degradation products shows Norrish I/ Norrish II at late stages	96
		Thermal Photooxidation (Energy of Activation for 25°C–250°C)	E_a 12.6–36.9 kcal mol^{-1} (N$_2$) E_a 15.7–33.1 kcal mol^{-1} (irradiated)	98
		Accelerated Aging	Thermooxidative (37°C–138°C) with/without 100% RH	94
		Thermal Hydrolysis (Energy of Activations)	^{17}O NMR characterization gives E_a = 87 kJ mol^{-1}	97
		Thermooxidation (recycled Nylon 6,6)	MALDI indicates high content of cyclic imides, pyridines, chain fragments, and cyclopentanones	99
		Thermooxidation	Formation of degradation resistant gel fractions	100
		γ radiolysis	Gas production trends H2 > CO > CO2 > CH4	101
		Thermo-nonoxidative	Review	103

References

1. Zimmerman J. In: Mark HF, et al, eds. *Encyclopedia of Polymer Science and Engineering.* 2nd ed. New York: John Wiley and Sons; 1985–1989.
2. Flory PJ. *Chem. Rev.* 1946;39:137.
3. Cologne J, Ficket E. *Bull. Soc. Chim.* 1955:412.
4. Griskey RG, Lee BI. *J. Appl. Polym. Sci.* 1966;10:105.
5. Pouchert CJ. *The Aldrich Library of FT-IR Spectra.* Aldrich Chemical, Milwaukee, 1985.
6. Noda I, Dowrey AE, Marcott C. In: Mark JE, ed. *Physical Properties of Polymers Handbook.* Woodbury, N.Y: AIP Press; 1996.
7. Okajima KC, Yamane, Ise F. In: Salamone J. *Polymeric Encyclopedia.* Boca Raton, Fla: CRC Press; 1996.
8. *Dupont Zytel product information sheet.* http://www.dupont.com.
9. Warfiled R. Kayser WEG, Hartmann B. *Makromol. Chem.* 1983;184:1927.
10. Pflüger R. In: Brandrup J, Immergut EH, eds. *Polymer Handbook.* 3rd ed. New York: John Wiley and Sons; 1989; V/109–115.

11. Wakelin JH, Sutherland A, Beck LR. *J. Polym. Sci.* 1960;42 (139):278.
12. Tautz H, Strobel L. *Koll. Z. f. Polym.* 1965;202 (1):33.
13. Griskey RG, Shou JKP. *Modern Plastics.* 1968;45:148.
14. Müller A, Pflüger R. *Kunststoffe.* 1960;50 (4):203.
15. Tobolsky AV. *Properties and Structures of Polymers.* New York: John Wiley and Sons; 1960.
16. Rigbi Z. *Polymer.* 1978;19:1229.
17. Hansen CM. *Skand. Tidskr. Faerg. Lack.* 1971;17:69.
18. Elias HG, Schumacher R. *Makromol. Chem.* 1964;76:23.
19. Threlkeld JO, Ende HA. *J. Polym. Sci., Part A-2.* 1966;4:663.
20. Saunders PR. *J. Polym. Sci.* 1962;57:131.
21. Wallach ML. *Polym. Prepr.* (Amer. Chem. Soc. Div. Poly. Chem.) 1965;6/1:53.
22. Schumacher R, Elias H-G. *Makromol. Chem.* 1964;76:23.
23. Saunders PR. *J. Polym. Sci.* 1965;A3:1221.
24. Beachell HC, Carlson DW. *J. Polym. Sci.* 1959;40:543.
25. Howard GJ. *J. Polym. Sci.* 1959;37:310.
26. Juilfs J. *Kolloid. J.* 1955;141:88.
27. Taylor GB. *J. Am. Chem. Soc.* 1947;69:638.
28. Howard GJ. *J. Polym. Sci.* 1963;A1:2667.
29. Morozov AG, et al. *Soviet Plast.* 1972;8:85.
30. Jacobi EH, Schuttenberg, Schultz RC. *Makromol. Chem. Rapid. Commun.* 1980;1:397.
31. Hughes AJ, Bell JP. *J. Polym. Sci., Polym. Phys.* 1978;16:201.
32. Ayers CW. *Abakyst* 1953;78:382.
33. Duveau N, Piguet A. *J. Polym. Sci.* 1962;57:357.
34. Burke JJ, Orofino TA. *J. Polym. Sci., Part A-2.* 1969;7:1.
35. Fox TG, Loshack S. *J. Appl. Phys.* 1955;26:1080.
36. Bueche F. *J. Chem. Phys.* 1952;20:1959; and 1956;26:599.
37. Heim E. *Faserforch. u. Texiltech.* 1960;11:513.
38. Flory PJ, Williams AD. *J. Polym. Sci., Part A-2.* 1967;5:399.
39. Saunders PR. *J. Polym. Sci. A-2.* 1964:3755.
40. Korshak VV, Frunze TM. *Synthetic Hetero-Chain Polyamides.* translated by N. Kaner. New York: Daniel Davey and Co; 1960.
41. Fowkes FM. *J. Phys. Chem.* 1962;66:382.
42. Echochard E. *J. Chim. Phys./Phys.-Chim. Biol.* 1946;43:113.
43. Itoh T. *Jap. J. Appl. Phys.* 1976;15:2295.
44. Starkweather HW Jr, Zoeller P, Jones GA. *J. Polym. Sci., Polym. Phys.* 1984;22:1615.
45. Bunn CW, Garner EV. *Proc. Roy. Soc.* 1947;A189:39.
46. Colclough ML, Baker R. *J. Mater. Sci.* 1978;13:2531.
47. Schaefgen JR. *J. Polym. Sci.* 1959;38:549.
48. Rybnikar F. *Collect. Czech. Chem. Commun.* 1959;24:2861.
49. Kirshenbaum I. *J. Polym. Sci.* 1965;A3:1869.
50. Haberkorn HH, Illers H, Simak P. *Polym. Bull. (Berlin)* 1979;1:485.
51. Van Krevelen DW, Hoftyzer PJ. *Properties of Polymers – Correlation with Chemical Structure.* 2nd ed. Amsterdam: Elsevier; 1976.
52. Starkweather HW Jr, et al. *J. Polym. Sci.* 1956;21:189.
53. Illers H-K, Haberkorn H. *Makromol. Chem.* 1971;146:267.
54. Starkweather HW Jr, Moynihan RE. *J. Polym. Sci.* 1956;22:363.
55. Magill J. H. *Polymer* 1965;6:367.
56. Boasson EH, Wostenenk JM. *J. Polym. Sci.* 1957;24:57.
57. Khoury F. *J. Polym. Sci.* 1958;33:389.

58. McLaren JV. *Polymer* 1963;4:175.

59. Burnett BB, McDevit WF. *J. Appl. Phys.* 1957;28:1101.

60. Lindegren CR. *J. Polym. Sci.* 1961;50:181.

61. Murayama T. *Polym. Eng. Sci.* 1982;22:788.

62. Birkinshaw CM, Buggy, Daly S. *Polym. Commun.* 1987;28:286.

63. Chung IE, Throckmorton, Chundury D. In: *Annual Technical Conference.* Society of Plastics Engineers, Brookfield Center, Conn; 1991;XXXVII:681.

64. Willbourn AH. *Trans. Faraday Soc.* 1950;54:717.

65. Wilhoit RC. *J. Phys. Chem.* 1953;57:14.

66. Dietrich W, Basch A. *Angew. Makromol. Chem.* 1974;38:40/41: 159.

67. Weisskopf K, Meyerhoff G. *Polymer* 1982;23:483.

68. Fendler HG, Stuart HA. *Makromol. Chem.* 1958;25:159.

69. Nasini AGC, Ambrosino, Trossarelli L. *Ricerca Sci.* (Int. Symp. Macromol. Chem., Milan-Turin, 1954) 1955;25:625.

70. Wu S. In: Brandrup J, Immergut EH, eds. *Polymer Handbook.* 3rd ed. New York: John Wiley and Sons; 1989:V1/421 (and references therein).

71. Fox HW, Zisman WA. *J. Phys. Chem.* 1954;58:503.

72. Owens DK, Wendt RC. *J. Appl. Polym. Sci.* 1969;13: 1741.

73. van Oss CJ, Good RJ, Busscher HJ. *J. Dispersion Sci. Technol.* 1990;11:75.

74. Pellon J, Carpenter WG. *J. Poly. Sci., Part A.* 1962;1:863.

75. Brandt WW. *J. Polym. Sci.* 1959;41:415.

76. Hartmann B, Jarzynski J. *J. Accoust. Soc. Am.* 1974;56:1469–1477.

77. Strauss LA, Wall J. *J. Res. Natl. Bur. Std.* 1959;63A:269; and 1958;60:280.

78. Peebles LH Jr., Huffman MW. *J. Polym. Sci., Part A-1.* 1971;9:1807.

79. Dawes K, Glover LC. In: Mark JE. *Physical Properties of Polymers Handbook.* Woodbury, N.Y: AIP Press; 1996:chap. 41.

80. Fuchs O. In: Brandrup J, Immergut EH, eds. *Polymer Handbook,* 3rd ed. New York: John Wiley and Sons; 1989; VII/393.

81. Lee, Stein Schreiber, Phillips, Paul J. *European Polymer Journal.* 2007;43(5):1933–1951.

82. Lee Stein Schreiber, Phillips, Paul J. *European Polymer Journal.* 2007; 43(5):1952–1962.

83. Qiu, Wulin, Habenschuss, Anton, Wunderlich, Bernhard. *Polymer* 2007;48(6):1641–1650.

84. Abastari, Sakai, Tetsuya, et al. *Polymer Degradation and Stability.* 2007;92(3):379–388.

85. Ramirez JM, Herrera, Bunsell AR, Colomban Ph. *Journal of Materials Science.* 2006;41(22):7261–7271.

86. Huang, Xiaohua Shi, Baoli Li, et al. *Polymer Testing.* 2006;25(7):970–974.

87. Zussman E, Burman M, Yarin AL, Khalfin R, Cohen Y. *Journal of Polymer Science, Part B: Polymer Physics.* 2006;44(10):1482–1489.

88. Feldman AY, Wachtel E, Vaughan GBM, Weinberg A, Marom G. *Macromolecules.* 2006;39(13):4455–4459.

89. Kim, Kap Jin, Dhevi D, et al. *Eun Kyung Polymer Degradation and Stability.* 2006;91(7):1545–1555.

90. Sengupta, Rajatendu, Sabharwal S, et al. *Journal of Applied Polymer Science.* 2006;99(4):1633–1644.

91. Markutsya S, Rapeaux M, Tsukruk VV. *Polymer.* 2005;46(18):7028–7036.

92. Brown L, Bui VT, Bonin HW. *Journal of Applied Polymer Science.* 2005; 97(6):2476–2487.

93. Cerruti, Pierfrancesco, Lavorgna, et al. *Luigi Polymer.* 2005;46(13):4571–4583.

94. Bernstein, Robert, Derzon, Dora K, Gillen, Kenneth T. *Polymer Degradation and Stability.* 2005; 88(3):480–488.

95. Goudeau, Sylvain; Charlot, Magali; Mueller-Plathe, Florian. *Journal of Physical Chemistry B.* 2004; 108(48):18779–18788.

96. Carroccio, Sabrina; Puglisi, Concetto; Montaudo, Giorgio. *Macromolecules.* 2004;37(16):6037–6049.

97. Alam, Todd M. *Polymer.* 2003;44(21):6531–6536.

98. Singh RP, Desai, Shrojal M, Pathak. *G Journal of Applied Polymer Science.* 2003;87(13):2146–2150.

99. Groning, Mikael, Hakkarainen. *Minna Journal of Applied Polymer Science.* 2002;86(13):3396–3407.

100. Puglisi, Concetto, Samperi, et al. *Polymer Degradation and Stability.* 2002;78(2):369–378.

101. Chang, Zheng, LaVerne, Jay A. *Journal of Physical Chemistry B.* 2002;106(2):508–551.

102. Cooper SJ, Coogan M, Everall N. *Priestnall, Polymer.* 2001;42(26):10119–10132.

103. Schaffer MA, Marchildon EK, McAuley KB, Cunningham MF. Journal of Macromolecular Science. *Reviews in Macromolecular Chemistry and Physics.* 2000;C40(4):233–272.

104. Davis RD, Jarrett WL, Mathias L. *J Polymer.* 2000;42(6):2621–2626.

Nylon 6,10

Z. Ahmad and M. I. Kohan

Acronyms, Trade Names PA 610, PA-610, Nylon-610, Amilan (Toray), Technyl D (Rhone Poulenc), Ultramid S (BASF)

Chemical Names Poly(hexamethylene sebacamide), poly(hexamethylene decanoamide), poly(iminohexamethylene-iminosebacoy1), poly[imino-1,6-hexariediylimino (1,10-dioxo-1,10-decanediyl)] (CAS Registry No. 9008-66-6)

Class Thermoplastic, aliphatic polyamides

Structure $-[NH(CH_2)_6NHCO(CH_2)_8CO]-$

Major Applications A general purpose medium viscosity grade polymer for injection molding and extrusion used for making hardware, industrial parts, and precision instruments. Generally used with glass fiber reinforcement.

Properties of Special Interest Relatively low melting point; resistance to solvents, particularly hydrocarbons, low water absorption; stiffness; abrasion resistance; dimensional stability.

Property	Units	Conditions	Value	Reference
Molecular weight	g mol^{-1}	Per amide group	141.21	
		Per repeat unit	282.42	
Typical molecular weight range	g mol^{-1}		10 000–20 000	(1)
Typical polydispersity index, M_m/M_n (M_w/M_n)			~ 2.0	
Volume molar	cm^{3}mol^{-1}	20°C crystalline (rods)	260	
		20°C amorphous	271	
Volume specific	cm^{3}g^{-1}	20°C amorphous	0.95	
		20°C crystalline	0.865	
		20°C melt	1.095	
Change on melting			0.02–0.03	
Density	g cm^{-3}	Crystalline, α, triclinic	1.156	(2)
		Crystalline	1.152	(3)
		Melt 270°C, 1 bar	0.913	(4)

Property	Units	Conditions	Value	Reference
		Melt 230–290°C	0.91–0.94	(1)
		Moldings, amorphous	1.05	(5)
		crystalline	1.09	(4, 5)
Correlation with crystallinity		Moldings	(990–1030)/ density	(5)
IR (characteristic absorption frequencies)	cm^{-1}	N-vic. CH$_2$ bend (a)	1474	(6)
		CH$_2$ bend	1466	
		CH$_2$ bend	1437	
		CO-vic. CH$_2$ bend (a)	1419	
		Amide III (?)	1284	
		(α)	1191	
		(γ, amorphous)	1180	
		(amorphous)	1133	
		C—CO stretch (α or $-\gamma$)	938	
		CH$_2$ wag	730	
		Amide V (α)	689	
		Amide VI (α)	583	
NMR				(7)
Coefficient of thermal expansion				
Linear	K^{-1}	Crystalline, 20°C	8–10 × 10^{-5}	(8)
Volume	K^{-1}	Crystalline 20°C	3.8 × 10^{-4}	
Compressibility of the melt	Pa^{-1} (bar^{-1})		~5 (~5 × 10^{-5})	(1)
PVT curves				(9)
Reduction temperature T^*	K		8240	
Reduction pressure P^*	MPa		661	
Reduction volume $v^.$	cm^3 g^{-1}		0.845	
Solvents		25°C	Concentrated sulfuric acid, *m*-cresol	
		Redissolution, 156°C	Ethylene glycol	(10)
		Redissolution, 139°C	Propylene glycol	(10)
Mark–Houwink parameters: *K* and *a*	$K =$ cm^3 g^{-1} $a =$ none	*m*-Cresol, 25°C, for M_n = 8000–24 000	$K = 13500$ $a = 0.96$	(11)
Polymers with which compatible				(12)
Unit cell dimensions		α-Triclinic	$a = 4.95, b = 5.4,$ $c = 22.4$	(2,13)
		β-Triclinic	$a = 4.9, b = 8.0,$ $c = 22.4$	
Unit cell angles	Degrees	α-Triclinic	$\alpha = 49, \beta = 76.5,$ $\gamma = 63.5$	(2,13)
		α-Triclinic	$\alpha = 90, \beta = 77,$ $\gamma = 67.5$	

Nylon 6,10

Property	Units	Conditions	Value		Reference
Units in cell		α-Triclinic	1		(2,13)
		β-Triclinic	2		
Degree of crystallinity	%	Range, injection molded	25–45		(14,15)
Activation energy of crystallization	kJ mol^{-1}		53.6		(16)
Heat of sorption	J mol^{-1}	Melt	-58.5×10^3		(17)
Heat of fusion (per repeat unit)	kJ mol^{-1} (kJ kg^{-1})	Crystalline, from ΔH_m, DTA	56.8 (201)		(18)
		Crystalline, from ΔH_m, DTA	54.6 (193)		(19)
		Crystalline, from sp. ht.	53.2 (188)		(20)
Entropy of fusion (per repeat unit)	J K^{-1} mol	Crystalline	110–114		(16)
Glass transition temperature	K	Dry, mech. loss peak	340		(21)
		Dry, flex. mod. vs. temp	343		(21)
		Dry, DTA	315		(22)
		50% RH, mech. loss peak	313		(21)
		100% RH, mech. loss peak	283		(21)
Melting point	K	X-ray	500		(23)
		DTA			(23)
		Start	494		
		Peak	497		
		End	499		
		Equilibrium	>510		(24)

			Range	Average	
		Fisher–Johns	489–496	492	(23)
		Capillary	485–494	490	(23)
		Kofler hot stage	485–503	493	(23)

Property	Units	Conditions	Value	Reference
Heat capacity (per repeat unit)	J K^{-1} mol^{-1}		502	(8)
Deflection temperature	K	ASTM D 648 = DIN53461 = ISO 75		(25, 26)
		Dry		
		455 kPa	430–448	
		1820 kPa	339	
		50% RH		
		455 kPa	433	
		1820 kPa	333	
Flash Ignition Temperature	K	ASTM D 1929–1979	698	
Ignition resistance	HB	D 635	HB@1.5mm	
Flammability				
Limiting Oxygen Index	%	ASTM D 2863, Dry	24	(4)
Flame Rating (UL 7 mm)		UL 94	HB	

Property	Units	Conditions	Value	Reference
Enthalpy	kJ kg^{-1}	Temperature T (°C) (reference to 20°C)		
		60	80	
		100	160	
		200	400	
		250	580	
		300	700	
Tensile properties		ASTM D 638 = DIN 53455 = ISO 527		
Tensile modulus	MPa	23°C		(25, 26)
		Dry	2400	
		50% RH	1500	
Tensile strength	MPa	−40°C		(25, 26)
		Dry	83	
		50% RH	83	
		23°C		
		Dry	59	
		50% RH	49	
		77°C		
		Dry	37	
		I 50% RH	37	
Yield stress	MPa	−40°C		(25, 26)
		Dry	83	
		50% RH	83	
		23°C		
		Dry	60	
		50% RH	50	
		77°C		
		Dry	37	
		50% RH	37	
Yield strain $(L/L_0)\gamma$	%	−40°C		(25, 26)
		Dry	10	
		50% RH	13	
		23°C		
		Dry	10	
		50% RH	30	
		77°C		
		Dry	30	
		50% RH		
Maximum extensibility	%	−40°C		(25, 26)
		Dry	20	
		50% RH	30	
		23°C		
		Dry	70–100	
		50% RH	> 150	

Property	Units	Conditions	Value	Reference
		77°C		
		Dry	300	
		50% RH	—	
Flexural modulus	MPa	ASTM D 790 = DIN 53457 = ISO 178		(25, 26)
		−40°C		
		Dry	2240	
		50% RH	2520	
		23°C		
		Dry	2000	
		50% RH	1100	
		100% RH	690	
		77°C		
		Dry	480	
Bulk modulus	MPa	25°C	2300	(27)
Shear strength	MPa	ASTM D 732, 23°C, dry	58	(26)
Impact strength	$J\,m^{-1}$	ASTM D 256, DIN 53453, N ISO 179 Notched Izod		(25, 26)
		Dry	50	
		50% RH	200	
	$kJ\,M^{-2}$	Charpy, 20°C		(26)
		Dry	410	
		65% RH, 4 months	13–15	
Hardness		ASTM D 785; 23°C		
	M scale	Dry	75	(25, 26)
	M scale	50% RH	60	(25, 26)
	R scale	Dry	110–111	(25)
Poisson ratio		20°C, moldings	0.3–0.4	(4)
		100°C	0.47	
		Melt	0.50	
Abrasion resistance, Taber	$mg\,kHz^{-1}$	C17 wheel, 1 kg	5–6	
Index of refraction		25°C, molded, un-drawn	1.532	
		Isotropic	1.52	
		Parallel	1.57	
		Perpendicular	1.52	
Dielectric constant		ASTM D 150, IEC 250		(26)
		Dry		
		50–100 Hz	3.9	
		1 k Hz	3.6	
		1 MHz	3.3	
		−30.0°C; 100 Hz–1 GHz	3.0	(4)
		30°C		
		100 Hz–1 kHz	3.2	
		1 MHz–1 GHz	3.0	

Property	Units	Conditions	Value	Reference
		60°C		(4)
		100 Hz	4.6	
		1 kHz	4.2	
		1 MHz	3.4	
		1 GHz	3.0	
		90°C		(4)
		100 Hz	13	
		1 kHz	10.5	
		1 MHz	5.2	
		1GHz	3.1	
		20°C, 65% RH		(4)
		100 Hz	6.5	
		1 kHz	5.4	
		1 MHz	3.5	
		1 GHz	3.0	
Dissipation factor, dielectric loss		ASTM D150, IEC 250		
		Dry		(26)
		50–100 Hz	0.04	
		1 kHz–1 MHz	0.03	
		–30°C		(4)
		100 Hz	0.012	
		1 kHz	0.011	
		1 MHz	0.015	
		1 GHz	0.006	
		0°C		(4)
		100 Hz–1 kHz	0.013	
		1 MHz	0.017	
		1GHz	0.010	
		30°C		(4)
		100 Hz	0.010	
		1 kHz	0.015	
		1 MHz	0.021	
		1 GHz	0.013	
		60°C		(4)
		100 Hz	0.090	
		1 kHz	0.065	
		1 MHz	0.054	
		1GHz	0.025	
		90°C		(4)
		100 Hz	0.250	
		1 kHz	0.170	
		1 MHz	0.190	
		1GHz	0.035	
		20°C; 65% RH		(4)
		100 Hz	0.200	
		1 kHz	0.150	
		1 MHz	0.080	
		1GHz	0.020	

Property	Units	Conditions	Value	Reference
Dielectric strength		VDE 0303, part 2; IEC-243, electrode K20/P50		
		dry	100×10^4	
		moist ISO -1110	60×10^4	
Volume resistivity		ASTM D 257, IEC 93		
		Dry		(4, 26)
		20°C	10^{15}	
		60°C	5×10^{11}	
		100°C	5×10^8	
		20°C		(4)
		50% RH	2×10^{12}	
		100% RH	3×10^{10}	
Surface tension	mN m^{-1}	Melt, 265°C	37	(28)
Thermal conductivity	W m^{-1} K^{-2}		0.23	(26)
		Amorphous, moist, 30°C	0.35	(27, 29)
		Dependence on pressure, 25°C λ(25 kbar)/λ (atm. pressure)	1.90	(27, 29)
Thermal diffusivity	cm^{2}s^{-1}		1.4×10^{-3}	(30)
Melt viscosity	Pa s	Commercial injection molding grade resin, 280°C		(31)
		10 s^{-1}	37	
		10^2 s^{-1}	34	
		10^3 s^{-1}	27	
		10^4 s^{-1}	14	
Activation energy of viscous flow	kJ mol^{-1}		60	(32)
Coefficient of friction		Thrust washer, 275 kPa, 0.25 ms^{-1}		(33)
		Static	0.23	
		Dynamic	0.31	
Limiting PV against steel	kPa m s^{-1}	0.5 m s^{-1}	70	(33)
Absorption	%			
Water		50% RH	1.4–1.5	
		100% RH	33 ± 03	(4)
Organic solvents		Ethanol, 20°C, saturation	8–13	(4)
		Butanol, 20°C, saturation	8–12	
		Glycol, 20°C, saturation	2–4	
		Methanol, 20°C, saturation	16	
		Propanol, 20°C, saturation	10	
Dew point	K		255	

References

1. ' "Ultramid" S Processing Properties.' BASF Tech. Bulletin, July 1969.
2. Bunn CW, Garner KV. *Proc. Roy. Soc. (London) A.* 1947;189:39.
3. Starkweather HW Jr, Moynihan RE. *J. Polym. Sci.* 1956;22:363.
4. Pfluger R. In: Brandrup J, Immergut EH, eds. *Polymer Handbook.* 3rd edition. New York: Wiley Interscience; 1989:V/109–116.
5. Muller A, Pflunger R, Kunstoffe. 1960;50(4):S. 203.
6. Sibilia JP. et al. In: Kohan MI. *Nylon Plastics Handbook.* Cincinnati: Hanser/Gardner Publishers; 1995:88.
7. ibid, pp. 90–97.
8. Warfield RW, Kayser EG, Hartmann B. *Makromol. Chem.* 1983;184:1927.
9. Walsh DJ. In: Kohan MI. *Nylon Plastics Handbook.* Cincinnati: Hanser/Gardner Publishers; 1995:165–171.
10. Johnson FR, Weadon EJ. *Tex. Inst. Trans.* 1964;55:T162.
11. Morgan PW, Kwolek SL. *Polym. Set, Part A.* 1963;1:1147–1162.
12. Ellis TS. In: Kohan MI, ed. *Nylon Plastics Handbook.*Cincinnati: Hanser/Gardner Publishers; 1995:268–277.
13. Jones NA, Atkins EDT, Hill MJ. *J Polym. Sci., Polym. Phys.* 2000;38:1209.
14. Ramesh C. *Macromolecules.* 1999;32:3721.
15. Bonner RM, et al. In: Kohan MI, ed. *Nylon Plastics.* New York: Wiley-Interscience; 1973: 327–407.
16. Van Krevelen DW, Hoftyzer PJ. *Properties of Polymers: Correlation with Chemical Structure.* 2nd ed. Amsterdam: Elsevier; 1976.
17. Ogata N. *Makromol. Chem.* 1960;42:52.
18. Inoue MJ. *Polym. Sci., Part A.* 1963;1:2697–2709.
19. Ke B, Sisko AW. *J. Polym. Sci.* 1961;50:87–98.
20. Dole M, Wunderlich B, Makromol. *Chem.* 1959;34:29.
21. Kohan MI, ed. *Nylon Plastics.* Wiley-Interscience; 1973:330.
22. Gordon GA. *J. Polym. Sol., Part A-2.* 1971;9:1693.
23. Starkweather HW Jr. In: Kohan MI. *Nylon Plastics.* New York: Wiley-Interscience; 1973:308.
24. Mandelkern L, Jain NL, Kim H. in: *J. Polym. Sci., Part A-2.* 1968;6:165–180.
25. 'Nylon Resin 610.' Monsanto Bulletin. In: Kohan M I, ed. (Cited in *Nylon Plastics Handbook.* Cincinnati: Hanser/Gardner Publishers; 1995:557.
26. Willams JCL, Watson S, Boydell. In: Kohan MI, ed. *Nylon PlasticsHandbook.* Cincinnati: Hanser/Gardner, Publishers; 1995: 293–360.
27. Anderson P, Makromol. *Chem.* 1976;177:271.
28. Hybart FJ, White TR. *J. Appl. Polym. Sci.* 1960;3(7):118–121.
29. Hellwege K, Hoffmann HR, Knappe W, Kolloid-Z. *Polymere* 1968;226(2):109–115.
30. Dietz W, Colloid. *Polym. Sci.* 1977;255:755.
31. Kohan MI, ed. In: *Nylon Plastics.* New York: Wiley-Interscience; 1973:115–153.
32. Kohan MI, ed. Estimated from data on PA-6 and PA-MXD6 in *Nylon Plastics Handbook.* Cincinnati: Hanser/Gardner Publishers; 1995:177, 568 and Laun, M.H. Rheol. Acta 18 (1979): 478.
33. LNP Internally Lubricated Reinforced Plastics. LNP Corp. Bulletin. 1978:254–278.

Nylon 6,12

Gus G. Peterson, W. Brooke Zhao and R. Abouhussein

Alternative Names Poly[imino-1,6-hexanediylimino(1,12-dioxo-1,12-dedecanediyl)]

Class Aliphatic polyamides
Repeat Unit $C_{18}H_{34}O_2N_2$

Structure

$$-[\ \overset{\overset{\displaystyle H}{|}}{N}-(CH_2)_6-\overset{\overset{\displaystyle H}{|}}{N}-\overset{\overset{\displaystyle O}{\|}}{C}-(CH_2)_{10}-\overset{\overset{\displaystyle O}{\|}}{C}\ ]-$$

Molecular weight of repeat unit: 310.48 g/mol.

Preparative Techniques Polycondensation of hexamethylenediamine and

Major Applications Engineering resin, fibers for textiles, brushes and sutures, cable sheathing and tubing and bearings, cams, gears, and casings for tools and appliances.

Properties of Special Interest Low water-absorbing nylon; heat , chemical, and wear resistance and lubricity; excellent toughness and strength.

Property	Units	Conditions	Value	Reference
Molecular weight (of repeat unit)	$g\ mol^{-1}$	Size-exclusion chromatography	$25\ 700 \pm 700$	(1)
IR (characteristic absorption frequencies)	cm^{-1}	N–H stretching	3050	(2)
		N–H	3300	(3)
		N–H	3299	(4)
		C–O stretching (amide I band)	1650–1634 1540–1642	(4)
NMR (15N)	ppm	32°C	119.8	(6)
		36°C	119.8	
		42°C	119.8	
		49°C	119.7	
		56°C	119.6	
NMR $^{13}C\ (\alpha\ C)$		–	36.8, 35.8	(4)
Thermal expansion coefficient	K^{-1}	Linear	9×10^{-5}	(2)

Property	Units	Conditions	Value	Reference
Density	g cm^{-3}	–	1.06	(7)
			1.065	(3)
			1.04	(5)
Reduced viscosity	g dl^{-1}	298 K , 97%	1.6850	(4)
		sulfuric acid	1.5856	(3)
Common solvents		Phenols formic acid, chloral hydrate, fluorinated alcohols, mineral acids		(2)
Contact angle	Degrees	c-Hex	113.9 ± 1.0	(8)
		i-Oct	109.0 ± 0.8	
Equilibrium heats of fusion ΔH_f^0	kJ mol^{-1}	–	80.1	(9)
Glass transition temperature T_g	K	–	319	(9)
Melting temperature T_m	K	–	520–480	(9)
			491	(5)
Heat capacity	kJK^{-1}mol^{-1}	230°C	0.382	(10)
		300°C	0.494	
		400°C	0.771	
		600°C	0.981	
Defection temperature	K	0.455 MPa	453	(2)
		1.82 MPa	363	
Brittleness temperature	K	–	164	(2)
Specific heat	kJK^{-1}mol^{-1}	–	0.525	(2)
Tensile strength	MPa	–	60.7	(2)
Yield stress	MPa	–	51.0	(2)
Elongation at break	%	–	≥ 300	(2)
Elongation at yield	%	–	25	(2)
Shear strength	MPa	Dry	55.8	(2)
Flexural modulus	MPa	–	1241	(2)
			1350	(3)
Izod impact strength	Jm^{-1}	–	75	(2)
Dielectric constant ε'	–	–	5.3×10^3	(2)
Volume resistivity	ohm cm	–	10^{13}	(2)
Dissipation factor	–	1000 Hz	0.15	(2)
Thermal activation energy of electrical conduction	eV	–	1.58	(11)
Dispersion force component of surface free energy γ_S^d	mJm^{-2}	–	62 ± 9	(8)
Nondispersive interaction free energy between solid and water I_{SM}^n	mJm^{-2}	–	30.7 ± 0.4	(8)
Polar surface free energy γ_S^p	mJm^{-2}	–	4.7	(8)
Surface free energy γ_S	mJm^{-2}	–	67	(8)
Thermal conductivity	Wm^{-1}K^{-1}	–	0.22	(2)

Nylon 6,12

Property	Units	Conditions	Value	Reference
Intrinsic viscosity	dlg^{-1}	–	1.45	(12)
Water absorption	%	At saturation	3.0	(2)
		–	2.02	(4)
Flammability, oxygen index	–		28	(2)
Diffusion coefficients	$D \times 10^{-6}$ cm^2/s	CO_2	0.2	(5)
		O_2	0.04	(5)
		H_2	0.53	(5)

References

1. Mourey TH, Bryan TG. *J. Chromatography* 1994;679:201.
2. Zimmerman J. In: Mark HF, et al, eds. *Encyclopedia of Polymer Science and Engineering.* Vol. 11. New York: John Wiley and Sons; 1989:315.
3. Rusu GK, Ueda E. Rusu, Rusu M. *Polymer.* 2001;42:5669.
4. Rusu G, Rusu E. *High Perfor. Polym.* 2004;16:569.
5. Zheng Chang, Jay A, LaVerne. *J. Phys. Chem. B.* 2002;106:508.
6. Holmes BSGC, Chingas WB, Moniz, Ferguson RC. *Macromolecules* 1981;14:1785.
7. Deanin RD. In: Salamone JC, ed. *Polymeric Materials Encyclopedia.* 2nd ed. Vol 3. New York: CRC Press; 1996:2080.
8. Matsunaga TJ. *Appl. Polym. Sci.* 1977;21:2847.
9. Xenopoulos A, Wunderlich BJ. *Polym. Sci., Part B Polym. Phys.* 1990;28:2271.
10. Wen J. In: Mark JE, ed. *Physical Properties of Polymers Handbook.* New York: American Institute of Physics; 1996.
11. Rusu GE, Rusu L, Leontie, Rusu G. *J. Polym. Sci.: Part B:Polym Phys.* 45(2007): 794.
12. Yeung MW-Y, Williams HL. *J. Appl. Polym. Sci.* 1986;32:3695.

Nylon 11

George Apgar and Mohammad Hassan

Acronyms, Trade Name Polyamide 11, PA-11, Rislan® B (Elf Atochem)

Class Aliphatic polyamides

Structure [—C=O—(CH$_2$)$_{10}$—NH—]

Major Applications Tubing, hoses, and pipes for automotive, trucking, industrial, and petroleum applications. Examples are heavy truck airbrake tubing, automotive fuel lines, and submarine flexible pipes for offshore oil production. Thermoplastic powder coatings for industrial, transportation, and retail items are prepared in a Nylon 11 base. Nylon 11 has been used in a variety of food-contact applications including sausage casting, beverage tubing, and reusable kitchen devices.

Preparative Techniques Nylon 11 is prepared by a condensation polymerization reaction. The commercial monomer is 11, aminoundecanoic acid. This aminoacid is unique among the nylon monomers because it is made from castor oil, a renewable, agricultural raw material. The 18-carbon ricinoleicacid is thermally cracked to 7-carbon and 11-carbon fractions. The 11-carbon portion has an omega unsaturation, which is hydrobrominated and then aminated to the aminoacid monomer.[1]

Property	Units	Conditions	Value	Reference
Common form	–	–	α, triclinic	(2)
Unit cell dimension	Å	a axis	4.9	(2)
		b axis	5.4	
		c axis	14.9	
Angles	Degrees	Alpha	40	(2)
		Beta	77	
		Gamma	63	
Density, crystalline	g cm^{-3}	–	1.15	(2)
Density, amorphous	g cm^{-3}	25% crystallinity is typically after melt processing	1.01	(2)
Specific gravity	–	23°C		(2)
		Unmodified	1.03	
		Plasticized	1.05	
		43% glass fiber	1.36	

Property	Units	Conditions	Value	Reference
Refractive index	–	Isotropic	1.52	(3)
		Parallel to the optical axis	1.55	
		Perpendicular to the optical axis	1.51	
		Calculated using atomic refractivities for a density of 1.04 g cm^{-3}	1.53	
Solubility	–	In 4.2 M HCl		(4)
		Room temperature	Insoluble	
		Boiling	Insoluble	
		In 90% formic acid		
		Room temperature	Insoluble	
		In boiling propylene glycol, redissolution at 145°C	Soluble	
		In conc. H$_2$SO$_4$	Soluble	
		In m-cresol	Soluble	
Water absorption	wt%	Equilibration at		(2)
		23°C, 65% RH	1.1	
		23°C, 100% RH	1.9	
		100°C, 65% RH	3.0	
		70°C, 100% RH	2.3	
Heat of fusion	J g^{-1}	24% crystallinity	39	(2)
Specific heat	J g^{-1} K^{-1}	23°C	1.26	(5)
Glass transition temperature	K	–	315	(2)
Melting point	K	Unmodified	461	(2)
		Plasticized	457	
		43% glass fiber	461	
Thermal conductivity	W m^{-1} K^{-1}	–	0.19	(5)
Heat deflection temperature	K	Unmodified	320	(2)
		Plasticized	313	
		43% glass fiber	452	
Coefficient of linear thermal expansion	K^{-1} ($\times 10^{-5}$)	−30–50°C		(2)
		Unmodified	8.5	
		Plasticized	11	
		43% glass fiber	7	
		50–120°C		
		Unmodified	15	
		Plasticized	21	
		43% glass fiber	13	
Mark–Houwink parameters: K and a	K = cm^3/g a = None	For PA-11; mol.wt. = 1.8– 9 × 10^4 at 30°C in m-Cresol	K = 0.091 a = 0.69	(6)
Melt viscosity	Poise	For commercial grades of PA-11; 240°C; 500 s^{-1} shear rate	1000–7000	(7)

Property	Units	Conditions	Value	Reference
Yield stress	MPa	23°C		(2)
		Unmodified	36	
		Plasticized	21	
Yield elongation	%	23°C		(2)
		Unmodified	22	
		Plasticized	26	
Break stress	MPa	−40°C		(2)
		Unmodified	72	
		Plasticized	76	
		23°C		
		Unmodified	68	
		Plasticized	62	
		43% glass fiber	145	
		80°C		
		Unmodified	60	
		Plasticized	48	
Break elongation	%	−40°C		(2)
		Unmodified	160	
		Plasticized	220	
		23°C		
		Unmodified	360	
		Plasticized	380	
		43% glass fiber	8	
		80°C		
		Unmodified	420	
		Plasticized	420	
Tensile strength	MPa	Dry		Ref. 5, p. 299
		−40°C	68	
		23°C	59	
		77°C	42	
		50% RH		
		23°C	54	
Flexural modulus	MPa	−40°C		(2)
		Unmodified	1586	
		Plasticized	2275	
		23°C		
		Unmodified	1269	
		Plasticized	310	
		43% glass fiber	8480	
		80°C		
		Unmodified	255	
		Plasticized	159	
Izod impact strength	J m^{-1}	−40°C		(2)
		Unmodified	27	
		Plasticized	21	

Property	Units	Conditions	Value	Reference
		23°C		
		Unmodified	99	
		Plasticized	No break	
		43% glass fiber	247	
		80°C		
		Unmodified	No break	
		Plasticized	No break	
Compressive modulus	MPa	Dry	1280	Ref. 5, p. 304
Shear strength	MPa	Conditioned for 15 days at 20°C and 50% RH	42	Ref. 5, p. 306
Brittleness temperature	°C	Dry	<-70	Ref. 5, p. 311
Taber abrasion[a]	mg loss/1000 cycles	CS-17 wheel, 1000 g load	5	(8)
Rockwell hardness	–	Unmodified	R108	(2)
		Plasticized	R75	
		43% glass fiber	R111	
Hardness	Shore D values	23°C		(2)
		Unmodified	72	
		Plasticized	63	
Dielectric constant	–	Dry, 10^6 Hz	3.1	Ref. 8, p. 346
Dissipation factor	–	Dry, 10^6 Hz	0.04	Ref. 8, p. 346
Volume resistivity	ohm.cm	500 VDC; 20°C		(2)
		Unmodified	10^{14}	
		Plasticized	10^{11}	
		43% glass fiber	10^{14}	
Surface resistivity	ohm	20°C		(2)
		Unmodified	10^{14}	
		Plasticized	10^{11}	
		43% glass fiber	10^{14}	
Dielectric strength	kV mm^{-1}	20°C		(2)
		Unmodified	30	
		Plasticized	24	
		43% glass fiber	45	
Gas permeability		0.1 mm thick films, 23°C		Ref. 8, p. 357
	cm^3/m^2.bar. day	Oxygen at 0% RH	120	
	gm/m^2.day	Water vapor at 85% RH	3.7	
Weather resistance	–	% Tensile strength retained after 12 years outdoor exposure, carbon black containing		Ref. 5, p. 341
		Hot/wet (Australia)	91	
		Hot/dry (Australia)	78	
		Temperature (UK)	98	

Property	Units	Conditions	Value	Reference
Relative thermal index (RTI)[b]	°C	Air oven aging, 3 mm thick		Ref. 8, p. 337
		Not stabilized:		
		Electrical	65	
		Mechanical with impact	65	
		Mechanical without Impact	65	
		Heat-stabilized:		
		Electrical	105	
		Mechanical with impact	90	
		Mechanical without impact	105	
Solvents absorption	wt%	Low crystallinity, equilibrated at room temperature		Ref. 3, p. 375
		Water	1.8	
		Methyl alcohol	9.5	
		Ethyl alcohol	10.5	
		Methylene chloride	19	
		Chloroform	33	
		Perchlorethylene	11	
		Acetone	4.5	
		Carbon tetrachloride	4.5	
		Methyl acetate	5.5	
		Benzene	7.5	
		Aliphatic hydrocarbons	1	
		Toluene	6.8	
		Cyclohexane	1	
Modifications		Reinforcement with 30% short glass fiber		(9)
Tensile strength	MPa	Dry	110	
Tensile elongation	MPa	Dry	6	
Tensile modulus	MPa	Dry	3.4	
Flexural modulus	GPa	Dry	6	
Flexural strength	MPa	Dry	140	
Compressive strength	MPa	Dry	90	
Notched izod impact	J/m	Dry, 6.4 mm	117	
Heat deflection temperature	°C	Dry	171	
Melting point	°C	Dry	186	
Thermal expansion coefficient	C^{-1} $(\times 10^{-5})$	Dry, machine direction	3	
Volume resistivity	ohm.cm	Dry	10^{14}	
Surface resistivity	ohm	Dry	10^{14}	
Dissipation factor	–	Dry, 10^3 Hz	0.03	

Property	Units	Conditions	Value	Reference
Specific gravity	–	Dry	1.26	
Water absorption	wt%	23°C, 50% RH	0.55	
		Saturation	1.4	

* All properties measured in a dry, as-molded state.

ᵃ Taber abrasion is a test that has been traditionally used to measure the abrasion resistance of nylons. The results of this test are expressed as milligrams of sample lost in 1000 cycles with conditions specified as a CS-17 wheel with a 1000 g load.

ᵇ RTI is a measure of a material's thermal endurance, and although relative, typically corresponds to the temperature at which 50% of the property is retained for 20 000–100 000 h in an air oven aging test.

References

1. Apgar G, Koskoski M. In: Seymour RB, Kirshenbaum GS, eds. *High Performance Polymers: Their Origin and Development.* New York: Elsevier; 1986:55–65.
2. Apgar G. In: Kohan MI, ed. *Nylon Plastics Handbook.* Munich: Hanser; 1995:576–582.
3. Bonner RM, Kohan MI, Lacey EM, Richardson PN, Roder TM, Sherwood LT. In: Kohan MI, ed. *Nylon Plastics.* New York: Wiley-Interscience; 1973:396.
4. Lacey EM. In: Kohan MI, ed. *Nylon Plastics.* New York: Wiley-Interscience; 1973:88–89.
5. Williams JCL. In: Kohan MI, ed. *Nylon Plastics Handbook.* Munich: Hanser; 1995:344.
6. Sibila JP, Murthy NS, Gabriel MK, McDonnell ME, Bray RG, Curran SA. In: Kohan MI, ed. *Nylon Plastics Handbook.* Munich: Hanser; 1995:81.
7. *Technical literature.* Elf Atochem, Paris and Philadelphia.
8. Watson SJ. In: Kohan MI, ed. *Nylon Plastics Handbook.* Hanser, Munich, 1995:323.
9. Scheetz HA. In: Kohan MI, ed. *Nylon Plastics Handbook.* Munich: Hanser; 1995:398–399.

Nylon 12

H. Ulf W. Rohde-Liebenau

Acronyms, Trade Names PA 12, polyamide 12, polydodecanolactam, polylaurolactam; Daiamid® (Daicel Chemical Industries); Grilamid® (EMS Chemie); Rilsan® A (Elf Atochem); UBE Nylon 12® (UBE Industries); Vestamid® (Creanova)

Class Aliphatic polyamides

Structure $-[NH_2-(CH_2)_{11}-CO]_p-$

Properties of Special Interest Hydrolytic polycondensation at 260–300°C. Very low monomer content in melt-equilibrium. Activated anionic polymerization = monomer casting (small market volume). PA 12 crystallizes in pseudo-hexagonal modification. Combination of typical nylon and polyolefin properties. Low moisture absorption and density, chemical resistance similar to other nylons, not sensitive to stress cracking. Good to excellent impact strength, in dry state or at low temperatures. Engineering plastic, can be modified by glass or carbon fiber reinforcement, plasticizer, or other additives. PA 12 copolymers with PTHF: polyether block amides (PEBA)–see below.[1]

Major Applications Multiplicity of applications in technical engineering, especially in automotive and electrical industries. Antistatic parts. Precision molding. Sports and leisure goods. Coatings by extrusion, fluidized bed, or electrostatic process.

General Information Most properties were determined by relevant ISO and IEC standards in accordance with CAMPUS®. Three grades from the vast range of grades were selected: (1) unmodified extrusion, (2) with ~13% plasticizer, and (3) 30% glass fiber modified grade. (See ISO 1874-2 for a list of relevant standards.)

Properties	Unit	Conditions	Value			Reference
			Unmodified	Plasticized	30% Glass Fiber	
Density	g cm^{-3}	Standard: ISO 1183				
		At 23°C	1.01–1.02	1.03	1.24	(2–4)
		Annealed at 160°C	1.028	Monomer casting	–	(2–4)
		At 260°C (melt)	~0.86	~0.88	~1.04	(5)
Moisture absorption	%	Standard: DIN 53495				(2–4)
		23°C, 50% RH	0.8	0.7	0.4–0.5	
		23°C, immersed	1.5	1.4	1.1	
Melting range	K	Polarization microscopy	448–453			(2–4)

Properties	Unit	Conditions	Value			Reference
			Unmodified	Plasticized	30% Glass Fiber	
Heat deflection temperature	K	Standard: ISO 75; load = 0.45 MPa	388	363	448	(2–4)
Vicat softening point	K	Standard: ISO 306; load = 10 N	443	433	448	(2–4)
Glass transition temperature	K	Standard: ISO 537; tan δ by torsional pendulum				(2–4)
		Dry as molded	328			
		50% RH ($=0.7\%$ H_2O)	318			
Thermal expansion coefficient	K^{-1} ($\times 10^{-4}$)	Standard: DIN 53752; for 23–80°C				(2–4)
		In flow direction	1.5	1.8	0.6	
		Perpendicular direction	1.1	1.5	–	
Specific heat	$J\,g^{-1}\,K^{-1}$	Solid (23–60°C)	2.0	–	1.6	(3)
		Melt (250°C)	2.9	3.0	2.5	
Heat of fusion	$J\,g^{-1}$	–	65–75[a]	–	35–40[b]	(3)
Thermal conductivity	$W\,m^{-1}\,K^{-1}$	20–100°C	0.24	0.23	0.29	(3)
Melt volume index	$ml\,(10\,min)^{-1}$	275°C (5 kg load)$^{-1}$	~36	~60	~30	(5)
Maximum use temperature	K	Standard: UL 746B	358	353	378	(UL 746)
Flammability		Most PA 12 grades are slow burning (HB acc. UL 94), but there are self-extinguishing grades				(UL 94)
Oxygen index	%	Unmodified PA 12	21–22			(5)
Tensile modulus	MPa	Standard: ISO 527; equilibrated to 50% RH	1,450	400	6,500	(2–4)
Yield stress	MPa	Standard: ISO 527; equilibrated to 50% RH	46	26	130	(2–4)
Strain at yield	%	Standard: ISO 527; equilibrated to 50% RH	5	30	5	(2–4)
Strain at break	%	Standard: ISO 527; equilibrated to 50% RH	>200	>200	5–6	(2–4)
Notched impact strength (Izod)	$kJ\,m^{-2}$	Standard: ISO 180/1A; equilibrated to 50% RH				(2–4)
		At 23°C	20	No break	24	
		At −30°C	7	6	20	

Properties	Unit	Conditions	Value			Reference
			Unmodified	Plasticized	30% Glass Fiber	
Notched impact strength (Charpy)	kJ m^{-2}	Standard: ISO 179; equilibrated to 50% RH	A$^{(c)}$	B$^{(d)}$		(2–4)
		At 23°C	6	20		
		At −30°C	5	7		
Dielectric constant ε'	–	Standard: IEC 250; 1 MHz; equilibrated to 50% RH	3.0	3.8	3.4	(4)
Dielectric loss ε''	–	Standard: IEC 250; 1 MHz; equilibrated to 50% RH	280×10^{-4}	$1,500 \times 10^{-4}$	230×10^{-4}	(4)
Dielectric strength	kV mm^{-1}	Standard: IEC 243; equilibrated to 50% RH	26	31	44	(4)
Surface resistivity R_{OA}	ohm	Standard: IEC 93; equilibrated to 50% RH	10^{13}	10^{12}	10^{13}	(4)
Volume resistivity	ohm cm	Standard: IEC 93; equilibrated to 50% RH	10^{15}	10^{12}	10^{15}	(4)
Comp. tracking index	–	Standard: IEC 112; equilibrated to 50% RH	600	600	>600	(4)
Molecular mass	g mol^{-1}	–	$M_n = 1.4\text{--}3.0\ (\times 10^4)$ $M_w = 3.5\text{--}10.5\ (\times 10^4)$			(6–8)
Typical polydispersity index (M_w/M_n)	–	–	2.5–3.5			(6–8)
Mark-Houwink parameters: K and a	$K = $ ml g^{-1} $a = $ None	–	$K = 524 \times 10^{-4}$ $a = 0.73$			(6–8)
Degree of crystallinity	%	Cooled After annealing at 150°C	∼0.3 0.35–0.40			–
Unit cell dimensions		Pseudohexagonal gamma-modification with unit cell dimensions				(2, 3)
Lattice	–	–	Pseudohexagonal			(2, 3)
Unit cell content (number of repeat units)	–	–	4			(9)
Cell dimensions	nm	–	$a = 0.479, b = 3.19, c = 0.958$			(9)
Cell angle	Degrees	–	$\beta = 120$			(9)

Properties	Unit	Conditions	Value			Reference
			Unmodified	Plasticized	30% Glass Fiber	
Density (crystalline)	g cm^{-3}	Also unstable monoclinic α modification	1.106[e]			(10)
Index of refraction $n_D{}^{25}$	–	Only film and thin quenched parts are transparent	1.52–1.53			(5)

[a] Range = 160–195°C. [b] Range = 155–185°C.
[c] A = low molecular weight/injection molding. [d] B = high molecular weight/extrusion.
[e] Some sources give the crystalline density as 1.03 to 1.05 g cm^{-3}, which is too low. If one extrapolates data from reference (11) or if a parallel for nylon 12 is drawn to the line of density vs. crystallinity for nylon 11 from reference (12), then one can derive the approximate crystalline density of 1.10 g cm^{-3}.

Polyether block amides (PEBA) are internally plasticized by copolycondensation of PA 12 and PTHF block segments. The grades are differentiated by Shore hardness D as a measure of flexibility. In addition to typical PA 12 application ranges, PEBA are used for seals, gaskets and in medical devices. (Trade name of these grades of Elf Atochem is Pebax®)

Property	Units	[Standard]/ Conditions	Shore D Hardness*				PA 12	Reference
			35	47	55	62		
Density	g cm^{-3}	[ISO 1183]	1.01	1.02	1.03	1.03	1.01–1.02	(1–3)
Tensile modulus	MPa	[ISO 527]	–	120	230	370	1,450	(1–3)
Yield stress	MPa	[ISO 527]	–	–	–	24	47	(1–3)
Tensile strength	MPa	[ISO 527]	17	23	32	–	–	(1–3)
Strain at break	%	[ISO 527]	>200	>200	>200	>200	>200	(1–3)
Notched impact strength (Izod)	kJ m^{-2}	[ISO 180/1A] At 23°C At −30°C	No break No break	No break No break	No break 22	No break 8	20 7	(1–3)
Heat deflection temperature	K	[ISO 75]; load 0.45 MPa	328	338	363	373	393	(1–3)
Vicat softening point	K	[ISO 306]; load 10 N	398	413	433	438	443	(1–3)

* Standard: ISO 868.

Suppliers

EMS Chemie AG, Domat, Switzerland
Elf Atochem S.A., Paris, France
UBE Industries, Tokyo, Japan
Creanova GmbH., Division of Degussa-Hüls AG., Marl, Germany

References

1. Apgar GB, Koskoski MJ. In: Seymour RB, Kirshenbaum CS, eds. *High Performance Polymers: Their Origin and Development.* New York: Elsevier Science Publishing; 1986:55–65.
2. Kohan MI, ed. *Nylon Plastics Handbook,* Hanser Publishers, Munich (Hanser/Gardner Publications, Cincinnati), 1995 (and references therein).
3. Bottenbruch L, Binsack R, eds. *Kunststoff Handbook, Vol. 3–4, Polyamide.* Carl Hanser Verlag, Munich and Vienna; 1998:sec. 4 (and references therein).
4. Technical literature and CAMPUS® data bank from Daicel; EMS; Elf Atochem; Hüls (see suppliers above).
5. Unpublished data from Hülls AG.
6. Scholten H, Feinauer R. *Agnew. Makromol. Chem.* 1973;21:187.
7. Hammel R, Gerth C. *Makromol. Chem.* 1973;34: 2697.
8. Griehl W, Zarate J. *Plastverarb* 1967;18:527.
9. Gogolewski SK, Czerniawska, Gasiorek M. *Colloid and Polym. Sci.* 1980;258:1130.
10. Cojazzi G, et al. *Makromol. Chem.* 1973;168:289.
11. Müller A, Pflüger R. *Kunstst.* 1960;50:203.
12. Kohan MI, ed. *Nylon Plastics.* New York: Wiley-Interscience; , 1973:332.

Nylon MXD6

Seung Woo Lee and Gil S. Sur

Trade Name Reny (Mitsubishi Gas Chemical Co.)

Class Partially aromatic polyamide

Structure

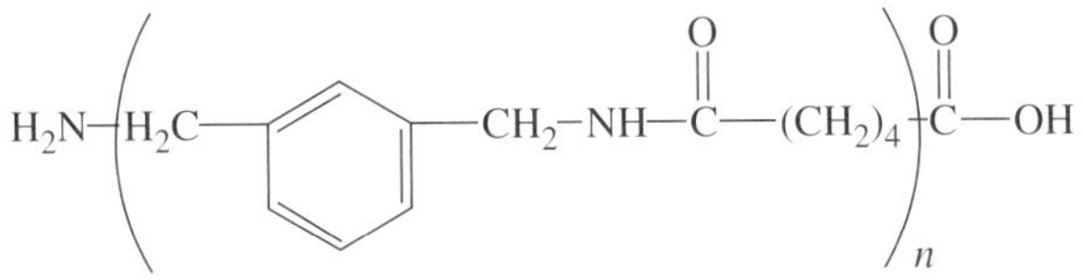

Major Applications Blow molded bottles. Extruded film and sheets for food packing, including blend, multilayer, and laminate with nylon 6, PET, and polyolefins. Monofilament for bristle and filter cloth. Glass fiber reinforced injection molding materials used to make parts for automotive, machine, electrical /electronic, civil engineering, sports, and other industries as a metal substitute.

Properties of Special Interest Relatively low cost. High mechanical strength, modulus, and heat resistance. Very low oxygen permeability in humid atmosphere.

Type of Polymerization Polycondensation in melt or solid phase.

Typical Comonomers *p*-Xylylenediamine

Property	Units	Conditions	Value	Reference
Appearance	–	Room temp.	Pellet	(1)
Molecular weight (of repeat unit)	$g\,mol^{-1}$	–	246.31	(1)
Typical molecular weight range	$g\,mol^{-1}$	End group titration	$(1.6\text{–}4.0) \times 10^4$	(1)
IR	cm^{-1}	Ref. KBr tablet	1650; 1550; 1440; 1030; 790; 700	(1)
UV	nm	Ref. 96% H_2SO_4	260	(6)
^{1}H-NMR	ppm	Formic acid solution	1.8, 2.5, 4.5, 7.3	(1)
^{13}C-NMR	ppm	Formic acid solution	25.7, 36.3, 44.7, 127.7, 130.0, 138.7, 177.7	(1)

Property	Units	Conditions	Value	Reference
Thermal expansion coefficient	K^{-1}	ASTM D696	5.1×10^{-5}	(1)
Specific gravity	–	ASTM D792	1.22	(1)
Density (amorphous)	$g\ cm^{-1}$	296 K	1.19	(1)
Density (bulk)	$g\ cm^{-1}$	–	0.6–0.8	(1)
Solvent	–	Room temp.	Sulfuric acid, formic acid, trifluoroacetic acid, m-cresol, o-cresol, phenol/ethanol (4:1 by vol), hexafluoroisopropanol	(1)
Solvent	–	433 K	Benzyl alcohol, ethylene glycol	
		473 K	Diethylene glycol, triethylene glycol	
Nonsolvent	–	Room temp.	Water, n-butanol, n-heptane	
Crystalline state	–	Lattice	Triclinic	(5)
		Space group	C_i^{1-P1}	
		Chain conformation	Planes incline to the c axis by a few degrees from planar zigzag	
Unit cell dimension	Å	–	$a = 12.01, b = 4.84, c = 29.8$	(5)
Unit cell angle	Degrees	–	$\alpha = 75.0, \beta = 26.0, \gamma = 65.0$	(5)
Unit cell contents	–	–	2	(5)
Degree of crystallinity	%	Solid phase polymerized, DSC	35	(1)
Heat of fusion	$kJ\ mol^{-1}$	DSC	37	(1)
Density (crystalline)	$g\ cm^{-3}$	–	1.25	(5)
Glass transition temperature	K	DSC	358	(1)
Melting point	K	DSC	510	(1)
Flash point	K	–	608	(1)
Heat capacity	$J\ K^{-1}\ g^{-1}$	DSC		(1)
		313K	1.31	
		533K	2.51	

Property	Units	Conditions	Value	Reference
Deflection temperature	K	ASTM D648, 1.8 MPa	1.31	(1)
Tensile modulus	MPa	ASTM D638 dry	4700	(1)
Tensile strength	MPa	ASTM D638 dry	99	(1)
Maximum extensibility (L/L_0)	%	ASTM D638 dry	2.3	(1)
Flexural modulus	MPa	ASTM 790 dry	4400	(1)
Flexural strength	MPa	ASTM 790 dry	160	(1)
Impact strength	$J\,m^{-1}$	ASTM 256 dry, notched	20	(1)
Hardness	Rockwell M	ASTM D785 dry	108	(1)
Abrasion resistance	$g\,kcycles^{-1}$	ASTM D1044	19×10^{-3}	(2)
Index of refraction n	–	ASTM D542, amorphous	1.582	(1)
Dielectric constant ε'	–	ASTM D150, 110 and 10^3 MHz	3.9	(2)
Dielectric loss constant ε''	–	ASTM D150, 110 and 10^3 MHz	0.039	(2)
Resistivity	ohm cm	ASTM D257	1.2×10^{16}	(2)
Permeability coefficient	$m^3(STP)\,ms^{-1}$ $m^{-2}\,Pa^{-1}$	O_2, 296 K, 60%RH	5.7×10^{-21}	(1)
Transparency coefficient	$cc\,mm\,m^{-2}$ $day^{-1}\,atm^{-1}$	OTR 23°C 60% RH CO$_2$TR23°C 60% RH WVTR23°C 60% RH	0.09 0.30 1.36	(8)
Semi-crystallization time	s		100	(7)
Thermal conductivity	$W\,m^{-1}\,K^{-1}$	–	0.38	(2)
Melt viscosity	Pa s	543 K, shear stress 24.5 kPa $M_n = 16\,000$ $M_n = 19\,000$ $M_n = 25\,000$ $M_n = 39\,000$	140 280 730 2400	(1)
	Pa s	Nozzle D: 1mm, Nozzle L: 1 mm Shear rate 100 s^{-1} 260°C 270°C 280°C	150 120 100	(7)
Melt index	g	$M_n = 16\,000$ $M_n = 19\,000$ $M_n = 25\,000$ $M_n = 39\,000$	7 4 2 0.5	(5)
Decomposition temperature	K	TGA	653	(6)
Water absorption	%	293 K, equilibrium	5.8	(1)
Important patents	–	–	–	(3, 4)

Property	Units	Conditions	Value	Reference
Cost	US$ kg^{-1}	–	4–6	
Suppliers	Mitsubishi Gas Chemical Co., Inc., Tokyo, Japan			
	Solvay & Cie , Brussels, Belgium			
	Toyobo Co. Ltd., Osaka, Japan			
	Du pont-Toray Co., Ltd., Tokyo, Japan			

References

1. Mitsubishi Gas Chemical Catalog. *Polyamide MXD6.*
2. Mitsubishi Gas Chemical Catalog. Reny, *Engineering Plastics.*
3. Miyamoto A. et al. *U.S. Patents 4 433 136 and 4 438 257(1984); European Patents 0 071 000 and 0 084 661 (1986).*
4. Miyamoto, A. et al. *U.S. Patents 3 962 524 and 3 968 071 (1976).*
5. Ota TM, Yamashita O, Yoshizaki, Nagai E. *J. Polymer Sci., Part A-2.* 1996;4:959.
6. Tsukamoto AH, Nagai K, Eto, Fujimoto N. *Kobunshi Kagaku.* 1973;30:339.
7. Nylon-Nanocomposite Formulations For Improved Barrier PET Container Application. Kingsport, TN, USA, Nova Pak Americas: Eastman Chemical Company; 2000: Proceeding, pp. 243–260.
8. Hu YSS, Mehta DA, Schiracdj A, Hiltner, Baer E. *J. Polymer Sci., Part B.* 2005;43:1365.

Perfluorinated Ionomers

Richard E. Fernandez, Ralph B. Lloyd, R. Daniel Lousenberg and
Stephen Mazur

Manufacturers (Trade Names) DuPont (Nafion®), Asahi Glass (Flemion®), Asahi Kasei (Aciplex®), 3M, Solvay-Solexis (Hyflon®)

Class Chemical copolymers

Preparative Techniques Free radical copolymerization of highly fluorinated vinyl ether monomers, usually with tetrafluoroethylene, in fluorocarbon solvents or by aqueous emulsion.

Typical Vinly Ether Monomers

Sulfonyl Fluoride Monomers	x	y
Nafion®, Flemion®	1	2
Aciplex®	1	3
Hyflon®	0	2
3M	0	4

$$CF_2 \!=\! CF \!-\! (OCF_2CF)_x \!-\! O(CF_2)_y SO_2F$$
$$\underset{\displaystyle CF_3}{|}$$

Carboxyl Monomers	x	y
Nafion®, Aciplex®	1	2
Flemion®	0	3

$$CF_2 \!=\! CF \!-\! (OCF_2CF)_x \!-\! O(CF_2)_y CO_2CH_3$$
$$\underset{\displaystyle CF_3}{|}$$

Copolymer Structures after Hydrolysis and Acid Exchange

Perfluorosulfonic acid (PFSA) ionomer

$$-(CF_2CF_2)_n - CF - (OCF_2CF)_x - O(CF_2)_y SO_3H$$

Carboxylate

$$-(CF_2CF_2)_n - CF - (OCF_2CF)_x - O(CF_2)_y CO_2H$$

(For commercial materials n varies from about 5–11)

Major Producers Nafion® is the DuPont trademark for its commercial line of perfluorinated ionomers, available as resins, membranes, and dispersions. Asahi Chemical Industry Company produces Aciplex® and Asahi Glass Company, Ltd., Japan, produces Flemion®; both are competitive products to Nafion® in form and function. 3M and Solvay-Solexis are also making perfluorinated ionomers for fuel cell and other applications.

Major Applications Perfluorinated ionomers are used in a variety of applications, the largest of which is as ion exchange membrane separators in the commercial electrolysis of brine to produce caustic and chlorine. Nafion® membranes are also used in low temperature (<100°C) polymer membrane fuel cells, for water electrolysis (oxygen and hydrogen generation), ozone generation, and gas drying and humidification applications. Nafion® is also used as a heterogeneous super acid catalyst in supported, cubed, or powdered form.

Manufacturing and Processing Perfluorinated ionomers are made via either emulsion or solution polymerization processes, the latter being produced in expensive fluorocarbon solvents. The sulfonate ionomers are made using monomers with the sulfonyl fluoride precursor functional site; the carboxylate ionomers are likewise made using monomers with the methyl ester (or other alkyl ester) precursor functional site. The resulting polymer precursors are thus able to be melt processed using conventional melt processing techniques into pellets, films, fibers, and other shapes. Usually the most economical process for making a particular shape, for example a 5 mil film, is to first melt extrude the polymer precursor into the desired shape and then perform the subsequent hydrolysis and acid exchange steps to convert the precursor to the final ionomer form.

As noted, the predominant application for perfluoroionomers is as barrier membranes in caustic-chlorine production by brine electrolysis. In these applications bi-layer membranes with both sulfonate and carboxylate perfluoroionomer resin layers are required, as layers with both functional sites are needed to achieve the desired properties of high current efficiency and density and low voltage; the carboxylate layer is typically much thinner than the sulfonate layer and gives the needed anion rejection. Membranes made for use in caustic-chlorine production are also usually reinforced with a Teflon®-based fabric. The ability to melt process the films in the polymer precursor form aids in the manufacture of these specialized composite membranes.

After the polymer is in the desired form, whether film, pellet, or other, the precursor functional site must then be hydrolyzed to obtain the ionomer form. This is achieved by exposing the precursor film to strong base solutions, usually sodium or potassium hydroxide. Time, temperature, and concentration—all affect the speed of hydrolysis. Once hydrolyzed, the polymer is now in the ionic form, but it may be desirable to exchange the cation of the hydrolysis solution with some other cation. While the chloralkali membranes are used in the salt form, many of the commercially available membranes are used in the acid or proton form. Therefore, the sodium or potassium cation that results from the hydrolysis step must be ion-exchanged to the proton form. This is readily achieved by exposing the polymer to a strong acid solution, such as sulfuric, nitric, or hydrochloric acid. The membrane is then rinsed, dried, and wound up on a roll.

Dispersions of the acid form sulfonate polymer are also manufactured by techniques described in US Patent 4,433,082. These dispersions (described in more detail below) are made in mixtures of water and short-chain alcohols at temperatures in excess of 200°C. Batch processing is typically used due to the high pressures and relatively small volumes. Once in dispersion, the solvent can then be exchanged to make dispersions in other solvents, usually by simple distillation. The dispersions have become an important component of membrane electrode assemblies, where they are used for catalyst layer formulations and can be cast into thin membranes (1).

Physical Properties—Phase Segregation The equivalent weight (EW) is a key indicator of the polymer and is defined as the grams of polymer per mole of exchange sites, that is, —SO_3H or —CO_2H groups. The physical properties of perfluorinated ionomers are strongly influenced by equivalent weight (EW), acid type (sulfonic versus carboxylic), and nature of the counter ion (2–4). The present summary focuses only on Nafion® perfluorosulfonic acid

(PSFA) ionomers with EW from 1000 to 1100, an important commercial product and the subject of many scientific studies. The acid side chains in EW 1000 Nafion® represent 14% of the monomer units, occupy about 40% of the bulk volume, and are segregated into hydrophilic phase domains whose characteristic dimensions ($\sim$5 nm) have been determined by small-angle X-ray and neutron scattering. Several studies have examined the effects of water sorption (2,5), counter ions (2,6), mechanical orientation (2,7), and thermal relaxation (4) on these domains. In addition, wide angle X-ray diffraction indicates a small crystalline component within the hydrophobic PTFE phase domains (8). Nevertheless, understanding of the global structure remains incomplete because the scattering data is equally compatible with different alternative models. Molecular relaxations in Nafion®, as manifest in differential scanning calorimetry, dynamic mechanical, dielectric, and NMR spectroscopies reveal several distinct features which have been assigned to either hydrophilic or hydrophobic phase domains. But an important consequence of the small size and intimate connections between these domains is that relaxations in each domain influence mobility in its neighbors and contribute to the mechanical properties and processing characteristics of the material.

Physical Properties—Thermomechanical Relaxations Film, fibers, tubing, and complex shapes are fabricated by extrusion or injection molding of the sulfonyl fluoride resin followed by hydrolysis with sodium hydroxide and proton exchange to the sulfonic acid form. Thermal analysis of EW 1000 acid indicates a melting transition near 240°C, but chemical decomposition above this temperature precludes melt processing in the acid form. An important consequence of the molecular mobility which develops above 240°C is the ability of the bulk PFSA resin to spontaneously disperse in water or aqueous alcohol to form charge-stabilized colloidal particles as described in the following section. At lower temperatures, two other thermal relaxation processes are important for the fabrication and applications of membranes. Above 170°C, while not truly fluid, the elastic modulus decreases to approximately 1 kPa and molecular mobility is sufficient for the decay of residual orientation. At still lower temperatures Nafion® behaves as a viscoelastic solid whose mechanical properties vary with temperature, time, and water content. For thoroughly dried acid resin, the dynamic storage modulus at 10 Hz decreases 100-fold at 70°C, but the temperature of this relaxation increases to 100°C with hydration. This phenomenon is identified with ionic interactions and molecular mobility within the hydrophilic domains and is strongly influenced by replacing the acid proton with alkali metal or organic cations. At ambient temperatures Nafion® exhibits significant long-time viscoelastic behavior. Thus, at 25°C the initial tensile modulus of a water-saturated membrane (249 MPa) decays by about 50% within 3 h to an equilibrium value.

PFSA Ionomer Dispersions PFSA ionomers in the acid form are effectively 'dispersed' into aqueous alcoholic liquids (9,10) at temperatures at or above 230°C and into water (11) at 270°C. However, the PFSA ionomers do not form true solutions in water or any polar-organic liquid medium. The ionomer molecules form phase-separated colloidal aggregates with a fluorocarbon core and the ionic groups at the polymer–liquid interface (12). A significant equilibrium between the aggregates and discrete polymer molecules does not appear to exist at ambient conditions (13). The ionomer dispersions can be relatively stable with the ionic pendant groups effectively functioning as surfactants. Small and ultra small angle X-ray scattering experiments have suggested elongated or anisotropic structures of indeterminate lengths but with a mean diameter of 4–5 nm for Nafion® PFSA ionomer dispersions (14–16). Commercially available Nafion® PFSA ionomer dispersions with concentrations between 5 and 20% have viscosity-shear rate profiles that are predominantly Newtonian. However, high viscosity dispersions can show shear-thinning behavior.

Size exclusion chromatography and light scattering detection were used to measure the molecular weight distributions of experimental aqueous dispersions that had high viscosities (13,17). Bimodal distributions were observed with a main peak at approximately 5×10^5 g/mol and a higher molecular weight shoulder at 4×10^6 g/mol. On heating the aqueous dispersions to 270°C, these peaks disappeared and new peaks approaching log-normal distributions had formed. The weight average molecular weights were reduced to 2.5×10^5 g/mol and the polydispersity was 1.8. Heating at high temperatures had caused aggregate breakup to reveal the discrete molecular weight distributions of individual polymer molecules and significantly reduced viscosity. Limited two-angle dependant light scattering measurements also indicated a linear relationship between the radius of gyration and molecular weight for the initial high molecular weight shoulder and was consistent with an elongated aggregate structure (17). The initial peak and high molecular weight aggregate shoulder contained approximately 4 and 30 molecules, respectively.

Property	Units	Conditions	Value	Reference
Average molecular weight (of repeat unit)	–	Defines equivalent weight	–	–
Head-to-head contents	%	–	Unknown	–
Degree of branching	%	–	0	–
Typical molecular weight (M_n)	g mol^{-1}	Size exclusion chromatography	1.5–2.0×10^5	(13)
Typical polydispersity index (M_w/M_n)	– Dimensionless	–	1.8–2.0	(13)
Morphology	Structure of hydrolyzed membranes is generally believed to be of a reverse micelle type, 30–50 Å in size, containing the aqueous ions, acid, and/or salt groups embedded in a continuous fluorocarbon phase.			(18–22)
IR	–	–	–	(23, 24)
UV	Transparent down to 200 nm		–	–
NMR	–	–	–	(25–29)
Solvent	For hydrolyzed sulfonic polymer, aqueous or alcoholic solutions can be made by dissolving the acid form of the polymer between 150 and 300°C.			(30)
	For hydrolyzed carboxylic polymer, the lithium ion form is preferred and degradation can occur between 250 and 300°C.			(31)
Swelling	As a function of the solvent, counter ion, EW, and temperature			(32, 33)
Solubility parameter	As a measure of the intermolecular forces present			(34)
Solvent effects on molecular motion	–		–	(35)
Heat of fusion	Jg^{-1}	Depends on EW	5–25	–
Density	gcm^{-3}	Unhydrolyzed Hydrolyzed	2 1.4–2.05	(36)

Property	Units	Conditions	Value	Reference
Glass transition temperature	°K	Sulfonyl fluoride	~273	–
		—SO$_3$H form	376	(37)
		—SO$_3$Li form	489	(37)
		—SO$_3$Na form	508	(37)
		—SO$_3$K form	498	(37)
		—SO$_3$Cs form	483	(37)
Melting point (depends on EW)	°K	Sulfonyl fluoride 1050EW	523 (typical)	–
Other thermal transitions	–	–	–	(38)
Mechanical properties	–	Sulfonate membranes	–	(39–41)
		Carboxylate membranes	–	(42)
		Both types	–	(43)
Dielectric properties	–	–	–	(44, 45)
Electronic conductivity	–	–	–	(46–48)
Permeability coefficient		For oxygen permeation through 700–800 EW Flemion carboxylate membranes		(49)
		Oxygen and hydrogen permeation through Nafion 117 membranes		(50)
Ion and water transport	–	–	–	(51–64)
Water transport	–	–	–	(65–67)
Proton Transport		–	–	(68–71)
		For Dow membrane (same as Hyflon®)	(72)	
Melt index	g	10 min at 270°C using a 1200 g weight in nonhydrolyzed form	5–15 (typically)	–
Biodegradability, effective microorganisms		–	None known	
Maximum use temperature	°K	Atmospheric cell pressure	353–363 (typically)	–
Decomposition temperature	°K	Sulfonate in Na$^+$ form	~673	–
		Carboxylate	~573	
Water absorption	%	Sulfonate in Na$^+$ form	15–25	(73)
		Sulfonic in H$^+$ form (depending on EW for both forms)	40	
Flammability, flame propagation rate			None	
Cost	US\$ kg^{-1}	Sulfonic resin	2000	
Availability	Commercially available			
Suppliers	Asahi Chemical Industry Company Asahi Glass Company, Ltd., Japan E. I. DuPont de Nemours and Company, Inc.			

Important Patents

5310765	'Process for Hydrolysis of ion exchange membranes'
5281680	'Polymerization of fluorinated copolymers'
4666574	'Ion-exchange membrane cell and electrolytic process'
4591439	'Ion exchange process and apparatus'
4437951	'Membrane, electrochemical cell, and electrolysis process'
4030988	'Process for producing halogen and metal hydroxides with cation exchange membranes of improved permselectivity'
4026783	'Electrolysis cell using cation exchange membranes of improved permselectivity'

Review Articles

Mauritz KA, Moore RB. *Chem. Rev.* 2004;104:4535.

Eisenberg A, Bailey F, eds. Coulombic Interactions in Macromolecular Systems. *ACS Symp. Ser. 302.* Washington, DC: American Chemical Society; 1986.

Eisenberg A, King M. *Ion-Containing Polymers.* New York, NY: Academic Press; 1977.

Eisenberg A, Yeager H, eds. *Perfluorinated Ionomer Membranes. ACS Symp. Ser.* 180. Washington, DC: American Chemical Society; 1982.

Heitner-Wirguin C. *J. Membrane Science.* 1996;120:1–33.

Lloyd D,ed. *Material Science of Synthetic Membranes.* ACS *Symp. Ser.* 269. Washington, DC: American Chemical Society; 1985.

Schlick S, ed. *Ionomers.* Boca Raton, Fla: CRC Press; 1996.

Sondheimer SN, Bunce, Fyfe C. *J. Macromol, Sci., Rev. Macromol, Chem. Phys.* 1986;C26:353.

Tant MK, Mauritz, Wilkes G, eds. *Ionomers,.* London: Blackie; 1997.

References

1. Banerjee S, Curtin DE. *Journal of Fluorine Chemistry.* 2004;125: 1211–1216.
2. Mauritz KA, Moore RB. *Chem. Rev..* 2004;104: 4535.
3. Page KA, Cable KM, Moore RB. *Macromol.* 2005;38:6472.
4. Page KA, Landis FA, Phillips AK, Moore RB. *Macromol.* 2006;39:3939.
5. Elliott JA, Hanna S, Elliott AMS, Cooley GE. *Macromol.* 2000;33:4171.
6. Rollet A-L, Diat O, Gebel G. *J. Phys. Chem. B.* 2002;106: 3033.
7. Barbi V, Funari SS, Gehrke R, Scharnagl N, Stribeck N. *Polymer.* 2003;44: 4853.
8. Gebel G, Aldebert P, Pineri M. *Macromol.* 1987;20:1425.
9. Grot WG. United States Patent. 1984:4433082.
10. Martin CR, Moore RB. United States Patent. 1988:4731263.
11. Curtin DE, Howard EG. United States Patent 2000:6150426.
12. Jiang SK, Xia, Xu G. *Macromolecules,.* 2001;34:7783–7788.
13. Lousenberg RD. *J. Polymer Sci.: Part B: Polym. Phys.* 2005;43:421–428.
14. Aldebert PB, Dreyfus G, Gebel N, Nakamura M, Pineri, Volino F. *J. Phys. France.* 1988;49:2101–2109.
15. Loppinet BG, Gebel, Williams CE. *J. Phys. Chem. B.* 1997;101:1884–1892.
16. Rubatat LAL, Rollet G, Gebel, Diat O. *Macromolecules.* 2002;35:4050–4055.
17. Curtin DR, Lousenberg T, Henry P, Tangeman M, Tisak. *J. Power Sources.* 2004;131:41–48.

18. Gierke T, Hsu W. In: Eisenberg A, Yeager H, eds. *Perfluorinated Ionomer Membranes*. Washington, DC: ACS Symp.Ser. 180. American Chemical Society; 1982.

19. Rodmacq BJ, Coey, Pineri M. In: Eisenberg A, Yeager H, eds. *Perfluorinated Ionomer Membranes*. Washington, DC: ACS Symp. Ser. 180 American Chemical Society; 1982.

20. Gierke TG, Munn, Wilson F. In: Eisenberg A, Yeager H, eds. *Perfluorinated Ionomer Membranes*. Washington, DC: ACS Symp. Ser. 180. American Chemical Society; 1982.

21. Hashimoto TM, Fujimura, Kawai H. In: Eisenberg A, Yeager H, eds. *Perfluorinated Ionomer Membranes*. Washington DC: ACS Symp. Ser. 180. American Chemical Society; 1982.

22. Gierke T, Hsu W. In: Eisenberg A, Yeager H, eds. *Perfluorinatd Ionomer Membranes*. Washington, DC: ACS Symp.Ser. 180. American Chemical Society; 1982.

23. Sondheimer SN, Bunce, Fyfe C. *J. Macromol. Sci., Rev. Macromol*, Chem. Phys. 1986;C26:353.

24. Falk M. In: Eisenberg A, Yeager H, eds. *Perfluorinated Ionomer Membranes*. Washington, DC: ACS Symp. Ser. 180. American Chemical Society; 1982.

25. Duplessix R, et al. In: *Adv. Chem. Ser.* 187, Chapter 28. Washington, DC: American Chemical Society; 1982.

26. Boyle NV, McBrierty, Douglass D. *Macromolecules*. 1983;16:80.

27. Boyle NV, McBrierty, Eisenberg A. *Macromolecules*. 1983:16:75.

28. Boyle N, et al. *Macromolecules* . 1984;17:1331.

29. Komoroski R, Mauritz K. In: Eisenberg A, Yeager H, eds. *Perfluorinated Ionomer Membranes*. Washington, DC: ACS Symp. Ser. 180. American Chemical Society; 1982.

30. Grot W, Chadds C. *European Pat.* 0066369.

31. Martin CT, Rhoades, Ferguson J. *Anal. Chem,*. 1982;5:1639.

32. Gebel GA, Aldebert, Pineri M. *Polymer* . 1993;34:333.

33. Yeo R. *J. Appl, Poly. Sci.* 1986;32:5733.

34. Yeo R. In: Eisenberg A, and Yeager H, eds. *Perfluorinated Ionomer Membranes*. Washington, DC: ACS Symp. Ser. 180. American Chemical Society; 1982.

35. Miura Y, Yoshida H. *Thermochim. Acta* . 1990;163:161.

36. Zook LA, Leddy J. *Anal. Chem.* 1996;68:3793.

37. Yeo SC, Eisenberg A. *J. Appl. Polym.* Sci. 1977;21(4):875.

38. Moore RB, Cable KM. *Polym. Prepr.* (American Chemical Society,Division of Polymer Chemistry). 1997;38(1):272.

39. Kyu T, Eisenberg A. In: Eisenberg A, Yeager H, eds. *Perfluorinated Ionomer Membranes*. Washington, DC: ACS Symp. Ser. 180. American Chemical Society; 1982.

40. Deng Z, Mauritz K. *Macromolecules*. 1992;25:2369.

41. Perusich SP, Avakian, Keating M. *Macromolecules*. 1993;26:4756.

42. Nakano Y, MacKnight W. *Macromolecules*. 1984;17:1585.

43. Kirsh YS, Smirov Y, Popkov, Timashev S. *Russian Chemical Reviews*. 1990;59:560.

44. Su S, Mauritz K. *Polym. Mater. Sci. Eng.* 1993;70:388.

45. Su S, Mauritz K. *Macromolecules* . 1994;27 (8):2079.

46. Narebski A, Koter S. *Electrochim. Acta.* 1987;32:449.

47. Koter S, Narebski A. *Electrochim. Acta.* 1987;32:455.

48. Halim J, et al. *Electrochim. Acta.* 1994;39:1303.

49. Inaba Mk, et al. *Electrochim. Acta.* 1993;38(13):1727–1731.

50. Broka K, Ekdunge P. *J. Appl. Electrochem..* 1997;27:117.

51. Yeager HZ, Twardowski, Clarke L. *J. Electrochem. Soc.* 1982;129:324.

52. Twardowski ZH, Yeager, O'Dell B. *J. Electrochem. Soc.* 1982;129:328.

53. Steck A, Yeager H. *J. Electrochem. Soc.* 1983;130:1297.

54. Hsu W, Gierke T. *J. Membrane Sci.* 1983;13:307.

55. Herrera A, Yeager H. *J. Electrochem. Soc.*. 1987;134:2446.

56. Kujawski W, Narebska A. *J. Membrane Sci.* 1991;56:99.

57. Narebski A, Koter S. *J. Membrane Sci.* 1987;30:141.

58. Narebski AW, Kujawski, Koter S. *J. Membrane Sci.*. 1987;30:125.

59. Narebski AS, Koter, Kujawski W. *J. Membrane Sci.*. 1985;25:153.

60. Pourcelly GA, Lindheimer, Gavach C. *J. Electroanal. Chem.* 1991;305:97.

61. Verbrugge, M., and R. Hill. *J. Electrochem. Soc.* 137 (1990): 886.

62. Verbrugge M, Hill R. *J. Electrochem. Soc.* 1990;137:893.

63. Verbrugge M, Hill R. *J. Electrochem. Soc.* 1990;137:1131.

64. Verbrugge M, Hill R. *Electrochem. Acta.* 1992;37:221.

65. Fuller T, Newman J. *J. Electrochem. Soc.*. 1992;139:1332.

66. Zawodzinski T Jr, et al. *J. Electrochem. Soc.* 1993;140:1041.

67. Zawodzinski T Jr, Gottesfeld S, Shoichet S, McCarthy T. *J. Appl. Electrochem.* 1993;23:86.

68. Chen Y, Chou T. *Electrochim. Acta.* 1992;38:2171.

69. Cahan B, Wainright J. *J. Electrochem. Soc.* 1993;140:L185.

70. Cappadonia MJ, Erning, Stimming U. *J. Electroanal. Chem.* 1994;376:189.

71. Kreur KT, Dippel W, Meyer, Maier J. . *Res. Soc. Symp. Proc.* 1993;293:273.

72. Tsou YM, Kimble, White R. *J. Electrochem Soc.* 1992;139:1913.

73. Pushpa KD, Nandan, Iyer R. *J. Chem. Soc. Faraday Trans.* 1988;184(6):2047–2056.

Phenolic Resins

Alexander B. Morgan and Milind Sohoni

Alternative Names Novolacs, resoles

Trade Name Bakelite (Georgia Pacific Resins, Inc.) Durite (Hexion)

Class Thermoset polymers; chemical copolymers

Typical Comonomers Phenols, substituted phenols, formaldehyde

Polymerizations Condensation

Major Applications Construction materials, electronics, aerospace, molded parts, insulating varnishes, laminated sheets, industrial coatings, wood bonding, fiber bonding, and plywood adhesives.

Properties of Special Interest Toughness, temperature resistance, low void content, chemical resistance, corrosion inhibition, and low flammability.

Substituted phenols used for phenolic resins[1]

Substituted Phenol	Resin Application
Cresol (*o*-, *m*-, *p*-)	Coatings, epoxy hardners
p-t-Butylphenol	Coatings, adhesives
p-Octylphenol	Carbonless paper, coatings
p-Nonylphenol	Carbonless paper, coatings
p-Phenylphenol	Carbonless paper
Bisphenol A	Low color molding compounds, coatings
Resorcinol	Adhesives
Cashew nutshell liquid	Friction particles

Forms of formaldehyde used in phenolic resin synthesis[1]

Type	Chemical Formula	Resin Preparation	
		Advantages	Disadvantages
Gaseous formaldehyde Formalin	CH_2O	–	Unstable
36%	$HO(CH_2O)_nH, n \approx 2$	Easy handling, moderate reactivity, stable at RT	High water content
50%	$HO(CH_2O)_nH, n \approx 3$	Increased capacity	Elevated temp, storage, formic acid formation
Paraformaldehyde	$HO(CH_2O)_nH, n \approx 20\text{–}100$	Increased capacity, water free	Dangerously high reactivity, solids handling

Type	Chemical Formula	Resin Preparation	
		Advantages	Disadvantages
Trioxane	$(CH_2O)_3$	Water-free	Catalyst requirements, cost
Hexamethylenetetramine	$(CH_2)_6N_4$	Autocatalytic	Amine incorporation

Relative rate constants for methylolation of phenol

Rate Constant	Ref. (2)	Ref. (3)	Ref. (4)
	1.00	1.00	1.00
	1.18	1.09	1.46
	1.66	1.98	1.75
	1.39	1.80	3.00
	0.71	0.79	0.85
	1.73	1.67	2.04
	7.94	3.33	4.36

Methylene group distribution, % in resoles[1]

Methylene Group	Catalyst	
	NaOH	Hexamethylene Tetramine (6 pph)
$2\text{-}CH_2OH$	30	24
$2\text{-}CH_2OCH_2OH$	24	1
$2\text{-}CH_2OR$	2	4
$4\text{-}CH_2OH$	12	9
$4\text{-}CH_2OCH_2OH$	16	0
$4\text{-}CH_2OR$	2	4
$2, 2'\text{-}CH_2$	0	0
$2, 4'\text{-}CH_2$	7	12
$4, 4'\text{-}CH_2$	7	10
$2\text{-}CH_2N$	0	27
$4\text{-}CH_2N$	0	7
Benzoxazine	0	2

Proton NMR chemical shifts of methylene groups in phenolic resins[5]

Methylene Group	Chemical Shift* (ppm)
2-CH$_2$OH	5.1
2-CH$_2$OR	5.0
4-CH$_2$OH	4.8
4-CH$_2$OR	4.7
2,2'-CH$_2$	4.2
2,4'-CH$_2$	4.1
4,4'-CH$_2$	3.8
2-CH$_2$N	4.0
4-CH$_2$N	3.5

* 10% concentration in d_5-pyridine.

Chemical shifts of methylene carbons in liquid resoles[1]

Structure*	Chemical Shift† (ppm)
Methylol C in	61.3
	(a) 65.4 (b) 88.0
Benzyl C in	68.9
Methylol C in	63.8
	(a) 68.5 (b) 88.0
	71.5
Methylene C in	31.5
Methylene C in	35.0
Methylene C in	40.4

* Designated carbon is shown underlined or described.

† From tetramethylsilane in d_6-acetone solution.

Phenolic resins used in coatings[1]

Property	Unsubstituted Phenol		Substituted Phenol	
	Heat-reactive	Non-heat-reactive	Heat-reactive	Non-heat-reactive
Type	Phenol	Phenol	Cresol *p-t*-Butyl phenol Bisphenol A	Cresol *p-t*-Butyl phenol Bisphenol A
Formaldehyde ratio	F > P	P > F	F > P	P > F
Catalyst	Alkaline	Acid	Alkaline	Acid
Stability	Low	High	Low	High
Softening point	Low	High	Low	High

Strength properties of phenolic-carbon-fiber composites[1]

Property	Units	Resin (%)		
		Phenolic		Epoxy Novolak, 27
		40	35	
Tensile strength	MPa*	115	63	64
Flexural strength	MPa*	183	126	110
Flexural modulus	GPa†	15.8	6.3	6.4

* To convert MPa to psi, multiply by 145.
† To convert GPa to psi, multiply by 145,000.

Functionality versus number of phenol alcohols[6]

Phenol	Functionality of Phenol	Number of Mono-alcohols	Number of Di-alcohols	Number of Tri-alcohols	Number of Tetra-alcohols	Total Number of Alcohols
2,4-Dimethylphenol	1	1	–	–	–	1
2,6-Dimethylphenol	1	1	–	–	–	1
p-Cresol	2	1	1	–	–	2
o-Cresol	2	2	1	–	–	3
2,3-Dimethylphenol	2	2	1	–	–	3
2,5-Dimethylphenol	2	2	1	–	–	3
3,4-Dimethylphenol	2	2	1	–	–	3
3,5-Dimethylphenol	3	2	2	1	–	5
Phenol	3	2	2	1	–	5
Resorcinol	3	2	2	1	–	5
m-Cresol	3	3	3	1	–	7
Hydroquinone	4	1	3	1	1	6
Catechol	4	2	3	2	1	8

First-order rate constants and comparative rates of reaction for various phenols[7]

Phenol	Apparent First-order Rate Constant	Relative Reactivity
3,5-Xylenol	0.0630	7.75
m-Cresol	0.0233	2.88
2,3,5-Trimethylphenol	0.0121	1.49
Phenol	0.00811	1.00
3,4-Xylenol	0.00673	0.83
2,5-Xylenol	0.00570	0.71
p-Cresol	0.00287	0.35
Saligenin	0.00272	0.34
o-Cresol	0.00211	0.26
2,6-Xylenol	0.00130	0.16

Properties of phenol-formaldehyde molding compounds[8]

Property	Units	Phenol-formaldehyde, Wood Flour and Cotton Floe
Pigmentation and coloring possibilities	–	Limited
Appearance	–	Opaque
Molding qualities	–	Excellent
Type of resin	–	Thermosetting
Molding temperature	°F (°C)	290–380 (143–193)
Molding pressure	psi	2,000–4,000
Mold shrinkage	in in^{-1}	0.004–0.009
Specific gravity	–	1.32–1.45
Tensile strength	psi	6.5–9 $\times 10^3$
Flexural strength	psi	8.5–12 $\times 10^3$
Notched Izod impact strength	ft-lb in^{-1}	0.24–0.6
Rockwell hardness	–	M 96–M 120
Thermal expansion	°C^{-1}	3.0–4.5 $\times 10^{-5}$
Deflection temperature under load	°F	260–340
Dielectric strength, short time, 0.125 in thickness	V mil^{-1}	200–425
Dielectric constant	–	4.0–7.0
Dissipation factor	–	0.03–0.07
Arc resistance	s	Tracks
Cold-water absorption, room temperature		
24 h, 0.125 inch thickness	%	0.3–1.0
7 days	mg (100 cm^2)$^{-1}$	200–750
Boiling water test, 10 min, 100°C	%	0.4–1.0
Burning rate	–	Very low
Effect of sunlight	–	General darkening

Properties of phenol-formaldehyde laminates[8]

Properties	Units	Phenol-formaldehyde Laminate	
		Paper-base Filler	Glass Fabric Base
Coloring possibilities	–	Limited	Limited
Appearance	–	Opaque	Opaque
Laminating temperature	°F	275–350	275–350
Laminating pressure	psi	1,000–1,800	1,500–2,000
Specific gravity	–	1.28–1.4	1.4–1.9
Tensile strength	psi	$8–20 \times 10^3$	$9–50 \times 10^3$
Flexural strength	psi	$10.5–30 \times 10^3$	$16–80 \times 10^3$
Notched Izod impact strength	ft-lb in^{-1}	0.3–1.0	4–18
Rockwell hardness	–	M 70–M 120	M 105–M 110
Water absorption, 24 h, room temperature, 0.125 inch thickness	%	0.2–4.5	0.3–1.5
Effect of sunlight	–	General darkening and lower surface resistance	General darkening and lower surface resistance
Machining qualities	–	Fair to excellent	Fair to good
Thermal expansion	$°C^{-1}$	$1.4–3.0 \times 10^{-5}$	$1.5–2.5 \times 10^{-5}$
Resistance to heat (continuous)	°F	225–250	250–500
Heat-distortion temperature	°F	250–over 320	Over 320
Burning rate	–	Very low	Nil
Dielectric strength, short time	V mil^{-1}	300–1,000	300–700
Dielectric constant, at 10^6 cps	–	3.6–6.0	3.7–6.0
Dissipation factor, at 10^6 cps	–	0.02–0.08	0.005–0.05
Arc resistance	s	Tracks	Tracks

Properties of epoxy/phenolic foams[9]

	Flammability Data Cone Calorimeter (ASTM E-1354)				Compression Test Data (ASTM D1621) Foam Density: 200 kg/m^2				Shear Test Data (ASTM C273) Foam Density: 200 kg/m^2	
	50 kW/m^2 Heat flux				Parallel Loading Direction		Perpendicular Loading Direction		Perpendicular Loading Direction	
	Peak Heat Release Rate	Average Heat Release Rate@ 300 seconds	Time to Ignition	Mass Lost	Modulus	Strength	Modulus	Strength	Modulus	Strength
Epoxy/Phenolic Foam Formulation	(kW/m^2)	(kW/m^2)	(s)	(%)	(MPa)	(MPa)	(MPa)	(MPa)	(MPa)	(MPa)
Epoxy (100%)	314	195	2.3	100	48.01	1.89	21.12	1.69	11.9	1.07
Epoxy/Phenolic (50%/50%)	258	179	4	89	44.32	2.64	24.22	2.05	7.9	0.82
Epoxy/Phenolic (35%/65%)	253	174	3.8	84	72.23	2.87	52.6	2.28	14.2	1.7
Epoxy/Phenolic (20%/80%)	247	170	3.6	91	8.4	0.62	–	–	5.59	0.33
Phenolic (100%)	106	54	6	84	106.05	3.39	79.87	3.27	7.1	0.84

Inherent flammability of epoxy novolac phenolic resin[10]

ASTM D-7907: Micro combustion calorimetry

	Heat Release Capacity (J/gK)	Total Heat Release (KJ/g)	Char (%)
Epoxy Novolac, catalytic Cure	246	18.9	15.9

Phenolic (Resol/Novolac) Polymer molecular weights, polydispersity, and polymerization temperatures.[11]

Formaldehyde to Phenol Molar Ratio	Polymer Characteristics					Polymerization Temperature as Function of Heating Rate by Differential Scanning Calorimetry					Energy of Activation
	Mn	Mw	Mn/Mw	Nonvolatile Content (%)	Viscosity (mPa s)	2 C/min	5 C/min	10 C/min	20 C/min	40 C/min	(kJ/mol)
Resol 1.3	240	460	1.92	59.2	157.0	110.0	131.0	145.2	153.1	160.6	17.65
Resol 1.9	380	630	1.66	53.5	445.0	103.8	106.7	107.8	119.4	129.5	17.91
Resol 2.5	400	670	1.67	50.4	1390.0	91.7	106.7	124.9	132.7	143.9	15.23
Novolac 0.5	380	540	1.44	46.6	22.6	105.0	106.7	117.4	130.9	136.6	22.93
Novolac 0.7	390	560	1.42	45.4	25.7	116.5	112.8	114.2	138.6	136.8	21.34
Novolac 0.9	390	580	1.49	45.9	29.8	104.9	110.7	120.0	139.9	154.4	16.02

References

1. Kopf PW. In *Encyclopedia of Polymer Science and Engineering*. Vol 11. New York: John Wiley and Sons; 1988:45.

2. Freeman JH, Lewis C. *J. Am. Chem. Soc.* 1954;76:2080.

3. Zsavitsas A, Beaulieu A. *Am. Chem. Soc. Div. Org. Coat. Plast. Chem. Pap.* 1967;27:100.

4. Eapen K, Yeddanapalli L. *Makromol. Chem.* 1968;4:119.

5. Kopf PW. In: Kroschwitz JI, ed. *Kirk-Othmer Encyclopedia of Chemical Technology*. 4th ed. Vol 18. New York: John Wiley and Sons; 1996:603.

6. Martin RW. *The Chemistry of Phenolic Resins*. New York: John Wiley and Sons; 1956:12.

7. Martin RW. *The Chemistry of Phenolic Resins*. New York: John Wiley and Sons; 1956:262.

8. Widmer G. In: Mark HF, ed. *Encyclopedia of Polymer Science and Technology*. Vol. 2. New York: John Wiley and Sons; 1965:54.

9. Auad MLL, Zhao H, Shen SR, Nutt, Sorathia U. "Flammability properties and mechanical performance of epoxy modified phenolic foams" *J. App. Polym. Sci.* 2007;104:1399–1407.

10. Walters RN, Lyon RE. "Molar group contributions to polar flammability" *J. App. Polym. Sci.* 2003;87:548–563.

11. Lee Y-KD-J, Kim H-J, Kim T-S, Hwang M. Rafailovich, Sokolov J. "Activation energy and curing behavior of resol- and novolac-type phenolic resins by differential scanning calorimetry and thermogravimetric analysis". *J. App. Polym. Sci.* 2003;89:2589–2596.

Polyacetylene

Shuhong Wang and Ping Xu

Class Conjugated and other unsaturated polymers

Structure *cis*-Polyacetylene

$$
\begin{array}{c}
\quad\text{H}\ \ \text{H}\qquad\quad\ \text{H}\ \ \text{H}\\
\quad|\ \ \ |\qquad\qquad\ |\ \ \ |\\
\text{C}=\text{C}-\text{C}=\text{C}-\text{C}=\text{C}-\text{C}=\text{C}\\
\qquad\ \ |\ \ \ |\qquad\qquad\ |\ \ \ |\\
\qquad\ \ \text{H}\ \ \text{H}\qquad\qquad\ \text{H}\ \ \text{H}
\end{array}
$$

trans-Polyacetylene

$$
\begin{array}{c}
\quad\text{H}\qquad\ \text{H}\qquad\ \text{H}\qquad\ \text{H}\\
\quad|\qquad\quad|\qquad\quad|\qquad\quad|\\
\text{C}=\text{C}-\text{C}=\text{C}-\text{C}=\text{C}-\text{C}=\text{C}\\
\ \ |\qquad\quad|\qquad\quad|\qquad\quad|\\
\ \ \text{H}\qquad\ \text{H}\qquad\ \text{H}\qquad\ \text{H}
\end{array}
$$

Major Applications Power cable sheathing, prime conductor, energy load leveling systems, batteries, and signal processing devices.

Properties of Special Interest Insulating, semiconducting, conducting, and nonlinear optical properties.

Polymerization Solvent evacuation (SE) method and intrinsic nonsolvent (INS) method.

Thermal behavior[1]
  Cis isomer 1. *Cis* to *trans* isomerization at 145°C
         2. Molecular rearrangement at 325°C
         3. Thermal decomposition at 420°C

Unit cell dimensions

Isomer	Lattice	Cell Dimensions (Å)			Reference
		a	*b*	*c*	
Cis	Orthorhombic	7.61	4.47	4.39	(2–5)
Trans	Orthorhombic	7.32	4.24	2.46	(6–8)

Property	Units	Conditions	*Cis* Value	*Trans* Value	Reference
Tensile strength	MPa	SE polyacetylene	600	900	(9)
		INS polyacetylene	800	2,100	
Tensile elongation	%	SE polyacetylene	6–8	–	(9)
		INS polyacetylene	6–9	–	

Property	Units	Conditions	*Cis* Value	*Trans* Value	Reference
Tensile modulus	MPa	SE polyacetylene	30–40	100	(9)
		INS polyacetylene	28	40	
Cis content	%	SE polyacetylene	70–90	–	(9)
		INS polyacetylene	85–95	–	
Density	g cm^{-3}	SE and INS polyacetylene	1.0–1.15	1.0–1.15	(9)
Magic angle spinning ^{13}C NMR	ppm	Solid-state	127–128	136–137	(10)
Linear absorption coefficient	cm^{-1}	Reflection method: *cis* at 18,500 cm^{-1}; *trans* at 15,400 cm^{-1}	1.4×10^5	1.5×10^5	(11)
Absorption edge	eV	–	1.90	1.35	(12)
Thermal activation energy	eV	–	0.6	0.3	(12)
Dark conductivity	(W cm)$^{-1}$	–	2×10^{-9}	5×10^{-6}	(12)
Electrical conductivity	S cm^{-1}	Doping species			(13)
		None	1.9×10^{-9}	4.4×10^{-5}	
		I$_2$	360	160	
		AsF$_5$	560	400	
		IBr	400	120	
		NaC$_{10}$H$_8$	25	80	
		MoCl$_5$	200	–	
		WCl$_6$	200	–	
		PtCl$_4$	134	–	
		RhCl$_3$	6×10^{-4}	–	
		CuCl$_2$	2×10^{-3}	–	
		InCl$_3$	600	–	
		LiAlH$_4$	–	6	

References

1. Ito T, Shirakawa H, Ikeda S. *J. Polym. Sci. Polym. Chem. Ed.* 1975;13:1943.
2. Baughmann RH, Hsu SL, Pez GP. Signorelli AJ. *J. Chem. Phys.* 1972;68:5405.
3. Akasimi T, et al. *J. Polym. Sci. Polym. Phys. Ed.* 1980;18:745.
4. Fincher CR, et al. *Phys. Rev. Lett.* 1982;48:100.
5. Robin P, et al. *Phys. Rev. Sect.* 1983;B27:3,938.
6. Shimamura K, Karasz FE, Hirsch J, Chien JCW. *Makromol. Chem. Rapid Commun.* 19812:473.
7. Bolognesi A, et al. *Makromol. Chem. Rapid Commun.* 1983;4:403.
8. Robin P, et al. *Polymer.* 1983;24:1558.
9. Akagi K, Shirakawa H. In: Salamone JC, ed. *The Polymer Materials Encyclopedia.* Boca Raton, Fla: CRC Press; 1996.
10. Maricq MM, et al. *J. Am. Chem. Soc.* 1978; 100:7729.

11. Fujimoyo HK, Kamiya M, Tanaka, Tanaka J. *Synth. Met.* 1985;10:367.
12. Kanicki J. In: Skotheim TA, ed. *Handbook of Conducting Polymers.* Vol 1. New York: Marcel Dekker; 1986.
13. Gibson HW, Pochan JM. In: Kroschwitz JI, ed. *Encyclopedia of Polymer Science and Engineering.* 2nd ed. Vol 1. New York: John Wiley and Sons; 1985.

Polyacrylamide

Robert A. Orwoll and Yong S. Chong

Acronym; Chemical Abstracts Name and Number; Trade Name PAAm;
2-propenamide homopolymer [9003-05-8]; Cyanamer (American Cyanamid)

Class Vinyl polymers

Structure $[-CH_2-CH-]$
           $|$
         $CONH_2$

Major Applications Flocculants in water treatment, paper manufacture, mining, and oil recovery; absorbents; gels for electrophoresis; thickening and binding agents; lubrication.

Properties of Special Interest Amorphous. High affinity for water and completely miscible in water. Low toxicity. Low cost. Very high molecular weights attainable.

Polymerization Conditions Most commonly, free-radical polymerizations of acrylamide in aqueous or aqueous salt solutions; but also precipitation polymerization in organic solvents; and emulsion polymerizations.[26]

Property	Units	Conditions	Value	Reference
Molecular weight (of repeat unit)	$g\ mol^{-1}$	–	71.08	–
Tacticity (stereoregularity)	–	Reaction conditions: temp. = 70°C; monomer conc. = 16 wt% in water; initiator = $(NH_4)_2S_2O_8$; chain-transfer agent = isopropanol	Probability meso $P_m = 0.43$	(1)
Head-to-head contents	–	Reaction conditions: temp. = 25°C; monomer conc. = 10% in water; initiators (25 mg/100 ml) = $K_2S_2O_8$, $Na_2S_2O_5$	Head-to-head units = 4.5%	(2)
IR spectrum	–	–	–	(3, 4, 27, 28, 30)
Raman spectrum	–	–	–	(5, 28)
NMR	–	^{13}C spectrum, 100 MHz	–	(1)
X-ray diffraction	–	–	–	(27, 28)
Solvents	Water aqueous sodium chloride, ethylene glycol, formamide, hydrazine			(6)

Property	Units	Conditions			Value	Reference
Nonsolvents		Methanol, hydrocarbons, and other common organic liquids				(6)
Partial specific volume $(\partial V / \partial m_2)$	$cm^3\,g^{-1}$	20°C, water			0.696	(7)
		25°C, water			0.716	(8)
		25°C, water			0.693 ± 0.002	(9)
		25°C, water			0.674	(10)
		20°C, water/methanol (3:2 v/v)			0.655	(10)
		0.1 M NaCl (aq.)			0.702	(11)
Apparent adiabatic compressibility in solution	$cm^3\,bar^{-1}\,g^{-1}$	25°C, water			-4.2×10^{-6}	(8)
Theta temperature Θ	K	Water (extrapolated value)			235	(12)
		Water/methanol (3:2 v/v), $0.33 < M_w \times 10^{-4} < 81$			293	(10)
		Water/methanol (59:41 v/v), $92 < M_w \times 10^{-4} < 820$			294	(13)
		Water/methanol (59:41 v/v), $43 < M_w \times 10^{-4} < 1{,}000$			298	(14)
		0.1 M Nacl (aq.), $M_w \times 10^{-4} = 670$			265	(31)

Interaction parameter χ	–	Solvent	Temp. (°C)	$M \times 10^{-6}$ $(g\,mol^{-1})$		
		Water	25	0.71	0.44	(12)
		Water	60	0.71	0.42	(12)
		Water	25	0.107	0.495	(9)

Enthalpy parameter χ_H	–	Solvent	Temp. (°C)	$M \times 10^{-6}$ $(g\,mol^{-1})$		
		Water	25	0.71	0.22	(12)
		Water	60	0.71	0.20	(12)
		Water	25	0.107	0.08 ± 0.008	(9)

Second virial coefficient A_2	$mol\,cm^3$ $g^{-2}\,(\times 10^4)$	Solvent	Temp. (°C)	$M \times 10^{-6}$ $(g\,mol^{-1})$		
		Water	20	0.25	3.1	(7)
		Water	20	2.4	2.9	(7)
		Water	20	11	2.2	(7)
		Water	25	0.43	4.4	(14)
		Water	25	4.7	0.64	(15)
		Water	25	0.5–6	4 ± 2	(16)
		Water	25	0.11	1.4	(9)
		Water	25	10	1.7	(14)
		0.1 M NaCl (aq.)	–	6	2.5 ± 0.4	(11)
		0.1 M NaCl (aq.)	4	0.033	3.3	(31)

Polyacrylamide

Property	Units	Conditions			Value	Reference
				$M \times 10^{-6}$		
		Solvent	Temp. (°C)	(g mol^{-1})		
		0.1 M NaCl (aq.)	60	0.033	6.8	(31)
		0.1 M NaCl (aq.)	4.5	0.22	2.0	(31)
		0.1 M NaCl (aq.)	55.5	0.22	3.2	(31)
		0.1 M NaCl (aq.)	2.5	6.7	0.3	(31)
		0.1 M NaCl (aq.)	30	6.7	1.8	(31)
		1 M NaCl (aq.)	–	5.5	2.7	(11)
		4 M NaCl (aq.)	–	5.5	2.9	(11)
		0.1 M LiCl (aq.)	–	6.8	1.9	(11)
		Water/methanol (3:2 v/v)	20	0.77	0.008	(10)
		Ethylene glycol	25	0.5–5	0.27 ± 0.08	(16)
		Formamide	–	6.8	1.3	(11)

Mark-Houwink parameters: K and a

Solvent	Temp. (°C)	$M \times 10^{-6}$ (g mol^{-1})	$K \times 10^2$ (with [η] in ml g^{-1})	a	Reference
Water	20	0.25–3	3.09	0.67	(7)
Water	25	0.5–6	0.49	0.8	(16)
Water	25	0.038–9	1.00	0.755	(6)
Water	25	0.01–0.36	6.8	0.66 ± 0.05	(17)
Water	25	0.003–0.8	1.83	0.72	(18)
Water	25	0.43–10	0.742	0.775	(14)
Water	30	0.02–0.5	0.631	0.80	(19)
Water	30	0.04–1.3	0.65	0.82	(20)
0.1 M NaCl (aq.)	–	0.2–8	0.933	0.75	(11)
0.2 M NaCl (aq.)	20	0.25–3	3.02	0.68	(7)
0.5 M NaCl (aq.)	25	0.5–6	0.719	0.77	(16)
1.0 M NaCl (aq.)	20	0.25–3	2.88	0.69	(7)
10% NaCl (aq.)	25	0.43–10	0.81	0.78	(14)
1.0 M NaNO$_3$ (aq.)	30	0.5–3	3.73	0.66	(6)
Water/methanol (3:2 v/v)	20	0.006–0.8	0.127	0.50	(10)
Water/methanol (59:41 v/v)	25	0.43–10	15	0.50	(14)
0.1 M NaCl (aq.)/methanol (3:2 v/v)	21	0.05–1.7	11	0.51	(31)
Ethylene glycol	25	0.5–5	13.6	0.54	(16)
Formamide	25	0.5–6	1.27	0.74	(21)

Property	Units	Conditions	Value	Reference
Huggins constant k'	–	0.5 M NaBr (aq.)	0.46	(22)
		20°C	See table below	(7)

$M_w \times 10^{-6}$ (g mol^{-1})	Huggins Constant k'		
	Water	0.2 M NaCl (aq.)	1.0 M NaCl (aq.)
0.26	0.41	0.38	0.38
0.62	0.40	0.41	0.37
1.0	0.28	0.40	0.36
2.4	0.17	0.34	0.37
2.8	0.39	0.38	0.39
11	0.37	0.40	0.35

Property	Units	Conditions			Value	Reference
Sedimentation constant S_0	s^{-1} ($\times 10^{13}$)			$M \times 10^{-6}$		(16)
		Solvent	Temp. (°C)	(g mol^{-1})		
		0.5 M NaCl (aq.)	20	0.8–6	0.009 $M_w^{0.32}$	
Characteristic ratio $\langle r^2 \rangle / nl^2$ (l = 0.154 nm)				$M \times 10^{-6}$		
		Solvent	Temp. (°C)	(g mol^{-1})		
		Water	25	0.5–6	3.6 $M_w^{0.18}$	(16)
		0.1 M NaCl (aq.)	–	0.8–8	4.9 $M_w^{0.28}$	(11)
		Water/ methanol (3:2 v/v), Θ solvent	20	0.08–0.8	9.3	(18)
		Water/methanol (59:41 v/v), Θ solvent	25	0.43–10	11.3	(14)
		Salt/water/ methanol (?:59:41v/v), Θ solvent	21	0.9–8	14	(13)
		Ethylene glycol	25	0.5–6	21 $M_w^{0.02}$	(16)
Glass transition temperature T_g	K	–			457–469	(22, 28, 29)
Softening temperature	K	–			481	(23)

Refractive index increment dn/dc

Solvent	Temp (°C)	dn/dc (cm^3 g^{-1})			Reference
		$\lambda = 436$ nm	$\lambda = 546$ nm	λ not Reported	
Water	20	0.185	0.182	–	(7)
Water	25	–	0.187	–	(16)
Water	25	–	0.189	–	(14)
Water	20–60	–	0.149	–	(12)
Water	–	–	–	0.165	(11)

Polyacrylamide

Solvent	Temp (°C)	dn/dc (cm³ g⁻¹)			Reference
		$\lambda = 436$ nm	$\lambda = 546$ nm	λ not Reported	
0.1 M LiCl (aq.)	–	–	–	0.164	(11)
0.1 M NaCl (aq.)	–	–	–	0.165	(11)
0.2 M NaCl (aq.)	20	0.186	0.182	–	(7)
1 M NaCl (aq.)	–	–	–	0.159	(11)
1 M Mg(ClO$_4$)$_2$ (aq.)	25	–	0.174	–	(10)
Ethylene glycol	25	–	0.095–0.105	–	(16)
Formamide	–	–	–	0.095	(11)

Property	Units	Conditions	Value	Reference
Critical surface tension γ_c	mN m⁻¹	20°C, contact angle method	52.3	(24)
Water absorption (residual wt% water)	%	Dried under vacuum at 20°C	15	(25)
		Dried overnight under vacuum at 60–80°C	7–11	(7)
		Dried overnight under vacuum at 60–80°C, then 4 h at 120°C	~0	(7)
		Dried under vacuum for 24 h at 25°C	3	(16)
		Dried under vacuum for 24 h at 25°C, then 9 h at 50°C	0.9	(16)
		Dried under vacuum for 24 h at 25°C, then 9 h at 50°C, then 7 h at 110°C	~0	(16)

References

1. Lancaster JE, O'Connor MN. *J. Polym. Sci., Polym. Lett. Ed.* 1982;20:547.
2. Sawant S, Morawetz H. *Macromolecules.* 1984;17: 2427.
3. Kulicke W-M, Siesler HW. *J. Polym. Sci., Polym Phys. Ed.* 1982;20:553.
4. Pouchert CJ. *The Aldrich Library of Infrared Spectra.* 3d ed. Milwaukee: Aldrich Chemical Company; 1981:1592, spectrum A.
5. Gupta MK, Bansil R. *J. Polym. Sci., Polym. Phys.Ed.* 1981;19:353.
6. Thomas WM, Wang DW. In: Mark HF, et al, eds. *Encyclopedia of Polymer Science and Engineering.* 2nd ed. Vol 1. New York: John Wiley and Sons; 1985:169–211.
7. Munk P, et al. *Macromolecules,* 1980;13:871.
8. Roy-Chowdhury P, Kale KM. *J. Appl. Polym. Sci.* 1970;14:2937.
9. Day JC, Robb ID. *Polymer.* 1981;22:1530.
10. Bohdanecky MV, Petrus, Sedlácek B. *Makromol. Chem.* 1983;184:2061.
11. François J, et al. *Polymer.* 1979;20:969.
12. Silberberg A, Eliassaf, J, Katchalsky A. *J. Polym. Sci.* 1957;23:259.
13. Schwartz T, Sabbadin, J, François J. *Polymer.* 1981;22:609.
14. Izyumnikov AL, et al. *Vysokomol. Soedin, Ser. A.* 1988;3:1030; *Polym. Sci. U.S.S.R.* 1988;30:1062.
15. Lechner MD, Nordmeier E, Steinmeier DG. In: Brandrup J, Immergut EH, Grulke EA, eds. *Polymer Handbook.* 4th ed. New York: John Wiley and Sons; 1999:VII/170.

16. Klein J, Conrad K-D. *Makromol. Chem.* 1980;18: 227.
17. Collinson E, Dainton FS, McNaughton GS. *Trans.Faraday Soc.* 1957;53:489.
18. Calculated from data in reference (10).
19. Scholtan W. *Makromol.Chem.* 1954;14:169.
20. Misra GS, Bhattacharya SH. *Eur. Polym. J.* 1979;1:125.
21. Klein J, Hannemann G, Kulicke W-M. *Colloid. Polym. Sci.* 1980;258:719.
22. Klein J, Heitzmann R. *Makromol. Chem.* 1978;179: 1895.
23. Miller ML. *Can. J. Chem.* 1958;36:309.
24. Kitazaki Y, Hata T. *J. Adhesion Soc. Japan.* 1971;8:131; as recorded in Wu S. In: Brandrup J, Immergut EH. *Polymer Handbook.* 3rd ed. New York: John Wiley and Sons; 1989:VI/416.
25. Sawant S, Morawetz H. *J. Polym. Sci, Polym. Lett. Ed.* 1982;20:385.
26. Kurenhyov VF, Myagchenkov VA. In: Joseph C, Salamone, eds. *Polymeric Materials Encyclopedia.* Boca Raton: CRC Press; 1996:47–54.
27. Xiao C, et al. *J. Macromol. Sci., Pure and App. Chem.* 2001;A38:761.
28. Yan F, et al. *J. Polym. Sci., Polym. Chem. Ed.* 1998;6:747.
29. Janigova I, et al. *Macromol. Chem. Phys.* 1994;195:3609.
30. Guha S, Ray B, Mandal BM. *J. Polym. Sci., Polym. Chem. Ed.* 2001;39:3434.
31. Kanda A, et al. *Polymer.* 1985;26:406.

Poly(acrylic acid)

Robert A. Orwoll and Yong S. Chong

Acronyms; Chemical Abstracts Number; Trade Names PAA, PAAc; [9003-01-4]; Acrysol, Acumer, Acusol, Duolite (Rohm & Haas); Alcogum, Alcosperse, Aquatreat (Alco); Carbopo, Good-ritel (B F Goodrich); Sokalan (BASF)

Class Vinyl polymers

Structure $[-CH_2-CH-]$
$\qquad\qquad\ \ \ \ \ $ COOH

Major Applications Thickening and suspension agents for petroleum recovery, pigment dispersements in paint, ion exchange resins (with cross-linking), flocculating agents for particles suspended in water, adhesives, dental cements, drag reduction in fluid flow through pipes. Many applications involve copolymers of acrylic acid.

Properties of Special Interest Amorphous polymers.

Property	Units	Conditions	Value	Reference
Molecular weight (of repeat unit)	$g\,mol^{-1}$	–	72.06	–
IR spectrum	–	–	–	(1)
	cm^{-1}	OH stretching	3,101 (broad)	(13)
		CH_2 or CH stretching	2,947	(13)
		overtones and combinations	2,700–2,500	(13)
		C=O stretching	1,711 (strong)	(13)
		CH_2 deformation	1,451	(13)
		C—O stretching and OH bending	1,413; 1,248; 1,178	(13)
		C—CH_2 stretching	1,112	(13)
		OH out of plane bending	902 (broad)	(13)
		CH_2 twisting and C—COOH stretching	804	(13)
Raman	cm^{-1}	CH_2 or CH stretching	2,927; 2,877	(13)
		C=O stretching	1,698	(13)
		CH_2 deformation	1,457	(13)
		CH_2 twisting	1,333	(13)
		C—O stretching and OH bending	1,191	(13)
		C—CH_2 stretching	1,105	(13)
		CH_2 rocking	1,026	(13)
		C—COOH stretching	846	(13)
Density	$g\,cm^{-3}$	–	1.22	(2)

Property	Units	Conditions	Value	Reference
Solvents		Water, dioxane, ethanol, dimethylformamide, methanol		(3)
Nonsolvents		Acetone, diethyl ether, benzene, aliphatic hydrocarbons		(3)
Partial specific volume	$cm^3 g^{-1}$	Water, 25°C	0.648	(4)
Apparent adiabatic compressibility in solution	cm^3 $bar^{-1} g^{-1}$	25°C, water	1.2×10^{-6}	(4)
		25°C, PAAc 25% neutralized with NaOH, water	-18×10^{-6}	
		25°C, PAAc 100% neutralized with NaOH, water	-54×10^{-6}	
		25°C, PAAc 25% neutralized with NaOH, 1.0 M NaCl (aq.)	-53×10^{-6}	
Theta temperature Θ	K	Dioxane	303 ± 1 (LCST)	(5)
		Water, 1.245 M in NaCl, and enough NaOH to neutralize 1/3 of acid groups	305 ± 3 (UCST)	(5)
		0.2 M HCl (aq.)	287	(6)
Interaction parameter χ	–	0.2 M HCl (aq.); $M_v = 0.43 \times 10^6$ g mol^{-1}		(6)
		20°C	0.498	
		68°C	0.490	
Enthalpy parameter χ_H	–	Water; $M = 0.43 \times 10^6$ g mol^{-1}		(6)
		20°C	0.0631	
		68°C	0.0542	

Second virial coefficient A_2

Solvent	Ionic Strength (mol L^{-1})	Degree of Ionization	Temp (°C)	$M_w \times 10^{-5}$ (g mol^{-1})	$A_2 \times 10^4$ (cm^3 mol g^{-2})	Reference
0.2 M HCl			20–68	11	$49.9 \times (1-287/T)$; T in K	(6)
1,4 Dioxane			25	2.3–56	−0.1 to 0.3	(16)
NaCl (aq.)	0.10	0.102	27.5	7.7	2	(17)
NaCl (aq.)	0.10	0.335	27.5	7.7	22.2	(17)
NaCl (aq.)	0.01	0.344	27.5	7.7	69.5	(17)
NaCl (aq.)	1.00	0.947	27.5	7.7	10.0	(17)
NaCl (aq.)	0.10	0.959	27.5	7.7	43.9	(17)
NaCl (aq.)	0.01	0.994	27.5	7.7	196	(17)

Poly(acrylic acid)

Property	Units	Conditions	Value	Reference

Mark-Houwink parameters: K and a

Solvent	$M \times 10^{-6}$ (g mol^{-1})	Temp (°C)	$K \times 10^2$ (with $[\eta]$ in ml g^{-1})	a	Reference
1,4 Dioxane	0.13–0.82	30	8.5	0.50	(7)
1,4 Dioxane	0.23–5.6	25	6.63	0.50	(16)
Na salt of PAA in 0.05 M NaCl (aq.)	0.09–0.62	25	0.735	0.88	(15)
Na salt of PAA in 0.3 M NaCl (aq.)	0.09–0.62	25	1.69	0.75	(15)
Na salt of PAA in 1.0 M NaCl (aq.)	0.09–0.62	25	4.15	0.63	(15)
Na salt of PAA in 1.5 M NaBr (aq.)	0.09–1.50	15	14.5	0.5	(15)

Property	Units	Conditions	Value	Reference
Huggins constant k'	–	1,4-Dioxane, 30°C	0.25–0.30	(3)
		0.5 M NaBr(aq.)	0.30	(8)
Characteristic ratio $<r^2>_o^{1/2}/M_w$	nm·mole$^{1/2}$·g$^{-1/2}$	$M_w = (1.3 \times 10^5 - 1.2 \times 10^6)$; 1,4-dioxane; 30°C	0.077 ± 0.003	(7)

Root-mean square radius of gyration $<s^2>^{1/2}$

Solvent	Ionic Strength (mol L^{-1})	Degree of Ionization	Temp (°C)	$M_w \times 10^{-5}$ (g mol^{-1})	$<s^2>^{1/2}$ (nm)	Reference
1,4 Dioxane			25	2.3	21	(16)
1,4 Dioxane			25	4.2	29	(16)
1,4 Dioxane			25	8.9	44	(16)
1,4 Dioxane			25	15.1	53	(16)
1,4 Dioxane			25	24.4	54	(16)
1,4 Dioxane			25	56	67	(16)
NaCl (aq.)	0.10	0.102	27.5	7.7	56	(17)
NaCl (aq.)	0.10	0.335	27.5	7.7	72	(17)
NaCl (aq.)	0.01	0.344	27.5	7.7	108	(17)
NaCl (aq.)	1.00	0.947	27.5	7.7	67	(17)
NaCl (aq.)	0.10	0.959	27.5	7.7	75	(17)
NaCl (aq.)	0.01	0.994	27.5	7.7	124	(17)

Property	Units	Conditions	Value	Reference
Optical configuration parameter $\Delta\alpha$	Å^3	Dioxane	-0.5	(11)
Glass transition temperature T_g	K	–	376	(9)
			379 ± 2	(10)
			399	(8)
			389	(12)

Poly(acrylic acid)

Property	Units	Conditions	Value	Reference
Ionization pK	–	10^{-3} M NaCl (aq.)	5	(14)
Refractive index increment dn/dc	$cm^3\,g^{-1}$	1,4-Dioxane, 25°C, $\lambda = 436$ nm	0.089	(7)
		1,4-Dioxane, 25°C, $\lambda = 546$ nm	0.070	(16)
		0.2 M HCl (aq.), 20–60°C, $\lambda = 546$ nm	0.146	(6)
		10^{-3} M NaCl (aq.), pH = 5.5 with HCl (aq.), $\lambda = 589$ nm	0.201	(14)
		Na salt of PAA in 0.3 M NaCl (aq.), 25°C, $\lambda = 488$ nm	0.159	(15)
		Na salt of PAA in 1.0 M NaCl (aq.), 25°C, $\lambda = 488$ nm	0.152	(15)
Water absorption (wt% water)	%	30°C, 32% relative humidity	4.8	(10)
		30°C, 54% relative humidity	7.7	
		30°C, 69% relative humidity	13.7	

References

1. Pouchert CJ. *The Aldrich Library of Infrared Spectra.* 3d ed. Milwaukee: Aldrich Chemical Company; 1981:1580, spectra A and B.
2. Welsh WJ. In: Mark JE. *Physical Properties of Polymers Handbook.* Woodbury, N.Y: AIP Press; 1996:401–407.
3. Nemec JW, Bauer W Jr. In: Mark HF, et al, eds. *Encyclopedia of Polymer Science and Engineering.* 2nd ed. Vol 1. New York: John Wiley and Sons; 1985:211–234.
4. Roy-Chowdhury P, Kale KM. *J. Appl. Polym. Sci.* 1970;14:2937.
5. Flory PJ, Osterheld JE. *J. Phys. Chem.* 1954;58:653.
6. Silberberg AJ, Eliassaf, A, Katchalsky. *J. Polym. Sci.* 1957;23:259.
7. Newman S, et al. *J. Polym. Sci.* 1954;14:451.
8. Klein J, Heitzmann R. *Makromol. Chem.* 1978;179:1895.
9. Eisenberg A, Yokoyama T, Sambalido E. *J. Polym. Sci., Part A-1, 7.* 1969:1717.
10. Hughes LJT, Fordyce DB. *J. Polym. Sci.* 1956;22:509.
11. Tsvetkov VN, Ya S, Lyubina TV, Barskaya. *Vysokomol. Soedin.* 1964;6: 806.
12. Li B, et al. *Macromolecules.* 2007;40:5776.
13. Dong J, Ozaki Y, Nakashima K. *Macromolecules.* 1997;30:1111.
14. Buron CC, et al. *Coll. Surf, A.* 20062;89:163.
15. Kato T, Tokuya T, Nozaki, T, Takahashi A. *Polym.* 1984; 25:218.
16. Penzel E, Goetz N. *Angew. Makromol. Chem.* 1990;178:191.
17. Orofino TA, Flory PJ. *J. Phys. Chem.* 1959;63:263.

Poly(acrylonitrile)

Anthony L. Andrady, Taner Z. Sen
and M. Göktuğ Ahunbay

Trade Name Barex (copolymer)

Class Acrylic polymers

Structure $[-CH_2CHCN-]$

Major Applications Acrylonitrile copolymers are used extensively in textile fiber manufacture and in nitrile rubber. Copolymers are used in gaskets, grommets, hoses, printing roll surfaces, diaphragms, and in plumbing accessories. They also are used in coating applications.

Polymer	Units	Conditions	Value	Reference
Preparative techniques	Radical polymerization: Bulk polymerization using conventional initiators (AIBN, peroxides) at <100°C Continuous slurry process Emulsion polymerization			(1) (2) (3)
Typical comonomers	Vinylidene chloride, 4-vinyl pyridine, styrene, butadiene and styrene			(4)
Molecular weight (of repeat unit)	$g\,mol^{-1}$	–	53.06	–
IR	FTIR study of the homopolymer and its thermal degradation			(5–7)
NMR	^{13}C NMR of homopolymer in 20 wt% DMSO at 50°C			(8) (9, 10)
X-ray	WAXS study of the homopolymer and its methylacrylate copolymer			(11)
	X-ray photoelectron spectroscopy study for the evaluation of microstructure and characterization of the homopolymer			(12)
Density	$g\,cm^{-3}$		1.184	(13)
Solvents	Dioxanone, ethylene carbonate, DMSO, chloroacetonitrile, dimethyl phosphite, dimethyl sulfone, sulfuric acid, nitric acid, DMF			(14–18)
Nonsolvents	Hydrocarbons, chlorinated hydrocarbons ketones, diethyl ether, acetonitrile			(15, 16)

Poly(acrylonitrile)

Polymer	Units	Conditions		Value		Reference
Second virial coefficient A_2	mol cm^3 g^{-2} ($\times 10^4$)	Temp. (°C)	M_n			
		20	98–120	22.9–21.4		(19, 20)
			9–69	32.3–7.0		(19)
		25	43–298	21		(19, 21)
			27–159	16–20		(19, 22)
		25–40	35–101	19.1		(19, 23)
Mark-Houwink parameters: K and a	$K = $ ml g^{-1} $a = $ None	Butyrolactone		$K \times 10^3$	a	(24)
		20°C		34.3	0.730	
		30°C		57.2	0.67	
		30°C		34.2	0.70	
		30°C		40.0	0.69	
		DMF, 20°C		30.7	0.76	(25)
Unit cell dimensions	Å	Orthorhombic		$a = 10.55$, $b = 5.8$, $c = 5.08$		(26)
				$a = 21.2$, $b = 11.6$, $c = 5.04$		(27)
				$a = 18.1$, $b = 6.12$, $c = 5.00$		(28)
Heat of fusion	kJ mol^{-1}	–		5.021		(29, 30)
Entropy of fusion	kJ mol^{-1}	–		0.0085		(29, 30)
Glass transition temperature	K	Dielectric, 1 Hz		398		(31)
		Calorimetry		370		(32)
				361		(33)
		Free radically prepared: in dimethyl sulfoxide		374		(34)
		bulk		363		
		in water		364		
Melting transition temperature	K	Calorimetry		593		(35)
		Calorimetry (40°C min^{-1} heating rate)		599		(36)
Heat capacity	kJ K^{-1} mol^{-1}	100°C		0.0302		(37)
		200°C		0.0493		
		300°C		0.0688		
		370°C		0.0862		
Tensile strength	MPa	Styrene-acrylonitrile copolymers: % Acrylonitrile				(38)
		27		72.47		
		21		63.85		
		14		57.37		
		9.8		54.61		
		5.5		42.27		

Poly(acrylonitrile)

Polymer	Units	Conditions	Value	Reference
		Methyl acrylate-acrylonitrile (10:90)	57.0	(39)
		Acrylonitrile homopolymer	55.3	(40)
Tensile modulus	MPa	Methyl acrylate-acrylonitrile(10:90)	2700	(39)
		Acrylonitrile homopolymer	2370	(40)
Toughness	MPa	Acrylonitrile homopolymer	2.97	(40)
Elongation	%	Styrene-acrylonitrile copolymers:		(38)
		% Acrylonitrile		
		27	3.2	
		21	2.5	
		14	2.2	
		9.8	2.1	
		5.5	1.6	
		Methyl acrylate-acrylonitrile(10:90)	22.3	(39)
		Acrylonitrile homopolymer	11.6	(40)
Dielectric constant (film)	–	Frequency (Hz)		(41)
		10^6	4.2	
		10^3	5.5	
		60	6.5	
Dissipation factor	–	Frequency (Hz)		(41)
		10^6	0.033	
		10^3	0.085	
		60	0.113	
Permeability coefficient P	$m^3(STP)m\ s^{-1}\ m^{-2}$ $Pa^{-1}\ (\times 10^{-9})$	Unplasticized film, 25°C		(42)
		O_2	0.00015	
		CO_2	0.00060	
		H_2O	230	
Permeability rate		Vapor permeation of $LiPF_6$ in ethylene carbonate/diethyl carbonate/ propylene carbonate/ethylmethyl carbonate solvent through acrylonitrile film		(43)
Pyrolyzability		Thermal degradation and cyclization of homopolymer and copolymers		(5, 44)
Thermal conductivity	$Wm^{-1}\ K^{-1}$	293°C	0.26	(45, 46)
Coefficient of thermal expansion	$°C^{-1}$	35–91 °C (Tg)	5.7×10^{-5}	(39)
		91 (Tg)–150°C	10.8×10^{-5}	
Viscosity	$cm^3\ g^{-1}$	in DMF at 25°C	187	(47)
Refractive Index			1.52	(48)
Functional free volume	$cm^3\ g^{-1}$	Park–Paul method	6.76×10^{-2}	(49)
Cohesive energy	J/cm3	Hoftyzer–Van Krevelen method	1116	(49)

References

1. Garcia-Rubio LH, Hamielec AE, MacGregor JF. *J. Appl. Polym. Sci.* 1979;23(5):1413.
2. Mallison WC. *U.S. Patent* 2847405 (12 Aug. 1958), to American Cyanamid.
3. Brubaker MM. *U.S. Patent* 2462354 (22 Feb. 1949), to E.I du Pont de Nemours and Co.
4. Peng FM. In: Mark HF, et al, eds. *Encyclopedia of Polymer Science and Engineering*, 2nd ed. Vol 1. John Wiley and Sons; 1987:426.
5. Coleman MM, Petcavich RJ. *J. Polym. Sci., Polym. Phys. Ed.* 1978;16(5):821.
6. Tadokoro H, et al. *J. Polym. Sci., Part A-1.* 1963:3029.
7. Grassie N, Hay JN. *J. Polym. Sci.* 1962;56:189.
8. Inoue Y, Nishioka A, Chujo R. *J. Polym. Sci., Polym. Phys. Ed.* 1973;11:2237.
9. Yoshino J. *J. Polym. Sci.* 1967;B5:703.
10. Svegliado G, Talamini G. *J. Polym. Sci., Part A-1.* 1967;5:2875.
11. Godshall DP, Rangarajan DG, Baird GL, Wilkes VA, Bhanu, McGrath JE. *Polymer.* 2003;44:4221–4228.
12. Liu T-Y. *Biomaterials.* 2004;26:1437–1444.
13. Chase GG. *Fluid/Particle Sep. J.* 2004;16:105–117.
14. Kurata M, Stockmeyer WH. *Adv. Polymer Sci.* 1963;3:196.
15. Moyer WW, Grev DA. *J. Polym. Sci.* 1963;B1:29.
16. Ham GE. *Ind. Eng. Chem.* 1954;46:390.
17. Thinius K. *Analytische Chemie der Plaste.* Berlin: Springer Verlag; 1963.
18. Nitsche R, Wolf KA. *Struktur und Physikalisches Verhalten der Kunslstoffe.* Vol 1. Berlin: Springer Verlag; 1961.
19. Brandrup J, Immegut EH, eds. *Polymer Handbook.* 3rd ed. New York: John Wiley and Sons; 1989.
20. Kamide K. *Chem. High Polym. (Tokyo).* 1967;24:679.
21. Onyon PE. *J. Polym. Sci.* 1959;37:315.
22. Onyon PE. *J. Polym. Sci.* 1956;22:13.
23. Krigbaum WR, Kotliar AM. *J. Polym. Sci.* 1958;32:323.
24. Inagaki H, Hayashi K, Matsuo T. *Makromol. Chem.* 1945;84:80.
25. Fujisaki Y, Kobayashi H. *Kobunshi Kagaku (Chem. High Polym., Tokyo).* 1962;19:7381.
26. Koboyashi H. *J. Polym. Sci.* 1963;B1:209.
27. Klement JJ, Geil PH. *J. Polym. Sci., Part A-2.* 1966;6:1381.
28. Menzcik Z. *Vysokomol. Soedin.* 1960;2:1635.
29. Krigbaum WR, Takita N. *J. Polym. Sci.* 1960;43: 467.
30. Natta G, Moraglio G. *Rubber Plastic Age.* 1963;44: 42.
31. Gupta AK, Vhand N. *J. Polym. Sci., Polym. Phys. Ed.* 1980;18(5):1125.
32. Park HC, Mount EM. In: Mark HF, et al, eds. *Encyclopedia of Polymer Science and Engineering.* 2nd ed. Vol 7. New York: John Wiley and Sons; 1987:89.
33. Liu T-Y, *Polymers Adv. Tech.* 2005;16:413–419.
34. Minagawa M, Kanoh H, Tanno S, Nishimoto Y. *Macromol. Chem. Phys.* 2002;203:2475–2480.
35. Hinrichsen G. *Angew Makromol. Chem.* 1974;20:121.
36. Dunn P, Ennins BC. *J. Appl Polym. Sci.* 1970;14:1759.
37. Gaur U, Lau SF, Wunderlich BB. *J. Phys. Chem. Ref. Data.* 1982;11:1065.
38. Hanson AW, Zimmerman RI. *Ind. Eng. Chem.* 1957;49(11):1803.
39. Guo H. *Polymer.* 2005;46:3001–3005.
40. Chae DW. *J. Appl. Polym. Sci.* 2006;99:1854–1858.
41. Harris M. *Handbook of Textile Fibers.* Washington, D.C: Harris Research Laboratories; 1954.

42. Salame M. *J. Polym. Sci. Symp.* 1973;41:1.

43. Jansen AN. *JOM* 2002;54:29–32.

44. Grassie N. *Dev. Polym. Deg.* 1977;1:137.

45. Thompson EV. In: Mark HF, et al, eds. *Encyclopedia of Polymer Science and Engineering.* Vol 16. New York: Wiley-Interscience; 1985:711–737.

46. Harper CA, ed. *Handbook of Plastics, Elastomers, and Composites.* New York: McGraw-Hill; 1992.

47. Pekel N. *Macromol. Chem. Phys.* 2004;205:1088–1095.

48. Holder AJ, Harris CD, Eick JD, Chappelow CC. *THEOCHEM.* 2000;507:265–275.

49. Kawaguchi Y. *J. Appl. Polym. Sci.* 2005;96:1306–1312.

Poly(L-alanine)

Douglas G. Gold, Wilmer G. Miller,
Taner Z. Sen and Andrzej Kloczkowski

Class Polypeptides and proteins

Structure

$$\left[\!-\!NH\!-\!\underset{\underset{CH_3}{|}}{CH}\!-\!\overset{\overset{O}{\|}}{C}\!-\!\right]_n$$

Major Applications Serves as a model for various proteins.

Properties of Special Interest Two crystalline forms of poly(L-alanine), the α-helix and β-sheet, have been observed.[1]

Synthesis Similar to the synthesis of poly(γ-benzyl-L-glutamate) (see the entry on *Poly(γ-benzyl-L-glutamate$)$* in this handbook); involves the conversion of the amino acid to the N-carboxyanhydride (NCA) monomer by reaction with phosgene gas followed by polymerization of the NCA with an appropriate initiator (e.g., *n*-butyl amine). Typical comonomers include other amino acid NCAs.

Primary amine-initiated polymerization of alanine-NCA that gives high yields was recently reported.[2]

Property	Units	Conditions	Value	Reference
Molecular weight (of repeat unit)	g mol^{-1}	–	71	–
Typical molecular weight range	g mol^{-1}	–	<50,000	–
IR (characteristic absorption frequencies)	cm^{-1}	α-helix β-sheet	1,657; 2,930; 2,985; 3,293 1,634; 2,930; 2,985; 3,283	(3)
NMR	–	–	–	(4)
Solvents	–	25°C	Dichloroacetic acid (DCA), tri-fluoroacetic acid (TFA), phosphoric acid, mixed solvents containing TFA	(5, 6)

Poly(L-alanine)

Property	Units	Conditions	Value		Reference
Nonsolvents	–	–	Water		–
Second virial coefficient	$\mathrm{mol\ cm^3\ g^{-2}}$	DCA, 25°C, 1.6×10^4	5×10^{-3}		(7)
Characteristic ratio	–	DCA, 25°C	5.3–5.6		(7)
Persistence length	Å	DCA, 25°C	44		(8)
Density (crystalline)	$\mathrm{g\ cm^{-3}}$	α-helix	1.25		(9)
		β-sheet	1.34–1.37		
Optical activity $[m']_D$	–	99% CHCl$_3$, 1% DCA	$[m']_D + 21$		(9)
		TFA	$[m']_D - 90$		
Surface tension	$\mathrm{mN\ m^{-1}}$	20°C	45.2		(9)
Helix pitch	Å	α-helix	5.41		(1)
Axial translation per residue	Å	α-helix	1.496		(1)
Conduction and valence bands of PolyAla(8)	eV	k-values in parantheses	CB$_{max}$	0.41(8)	(10)
			CB$_{min}$	−0.20(0)	
			CB$_{width}$	0.61	
	eV		VB$_{max}$	−7.24(0)	
			VB$_{min}$	−7.38(5)	
			VB$_{width}$	0.14	
Fundamental gap value	eV		7.07		(10)
Isotropic ^{13}C chemical shift	ppm	Nucleus: C=O	α-Helix	β-Sheet	(11)
			176.2	171.6	
	ppm	Nucleus: C$_\alpha$	α-Helix	β-Sheet	(11)
			53.0	48.7	
	ppm	Nucleus: C$_\beta$	α-Helix	β-Sheet	(11)
			15.5	21.0	

References

1. Fraser RDB, MacRae TP. *Conformation in Fibrous Proteins and Related Synthetic Polypeptides.* New York: Academic Press; 1973.
2. Kricheldorf HR, von Lossow C, Schwarz G. *Macromol. Chem. Phys..* 2004;2005:918
3. Elliott A. *Proc. Roy. Soc. A.* 1954;226:408.
4. Ferretti JA, Paolillo L. *Biopolymers.* 1969;7:155.
5. Sober HA, ed. *Handbook of Biochemistry: Selected Data for Molecular Biology.* 2nd ed. Cleveland: CRC Company; 1970.
6. Bamford CH, Elliott A, Hanby WE. *Synthetic Polypeptides: Preparation, Structure, and Properties.* New York: Academic Press; 1956.
7. Nakajima A, Murakami M. *Biopolymers.* 1972;11:1295.

8. Brumberger H, Anderson LC. *Biopolymers.* 1972:11:679.
9. Brandrup J, Immergut EH, eds. *Polymer Handbook.* 3rd ed. New York: John Wiley and Sons; 1989.
10. Bogar F, Ladik J. *Int. J. Quant. Chem.* 2004;99:47
11. Henzler Wildman KA, Lee DK, Ramamoorthy A. *Biopolymers.* 2002;64:246

Poly(amide imide)

Loon-Seng Tan

Acronyms, Trade Names PAI, Torlon, Torlon 4000T, Torlon-AI-10, AMOCO-AI-10, AI polymer

Class Engineering thermoplastic

Structure

Ar, Ar$'$ = bridging groups; n and m are molar fractions, and $n + m = 1.0$

Synthesis Poly(amide imides) can be prepared via the following methods: (a) two-step polycondensation of 4-chloroformylphthalic anhydride (trimellitic anhydride chloride) and aromatic amines[1,2]; (b) low temperature polymerization of trimellitic anhydride-based diacid chlorides and aromatic amines[3], and (c) polycondensation of trimellitic anhydride or dicarboxylic acids derived from trimellitic anhydride with aromatic diisocyanates[4]; (d) direct polycondensation of dicarboxylic acids derived from trimellitic anhydride and aromatic amines via Yamazaki–Higashi reaction[5] or with the aid of thionyl chloride[6]; (e) interfacial polymerization (methylethyl ketone/water medium)[1,7] or (alkaline/CH_2Cl_2 medium).[8] A recent report[9] revealed that Torlon 4000T is a copolymer derived from trimellitic anhydride chloride:oxydianiline:m-phenylenediamine in the following molar ratio, 1.0:0.7:0.3.

Major Applications A wide variety of injection-molded automotive parts such as housings, connectors, switches, relays, thrust washers, spline liners, valve seats, bushings, piston rings and seals, wear rings, ball bearings, rollers, thermal insulators etc.; laminated parts such as printed circuit boards, honeycomb core, radomes, etc.; coating for magnet wires; mechanical parts for business machines; aerospace applications such as jet-engine components, compressor and generator parts, and electronic devices; hydraulic/pneumatic parts such as bushings, seals, vanes, and flow-control parts.[10]

Properties of Special Interest Commercial poly(amide imide) (Torlon®) is melt-processable, and injection-moldable either in neat form or with reinforcing fillers such as glass fiber, graphite fiber, and combination of these with polyfluorocarbon and with TiO_2. Parts fabricated from Torlon® have excellent frictional properties, and can be used without lubrication in many applications. Additional features of Torlon family is their inherent ignition resistance without the aid of flame retardants and capability to perform throughout a wide temperature range; in particular, it retains its high strength at very low temperatures.[11,12] Their thin films have also been considered for integrated optics/microdevice applications.[13]

Commercial poly(amide imide) products

Product Names	Product Description	Supplier
Torlon 4000T[14]	Unfilled poly(amide imide) powder for adhesive and coating applications; available in four powder grades, **4000T-LV, 4000T-MV, 4000T-HV,** and a finer particle size version, **4000TF,** with a particle size range of 0–150 μm, and 95% of the particles are less than 75 μm	Solvay Advanced Polymers, LLC. 4500 McGinnis Ferry Road Alpharetta, GA30202-3914 http://www.solvayadvancedpolymers.com/
Torlon high-strength grades	High-strength grades perform more like metals at elevated temperature and are recommended for repetitively used precision mechanical and load-bearing parts; **4203L** contains 3% TiO_2 and 0.5% fluorocarbon; **5030** contains 30% glass fiber and 1% fluorocarbon; **7130** contains 30% graphite fiber and 1% fluorocarbon	
Torlon wear resistant grades	**4301** contains 12% graphite powder and 3% fluorocarbon (PTFE); **4275** contains 20% graphite powder and 3% fluorocarbon (PTFE); **4435,** proprietary composition **4601,** proprietary composition **4630** contains graphite powder and PTFE; **4645** contains carbon fiber and PTFE	
Torlon AI-10[15,16]	Poly(amide imide) powder composed of about 50% of amic acid form	
Torlon AI-30	Water-soluble analog of **Torlon AI-10,** consisting of 35% solids, 63% water, and 2% NMP; made water soluble by forming an ammonium salt with the amic acid groups on the polymer backbone	
Torlon AI-50	Water-soluble analog of **4000TF;** similar to **Torlon AI-30** in solid composition but with a lower viscosity	
AI polymer	Poly(amide imide) with undisclosed composition	Mitsubishi Gas Chemical Company, Inc. (MGC) Mitsubishi Building 5-2, Marunouchi 2-chome Chiyoda-ku, Tokyo 100-8324 http://www.mgc.co.jp/eng/company/ specialty/products/enpla/aipolym/index. html
MS series	High heat resistance; excellent melt flow properties MS-3 grade contains GF40w% of graphite fibers	
MP series	Enhanced moldability. MP-4 grade contains GF40w% of graphite fibers	
MI-D series	Unfilled poly(amide imide) resins	

Poly(amide imide)

Typical mechanical, thermal, electrical, and other properties of unfilled Torlon® 4000T*

Property	Units	Conditions	Value	Reference
Density	gcm^{-3}	ASTM D792	1.380	17
			1.412	18
Tensile strength, break	MPa	ASTM D638 23°C	117.2	10
Tensile modulus	GPa	ASTM D638 23°C	5.2	10, 17
Elongation, break	%	ASTM D638 23°C	10–18	10, 17
Flexural strength, yield	MPa	ASTM D790 23°C	189.0	10, 17
Flexural modulus	GPa	ASTM D790	3.59	10, 17
Compressive strength	MPa	ASTM D695 23°C	241.4	10, 17
Impact strength Notched Izod	$J\,m^{-1}$	ASTM D256 23°C, 3.2 mm	136	10, 17
Impact Strength Unnotched Izod	$J\,m^{-1}$	ASTM D256 23°C, 3.2 mm	1088	17
Fracture Energy (G_{IC})	$J\,m^{-2}$	Compact tension specimen	3400	14
Hardness (Rockwell)	–	–	E78	10
Glass transition temperature	°C	DSC	274.7	19
		Modulated DSC	276.4	
Loss modulus	°C	DMA	281.54	19
Tan δ	°C	DMA	291.6	19
Thermal Conductivity	$W\,m^{-1}\,°C^{-1}$	ASTM C177	0.24	10
Linear coefficient of thermal expansion	$10^{-5}\,°C^{-1}$	ASTM D696 (cm/cm)	3.60	10, 17
Deflection temp.		ASTM D648 @1.81 MPa	252–260	10, 17
Volume resistivity	ohm m	ASTM D257	3.0×10^{13}	10
Heat release capacity	$J\,g^{-1}\,°C^{-1}$	Pyrolysis-combustion flow calorimetry	33	20
Surface resistivity	ohm	ASTM D257	$>1.0 \times 10^{17}$	10
Dielectric strength	$kV\,mm^{-1}$	ASTM D149	17.3	10
Dielectric constant	–	ASTM D150 @ 106 Hz	4.0	10
Dissipation factor	–	ASTM D150 @ 106 Hz	0.009	10
Apparent affinity constant toward adenosine	$10^4\,mol^{-1}\,dm^3$	Surface Plasmon resonance spectroscopy; spin-coated film containing 9-ethylaldenine on pre-treated gold-deposited glass plate	2.73–7.86	21
Vapor Permselectivity toward water α (H_2O/iPrOH)	–	Membrane thickness 14–18 μm; cast from DMSO; fraction of vapor in feed, ca. 3.0×10^{-3}; downstream pressure, 133.3 Pa (1.0 mmHg); ca. 82°C.	~100	22

* CAS Registry No: 42955-03-3.

Poly(amide imide)

Mechanical properties of Torlon 4203L®*

Property	Units	Conditions	Value	Reference
Tensile strength	MPa	ASTM D1708		23, 24
		−196°C	218	
		23°C	192	
		135°C	117	
		232°C	66	
Tensile elongation	%	ASTM D1708		23, 24
		−196°C	6	
		23°C	15	
		135°C	21	
		232°C	22	
Tensile modulus	GPa	ASTM D1708 23°C	4.9	23, 24
Flexural strength	MPa	ASTM D790		23, 24
		−196°C	287	
		23°C	244	
		135°C	174	
		232°C	120	
Flexural modulus	GPa	ASTM D790		23, 24
		−196°C	7.9	
		23°C	5.0	
		135°C	3.9	
		232°C	3.6	
Compressive strength	MPa	ASTM D695 23°C	220	23, 24
Compressive modulus	GPa	ASTM D695 23°C	4.0	23, 24
Shear strength	MPa	ASTM D732 23°C	128	23, 24
Impact strength Izod, notched	J m^{-1}	ASTM D256 23°C 3.2 mm	142	23, 24
Impact strength Izod, unnotched	J m^{-1}	ASTM D256 23°C 3.2 mm	1062	23, 24
Poisson's ratio	–	–	0.45	23, 24

* Torlon 4203L (filler contents: 3% TiO_2; 0.5% fluorocarbon).

Thermal properties of Torlon 4203L®*

Property	Units	Conditions	Value	Reference
Deflection temperature	°C	ASTM D648 @1.8 MPa	278	23, 24
Coefficient of linear thermal expansion	10^{-6} °C^{-1}	ASTM D696 (cm/cm)	30.6	23, 24
Thermal conductivity	W m^{-1} °C^{-1}	ASTM C177	0.26	23, 24

* Torlon 4203L (filler contents: 3% TiO_2; 0.5% fluorocarbon).

Flammability data of Torlon 4203L®*

Property	Units	Conditions	Value	Reference
Limiting Oxygen Index	%	ASTM D2863	45	23, 24
FAA Smoke Density; minimum light transmittance	%	National Bureau of Standards, NFPA 258 specimen thickness 1.3–1.5 mm	92 (smoldering); 6 (flaming)	23, 24

Poly(amide imide)

Property	Units	Conditions	Value	Reference
Maximum specific optical density (D_m)	–		5 (smoldering) 170 (flaming)	23, 24
Time to 90% D_m	min		18.5 (smoldering) 18.6 (flaming)	23, 24
Flash ignition temperature	°C	ASTM D1929	570	23, 24
Self ignition temperature	°C	ASTM D1929	620	23, 24
Flammability	–	UL-94	94V-O	23, 24
Heat of combustion	kJ g^{-1}	Oxygen consumption calorimetry	24.31	25, 25

* Torlon 4203L (filler contents: 3% TiO_2; 0.5% fluorocarbon).

Electrical properties of Torlon 4203L®*

Property	Units	Conditions	Value	Reference
Dielectric constant	–	ASTM D150		23, 24
		10^3 Hz	4.2	
		10^6 Hz	3.9	
Dissipation factor	–	ASTM D150		23, 24
		10^3 Hz	0.026	
		10^6 Hz	0.031	
Volume resistivity	ohm m	ASTM D257	2×10^{15}	23, 24
Surface resistivity	ohm	ASTM D257	5×10^{18}	23, 24
Dielectric Strength	kV mm^{-1}	ASTM D149 1 mm	23.6	23, 24

* Torlon 4203L (filler contents: 3% TiO_2; 0.5% fluorocarbon).

Other physical properties of Torlon 4203L®*

Property	Units	Conditions	Value	Reference
Density	g cm^{-3}	ASTM D792	1.42	23, 24
Hardness, Rockwell E	–	ASTM D785	86	23, 24
Water absorption	%	ASTM D570	0.33	23, 24
Wettability	°	Contact angle measurement	55	26
Friction coefficient	–	ASTM 99G 10 000 cycles (PEEK 450G disk)		26
		Dry:	0.25	
		In water:	0.50	
Friction coefficient	–	ASTM 99G 10 000 cycles (UHMWPE disk)		26
		Dry:	0.22	
		In water:	0.15	

Property	Units	Conditions	Value	Reference
Wear	mm	ASTM 99G 10 000 cycles (PEEK 450G disk)		26
		Dry:	0.020	
		In water:	0.070	
Wear	mm	ASTM 99G 10 000 cycles (UHMWPE disk)		26
		Dry:	0.015	
		In water:	0.035	

* Torlon 4203L (filler contents: 3% TiO_2; 0.5% fluorocarbon).

Glass transition temperatures of polyamide-imides derived from trimellitic anhydride:

Ar	Unit	Conditions	Value	Reference
Torlon®	°C	–	277	27
	°C	TMA in air at heating rate of 10°C/min	260	28
	°C	Dielectric constant and dissipation factor measurements	285	29
(Torlon AI-10* after cure)	°C	–	272	30, 31
	°C	TMA in air at heating rate of 10°C/min	330	2

* Previously known as Amoco-AI-10.

Glass transition and secondary-relaxation temperatures and associated activation energy values of poly(amide-imide) (Torlon 4203L®)

Conditions	T_g (°C)	E_a (kJ mol^{-1})	T_β (°C)	E_a (kJ mol^{-1})	T_γ (°C)	E_a (kJ mol^{-1})	Reference
Forced oscillation dynamic mechanical analysis @1 Hz	276	–	65	117	−69	–	32, 33

Poly(amide imide)

Film (thickness 33 μm) properties of poly[N,N′-(4,4′-diphenylether)4-aminophthalimide]

Property	Units	Conditions	Value	Reference
Density	gcm^{-3}	ASTM D792	1.39	34
Tensile strength, break	MPa	ASTM D882		34
		25°C	103–145	
		200°C	54–64	
Tensile modulus	GPa	ASTM D882		34
		25°C	2.41–3.45	
		200°C	2.21	
Elongation, break	%	ASTM D882		34
		25°C	10–60	
		200°C	10–25	
Fold endurance	cycles	ASTM D2176 1 kg	20 000–500 000	34
		25°C		
Water absorption	%	24 h, 37.8°C, 100% R.H.	2.5	34
Tear strength	$g\,mm^{-1}$	ASTM D1004	0.018	34
Volume resistivity (500 volts)	ohm m	ASTM D257		34
		25°C	10^{14}	
		200°C	10^{10}	
Dielectric strength	$kV\,mm^{-1}$	ASTM D149 @ 60 Hz		34
		25°C	163–265	
		200°C	163	
Dielectric constant	–	ASTM D150 @ 1 KHz		34
		25°C	3.5	
		200°C	4.0	
Dissipation factor	–	ASTM D150 @ 1 KHz		34
		25°C	0.008	
		200°C	0.018	

Poly(amide imide)

Typical properties of cured Torlon-AI-10* Film

Property	Units	Conditions	Value	Reference
Glass transition temperature	°C	–	272	16, 31
Refractive index	–	–	1.656	16, 31
Tensile strength	MPa	ASTM D638 23°C	117	16, 31
Tensile modulus	GPa	ASTM D638 23°C	3.03	16, 31
Tensile elongation	%	ASTM D638 23°C	23	16, 31

* Previously known as Amoco-AI-10.

References

1. Alvino WM. *J. Appl. Polym. Sci.* 1975;19:65.
2. Imai Y, Maldar NN, Kakimoto. *J. Polym.Sci. Polym. Chem. Ed.* 1985;23:2077.
3. (a) Wrasilo W, Augl JM. *J. Polym. Sci. Polym. Chem. Ed.* 1969;7:321.
   (b) Ray A, Rao YV, Bhattacharya VK, Maiti S. *Polm. J.* 1983;15:169.
   (c) Das S, Maiti S. *Makromol. Chem. Rapid Commun.* 1980;1:403.
   (d) Ray A, Das S, Maiti S. *Makromol. Chem. Rapid Commun.* 1981;2:333
   (e) Mauti S, Ray A. *Makromol. Chem.Rapid Commun.* 1981;2:649.
   (f) de Abajo J, Gabarda JP, Fontan J. *Angew. Makromol. Chem.* 1978;71:143.
4. (a) Nieta JL, de la Campa JG, de Abajo J. *Makromol. Chem.* 1982;183:557.
   (b) de la Campa JG, de Abajo J, Nieta JL. *Makromol. Chem.* 1982;183:571.
   (c) Kakimoto M, Akiyama R, Negi YS, Imai Y. *J. Polym. Sci. Polym. Chem. Ed.* 1988;26:99.
5. Yang C-P, Lin J-H. *J. Polym. Sci. Part A: Polym. Chem.* 1994;32:2653.
6. (a) Choi KY, Hong YT. *Macromol. Symp.* 1997;118:163.
   (b) Hong YT, Jin MY, Suh DH, Lee JH, Choi KY. *Angew. Makromol. Chem.* 1997;248:105.
7. Imai Y, Hiroshi Uchiyama. *J. Polym. Sci.Polym. Lett..* 1970;8:559.
8. Buch PR, Mohan DJ, Reddy AVR. *Polym. Int..* 2006;55:391.
9. Robertson GP, Guiver MD, Yoshikawa M, Brownstein S. *Polymer.* 2004;45:1111.
10. Cekis GV. *Modern Plastics.* mid-October Encyclopedia issue, 1990:32–33.
11. Barucci M, Olivieri E, Pasca E, Risegari L, Ventura G. *Cryogenics.* 2005;45:295.
12. Chagovets VK, Rudavskii EY, Taubenreuther KU, Eska G. *Physica B (Amsterdam).* 2000;2045:284–288.
13. Bryce RM, Nguyen HT, Nakeeran P, et al. *Thin Solid Films.* 2004;458:233.
14. 'Torlon Engineering Polymers Design Manual', Amoco Performance Products, Inc., Atlanta, Georgia., see Solvay Advanced Polymer website for recent technical data. 'Torlon Design Guide ver.2.1' http://www.solvayadvancedpolymers.com/static/wma/pdf/9/9/7/TORLON_Design_Guide.pdf
15. 'TORLON AI Powders for Coatings,' Product Bulletin, Solvay Advance Polymers, Alpharetta, Georgia, USA.
16. Previously known as Amoco-AI-10. Application Bulletin AMOCO-AI-10 Polymer. Atlanta, Georgia: Amoco Performance Products, Inc; 1997.

17. 'Plastic: A Desk-Top Data Bank,' Book B. 5th ed. San Diego, California: The International Plastic Selector, Inc., a subsidiary of Cordura Publications, Inc; 1980:B-396.

18. Benabdallah HS. *J. Tribol.* 2006;128:96–102.

19. Wang Y, Goh SH, Chung T-S. *Polymer.* 2007;48:2901.

20. Lyons RE, Walters RN. *J. Aanl. Appl. Pyrolysis.* 2004;71:27.

21. Yoshikawa M, Guiver MD, Robertson GP. *J. Mol. Struct.* 2005;739:41.

22. Yoshikawa M, Higuchi A, Ishikawa M, Guiver MD, Robertson GP. *J. Membr. Sci.* 2004;243:89.

23. *Torlon Engineering Polymers Design Manual.* Atlanta, Georgia.: Amoco Performance Products, Inc;

24. Sroog CE. Chemistry and Properties of Addition Polyimides. In: Wilson D, Stenzenberger HD, Hergenrother PM, eds. *Polyimides.* New York: Chapman & Hall; 1990:270.

25. Walters RN, Hackett SM, Lyon RE. *Fire Mater.* 2000;24:245.

26. Borruto A, Marrelli L, Palma F. *Tribology Lett.* 2005;20:1.

27. Bicerano J. *Prediction of Polymer Properties.* New York: Marcel Dekker, Inc; 1993:157.

28. Imai Y, Maldar N, Kakimoto M-A. *J. Polym. Sci., Polym. Chem. Ed.* 1985;23:2077.

29. Alvino WM. *J. Appl. Polym. Sci.* 1975;19:665.

30. Lee H, Stoffey D, Neville K. *New Linear Polymers.* New York: McGraw-Hill; 1967:Chapter 7. 171.

31. *'Torlon AI-10 Coatings' Technical Bulletin, Solvay Advanced Polymers.* Georgia: Alpharetta;

32. Fried JR. 'Sub-T_g Transitions'. In: Mark JE, ed. *Physical Properties of Polymers Handbook.* Woodbury, New York: American Institute of Physics; 1996:Chapter 13 166–167.

33. Dallas G, Ward T. *Eng. Plast.* 1994;7:329.

34. Alvino WM. *J. Appl. Polym. Sci.* 1975;19:665.

Poly(amidoamine) Dendrimers

Donald A. Tomalia and Linda S. Nixon

Acronym, Trade Names PAMAM dendrons and dendrimers, Starburst® dendrons and dendrimers

Class Dendritic polymers; dendrons; dendrimers

Structure Dendrimers/dendrons were first synthesized in 1979 by Tomalia et al.[1,2] They are well-defined three-dimensional macromolecules that are considered to be subclasses of the fourth major architectural polymer class referred to as *dendritic polymers*.[3,4] Dendrimers consist of three major architectural components; namely: (I) a *core*, (II) an *interior (branch cells)*, and (III) *terminal groups*. They are produced in well-defined nanoscale shapes and sizes depending on the nature of the core and the extent of generational growth from the core. Generational growth involves the covalent concentric attachment of repeat units called *branch cells* (e.g. $-N(H)CH_2CH_2N[CH_2CH_2C(O)]_2$) around the initiator cores. This leads to amplification of the branch cells and terminal groups according to dendritic rules and principles. Generational amplification of branch cells and terminal groups, respectively, occur as described in Figure 1; wherein, N_c = multiplicity of core; N_b = multiplicity of branch cell, and Z = terminal groups (i.e. $-OCH_3$; $-NH-(CH_2)_2-NH_2$; $-NH-C-(CH_2OH)_3$; or $-NH-(CH_2)_2-OH$. Precise nanoscale macromolecules (i.e. diameters between 1 and 15 nm) may be synthesized as a function of generation. The cores may be nonfunctional and spherical, if grown from a point-like core such as NH_3, ellipsoidal; if grown from α,ω-alkylendiamines (e.g. $NH_2-(CH_2)_6-NH_2$), or rod-like if amplified from *linear*-poly(ethyleneimine).[5] Alternatively, the core may be functional (cleavable) and produces an ellipsoidal dendrimer if grown from cystamine. Mild reduction of such dendrimers yield focal point thiol-functionalized dendrons with conical shapes.[6]

Molecular properties for the cystamine core, poly(amidoamine), amine-terminated dendrimer family are as described in Figure 2. The accepted shorthand nomenclature for a typical, cystamine core, generation four, poly(amidoamine) dendrimer terminated with amine groups utilizes the architectural components as follows; [*core*; cystamine];$(G = 4.0)$; *dendri*-PAMAM$-(NH_2)_{64}$.[3]

Major Applications Dendrimers are ideal macromolecular standards for use in size exclusion chromatography,[7] membrane porosity evaluations, Newtonian viscosity applications,[8] and electron microscopy.[9] The mathematically defined surface functionality (Z may range from 2, 3, or 4 to several thousand) manifested by dendrimers (Figure 2) provides well-defined nanoscale building blocks for constructing complex nanostructures based on either covalent bonding or self-assembly-type processes.[10] On the other hand, PAMAM dendrimers may function as hosts for the encapsulation of pharmaceuticals, agricultural chemicals, metals, fragrances, etc. once they have been grown to an appropriate congestion state that exhibits nanocontainer properties as shown in Figure 3.

Dendrimer/Dendron Shapes	Core	Branch Cells	Terminal Groups

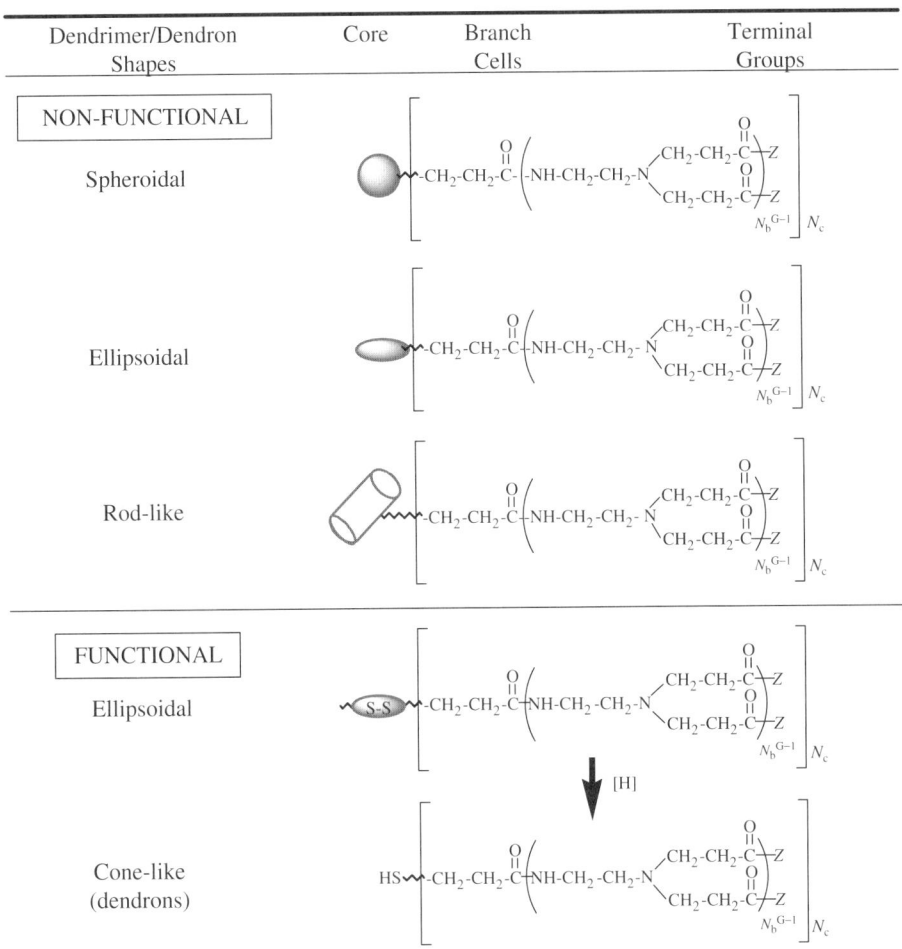

Figure 1. PAMAM dendrimers as a function of shape, core, branch cell, and terminal groups, *where*: N_b = branch cell multiplicity, N_c = core multiplicity, and G = generation.

In the biomedical field, dendrimers have been used for drug delivery,[11–13] *in vitro* gene transfection,[14] antibody conjugates, (diagnostics)[3,15,16] MRI contrast agents,[17,18] antimicrobials[19] and as topical HIV preventatives.[20–22] A recent overview describes a wide range of dendrimer-based nanomedicine and biotechnology applications.[23] In the materials science area, dendrimers have been used for adhesive tie coats to glass, metal, carbon, or polymer surfaces, additives for polymer resins and composites, printing inks,[24] surfactants, cross-linking agents, electrically conductive nanodevices,[25] flow regulators, processing aids, and chemical sensors.[26]

Properties of Special Interest A unique dendrimer property not found in traditional linear macromolecular architecture, as well as the other two dendritic subclasses (i.e. random hyperbranched and dendrigrafts) is the distinct parabolic intrinsic viscosity curve with a maximum as a function of molecular weight; that is exhibited at approximately generation four in the PAMAM series.

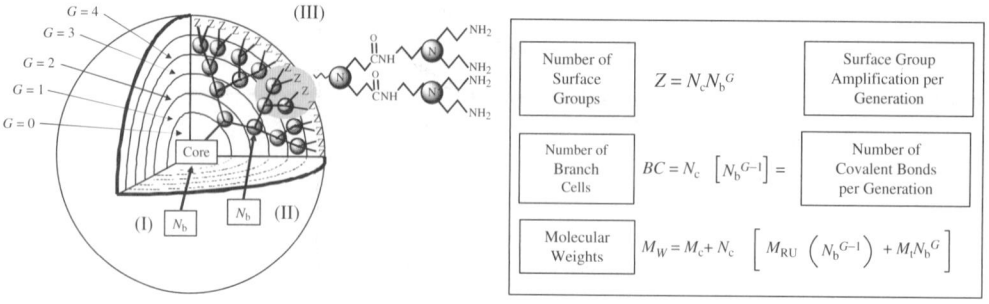

Gen	No. of NH$_2$ Surface Groups	Molecular Formula	M_W	Hydrodynamic Diameter ~(nm)
0	4	$C_{24}H_{52}N_{10}O_4S_2$	609	1.5
1	8	$C_{64}H_{132}N_{26}O_{12}S_2$	1522	2.2
2	16	$C_{144}H_{292}N_{58}O_{28}S_2$	3348	2.9
3	32	$C_{304}H_{612}N_{122}O_{60}S_2$	7001	3.6
4	64	$C_{624}H_{1252}N_{250}O_{124}S_2$	14 307	4.5
5	128	$C_{1264}H_{2532}N_{506}O_{252}S_2$	28 918	5.4
6	256	$C_{2544}H_{5092}N_{1018}O_{508}S_2$	58 140	6.7
7	512	$C_{5104}H_{10212}N_{2042}O_{1020}S_2$	116 585	8.1

Z = monomer-shell-saturation level, N_c = core (cystamine) multiplicity, N_b = branch-cell (BC) multiplicity, G = generation.

Figure 2. Mathematical expressions for calculating the theoretical number of surface groups (Z), branch cells (BC), and molecular weights (M_W) for [cystamine core]–PAMAM dendrimers as a function of generation. Approximate hydrodynamic diameters (Gen0–7) based on gel electrophoretic comparison with the corresponding [ethylenediamine core]–PAMAM dendrimers.

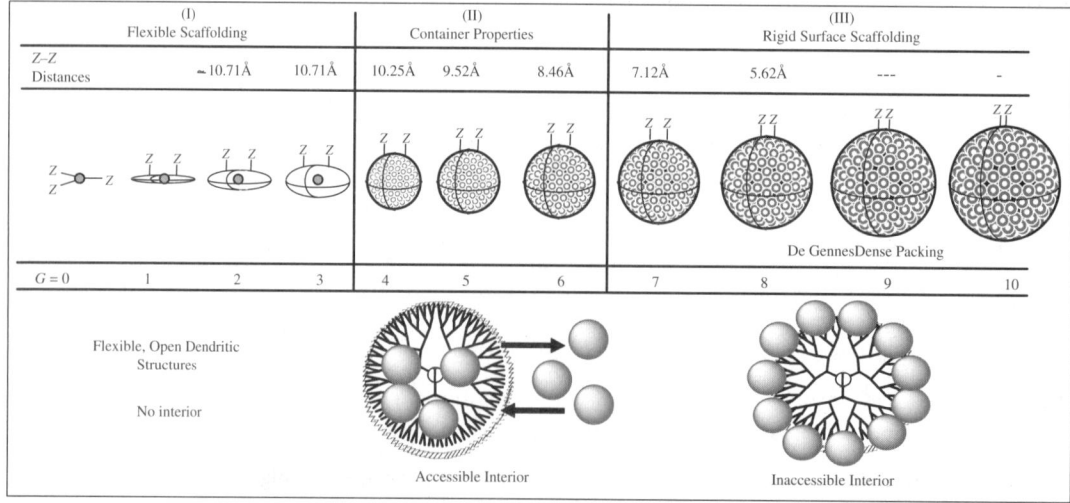

Figure 3. Periodic properties of PAMAM dendrimers as a function of generation. Various chemophysical dendrimer surfaces amplified according to Z, N_c, N_b, and G; where N_c, core multiplicity; N_b, branch-cell multiplicity; and G, generation.

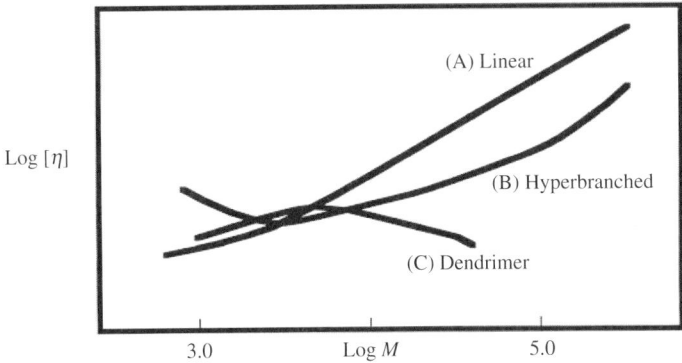

Figure 4. Comparison of intrinsic viscosities (log (η)) versus molecular weight (log M) for (A) linear, (B) random hyperbranched, (C) dendrimers, and (D) dendrigraft topologies. Data for A, B, C is adapted from Fréchet et al.[27]

Other distinguishing features include (1) very monodispersed sizes and shapes (i.e. M_w/M_n routinely below 1.1 even at high molecular weights); (2) *exo* presentation of exponentially large numbers of surface groups; (3) selective synthesis of specific chemical functionalities of surface groups; (4) a *dense-shell*-type surface with a soft, spongy interior;[8] (5) typical Newtonian-type rheology even at molecular weights exceeding 50 000 g mol^{-1}; and (6) functional (i.e. cleavable) disulfide core dendrimers (e.g. $H_2N-CH_2-S-S-CH_2-NH_2$) that allow the facile preparation of thiol-functionalized PAMAM dendrons. In the PAMAM series, over 100 different surface group modifications have been reported.[4] Surface functionalities reside in three main categories; namely: (a) cationic (i.e. ethylenediamine and aminoethyl ethanolamine) terminated dendrimers; (b) anionic (i.e. sodium carboxylate surfaces); (c) neutral surface, which include polyethylene glycol (PEG), 3-carbomethoxy-pyrrolidinone or hexylamide surfaces.[28]

Preparative Techniques PAMAM dendrimers are synthesized by the divergent method starting from either: $H_2N-CH_2-CH_2-NH_2$ (EDA) ($N_c = 4$); $H_2N-(CH_2)_4-NH_2$ (DAB) ($N_c = 4$) or $H_2N-CH_2-CH_2-S-S-CH_2-CH_2-NH_2$ (cystamine) initiator core reagents.[29,30] They are amplified by progressing through a reiterative sequence consisting of (a) a double Michael addition of methyl acrylate to a primary amino group followed by (b) amidation of the resulting carbomethoxy intermediate with a large excess of ethylenediamine (EDA). Products up to generation 10 (i.e. molecular weight of over 930 000 g mol^{-1}) have been obtained. Higher generation carbomethoxy intermediates can be amidated with primary amine molecules to obtain nonreactive surface groups (i.e. tris(hydroxymethyl) aminomethane or ethanolamine). Reactions are performed between room temperature and about 50°C in methanol. Samples are available in methanol or in water solutions. Dendrimers soluble in organic solvents (e.g. toluene or chloroform) can be readily prepared by modification of amine-terminated dendrimers with hydrophobic reagents.

Supplier: Dendritic Nanotechnologies, Inc., 2625 Denison Drive, Mt. Pleasant, Michigan 48858.

Molecular properties of (*core*; ethylenediamine);($G = 0$–10); *dendri*-(PAMAM)—$(NH_2)_z$ dendrimers

Generation	Number of Terminal Groups[a]	Molecular Weight (g mol^{-1})[a]	Hydrodynamic Diameters (nm)[b]		Hydrodynamic Volumes (nm^3)[e]	
			SEC[c]	DSV[d]	SEC	DSV
0	4	517	1.52	–	183.8	–
1	8	1430	2.17	2.02	534.8	431.4
2	16	3256	2.86	2.88	1224.3	1250.1
3	32	6909	3.57	3.89	2381.1	3080.5
4	64	14 215	4.48	5.0	4705.6	6541.7
5	128	28 826	5.44	6.58	8425.1	14 909.3
6	256	58 048	6.74	–	16 023.5	–
7	512	116 493	8.1	–	27 812.1	–
8	1024	233 383	9.7	–	47 763.2	–
9	2048	467 162	11.4	–	77 534.1	–
10	4096	934 720	13.5	–	128 759.6	–

[a] Theoretical values.
[b] At 25°C; 0.1 M citric acid in water; pH = 2.7.
[c] Size exclusion chromatography; relative to linear PEO standards.
[d] Dilute solution viscometry.
[e] Calculated from hydrodynamic diameters assuming ideal sphericity.

Commercial availability

Property	Conditions	Value
Availability	Gold standard to diagnostic grade: low defect levels, biomedical applications	Units: 100 mg; 500 mg; g
	Technical grade: higher defect levels, reduced regularity, materials applications	Units: kg
Suppliers	Gold standards to diagnostics grade (mg); technical grade (kg)	Dendritic Nanotechnologies, Inc. 2625 Denison Drive Mt. Pleasant, MI 48858 USA www.dnanotech.com
	Primary amine, sodium carboxylate, and certain hydroxyl surface groups are available	Aldrich Chemical Company, Inc. 1001 West St. Paul Avenue Milwaukee, Wisconsin 53233 USA
Property Significant Patents	U.S. Patent 5,338,532 U.S. Patent 5,527,524 U.S. Patent 5,714,166 U.S. Patent 5,731,095 U.S. Patent 6,020,457 U.S. Patent 6,475,994 U.S. Patent 6,585,956 U.S. Patent 6,635,720 U.S. Patent 6,790,437 U.S. Patent 7,005,124	

STARBURST® is a registered trademark of Dendritic Nanotechnologies, Inc.

Acknowledgements We would like to thank Dr. Ryan Hayes and Mr. Joe Heinzelmann and the DNT staff for many helpful discussions.

References

1. Tomalia DA, Fréchet JMJ. *J. Polym. Sci. Part A: Polym. Chem.* 2002;40:2719–2728.
2. Tomalia DA, Naylor AM, Goddard WA III. *Angew.Chem. Int. Ed. Engl.* 1990;29(2):138.
3. In: Tomalia DA, . Fréchet JMJ. *Dendrimers and Other Dendritic Polymers.* Chichester:Wiley & Sons Ltd; 2001.
4. (a) Tomalia DA. *Aldrichimica Acta.* 2004;37(2):39–57.
   (b) Tomalia DA. *Prog. Polym. Sci.* 2005;30:294–324.
5. Yin R, Zhu Y, Tomalia DA. *J. Am. Chem. Soc.* 1998;120:2678.
6. Tomalia DA, Huang B, Swanson DR, Brothers H.M II, Klimash JW. *Tetrahedron.* 2003;59:3799–3813.
7. (a) Dubin PL, et al. *Analytical Chemistry.* 1992;64:2344.
   (b) Dubin PL, Edwards SL, Mehta MS. *Journal of Chromatography.* 1993;635:51.
8. Uppuluri S, Keinath SE, Tomalia DA, Dvornic PR. *Macromolecules.* 1998;31:4498–4510.
9. Jackson CL, Chanzy HD, Booy FB, et al. *Macromolecules.* 1998;31:6259–6265.
10. Tomalia DA, Henderson SA, Diallo MS. In: Goddard WA III, Brenner DW, Lyshevski SE, Irafrate GJ, eds. *Handbook of Nanoscience, Engineering and Technology.* 2nd ed. Chapter 24, Boca Raton, Florida: CRC Press, Taylor and Francis; 2007:24.1–24.47.
11. Svenson S, Tomalia DA. *Advanced Drug Delivery Reviews.* 2005;57:2106–2129.
12. Esfand R, Tomalia DA. *Drug Discovery Today.* 2001;6(8):427–436.
13. Tomalia DA, Reyna LA, Svenson S. *Biochemical Society Transactions.* 2007;35(1):61–67.
14. Kubasiak LA, Tomalia DA. In: Amiji MM, ed. *Polymeric Gene Delivery: Principles and Applications.* Chapter 9, Boca Raton, Florida: CRC Press; 2004:133–157.
15. Singh P. *Bioconjugate Chem.* 1998;9(1):54.
16. Singh P, et al. *Clinical Chemistry.* 1996;42(9):1567.
17. Wiener EC, et al. *Magnetic Resonance in Medicine.* 1994;31:1.
18. Xu H, Regino CAS, Koyama Y, et al. *Bioconjugate Chem.* 2007;18:1474–1482.
19. Balogh L, Swanson DR, Tomalia DA, Hagnauer GL, McManus AT. *NanoLetters.* 2001;1(1):18–21.
20. Gong Y, Matthews B, Cheung D, et al. *Antiviral Res.* 2002;55:319.
21. Bernstein DI, Stanberry LR, Sacks S, et al. *Antimicrob. Agents Chemother.* 2003;47:3784.
22. Bourne N, Staberry LR, Kern ER, Holan B, Matthews B, Bernstein DI. *Antimicrob. Agents Chemother.* 2000;44:2471.
23. Boas U, Christensen JB, Heegaard PMH. *Dendrimers in Medicine and Biotechnology.* Cambridge, UK: RSC Publishing; 2006.
24. Tomalia DA. *Materials Today.* March (2005):34–46.
25. Miller L, et al. *J. Am. Chem. Soc.* 1997;119:1005.
26. Crooks RM, Ricco AJ. *Acc. Chem. Res.* 1998;31:219.
27. Tomalia DA, Fréchet JMJ. In: Tomalia DA, Fréchet JMJ, eds. *Dendrimers and Other Dendritic Polymers.* Chapter 1, Chichester: J. Wiley & Sons Ltd; 2001:3–44.
28. Tomalia D.A. In: James E, Mark, ed. *Physical Properties of Polymers Handbook.* Chapter 42, New York: Springer; 671–692, 2007.

29. Esfand R, Tomalia DA. In: Tomalia DA, Fréchet JMJ, eds. *Dendrimers and Other Dendritic Polymers*. Chapter 25, Chichester: J. Wiley & Sons Ltd; 2001:587–604.
30. Tomalia DA. In: Davis FJ, eds. *Polymer Chemistry – A Practical Approach*. Chapter 7, New York:Oxford University Press; 2004:188–200.

Polyaniline

Stephen S. Hardaker, Abhijit V. Jadhav, Anna D. Gudmundsdottir
and Richard V. Gregory

Acronym, Alternative Names, Trade Names PANI, emeraldine, leucoemeraldine, pernigraniline, Ormecron (Zipperling Kessler and Co.), Zypan (Dupont)

Class Conjugated and other unsaturated polymers; electrically conductive polymers

Structure Polyaniline base of variable oxidation state

$x = 1 - y$
$y = 0$: Leucoemeraldine base (LEB)
$y = 0.5$: Emeraldine base (EB)
$y = 1$: Pernigraniline base (PNB)

Emeraldine Salt (ES)

Major Applications Polyaniline is finding widespread use in novel organic electronic applications such as light emitting diodes (LED), electroluminescence, metallic corrosion resistance, organic rechargeable batteries, biological and environmental sensors, composite structures, textile structures for specialized applications or static dissipation, membrane gas-phase separation, actuators, EMI shielding, organic semiconductor devices for circuit applications, blends with insulative host polymers to impart a slight electrical conductivity, bioelectronic medical devices, and a variety of other applications where tunable conductivity in an organic polymer is desirable.

Properties of Special Interest Electrical conductivity is in the range of 10^{-8}–$400 \ \mathrm{Scm^{-1}}$. Stretch alignment of polyaniline films can lead to conductivities around $1000 \ \mathrm{Scm^{-1}}$. The conductivity will increase as better processing methods are developed for reducing structural defects. The conductivity can be tuned to specific end uses for a variety of applications. Polyaniline is reasonably stable under ambient conditions and, with proper selection of dopants, retains its conductivity over long periods of time (i.e. 5 years and longer). Polyaniline easily switches from the conductive form (emeraldine salt) to the insulative form (emeraldine base) as a function of pH. Under acidic conditions the polymer dopes and becomes conductive. When exposed to higher pH levels the polymer switches to the insulating form. This facile switching can be cycled many times.

Polyaniline

Property	Units	Conditions	Value	Reference
Synthesis		Oxidative polymerization by $(NH_4)_2S_2O_8$		
		Chemical synthesis—Emeraldine base		1, 2
		Electrochemical synthesis—Emeraldine base		3
		Substituted polyanilines:		
		Alkyl, alkoxyl		4
		Azide		5
		Azobenzene		6
		Boronic acid		7
		o-Toluidine, *m*-toluidine, *o*-Ethylaniline		8
		Sulfonic acid		9
		Tertbutoxycarbonyl		10
Molecular weight	g/mol	*Synthesis at room temperature:*		
		Emeraldine base	$M_w{}^* = 42\ 206$ $P.I.^{**} = 2.3$	11
		Synthesis at $0°\,C$:		
		Emeraldine base	$M_w = 82296$ $P.I. = 2.4$	11
		Synthesis at $-30°\,C$:		12
		Emeraldine base	$M_w = 440000$ $P.I. = 3.46$	
		$^*M_w =$ Weight average molecular weight $^{**}P.I. =$ Polydispersity index $= M_w/M_n$ $M_n =$ Number average molecular weight		
UV-Vis spectroscopy	nm	*Emeraldine base in N-methyl pyrollidinone*		13, 14
		Electron transitions within quinoid region	620	
		Electron transitions within benzenoid region	330	
Infrared spectroscopy	cm^{-1}	*Emeraldine base films cast from N-methyl pyrollidinone*		15
		N—H (2° amine)	3390	
		C—N (Iminoquinone)	1595	
		C—C (Aromatic)	1508	
		C—N (2° aryl amine)	1310	
		C—H (Aromatic, para disubstitution)	831, 1009, 1108, 3046	

Property	Units	Conditions	Value	Reference
NMR spectroscopy	ppm	*Solid state*^{13}C NMR		16
		Quinoid carbons:		
		Quarternary carbons	~157	
		Nonquaternary carbons	~135	
		Benzenoid carbons:		
		Quaternary carbons attached to quinoid ring	~145	
		Quaternary carbons attached to 2° amine	~141	
		Nonquaternary carbons on quinoid side	~123	
		Nonquaternary carbons	~115	

Unit cell dimensions

Form	a (Å)	b (Å)	c (Å)	Lattice	Comments	Reference
EB-II	7.80	5.75	10.05	Orthorhombic	NMP-cast, stretched film	17
	7.65	5.75	10.20	Orthorhombic	THF/NMP–extracted powder	17
	7.65	5.65	10.40	Orthorhombic	Powder from THF-extracted solution	17
	21.5	3.5	18.6	Orthorhombic	CSA-doped/*m*-cresol-cast	18
ES-II	7.1	7.9	10.4	Orthorhombic	NMP-Cast, stretched film, HCl dopant	17
	7.0	8.6	10.4	Orthorhombic	THF/NMP–extracted powder, HCl dopant	17
ES-I	4.3	5.9	9.6	Pseudoorthorhombic	As synthesized, HCl dopant	17

Solubility parameters of polyaniline and several solvents

Compound	δ (MPa$^{1/2}$)	δ_d (MPa$^{1/2}$)	δ_p (MPa$^{1/2}$)	δ_h (MPa$^{1/2}$)	Comment	Reference
Emeraldine base	22.2	17.4	8.1	10.7	Empirical	19
Emeraldine salt	23.6	17	8.9	13.7	Empirical	19
Leucoemeraldine base	23–25	21.1	5.6	7.3	Empirical	19
1-Methyl-2-pyrrolidinone (NMP)	23.7	16.5	10.4	13.5	Calculated	19
N,N′-dimethyl propylene urea (DMPU)	22.3	16.4	11.3	10.0	Calculated	20
m-Cresol	22.7	18.7	4.8	13.5	Calculated	19

Mechanical properties of polyaniline fibers and films

Fiber process[a]	Base			Doped				Conductivity (S cm^{-1})	Reference
	Tenacity	Modulus	Elongation (%)	Dopant	Tenacity	Modulus	Elongation (%)		
PANI-CSA/ *m*-cresol[c]	n/a	n/a	n/a	CSA	0.2 gpd[(b)]	7.3 gpd	8.4	203	21
PANI-EB/H$_2$SO$_4$	n/a	n/a	n/a	H$_2$SO$_4$	1.8 gpd	39.3 gpd	25.4	6.3	21
PANI-EB/NMP[d] drawn	3.9	–	–	HCl	1.4	–	–	160	22

Fiber process[a]	Base			Dopant	Doped			Conductivity (S cm^{-1})	Reference
	Tenacity	Modulus	Elongation (%)		Tenacity	Modulus	Elongation (%)		
PANI-EB/NMP as-spun	n/a	0.51 GPa	n/a	HCl	n/a	n/a	n/a	0.14	23
PANI-EB/NMP 2 × drawn		1.02 GPa	n/a	n/a	n/a	n/a	n/a	n/a	23
PANI-EB/DMPU as-spun	0.2–0.6	27	7	CH$_3$SO$_3$H	<0.2 gpd	–	–	10–32	24
PANI-EB/DMPU 4× drawn	2.4 gpd	56 gpd	13	CH$_3$SO$_3$H	<1.0 gpd	–	–	15	24
PANI-LEB/DMPU as-spun	1.1	57	51	CH$_3$SO$_3$H	0.8 gpd	–	–	15	25
PANI-LEB/DMPU 2× drawn	3.6	89	15	HCl	1.9 gpd	–	–	140	25
Film process[a]									
PANI-EB/NMP	n/a	4.3 GPa	n/a	–	n/a	n/a	n/a	n/a	23
PANI-EB/NMP 4 × drawn	n/a	14 GPa	n/a	1 M HCl	n/a	n/a	n/a	45	23

[a] Fiber/Film process is designated as: polyaniline form/solvent, post process;
[b] gpd ≡ g denier^{-1}. Denier is linear density: 1 denier = 1 g (9000 m)$^{-1}$;
[c] Mixture of emeraldine base and (±)-camphor sulfonic acid dissolved in *m*-cresol;
[d] Solution also contained a gel inhibitor.

Property	Units	Conditions	Value	Reference
Permeability	m^3 (STP) m s^{-1} m^{-2} Pa^{-1}	Gas		26
		H$_2$	3580	
		CO$_2$	586	
		O$_2$	123	
		N$_2$	13.4	
		CH$_4$	3.04	
		Ar		
Huggins parameter: k'	–	Form/Solvent		27
		EB/NMP	0.384	
		EB/DMPU	0.371	
Storage modulus	MPa	EB form; EB film cast from NMP; DMTA, 1 Hz, 25°C	2000	1
		ES-HCl form; EB film cast from NMP then doped with HCl; DMTA, 1 Hz, 25°C	2300	
Loss modulus	MPa	EB form; EB film cast from NMP; DMTA, 1 Hz, 25°C	256	1
		ES-HCl form; EB film cast from NMP then doped with HCl; DMTA, 1 Hz, 25°C	218	
Room temperature conductivity	Scm^{-1}	CSA dopant; film cast from *m*-cresol;	400	28
		PANI–CSA complex formed in solution	967 ± 30	29

Property	Units	Conditions	Value		Reference
		Above film 120% elongated HCl dopant; film cast from EB/NMP	60 350		30
		Above film 650% elongated CSA dopant; as-spun fiber from m-cresol; EB and CSA mixed as powder	203		21
		CSA dopant; film cast from 30/70 chloroform/m-cresol; EB and CSA mixed as powder	70		21
		CH_3SO_3H/ acetic acid dopant; film cast from EB/DMPU then doped	60		27
		I_2 dopant; spin coated from LEB/DMPU then doped	11.4		31
		H_2SO_4 dopant	6.31		32
Apparent bandgap (absorption)	eV	Polyaniline form LEB EB PNB	*Onset* 3.2 1.6(3.0) 1.8	*Peak* 3.6 2.0(3.8) 2.3	28
Melting temperature	K	LEB film and fiber from DMPU; DSC, 20°C min^{-1}	658		33
Glass transition temperature	K	LEB fiber spun from DMPU; DSC, 5°C min^{-1}, N_2	474		34
		EB film cast from NMP; DMTA, 5°C min^{-1}, 1 Hz	493		1
Sub-Tg transition temperature	K	EB film cast from NMP; DMTA, 3°C min^{-1}, 1 Hz; assigned to phenyl ring twisting	193		35
Thermal stability	K	Cross-linking reaction: EB film cast from NMP; DMTA, 3°C min^{-1}, 1 Hz	453		35
		Decomposition (LEB) reaction; EB film cast from NMP; TGA, 20°C min^{-1}, N_2	780		34
		Decomposition (LEB) reaction; EB film cast from NMP; TGA, 20°C min^{-1}, N_2	673		1
Index of refraction n	–	EB spin coated from DMPU, average, 1550 nm	1.85		36
Zero-T dielectric constant $\varepsilon_{mw}(T \to 0)$	–	PANI-CSA cast from chloroform	~30		21
Dielectric constant ε_{mw}	–	PANI-CSA cast from m-cresol, 300 K, 6.5 GHz	-4.5×10^{-4}		21
Plasma frequency ω_p	eV	PANI-CSA cast from m-cresol, 300 K	0.016		21
Dielectric relaxation time τ	s	PANI-CSA cast from m-cresol, 300 K	1.1×10^{-11}		21
Electroluminescence emission peak	Nm	Porous Si/ PANI-CSA (m-cresol), 0.5 A cm^{-2} current density	800		37

Polyaniline

Surface energies[38]

Form	Surface Energy			Comments
	γ (erg cm^{-2})	γ^d (erg cm^{-2})	γ^p (erg cm^{-2})	
EB	44.6	36.9	7.7	NMP cast film
PANI-HCl	63.5	38.7	24.8	NMP cast EB; doped with HCl (pH = 0)

Electrochemical potentials of redox processes in polyaniline

Redox couple	Potential* (V)	Conditions	Reference
Leucoemeraldine/emeraldine	0.15	vs. Cu/CuF$_2$ in NH$_4$F + 2.3HF	39
Emeraldine/pernigraniline	0.80	vs. Cu/CuF$_2$ in NH$_4$F + 2.3HF	39
Leucoemeraldine/emeraldine	0.115	vs. SCE in 1.0M HCl	40
Emeraldine/pernigraniline	0.755	vs. SCE in 1.0M HCl	40

References

1. Wei Y, et al. *Polymer*. 1992;33(2):314.
2. Adams PN, Laughlin PJ, Monkman AP, Kenwright AM. *Polymer*. 1996;37(15):3411.
3. Manohar SK, MacDiarmid AG, Epstein AJ. *Synth. Met*. 1991;41–43:71.
4. Liao YH, Angelopoulos M, Levon K. *J. Poly. Sci., Part A: Poly. Chem*. 1995;33:2725.
5. Jadhav AV, Gulgas CG, Gudmundsdottir AD. *Eur.Poly. Journal*. 2007;43:2594.
6. Huang K, Qui H, Wan M. *Macromolecules*. 2002;35:8653.
7. English JT, Deore BA, Freund MS. *Sens.Act B*. 2006;115:666.
8. Wei Y, et al. *J. Phy. Chem*. 1989;93:495.
9. Yue J, Epstein AJ. *J. Am. Chem. Soc*. 1990;112:2800.
10. Lee C, Seo Y, Lee S. *Macromolecules*. 2004;37:4070.
11. Kolla HS, et al. *J. Am. Chem. Soc*. 2005;127:16770.
12. Oh EJ, et al. *Synth. Met*. 1993;55–57:977.
13. Wei Y, et al. *J. Phys. Chem*. 1989;93(1):495.
14. Stejskal J, Kratochvil P, Radhakrishnan N. *Synth.Met*. 1993;61(3):225.
15. Monkman AP, Adams. *Synth. Met*. 1991;41–43:891.
16. Kaplan S, Conwell EM, Richter AF, McDiarmid AG *J. Am. Chem. Soc*. 1988;110(23):7647.
17. Pouget JP, et al. *Macromolecules*. 1991;24:779.
18. Djurado D, Nicolau YF, Dalsegg L, Samuelsen EJ. *Synth. Met*. 1997;84(1–3):121.
19. Shacklette LW, Han CC. *Mat. Res. Soc. Symp. Proc*. 1994. 328:157.
20. Ou R. *Private Communication*. 1998.
21. Hsu CH, Epstein AJ. *SPE ANTEC'96*. 1996;54(2):1353.
22. Hsu CH, Cohen JD, Tietz RF. *Synth. Met*. 1993;59:37.
23. Wang H-L, Romero RJ, Mattes BR, Zhu Y, Winokur MJ. *J. Poly. Sci., Part B: Poly. Phy*. 2000;38(1):194.
24. Hardaker SS, et al. *SPE ANTEC'96*. 1996;54(2):1358.
25. Chacko AP, Hardaker SS, Gregory RV. *Polymer Preprints*. 1997;37(2):743.
26. Conklin JA, et al. In: Skotheim TA, Elsenbaumer RL, Reynolds JR, eds. *Handbook of Conducting Polymers*. New York: Marcel Dekker; 1998:945.

27. Gregory RV. In: Skotheim TA, Elsenbaumer RL, Reynolds JR, eds. *Handbook of Conducting Polymers*. New York: Marcel Dekker; 1998:437.

28. Kohlman RS, Epstein AJ. In: Skotheim TA, Elsenbaumer RL, Reynolds JR, eds. *Handbook of Conducting Polymers*. New York: Marcel Dekker; 1998: 85.

29. Abell L, Pomfret SJ, Adams PN, Middleton AC, Monkman AP. *Synth. Me.* 1997;84(1–3):803.

30. Monkman AP, Adams P. *Synth. Met.* 1991;40(1):87.

31. Hardaker SS, et al. *Mat. Res. Soc. Symp Proc.* 1998;488:365.

32. Menon R, et al. In: Skotheim TA, Elsenbaumer RL, Reynolds JR, eds. *Handbook of Conducting Polymers*. New York: Marcel Dekker; 1998:27.

33. Chacko AP, et al. *Polymer.* 1998;39(14):3289.

34. Chacko AP, et al. *Synth. Met.* 1997;84:41.

35. Milton AJ, Monkman AP. *J. Phys. D: Appl. Phys.* 199326:1468.

36. Cha C, et al. *Synth. Met.* 1997;84:743.

37. Halliday DP, et al. *ICSM'96.* New York: Elsevier Science Publishers; 1997:1245.

38. Liu MJ, Tzou K, Gregory RV. *Synth. Met.* 1994;63:67.

39. Genies EM, Lapkowski M. *Electroanal. Chem.* 1987;236:189.

40. Focke WW, Wnek GE, Wei Y. *J. Phys. Chem.* 1987;91:5813.

Poly(aryloxy)thionylphosphazenes

Joseph H. Magill

Acronyms PATP, PTP

Class Polyphosphazenes; poly(thionylphosphazenes)

Structure $-[(NSOX)(NP(OAr_2)_2)]_n-$ (X = Cl in this context)

$$\left[\begin{array}{ccc} O & OR & OR \\ \| & | & | \\ -S=N-P=N-P=N- \\ \| & | & | \\ X & OR & OR \end{array} \right]_n$$

Major Applications Experimental specimens have considerable potential interest. There are ongoing evaluations and development of these new types of noncrystalline polymers.[1−6]

Properties of Special Interest Film-forming elastomers, potential oxygen sensors for biomedical and aerospace are among this class of poly(thionylphosphazenes) depending upon substituents present.[3,4,6]

Synthesis Techniques and Types of Structures Thermal ring-opening polymerization of cyclic thiophosphazene – comprehensive reviews on the chemistry of halogen side-group replacement reactions in cycloheterophosphazenes have been published by van de Grampel[2] – to produce a linear polymer intermediate.[2] Upon reaction of this halo side group intermediate – these elastomeric materials are hydrolytically sensitive as are other halogenated polymer intermediates – with organic nucleophiles poly(thiophosphazenes) are produced, whose properties depend on the nature of the substituents. A wide variety of material properties are anticipated following these procedures. To date only amorphous polymers have been synthesized and characterized by conventional analytical methods.

Property	Units	Conditions	Value	Reference
Chemical structure and properties	–	Depends upon substituents	Variable	(5, 6)
Molecular mass (of repeat unit)	$g\,mol^{-1}$	–	569.64	–
Typical molecular weight	$g\,mol^{-1}$	GPC	$M_w = 1.4 \times 10^5$ $M_n = 5.1 \times 10^4$	–
Typical polydispersity M_w/M_n	–	–	<3	–
Solvents		Generally THF, toluene, chlorinated hydrocarbons such as CH_2Cl_2, etc.		(7)
Nonsolvents		Nonpolar (hexanes) or highly polar H-bonded liquids such as H_2O or MeOH		(7)

Poly(aryloxy)thionylphosphazenes

Spectroscopic properties

Property	Units	Conditions*	Value	Reference
UV-visible spectrum	Nm	Unresolved peaks	252; 272	(6)
IR-spectrum	cm^{-1}	Thin films cast on KBr disks		–
		S=O	1,307	
		P=N	1,203	
		C—O	1,165	
		S=O	1,148	
		P—O	967	
NMR-spectrum (solution)	ppm	^{1}H in CDCl$_3$	7.1 (m)	(6)
		^{31}P in CHCl2	−20.9	
		^{31}C in CDCl3		
		o-Ph	121.3	
		m-Ph′	126.9	
		p-Ph′	127.2	
		o-Ph′	128.0	
		m-Ph′	128.8	
		p-Ph	138.1	
		ipso Ph′	139.9	
		ipso Ph′	149.9	
		ipso Ph, all s	149.9	

* Ph denotes the phenyl ring closest to the polymer backbone; Ph′ refers to the ring furthest away.[6] *Ab initio* molecular orbital calculations depicting conformational parameters are compiled in reference (4).

Transition temperatures

Property	Units	Conditions	Value	Reference
Glass transition temperature*	K	DSC method ($10°$ min^{-1} heating rate)	328	(6)
Mesophase transition	–	DSC method ($10°$ min^{-1} heating rate)	None reported	–
Melting temperature	–	DSC method ($10°$ min^{-1} heating rate)	None reported	–

* Values are reported for various poly(thiophosphazenes) ranging from 217 to 330 K, depending on the side groups and molecular weight.[4,6,8]

Solution properties

Property	Units	Conditions	Value	Reference
Solvents	–	–	THF, CH$_2$Cl$_2$, dioxane	(7)
Theta temperature Θ	K	THF solution, $M_w = 6.4 \times 10^4$*	295	(6)
Hydrodynamic Stokes	Å	THF, 295 K, $M_w = 6.4 \times 10^4$	59 (radius, R_h, eff.)	(6)
Diffusion coefficient	cm$_2$ s^{-1}	THF, 295 K, $M_w = 6.4 \times 10^4$	7.75×10^{-7}	–

Poly(aryloxy)thionylphosphazenes

Property	Units	Conditions	Value	Reference
Second virial coefficient	$mol\ cm^3\ g^{-2}$	THF, 295 K, $M_w = 6.4 \times 10^4$	~0.0	–
Refractive index increment dn/dc	$ml\ g^{-1}$	THF, 295 K, $M_w = 6.4 \times 10^4$	0.208	–

* There is an apparent discrepancy in molecular weights measured by GPC and low-angle laser light scattering (LALLS) techniques.[6] GPC overestimates it by about it 30% unless corrections are made to coil size between the phosphazene and the polystyrene calibrant.

Stabilities

Conditions	Value
In air (years)	Stable
In hot solution, NaOH or Na aryloxide	Rapidly decomposes by nucleophilic attack at the S in backbone

References

1. Dodge JA, et al. *J. Amer. Chem. Soc.* 1990;112:1268.
2. van de Grampel HC. *Coordination Chem. Revs.* 1992;112:247.
3. Liang M, Manners I. *J. Amer. Chem. Soc.* 1991;113:4044;
   Gates DP, Manners I. *J. Chem. Soc., Daltons Trans.* 1997:2525.
4. Jaeger R, et al. *Macromolecules.* 1995;28:539.
5. Ni Y, et al. *Macromolecules.* 1996;29:3401.
6. Ni Y, et al. *Macromolecules.* 1992;25:7119.
7. Manners I. *Private communication.*
8. Manners I. *Polymer News.* 1993;18:133.

Poly(p–benzamide)

Guru Sankar Rajan

Acronyms PBA, PPBA

Registry PBA (SR) 24991–08–0; PBA (homopolymer) 25136–77–0

Structure[1–9]

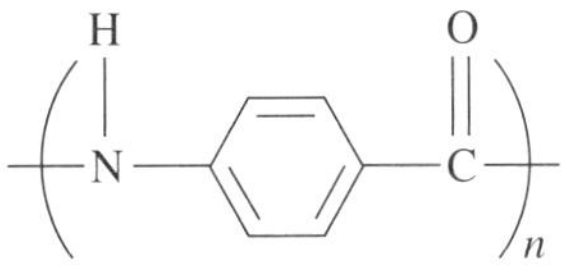

General Information PBA is the first nonpeptide, synthetic condensation polymer reported to form liquid–crystalline solution.[1,4] There are several routes to obtain PBA.[1–9] PBA forms liquid–crystalline solutions because of an inherently extended rigid chain structure produced by a combination of the carbon–nitrogen bond in predominantly *trans* amide linkages.[1,4] The molecular conformation is TCTC(T = *trans*, C = *cis*), where the internal rotation angles about the N—C bond of the amide group and about the virtual bond of N—phenyl—C are T and C conformations, respectively.[7] The chain of all amide groups is in the *head-to-tail* order for PBA.[8]

Major Applications The dopes of PBA can be utilized for the preparation of films, filaments, fibrids, and coatings. Wet extruded, tough, clear, flexible films can be applied to substrates like glass, ceramics, metals, concrete, and polymeric materials.[1,4] The high temperature resistance of the polyaramids makes them suitable for asbestos replacement in heat-resistant work wear. PBA has been superseded by poly(p–phenylene terephthalamide).

Synthesis

Scheme 1[1,4]

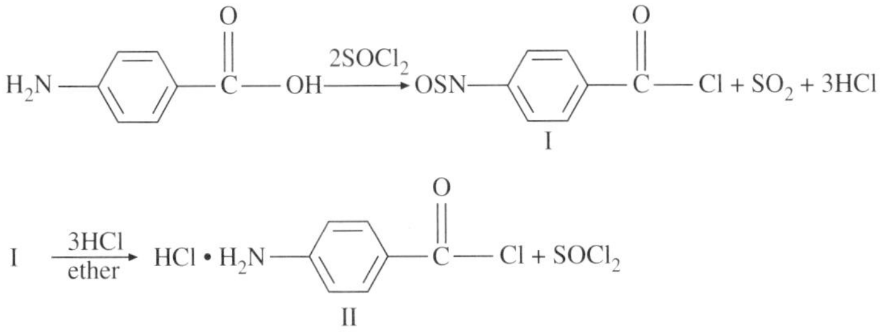

Poly(*p*–benzamide)

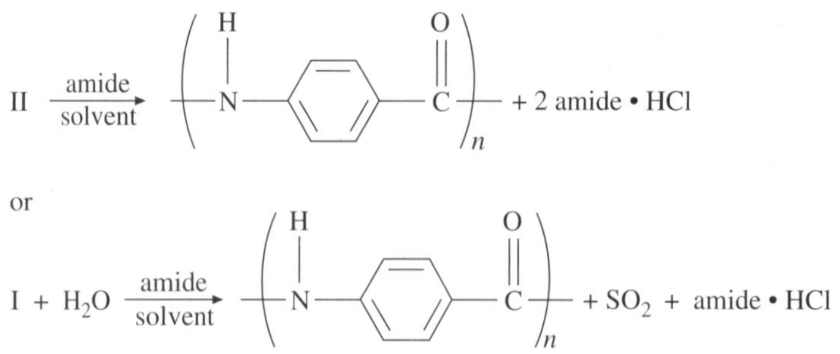

$$II \xrightarrow[\text{solvent}]{\text{amide}} \left(N \overset{H}{|} \text{—} \bigcirc \text{—} C \overset{O}{\|} \right)_n + 2 \text{ amide} \cdot HCl$$

or

$$I + H_2O \xrightarrow[\text{solvent}]{\text{amide}} \left(N \overset{H}{|} \text{—} \bigcirc \text{—} C \overset{O}{\|} \right)_n + SO_2 + \text{amide} \cdot HCl$$

Scheme 2[2]

$$HO \text{—} \overset{O}{\underset{\|}{C}} \text{—} \bigcirc \text{—} NH_2 + CS_2 \underset{\longleftarrow}{\overset{B}{\rightleftharpoons}} \left[BH^+ {}^-OC \text{—} \bigcirc \text{—} NHC \overset{S}{\|} \text{—} S^- HB^+ \right]$$

$$\left[BH^+ {}^-OC \text{—} \bigcirc \text{—} NCS \right]$$

$$+[HS^- HB^+ \rightleftharpoons H_2S\uparrow + B]$$

$$\left[BH^+ {}^-OC \text{—} \bigcirc \text{—} \left(NHC \overset{S}{\|} \text{—} O \text{—} \overset{O}{\underset{\|}{C}} \text{—} \bigcirc \right)_n NCS \right]$$

$$\xrightarrow{-COS} BH^+ {}^-OC \text{—} \bigcirc \text{—} \left(NH \text{—} \overset{O}{\underset{\|}{C}} \text{—} \bigcirc \right)_n NCS \xrightarrow[\Delta \text{ solid state}]{-COS} PBA$$

B =

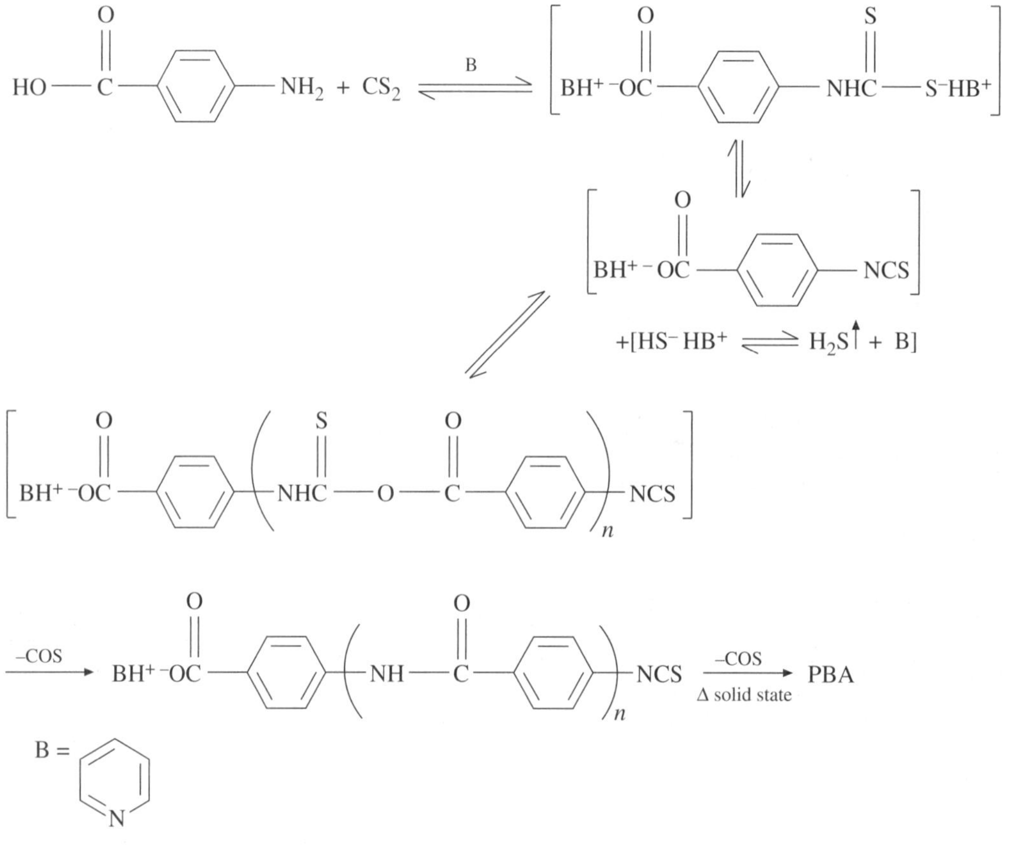

Poly(p–benzamide)

Intrinsic viscosities, molecular weights, (M_w and M_n), and M_w/M_n[10]

$[\eta]^*$(dl/g)	M_w(g/mol)	M_n(g/mol)	M_w/M_n
0.46	5300*	2700	2.0
0.88	6800*	4400	1.5
1.21	7250*	4250	1.7
1.15	8700*	–	–
1.62	10 400*	5750	1.8
2.08	11 700*	6760	1.7
2.10	12 400*	–	–
2.20	13 800*	–	–
2.43	14 200*	–	–
5.20	26 000#	–	–
9.00	51 000#	–	–
12.5	64 400*	–	–

* In concentrated sulfuric acid.

In chlorosulfonic acid, 0.1 N LiClSO$_3$.

Mark–Houwink parameters[10–13]

Molecular Weight Range	K (dl/g)	a	Ref
<12 000*	1.9×10^{-7}	1.70	10
<12 000*	7.8×10^{-5}	1.08	10
$3100 \leq M_w \leq 13\,000$*	2.14×10^{-5}	1.20	11
$5300 \leq M_w \leq 51\,000$#	1.67×10^{-5}	1.23	10
$7140 \leq M_w \leq 23\,000$#	2.69×10^{-5}	1.41	12

* In sulfuric acid.

In dimethyl acetamide + 3% LiCl.

Average molecular weight and molecular weight distribution using different methods of data analysis[8]

Method	Sample	$10^{-4}M_n$	$10^{-4}M_w$	$10^{-4}M_z$	M_w/M_n	M_z/M_w
CONTIN	4	2.63	4.77	11.8	1.82	2.47
	5	1.81	3.00	6.69	1.66	2.23
MSVD	4	2.30	3.84	10.8	1.67	2.80
	5	2.20	3.28	8.67	1.49	2.64

Observed band frequencies for the *cis–trans* conformation by infrared spectroscopy[14]

Intensity	Frequency (cm^{-1})
Very strong	1507, 1319, 1238
Strong	3346, 1662, 1605, 1593, 1527, 1410, 1402, 1272, 1186, 847
Medium strong	1091
Medium	1127, 1019, 807, 764
Weak	3062, 3036, 701, 691, 631, 539, 486
Very weak	972, 950, 601

Relaxation times of selected ^{13}C resonances*[15]

Site	At ppm	T_1(ms)*	T_2 (ms)*
Protonated aromatic, isotropic	126	130 ± 10	1.20 ± 0.30
	131	140 ± 10	1.25 ± 0.50
Protonated aromatic, nematic	154	120 ± 20	0.90 ± 0.20
	159	110 ± 20	1.15 ± 0.40
Carbonyl, isotropic	173	470 ± 70	1.90 ± 0.70
	205	450 ± 100	1.60 ± 1.00

* Under conditions of nematic/isotropic coexistence (inversion recovery and spin echo data fits in 12.4% w/w PBA/H_2SO_4 solutions at 45°C).

Shift of fluorescence peak wavelength[16]

λ_{ex} (nm)	380	390	400	410	420	465
Peak λ_{ex}(nm)	431	431	460	478	496	513

Standard values for bond lengths and bond angles[7]

Bond	Bond Length (Å)	Bond	Bond Angle (Degree)
C1=O	1.24	$O'-C1'-C7$	120.9
C1—C7	1.50	$C7-C1'-N'$	116.3
C1—N	1.35	O—C1—N	122.8
N—H1	0.96	C1—N—H1	117.9
C—C (phenyl)	1.395	C1—N—C2	124.5
C—H (phenyl)	1.084	H1—N—C2	117.6

Average thermal expansion coefficients[17,18]

Expansion Coefficient (10^{-5} K^{-1})	300–500 K	
	Calculated	Experimental
α_1	7.7	7.0
α_2	4.6	4.1
α_3	−0.84	−0.77

Solvents and nonsolvents[1,19]

Solvents	Tetramethylurea (TMU), *N,N*-dimethylacetamide (DMAc), *N,N*-diethylacetamide, *N,N*-dimethyl ethylene urea, *N*-methylpyrrolidone (NMP), *N*-ethylpyrrolidone, *N*-acetylpyrrolidone, *N,N*-dimethylpropionamide,*N*, *N*-dimethylbutyramide, *N*, *N*-dimethylisobutyramide. (Li or Ca chloride can increase the solubility of sparingly soluble rod-like polyamides such as PBA in DMAc, TMU, or NMP.)
Nonsolvents	Water bath (65–90°C), ethylene glycol, mixtures of TMU and water, mixtures of alcohol and water, and aqueous salt baths (preferably maintained at 40–45°C or above).

Poly(p–benzamide)

Values of two phases of a solution of PBA in TMU–LiCl[1,4] with p-aminobenzoic acid as chain terminator ($\eta_{inh} = 0.81$; 0.5g/100 ml H_2SO_4 at 30°C)

Property	Units	Isotropic	Optically Anisotropic Phase
Proportion	% by volume	31	69
Density	g/cm^3	1.0598	1.0664
LiCl content	g/cm^3	0.085	0.082
Polymer content	g/cm^3	0.105	0.121
Polymer η_{inh}	dl/g	0.59	0.88
Bulk viscosity	cP	~6000	~3000

Static and dynamic properties of PBA in DMAc+3% (g/cm^3) LiCl[8]

Sample	$10^{-4} M_w$	Second Virial Coefficient $A_2 10^{-3}$ (cm^3.mol/g^2)	Radius of Gyration R_g (nm)	Persistence Length (nm)	Anisotropy δ
1	5.17	1.63	32	75 ± 3	0.40
2	5.52	0.31	36	75 ± 3	0.40
3	4.41	–	–	75 ± 3	0.41
4	2.88	–	–	75 ± 3	0.51

Persistence length and Kuhn segment[20–23]

Method	Kuhn Segment (nm)	Persistence Length (nm)	Method	Kuhn Segment (nm)	Persistence Length (nm)
Sedimentation (in DMAc + LiCl)	38–39	19–19.5	Flow birefringence (in H_2SO_4)	210	105
Light scattering (in H_2SO_4)	–	40	–	196	98
Viscosity (in H_2SO_4)	–	18–24	–	100	50

Constants* for the clearing temperatures[24]

M_w	A	α
10 000	41 (6)	0.92 (0.06)

* A and α are from least squares fits of the clearing temperature (T_{ni}) vs. concentration (c) measurements to a relation of the form $T_{ni} = Ac^{\alpha}$. Standard deviations are shown in parentheses.

Thermal properties[25−27]

Property	Units	Value
Glass transition temperature*	K	>503
Crystal transformation temperature	K	487 (modification I to II)
		748 (II)
Crystal–nematic transition temperature#	K	817 (III)

* DSC thermogram of the PBA crystalline solvate exhibits two broad endotherms in the 120–230°C temperature range, which disappears when the sample anneals above 230°C, but a different crystal form appears.

By heating II above 475°C and cooling, or by washing with water, or washing I with water and annealing, III can be obtained.

Crystallographic data

Parameters	Units	Ref 7	Ref 25	Ref 28
Crystal system			Orthorhombic	
Space group		$P2_12_12_1-D_2^4$	$P2_12_12_1-D_2^4$	
Lattice constant				
a	Å	7.75	7.71	8.06
b		5.30	5.14	5.13
c(fiber axis)		12.87	12.8	12.96
Density				
Observed	g/cm^3	1.48	1.48	1.48
Calculated		1.50	1.54	1.48

Isothermal liquid crystallization data[29]

Property	Units	PBA/H$_2$SO$_4$ system			
Liquid crystallization temperature	K	318	323	328	333
Avrami exponent n		1.35	1.20	1.15	0.95
Half-time for the liquid crystallization $t_{1/2}$	s	8.8	11.4	16.4	22.7

Effect of anisotropy on fiber properties ($\eta_{inh} = 2.1$, in H$_2$SO$_4$)[4]

Anisotropic Phase in Spin Dope	Spin Dope*		As Extruded Filaments			
	η (cP)	Wt% of Polymer	Tenacity N/Tex	Elongation (%)	Initial Modulus (N/Tex)	Orientation Angle (Degree)
None	14 000	4.6	0.39	10.9	16.1	33
Small amount	5600	5.8	0.75	9.7	29.1	20
Larger amount	1800	6.7	0.86	8.3	37.4	16

* Spin dope = 13% in tetramethylurea–lithium chloride (6.54%).

Poly(p–benzamide)

Mechanical properties of undrawn fiber by dry spinning ($\eta_{inh} = 1.48$, in H_2SO_4)[4]

Spin Stretch Factor*	Tex Per Filament	Tenacity N/Tex	Elongation (%)	Initial Modulus (N/Tex)	Orientation Angle (Degree)
Free fall	0.67	0.28	3.7	12.4	39
1.90	0.68	0.34	2.7	20.7	37
2.42	0.53	0.37	3.4	19.4	38
2.56	0.50	0.49	3.3	24.7	26
3.83	0.34	0.61	2.9	34.4	22
5.11	0.26	0.76	3.0	41.5	22
6.39	0.21	0.71	2.8	38.0	19

* Spin dope = 13% in tetramethylurea–lithium chloride (6.54%).

Effect of spinning method on fiber properties[30,31]

η_{inh} (dl/g)[a]	Spun From[b]	Spinning Method[c]	Tex Per Filament	Tenacity N/Tex	Elongation (%)	Initial Modulus (N/Tex)
1.67	O(−)	D	–	0.72	3.1	44.9
2.36	O(A)	W	0.54[d]	0.64	8.1	25.0
Same dope	O(I)	W	2.53[e]	0.11	9.0	5.6
3.7	Acid (A)	DJ–W	0.11	1.7	4.0	50.3

[a] In H_2SO_4; [b] O = organic solvent; Acid = acid (H_2SO_4); I = isotropic dope; A = anisotropic dope; [c] D = dry spun; W = wet spun; DJ–W = dry jet wet spun; [d] Spun from anisotropic layer of dope; [e] Spun from isotropic dope.

Tensile properties before and after annealing[32]

	Units	Dry Spinning*	Annealing[#]
Tensile modulus	MPa	65 000	137 000
Tensile strength	MPa	1050	2200
Elongation	%	3.1	1.9

* Spin stretch factor of 3.2.
# Annealing for a few seconds at 525°C under nitrogen.

Calculated elastic constants (GPa)[28]

E_x	12.8	G_{xy}	8.90	μ_{yx}	0.872
E_y	60.8	G_{yz}	14.0	μ_{zy}	0.259
E_z	322.3	G_{xz}	0.90	μ_{zx}	0.331

Poly(*p*–benzamide)	Reference
Patents	33–41
Synthesis	1–9, 75, 76
Crystal structure	3, 7–9, 17, 18, 25, 27, 28, 45
Phase diagram, composition, and fractionation	3, 43, 47, 48, 50–52, 54, 70
Static and dynamic properties	8
Mechanical properties	1, 30–32
Optical properties	69, 81, 82
Viscosity–molecular weight relationships	10–13
IR, NMR, chemical shift tensor parameters, fluorescence spectra, EPR	2, 9, 14–16, 58, 59, 83
Moments of end-to-end vectors, order parameter, molecular entanglements, conformation	25, 42, 46, 55
Thermal expansion and isothermal elastic stiffness constants	17, 18, 25
Solubility, persistence length, effect of solvent on the structure and property	1, 2, 4, 19, 20–23, 53, 80
Thermal transition, thermal behavior, clearing temperatures	9, 24–27
Orientation under magnetic field	25, 78, 79
Vapor permeation	49
Kinetics of liquid crystallization	29
Dynamic birefringence, rigidity	12, 45, 53
Thermomechanical and ultrasonic properties	44, 84
Molecular simulation	28
PBA block copolymers	62, 63–65, 67
Shear-induced structure transition, small angle light scattering, rheological properties	56, 77, 81
Alkylated PBA	57, 66
Molecular composites	60
Nanocomposites	61, 68, 72
Condensation reaction between chain ends	71
Dynamic characterization of liquid crystalline polymers, liquid crystallinity	73, 74
Effect of molecular weight on fiber, some problems	85, 86
XRD	1, 7, 9

References

1. Kwolek SL. *U. S. Patent* 3,600,350 (17 August 1971); *U. S. Patent* 3,671,542 (20 June 1972).
2. Memeger W. *Macromolecules.* 1976;9:195.
3. Panar M, Beste LF. *Macromolecules.* 1976;10:1401.
4. Kwolek SL, et al. *Macromolecules.* 1977;10:1390.
5. Krigbaum WR, et al. *J. Polym. Sci., Polym. Chem. Ed.* 1984;22:4045.
6. Krigbaum WR, et al. *J. Polym. Sci., Polym Chem. Ed.* 1985;23:1907.
7. Takahashi Y, et al. *J. Polym. Sci., Polym. Phys. Ed.* 1993;31:1135.
8. Ying Q, Chu B. *Macromolecules.* 1987;20:871.
9. Bao H, et al. *J. Macromol. Sci. – Physics.* 2001;B40:869.
10. Schaefgen JR, et al. *Polym. Prepr. Am. Chem. Soc. Div. Polym. Chem.* 1976;17:69.
11. Aharoni SM. *Macromolecules.* 1987;20:2010.

12. Tsvetkov VN, et al. *European Polymer J.* 1976;12:517.

13. Ying Q, et al. *Polym. Mater. Sci. Eng.* 1986;54:546.

14. Yang X, et al. *Polymer.* 1993;34:43.

15. Zhou M, Frydman V, Frydman L. *Macromolecules.* 1997;30:5416.

16. Bai F, et al. *Macromol. Chem. Phys.* 1994;195:969.

17. Lacks DJ, Rutledge GC. *Macromolecules.* 1994;27:7197.

18. Ii T, et al. *Macromolecules.* 1986;19:1772.

19. Orwall RA. In: Mark HF, et al. eds. *Encyclopedia of Polymer Science and Engineering.* 2nd ed. Vol 15. New York: John Wiley and Sons; 1989:399.

20. Arpin M, Strazielle C. *Polymer.* 1977;18:591.

21. Papkov SP. *Advances in Polymer Science.* 1984;59:75.

22. Ciferri A, ed. In: *Liquid Crystallinity in Polymers: Principles and Fundamental Properties.* New York: VCH Publishers; 1991.

23. Blumstein A, ed. In: *Liquid Crystalline Order in Polymers.* New York: Academic Press; 1978.

24. Picken SJ. *Macromolecules.* 1989;22:1766.

25. Tashiro K, et al. *Macrmolecules.* 1977;10:413.

26. Takase M, et al. *J. Polym. Sci., Polym Phys. Ed.* 1986;24:1115.

27. Takase M, et al. *J. Polym. Sci., Polym Phys. Ed.* 1986;24:1675.

28. Yang X, Hsu SL. *Macromolecules.* 1991;24:6680.

29. Lin J, et al. *Polymer International.* 1994;34:141.

30. Preston J. In: Mark HF, et al. eds. *Encyclopedia of Polymer Science and Engineering.* 2nd ed. Vol 11. New York: John Wiley and Sons; 1989:392.

31. Conio G, et al. *Polymer J.* 1987;19:757.

32. Collyer AA. *Materials Science and Technology.* October 1990;6:981.

33. Huffmann WA, Smith RW. *U. S. Patent.* 3,203,933 (31 August 1965).

34. Preston J, Smith RW. *U. S. Patent.* 3,225,011 (21 December 1965).

35. Stephens CW. *U. S. Patent.* 3,472,819 (14 October 1969).

36. Pikl J. *U. S. Patent.* 3,541,056 (17 November 1970).

37. Hoegger EF, et al. *U. S. Patent.* 3,575,933 (20 April 1971).

38. Kwolek SL. *U. S. Patent.* 3,819,587 (25 June 1974).

39. Kwolek S. *U. S. Patent.* 3,888,965 (10 June 1975).

40. Krigbaum WR, et al. *U. S. Patent.* 4,412,059 (25 October 1983).

41. Cohen A, et al. *U. S. Patent.* 5,788,888 (4 August 1998).

42. Erman B, et al. *Macromolecules.* 1980;13:484.

43. Conio G, et al. *Macromolecules.* 1981;14:1084.

44. Ii T, et al. *Macromolecules.* 1986;19:1809.

45. Tsvetkov VN, Shtennikova IN. *Macromolecules.* 1978;11:306.

46. Sartirana ML, et al. *Macromolecules.* 1986;19:1176.

47. Balbi C, et al. *J. Polym. Sci., Polym. Phys. Ed.* 1980;18:2037.

48. Krigbaum WR, et al. *J. Polym. Sci., Polym. Phys. Ed.* 1987;25:1043.

49. Ikeda RM, Gay FP. *J. Appl. Polym. Sci.* 1973;17:3821.

50. Sato T, et al. *Polymer.* 1989;30:311.

51. Bianchi E, et al. *Polymer J.* 1988;20:83.

52. Bianchi E, et al. *Macromol. Chem. Phys.* 1997;198:1239.

53. Krigbaum WR, et al. *Macromolecules.* 1991;24:4142.

54. Bianchi E, et al. *Macromolecules.* 1982;15:1268.

55. Aharoni S. *Macromolecules.* 1983;16:1722.

56. Siddiquee SK, van Egmond JW. *J. Rheol.* 2002;46:367.

57. Tanatani A, et al. *J. Am. Chem. Soc.* 2005;127:8553.

58. Grinshtein J, et al. *J. Chem. Phys.* 2001;114:5415.
59. McElheny D, et al. *J. Magn. Reson.* 2001;148:436.
60. Takayanagi M. *Pure Appl. Chem.* 1983;55:819.
61. Davis VA, et al. *Macromolecules.* 2004;37:154.
62. Tsutomu Y, et al. *Makromol. Chem. Macromol. Symp.* 2002;199:187.
63. Chavan NN, et al. *Macromol. Chem. Phys.* 1996;197:2415.
64. Cavalleri P, et al. *Macromol. Chem. Phys.* 1998;199:2087.
65. Li X-G. *Polym. Int.* 1999;48:1277.
66. Shi H, et al. *Polymer.* 2004;45:6299.
67. Tsutomu Y, et al. *J. Amer. Che. Soc.* 2002;124:15158.
68. Ruckenstein E, Yuan Y. *Polymer.* 1997;38:3855.
69. Lin J. Li S. *Euro Polym J.* 1994;30:671.
70. Salaris F, et al. *Makromol. Chemie.* 1976;177:3073.
71. Sugamiya K. et al. *Polym. J.* 1978;10:275.
72. Marsano E. et al. *Macromol. Symp.* 2006;234:33.
73. Nemtsov VB. et al. *J. Engg. Phys. Thermophysics.* 2003;76:540.
74. Ciferri A. *Liq. Cryst.* 2004;31:1487.
75. Wu G-C, et al. *Polym. J.* 1982;14:571.
76. Ali S, El-Sabbah MMB. *Acta Polymerica.* 1980;31:638.
77. Papkov SP, et al. *J. Polym. Sci., Polym. Phys. Ed.* 1974;12:1753.
78. Platonov VA, et al. *Fibre Chemistry.* 1976;7:393.
79. Platonov VA, et al. *Fibre Chemistry.* 1976;7:577.
80. Arfe'ev NM, et al. *Fibre Chemistry.* 1981;13:226.
81. Khanchich OA, et al. *Fibre Chemistry.* 1976;7:574.
82. Khanchich OA. *Fibre Chemistry.* 1987;18:241.
83. Dovbii EV. *Fibre Chemistry.* 2002;34:206.
84. Kalashnik AT, et al. *Fibre Chemistry.* 1998;30:163.
85. Volokhina AV. *Fibre Chemistry.* 2002;34:1.
86. Iovleva MM. *Fibre Chemistry.* 1992;23:341.

Poly(benzimidazole)

Vladyslav Kholodovych and William J. Welsh

Acronyms, Alternative Name PBI, PBZI, poly[2,2′-(*m*-phenylene)-5,5′- bibenzimidazole]

Class Rigid-rod polymers

Structure

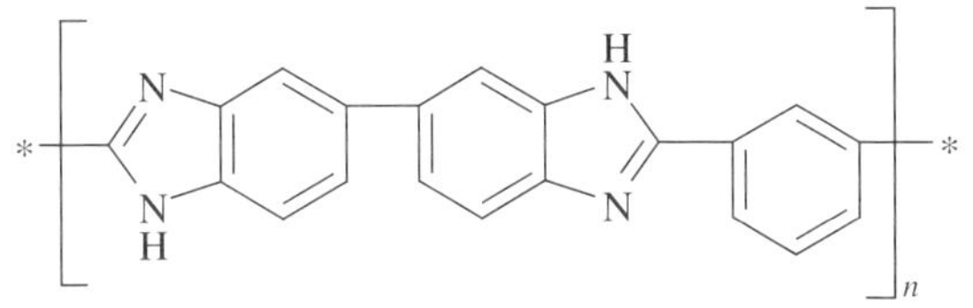

Major Applications Fire-resistant material, replacement for asbestos, thermal-protective clothing, ion-exchange resins, microporous absorbent beads, membrane applications.

Properties of Special Interest High-temperature stability, nonflammability, unusual resistance to organic solvents, excellent mechanical properties, interesting electrical and nonlinear optical properties.

Synthesis Condensation polymerization of 3,3′,4,4′-tetraaminobiphenyl (TAB) and diphenyl isophthalate (DPIP) in poly(phosphoric) acid, [1] or in a hot molten nonsolvent such as sulfolane or diphenyl sulfone [2].

Property	Units	Conditions	Value	Reference
Density	g cm^{-3}	Fiber, stabilized	1.43	3
		Fiber, unstabilized	1.39	
		Fiber-grade ®lm		
		Untreated	1.2	
		Annealed	1.3	
		Plasticized	1.4	
Young's modulus	N/tex	Fiber, stabilized	39.6	3
		Fiber, unstabilized	79.2	
	MPa	Fiber-grade film	2750	
		Untreated	3790	
		Annealed	2270	
		Plasticized		
		High MW film	3170	
		Untreated	2820	
		Plasticized		

Property	Units	Conditions	Value	Reference
Tensile strength (tenacity)	N/tex	Fiber, stabilized	2.3	3
		Fiber, unstabilized	2.3	
	MPa	Fiber-grade film		
		Untreated	117	
		Annealed	186	
		Plasticized	103	
		High MW film		
		Untreated	96	
		Plasticized	96	
Elongation at break	%	Fiber	30	3
		Fiber-grade film		
		Untreated	14	
		Annealed	24	
		Plasticized	20	
Glass transition temperature T_g	K	–	~700	3
		After annealing	773	
Thermal decomposition onset	K	–	~873	3
Flame-test shrinkage	%	Fiber, stabilized	6	3
		Fiber, unstabilized	50	
Moisture content	%	Fiber, stabilized, 21°C		3
		65% relative humidity	15	
		Fiber-grade film		
		Untreated	10	
		Annealed	5	
		Plasticized	12	
		High MW film		
		Untreated	10	
		Plasticized	12	
Surface resistivity	ohm·sq^{-1}	Film	10^{11}	3
Volume resistivity	ohm-cm		10^{13}	3
Dielectric constant	–	Film, at 100 Hz		3
		25°C	5.4	
		250°C	3.7	
Dielectric strength	V m^{-1}	Film, at 100 Hz		3
		25°C	3900	
		250°C	2500	
Dissipation factor	–	Film, at 100 Hz		3
		25°C	0.013	
		250°C	0.021	
Bulk protonic conductivity	ohm^{-1}·cm^{-1}	Film, at 100% relative humidity	8×10^{-5}	3
Characteristic peaks	cm^{-1}	FTIR, dry polymer film		4
		Aromatic C–H stretch	3150	
		Imidazole free N–H stretch	3420	

Poly(benzimidazole)

Property	Units	Conditions	Value	Reference
		FTIR, wet polymer		
		Aromatic C–H stretch	3150	
		Imidazole free N–H stretch	3420	
		Water O–H stretch	3620	

References

1. Iwakura Y, Uno K, Imai Y. *J. Polym. Sci.* 1948;12:2605.
2. Hedberg FL, Marvel CS. *J. Polym. Sci.* 1974;12:1823.
3. Buckley A, Stuetz DE, Serad GA. In: Mark HF, et al. eds. *Encyclopedia of Polymer Science and Engineering.* Vol 11. New York: John Wiley and Sons; 1988:572. (and references therein).
4. Brooks NW, et al. *Polymer.* 1993;34:4038.

Poly(benzobisoxazole)

Vladyslav Kholodovych and William J. Welsh

Acronym, Alternative Names PBO, poly(*p*-phenylene-2,6-benzoxazolediyl), poly[(benzo[1,2-d:5,4-d′]bisoxazole-2,6-diyl)-1,4-phenylene]

Class Rigid-rod polymers

Structure

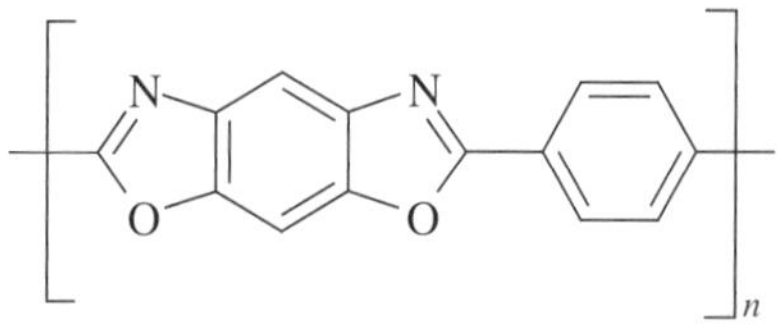

Major Applications High-performance films, fibers, and coatings.

Synthesis Polycondensation of a terephthalic acid with 4,6-diamino-1,3-benzenediol dihydrochloride in poly(phosphoric acid). Processing is primarily limited to variations of wet extrusion. [1,2]

Properties of Special Interest High-temperature resistance, unusual resistance to organic solvents, excellent mechanical properties, interesting electrical and nonlinear optical properties.

Solubility (a) Protonic sulfonic acids RSO_3H, where R $=$ $-OH$, $-CH_3$, $-Cl$, $-CF_3$, $-C_6H_5$, etc., polyphosphoric acid (PPA), *m*-cresol/dichloroacetic acid (70/30), dichloroacetic acid/MSA (90/10) [1] (b) Aprotic organic solvents (e.g. nitroalkanes) containing metal halide Lewis acids (e.g. $AlCl_3$, $GaCl_3$, $FeCl_3$) up to 7.5% polymer. [3]

Unit cell dimensions

Lattice	Monomers Per Unit Cell	Cell Dimensions (Å)			Cell Angles (degrees)			Reference
		a	*b*	*c* (Chain Axis)	*α*	*β*	*γ*	
Monoclinic	2	11.20	3.54	12.05	90	90	101.3	6,20
Monoclinic	1	5.651	3.57	12.05	90	90	101.48	20,21
Monoclinic	1	5.65	3.58	11.74	90	90	102.5	7
Monoclinic	1	5.598	3.540	12.05	90	90	102.5	8

Poly(benzobisoxazole)

Property	Units	Conditions	Value	Reference
Density	g cm^{-3}	As-spun fiber	1.50	4
		Fiber	1.58	5
		X-ray diffraction data	1.50	6
Torsion angle between the benzobisoxazole and phenyl rings	degrees		12	6,20
			25.7	20,21
			0	20,22
d-spacing of the first equatorial Bragg reflection (side-by-side packing)	Å		5.8	6,20
d-spacing of the second equatorial Bragg reflection (face-to-face packing)	Å		3.5	6,20
Repeat unit distance	Å		12.0	6,20
Young's (tensile) modulus	g·denier^{-1}	Fiber, as-spun	502	9
		Fiber, heat-treated	711	9
	GPa	Ribbon	7.6	9
		Along fiber or draw direction	85	10
		Perpendicular to fiber or draw direction	6.5	10
		Fiber, heat-treated (value depends on MW)	221–304	4
		Fiber, as-spun	144	11
		Heat-treated (600°C)	250	
		Heat-treated (650°C)	262	
		Fiber	200–360	5
		Fiber	370	7
		Fiber	317, 365	12
		Fiber, as-spun		
		Heat-treated (600°C)	318	
		Heat-treated (650°C)	290	
		Fiber, as-spun	144 ± 23	
		Heat-treated (600°C)	250 ± 20	
		Heat-treated (650°C)	262 ± 25	
X-ray modulus	–	Fiber, as-spun	387	13
		Heat-treated (600°C)	477	
		Heat-treated (650°C)	433	
Compressive modulus	GPa	Fiber	240	15
Tensile strength	g·denier^{-1}	Fiber, as-spun	4.2	2
		Fiber, heat-treated	4.8	2
	GPa	Ribbon	0.103	3
		Fiber	4.9, 5.8	12
		Fiber, as-spun	2.31	4
		Fiber, as-spun	4.6	11
		Heat-treated (600°C)	5.1	
		Heat-treated (650°C)	3.4	
		Fiber	3.0–5.7	5
		Fiber	3.6	7

Property	Units	Conditions	Value	Reference
		Fiber, heat-treated		4
		(value depends on MW)	2.2–4.7	
		Fiber, as-spun	4.6	13
		Heat-treated (600°C)	4.9	
		Heat-treated (650°C)	3.0	
		Fiber, as-spun	4.6 ± 0.5	14
		Heat-treated (600°C)	5.1 ± 0.6	
		Heat-treated (650°C)	3.4 ± 0.5	
Elongation at break	%	Fiber, as-spun	1.4	2
		Fiber, heat-treated	0.7	2
		Ribbon	0.8	2
		Fiber	1.7, 1.6	12
		Fiber, as-spun	2.1	4
		Fiber, as-spun	3.2	11
		Heat-treated (600°C)	1.9	
		Heat-treated (650°C)	1.3	
		Fiber	1.9	7
		Fiber, heat-treated		4
		(value depends on MW)	1.1–18	
		Fiber, as-spun	2.8	13
		Heat-treated (600°C)	1.7	
		Heat-treated (650°C)	1.2	
		Fiber, as-spun	3.2 ± 0.4	14
		Heat-treated (600°C)	1.9 ± 0.3	
		Heat-treated (650°C)	1.3 ± 0.3	
Compressive strength	GPa	Fiber	0.2–0.3	5
		Fiber	0.68	4
		Fiber	0.300 ± 0.035	15
Torsional modulus	GPa	Fiber	1.0	5
Persistence length Q	nm	300°C	20–30	16
Elastic moduli	GPa	C_{11}	16.33	5
		C_{12}	16.64	
		C_{13}	−0.49	
		C_{15}	−42.19	
		C_{22}	84.0	
		C_{23}	0.69	
		C_{25}	2.01	
		C_{33}	0.49	
		C_{35}	19.11	
		C_{44}	3.79	
		C_{46}	−4.18	
		C_{55}	14.10	
		C_{66}	10.34	
Coefficient of thermal expansion	ppm·K^{-1}	Fiber	−7 to −10	5

Poly(benzobisoxazole)

Property	Units	Conditions	Value	Reference
Degradation temperature	K	Film, uniaxial	>873	17
Fiber flammability-critical oxygen concentration (COC)	–	Fiber	36.1 (top) 22.8 (bottom)	2
Apparent activation energy of polymerization	kcal·mol^{-1}	–	7.16	18
Index of refraction		n_r	1.663	4
		n_t	1.589	
		n_2	>3.0 (est.)	
Third-order nonlinear optical susceptibility $\chi^{(3)}$	esu	Nonresonant ($\lambda = 602$ nm)	$\sim 10^{11}$	19
Raman characteristic frequencies	cm^{-1}	–	1615 1540 1280	14

References

1. Wolfe JF, Arnold FE. *Macromolecules*. 1981;14:909.
2. Choe EW, Kim SN, *Macromolecules*. 1981;14:920.
3. Jenekhe SA, Johnson PO, Agrawal AK. *Macromolecules*. 1989;22:3216.
4. Northolt MG, Sikkema DJ. In: Collyer AA, ed. *Liquid Crystal Polymers: From Structures to Applications*. London and New York: Elsevier Applied Science; 1992:273.
5. *Polymeric Materials Encyclopedia*. Vol 10. 1996.
6. Fratini AV, et al. The Materials Science and Engineering of Rigid-Rod Polymers. In: Adams WW, Eby RK, McLemore DE. eds. *Mat. Res. Soc. Symp. Proc.* Vol 134. Pittsburgh: Materials Research Society; 1989:431.
7. Krause SJ, et al. *Polymer*. 1988;29:1354.
8. Adams WW, et al. *Polymer Commun*. 1989;30:285.
9. Choe EW, Kim SN. *Macromolecules*. 1981;14:920.
10. Rao DN, et al. *Macromolecules*. 1989;22:985.
11. Young RJ, Day RJ, Zakikhami M. *J. Mater. Sci*. 1990;25:127.
12. Ledbetter HD, Rosenberg S, Hurtig CW. In: Adams WW, Eby RK, McLemore DE. ed. *Mat. Res. Soc. Symp. Proc.* Vol 134. Pittsburgh: Materials Research Society; 1989:253.
13. Lenhert PG, Adams WW. The Materials Science and Engineering of Rigid-Rod Polymers. In: Adams WW, Eby RK, McLemore DE. eds. *Mat. Res. Soc. Symp. Proc.* Vol 134. Pittsburgh: Materials Research Society; 1989:329.
14. Young RJ, Day RJ, Zakikhami M. The Materials Science and Engineering of Rigid-Rod Polymers. In: Adams WW, Eby RK, McLemore DE. eds. *Mat. Res. Soc. Symp. Proc.* Vol 134. Pittsburgh: Materials Research Society; 1989:351.
15. Fawaz SA, Palazotto AN, Wang CS. The Materials Science and Engineering of Rigid- Rod Polymers. In: Adams WW, Eby RK, McLemore DE. eds. *Mat. Res. Soc. Symp. Proc.* Vol 134. Pittsburgh: Materials Research Society,1989:381.
16. Roitman DB, McAdon M. *Macromolecules*. 1993;26:4381.
17. Wolfe JF, Loo BH, Arnold FE. *Macromolecules*. 1981;14:909.
18. Cotts DB, Berry GC. *Macromolecules*. 1981;14:930.

19. Prasad PN. The Materials Science and Engineering of Rigid-Rod Polymers. In: Adams WW, Eby RK, McLemore DE. eds. *Mat. Res. Soc. Symp. Proc.* Vol 134. Pittsburgh: Materials Research Society; 1989:635.

20. Park SY, Moon SC, Venkatasubramanian N, Dang TD, Lee JW, Farmer BL. *J. Polym. Sci., Part B: Polym. Phys.* 2006;44:1948.

21. Takahashi Y, *Macromolecules.* 1999;32:4010.

22. Tashiro K, Yoshino J, Kitagawa T, Murase H, Yabuki K. *Macromolecules.* 1998;31:5430.

Poly(benzobisoxazole), Naphthalene Derivatives

N. Venkatasubramanian, Thuy D. Dang, Soo-Young Park,
Vladyslav Kholodovych and William J. Welsh

Acronyms, Alternative Names Poly(2,6-naphthalenebenzobisoxazole) (Naph-2,6-PBO),
poly(2,6-naphthalene-2,6-benzobisxazolediyl); Poly(1,5-naphthalenebenzobisoxazole)
(naph-1,5-PBO), poly(1,5-naphthalene-2,6-benzobisoxazolediyl)

Class Rigid-rod polymers

Structure

where Ar =

Major Applications of Interest High performance fibers, films, and coatings.

Synthesis Polycondensation of 2,6-naphthalenedicarboxylic acid or
1,5-naphthalenedicarboxylic acid with 4,6-diaminoresorcinol dihydrochloride in 83%
polyphosphoric acid (PPA)[1−4] using the 'P$_2$O$_5$ adjustment method'.[5] Although
Naph-2,6-PBO composition is mentioned in the Japanese patent (reference 3), the properties
of this rigid-rod polymer have not been described in the literature.

Processing Isotropic films are mainly cast from dilute methanesulfonic acid solutions.
Fabrication of the polymeric fibers is from the lyotropic liquid crystalline phase of the
polymer in PPA; polymer fibers are spun from the dope by a dry-jet wet spinning technique,
a variation of wet extrusion,[6,7] at 90°C with draw ratios in the range 35–45.

Properties of Special Interest High thermal and thermo-oxidative stabilities, exceptional
resistance to organic solvents, good mechanical properties; lower geometrical symmetry and
extended planar packing relative to PBO,[8] influence chain conformation, packing and
crystallinity, as well as the mechanical properties of the Naph-PBO system in a manner
analogous to the well-known PEN and PET systems.[9]

Solubility Strong, corrosive protonic acid media such as RSO$_3$H where R = —CH$_3$,
—CF$_3$, or —Cl as well as PPA; potentially soluble in polar aprotic organic solvents
(e.g. nitroalkanes) containing metal halide Lewis acids (e.g. AlCl$_3$, GaCl$_3$, and FeCl$_3$).[10,11]

Poly(benzobisoxazole), Naphthalene Derivatives

Polymer	Property	Units	Conditions	Value	Reference
Naph-2,6-PBO	Intrinsic viscosity	dl g^{-1}	Methanesulfonic acid, 30°C	26.4	(4)
Naph-1,5-PBO	Intrinsic viscosity	dl g^{-1}	Methanesulfonic acid	9.6	(2, 4)
Naph-2,6-PBO	Thermal stability	°C	Bulk polymer in helium (TGA)	650	(4)
	Thermo-oxidative stability	°C	Bulk polymer in air (TGA)	580	(4)
Naph-1,5-PBO	Thermal stability	°C	Bulk polymer in helium (TGA)	600	(4)
	Thermo-oxidative stability	°C	Bulk polymer in air (TGA)	550	(4)
Naph-2,6-PBO	Tensile modulus	GPa	Fiber, as-spun	35.4	(4)
			Fiber, heat-treated (350°C)	60	
	Tensile strength	GPa	Fiber, as-spun	2.3	(4)
			Fiber, heat-treated (350°C)	2.5	(4)
	Elongation at break	%	Fiber, as-spun	6.4	(4)
			Fiber, heat-treated (350°C)	3.8	(4)
Naph-1,5-PBO	Tensile modulus	GPa	Fiber, as-spun	11.7	(4)
			Fiber, heat-treated (350°C)	32.5	(4)
	Tensile strength	GPa	Fiber, as-spun	0.93	(4)
			Fiber, heat-treated (350°C)	1.2	(4)
	Elongation at break	%	Fiber, as-spun	8.0	(4)
			Fiber, heat-treated (350°C)	3.7	(4)
Naph-2,6-PBO	chain conformation		Molecular modeling	Trans	(4)
	Repeat unit distance	Å	Fiber diffraction	14.15	(4)
	d-spacing of the first equatorial reflection (side-by-side packing)	Å	Fiber diffraction	6.01	(4)
	d-spacing of the second equatorial reflection (face-to-face packing)	Å	Fiber diffraction	3.40	(4)
Naph-1,5-PBO	chain conformation		Molecular modeling	Trans	(4)
	Repeat unit distance	Å	Fiber diffraction	12.45	(4)
	d-spacing of the first equatorial reflection (side-by-side packing)	Å	Fiber diffraction	6.88	(4)
	d-spacing of the second equatorial reflection (face-to-face packing)	Å	Fiber diffraction	3.50	(4)

Poly(benzobisoxazole), Naphthalene Derivatives

Crystal parameters from X-ray (fiber diffraction) results

Dimension	Units	Naph-2,6-PBO	Naph-1,5-PBO	Reference
a^*	(1/Å)	1/6.01	1/6.88	(4)
b^*	(1/Å)	1/3.40	1/3.50	(4)
β^*	(°)	59.7	67.2	(4)
c	(Å)	14.15	12.45	(4)
$(c/\Delta c)^1$	–	0.25	0.23	(4)

[1] Staggering ratio between adjacent chains on the ac plane of the crystal structure.

References

1. Dang TD, Venkatasubramanian N, Talicsa A, Park S-Y, Arnold FE. *Polymer Preprints (American Chemical Society)*. 2002;43(1):660.

2. Dang TD, Venkatasubramanian N, Lee J-W, Park S-Y, Arnold FE, Farmer BL. *US Patent*. 2006;7041779.

3. Matsuoka T, Kuboto F. *Jpn Kokai Tokkyo Koho*. 1998;10158213.

4. Park S-Y, Moon S-C, Venkatasubramanian N, Dang TD, Lee J-W, Farmer BL. *J. Poly. Sci. (Polym. Phys.)*. 2006;44:1948.

5. Wolfe JF. *Encyclopedia of Polymer Science and Technology*. New York: Wiley Interscience; 1985;11:601.

6. Choe EW, Kim SN. *Macromolecules*. 1981;14:920.

7. Allen SR, Filippov AG, Farris RJ, et al. *Macromolecules*. 1981;14:1135.

8. For a description of properties of PBO, please see Welsh, WJ. In: James E Mark. ed. *Polymer Data Handbook*. 1999:291.

9. Tonelli AE. *Polymer*. 2002;43:637.

10. Jenekhe SA, Johnson PO, Agrawal AK. *Macromolecules*. 1989;22:3216.

11. Connolly JW, Dudis DS, Kumar S, Gelbaum LT, Venkatasubramanian N. *Chem. Mater.* 1996;8(1):54.

Poly(benzobisthiazole)

Vladyslav Kholodovych and William J. Welsh

Acronyms, Alternative Names PBT, PBZT, poly(*p*-phenylene-2,6-benzobisthiazolediyl), poly[(benzo[1,2-d:4,5-d′]bisthiazole-2,6-diyl)-1,4-phenylene]

Class Rigid-rod polymers

Structure

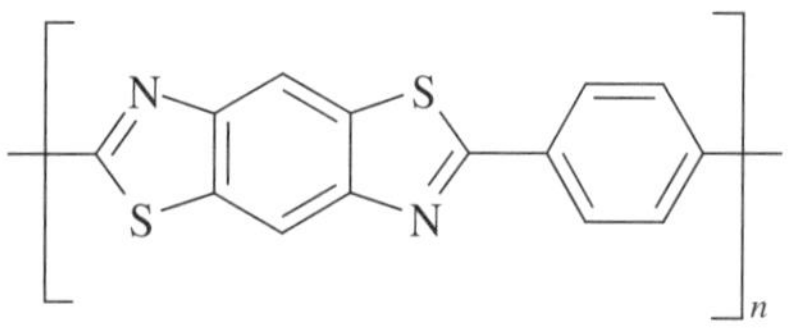

Major Applications High-performance films, fibers, and coatings.

Properties of Special Interest High-temperature resistance, unusual resistance to organic solvents, excellent mechanical properties, interesting electrical and nonlinear optical properties.

Preparative Techniques Polycondensation of terephthalic acid with 2,5-diamino-1,4-benzenedithiol dihydrochloride in poly(phosphoric acid). Processing is primarily limited to variations of wet extrusion. [1]

Solubility Protonic sulfonic acids RSO_3H, where R = —OH, —CH_3, —Cl, —CF_3, —C_6H_5, etc., polyphosphoric acid (PPA). [1] Aprotic organic solvents (e.g. nitroalkanes) containing metal halide Lewis acids (e.g. $AlCl_3$, $GaCl_3$, $FeCl_3$) up to 7.5% polymer. [2]

Unit cell dimensions

Lattice	Monomers Per Unit Cell	Cell Dimensions (Å)			Cell Angles (degrees)			Reference
		a	*b*	*c* (chain axis)	α	β	γ	
Monoclinic	2	11.79	3.54	12.51	90	90	94	10, 31
Oblique		11.60	3.59	12.51			92	31–33
Monoclinic	1	5.83	3.54	12.35	90	90	96	6
Monoclinic	2	7.10	6.65	12.35	90	90	63	6
Monoclinic	2	11.957	3.555	12.35	90	90	100.9	7
Monoclinic	1	6.55	3.56	12.35	90	90	116.4	7
Monoclinic	2	11.79	3.539	12.514	90	90	94	10

Poly(benzobisthiazole)

Property	Units	Conditions	Value	Reference
Density	g·cm^3	As-spun fiber	1.47–1.53	3
		Heat treated fiber	1.54–1.60	3
		Model compound	1.44	4
		Film, uniaxial	1.56	5
		Film, balanced biaxial(quasi-isotropic)	1.56	5
		X-ray diffraction data	1.69	6
		X-ray diffraction data	1.713	7
		Microfibrils	1.46	8
		Fiber	1.58	9
Torsion angle between the benzobisthiazole and phenyl rings	degrees	X-ray fiber pattern	46	10,31
			23	4,31
			20.5	31–33
		Calculated	27	31, 34
		Calculated	20	31, 35
		Calculated	21	31, 36
		Calculated	29	31, 37
d-spacing of the first equatorial Bragg reflection (side-by-side packing)	Å		5.5	31, 39
d-spacing of the second equatorial Bragg reflection (face-to-face packing)	Å		3.5	31, 39
Repeat unit distance	Å		12.5	31, 39
Tensile modulus	GPa	Fiber heat-treated spun from PPA	250	38, 39
Elongation at break	%	Fiber heat-treated spun from PPA	1.5	38, 39
Intrinsic viscosity	dl/g	in MSA at 30°C	18.3	40
Intermolecular interaction energy between two chains packed face-to-face	kJ/mol	Calculated	123	41
Glass transition temperature T_g	°C	Pristine PBT	≥500	41
		GaCl$_3$-complex	26	
Young's modulus	GPa	Fibers (25 mm)	18–331	3
		As-spun fiber	110	11
		Heat-treated fiber	280	11
		Ribbon	40	12
		Fiber	186	13
		Fiber	310	14
		Filaments, as-spun	17–159	15
		Filaments, heat-treated	303–331	15
		Fiber	200–330	9
		Fiber	320	16
		Film, uniaxial	270	5
		Film, balanced biaxial (quasi-isotropic)	34	5

Property	Units	Conditions	Value	Reference
Tensile strength	GPa	Fibers (25 mm)	2.35–4.19	3
		As-spun fiber	1.1	11
		Heat-treated fiber	2.7	11
		Ribbon	0.5	12
		Fiber	1.518	13
		Filaments, as-spun	2.28–2.35	15
		Filaments, heat-treated	3.49–4.19	15
		Fiber	3.0–4.2	9
		Fiber	3.1	16
		Film, uniaxial	2.0	5
		Film, balanced biaxial (quasi-isotropic)	0.55	5
		Fiber		
		heat treated spun from PPA	2.4	38,39
Elongation at break	%	Fibers (25 mm)	1.3–7.1	3
		Filaments, as-spun	2.4–7.1	15
		Filaments, heat-treated	1.3–1.4	15
		Fiber	1.1	16
		Film, uniaxial	0.88	5
		Film, balanced biaxial (quasi-isotropic)	2.5	5
Compressive strength	GPa	–	0.3	8
		–	0.68	3
		Fiber	0.2–0.4	9
Torsional modulus	GPa	Fiber	1.2	9
Persistance length Q	nm	300°C	55–80	17
		CSA solvent	64.0 ± 0.9	18
Coefficient of thermal expansion	ppm·K^{-1}	Film, uniaxial	−10	5
		Film, balanced biaxial (quasi-isotropic)	−5	5
Degradation temperature	K	Film, uniaxial	>873	5
		Film, balanced biaxial (quasi-isotropic)	>873	5
		Fiber	~873	19
Fiber flammability-critical oxygen concentration (COC)	–	Fiber	35.7 (top) 22.6(bottom)	20
Dielectric constant ε	–	Film, uniaxial	2.8	5
		Film, balanced biaxial (quasi-isotropic)	2.8	
Dissipation factor	–	Film, uniaxial	0.005	5
		Film, balanced biaxial (quasi-isotropic)	0.005	

Poly(benzobisthiazole)

Property	Units	Conditions	Value	Reference
Dielectric strength	volt·mil^{-1}	Film, uniaxial	8,900	5
		Film, balanced biaxial (quasi-isotropic)	8,900	
Electrical conductivity	ohm^{-1}·cm^{-1}	Electrochemically doped	~20	21
		Undoped	~10^{12}	
Cathodic peak	volts	Versus SCE	−1.70	21
Anodic peak			−1.23	
Energy band gap	eV	Band edge at ~500 nm	2.48	22
Index of refraction	–	Film ($\lambda = 602$ nm)	2.16	23
Optical loss α	cm^{-1}	Film	5.2×10^3	23
Third-order nonlinear optical susceptibility $\chi^{(3)}$	esu	Nonresonant ($\lambda = 602$ nm)	4.5	23
		–	~10^{-11}	24
		1.3 μm	8.31 ± 1.66 $(\times 10^{-11})$	25
Quantum efficiency	%	Solid state	6	26
IR characteristic frequencies (intensity)	cm^{-1}	Highly oriented film	3076 (w); 3076 (w); 3027 (w); 1605 (w); 1532 (m); 1500 (sh); 1485 (vs); 1428 (m); 1410 (s); 1401 (s); 1314 (vs); 1252 (s); 1211 (w); 1113 (m); 1056 (m); 1017 (w); 960 (vs); 860 (s); 837 (s); 732 (w); 705 (m); 689 (s); 627 (w); 605 (s); 488 (m)	27
Raman characteristic frequencies (intensity)	–	–	1605 (s) 1481 (s) 1160–1300 (m)	28
Wavelength at maximum of band	nm	UV-vis absorption in MSA	440	29
Birefringence	cm^{-1}	IR region	0.88 ± 0.04	30

References

1. Wolfe JF, Arnold FE. *Macromolecules.* 1981;14:915.
2. Jenekhe SA, Johnson PO, Agrawal AK. *Macromolecules.* 1989;22:3216.
3. Northolt MG, Sikkema DJ. In Collyer AA, ed. *Liquid Crystal Polymers: From Structures to Applications.* London and New York: Elsevier Applied Science; 1992:273.
4. Wellman MW, et al. *Macromolecules.* 1981;14:935.
5. Lusignea RW. The Materials Science and Engineering of Rigid-Rod Polymers. In: Adams WW, Eby RK, McLemore DE, eds. *Mat Res. Soc. Symp. Proc.* Vol 134. Pittsburgh: Materials Research Society; 1989:265.
6. Roche EJ, Takahashi T, Thomas EL. Fibre Diffraction Methods. In: French AD, Gardner KH, eds. *ACS Symp.* Ser. 141. Washington, D.C: American Chemical Society; 1980:303.
7. Odell JA, et al. *J. Mat. Sci.* 1981;16:3309.
8. Cohen Y, Thomas EL. *Macromolecules.* 1988;21:433.
9. Kumar S. In: *Polymeric Materials Encyclopedia.* Vol 10. Boca Raton, Fla: CRC Press; 1996:7512.
10. Fratini AV, et al. The Materials Science and Engineering of Rigid-Rod Polymers. In: Adams WW, Eby RK, McLemore DE, eds. *Mat. Res. Soc. Symp. Proc.* Vol 134. Pittsburgh: Materials Research Society; 1989:431.
11. Allen SR, et al. *J. Appl. Polymer Sci.* 1981;26:291.
12. Minter JR, Shimamura K, Thomas EL. *J. Mat. Sci.* 1981;16:3303.
13. Critchley JP. *Die Angewandte Makromolekulare Chemie.* 1982;41:109–110.
14. Hwang WF, et al. *Polym. Eng. Sci.* 1983;23:784.
15. Wolfe JF. In: Mark HF, et al. eds. *Encyclopedia of Polymer Science and Engineering.* Vol 11. New York: John Wiley and Sons; 1988:572.
16. Krause SJ, et al. *Polymer.* 1988;29:1354. (see reference 14 therein).
17. Roitman DB, McAdon M. *Macromolecules.* 1993;26:4381.
18. Crosby CR, et al. *J. Chem. Phys.* 1981;75:4298.
19. Wolfe JF, Loo BH, Arnold FE. *Macromolecules.* 1981;14:915.
20. Choe EW, Kim SN. *Macromolecules.* 1981;14:920.
21. DePra PA, Gaudiello JG, Marks TJ. *Macromolecules.* 1988;21:2295.
22. Jenekhe SA, Johnson PO, Agrawal AK. *Macromolecules.* 1989;22:3216.
23. Lee CYC, et al. *Polymer.* 1991;32:1195.
24. Rao DN, et al. *Appl. Phys. Lett.* 1986;48:1187. (Note: The lower value than given in reference 23 may be due to poor film quality.)
25. Jenekhe SA, et al. *Polym. Prepr.* 1991;32:140.
26. Osaheni JA, Jenekhe SA. *Macromolecules.* 1995;28:1172.
27. Shen DY, Hsu SL. *Polymer.* 1982;23:969. (supplement).
28. Osaheni JA, et al. *Macromolecules.* 1992;25:828.
29. Shen DY, et al. *J. Polym. Sci., Polym. Phys. Ed.* 1982;20:509.
30. Chang C, Hsu SL. *J. Polym. Sci., Polym. Phys. Ed.* 1985;23:2307.
31. Park S-Y, Moon S-C, Dang TD, Venkatasubramanianc N, Lee J-W, Farmer BL. *Polymer.* 2005;46:5630.
32. Takahashi Y. *Macromolecules.* 1999;32:4010.
33. Takahashi Y. *Macromolecules.* 2001;34:2012.
34. Welsh WJ, Mark JE, Yang Y, Das GP. *Mater Res Soc Symp Proc.* 1989;134:621.; Yang Y, Welsh WJ. *Macromolecules.* 1990;23:2410.
35. Welsh WJ, Mark JE. *J. Mater Sci.* 1983;18:1119.
36. Welsh WJ, Yang Y. *Comp Polym Sci.* 1991;1:139.

37. Farmer BL, Wierschke SG, Adams WW. *Polymer.* 1990;31:1631.

38. Allen SR, Filippov AG, Farris RJ, et al. *Macromolecules.* 1981;14:1135.

39. Park S-Y, Moon S-C, Dang TD, Venkatasubramanian N, Lee J-W, Farmer BL. *Macromolecules.* 2005;38:1711.

40. Jenekhe SA, Johnson PO. *Macromolecules.* 1990;23:4419.

41. Jenekhe SA, Roberts MF. *Macromolecules.* 1993;26:4981.

Poly(benzobisthiazole), Naphthalene Derivatives

N. Venkatasubramanian, Thuy D. Dang, Soo-Young Park, Vladyslav Kholodovych and William J. Welsh

Acronyms, Alternative Names Poly(2,6-naphthalenebenzobisthiazole) (Naph-2,6-PBT), poly(2,6-naphthalene-2,6-benzobisthiazolediyl); Poly(1,5-naphthalenebenzobisthiazole) (naph-1,5-PBT), poly(1,5-naphthalene-2,6-benzobisthiazolediyl).

Class Rigid-rod polymers

Structure

where Ar =

Major Applications of Interest High performance fibers, films, and coatings

Synthesis Polycondensation of 2,6-naphthalenedicarboxylic acid or 1,5-naphthalenedicarboxylic acid with 2,5-diamino-1,4-benzenedithiol dihydrochloride in 83 % polyphosphoric acid (PPA)[1−4] using the 'P$_2$O$_5$ adjustment method'.[5]

Processing Isotropic films are mainly cast from dilute methanesulfonic acid solutions. Fabrication of the polymeric fibers is from the lyotropic liquid crystalline phase of the polymer in PPA; polymer fibers are spun from the dope by a dry-jet wet spinning technique, a variation of wet extrusion,[6,7] at 90°C with draw ratios in the range 20–40.

Properties of Special Interest High thermal and thermo-oxidative stabilities, exceptional resistance to organic solvents, good mechanical properties; lower geometrical symmetry and extended planar packing relative to PBT,[8] influence chain conformation, packing and crystallinity, as well as the mechanical properties of the Naph-PBT system in a manner analogous to the well-known PEN and PET systems.[9]

Solubility Strong, corrosive protonic acid media such as RSO$_3$H where R = —CH$_3$, —CF$_3$, or —Cl as well as PPA; potentially soluble in polar aprotic organic solvents (e.g. nitroalkanes) containing metal halide Lewis acids (e.g. AlCl$_3$, GaCl$_3$, and FeCl$_3$).[10,11]

Poly(benzobisthiazole), Naphthalene Derivatives

Polymer	Property	Units	Conditions	Value	Reference
Naph-2,6-PBT	Intrinsic viscosity	dl g^{-1}	Methanesulfonic acid, 30°C	27	(1,4)
Naph-1,5-PBT	Intrinsic viscosity	dl g^{-1}	Methanesulfonic acid, 30°C	13.2	(2,3)
Naph-2,6-PBT	Density	g cm^{-3}	Fiber (flotation method)	1.56	(4)
			Calculated	1.68	(4)
Naph-1,5-PBT	Density	g cm^{-3}	Fiber (flotation method)	1.52	(3)
			Calculated	1.68	(3)
Naph-2,6-PBT	Thermal stability	°C	Bulk polymer in helium (TGA)	690°C	(4)
	Thermo-oxidative stability	°C	Bulk polymer in air (TGA)	595°C	(4)
Naph-1,5-PBT	Thermal stability	°C	Bulk polymer in helium (TGA)	610°C	(3)
	Thermo-oxidative stability	°C	Bulk polymer in air (TGA)	550°C	(3)
Naph-2,6-PBT	Tensile modulus	GPa	Fiber, as-spun	20.3	(4)
			Fiber, heat-treated (350°C)	32.8	
	Tensile strength	GPa	Fiber, as-spun	1.42	(4)
			Fiber, heat-treated (350°C)	1.45	(4)
	Elongation at break	%	Fiber, as-spun	7.1	(4)
			Fiber, heat-treated (350°C)	4.5	(4)
Naph-1,5-PBT	Tensile modulus	GPa	Fiber, as-spun	13.4	(3)
			Fiber, heat-treated (350°C)	25.1	(3)
	Tensile strength	GPa	Fiber, as-spun	0.93	(3)
			Fiber, heat-treated (350°C)	1.1	(3)
	Elongation at break	%	Fiber, as-spun	7.7	(3)
			Fiber, heat-treated (350°C)	5.4	(3)
Naph-2,6-PBT	d-Spacing of the first equatorial Bragg reflection (side-by-side packing)	Å	Fiber diffraction	6.14	(4)
	d-Spacing of the second equatorial Bragg reflection (face-to-face packing)	Å	Fiber diffraction	3.45	(4)
	Repeat unit distance	Å	Fiber diffraction	14.6	(4)
	Torsion angle between the naphthalene and heterocyclic rings	degrees	Molecular modeling	23	(4)
	Angle of rotation of the naphthalene ring from the *ac* plane	degrees	Molecular modeling	-9 ± 3	(4)
Naph-1,5-PBT	d-Spacing of the first equatorial Bragg reflection (side-by-side packing)	Å	Fiber diffraction	6.6	(3)
	d-Spacing of the second equatorial Bragg reflection (face-to-face packing)	Å	Fiber diffraction	3.7	(3)

Poly(benzobisthiazole), Naphthalene Derivatives

Polymer	Property	Units	Conditions	Value	Reference
	Repeat unit distance	Å	Fiber diffraction	12.7	(3)
	Torsion angle between	degrees	Molecular modeling		(3)
	the naphthalene and		toward nitrogen atom	138	
	heterocyclic rings		toward sulfur atom	50	(3)
	Angle of rotation of the	degrees	Molecular modeling	4.4 ± 4.6	(3)
	naphthalene ring from the				
	ac plane				

Unit cell dimensions from (annealed) fiber diffraction

Polymer	Lattice	Monomers Per Unit Cell	Cell Dimensions (Å)			Cell Angles (°)			Reference
			a	*b*	*c*	α	β	γ	
Naph-2,6-PBT	Triclinic	1	6.78	3.46	14.61	88.0	114.7	94.8	(4)
Naph-2,6-PBT	Triclinic	1	7.67	3.47	14.61	86.7	126.6	95.5	(4)
Naph-1,5-PBT	Triclinic (metrically monoclinic)	1	7.31	3.68	12.68		115.1		(3)

Naph-2,6-PBT: The indices and the observed and calculated d-spacings based on the triclinic unit cell parameters of $a = 6.78$ Å, $b = 3.46$ Å, $c = 14.61$ Å, $\alpha = 88.0°$, $\beta = 114.7°$, and $\gamma = 94.8°$ (reference 4)

h	k	l	d_o (Å)	d_c (Å)
1	0	0	6.14	6.14
0	1	0	3.45	3.45
1	1	0	3.11	3.11
2	0	0		3.07
0	0	2	6.64	6.64
−1	0	2	5.87	5.89
			4.84[a]	
0	0	3	4.07	4.43
−1	0	4	3.64	3.62
0	0	4	3.32	3.32
−2	0	4	2.99	2.95
0	0	5	2.72	2.66
−1	1	5	2.21	2.19
1	0	5		2.14
0	0	6	2.20	2.21

[a] Cannot be indexed with these parameters.

Naph-1,5-PBT : The indices and the observed and calculated d-spacings based on the triclinic unit cell parameters of $a = 7.31$ Å, $b = 3.68$ Å, $c = 12.68$ Å, and $\beta = 115.1°$ (reference 3)

h	k	l	d_o (Å)	d_c (Å)
1	0	0	6.62	6.62
0	1	0	3.68	3.68
2	0	0	3.33	3.31
1	1	0		3.21
0	0	1	11.49	11.49
−1	0	1	7.13	7.20
1	0	1	4.90	4.91
1	1	1	2.87	2.88
−1	1	−1		
2	0	1		
−1	0	2	5.78	5.74
0	0	2		
−2	0	3	3.26	3.29
0	0	3	3.90	3.83
−1	1	3	2.84	2.84
1	1	−3		
1	0	3		
−2	0	4	2.87	2.87
0	0	4		
−1	1	4	2.31	2.30
1	1	−4		
1	0	4		
0	0	5	2.30	2.30
1	−1	−5	2.08	2.10
−1	1	5		
−3	0	5		

References

1. Dang TD, Venkatasubramanian N, Talicsa A, Park S-Y, Arnold FE. *Polymer Preprints (American Chemical Society)*. 2002;43(1):660.
2. Dang TD, Venkatasubramanian N, Lee J-W, Park S-Y, Arnold FE, Farmer BL. *US Patent*. 2006;7041779.
3. Park S-Y, Moon S-C, Dang TD, Venkatasubramanian N, Lee J-W, Farmer BL. *Macromolecules*. 2005;38:1711.
4. Park S-Y, Moon S-C, Dang TD, Venkatasubramanian N, Lee J-W, Farmer BL. *Polymer*. 2005;46(15):5630.
5. Wolfe JF. *Encyclopedia of Polymer Science and Technology*. New York: Wiley Interscience; 1985;11:601.
6. Choe EW, Kim SN. *Macromolecules*. 1981;14:920.
7. Allen SR, Filippov AG, Farris RJ, et al. *Macromolecules*. 1981;14:1135.

8. For a description of properties of PBT, please see Welsh WJ. In: James E. Mark, ed. *Polymer Data Handbook*. 1999:295.
9. Tonelli AE. *Polymer*. 2002;43:637.
10. Jenekhe SA, Johnson PO, Agrawal AK. *Macromolecules*. 1989;22:3216.
11. Connolly JW, Dudis DS, Kumar S, Gelbaum LT, Venkatasubramanian N. *Chem. Mater.* 1996;8(1):54.

Poly(γ-benzyl-L-glutamate)

Douglas G. Gold, Wilmer G. Miller, Taner Z. Sen
and Andrzej Kloczkowski

Acronym PBLG

Class Polypeptides and proteins

Structure

Major Applications Modeling of conformational changes of biopolymers and modeling of α-helical polypeptides. Used in chromatography as a stationary phase for the resolution of racemic materials. Microencapsulation of pharmaceutically active hydrophobic liquids. Improves shatter resistance of plastics when blended with poly(vinyl chloride), poly(vinyl acetate), or their copolymers.

Photo-induced energy transfer switches containing chromophores at both ends of oligopeptide chains.[1]

Properties of Special Interest Exists in a highly ordered, well-defined, α-helical conformation held intact by intramolecular hydrogen bonds. The α-helical structure renders the polymer as a relatively stiff rigid rod and is retained when the polymer is dissolved in many solvents. In these helicogenic solvents, PBLG exists as a single isotropic phase at low concentration. At higher concentrations a liquid-crystalline cholesteric phase is present.

α-Helical peptide microcapsulation formed by emulsion-templated self-assembly of amphiphilic PBLG.[2]

Some PBLG solutions form a self-supporting gel. Blending between rigid rodlike PBLG polymer and benzyl methacrylate solvent allows for morphology control at the nanoscale.[3]

Common Solvents and Nonsolvents α-helical conformation when dissolved in solvents such as dimethylformamide, benzene, toluene, methylene chloride, and chloroform. Random coil conformation in trifluoroacetic acid (TFA) and dichloroacetic acid (DCA), and in mixed solvents containing TFA and DCA. Nonsolvents include water and methanol.

Synthesis The first step involves the synthesis of the amino acid γ-benzyl-L-glutamate by a standard Fischer esterification reaction of L-glutamic acid with benzyl alcohol in the presence of strong acid. The amino acid is subsequently converted to the N-carboxyanhydride (NCA) monomer by reaction with phosgene gas,[4] or by reaction with the less hazardous compound triphosgene.[5] The NCA is polymerized by initiation with a variety compounds such as primary and secondary amines, and alkoxides.[4] Typical comonomers include other amino acid NCAs.

A novel highly efficient method (yield 84%) of using di-*tert*-butyltricarbonate instead of highly toxic phosgene or triphosgene has been recently proposed.[6] Pyridine-initiated polymerization yields cyclic polypeptides.[7]

Fractionation Fractionation has been accomplished using the following solvent/nonsolvent combinations: dichloroethane/petroleum ether, dioxane/ethanol, methylene chloride/methanol.[8]

Property	Units	Conditions	Value	Reference
Molecular weight (of repeat unit)	g mol^{-1}	–	219	–
Typical molecular weight range	g mol^{-1}	–	$10^4 - 3 \times 10^5$	–
Typical polydispersity index (M_w/M_n)	–	–	1.2	–
IR (characteristic absorption frequencies)	cm^{-1}	–	3,291; 1,733; 1,652; 1,550; 1,167	(4)
UV (characteristic absorption frequencies)	cm^{-1}	–	61,000; 53,800; 51,000; 47,800; 45,700	(4)
NMR	–	–	–	(4, 9)
Thermal expansion coefficients	K^{-1}	$T < T_g \approx 15°C$, buoyant-weight technique	2.3×10^{-4}	(10)
		$T > T_g \approx 15°C$, buoyant-weight technique	4.5×10^{-4}	
Second virial coefficient	mol cm^3 g^{-2}	Dry DMF, 5–75°C, $M_w \sim 10^5$	4×10^{-4}	(11)

Property	Units	Conditions	K	a	Reference
Mark-Houwink parameters: K and a	$K = $ ml g^{-1} $a = $ None	Dimethylformamide, 25°C, helical, 70,000–340,000	2.9×10^{-7}	1.7	(8)
		Dimethylformamide, 25°C, 60,000–570,000	5.6×10^{-6}	1.45	
		Dichloroacetic acid, 25°C, random coil, 20,000–340,000	2.78×10^{-3}	0.87	
		Dichloroacetic acid, 25°C, 60,000–570,000	8.8×10^{-3}	0.77	
Characteristic ratio	–	Dichloroacetic acid, 25°C, random coil	10.3		(8)
		m-Cresol, helical	400–622		(12)
Persistence length	Å	Helicogenic solvents	$1,100 \pm 500$		(12–14)
Theta temperature	K	Dichloroethane/diethylene glycol (80:20)	298		(8)
Density (crystalline)	g cm^{-3}	–	1.26–1.30		(8)
T_g-like transition temperature	K	Onset of side-chain rotation	288–293		(10, 15)

Poly(γ-benzyl-L-glutamate)

Property	Units	Conditions	Value	Reference
Shear modulus	MPa	25°C	1,000	(10)
		−40°C	7,000	
Storage modulus	MPa	0°C, 0.1 Hz	1,000	(15)
		25°C, 0.1 Hz	100	
Loss modulus	MPa	0°C, 0.1 Hz	100	(15)
		25°C, 0.1 Hz	30	
WLF parameters: C_1 and C_2	°C (C_2)	–	$C_1 = -8.86$	(15)
			$C_2 = 101.6$	
Refractive index increment dn/dc	ml g^{-1}	Dichloroacetic acid, 25°C	0.085	(8)
		Dioxane, 25°C	0.114	(4, 8)
		Dimethylformamide, 25°C, λ variable	0.118–0.127	(11)
Optical activity $[\alpha]_D$	–	Chloroform	$[\alpha]_{546} + 14$	(8)
		dichloroacetic acid	$[\alpha]_{546} - 15$	
Electronic band gap	eV	–	2.07	(4)
Conductance	ohm^{-1} cm^{-1}	–	2×10^{-17}	(4)
Piezoelectric coefficient	pCN^{-1}	–	−0.4	(4)
Magnetic susceptibility	emu g^{-1}	–	-0.52×10^{-6}	(4)
Surface tension	mN m^{-1}	20°C	39.2	(8)
Decomposition temperature	K	–	473	(4)
Helix pitch	Å	–	5.42	(4)
Axial translation per residue	Å	–	1.505	(4)
Residues per turn	–	–	3.6	(4)
Peak-to-peak intensity for vibrational circular dichroism in the amid I band	–	Mol. Weight		
		MW = 22000	2.10×10^{-4}	
		MW = 37000	2.26×10^{-4}	(16)
		MW = 260000	2.76×10^{-4}	
Film thickness	nm	chemisorbed	20	(17)
		physisorbed	75	
Rms roughness	nm	chemisorbed	1	(17)
		physisorbed	10	
Contact angle	degree	chemisorbed	advancing/receding $87 \pm 2/78 \pm 3$	(17)
		physisorbed	$84 \pm 1/75 \pm 4$	

References

1. Kishimoto A, Mutai T, Araki K. *Chem. Commun.* 2003;742.
2. Morikawa M, Yoshihara M, Endo T, Kimizuka N. *Chem. Eng. J.* 2005;11:1574.
3. van Hooy-Corstjens CSJ, Rastogi S. *Biomacromolecules.* 2006;7:1542.

4. Block H. *Poly(γ-benzyl-glutamate) and Other Glutamic Acid Containing Polymers*. New York: Gordon and Breach Science Publishers; 1983.

5. Daly WH, Poche D. *Tetrahedron Lett.* 1988;29:5859.

6. Nagai A, Sato D, Ishikawa J, Ochiai B, Kudo H, Endo T. *Macromolecules*. 2004;37:2332.

7. Kricheldorf HR, von Lossow C, Schwarz G. *J. Pol. Sci: Part A: Pol. Chem.* 2006;44:4680.

8. Brandrup J, Immergut EH, eds. *Polymer Handbook*. 3rd ed. New York: John Wiley and Sons; 1989.

9. Bovey FA. *Polymer Conformation and Configuration*. New York: Academic Press; 1969.

10. McKinnon AJ, Tobolsky AV. *J. phys. chem.* 1968;72(4):1157.

11. DeLong LM, Russo PS. *Macromolecules*. 1991;24:6139.

12. Aharoni SM. *Macromolecules*. 1983;16:1722.

13. Schmidt M. *Macromolecules*. 1984;17:553.

14. Iwata K. *Biopolymers*. 1980;19:125.

15. Yamashita Y, et al. *Polymer Journal*. 1976;8(1):114.

16. Buffeteau T, Lagugnè-Labarthet F, Sourisseau C. *Applied Spectroscopy*. 2005;59:732.

17. Lee NH, Christensen LM, Frank CW. *Langmuir*. 2003;19:3225.

Poly(1,3-bis-*p*-carboxyphenoxypropane anhydride)

Abraham J. Domb and Robert Langer

Acronyms, Trade Names BIODEL-CPP, Poly(CPP), Poly(CPP-SA)

Class Polyanhydrides

Structure $[-CO-C_6H_4-O-CH_2-CH_2-CH_2-O-C_6H_4-COO-]$

Major Applications Biodegradable polymer for controlled drug delivery in a form of implant or injectable microspheres (e.g., Gliadel™-BCNU-loaded wafer for the treatment of brain tumors).

Properties of Special Interest Anhydride copolymers of 1,3-bis-*p*-carboxyphenoxypropane (CPP) with aliphatic diacids such as sebacic acid (SA) degrade in a physiological medium to CPP and SA. Matrices of the copolymers loaded with dissolved or dispersed drugs degrade in vitro and in vivo to constantly release the drugs for periods from 1–10 weeks.

Property	Units	Conditions	Value	Reference
Molecular weight		P(CPP-SA)		
	10^4 g mol^{-1}	GPC-polystyrene standards	M_w = 3–20, M_n = 0.5–3	–
	dl g^{-1}	Viscosity 25°C, dichloromethane	η_{sp} = 0.2–0.9	–
IR (characteristic absorption frequencies)	cm^{-1}	Film on NaCl pellet PSA P(CPP-SA) P(CPP)	1,750, 1,810 1,740, 1,770, 1,810 1,712, 1,773	(1)
Raman	cm^{-1}	Film on NaCl pellet PSA P(CPP-SA) P(CPP)	 1,739, 1,803 1,723, 1,765, 1,804 1,712, 1,764	(1)
UV (characteristic absorption wavelength)	nm	P(CPP-SA), dichloromethane CPP monomer, 1 N NaOH solution	265 265	–
Optical rotation	–	Dichloromethane	No optical rotation	–

Poly(1,3-bis-p-carboxyphenoxypropane anhydride)

Property	Units	Conditions	Value				Reference
Solubility	mg ml^{-1}		P(CPP-SA), 0–60 mol% CPP	P(CPP-SA), 70–100 mol% CPP			(2)
		Chloroform	>300	<1			
		Dichloromethane	>300	<1			
		Tetrahydrofuran	20	<1			
		Ketones	1	<1			
		Ethyl acetate	<1	<1			
		Alkanes and arenes	<1	<1			
		Ethers	<1	<1			
		Water	<1	<1			
Mark-Houwink parameters: K and a	ml g^{-1} None	CHCl$_3$, 23°C	$K = 3.88$ $a = 0.658$				(3)
Thermal properties		P(CPP-A), DSC, 10°C min^{-1}	0:100	22:78	46:54	100:0	(3)
	K	T_m	359.0	339.0	458.0	513.0	
	K	T_g	333.1	320.0	274.8	369.0	
	kJkg^{-1}	ΔH	150.7	64.0	13.0	110.9	
Crystallinity	%	P(CPP-SA), powder, X-ray diffraction	0:100	22:78	46:54	100:0	(3)
		X_c	–	30.0	6.1	–	
		W_c	66.0	35.0	14.2	61.4	
Comonomer sequence distribution		P(CPP-SA), ^{1}H-NMR, CDCl$_3$	8:92	22:78	59:41	49:51	(3)
		Probability for SA-SA	0.86	0.61	0.36	0.24	
		Probability for SA-CPP	0.14	0.34	0.47	0.49	
		Average block length L(SA)	12.3	4.6	2.5	2.0	
		Degree of randomness	0.3	0.7	0.9	1.0	
Stability in chloroform solution (decrease in M_w) (anhydride interchange depolymerization)			P(CPP-SA)				(4)
			0:100	20:80	40:50		
Depolymerization rate constant	t^{-1}	37°C	0.1325	0.1535	0.0743		
Activation energy	kcal mol^{-1} K^{-1}		8.08	8.27	7.27		
Erosion rate	mg h^{-1}	P(CPP-SA), 14 × 1.2 mm disc, 0.1 M phosphate buffer, pH 7.4, 37°C	0:100	22:78	49:51	100:0	(5)
		SA	2.3	1.8	0.4	–	
		CPP	–	0.5	0.3	<0.01	
Erosion front	µm day^{-1}	0.1 M phosphate buffer, pH 7.4, 37°C					(6)
		P(CPP-SA), 20:80	106 ± 5				
		P(CPP-SA), 50:50	118 ± 18				

Poly(1,3-bis-*p*-carboxyphenoxypropane anhydride)

Property	Units	Conditions	Value		Reference
Elimination in vivo	%		CPP	SA	(7)
		7 days in rat brain	2	95	
		21 days in rat brain	64	100	
Drug release in vitro	% day^{-1}	P(CPP-SA), 20:80			(7)
		3.8% BCNU in disc	30		(6)
		5% indomethacin in disc	9		
Drug release in vivo	% day^{-1}	3.8% BCNU disc implanted in rat brain	16		(7)
Biocompatibility		Compatible with human brain			(8)
		Compatible with rabbit brain, cornea, muscle, subcutane			
Supplier		Guilford Pharmaceuticals, Inc., Baltimore, Maryland, USA			

References

1. Tudor AM, et al. *Spectrochimica Acta.* 1991;9/10:1335–1343.
2. Domb AJ, Maniar M. *J. Polym. Sci.* 1993;31:1275–1285.
3. Ron E, et al. *Macromolecules.* 1991;24:2278–2282.
4. Domb AJ, Langer R. *Macromolecules.* 1989;22:2117–2122.
5. Tamada JA, Langer R. *Proc. Natl. Acad. Sci. USA.* 1993;90:552.
6. Gopferich A, Karydas D, Langer R. *Eur. J. Pharm. Biopharm.* 1995;41:81–87.
7. Domb AJ, et al. *Biomaterials.* 1994;15:681–688.
8. Domb AJ, Amselem S, Langer R, Maniar M. In: Shalaby S, ed. *Designed to Degrade Biomedical Polymers*, Carl Hauser Verlag; 1994:69–96.

Poly(bis maleimide)

Loon-Seng Tan

Acronyms, Trade Names BMI, Compimide®, CYCOM®, Matrimid®, Xponent[TM], Kerimid, Desbimid.

Class Thermoset resins, addition polyimides, composite matrix resins

Structure

X = aromatic, heterocyclic, or
aliphatic bridging groups

Major Applications Printed circuit boards, laminating powder, carbon-fiber composites for aero-engines and military aircraft parts such as flap inboard cover, forward nozzle, gun pack, ammunition pack, blade choke, deep choke, speed brake, and ventral fin[1].

Properties of Special Interest BMI resins are generally brittle. They can be toughened with additives such as aromatic diamines (chain extension via Michael Addition reaction), divinylbenzene, or bis(allylphenyl) compounds (chain extension via Diels-Alder reaction and ene reaction, respectively), benzocyclobutene derivatives (chain extension via Diels-Alder reaction),[2] low molecular weight rubber,[3] and thermoplastics (linear[4] and hyperbranched[5] polymers). Certain bismaleimides are liquid-crystalline[6,7].

Synthesis Bismaleimides are generally prepared from the two-step reaction of maleic anhydride and diamines in the presence of acetic anhydride and catalytic amounts of nickel acetate and triethylamine[8,9]. Sodium acetate may be substituted for nickel acetate[10]. Poly(bismaleimides) are highly cross-linked polymers formed from thermally cured bismaleimides. Thermal curing can be promoted by either a radical-type initiator (peroxides or azo compounds)[11] or an ionic-type initiator such as 1,4-diazabicyclo-[2.2.2.]octane (DABCO), 2-methylimidazole[12], and triphenylphosphine[13]. Radiation curing of BMI resin with e-beam[14], UV light[15], or microwave[16] has been reported.

Poly(bis maleimide)

Thermal properties of bismaleimides

Linking Group (X)	$T_m{}^a$ (°C)	$T_{poly, max}{}^b$ (°C)	$\Delta H_{poly}{}^c$ (kJ mol^{-1})	Reference
	363 (dec.)*	–	–	17
	202–203	–	56.3	18
	241	–	–	19
	174–176	–	50.8	18
	146–150	–	54.1	18
	174–175	–	61.1	18
	155–157	235	86.0 70.9	18 20
	195–196	–	–	21
	164–165	–	–	21
	210–212	–	–	20, 22

413

Poly(bis maleimide)

Linking Group (X)	$T_m{}^a$ (°C)	$T_{poly, max}{}^b$ (°C)	$\Delta H_{poly}{}^c$ (kJ mol^{-1})	Reference
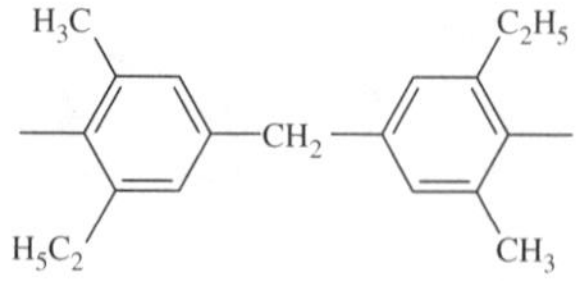 (structure: diaryl methane, ring substituents H₃C, C₂H₅, H₅C₂, CH₃)	150–154	298	85.0 / 82.7	18 / 20
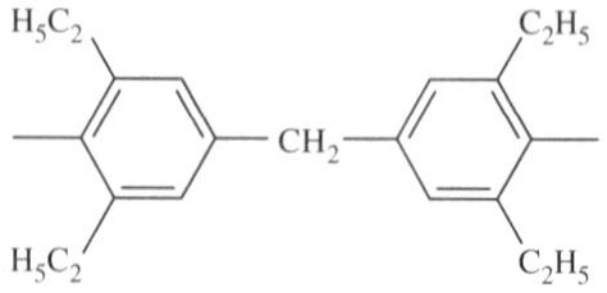 (structure: diaryl methane, ring substituents H₅C₂, C₂H₅, H₅C₂, C₂H₅)	149–151	328	132 / 96.9	18 / 20
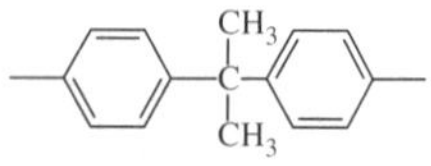 (structure: isopropylidene diphenyl, C(CH₃)₂)	235	290	83.5	20
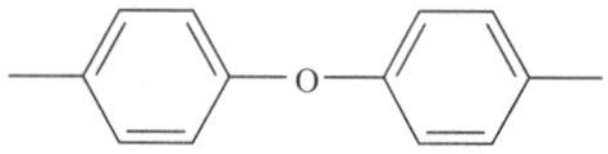 (structure: 4,4′-diphenyl ether, O)	172–178	286	57.6	18, 20
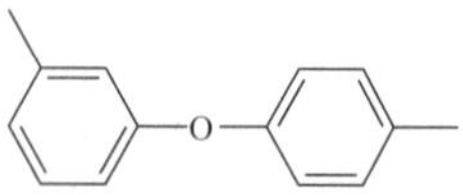 (structure: 3,4′-diphenyl ether, O)	212	236	–	19
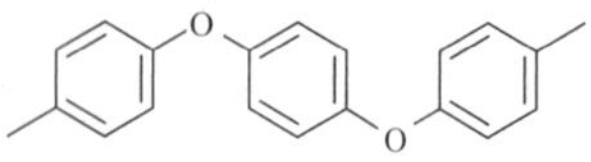 (structure: bis(phenoxy)phenylene, O...O)	239	252	84.6	20
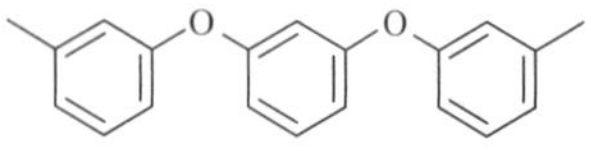 (structure: bis(phenoxy)phenylene, meta linkages, O...O)	163 / 116	254 / 277	115 / 77	20 / 16
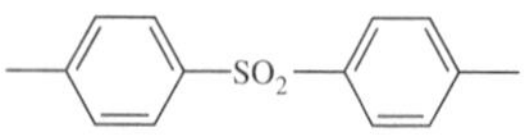 (structure: 4,4′-diphenyl sulfone, SO₂)	252–255	264	77.6 / 60.8	18 / 20
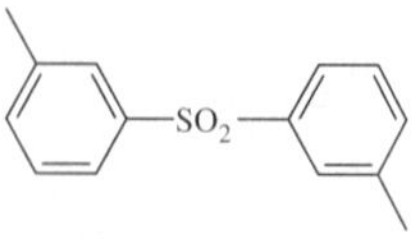 (structure: 3,3′-diphenyl sulfone, SO₂)	210–211	217	85.8 / 76.4	18 / 20
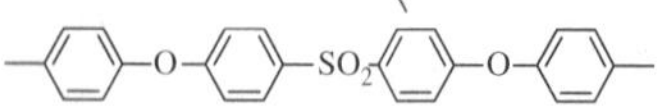 (structure: bis(phenoxyphenyl) sulfone, O...SO₂...O)	250	–	–	20

414

Poly(bis maleimide)

Linking Group (X)	$T_m{}^a$ (°C)	$T_{poly, max}{}^b$ (°C)	$\Delta H_{poly}{}^c$ (kJ mol^{-1})	Reference
(structure: $-\text{O}-\text{C}_6\text{H}_4-\text{SO}_2-\text{C}_6\text{H}_4-\text{O}-$ linkage)	80–92	295	114	20
(structure: P(=O)(CH$_3$) linked diaryl)	195	250	80	19, 23, 24
(structure: triaryl phosphine oxide)	148	214	100.3	25
(structure: $-\text{O}-\text{C}_6\text{H}_4-$P(=O)(C$_6H_5$)$-\text{C}_6\text{H}_4-\text{O}-$)	92	210	98.2	26
(structure: isopropylidene bis(phenoxy))	83	200	99.0	27
	142	270	56.6	28
	155	290	–	29
(structure: aryl ethyl phosphate, CH$_2$CH$_3$)	137	~280	–	30
(structure: aryl phenyl phosphate)	205	~280	–	30
(structure: hexafluoroisopropylidene bis(phenoxy), CF$_3$)	136	281	69.4	27
(structure: CF$_3$-substituted hexafluoroisopropylidene bis(phenoxy))	112	323	66.5	27
(structure: tetramethyl norbornane-linked bis(phenoxy), H$_3$C/CH$_3$)	~150	~250	–	56

Linking Group (X)	$T_m{}^a$	$T_{poly, max}{}^b$ (°C)	$\Delta H_{poly}{}^c$ (kJ mol^{-1})	Reference
(structure: H_3C/CH_3 substituted bisphenol ether)	144	252	–	31
(structure: oxadiazole–C_6H_4–Si(CH_3)$_2$–C_6H_4–oxadiazole)	263	281	–	32
(structure: bis-oxadiazole with H_3C–C–CH_3/CH_3 tert-butyl)	288	290	–	32
(structure: diphenoxy benzonitrile, CN)	205	–	109	34
(structure: 2,6-diphenoxy pyridine, N)	137	258	186	33
(structure: ether–C_6H_4–CO–C_6H_4–ether)	239	250	89–100	20
(structure: meta ether–C_6H_4–CO–C_6H_4–ether)	85–91	304	123	20
(structure: bis-benzoyl, two C=O with central ring)	226	285	– 74.6	20 34
(structure: bis-benzoyl para)	293	–	64.1	34
(structure: bis-benzoyl meta-terminal)	209	–	104	34
(structure: bis-benzoyl with central meta ring)	60–65 185	314	87.2	20 34
(structure: C_6H_4–O–CH_2CH_2–O–C_6H_4)	230	245	–	35
(structure: C_6H_4–O–$(CH_2CH_2)_3$–O–C_6H_4)	176	274	–	35

Poly(bis maleimide)

Linking Group (X)	$T_m{}^a$	$T_{poly, max}{}^b$ (°C)	$\Delta H_{poly}{}^c$ (kJ mol^{-1})	Reference
	160	217	–	36
	150	265	–	36
	125	295	–	36
	180	235	–	36
	281–283	~283	–	37
	90–100	203	39.4	9, 20
	315	320	–	38
	150	226 (onset)	49.1	39
	181	274	–	40
	324–327	317	–	41
	237–240	286	–	41
	232–236	300	–	41

Poly(bis maleimide)

Linking Group (X)	$T_m{}^a$	$T_{poly, max}{}^b$ (°C)	$\Delta H_{poly}{}^c$ (kJ mol^{-1})	Reference
	109–113	229	–	41
MW(PMDS) = 900	Viscous liquid	268.6	35.73	42
MW(PMDS) = 1680	Viscous liquid	294.2	27.16	42
MW(PMDS) = 3000	Viscous liquid	295.3	24.40	42
MW(PMDS) = 4600	Viscous liquid	318.3	30.13	42
	181–182	300	–	19, 43
	255–257	–	–	42
	185–186	–	–	42
	70–130	–	–	19
$-(CH_2)_2-$	191–192	–	–	44
$-(CH_2)_6-$	140–141.5	–	–	44

Poly(bis maleimide)

Linking Group (X)	$T_m{}^a$	$T_{poly, max}{}^b$ (°C)	$\Delta H_{poly}{}^c$ (kJ mol^{-1})	Reference
—(CH$_2$)$_8$—	123	–	–	45
	120–122			44
—(CH$_2$)$_{10}$—	113.5–115	–	–	44
—(CH$_2$)$_{12}$—	110–112	–	–	44

^{a} T_m = normal melting temperature; ^{b} $T_{poly, max}$ = maximum of polymerization exotherm; ^{c} ΔH_{poly} = enthalpy of polymerization.
* Decomposition temperature.

Typical physical properties of bis(4-maleimidophenyl)methane*

Property	Units	Conditions	Value	Reference
Physical form	–	–	Fine powder	19
Color	–	–	Yellow	19
M.P.	°C	DSC	148–158	19
Exotherm peak temperature	°C	DSC	260	19
Polymerization energy	J g^{-1}	DSC, heating rate at 20°C min^{-1}	>190	19

* CAS Registry: 13676–54–5.

Typical properties of cured bis(4-maleimidophenyl)methane

Property	Units	Conditions	Value	Reference
Glass transition temperature	°C	DSC, heating rate at 20°C min^{-1}	230–290	19
Tensile strength	MPa	23°C	41–83	19
Tensile modulus	GPa	23°C	4–5	19
Flexural strength	MPa	23°C	76–145	19
Flexural modulus	GPa	23°C	3.4–4.8	19
Flexural strain to failure	%	23°C	1.3–2.3	19
Fracture energy (G_{IC})	J m^{-2}	23°C	24–33	19

Compimide 353* property data

Property	Units	Conditions	Value	Reference
Softening point	°C	50–70		46
Gel time	minutes	at 170°C	>35	18, 46
Viscosity	mPa.s	at 110°C	400–1400	18, 46
Flash point	°C	–	200	46
Polymerization exotherm, onset temperature	°C	DSC, heating rate at 20°C min^{-1}	193 ± 10	18
Exotherm peak temperature	°C	DSC, heating rate at 20°C min^{-1}	274	18
Polymerization energy	J g^{-1}	DSC, heating rate at 20°C min^{-1}	220 ± 40	18

* A brown mixture (resolidified melt) comprising 50 % of 1,4-bis(maleimido)diphenylmethane, 40% of 2,5- bis(maleimido)toluene, and 10% 1,6- bis(maleimido)-2,2-dimethyl-4-methyl-hexane.

Poly(bis maleimide)

Properties of compimide 353* neat resin castings

Property	Units	Conditions	Value	Reference
T_g	°C		>300	18
Flexural strength	MPa	At 25°C	60	18
	MPa	At 250°C	50	18
Flexural modulus	GPa	At 24°C	5.5	18
	GPa	At 250°C	3.4–3.5	18
Fracture energy, G_{IC}	J m^{-2}	At 25°C,	~25	18

* A eutectic mixture comprising 50% of 1,4-bis(maleimido)diphenylmethane, 40% of 2,5-bis(maleimido)toluene, and 10% 1,6-bis(maleimido)-2,2-dimethyl-4-methyl-hexane.

Range of properties of bismaleimide resins

Property	Units	Conditions	Value	Reference
Tensile strength	MPa	24°C	332–617	47
		200°C	275–497	
Tensile strain to failure	%	24°C	1.2–3.6	47
		150°C	2.6	
Flexural strength	MPa	24°C	121.3–166.8	47
Fracture toughness, G_{IC2}	J m^{-1}	At 24°C,	30–389	47
Glass transition temperature	°C		205–320	47
Density	g cm^{-3}	24°C	1.22–1.30	47
Moisture absorption	wt.%		1.0–4.8	47

Fracture toughness (G_{IC}) of some BMI resins (cured)

Material	Units	Conditions	Value	Reference
Kerimid 601	J m^{-2}	23°C	34	48
Kerimid 70003	J m^{-2}	23°C	82	48
Kerimid or Compimide 353	J m^{-2}	23°C	25	48
Compimide 795, 766, 800, 183	J m^{-2}	23°C	40–180	48
Modified Compimide 353	J m^{-2}	23°C	389	48
Desbimid	J m^{-2}	23°C	470	49
Ciba-Geigy Matrimid 5292 (XU292)	J m^{-2}	23°C	210, 259	48

Poly(bis maleimide)

Fracture toughness (K_{IC}) of some BMI resins (cured)

Material	Units	Conditions	Value	Reference
Matrimid 5292 A/B	MPa.m$^{1/2}$	ASTM E399-86 Compact tension	0.42 ± 0.1	50
Matrimid 5292 A/B	MPa.m$^{1/2}$	ASTM E-399	0.67 ± 0.01	51, 4
			0.64 ± 0.04	
Matrimid 5292 A/B	MPa.m$^{1/2}$	ASTM E399	0.5 ± 0.2	52
Kerimid 736	MPa.m$^{1/2}$	ASTM E399 Compact tension	~ 0.55	53, 54
BPA-BMI	MPa.m$^{1/2}$	ASTM D5045-99	~ 0.48	55

BPA-BMI =

Range of mechanical properties of graphite fiber/BMI composites (unidirectional)

Property	Units	Conditions	Value	Reference
Density	g cm^{-3}	60 vol.% fiber and 40 vol.% BMI	1.5–1.6	47
Flexural strength	MPa	24°C, dry	1916–2047	47
		24°C, wet	1930–2041	
Flexural strength	MPa	177°C, dry	1930	47
		177°C, wet	1378	
		232°C, dry	1234–1378	
Flexural modulus	GPa	24°C, dry	124.1–144.1	47
		24°C, wet	142.0	
Glass transition temperature	°C	DSC	205–320	47

Dielectric constant values of some cured BMI

X	Conditions	Value	Reference
	1 MHz, r.t.	4.03	31
	1 GHz, r.t.	3.89	
	1 MHz, r.t.	3.2	27
	1 MHz, r.t.	3.0	27
	1 MHz, r.t.	2.8	27

X	Conditions	Value	Reference
	1 MHz, r.t.	3.01	31, 56
	1 GHz, r.t.	2.98	
	1 MHz, r.t.	3.03	31
	1 GHz, r.t	2.99	

Thermal properties of liquid crystalline, bismaleimide-based ester monomers

Molecular Structure	Conditions	Transition Temperatures			Reference
		Crystal to LC	LC to Isotropic	Thermosetting/Solidification	
R = H; $m = 0$, $n = 0$	(a)	282 (k → n)	Not observed	293	6
R = CH$_3$; $m = 0$, $n = 0$	(a)	245 (k → n)	Not observed	280	6
R = Cl; $m = 0$, $n = 0$	(a)	215 (k → n)	Not observed	270	6
R = H; $m = 1$, $n = 6$	(b)	154–169 (k → d. s.)	Not observed	299	7
R = H; $m = 1$, $n = 8$	(b)	218–225 (k → d. s.)	Not observed	299	7

Note: LC = liquid crystalline phase; k = crystal; n = nematic; d.s. = disordered smectic.

Conditions: Transition temperatures were determined by (a) hot-stage, polarized light microscopy or with a heating rate of approximately (a) 20°C min^{-1} or (b) 5°C min^{-1}.

Commercial BMI products

Product Names	Product Descriptions	Suppliers
Compimide MDAB	A BMI building block based on 4,4′-bismaleimidodiphenylmethane	Degussa AG Paul-Baumann-Str. 1 Building
Compimide TDAB.	A BMI building block based on 2,4-bismaleimidotoluene	1043 Marl, D-45764 Germany
Compimide 353, 353A, 796	A basic hot-melt type, eutectic mixture of BMI based on methyldianiline and aliphatic diamines.	Degussa Corporation P O Box 677 379 Interpace Parkway, Building C Parsippany,
Compimide 353RTM.	Heat-curable BMI resin supplied as high viscosity or resolidified melts; suitable for resin-transfer molding	New Jersey, 07054 USA

Poly(bis maleimide)

Product Names	Product Descriptions	Suppliers
Compimide 1206R55, 1251RH60	A 53–60% by weight solution (in DMF or NMP) of BMI resin for prepregging, recommended for the manufacture of printed wiring board (PWB)	Degussa Japan Co., Ltd. 2-3-1, Nishi-Shinjuku Shinjuku-ku P O Box 7016 Tokyo, 163-0938 Japan
Compimide 65FWR	A filament windable BMI resin	
Compimide P500	Heat-curable BMIe resin supplied as a powder, generating no volatiles during cure; used for the production of high performance fiber-reinforced composites	Web: http://www.degussa-bk.com Web: http://www.degussa-ec.com Web: http://www.degussa-ec.com/ ec2/standard_solutions/product_story.htm
Cycom 5250-2, 5250-3 and 5250-4 (formerly, Narmco5250-2, 5250-3 and 5250-4)	Proprietary BMI prepreg resins; Cycom-5250-2 has the highest T_g and Cycom-5250-4 is the toughest of the series	Cytec Industries Inc. 5 Garret Mountain Plaza West Paterson, New Jersey 07424 USA Web: http://www.cytec.com/index.htm
Matrimid 5292 A,B	A two-component resin system consists of 4,4′-bismaleimidodiphenylmethane (M5292A) and diallylbisphenol-A (M5292B)	Huntsman Advanced Materials Everslaan 45, B-3078 Everberg Belgium. Web: http://www.huntsman.com/ advanced_materials/
HexPly F65	Toughened BMI system	Hexcel Corporation,
HexPly F650	A high T_g BMI resin system, for prepreg uses	11711 Dublin Blvd. Dublin, California 94586-2832
HexPly F 652	A tough, controlled-flow BMI resin formulation for prepreg uses	USA Web: http://www.hexcel.com/
HexTOOL™/M61	Multi-axial high temperature prepreg tooling compound	
HexPly F 655	A controlled-flow, tough BMI system (for intermediate modulus fibers) for prepreg uses	
RTM 651	One-component modified bismaleimide resin transfer molding (RTM) compound	
Hysol EA-9673	BMI-based film adhesives	Henkel Corporation Aerospace Group
Hysol EA 9833.1	Modified BMI foaming core splice adhesive	2850 Willow Pass Road P.O. Box 312, Bay Point, California 94565 USA Web: http://www.aerospace.henkel.com/

Poly(bis maleimide)

Product Names	Product Descriptions	Suppliers
BT resins	Blends of bismaleimide (B) and triazine (T) resins, primarily for printed circuit board applications	Mitsubishi International Corporation (MIC) 655 Third Avenue New York, NY 10017 USA Web: http://www.micchem.com/products/BtResin.htm
PX-300, PX-307	Filler-free, solvent-free, hot-curing, 1-part bismaleimide adhesive film	Polymerics GmbH Landsberger Allee 378 D-12681 Berlin Germany Web: http://www.polymerics.de
RS-8, RS-8HT, RS-8M (Xponent™)	**RS-8**, a BMI resin with excellent elevated temperature properties and processability; evaluated and qualified in the areas of satellite and airframe/missile structures **RS-8HT**, a high service temperature formulation; **RS-8M**, a high service temperature, controlled flow formulation	YLA Inc., 2970 C Bay Vista Court, Benicia, California, 94510 USA Web: http://www.ylainc.com
SF-4: 400°F Cure, BMI syntactic film	A BMI syntactic film with excellent elevated temperature properties; available in unsupported or supported continuous film form in thicknesses from 0.010" to 0.300"; evaluated and qualified in aerospace structures	
BMI-1, BMI-2	Developmental, proprietary formulations	Fibraplex Corp 120 Industrial Park Lane Celina, Tennesse 38551 USA Web: http://www.fibraplex.com/

References

1. Wilson D, Polyimides as Resin Matrices for Advanced Composites. In: Wilson D, Stenzenberger HD, Hergenrother PM, eds. *Polyimides.* New York: Chapman & Hall; Chapter 7. 1990:190–198.
2. (a) Tan L-S, Arnold FE, Soloski EJ. *J. Polym. Sci., Part A: Polym. Chem.* 1988;26:3103. (b) Denny LR, Goldfarb IJ, Farr MP. *ACS Symp. Ser.* 1988;367:366. (c) Bruza KJ, Bell KA, Bishop MT, Woo EP. *Polym. Prepr (Am. Chem. Soc., Div. Polym. Chem).* 1994;35:373. (d) Bishop MT, Bruza KJ, Laman SA, Lee WM, Woo EP. *Polym. Prepr. (Am. Chem. Soc., Div. Polym. Chem).* 1992;33:362.
3. (a) Abbate M, Martuscelli E, Musto P, Ragosta G. *Angew. Makromol. Chem.* 1997;246:23. (b) Pritchard G, Swan M, Rose RG. *Polym. Int.* 1995;36:1. (c)Takeda S, Kakiuchi H. *J. Appl. Polym. Sci.* 1988;35:1351. (d)Shaw SJ, Kinloch AJ. *Int. J. Adhes. Adhes.* 1985;5:123.

4. Iijima T, Hayashi N, Oyama T, Tomoi M. *Polym. Int.* 2004;53:1417. and references therein.

5. (a) Gopala A, Wu H, Xu J, Heiden P. *J. Appl. Polym. Sci* 1999;71:1809. (b)Qin H, Mather PT, Baek J-B, Tan L-S. *Polymer.* 2006;47:2813.

6. Hoyt AE, Benicewicz BC. *J. Polym. Sci.: Part A, Polym. Chem.* 1990;28:3417.

7. Kallal-Bartolomeo K, Milano J-C, Vernet J-L, Gallot B. *Macromol. Chem. Phys.* 2000;201:2276.

8. Searle NE. *U. S. Patent.* 2444536.; *Chem Abst.* 1948;42:1340.

9. (a) Lee BH, Chaudhari MA, Galvin T. *Proc. 17 National SAMPE Tech. Conf.*1985;172–178.; (b) Barrett KA, Chaudhari MA, Lee BH. *Proc. 33rd International SAMPE Symp.* 1988;33:398.

10. Cole N, Grubber WF. *U.S. Patent.* March, 1964;3127414.

11. Cubbon RCP. *Polymer.* 1965;6:419.

12. Stenzenberger HD, Herzog M, Romer W, Scheiblich R, Pierce S, Canning M. *Proc. 30th National. SAMPE Symp.*,. 1985;30:1568.

13. Shibahara S, Enoki T, Yamamoto T, Motoyoshiya J, Hayashi S. *Polymer J.* 1996;28:752.

14. Li Y, Organ Roger J, Tschen F, Sue H-J, Lopata V. *J. Appl. Polym. Sci.* 2004;94:2407.

15. Abadie MJM, Xiong Y, Boey FYC. *Eur. Polym. J.* 2003;39:1243.; (b)Decker C, Bianchi C, Joensson S. *Polymer.* 2004;45:5803.

16. (a) Zainol I, Day R, Frank Heatley, *J. Appl. Polym. Sci.* 2003;90:2764.; (b) Liptak SC, Wilkinson SP, Hedrick JC, Ward TC, McGrath JE. *Am. Chem. Soc. Symp. Ser.* 1991;475:364.; (c) Mijovic J, Corso WV, Nicolais L, d'Ambrosio G, *Polym. Adv. Technol.* 1988;9:231.

17. Crivello JV. *J. Polym. Sci. Polym. Chem. Ed.* 1973;11:1185.

18. Stenzenberger HD. In: Kinloch AJ, ed. *Structural Adhesives: Developments in Resins and Primers.* New York: Elsivier Applied Science Publishers; Chapter 4. 1986:77–126.

19. Lin S-C, Pearce EM. *High-Performance Thermosets: Chemistry, Properties and Applications.* Munich: Hanser Publishers; Chapter 2. 1994:13–63.

20. Stenzenberger HD. Chemistry and Properties of Addition Polyimides. In: Wilson D, Stenzenberger HD, Hergenrother PM, eds. *Polyimides.* New York: Chapman & Hall; Chapter 4. 1990:79–128.

21. Bell VL, Young PR. *J. Polym. Sci., Polym. Chem. Ed.* 1986;24:2647–2655.

22. Kraiman EA. *U. S. Patent,* 2890206. 1959.; (b) *idem U. S. Patent.* 2890207.; (c)*Chem. Abst.* 1959;53:17572.

23. Varma IK, Fohlen G, Parker JA. *U. S. Patent.* 4276344. 1981.; (b) Varma IK, et al. In: Labana SS, ed. *Chemistry and Properties of Crosslinked Polymers.* New York: Academic Press; 1977:115.

24. Tsai P-F, Shau M-D, Chen G-F. *J. Appl. Polym. Sci..* 2006;100:1.

25. Liu YL, Liu YL, Jeng RJ, Chiu Y-S. *J. Polym. Sci., Part A: Polym. Chem.* 2001;39:1716.

26. Heisey C, Wood PA, McGrath JE, Wightman JP. *Polym. Mater. Sci. Eng.* 1992;67:28.

27. Baek J-B, Qin H, Mather PT, Tan L-S. *Macromolecules.* 2002;35:4951.

28. Nagai A, Takahashi A, Sizuki M, Mukoh A. *J. Appl. Polym. Sci.* 1992;44:159.

29. Nalwa HS, Suzuki M, Takahashi A, Kageyama A. *Appl. Phys. Lett.* 1998;72:1311.

30. Shu W-J, Peng L-S, Chin W-K. *J. Appl. Polym. Sci.* 2002;83:1919.

31. Hwang H-J, Li C-H, Wang C-S. *Polym. Int.* 2006;55:1341.

32. Tang H, Song N, Gao Z, Chen X, Fan X, Xiang Q, Zhou Q. *Polymer.* 2007;48:129–138.

33. Pascal T, Mercier R, Sillion B. *Polymer.* 1989;30:739.

34. Stenzenberger HD. *Adv. Polym. Sci.* 1994;117:165.

35. Takeda S, et al. *J. Appl. Polym. Sci.* 1988;35:1341.

36. Liu Y-L, Chen Y-J. *Polymer.* 2004;45:1797.

37. Wang CS, Lin CH. *Polymer.* 1999;40:5665.

38. Wang C-S, Leu TS, Hsu K-R. *Polymer.* 1998;39:2921.

39. Wang C-S, Leu T-S. *J. Appl. Polym. Sci.* 1999;73:833.

40. Wang C-S, Hwang HJ. *J. Polym. Sci. Part A: Polym. Chem.* 1996;34:1493.

41. Sava M. *J. Appl. Polym. Sci.* 2006;101:567.

42. Wang C-S, Leu T-S. *Polymer.* 1999;40:5407.

43. Sergeyer VA, et al. *Vysokomol. Soyed.* 1986;28(9):1925.

44. Stenzenberger HD, Heinen KU, Hummel DO. *J. Polym. Sci., Polym. Chem. Ed.* 1976;14:2911.

45. White JE, Scaia MD, Snider DA. *J. Appl. Polm. Sci.* 1984;29:891.

46. 'Compimide 353 A', Product Information ID 3390, Degussa.

47. Scola DA. In: Lee SM, ed. *International Encyclopedia of Composites.* Vol 6. New York: VCH Publishers; 1991:34.

48. Scola DA. Polyimide Resins. In *Engineered Materials Handbook, Composites.* Metals Park, Ohio: ASM International; 1987:78–89.

49. Scholle KFMGJ, Winter H. *Proc. 33rd International SAMPE Symp.* 1988;33:1109.

50. Gopala A, Wu H, Heiden P. *J. Appl. Polym. Sci.* 1998;70:943.

51. Iijima T, Nishina W, Fukuda, Tomoi M. *J. Appl. Polym. Sci.* 1996;60:37.

52. Wilkinson SP, Ward TC, McGrath JE. *Polymer.* 1993;34:870.

53. Voit JF, Giraud Y, Camberlin Y, Lopez P, Meissonnier J. *Polym. Compos.* 1989;5:367.

54. Regnier N, Fayos M,Lafontaine E. *J. Appl. Polym. Sci.* 2000;78:2379.

55. Qin H, Mather PT, Baek J-B, Tan L-S. *Polymer.* 2006;47:2813.

56. Hwang H-J, Li C-H, Wang C-S. *Polymer.* 2006;47:1291.

1,2-Polybutadiene

Rahul D. Patil

Class Diene elastomers; diene polymers

CAS Registry Number [26160-98-5]

Structure

Major Applications With one chiral center, 1,2-polybutadiene can exist in the amorphous atactic form and two crystalline forms: isotactic and syndiotactic. In the formation of 1,2-polybutadiene, it is believed that the syn p-allyl form yields the syndiotactic structure, while the anti p-allyl form yields the isotactic structures. The equilibrium mixture of syn and anti p-allyl structures yields heterotactic polybutadiene. The two stereo-isomers that are most used commercially are the syndiotactic and heterotactic structures[1].

Commercial Use Syndiotactic 1,2-polybutadiene is used in films, footwear soles, tubes, and hoses; atactic 1,2-polybutadiene is extensively used in the rubber and tire industry[1,2].

Properties of Special Interest Syndiotactic 1,2-polybutadiene is a reactive thermoplastic resin, which has characteristics of both a thermoplastic and an elastomer[1,2].

Preparation The preparation of amorphous high (99%) 1,2-polybutadiene was first reported in 1981[3]. Several reports in the literature describe the preparation of low, medium, and high vinyl 1,2-polybutadienes[1,4–7]. Syndiotactic 1,2-polybutadiene can be prepared using various cobalt catalysts[1,6–11].

Low molecular weight, low crystalline syndiotactic 1,2-polybutadiene has been commercially developed by JSR Corporation[12]. It has an average molecular weight of 120 000 and is 15–30 % crystalline.

Property	Units	Conditions	Value	Reference
Average molecular weight M_w	g mol^{-1}	Syndiotactic	100 000	(13)
Specific gravity	g cm^{-3}	92% 1,2 content	0.902	(14)
Melting temperature	K	Syndiotactic	429	(15)
		Isotactic	399	
		Atactic	–	

Property	Units	Conditions	Value	Reference
Glass transition temperature	K	Syndiotactic	245	(15)
		Isotactic	–	
		Atactic	269	
Solubility parameter	$(MPa)^{1/2}$	90% 1,2-units	17.4	(16)
Mark–Houwink parameters: K and a	$K = ml\,g^{-1}$ $a = $ none	In toluene	$K = 9.05 \times 10^{-5}$ $a = 0.81$	(17)
Intrinsic viscosity	$dl\,g^{-1}$	In benzene at 30°C	1.52	(18)
Infrared absorption coefficients	$(dl\,cm^{-1}\,mg^{-1}) \times 10^{-3}$	Wavelength (mm)		(19)
		10.35	0.828	
		10.95–10.98	26.7	
		13.5–13.65	0.231	
Infrared absorption coefficients	$mol^{-1}\,cm^{-1}$	Wavelength (mm)		(20)
		10.3	6.7	
		11.0	184	
		12–15.75	4.7	
Water contact angle	Degrees	At pH 1	95	(13)
		At pH 12	97	
Critical surface tension	$(Nm^{-1}) \times 10^3$	–	25	(21)
Refractive index	–	–	1.51	(1)
Solubility	–	In THF at 25°C	Soluble	(17)
		In toluene at 25°C	Soluble	(17)

Unit cell dimensions[22–25]

Isomer	Lattice	Monomers Per Unit Cell	Cell Dimensions (Å)			Cell Angles (degrees)		
			a	b	c (Chain Axis)	α	β	γ
Isotactic (99%)	Rhombohedral	18	17.3	17.3	6.5	90	90	120
Syndiotactic (98%)	Orthorhombic	4	10.98	6.60	5.14	90	90	90

Parameters of internal rotation[16]

1,2 content (%)	C	s	U	U_0 (J mol^{-1})	ε
30	2.32	1.67	1.40	2385	0.83
50	2.34	1.83	1.55	2427	0.97
66	2.41	1.89	1.55	2469	1.06
90	2.41	2.05	1.79	2469	1.20

1,2-Polybutadiene

Molecular weight and intrinsic viscosity *[16]

$[\eta]$ (dl g^{-1})	$M_{\mathrm{w}} \times 10^{-4}$
1.40	20.12
1.98	30.94
2.45	41.66
2.74	49.20
3.40	68.41
4.27	92.92

* For 90% 1,2- and 10% trans, intrinsic viscosity
   was measured in toluene at 30°C $\pm$ 0.05°C.

Microstructure and properties of syndiotactic 1,2-polybutadiene[26]

MP (°C)	Heat of Fusion (J g^{-1})	Crystallinity (%)	$[\eta]$ (dl g^{-1})	^{1}H NMR 1,2 Content (%)	^{13}C-NMR (%) 1,2 Content	Syndiotactic
210	78.7	77.5	6.06	99.72	99.0	99.6
208	77.4	79.7	5.08	99.74	99.2	99.4
206	75.7	81.7	2.00	99.02	98.8	98.8
202	74.5	77.2	1.94	97.75	96.8	97.9
200	79.5	78.3	1.11	97.28	96.0	97.8
192	76.6	72.2	0.46	95.35	93.6	95.1
156	45.2	55.6	0.12	86.27	83.2	87.7

Mechanical properties[14]

Property	Units	Conditions	Value
Tensile strength	MPa	–	11.2
Elongation	%	–	650
M300 stress at 300°C	MPa	–	6.9
Tear strength	kN m^{-1}	–	68.8
Yield stress	MPa	–	5.6
Hardness	Shore D	–	35
Impact strength	(J m^{-1}) $\times 10^{-3}$	–	5.0
Tension set	%	At 100% elongation	22
		At break	145
Hysteresis	–	At 30% strain	0.177
		At 300% strain	0.772

Properties of syndiotactic-PB fibers[27]

MP (°C)	Stretching Temp. Ratio	Diameter (μm)	Initial Modulus (t cm^{-2})	Tenacity (t cm^{-2})	Elongation (%)	Birefringence $\Delta \times 10^3$
187	60 × 1.8	14.5	11.4	1.50	68	−11.5
185	60 × 2.1	14.4	16.7	2.25	19	−12.1
192	60 × 2.3	14.0	16.9	1.61	18	−13.6

Literature available

1,2-Polybutadiene	Reference
Conformational properties	(28, 29)
Effects of fillers on mechanical and viscoelastic properties	(14)
Infrared laser-induced reactions with difluorovinylidene	(30)
Adhesion and wettability studies	(21, 31)
Crystallization behavior and kinetics	(32–34)
Radiation-induced addition of carbon tetrachloride	(35)
Hydrogenation	(36, 37)

Suppliers

1. JSR Corporation, 5-6-10 Tsukiji Chuo-ku, Tokyo, 104–8410 Japan.
2. JSR AMERICA, INC. 312 Elm St. Suite 1585, Cincinnati, OH 45202.
3. Scientific Polymer Products, Inc., 6265 Dean Parkway, Ontario, New York 14519–8997, USA.
4. Acros Organics USA, 711 Forbes Avenue, Pittsburgh, Pennsylvania 15219–4785, USA.
5. Niss America, Inc., 220E 42nd Street, Suite 3002, New York, New York 10017, USA.

References

1. Halasa AF, Massiein JM. In: Kroschwitz JI, ed. *Kirk-Othmer Encyclopedia of Chemical Technology.* 4th ed. Vol 8. New York: John Wiley and Sons; 1989.
2. Tate DP, Bethea TW. In: Mark HF, et al. eds. *Encyclopedia of Polymer Science and Engineering.* 2nd ed. Vol 2. New York: John Wiley and Sons; 1989.
3. Halasa AF, Lohr DF, Hall JE. *J. Polym. Sci., Poly. Chem. Ed.* 1981;19:1347.
4. Tobolsky AV, Kelley DI, Hsieh HJ. *J. Polym. Sci.* 1957;26:240.
5. Binder JL. *Anal. Chem.* 1954;26:1877.
6. Monteil V, Bastero A, Mecking S. *Macromolecules.* 2005;38:5393.
7. Lu J, et al. *J. Appl. Polym. Sci.* 2006;100:4265.
8. Natta G, Corradin P. *J. Polym. Sci.* 1956;20:251.
9. Susa E. *J. Polym. Sci., Part C.* 1964;4:399.
10. Longiave C, Castelli R. *J. Polym. Sci., Part C.* 1964;4:387.
11. Ashitaka H, Jinda K, Ueno H. *J. Polym. Sci., Poly. Chem. Ed.* 1983;21:1951.
12. http://www.jsr.co.jp/jsr_e/pd/images/rb_sis_3e.pdf
13. Carey DH, Ferguson GS. *Macromolecules.* 1994;27:7254.
14. Bhagawan SS, Tripathy DK, De SK. *J. Appl. Polym. Sci.* 1987;34:1581.
15. Lee WA, Rutherford RA. In: Brandrup J, Immergut EH, eds. *Polymer Handbook.* 2nd ed. New York: John Wiley and Sons' 1975:III–139.
16. He T, Li B, Ren S. *J. Appl. Polym. Sci.* 1986;31:873.
17. Anderson JN, Barzan ML, Adam HE. *Rubber Chem. and Tech.* 1972;45:1270.
18. Liaw DJ, Lin LL. *J. Appl. Polym. Sci.* 1989;37:1993.
19. Morero P, et al. *Chim. Ind. (Milan).* 1959;41:758.
20. Silas RS, Yates J, Thornton V. *Anal. Chem.* 1959;31:529.
21. Lee LH. *J. Poly. Sci., Part A-2.* 1967;5:1103.
22. Stephens HL. In: Brandrup J, Immergut EH, eds. *Polymer Handbook* 3rd ed. New York: John Wiley and Sons; 1989:III–139.
23. Natta G, Corradini P. *Nuovo Cimento.* 1960;15(Suppl. 1):9.

24. Natta G, et al. *Atti. Accad. Nazl. Lincei Rend.* 1956;20:560.
25. Natta G, Corradini P. *Rubber Chem. and Tech.* 1960;33:732.
26. Ashitaka H, Inaishi K, Ueno H. *J. Polym. Sci., Poly. Chem. Ed.* 1983;21:1973.
27. Ashitaka H, et al. *J. Appl. Polym. Sci.* 1984;29:2763.
28. Ma H, Zhang L. *Polymer Journal.* 1994;26:121.
29. Roland CM, et al. *Macromolecules.* 2003;36:4954.
30. Thomsen MW, Kimmich BF. *Macromolecules.* 1991;24:6343.
31. Friedmann G, Brossas J. *J. Appl. Polym. Sci.* 1985;30:755.
32. Chen Y, et al. *Polymer.* 2006;47:1667.
33. Ren MQ, et al. *J. Polym. Sci., Poly. Phys. Ed.* 2005;43:553.
34. Cai JL, et al. *Polym. Int.* 2004;53:1127.
35. Okamoto H, Adachi S, Iwai T. *J. Polym. Sci., Poly. Chem. Ed.* 1979;17:1267.
36. Schulz DN. In: Mark HF, et al. eds. *Encyclopedia of Polymer Science and Engineering.* 2nd ed. Vol 7. New York: John Wiley and Sons; 1989.
37. Jones RV, Marberly CW, Reynolds WB. *Ind. Eng. Chem.* 1953;45:1117.

Cis-1,4-Polybutadiene

Gui Lin and M. A. Sharaf

Acronyms, Alternative Names, Trade Name PBD, BR

Class Diene elastomers

Structure Linear and branched *cis*-1,4-polybutadiene

Major Applications Tires and tire products, sealants, belts, gaskets, hoses, automotive molded articles, rubber bands, gloves, footwear, sporting goods, and rubber sheeting. Block copolymers with styrene are used for adhesives and footwear.

Properties of Special Interest High green strength, tack, can be compounded with fillers and other polymers, can form block copolymers for specialty applications, high tensile strength owing to strain-induced crystallization.

Producers and/or Suppliers Anic; Bayer; Bunawerke Huels; Bridgestone/Firestone Tire and Rubber Company; Goodyear Tire and Rubber Company; Michelin; Dow Chemical Company; Aldrich; PhilipsPetroleum Co.

Preparative techniques

Catalyst Systems	Microstructure (%)			Reference
	Cis-1.4	*Trans*-1.4	1,2-Vinyl	
Ziegler–Natta				
AlR_3/TiI_4	92	3	5	(1–4)
$AlR_3/TiI_4O(i\text{-}C_3H_7)_2$	92–94	2–3	5	(5)
$AlR_3/TiCl_4/TiI_4$	93–94	2–3	4	(6,7)
$AlR_2Cl/Co(acac)_2/2H_2O$	98	≈ 1	≈ 1	(8–15)
$AlR_3/Ni(OCOR)_2/BF_3O(C_2H_5)_2$	97	≈ 1.5	≈ 1.5	(16,17)
$Al(C_2H_5)_2Cl/Nd(OCOR)_3/AlR_3$	98–99	–	≈ 1	(18–21)
$AlR_3/NdCl_3/Donor$	99	–	≈ 1	(20,21)
$AlEt_2Cl/NdV_4/Al(iBu)_3$	99			(22)
$CoR/NR_3/C_3HF_6OH$				(23)
$AlR_3/Cp^*Ti(OBz)_3//NR_3$	61.8–63.0	10.2–12.6	24.7–28.0	(24)
$AlR_3/CpTiCl_3$				(25)

Catalyst Systems	Microstructure (%)			Reference
	Cis-1.4	*Trans*-1.4	1,2-Vinyl	
Al(i-C$_4$H$_9$)$_3$/T iCl$_4$	75–78			(26)
AlR$_3$/Ti(η^5-C$_5$H$_5$) (η^2-MBMP)Cl				(27)
Al(i-C$_4$H$_9$)$_2$H/NdV$_3$/tert-C$_4$H$_9$Cl	98–99%			(28)
AlBu$_3$/[(C$_5$Me$_5$)$_2$Ln][B(C$_6$F$_5$)$_4$](Ln = Pr, Nd, or Gd)	99.9%			(29)
[[2,6-[2,6-(i-Pr)$_2$C$_6$H$_3$] NCd(CH$_2$)]$_2$(C$_5$H$_3$N)]Nd(THF)(μ-Cl)$_2$[Li(THF)$_2$] 0.5(hexane)	97.0	2.3	0.7	(30)
[[2,6-[2,6-(i-Pr)$_2$C$_6$H$_3$]NCd(CH$_2$)]$_2$(C$_5$H$_3$N)]Nd[CH$_2$Si(CH$_3$)$_3$](THF)	61.8	36.4	1.8	(30)
[[2,6-[2,6-(i-Pr)$_2$C$_6$H$_3$]NCd(CH$_2$)]$_2$(C$_5$H$_3$N)]Nd(i-CH$_3$)$_2$[Li(THF)$_2$]	95.9	3.3	0.8	(30)
[[2,6-[2,6-(i-Pr)$_2$C$_6$H$_3$]NCd(CH$_2$)]$_2$(C$_5$H$_3$N)]Nd(μ-Cl)(μ-X)[Li(THF)$_2$]	97.0	2.5	0.5	(30)
[[2,6-[2,6-(i-Pr)$_2$C$_6$H$_3$]NCd(CH$_2$)]$_2$(C$_5$H$_3$N)]Nd(μ-Cl)(μ-X)[Li(THF)$_2$]	95.0	4.5	0.5	(30)
[[2,6-[2,6-(i-Pr)$_2$C$_6$H$_3$]NCd(CH$_2$)]$_2$(C$_5$H$_3$N)]Nd(μ-Cl)(μ-X)[Li(THF)$_2$]	96.1	4.5	3.4	(30)
η^3-Allyl derivatives of transition metals				
(η^3-C$_4$H$_7$)CrOCOCl$_3$	93	4	3	(31,32)
(η^3-C$_4$H$_7$)CrCl$_2$	90	5	5	(31,32)
(η^3-C$_4$H$_7$)CoCl	91	2	7	(31,32)
(η^3-C$_4$H$_7$)NiCl$_2$	85–90	5–10	-	(33,34)
(η^3-C$_4$H$_7$)NiOC$_6$H$_2$(NO$_2$)$_3$	97	3	-	(35)
(η^3-C$_4$H$_7$)NiOCOCF$_3$	91–98	1-8	1	(35,36)
(η^3-C$_{12}$H$_{19}$)NiOCOCF$_3$	98	2	-	(37)
η^5:η^3-C$_5$H$_4$(CH$_2$CH$_2$OMe)TiCl$_3$/MAO	97.7	0.8	1.2	(38)
	99.2	0	0.8	(38)
η^5-C$_5$H$_4$(CH$_2$CH$_2$CH$_2$OMe)TiCl$_3$/MAO	78.6	4.6	16.8	(38)
	83.7	3.0	13.3	(38)
η^5:η^1-C$_5$H$_4$(CH$_2$COOMe)TiCl$_3$/MAO	93.0	2.0	5.0	(38)
	95.0	0	5.0	(38)
[η^5:η^1-C$_5$H$_4$(CH$_2$CH$_2$)O]TiCl$_2$]$_2$/MAO	87.7	1.5	11.8	(38)
	88.2	2.4	9.4	(38)

Property	Units	Conditions	Value	Reference
Heat of polymerization	KJ mol^{-1}	*cis*-1,4-addtion at 25°C	78	(39)
Entropy of polymerization	J K^{-1}mol^{-1}	*cis*-1,4-addtion at 25°C	84	(40)
Activation energy of thermal depolymerization	KJ mol^{-1}	–	259	(41)
Major infrared bands	cm^{-1}	In phase out-of-plane CH wag CH$_2$ wag C=C stretching	730 1310 1655	(42)
Infrared absorption coefficients	dl cm^{-1}mg^{-1} ($\times 10^{-3}$)	In phase out-of-plane CH wag, 740 cm^{-1} (shift to 725 cm^{-1} for lower content of *cis*-1,4-units)	5.73	(43–47)
Infrared molar absorptivities	mol^{-1} cm^{-1}	In phase out-of-plane CH wag, 740 cm^{-1} (shift to 725 cm^{-1} for lower content of *cis*-1,4-units)	10.1	(42–48)

Property	Units	Conditions	Value	Reference
High resolution ^{1}H NMR resonance lines				
Proton resonance lines	ppm	250 MHz or greater		(49,50)
		1,2 methylene	1.3	
		1,4 methylene (occurs as doublet corresponding to *cis–trans* units)	2.0	
		1,2 terminal vinyl	4.8	
		1,4 olefinic (occurs as doublet corresponding to *cis–trans* units)	5.4	
		1,2 nonterminal vinyl	5.6	
Solid state proton resonance lines	ppm	1,4 methylene	25	(51)
		1,4 olefinic		
^{13}C resonance lines	ppm	In CDCl$_3$ at 40°C		(52–56)
		1,4 methylene	≈ 27.5	
		1,4 olefinic	129	
^{13}C T_1	s	In CDCl$_3$ at 54°C	3	
^{13}C correlation time	ns	In CDCl$_3$ at 54°C	0.01–0.016	(57,58)
Schaefer width parameter p		Methylene carbon, $T = 40$–45°C	9	(57,58)
J-coupling constant (^{13}C satellite signal of olefinic protons)	Hz	55°C	10.7	(59)
Neutron scattering length density	10^{14} m^{-2}	23°C	0.41	(60)
Density	g cm^{-3}	1,4-*cis*	0.915	(61)
		1,4-*cis* (98-99%), 5°C	1.01	(62,63)
Reducing parameters for the Simha-Somcynacky equation of state: T^*, V^*, P^*	$T^* =$ K $V^* =$ cm^{3}g^{-1} $P^* =$ MPa	T-range = 5–55°C, P-range = 0–300 MPa	$T^* = 9644$ $V^* = 1.0861$ $P^* = 771.4$	(64)

Theta solvents and temperatures

Cis units (%)	Solvent	Θ Temp. (K)	Method	Reference
97	n–Heptane	272	Phase equilibrium	(65)
97	n–Propyl actate	308.5	Virial coefficient, viscosity	(65)
97	5-Methyl-2-hexanone/2-pentanone (1/3 V)	319.2	Phase equilibrium	(66)
97	5-Methyl-2-hexanone/2-pentanone (1/1 V)	305.7		(66)
97	5-Methyl-2-hexanone/2-pentanone (3/1 V)	295.3		(66)
97	3-Pentanone	283.3		(66)
97	3-Pentanone/2-pentanone (3/2 V)	303.0		(66)
94	Diethyl ketone	486		(67)
94	Ethyl propyl ketone	513		(67)
94	Propylene oxide	419		(67)
93	Diethyl ketone	287		(67)

Cis units (%)	Solvent	Θ Temp. (K)	Method	Reference
93	Diethyl ketone	481		(67)
93	Ethyl propyl ketone	251		(67)
93	Ethyl propyl ketone	510		(67)
93	Propylene oxide	308		(67)
93	Propylene oxide	414		(67)
90	i-Butyl acetate	293.5	Virial coefficient, viscosity	(68)
90	n-Heptane/n-hexane (50/50 V)	278	Phase equilibria	(69,70)
90	n-Heptane/n-hexane (25/75 V)	293		(69)
90	5-methyl-2-hexaneone	285.6		(71)
90	2-Pentanone	332.7		(71)
90	3-Pentanone	283.6		(71)

Second virial coefficient (72)

Solvent	Temp. (°C)	$M \times 10^{-5}$ (g mol^{-1})	Condition	Method	$A2 \times 10^{10}$ (mol m^3 g^{-2})
Benzene	28.6	0.6–2.93		Osmometry	15.3
		1.38			27.9
Cyclohexane	28.6	8.4–43.5	Unfractionated sample	Light scattering	2.92
		1.43–1.64	Fractionated sample		1.63–7.5

Property	Conditions	Value	Reference
Solvents	Hydrocarbons tetrahydrofuran, higher ketones, higher aliphatic esters		(73,74)
Nonsolvents	Alcohol, lower ketones, lower esters, nitromethane, proponitrile, water, dilute acids, dilute alkalies, hypochlorite solutions		(73,74)
Solvent/nonsolvent mixtures	Benzene/actone, benzene/n-butanol, benzene/n-methanol, chloroform/acetone, dichloroethane/2-butanone, toluene/n-butanol, toluene/methanol	Fractional precipitation	(74)
	Amyl acetate/2-ethoxy ethanol, benzene/methanol, acetone, (acetone, n-hexane)	Fractional solution	
	Benzene/methanol, carbon tetrachloride/n-butanol	Turbidimetric titration	

Supercritical fluids (75,76)

Mol. Wt. (g mol^{-1})	Solvent	Temp. (°C)	Pressure (MPa)	Conc. (wt%)	Method	Conditions
5000	CO_2	25	19.3	0.27	Cloud point	*cis*-1,4

Unit cell dimensions (61,77–81)

Isomer	Lattice	Space Group	Monomer Per Unit Cell	Density, Cryst. (g cm^{-3})	Melting Point T_m (°C)	Unit Cell Dimensions (Å)			Cell Angles (°)		
						a	b	c (Chain Axis)	α	β	γ
1,4-*cis* (98–99%)	Monoclinic	CS-4	4	1.012	1	4.6	9.5	8.6	90	109	90

Property	Units	Conditions	Value	Reference
Heat of fusion	kJ mol^{-1}	1,4-*cis*	2.51	(82)
			9.2 ± 0.5	(83)
Entropy of fusion	J K^{-1} mol^{-1}	1,4-*cis* (98%)	33.5	(84)
Glass transition temperature	K	1,4-*cis*	167	(82)
		1,4-*cis*	171	(84)
		1,4-*cis* (98–99%)	178	(85)
Melting temperature	K	1,4-*cis* (98–99%)	275	(61,86)
		1,4-*cis* (98.5%)	1.5	(88)
		–	285	(61,77–81)
Thermal conductivity	W m^{-1} K^{-1}	Unspecified microstructure, $T = 20°C$	0.22	(61,86)
Coefficient of thermal expansion	K^{-1} × 10^{-4}	Unspecified microstructure, $T = 25°C$	6.7	(87)
			1.5	(88)
			7.25	(87)
Cohesive energy density	(J m^{-3})$^{1/2}$ × 10^{-3}	1,4-*cis*	16.85	(89)
Solubility parameters δ	(MPa)$^{1/2}$	Polybutadiene, unspecified microstructure	16.57	(90)
			17.08	(91)
			14.65–17.6	(92)
		Different solvents	–	(90,92)
Huggins coefficients: [η] and k'	–	*cis*; toluene; 25°C	[η] = 2.52 $k' = 0.33$	(93)
Heat capacity C_p	J K^{-1} mol^{-1}	94% *cis*, 3% *trans*, 3% vinyl 1,2	See table below	(94,95)
dT/dP	K MPa^{-1}	92% *cis*, 4% *trans*, 4% vinyl 1,2 98.5% *cis*, 1.1% trans, 0.4% vinyl 1,2	0.15	(96)

Temp. (K)	C_p	Condition	T_g (K)	$(\Delta C_p)_{Tg}$	T_m(K)	$(\Delta C_p)_{Tm}$
10	1.18	Solid	171	29.1	284	86.78–103.1
50	19.98	Solid				
100	34.63	Solid				
150	48.5	Solid				
300	106.00	Melt				
350	114.9	Melt				

Cis-1,4-Polybutadiene

Polymer–solvent interaction parameter x

Solvent	v_2*	x	Conditions	Method	Reference
n-Heptane	–	$0.45 + 0.35 v_2$	–	–	(97)
Benzene	–	0.21	98% *cis* units	Vapor sorption	(98)
		0.325	Unspecified *cis* content	–	(99)
Benzene	0.0995	0.253	$v_{eff} = 136$	Calculated from stress–strain and swelling measurements on networks	(100)
Decane	0.256	0.477			
Benzene	0.0679	0.275	$v_{eff} = 64^\dagger$	–	(100)
Decane	0.191	0.477			
Benzene	0.08	0.292	$v_{eff} = 79^\dagger$	–	(100)
Decane	0.186	0.445			
Hexadecane	0.304	0.545			
Benzene	0.0588	0.28	$v_{eff} = 48.2^\dagger$		(100)
Decane	0.155	0.453			
Hexadecane	0.256	0.538			

* v_2 = Volume fraction of polymer.

$\dagger$ v_{eff} = Number density of elastically effective chains (mol/m^3).

Mark–Houwink parameters: K and a

Microstructure			Solvent	Temp. (°C)	Mol. Mass Range ($M \times 10^{-5}$)	Conditions	Method	K (ml g^{-1})	a	Reference
% *Cis*	%, *Trans*	%, 1,2-vinyl								
98		2	Benzene	30	5		Osmometry	0.0337	0.715	(68)
			Isobutyl acetate	20.5	0	Θ solvent	Osmometry	0.185	0.5	(68)
			Toluene	30	5		Osmometry	0.0305	0.725	(68)
95	1	4	Benzene	30	5		Light scattering	0.0085		(101)
			Cyclohexane	30	5		Light scattering	0.0112		(101)
			5-methyl-2-hexanone	12.6	–	Θ solvent	Light scattering	0.15		(102)
			3-pentanone	10.3	–	Θ solvent	Light scattering	0.152		(102)
			Toluene	30	–		Osmometry	0.0339		(103)
94	4	2	Benzene	–	–		Osmometry	0.0414		(104)
			Dioxane	20.2	12	Θ solvent	Osmometry	0.205		(104)
92	3	5	Benzene	32	16	Θ solvent	Light scattering	0.01	0.77	(72)

Unperturbed dimension, $<r>_0/M^{1/2}$, and characteristic ratio, C

Microstructure			Solvent	Temp. (°C)	Conditions	Method	$<r>_0/M^{1/2}$ $\times 10^4$ (nm)	C	Reference
% *Cis*	%, *trans*	%, 1,2-vinyl							
100			Θ solvent	20.2	Θ solvent	Viscosity	920	5.15	(104)
98	2			20.5		Viscosity	880	4.75	(68)
95	5			59.7		Viscosity	835	4.3	(103)
				10.3		Viscosity	825	4.2	(103)

Temperature coefficients of unperturbed mean squared end-to-end distance, $d\ln<r>_0/dT$; and energetic contribution to total elastic force, f_e/f

% *Cis* Units	Solvent	Temp. (°C)	Deformation α	$d\ln<r>_0/dT$	f_e/f	Conditions	Method	Reference
94	Undiluted	50–90	1.25	0.4	0.12	Peroxide	Stress–	(105)
			1.34	0.12	0.12	cure	temperature	
			1.35	0.24	0.08	–	elongation	
						β–Radiation		(105)
–	Undiluted	25–65	0.95	0.43	0.14	γ–Radiation;	Stress–	(105)
			0.9	0.45	0.14	cross-linking	temperature,	
			0.85	0.48	0.15	in solution	compression	
						of toluene		
	1-Chloronaphthalene		0.9	0.41	0.13			(105)
			0.85	0.44	0.14			
			0.8	0.47	0.15			
			0.75	0.51	0.16			
96				0.31	0.1			(106)
96				0.41	0.12			(107)

Rotational isomeric state (RIS) parameters and matrices (108)

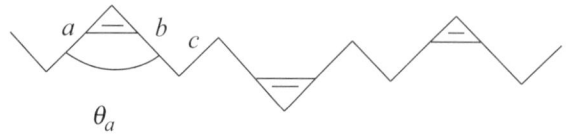

	u_i	u_0	E_0 (KJ mol^{-1})	Conditions
γ	10	1	−6.7	343 K
σ	1.4	1	−0.8	–

Cis-1,4-Polybutadiene

$U_a =$

	t	s^+	g^+	c	g^-	s^-
t	1	1	0	0	0	1
s^+	0	0	0	0	0	0
g^+	1	1	0	0	0	1
c	0	0	0	0	0	0
g^-	1	1	0	0	0	1
s^-	0	0	0	0	0	0

$U_b =$

	t	s^+	g^+	c	g^-	s^-
t	0	1	0	0	0	1
s^+	1	γ	0	0	0	γ
g^+	0	0	0	0	0	0
c	0	0	0	0	0	0
g^-	0	0	0	0	0	0
s^-	0	γ	0	0	0	γ

$U_c =$

	t	s^+	g^+	c	g^-	s^-
t	1	0	σ	0	γ	0
s^+	1	0	γ	0	γ	0
g^+	0	0	0	0	0	0
c	0	0	0	0	0	0
g^-	0	0	0	0	0	0
s^-	1	0	γ	0	γ	0

Further RIS work on *cis*-1,4-polybutadiene can be found in references (109–112).

Property	Units	Conditions	Value		Reference
Refractive index	–	*cis*-1,4, 25°C	1.526		(113)
			1.5		(114)
Refractive index increment dn/dc	ml g^{-1}	1,4-units, 25°C, cyclohexane			(115)
		$\lambda = 436$ nm	0.121		
		$\lambda = 54$ nm	0.113		
Molar polarizability α	m$^3 \times 10^{-31}$	*cis*-1,4	71.4		(114)
Directional polarizabilities	m$^3 \times 10^{-31}$	96% *cis*, undiluted	93.52		(116)
		b_{xx}	77.19		
		b_{zz}	50.90		
Optical segmental anisotropy ($\alpha_1 - \alpha_2$) and optical anisotropy of monomer units ($b_1 - b_2$)	m$^3 \times 10^{-31}$	Swollen polymer networks	$(\alpha_1 - \alpha_2)$	$(b_1 - b_2)$	
		Benzene	61.3–63	30.8	(105, 117, 118)
		Carbon tetrachloride	53.5	31.7	(118)
		Carbon tetrachloride	55.2	–	(105,117,118)
		Cyclohexane	57.3	33.9	(118)
		Toluene	72	42.6	(118)
		p-Xylene	86.9	51.4	(118)
		undiluted	77.7	29.2	(116)
Temperature coefficient of segmental anisotropy, $R\, d\ln(\alpha_1 - \alpha_2)/d(1/T)$	KJ mol^{-1}	96% *cis* units	0.355		(116)
Surface tension	mN m^{-1}	*Cis*-1,4	32		(119)

Sedimentation coefficient, s_0, and diffusion coefficient, D_0

% *cis* units	Solvent	Temp. (°C)	Mol. mass, $(M \times 10^{-5})$ (g mol^{-1})	$s_0 \times 10^{13}$ (s^{-1})	$D_0 \times 10^{11}$ (m^{2}s^{-1})	Conditions	Reference
94	Diethylketone	10.3	–	$0.53 \times 10^{-15} M^{0.5}$		Θ solvent	(120)
90	Hexane/heptane (1:1)	20	0.5–10.8	$2.80 \times 10^{-15} M^{0.48}$		TiI$_4$ catalyst system	(69)
90	Hexane/heptane (1:1)	20	0.35–10.4	$2.33 \times 10^{-15} M^{0.5}$	–	CoCl$_2$ catalyst system	(69)
High	Hexatriacontane	80			4.78		(121)
High	Dodecane	80			23.8		(121)
High	Hexaflourobenzene	80			16.3		(121)

Property	Units	Condition			Value	Reference
Transport of gases		Permeant	Temp. (°C) (°C)	Temp. range (°C)		(88, 122, 123)
Permeability coefficient P	m^3(STP) m s^{-1} m^{-2} Pa$^{-1} \times 10^{-17}$	N$_2$	25	25–50	4.84	
		O$_2$	–	25–50	14.3	
		CO$_2$		25–50	104	
		He	24	0–45	24.5	
		Ne		0–45	14.4	
		Ar		0–45	30.8	
		N$_2$		0–45	14.4	
Preexponential factor P_0	m^3(STP) m s^{-1} m^{-2} Pa$^{-1} \times 10^{-17}$	N$_2$	25	25–50	4.91	
		O$_2$	–	25–50	2.27	
		CO$_2$		25–50	0.683	
		He	24	0–45	0.0855	
		Ne		0–45	0.096	
		Ar		0–45	0.084	
		N$_2$		0–45	0.078	
Transport of gases Diffusion coefficient D	m^{2}s$^{-1} \times 10^{-10}$	N$_2$	25	25–50	1.1	
		O$_2$	–	25–50	1.5	
		CO$_2$		25–50	1.05	
		He	24	0–45	15.7	
		Ne		0–45	6.55	
		Ar		0–45	4.06	
		N$_2$		0–45	2.96	
Gas permeability	cm^3, cm/cm^2. s.atm	air		40	27.7×10^{-8}	
				60	44.1×10^{-8}	
				80	65.5×10^{-8}	
Solubility coefficient S	m^3(STP) m^{-3} Pa$^{-1} \times 10^{-6}$	N$_2$	25	25–50	0.444	
		O$_2$	–	25–50	0.957	
		CO$_2$		25–50	9.87	

Property	Units	Conditions			Value	Reference
		He	24	0–45	0.156	
		Ne		0–45	0.220	
		Ar		0–45	0.758	
		N_2		0–45	0.488	
Activation energy of permeation E_p	KJ mol^{-1}	N_2	25	25–50	34.3	
		O_2	–	25–50	29.7	
		CO_2		25–50	21.8	
		He	24	0–45	20.3	
		Ne		0–45	21.8	
		Ar		0–45	19.4	
		N_2		0–45	21.3	
Activation energy of diffusion E_D	KJ mol^{-1}	N_2	25	25–50	30.1	
		O_2	–	25–50	28.5	
		CO_2		25–50	30.6	
		He	24	0–45	17.3	
		Ne		0–45	17.4	
		Ar		0–45	21.3	
		N_2		0–45	25.0	
Heat of solution Es	KJ mol^{-1}	N_2	25	25–50	4.2	
		O_2	–	25–50	1.2	
		CO_2		25–50	–8.8	
		He	24	0–45	2.9	
		Ne		0–45	4.4	
		Ar		0–45	–1.9	
		N_2		0–45	–3.7	
Heat of solution Es	KJ mol^{-1}	–			2.9	(123)

Property	Units	Conditions	Value	Reference
Activation energy of viscous flow	KJ mol^{-1}	Unspecified microstructure	33.4–41.8	(88)
Rheological properties		96% *cis*-units, $T = 25°C$		(124)
Plateau modulus $G^o{}_N$	MPa		0.76 0.73 (calculated)	
Entanglement molecular mass Me	g mol^{-1}		2300 2400 (calculated)	
Tube diameter dt	nm		430 (calculated)	
Packing length p	nm		24.3	
WLF parameters: $C1$ and $C2$	K	Microstructure = 96% *cis*, 2% trans, and 2% vinyl 1,2; reference temp. $T_0 = 29.8°C$; $T_g = 161°C$; shift factor $a_{T,S}$ of the softening dispersion	$C1 = 3.44$ $C2 = 196.6$	(125)

Property	Units	Conditions	Value	Reference
Effect of radiation: G-factors for cross-linking and scission, $G(x)$ and $G(s)/G(x)$		*Cis*-1,4	$G(x)$ = 5.3 $G(s)/G(x)$ = 0.1	(126, 127)
Thermal oxidative stability	K	Unspecified microstructure $T_{1/2}$ = temperature at which the polymer loses half its mass when heated in vacuum for 30 min	680	(128–131)
		Th = upper use temperature	373	(132)
		Td = lowest temperature reported for thermal decomposition	598	(133)
Biodegradability		*cis*-1,4, Mn = 650; method = molecular mass measurements and biomass; inoculum = acinetobacter		(132, 133)
Conductivity	S/cm	Doped with iodine	10^{-6}	(134, 135)

Catalyst System Metal	Condition				Tan δ		Reference
	cis (%)	Trans (%)	Vinyl (%)	Tg (°C)	0°C	50°C	
Co	97.1	0.9	2.0	−108	0.188	0.173	(136, 137)
	98.1	0.9	1.0	−109			
	98.1	0.9	1.0	−109			
Ni	97.4	1.5	1.1	−106	0.193	0.178	
	97.3	1.5	1.2	−109			
	97.7	1.5	0.8	−109			
Nd	98.0	1.7	0.3	−109	0.174	0.155	
	98.2	1.4	0.4	−109			

References

1. Moyer PH, Lehr MH. *J. Polym. Sci.* 1965;A3:217.
2. Saltman WM, Link TH. *Ind. Eng. Chem. Prod. Res. Dev.* 1964;199.
3. Zelinski RP, Smith DR. *U.S. Patent.* 1956;3050513.
4. Moyer PH. *J. Polym. Sci.* 1965;A3:209.
5. Henderson JF. *J. Polym. Sci.* 1963;C4:233.
6. Marconi W, Mazzei A, Araldi M, De Malde M. *J. Polym. Sci.* 1965;A3:735.
7. Mazzei A, Araldi M, Marconi W, De Malde M. *J. Polym. Sci.* 1965;A3:753.
8. Longiave CRC, Croce GF. *Chim. Ind. (Milan).* 1961;43:625.
9. Gippin M. *Ind. Eng. Chem. Prod. Res. Dev.* 1965;4:160.
10. Takahashi A, Kambara S. *J. Polym. Sci.* 1965;B3:279.

11. Racanelli P, Porri L. *Eur. Polym. J.* 1970;6:751.
12. van de Kamp FP. *Makromol. Chem.* 1966;93:202.
13. Bawn CEH. *Rubber Plast. Age.* 1965;46:510.
14. Porri L, Di Corato A, Natta G. *J. Polym. Sci.* 1967;B5:321.
15. Longiave C, Castelli R, Croce GF. *Belgium Patent.* 573680.
16. Sakata R, Hosono J, Onishi A, Ueda K. *Makromol. Chem.* 1970;139:73.
17. Beebe DH, et al. *J. Polym. Sci., Polym. Chem. Ed.* 1978;16:2285.
18. Shen Z, et al. *J. Polym. Sci., Polym. Chem. Ed.* 1980;18:3345.
19. Witte JA. *Makromol. Chem.* 1981;94:119.
20. Hsieh HL, Yeh YC. *Rubber Chem. Technol.* 1985;58:117.
21. Yang JH, Tsutsui M, Chen Z, Bergbreiter DE. *Macromolecules.* 1982;15:2303.
22. Kwag G, Kim P, Han S, Choi H. *Polymer.* 2005;46:3782–3788.
23. Castner KF. Synthesis of *cis*-1,4-polybutadiene rubber in presence of cobalt containing catalyst system. In: Patent US United States of America: The Goodyear Tire & Rubber Company; 1998.
24. Zhu F, Huang D, Lin S. *J. Appl. Polym. Sci.* 2004;93:2494–2500.
25. Ban HT, Tsunogae Y, Shiono T. *J Polym Sci Part A: Polym Chem.* 2004;42:2698–2704.
26. Mingaleev VZ, Zakharov VP, Monakov YB. *Rus. J. Appl. Chem.* 2007;80:1130–1134.
27. Cuomo C, Serra MC, Maupoey MG, Grassi A. *Macromolecules.* 2007;40:7089–7097.
28. Mello IL, Coutinho FMB, Nunes DSS, Soares BG, Costa MAS, de Santa Maria LC. *Europ. Polym. J.* 2004;40:635–640.
29. Kaita S, Hou Z, Nishiura M, et al. *Macromol Rap Comm.* 2003;24:179–184.
30. Sugiyama H, Gambarotta S, Yap GPA, Wilson DR, Thiele SKH. *Organometallics.* 2004;23: 5054–5061.
31. Oreshkin IA, Tinyakova EI, Dolgoplosk BA. *Vysokomol. Soedin.* 1969;A11:1840.
32. Dolgoplosk BA, Oreshkin IA, Tinyakova EI, Yakovlev VA, Akad Izv. *Nauk SSSR, Ser. Khim.* 1967;2130.
33. Kormer VA, Babitskii BD, Lobach MI, Chesnokova NN. *J. Polym. Sci.* 1969;C16:3451.
34. Porri L, Natta G, Gallazzi MC. *J. Polym. Sci.* 1967;C16:2525.
35. Dawans F, Durand JP, Teyssie Ph. *J. Polym. Sci.* 1972;B10:493.
36. Dawans F, Teyssie Ph. *J. Polym. Sci.* 1969;B7:11.
37. Durand JP, Dawans F, Teyssie Ph. *J. Polym. Sci.* 1968;B6:757.
38. Miyazawa A, Kase T, Hashimoto K, Choi J, Sakakura T, Ji-zhu J. *Macromolecules.* 2004;37: 8840–8845.
39. Roberts DE. *J. Res. Natl. Bur. Std.* 1950;44:221.
40. Dainton FS, Evans DM, Hoare FE, Melia TP. *Polymer.* 1962;3:297.
41. Sawada H. In *Thermodynamics of Polymerization.* New York: Marcel Dekker; 1976.
42. Colthup NB, Daly LH, Wiberly SE. In: *Introduction to Infrared and Raman Spectroscopy.* 2nd ed. New York: Academic Press; 1975.
43. Morero P, Santambrogio A, Porri L, Ciampelli F. *Chim. Ind. (Milan).* 1959;41:758.
44. Binder JL. *Anal. Chem.* 1954;26:1877.
45. Binder JL. *J. Polym. Sci., Part A.* 1965;3:1587.
46. Hampton RR. *Anal. Chem.* 1949;21:923.
47. Silas RS, Yates J, Thornton V. *Anal. Chem.* 1959;31:529.
48. Amram B, Bokobza L, Monnerie L. *Polymer.* 1988;29:1155.
49. Santee ERJ, Chang R, Morton M. *J. Polym. Sci., Polym. Lett. Ed.* 1973;11:449.
50. Santee ERJ, Mochel VD, Morton M. *J. Polym. Sci., Polym. Lett. Ed.* 1973;11:453.
51. English AD, Dybowski CR. *Macromolecules.* 1984;17:446.
52. Bovey FA, Jelinski LW. In: Mark HF, et al. eds. *Encyclopedia of Polymer Science and Engineering.* 2nd ed. New York: John Wiley and Sons; 1988:254.

53. Duch MW, Grant DM. *Macromolecules.* 1970;3:165.

54. Elgert KF, Quack G, Stuetzel P. *Polymer.* 1974;15:612.

55. Elgert KF, Quack G, Stuetzel P. *Polymer.* 1975;16:154.

56. Guerry JL. *Rev. Gen. Caoutch. Plast.* 1977;103.

57. Komoroski RA, Mandelkern L. In: Brame EG, ed. *Applications of Polymer Spectroscopy.* New York: J. Academic Press; 1978:57.

58. Grunski W, Murayama N. *Makromol. Chem.* 1976;177:3017.

59. Katoh T, Ikura M, Hikichi K. *Polymer J.* 1988;20:185.

60. Wignall GD. In: Mark JE, ed. *Physical Properties of Polymers Handbook.* Woodbury, New York: AIP Press; 1996:299.

61. Natta G. *Science.* 1965;147:269.

62. Sharaf MA, Mark JE, Cesca S. *Polym. Eng. Sci.* 1986;26:304.

63. Sharaf MA. *Rubber Chem. Tech.* 1994;67:88.

64. Jaine RK, Simha R. *Polym. Eng. Sci.* 1979;19:845.

65. Moraglio G. *Europ. Polym. J.* 1965;1:103.

66. Abe M, Fujita H. *J. Phys. Chem.* 1965;69:3263.

67. Cowie JMG, McEwen IJ. *Polymer.* 1975;16:933.

68. Danusso F, Moraglio G, Gianott G. *J. Polym. Sci.* 1961;51:475.

69. Poddubnyi IY, Grechanovskii VA, Mosevitskii MI. *Vysokomol. Soedin.* 1964;5:1049.

70. Poddubnyi IY, Grechanovskii VA, Mosevitskii MI. *Polym. Sci. USSR* 1964;5:105.

71. Abe M, Fujita H. *Reports Prog. Polym. Phys. (Japan).* 1964;7:42.

72. Cooper WG, Vaughan G, Eaves DE, Madden RW. *J. Polym. Sci.* 1961;50:159.

73. Fuchs O. In: Brandrup JB, Immergut EH, eds. *Polymer Handbook.* 3rd ed. New York: Wiley-Interscience; 1988:VII–379.

74. Bello A, Barrales-Rienda JM, Guzman GM. In: Brandrup, J, Immergut EH, ed. *Polymer Handbook.* 3rd ed. New York: Wiley-Interscience; 1988:VII/233.

75. Heller JP, Dandge K. *Soc. Petroleum Engineers Preprints.* 1983;11:173.

76. Shine AD. In: JE, ed. *Physical Properties of Polymers Handbook.* Woodburry, New York: MarkAIP Press; 1996:249.

77. Natta G. *Rubber Plastics Age.* 1957;38:495.

78. Natta G, Corradini P. *Angew. Chem.* 1956;68:615.

79. Natta G, Corradini P. *J. Polym. Sci.* 1956;20:251.

80. Natta G, Corradini P. *Rubber Chem. Technol.* 1960;33:732.

81. Wunderlich B. In: *Macromolecular Physics.* New York: Academic Press; 1973.

82. Bahary WS, Sapper DI, Lane JH. *Rubber Chem. Technol.* 1967;40:1529.

83. Natta G, Moraglio G. *Makromol. Chem.* 1963;66:218.

84. Trick GS. *J. Appl. Polym. Sci* 1960;3:253.

85. Baccaredda M, Butta E. *Chim. Ind. (Milan).* 1960;42:978.

86. Berger M, Buckley DJ. *J. Polym. Sci.* 1963;A1:2945.

87. DiBenedetto AT. *J. Polym. Sci.* 1963;A1:3459.

88. Thompson EV. In: Mark HF, et al. ed. *Encyclopedia of Polymer Science and Engineering.* New York: Wiley-Interscience; 1985:737.

89. Stephens HL. In: Brandrup J, Immergut EH, ed. *Polymer Handbook.* 3rd ed. New York: Wiley-Interscience; 1988.

90. Tobolsky AV. In *Properties and Structure of Polymers.* New York: John Wiley and Sons; 1960:66.

91. Scott RL, Magat M. *J. Polym. Sci.* 1949;4:555.

92. Grulke EA. In: Brandrup J, Immergut EH, ed. *Polymer Handbook.* 3rd ed. New York: Wiley-Interscience, 1988:VII–519.

93. Bristow GM. *J. Polym. Sci.* 1962;62:S168.

94. Gaur U, Lau SF, Wunderlich BB, Wunderlich B. *J. Phys. Chem. Ref. Data.* 1983;12:29.

95. Grebowicz J, Avcock W, Wunderlich B. *Polymer.* 1986;27:575.

96. Kirpatch A, Adolf DB. *Macromolecules.* 2004;37:1576–1582.

97. Burton ASM. In: *Handbook of Solubility Parameters.* Boca Raton, Fla: CRC Press; 1983.

98. Saeki S, Holste JC, Bonner DC. *J. Polym. Sci., Polym. Phys. Ed.* 1982;20:793.

99. Jessup RS. *J. Res. Nat. Bur. Stand.* 1958;60:47.

100. Brotzman RW, Flory PJ. *Macromolecules.* 1987;20:351.

101. Kurata M, Tsunashima Y. In: Brandrup J, Immergut EH, ed. *Polymer Handbook.* 3rd ed. New York: Wiley-Interscience; 1989:VII–1.

102. Abe M, Murakami Y, Fujita H. *J. Appl. Polym. Sci.* 1965;9:2549.

103. Takeda M, Endo R. *Rep. Prog. Polym. Phys. Japan.* 1963;6:37.

104. Poddubnyi IY, Erenburg YeG, Yeremina MA. *Vysokomol. Soedin.* 1968;10:1381.

105. Becker RH, Yu CU, Mark JE. *Polymer J.* 1975;8:234.

106. Shen M, Chen TY, Cirlin EH, Gebhard HM. In: Chomff AJ, Newman S, ed. *Polymer Networks: Structure and Mechanical Properties.* New York: Plenum Press; 1971.

107. Price C, Yoshimura N. *Polymer.* 1975;16:261.

108. Mark JE. *J. Am. Chem. Soc.* 1966;88:4354.

109. Rehahn M, Mattice WL, Suter UW. In *Rotational Isomeric State Models in Macromolecular Systems.* New York: Springer-Verlag; 1997:V001.

110. Ishikawa T, Nagai K. *J. Polym. Sci., Part A-2.* 1969;7:1123.

111. Abe Y, Flory PJ. *Macromolecules.* 1971;4:219.

112. Tanaka S, Nakajima A. *Polymer J.* 1972;3:500.

113. Furukawa J, Yamashita S, Kotani T, Kawashima M. *J. Appl. Polym. Sci.* 1969;13:2527.

114. Seferis JC. In: Brandrup J, Immergut EH, ed. *Polymer Handbook.* 3rd ed. New York: Wiley-Interscience; 1988:VI–451.

115. Kratochvil R, Strakova D, Schmidt P. *Angew. Makromol. Chem.* 1972;23:169.

116. Morgan RJ, Treloar LRG. *J. Polym. Sci., Part A-.* 1972;10:51.

117. Ishikava T, Nagai K. *Polym. J.* 1970;1:116.

118. Fukuda M, Wilkes GL, Stein RS. *J. Polym. Sci., Part A.* 1971;2:1417.

119. Lee L, Lee H. *J. Polym. Sci., Part A.* 1967;5:1103.

120. Cerny LC, Graham RC, James H Jr. *J. Appl. Polym. Sci.* 1967;11:1941.

121. Ferguson RD, von Meerwall E. *J. Polym. Sci., Polym. Phys. Ed.* 1980;18:1285.

122. Amerongen GJ. *J. Polym. Sci.* 1950;5:307.

123. Paul DR, DiBenedetto AT. *J. Polym. Sci., Part C.* 1965;10:17.

124. Fetters LJ, Lohse DJ, Colby RH. In: Mark JE, Ed. *Physical Properties of Polymers Handbook.* Woodbury, New York: AIP Press; 1996:335.

125. Nagai KL, Plazek DJ. In: Mark JE, ed. *Physical Properties of Polymers Handbook.* Woodbury, New York: AIP Press; 1996:341.

126. Kozlov VT, Yevseyev AG, Zubov PI. *Vysokomol. Soed.* 1969;A11:2330.

127. Bohm GG, Tveekrem JO. *Rubber Chem. Technol.* 1982;55:575.

128. van Krevelen DW, Hoftyzer PJ. In *Properties of Polymers.* Amsterdam: Elsevier Scientific; 1976:459.

129. Billmeyer FWJ. In *Textbook of Polymer Science.* New York: Wiley-Interscience; 1984:143.

130. Welsh WJ. In: Mark JE, ed. *Physical Properties of Polymers Handbook.* Woodbury, New York: AIP Press; 1996.

131. Grassie N. In: Brandrup J, Immergut EH, eds. *Polymer Handbook.* 3rd ed. New York: Wiley-Interscience; 1988:II–365.

132. Tsuchii A, Suzuki T, Takahara Y. *Agribiol. Chem.* 1978;42:1217.

133. Andrady AL. In: Mark JE, Ed. *Physical Properties of Polymers Handbook.* Woodburry, New York: AIP Press; 1996:625.

134. Dai L. *Macromolecules.* 1996;29:282.

135. Yildirm P, Kucukyavuz Z, Erman B. *Synth. Metal.* 1997;88:231–235.

136. Pires NMT, Ferreira AA, Lira CHD, et al. *J. Appl. Polym. Sci.* 2006;99:88–99.

137. Kwag G, Kim P, Han S, Lee S, Choi H, Kim S. *J. Appl. Polym. Sci.* 2007;105:477–485.

trans-1,4-Polybutadiene

Zhengcai Pu

Acronyms PBD, BR

Class Diene elastomers

Structure

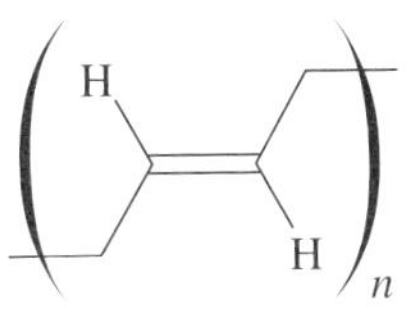

Major Applications Tire treads, carcass, belts, hoses, gaskets, seals, and protective coatings; component in other synthetic rubbers and blends.[1]

Properties of Special Interest Good low-temperature properties and adhesion to metals; good resilience, durability, and abrasion resistance.[1−3]

Producers and/or Suppliers Anic; Bayer; Bunawerke Huels; Bridgestone/Firestone Tire and Rubber Company; Goodyear Tire and Rubber Company; Michelin.[1]

Property	Units	Conditions	Value		Reference
Anistropy of segment	cm^{-3}		$\alpha_1 - \alpha_2$	$\alpha_{\|\|} - \alpha_{\perp}$	(4)
		Benzene	$+71 \times 10^{-25}$	$+37.4 \times 10^{-25}$	
		CCl_4	$+61.1 \times 10^{-25}$	$+36.3 \times 10^{-25}$	
		Cyclohexane	$+57.3 \times 10^{-25}$	$+33.1 \times 10^{-25}$	
		Toluene	$+81.6 \times 10^{-25}$	$+48.6 \times 10^{-25}$	
		p-Xylene	$+101 \times 10^{-25}$	$+60.2 \times 10^{-25}$	
Coefficient of thermal expansion	K^{-1}	Cubicle	6.75×10^{-4}		(6)
Critical surface tension of spreading γ_c	$N\,m^{-1}$		0.031		(7)
Crystallographic interplanar spacing	nm		0.396–0.400		(16)
			0.395–0.397		(17)
Decomposition temperature	K	Initial decomposition	598		(5)
		Half decomposition; heated in vacuum for 30 min	680		

447

Property	Units	Conditions	Value	Reference
Density	g cm^{-3}		0.93–0.97	(4)
Dielectric constant		1 MHz	2.51	(8)
			3.3	(6)
Dielectric loss factor			0.002	(8)
Dielectric strength	kV m^{-1}		400–600	(6)
Enthalpy of fusion	J mol^{-1}	Modification I	13 807	(5)
		Modification II	4602	
Enthalpy of melting	J g^{-1}	Crystalline rubber with 82% of trans	0.48–7.51	(14)
Enthalpy of transition	J g^{-1}	Crystalline rubber with 82% of trans	31.36–46.05	(14)
Entropy of fusion	J K^{-1} mol^{-1}	Modification I	26.8, 37.4	(2, 4, 5)
		Modification II	11.3, 10.9	
Glass transition temperature	K		166, 171	(4,5,9)
Heat of fusion	kJ mol^{-1}	Modification I	4.184	(10)
		Modification II	4.184–4.6	(11, 12)

Fractionation systems[4]

Method of Fractionation	Solvent or Solvent/Nonsolvent Mixture
Fractional precipitation	Benzene/acetone
	Benzene/methanol
	Benzene/n-butanol
	Pentane/methanol
	Tetrahydrofuran/water
	Toluene/ethanol
	Toluene/n-butanol
	Toluene/methanol
Fractional solution	Benzene/methanol
	Chloroform/methanol
	Ethyl ether
Fractional crystallization	Heptane
Gel permeation chromatography	Chloroform
	o-Dichlorobenzene
	Dichloromethane
	Tetrahydrofuran
	Toluene
	1,2,4-Trichlorobenzene
	Trichloroethylene

trans-1,4-Polybutadiene

Heat capacity[4,5]

Solid		Melt	
Temperature (K)	Heat Capacity (kJ K^{-1} mol^{-1})	Temperature (K)	Heat Capacity (kJ K^{-1} mol^{-1})
30	0.00970	360	0.1166
40	0.01439	370	0.1184
50	0.01874	380	0.1202
60	0.02256	390	0.1220
70	0.02589	400	0.1237
80	0.02900	410	0.1255
90	0.03202	420	0.1273
100	0.03498	430	0.1291
110	0.03786	440	0.1309
120	0.04070	450	0.1326
130	0.04353	460	0.1344
140	0.04626	470	0.1362
150	0.04899	480	0.1380
160	0.05171	490	0.1390
170	0.05452	500	0.1415
180	0.05788		

Intersurface tension[4]

Surface 1	Surface 2	Interfacial Tension γ_{12} (N m^{-1})			$-d\gamma_{12}/dT$ (N m^{-1} K^{-1})
		293 K	423 K	473 K	
PBD, $M_n = 960$	PDMS, $M_n = 3900$	0.00248	0.00207	0.00191	3.22×10^{-6}
PBD, $M_n = 2350$	PDMS, $M_n = 3900$	0.00386	0.00270	0.00225	8.95×10^{-6}
PBD, $M_n = 960$	PDMS, $M_n = 5200$	0.00398	0.00285	0.00242	8.65×10^{-6}
PBD, $M_n = 2350$	PDMS, $M_n = 5200$	0.00258	0.00225	0.00213	2.50×10^{-6}

Mark–Houwink parameters: K and a[4]

Solvent	Temperature (K)	Molecular Weight (kg mol^{-1})	K (ml g^{-1})	a
Cyclohexane	313	170	0.0282	0.70
Toluene	303	160	0.0294	0.753

Property	Units	Conditions	Value	Reference
Melting point	K	Modification I	370	(2,4)
		Modification II	418	
		$M_v = 5.6 \times 10^4$, 91.5 % of trans	362	(15)
		$M_v = 8.0 \times 10^5$, 96.5 % of trans	408.6	

Property	Units	Conditions	Value	Reference
Nonsolvents			Alcohol, lower ketones and esters, nitromethane, propionitrile, water, diluted acids, diluted alkalies, hypochlorite solutions	(4)
Phase transition temperature	K	$M_v = 5.6 \times 10^4$, 91.5 % of trans	321	(15)
		$M_v = 8.0 \times 10^5$, 96.5 % of trans	346.8	
Refractive index	–	298 K	1.515	(4)
Refractive index increment	$l\,kg^{-1}$	303 K, in cyclohexane	0.110	(4)
Scattering length density of neutron scattering	cm^{-2}	–	4.1×10^{-11}	(5)

Second virial coefficient A_2	$mol\,m^3\,kg^{-2}$ ($\times 10^{-4}$)	Solvent	Temp. (K)	M.W. $(kg\,mol^{-1})$		(4)
		–	307	17.4–370	18.3–9.48	
		Dioxane	307	17.9–434	3.81–1.49	
		Toluene	307	18.2–466	19.4–11.5	

Sedimentation coefficient at zero concentration $S_0(s)$	–	Diethyl ketone, 283 K Mw $(kg\,mol^{-1})$		(4)
		60	1.76×10^{-13}	
		187	2.76×10^{-13}	
		350	3.45×10^{-13}	
		436	4.28×10^{-13}	
		778	4.52×10^{-13}	
		1380	5.15×10^{-13}	

Property	Units	Conditions	Value	Reference
Service temperature	K	–	172–366	(6)
Solubility parameter	$(MPa)^{1/2}$	–	14.6–17.6	(4,5)
Solvents			Hydrocarbons, tetrahydrofuran, higher ketones, higher aliphatic esters	(4)

Surface tension[4]

M_n (g mol^{-1})	End Group	Surface Tension γ (N m^{-1})			$-d\gamma/dT$ (N m^{-1} K^{-1})
		293 K	423 K	473 K	
5400	Carboxylic acid	0.0486	0.0299	0.0227	1.440×10^{-4}
5400	Methyl ester	0.0431	0.0288	0.0233	1.098×10^{-4}

trans-1,4-Polybutadiene

Property	Units	Conditions	Value	Reference
Tensile compression set	%	–	10–30	(6)
Tensile elongation	%		450	(6)
Tensile modulus	MPa	–	2–10	(6)
Tensile resilience	%	ASTM 945	50–90	(6)
Tensile strength	MPa	–	14–17	(6)
Thermal conductivity	$W\,m^{-1}\,K^{-1}$	293 K	0.22	(5)
Theta temperature	K	Diethyl ketone Ethyl propyl ketone, $M_n = 47$–193 kg mol^{-1} Propylene oxide	486 513 419	(5)
Transition temperature	K	Modification I to Modification II Crystal to meso phase Meso phase to melt	348 356 437	(4) (17)
Unperturbed dimension $\sigma = r_0/r_{0f}$ $C_\infty = r_0^2/nl^2$	–	328 K, in decalin	1.23 5.8	(4)

Unit cell data$^{(3,4)}$

Property	Modification I		Modification II	
Crystallographic system	PHEX	MONO	PHEX	HEX
Space group	–	C_{2h}^5	–	–
Cell dimension				
a_0 (Å)	4.54	8.63	4.88	4.95
b_0 (Å)	4.54	9.11	4.88	4.95
c_0 (Å)	4.9	4.83	4.68	4.66
α (°)	–	–	–	–
β (°)	–	114	–	–
γ (°)	–	–	–	–
Crystal density (g cm^{-3})	1.03	1.036	0.930	0.908
Repeat distance	0.485	–	0.465	–
Repeat unit per unit cell	1	4	1	1

References

1. Ulrich H. *Introduction to Industrial Polymers*. 2nd ed. Munich: Hanser Publishers; 1993.
2. Salamone JC. *Polymer Materials Encyclopedia*. Vol 8. Boca Raton, Fla: CRC Press; 1996.
3. Mark HF, et al. eds. *Encyclopedia of Polymer Science and Engineering*. Vol 6. New York: John Wiley and Sons; 1996.
4. Brandrup J, Immergut EH. *Polymer Handbook*. 3rd ed. New York: Wiley-Interscience; 1989.
5. Mark, JE. ed. *Physical Properties of Polymers Handbook*. New York: AIP Press; 1996.
6. Pentone Publication. *Materials Engineering*. 1989;106:178.

7. Anderson JN, Barzan ML, Adams HE. *Rubber Chem. Technol.* 1972;45:1281.
8. Pegoraro M, Mitoraj K. *Makromol. Chem.* 1963;61:132.
9. Bahary WS, Sapper DI, Lane JH. *Rubber Chem. Technol.* 1967;40:1529.
10. Berger M, Buckley DJ. *J. Polym. Sci.* 1963;A1:2945.
11. Natta G, Moraglio G. *Makromol. Chem.* 1963;66:218.
12. Mandeldern L, Quinn FA Jr. *J. Polym. Sci.* 1956;19:77.
13. Hummel DO. *Infrared Spectra of Polymers in the Medium and Long Wavelength Regions.* New York: Interscience Publishers; 1966.
14. Rivera M, Najera RH, Tapia JJB, Guerrero LR. *J. Elastomers and Plastics.* 2005;37:267.
15. Yang XN, Cai JL, Kong XH, Dong WM, Li G, Zhou EL. *Macromol. Chem. Phys.* 2001;202:1166.
16. Antipov EM, Mushina EA, Stamm M, Fischer EW. *Macromol. Chem. Phys.* 2001;202:73.
17. Antipov EM, Schklyaruk BF, Stamm M, Fischer EW. *Macromol. Chem. Phys.* 2001;202:82.

Poly(butene-1)

R. P. Patki, D. R. Panse and P. J. Phillips

Acronyms and Trade names Polybutylene, polybutene, PB, Butuf, Vestolen BT

Class α-Olefins

Structure of the Repeat Unit $-[-CH_2-CH(C_2H_5)-]-$

Major Applications Hot water and high pressure piping, films, adhesives, comonomer for ethylene polymer, atactic polymer used in sealant, cosmetic (lip) sticks, used in skin cleanser formulations.

Properties of Special Interest The polymer exhibits excellent creep properties. It is tough, retains strength at elevated temperatures, and is resistant to environmental stress cracking and abrasion.

Preparative techniques

Type of Polymerization	Conditions	Ref.
Ziegler–Natta	Nickel-based catalysts, 80–120°C, 7–15 MPa	(1–5)
	Other typical catalytic systems:	
	$TiCl_3 + Et_2AlCl$, $TiCl_3$+cocatalyst+$MgCl_2$,	
	chiral racemic zirconium dichloride+methyl aluminoxane	
Metallocene	rac-(dimethylsilyl) bis(4,5,6,7-tetrahydro-l-indenyl)zirconiumdichloride	(6)
	and methylaluminoxane (MAO) catalysts, 100°C	
	(n^5-pentamethylcyclopentadienyl) titanium	(7)
	tricinnamyloxide [$Cp^*Ti(OCH_2–CH=CHC_6H_5)_3$]	
	with MAO cocatalyst, 32 kPa	
	Cp^* = pentamethylcyclopentadienyl	

Typical comonomers used: Ethylene, propylene, and 1-pentene

Structure and morphology

Property	Units	Condition	Value	Ref.
Molecular weight of repeat unit	gmol^{-1}	–	56.11	
Stereoregularity		Type of polymerization/catalyst system:	84% isotactic	(6)
		(1) Metallocene/zirconium dichloride and methylaluminoxane (MAO) catalysts		

Property	Units	Condition	Value	Ref.
		(2) Metallocene/titanium tricinnamyloxide with MAO cocatalyst	Atactic, high molecular weight	(7)
		(3) Ziegler–Natta/lithium diphenyl-phosphide + Et_2AlCl + $TiCl_3$-AA	94.7% Isotactic	(8)
		(4) Ziegler–Natta/DMH + hydrogen + Et_2AlCl + $TiCl_3$-AA	97.7% Isotactic	(9)
		(5) Ziegler–Natta/Et_2AlCl + $TiCl_3$	84.5% isotactic	(9)
		(6) Metallocene/$Cp_2^*MCl_2$/MAO catalysts. Cp^* = pentamethylcyclopentadienyl; M = Zr, Hf; MAO = methylaluminoxane	Predominantly syndiotactic	(10)
Typical molecular weight range	gmol^{-1}	Metallocene (Weight avg.) Ziegler–Natta:	530 000–960 000	(7) (11)
		(1) Number avg.	70 000–75 000	
		(2) Weight avg.	725 000–750 000	
		(3) Z-avg.	$2.5*10^6$–$3.0*10^6$	
Typical polydispersity index		Metallocene	1.2–1.4	(7)
		Ziegler–Natta	10–11	(11)
		Anionic	1.02	(12)

Equation of state properties

Property	Units	Condition	Value	Ref.
Thermal expansion coefficient	K^{-1}	Temperature range: 140–240°C	$6.7*10^{-4}$	(13)
Compressibility coefficient	bar^{-1}	140–240°C	$10.1*10^{-5}$–$15.8*10^{-5}$	
Reducing temperature	K		10808	(14)
Reducing pressure	Pa		$608.5*10^6$	
Reducing volume	cm^3 g^{-1}		1.1635	
Density	g cm^{-3}	23°C	0.859	(15)

Solution properties

Solvents	Nonsolvents	Ref.
Above 100°C: benzene, toluene, decalin, tetralin, chloroform, chlorobenzene	Organic solvents at room temperature	(16)

Poly(butene-1)

Equation of state properties

Property	Units	Condition	Value	Ref.
Theta temperature	K	(1) Atactic polymer		
		Solvent/method used:		
		anisole/VM*	356	(17)
		anisole/PE*	359.2	(18)
		diphenyl ether/PE, VM	414	(19)
		phenetole/VM	334	(17)
		i-amyl acetate/PE, VM	296	(19)
		(2) Isotactic polymer		
		toluene/PE, VM	227	(19)
		anisole/PE	362.1	(18)
		anisole/PE, VM	362	(19)
		cyclohexane and n-propanol (69/31 by volume)/VM	308	(20)
		diphenyl ether/PE, VM	421	(19)
		phenetole/PE, VM	337.5	(19)
Polymer–solvent interaction parameter χ		Solvent/Temperature (°C)		(21)
		n-heptane/115–135	0.38	
		n-octane/115–135	0.36	
		n-nonane/115–135	0.32	
		n-decane/115–135	0.30	
		benzene/135	0.49	
		cyclohexane/135	0.20	
		2,5-dimethylhexane/115–135	0.36	
		2,4-dimethylpentane/115–135	0.40	
		2,3-dimethylpentane/115–135	0.35	
		3-ethylpentane/115–135	0.34	
		2-methylhexane/115–135	0.39	
		3-methylhexane/115–135	0.38	
		toluene/135	0.47	
Second virial coefficient	$mol\ cm^3\ g^{-2}$	Solvent/temperature/molecular weight		(18)
		(1) Atactic polymer	$2.4*10^{-4}$	
		n-nonane/35°C/44.1–1300*10^3 gmol^{-1}		
		toluene/45°C/26.3–558*10^3 gmol^{-1}	$10.8–4.1*10^{-4}$	
		(2) Isotactic polymer	$6.05–1.05*10^{-4}$	
		n-nonane/80°C/105–935*10^3 gmol^{-1}		
		toluene/45°C/90.1–775*10^3 gmol^{-1}	$6.57–3.73*10^{-4}$	
Characteristic ratio $C\infty$		Solvent/temperature(°C)/method	15.1	(18)
		nonane/35/LS*		
		—/25 to 223/SANS*	5.1–5.5	(12)
		—/5 to 53/VM*	5.3–5.9	(8)

* PE-Phase equilibria, VM-intrinsic viscosity/molar mass, LS-light scattering, SANS-small angle neutron scattering.

Poly(butene-1)

Mark–Houwink parameters

Isotactic polymer

Solvent/Method	Temperature°C	Mol. Wt.*10^{-4}	$k*10^3$ (ml g^{-1})	a	Ref.
Anisole/LS	89	57	111	0.5	(19)
Decalin/LS	115	90	9.49	0.73	(22)
Ethylcyclohexane/LS	70	94	7.34	0.80	(18)
Heptane/LS	35	90	4.73	0.80	(22)
Heptane/LS	60	90	15	0.69	(22)
Nonane/LS	80	94	5.85	0.80	(18)
Phenetole/OS*	64.5	57	113	0.5	(19)
Phenylether/OS	148	57	103	0.5	(19)
1,2,4-trichlorobenzene/GPC	135	–	11.8	0.729	(23)
Cryclohexane and propanol (80/20)/LS	35	73	102	0.59	(24)

* OS-Osmotic pressure.

Atactic polymer

Solvent/Method	Temperature°C	Mol. Wt.*10^{-4}	$k*10^3$ (ml g^{-1})	a	Ref.
Anisole/LS	86.2	130	123	0.5	(18)
Benzene/EG*	30	0.5	22.4	0.72	(25)
Ethylcyclohexane/LS	70	130	7.34	0.80	(18)
Phenylether/OS	141	66	104	0.5	(19)

* EG-End group titration.

Crystalline state properties

Isotactic polymorphs

Property	I	II	III	Ref.
Lattice	Trigonal	Tetragonal	Orthorhombic	(26–28)
Unit cell dimensions (Å)				
a	17.7	14.85	12.38	
b	17.7	14.85	8.92	(26–28)
c	6.5	20.6	7.45	
Unit cell angles (°)	All 90	All 90	All 90	(26–28)
Monomers per unit cell	18	44	8	(26–28)
Helix conformation	3_1	11_3	4_1	(26–28)
Space group	D3D-6	-	-	(26)
Crystalline density@ 23°C	0.951	0.902	0.905	(26–28)
Growth kinetics coefficients				
K (K^2)	$3.17*10^5$	$9.54*10^5$		
G_o (μm s^{-1})	$4.32*10^5$	$1.05*10^4$		(29)
Lateral surface energy σ	$15.0*10^{-3}$	$4.3*10^{-3}$		
Fold surface energy σ_e	$53.9*10^{-3}$	$57*10^{-3}$		

Poly(butene-1)

Syndiotactic polymorphs

Property	I	II	Ref.
Lattice	Orthorhombic	Monoclinic	(30,31)
Unit cell dimensions (Å)			
a	16.81	15.45	
b	6.06	14.36	
c	7.73	20	
Unit cell angles (°)	All 90	$\gamma = 116$	
Helix conformation	2_1	5_3	
Space group	$C222_1$	$P2_1/a$	

Property	Units	Condition	Value	Ref.
Degree of crystallinity	%	Form I (isotactic) from extrusion	48–55	(11)
Heat of fusion	kJ		6.318	(32,33)
	mol^{-1}		6.276	
			6.485	
		Clapeyron equation	7.782	(34,35)
			7.531	
Entropy of fusion	$kJ\,K^{-1}\,mol^{-1}$		$15.5*10^{-3}$	(32,33)
		Clapeyron equation	$15.8*10^{-3}$	(34,35)
			$19.2*10^{-3}$	
Avrami exponent		Compression molded, cooled @ 40°C/min from 180°C	0.9–1.07	(36)
		Blown film samples, draw ratio between 1 and 6	0.32–0.74	

Transition temperatures

Property	Units	Condition	Value	Ref.
Glass transition	K	DMA	256–256	(37)
Melting temperature	K	(1) Isotactic polymer		(16)
		I	411–415	
		II	393–403	
		III	374–383	
		(2) Syndiotactic polymer		
		I	323	
		II	323	
Sub T_g transition	K	Nature of transition: onset of local motion of side groups	115	(37)

Poly(butene-1)

Thermodynamic and other properties

Property	Units	Condition	Value	Ref.
Heat capacity	kJ mol^{-1} K^{-1}	Temperature		(38)
		100 K	0.0377	
		200 K	0.0684	
		300 K	0.1170	
		600 K	0.1720	
Polymers with which compatible		All proportions	Polypropylene	(39)
		Random copolymer	Polyethylene	(40)

Mechanical properties

Property	Units	ASTM Method	Value	Ref.
Tensile modulus	MPa	D638	290–295	(11)
Tensile strength at yield	MPa	D638	16–18	(11)
Tensile strength at break	MPa	D638	32–35	(11)
Elongation at break	%	D638	275–320	(11)
Flexural modulus	MPa	D790	375–380	(11)
Notched Izod impact strength	J m^{-1}	D256	640–800	(11)
Hardness, Shore D		D2240	55–65	(11)
Poisson's ratio		At 25°C	0.47	(41)
Dart impact strength	g	D1709 (for film thickness = 50.8 μm)	350	(11)
Elmendorf tear strength MD	kN m^{-1}	D1922 (50.8 μm thick film)	425	(11)
TD	kN m^{-1}	D1922 (50.8 μm thick film)	386	(11)

Electro-optical properties

Property	Units	Condition	Value	Ref.
Index of refraction n		Isotactic polymer	1.5125	(11)
Birefringence		Polymorphs		
		I	0.034	(42)
		II	0.013	(42)
Refractive index increment dn/dc	ml g^{-1}	Solvent/temperature (°C)		
		n-Nonane/35	0.092	(43)
		n-Nonane/80	0.108	(44)
		1-Chloronaphthalene/135	−0.206	
		Cyclohexane/25	0.074	
Dielectric constant		103–106 Hz	2.53	(11)
Dissipation factor		103–106 Hz	0.0005	(11)

Poly(butene-1)

Transport properties

Property	Units	Condition	Value	Ref.
Thermal conductivity	W m^{-1} K^{-1}	C177	0.22	(11)
Melt Index	g/10 min	ASTM D 1238 (E)	0.4	(11)
Oxygen permeability	m^3(STP) m s^{-1} m^{-2} Pa^{-1}	100 mil film	1.74*10^{-15}	(45)

Stabilities

Property	Units	Condition	Value	Ref.
Heat deflection temperature	K	@ 1.82 MPa, ASTM D648	327–333	(11)
	K	@ 0.46 MPa, ASTM D648	375–386	
Brittleness temperature	K	ASTM D746	255	(11)
Water absorption	%	24 h, ASTM D570	<0.03	(11)

References

1. Keim W. *Makromol. Chem. Macromol. Symp.* 1993;66:225.

2. Morris DD, Roberts M. *Chem. Week.* Nov. 18 1992;43.

3. Eur. Pat. Appl. 2522 (1979) to: Phillips Petroleum Co.

4. Kashiwa N, Yoshitake J. *Polym. Bull.* 1985;11:485.

5. Kaminsky W. *Angew. Chem., Int. Ed.* England, 1985;24:507.

6. Rossi A, Odian G, Zhang J. *Macromolecules.* 1995;28:1739.

7. Huang Q, Wu Q, Zhu F, Lin S. *J. Polym. Sci., Pt. A: Polym. Chem.* 2001;39:4068.

8. Eur. Pat. Appl. 6968 (1980) to: Conoco Inc.

9. U.S. Pat. 3907761 (1975) to: Ethylene Plastique, France.

10. Resconi L, Abis L, Franciscono G. *Macromolecules.* 1992;25:6814.

11. Chaterjee AM. In: Kroschwitz JI. ed. *Encyclopedia of Polymer Science and Engineering.* 2nd ed. Vol 2. New York: John Wiley and Sons Inc; 1985.

12. Zirkel A, Urban V, Richter D. *Macromolecules.* 1992;25:6148.

13. Zoller P. *J. Appl. Polym. Sci.* 1979;23:1057.

14. Zoller P. *J. Polym. Sci., Pt. B: Polym. Phys.* 1978;16:1491.

15. Wunderlich B. *Macromolecular Physics.* Vol 3. New York: Academic Press; 1980.

16. Olefin Polymers (Higher Olefins). In: Kirk-Othmer, ed. *Encyclopedia of Chemical Technology.* John Wiley & Sons; 1996.

17. Moraglio G, Gianotti G, Danusso F. *Europ. Polym. J.* 1967;3251.

18. Krigbaum WR, Kurz JE, Smith P. *J. Phys. Chem.* 1961;65:1984.

19. Moraglio G, Gianotti G, Zoppi F, Bonicelli U. *Europ. Polym. J.* 1971;7:303.

20. Sastry S, Patel RD. *Europ. Polym. J.* 1975;11:881.

21. Charlet G, Ducasse R, Delams G. *Polymer.* 1981;22:1190.

22. Stivala SS, Wales RJ, Levi DW. *J. App. Polym. Sci.* 1963;797.

23. Constantin D. *Europ. Polym. J.* 1977;13:907.

24. Katime I, Garro P, Teijon JM. *Europ. Polym. J.* 1975;11:881.

25. Endo R, Iimura K, Takeda M. *Bull. Chem. Soc.* Japan, 1964;37:950.

26. Natta G, Corradini P, Bassi IW. (A) *Makromol. Chem.* 1956;21:240. (B) *Nuovro Cimento Suppl.* 1960;15:52.

27. Turner-Jones A. *J. Polym. Sci. Part B.* 1963;1:455.

28. Fischer EW, Kloos F, Lieser G. *J. Polym. Sci. Part B.* 1969;7:845.

29. Yamashita M, Hoshino A, Kato M. *J. Polym. Sci., Pt. B: Polym. Phys.* 2007;45:684.

30. DeRosa C, Venditto V, Guerra G, Corradini P. *Makromol. Chem.* 1992;193:1351.

31. DeRosa C, Scaldarella D. *Macromolecules.* 1997;29:471.

32. Danusso F, Gianotti G. *Makromol. Chem.* 1963;61:139.

33. Wilski H, Grewer T. *J. Polym. Sci., Polym. Symp.* 1964;6:33.

34. Starkweather HW Jr, Jones GA. *J. Polym. Sci., Polym. Phys. ed.* 1986;24:1509.

35. Leute U, Dollhopt W. *Colloid. Polym. Sci.* 1983;261:299.

36. Kwang-Bum Hong, Joseph E. Spruiell. *J. Phys.Chem.* Ref. data, 1983;12:29.

37. Choy CL, Luk WK, Chen FC. *Polymer.* 1981;22:543.

38. Gaur U, Lau SF, Wunderlich BB, et al. *J. Phys. Chem.* 1972;27:215.

39. Foglia AJ. *App. Polym. Symp.* 1969;11:1.

40. Morton M, Fetters LJ. *Rubber Chem. Technol.* 1975;48:359.

41. Warfield RW, Robert Barnet F. *Angew. Makromol. Chem.* 1972;27:215.

42. Tanaka A, Sugimoto N, Asada T, Onogi S. *Polym. J.* 1975;7:529.

43. Horska J, Stejskal J, Kratochril P. *J. Appl. Polym. Sci.* 1983;28:3873.

44. Chandra R, Singh RP. *Makromol. Chem.* 1980;181:1637.

45. Luciani L, Seppala J, Lofgren B. *Prog. Polym. Sci.* 1988;13:37.

Poly[(*n*-butylamino)thionylphosphazene]

Ian Manners

Class Inorganic and semi-inorganic polymers

Structure [NSO(HNnBu){NP(NHnBu)$_2$}$_2$]$_n$

Major Applications matrix for gas sensing (3)

Properties of Special Interest Low glass transition temperature

Property	Units	Condition	Value	Reference
UV-Vis Absorption ($\lambda_{\text{max.}}$)	nm	THF Solution		
UV-Vis Absorption Coefficient (ε)	M^{-1}cm^{-1}	THF Solution		
Glass Transition Temperature	°C	DMA Experiment		
Glass Transition Temperature	°C	DSC Experiment	−16	1
Melting Temperature	°C	DSC Experiment		

Poly[(*n*-butylamino)thionylphosphazene]

References

1. Ni Y, Park P, Liang M, Massey J, Walding C, Manners I. *Macromolecules.* 1996;29:3401.
2. van Bolhuis F, van de Grampel JC. *Acta Crystallogr.* 1976;B32:1192.
3. Wang Z, McWilliams AR, Evans CEB, et al. *Adv. Func. Mater.* 2002;12:415.

Poly(butylene terephthalate)

Jude O. Iroh

Acronyms, Trade Names PBT

Class Linear and flexible aromatic polyester, thermoplastic

Structure

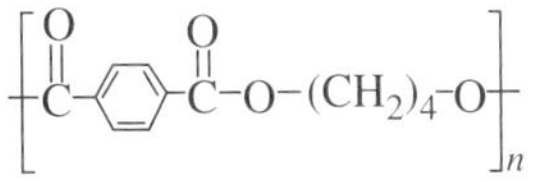

Synthesis

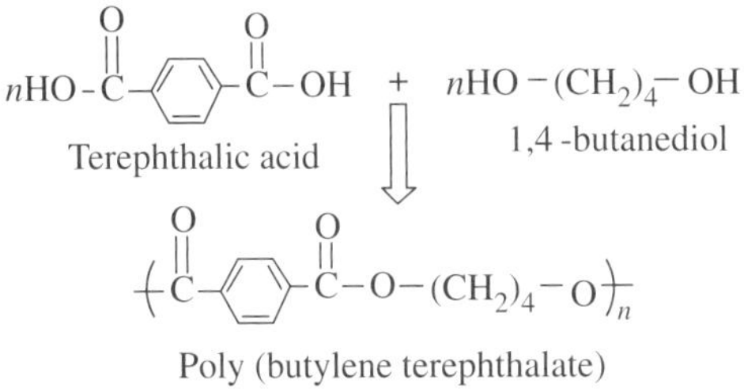

Terephthalic acid + 1,4 -butanediol

Poly (butylene terephthalate)

Molecular Conformation Nearly planar

Major Applications Molding plastic, molecular component of polyether ester thermoplastic block copolymer elastomer, fiber and plastic forming, used in tooth and paint brush and in bristles and filler fabrics.

Properties of Special Interest Mostly synthesized as flexible semi-crystalline thermoplastic, PBT has outstanding resilience and toughness. High toughness and resilience is due to improved chain flexibility derived from the four methylene units. Used in thermoplastic matrix composites for gears, machine parts, small pump housings, and insulators.

Preparative Techniques Synthesized by step growth polymerization between butylene glycol and terephthalic acid. PBT is often synthesized by ester exchange polymerization using weak basic catalysts such as alkanoates, hydrides, and alkoxides of sodium, lithium, zinc, calcium, magnesium, titanium etc. PBT is formed by the reaction of dimethyl terephthalate with 1,4-butanediol at 0.020 atm and 160–230°C. Final reaction occurs in the range of 260–300°C under vacuum at 0.001 atm. [1–4, 18–22]

Poly(butylene terephthalate)

Property	Units	Conditions	Value		Reference
Unit cell	–	–	Triclinic/allomorphs		[4]
Lattice constants	Degree	X-ray diffraction	α	β	
			$a = 4.86$	4.72	[5,6]
			$b = 5.96$	5.79	[5,6]
			$c = 1.165$	1.300	[5,6]
			$\alpha = 99.70$	102.70	[5,6]
			$\beta = 116$	120.2	[5,6]
			$\gamma = 110.8$	103.7	[5,6]
Unit cell volume	nm^3	X-ray diffraction	0.2615	0.2729	[5,6]
No of chains per unit cell	–	–	1		[6]
Unit cell density	g/cm^3	X-ray diffraction	1.397	1.338	[6]
Measured density	g/cm^3	–	1.34	1.33	[6]
Number of chains	–	–	1		[5,6]
No of monomers	–		1		[5,6]
Glass transition temperature (T_g)	°C	ASTM D3418	30–60		[5,7,8]
Melting temperature (T_m)	°C	ASTM D3418	222–232°C		[3,5 8,9]
Heat of fusion (ΔH)	KJ/mole	DSC	21.2		[5,7]
Mechanical properties					
Breaking strength, (σ_B)	MPa	ASTM D638	55		[5,8,10–14]
Tensile (Young's) modulus, E	GPa	ASTM D638	2.6		[5,8,10–14]
Flexural modulus (rigidity), E	GPa	3-point flexure ASTM D790	2.3		[5,8,10–14]
Ultimate strain, ε_B	%	ASTM D638	200–300		[5,8,10–14]
Yield strain, ε_Y	%	ASTM D638	4		[5,8,10–14]
Yield strength, σ_Y	MPa	ASTM D638	52		[15]
Impact strength	J/m	ASTM D256–86	53		[5,8,10–14]
Thermal					
Linear coefficient of thermal expansion, α	mm/mm/°C	ASTM D696	7.4×10^{-5}		[5,8,13,14]
Specific heat	kJ/kg.K	–	223		[5,8,13,14]
Thermal conductivity	W/mK	–	1.35		[5,8,13,14]
Thermal deflection	°C	ASTM D648	@ 264 psi = 54		[5,8,13,14]
			@ 66 psi = 154		[5,8,13,14]
Outdoor weathering	%	ASTM D1435	Good resistance		[5,8,13,14]

Poly(butylene terephthalate)

Property	Units	Conditions	Value	Reference
Electrical				
Volume resistivity,	Ωcm $\times 10^{16}$	ASTM D257	0.1	[5,8,13,14]
Dissipation factor				
100 Hz	100 Hz	D150	0.005	[5,8,13,14]
10^6 Hz	10^6 Hz	D150	0.012	[5,8,13,14]
Dielectric strength	kV/mm	ASTM D149	15.8	[5,8,13,14]
Dielectric constant	10^6 Hz	ASTM D150	3.24	[5,8,13,14]
Molecular weight of repeat unit	g/mole	–	220	–
Mark–Houwink constants	–	Solution viscometry (30°C)	$K = 1.17 \times 10^{-4}$ $a = 0.87$	[5,16]
Weight average molecular weight	g/mole	Light scattering	30 000–80 000	[5,17]

References

1. Whinfield JR, Dickson JT. US Patent. 2,465,319 (1940) to E.I. du Pont de Nemours & Co., Inc.
2. Jaquiss DBG, Borman WFH, Campbell RW. In: Grayson M, ed. *Encyclopedia of Chemical Technology.* Vol 18. New York: John Wiley and Sons, Inc; 1982:549.
3. Rodriguez F. *Principles of Polymer Systems.* 2nd ed. London: MacGraw Hill, International Student Edition; 1983:432, 435.
4. Wilfong RE. *J. Polym. Sci.* 1961;54:385.
5. Mark HF, Bikales NM, Overberger CG, Menges G, Kroschwitz JI, eds. *Encyclopedia of Polymer Science and Engineering.* Vol 12. New York: Wiley; 1985:226.
6. Hall IH. *Structure of Crystalline Polymers.* Barking, UK: Elsevier Applied Science Publishers, Ltd; 1984:39.
7. Illers KH, *Colloid Polym. Sci.* 1980;258:117.
8. Rubin II, ed. *Handbook of Plastics Materials and Technology.* New York: John Wiley and Sons, Inc, 1990:634.
9. Rosen SL, *Fundamental Principles of Polymeric Materials.* 2nd ed. New York: John Wiley and Sons Inc; 1993:111.
10. Theberge JE, Crosby J, Hutchins M. *Mach. Des.* 1985;57930:67.
11. Theberge JE. Polym. Plast. Technol. Eng. 1981;16(1):41.
12. U.S Pat. 3,764,576 (Oct. 8, 1973), Russo R. V. (to Celanese Corp.).
13. Jaquiss DBG, Borman WFH, Campbell RW. In: Grayson M, ed. *Encyclopedia of Chemical Technology.* Vol 18. New York: John Wiley & Sons, Inc; 1982:549.
14. Brozenick NJ. *Modern Plastics Encyclopedia.* New York: McGraw-Hill, Inc; 1986–1987:464.
15. Mark HF, Bikales NM, Overberger CG, Menges G, Kroschwitz JI, eds. *Encyclopedia of Polymer Science and Engineering.* Vol 12. New York: Wiley; 1985:23.
16. Borman WF. *J. Appl. Polym. Sci.* 1978;22:2119.
17. Brit. Pat. 1,320, 520 (1973), Jackson WJ Jr, Kuhfuss HF, Caldwell JR. (to Eastman Kodak Co.).

Poly(butylene terephthalate)

18. Colonna M, et al. *Polymer.* 2003;44:4773.
19. Banach TE, et al. *Polymer.* 2001;42:7511.
20. Banach TE, Colonna M. *Polymer.* 2001;42:7517.
21. Lotti N, Finelli L, Righetti MC, Munari A. *Polymer.* 2000;41:5297.
22. Chang J-H, Mun MK. *J Appl.* 2006;1247.

Poly (*n*-butyl isocyanate)

Chandima Kumudinie Jaysuriya and Jagath K. Premachandra

Acronym PBIC

Class poly (isocyanates); N-substituted 1-nylons

Structure

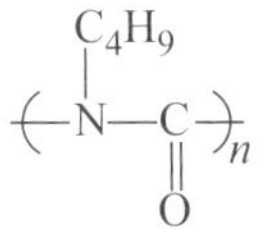

$$\left(\text{N} - \underset{\underset{\text{O}}{\|}}{\text{C}} \right)_n \quad \overset{\text{C}_4\text{H}_9}{|}$$

Major Application An ideal example of a polymer model for a rigid-rod macromolecular chain material amenable to physical studies.

Properties of Special Interest Hydrodynamic rigid-rod molecule, unusual chain stiffness, helical conformation[1].

Other Polymers Showing this Special Property Rigid-rod molecule, helical conformation: Poly (*n*-hexyl isocyanate), poly (γ-benzyl-L-glutamate).

Preparative techniques

Polymerization Process	Solvent	Temp. (°C)	Catalyst	Reference
Anionic	Benzene	−55	NaCN in dimethylformamide	(2)
Anionic	Toluene-THF	–	*n*-Butyllithium	(3)
Anionic	Toluene	−78	*n*-Butyllithium	(4)
Anionic	Toluene	−78	Fluorenyl sodium	(4)
Anionic	Toluene	−78	Ethyllithium in benzene	(5)
Anionic	Toluene	−78	$(C_2H_5)_2Be$	(5)
Anionic	–	−40 to −70	NaCN in dimethylformamide	(6)
Anionic	CH_2Cl_2	−78	*n*-Octyl sodium	(5)
Anionic	THF	−78	Ethyllithium in benzene	(5)
Anionic	Acetone	−78	$LiOC_4H_9$	(5)
Anionic	CH_2Cl_2	−78	Ethyllithium in benzene	(5)
Cationic*	–	–	–	(5)

* No polymers were obtained.

Poly (*n*-butyl isocyanate)

Property	Units	Conditions	Value	Reference
Molecular weight of repeat unit	g mol^{-1}	–	99	–
Typical molecular weight range of polymer	g mol^{-1}	M_w	$(0.14–2.3) \times 10^6$	(1)
		M_w	$(1.75–2.4) \times 10^5$	(7)
		M_w	$(0.23–5.3) \times 10^5$	(8)
Typical polydispersity index (M_w/M_n)	–	–	1.0–1.2	(9)
		–	2.5	(9)
		–	1.07–1.20	(10)
		Light scattering and osmometry	1.15–1.30	(11)
		Dielectric measurements	1.06–1.44	(11)
		In chloroform, initiator: NaCN in dimethylformamide	1.1–1.4	(4)
		In chloroform, initiator: Fluorenyl sodium	1.1–9.5	(4)
		In chloroform, initiator: n-butyl lithium	1.3–8.4	(4)
		In benzene, dielectric measurements	3.2–4.4	(4)
IR (characteristic absorption frequencies)	cm^{-1}	Carbonyl, solid state	~1700	(2,5,12–14)
		Carbonyl, dilute solution	1690–1695	(13,14)
		Disubstituted amide, solid state	1282 and 1390	(2)
UV	nm	Absorption maxima at the high wavelength band, λmax	254	(15,16)
	1 mol^{-1} cm^{-1}	Extinction coefficient	3.7×10^3	(16)
NMR		^{13}C NMR, 125.7-MHz spectrometer, in CDCl$_3$ at 55°C		(16)
		^{13}C NMR, 20-MHz spectrometer		(13,17)
		^{1}H NMR, poly[(S)−(+)_2- methyl butyl isocyanate], 220-MHz spectrometer		(18)
		^{1}H NMR		(13,17)
Solvents		Aromatic and chlorinated hydrocarbons		(2)
		Nonpolar solvents such as benzene and toluene		(1)
		THF, benzene, toluene		(3,7)
		CCl$_4$,CHCl$_3$, 1,2,4-trichlorobenzene-chloroform (75/25 v/v)		(7)
Nonsolvents		Methanol		(2,19)
		Acetone, ethyl acetate, dimethylformamide, methylethylketone		(3)
Second virial coefficient	mol cm^3 g^{-2}	In toluene at 37°C, $M_n = (0.23–5.3) \times 10^5$ gmol^{-1}, osmometry	$(1.85–2.36) \times 10^{-3}$	(8)
		In toluene at 37°C, $M_n = (0.23–5.2) \times 10^5$ g mol^{-1}, osmometry	$(1.8–2.5) \times 10^{-3}$	(4)
		In toluene, osmometry, arithmetic mean	2.13×10^{-3}	(4)
		In chloroform, $M_w = 6.1 \times 10^4$ g mol^{-1}	3×10^{-3}	(4)

Poly (*n*-butyl isocyanate)

Property	Units	Conditions	Value	Reference
		In chloroform, $M_w = 1.0 \times 10^7$ g mol^{-1}	2.1×10^{-3}	(4)
		In chloroform, $M_w = 1.03 \times 10^5$ g mol^{-1}	1.5×10^{-3}	(4)
		In chloroform, $M_w = 5.40 \times 10^5$ g mol^{-1}	2.5×10^{-3}	(4)
		In chloroform, $M_w = 1.65 \times 10^5$ g mol^{-1}	2.0×10^{-3}	(4)
		In chloroform, $M_w = 2.45 \times 10^6$ g mol^{-1}	2.4×10^{-3}	(4)
		In chloroform, light scattering, arithmetic mean	2.51×10^{-3}	(4)
Mark–Houwink parameters: K and a	–	In tetrahydrofuran	$a = 1.18$	(20)
		In carbon tetrachloride	$K = 3.16 \times 10^{-4}$ $a = 1.2$	(21, 22)
Huggins constant	–	In carbon tetrachloride, at 22°C	0.27–0.97	(10)
Radius of gyration	Å	In chloroform, $M_w = 6.1 \times 10^4$ g mol^{-1}	320	(4)
		In chloroform, $M_w = 1.0 \times 10^7$ g mol^{-1}	5200	
		In chloroform, $M_w = 1.03 \times 10^5$ g mol^{-1}	420	
		In chloroform, $M_w = 5.40 \times 10^5$ g mol^{-1}	1250	
		In chloroform, $M_w = 1.65 \times 10^5$ g mol^{-1}	625	
		In chloroform, $M_w = 2.45 \times 10^6$ g mol^{-1}	2800	
Monomer projection length		Translation diffusion	2.0	(21)
		Rotatory diffusion	2.7	(21)
		Viscosity	2.0	(21)
		Viscosity	1.50–1.80	(10)
		Relaxation time measurements	1.19–1.49	(10)
		Viscosity sedimentation-diffusion	2.1	(23, 24)
		Dielectric measurements	1.1	(25)
		Dielectric and viscoelastic relaxation	0.6–1.1	(26)
		In chloroform, light scattering	1.8–2.1	(4, 27)
		In chloroform	2	(28)
		X-ray diffraction	1.94	(29)
Chain diameter	Å	In toluene at 37°C, osmometry	10	(27)
			11	(9)

Property	Units	Conditions	Value	Reference
Number of monomer units in segment of molecular chain		Translatory diffusion	600	(21)
		Rotatory diffusion	500	
		Flow birefringence	370	
		Birefringence in an electric field	700	
Persistence length	Å	In carbon tetrachloride	600	(21)
		In chloroform	500–600	(27)
		In chloroform	300	(28)
		Viscosity sedimentation-diffusion	1300	(23, 24)
		Dielectric measurements	880	(25)
		Dielectric and viscoelastic relaxation	>400	(26)
		Light scattering	500–600	(4)
Space group		Triclinic with the axes of the molecule at 1/3,1/6,z and 2/3,5/6,z		(29)
Chain conformation		Rigid rod, nonpolar, and possibly helical		(1)
		Helical structure with a translation of 1.94 Å and rotation of 135° per monomeric unit (8_3 helix)		(29)
		Rigid rod up to degree of polymerization, DP~600, with an onset of flexibility at higher DP		(30)
		The conformation of the polymer is same in the solid phase and in solution		(29)
		The onset of flexibility occurs at $M_w = (0.73–1.33) \times 10^5$ g mol^{-1}		(15)
		Flexibility is encountered when $M_w > 5.0 \times 10^4$ g mol^{-1}		(9)
		Low molecular weight molecule, $M_w < 8.0 \times 10^4$ is rodlike and helical. At high molecular weight, $M_w > 1.0 \times 10^6$, the polymer chain conformation is random coil		(25)
Unit cell dimensions				(29)
Lattice	–	–	Pseudohexagonal	
Monomers per unit cell	–	–	2	
Cell dimensions	Å	–	$a = b = 13.3, c = 15.4$	
Degree of crystallinity	%	$[\eta] = 4.2$ dl g^{-1} in benzene	44.5	(5)
		$[\eta] = 7.3$ dl g^{-1} in benzene	36.4	(5)
		$[\eta] = 10.6$ dl g^{-1} in benzene	22.7	(5)
		$[\eta] = 11.6$ dl g^{-1} in benzene	19.1	(5)
		Relatively high and depends on the catalyst used in polymerization	–	(5)
		Depends on the post treatment of the polymer	–	(31)
		Higher degree of crystallinity of the polymer prepared with C_2H_5Li than with NaCN in dimethylformamide	–	(5)

Poly (*n*-butyl isocyanate)

Property	Units	Conditions	Value	Reference
Density	g cm^{-3}	–	0.97	(29)
		–	1.25	(7)
		Highly crystalline	1.071	(25)
		Calculated	1.10	(29)
Glass transition temperature	K	DSC	None observed	(7)
		DTA-DSC, at 20°C min^{-1}: Ar atmosphere	258	(13)
Melting point	K	DTA-DSC; Ar atmosphere; heating rate; 20°C min^{-1}	~458	(13)
		^{13}C NMR thermal cycling in the buck	463	(17)
		Polymerization catalyst: n-octyl sodium; solvent: CH$_2$Cl$_2$	438	(5)
		Polymerization catalyst: C$_2$H$_5$Li in benzene: solvent: THF	438	(5)
		By hot-stage microscopy	487	(32)
			471–473	(31)
			472–474	(31)
			482	(33)
Mesomeric transition temperatures	K	Crystalline–nematic transition, by hot-stage microscopy	484	(32)
Sub-T_g transition temperature	K	Relaxations, by hot-stage microscopy	444	(32)
Softening temperature	K	–	453	(2,33)
Enthalpy of propagation	kJ mol^{-1}	–	5.7	(6)
Tensile modulus	MPa	At room temperature, η_{sp}/C in CHCl$_3$ = 8.1 dl g^{-1}	~810	(13)
		At 23°C, strain rate = 8.3 × 10^{-4} s^{-1}	~500	(34)
		At 20°C, strain rate = 6.7 × 10^{-4} s^{-1}	~1.570	(7)
		At 80°C, strain rate = 6.7 × 10^{-4} s^{-1}	~2.940	(7)
		At 0°C, strain rate = 6.7 × 10^{-4} s^{-1}	~1.765	(7)
Tensile strength	MPa	At room temperature, η_{sp}/C in CHCl$_3$ = 8.1 dl g^{-1}	~45	(13)
		At 23°C, strain rate = 8.3 × 10^{-4} s^{-1}	~52	(34)
		At –20°C, strain rate = 6.7 × 10^{-4} s^{-1}, 30–40% crystalline, η = 7.25 dl g^{-1}	~58	(7, 35)

Property	Units	Conditions	Value	Reference
		At 0°C, strain rate = 6.7×10^{-4} s^{-1}, 30–40% crystalline, $\eta = 7.25$ dl g^{-1}	~50	(7, 35)
		At 23°C, strain rate = 6.7×10^{-4} s^{-1}, 30–40% crystalline, $\eta = 7.25$ dl g^{-1}	~34	(7, 35)
Maximum extensibility	%	At room temperature, η_{sp}/C in $CHCl_3 = 8.1$ dl g^{-1}	32	(13)
		At 23°C, strain rate = 8.3×10^{-4} s^{-1}	~33	(34)
		At –20°C, strain rate = $6.7 \times 10^{-4} s^{-1}$, 30–40% crystalline, $\eta = 7.25$ dl g^{-1}	~15	(7, 35)
		At 0°C, strain rate = 6.7×10^{-4} s^{-1}, 30–40% crystalline, $\eta = 7.25$ dl g^{-1}	~23.5	(7, 35)
		At 23°C, strain rate = 6.7310^{-4} s^{-1}, 30–40% crystalline, $\eta = 7.25$ dl g^{-1}	~39	(7, 35)
Specific refractive index increment, *dn/dc*	ml g^{-1}	In chloroform, $\lambda_o = 5460$Å	0.054	(1, 11)
		In tetrahydrofuran, $\lambda_o = 4358$Å	0.99	(27)
Relative electrical birefringence	–	In carbon tetrachloride, frequency $< 5 \times 10^3$ Hz, $M_w = 2.46 \times 10^5$ g mol^{-1}	~1–0.03	(21)
		In carbon tetrachloride; frequency range: $< 10^3$–10^4 Hz; $M_w = 2.0 \times 10^4$ g mol^{-1}	~1–0.75	
Dielectric constant ε'	–	In dilute benzene solution (10^{-4} g cm^{-3}), at 22.5°C; frequency range: 10^{-1} to 2×10^6 Hz; $M_w = 1.84 \times 10^6$ g mol^{-1}	~2.28–2.76	(1)
		In carbon tetrachloride at 22.9°C; frequency range: 10^{-1}–10^6 Hz; $M_w = 2.35 \times 10^5$ g mol^{-1}	2.35–2.37	(25)
Dielectric loss ε''	–	In dilute benzene solution (10^{-4} g cm^{-3}), at 22.5°C; frequency range: 10^{-1} to 2×10^6 Hz; $M_w = 1.84 \times 10^6$ g mol^{-1}	~0.01–0.10	(1)
		In carbon tetrachloride at 22.9°C; frequency range: 10^{-1}–10^6 Hz; $M_w = 2.35 \times 10^5$ g mol^{-1}	~0.003–0.053	(11, 25)

Poly (*n*-butyl isocyanate)

Property	Units	Conditions	Value	Reference
Dielectric critical frequency	Hz	In dilute benzene solution (10^{-4} g cm^{-3}), at 22.5°C; frequency range: 10^{-1} to 2×10^6 Hz; $M_w = (0.14-2.3) \times 10^6$ g mol^{-1}	100 000–32	(1)
		In carbon tetrachloride at 22.9°C; frequency range: 10^{-1}–10^6 Hz; $M_w = 2.35 \times 10^5$ g mol^{-1}	~540	(11, 25)
Relaxation time	μs	In dilute benzene solution (10^{-4} g cm^{-3}), at 22.5°C; frequency range: 10^{-1} to 2×10^6	1.6–5000	(1)
		Hz; $M_w = (0.14-2.3) \times 10^6$ g mol^{-1}	17.8	(9)
		In tetrahydrofuran, $M_w = 3 \times 10^5$ g mol^{-1}	20.0	(9)
		In chloroform, $M_w = 3 \times 10^5$ g mol^{-1}	25.0	(9)
		In benzene, $M_w = 3 \times 10^5$ g mol^{-1}	30.0	(9)
		In benzene, $M_w = 1.33 \times 10^5$ g mol^{-1}	3.3	(9)
		In benzene, $M_w = 6.0 \times 10^4$ g mol^{-1}	0.42	(9)
		In benzene, $M_w = 2.0 \times 10^4$ g mol^{-1}	0.40–1260	(10)
		In carbon tetrachloride at 22°C; $M_w = (0.2-3.8) \times 10^5$ g mol^{-1}	~0.40–20	(25)
		In carbon tetrachloride at 22.9°C; $M_w = (2-7.3) \times 10^4$ g mol^{-1}	~45–1995	(25)
		In carbon tetrachloride at 22.9°C; $M_w = (1.03-5.4) \times 10^5$ g mol^{-1}	~4467–89 125	(25)
		In carbon tetrachloride at 22.9°C; $M_w = (0.12-1.0) \times 10^7$ g mol^{-1}		
Dipole moment D	–	Net dipole moment is parallel to the major axis of the molecule		(1)
		$M_w = 1.7 \times 10^6$ g mol^{-1}	4500	(36)
		$M_w = 3.8 \times 10^5$ g mol^{-1}	2120	(36)
		In benzene, $M_n = 1.2 \times 10^5$ g mol^{-1} specific volume $= 0.8$ cm^3 g^{-1}	1224	(9)
		In benzene, $M_n = 5 \times 10^4$ g mol^{-1} specific volume $= 0.8$ cm^3 g^{-1}	481	(9)

Property	Units	Conditions	Value	Reference
		In benzene, $M_n = 2 \times 10^4$ g mol^{-1} specific volume $= 0.8$ cm^3 g^{-1}	179	(9)
		In carbon tetrachloride at 22.9°C; $M_w = (2–7.3) \times 10^4$ g mol^{-1}	226–726	(25)
		In carbon tetrachloride at 22.9°C; $M_w = (1.03–5.4) \times 10^5$ g mol^{-1}	806–2561	(25)
		In carbon tetrachloride at 22.9°C; $M_w = (1.2–3.5) \times 10^6$ g mol^{-1}	3768–6277	(25)
		In carbon tetrachloride at 22.9°C; $M_w = (0.54–1.0) \times 10^7$ g mol^{-1}	2561–10 954	(25)
Anisotropy of the monomer unit	–	–	1.1×10^{-24}	(21)
			1.46×10^{-24}	(37)
Anisotropy of the Kuhn statistical segment	–	–	3.6×10^{-22}	(21)
Kerr constant	–	–	$(0.25–24) \times 10^{-7}$	(21)
Optical activity: $[\alpha]_D$	Degree	Poly[(S)-(+)-2-methylbutyl isocyanate, in chloroform	+160	(18)
Reduced dichorism	–	In carbon tetrachloride at 22.9°C; $M_w = 73\,000$ g mol^{-1}, field strength $\sim(1.3–2.3) \times 10^4$ V cm^{-1}	$\sim$0.24–0.068	(15)
Diffusion coefficient D	–	In carbon tetrachloride, $M_w = 10^3–10^6$ g mol^{-1}	$(0.075–1.6) \times 10^{-6}$	(21)
		$M = (0.10–3.0) \times 10^5$ g mol^{-1}	1.07×10^{-3} M$^{-0.8}$	
Rotatory diffusion coefficient	S^{-1}	In tetrahydrofuran, $M_w = 3 \times 10^5$ g mol^{-1}	2.8×10^4	(9)
		In chloroform, $M_w = 3 \times 10^5$ g mol^{-1}	2.5×10^4	
		In benzene, $M_w = 3 \times 10^5$ g mol^{-1}	2.0×10^4	
		In benzene, $M_w = 1.33 \times 10^5$ g mol^{-1}	1.66×10^4	
		In benzene, $M_w = 6.0 \times 10^4$ g mol^{-1}	1.5×10^5	
		In benzene, $M_w = 2.0 \times 10^4$ g mol^{-1}	1.20×10^6	

Poly (*n*-butyl isocyanate)

Property	Units	Conditions	Value	Reference
Index of sedimentation $[S]$	–	In carbon tetrachloride, $M_w = 10^3 - 10^6$ g mol^{-1}	$\sim(1.26-1.38) \times 10^{-13}$	(21)
		$M = (0.10-3.0) \times 10^5$ g mol^{-1}	$1.97 \times 10^{-14} M^{0.2}$	(21)
Intrinsic viscosity $[\eta]$	dl g^{-1}	In benzene, $C = 0.2$ g dl^{-1}	1.2	(31)
		In carbon tetrachloride, $M_w = 10^3 - 10^6$ g mol^{-1}	$\sim 12-2370$	(21)
		$M = (0.10-3.0) \times 10^5$ g mol^{-1}	$3.16 \times 10^{-4} M^{1.2}$	(21)
Decomposition temperature	K	At polymer melting point	482	(2)
		TGA, heating rate: 10°C min^{-1}	~ 423	(13)
		TGA, heating rate: 20°C min^{-1} nitrogen atmosphere	458	(38)
		By hot-stage microscopy	~ 492	(32)

Pyrolyzability[38]

Conditions	Observation
Direct pyrolysis mass spectrometry	Cyclic trimer of n-butyl isocyanate as the principle decomposition product,and trace amounts of monomer

References

1. Yu H, Bur AJ, Fetters LJ. *J. Chem. Phys.* 1966;44:2568.
2. Shashoua VE, Sweeny W, Tietz RF. *J. Am. Chem. Soc.* 1959;82:866.
3. Godfrey RA, Miller GW. *J. Polym. Sci., Part A-1.* 1969;7:2387.
4. Fetters LJ, Yu H. *Macromolecules.* 1971;4:385.
5. Natta G, Dipietro J, Cambini M. *Macromol. Chem.* 1962;56:200.
6. Eromosele IC, Pepper DC. *J. Polym. Sci., Polym. Chem., Ed.* 1987;25:3499.
7. Owadh AA, Parsons IW, Hay JN, Haward RN. *Polymer.* 1978;19:386.
8. Fetters LJ, Yu H. *Polym. Prepr.* 1966;7:443.
9. Jennings BR, Brown BL. *Eur. Polym. J.* 1971;7:805.
10. Bur AJ, Fetters LJ, Yu H. *Macromolecules.* 1973;6:874.
11. Bur AJ. *J. Chem. Phys.* 1970;52:3813.
12. Iwakura Y, Uno K, Kobayashi N. *J. Polym. Sci., Part A-1.* 1968;6:1087.
13. Aharoni SM. *Macromolecules.* 1979;12:94.
14. Aharoni SM, Sibilia JP. *Polym. Prepr.* 1979;20:118.
15. Milstein JB, Charney E. *Macromolecules.* 1969;2:678.
16. Green MM, Gross RA, Crosby C III, Schilling FC. *Macromolecules.* 1987;20:992.
17. Aharoni SM. *Polym. Prepr.* 1980;21(1):209.
18. Goodman M, Chen S. *Macromolecules.* 1971;4:625.
19. Berger MN, Tidswell BM. *J. Polym. Sci.* 1973;42:1063.
20. Burchard W. *Macromol. Chem.* 1963;67:182.
21. Tsvetkov VN, Shtennikova IN, Rjumtsev EI, Yu P. Getmanchuk. *Eur. Polym. J.* 1971;7:767.
22. Tsvetkov VN, Rjumtsev YeI, Shtennikova IN. *Polym. Sci ,USSR.* 1971;13:579.
23. Tsvetkov VN, et al. *Polym. Sci ,USSR.* 1968;10:2482.

24. Tsvetkov VN. *Eur. Polym. J., Suppl.* 1969;23:7.
25. Bur AJ, Roberts DE. *J. Chem. Phys.* 1969;51:406.
26. Dev SB, Lochhead RY, North AM. *Discuss. Faraday Soc.* 1970;49:244.
27. Fetters LJ, Yu H. *Polym. Prepr.* 1970;2:1093.
28. Rubingh DN, Yu H. *Polym. Prepr.* 1973;14(2):1118.
29. Shmueli U, Traub W. *J. Polym. Sci., PartA-2.* 1969;7:515.
30. Troxell TC, Scheraga HA. *Macromolecules.* 1971;4:528.
31. Iwakura Y, Uno K, Kobayashi N. *J. Polym. Sci., Part A-2.* 1966;4:1013.
32. Aharoni SM. *J. Polym. Sci., Polym. Phys. Ed.* 1980;18:1303.
33. Ulrich H. In: Mark HF, et al., eds. *Encyclopedia of Polymer Science and Engineering.* Vol 8. New York: John Wiley and Sons; 1987:448–462.
34. Aharoni SM. *Polymer.* 1981;22:418.
35. Owadh AAJ, Parsons IW, Hay JN, Haward RN. *Polymer.* 1976;17:926.
36. Baijal MD, Diller RM, Pool FR. *Polym. Prepr.* 1969;10:1464.
37. Tsvetkov VN, Andreeva LN. *Adv. Polym. Sci.* 1981;39:95.
38. Durairaj B, Dimock AW, Samulski ET, Shaw MT. *J. Polym. Sci., Part A, Polym. Chem.* 1989;27:3211.

Poly(ε-caprolactone)

Jude O. Iroh

Acronyms, Trade Names PCL

Class Linear aliphatic flexible polyester, thermoplastic

Structure

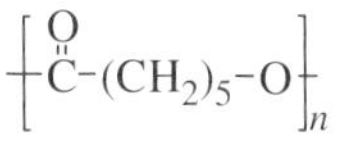

$$\left[\!\!\!\begin{array}{c} O \\ \| \\ C\text{-}(CH_2)_5\text{-}O \end{array}\!\!\! \right]_n$$

Poly(ε-caprolactone)

Molecular Conformation Nearly planar

Major Applications Films, formulation of copolymers, biodegradable polyesters, formulation of elastomeric block copolyesters, formation of diol for extension by diisocyanate

Properties of Special Interest Mostly synthesized as semi-crystalline thermoplastic. PCL is a clear and flexible polyester with elastomeric properties.

Preparative Techniques Synthesized by ring opening addition polymerization of ε-caprolactone at 170°C in nitrogen atmosphere using dibutyl stanneous oxide (Bu_2SnO) as the catalyst. A wide range of initiators such as organometallic catalysts and alkanolamine can be used [1–7, 14]. The copolymer poly(ε-caprolactone-co-ethylene glycol) is used as diol extension for polyurethane. PCL has also been synthesized by microwave irradiation polymerization [15].

Synthesis

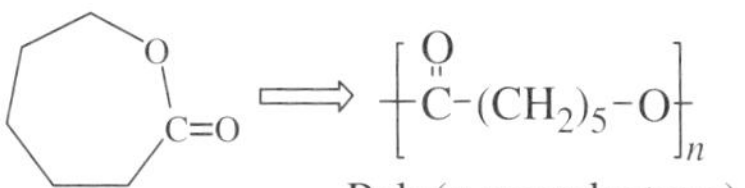

ε-Caprolactone $\implies$ Poly(ε-caprolactone)

Property	Units	Conditions	Value	Reference
Molecular weight of repeat unit	g/mol	–	114	–
Weight average molecular weight	g/mol	GPC	74 000	[8]
Number average molecular weight	g/mol	GPC	25 000	[11]

Property	Units	Conditions	Value	Reference
Intrinsic viscosity	cm^3/g	Dilute solution viscometry	0.9	[10]
Solvents	–	Dimethylacetamide (DMAc), benzene, chloroform	–	[9, 13]
Thermodynamic polymerization data (@ 25°C and 1 atm)				
Enthalpy of polymerization	kJ/mol	**25°C and 1 atm**	–28.8	[8]
Entropy of polymerization	kJ/mol	**25°C and 1 atm**	–53.9	[8]
Gibbs free energy of polymerization	kJ/mol	**25°C and 1 atm**	–12.8	[8]
Physical state	–	Semi-crystalline	–	–
Degree of crystallinity	%	DSC	69	[11]
Unit cell	–	X-ray diffraction	Orthorhombic: 2θ peaks at 22° and 24°	[9–12]
Lattice constants	Angstrom, Å	X-ray diffraction	$a = 7.45$ $b = 4.98$ $c = 17.05$	[10] [10] [10]
Number of repeating units per unit cell	–	–	4	[10]
Unit cell density	g/cm^3	X-ray diffraction	1.20	[10]
Measured density	g/cm^3	X-ray diffraction	1.094–1.2000	[8, 10, 12]
Elongation	%	–	700	[8]
Glass transition temperature (T_g)	°C	DSC	–72	[8, 12]
Melting temperature (T_m)	°C	DSC	58	[8, 12]
Heat of fusion (ΔH_f)	kJ/mol	DSC	8.9	[12]

References

1. Lundberg RD, Koleske JV, Wischmann KB. *J. Polym. Sci. Part A-1.* 1969;7:2915.
2. Mazier C, Douillard C, Merle G, Pascault JP. *Eur. Polym. J.* 1980;16:773.
3. U.S. Pat. 3,274,143 (Sept. 10, 1966), Hostettler F, Young DM. (to Union Carbide Corp.).
4. Schindler A, Jeffcoat R, Kimmel GL, Pitt CG, Wall ME, Zweidinger R. In: Pearce EM, Schaefgen JR. eds. *Contemporary Topics in Polymer Science.* Vol 2. New York: Plenum Publishing Corp; 1977:251–289.
5. Teyssie P, Bioul JP, Hamitou A, et al. *Polym. Prepr. Am. Chem. Soc. Div. Polym. Chem.* 1977;18:65.
6. Ito K, Hashizuka Y, Yamashita Y. *Macromolecules.* 1977;10:821.
7. Mark HF, Bikales NM, Overberger CG, Menges G, Kroschwitz JI, eds. *Encyclopedia of Polymer Science and Engineering.* Vol 12. New York: Wiley; 1985:1–37.
8. Shuster M, Narkis M. *Polym. Eng. & Sci.* 1994;34(21):1613.

9. Nishio Y, Manley R. *St. J., Polym. Eng. & Sci.* 1990;30(2):71.

10. Chatani Y, Okita Y, Tadokoro H, Yamashita Y. *Polym. J.* 1970;1:555.

11. Ong CJ, Price FP. *J. Polym. Sci. Polym. Symp.* 1978;63:45.

12. Huarng JC, Min K, White JL. *Polym. Eng. & Sci.* 1988;28(24):1590.

13. Mark HF, Bikales NM, Overberger CG, Menges G, Kroschwitz JI, eds. *Encyclopedia of Polymer Science and Engineering.* Vol 12. New York: Wiley; 1985:50.

14. Gorrasi GM, et al. *Polymer.* 2003;44:2271.

15. Barbier-Baudry D, et al. *Environ Chem Lett.* 2003;1:19.

Polycarbonate

Robert R. Gallucci and James L. DeRudder

Acronyms, Alternate Names, Trade Names PC, BPA-PC, bisphenol-A polycarbonate, Lexan® (SABIC Innovative Plastics), Makrolon® (Bayer), Calibre® (Dow), Panlite® (Teijin), Iupilon® (Mitsubishi Eng.), Xantar® (DSM).

Class Polyesters

Synthesis Polycondensation. Polymerized in solution using interfacial polymerization of bisphenol-A (BPA) and phosgene, also made in the melt by transesterification of diphenyl carbonate and BPA [1–3].

Structure

Major Applications Polycarbonate resins are used in a very wide variety of applications primarily for the production of durable goods [1–3]. It is a naturally clear and colorless polymer but can be provided in many transparent, translucent, or opaque colors. The polymer has an excellent balance of high heat resistance, stiffness, strength, dimensional stability, low creep, ignition resistance, and exceptional impact strength.

Well-known applications are optical recording media, for example CDs and DVDs, where its transparency and dimensional stability are important. In automotive lighting, especially headlamp lenses, its high heat and impact strength are vital. Its clarity, stiffness, and impact are also utilized in large and small water bottles, kitchenware, and containers. Computer enclosures and cell phones use its good flame resistance, toughness, and electric insulating capabilities. Architectural glazing, outdoor signs, and films use its toughness, high gloss, and water resistance. Films for appliance control panels, including LCD displays, use its optical properties, high heat, flexibility, and toughness. Polycarbonate is used in medical devices where it provides resistance to sterilization techniques. It is additionally used in eyeglass lens and frames, helmets, face shields, safety equipment, airplane windows and canopies, as well as bulletproof glazing. The exceptionally high impact strength of polycarbonate is an essential feature of many automotive and transportation market components, where it is used in exterior and interior components.

Polycarbonate is also used as a major component in many polymer blends, notably with ABS, ASA, PEI, and thermoplastic polyesters, such as PET, PBT, or PCT. Many fillers, such as glass fiber, carbon fiber, talc, clay, and PTFE are used to make commercial products.

Polycarbonate

Processing Polycarbonate is a thermoplastic that is usually processed in the melt by injection molding. It can also be processed by foam molding, gas assist molding, extrusion, blow molding, compression molding, or rotomolding. Polycarbonate should be thoroughly dried prior to melt processing to prevent degradation of the resin by hydrolysis. Polycarbonate can also be processed as a polymer solution in solvents such as chloroform or methylene chloride. Polycarbonate is also available in sheet, film, slabs, and rod stock that may be thermoformed, vacuum formed, pressure formed, sawn, sheared, or machined into final products.

Properties of Special Interest Polycarbonate engineering thermoplastics are amorphous clear polymers that exhibit superior dimensional stability, good electrical insulating properties, and outstanding impact strength. The balance of clarity, high heat resistance and impact with high modulus, as well as resistance to ignition make them useful in many applications.

PC is easily welded or solvent bonded. The limited resistance of PC to polar organic solvents, such as acetone, facilitates painting, hard coating for improved scratch resistance, decoration, and adhesion to other substrates. Strong bases such as caustic, ammonia, and ethylamine attack polycarbonate resins.

General references to polycarbonate: crystallinity, thermal properties, fracture toughness, optical properties, viscoelastic behavior and critical molecular weight, solvents, gas permeation as well as photo aging and degradation are shown in references 4–11 respectively.

Note that that there are many specific grades of polycarbonate having a wide range of performance features. The properties presented here are average values or ranges for standard PC resins with minimal additives. Check with the manufacturers for detailed information about a specific grade and color combinations. This data is not provided for specification and provides no license or warranty. The fitness for use of a resin in a specific application must be determined by end use testing of the device in the actual application.

Property	Value	Method	Reference
Thermal			[4,5]
T_g Glass transition temp. °C	140–151°C	ASTM D3418	SABIC Innovative Plastics Data
Vicat 10N	154°C	ASTM D1525	SABIC Innovative Plastics Data
Vicat 50N	150°C	ASTM D1525	SABIC Innovative Plastics Data
HDT 66 psi	140°C	ASTM D648	SABIC Innovative Plastics Data
HDT 264 psi	134°C	ASTM D648	SABIC Innovative Plastics Data
Specific heat	1.25 J/g°C	ASTM C351	SABIC Innovative Plastics Data
Thermal conductivity	0.29 W/m-°C	ASTM C177	SABIC Innovative Plastics Data
Coefficient of Linear Thermal Expansion (CTE) m/m C (−40–95°C)	6.75×10^5	ASTM E831	SABIC Innovative Plastics Data
Annealing temperature, degrees below T_g, °C	30°C		SABIC Innovative Plastics Data
Tm, Melting point, solvent-induced crystallinity	220–260°C		[4]

Polycarbonate

Property	Value	Method	Reference
Mechanical			
Tensile strength (Y)	62 Ma	ASTM D638	SABIC Innovative Plastics Data
Elongation (Y)	5–7%	ASTM D638	SABIC Innovative Plastics Data
Elongation (B)	100–150%	ASTM D638	SABIC Innovative Plastics Data
Flex modulus	2340 MPa	ASTM D790	SABIC Innovative Plastics Data
Flex strength	97 MPa	ASTM D790	SABIC Innovative Plastics Data
Compressive strength	86 MPa	ASTM D695	SABIC Innovative Plastics Data
Compressive modulus	2400 MPa	ASTM D695	SABIC Innovative Plastics Data
Shear strength (Y)	41 MPa	ASTM D732	SABIC Innovative Plastics Data
Shear strength (B)	69 MPa	ASTM D732	SABIC Innovative Plastics Data
Shear modulus	785 MPa	ASTM D732	SABIC Innovative Plastics Data
Poisson's ratio	0.38		SABIC Innovative Plastics Data
Notched Izod impact 23°C	600–900 J/m	ASTM D256	SABIC Innovative Plastics Data
Unnotched Izod impact 23°C	3200 J/m	ASTM D256	SABIC Innovative Plastics Data
Tensile impact kJ/m^2 (S type)	470–762 kJ/m^2	ASTM D1822	SABIC Innovative Plastics Data
Multi axial impact J	60–65 J	ASTM D3763	SABIC Innovative Plastics Data
Fracture toughness (MPa/m$^{0.5}$)K_{1c} brittle	4.2		[6]
Optical			[7]
Abbe Number @ 20°C	27.0	ASTM C1648	SABIC Innovative Plastics Data
Refractive Index 435.8 nm	1.6115	ASTM D542	SABIC Innovative Plastics Data
Refractive Index 480.0 nm	1.6011	ASTM D542	SABIC Innovative Plastics Data
Refractive Index 546.1 nm	1.5906	ASTM D542	SABIC Innovative Plastics Data
Refractive Index 587.6 nm	1.5860	ASTM D542	SABIC Innovative Plastics Data
Refractive Index 643.8 nm	1.5813	ASTM D542	SABIC Innovative Plastics Data
% Transmission 3.2 mm	89%	ASTM D1003	SABIC Innovative Plastics Data
Physical			
Density	1.19	ASTM D792	SABIC Innovative Plastics Data
Specific volume	0.83 cm^3/g	ASTM D792	SABIC Innovative Plastics Data
Water absorption 24 h immersion 23°C	0.15%	ASTM D570	SABIC Innovative Plastics Data
Water absorption equilibrium immersion 23°C	0.35%	ASTM D570	SABIC Innovative Plastics Data
Water absorption equilibrium immersion 100°C	0.58%	ASTM D570	SABIC Innovative Plastics Data
Coefficient of friction, static, steel	0.33	ASTM D1894	SABIC Innovative Plastics Data

Polycarbonate

Property	Value	Method	Reference
Coefficient of friction, dynamic, steel	0.39	ASTM D1894	SABIC Innovative Plastics Data
K factor ($\times 10^7$ in/h @zpv = 2000 psi ft/min)	13		SABIC Innovative Plastics Data
Mold shrinkage	~6 in/in-C	ASTM D955	SABIC Innovative Plastics Data
Hardness Rockwell M	70 M	ASTM D785	SABIC Innovative Plastics Data
Hardness Rockwell R	118 R	ASTM D785	SABIC Innovative Plastics Data
Fatigue limit 2.5 MM cycles (MPa)	6.9	ASTM D671	SABIC Innovative Plastics Data
Fatigue life @ 20 MPa, Uniaxial tensile fatigue	10^5–10^7 cycles		SABIC Innovative Plastics Data
Deformation under load 27 MPa 23°C	0.20%	ASTM D621	SABIC Innovative Plastics Data
Deformation under load 27 MPa 70°C	0.50%	ASTM D621	SABIC Innovative Plastics Data
Molecular weight			
Melt flow index @ 300°C 1.2 kg	2–60 g/10 min	ASTM D1238	SABIC Innovative Plastics Data
Mw, Absolute	15 000–40 000	ASTM D5296	SABIC Innovative Plastics Data
Mn, Absolute	7000–19 000	ASTM D5296	SABIC Innovative Plastics Data
Disp.	~2.1	ASTM D5296	SABIC Innovative Plastics Data
Mw, Relative to Polystyrene	28 000–85 000	ASTM D5296	SABIC Innovative Plastics Data
Mn, Relative to Polystyrene	13 000–40 000	ASTM D5296	SABIC Innovative Plastics Data
Molecular weight of a repeat unit	254		SABIC Innovative Plastics Data
Critical molecular weight, Me	2400 g/mol		[8]
Critical molecular weight, Mc	4800 g/mol		[8]
Solvents, (Solubility @ 23°C)			[9]
Methylene chloride	~22 wt%		SABIC Innovative Plastics Data
Chloroform	~15 wt%		SABIC Innovative Plastics Data
Dioxane	~5 wt%		SABIC Innovative Plastics Data
1,1,2,2 -Tetrachloroethylene	~15%		SABIC Innovative Plastics Data
1,2-Dichloroethane	~2%		SABIC Innovative Plastics Data
Pyridine	~16%		SABIC Innovative Plastics Data
Flammability			
Oxygen index	26	ASTM D2863	SABIC Innovative Plastics Data
UL 94 Flammability	HB to V0, 5VB	UL 94	SABIC InnovativePlastics Data
Continuous use temp.	140°C	UL 746 A,B,C	SABIC Innovative Plastics Data
Hot wire ignition	Class 2	UL 746A	SABIC Innovative Plastics Data
High voltage arc track rate	Class 2	UL 746A	SABIC Innovative Plastics Data
High Ampere arc ignition, surface	Class 1	UL 746A	SABIC Innovative Plastics Data
Comparative tracking index	Class 2	UL 746A	SABIC Innovative Plastics Data

Property	Value	Method	Reference
Electrical			
Volume resistivity ohm-cm	$> 1.0 \times 10^{17}$	ASTM D257	SABIC Innovative Plastics Data
Dielectric strength v/mil (in air)	380–425	ASTM D149	SABIC Innovative Plastics Data
Dielectric constant 60 Hz	3.17	ASTM D150	SABIC Innovative Plastics Data
Dielectric constant 1 Hz	2.96	ASTM D150	SABIC Innovative Plastics Data
Dissipation factor 60Hz	0.0009	ASTM D150	SABIC Innovative Plastics Data
Dissipation factor 1Hz	0.01	ASTM D150	SABIC Innovative Plastics Data
Gas permeation @ 25°C			[10]
Diffusion constant (D) cm^2/s			
H_2	640×10^{-9}		[10]
H_2O	68×10^{-9}		[10]
O_2	21×10^{-9}		[10]
Argon	15×10^{-9}		[10]
CO_2	5×10^{-9}		[10]
Permeation rate (P) cm^2 gas (STP)/s			
H_2	120×10^{-10}		[10]
H_2O	$14\,000 \times 10^{-10}$		[10]
O_2	14×10^{-10}		[10]
Argon	8×10^{-10}		[10]
CO_2	80×10^{-10}		[10]
Solubility (S)			
H_2	0.14		[10]
H_2O	169.0		[10]
O_2	0.51		[10]
Argon	0.42		[10]
CO_2	12.6		[10]
Infrared spectra (frequency cm^{-1})			
Para out-of-plane aromatic CH wag 2 adjacent H	830		SABIC Innovative Plastics Data
Para in-plane aromatic CH bend	1015		SABIC Innovative Plastics Data
CH_3 rock/C—C stretch	1080		SABIC Innovative Plastics Data
Carbonate C—O stretch	1160 and 1193		SABIC Innovative Plastics Data
Carbonate aryl-O-aryl-C—O stretch	1230		SABIC Innovative Plastics Data
CH_3 sym (umbrella) deformation	1362		SABIC Innovative Plastics Data
Para aromatic ring semicircle stretch	1405 and 1505		SABIC Innovative Plastics Data
Para aromatic ring quadrant stretch	1600		SABIC Innovative Plastics Data
Carbonate C–O stretch	1775		SABIC Innovative Plastics Data
CH_3 symmetric stretch	2875		SABIC Innovative Plastics Data
CH_3 asymmetric stretch	2970		SABIC Innovative Plastics Data

Polycarbonate

Property	Value	Method	Reference
Aromatic C—H stretches	3000–3200		SABIC Innovative Plastics Data
Ultra Violet (UV) cut-off	280 nm		SABIC Innovative Plastics Data
Raman spectra (frequency cm^{-1})			
Phenyl ring/diagonal breathing	634		SABIC Innovative Plastics Data
Phenyl ring breathing	702 and 732		SABIC Innovative Plastics Data
Methyl group stretching	885		SABIC Innovative Plastics Data
Phenyl ring lateral stretch	1108 and 1177		SABIC Innovative Plastics Data
Carbonyl stretching	1232		SABIC Innovative Plastics Data
Quaternary carbon stretch	1600		SABIC Innovative Plastics Data
Proton NMR spectra (CDCl$_3$) ppm			
Methyl group hydrogens	1.5		SABIC Innovative Plastics Data
Protons alpha to oxygens A2B2 pattern	7.0–7.2		SABIC Innovative lastics Data
Protons alpha to quaternary carbon A2B2 pattern	7.0–7.2		SABIC Innovative Plastics Data
Carbon NMR spectra (CDCl$_3$) ppm			
Methyl group carbons	31.6		SABIC Innovative Plastics Data
Quaternary carbon	42.6		SABIC Innovative Plastics Data
Phenyl carbons alpha to oxygen	120.1		SABIC Innovative Plastics Data
Phenyl carbons alpha to quaternary carbon	130.4		SABIC Innovative Plastics Data
Phenyl carbon attached to quaternary carbon	148.0		SABIC Innovative Plastics Data
Phenyl carbon attached to oxygen	149.6		SABIC Innovative Plastics Data
Carbonyl carbon	152.8		SABIC Innovative Plastics Data
Injection molding parameters (representative)*			
Drying temperature	120°C		SABIC Innovative Plastics Data
Drying time	3–4 h		SABIC Innovative Plastics Data
Melt temp	300–340°C		SABIC Innovative Plastics Data
Nozzle temp	305–325°C		SABIC Innovative Plastics Data
Front zone	310–330°C		SABIC Innovative Plastics Data
Middle zone	300–320°C		SABIC Innovative Plastics Data
Rear zone	290–310°C		SABIC Innovative Plastics Data
Mold temp	80–120°C		SABIC Innovative Plastics Data
Back pressure	0.3–0.7 Mpa		SABIC Innovative Plastics Data
Screw speed	40–70 rpm		SABIC Innovative Plastics Data
Shot to cylinder size	40–60%		SABIC Innovative Plastics Data
Vent depth	0.025–0.076 mm		SABIC Innovative Plastics Data

* Check with manufacturers for specific processing condition for each grade of PC.

Acknowledgement Thanks to John Bendler for enlightening discussions and references regarding critical molecular weight.

References

1. DeRudder J, Rosenquist N, Sapp B, Sybert P. Polycarbonates. In: Margolis J, ed. *Engineering Plastics Handbook.* McGraw Hill Publishers; Chapter 14. 2005;327–383.

2. Grigo U, Kircher K, Miller PR. Polycarbonates. In: Bottenbruch L, ed. *Engineering Thermoplastics; Polycarbonates, Polyacetals, Polyesters and Cellulose Esters.* Hanser Publishers; Chapter 3. 1996:112–287.

3. *Handbook of Polycarbonate Science and Technology.* LeGrand DG, Bendler JT, eds. 2000; Marcel Dekker Publishers.

4. Alizadeh A, et al. Crystalline PC. *Macromolecules.* 2001;34:4066–4078. Jonza JM, Porter RS. *J. Poly. Sci. Part B; Polymer Physics.* 1986;24:2459–2472.

5. Steere RC. Thermal Properties. *J. Appl. Poly. Sci.* 1966;10:1673–1685.

6. Kim A, Bosnyak CP, Chudnovsky A. Fracture Toughness. *J. Appl. Poly. Sci.* 1994;51:1841–1848.

7. Philipp HR, LeGrand DG, Cole HS, Liu YS. Optical Properties. *Polym. Eng. Sci.* 1987;27:1148–1155.

8. Me Mercier JP, Aklonis JJ, Litt M, Tobolsky AV. Viscoelastic Behavior and Critical Molecular Weight. *J. Appl. Poly. Sci.* 1965;9:447–459. Mc, Wool RP. *Macromolecules.* 1993;26:1564–1575. Leon S, van der Vegt N, Della Site L, Kremer K. *Macromolecules.* 2005;38:8078–8092. Fetters LJ, Lohse DJ, Milner ST. *Macromolecules.* 1999;32:6847–6851.

9. Caird DW, Burkinshaw LD. Solvents. In: Sweeting OJ, ed. *Science and Technology of Polymer Films.* Interscience Publishers; 1968.

10. Norton FJ. Gas Permeation. *J. Applied Polym. Sci.* 1963;7:1649–1659.

11. Factor A, Chu ML. Photo Aging and Degradation. *Polym. Deg. Stab.* 1980;2:203–223.; Factor A, Ligon WV, May RJ. *Macromolecules.* 1987;20:2461–2468.

Polychloral

Jagath K. Premachandra, Chandima Kumudinie Jayasuriya and Junzo Masamoto

Class Polyacetals; poly(aldehydes)

Structure

$$\begin{array}{c} CCl_3 \\ | \\ \!-\!\!\left(CH\!-\!O\right)_{\!\overline{n}} \end{array}$$

Major Applications Potential material for engineering plastics with rodlike backbone. Possible use as a packing material for high-performance liquid chromatography.[1]

Properties of Special Interest Crystalline material with mechanical properties comparable to engineering plastics with rigid backbone of 4_1 helix similar to that of isotactic polyacetaldehyde, completely isotactic nature, and clean degradation to monomer at elevated temperature. Optical activity of polychloral based on macromolecular asymmetry is a new development; the values of optical activity of these polymers is in the thousands of degrees.

Other Polymers Showing this Special Property Poly(n-hexyl isocyanate): rigid polymer backbone with bulky side chain. Triarylmethyl methacrylates: high value of optical activity with helix polymer structure.

Preparative techniques*

Polymerization Process	Conditions	Reference
Anionic	Anionic polymerization is widely used. Initiators: alkali metal oxides, tertiary amines, tertiary phosphines, organometallic compounds, etc. Chloral is mixed with an anionic initiator above the threshold temperature, for bulk polymerization, and the mixture is then cooled (usually to 0°C under quiescent conditions). Thus, polychloral pieces of desired shape can be prepared.	(2–4)
Anionic	With lithium alkoxide of cholesterol at 0°C in hexane (0.2 mol% initiator relative to chloral)	(1)
Anionic	Oligomerization of chloral with lithium t-butoxide	(5, 6)
Anionic	Cooligomerization of chloral and bromal with lithium t-butoxide or bornyl oxide followed by acetate end-capping	(7)
Anionic	At −78°C using a series of organometallic catalysts, (3–45%) yield	(8)
Anionic	Initiator: lithium alkoxide of (−)borneol	(9)
	Initiators: H_2SO_4 and pyridine	(10)
Cationic	Initiators: BF_3, CH_3SO_3H, H_2SO_4 etc.	(2–4)

* For synthesis of the monomer, chloral (trichloro acetaldehyde), see reference (2).

Property	Units	Conditions	Value	Reference
Ceiling temperature	K	In toluene	282	(2, 11)
		In tetrahydrofuran	284	
Enthalpy of polymerization	kJ mol^{-1}	From solution in toluene to partially crystalline polymer	37.8	(2, 11)
		From solution in tetrahydrofuran to polymer	14.6	
Entropy of polymerization	J mol^{-1} K^{-1}	From solution in toluene to partially crystalline polymer	134	(2, 11)
		From solution in tetrahydrofuran to polymer	52	
Typical comonomers		Most thoroughly studied family of polychloral copolymers are: chloral with isocyanates		(4, 12, 13, 14)
		Monochloroacetaldehyde and dichloroacetaldehyde		(15)
		Ketenes, formaldehyde, trioxane, etc.		(2)
Molecular weight (of repeat unit)	g mol^{-1}	–	147.5	–
Tacticity	–	X-ray diffraction, magic angle ^{13}C NMR and ^{35}Cl-NQR	Completely isotactic structure because of bulkiness of the trichloromethyl group	(2, 16)
		X-ray single crystal analysis, linear oligomers of chloral	Isotactic	(17)
Degree of polymerization	–	Catalyst AlBr$_3$, solvent CH$_2$Cl$_2$, at −30°C for 5 h	190	(10)
		Catalyst (C$_4$H$_9$)$_3$CH$_3$NI, no solvent, at 0°C for 0.5 h	380	
		Catalyst 2,6-dimethoxyphenyllithium, solvent propylene, at −48°C for 1 h	600	
IR (characteristic absorption frequencies)	cm^{-1}	CH bending	1,360	(18)
		C–O stretching	1,125	
		C–Cl stretching	682	
NMR		^{13}C NMR spectroscopy, solid-state at 100.5 MHz		(19)
		Cross polarization/magic angle spinning (CP/MAS) measurement		(19)

Property	Units	Conditions	Value	Reference
		^{13}C NMR, Fourier transform NMR spectrometer at room temperature and at 25 MHz. Internal standard: tetramethylsilane		(1)
		^{1}H and ^{13}C and 2D-NMR, at 35°C in CDCl$_3$ under a nitrogen atmosphere. Internal standard: tetramethylsilane		(5)
		^{1}H NMR, mixture of acetylated chloral addition products		(20)
		^{1}H and ^{13}C NMR, chloral oligomers prepared by lithium t-butoxide initiation		(21)
Mass spectrometry		Cooligomers of chloral and bromal (up to pentamers)		(7)
		Chloral oligomers by GC		(5)
		K+ ionization of desorbed species (K+IDS) spectrometry, linear t-butoxide-initiated, acetate-capped chloral oligomers, ion source pressure <10^{-6} Torr, source temperature = 200°C		(6)
		Amine-initiated chloral oligomers, K+IDS mass spectrometry and capillary GC		(22)
Thermal expansion coefficients	K^{-1}	–	4×10^{-5}	(2, 13)
Solvents		Completely insoluble in any organic solvent		(2)
		Copolymers of chloral with monochloroacetaldehyde is dissolved in CHCl$_3$		
Nonsolvents		Any organic solvent		(2, 23)
		Conventional organic solvents		(24)
		Any solvent		(1, 10)
		Chloral monomer		(2)
Lattice	–	–	Tetragonal	(25)
Space group	–	–	$I\,4_1/a$	(25)
Chain conformation	–	Helical, by IR spectroscopy		(10)
		4_1 helix similar to polyacetaldehyde, electron microscopy shows no evidence of chain folding or lamellar structure and small angle scattering suggests rodlike polymer		(2, 24, 25)
		4_1 helix, axis of helix is parallel to the c-axis of the crystal, by X-ray single crystal analysis, linear oligomers of chloral		(17)
		4_1 helix by X-ray studies on cold-rolled film samples		(25)
		4_1 helix by X-ray studies on film samples that were drawn over a hot pin at 180–210°C		(13)
		Helix symmetry 4_1, helix pitch (c-axis) 5.2 Å, monomer repeat (c-axis) 1.3 Å, and backbone atoms in monomer unit 2		(23)
Unit cell dimensions	Å	–	$a = 17.38, b = 6.45$	(25)
			$c = 5.2$	(2, 13, 23)
			$c = 4.81$	(17)
Unit cell contents (number of repeat units)			16	(25)

Property	Units	Conditions	Value	Reference
Degree of crystallinity	%	Wide angle X-ray diffraction of an oriented fiber	20–30	(2, 13)
		Depends upon the method of preparation	–	(10)
Density (crystalline)	g cm^{-3}	Theoritical density	2.012	(13, 25)
		Observed density	1.9	(2, 13)
Glass transition temperature T_g	K	No T_g was observed between 123 and 473, and over 493		(13)
Melting point	K	Hypothetical melting point	733	(2, 13)
			>533	(23)
Vicat softening temperature	K	Onset of decomposition	473	(2, 13)
Tensile modulus	MPa	–	2,500–3,500	(2, 13)
		Chloral/dichloroacetaldehyde (DCA) copolymer, 28 mol% DCA	~1,100	(10)
		Chloral/DCA copolymer, 45 mol% DCA	~1,450	(10)
		Chloral/ aromatic isocyanate (10%) copolymer	1,700–2,800	(13)
Shear modulus	MPa	Room temperature	850	(2, 13)
		−18°C	2,000	(2)
Tensile strength	MPa	–	35–50	(2, 13)
		Polychloral fibers	40	(2)
		Chloral/DCA copolymer, 28 mol% DCA	41.4	(10)
		Chloral/DCA copolymer, 45 mol% DCA	42	(10)
		Chloral/aromatic isocyanate (10%) copolymer	42–63	(13)
Yield stress	MPa	–	38	(2, 13)
Yield strain $(L/L_0)_y$	%	–	5	(13)
Maximum extensibility $(L/L_0)_r$	%	–	12–20	(2, 13)
		Chloral/DCA copolymer, 28 mol% DCA	12	(10)
		Chloral/DCA copolymer, 45 mol% DCA	12	(10)
		Chloral/aromatic isocyanate (10%) copolymer	15–45	(13)
Flexural modulus	MPa	–	2,200	(13)

Property	Units	Conditions	Value	Reference
Compressive strength	MPa	–	10	(13)
Notched Izod impact strength	J m^{-1}	–	60–80	(2, 13)
Hardness	–	Rockwell hardness	R10, M50	(13)
Index of refraction n	–	25°C	1.58	(2, 13)
Dielectric constant ε'	–	Room temperature	2.8	(2, 13)
Dissipation factor	–	–	0.003	(2, 13)
Optical activity: specific rotation $[\alpha]_d$	Degree	Using chiral lithium alkoxides as initiators	Thousands	(2)
			4,000 to −4,000	(26)
		Initiated with tetramethylammonium (−) acetyl mandelate, holding time = 10–50 min	+84 to +310 (a linear increase in specific rotation with time was observed)	(27)
		Initiated with tetramethylammonium (−) α-methoxy mandelate	Initial increase followed by a leveling off	(27)
		Holding time = 10–30 min	+136 to +163	(27)
		Holding time = 50 min	+114	(27)
		Initiated with lithium $R(-)$-2-octanoxide, holding temperature = 65°C, holding time = 10–70 min	∼(4,300–1,000), higher holding times results in lower rotation values	(28)
		Initiated with lithium $R(-)$-2-octanoxide, holding temperature = 85°C, holding time = 10–70 min	∼(2,300–500), higher holding temperatures results in lower rotation values	(28)
		Initiated with lithium $S(+)$-2-octanoxide, holding temperature = 85°C, holding time = 10–70 min	∼(−2,000 to −800)	(28)
		Initiated with lithium (±)-2-octanoxide, holding temperature = 85°C, holding time = 10–70 min	Optically inactive	(28)
		Initiated with lithium cholestenoxide, holding time = 10–70 min		
		At 65°C	∼(3,400–1,000)	(28)
		At 75°C	∼(2,500–600)	(28)
		At 85°C	∼(2,500–250)	(28)
		Using chiral initiators, solid films	Up to 5,000	(23)

Property	Units	Conditions	Value	Reference
Resistivity	ohm cm	–	4×10^{15}	(2, 13)
Coefficient of sliding friction μ	–	Both static and dynamic	0.55	(2, 13)
Pyrolyzability, nature of product	–	–	Clean degradation to chloral	(2, 18)
Pyrolyzability, amount of product	–	–	Clean degradation to chloral	(2, 18)
Decomposition temperature	K	Usually start to degrade by depolymerization to monomer	493	(2, 29, 30)
		$20°C \ min^{-1}$, properly end-capped and stabilized polymer under nitrogen atmosphere	523	(2, 29)
		$20°C \ min^{-1}$, maximum rate of degradation of properly end-capped and stabilized polymer under N_2 atmosphere	613	(2, 29)
Relative thermal decomposition rate	$\% \ min^{-1}$	Acetate end-capped polymer	0.04	(10)
		Unend-capped polymer	1.48	
Flammability	–	–	Nonflammable	(13, 24)

References

1. Hatada K, Kitayama T, Shimizu S, Yuki H. *J. Chromatography*. 1982;248:63.
2. Vogl O. In: Jacqueline K, ed. *Encyclopedia of Polymer Science and Engineering*. Vol 1. New York: Wiley; 623–643.
3. Vogl O. *Macromolecules*. 1972;5:658.
4. Kubisa P, Vogl O. *Macromol. Synth*. 1977;6:49.
5. Jaycox GD, et al. *Polym. Prepr*. 1989;30(2):167.
6. Simonsick WJ Jr, Hatada K, Xi F, Vogl O. *Macromolecules*. 1991;24:1720.
7. Kruger FWH, et al. *Polym. Prepr*. 1992;33(1):1012.
8. Furukawa J, Saegusa T, Fujii H. *Makromol. Chem*. 1961;44:398.
9. Zhang J, Jaycox GD, Vogl O. *Polym. J*. 1987;19:603.
10. Rosen I. *Polym. Prepr*. 1966;7:221.
11. Kubisa P, Vogl O. *Polymer*. 1980;21:525.
12. Kubisa P, et al. *Macromol. Chem*. 1980;181:2267.
13. Kubisa P, et al. *Polym. Eng. Sci*. 1981;21:829.
14. Odian G, Hiraoka LS. *J. Polym. Sci., Part A-1*. 1970;8:1309.
15. McCain GH, Hudgin DE, Rosen I. *Polym. Prepr., ACS., Div. Polym. Chem*. 1965;6:659.
16. Brame EG, et al. *Polym. Bull. Berlin*. 1983;10:521.
17. Vogl O, et al. *Macromolecules*. 1989;22:4660.
18. Corley LS, Vogl O. *J. Macromol. Sci.-Chem*. 1980;A14:1105.
19. Hatada K, et al. *Polymer J*. 1994;26:267.

20. Zhang J, Jaycox GD, Vogl O. *Polymer.* 1988;29:707.
21. Hatada K, et al. *Makromol. Chem.* 1989;190:2217.
22. Bartus J, Simonsick WJ Jr, Hatada K, Vogl O. *Polym. Prepr.* 1992;33(2):114.
23. Vogl O, Jaycox GD. *Polymer.* 1987;28:2179.
24. Abe A, Tasaki K, Inomata K, Vogl O. *Macromolecules.* 1986;19:2707.
25. Wasai T, et al. *Kogyo Kagaku Zasshi J. Ind. Chem. Jpn.* 1964;67:1920.
26. Jaycox GD, Vogl O. *Polym. J.* 1991;23:1213.
27. Harris WJ, Vogl O. *Polym. Prepr.* 1981;22(2):309.
28. Jaycox GD, Vogl O. *Polym. Prepr.* 1989;30(1):181.
29. Corley LS, Vogl O. *Makromol. Chem.* 1980;181:2111.
30. Ilyina DE, Krentsel BA, Semenido GE. *J. Polym. Sci., Part C.* 1964;4:999.

Polychloroprene

Vassilios Galiatsatos

Alternative Names, Acronyms, Trade Names Poly(1-chloro-1-butenylene), poly(2-chloro-1,3-butadiene), chloroprene rubber (CR), GR-M, Baypren, Butaclor®, Neoprene, Perbunan C, Skyprene

Class Diene elastomers

Structure $-CH_2-Cl-C=CH-CH_2-$

Major Applications Aerospace industry (gaskets, seals, deicers); automotive industry (timing belts, window gaskets, fuel-hose covers, cable jacketing, sparkplug boots, hoses, and joint seals); industrial applications (pipeline pigs, gaskets, hoses, power transmission belts, conveyor belts, escalator handrails); and electronics (wire and cable jacketing). Also for sponge shoe soles and foam cushions. Specialty items made by dipping or electrophoresis from latex. Adhesive tape to replace metal fasteners for automotive accessories such as seat cushions, sealants, and carpet backing.

Main Processing Methods Vulcanization, dip coating, sheeting, calendering, extrusion.

Property	Units	Conditions	Value	Reference
Type of polymerization	–	–	Emulsion polymerization	–
Typical initiator	–	–	$K_2S_2O_8$	–
Typical regulator	–	–	n-Dodecyl mercaptan	–
Typical comonomer	–	–	Sulfur	–
Molecular weight (of repeat unit)	g mol^{-1}	–	88.54	–
Typical molecular weight range of polymer	g mol^{-1}	–	1×10^5 to $>1 \times 10^6$	–
Tacticity		Major isomeric form is the trans-1,4 unit, which varies between 70 and 90% depending on temperature of polymerization. Remaining units are cis-1,4 and -1,2 types.		–
Head-to-head content	%	–	10–15	–
Mark–Houwink parameters	$K =$ dlg^{-1} $a =$ None		$K \times 10^5$ a	(1–4)
		Polychloroprene, toluene at 25°C	50 0.615	
		Linear polychloroprene, THF at 30°C	4.18 0.83	
		Neoprene CG, benzene	2.02 0.89	
		Neoprene GN, benzene	14.6 0.73	
		Neoprene W, benzene	15.5 0.71	

Property	Units	Conditions	Value	Reference
Glass transition temperature	K	–	228–234	(5)
		Cooling rate = 0.3°C min⁻¹ for the all *trans* polymer	228.5	
		1,4-*cis* polymer	253	
Melting temperature	K	Polymerization temperature range = −40 to 40°C	318–348	(6)
		1,4-*cis* polymer	343	
		All *trans* form	388, 380, and 353	
		Polymer prepared at −150°C	651	
Heat of fusion	kJ mol⁻¹	–	8.37	–
Unit cell dimensions	nm	Orthorhombic	$a = 0.884$, $b = 1.024$, $c = 0.48$	(7)
Unit cell content (number of repeat units)	–	–	4	–

^{13}C-NMR analysis of polychloroprenes*(8,9)

Polymerization Temp. (°C)	Total (%) (1,4-trans)	Inverted (%) (1,4-trans)	1,2	1,2 Isomerized	3,4	*Cis*-1,4
90	85.4	10.3	2.3	0.6	4.1	7.8
40	90.8	9.2	1.7	0.8	1.4	5.2
20	927	8.0	1.5	0.9	1.4	3.3
0	95.9	5.5	1.2	1.0	1.1	1.8
−20	97.1	4.3	0.9	0.6	0.5	0.8
−40	97.4	4.2	0.8	0.6	0.5	0.7

* Resulting structure depends on polymerization temperature.

Property	Units	Conditions	Value	Reference
Mooney viscosity	°ML	–	47	–
Specific gravity	–	Neoprene WM1 (DuPont)	1.23	–
Thermal conductivity	W m⁻¹ K⁻¹	20°C	0.19	–
Specific heat capacity	J K⁻¹ kg⁻¹	–	2175	–
Relative gas permeability and selectivity	–	Polychloroprene film in the range of 23–25°C		–
		Helium vs. methane	5.0	
		Oxygen vs. nitrogen	3.64	
		Hydrogen vs. methane	7.69	
		Carbon dioxide vs. methane	8.5	
Dose required to reduce the elongation at break to 50% of original	Gy	Low-dose rate conditions in air	3×10^5	(10)
		High-dose rate or inert atmospheric conditions	5×10^5	

Property	Units	Conditions		Value	Reference
Flow behavior index n'	–	80°C		0.15	(11)
		100°C		0.11	
		120°C		0.07	
Consistency of flow K'	–	80°C		225.1	(11)
		100°C		257.6	
		120°C		279.6	
Shear viscosity	kPa s	Temp. (°C)	Shear rate (s^{-1})		(11)
		80	122.6	1.6	
		80	245.2	0.9	
		80	490.4	0.48	
		80	735.6	0.035	
		100	122.6	1.10	
		100	245.2	0.06	
		100	490.4	0.032	
		100	735.6	0.02	
		120	122.6	0.075	
		120	245.2	0.042	
		120	490.4	0.022	
		120	735.6	0.015	
Dynamic extensional viscosity	MPa s	Temp. (°C)	Shear rate (Hz)		(11)
		80.5	110	0.004	
		80.5	35	0.003	
		80	11	0.027	
		80	3.5	0.046	
		100	110	0.005	
		100	35	0.0025	
		100	11	0.038	
		100	3.5	0.042	
		120	110	0.002	
		120	35	0.004	
		120	11	0.03	
		120	3.5	0.025	
Scorch time	min	Mooney viscosity measured at 140°C, Δ5°ML		30	(12)
Minimum plasticity	°ML	–		43	(12)
Shrinkage on calendering	%	50°C		331	(12)
Tensile strength	kg cm^{-2}	Vulcanization time			(12)
		5 min		179	
		10 min		210	
		15 min		190	
Elongation at break	%	Vulcanization time			(12)
		5 min		1020	
		10 min		930	
		15 min		830	

Polychloroprene

Property	Units	Conditions	Value	Reference
300% modulus	kg cm^{-2}	Vulcanization time		(12)
		5 min	10	
		10 min	14	
		15 min	16	
500% modulus	kg cm^{-2}	Vulcanization time		(12)
		5 min	21	
		10 min	26	
		15 min	32	
Permanent set	%	Vulcanization time		(12)
		5 min	13	
		10 min	8	
		15 min	6	
Dielectric loss peaks	K	300 Hz		(12)
		α-relaxation	200 (in the glassy state)	
		β-relaxation	251 (above T_g)	
Extension dependence	K	Of the glass transition temperature in the dilatometric time scale	1.05–1.25	–
Anisotropy of segments and monomer units	$(\alpha_1 - \alpha_2)$ cm^3	α-Bromonaphthalene	+110	(13)
		Carbon tetrachloride	+33	
		Chlorobenzene	+64	
		Dichloroethane	+39	
		α-Methylnapthalene	+99	
		Tetrachloroethylene	+46	
		Toluene	+67	
		p-Xylene	+88	

Environmental properties Polychloroprene has very good weathering characteristics due to the presence of the electron-withdrawing chlorine atom. It is resistant to ozone and oil. It is heat resistant up to 100°C. Polychloroprene has low permeability to air and water vapors. Accelerated ageing studies are widely available. (14–16)

Formulation Fillers may be carbon black, zinc oxide, or magnesium oxide. For EMI shielding fillers may be silver-plated aluminum or nickel or copper or glass spheres. (17)

Suppliers

Trade Name	Supplier
Baypren	LANXESS, Germany - http://techcenter.lanxess.com
Neoprene	DuPont Performance Elastomers, Wilmington, Delaware, USA - www.dupontelastomers.com
Skyprene	Tosoh Corporation, Tokyo, Japan - www.tosoh.com

References

1. Coleman MM, Fuller RE. *J. Macromol. Sci. Phys.* 1975;11(3):419.
2. Mochel WE, Nichols JB. *J. Am. Chern. Soc.* 1949;71:3425.
3. Mochel WE, Nichols JB, Mighton CJ. *J. Am. Chem. Soc.* 1948;70:2185.
4. Mochel WE, Nichols JB. *Ind. Eng. Chem.* 1951;43:154.
5. Aufdermarsh CA, Pariser R. *Polym. Sci., Part A.* 1964;2:4727.
6. Garrett RR, Hargreaves CA II, Robinson DN. *J. Macromol. Sci. Chem.* 1970;4(8):1679.
7. Bunn CW. *Proc. R.S. London Ser. A.* 1942;180:40.
8. Coleman MM. Brame EG. *Rubber Chem. Techmol.* 1978;51:668.
9. Coleman MM, Tabb DL, Brame EG Jr. *Rubber Chem. Technol.* 1977;50:49.; Ebdon JR. *Polymer.* 1978;19:1232.
10. Gillen KT, Clough RL. *Radiat. Phys. Chem.* 1981;18:679.
11. Kundu PP, Bhattacharya AK, Tripathy DK. *J. Appl. Polym. Sci.* 1997;66:1759.
12. Nakajima K, Naoki M, Nose T. *Polym. J.* 1978;10(3):307.
13. Brandrup J, Immergut EH, Edmund H, Grulke EA, Abe A, Bloch DR,eds. *Polymer Handbook.* 4th ed. New York: John Wiley and Sons; (1999; 2005).
14. Speight JG, ed. *Lange's Handbook of Chemistry.* 16th ed. McGraw-Hill; 2005.
15. Gent AN, ed. *Engineering with Rubber—How to design Rubber Components.* 2nd ed. Hanser Publishers; 2001.
16. Brown RP, Butler T, Hawley SW. *Ageing of Rubber—Accelerated Heat Ageing Results.* Rapra; 2001.
17. Wypych G, ed. *Handbook of Fillers—A Definite User's Guide and Databook.* 2nd ed. 2001.

Poly(*p*-chlorostyrene)

Anand K. Patel, Jonathan H. Laurer and Richard J. Spontak

Acronyms *p*-CST, PCS, *p*-ClST, *p*-ClSt

Class Vinyl polymers; *p*-halostyrenes

CAS Number 24991-47-7

Structure

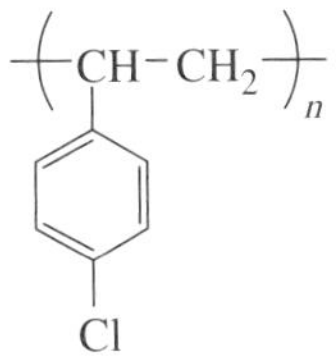

Major Applications High-contrast negative e-beam resist.[1] Comonomer in numerous block copolymer systems.[2,3]

Property	Units	Conditions	Value	Reference
Density	g cm^{-3}	$\overline{M}_w = 650\,000$ g mol^{-1}	1.246	(4)
Specific free volume	cm^3 g^{-1}	$\overline{M}_w = 650\,000$ g mol^{-1}	0.124	(4)
^{13}C NMR chemical shift δ	ppm	For β-carbon, δ	114.4	(5)
Thermal expansion parameter α	K^{-1}	–	0.582	(6)
Ionization potential I_p	eV	–	8.45	(7)
Heat capacity increase ΔCp	J K^{-1} mol^{-1}	At T_g	31.1	(8)
		300 K to T_g	$-4.20 + 0.4866\,T$	
		T_g to 550 K	$112.57 + 0.2775\,T$	
γ-Loss peak	K	–	143	(9)

Dynamic elastic shear modulus* compared to linear polystyrene[9]

T(°C)	Poly(*p*-chlorostyreme) G′(MPa)	Linear polystyrene G′(MPa)
−78	1400	1260
−98	1420	1280
−123	1420	1320
−148	1450	1350
−173	1500	1350

* Calculated from the frequencies of free vibration, moment of inertia of the system, and ambient-temperature polymer dimensions. Measurements performed on a torsion pendulum at a constant frequency of 1 Hz.

Poly(p-chlorostyrene)

Property	Units	Conditions		Value	Reference
Solubility parameters δ	$(MPa)^{1/2}$	$T = 25°C$		19.7	(10)
		$T = 25°C$; (group contribution calculation)		20.0	(6)
		$T = 160°C$; (experimental)		12.3	(6)
		Small method[11]		19.5	(12)
		Hoy method[13]		20.2	(12)
		Hoftyzer and van Krevelen method[14]		19.9	(12)
		Cohesive energy method[12]		19.3	(12)
Theta temperature θ	K	Composition (v/v) by phase equilibria			
		Benzene		281	(15)
		Benzene/Methanol (4.5/1)		314.6	(15)
		Benzene/Methanol (5.0/1)		305.4	(15)
		Benzene/Methanol (5.5/1)		299.7	(15)
		Carbon tetrachloride		323.7	(16)
		Carbon tetrachloride/Toluene (1/1)		286.6	(17)
		Carbon tetrachloride/Toluene (5/2)		305.2	(17)
		Cumene		332	(18)
		Ethyl acetate		284.9	(19)
		Ethyl benzene		258.3	(16)
		Ethyl carbitol		300.8	(16)
		Ethyl chloroacetate		271.2	(16)
		Isopropyl acetate		348.5	(16)
		Isopropyl benzene		332	(16)
		Isopropyl chloroacetate		264.8	(16)
		Methyl chloroacetate		337.6	(16)
		n-Butyl acetate		306.9	(19)
		n-Butyl carbitol		323.1	(16)
		t-Butyl acetate		338.4	(16)
		Tetrachloroethylene		317.4	(16)
Second virial coefficient A_2	$mol\ cm^3\ g^{-2}$	Solvent			(10)
		Toluene		1.20	
		Methylethylketone		1.72	
		Cumene		−0.20	
Flory–Huggins interaction parameter χ	–	Solvent	T (°C)		
		t-Butyl acetate	30	0.462	(16)
		n-Butyl carbitol	30	0.468	(16)
		Ethyl carbitol	30	0.505	(16)
		Ethyl chloroacetate	30	0.464	(16)
		Isopropyl acetate	30	0.446	(16)
		Isopropyl chloroacetate	30	0.460	(16)
		Toluene	30	0.460	(10)
			150	0.294	(6)
			160	0.270	(6)
			170	0.233	(6)
		Methylethylketone	30	0.447	(10)
		Methyl chloroacetate	30	0.549	(16)
		Cumene	55	0.505	(10)

Poly(p-chlorostyrene)

Property	Units	Conditions		Value	Reference
Flory–Huggins interaction parameter χ	–	Solvent	$T\ (^\circ C)$		
		Carbon tetrachloride	30	0.528	(16)
		n-Pentane	150	0.600	(6)
			160	0.540	(6)
			170	0.470	(6)
		n-Hexane	150	0.755	(6)
			160	0.694	(6)
			170	0.641	(6)
		n-Heptane	150	0.843	(6)
			160	0.798	(6)
			170	0.771	(6)
		Benzene	150	0.305	(6)
			160	0.271	(6)
			170	0.225	(6)
		Ethyl benzene	30	0.474	(16)
		Isopropyl benzene	30	0.478	(16)
			150	0.491	(6)
			160	0.473	(6)
			170	0.455	(6)
		n-Propyl benzene	150	0.497	(6)
			160	0.472	(6)
			170	0.429	(6)
		Tetrachloroethylene	30	0.618	(16)
	–	Monomer or polymer	$T\ (^\circ C)$		
		Phenylene oxide	200	0.030	(2)
		Phenylsulfonylated phenylene oxide	200	0.017	(2)
		o-Chlorostyrene	150	0.0915	(3)
			200	0.0940	(2,3)
			250	0.109	(3)
			300	0.135	(3)
		Styrene	150	0.0720	(3)
			200	0.0792	(3)
			250	0.0927	(3)
			300	0.111	(3)
Interaction parameter χ_{12}^* based on hard-core volumes	–	Solute	$T\ (^\circ C)$		(6)
		n-Pentane	150	0.965	
			160	0.951	
			170	0.941	
		n-Hexane	150	1.047	
			160	1.016	
			170	0.996	
		n-Heptane	150	1.085	
			160	1.064	
			170	1.061	
		Toluene	150	0.478	
			160	0.466	
			170	0.451	

Property	Units	Conditions		Value	Reference
Interaction parameter χ_{12}^* based on hard-core volumes	–	Solute	T (°C)		(6)
		Benzene	150	0.517	
			160	0.503	
			170	0.479	
		Isopropyl benzene	150	0.640	
			160	0.635	
			170	0.632	
		n-Propyl benzene	150	0.641	
			160	0.629	
			170	0.608	
Glass transition temperature T_g	K	$\overline{M}_w \times 10^{-4}$ (g mol^{-1})	$\overline{M}_n \times 10^{-4}$ (g mol^{-1})		
		8.7	16.0	405	(2)
		16.4	8.0	402	(7)
		25.3	6.2	388	(20)
		26.7	14.0	406	(8)
		65.0	–	399	(4)

Interaction parameter in toluene at 21–22°C[21]

Polymer Volume Fraction	χ^*
0.20	0.528
0.24	0.538
0.28	0.554
0.32	0.572
0.36	0.579
0.40	0.581
0.44	0.518
0.48	0.573
0.52	0.568
0.56	0.563
0.60	0.553
0.64	0.543

* Determined from excess free energies of solution.

Kuhn–Houwink parameters in various solvents

	$K \times 10^{-5}$	a	Reference
Benzene	30.6	0.56	(22)
Chlorobenzene	2.19	0.80	(23)
Chloroform	14.8	0.65	(22)
Dioxane	17.6	0.62	(22)
Methylethylketone	16.6	0.62	(22)
	3.52	0.75	(23)
	12.0	0.64	(24)
	29.0	0.59	(25)

Poly(p-chlorostyrene)

	$K \times 10^{-5}$	a	Reference
Toluene	5.37	0.71	(23)
	39.0	0.59	(24)
	13.0	0.64	(25)
Xylene	30.0	0.56	(22)

Intrinsic viscosity $[\eta]$; and partial specific volume v_p

Solvent	$\overline{M} \times 10^{-4}$ (g mol^{-1})	T(°C)	v_p(cm^3 g^{-1})	$[\eta]$ (dl g^{-1})	Reference
Toluene	33.0*	30	0.778	0.557	(10)
Methylethylketone	33.0*	30	0.770	0.603	(10)
Cumene	33.0*	55	0.788	0.354	(10)
Chloroform	14.04*	25	–	0.79	(8)
Benzene	179.9†	27	–	0.963	(15)
	129.7†	27	–	0.793	(15)
	81.7†	27	–	0.617	(15)
	48.9†	27	–	0.466	(15)
	34.1†	27	–	0.375	(15)
	179.9†	32	–	1.020	(15)
	129.7†	32	–	0.825	(15)
	81.7†	32	–	0.626	(15)
	48.9†	32	–	0.481	(15)
	34.1†	32	–	0.383	(15)
	179.9†	42	–	1.059	(15)
	129.7†	42	–	0.863	(15)
	81.7†	42	–	0.649	(15)
	48.9†	42	–	0.484	(15)
	34.1†	42	–	0.390	(15)
Experimental fits	–				
Toluene	–	30	–	$(12.3 \times 10^{-5})\,\overline{M}_w^{0.65}$	(18)
Ethyl benzene		30	–	$(21.7 \times 10^{-5})\,\overline{M}_w^{0.60}$	(18)

* Number-average molecular weight.

† Viscosity-average molecular weight.

Solution properties at 30°C in benzene[22]

$[\eta]$ (dl g^{-1})	$\overline{M}_n \times 10^{-5}$ (g mol^{-1})	$\overline{M}_w \times 10^{-5}$ (g mol^{-1})	$A_2 \times 10^{-4}$ (ml mol g^{-2})
0.950	6.41	17.83	0.39
0.600	5.01	8.02	0.97
0.515	4.05	6.38	1.06
0.485	3.37	5.82	1.04
0.400	2.83	3.70	1.20
0.290	1.56	2.51	1.60
0.240	1.09	1.37	2.20
0.155	0.59	0.78	1.80

Solution properties at 30°C in toluene[26]

$[\eta]$ (dl g^{-1})	$\overline{M}_n \times 10^{-4}$ (g mol^{-1})	$\overline{M}_w \times 10^{-4}$ (g mol^{-1})	$A_2 \times 10^{-4}$ (ml mol g^{-2})
1.25	—	137	0.57
0.869	66.9	79.7	0.75
0.670	—	58.1	0.75
0.617	40.2	49.9	0.79
0.545	33.2	40.1	0.83
0.377	18.5	21.4	0.89
0.355	—	20.6	0.82

Second virial coefficient and intrinsic viscosity at 30°C in various solvents[27]

$\overline{M}_n \times 10^{-4}$ (g mol^{-1})	$\overline{M}_w \times 10^{-4}$ (g mol^{-1})	$A_2 \times 10^{-4}$ (ml mol g^{-2})	$[\eta]$ (dl g^{-1})		
			Toluene	Butanone	Chlorobenzene
50.7	115	—	—	1.114	1.396
47.2	99	0.96	0.850	0.915	1.247
30.9	75	1.47	0.658	0.735	0.922
31.9	66	1.42	0.652	0.712	0.918
27.2	47	1.54	0.556	0.594	0.753
19.2	31	1.69	0.395	0.435	0.537

Intrinsic viscosity, thermal expansion factor and polymer–solvent interaction parameter in diethylene glycol monobutyl ether[28]

T (°C)	$[\eta]$ (cm^3 g^{-1})	α	χ
25	26.06	1.36	0.35
30	19.72	1.24	0.41
35	19.21	1.23	0.42
40	18.38	1.21	0.42
45	16.06	1.16	0.45
50	10.33	1.00	0.50
55	8.78	0.95	0.51
60	5.12	0.79	0.55
65	8.82	0.95	0.51
70	6.18	0.84	0.54
75	6.72	0.87	0.53
80	4.05	0.73	0.56
85	5.27	0.80	0.55

Poly(p-chlorostyrene)

Osmotic pressure ($\overline{M}_n = 47.2 \times 10^4$ g mol^{-1}, $\overline{M}_w = 99 \times 10^4$ g mol^{-1})[27]

Solvent	T(°C)	$c \times 10^2$ (g cm^{-3})	π (g cm^{-2})
Toluene	30	0.200	0.120
		0.398	0.253
		0.601	0.428
		0.801	0.610
		0.989	0.804
		1.751	1.855
		2.513	3.418
	60	0.178	0.112
		0.392	0.274
		0.570	0.436
		0.792	0.686
		1.016	0.988
		1.751	2.250
		2.504	4.129
Chlorobenzene	30	0.224	0.157
		0.426	0.360
		0.579	0.549
		0.799	0.900
		1.013	1.320
		1.408	2.402
	60	0.196	0.154
		0.421	0.394
		0.605	0.666
		0.820	1.044
		1.010	1.474
		1.511	3.036

Property	Units	Conditions		Value	Reference
Permeability coefficient P	cm^3(STP) cm (cm^2 s cm Hg)$^{-1}$	$\overline{M}_w = 650\,000$ g mol^{-1}, measured at 1 atm, 35°C			(4)
		Gas	He	16.4×10^{-10}	
			CH$_4$	2.6×10^{-11}	
			O$_2$	1.2×10^{-10}	
			N$_2$	2.3×10^{-11}	
			CO$_2$	4.3×10^{-10}	
Diffusion coefficient D	cm^2 s^{-1}	$\overline{M}_w = 650\,000$ g mol^{-1}, measured at 1 atm, 35°C			(4)
		Gas	CH$_4$	6.1×10^{-9}	
			O$_2$	7.8×10^{-8}	
			N$_2$	2.4×10^{-8}	
			CO$_2$	2.2×10^{-8}	

Property	Units	Conditions		Value		Reference
Refractive index (n) increment, and dielectric constant (ε) increment, ($c = p$-ClSt weight fraction in solvent)	–	Solvent	T (°C)	dn^2/dc	$d\varepsilon/dc$	(29)
		Benzene	15	0.271	2.237	
			25	2.277	1.869	
			35	0.283	1.757	
			50	0.290	1.712	
			60	0.291	1.722	
			70	0.296	1.668	
		Isopropyl benzene	30	0.294	1.749	
			40	0.295	1.735	
			50	0.296	1.685	
			60	0.297	1.665	
			70	0.300	1.588	
			80	0.301	1.534	
			90	0.301	1.438	
		n-Propyl benzene	20	0.291	1.644	
			30	0.292	1.600	
			40	0.296	1.597	
			50	0.297	1.569	
			60	0.298	1.552	
			70	0.298	1.513	
			80	0.299	1.453	
			90	0.302	1.392	
Mean square dipole moment ratio $D_\infty = <\mu^2>/x\mu_0^2$	–	Solvent	T (°C)			(29)
		Benzene	15	0.599		
			25	0.513		
			35	0.501		
			50	0.521		
			60	0.556		
			70	0.562		
		Isopropyl benzene	30	0.493		
			40	0.513		
			50	0.521		
			60	0.540		
			70	0.534		
			80	0.537		
			90	0.520		
		n-Propyl benzene	20	0.436		
			30	0.445		
			40	0.465		
			50	0.478		
			60	0.496		
			70	0.505		
			80	0.503		
			90	0.499		

Poly(p-chlorostyrene)

Dipole moment in solution

Solvent	T (°C)	$\langle\mu^2\rangle/ \times \mu_0$	$\langle\mu^2\rangle/ \times (D^2)$	Remarks	Reference
Benzene	25	0.64		$\mu_0 = 2.00$ D	(30)
	30	0.56	2.10	$\mu_0 = 1.93$ D	(31)
	30		1.49		(32)
	30		1.49	Isotactic	(32)
	15–70	0.60–0.56	2.37–2.22		(29)
Carbon tetrachloride	25	0.57		$\mu_0 = 1.88$ D	(33)
	50		1.89		(34)
Cumene	25–65		1.87–1.73		(35)
Isopropyl benzene	30	0.53	1.50	$\mu_0 = 1.68$ D	(36)
	30–90	0.73–0.70	2.06–1.98	$\mu_0 = 1.68$ D	(37)
	30–90	0.49–0.52	1.85–1.96		(29)
n-Propyl benzene	20–90	0.44–0.50	1.64–1.88		(29)
p-Dioxane	20–50		2.11–2.46		(34)
Toluene	25		1.82		(35)
	25		1.65		(38)
	25	0.60	1.59	$\mu_0 = 1.63$ D	(39)
	30	0.77	2.17	$\mu_0 = 1.68$ D	(36)
	30–60	0.74–0.75	2.09–2.12	$\mu_0 = 1.68$ D	(37)
Xylene	30	0.71	2.00	$\mu_0 = 1.68$ D	(36)
	30–50	0.42–0.56			(40)

Dielectric loss tangents ($\times 10^3$) in benzene[41]

$\overline{M}_w \times 10^{-3}$ (g mol^{-1})	Concentration (wt%)	Frequency, f (MHz)						$\log f_{max}$ (Hz)
		10	30	48.5	101	150	300	
2.03	3.01	1.32	2.39	3.51	4.11	4.15	3.89	8.15
4.60	2.99	3.30	5.04	5.86	5.96	5.52	3.93	7.86
11.1	3.11	3.85	5.50	5.96	5.42			7.73
21.1	3.02	4.31	6.31		5.33	5.05	3.65	7.63
	1.51	2.75	4.30	4.66	3.44			7.62
	0.75	1.77	2.93	3.21	2.30			7.62
38.8	3.02	4.90	5.76	5.96	5.38	4.55	3.03	7.58

Temperature dependence of the dipole moment in various solvents

Solvent	T Range (°C)	$d \ln\langle\mu^2\rangle/dT$	Reference
Benzene	10–40	-6×10^{-3}	(42)
Benzene	10–35	-8×10^{-3}	(33)
Benzene	35–70	-6×10^{-3}	(33)
Benzene	15–35	-6×10^{-3}	(29)
Benzene	35–70	4×10^{-3}	(29)
Benzene	10–70	-0.5×10^{-3}	(43)
Benzene	10–70	-0.9×10^{-3}	(44)

Poly(p-chlorostyrene)

Solvent	T Range (°C)	$d \ln <\mu^2>/dT$	Reference
Isopropyl benzene	30–60	3×10^{-3}	(29)
Isopropyl benzene	60–90	-1×10^{-3}	(29)
n-Propyl benzene	20–50	3×10^{-3}	(29)
n-Propyl benzene	50–90	~ 0	(29)
Toluene	20–90	-4×10^{-3}	(42)

Pressure and temperature dependence of specific volume $(cm^3\ g^{-1})^{(45)}$

$T(°C)$	Pressure (kg cm^{-2})										
	0	200	400	600	800	1000	1200	1400	1600	1800	2000
29.7	0.8211	0.8161	0.8119	0.8080	0.8044	0.8009	0.7977	0.7945	0.7914	0.7884	0.7855
39.1	0.8227	0.8174	0.8131	0.8091	0.8054	0.8019	0.7984	0.7952	0.7922	0.7891	0.7863
50.1	0.8243	0.8189	0.8146	0.8104	0.8067	0.8031	0.7996	0.7963	0.7932	0.7901	0.7870
60.4	0.8256	0.8203	0.8160	0.8116	0.8079	0.8042	0.8007	0.7973	0.7940	0.7907	0.7875
70.6	0.8272	0.8217	0.8171	0.8128	0.8088	0.8051	0.8015	0.7980	0.7947	0.7916	0.7885
81.4	0.8289	0.8231	0.8183	0.8141	0.8101	0.8063	0.8026	0.7989	0.7956	0.7922	0.7892
91.0	0.8306	0.8246	0.8197	0.8152	0.8111	0.8071	0.8033	0.7998	0.7963	0.7931	0.7899
100.8	0.8317	0.8258	0.8209	0.8163	0.8120	0.8079	0.8042	0.8007	0.7972	0.7939	0.7905
110.1	0.8338	0.8274	0.8222	0.8176	0.8132	0.8089	0.8052	0.8015	0.7979	0.7946	0.7912
120.3	0.8364	0.8292	0.8235	0.8184	0.8139	0.8098	0.8059	0.8022	0.7987	0.7952	0.7920
130.8	0.8408	0.8321	0.8253	0.8196	0.8149	0.8104	0.8063	0.8024	0.7989	0.7955	0.7921
140.1	0.8458	0.8358	0.8277	0.8211	0.8158	0.8110	0.8068	0.8027	0.7991	0.7955	0.7920
149.7	0.8507	0.8401	0.8316	0.8241	0.8177	0.8124	0.8075	0.8034	0.7995	0.7957	0.7924
159.5	0.8556	0.8445	0.8356	0.8278	0.8207	0.8145	0.8092	0.8045	0.8003	0.7964	0.7928
169.6	0.8608	0.8491	0.8399	0.8317	0.8246	0.8178	0.8116	0.8064	0.8016	0.7974	0.7934
178.8	0.8657	0.8538	0.8442	0.8357	0.8282	0.8214	0.8152	0.8094	0.8040	0.7992	0.7949
189.8	0.8712	0.8588	0.8488	0.8400	0.8324	0.8252	0.8189	0.8132	0.8076	0.8022	0.7973
199.8	0.8771	0.8640	0.8537	0.8445	0.8366	0.8294	0.8227	0.8167	0.8110	0.8057	0.8009
209.9	0.8812	0.8677	0.8570	0.8476	0.8394	0.8320	0.8252	0.8190	0.8133	0.8079	0.8027
219.9	0.8863	0.8719	0.8608	0.8511	0.8427	0.8351	0.8282	0.8218	0.8157	0.8104	0.8055
229.7	0.8910	0.8762	0.8647	0.8547	0.8459	0.8380	0.8311	0.8246	0.8186	0.8133	0.8079
240.8	0.8963	0.8810	0.8689	0.8587	0.8496	0.8415	0.8344	0.8278	0.8217	0.8159	0.8106
250.5	0.9020	0.8855	0.8731	0.8623	0.8531	0.8447	0.8373	0.8305	0.8242	0.8184	0.8131
260.3	0.9072	0.8901	0.8772	0.8660	0.8566	0.8481	0.8405	0.8336	0.8272	0.8211	0.8156
269.8	0.9133	0.8956	0.8822	0.8708	0.8610	0.8522	0.8443	0.8372	0.8307	0.8246	0.8190
280.4	0.9191	0.9003	0.8864	0.8746	0.8644	0.8553	0.8473	0.8400	0.8333	0.8271	0.8214
289.7	0.9250	0.9052	0.8908	0.8787	0.8681	0.8590	0.8508	0.8433	0.8365	0.8301	0.8246

* The specific volume at ambient conditions is 0.8183 cm^3 g^{-1}.

Poly(p-chlorostyrene)

Preparative techniques

Living cationic polymerization

Initiating system	1-Phenylethyl chloride/SnCl$_4$
Solvent	Methylene chloride
Temperature	0°C
Reagent concentrations	[p-ClSt] = 1.0 M, [1-phenylethyl chloride] = 20 mM, [SnCl$_4$] = 100 mM
% Conversion	80% in 2 h
Molecular weight range	10^3–10^4 g mol^{-1}
$\overline{M}_w/\overline{M}_n$	1.1
Reference/Note	(5)/Living polymerization is also reported at 25°C

Living carbocationic polymerization

Initiating system	Cumyl methyl ether/BCl$_3$
Solvent	Methylene chloride
Temperature	−60°C
Reagent concentrations	[Cumyl methyl ether] = 4.92 × 10^{-3} M, [BCl$_3$] = 0.104 M
% Conversion	66–72
Molecular weight range	4190–12 390 g mol^{-1}
$\overline{M}_w/\overline{M}_n$	1.40–1.79
Reference/Note	(46)/High $\overline{M}_w/\overline{M}_n$ values are attributed to slow cationation

Living carbocationic polymerization

Initiating system	2-Chloro-2,4,4-trimethylpentane (TMPCl)/TiCl$_4$ + dimethylacetamide (DMA) (electron donor) + 2,6-*di-t*-butylpyridine (DtBP) (proton trap)
Solvent	Methyl chloride/methylchlorohexane
Temperature	−80°C
Reagent concentrations	[TMPCl] = 5.4 mM, [TiCl$_4$] = 0.086 M, [DMA] = 4.3 mM, [DtBP] = 3.6 mM
% Conversion	94–100%
Molecular weight range	3500–7670 g mol^{-1}
$\overline{M}_w/\overline{M}_n$	1.26–1.51
Reference/Note	(46)/TiCl$_4$ is moisture sensitive compared to BCl$_3$

Photochemical polymerization

Initiating system	Polymethylphenylsilane (PMP): UV with λ = 300–400 nm
Solvent	In bulk
Temperature	30°C
Reagent concentrations	[p-ClSt] = 7.96 M, [PMPS] = 0.39–1.50 M
Molecular weight range	50 000–85 500 g mol^{-1}
$\overline{M}_w/\overline{M}_n$	2.09–2.63
Reference/Note	(47)

Comonomers used in copoylmerizations

Comonomer	Reference
Citraconic anhydride	(48)
Styrene	(7,49–51)
Maleic anhydride	(52)
o-Chlorostyrene	(2,3)
Hexyl methacrylate	(53)
o-Methylstyrene	(51)
p-Methylstyrene	(51)
m-Methylstyrene	(51)
4-*t*-Butylstyrene	(51)
o-Fluorostyrene	(54)

References

1. Liutkus J, Hatzakis M, Shaw J, Paraszczak J. *Polym. Engr. Sci.* 1983;23:1047.
2. Vukovic R, et al. *J. Polym. Sci. B: Polym. Phys.* 1994;32:1079.
3. Cimmino S, Karasz FE, MacKnight WJ. *J. Polym. Sci. B: Polym. Phys.* 1992;30:49.
4. Puleo AC, Muruganandam N, Paul DR. *J. Polym. Sci.: Polym. Phys. Ed.* 1989;27:2385.
5. Kanaoka S, Eika Y, Sawamoto M, Higashimura T. *Macromolecules* 1996;29:1778.
6. Yilmaz F, Cankurtaran OEG, Baysal BM. *Polymer* 1992;33:4563.
7. Gustafsson A, Wiberg G, Gedde UW. *Polym. Engr. Sci.* 1993;33:549.
8. Judovits L, . Bopp RC, Gaur U, Wunderlich B. *J.Polym. Sci.: Polym. Phys. Ed.* 1986;24:2725.
9. Illers KH, Jenckel E. *J. Polym. Sci.* 1959;41:528.
10. Ogawa E, Yamaguchi N, Shima M. *Polym. J.* 1986;18:903.
11. Small PA. *J. Appl. Chem.* 1953;3:71.
12. Ahmad H, Yaseen M. *J. Oil Col. Chem. Assoc.* 1977;60:99.
13. Hoy KL. *J. Paint Technol.* 1970;42:76.
14. van Krevelen DW. *Properties of Polymers: Their Estimation and Correlation with Chemical Structure*, 3rd ed. Amsterdam: Elsevier; 1997.
15. Hernaândez-Fuentes I, Prolongo MG. *Eur. Polym. J.* 1979;15:571.
16. Izumi Y, Miyake Y. *Polym. J.* 1972;3:647.
17. Kubo K, Ogino K, Nakagawa T. *Rep. Prog. Polym.Phys. Jpn.* 1967;10:7.
18. Izumi Y, Miyake Y. *Polym. J.* 1973;4:205.
19. Kubo K, Ogino K. *Rep. Prog. Polym. Phys. Jpn.* 10(1967): 111.
20. Malhotra SL, Lessard P, . Blanchard LP. *J. Macromol. Sci.: Chem.* 1981;A15:279.
21. Corneliussen R, Rice SA, Yamakawa H. *J. Chem. Phys.* 1963;38:1768.
22. Mohite RB, Gundiah S, Kapur SL. *Makromol. Chem.* 1968;2759:280.
23. Kuwahara N, et al. *J. Polym. Sci. A.: Polym. Chem.* 1965;3:985.
24. Saito T. *Bull. Chem. Soc. Jpn.* 1962;35:1580.
25. Kotera A, Saito T, Matsida H, Kamata R. *Rep. Prog. Polym. Phys. Jpn.* 1960;3:51.
26. Noguchi Y, Aoki A, Tanaka G, Yamakawa H. *J. Chem. Phys.* 1970;52:2651.
27. Takamizawa K. *Bull. Chem. Soc. Jpn.* 1966;39:1186.
28. Şakar D. *Monatsh. Chem.* 2006;137:919.
29. Yilmaz F, Baysal BM. *J. Polym. Sci. B: Polym. Phys.* 1992;30:197.

30. Baysal B, Yu H, Stockmayer WH. In: Karasz FE, ed. *Dielectric Properties of Polymers.* New York: Plenum Press; 1971.

31. Debye P, Bueche F. *J. Chem. Phys.* 1951;19:589.

32. Pohl HA, Zabusky HH. *J. Phys. Chem.* 1962;66:1390.

33. Baysal BM, Aras L. *Macromolecules.* 1985;16:1693.

34. Tonelli AE, Belfiore LA. *Macromolecules.* 1983;16:1740.

35. Yamaguchi N, Sato M, Ogawa E, Shima M. *Polymer.* 1981;22:1464.

36. Burshtein LL, Stepenova TP. *Polym. Sci. USSR.* 1969;11:2885.

37. Blasco F, Riane E. *J. Polym. Sci., Part B: Polym. Phys.* 1983;21:835.

38. Kücükyavuz K, Baysal BM. *Polymer.* 1991;32:3481.

39. Roig A, Hernaândez-Fuentes I. *An. Quím.* 1974;70:668.

40. Ptitsyn OB. *Soviet Phys.-Usp.* (English transl.) 1960;2:797.

41. Stockmayer WH, Matsuo K. *Macromolecules.* 1972;5:766.

42. Baysal B, Lowry BA, Yu H, Stockmayer WH. In: Karasz FE, ed. *Dielectric Properties of Polymers.* New York: Plenum Press; 1971.

43. Saiz E, Mark JE, Flory PJ. *Macromolecules.* 1977;10:967.

44. Bahar I, Baysal BM, Erman B. *Macromolecules.* 1986;19:1703.

45. Zoller P, Walsh DJ. In: *Standard Pressure-Volume-Temperature Data for Polymers.* Lancaster, Pennsylvania: Technomic Publishing Co., Inc; 1995.

46. Kennedy JP, Kurian J. *Macromolecules.* 1990;23:3736.

47. Chen HB, Chang TC, Chiu YS, Ho SY. *J. Polym. Sci. A: Polym. Chem.* 1996;34:679.

48. Brown PG, Fujimori K. *Macromol. Chem. Phys.* 1994;195:917.

49. Ramelow U, Baysal BM. *J. Appl. Polym. Sci.* 1986;32:5865.

50. Grassi A, Longo P, Proto A, Zambelli A. *Macromolecules* 1989;22:104.

51. Jones RG, Tate PCM, Brambley DR. *Polymer.* 1993;34:1768.

52. Rungaphinya W, Fujimori K, Craven IE, Tucker DJ. *Polym. Int.* 1997;42:17.

53. Sato T, Ikeda M, Sugawara A. *Polym. Int.* 2004;53:951.

54. Vukovic R, et al. *J. Polym. Sci.* 1985;30:317.

Poly(chlorotrifluoroethylene)

Anthony L. Andrady, Taner Z. Sen and M. Göktuğ Ahunbay

Acronym, Trade Name PCTFE, Kel-F 81 (3M Company)

Class Vinylidene polymers

Structure ($-CF_2CFCl-$)

Major Applications Used to mold equipment parts, seals, and gaskets, particularly in chemical process equipment and in cryogenic systems. Also used as barrier packaging in pharmaceutical industry. Elastomeric homo- and copolymers used in o-rings, gaskets, and diaphragms.

Property	Units	Conditions	Value	Reference
Radical polymerization	–	Bulk polymerization with the initiators		
		Trichloroacetyl peroxide		(1)
		Dichlorotrifluoropropionyl peroxide		(2)
Typical comonomers		Vinylidene fluoride (Kel-F 800, 3M Company)		
		Ethylene (Halar, Aclar, Allied Corporation)		–
Molecular weight (of repeat unit)	g mol^{-1}	–	116.47	–
Typical molecular weight range of polymer	g mol^{-1}	–	7–40×10^4	(3)
Solvents		Cyclohexane (235°C), benzene (200°C), toluene (142°C), 1,1,1-trichloroethane (120°C), carbon tetrachloride (114°C)		(4–7)
Nonsolvents		Hydrocarbons, alcohols	–	
Crystalline structure	nm	Pseudohexagonal structure	$a = 0.644$, $c = 4.15$	(8)
Degree of crystallinity	%	Commercial polymer	40–80	(3)
Heat of fusion	kJ mol^{-1}	–	5.021	(9)
			1.747–2.329	(10)
Entropy of fusion	kJ mol^{-1}	–	0.0104	(9)

Poly(chlorotrifluoroethylene)

Property	Units	Conditions	Value	Reference
Density (crystalline)	g cm^3	Estimate for completely crystalline polymer	2.187	(10)
		Estimate for completely amorphous polymer	2.077	
Avrami exponent	–	Dilatometry 180–196°C	3	(12)
Glass transition temperature	K	Sample with ∼80% crystallinity (dynamic mechanical at 1 Hz)	423	(13)
			325	(11)
			337	(14)
			348	(15)
		Depending on crystallinity	320–350	(10)
Melting transition temperature	K	Differential thermal analysis	483–488	(16)
		DSC	481–489	(10)
Sub-T_g transitions	K	Sample with ∼80% crystallinity Dynamic mechanical (1 Hz)		(13)
		β transition	363	
		γ transition	236	
		Mechanical loss		(17)
		β transition	368	
		γ transition	230	
Heat capacity	kJ K^{-1} mol^{-1}	100°C	0.0363	(18)
		200°C	0.0575	
		250°C	0.0690	
		−270.5°C–97°C by Adiabatic Calorimetry and −23°C–347°C by DSC	0.13×10^{-3}–0.1267	(10)
Young modulus GPa		Dynamic mechanical analysis	3.66	(19)
Shear modulus GPA			1.31	(19)
Bulk modulus GPA			5.80	(19)
Tensile modulus	MPa	−196°C	7,660	(8)
		25°C	14,000	(16)
Tensile strength	MPa	−196°C	119–173	(8)
		25°C	40	(16)
		125°C	4	–
Elongation	%	−196°C	2–4	(8)
		25°C	150	(16)
		125°C	400	–
Flexural modulus	MPa	ASTM D790		(16)
		23°C	1,250	
		−196°C	14,420	

Property	Units	Conditions	Value	Reference
Flexural strength	MPa	–	74	(16)
		$-190°C$	400	(20)
Permeability coefficient P	m^3 (STP) m s^{-1} m^{-2} Pa^{-1} ($\times 10^{-17}$)	Unplasticized film, $\sim$30% crystalline	0.705	(20)
		H_2, 20°C	0.0038	(21)
		N_2, 25°C		
		O_2, 40°C	0.03	(21)
		CO_2, 40°C	0.158	(21)
		H_2O, 25°C	0.218	(22)
		Unplasticized film, $\sim$80% crystalline		
		N_2, 25°C	0.0023	(21)
		O_2, 40°C	0.0188	(21)
		CO_2, 40°C	0.036	(21)
Pyrolyzability	The main decomposition product is monomer (26% by weight). Halocarbon waxes also formed			(23)
Thermal decomposition temperature	K	–	623–643	(23)
Dielectric constant			2.3–2.8	(24)

References

1. Jewel JW. U.S. Patent 3,014,015 (to 3M Company), 19 December 1961.
2. Dittman AL, Wrightson JM. U.S. Patent 2,705,706 (to M. W. Kellog Company), 5 April 1955.
3. Chandrasekaran S. In: Mark HF, Bikkales NM, Overberger CG, Menges G, eds. *Encyclopedia of Polymer Science and Engineering.* 2nd ed. Vol 3. New York: John Wiley and Sons; 1987:466.
4. Thinius K. *Analytische Chemie der Plaste.* Berlin: Springer-Verlag; 1963.
5. Nitsche R, Wolf KA. *Struktur und Physikalisches Verhalten der Kunstoffe.* Vol 1. Berlin: Springer-Verlag; 1961.
6. Kurata M, Stockmeyer WH. *Adv. Polymer Sci.* 1963;3:196.
7. Hall HT. *J. Am. Chem. Soc.* 1952;74:68.
8. Mencik Z. *J. Polym. Sci., Polym. Phys. Ed.* 1973;11:1585.
9. Bueche AM. *J. Am. Chem. Soc.* 1952;74:65.
10. Chang SS, Weeks JJ. *J. Res. Natl. Inst. Stand. Technol.* 1992;97:341.
11. Hoffman JD, Weeks JJ. *J. Res. Natl. Bur. Stand.* 1958;60:465.
12. Rybnikar F. *Coll. Czech. Commun.* 1962;27:449.
13. Scott AH, et al. *J. Res. Natl. Bur. Stand.* 1962;66A:269.
14. Privalko VP, et al. *Polym. Sci. USSR.* 1985;27:642.
15. Khanna YP, Kumar R, *Polymer.* 1991;32:2010.
16. Brandup J, Immergut EH. *Polymer Handbook.* 3rd ed. New York: John Wiley and Sons; 1989:V-54.
17. McRum NG. *J. Polym. Sci.* 1962;60:53.
18. Gaur U, Lau SF, Wunderlich BB. *J. Phys. Che. Ref. Data.* 1983;12:29.

19. Brown EN, Rae PJ, Orler EB. *Polymer*. 2006;47:7506–7518.

20. Ito Y. *Kobunshi Kagaku*. 1961;18:124.

21. Myers AW, Tammela V, Stannett V, Szwarc M. *Mod. Plastics*. 1960;37(10):139.

22. Rust G, Herrero F. *Materialprufung*. 1969;11(5):166.

23. Mardosky SL, Straus S. *J. Res. Natl. Bur. Stand*. 1955;55:223.

24. Brown BA. WO 2002045184 A1 2002.

Poly(cyclohexyl methacrylate)

Jianye Wen

Acronyms, Trade Names PCHMA

CAS number 25768-50-7

Class Vinylidene polymers

Structure

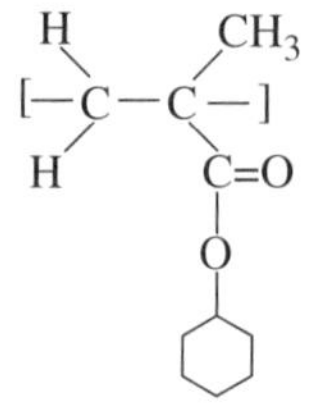

Major Applications Adhesives and binders, coatings, optical waveguides, blends with other polymers for various applications.

Properties of Special Interest Hard, T_g similar to poly(methyl methacrylate) but much higher than its n-hexyl isomer due to the bulkiness of cyclohexyl group.

Property	Units	Conditions	Value	References
^{13}C-NMR spectra	ppm	Selected group assignments		2
		Carbonyl **C**=O	177	
		OCH attached to the ring	74	
		Secondary CH$_2$ groups	26–56	
		CH$_3$ group	22	
Coefficient of thermal expansion	10^{-4} K^{-1}			
		at 20°C	2.4	3
		at 40°C	2.5	3
		at 60°C	2.5	3
		at 80°C	2.5	3
		at 120°C	5.6	3
		at 140°C	6.0	3
		at 160°C	6.2	3
		at 180°C	6.3	3
		at 200°C	6.4	3

Poly(cyclohexyl methacrylate)

Property	Units	Conditions	Value	References
Cohesive energy E_{coh}	J/mol		59978	4
Compressibility	$cm^3/(gkPa)$			
		at 20°C, Glassy state	2.5	3
		at 40°C, Glassy state	2.7	3
		at 60°C, Glassy state	2.8	3
		at 140°C	6.3	3
		at 160°C	7.0	3
		at 180°C	7.7	3
		at 200°C	8.6	3
Decomposition temperature	°C	Initial	190	5
			200	6
		50%	270	5
		max	284, 356, 446	6
			350	5
		at 160°C	1.041	3
		at 180°C	1.028	3
		at 200°C	1.015	3
		110–199°C	$1.1394 - 5.90 \times 10^{-4} t - 0.163 \times 10^{-6} t^2$	3
Density	g/cm^3			
		at 20°C	1.100	1(a),7,8
		at 40°C	1.095	3
		at 80°C	1.084	3
		at 120°C	1.066	3
		at 140°C	1.054	3
FTIR spectra	cm^{-1}	Selected group assignments		9
		C=O	1725	
		C—CH3 bending	1450	
		C—O—C stretch	1261, 1161	
		Aliphatic C—H stretches	2936	
Glass-transition temp. (T_g)	°C		111	6
		Atactic	104	7,10,11
		Syndiotactic	163	7,10
		Isotactic	51	7,10

Mark–Houwink constants

Solvent	Temp. (°C)	Mol. Wt. range ($M \times 10^{-4}$)	$K \times 10^3$ (ml/g)	a	References
Benzene	30	−200	8.4	0.69	12
	25	−419	3.54	0.77	13
Butanol	23 (θ)	−445	33.7	0.50	14
	22.5 (θ)	−125	45.2	0.50	15
	25	−125	31.8	0.533	15
Butanone	25	−560	5.79	0.68	14
	30	−200	7.0	0.66	12
Cyclohexane	25	−418	8.8	0.67	13

Poly(cyclohexyl methacrylate)

Property	Units	Conditions	Value	References
Molar refraction (R)	cm^3		45.05	16
Molar volume (V_m)	cm^3/mole		150.80	16
Molecular weight of repeat unit	g/mol		168.24	
Rate of polymerizarion	$\times 10^4 s^{-1}$	(apparent propagation constant)		17
		Bulk	11.49	
		Toluene	5.42	
		Diphenyl ether	4.93	
		Benzonitrile	3.99	
Refractive index	n_{25}^D		1.50645	1(a),8,18
			1.5066	19
Solubility parameter	(J/cm^3)$^{1/2}$		19.8	20
Theta solvent				
		n-butanol	22.5°C	21
			23°C	14
		n-decane	93.6°C	22
			120.0°C	23
		n-decanol	23.0°C	22
		n-dodecane	97.5°C	22
			129.9°C	23
		n-hexanol	9.2°C	22
		n-nonanol	20.2°C	22
		n-octanol	17.9°C	22
		n-octane	83.4°C	22
			112.1°C	23
		n-propanol	39.5°C	22
Solvents	Tetrachloromethane, toluene, tetrahydrofuran, benzene, chloroform, methyl ethyl ketone, cyclohexanone			5
Nonsolvents	Hexane, dimethylformamide, methanol			5

Unperturbed dimension

Solvent	Temp. (°C)	$[0]/M^{1/2} \times 10^4$	References
n-propanol	39.2	4.88	23
n-butanol	22.7	4.58	23
n-hexanol	9.2	4.37	23
n-ocanol	17.0	4.51	23
n-nonanol	19.8	4.56	23
n-decanol	23.0	4.66	23
n-octane	112.1	4.08	23
n-decane	120.0	4.18	23
n-dedecane	129.9	4.36	23

References

1. General references:
   (a) Brandrup J, Immergut EH. , *Polymer Handbook*. 3rd ed. New York: Wiley-Interscience; 1989.
   (b) Daniels W, Bikales NM, eds. *Encyclopedia of Polymer Science & Technology*. Vol 1. New York: Wiley-Interscience; 1987:234;
   (c) Mark JE, ed., *Physical Properties of Polymers Handbook*. New York: AIP Press, Woodbury; 1996.

2. Chang L, Woo EM, *J. Polym. Sci. Part B*. 2003;41:772.

3. Olabisi O, Simha R. *Macromolecules*. 1975;8:206.

4. Yu X, Wang X, Li X, Gao J, Wang H. *J. Polym. Sci. Part B*. 2006;44:409.

5. Ryttel A. *J. Appl. Polym. Sci*. 1995;57:863.

6. Matsumoto A, Mizuta K, Otsu T. *J. Polym. Sci., Polym. Chem*. 1993;31:2531.

7. Van Krevelen DW. *Properties of Polymers*. Amsterdam: Elsevier Publishing Co; 1976.

8. The Aldrich library of FT-IR spectra , ed. C. J. Charles, Milwaukee I, Wis. Aldrich Chemical Co, V2, P1186,1985.

9. Lewis OG. *Physical Constants of Linear Homopolymers*. New York: Springer-Verlag; 1968.

10. Crawford JWC. *J. Soc. Chem. Ind. London*. 1949;68:201.

11. Novak RW, Lesko PM. In: *Kirk-Othmer Encyclopedia of Chemical Technology*. 4th ed. Vol 16. New York: Wiley- Interscience; 1995:506.

12. Cohn ES, Scogna IL, Orofino TA. unpublished work cited in S. Krause, *Dilute Solution Properties of Acrylic and Methacrylic Polymers*. Philadephia, Pennsylvania: Part I, Revison 1, Rohm & Haas Co; 1961.

13. Hadjichristidis N, Devaleriola M, Desreux V. *Eur. Polym. J*. 1972;8:1193.

14. Hakozaki J. *Nippon Kagaku Zasshi J. Chem. Soc. Japan, Pure Chem. Sec*. 1961;82:158.

15. Katime Amashta IA, Sanchz G. *Eur. Polym. J*. 1975;11:223.

16. Patel MP, Davy KWM, Braden M, *Biomaterials*. 1992;13:643.

17. Muñoz-Bonilla A, Madruga EL, Fernández-García M. *J. Polym. Sci. Part A*. 2005;43:7.

18. Wiley RH, Braver GM. *J. Polym. Sci*. 1948;3:455.

19. Seferis JC, Brandrup J, Immergut EH. In: *Polymer Handbook*. 3rd ed. New York: Wiley-Interscience; 1989:VI–451.

20. Yu X, Wang X, Li X, Gao J. *QSAR Comb. Sci*. 2006;25:156.

21. Amashta K, Sanchez G. *Eur. Polym. J*. 1975;11:223.

22. Sedlak K, Lath D. *Makrotest Sb. Prednasek, Celustatni Konf*. 1978;5:71.

23. Lath D, Sedlak K, Florian S, Lathova E. *Polym. Bull*. 1986;16:453.

Poly(di-*n*-butylsiloxane)

Yuli K. Godovsky and Vladimir S. Papkov

Acronym PDBuS

Class Polysiloxanes

Structure [—C$_4$H$_{10}$)$_2$SiO—]

Properties of Special Interest Low glass transition temperature, mesophase behavior.

Property	Units	Conditions	Value	Reference
Preparative techniques		Anionic ring opening polymerization of hexabutylcyclotrisiloxane.		(1–4)
Molecular weight (repeat unit)		–	158.31	–
Typical molecular weight range of polymer	g mol^{-1}	–	10^4–10^5	–
NMR spectroscopy		Solid state ^{1}H, ^{13}C, ^{29}Si		(3, 4)
Heat of fusion	kJ mol^{-1}	High temperature crystal 2 to mesophase	0.9–1.1	(3–5)
Entropy of fusion	J mol^{-1}K^{-1}		3.9	(3–5)
Glass transition temperature	K	DSC	157	(3)
Melting temperature	K	High temperature crystal 2 to mesophase	254	(3–5)
Polymorphs		DSC, X-ray data		
		Low temperature crystal 1		(3–6)
		High temperature crystal 2		(3–6)
		Mesophase		(3–6)
Transition temperature	K	Crystal 1–crystal 2, DSC	229	(3–5)
Heat of transition	kJ mol^{-1}	Crystal 1–crystal 2	3.6	(3–5)
Isotropization temperature	K	Polarization microscopy		
		Strong MW dependence		
		2.8×10^4	489	(3)
		1.28×10^5	572	(3)

Poly(di-*n*-butylsiloxane)

References

1. Moeller M, Siffrin S, Koegler G, Oelfin D. *Makromol. Chem., Macromol. Symp.* 1990;34:171.
2. Moeller M, Siffrin S, Out GJJ, Boileau S. *ACS Polym. Prep.* 1992;33(1):176.
3. Out GJJ, Schloessels F, Turetskii AA, Moeller M. *Makromol. Chem. Phys.* 1995;196:2035.
4. Out GJJ, Siffrin S, Frey H, Oelfin D, KoeglerG, Moeller M. *Polym. Adv. Technol.* 1994;5:796.
5. Out GJJ, Turetskii AA, Moeller M. *Macromol. Rapid. Commun.* 1995;16:107.
6. Out GJJ, Turetskii AA, Moeller M, Oelfin D. *Macromolecules.* 1994;27:3310.

Poly(diethylsiloxane)

Yuli K. Godovsky and Vladimir S. Papkov

Acronym PDES

Class Polysiloxanes

Structure $[-C_2H_5)_2SiO-]$

Major Applications Comonomer for low-temperature silicone rubbers: for example, poly(dimethyl-diethyl)siloxane. Low molecular weight PDES is the basis for low- and high-temperature silicone oils, greases, and lubricants.[1,2]

Properties of Special Interest Low glass transition temperature, mesophase behavior including reversible stress-induced mesophase formation accompanied by necking–denecking phenomena in cyclic deformation of elastomers.[3–8]

Property	Units	Conditions	Value	Reference
Preparative technique		Anionic nonequilibrium ring-opening polymerization of hexaethylcyclotrisiloxane (initiators: KOH, CF_3SO_3H, sec-BuSi(Et2)OLi , nBuLi-(nBuO)$_3$PO, NaOH/12-crown-4 (1,4,7,10 tetraoxacyclododecane) complex)		(9–15)
Enthalpy of polymerization	kJ mol^{-1}	$T_{pol} = 443$ K	15.1 ± 0.4	(16)
Molecular weight (of repeat unit)	g mol^{-1}	–	102.21	–
Typical molecular weight range of polymer	g mol^{-1}	–	$2 \times 10^3 - 1 \times 10^6$	–
Typical polydispersity index (M_w/M_n)	–	–	1.1–2.0	(11–15)
Raman spectroscopy		Temperature range: 129–293 K		(17)
IR spectroscopy		Temperature range: 129–303 K		(5)
NMR spectroscopy		Solid state ^1H, ^2H, ^{13}C, ^{29}Si, ^{17}O		(18–24) (25)
Density (amorphous)	g cm^{-3}	293 K	0.99	(26, 27)

Poly(diethylsiloxane)

Property	Units	Conditions	Value	Reference
Thermal expansion coefficient	K^{-1}	293–363 K(dilatometry); pressure $P = 0$–0.1 GPa (mesophase state and isotropic melt)	$(6.3–17P) \times 10^{-4}$	(27)
		$M_n = 3 \times 10^3$ 293–360 K (amorphous)	6.1×10^{-4}	
		$M_n = 3.27 \times 10^5$ 293–305 K (mesophase) 330–350 K (isotropic melt) (from refractometry data)	4.8×10^{-4} 5.9×10^{-4}	(28)
		$M_n = 1.12 \times 10^5$; $M_w/M_n = 2.5$ 25°C	5.75×10^{-4}	(29)
Refractive index	–	$M_n = 3.27 \times 10^5$; $M_w/M_n = 2.2$ $n_D{}^{20}$		(28)
		Mesophase	1.4537	
		Isotropic melt (extrapolated)	1.4519	
		dn_D/dT		(28)
		293–305 K (mesophase)	−2.52	
		330–350 K (isotropic melt)	−2.96	
Mark–Houwink parameters: K and a	$K = $ ml g^{-1} $a = $ None	$MW = (1.15–3.7) \times 10^5$ Toluene, 25°C,	$K = 2.71 \times 10^{-2}$ $a = 0.636$	(30)
		$MW = (5.1–192) \times 10^3$ Toluene, 25°C	$K = 1.15 \times 10^{-4}$ $a = 0.704$	(31)
		THF, 30°C	$K = 1.37 \times 10^{-4}$ $a = 0.715$	
Characteristic ratio $\langle r^2 \rangle_0/nl^2$		Toluene, 25°C (osmometric and viscosimetric measurements)	7.7 ± 0.8	(30)
		MD simulation	∼7	(32)
Temperature dependence of the unperturbed dimensions of chains $d\ln\langle r^2 \rangle/dT$		Stretching calorimetry, 25–100°C	-0.80×10^3	(33)
Energy contribution f_e/f			−0.25	(33)
PDES–solvent interaction parameter at infinite dilution χ_0		Stretching calorimetry, 25–100°C		(31)
		Toluene, 30°C	0.467	
		THF, 30°C (viscosity measurement)	0.412	
		Toluene,		
		60°C	0.72	
		100°C	0.70	
		130°C	0.71	

Property	Units	Conditions	Value	Reference
		Benzene,		
		60°C	0.82	
		100°C	0.77	
		130°C	0.77	
		n-Heptane		
		60°C	0.53	
		100°C	0.55	
		130°C	0.60	
		(gas–liquid chromatography measurement)		
Diffusion coefficient of PDES chains, D	cm^2s^{-1}	In mesophase	$\approx 10^{-8}$	(24)
		In amorphous phase (pulse field-gradient stimulated-echo) ^{13}C NMR measurement at 20–30°C)	$\approx 10^{-(10-11)}$	

Polymorph	Lattice	Monomers Per Unit Cell	Cell Dimensions (A)			Cell Angles (Degrees)		
			a	b	c (Chain Axis)	α	β	γ
α_1	Monoclinic[26]	2	14.45	8.75	4.72	90	90	29.8
α_2	Monoclinic[26]	2	14.59	8.90	4.75	90	90	29.7
β_1	Tetragonal[26]	2	7.83	7.83	4.72	90	90	90
β_2	Tetragonal[26]	2	7.90	7.90	4.72	90	90	90
β_2	Monoclinic[34]	2	8.32	8.12	5.02	90	90	108
μ (mesophase)	Monoclinic[26] (close to pseudohexagonal)	2	14.75	8.89	4.88	90	90	31.2
μ (mesophase)	Monoclinic[34]	2	8.39	8.77	4.91	90	90	65.5

Property	Units	Conditions	Value	Reference
Degree of crystallinity	%	Cold crystallization after quenching amorphous sample	≈ 30	(10, 26, 35)
		Crystallization from mesomorphic state	≈ 90	
Heat of fusion	$kJmol^{-1}$	$\alpha_2 \rightarrow \mu$	1.72	(10)
			1.51	(13)
		$\beta_2 \rightarrow \mu$	2.14	(10)
			2.27	(13)
Heat of isotropization	$kJmol^{-1}$	$\mu \rightarrow$ isotropic melt	0.31	(5, 36)
			0.34	(13)
Entropy of fusion	$Jmol^{-1}K^{-1}$	$\alpha_2 \rightarrow \mu$	6.14	(37)
			5.4	(13)
		$\beta_2 \rightarrow \mu$	7.38	(37)
			7.8	(13)

Poly(diethylsiloxane)

Property	Units	Conditions	Value	Reference
Entropy of isotropization	$Jmol^{-1}K^{-1}$	$\mu \rightarrow$ isotropic melt	0.92	(37)
			1.07	(13)
Density (crystalline)	$g\,cm^{-3}$	From X-ray data		
		α_1 <212 K	1.17	
		α_2, 223 K	1.10	
		β_1, 193 K	1.17	(26)
		β_2, 223 K	1.14	
		β_2, 273 K	1.05	
Density (mesophase)	$g\,cm^{-3}$	From X-ray data	1.02	
		Observed, 293 K	1.015	(26)
Glass transition temperature	K	DSC	134	(5)
		Adiabatic calorimetry	130	(8, 22)
		Dielectric, 1 KHz	146	(11)
		Dielectric, 100 Hz	133	(23)
		NMR, T_1 and T_2	138–140	(13, 22)
		DMA, 1 Hz	134	(11)
Melting temperature	K	$\alpha_2 \rightarrow \mu$ $MW = 3 \times 10^4$–6×10^5	277–282	(10, 13)
		$\beta_2 \rightarrow \mu$	284–291	
Transition temperature	K	$\alpha_1 \rightarrow \alpha_2$ $MW = 1.6 \times 10^5$	214	(10)
		$\beta_1 \rightarrow \beta_2$	206	
Heat of transition	$kJmol^{-1}$	$\alpha_1 \rightarrow \alpha_2$	2.86	(10)
		$\beta_1 \rightarrow \beta_2$	2.65	
Isotropization temperature	K	$MW(\times 10^3) =$		
		765	326	
		425	325	
		172	319	
		100	307	(10, 13, 36)
		55	301	
		42	292	
		$MW = 4 \times 10^5$; Under hydrostatic pressure		
		0–170 MPa	325–360 K	
		170–265 MPa	360–347 K	(38)
Avrami exponent n	–	Crystallization at 276 K from mesophase	≈ 2	(36)
		Formation of mesophase from the melt in the range of 293–306 K	1.75	(36)
Morphology		Crystalline and mesomorphic lamellae and their aggregates		(13, 39, 40)
Heat capacity	$J\,K^{-1}mol^{-1}$ (of repeat units)	β_1 polymorph		
		10 K	2.289	(16, 41, 42)
		20 K	10.97	
		50 K	39.60	
		100 K	75.40	
		200 K	123.0	

Property	Units	Conditions	Value	Reference
		β_2 polymorph		
		250 K	140.5	
		330 K (melt)	174.0	(16,41,42)
		ΔC_p at T_g	36.0	(10,16)
Dielectric constant ε'	–	83–123 K	2.60–2.70	(43)
Loss factor tan δ Rheological properties	–	200–300 K	0.015	(5)
Effective viscosity	Pa s	$MW = 7.27 \times 10^5$, shear rate 10^{-1} s^{-1} Mesophase (20°C) Melt (60°C)	10^5 10^4	(44,45)
Flow index, n	–	Mesophase (20°C) Melt (60°C)	0.2 0.72	
Activation energy of flow, E_τ	kJ mol^{-1}	Melt (60–120°C)	16	
Dynamic elastic module, G'	dyn cm^2	$MW = 1.92 \times 10^{-3}$ Frequency 10^{-1}–4×10^2 rad s^{-1} (mesophase, 24°C) (melt, 60°C)	4.7×10^6– 7.1×10^6 2.8×10^2– 6.2×10^5	(46)
Dynamic loss module, G''	dyn cm^2	Frequency 10^{-1}–4×10^2 rad s^{-1} (mesophase, 24°C) (melt, 60°C)	1.1×10^6– 7.1×10^5 4.7×10^6– 7.1×10^5	

References

1. Sobolevskii M, Skorokhodov I, Grinevich K. *Oligoorganosiloxanes.* Moscow: Khimiya; 1985.
2. SILICONE FLUIDS. Stable, Inert Media, Gelest, Inc; 2004:22. http://gelest.com/company/pdfs/siliconefluids.pdf.
3. Beatty CL, et al. *Macromolecules.* 1975;8:547.
4. Papkov VS, et al. *Polymer Science USSR.* 1989;A31:1729.
5. Papkov VS, Yu P. Kvachev. *Progr. Colloid Polym. Sci.* 1989;80:221.
6. Godovsky Yu, Angew K. *Makromol. Chem.* 1992;187:202–203.
7. Papkov VS, Turetski A., Out GJ, Moeller M. *Int. J. Polymeric Materials.* 2002; 51:369.
8. Hedden RC, Tachibana H, Duncan TM, Cohen C. *Macromolecules.* 2001;34:5540.
9. Lee CL, et al. *ACS Polym. Preprints.* 1969;10(2):1319.
10. Papkov VS, et al. *J. Polym. Sci., Polym. Chem. Ed.* 1984;22:3617.
11. Koegler G, Hasenhind A, Moeller M. *Macromolecules.* 1989;22:4190.
12. Zavin BG, et al. *Polym. Sci.* 1995;A37:355.
13. Molenberg A, Moeller M. *Macromolecules.* 1997;30:8332.
14. Tsuchihara K, Fujishige S. *Polym. Bull.* 1999;43:129.

15. Hedden RC, Cohen C. *Polymer.* 2000;41:6975.

16. Lebedev B, et al. *Vysokomol. soedin.* 1984;26:2476.

17. Friedrich J, Rabolt JF. *Macromolecules.* 1987;20:1975.

18. Froix MF, et al. *J. Polym. Sci., Polym. Phys. Ed.* 1975;13:1269.

19. Litvinov V, et al. *Vysokomol. Soedin.* 1985;27A:1529.

20. Moeller M, et al. *Makromol. Chem., Macromol. Symp.* 1990;34:171.

21. Litvinov V, et al. *Colloid Polym. Sci.* 1989;267:681.

22. Litvinov VM, Macho V, Spiess HW. *Acta Polym.* 1997;48:471.

23. Kimmich R, Anoardo E. *Prog. Nucl. Magn. Reson. Spectroscopy.* 2004;44:257.

24. Kanesaka S, et al. *Macromolecules.* 2004;37:453.

25. Kimura H. *Magn. Reson. Chem.* 2005;43:209.

26. Tsvankin D, Ya, et al. *J. Polym. Sci., Polym. Chem. Ed.* 1985;23:1043.

27. Pechhold W, Schwarzenberger P. In: Hochheimer HD, Etters D, eds. *Frontiers of High-pressure Research.* New York: Plenum Press; 1991:58–71.

28. Chalykh AE, et al. *Polymer Science.* 2002;A44:1048.

29. Galin M. *Macromolecules.* 1977;10:1239.

30. Mark JE, Chin DS, Su TK. *Polymer.* 1978;19:407.

31. Hedden RC, Saxena H, Cohen C. *Macromolecules.* 2000;33:8676.

32. Neuburger N, Bahar IT, Mattice WL. *Macromolecules.* 1992;25:2447.

33. Godovsky Yu. Synthesis K. In: Aharoni SM, ed. *Characterization and Theory of Polymeric Networks and Gels.* New York: Plenum Press; 1992:127–145.

34. Inomata K, Yamamoto K, Nose T. *Polymer Journal.* 2000;32:1044.

35. Out G, et al. *Polymer.* 1995;36:3213.

36. Papkov VS, et al. *J. Polym. Sci.: Part B: Polym. Phys.* 1987;25:1859.

37. Godovsky YuK, Papkov VS. *Makromol. Chem., Macromol. Symp.* 1986;4:71.

38. Reck T, et al. *Rev. Sci. Instrum.* 1998;69:1823.

39. Obolonkova ES, Papkov VS. *Vysokomol. Soed.* 1991;B31:691.

40. Godovsky YuK, Papkov VS, Magonov SN. *Macromolecules.* 2001;34:976.

41. Beatty CL, Karasz FE. *J. Polym. Sci., Polym. Phys. Ed.* 1975;13:971.

42. Varma-Nair M, Wesson JP, Wunderlich B. *J. Thermal Anal.* 1989;35:1913.

43. Pochan JM, Beatty CL, Hinman DF. *J. Polym. Sci., Polym. Phys. Ed.* 1975;13:977.

44. Papkov VS, et al. *Polymer Science.* 2001;A43:200.

45. Chaika EM, Vasil'ev VG, Papkov VS. *Polymer Science.* 2007;A49:337.

46. Saxena H, Hedden RC, Cohen C. *J. Rheol.* 2002;46:1177.

Poly(di-*n*-hexylsiloxane)

Yuli K. Godovsky and Vladimir S. Papkov

Acronym PDHeS

Class Polysiloxanes

Structure $[-C_6H_{13})_2SiO-]$

Properties of Special Interest Low glass transition temperature, mesophase behavior.

Property	Units	Conditions	Value	Reference
Preparative techniques		Anionic ring opening polymerization of hexahexylcyclotrisiloxane.		(1, 2)
Molecular weight (repeat unit)		–	214.41	–
Typical molecular weight range of polymer	$g\,mol^{-1}$	–	10^4–10^6	–
NMR spectroscopy		Solid state ^{29}Si		(2, 3)
Mark–Houwink parameters K and a	$K = ml\,g^{-1}$ a = none	Toluene, 298 K	$K = 0.275$ $a = 0.463$	(3)
Heat of fusion	$kJ\,mol^{-1}$	High temperature crystal 2 to mesophase	1.8–2.2	(3–5)
Entropy of fusion	$J\,mol^{-1}K^{-1}$		7.2	(3–5)
Melting temperature	K	High temperature crystal 2 to mesophase	296	(3–5)
Polymorphs		DSC, X-ray data Low temperature crystal 1 High temperature crystal 2 Mesophase		(3–5) (3–5) (3–5)
Transition temperature	K	Crystal 1–crystal 2, DSC	246	(3–5)
Heat of transition	$kJ\,mol^{-1}$	Crystal 1–crystal 2	6.7	(3)
Isotropization temperature	K	Polarization microscopy	603	(3)

Poly(di-*n*-hexylsiloxane)

References

1. Moeller M, Siffrin S, Out GJJ, Boileau S. *ACS Polym. Prep.* 1992;33(1):176.
2. Out GJJ, Turetskii AA, Moeller M. *Macromol. Rapid. Commun.* 1995;16:107.
3. Out GJJ, Turetskii AA, Moeller M, Oelfin D. *Macromolecules.* 1994;27:3310.
4. Out GJJ. *Dissertation.* The Netherlands: Universiteit Twente; 1994.
5. Molenberg A. *Dissertation.* Germany: University of Ulm; 1997.

Poly(di-*n*-hexylsilylene)

Robert West

Acronym, Alternative Name PDHS, polydi-*n*-hexylsilane

Class Polysilanes

Structure $[-nC_6H_{13}-Si-nC_6H_{13}-]$

Major Applications None

Properties of Special Interest Transition from crystalline phase to columnar mesophase at 42°C. That is, PDHS is crystalline with an all-*trans* arrangement of the polysilane chain below the disordering temperature of 42°C; above this temperature the polymer exists in a hexagonal columnar liquid crystalline phase.

For general information about polysilane polymers see the entry for *Poly(methylphenylsilylene)* in this handbook.

Preparative techniques

Reactants	Temp. (°C)	Yield (%)	M_w	M_w/M_n	Reference
n-Hex$_2$SiCl$_2$, Na, toluene, 15-crown-5	110	24	–	–	(1)
n-Hex$_2$SiCl$_2$, Na, toluene, 15-crown-5, ultrasound	20	50	67,000	–	–
n-Hex$_2$SiCl$_2$, Na, toluene (25% diglyme)	24	–	45,000	1.73	(2)
Same as above with 3% 18-crown-6	–	63.4	9,200	3.6	(2)
n-Hex$_2$SiCl$_2$, Na, Et$_2$O, 15-crown-5	35	22	23,800	–	(3)
n-Hex$_2$SiCl$_2$, Na, toluene, 2% EtOAc	110	12.6	1,300,000	1.8	(4)

Property	Units	Conditions	Value	Reference
Repeat unit	g mol^{-1}	(C$_6$H$_{13}$)$_2$Si	198	–
Infrared spectrum	cm^{-1}	–	2,961, 2,924, 2,871, 2,854, 1,471, 1,417, 1,379, 1,233, 1,175, 1,110, 977, 899, 727, 673	(5)
UV absorption	λ_1 (nm)	Hexane, $\varepsilon = 9,700$	318	(2)
Emission spectrum	λ_1 (nm)	Hexane, $\phi = 0.42$, $\tau = 150$ ps	342	(6)

Property	Units	Conditions			Value	Reference
NMR spectra	δ (ppm)	Nucleus	Conditions	Temp. (°C)		
		^{29}Si	Solution, trichlorobenzene-dioxane-d_8	25	−24.8	(7)
		^{29}Si	Solid	25	−20.8	(8)
				44.3	−24.1	(8)
		^{13}C	Solution, trichlorobenzene-dioxane-d_8	25	14.45	(7)
					23.49	
					32.55	
					35.19	
					28.41	
					15.92	
		^{1}H	Solution, CDCl$_3$	25	0.6–1.7	(2)
Solvents	THF, toluene, CH$_2$Cl$_2$, hexane					
Nonsolvents	Ethanol, 2-propanol					
Properties from light scattering study						(9)
M_w	g mol^{-1}	Hexane			2.2×10^6	
M_w/M_n	–	–			2.3	
dn/dc	ml g^{-1}	–			0.138	
A_2	ml mol g^{-2}	–			1.62×10^{-4}	
R_g	nm	–			10^8	
$\lambda_{persist}$	nm	–			6–7	
M_w	g mol^{-1}	THF			2.2×10^6	
M_w/M_n	–	–			2.3	
dn/dc	ml g^{-1}	–			0.177	
A_2	ml mol g^{-2}	–			1.14×10^{-1}	
R_g	nm	–			92	
l/k	nm	–			5.4	
C_∞	nm	–			19	
Crystalline properties	Monoclinic lattice below transition temperature of 42°C					
	Hexagonal columnar mesophase above transition temperature					
Crystalline phase	Å	25°C			$a = 13.75,$ $b = 21.82,$ $c = 4.07$	(10, 11)
	Degrees	25°C			$\gamma = 88$	(10, 11)
Hexagonal columnar mesophase, a	Å	>42°C			1.56	(10, 11)
Surface tension	mN m^{-1}	–			29.9	(12)
Scission, quantum yield, ϕ_s	mol Einstein^{-1}	Toluene solution, $\lambda = 353$ nm			0.6	(6)

Poly(di-n-hexylsilylene)

Property	Units	Conditions	Value	Reference
Cross-linking, quantum yield, ϕ_x	mol Einstein^{-1}	Toluene solution, $\lambda = 353$ nm	0	(6)
Suppliers		Gelest, Inc., 612 William Leigh Drive, Tullytown, PA 19007-6308, USA		

Nonlinear optical properties$^{(13)}$

M_w (g mol^{-1})	Temp. (°C)	λ (nm)	Lp (thickness, nm)	χ^{131} (esu$^{(\times 10-12)}$)
>300,000	23	1,064	50	11
	21	1,064	120	5.5
	50	1,064	120	2.0
	23	1,064	240	4.6
		1,907	240	1.3
		1,907	240	0.9

References

1. Miller RD, Thompson D, Sooriyakumaran R, Fickes GN. *J. Polym. Sci, Polym. Chem. Ed.* 1991;29:813.
2. Matyjazewski K, Greszka D, Hrkach JS, Kim HK. *Macromolecules.* 1995;28:59.
3. Cragg RH, Jones RG, Swain AC, Welsh SJ. *J. Chem. Soc., Chem. Commun.* 1990;1147.
4. Miller RD, Jenker PK. *Macromolecules.* 1994;27:5921.
5. Rabolt JF, Hofer D, Miller RD, Fickes GN. *Macromolecules.* 1986;19:611.
6. Miller RD, Michl J. *J. Chem. Rev.* 1989;89:1359.
7. Schilling FC, Bovey FA, Zeigler JM. *Macromolecules.* 1986;19:2309.
8. Gobbi GC, Fleming WW, Sooriyakumaran R, Miller RD. *J. Am. Chem. Soc.* 1986;108:5624.
9. Shukla P, et al. *Macromolecules.* 1991;24:5606.; Cotts PM, Ferline S, Dalgi G, Pearson JC. *Macromolecules.* 1991;24:6730.
10. Kuzmany H, Rabolt JF, Farmer BL, Miller RD. *J. Chem. Phys.* 1986;85:7413.
11. Weber P, Guillon D, Skoulios A, Miller RD. *Liq. Cryst.* 1990;8:825.
12. Fujisaka T, West R, Murray C. *J. Organometal. Chem.* 1993;449:105.
13. Baumert JC, et al. *Appl. Phys. Lett.* 1988;53:1147.

Poly(dimethylferrocenylethylene)

Ian Manners

Class Inorganic and semi-inorganic Polymers

Structure $[(C_5H_3Me)Fe(C_5H_3MeCH_2CH_2)]_n$

Properties of Special Interest Low cost; ease of synthesis; interesting optical, magnetic, and electrical properties.

Property	Units	Conditions	Value	Reference
UV-Vis absorption ($\lambda_{max.}$)	nm	THF solution	440	(1)
UV-Vis absorption coefficient (ε)	$M^{-1}cm^{-1}$	THF solution	190	(1)
Glass transition temperature	°C	DMA experiment		
Glass transition temperature	°C	DSC experiment	65	(1)
Melting temperature	°C	DSC experiment		

For Monomer $(C_5H_4)_2FeCH_2CH_2$[6]

Unit cell dimensions

Lattice	Monomers per unit cell	Cell dimensions			Cell angles		
		a	b	c	α	β	γ
Orthorhombic	8	7.421(2)	12.305(2)	19.839(4)	90	90	90

Reference

1. Nelson JM, Nguyen P, Petersen R, et al. *Chem. Eur. J.* 1997;3:573.

Poly(2,6-dimethyl-1,4-phenylene oxide)

Edward N. Peters

Acronym PPE

Trade Names PPO (Sabic Innovative Plastics.), Noryl (Sabic Innovative Plastics.), Noryl GTX (Sabic Innovative Plastics.), Prevex (Sabic Innovative Plastics.), Xyron (Ashahi Chemical), Iupiace (Mitsubishi Engineering Plastics), Lemalloy (Mitsubishi Engineering Plastics), Artley (Sumitomo Chemical Co. ltd.), Blue Star (Blue Star)

Structure

Class Polyaryl ether, engineering thermoplastic

General Information PPE polymers have found broad use in polymer blends or alloys.[1−5] PPE forms rather rare miscible, single-phase blends with polystyrene over the entire range of composition. PPE blends with impact polystyrene are often referred to as 'modified PPE' resins and are characterized by ease of processing, high impact strength, outstanding dimensional stability at elevated temperatures, long-term stability under load (creep resistance), and excellent electrical properties over a wide range of frequencies and temperatures. Another unique feature of modified PPE resins is their ability to make chlorine/bromine-free flame retardant (ECO-label compliant) grades. PPE alloys with polyamides and polypropylene are available which exhibit chemical resistance and dimensional stability at elevated temperatures. PPE macromonomers are available for enhancement of the performance of thermoset resins.[6]

Major Applications Applications include automotive (instrument panels, trim, spoilers, under the hood components, grilles), telecommunication equipment (TV cabinets, cable splice boxes, wire board frames, electrical connectors, structural and interior components in electrical/electronic equipment), plumbing/water handling (pumps, plumbing fixtures), consumer goods (microwavable food packaging, and appliance parts) medical, building and construction.[2−5]

Properties of Special Interest PPE is an amorphous polymer which has an extremely high glass transition temperature, outstanding electrical properties over a wide frequency and temperature range, low density and very low moisture absorption relative to other engineering thermoplastics, good inherent resistance to flames, and excellent hydrolytic stability.[1−5]

Poly(2,6-dimethyl-1,4-phenylene oxide)

Preparative Techniques PPE is produced by the oxidative coupling polymerization of 2,6-dimethyl phenol.[7–9]

Property	Units	Conditions	Value	Reference
Molecular weight of repeat unit	g mol^{-1}	–	120.15	(1, 7–9)
Typical molecular weight	g mol^{-1}	M_w	40 000	(8)
	g mol^{-1}	M_n	18 000	
Typical polydispersity index (M_w/M_n)	–	–	2.2	(8, 10)
NMR		^{1}H-NMR (CDCl$_3$)		(11)
		^{13}C-NMR (CDCl$_3$)		(12)
		^{31}P-NMR (hydroxyl end group)		(13)
Density	g cm^{-3}	296 K	1.06	(10)
		Melt	0.958	(8)
		Crystalline	1.16	(14)
Solvents	–	–	Benzene	(10, 15)
			Toluene	
			Xylene	
			Chlorobenzene	
			Chloroform	
			Carbon tetrachloride	
Nonsolvents	–	–	Alcohols, ketones	(10)
Solubility parameter	MPa$^{1/2}$		9.5–10.21	(16)
Theta temperature	K	Methylene chloride	342.4	(17)
Second virial coefficient	A$^2 \times 10^4$mol cm^3 g^{-3}			(15)
		Toluene, 298 K, M_w 111 000	9.4	
		Benzene, M_w 130 000	11.1	
		Xylene, M_w 106 000	8.8	
		Chloroform, M_w 130 000	14.1	
		Dioxane, M_w 85 000	2.0	
Intrinsic viscosity	dl g^{-1}	CHCl$_3$	0.30–0.58	(2, 8)
Mark–Houwink parameters K and a	K = g mol^{-1} a = none	Benzene, 298 K 26 500 < M_w < 415 000	$K = 260$ $a = 0.69$	(15)
		Toluene, 298 K 26 500 < M_w < 415 000	$K = 285$ $a = 0.68$	(18)
		Chlorobenzene, 298 K 26 500 < M_w < 415 000	$K = 378$ $a = 0.66$	(18)
		Chloroform, 298 K 26 500 < M_w < 415 000	$K = 483$ $a = 0.64$	(18)
		Carbon tetrachloride, 298 K 26 500 < M_w < 415 000	$K = 744$ $a = 0.58$	(15)
Interaction parameter χ				(19)

Poly(2,6-dimethyl-1,4-phenylene oxide)

Property	Units	Conditions	Value	Reference
Persistence length	Å	Benzene	0.84	(15)
		Carbon tetrachloride	0.86	
Lattice	–	–	Monoclinic	(20)
Chain conformation	–	–	(4/1) helix	(20)
Unit cell dimensions	Å	α-Pinene	$a = 11.92$	(20)
			$b = 17.10$	
Unit cell angles	Degree	–	97	(20)
			91.02	(21)
Unit cell contents (number of repeat units)			6	(20)
Glass transition temperature	K	DSC (5 K h^{-1})	498	(22)
Melting point	K	DSC (5 K h^{-1})	535	(23)
Degree of crystallinity	%	X-ray	25	(22)
		Cooling from melt at 12 K h^{-1}	0	(22)
		Exposure to 2-butanone	~30	(24)
Heat of fusion (of repeat units)	kJmol^{-1}	DSC	1.80 ± 0.36	(22)
		Methylene chloride	5.88	(17)
		1-Chloronaphthalene	3.77	(25)
Entropy of fusion (of repeat units)	kJ K^{-1}mol^{-1}	DSC	9.5×10^{-3}	(22)
Heat capacity	kJ K^{-1}mol^{-1}	400–482 K	$(0.3428T + 53.86) \times 10^{-3}$	(26)
		482–540 K	$(0.2279T + 141.09) \times 10^{-3}$	
Avrami exponent	–	Melt-crystallization	1.6	(27)
Coefficient of thermal expansion	K^{-1}	–	5.2×10^{-5}	(8)
Oxygen Index	%	–	29	(28)
Heat deflection temperature	K	At 0.46 MPa	452	(29)
		At 1.82 MPa	447	
Tensile modulus	MPa	ASTM D638		(8)
		296 K	2690	
		366 K	2480	
Shear modulus	MPa	ASTM D638		(8)
		296 K	2690	
		366 K	2480	
Tensile strength	MPa	ASTM D638		(8)
		296 K	80	
		366 K	55	
Elongation at break	%	ASTM D638		(8)
		296 K	30	
		366 K	50	

Poly(2,6-dimethyl-1,4-phenylene oxide)

Property	Units	Conditions	Value	Reference
Flexural modulus	MPa	ASTM D790		(8)
		256 K	2650	
		296 K	2590	
		366 K	2480	
Flexural strength	MPa	ASTM D790		(8)
		256 K	134	
		296 K	114	
		366 K	87	
Izod impact strength	Jm^{-1}	ASTM D790		(8)
		Notched, 256 K	53	
		Notched, 296 K	64	
		Notched, 366 K	91	
		Unnotched, 296 K	>2000	
Hardness	–	Rockwell	M78	(8)
Refractive index	–	20°C	1.575	(30)
	–	–	1.567	(31)
Dielectric constant	–	60 Hz, 78 K	2.57	(2, 8)
		60 Hz, 273 K	2.56	
		60 Hz, 298 K	2.55	
		60 Hz, 323 K	2.55	
		60 Hz, 348 K	2.55	
		60 Hz, 373 K	2.54	
		60 Hz, 453 K	2.52	
		1 MHz, 298 K	2.56	
		1 GHz, 298 K	2.57	
Dissipation factor	–	60 Hz, 78 K	0.0010	
		60 Hz, 273 K	0.0027	
		60 Hz, 298 K	0.0032	
		60 Hz, 323 K	0.0029	
		60 Hz, 348 K	0.0028	
		60 Hz, 373 K	0.0031	
		60 Hz, 453 K	0.0043	
		1 MHz, 298 K	0.0040	
Volume resistivity	ohm cm	60 Hz, 78 K	4.4×10^{15}	(2, 8)
		60 Hz, 273 K	8.2×10^{16}	
		60 Hz, 298 K	1.3×10^{17}	
		60 Hz, 323 K	2.6×10^{17}	
		60 Hz, 348 K	2.7×10^{17}	
		60 Hz, 373 K	2.5×10^{17}	
		60 Hz, 453 K	7.5×10^{16}	
Surface resistivity	ohm	ASTM D257	4.3×10^{17}	(2, 8)
Dielectric strength	$kV\ mm^{-1}$	ASTM D149	20	(2, 8)
Arc resistance	s	ASTM D195	75	(2, 14)

Poly(2,6-dimethyl-1,4-phenylene oxide)

References

1. Hay AS, Ding Y. In: Mark JE, ed. *Polymer Data Handbook.* New York: Oxford University Press; 1999:406–410.
2. Peters EN. In: Margolis J, ed. *Engineering Plastics Handbook Thermoplastics, Properties, and Applications.* New York: McGraw-Hill; 2005.
3. Peters EN. In: Kutz M, ed. *Mechanical Engineers' Handbook.* 3rd ed. New York: Wiley-Interscience; 2005.
4. Peters EN, Arisman RK. In: Craver CD, Carraher CE, eds. *Applied Polymer Science – 21st Century.* New York: Elsevier; 2000.
5. Peters EN, Parthasarathy M. *Kunstoffe.* October 2005:108–113.
6. Peters EN, Kruglov A, Delsman E, Guo H, Carrillo A, Rocha G. *SPE ANTEC 2007.*
7. Hay AS, Blanchard HS, Endres GF, Eustance JW. *J. Am. Chem. Soc.* 1959;81:6335.
8. Aycock D, Abolins V, White DM. In: Mark HF, Bikales NM, Overberger CG, Menges G. eds. *Encyclopedia of Polymer Science Engineering.* Vol 13. New York: John Wiley and Sons; 1988:1.
9. Hay AS, Blanchard HS, Endres GF, Eustance JW. *Macromol. Synth.* 1963;1:75.
10. Hay AS, et al. In: Mark HF, ed. *Encyclopedia of Polymer Science Engineering.* 1st ed. Vol 10. New York: John Wiley and Sons; 1969:92.
11. White DM. *J. Org. Chem.* 1969;34:297.
12. White DM, Nye SA. *Macromolecules.* 1990;23:1318.
13. Chan KP, et al. *Macromolecules.* 1994;27:6371.
14. Magré EP, Boon J. *Communication presented at the Microsymposium on Structure of Organic Solids.* Czechoslovakia: Prague; 16–19 September 1968.
15. Akers PJ, Allen G, Bethell MJ. *Polymer.* 1968;9:575.
16. Krause S. In: Paul DR, Newman S, eds. *Polymer Blends.* Vol I. Orlando, Florida: Academic Press; 1978:15–113.
17. Shultz AR, McCullough CR. *J. Polym. Sci. Part A-2.* 1969;7:1577.
18. Barrales-Rienda JM, Pepper DC. *Polym. Lett.* 1966;4:939.
19. Barton AFM. *CRC Handbook of Polymer-Liquid Interaction Parameters and Solubility Parameters.* Boca Raton, Florida: CRC Press; 1990.
20. Horikiri S. *J. Polym Sci., Part A-2.* 1972;10:1167.
21. Butte WA, Price CC, Hughes RE. *J. Polym. Sci..* 1962;61:528.
22. Karasz FE, Blair HE, O'Reilly JM. *J. Polym. Sci., Part A-2.* 1968;6:1141.
23. Su AC, Fried JR, Lorenz T. *Polym. Mater. Sci. Eng.* 1984;51:275.
24. Wenig W, Hammel R., Macknight WJ, Karasz FE. *Macromolecules.* 1976;9:253.
25. Karasz FE, O'Reilly JM, Bait HE, Kluge RA. *Polym. Prep.* 1968;9:822.
26. Gaur U. Wunderlich B. *J. Phys. Chem. Ref. Data.* 1981;10:1005.
27. Packter A, Sharif KA. *Polym. Lett.* 1971;9:435.
28. Johnston NW, Joesten BL. *J. Fire & Flamm.* 1972;3:274.
29. Heijboer UJ, *Polym. Sci., Polym. Symp.* 1968;16:3755.
30. Seferis JC, In: Brandrup J, Immergut EH, eds. *Polymer Handbook.* 3rd ed. Section VI. New York: Wiley-Interscience; 1989:451.
31. Koenhen DM, Smolders CA. *J. Appl. Polym. Sci.* 1975;19:1163.

Poly(dimethylsiloxane)

Alex C. M. Kuo

Acronym, Alternate Names, Trade names PDMS poly[oxy(dimethylsilylene)]; dimethicone; methylsilicone oil; Dow Corning 200® fluid; Wacker-Belsil® DM fluid; SF96® fluid.

Class Polysiloxane; di-methyl silicones and siloxanes.

Structure $-[(CH_3)_2Si-O-]_n$

Major Applications Release agents, rubber molds, sealants and gaskets, surfactants, water repellents, masonry coating, adhesives, process aids, foam control agents, biomedical devices, personal care and cosmetics, medicinal, pharmaceuticals, dielectric encapsulation, greases, lubricants, hydraulic fluids, heat transfer fluids, fuser oil, polish and coatings, paper coating.

Properties of Special Interest Thermal stability, low temperature performance, and minimal temperature effect. Good resistance to UV radiation. Excellent release properties and surface activity. High permeability to gases. Good damping behavior, anti-friction, and lubricity. Hydrophobic and physiological inertness. Shear stability, weak intermolecular forces, and excellent dielectric strength. Low volatility at high molecular weight and high volatility at low molecular weight.

Shorthand notation for siloxane polymeric buildings blocks

MDTQ formula: M (monofunctional)   D (difunctional)   T (trifunctional)   Q (tetrafunctional)

Formula: $(CH_3)_3SiO_{0.5}$ $(CH_3)_2SiO$ $(CH_3)SiO_{1.5}$ SiO_2

End-group and structure of certain dimethylsiloxanes

End Group	Structure	MDTQ Formula	CAS Reg. No.
Methyl	$(CH_3)_3Si-O-[(CH_3)_2Si-O-]_m-Si(CH_3)_3$	MD_nM	63148-62-9
Hydroxyl	$HO-(CH_3)_2Si-O-[(CH_3)_2Si-O-]_m-Si(CH_3)_2-OH$	$M^{OH}D_nM^{OH}$	70131-67-8
Vinyl	$CH_2{=}CH-(CH_3)_2\,Si-O-[(CH_3)_2Si-O-]_m-Si(CH_3)_2-CH{=}CH_2$	$M^{vi}D_nM^{vi}$	68083-19-2
Hydrogen	$H-(CH_3)_2Si-O-[(CH_3)_2Si-O-]_m-Si(CH_3)_2-H$	$M^HD_nM^H$	70900-21-9
None	Cyclic trimer; $[(CH_3)_2Si-O-]_3$	D_3	541-05-9
Methyl	$[(CH_3)_3Si-O-]_3SiH$	M_3T^H	1873-89-8

Poly(dimethylsiloxane)

Product form and properties[1]

Form	Structure and Properties
Fluids	Linear polymer. Liquid at low molecular weights and solid gum at high molecular weights
Elastomers	Cross-linked solids. Reinforcement necessary for property performance.
Resins	Highly branched cross-linked solids or fluids with high % T and Q in structure

Branched polymers[1] Silicone resins and elastomers are cross-linked polymers with branched polymer chains containing M (monofunctional), D (difunctional), T (trifunctional), and Q (tetrafunctional) units. Slightly branched polymers made from D, T, and Q structures have lower bulk and intrinsic viscosity than linear polymers of the same average molecular weight.

Infrared characteristic absorption[2,3]

Group	Absorption, Wave Number (cm^{-1})
$-Si(CH_3)_2-O-Si(CH_3)_2-$	2905–2960; 1020; 1090
$Si(CH_3)_3$	2905–2960; 1250; 840; 765
$Si(CH_3)_2$	2905–2960; 1260; 855; 805
$Si-(CH_3)$	2905–2960; 1245–1275; 760–845
$Si-H$	2100–2300; 760–910
$Si-OH$	3695; 3200–3400; 810–960
$Si-CH=CH_2$	1590–1610; 1410; 990–1020; 940–980

^{29}Si Nuclear magnetic resonance spectroscopy for typical structural building units in Dimethylsiloxanes[4–6]

Structure	MDTQ Formula	Chemical Shifts (ppm Down-Field from TMS)
$-O-Si(CH_3)_3$	M	6.6–7.3
$-Si(CH_3)_2-(C_6H_5)$	M^{ph}	-1
$-Si(CH_3)_2-CH=CH_2$	M^{vi}	-4
$-Si(CH_3)_2-H$	M^H	-7
$-Si(CH_3)_2-OH$	M^{OH}	-12
$-[O-Si(CH_3)_2-]$	D	-19 to -23
$[O-Si(CH_3)_2-]_3$	D_3	-9.1
$[O-Si(CH_3)_2-]_4$	D_4	-19.5
$[O-Si(CH_3)_2-]_5$	D_5	21.93
$[O-Si(CH_3)_2-]_6$	D_6	22.48
$(-O_{0.5}-)_3Si-CH_3$	T	-63 to -68
$(-O_{0.5}-)_4Si$	Q	-105 to -115

X-ray photoelectron spectroscopy elemental analysis[7]

Element Identification	Binding Energy	Atomic Composition
Si-2p	102.6	25.0
C-1s	285.0	50.0
O-1s	532.6	25.0

Poly(dimethylsiloxane)

Preparative techniques

Polymerization Process	Monomers	Major Catalysts	Reference
Hydrolysis	Dichlorodimethylsilane and dialkoxydimethylsilane	Acids, alkalis, and polychlorophosphazenes	(1, 8, 9)
Condensation	Oligomeric dimethylsiloxane-diol	H_2SO_4, HCl, tin dicarboxylates, hydroxides of alkali metals or zeolites, phosphonitrilic chlorides, quaternary phosphonium compounds	(1, 10–13)
Base ring-opening	Cyclic dimethylsiloxanes; dimethylsiloxane-diol	Hydroxides, silanolates, and alcoholates of alkali metals, quaternary ammonium or phosphonium compounds	(1, 10, 14–18)
Acid ring-opening	Cyclic dimethylsiloxanes; dimethylsiloxane-diol	Strong protic acids (sulfonic acid and triflic acid) and their derivatives	(1, 10, 18–20)
Emulsion	Silanol-ended oligomer or cyclic dimethylsiloxanes	Sodium silicate, tin dicarboxylates, acid salt hydroxides of alkali metals	(21–23)
Radiation	Cyclic dimethylsiloxanes	γ (^{60}Co)	(10)

Property	Units	Conditions	Value	Reference
Enthalpy of polymerization $-\Delta H_p$	kJ mol^{-1}	D_3 at 25°C	2.79	(24)
		D_3 at 77°C	23.4	(24)
		D_4 at 25°C	−6.4	(24)
		D_4 at 77°C	−13.4	(24)
Entropy of polymerization ΔS_p	J K^{-1}mol^{-1}	D_3 at 25°C	51.0	(24)
		D_3 at 77°C	−3.03	(24)
		D_4 at 25°C	194.4	(24)
		D_4 at 77°C	190	(24)
Solvents		Benzene, toluene, xylene, diethyl ether, methyl ether, chloroform, carbon tetrachloride, ethyl acetate, butanone, perchloroethylene, methylene chloride, trichloroethylene, gasoline, kerosene, Naptha VM&P		(25, 59)
Partially soluble solvents		Acetone, ethanol, isopropanol, butanol, dioxane, ethyl phenyl ether, heptadecanol		(25, 59)
Nonsolvents		Water, methanol, cyclohexanol, ethylene glycol, 2-ethoxy ethanol, dimethyl phthalate, aniline, 2-ethoxyethanol, 2-(2-ethoxyethoxy)ethanol, bromobenzene, propylene glycol, paraffine oil		(25, 59)
Solubility parameter δ	(MPa)$^{1/2}$	Average range (PDMS 100–60000 cs)	14.9–15.59	(26)
		Static vapor sorption for PDMS ($M_n = 89\,000$)	14.9	(27)

Property	Units	Conditions	Value	Reference
		Gas chromatography method for PDMS ($M_n = 2140$–218 000) at 298 K	15.1	(28)
		Gas chromatography method for PDMS ($M_n = 2140$–218 000) at 363 K	13.4	(28)
Theta temperature Θ	K	Bromobenzene	351.7	(29)
		Bromocyclohexane	300.6	(30)
		Ethyl iodide	275.1	(29)
		Phenetole	356	(31)
		Butanone	293	(31, 32)
		Benzene	293	(32)
		Mixture of C_8F_{18} and $C_2Cl_4F_2$ (1:2)	295.5	(32)
Second virial coefficient A_2	mol cm^{3}g^{-2}	PDMS ($M_w = 8.4 \times 10^5$) in benzene at 20°C	2.1×10^{-4}	(32)
		PDMS ($M_w = 1.2 \times 10^6$) in benzene at 20°C	1.84×10^{-4}	(32)
		PDMS ($M_w = 5.78 \times 10^5$)		
		in toluene at 27°C	4.5×10^{-4}	(30)
		in benzene at 27°C	2.95×10^{-4}	(30)
		in chlorobenzene at 30°C	1.04×10^{-4}	(30)
A_2 vs. pressure		In bromocyclohexane at 27°C/0 kg/cm^2	-1.58×10^{-6}	(30)
		27°C/100 kg/cm^2	2.58×10^{-6}	(30)
		27°C/200 kg/cm^2	5.38×10^{-6}	(30)
		27°C/300 kg/cm^2	7.82×10^{-6}	(30)
		27°C/400 kg/cm^2	9.95×10^{-6}	(30)
A_2 vs. temp.		In bromocyclohexane at 36.5°C	3.62×10^{-5}	(30)
		In bromocyclohexane at 47.2°C	6.57×10^{-5}	(30)
		In bromocyclohexane at 56.2°C	9.54×10^{-5}	(30)

Mark–Houwink parameters K and a

Solvents	Temp. (°C)	$K \times 10^3$ (ml g^{-1})	a	Reference
Butanone	20	81.5	0.5	(31)
Butanone	20	89	0.5	(33)
Butanone	20	78.3	0.5	(32)
Ethyl phenyl ether	83	77	0.5	(31)
Toluene	25	20	0.66	(34)
Toluene	25	8.28	0.72	(32)
Toluene	25	11	0.92	(33)
Benzene	20	12	0.68	(32)
Mixture of C_8F_{18} and $C_2Cl_4F_2$ (1:2)	22.5	105.7	0.5	(32)
Bromobenzene	78.7	76	0.5	(29)
Ethyl iodide	2.1	70	0.5	(29)
Bromocyclohexane	29	74	0.5	(29)

Property	Units	Conditions		Value	Reference
Characteristic ratio $C_\infty = <r^2>_0/nl^2$		Mixture of C_8F_{18} and $C_2Cl_4F_2$ (1:2) at 22.5°C		7.7	(32)
		Butanone at 20°C		6.3	(32)
		Calculation based on Ising lattice method ($l = 1.64$ Å, $\theta_1 = 110°$, θ_2 - 143°)		3.32–5.28	(35)
Root-mean-square end-to-end chain length $(<r^2>_0/M)^{1/2}$	nm mol$^{1/2}$ g$^{-1/2}$	PDMS in various theta solvents		2.5×10^{-2}	(29)
		PDMS in butanone at 20°C		7.30×10^{-2}	(31)
		Free rotation value calculated at 20°C for $l = 1.65$ Å, $\theta_1 = 110°$, $\theta_2 = 130°$		4.56×10^{-2}	(31)
		Free rotation value calculated at 20 °C for $l = 1.65$ Å, $\theta_1 = 110°$, $\theta_2 = 160°$		5.30×10^{-2}	(31)
Root-mean square radii of gyration $R_g = <s^2>_z^{1/2}$	Å	Blend of PDMS and preduterated PDMS ($M_n = 3000$–25 000)		41	(36)
		PDMS in benzene-d$_6$			
		$M_z = 4990$		18.6	(37)
		$M_z = 8670$		25.2	(37)
		$M_z = 12\,890$		33.8	(37)
		$M_z = 20\,880$		49.4	(37)
		Elastomer prepared by PDMS and preduterated PDMS		39	(36)
Z-average square radii of gyration $<s^2>_{z,linear}/<s^2>_{z,ring}$	–	Linear and cyclic PDMS in diluted benzene-d$_6$		1.9 ± 0.2	(37)
Interaction parameter of PMDS in organic solvents χ_{12}	–	Organic solvent	Conditions		
		Pentane	Swelling at 25°C	0.43	(38)
		Toluene	Swelling at 25°C	0.465	(38)
		Nitrobenzene	Swelling at 25°C	2.2	(38)
		Ethyl ether	Swelling at 25°C	0.43	(38)
		Cyclohexane	Swelling at 25°C	0.44	(38)
		Hexane	Swelling at 25°C	0.40	(38)
		Carbon tetrachloride	Swelling at 25°C	0.45	(38)
		Ethyl iodide	Swelling at 25°C	0.58	(38)
		Dioxane	Swelling at 25°C	0.61	(38)
		2,3-dimethylpentane	Swelling at 25°C	0.392	(39)
		2,2,4-trimethylpentane	Swelling at 25°C	0.38	(39)
		Chlorobenzene	Osmotic measurement at 20°C	0.477	(40)
		Cyclohexane	Osmotic measurement at 25°C	0.429	(40)
		Benzene	Osmotic measurement at 25°C	0.481	(40)
		Pentane	Gas chromatography at 100°C	0.311	(41)
		Toluene	Gas chromatography at 100°C	0.594	(41)

Property	Units	Conditions		Value	Reference
		Organic solvent	Conditions		
		Cyclohexane	Gas chromatography at 100°C	0.351	(41)
		Hexane	Gas chromatography at 100°C	0.296	(41)
		Chloroform	Gas chromatography at 100°C	0.60	(41)
		Benzene	Gas chromatography at 100°C	0.577	(41)
		Chlorobenzene	Gas chromatography at 100°C	0.764	(41)
		Dioxane	Gas chromatography at 100°C	1.064	(41)
		n-Butanol	Gas chromatography at 100°C	1.908	(41)
		Ethanol	Gas chromatography at 100°C	2.571	(41)

Interaction parameter of PDMS with other siloxanes, χ_{12}

Materials/ Condition	Temp. (K)	Method	χ_{12}	Reference
PDMS network/PDMS ($M = 422$–875)	298	Swelling	0.19–0.25	(38)
PDMS network/PDMS ($M = 700$–$26\,400$)	298	Swelling	−0.017 to 0.006	(42)
PDMS network/$[(CH_3)_2\,Si{-}O{-}]_5$	298	Swelling	0.247	(43)
PDMS/$[(CH_3)_2\,Si{-}O{-}]_4$	298	Osmotic measurement	0.298	(44)
PDMS network/$MD_3^{ph}M$	298	Swelling	0.345	(45)
PDMS network/$MD_2^{ph}M$	298	Swelling	0.438	(45)
PDMS network/$MD^{ph}M$	298	Swelling	0.356	(45)
PDMS/$MD_{23}^{ph}M$	518	Light scattering	0.122	(46)
PDMS/$MD_{28}^{ph}M$	458	Light scattering	0.111	(47)
PDMS/D_4^{ph} cycloisomers	360–371	Light scattering	0.300	(43)
PDMS/Polyethylmethylsiloxane ($M_n = 30\,300$)	332.5	Light scattering	0.00664–0.0077	(48)
PDMS/ poly(ethylene oxide)	343–373	Gas chromatography	0.4–1.1	(49)

Parameters for the equation-of-state

PDMS	Method	T^* (K)	v_{sp}^* (cm^3 g^{-1})	P* (MPa)	Reference
$M_v = 1 \times 10^5$	Flory–Orwoll and Vrij theory from 25°C	5528	0.8395	341	(50)
$M_n = 162.4$	Flory–Orwoll and Vrij theory from 25°C	4468	0.9995	325.3	(51)
$M_n = 340$	Modified Flory–Orwoll and Vrij theory from 40 to 73°C	3726.5	0.94877	373.9	(52)
$M_n = 958$	Flory–Orwoll & Vrij theory from 25°C	5288	0.8694	313.3	(51)
$M_n = 7860$	Flory–Orwoll & Vrij theory from 25°C	5554	0.8403	311.5	(51)
$M_n = 187\,000$	Modified Flory–Orwoll & Vrij theory from 42 to 93°C	4386.7	0.88085	382.6	(52)
$M_n = 47\,200$	Ising fluid model from 25 to 70°C	476	0.9058	302	(53)

Poly(dimethylsiloxane)

Morphology in multiphase systems

System A/B	Microstructure	Architecture	Reference
Polybutadiene/PDMS	Cylinders/spheres	A-B diblock	(54)
Polystyrene/PDMS	Spheres/lamellae/cylinders	A-B diblock	(55)
Polydiphenylsiloxane/PDMS	Lamellae	A-B-A triblock or star-block	(56)
Poly(methyl styrene)/PDMS	Spheres /lamellae	A-B diblock and A-B-A triblock	(57)
Poly(ethylene oxide)/PDMS	Lamellae/ cylinders	B-A-B triblock	(49)
Poly(methyl methacrylate)/PDMS	Spheres /cylinders	A-g-B graft	(58)

Physical properties of trimethylsiloxy terminated polydimethylsiloxane vs. viscosity[59,60]

Properties	Units	PDMS viscosity at 25°C (cs)						
		0.65	2.0	10	100	1000	12 500	60 000
Molecular weight (estimated)	g mol^{-1}	162	410	1250	5970	28 000	67 700	116 500
Flash point	K	269.7	352	484	>599	>599	>599	>599
Boiling point at 760 mm	K	373	503	–	–	–	–	–
Pour point	K	205	173	173	208	223	227	232
Specific gravity at 25°C	–	0.760	0.872	0.935	0.964	0.970	0.974	0.977
Viscosity-temperature coefficient $[1-(\eta_{273K}/\eta_{311K})]$	–	0.31	0.48	0.56	0.60	0.61	0.61	0.61

Property	Units	Conditions	Value	Reference
Density ρ	g cm^{-3}	PDMS(1000–12 500 cs)	0.97	(59,60)
		Theoretical of crystallized PDMS	1.07	(61)
ρ vs. temperature	g cm^{-3}	PDMS ($M_v = 1 \times 10^5$) from 20–207°C	$\rho = 0.9919 - (8.925 \times 10^{-4})t + (2.65 \times 10^{-7})t^2 - (3.0 \times 10^{-11})t^3$	(50)
Specific volume $v_{sp} v_{sp}$ vs. temp.	cm^3 g^{-1} cm^3 g^{-1}	20–90°C 90–170°C	$v_{sp} = 1.0265 + (9.7 \times 10^{-4})(t - 20)$ $v_{sp} = 1.0944 + (10.3 \times 10^{-4})(t - 90)$	(62) (62)
Thermal expansion coefficient α	K^{-1}	PDMS ($M_v = 1 \times 10^5$) at 298 K	9.07×10^{-4}	(52)
		PDMS ($M = 1.5 \times 10^4$) at 303 K	9.0×10^{-4}	(63)
		PDMS (from 100 to 60 000 cs)	9.6×10^{-4}	(59)
α vs. temperature		PDMS ($M_v = 1 \times 10^5$) from 20 to 207°C	$\alpha = 0.90 \times 10^{-3} + (2.76 \times 10^{-7})t + (1.0 \times 10^{-10})t^2$	(50)
Thermal pressure coefficient, γ, vs. temperature	bar K^{-1}	PDMS ($M_v = 1 \times 10^5$) from 24 to 161°C	$\gamma = 8.71 + (4.74 \times 10^{-2})t + (9.3 \times 10^{-5})t^2$	(50)

Property	Units	Conditions	Value	Reference
Water solubility	ppm	MDM at 296 K nonturbulent measurement	3.45×10^{-2}	(64)
	ppm	MD$_3$M at 296 K nonturbulent measurement	7.04×10^{-5}	(64)
	ppm	PDMS ($M = 1200$) at 298 K water elution measurement	1.6	(65)
	ppm	PDMS ($M = 6000$) at 298 K water elution measurement	0.56	(65)
	ppm	PDMS ($M = 25\,000$) at 298 K water elution measurement	0.17	(65)
	ppm	PDMS ($M = 56\,000$) at 298 K water elution measurement	0.076	(65)

Compressibility of polydimethylsiloxane vs. viscosity[66]

Pressure (kgf cm^{-2})	Viscosity of PDMS (cs)						
	0.65	1	2	100	350	1000	12 500
	Volume Reduction %						
0	0	0	0	0	0	0	0
500	6.34	5.36	4.85	4.49	4.47	4.58	4.46
1000	10.04	8.84	8.21	–	7.42	7.36	7.29
25 000	16.33	15.08	14.34	12.71	12.78	12.74	12.53
50 000	Gel	20.66	20.07	17.43	17.96	17.87	17.71
30 000	–	34.57	34.56	–	32.94	31.31	31.25

X-ray diffraction pattern[67]

Condition	2θ	Reflection
PDMS rubber at $-50°$C for 6 h	11°40′ (amorphous halo)	110 + 001
	19°30′	110 + 020
	23°20′	021 + 112

Unit cell dimensions[61]

Lattice	Monomer Per Unit Cell	Unit Cell Dimension (Å)			Cell Angle (deg)			Theoretical Density (g cm^{-3})	
		a	c	b	α	γ	β	Crystal	Amorphous
Monoclinic	6	13.0	7.75	8.3	90	90	60	1.07	0.98

Crystalline-state properties

Property	Units	Conditions	Value	Reference
Si—C bond length	Å	[(CH$_3$)$_2$SiO]$_4$ at $-50°$C	1.92	(68)
Si—O bond length	Å	[(CH$_3$)$_2$SiO]$_4$ at $-50°$C	1.65	(68)
Si—C bond energy	kJ mol^{-1}		326	(69)
Si—O bond energy	kJ mol^{-1}		443	(69)
O—Si—O bond angle	Degree	[(CH$_3$)$_2$SiO]$_4$ at $-50°$C	109	(68)
		Conformation analysis	112	(70)

Poly(dimethylsiloxane)

Property	Units	Conditions	Value	Reference
Si—O—Si bond angle	Degree	X-ray diffraction analysis	140 ± 10	(61)
		Conformation analysis for hexamethyldisiloxane	145–150	(71)
		$[(CH_3)_2SiO]_4$ at $-50°C$	142.5	(68)
C—Si—C bond angle	Degree	$[(CH_3)_2SiO]_4$ at $-50°C$	106	(68)
Degree of crystallinity	%	X-ray measurement for 17% silica-filled PDMS elastomer at $-60°C$	42	(72)
		X-ray measurement for 17% silica-filled PDMS elastomer at $-80°C$	59	(72)
		DSC measurement for PDMS ($M_n = 1.11 \times 10^5$) at T_g using a cooling rate $=10$ K min^{-1}	58.8	(73)
		Calorimeter measurement for PDMS ($M \sim 6 \times 10^5$)	67	(24)
		DSC measurement for PDMS at T_g using a cooling rate $= 2.1$ K min^{-1}	79	(74)

Avrami parameters

Conditions	Crystallization Temp. T_c (°C)	$k \times 10^3$	N	$\tau_{0.5}$ (min)	Reference
Isothermal crystallization of PDMS ($M = 4 \times 10^5$)	-55.6	1.905	2.19	15	(67)
	-57.5	7.0	2.2	9	(75)
	-60.5	120	1.75	2.5	(75)
Isothermal crystallization of PDMS ($M = 1 \times 10^5$)	-60.5	1.0	2.5	13.5	(75)
	-65.0	7.35	2.2	8.2	(75)
	-71.0	23	2.25	4.8	(75)
NMR measurement for PDMS ($M_n = 7.4 \times 10^5$)	-58.8	–	3.1	–	(76)

Property	Units	Conditions	Value		Reference
Glass transition temperature T_g	K	Measured by DSC	150		(74)
			123.3–149.9		(77)
Melting point T_m	K	Measured by DSC	T_{m1}	T_{m2}	
			226–232	232	(74)
			217.8–228.3	235.3–235.6	(77)
Crystallization temperature T_c	K	Measured by DSC	173–183		(74)
			181.4–196.8		(77)
Enthalpy of fusion, ΔH_u	kJ mol^{-1}	Calculation by melting temperature depression of PDMS in toluene solution	1.36		(74)
	kJ mol^{-1}	Calorimeter measurement for a PDMS with 67% crystallinity	3.010		(24)
Entropy of fusion, ΔS	kJ K^{-1}mol^{-1}	Calculation by melting temperature depression of PDMS in toluene solution	5.78×10^{-3}		(74)

Property	Units	Conditions	Value	Reference
Specific heat C_p	kJ kg^{-1}K^{-1}	PDMS (2–1000 cs) at 298 K	1.35–1.51	(60)
		PDMS (350 cs) at 298 K	1.464	(59)
		PDMS (1000 cs) at 298 K	1.461	(59)
		PDMS ($M = 400\,000$)	1.552	(59)
Specific heat, C_p, effect of temperature		PDMS ($M_n = 1.11 \times 10^5$) at 120 K	0.660	(73)
		PDMS ($M_n = 1.11 \times 10^5$) at 140 K	0.824	(73)
		PDMS ($M_n = 1.11 \times 10^5$) at 250 K	1.439	(73)
		PDMS ($M_n = 1.11 \times 10^5$) at 300 K	1.532	(73)
Viscosity vs. molecular weight	cs	PDMS ($M_n > 2500$) at 25°C	$\log \eta = 1.00 + 0.0123 M^{0.5}$	(34)
Energy of vaporization E_{vap}	kJ mol^{-1}	MD$_2$M	43.55	(78)
		MD$_9$M	90.45	(78)
Energy of activation for viscous flow E_{visc}	kJ mol^{-1}	MD$_9$M	13.74	(78)
		PDMS ($M = 4.7 \times 10^3$ to 4.8×10^5)	14.6	(79)
Critical molecular weight for entanglement, M_c	g mol^{-1}	Linear PDMS	21 000	(80)
		Linear PDMS	29 000	(34, 81)
		Linear PDMS	30 000	(79)
		Linear PDMS	33 000	(82)
		Trifunctional branched PDMS	98 000	(82)
		Tetrafunctional branched PDMS	110 000	(82)
Color	APHA	PDMS (200 ®fluids)	5	(59)

Monolayer properties of force vs. area isotherm for PDMS on water surface[83]

Property	Units	Materials	Value
Area per monomer unit, A_0	Å^2	MD$_{14}$M	22
Film pressure, F, at 7 Å^2	dyne/cm	MD$_{14}$M	10
Surface electrostatic potential difference, ΔV, at 7 Å^2	mV	MD$_{14}$M	150
Apparent dipole moment per mole per monolayer, μ_p, at 7 Å^2	mD	MD$_{14}$M	30

Property	Units	Conditions	Value	Reference
Water contact angle θ	Degrees	PDMS (500 cs) film on soda-lime glass after a 15-min treatment		
		At 25°C	54	(84)
		At 100°C	70	(84)
		At 200°C	102	(84)
		At 300°C	110	(84)

Property	Units	Conditions	Value	Reference
		At 400°C	103	(84)
		At 500°C	85	(84)
		At 525°C	0	(84)
		Unfilled PDMS elastomer	105	(85)
		PDMS elastomer coating	95–103	(85)
		Filled PDMS elastomer with PDMS migration	113	(85)
		Silica-filled PDMS elastomer treated by RF air plasma		
		At 150 mTorr, 2 mim	6.8	(86)
		At 300 mTorr, 2 min	8.8	(86)
		At 500 mTorr, 2 min	7.4	(86)
		At 300 mTorr, then 3-h exposure to air	20	(86)
		At 300 mTorr, then 30-day exposure to air	60	(86)
		Silica-filled PDMS elastomer, untreated	113	(86)
Methylene Iodide θ contact angle	Degree	PDMS fluid, cross-linked PDMS paper coating, and unfilled PDMS elastomer	67–77	(85)
n-Hexadecane contact angle θ	Degree	Surface of cross-linked PDMS sheet	40	(87)
Perfluorodecalin contact angle θ	Degree	PDMS elastomer vs. perfluorocarbon monolayer on mica surface	37	(88)
Critical surface tension of wetting γ_c	mN m^{-1}	Silica-filled PDMS rubber at 20°C	20–23	(89)
		Dimethylsiloxane dimer at 20°C	15.7	(90)
		Dimethylsiloxane trimer at 20°C	16.96	(90)
		Dimethylsiloxane tetramer at 20°C	17.60	(90)
		Dimethylsiloxane heptamer at 20°C	18.60	(90)
		Dimethylsiloxane dodecamer at 20°C	19.56	(90)
		Dimethylsiloxane heptadecamer at 20°C	19.87	(90)
		PDMS (35 cs) at 20°C	19.9	(90)
		PDMS (70 cs) at 20°C	20.3	(90)
		PDMS (6×10^4 cs) at 20°C	20.4	(62)
		PDMS (6×10^4 cs) at 140°C	14.1	(91)
		PDMS (6×10^4 cs) at 150°C	13.6	(62)
		PDMS (6×10^4 cs) at 180°C	12.2	(91)
Temperature coefficient of surface tension $-d\gamma_{12}/dT$	mN m^{-1} K^{-1}	PDMS melt (10^6 and 6×10^4 cs) at 150°C PDMS (35 cs) at 20°C	0.048 0.067	(62) (90)
Interfacial tension against water γ_{1w}	mN m^{-1}	PDMS (0.65 cs) at 20°C	39.9	(90)
		PDMS (1.0 cs) at 20°C	42.5	(90)
		PDMS (5.0 cs) at 20°C	42.2	(90)
		PDMS (10.0 cs) at 20°C	39.9	(90)
		PDMS (35 cs) at 20°C	43.1	(90)
Polarity x^p		Form measurement of interfacial tension of PDMS (6×10^4 cs)	0.04	(91)
Surface shear viscosity	μN s m^{-1}	PDMS ($M = 500$–105 000)	~1	(92)

Williams–Landel–Ferry (WLF) parameters measurement for trimethylsiloxy-terminated PDMS[93]

PDMS, M_n (g mol^{-1})	Reference Temperature, T_0 (K)	C_1	C_2/K	$T_{g,DSC}$ (K)
10 370	147	10.4	14.24	149.5
4160	145.5	14.22	23.84	149.3
2080	143.8	14.32	23.37	147.5
830	141.1	13.48	20.03	141.2
420	136.5	11.46	14.01	135.9

Interfacial tension of polymer blends

Polymer Pairs	γ_{12} (mN/m)	$-d\,\gamma_{12}/dT$ (mN/m-K)	References
PDMS/polypropylene	3.2 at 20°C	0.002	(94)
PDMS/poly(t-butyl methacrylate)	3.6 at 20°C	0.0025	(91)
PDMS/poly(isobutene)	4.9 at 20°C	0.006	(95)
PDMS/poly(isobutylene)	3.9 at 20°C	0.016	(96)
PDMS/polybutadiene	4.15 at 25°C	0.00865	(97)
PDMS/poly(n-butyl methacrylate)	4.2 at 20°C	0.0038	(91)
PDMS/polyethylene, branch	5.3 at 20°C	0.002	(92, 95)
PDMS/poly(oxytetramethylene)	6.4 at 20°C	0.0012	(91)
PDMS/polychloroprene	7.1 at 20°C	0.0050	(91)
PDMS/poly(vinyl acetate)	8.4 at 20°C	0.0081	(91, 96)
PDMS/polyethylene	5.08 at 150°C	0.0016	(98)
PDMS/polyvinyl acetate	7.43 at 150°C	0.0087	(98)
PDMS/poly(oxyethylene)	9.85 at 150°C	0.0078	(98)
PDMS/poly(tetrahydrofuran)	6.26 at 150°C	0.0004	(98)

Solubility of gases in PDMS membrane at 25°C /760 mmHg

Gas	Solubility (ml g^{-1}) [66]	Solubility (ml g^{-1})[99]	Diffusion Rate, $D \times 10^5$ (cm^2 s^{-1})[99]
He	0.010	0.045	60
Ar	0.301	0.33	14
Air	0.168	–	–
O_2	0.258	0.31	16
N_2	0.166	0.15	15
CO_2	1.497	2.2	11
CH_4	0.543	0.57	12.7
SF_6	0.996	–	–
C_3F_8	1.041	–	–
H_2	–	0.12	–

Poly(dimethylsiloxane)

Gas permeability from PDMS membranes filled with 33% Silica, $cm^3(STP)$ $cm/(s$ cm^2 cm $Hg)$[99]

Gas	$P_r \times 10^9$	Gas	$P_r \times 10^9$	Gas	$P_r \times 10^9$
H_2	65	N_2O	435	$n\text{-}C_6H_{14}$	940
He	35	NO_2	760	$n\text{-}C_8H_{18}$	860
NH_3	590	SO_2	1500	$n\text{-}C_{10}H_{22}$	430
H_2O	3600	CS_2	9000	HCHO	1110
CO	34	CH_4	95	CH_3OH	1390
N_2	28	C_2H_6	250	$COCl_2$	1500
NO	60	C_2H_4	135	Acetone	586
O_2	60	C_2H_2	2640	Pyridine	1910
H_2S	1000	C_3H_8	410	Benzene	1080
Ar	60	$n\text{-}C_4H_{10}$	900	Phenol	2100
CO_2	325	$n\text{-}C_5H_{12}$	2000	Toluene	913

Temperature effect of oxygen permeability and solubility from PDMS membrane[99]

Temperature ($°C$)	$P_r \times 10^9$, $cm^3(STP)$ $cm/(s$ cm^2 cm $Hg)$	Solubility (ml/g)
28	62	0.31
−40	20	0.39
−75	0.74	47

Property	Units	Conditions	Value		Reference
Thermal conductivity	W $m^{-1}K^{-1}$	PDMS (1000–60 000 cs) at 50°C	0.1591		(59)
		PDMS (2.0 cs) at 80°C	0.1055		(60)
		PDMS (20 cs) at 80°C	0.1398		(60)
		PDMS (100 cs) at 80°C	0.1511		(60)
		PDMS (12 500 cs) at 80°C	0.1520		(60)
Load-bearing capacity	kg	PDMS fluid	50–150		(1)
Lubricity	mm	Shell four ball test, wear scar, steel on steel; PDMS(100 cs) at 1 h/600 rpm/50 kg load/ambient temperature	1.91		(100)
		Shell four ball test, wear scar, steel on bronze; PDMS(100 cs) at 1 h/600 rpm/50 kg load/ambient temperature	2.0		(100)
Speed of sound longitudinal velocity	m s^{-1}	Viscosity	30°C	50.7°C	
		PDMS (0.65 cs)	873.2	795.3	(101)
		PDMS (1.0 cs)	901.3	828.5	(101)
		PDMS (50 cs)	981.6	925.3	(101)
		PDMS (1000 cs)	987.3	933.3	(101)
Anomalous longitudinal velocity	m s^{-1}	PDMS (2×10^5 cs) cooling at $T < 205$ K	1850		(102)
		PDMS (2×10^5 cs) warming at $T > 235$ K	1200		(102)

Property	Units	Conditions	Value	Reference
Refractive index n_D^{25}		PDMS (0.65–10 cs) at 25°C	1.375–1.399	(60)
		PDMS (100–125 000 cs) at 25°C	1.403–1.4035	(60)
Diamagnetic susceptibility X_m	$cm^3\ g^{-1}$	PDMS ($M = 1200$)	0.62×10^{-6}	(103)
		MD_5M	0.658×10^{-6}	(104)
		D_3 and D_4	0.632×10^{-6}	(104)
Verdet constant of magnetic rotary power	min gauss^{-1} cm^{-1}	PDMS (0.65–1000 cs) at 25°C and 5893Å	$(1.623–1.693) \times 10^{-2}$	(105)
Dipole moment μ	D	Hydroxy-terminated PDMS ($M = 20\ 000$) in cyclohexane at 25°C	11.54	(106)
		Hydroxy-terminated PDMS ($M = 70\ 230$) in cyclohexane at 25°C	21.48	(106)
		Trimethylsiloxy-terminated PDMS ($M = 78\ 500$) in cyclohexane at 25°C	22.24	(106)
		Trimethylsiloxy-terminated PDMS ($M = 515\ 000$) in cyclohexane at 25°C	83.42	(106)
Dipole moment ratio, per repeat unit $\mu/n^{1/2}$	D	Trimethylsiloxy-terminated PDMS in cyclohexane	0.697	(106)
		Hydroxy-terminated PDMS in cyclohexane	0.666	(106)
Root-mean-square dipole moment ratio $<\mu_0>/nm^2$	–	PDMS (DP = 194–2076) undiluted at 25°C	0.30	(107)
		PDMS (DP = 194–2076) in cyclohexane at 25°C	0.40	(107)
		PDMS (DP = 2–4940) in cyclohexane at 25°C	0.29	(108)

Dielectric properties of trimethylsiloxy-terminated PDMS at various viscosities[59]

Viscosity at 25°C (cs)	0.65	2.0	10	100	1000	12 500	60 000
Dielectric constant at 10^2–10^4 Hz	2.2	2.45	2.72	2.75	2.75	2.75	2.75
Dielectric strength at 25°C kV cm^{-1}	118	138	148	158	158	158	158
Volume resistivity at 25°C ohm cm	1×10^{14}	5.0×10^{14}	1.0×10^{15}	1.0×10^{15}	1.0×10^{15}	1.0×10^{15}	1.0×10^{15}

Dielectric data for PDMS (440 cs) at various temperatures[1]

Properties	Sample	20°C	100°C	200°C
Dielectric constant ε	PDMS (440 cs)	2.8	2.5	2.3
Dissipation factor, tan δ, at 800 Hz	PDMS(440 cs)	1.2×10^{-4}	1.3×10^{-4}	1.5×10^{-4}
Volume resistivity (ohm-cm)	PDMS (440 cs)	4×10^{15}	6×10^{14}	1×10^{14}
Dielectric strength (kV cm^{-1}), 5 cps, 6 kV s^{-1}	PDMS (440 cs)	120	100	95

Poly(dimethylsiloxane)

Property	Units	Conditions	Value	Reference
Autoignition temperature	K	ASTM D286-30, PDMS (1 cs)	691	(25)
		ASTM D286-30, PDMS (10 cs)	725	(25)
		ASTM D286-30, PDMS (100 cs)	>763	(25)
Limiting oxygen index, LOI	%	General silicone (VMQ) rubber	26–42	(109)
Arc resistance	s	General silicone rubber	250	(109)
Corona resistance	kV	General silicone rubber	40	(109)
Anisotropy, optical configuration parameter Δa or $(\alpha_1 - \alpha_2)$	cm^3	PDMS ($M = 1.8 \times 10^6$) in petroleum ether	4.7×10^{-25}	(110)
		Cross-linked PDMS at 20°C	4.5×10^{-25}	(111)
		Cross-linked PDMS at 70°C	8.1×10^{-25}	(112)
		Cross-linked PDMS swelled in decalin at 70°C	5.1×10^{-25}	(112)
		Cross-linked PDMS swelled in cyclohexane at 70°C	3.8×10^{-25}	(112)
		Cross-linked PDMS swelled in CCl_4 at 70°C	1.8×10^{-25}	(112)
Stress-optical coefficient C	$m^2\,N^{-1}$	PDMS		
		At 200°C	1.35×10^{-10}	(113)
		At 22/25°C	$1.35/1.75 \times 10^{-10}$	(114)
		At 105/190°C	$1.9/2.65 \times 10^{-10}$	(114)
Shear modulus	Pa	Unfilled PDMS elastomer ($M_n = 10\,000$)	2.03×10^5	(115)
		Trifunctional PDMS cured elastomers	2.32×10^5	(116)

Fire parameters (cone calorimeter test)[117]

Samples	External Heat Flux $(kW\,m^{-2})$	Peak Rate of Heat Release $(kW\,m^{-2})$	Specific Extinction Area $(m^2\,kg^{-1})$
MM	30	2800	–
MD_2M	60	2200	–
MD_3M	60	1750	–
MD_8M	60	750	–
10 cs PDMS	60	175	–
50 cs PDMS	60	140	600
6×10^5 cs PDMS	60	105	550
1×10^7 cs PDMS	60	95	550
Elastomers/silica filled	60	80–110	1300–1700

Thermochemical parameters[117]

Viscosity of PDMS (cs)	Heat of Gasification (MJ kg^{-1})	Heat of Combustion (MJ kg^{-1})	Flame Heat Radiated to Surface (kW m^{-2})
0.65	0.327	36.1	–
2.0	0.492	30.0	–
10	3.0–3.6	26.8	26
10 000 000	3.0–3.6	26.8	26

Degradation behavior

End-Group of PDMS	Depolymerization Conditions	Activation Energy (kJ mol^{-1})	Reference
Trimethylsiloxy-terminated	Random scission thermal depolymerization from 420 to 480°C	176	(118)
Trimethylsiloxy-terminated	Thermal oxidation depolymerization from 350 to 420°C	126	(118)
Hydroxyl-terminated	Unzipping in vacuum at $T > 250$°C	35.6	(118)
Hydroxyl-terminated	0.01% NaOH or 0.01% H_2SO_4 catalyzed depolymerization from 170 to 300°C	58.6	(119)
Hydroxyl-terminated	Stress relaxation measurement in anhydrous argon 150 to 260°C	95.4	(120)
Hydroxyl-terminated	0.01% KOH catalyzed reaction from 60 to 140°C	21.4	(117)
Trimethylsiloxy-terminated ^{14}C-PDMS	Degradation occurred in soil with <3% moisture and formed volatilized dimethylsilane 1,1-diol.	–	(121, 122)
Trimethylsiloxy-terminated PDMS measured by GPC-ICP or HPLC-ICP	Soil, sludge, sediment from wastewater treatment plant found PDMS effluent, PDMS silanol, and dimethylsilane 1,1-diol	–	(123)

Decomposition products[124]

Thermal Decomposition Products (100 cs PDMS)	% at 475°C	Thermal-Oxidative Decomposition Product	% at 430°C (approx.)
D_3	45	Cyclic siloxanes	81
D_4	19	HCHO	13
D_5	5	CO_2	3
D_6	11	CO	2
D_7	7	CH_3OH	1.5
D_8	2	HCO_2H	0.2

Poly(dimethylsiloxane)

Acute oral toxicity

Species	PDMS Viscosity (cs)	Result or Hazard Rating, LD_{50} (mg/kg)	Reference
Rat	10	>4990	(121)
Guinea pig	50	>47 750	(121)
Rat	100	>4800	(121)
Rabbit/dog/cat	140	>9800	(121)
Rat	350	>48 600	(121)
Rat	1000	>4800	(121)
Rat	350 (SF96® 350)	>5000	(125)

Acute dermal toxicity

Species	PDMS Viscosity (cs)	Result or Hazard Rating, LD_{50} (mg/kg)	Reference
Rabbit (male New Zealand)	350	No adverse effect at 24 h LD_{50} is > 19 400 mg kg^{-1} bw	(121)
Rat	50, 500, and 1000	LD_{50} is > 2000 mg kg^{-1} bw	(121)
Rabbit	350 (SF96® 350)	LD_{50} is > 10 000 mg kg^{-1}	(125)
Rat	0.65, 10, 350 up to 500 000	LD_{50} is > 2008 mg kg^{-1}	(126)

Inhalation toxicity

Species	PDMS Materials	Result and Hazard Rating, LC_{50} (mg/kg)	Reference
Wistar rat	PDMS (10 000 cs) aerosol in a 25% solution in white spirit	No observed adverse effect, $LC_{50} - 4$ h is > 11 582 mg m^{-3}	(121)
Rat	350 cs (SF96® 350)	LC_{50} is > 535 mg kg^{-1}	(125)
Wistar rat	Aerosol of 10 000 cs PDMS fluid 25% solution in dichloromethane	No observed adverse effect, $LC_{50} - 4$ h is > 695 mg m^{-3}	(121)

Skin irritation[1121]

PDMS Viscosity (cs)	Species	Volume (ml)	Type of Application	No. of Applications	Duration (days)	Effects
50	Rabbit	–	Semi occlusive (continuous application to intact skin)	10	14	Nonirritating
100	Rabbit	0.5	Applied to the ears under an occlusive dressing	1	1	Nonirritating
100	Guinea pig	0.5	Draize method, 10 times per day	10 (daily)	15	Nonirritating
1,000	Rabbit	0.5	Draize method, OECD Guideline 404	1	7	Nonirritating

Eco-toxicity in aquatic compartment

Species	Materials	Result or Hazard Rating	Reference
(Fresh water)			
Salmo gairdneri	PDMS (350 cs) 25% in food during 28 days; followed by 14-day observation period	No effect on behavior and growth with 10 mg PDMS fish^{-1}day^{-1}	(121)
Phoxinus phoxinus	PDMS (viscosity not specified)	LC$_{40}$ – 8 day = 3000 (mg l^{-1})	(121)
(Sea water)			
Pomatoschistus minutus Gasterosteus aculeatus	PDMS (100, 350, and 12 500 cs)	No mortality—96 h at saturation	(121)
Pleuronectes platessa	PDMS (50 cs)	Toxicity—96 h > 10 000 mg l^{-1} at the surface of water (5 mg l^{-1} in water)	(121)
Scorpaena porcus	PDMS (50 cs) 30% emulsion	LC$_{50}$ – 50 h = 700 (mg l^{-1})	(121)
Carassius auratus	PDMS (50 cs) 30% emulsion	LC$_{50}$ – 24 h = 3500 (mg l^{-1})	(121)
Marine Crab (Acartia tonsa)	PDMS (10 cs)	LC$_{50}$ – 48 h > 88 865 (mg l^{-1})	(126)

Ecotoxicity in terrestrial compartment[121]

Species	Materials	Result or Hazard Rating
Plant: Soybean	Soil containing a sewage sludge with ^{14}C-PDMS was examined as nutrients for plants from germination of the seed growth to grains during a 7-month period	No significant difference from controls was observed.
Insect activity: Acheta domesticus	PDMS (5–1000 cs) direct application of 5 μl to the ventral thorax of insect	The time of loss of righting reflex increased with the viscosity of the PDMS and the mortality at 48 h decreased 2 fold when the viscosity of PDMS increased 200 fold.
Birds: Anas platyrhynchos and Colinus virginatus	PDMS (100 cs) was used as feed for 5 days in the diet (5000 mg kg^{-1} food) and was kept on standard food for 3 additional days	No mortality and no other signs of toxicity occurred.

Silicone rubber preparation[109,127–130]

Method	PDMS Functional System	Chemistry	Major Applications
Room temperature vulcanizing rubber	Hydroxyl, alkoxyl, SiH, or vinyl group PDMS	Hydrosilylation, condensation	Sealant, adhesive, encapsulation, and mold making
Thermal vulcanizing rubber	Trimethylsiloxyl, vinyl, and SiH functional PDMS	Hydrosilylation or peroxide catalyzed free radical reaction	Injection- or extrusion-molded rubber, roll-coated thin-layer silicone release coating, adhesive, or encapsulation
Others	Acrylate or epoxy functional PDMS	Cure by ultraviolet (UV), electron beam, and gamma ray	Protective coating, release coating, encapsulation, and cable wire insulation

Poly(dimethylsiloxane)

Properties of commercialized PDMS rubber

Property	Units	Conditions	Value	Reference
Tensile strength	MPa	ASTM D412 DIE C	$3.2 \sim 11.8$	(131–133)
Specific gravity		ASTM D792	$1.06 \sim 1.47$	(131–133)
Hardness (shore A)	Points	ASTM D2240	$23 \sim 81$	(131–133)
Elongation at break	%	ASTM D412 DIE C	$140 \sim 978$	(131–133)
Tear strength	$kN\,m^{-1}$	ASTM D624 DIE B	$4 \sim 52$	(131–133)
Compression set	%	ASTM D395B, after 22 h/ 177°C	$10 \sim 49$	(131–133)
Dielectric strength	$V\,mil^{-1}$	IEC 60243	$450 \sim 600$	(131–133)
Volume Resistivity	ohm cm	IEC 60243	$6.1 \sim 9 \times 10^{15}$	(131–133)
Resilience, (Bashore)	%	ASTM 2632 reinforced PDMS rubber	$29 \sim 58$	(133)
Abrasion resistance	rev/0.254 cm	ASTM D1630-61 reinforced PDMS rubber	155–600	(133)
Tear propagation	cycles/1.27 cm	ASTM D813-59 reinforced PDMS rubber	$120 \sim 150\,000$	(133)

Major producers[134]

Company	America	Asia	Europe
Dow Corning Co.	✓	✓	✓
Momentive Performance Materials Inc. (former GE silicone)	✓	✓	✓
Wacker Chemie AG	✓	✓	✓
Blue Star Co. (former Rhodia)	✓	✓	✓
Shin-Etsu Chemical Co.	✓	✓	✓
Degussa AG-Goldschmidt		✓	✓
Chisso Co.		✓	
Penta-91 (Russia)			✓
Kumgang Korea Chemical Co.		✓	
United Chemical Technologies Inc.	✓		
NüSil Technology Co.	✓		

References

1. Noll W. *Chemistry & Technology of Silicone*. Chapters 5 and 6. New York: Academic Press; 1968.
2. Lipp ED, Smith AL. In: Smith AL, ed. *Analysis of Silicone*. 2nd ed. Chapter 11. New York: John Wiley and Sons; 1991.
3. Mayhan KG, Thompson LF, Magdalin CF. *J. Paint Tech*. 1972;44:85.
4. Harris RK, Robins ML. *Polymer*. 1978;19:1123.
5. Taylor RB, Parbhoo B, Fillmore DM. In: Smith AL, ed. *Analysis of Silicone*. 2nd ed. Chapter 12. New York: John Wiley and Sons; 1991.

6. Williams EA. In: Webb GA, ed. *Annual Reports on NMR Spectroscopy*. Vol 15. London: Academic Press; 1983:235.

7. Pertsin AJ, Gorelova MM, Levin VYu, Makarova LI. *J. Appl. Polym. Sci.* 1992;45:1195.

8. Chojnowski J. In: Clarson SJ, Semlyen JA, eds. *Siloxane Polymer*. Chapter 1. Prentice Hall, Englewood Cliffs, N.J: 1993.

9. Burkhardt J, et al. *European Patent*. EP 0,258,640 (1988).

10. Voronkov MG, Mileshkevich VP, Yuzhelevski YuA. *The Siloxane Bond*. New York: Consultants Bureau; 1978. Translation of Siloksanovaya Svyaz, Nauka, Novosybirsk, 1976 (and references cited therein).

11. Union Carbide Corp. *British Patent* GB 943,841 (1960).

12. Takahashi M, et al. U.S. Patent Application Pub. No.0225236 A1 (2003).

13. Rubinsztajn S. *European Patent*. EP 0779316 (2005).

14. Hyde JF. *U.S. Patent*. 2,490,357 (1949).

15. Hyde JF, Wehrly JR. *U. S. Patent*. 3,337,497 (1967).

16. Grzelka A, et al. *J. Inorg. Organometal. Polym..* 2004;14:85.

17. Eglin D, et al. *U.S. Patent*. 6,448,196 (2002).

18. Kendrick TC, Parbhoo BM, White JW. In: Allen G, et al. eds. *Comprehensive Polymer Science*. Vol.4. Oxford: Pergamon Press; 1989:459.

19. Sigwalt P. *Polym. J.* 1987;19:567.

20. Bordone C, et al. U.S. Patent Application Pub. No. 0109659 (2003).

21. Hyde JF, Wehrly JR. *U.S. Patent*. 2,891,920 (1959).

22. Graiver D, Huebner DJ, Saam JC. *Rubber Chem. Technol.* 1983;56:918.

23. De Gunzbourg A, Favier J-C, Hemery P. *Polym. Int.* 1994;35:179.

24. Lebedev BV, Mukhina NN, Kulagina TG. *Vysokomol. Soyed.* 1978;A20:1297.

25. Barry AJ, Beck HN. In: Stone FGA, Graham WAG, eds. *Inorganic Polymer*. New York: Academic Press; 1962.

26. Grulke EA. In: Brandrup J, Immergut EH, eds. *Polymer Handbook*. 2nd ed. New York: John Wiley and Sons; 1975:VII–557, (and references cited therein).

27. Ashworth AJ, Price GJ. *Macromolecules*. 1986;19:362.

28. Roth M. *J. Polym. Sci.: Part B, Polym. Phys.* 1990;28:2715.

29. Schulz GV, Haug A. *Z. Phys. Chem.(Frankfurt)*. 1962;34:328.

30. Kubota K, Kubo K, Ogino K. *Bull. Chem. Soc. Japan*. 1976;49:2410.

31. Flory PJ, Mandelkern L, Kinsinger JB, Shultz WB. *J. Am. Chem. Soc.* 1952;74:3364.

32. Crescenzi V, Flory PJ. *J. Am. Chem. Soc.* 1964;86:141.

33. Andrianov KA, et al. *Vysokomol. Soyed.* 1977;A19:2300.

34. Barry AJ. *J. Appl. Phys.* 1946;17:1020.

35. Flory PJ, Crescenzi V, Mark JE. *J. Am. Chem. Soc.* 1964;86:146.

36. Bleltzung M, Picot C, Rempp P, Herz J. *Macromolecules*. 1982;15:1594.

37. Higgins JS Dodgson K, Semlyen JA. *Polymer*. 1979;20:553.

38. Bueche AM. *J. Polym. Sci.* 1955;15:97.

39. Malone SP Vosburgh C, Cohen C. *Polymer*. 1993;34:5149.

40. Flory PJ, Shih H. *Macromolecules*. 1972;5:761.

41. Munk P, Hattam P, Du Q, Abdel-Azim AA. *J. Appl. Polym. Sci.: Appl. Polym. Symp.* 1990;45:289.

42. Mark JE, Zhang Z-M. *J. Polym. Sci.: Polym. Phys. Ed.* 1983;21:1971.

43. Kuo CM PhD. *Dissertation*. University of Cincinnati; 1991.

44. Shiomi T, Kohra Y, Hamada F, Nakajima A. *Macromolecules*. 1980;13:1154.

45. Clarson SJ, Galiatsatos V, Mark JE. *Macromolecules*. 1990;23:1504.

46. Kuo CM, Clarson SJ. *Macromolecules*. 1992;25:2192.

47. Kuo CM, Clarson SJ. *Eur. Polym. J.* 1993;29:661.

48. Meier G, Momper B, Fischer EW. *J. Chem. Phys.* 1992;97:5884.

49. Galin M, Mathis A. *Macromolecules.* 1981;14:677.

50. Shih H, Flory PJ. *Macromolecules.* 1972;5:758.

51. Lichtenthaler RN, Liu DD, Prausnitz JM. *Macromolecules.* 1978;11:192.

52. Dee GT, Ougizawa T, Walsh DJ. *Polymer.* 1992;33:3462.

53. Sanchez IC, Lacombe RH. *J. Polym. Sci. Polym. Lett. Ed.* 1977;15:71.

54. Li W, Huang B. *J. Polym. Sci.: Part B, Polym. Phys.* 1992;30:727.

55. Chu JH, Rangarajan P, Adams JL, Register RA. *Polymer.* 1995;36:1569.

56. Ibemesi J, Gvozdic N, Keumin M, Tarshiani Y, Meier DJ. *Mater. Res. Soc. Symp. Proc.* 1990;171:105.

57. Hartney MA, November AE, Bates SF. *J. Vac. Sci. Technol.* 1985;B3:1346.

58. Smith SD, et al. *Macromolecules.* 1992;25:2575.

59. Dow Corning® 200 Fluid, Product information about Dow Corning Silicone Fluid, Dow Corning Corp. Form No. 22-931A-90, 22-926D-93, 22-927B-90, 22-928E-94, 22-929A-90, 22-930A-90, 22-0069N-01.

60. Bates OK. *Ind. Eng. Chem.* 1949;41:1966.

61. Damaschun VG. *Kolloid Z.* 1962;180:65.

62. Roe RJ. *J. Phys. Chem.* 1968;72:2013.

63. Allen G, et al. *Polymer.* 1960;1:467.

64. Varaprath S, Frye CL, Hamelink J. *Environ, Toxicol. Chem.* 1996;15:1263.

65. Watanabe N, et al. *Sci. Total Environ.* 1984;38:167.

66. Tanimura M. *Silicone Materials Handbook.* Tokyo: Toray Dow Corning Silicone; 1993.

67. Andrianov KA, et al. *J. Polym. Sci. Part A-1.* 1972;10:1.

68. Steinfink H, Post B, Fankuchen I. *Acta Cryst.* 1955;8:420.

69. Cottrell TL. *The Strength of Chemical Bond.* 2nd ed. London: Butterworths; 1958.

70. Grigoras S, Lane TH. *J. Comput. Chem.* 1988;9:25.

71. Grigoras S, Lane TH. In: Zeigler JM, Fearon FWG, eds. *Silicone Based Polymer Science. Adv. Chem. Ser. 224.* Chapter 7. Washington, DC: American Chemical Society; 1990.

72. Ohlberg SM, Alexander LE, Warrick EL. *J. Polym. Sci.* 1958;27:1.

73. Wang B, Krause S. *Macromolecules.* 1987;20:2201.

74. Lee CL, et al. *Polymer Preprint Am. Chem. Soc. Polym. Chem. Div.* 1969;10(2):1311.

75. Slonimskii GL, Levin VYu. *Vysokomol. Soyed.* 1966;8:1936.

76. Feio G, Buntinx G, Cohen-addad JP. *J. Polym. Sci.: Part B, Polym. Phys.* 1989;27:1.

77. Clarson SJ, Dodgson K, Semlyen JA. *Polymer.* 1985;26:930.

78. Wilcock DF. *J. Am. Chem. Soc.* 1946;68:691.

79. Kataoka T, Ueda S. *J. Polym. Sci., Polym. Lett. Ed.* 1966;4:317.

80. Pethrick RA. In: Clarson SJ, Semlyen JS, eds. *Siloxane Polymer.* Chapter 10. Prentice Hall, Englewood Cliffs, N.J: 1993.

81. Bagley EB, West DC. *J. Appl. Phys.* 1958;29:1511.

82. Valles EM, Macosko CW. *Macromolecules.* 1979;12:521.

83. Bernett MK, Zisman WA. *Mcaromolecules.* 1971;4:47.

84. Hunter MJ, et al. *Ind. Eng. Chem.* 1947;39:1389.

85. Duel LA, Owen MJ. *J. Adhesion.* 1983;16:49.

86. Chen I-J, Lindner W. *Langmuir.* 2007;23:3118.

87. Chaudury MK, Whitesides GM. *Langmuir.* 1991;7:1013.

88. Chaudury MK. *J. Adhesion Sci. and Technol.* 1993;7:669.

89. Lee LH. *J. Adhes.* 1972;4:39.

90. Fox HW, Taylor PW, Zisman WA. *Ind. Eng. Chem.* 1947;39:1401.

91. Wu S. *J. Polym. Sci: Part C.* 1971;34:19.

92. Jarvis NL. *J. Phys. Chem.* 1966;70:3027.

93. Kirst KU, Kremer F, Pakula T, Hollingshurst J. *Colloid Polym. Sci.* 1994;272:1420.

94. Oda Y, Hata T. *Preprints from the 17th Annual Meeting High Polymer Society.* Japan; 1968:267.

95. Kitazaki Y, Hata T. *Preprints from the 18th Annual Meeting High Polymer Society.* Japan; 1969:478.

96. Wanger M, Wolf BA. *Macromolecules.* 1993;26:6498.

97. Anastasiadis SH, et al. *Polym. Eng. Sci.* 1986;26:1410.

98. Roe RJ. *J. Colloid Interface Sci.* 1969;31:228.

99. Robb WL. *Ann. N. Y. Acad. Sci.* 1968;146:119.

100. Meals RN, Lewis FM. *Silicone.* Chapter 2. New York: Reinhold Publishing; 1959.

101. Weissler A. *J. Am. Chem. Soc.* 1949;71:93.

102. Beattie AG. *J. Appl. Phys.* 1972;43:1448.

103. Bondi A. *J. Phys. Coll. Chem.* 1951;55:1355.

104. Mathur RM. *Trans. Faraday. Soc.* 1958;54:1477.

105. Lagemann R. *J. Polym. Sci.* 1948;3:663.

106. Nagy J, Ferenczi-Gresz S, Farkas R, Czuppon A. *Acta Chim. Acad. Sci.(Hung.).* 1976;91:351.

107. Liao SC, Mark JE. *J. Chem. Phys.* 1973;59:3825.

108. Yamada T, Yoshizaki T, Yamakawa H. *Macromolecules.* 1992;25:1487.

109. Tomanek A. *Silicone and Industry.* Hanser Velag, Munich: 1991.

110. Tsvetkov VN, Frisman EV, Boitsova NN. *Vysokomol. Soyed.* 1960;2:1001.

111. Mills NJ, Saunders DW. *J. Macromol. Sci. Phys.* 1968;B2:369.

112. Liberman MH, Abe Y, Flory PJ. *Macromolecules.* 1972;5:550.

113. Wales JLS. *The Application of Flow Birefringence to Rheological Studies of Polymer Melts.* Delft University Press: 1976.

114. Van Krevelen DW. *Properties of Polymer.* 2nd ed. Amsterdam: Elsevier; 1976.

115. Bleltzung M, Picot C, Herz J. *Macromolecules.* 1984;17:663.

116. Valles EM, Rost EJ, Macosko CW. *Rubber Chem. Technol.* 1984;57:55.

117. Buch RR. *Fire Safety Journal.* 1991;17:1.

118. Thomas TH, Kendrick TC. *J. Polym. Sci. A-2.* 1969;7:537.

119. Rodé VV, Verkhotin MA, Rafikov SR. *Vysokomol. Soyed.* 1969;A11:1529.

120. Osthoff RC, Bueche AM, Grubb WT. *J. Am. Chem. Soc.* 1954;76:4659.

121. Joint Assessment of Commodity Chemicals No. 26, *Linear Polydimethylsiloxanes (viscosity 10-100,000 centistokes),* European Centre for Ecotoxicology and Toxicology of Chemicals, Brussels, 1994 (and references cited therein).

122. Sabourin CL, et al. *Environ. Toxicol. Chem.* 1999;18:1913.

123. Fendinger NJ, et al. *Environ. Sci. Technol.* 1997;31:1555.

124. Hainline AN. Silicone Technology. In: Bruins PF, eds. *Applied Polymer Symposia, No. 14.* New York: John Wiley and Sons; 1970.

125. *MSDS of SF96® 350 Fluid.* Momentive Performance Materials Inc; 2006.

126. *MSDS of Wacker-Belsil® DM 10 fluid.* Wacker Chemie AG. 2007.

127. Koerner G, Schulze M, Weis J. *Silicone Chemistry and Technology.* Vulkan-Verlag, Essen: 1991.

128. Eckberg R, et al. *U.S. Patent.* 5,650,453(A) (1997).

129. Eckberg R. et al. *European Patent.* EP 0844267 (2004).

130. Döhler H, et al. *U.S. Patent.* 6,268,404(B1) (2001).

131. Silastic® Silicone Rubber Selection Guide. Dow Corning Corp. Form No. 45-1205-01 and 45-1206-01 (2003).

132. Silastic® Liquid Silicone Rubber Selection Guide. Dow Corning Corporation, Form No. 45-1406-01 (2004).

133. Polmanteer KE. *Rubber Chem. Technol.* 1987;61:470.

134. Will R, Lochner U, Yoneyama M. *Chemical Economics Handbook.* Menlo Park, Calif: SRI International; 2007:5830100A.

Poly(dimethylsilylene)

Robert West

Acronym, Alternative Name PDMS, polydimethylsilane

Class Polysilanes

Structure $[-(Me_2Si)_n-]$

Major Applications Precursor to silicon carbide ceramics via intermediate pyrolysis to polycarbosilane.[1]

Properties of Special Interest Relatively low cost, compared with other polysilanes. For general information about polysilane polymers see the entry for *Poly(methylphenylsilylene)* in this handbook.

Preparative techniques[2,3]

Reactants	Solvent	Temp. (°C)	Yield (%)
Me_2SiCl_2, Na	Toluene	110	80
	Octane	125	–

Property	Units	Conditions	Value	Reference
Typical comonomers for copolymerization			$PhMeSiCl_2$, Ph_2SiCl_2	
Repeat unit	$g\,mol^{-1}$	$(CH_3)_2Si$	58	–
IR absorption	cm^{-1}	–	2,950, 2,890, 1,905, 1,250, 835, 750, 695, 632	(2)
UV absorption	λ (nm)	Solid	340	(3)
NMR spectra	δ (ppm)	Solid; ^{29}Si nucleus	−34.45	(3)
Solvents	Fluorene (220°C), α-chloronaphthalene (238°C)			
Nonsolvents	Toluene, THF, hexane, 2-propanol, CH_2Cl_2, acetone			
Lattice	–	–	Monoclinic	(3)
Monomers per unit cell	–	–	2	(3)
Unit cell dimensions	Å	–	$a = 12.18, b = 8.00, c = 3.88$	(3)
Unit cell angles	Degrees	–	$\alpha = \beta = \gamma = 90$	(3)
Transition temperature	K	$2.5\,cal\,g^{-1}$	333	(3)
		$0.3{-}0.8\,cal\,g^{-1}$	499	

Poly(dimethylsilylene)

Property	Units	Conditions	Value	Reference
Density	g cm^{-3}	–	0.971	(2)
Electronic conductivity	S cm^{-1}	Undoped	$<10^{-12}$	(4)
		H$_2$SO$_4$	10^{-3}	
Suppliers		Nippon Soda Co. Ltd., 2-1, Ohtemachi 2-chome, Chiyoda-ku, Tokyo 100, Japan Gelest Inc., 612 William Leigh Drive, Tullytown, PA 19007-6308, USA.		

References

1. Yajima S, Okamura K, Hayashi J, Omori M. *J. Am. Ceramic Sci.* 1976;59:324.; Hayashi J, Omori M, Yajima S. *U.S. Patent.* 1979;4159259.
2. Wesson JP, Williams TC. *J. Polym. Sci., Polym. Chem. Ed.* 1979;17:2833.
3. Lovinger AJ, et al. *Macromolecules.* 1991;24:132.
4. Usuki A, Marase M. *Jpn. Kokai Tokkyo Koho.* JP 6959623.; *Chem. Abst.* 1987;107:218592u.

Poly(dimethylsilylene-*co*-phenylmethylsilylene)

Robert West

Acronym, Alternative Name PSS, polysilastyrene

Class Polysilanes

Structure $[-(Me_2Si)_n(PhMeSi)_m-]$ ($n, m = 0.5-2$)

Major Applications Precursor for silicon carbide ceramic.[1,2] Initiator for free-radical polymerization.[3]

Properties of Special Interest None. For general information about polysilane polymers see the entry for *Poly(methylphenylsilylene)* in this handbook.

General Information Polysilastyrene is a copolymer of Me_2Si units (58 g mol^{-1}) with PhMeSi units (120 g mol^{-1}). As ordinarily prepared by cocondensation of Me_2SiCl_2 and PhMeSiCl$_2$, it is somewhat blocklike, but a more ordered polymer is obtained from ClSiMe$_2$SiMePhCl. The polymer is atactic and amorphous. The molecular weight distribution is bi- or polymodal.

Preparative techniques

Reactants	Solvent	Temp. (°C)	Yield (%)	M_w	Reference
PhMeSiCl$_2$, Me$_2$SiCl$_2$, Na	Toluene	110	49	40,000 39,000 14,000	(4)
ClSiMePhSiMe$_2$Cl, Na	Toluene	110	45	600,000 4,000	(5)

Property	Units	Condition		Value	Reference
UV absorption	λ (nm)	THF, ε/repeat $= 8,000$		330	(4)
		THF, ε/repeat $= 6,000$		320	(5)
NMR spectra	δ (ppm)	Nucleus	Conditions		
		^{29}Si	Random polymer, C$_6$D$_6$ solution	−36.6 −37.4 −38.2 to −40.4	(5)
		^{29}Si	Ordered polymer	−35 to −39	(5)
		^{29}Si	Solid	−45	(6)

Poly(dimethylsilylene-*co*-phenylmethylsilylene)

Property	Units	Condition		Value	Reference
NMR spectra	δ (ppm)	Nucleus	Conditions		
		^{13}C	Random polymer, C_6D_6 solution	127.4	(4)
				134.7	
				137.0	
				−3.8	
				−5.3	
				−6.2	
				−6.9	
Solvents	THF, toluene, CH_2Cl_2				
Nonsolvents	Hexane, 2-propanol				
Electronic conductivity	S cm^{-1}	Undoped polymer		$<10^{-12}$	(1)
		AsF$_5$, 15 Torr		1.5×10^{-6}	
		SbF$_5$, 5 Torr		5×10^{-7}	
Suppliers	Nippon Soda Co. Ltd., 2-1, Ohtemachi 2-chome, Chiyoda-ku, Tokyo 100, Japan Gelest Inc., 612 William Leigh Drive, Tullytown, PA 19007-6308, USA				

References

1. West R, et al. *Ceramic Bull.* 1983;62:891.
2. Wolff AR, Nozue I, Maxka J, West R. *J. Polym. Sci., Polym. Chem. Ed.* 1988;26:701.
3. Welsh KJ, et al. *Polym Preprints.* 1983;24:131.
4. West R. *J. Organometal. Chem.* 1986;300:327.
5. Suganuma K, et al. *J. Materials Sci.* 1993;28:1175.
6. Wolff AR, West R. *Applied Organomet. Chem.* 1987;1:7.

Poly(1,3-dioxepane)

Evaristo Riande and Julio Guzmán

Acronym PDXP

Class Polyformals

Structure $[-CH_2-O-CH_2-CH_2-CH_2-CH_2-O-]$

Major Applications None known

Properties of Special Interest None known

Preparative Techniques Cationic polymerization of 1,3-dioxepane in solution or in bulk at temperatures normally lower than 25°C. Initiators: Lewis acids, oxonium salts, etc.[1−4]

Property	Units	Conditions	Value	Reference
Ceiling temperature	K	Initiator: ClO_4H in CH_2Cl_2	300	(5)
		Initiator: BF_3 in bulk	517	(6)
Typical comonomers	Isobutyl vinyl ether, 1,3-dioxolane			(7, 8, 9)
Specific volume	$cm^3\ g^{-1}$	Amorphous	$0.94 + (3.1 \times 10^{-4})T$	(10)
		Crystal	$0.813 + (2.4 \times 10^{-4})T$	
Solvents	Almost all the organic solvents			
Nonsolvents	Alkanes			
Solubility parameter	$(MPa)^{1/2}$	From viscosity measurements	18.81	(3)
Cohesive energy density	MPa	From viscosity measurements	353.6	(3)
Lattice	–	–	Orthorhombic	(11)
Space group	–	–	P2cn (C_{2v})	(11)
Monomers per unit cell	–	–	2	(11)
Unit cell dimensions	Å	Chain axis	$a = 8.50$	(11)
			$b = 4.79$	
			$c = 13.50$	
Degree of crystallinity	%	$M_n = 1.2 \times 105$	35	(10)
Heat of fusion	$kJ\ mol^{-1}$	DSC	14.454	(10)
			14.3	(12)

Poly(1,3-dioxepane)

Property	Units	Conditions	Value	Reference
Entropy of fusion	$kJ\,K^{-1}\,mol^{-1}$	DSC	0.0477	(10)
Avrami exponent	–	Dilatometry	3	(10)
Glass transition temperature T_g	K	DSC		
		$M_n = 1 \times 10^5$	192	(2)
		$M_n = 3{,}500$	179	(2)
		–	189	(12)
Equilibrium melting temperature $T_m{}^\circ$	K	Dilatometric data. Extrapolation T_m vs. T_c	303	(10)
Melting temperature T_m	K	DSC	296	(12)
			297	(11)
Heat capacity	$kJ\,K^{-1}\,mol^{-1}$	Crystal $(T_g < T < T_m)$	$(0.189 \times 10^{-3}) + (4.2 \times 10^{-6})T$	(12)
		Amorphous $(T > T_m)$	$(1.38 \times 10^{-3}) + (1.76 \times 10^{-6})T$	
Dipolar ratio $\langle \mu^2 \rangle_0 / nm^2$	–	30°C	0.158	(4)
$d\ln\langle \mu^2 \rangle_0 / dT$	K^{-1}	–	5.4×10^{-3}	(4)
Molecular conformation			$-CH_2-CH_2-CH_2-CH_2-O-CH_2-O-CH_2-CH_2-CH_2-CH_2-O-CH_2-O-$ T T T G T G G T T T G T G G	(4, 11)

References

1. Ivin KJ, Saegusa T. *Ring-Opening Polymerization*. Vol. 1, Ch. 6. New York: Elsevier; 1984.
2. Marco C, Garza J, Fatou JG, Bello A. *An. Quim.* 1981;77(2):250.
3. Marco C, Bello A, Fatou JG, Garza J. *Makromol. Chem.* 1986;187(1):177.
4. Riande E, Mark JE. *J. Polym. Sci., Polym. Phys. Ed.* 1979;17(11):2013.
5. Plesh PH, Westermann PH. *Polymer.* 1969;10:105.
6. Busfield WK, Lee RM. *Makromol. Chem.* 1973;169:199.
7. Okada M, Yamashita Y. *Makromol. Chem.* 1969;126:266.
8. Tüdos F, Kelen T, Turcsanyi B, Kennedy JP. *J. Polym. Sci., Polym. Chem. Ed.* 1981;19:1119.
9. Chwialkowska W, Kubisa P, Penczek S. *Makromol.Chem.* 1982;183:753.
10. Garza J, Marco C, Fatou JG, Bello A. *Polymer.* 1981;22:477.
11. Sasaki S, Takahashi Y, Tadokoro H. *Polym. J.* 1973;4:172.
12. Clegg GA, Melia TP. *Polymer.* 1970;11(5):245.

Poly(1,3-dioxolane)

Evaristo Riande and Julio Guzmán

Acronym PDXL

Class Polyacetals

Structure $[-CH_2-O-CH_2-CH_2-O-]$

Major Applications None known. Stabilizer of Delrin by copolymerization with trioxane.

Properties of Special Interest Possible use as a modifier for elastomers.

Preparative Techniques Cationic polymerization of 1,3-dioxolane in solution or in bulk at temperatures normally lower than 25°C. Initiators: Lewis acids, oxonium salts, etc.[1]

Property	Units	Conditions	Value	Reference
Ceiling temperature	K	Initiators		
		$HClO_4$ in CH_2Cl_2	274	(2)
		BF_3OEt_2 in benzene	265	(3)
		BF_3 in CDCl3	320	(4)
		$HClO_4$ in bulk	417	(5)
Typical comonomers	Cyclic ethers, cyclic acetals, diketene, lactones, styrene			(6, 7–10, 11, 12, 13)
Specific volume	$cm^3\ g^{-1}$	Amorphous	$0.796 + (7.64 \times 10^{-4})T$	(14)
		Crystal II	$0.6965 + (5.0 \times 10^{-4})T$	
		Crystal III	$0.7350 + (6.5 \times 10^{-4})T$	
Thermal expansion coefficient	K^{-1}	Liquid (dilatometry)	6.73×10^{-4}	(15)
		Glass (thermal mechanical analysis)	3.40×10^{-4}	
Solvents	Chlorinated solvents (methylene chloride, chloroform, etc.), aromatic (benzene, toluene, etc.), ketones, ethers			
Nonsolvents	Hydrocarbons (pentane, hexane)			
Solubility parameter	$(MPa)^{1/2}$	From viscosity measurements	20.67	(16)
Virial coefficient	$cm^3\ mol^{-1}\ g^{-2}$	In tetrahydrofuran at 25°C		(17)
		$M_n = 1.1 \times 10^5$	9.15×10^{-4}	
		$M_n = 9 \times 10^4$	9.40×10^{-4}	
		$M_n = 6.6 \times 10^4$	9.48×10^{-4}	
		$M_n = 3.55 \times 10^4$	9.78×10^{-4}	

Poly(1,3-dioxolane)

Property	Units	Conditions	Value	Reference
Characteristic ratio $C_\infty = \langle r^2 \rangle_0 / nl_2$	–	From virial coefficient	3.7	(17)
Temperature coefficient of unperturbed dimensions $d\ln\langle r^2\rangle_0/dT$	K^{-1}	Intrinsic viscosities	0.2×10^{-3}	(18)
Cohesive energy density	MPa	From viscosity measurements	427	(16)
Degree of crystallinity	%	$M_w = 1.2 \times 10^5$ $M_w = 8.8 \times 10^3$	55 80	(19)
Heat of fusion	kJ mol^{-1}	DSC	16.698 ± 0.32 15.49	(20) (21)
Entropy of fusion	kJ K^{-1} mol^{-1}	DSC	0.0423	(21)
Avrami exponent	–	DSC, crystallization between 0 and 21°C	2	(22)
		Dilatometry	3	(19)

Unit cell dimensions[23–24]

Lattice and Space Group	Monomers Per Unit Cell	Cell Dimensions (Å)			Cell Angles		
		a	b	c (Chain axis)	α	β	γ
Triclinic	15	12.32	4.66	24.7	–	–	–
Orthorhombic, Pbca-D$_{2H}^{15}$	8	9.07	7.79	9.85	–	–	–
Hexagonal	18	8.07	8.07	29.5	–	–	120

Property	Units	Conditions	Value	Reference
Glass transition temperature T_g	K	DSC	209 210	(20) (15)
Equilibrium melting temperature T_m	K	Dilatometric data. Extrapolation T_m vs. T_c Low molecular weight High molecular weight	352 358, 366	(22) (19, 22)
Melting temperature T_m°	K	DSC	333 325	(25) (20)
Heat capacity	KJ K^{-1} mol^{-1}	Crystal ($T_g < T < T_m$) Amorphous ($T > T_m$)	$(0.189 \times 10^{-3}) + (3.7 \times 10^{-6})T$ $(1.396 \times 10^{-3}) + (1.472 \times 10^{-6})T$	(20)

Poly(1,3-dioxolane)

Property	Units	Conditions	Value	Reference
Dipolar moment ratio $\langle\mu^2\rangle_0/nm^2$	–	30°C	0.17	(26)
$d\ln\langle\mu^2\rangle_0/dT$	K^{-1}	30–60°C	6.0×10^{-3}	(26)
Intrinsic viscosity $[\eta]$	dl g^{-1}	Chlorobenzene in tetrahydrofuran at 25°C ($3.55 \times 10^4 < M_n < 1.1 \times 10^5$)	$[\eta] = 0.002M^{0.5}$ $[\eta] = 1.7 \times 10^{-4}M_n^{0.73}$	(27) (17)
Molecular conformation	O—CH$_2$—O—CH$_2$—CH$_2$— G′ G′ T′ G′ G′ 79 74 173 − 63 − 94			(23)

References

1. Ivin KJ, Saegusa T. *Ring-Opening Polymerization*. Vol 1. Ch. 6. New York: Elsevier; 1984.
2. Plesch PH, Westermann PH. *J. Polym. Sci.* 1968;C16:3837.
3. Yamashita Y, Okada M, Suyama K, Kasahara H. *Makromol. Chem.* 1968;114:146.
4. Busfield WK, Lee RM, Merigold O. *Makromol. Chem.* 1972;156:183.
5. Binet R, Leonard J. *Polymer.* 1973;14:355.
6. Okada M, et al. *Makromol. Chem.* 1965;82:16.
7. Jaacks V. *Makromol. Chem.* 1967;101:33.
8. Kucera M, Pichler J. *Polymer.* 1964;5:371.
9. Yamashita Y, Asakura T, Okada M, Ito K. *Makromol. Chem.* 1969;129:1.
10. Gibas M, Jedlinsky Z. *Macromolecules.* 1981;14:102.
11. Okada M, Yokoyama Y, Sumitomo H. *Makromol. Chem.* 1972;162:31.
12. Yokoyama Y, Okada M, Sumitomo H. *Makromol. Chem.* 1974;175:2525.; 1975;176:2815, 3537.
13. Okada M, Yamashita Y, Ishii Y. *Makromol. Chem.* 1966;94:181.
14. Archambault P, Prud'Homme RE. *J. Polym. Sci.: Polym. Phys. Ed.* 1980;18:35.
15. Alamo R, Fatou JG, Guzmán J. *An. QuRm.* 1983;79:652.
16. Marco C, Bello A, Fatou JG, Garza J. *Makromol. Chem.* 1986;187:177.
17. Alamo R, Bello A, Fatou JG. *Polym. J.* 1983;15:491.
18. Rahalkar R, Mark JE, Riande E. *Macromolecules.* 1986;12:795.
19. Alamo R, Fatou JG, Guzmán J. *Polymer.* 1982;23:374, 379.
20. Clegg GA, Melia TP. *Polymer.* 1969;10:912.
21. Alamo RG, Bello A, Fatou JG, Obrador C. *J. Polym. Sci.: Part B, Polym. Phys. Ed.* 1990;28:907.
22. Neron M, Tardif A, Prud'Homme RE. *Eur. Polym. J.* 1976;12:605.
23. Brandrup J, Immergut EH, eds. *Polymer Handbook.* 2nd ed. Wiley, New York: 1975.
24. Sasaki S, Takahashi Y, Tadokoro H. *J. Polym. Sci.: Polym. Phys. Ed.* 1972;10:2363.
25. Prud'Homme RE. *J. Polym. Sci.: Polym. Phys. Ed.* 1977;15:1619.
26. Riande E, Mark JE. *Macromolecules.* 1978;11:956.
27. Pravinkova NA, Berman YB, Lyudvig YBL, Davtyan AG. *Polym. Sci. USSR.* 1970;12:653.

Poly(di-*n*-pentylsiloxane)

Yuli K. Godovsky and Vladimir S. Papkov

Acronym PDPeS

Class Polysiloxanes

Structure $[-C_5H_{11})_2SiO-]$

Properties of Special Interest Low glass transition temperature, mesophase behavior

Property	Units	Conditions	Value	Reference
Preparative techniques		Anionic ring-opening polymerization of hexapentylcyclotrisiloxane.		(1, 2)
Molecular weight (repeat unit)		–	186.36	–
Typical molecular weight range of polymer.	g mol^{-1}	–	10^4–10^6	–
NMR spectroscopy		Solid state ^{29}Si		(2, 3)
Mark–Houwink parameters K and a	K = ml g^{-1} a = none	Toluene, 298 K	$K = 0.141$ $a = 0.514$	(3)
Heat of fusion	kJ mol^{-1}	High temperature crystal 2 to mesophase	1.9	(3–5)
Entropy of fusion	J mol^{-1}K^{-1}		7.6	(3–5)
Glass transition temperature	K	DSC	167	(3)
Melting temperature	K	High temperature crystal 2 to mesophase	250	(3–5)
Polymorphs		DSC, X-ray data		
		Low temperature crystal 1		(3–5)
		High temperature crystal 2		(3–5)
		Mesophase		(3–5)
Transition temperature	K	Crystal 1–crystal 2, DSC	235	(3–5)
Heat of transition	kJ mol^{-1}	Crystal 1–crystal 2	9.0	(3)
Isotropization temperature	K	Polarization microscopy	603	(3)

Poly(di-*n*-pentylsiloxane)

References

1. Moeller M, Siffrin S, Out GJJ, Boileau S. *ACS Polym. Prep.* 1992;33(1):176.
2. Out GJJ, Turetskii AA, Moeller M. *Macromol. Rapid. Commun.* 1995;16:107.
3. Out GJJ, Turetskii AA, Moeller M, Oelfin D. *Macromolecules.* 1994;27:3310.
4. Out GJJ. *Dissertation.* The Netherlands: Universiteit Twente; 1994.
5. Molenberg A. *Dissertation.* Germany: University of Ulm; 1997.

Poly(diphenylsiloxane)

Dale J. Meier

Acronym PDPS

Class Polysiloxanes

Structure $[-Si(C_6H_5)_2O-]$

Major Applications PDPS is not a commercial polymer. Diphenylsiloxane is a component in various copolymers.

Properties of Special Interest Highly crystalline, high melting point, excellent thermal stability, mesomophic state at high temperatures.

Preparative Techniques	Conditions	Reference
Anionic	From hexaphenylcyclotrisiloxane	
	Li alkyl, bulk	(1)
	KOH, bulk	(2, 3)
	Li alkyl, solution	(4, 5)
Condensation	From diphensilanediol	(6)
Typical comonomer	Dimethylsiloxane	
	Random	(4, 7–9)
	Block	(1, 4, 5, 10)

Crystalline state properties

Lattice	Cell Dimensions (Å)			Cell Angles (Degrees)			Reference
	a	*b*	*c*	*α*	*β*	*γ*	
Pbn21, hexagonal pacxking in quasi-planar sequential configuration	20.145	9.820	4.944	90	90	90	(11)
Rhombic unit cell, 2 monomers per cell	20.1	10.51	10.24	–	–	–	(18)

Property	Units	Conditions	Value	Reference
Solvents	K	Diphenyl ether	>410	–
		1-Chloronaphalene	>410	–
		1,2,4 Trichlorobenzene	>410	–
		From quenched state: Chloroform, toluene	320	(4)

Property	Units	Conditions	Value	Reference
Density	g cm^{-3}	Experimental	1.22	(13)
		Unit cell	1.26–1.3	(11)
Melting temperature	K	To mesomorphic state	538	(16)
			545	(14)
			503	(15)
			553	(24)
		Oligomers	471, 481, 487	(19)
Transition temperature	K	To isotropic state	813	(16)
			833	(24)
Heat of fusion	J g^{-1}	To mesomorphic state	35.5	(14)
			20.4	(15)
			51.7	(24)
Entropy of fusion	J K^{-1} mol^{-1}	–	12.8	(14)
			7.98	(15)
Glass transition temperature	K	DSC	313	(16)
			322	(3)
Thermal stability	K	TGA, 10% weight loss 10°min^{-1} under N$_2$	784	(16)
Dielectric constant	–	MW = 1,500–2,600	3.5–2.2	(17)
Dielectric loss	–	MW = 1,500–2,600	0.004–0.5	(17)
Elastomer reinforcement		In dimethylsiloxane elastomers		(6)
Sequence distributions and crystallinity in copolymers with dimethysiloxane		Computer simulations		(20, 21)
Rheological properties		M$_w$ = 907,000 M$_w$/M$_n$ = 26.5		(23)
	K	T = 523–623 (in mesomorphic state)		
Thermodynamics	K	T = 5–620		(25)
Light emission	nm	KrF laser irradiation, 248 nm	340	(22)

References

1. Bosdic EE. *ACS Poly. Preprints.* 1969;10:877.
2. Buzin M, et al. *J. Poly. Sci., Part A: Polym. Chem.* 1997;35:1973.
3. Buzin MI, Kvachev YP, Svistunov VS, Psapkov VS. *Vysokomol. Soedin.* 1992;34, Series B:66.
4. Ibemesi J, et al. *ACS Poly. Preprints.* 1985;26:18.
5. Ibemesi J, et al. In: Mark JE, Schaefer DW, ed. *Polymer Based Molecular Composites.* Pittsburgh: Materials Research Society; 1989.
6. Wang S, Mark JE. *J. Materials Sci.* 1990;25:65.
7. Lee CL, Marko OW. *ACS Poly. Preprints.* 1978;19:250.
8. Babu GN, Christopher SS, Newmark RA. *Macromol.* 1987;20:2654.

9. Yang M-H, Chou C. *J. Poly. Research.* 1994;1:1.

10. Fritzsche AK, Price FP. In: Aggarwal SL, ed. *Block Copolymers.* New York: Plenum Press; 1970.

11. Grigoras S, et al. *Macromol.* 1995;28:7371.

12. Dubchak IL, et al. *Vysokomol. Soedin.* 1989;31, Series A:65.

13. Tsvankin DY, et al. *Poly. Sci. USSR English translation.* 1980;21:2348.

14. Govodsky YK, Papkov VS. *Adv. Poly. Sci.* 1989;88:129.

15. Falender JR, et al. *J. Poly. Sci.: Polymer Physics.* 1980;18:388.

16. Lee MK, Meier DJ. *Polymer.* 1993;34:4882.

17. Karavan YV, Gukalov SP. *Fiz. Elekron. (Lvov).* 1974;7:77.; CA 81: 121610.

18. Babchinitser TM, et al. *Polymer.* 1985;26:1527.

19. Harkness BR, Tachikawa M, Mita H. *Macromol.* 1995;28:1323.

20. Madkour TM, Mark JE. *Comput. Poly. Sci.* 1994;4:87.

21. Madkour TM, Mark JE. *ACS Poly. Preprints.* 1995;36:673.

22. Suzuki M, et al. *Material Sci. Eng.* 1997;B49:172.; CA 127: 332153.

23. Papkov VS, et al. *Poly. Sci., Ser. A Russia (English translation).* 2000;43:200.

24. Papkov VS, et al. *Macromol.* 2002;35:1079.

25. Lebedev BV, et al. *Poly. Sci. Ser. A Russia (English traslation).* 2000;42:1711.

Poly(di-*n*-propylsiloxane)

Yuli K. Godovsky and Vladimir S. Papkov

Acronym PDPrS

Class Polysiloxanes

Structure $[-(C_3H_7)_2Si\,O-]$

Properties of Special Interest Low glass transition temperature, mesophase behavior.

Property	Units	Conditions	Value	Reference
Preparative technique	Anionic ring-opening polymerization of hexapropylcyclotrisiloxane			(1–4)
Molecular weight (of repeat unit)	g mol^{-1}	–	130.26	–
Typical molecular weight range of polymer	g mol^{-1}	–	10^3–10^5	–
NMR spectroscopy	Solid state ^{1}H, ^{29}Si			(3, 5, 6)
Theta temperature	K	Toluene	283	(7)
		2-Pentanone	351	
Mark–Houwink parameters	$K=$ ml g^{-1} $a=$ None	MW $= (2.5$–$30) \times 10^5$		(7)
		Toluene, 25°C,	$K = 4.35 \times 10^{-2}, a = 0.58$	
		Toluene, 10°C	$K = 1.09 \times 10^{-1}, a = 0.5$	
		2-Pentanone, 78°C	$K = 8.71 \times 10^{-2}, a = 0.5$	
Characteristic ratio $(r^2)/nl^2$	–	(osmometric and viscosimetric measurements)	13 ± 1	(7–9)
		MD simulation	~ 7.5	(10)

Unit cell dimensions[6,11]

Polymorph	Lattice	Monomers Per Unit Cell	Cell Dimensions (Å)			Cell Angles (Degrees)		
			a	*b*	*c* (Chain Axis)	α	β	γ
High-temperature crystalline phase[11]	Tetragonal space group P4$_1$ or P4$_3$	4 Helix 4$_1$	9.52	9.52	9.40	90	90	90
High-temperature crystalline phase, α[6]	Pseudo-tetragonal monoclinic space group *C2/c* (No15)	4 *cis–trans* chain conformation	19.15	19.15	5.0	90	90	90

Poly(di-*n*-propylsiloxane)

Polymorph	Lattice	Monomers Per Unit Cell	Cell Dimensions (Å)			Cell Angles (Degrees)		
			a	*b*	*c* (Chain Axis)	α	β	γ
Low-temperature crystalline phase, β[6]	Monoclinic	4 *cis–trans* chain conformation	20.60	19.22	4.95	90	90	93.1

Property	Units	Conditions	Value	Reference
Heat of fusion	kJ mol^{-1}	α → μ (mesophase)	2.60–2.86	(2, 12, 13)
			3.13	(6)
Heat of isotropization	kJ mol^{-1}	μ → isotropic melt	0.26–0.42	(2, 12, 13)
Entropy of fusion	J mol^{-1}K^{-1}	α → μ (mesophase)	8.59	(2, 12, 13)
Entropy of isotropization	J mol^{-1}K^{-1}	μ → isotropic melt	0.88	(2, 12, 13)
Density (crystalline)	g cm^{-3}	α polymorph, 293 K	1.015	(11)
		From X-ray data	0.94 ± 0.02	(6)
		Experimental	0.95 ± 0.02	(6, 11)
Glass transition temperature	K	DSC	164	(2, 3, 6, 12)
Melting temperature	K	α → μ (mesophase)	333–335	(2, 6, 12)
Transition temperature	K	β → α	218	(2, 12, 13)
			228	(6)
Heat of transition	kJ mol^{-1}	β → α	2.04	(2, 12, 13)
			4.03	(6)
Isotropization temperature	K	MW (× 10^3) =		(14, 15)
		87	480	
		68	450	
		51	445	
		43	418	
		≈ 10	No mesophase	

References

1. Lee CL, et al. *ACS Polym. Preprints.* 1969;10(2):1319.
2. Godovsky, Yu-K, et al. *Makromol. Chem., Rapid Commun.* 1985;6:443.
3. Out GJJ, et al. *Polym. Adv. Technology.* 1994;5:796.
4. Molenberg A, et al. *Macromol. Symp.* 1996;102:199.
5. Moeller M, et al. *Makromol. Chem., Macromol. Symp.* 1990;34:171.
6. Gearba R, et al. *Macromolecules.* 2006;39:988
7. Lee CL, Emerson FA. *Polym. Sci. Part A-2.* 1967;5:829.
8. Mark JE. *Macromolecules.* 1978;11:627.
9. Stepto RFT. In: Clarson SJ, Semlyen JA, eds. *Siloxane Polymers.* Chapter 8. PTR Prentice Hall, Englewood Cliffs, N.J: 1993.

10. Neuburger N, Bahar IT Mattice WL. *Macromolecules.* 1992;25:2447.
11. Peterson DR, Carter DR, Lee CL: *Macromol. Sci. Phys.* 1969;B3:519.
12. Godovsky, Yu-K, Papkov VS. *Adv. Polym. Sci.* 1989;88:129.
13. Godovsky, Yu-K, Papkov VS. *Makromol. Chem. Macromol. Symp.* 1986;4:71.
14. Godovsky, Yu-K, et al. *Makromol. Chem., Rapid Commun.* 1985;6:797.
15. Molenberg A, Moeller M, Sautter E. *Progr. Polym. Sci.* 1997;22:1133.

Poly(epichlorohydrin)

Qingwen Wendy Yuan-Huffman

Acronyms PECH

Class Polyether

Structure $[-CH_2-CH(CH_2Cl)-O-]$

Mol. Wt. of Repeat Unit 92.5

Property	Unit	Conditions	Values	Ref.
Polymerization	None		Ring-opening polymerization	1, 2
Typical copolymers	None	Epichlorohydrin (EPI)–Ethylene Oxide (EO) copolymer EPI– Allyl Glycidyl Ether (AGE) copolymer EPI–EO–AGE terpolymer		3
Glass transition temperature	K	$n = 5000$–20 000 Heating rate = 20 K/min	258.5 251	2 3, 4
Tensile strength	MPa		17	5
Elongation	%		280	5
Engineering modulus	MPa	Elongation = 100% Elongation = 200%	5.1 12.6	5 5
Hardness	Shore A		72	5
Tear strength	kN/m		36	5
Compression set	%	70 h at 100°C 70 h at 150°C	26 57	5 5
Volume change	%	70 h, ASTM Fuel A, 20°C 70 h, ASTM Fuel C, 20°C 70 h, ASTM Oil #1, 150°C 70 h, ASTM Oil #3, 150°C	0 25 0 1	5 5 5 5
Surface tension	mN m^{-1}	$M = 1500$, $T = 293.5$ K	43.2	3
Fractionation		Extraction; precipitation	Acetone (cold), acetone/methanol, methanol/water	3

Poly(epichlorohydrin)

Crystalline-state properties

Crystl. Syst. (Lattice)	Space Group	Unit Cell Parameters			Monomer Per Unit	Density (g cm^{-3})	Ref.
		A (Å)	B (Å)	C (Å)			
Orthorhombic	D2-4 or C2V-9	12.14	4.90	7.07	4	1.461	3
Orthorhombic	C2V-9	12.16	4.90	7.03	4	1.467	3
Orthorhombic		12.24	4.92	6.96	4	1.466	3
Orthorhombic	D2-4	12.15	4.86	7.07	4	1.472	3

References

1. Odian G. *Principles of Polymerization.* 3rd ed. Wiley Interscience; 1991.
2. Rodriguez F. *Principles of Polymer Systems.* 4th ed. Taylor & Francis Publishers; 1996.
3. Brabdrup J, Immergut EH. *Polymer Handbook.* 3rd ed. New York: Wiley Interscience; 1989.
4. Blythe AR, Jeffs GM. *J. Macromol. Sci.* 1969;B3:141.
5. Mark HS, et al. *Encyclopedia of Polymer Science and Engineering.* Vol 16. Wiley Interscience; 1989.

Poly(erucic acid dimer anhydride)

Abraham J. Domb and Robert Langer

Acronyms, Trade Names BIODEL-EAD, Poly(EAD), Poly(EAD-SA)

Class Polyanhydrides

Structure

$$\left[CO-(CH_2)_7-CH-(CH_2)_8-CH_3 \right.$$
$$\left. CH_3-(CH_2)_8-CH-(CH_2)_7-COO \right]$$

Major Applications Biodegradable polymer for controlled drug delivery in a form of implant, film, or injectable microspheres (e.g., Septacin™–gentamicin-loaded linked beads for the treatment of chronic bone infections).

Properties of Special Interest Anhydride copolymers of erucic acid dimer (EAD) with aliphatic diacids such as sebacic acid (SA) degrade in a physiological medium to EAD and SA. Matrices of the copolymers loaded with dissolved or dispersed drugs degrade in vitro and in vivo to constantly release the drugs for periods from 1–12 weeks.

Property	Units	Conditions	Value		Reference
Molecular weight		P(EAD-SA)			(1)
	10^4 g mol^{-1}	GPC-polystyrene standards	M_w = 3–30, M_n = 1–3		
	dL g^{-1}	Viscosity 25°C, dichloromethane	η_{sp} = 0.2–1.4		
IR (characteristic absorption frequencies)	cm^{-1}	PSA, P(EAD-SA), or P(EAD) film on NaCl pellet	1,740, 1,810		(1)
UV (characteristic absorption wavelength)	nm	P(EAD-SA), EAD monomer dichloromethane	253		–
Optical rotation	–	Dichloromethane	No optical rotation		–
Solubility	mg ml^{-1}	25°C	P(EAD)	P(EAD-SA)	(2)
		Chloroform	>300	<300	
		Dichloromethane	>300	<300	
		Tetrahydrofuran	180	100	
		Ketones	80	50	
		Ethyl acetate	30	25	
		Ethers	5	3	
		Alkanes and arenes	<1	<1	
		Water	<1	<1	

Poly(erucic acid dimer anhydride)

Property	Units	Conditions	Value				Reference
Mark-Houwink parameters: K and a	$K = $ ml g^{-1} $a = $ None	CHCl$_3$, 23°C	$K = 3.46$ $a = 0.634$				(1)
Thermal properties	mol %	P(EAD-SA), DSC, 10°C min^{-1}	0:100	8:92	22:78	100:0	(2)
	K	T_m	359.0	348.0	337.0	293.0	
	K	T_g	333.1	<283.0	<283.0	<273.0	
	kJ kg^{-1}	ΔH	150.7	50.2	13.0	4.0	
Crystallinity	%	–	66	54	35	<5	
Stability in chloroform (decrease in M_w) (anhydride interchange depolymerization)			P(EAD-SA)				
			0:100	22:78			
Depolymerization rate constant	t^{-1}	37°C	0.1325	0.1233			
Tensile strength	MPa	Film by melt, P(EAD-SA) 22:78, $M_w = 1.55 \times 10^5$ g mol^{-1}	4.2				(1)
Tensile modulus	MPa	Film by melt, P(EAD-SA) 22:78, $M_w = 1.55 \times 10^5$ g mol^{-1}	45				(1)
Elongation yield	%	Film by melt, P(EAD-SA) 22:78, $M_w = 1.55 \times 10^5$ g mol^{-1}	14				(1)
Elongation at break	%	Film by melt, P(EAD-SA) 22:78, $M_w = 1.55 \times 10^5$ g mol^{-1}	85				(1)
Erosion rate, SA release	mg h^{-1}	14 × 2.7 mm P(EAD-SA) disc, 0.1 M phosphate buffer, pH 7.4, 37°C	0.3				(2)
Erosion front	mm day^{-1}	–	188				
Elimination time in vivo	days	Implant in dog bone	35				(3)
Drug release in vitro	% day^{-1}	From P(EAD-SA) 22:78					(1)
		Hydrophilic drugs (i.e. gentamicin, carboplatin)	3–6				
		Hydrophobic drugs (i.e., taxol, dexamethasone)	1–3				
Drug release in vivo	% day^{-1}	Beads of 20% gentamicin in rabbit bone	5				(3)
Biocompatibility		Compatible with human bone and muscle Compatible with rabbit brain, bone, muscle, subcutane					(1)
Supplier		Guilford Pharmaceuticals, Inc., Baltimore, Maryland, USA					

Poly(erucic acid dimer anhydride)

References

1. Domb AJ, Maniar M. *J. Polym. Sci.* 1993;31:1275–1285.
2. Shieh L, et al. *J. Biomat. Mater. Res.* 1994;28:1465–1975.
3. Shea J, et al. *Pharm. Res.* 1991;8:195.

Poly(ester-acrylate/amine) Dendrimers

Donald A. Tomalia and Linda S. Nixon

Acronym, Trade Names PEA dendrimers, Priostar® ester-acrylate dendrimers, Priostar® ester-amine dendrimers

Class Dendritic polymers; dendrons; dendrimers

Structure This dendrimer family is synthesized by the direct Michael addition of trimethylolpropane triacrylate (TMPTA) or pentaerythritol tetra-acrylate (PTA) type branch cell reagents to either alkyl amines or poly(alkylene) amines under carefully controlled conditions.[1] Various shaped (i.e. cone-like or ellipsoidal), high multiplicity, acrylate terminated, generation = 1, PEA dendrimers are obtained depending on the nature of the core. Secondary amines (i.e. morpholine) readily undergo Michael addition to these dendritic acrylates to produce morpholine-terminated dendrimers (Figure 1). Dendrimers derived from α, ω-alkylene diamine cores exhibited very low viscosities compared to equivalent molecular weight linear analogs or mixtures with reactive diluents[2] (i.e. Table 1 and Figures 2 and 3). Furthermore, under UV conditions, the PEA dendritic acrylates cured dramatically faster either neat or with reactive diluents compared to linear analogs.[2]

TMPTA   **PTA**

Dendrimer/ Dendron Shapes	Core	Branch Cell Reagents	Poly(ester) Acrylate/Morpholine Dendrimers	

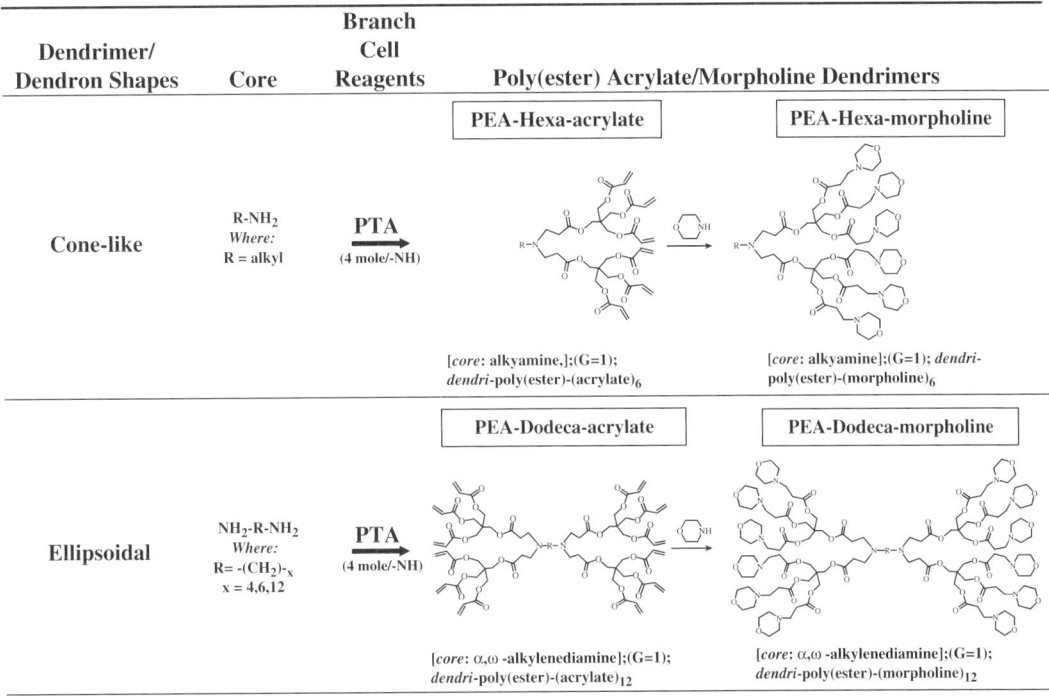

PEA-Hexa-acrylate → **PEA-Hexa-morpholine**

Cone-like · R-NH₂ *Where:* R = alkyl · **PTA** (4 mole/-NH)

[*core*: alkyamine,];(G=1); *dendri*-poly(ester)-(acrylate)₆

[*core*: alkyamine];(G=1); *dendri*-poly(ester)-(morpholine)₆

PEA-Dodeca-acrylate → **PEA-Dodeca-morpholine**

Ellipsoidal · NH₂-R-NH₂ *Where:* R= -(CH₂)-ₓ x = 4,6,12 · **PTA** (4 mole/-NH)

[*core*: α,ω -alkylenediamine];(G=1); *dendri*-poly(ester)-(acrylate)₁₂

[*core*: α,ω -alkylenediamine];(G=1); *dendri*-poly(ester)-(morpholine)₁₂

Figure 1. PEA dendrimers as a function of shape, core, branch cell, and terminal groups.

Table 1. Viscosity of various acrylate oligomers

Oligomer Name	Molecular Weight	Viscosity, mPa s
EBECRYL Resin 600	500[a]	3000 (H, 60°C)[a]
EBECRYL Resin 605	450[b]	7500 (H, 25°C)[a]
EBECRYL Resin 210	1500[a]	3900 (H, 60°C)[a]
EBECRYL Resin 270	1500[a]	3000 (H, 60°C)[a]
EBECRYL Resin 285	1200[a]	23 000 (H, 25°C)[a]
		47 000 (25°C)[c]
Priostar PEA: (Core:EDA)/acrylate[d]	1244[c]	5100 (25°C)[c]
75% Priostar PEA[d] with 25% TPGDA	1008[b]	2200(25°C)[c]

[a] Data came from product introduction of UCB Chemicals (Belgium). H refers to Höppler viscosity.

[b] Mean molecular weight calculated according to mass percent of two components (see Reference 2).

[c] Data obtained from experiments cited in Reference 2.

[d] Priostar hexa-acrylate derived from an ethylenediamine core which possesses only six acrylates vs. eight (see Reference 1 for details).

EBECRYL Resin 600 = bisphenol A epoxy diacrylate[(a)]

EBECRYL Resin 605 = EBECRYL Resin 600 with 25% (wt) TPGDA[(a)]

EBECRYL Resin 210 = aromatic urethane diacrylate[(a)]

EBECRYL Resin 270 = aliphatic urethane diacrylate[(a)]

EBECRYL Resin 285 = EBRCRYL Resin 270 with 25% (wt) TPGDA[(a)]

TPGDA= tripropylene glycol diacrylate.

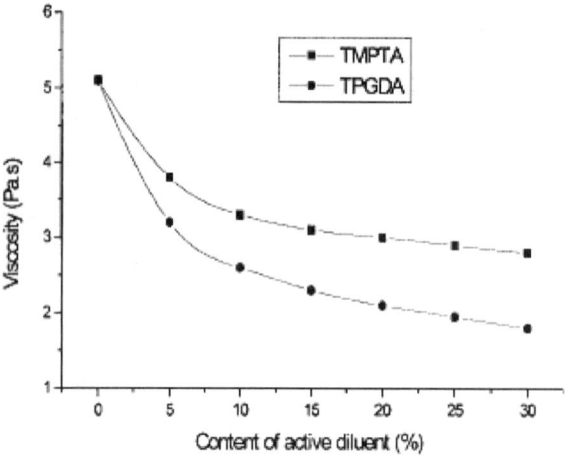

Figure 2. Viscosity of hexa-acrylate derived from EDA core (ref. 2) systems with different active diluents (25°C). Reprinted with permission of John Wiley & Sons, Inc.

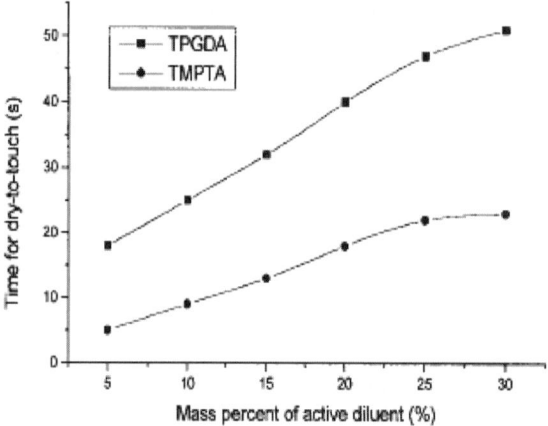

Figure 3. The curing time of hexa-acrylate derived from EDA core (ref. 2) vs. active diluent. Reprinted with permission of John Wiley & Sons, Inc.

Property	Conditions	Value
Availability	Priostar acrylate terminated dendrimers (4-methoxyphenol – inhibited); technical grade	Units: g-kg
	Priostar morpholine terminated dendrimers; technical grade	Units: g-kg

Poly(ester-acrylate/amine) Dendrimers

Property	Conditions	Value
Suppliers	Technical grade (g-kg)	Dendritic Nanotechnologies, Inc.
		2625 Denison Drive
		Mt. Pleasant, MI 48858
		www.dnanotech.com
Significant Patents:		PCT/US2005/013864
		PCT/US2005/047635

Priostar® is a registered trademark of Dendritic Nanotechnologies, Inc.

Acknowledgements

We would like to thank Dr. Ryan Hayes and the DNT staff for many helpful discussions.

References

1. Swanson DR, Huang B, Abdelhady HG, Tomalia DA. *New J. Chem.* 2007;31:1368–1378.
2. Xu DM, Zhang KD, Zhu XL. *J. Appl. Polym. Sci.* 2004;92:1018–1022.

Poly(ester-co-anhydride) of Ricinoleic Acid and Sebacic Acid

Boris Vaisman, Ariella Shikanov, Abraham J. Domb and R. Langer

Acronyms Poly(SA-RA), P(SA-RA)

Class Poly(ester-anhydride)s

Structure

Major Applications Biodegradable polymers for controlled drug delivery in a form of implant or injectable implant (e.g. for the treatment of solid tumors, delivery of local anesthetics, and delivery of antibiotics).

Properties of Special Interest Polyester-anhydride of ricinoleic acid (RA) with aliphatic diacids such as sebacic acid (SA) are viscous liquids at low contents of sebacic acid, <40% w/w, (pasty polymers). Pasty polymers that can be mixed with drugs at room temperature and injected into tissue as neat composition are advantageous as they allow simple preparation of delivery systems, particularly for heat-sensitive drugs. The formulation gels at the site of injection to form a depot of the drug protected in the hydrophobic gel polymer. The drug-loaded pasty polymer releases the drugs for a few weeks while being degraded in vitro and in vivo to short oligomers which future degrade to fatty acids.

Property	Units	Conditions	Value	Reference
Molecular weight	Da	P(SA-RA) GPC-polystyrene standards	$M_w = 3000$–$60\,000$, $M_n = 2000$–$21\,000$	(1)
IR (characteristic absorption frequencies)	cm^{-1}	Film on NaCl pellet	1740, 1810 1050–1300	(1)

Poly(ester-co-anhydride) of Ricinoleic Acid and Sebacic Acid

Property	Units	Conditions	Value				Reference
Solubility	mg ml^{-1}	P(SA-RA) 30:70					
		Chloroform	>300				
		Dichloromethane	>300				
		Tetrahydrofuran	200				
		n-Hexane	<1				
		Diethyl Ethers	<1				
		Water	<1				
Thermal properties		P(SA-RA) 30:70, DSC, 10°C min^{-1}					
	K	T_m	306.8				
	kJkg^{-1}	ΔH	21.5				(1)
Stability during storage (decrease in M_w)		P(SA-RA) 30:70, storage under nitrogen at different temperatures after 6 months GPC-polystyrene standards	Initial M_w	M_w at 298.15 K	M_w at 277.15 K	M_w at 256.15 K	(1)
			9000	2000	8400	9000	
Stability to γ-irradiation		P(SA-RA) 30:70, the polymers were exposed to γ-irradiation at a dose of 2.5 Mrad	Stable, no significant changes were observed				(1)
Degradation rate in vitro	mg h^{-1}	P(SA-RA) 30:70, 200 mg, diameter~5 mm, 0.1M phosphate buffer, pH 7.4, 37°C, with orbital shaking (100 rpm)	(1) Degradation to shorter oligomers				(1)
			Time, h				
			5	20	30	90	
			2.40	0.33	2.50	0.83	
			(2) Degradation of oligomers of ricinoleic acid				(2)
			RA oligomers M_w, Da				
			1000		3000		
			1.57		1.75		
Elimination in vivo (total weight loss from implantation site)	%	P(SA-RA) 30:70, M_w = 11600 Da					
		7 days in rat subcutane	38				
		21 days in rat subcutane	90				
Drug release in vitro % day^{-1}	% day^{-1}	P(SA-RA) 30:70 5% (w/w) cis-platin	0.45 (for the first 5 days), 9.36 (for the next 3 days), and 4.56 (for the following 9 days).				(1)

Property	Units	Conditions	Value	Reference
		5%, 10%, and 20% w/w Paclitaxel	Load dependent: (1) for 5% load: 0.75 (for the first 20 days), 0.35 (for the next 20 days), and 0.17 (for the following 60 days). (2) for 10% load: 0.40 (for the first 20 days), 0.35 (for the next 20 days), and 0.12 (for the following 60 days). (3) for 20% load: 0.10 (for the first 40 days), and 0.08 (for the following 60 days).	(3)
		10% w/w Gentamicin sulfate	1.8	(4)
		5%, 7%, and 10% w/w Bupivacaine free base	Load dependent: (1) for 5% and 7% load: 16 (for the first 3 days), 6.67 (for the next 6 days), and 1.0 (for the following 8 days). (2) for 10% load: 18.33 (for the first 3 days), 4.17 (for the next 6 days), and 0.62 (for the following 8 days).	(5)
Drug release in vivo % day^{-1}	% day^{-1}	P(SA-RA) 30:70 5% Paclitaxel (subcutaneous administration in mice)	7.5	(6)
Biocompatibility	Compatible with mice and rat subcutane, and mice muscles and peripheral nerves Compatible with rat muscles and brain			(3–5)

References

1. Krasko MY, et al. *J. Polym. Sci. Part A.* 2003;41:1059–1069.
2. Krasko MY, Domb AJ. *Biomacromolecules.* 2005;6:1877–1884.
3. Shikanov A, et al. *J. Biomed. Mat. Res.* 2004;69A:47–54.
4. Krasko MY, et al. *J. Control. Release.* 2007;117:90–96.
5. Shikanov A, Domb AJ, Weiniger CF. *J. Control. Release.* 2007;117:97–103.
6. Shikanov A, Ezra A, Domb AJ. *J. Control. Release.* 2005;105:52–67.

Polyesters of Ricinoleic Acid and Lactic Acid

Boris Vaisman, Ariella Shikanov, Abraham J. Domb and R. Langer

Acronyms Poly(LA-RA), P(LA-RA)

Class Polyesters, copolyesters

Structure

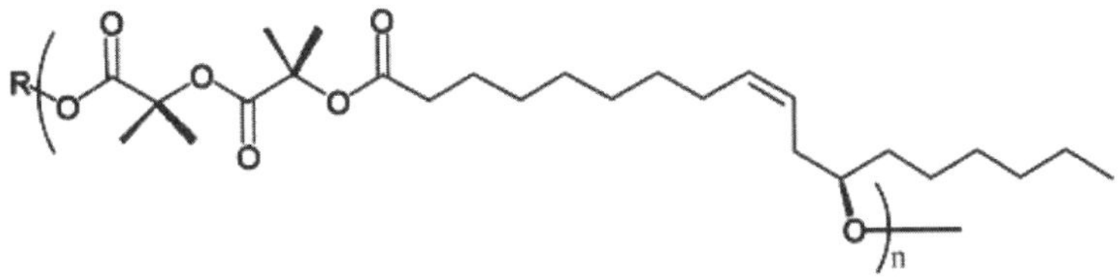

P(L-LA-RA)

Major Applications Biodegradable polymers for controlled drug delivery in a form of implant (e.g. for the delivery of antiinflammatory drugs or antineoplastic agents).

Properties of Special Interest Polyester copolymers of ricinoleic acid (RA) with lactic acid (LA) degrade in physiological medium giving an almost zero-order weight loss, with a 20–40% loss after 60 days of incubation. Lactic acid release to the degradation solution is proportional to weight loss of the polymer samples. The main decrease in molecular weight was observed during the first 20 days, followed by a slow degradation phase, which kept the number average molecular weight constant for another 40 days. These polymer matrices are suitable for delivery of both water-soluble and hydrophobic drugs.

Synthesis Copolyesters of lactic acid (LA) and ricinoleic acid (RA) with different LA/RA ratios may be prepared by three different synthetic approaches: (A) ring-opening polymerization; (B) random condensation of lactic acid (LA) and ricinoleic acid (RA); and (C) transesterification of poly(lactic acid) (PLA) with RA and repolyesterification (1).

Property	Units	Conditions	Value	Reference
Molecular weight	Da	P(LA-RA) GPC-polystyrene standards	$M_n = 4500$–$16\,000$, $M_w = 14\,000$–$35\,000$	(1)
IR (characteristic absorption frequencies)	cm^{-1}	Film on NaCl pellet	1748	(2)

Property	Units	Conditions	Value	Reference
Solubility	mg ml^{-1}	P(LA-RA) 60:40		
		Acetone	>300	
		Acetonitrile	≤50	
		Chloroform	>300	
		Dichloromethane	>300	
		Toluene	≥150	
		Ethers	<1	
		Ethanol	<1	
		Water	< 1	
Thermal properties	K	P(LA-RA) 60:40, DSC, 10°C min^{-1} T_m	366.15	(1)
Stability during storage		Storage under nitrogen at −253.15 K after 6 months, GPC, IR	Stable, no changes were observed	(2)
Stability to γ-irradiation		The polymers were exposed to γ-irradiation at a dose of 2.5 Mrad, GPC, IR	Stable, no changes were observed	(2)
Degradation rate in vitro	mg h^{-1}	P(LA-RA) 60:40 (3 × 3 × 3 mm, 70 mg in 10 ml of medium), 0.1 M phosphate buffer, pH 7.4, 37°C, with orbital shaking (100 rpm)	0.33	(1)
Drug release in vitro % day^{-1}	% day^{-1}	P(LA-RA) 60:40 5FU 10% w/w Triamcinolone 10% w/w	4.71 1.76	(1)

References

1. Slivniak R, Ezra A, Domb AJ. *Pharm. Res.* 2006;23:1306–1312.
2. Slivniak R, Domb. *Biomacromolecules.* 2005;6:1679–1688.

Polyesters, Unsaturated

Mee Y. Shelley and Jennifer L. Braun

Trade Names Dion, Hetron, Polylite, Advaco, Altek, Cargill, Cook, OCF, Pedigree, Pioester, etc.

Class Thermoset polymers (mixtures of polyester prepolymers with aliphatic unsaturation and a vinyl monomer)

Principal Components Prepolymers (oligomer): glycols (e.g., 1,3-propylene glycol), saturated acids (e.g., phthalic anhydride/acid), unsaturated acids (e.g., maleic anhydride/acid). Monomers: styrene, α-methylstyrene, methyl acrylate, methyl methacrylate, etc.

Other Ingredients Inhibitors to prevent premature cross-linking and to allow a suitable shelf life (e.g., hydroquinone). Initiators (catalysts): methyl ethyl ketone peroxides, benzoyl peroxides, etc. Accelerators: cobalt naphthenate, cobalt octanoate, etc.

Major Applications Laminates, coatings, art objects, insulation, construction (e.g., bath tubs, floor tiles, countertops, roofing, siding, skylights, fences, etc.), automobile parts, embedding of specimens (e.g., decorative, zoological), encapsulation of electronic assemblies, toys, playground equipment, furniture, pearl buttons, sports equipment (snow boards, skis, bowling balls, etc.), chemical storage tanks.

Properties of Special Interest Low cost, excellent wetting and surface quality, ease of moldability, versatility, processible over a wide temperature range, high impact resistance, good weathering resistance, high cure shrinkage.

Property	Units	Conditions*	Value	Reference
Linear mold shrinkage	Ratio	Unfilled	0.001–0.007	(1)
		Glass fiber-reinforced	0.0002–0.012	(2)
		SMC, glass fiber-reinforced	0.00002	(3, 4)
		BMC, glass fiber-reinforced		(3, 4)
		Compression	0.00001	
		Injection	0.00004	
		EMI shielding (conductive)	0.0002–0.001	(2)
		DMC, with filler and glass fiber	0.000–0.002	(13)
		SMC, with filler and glass fiber	0.000–0.002	(13)
Processing temperature	K	Glass fiber-reinforced		(2)
		Preformed, chopped roving, (compression)	350–430	
		Premix, chopped glass, (compression)	410–450	
		Woven cloth, (compression)	296–390	

Property	Units	Conditions*	Value	Reference
		Molding, glass fiber-reinforced		
		Compression	405–470	
		Injection	405–470	
		Transfer	405–450	
		EMI shielding (conductive)	405–470	
		Compression	405–460	
		Transfer	410–430	
Molding pressure	MPa	Glass fiber-reinforced		(2)
		Preformed, chopped roving	1.7–14	
		Premix, chopped glass	3.4–14	
		Woven cloth	2.1	
		SMC, glass fiber-reinforced	2.1–14	
		SMC, BMC, glass fiber-reinforced, low-density	3.4–14	
		SMC, glass fiber-reinforced, low pressure	1.7–5.5	
		SMC, glass fiber-reinforced, low shrink	3.4–14	
		BMC, TMC, glass fiber-reinforced	2.8–7.6	
		EMI shielding (conductive)	3.4–14	
Viscosity	Pa s	Brookfield model lvf #3 spindle at 60 rpm		(5)
		Cast, rigid	0.65–0.85	
		Cast, flexible	1.1–1.4	
Specific gravity	–	Cast, rigid	1.04–1.46	(2)
		Cast, flexible	1.01–1.20	
		Glass fiber-reinforced		
		Preformed, chopped roving	1.35–2.3	
		Premix, chopped glass	1.65–2.3	
		Woven cloth	1.5–2.1	
		SMC, glass fiber-reinforced	1.65–2.6	
		SMC, glass fiber-reinforced, low density	1.0–1.5	
		BMC, TMC, glass fiber-reinforced	1.72–2.1	
		EMI shielding (conductive)	1.75–1.85	
		PCT PET	1.32–1.41	(11)
		DMC, with filler and glass fiber	1.8–2.1	(13)
		SMC, with filler and glass fiber	1.7–2.1	(13)
Water absorption	%	1/8 in. thick sample, 24 h		
		Cast, rigid	0.15–0.6	(2)
		Cast, flexible	0.5–2.5	(2)
		SMC, glass fiber-reinforced	0.1–0.5	(3, 4)
		BMC, glass fiber-reinforced	0.20	(3, 4)
		Pultruded, glass fiber-reinforced	0.75	(3, 4)
		Spraying/lay-up, glass fiber-reinforced	1.30	(3, 4)
		Woven roving, lay-up, glass fiber-reinforced	0.50	(3, 4)

Property	Units	Conditions*	Value	Reference
Surface resistivity	Ohm	DMC, with filler and glass fiber	10^{10}–10^{12}	(13)
		SMC, with filler and glass fiber	10^{12}–10^{13}	(13)
Volume resistivity	ohm cm	Unspecified	10^{14}	(5)
		Glass fiber-reinforced SMC, compression	5.7×10^{14}	(3, 4)
		BMC, compression	27×10^{14}	(3, 4)
		Pultruded	10^{13}	(3, 4)
		Woven roving, lay-up	10^{14}	(3, 4)
		DMC, with filler and glass fiber	10^{11}–10^{13}	(13)
		SMC, with filler and glass fiber	$10^{12.5}$–10^{13}	(13)
Dielectric constant	–	At 1 MHz, cast	2.8–3.0	(1)
		At 1 MHz, molding	3.2–4.5	(1)
		Glass fiber-reinforced	5	(6)
Dissipation factor	–	At 1 MHz	0.02	(5)
Dielectric strength	MV/m	DMC, with filler and glass fiber	8–10	(13)
		SMC, with filler and glass fiber	8–10	(13)
Coefficient of thermal expansion	$°C^{-1}$	General range	0.000017–0.00015	(11)
		DMC, with filler and glass fiber	0.00002	(13)
		SMC, with filler and glass fiber	0.00002	(13)
Permittivity	–	At 1 MHz		
		DMC, with filler and glass fiber	4.0–5.5	(13)
		SMC, with filler and glass fiber	4.0–5.5	(13)
Insulation Resistance	Ohm	DMC, with filler and glass fiber	$>10^{13}$	(13)
		SMC, with filler and glass fiber	$>10^{13}$	(13)
Thermal conductivity	$W\,m^{-1}\,K^{-1}$	Glass fiber-reinforced		(3, 4)
		BMC, compression or injection	8.37	
		Pultruded	6.92	
		Spraying/lay-up	2.60	
Specific heat	$J\,kg^{-1}\,K^{-1}$	Glass fiber-reinforced		(3, 4)
		SMC or BMC	1.26	
		Pultruded	1.17	
		Spraying/lay-up	1.30	
Deflection temperature	K	Under flexural load, 0.45 MPa PCT PET	337–486	(11)
		Under flexural load, 1.80 MPa PCT PET	331–504	(11)
		Cast, rigid	330–480	(2)
		Blend (flexible:rigid = 30:70)	324	(5)
		Blend (flexible:rigid = 20:80)	331	(5)
		Blend (flexible:rigid = 10:90)	336	(5)
		Blend (flexible:rigid = 5:95)	358	(5)
		Glass fiber-reinforced	430–560	(2)
		EMI shielding (conductive)	470–480+	(2)

Property	Units	Conditions*	Value	Reference
Maximum resistance to continuous heat	K	Glass-reinforced	430	(6)
Arc resistance	s	SMC or BMC, glass fiber-reinforced	188–190	(3, 4)
		Pultruded, glass fiber-reinforced	80	
Flash point	K	Cast, rigid or flexible, Seta closed cup	305	(5)

* SMC = sheet molding compounds; BMC = bulk molding compounds; TMC = thick molding compounds; EMI = electromagnetic interference.

Resistance to chemicals[6]

Conditions	Satisfactory Resistance To:	Questionable Resistance To:
Glass-reinforced, 298 K	Nonoxidizing acids	Oxidizing acids
	Aqueous salt solutions	Aqueous alkalies
	Polar organic solvents	Nonpolar solvents
	Water	

Radiation resistance, half-value dose in air*

Conditions	Determined By:	Dose Rate (Gy h^{-1})	Value (M Gy)	Reference
Filled with glass fiber	Flexural strength	$\geq 10^5$	10–50	(7, 8)
Filled with mineral flour and glass fiber	Flexural strength	$\geq 10^5$	>30	(7, 9)
Filled with mineral flour and glass fiber	Impact strength	$\geq 10^5$	>10	(7, 9)
Filled with mineral flour and glass fiber (50% mineral flour)	Flexural strength	10	>1	(7, 9)
Filled with mineral flour and glass fiber (15% glass fiber)	Impact strength	10	>0.5	(7, 9)
Filled with mineral flour (82% quartz sand)	Flexural strength	500	>7	(7, 10)

* Defined as the absorbed dose that reduces the mechanical property in the second column to 50% of the initial value.

Property	Units	Conditions*	Value	Reference
Tensile strength at break	MPa	General range	27.6–344.7	(12)
		PCT PET	23.0–250	(11)
		DMC, with filler and glass fiber	30–70	(13)
		SMC, with filler and glass fiber	50–130	(13)
		Cast, rigid	4.1–90	(2)
		Cast, flexible	3.4–21	(2)
		Glass fiber-reinforced		
		Preformed, chopped roving	100–210	(2)
		Premix, chopped glass	21–69	(2)

Property	Units	Conditions*	Value	Reference
		Woven cloth	210–340	(2)
		Pultruded	207	(3, 4)
		SMC, glass fiber-reinforced	28–170	(2)
		BMC, TMC, glass fiber-reinforced	21–90	(2)
		EMI shielding (conductive)	28–55	(2)
Elongation at break	%	Cast, rigid	<2.6	(2, 5)
		Cast, flexible	40–310	(2)
		Blend (flexible: rigid = 30:70)	10	(5)
		Blend (flexible: rigid = 20:80)	4.8	(5)
		Blend (flexible: rigid = 10:90)	1.7	(5)
		Blend (flexible: rigid = 5:95)	1.3	(5)
		Glass fiber-reinforced	0–5	(2)
		PCT PET	2.0–150	(11)
Compressive strength	MPa	General range	100–350	(12)
(rupture or yield)		Cast, rigid	90–210	(2)
		Glass fiber-reinforced		
		Preformed, chopped roving	100–210	(2)
		Premix, chopped glass	140–210	(2)
		Woven cloth	170–340	(2)
		Pultruded	207	(3, 4)
		Glass fiber-reinforced	172	(6)
		Molding, glass fiber-reinforced	97–210	(2)
		EMI shielding (conductive)	120–170	(2)
Flexural strength	MPa	General range	70–550	(12)
(rupture or yield)		DMC, with filler and glass fiber	80–120	(13)
		SMC, with filler and glass fiber	125–225	(13)
		Cast, rigid	60–160	(2)
		Glass fiber-reinforced		
		Preformed, chopped roving	70–280	
		Premix, chopped glass	50–140	
		Woven cloth	280–550	
		Molding, glass fiber-reinforced	62–250	
		EMI shielding (conductive), SMC, TMC	120–140	
		EMI shielding (conductive), BMC	83	
Tensile modulus	MPa	General range	5,500–31,000	(12)
		DMC, with filler and glass fiber	8,000–10,000	(13)
		SMC, with filler and glass fiber	8,000–12,000	(13)
		Cast, rigid	2,100–4,400	(2)
		Glass fiber-reinforced	5,500–31,000	
Flexural modulus	MPa	Cast, rigid, 296 K	3,400–4,200	(2)
		Glass fiber-reinforced		
		Preformed, chopped roving, 296 K	7,000–21,000	(2)
		Premix, chopped glass, 296 K	7,000–14,000	(2)
		Woven cloth, 296 K	7,000–21,000	(2)

Property	Units	Conditions*	Value	Reference
		Woven cloth, 366 K	4,600	(2)
		Woven cloth, 394 K	3,000	(2)
		Woven cloth, 422 K	1,900	(2)
		Pultruded	11,000	(3, 4)
		SMC, 296 K	7,000–15,000	(2)
		SMC, low pressure, 296 K	7,000–150,000	(2)
		BMC, TMC, 296 K	10,000–12,000	(2)
		EMI shielding (conductive), 296 K	9,700–10,000	(2)
		PCT PET	1,560–3280	(11)
		DMC, with filler and glass fiber	9,000–15,000	(13)
		SMC, with filler and glass fiber	9,000–15,000	(13)
Impact strength, Izod	J m^{-1}	Cast, rigid	11–21	(2)
		Cast, flexible	>370	(2)
		Glass fiber-reinforced	80–1,600	(2)
		EMI shielding (conductive)	270–640	(2)
		PCT PET	14.0–80.4	(11)
Impact strength, Charpy	KJ m^{-2}	DMC, with filler and glass fiber	8–14	(13)
		SMC, with filler and glass fiber	40–50	(13)
Hardness	Rockwell	General range	M70–M120	(12)
	Barcol	General range	60–80	(12)
	Rockwell	PCT PET	88–120	(11)
	Rockwell	Glass-reinforced	M50	(6)
	Barcol	Glass fiber-reinforced	40–80	(2)
	Barcol	EMI shielding (conductive)	45–50	(2)
	Barcol	Cast, rigid	35–75	(2)
	Shore	Cast, flexible	D84–94	(2)
	Barcol	Blend (flexible:rigid = 30:70)	0–5	(5)
	Barcol	Blend (flexible:rigid = 20:80)	20–25	(5)
	Barcol	Blend (flexible:rigid = 10:90)	30–35	(5)
	Barcol	Blend (flexible:rigid = 5:95)	35–40	(5)

* SMC = sheet molding compounds; BMC = bulk molding compounds; TMC = thick molding compounds; DMC = dough molding compounds; EMI = electromagnetic interference.

References

1. *Plastics Digest, Thermoplastics and Thermosets.* 15th ed. Vol 1. Englewood: D.A.T.A. Business Publishing; 1994.
2. Kaplan WA, et al. eds. *Modern Plastics Encyclopedia '97.* New York: McGraw-Hill; Modern Plastics. Mid-November 1996.
3. Rosato D. In: Mark HF, et al. ed. *Encyclopedia of Polymer Science and Engineering.* Vol 14. New York: John Wiley and Sons; 1988:350–391.
4. *Fiberglas Plus Design: A Comparison of Materials and Processes for Fiber Glass Composites.* Owens-Corning Fiberglas Corp; July 1985.
5. Harper CA, ed. *Handbook of Plastics, Elastomer, and Composites.* 3rd ed. New York: McGraw-Hill; 1996.

6. Seymour RB. *Polymers for Engineering Applications.* Washington, D.C: ASM International; 1987.

7. Wündrich K. In: Brandrup J, Immergut EH, ed. *Polymer Handbook.* 3rd ed. New York: Wiley-Interscience; 1989:VI 463–474.

8. Schönbacher H, Stolarz-Izycka A. *CERN* 79-08 (1979).

9. Wilski H. *Europäisches Treffen der chemischen Technik (Achema).* Frankfurt; June 18 1970.

10. Rauhut K, Rösinger S, Wilski H. *Kunststoffe.* 1980;70:89.

11. ©1986–2007 IDES Inc. www.ides.com.

12. National Research Council Canada. CBD-159_Thermosetting Plastics-NRC-IRC. www.nrc-cnrc.qc.ca.

13. Hepworth Building Products. www.hepworthcomposites.co.uk.

Poly(ether ether ketone)

J. R. Fried

Acronym, Trade Name PEEK, Victrex® (ICI)

Class Polyketones

Structure

Major Applications General-purpose molding and extrusion polymer for high-performance applications, especially as resin for carbon fiber composites. Examples include chemical resistant tubing and electrical insulation, automotive bearings, pump and valve construction for corrosive applications, and compressor valve plates.

Properties of Special Interest Good abrasion resistance; low flammability and emission of smoke and toxic gases; low water absorption; resistance to hydrolysis, wear, solvents, radiation, and high-temperature steam; ease of processing and excellent thermal stability and mechanical properties at high temperatures.

Type of Polymerization Nucleophilic displacement of activated aromatic halides in polar solvents by alkali metal phenates or Friedel–Crafts processes; examples include polycondensation of the potassium salt of hydroquinone and 4,4′-difluorobenzophenone in DMSO at temperatures up to 340°C and the polycondensation of 4,4′-difluorobenzophenone and silylated hydroquinone in the range of 220–320°C.

Property	Units	Conditions	Value	Reference
Molecular weight (of repeat unit)	$g\,mol^{-1}$	–	288.31	–
Solvents		Very low or no solubility in ordinary solvents; concentrated sulfuric acid will dissolve and sulfonate PEEK; at high temperatures, dilute solutions can be obtained in hydrofluoric acid, trifluoromethanesulfonic acid, dichlorotetrafluoroacetone monohydrate, phenol-1,2,4-trichlorobenzene, and benzophenone		
Polymers with which compatible		Poly(ether ketone), poly(ether ether ether ketone), poly(ether ether ketone ketone), polyetherimide		
Characteristic ratio $\langle r^2 \rangle_0 / nl^2$	–	–	3.04	(1)
Compressibility coeffcient, isothermal	bar^{-1}	At T_m At 340°C	9.302×10^{-5} 9.4×10^{-5}	(2)

Property	Units	Conditions	Value	Reference
		At 360°C	10.2×10^{-5}	
		At 380°C	11.0×10^{-5}	
Continuous service temperature	K	–	473	(3)
Crystallinity	%	Typical	30–35	(4)
		Maximum	48	
Density	g cm^{-3}	Amorphous	1.263–1.265	(5–7)
		Crystalline	1.400–1.401	
Entropy of fusion	kJ K^{-1} mol^{-1}	PVT data	0.0758	(2)
		DSC data	0.0951	(5)
Maximum extensibility $(L/L_0)_r$	%	Annealed	42	(8)
		As molded	103	(8)
		ASTM D 638	150	(3)
Flexural modulus	MPa	At 23°C	3700	(9)
		At 100°C	3600	
		At 200°C	500	
		At 300°C	300	
Glass transition temperature	K	PVT data	425	(2)
		Quenched (DSC)	410	(8)
		Annealed (DSC)	415	(8)
Hardness	R scale	Rockwell	126	(9)
Heat capacity	kJ K^{-1} mol^{-1}	Amorphous, 350 K	0.367	(10)
		Amorphous, 400 K	0.415	
		Amorphous, 610 K	0.600	
		Amorphous, 660 K	0.623	
		32% crystalline, 350 K	0.366	
		32% crystalline, 400 K	0.425	
		32% crystalline, 450 K	0.484	
		32% crystalline, 500 K	0.529	
		32% crystalline, 550 K	0.559	
Heat deflection temperature	K	At 1.81MPa (D648)	433	(3)
Heat of fusion	kJ K^{-1} mol^{-1}	–	36.8	(2)
			37.5	(5)
Impact strength	J m^{-1}	Unnotched Izod	No break	(8)
		Notched Izod (D256)	84	
Index of refraction, n	–	–	1.671	(11)
Maximum use temperature	K	1 h exposure	673	(9)
Melt viscosity	Pa s	At 380°C and 1000 s^{-1}	100–300	(9)
Melting temperature	K	DSC	608–616	(5,6,8)
		Equilibrium	657–668	(5,12,13)
Persistence length	Å	97.4% H$_2$SO$_4$, 30°C	54	(14)

Property	Units	Conditions	Value	Reference
Plateau modulus G_N^0	MPa	At 623 K (calculated)	4.0	(15)
Reducing temperature T^*	K	Flory equation of state (0–500 bar)	9272	(16)
Reducing pressure P	MPa	Flory equation of state (0–500 bar)	726.6	(16)
Reducing volume ν^*	$cm^3\ g^{-1}$	Flory equation of state (0–500 bar)	0.6842	(16)
Solubility parameter	$(MPa)^{1/2}$	Calculated	21.2–22.6	(17)
Sub-T_g transition	K	1 Hz (DMTA)	205	(18)
		1 kHz (dielectric)	239	
Tensile modulus	GPa	D638	3.56	(8)
Tensile strength	MPa	At 23°C	92.0	(9)
		At 100°C	50	
		At 200°C	12.0	
		At 300°C	10.0	
Thermal conductivity	$W\ m^{-1}\ K^{-1}$	C177	0.25	(9)
Thermal expansion coefficient	K^{-1}	$30°C < T < 150°C$ at $P = 0$	1.610×10^{-4}	(2)
		Melt	6.690×10^{-4}	
Volume resistivity	$W\ cm^{-1}$	23°C	4.9×10^{16}	(9)
Water absorption	%	24 h at 40% RH	0.15	
WLF parameters: C_1 and C_2	$C_1 =$ None $C_2 =$ K	$T_0 = 412.9$ K	$C_1 = 29.96$ $C_2 = 53.74$	(19)
Yield stress	MPa	D638	91	(3)

Avrami parameters for isothermal crystallization

T_c (K)	n^*	$k(s^{-3})$	Reference
427.6	2.98	8.9×10^{-11}	(20)
429.6	2.81	7.1×10^{-10}	(20)
432.6	3.07	1.2×10^{-8}	(20)
643	3.4	2.6×10^{-3}	(21)
663	3.6	6.7×10^{-5}	(21)
683	3.8	2.9×10^{-5}	(21)

* At half-life for crystallization.

Property	Units	Conditions	Value				Reference
Gas evolution, G value (10^{-4}) of component gas	–	γ-irradiation (under vacuum)	H_2	CO	CO_2	CH_4	(10)
		Amorphous; quenched (7.4 MGy dose)	12	6.5	12	0.20	

Property	Units	Conditions	Value				Reference
		Crystalline (8.1 MGy dose)	6.3	12	5.5	0.14	
		Electron beam					(22)
		Amorphous (6 MGy dose)	12	5.2	16	0.22	
		Crystalline (5.8 MGy dose)	7.5	3.4	11.3	0.16	
Infrared spectrum (principal absorptions)	cm^{-1}	Assignment	Wavenumber				(23)
		In-place vibration of aromatic hydrogens	1160				
		Asymmetric stretch of diphenyl ether groups	1227 and 1190				
		Skeletal in-phase phenyl ring vibration	1599 and 1492				
		Carbonyl stretching	1655				
Permeability P	m^3 (STP) m s^{-1} m^{-2} Pa^{-1}	O_2, 7.8% crystallinity	6.2×10^{-16}				(24)
		CO_2, amorphous	6.0×10^{-18}				(25)
		CO_2, 26–30% crystallinity	2.4×10^{-18}				(25)
Lattice	–	–	Orthorhombic				(4,26)
Monomers per unit cell	–	–	2/3				(4,26)
Unit cell dimensions	Å	–	$a = 7.75–7.88$				(4,26)
			$b = 5.86–5.94$				(4,26)
			c (chain axis) $= 9.88–10.07$				
Important patent			J. Rose and P. Staniland (assigned to ICI Americas, Inc.) U.S. 4,320,224, 16 Mar. 1982.				

References

1. Roovers J, Cooney JD, Toporowski PM. *Macromolecules.* 1990;23:1611.
2. Zoller P, Kehl TA, Starkweather HW, Jones GA. *J. Polym. Sci.: Part B: Polym. Phys.* 1989;27:993.
3. Attwood TE, et al. *Polymer.* 1981;22:1096.
4. Nguyen HX, Ishida H. *Polym. Compos.* 1987;8:57.
5. Blundell DJ, Osborn BN. *Polymer.* 1983;24:953.
6. Dawson PC, Blundell DJ. *Polymer.* 1980;21:577.
7. Lu SX, Cebe P, Capel M. *Polymer.* 1996;37:2999.
8. Harris JE, Robeson LM. *J. Appl. Polym. Sci.* 1988;35:1977.
9. May R. In: Kroschwitz JI, ed. *Encyclopedia of Polymer Science and Engineering.* Vol 12. New York: John Wiley and Sons; 1990:313–320.
10. Hegazy E-SA, Sasuga T, Nishii M, Seguchi T. *Polymer.* 1992;33:2897.
11. Voice AM, Bower DI, Ward IM. *Polymer.* 1993;34:1154.

12. Hay JN, Kemmish DJ. *Plast. Rubber Process. Applic.* 1989;11:29.

13. Lee Y, Porter RS. *Macromolecules.* 1987;20:1336.

14. Bishop MT, Karasz FE, Russo PS, Langley KH. *Macromolecules,* 1985;18:86.

15. Fetter LJ, Lohse DJ, Colby RH. In: Mark JE, ed. *Physical Properties of Polymers Handbook.* Woodbury, N.Y: AIP Press; 1996:335–340.

16. Rodgers PA. *J. Appl. Polym. Sci.* 1993;48:1061.

17. Bicerano J. *Prediction of Polymer Properties.* 2nd ed. New York: Marcel Dekker; 1996:130.

18. Goodwin A, Marsh R. *Macromol. Rapid Comm.* 1996;17:475.

19. David L, Sekkat A, Etienne S. *J. Non-Cryst. Solids.* 1994;172–174:214.

20. Kemmish DJ, Hay JN. *Polymer.* 1985;26:905.

21. Lee Y, Porter RS. *Macromolecules.* 1988;21:2770.

22. Hegazy ESA, Sasuga T, Nishii M, Seguchi T. *Polymer.* 1992;33:2904.

23. Nguyen HX, Ishida H. *Polymer.* 1986;27:1400.

24. Orchard GA J, Ward IM. *Polymer.* 1992;33:4207.

25. De Candia F, Vittoria V. *J. Appl. Polym. Sci.* 1994;51:2103.

26. Dawson PC, Blundell DJ. *Polymer.* 1980;21:307.

Poly(ether imide)

Loon-Seng Tan

Acronyms, Trade Names PEI, Ultem®, PBNPI

Class Engineering thermoplastics

Structure

Synthesis Aromatic polyetherimides are usually prepared from: (a) bisphenoxide salts and aromatic dinitrobisimides via nucleophilic nitro-displacement reactions[1,2,3]; (b) bisphenoxide salt and a bis(N-chlorophthalimido) compound via nucleophilic chloro-displacement reactions in a solvent of low polarity (e.g. o-dichlorobenzene), and in the presence of a thermally stable, phase transfer catalyst stable at the polymerization temperatures, e.g. hexaalkylguanidinium chloride;[4] microwave irradiation accelerated polymerization process;[5] (c) two-step polycondensation of aromatic diamines and ether-dianhydrides in a polar aprotic solvent, followed by thermal[6] or chemical[7,8] cyclodehydration of the polyamic acid precursors, (d) one-step, high temperature solution polymerization of aromatic diamines and ether-dianhydrides in a phenolic solvent, removing water of condensation azeotropically[9]. Certain polyetherimides can also be synthesized via direct melt polymerization.[10]

Major Applications Printed circuit boards and hard disks for computers, under-the-hood automotive uses, reinforced composites for aerospace applications, flame-retardant mattress barrier, hot-gas filter, and machined parts for reusable medical devices, analytical instrumentation, electrical/electronic insulators (including many semiconductor process components).

Properties of Special Interest Commercial polyetherimide (Ultem®) is an amorphous thermoplastic with the following characteristics: high heat resistance, strength, and modulus; inherent flame resistance with low smoke evolution; high dielectric strength, stable dielectric constant and dissipation factor over a wide range of temperature and frequencies; transparency; amenable to conventional molding processes (injection, compression, or blow-molding).

Poly(ether imide)

Glass-transition and melting (T_m) temperatures (in °C, as determined by DSC;) of 4,4′-isomeric polyetherimides

Unless otherwise indicated, all the values are from references 2 and 9.

Ar$_1$ (bisphenolate)	Ar$_2$ (bisphthalimide)				
	242	–	255	–	199 330 (T_m)
	209	–	224	–	188
	229	–	247	–	205 343 (T_m)
	215	–	227	–	184
	212	226[11] 426 (T_m)	209 192[12]	257[11] 482 (T_m)	178
	260	–	265	–	219
	210	–	239	–	194
	223	–	215 217[13]	–	–

Poly(ether imide)

Glass-transition temperatures (in °C, as determined by DSC) of 3,3′-isomeric polyetherimides

All the values are from references 2 and 9

Ar$_1$ (bisphenolate)	Ar$_2$ (bisphthalimide)				
					$-(CH_2)_6-$
	263	259	–	214	–
	226	241	–	193	–
	277	275	–	224	–
	239	232	–	198	128
	234	231	–	202	–
	267	266	–	230	–
	–	248	–	216	
	235	236	230	–	135

Poly(ether imide)

Glass-transition and melting (T_m) temperatures (°C) of naphthalene-based polyetherimides determined by DSC

All the values are from reference 14

Ar (amine)	Naph (naphthalene bis anhydride)			
	1,5-naphthalene	2,3-naphthalene	2,6-naphthalene	2,7-naphthalene
m-phenylene	260	255	230 340 (T_m)	254
4,4′-oxydiphenylene	240	235	249	245
3,4′-oxydiphenylene (methyl)	235	226	225	221
bisphenol A diphenylene	265	–	236	247
tetramethyl terphenylene	253	–	246	246
1,3-bis(phenoxy)phenylene	–	229	–	228
1,2-bis(phenoxy)phenylene	–	208	–	Not found
bis(phenoxy)biphenylene	–	227	230	231

Ar (amine)	Naph (naphthalene bis anhydride)			
H₃C–/CH₂/–CH₃ structure (tetramethyl diphenylmethane)	281	265	–	272
tetramethyl biphenyl structure	Not found	308	–	Not found
3,5-dimethyl CF₃ benzene structure	250	250	–	235
bis(phenyl) C(CF₃)₂ structure	276	256	218	256

Product Names*	Product Description
Ultem 1000 Series	Unreinforced grade polyetherimide resins;

CAS Registry Number: 61128-24-3

Product Names*	Product Description
Ultem 2000 series	Glass reinforced resins (10–40% glass fillers)
Ultem 3000 series	Glass- and mineral-fiber reinforced polyetherimide resins for a balance of low warpage, dimensional stability, and low CTE
Ultem 4000 series	Polyetherimide containing internal lubricants
Ultem CRS5000 series	Copolyetherimide with improved chemical resistance
Ultem 7000 series	Carbon reinforced polyetherimide resins
Ultem 8000 and 9000 series aircraft resins	Polyetherimide resins to meet FAR 25.853 regulations for commercial aircraft interiors
Ultem LTX Series Resins	A PEI/polycarbonate blend with higher impact resistance
Ultem Healthcare Resins (HU, HAT, and MD series)	Ultem 1000 resins melt filtered to 40 microns

* Supplier: GE Plastics, Plastics Technology Center, One Plastics Ave, Pittsfield, MA 01201, USA. (Web site: http://www. geplastics.com or http://www.geplastics.com/gep/Plastics/en/ProductsAndServices/ProductLine/ultem.html

Mechanical properties of Ultem 1000

Property	Units	Conditions	Value	Reference
Tensile strength	MPa	ASTM D 638 yield, Type I, 3.2 mm	105	15
Tensile elongation	%	ASTM D 638 yield, Type I, 3.2 mm	7.0	15
Tensile elongation, ultimate	%	ASTM D 638 break, Type I, 3.2 mm	60	15
Tensile modulus	GPa	ASTM D 638 yield, Type I, 3.2 mm	3.0	15
Flexural strength	MPa	ASTM D 790 yield, 3.2 mm	150	15
Flexural modulus	GPa	ASTM D 790 yield, 3.2 mm	3.3	15
Compressive strength	MPa	ASTM D 695	150	15
Compressive modulus	GPa	ASTM D 695	3.3	15
Shear strength	MPa	ASTM D 732	100	15
Izod impact unnotched	$J\,m^{-1}$	ASTM D 256 3.2 mm, 23°C	1300	15
Izod impact notched	$J\,m^{-1}$	ASTM D 256 3.2 mm, 23°C	50	15
Gardner impact	J	ASTM D 3029, 23°C	37	15
Shear strength, ultimate	MPa		90–103	15
Rockwell hardness	M scale	ASTM D 785	109M	15
Taber abrasion	$mg\,(1000\ cycles)^{-1}$	ASTM D 1044 CS-17, 1 kg	10	15
Poisson's ratio	–	ASTM D638	0.36	15

Thermal properties of Ultem 1000

Property	Units	Conditions	Value	Reference
Vicat softening point, method B	°C	ASTM D1525 Rate B	219	15
Glass transition temperature	°C	DSC	225[16] 215[17] 210[18] 217.7[19] 218[20]	16–20
Heat deflection temperature	°C	ASTM D648 unannealed @ 0.45 MPa, 6.4 mm	210	15
	°C	ASTM D648 unannealed @ 1.8 MPa, 6.4 mm	200	15
Thermal conductivity	$W\,m^{-1}\,°C^{-1}$	ASTM C177	0.22	15
Coefficient of thermal expansion	$10^{-5}\,°C^{-1}$	ASTM E831 flow X E-5, from −20°C to 150°C	5.6 5.95	15 21
Storage modulus	MPa	Dynamic mechanical analysis (DMA)	179.3 (100°C) 116.6 (200°C)	21
$\tan \delta_{max}$	°C	Dynamic mechanical analysis (DMA)	225.8	21
Loss modulus (E'')	°C	Dynamic mechanical analysis (DMA)	223.3	21
Continuous service temp. index	°C	UL 756B	170	15
Heat release capacity	$J\,g^{-1}\,°C^{-1}$	Pyrolysis-combustion flow calorimetry	121	22

Poly(ether imide)

Glass transition and secondary-relaxation temperatures and associated activation energy values of poly(ether imide)(Ultem), adapted from reference 23

Conditions	T_g (K)	E_a (kJ mol^{-1})	T_β (K)	E_a (kJ mol^{-1})	T_γ (K)	E_a (kJ mol^{-1})	Ref.
Torsion pendulum; $\sim$ 1 Hz	485	–	343	–	168	–	24
Forced oscillation dynamic-mechanical analysis; 1 Hz	492	–	355	–	160	–	25
Forced oscillation dynamic-mechanical analysis; 1 Hz	501	330–1250	–	–	–	–	26
Forced oscillation dynamic-mechanical analysis; 35 Hz	–	–	379	–	186	–	27
Dielectric measurement; 1000 Hz	513	–	–	–	–	–	26
Dielectric measurement	–	–	–	–	–	43	28

Flammability of Ultem 1000

Property	Units	Conditions	Value	Reference
UL94V-O flame class rating	mm	UL 94	0.41	15
UL94-5VA flame class rating	mm	UL 94	1.9	15
Oxygen index (LOI)	%	ASTM D2863	47	15
NBS smoke density flaming mode	–	ASTM E662 Flaming, Ds @ 4 min	0.7	15
	–	ASTM E662 Flaming, Dmax @ 20 min	30	15
Heat of combustion	kJ g^{-1}	Oxygen consumption calorimetry[29]	28.17	30

Electrical properties of Ultem 1000

Property	Units	Conditions	Value	Reference
Volume resistivity	ohm-m	ASTM D2571 1.6 mm	1.0×10^{15}	15
Dielectric strength	kV mm^{-1}	ASTM D149 1.6 mm in air	33	15
	kV mm^{-1}	ASTM D149 1.6 mm in oil	28	15
	kV mm^{-1}	ASTM D149 3.2 mm in oil	20	15
Dielectric constant	–	ASTM D150 @100 Hz	3.15	15
	–	ASTM D150 @1kHz	3.15	15
Dissipation factor	–	ASTM D150 @100 Hz	0.0015	15
	–	ASTM D150 @1kHz	0.0012	15
	–	ASTM D150 @2450 MHz	0.0025	15
Arc resistance	s		128	15

Optical and spectroscopic properties of Ultem film

Property	Units	Conditions	Value	Reference
Refractive index	–	Measured with a prism	1.6525	31
In-plane refractive index	–	coupler at 632.8 nm;	1.6539	31
Out-of-plane refractive index	–	resolution +0.0005	1.6539	31
Birefringence (Δn)	–		0.0041	31

Poly(ether imide)

Property	Units	Conditions	Value	Reference
Binding energy	eV	X-ray photoelectron spectroscopy	285 (N1s) 400 (O1s) 533 (C1s)	32
Characteristic IR peaks	cm^{-1}	FTIR-attenuated total reflectance (ATR)	3068–3038 (sp^{2}C—H), 2968 (sp^{3}C—H) 1777 ν_{asym}(CO) & 1720 $_{sym}$(CO);mide I-band 1356 (C=N), imide II band) 1234 (sp^{2}C—O—sp^{2}C)	33
Characteristic Raman peaks	cm^{-1}	FT-Raman	$\sim$ 3080 (C—H, phenyl) $\sim$ 1780 (C=O)sym $\sim$1620 C=C, phenyl $\sim$ 1380 (C—N) $\sim$ 725 (imide ring)	34

Surface properties of Ultem^{TM} film

Property	Units	Conditions	Value	Reference
Contact angle with water (θ_{H_2O})	deg (°)	Measured with a goniometer (sessile drop method); average of 10 measurements	90	35
Contact angle with water (θ_{H_2O})	deg (°)	Measured with a goniometer	93.0	36
Contact angle with diiodomethane ($\theta_{CH_2I_2}$)	deg (°)	(sessile drop method); average of 6 measurements	27.0	36
Surface energy (dispersive component), γ_s^d	mJ m^{-2}	'Owen and Wendt' method[37]	46.7	36
Surface energy (polar component), γ_s^p	mJ m^{-2}		0.1	36
Surface energy γ_s	mJ m^{-2}		46.8	36
Contact angle with water (θ_{H_2O})	deg (°) θ_a = advancing contact angle θ_r = receding contact angle	Measured with a contact angle meter 14-inch horizontal beam comparator	θ_a = 86.1 θ_r = 69.3	38 38
Contact angle with ethylene glycol (θ_{EG})			θ_a = 50.2 θ_r = 38.1	
Surface energy (dispersive component),γ_s^d	mJ m^{-2}	'Harmonic mean expression' method[39]	21.6	38
Surface energy (polar component), γ_s^p	mJ m^{-2}		14.3	38
Surface energy, γ_s	mJ m^{-2}		36.0	38

Poly(ether imide)

Property	Units	Conditions	Value	Reference
Contact angle with water (θ_{H_2O})	deg (°)	Sessile drop method using the robust-shape-comparison technique[40]	84.2 ± 4.7	41
Contact angle with glycerol (θ_{gly})	deg (°)	Measured with a goniometer (sessile drop method); average of 6 measurements	69.8 ± 1.4	41
Contact angle with formamide (θ_{FAM})	deg (°)		56.1 ± 2.1	41
Contact angle with diiodomethane ($\theta_{CH_2I_2}$)	deg (°)		24.2 ± 1.2	41
Contact angle with ethylene glycol (θ_{EG})	deg (°)		45.0 ± 2.7	41
Contact angle with DMSO (θ_{DMSO})	deg (°)		13.6 ± 2.9	41
Contact angle with tricresyl phosphate (θ_{TP})	deg (°)		12.0 ± 1.9	41
Contact angle with water (θ_{H_2O})	deg (°)	Ultem 60—80 μm granules; measured with a goniometer (sessile drop method); average of 10 measurements	93 ± 1	42
Contact angle with diiodomethane ($\theta_{CH_2I_2}$)	deg (°)		27 ± 1	42
Surface energy γ_s	mJ m^{-2}	Ultem 60–80 μm granules; 'Owen and Wendt' method[37]	46.8 ± 1	42
Contact angle with water (θ_{H_2O})	deg (°)	Ultem film 75 μm thick (as-received); measured with a goniometer (sessile drop method); average of 6 measurements	75 ± 1	42
Contact angle with diiodomethane ($\theta_{CH_2I_2}$)	deg (°)		32 ± 1	42
Surface energy γ_s	mJ m^{-2}	Ultem film 75 μm thick (as-received); 'Owen and Wendt' method[37]	45.3 ± 1	42

Physical properties of Ultem 1000

Property	Units	Conditions	Value	Reference
Specific gravity	–	ASTM D792	1.27[17,25] 1.28[43]	17,25,43
Mold shrinkage	(m/m)	ASTM D955 flow, 3.2 mm	0.007	15
Water absorption	%	ASTM D570 @24 h, 23°C	0.25	15
	%	ASTM D570 equilibrium, 23°C	1.25	
O_2 permeability	barrer*	30°C and 1 bar pressure	0.50	44
N_2 permeability	barrer*	30°C and 1 bar pressure	0.07	44
CO_2 permeability	barrer*	30°C and 1 bar pressure	1.55	44
CO_2 permeability	barrer*	35°C and 10 atm pressure	1.33	45

Property	Units	Conditions	Value	Reference
Permselectivity for CO_2 and CH_4 (α)	–	35°C and 10 atm pressure	36.9	45
Permselectivity for H_2O and ethanol (α)	–	Extruded film (75 μm); vapor mixture of 4.4 wt% H_2O/95.6 wt% EtOH at total flow rate of 2–5 g/m^2 h and 40°C	850	42

* 1 Barrer = 10^{-10} cm^3 (STP) cm/(s cm^2 cmHg), where the standard temperature and pressure (STP) are 273.15 K and 1 atm (1.1013×10^{-5} Pa), respectively.

Diffusion coefficients (D; cm^2 s^{-1}) from 5 to 60°C and activation energy (E_a; kcal/mol) for water-uptake of Ultem® 1000 film (thickness 10–13 μm)

D					E_a	Reference
5°C	15°C	25°C	40°C	60°C		
5.1	7.8	12.8	22.3	26.0	5.69	31

References

1. Wirth JG, Heath DR. *U. S. Patent 3,730,940* (1973).
2. Takekoshi T, Wirth JG, Heath DR, Kochanowski JE, Manello JS, Webber MJ. *J. Polym. Sci., Polym. Chem. Ed*. 1980;18:3069.
3. White DM, Takekoshi T, Williams FJ, et al. *J. Polym. Sci., Polym. Chem. Ed*. 1981;19:1635.
4. (a) Brunelle DJ, Acar HY, Khouri FF, Guggenheim TL, Woodruff DW, Johnson NE. (General Electric Co., USA), *U.S. Pat. Appl. Publ.*, Cont.-in-part of U.S. Ser. No. 647,889. *Chem.Abst.* 2006;145:189365.
   (b) Brunelle DJ. (General Electric Co., USA), *U. S. Pat. 5,082,968.* , *Chem. Abst.* 1992;116:173582.
5. Gao C, Zhang S, Gao L, Ding M, *J. Appl. Polym. Sci..* 2004;92:2415.
6. Sroog CE, Endrey AL, Abramo SV, Berr CE, Edwards WM, Olivier KL. *J. Polym. Sci. Part A*. 1965;3:1373.
7. Vinogradova SV, Vygodski YS, Vorobiev VD, Shurakhina NA, Chudina LI, Spirina TN, Korshak VV. *Polym. Sci. USSR*. 1974;16:584.
8. Eastmond GC, Paprtny J, Webster I. *Polymer*. 1993;34:2865.
9. Takekoshi T, Kochanowski JE, Manello JS, Webber MJ. *J. Polym. Sci., Polym. Symp*. 1986;74:93.
10. (a) Takekoshi T, Kochanowski JE. (General Electric Co., USA), *U. S. Patent. 3,803,085*, *Chem. Abstr*. 1974;82:112468;
    (b) Nick RJ, Nelson ME, Caringi JJ, Williams DE. (General Electric Co., USA), *U. S. Patent* 6,066,74. *Chem. Abstr*. 2004;132:335031.
11. St. Clair TL, St. Clair AK. *J. Polym. Sci., Polm. Chem. Ed*. 1976;15:1529.
12. Hergenrother PM, Wakelyn NT, Havens SJ. *J. Polym. Sci. Part A, Polym. Chem*. 1987;25:1093.
13. Wang DH, Shen Z, Cheng SZD, Harris FW. Polymer. 2007;48:2572.
14. Eastmond GC, Paprotny J. *J. Mater. Chem*. 1996;6:1459.
15. ULTEM 1000 Technical Data Sheet, General Electric Company, http://www.ge.com
16. Giannotti MI, Mondragon I, Galante MJ, Oyanguren PA. *Polym. Int..* 2005;54:897.

17. Punsalan D, Koros WJ, *Polymer*. 2005;46:10214.

18. Peng M, Li H, Wu L, Chen Y, Zheng Q, Gu W. *Polymer*. 2005;46:7612.

19. Chiefari J, Dao B, Groth AM, Hodgkin JH. In: Mittel KL, ed. *Polyimides and Other High Temperature Polymers*. Netherlands: VSP International Science Publisher, the Netherlands; 2005;Part 1:3–13.

20. Sanner MA, May A. *Annu. Tech. Conf. - Soc. Plast. Eng.*. 2005;63:1810–1814.

21. Vora RH, Pallathadka PK, Goh SH, Chung T-S, Lim YX, Bang TK. *Macromol. Mater. Eng*. 2003;288:337.

22. Lyons RE, Walters RN. *J. Aanl. Appl. Pyrolysis*. 2004;71: 27.

23. Fried JR. 'Sub-T$_g$ Transitions,' In: Mark JE, ed. *Physical Properties of Polymers Handbook*. Woodbury, New York: American Institute of Physics; 1996;Chap. 13:166–167.

24. Harris JE, Robeson LM. *J. Appl. Polym. Sci*. 1988;35:1877.

25. Fried JR, Liu H-C, Zhang C. *J. Polym. Sci.: Part C, Polym. Lett*. 1989;27:385.

26. Biddlestone F, Goodwin AA, Hay JN, Mouledous GAC. *Polymer*. 1991;32:3119.

27. Pegoraro M, Landro LD. *Plast. Rubber Compos. Process. Appl*. 1992;17:269.

28. Schartel B, Wendorff JH. *Polymer*. 1995;36:899.

29. Lyon RE. In: Grand AF, Wilkie CA, eds. *Fire Retardancy of Polymeric Materials*. New York: CRC Press; 2000;Chapter 11:391–447.

30. Walters RN, Hackett SM, Lyon RE. *Fire Mater*. 2000;24:245.

31. Seo J, Lee C, Jang W, Sundar S, Han H. *J. Appl. Polym. Sci*. 2006;99:1692.

32. Kaba M, Essamri A, Mas A, et al. *J. Appl. Polym. Sci*. 2006;100:3579.

33. Albrecht W, Seifert B, Weigel T, et al. *Macromol. Chem. Phys*. 2003;204:510.

34. Devasahayam S, Hill DJT, Connell JW. *J. Appl. Polym. Sci*. 2006;101:1575.

35. Shen L-Q, Xu Z-K, Yang Q, Sun H-L, Wang S-Y, Xu Y-Y. *J. Appl. Polym. Sci*. 2004; 92:1709.

36. Kaba M, Romero RE, Essamri A, Mas A. *J. Fluorine Chem*. 2005;126:1476.

37. Owens DK, Wendt RC. *J. Appl. Sci*. 1969;13:1741.

38. Khayet M. *Appl. Surf. Sci*. 2004;238:269.

39. Wu S. *Polymer Interface and Adhesion*. New York: Marcel Dekker; 1982.

40. Song B. *Doctoral Thesis*. TU-Berlin, D83, 1994.

41. Bismarck A, Kumru ME, Springer J. *J. Colloid Interface Sci*. 1999;217:377.

42. Kaba M, Raklaoui N, Guimon MF, Mas A. *J. Appl. Polym. Sci.*. 2005;97:2088.

43. Liu T, Ozisik R, Siegel RW. *J. Polym. Sci., Part B: Polym. Phys*. 2006;44:3546.

44. López-González MM, Compañ V, Saiz E, Rainde E, Guzmán J. *J. Mmbrane Sci*. 2005;253:175.

45. Chern RT, Koros WJ, Yui B, Hopfenberg HB, Stannett VT. *J. Polym. Sci. Polym. Phys. Edn*. 1984;32:69.

Poly(ether ketone)

J. R. Fried

Acronym, Trade Name PEK, Kadel®

Class Polyketones

Structure

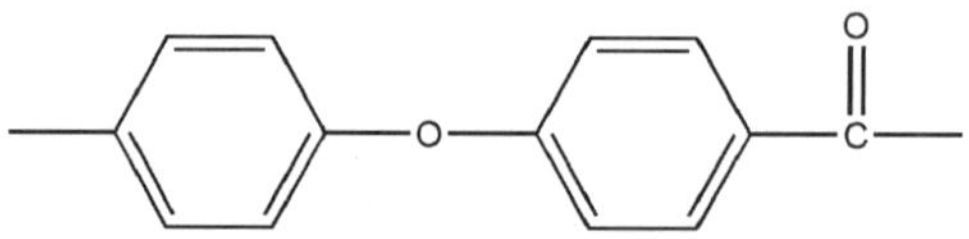

Type of Polymerization Nucleophilic displacement of activated aromatic halides in polar solvents by alkali metal phenates or Friedel–Crafts processes.

Property	Units	Conditions	Value	Reference
Molecular weight (of repeat unit)	g mol^{-1}	–	196.21	–
Density	g cm^{-3}	Amorphous	1.272	(1)
		Crystalline	1.430	(1, 2)
Maximum extensibility $(L/L_0)_r$	%	–	68	(3)
Impact strength	J m^{-1}	Notched Izod (D256)	59	(3)
Glass transition temperature	K	DSC	425	(1, 3)
			427	(2)
Melt flow	dg min^{-1}	At 400°C	1.5	(1)
Melting temperature	K	DSC	634–640	(1–3)
Tensile impact strength	kJ m^{-2}	–	168	(3)
Tensile modulus	GPa	D638	3.19	(3)
Tensile strength	MPa	D638	104.0	(3)
Unit cell dimensions	Å	–	$a = 7.63$	(1, 2)
			$b = 5.96$	
			c (chain axis) $= 10.0$	

References

1. Harris JE, Robeson LM. *J. Polym. Sci.: Part B: Polym. Phys.* 1987;25:311.
2. Dawson PC, Blundell DJ. *Polymer.* 198021:577.
3. Harris JE, Robeson LM. *J. Appl. Polym. Sci.* 1988;35:1977.

Poly(ether sulfone)

Sizhu Wu and Tarek M. Madkour

Acronym, Trade Names PES, Victrex 100P and 200P (ICI), Ultrason E6020P

Class Poly(ether sulfones)

Synthesis Polycondensation

Structure

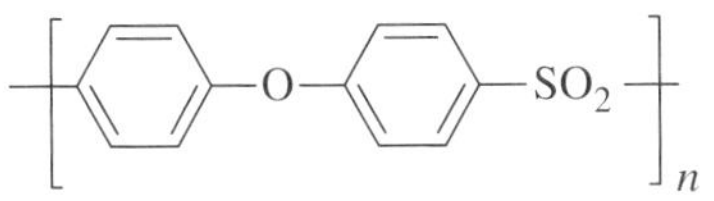

Major Applications Medical and household appliances that are sterilizable by hot air and steam such as corrosion-resistant piping. Also used in electric and electronic applications such as television components. Used as membranes for reverse gas streams and gas separation.

Properties of Special Interest High performance thermoplastic of relatively low flammability. Amorphous, high-creep resistance, and stable electrical properties over wide temperature and frequency ranges. Transparent with good thermal and hydrolytic resistance.

Property	Units	Conditions	Value	Reference
Molecular weight (of repeat unit)	g mol^{-1}		232.25	
Infrared bands (frequency)	cm^{-1}	Group assignments		(1, 2)
		SO$_2$ scissors deformation	560	
		SO$_2$ symmetric stretch	1151; 1175	
		SO$_2$ asymmetric stretch	1294; 1325	
		Aryl—O—aryl C—O stretch	1244	
		Aromatic CH stretches	3000–3200	
Thermal expansion coefficient	K^{-1}	Victrex 200P	5.5×10^5	(3)
		Victrex 430P (30% glass fiber)	2.3	
Density	g cm^{-3}	Victrex 200P	1.37	(3)
		Victrex 430P (30% glass fiber)	1.60	
Molar volume	cm^3 mol^{-1}	25°C	157	(4)
Solubility parameter δ	(MPa)$^{1/2}$	Calculated, 25°C	23.12	(4)
		Victrex 4800	22.9	(5)

Property	Units	Conditions	Value	Reference
Theta temperature	K	DMF/methanol (83/17)	298	(6)
		DMF/toluene (39/61)	303	
Glass transition temperature	K	Forced oscillation dynamic-mechanical analysis	498	(7)
Sub-T_g transition temperature	K	y-relaxation temperature	193	(7)
Heat capacity	kJ K^{-1}mol^{-1}		0.174	(8)
Heat deflection temperature	K	(1.82 MPa)	507	(9)
		30% glass fiber reinforced	476	
		30% carbon fiber reinforced	507	
Activation energy	kJmol^{-1}		290.7	(10)
Excess enthalpy loss	cal g^{-1}	Aging temperature at 160°	0.67	(11)
Specific surface area	m^{2}g^{-1}	Pure PES	28.13 ± 0.70	(12)
		30% activated carbon hybrid beads	89.63 ± 2.00	
Porosity	%	Pure ES	89.98 ± 2.00	(12)
		30% activated carbon hybrid beads	87.13 ± 2.00	
Pore volume	cm^3 g^{-1}	Pure PES	5.47 ± 0.10	(12)
		30% activated carbon hybrid beads	3.80 ± 0.10	
Conductivity	S cm^{-1}	50% HPS	0.08	(13)

Mechanical properties[3,9,14]

Property	Units	Resin		
		Neat Resin	30% Glass Fiber Reinforced	30% Carbon Fiber Reinforced
Tensile modulus	MPa	2413		
Tensile strength	MPa	82.8	146.9	211.7
Maximum extensibility $(L/L_0)_r$	%	40–80	2.57	1.11
Flexural modulus	MPa	2552	6987	13 973
Flexural strength	MPa	128	210	264
Notched Izod impact strength	Jm^{-1}	85.7	296	88
Unnotched Izod impact strength	J		1082	521
Hardness	Shore D	88	86	89

Property	Units	Conditions	Value	Reference
WLF parameters: C_1 and C_2			70.98 241.2	(15)
Refractive index n		20°C	1.545	(16)
Dielectric constant			3.5	(17)
Resistivity	ohm cm		1×10^{17}	(3)

Poly(ether sulfone)

Property	Units	Conditions	Value	Reference
Speed of sound, longitudinal	m		2260	(18)
Permeability coefficients	m^3 (STP) ms^{-1} m^{-2} Pa^{-1}	50°C and pressure difference of 10 bar		(19)
		Gas		
		He	7.95×10^{-17}	
		CO_2	3.15×10^{-17}	
		O_2	6.0×10^{-18}	
Thermal conductivity k	$Wm^{-1}K^{-1}$	–	0.18	(17)
Maximum use temperature	K	–	491	(17)
Water absorption	%	24 h	0.43	(3)

Dual-mode parameters[20]

Gas	Sorption parameters			Diffusion	
	k_D [m^3 (STP) m^{-3} atm^{-1}]	C_H [m^3 (STP) m^{-3}]	b (atm^{-1})	$D_D \times 10^{12}$ (m^2s^{-1})	$D_H \times 10^{12}$ (m^2s^{-1})
CO_2	0.807	16.310	0.398	2.792	0.441
C_2H_6	0.496	10.844	0.289	–	–
CH_4	0.240	6.445	0.109	0.151	0.128

References

1. Colthup N, Daly L, Wiberley S. *Introduction to Infrared and Raman Spectroscopy.* 2nd ed. New York: Academic press; 1975.
2. Pouchert C. *The Aldrich Library of FT-IR Spectra.* Milwaukee: Aldrich Chemical; 1985.
3. Elias H, Vohwinkel F. *New Commercial Polymers 2.* New York: Gordon and Breach Science Publishers; 1986, Chapter 8.
4. Bucknall C, Partridge I. *Polym. Eng. Sci.* 1986;26:54.
5. Wang D, Li K, Teo W. *J. Membr. Sci.* 1996;115:85.
6. Park Y, Lee D. *Polymer (Korea).* 1988;12:749.
7. Aitken C, McHattie J, Paul D. *Macromolecules.* 1992;10:2910.
8. In: Mark JE, ed. *Physical Properties of Polymers Handbook.* Woodbury, N.Y: AIP Press; 1996.
9. Ma C. In: *Proc. of the Natl. SAMPE Symp. Exhib.*, 30 (Adv. Technol. Mater. Processes), 1985:543.
10. Zhou,XM, Jiang ZH. *J. Appl. Polym. Sci.* 2006;102:530–534.
11. Shi Ru Jong, Tzyy Lung Yu. *Macromol. Chem. Phys.* 1999;200:87–94.
12. Deng XP, Wang T, Zhao F, Li LJ, Zhao CSJ. *Appl. Polym. Sci.* 2007;103:1085–1092.
13. Krishnan NN, Kim H-J, Prasanna M, et al. *Power Sources.* 2006;158:1246–1250.
14. Hisue E, Miller R. In: *Proc. of the Natl. SAMPE Symp. Exhib.*, 30 (Adv. Technol. Mater. Processes). 1985:1035.

15. David L, Sekkat A, Etienne S. *J. Non-Cryst. Solids.* 1994;214:172.

16. In: Brandrup J, Immergut EH, eds. *Polymer Handbook.* 3rd ed. New York: John Wiley and Sons; 1989.

17. In: Mark H, et al, eds. *Kirk-Othmer: Encyclopedia of Chemical Technology.* 3rd ed. New York: WileyInterscience; 1984.

18. Phillips D, North A, Pethrick R. *J. Appl. Polym. Sci.* 1977;21:1859.

19. Wang D, Li K, Teo W. *J. Membrane Sci.* 1995;105:89.

20. Reimers M, Barbari T. *J. Polym. Sci. Polym. Phys.* 1994;32:131.

Poly(ethyl acrylate)

Jianye Wen

Acronyms, Trade Names PEA

CAS Number 9003-32-1

Class Vinyl polymers

Structure

$$[-CH_2 - CH-]$$
$$|$$
$$COOC_2H_5$$

Major Applications Coatings, textile finishing, paper saturants, leather finishing, oil-resistant and high temperature resistant elastomers

Properties of Special Interest A rubber-like, considerably softer, and more extensible polymer compared to poly(methyl acrylate); have superior resistance to degradation and show remarkable retention of their original properties under use conditions

Property	Units	Conditions	Value	Reference
^{13}C-NMR spectra	ppm	Selected group assignments		2
		Carbonyl C=O	174.31	
		Methane		
		$CH_2 - CHCOOCH_2CH_3$	41.23	
		Methylene		
		$CH_2 - CHCOOCH_2CH_3$	34.79	
		$CH_3 - CH_2OOCCH_2CH_3$	60.33	
		Methyl, CH_3	13.99	
^{1}H-NMR spectra	ppm	Selected group assignments		3
		main-chain Methylene		
		mmm	1.62, 2.21	
		mrm	1.80	
		mmr	1.56, 2.11	
		mrr	1.78	
		rrr	1.75	
		rmr	1.48, 2.10	
		main-chain Methine	2.58	
		OCH$_2$ proton	~4	
		Methyl, CH_3	0.8–1.0	

Property	Units	Conditions	Value	Reference
Cohesive energy E_{coh}	J/mol		36 619	4
Density	g/cm^3	25°C	1.12	5–7
FTIR spectra	cm^{-1}	Selected group assignments		8
		C=O	1733	
		C—CH$_3$ bending	1382, 1447	
		C—O—C stretch	1257, 1159	
		Aliphatic C—H stretches	2982	
Glass transition temp.	°C	Inherent viscosity 0.05 g polymer in 100 ml toluene		
		0.25	−42	9
		0.6	−43	9
		0.9	−35	9
		5.0	−27	9
		7.0	−28	9
		Conventional	−24	9–13
		Syndiotactic	−24	14,15
		Isotactic	−25	16,17
Heat capacity	J/g.K	−183°C	0.5792	18
		−73°C	1.0301	18
		27°C	1.7867	18
		227°C	2.2189	18
		ΔC_p	45.60/100.12	18
Interaction parameter χ, with;		butane, 70–90°C	1.318 to 1.232	19
		hexane, 70–110°C	1.483 to 1.296	19
		heptane, 70–110°C	1.585 to 1.345	19
		decane, 70–110°C	1.926 to 1.645	19
		cyclohexane, 70–110°C	1.148 to 0.974	19
		benzene, 70–110°C	0.183 to 0.188	19
		toluene, 70–100°C	0.289 to 0.301	19
		chloroform, 70–110°C	−0.478 to −0.322	19
		carbon tetrachloride, 70–110°C	0.384 to 0.365	19
		acetone, 70–110°C	0.507 to 0.411	19
		methyl ethyl ketone, 70–110°C	0.400 to 0.218	19
		tetrahydrofuran, 70–100°C	0.215 to 0.191	19
		dioxane, 70–100°C	0.239 to 0.255	19
		methyl acetate, 70–110°C	0.402 to 0.394	19
		ethyl acetate, 70–110°C	0.365 to 0.363	19

Mark–Houwink coefficients

Solvent	Temp. (°C)	Mol. Wt. range ($M \times 10^4$)	$K \times 10^3$ (ml/g)	a	Refs.
Acetone	25	−450	51	0.59	20
	30	−50	20.0	0.66	21

Poly(ethyl acrylate)

Property	Units	Conditions		Value	Reference

Solvent	Temp. (°C)	Mol. Wt. range ($M \times 10^4$)	$K \times 10^3$ (ml/g)	a	Refs.
Benzene	30	−67	27.7	0.67	22
Butanone	30	−700	2.68	0.80	23
Chloroform	30	−54	31.4	0.68	22
Ethyl acetate	30	−54	26.0	0.66	22

Property	Units	Conditions	Value	Reference
Mechanical properties				
Tensile strength	MPa	−	0.2	24, 25
Elongation at break	%	−	1800	24, 25
Refractive index, n_D^{25}	−	−	1.464	1(b)
			1.4685	26
			1.3975	27
2nd virial coefficient	$A_2 \times 10^4$ $(mol.cm^3/g^2)$	acetone 20°C $M \times 10^3 = 55$–860	5.0–3.1	28
		28°C $M \times 10^3 = 320$–8000	10.52	29
		30°C $M \times 10^3 = 145$–191	14.6	30
Solubility parameter	$(J/cm^3)^{1/2}$	−	19.3	31
			20.5	32
			20.6	
Solvents		Aromatic hydrocarbons, chlorinated hydrocarbons, tetrahydrofuran, esters, ketones, methanol, butanol, glycol ether		33
Nonsolvents		Aliphatic hydrocarbons, hydrogenated naphthalenes, diethyl ether, aliphatic alcohols (C > 5), cyclohexanol, tetrahydrofurfuryl alcohol		33
Surface tension	mN/m	$M_w = 28\,000$		34
		at 20°C	37.0	
		at 150°C	27.0	
		at 200°C	23.2	
	mN/m.K	$-d\gamma/dT$	0.070	
	χ^P	polarity	0.174	
Thermal conductivity k	W/mK	at 310.9 K	0.213	35
		at 422.1 K	0.230	35
		at 533.2 K	0.213	35
Theta solvent	−	n-butanol	44.9°C	36
		ethanol	37.4°C	
		methanol	20.5°C	
		n-propanol	39.5°C	

Property	Units	Conditions	Value	Reference

Unperturbed dimension[a]

Conditions	$r_0/M^{1/2} \times 10^4$ (nm)	$r_{of}/M^{1/2} \times 10^4$ (nm)	$\sigma = r_0/r_{of}$	$C_\infty = r_0^2/nl^2$	Refs.
Acetone; methanol 30°C	720 ± 30	308	2.34 ± 0.10	10.9	37
Acetone, 25°C	856	308	2.78	15.4	38
Undiluted, 60°C	$dlnr_0^2/dT = -0.2 \times 10^{-3}$ $[\mathrm{deg}^{-1}]$	–	–	–	39

[a] See reference 1 for details.

References

1. General references:
   (a) In: Brandrup J, Immergut EH, eds. *Polymer Handbook*, 3rd ed. New York: Wiley-Interscience; 1989;
   (b) Kine BB, Novak RW. In: Bikales NM, ed. *Encyclopedia of Polymer Science & Technology*. Vol 1. New York: Wiley-Interscience; 1987:234;
   (c) In: Mark JE, ed. *Physical Properties of Polymers Handbook*. Woodbury, New York: AIP Press; 1996.
2. Mathakiya I, Rakshit AK. *International J. Polym. Anal. Charact.* 2003;8:339.
3. Suchopárek M, Spěváček J, Masař B. *Polymer.* 1994;35(16):3389.
4. Yu X, Wang X, Li X, Gao J, Wang H, *J. Polym. Sci. Part B.* 2006;44:409.
5. Van Krevelen DW, *Properties of Polymers*, Amsterdam: Elsevier Publishing Co; 1976.
6. Shetter JL. *J. Polym. Sci., Part B.* 1963;1:209.
7. Kine BB, Novak RW. In: Bikales NM, ed. *Encyclopedia of Polymer Science and Technology.* Vol 1. New York: Wiley-Interscience; 1989:257.
8. The Aldrich library of FT-IR spectra , ed. C. J. Charles, 1 Milwaukee, Wis. Aldrich Chemical Co, V2, P1186,1985.
9. Wiley RH, Braver GM. *J. Polym. Sci.* 1948;3:647.
10. Van Krevelen DW. *Properties of Polymers.* Amsterdam: Elsevier Publishing Co; 1976.
11. Crawford JWC. *J. Soc. Chem. Ind. London.* 1949;68:201.
12. Reding FP, Faucher JA, Whitman RD. *J. Polym. Sci.* 1962;5:483.
13. Hughes LJ, Brown GL. *J. Appl. Polym. Sci.* 1961;5:580.
14. Rehberg CE, Fisher CH. *Ind. Eng. Chem.* 1948;40:1429.
15. Shetter JA. *J. Polym. Sci.,B.* 1963;1:209.
16. Mikhailov GP, Shevelev VA. *Polym. Sci. USSR.* 1967;9:2762.
17. Lawler J, Chalmers DC, Timar J, *ACS Div. Rubber Chem.* Spring Meeting, May 1967: paper 42. (10)
18. Gaur U, Lau SF, Wunderlich B, et al. *J. Phys. Chem. Ref. Data.* 1982;11:1065.
19. Tian M, Munk P. *J. Chem. Eng. Data.* 1994; 39:742.
20. Giurgea M, Ghita C, Baltog I, Lupu A. *J. Polym. Sci.* 1966;A2, 4:529.
21. Sumitimo H, Hachihama Y. *Kobunshi Kagaku (Chem. High. Polym. (Tokyo)).* 1953;10:544.

22. Sumitimo H, Hachihama Y, *Kobunshi Kagaku (Chem. High. Polym. (Tokyo))*. 1955;12:479.

23. Mangaraj D, Patra SK. *Makromol. Chem.* 1967;107:230.

24. Brendley WH Jr. *Paint Varn. Prod.* 1973;63:19.

25. Craemer AS. *Kunststoffe.* 1940;30:337.

26. Holder AJ. *QSAR Comb. Sci..* 2006;25(10):905.

27. Garner AY. *J. Org. Chem..* 1959;24:532.

28. Wunderlich W. *Angew. Makromol. Chem.* 1970;11:189.

29. Hansen JE, McCarthy MG, Dietz TJ. *J. Polym. Sci.* 1951;7:77.

30. Hachihama Y, Sumitomo H. *Tech. Rept. Osaka Univ.* 1953;3:385.

31. Gardon JL. In: Bikales NM, ed. *Encyclopedia of Polymer Science and Technology.* Vol 3. New York: Wiley-Interscience; 1965:833.

32. Yu X, Wang X, Li X, Gao J. *QSAR Comb. Sci.* 2006;25:156.

33. Fuchs O. In: Brandrup J, Immergut EH, eds. *Polymer Handbook.* 3rd ed. New York: Wiley-Interscience; 1989:VII379.

34. Partington JR. *An Advanced Treatise of Physical Chemistry: Physico-Chemical Optics.* Vol IV. London: Longmans, Green and Co; 1960.

35. In: Touloukian YS, Powell RW, Ho CY, Klemens PG, eds. *Thermal Conductivity, Nonmetallic Solids, Vol. 2 of Thermophysical Properties of Matter.* New York: IFI/Plenum; 1970.

36. Liopis J, Albert A, Usobiaga P, *Eur. Polym. J.* 1967;3:259.

37. Kurata M, Stockmayer WH. *Fortschr. Hochpolymer. Forsch.* 1963;3: 196.

38. Giurgea M, Ghita C, Baltog I, Lupu A. *J. Polym. Sci.* 1966;A2, 4:529.

39. Tobolsky AV, Carlson D, Indictor N. *J. Polym. Sci.* 1961;54:175.

Polyethylene, Elastomeric (very highly branched)

A. Prasad

Acronym, Alternative Names POE, polyolefin elastomer, ultra-low-density ethylene copolymer

Class Poly(α-olefins)

Structure —[CH$_2$—CH$_2$—CHR—CH$_2$]$_n$— (R = α-olefins)

Introduction POE is a new family of ethylene α-olefin copolymers produced using a metallocene catalyst.[1-5] The metallocene catalyst selectively polymerizes the ethylene and comonomer sequence and on increasing the comonomer content, produces chains of higher elasticity due to efficient incorporation of the comonomer that effectively disrupts the polymer crystallinity. The uncross-linked polymers referred to in this chapter are known to have only moderate elastomeric recovery properties (up to 96%). These copolymers are characterized by a narrow molecular weight distribution (MWD) ($M_w/M_n = 2$–2.5) and homogeneous comonomer distribution.[1-24] The control of chain microstructure by the use of metallocene catalyst makes it possible to produce poly(α-olefin) copolymers with considerably lower density, which has not been possible before using the conventional Ziegler–Natta catalyst. Some of the highly branched ethylene copolymers presented in the entry on Polyethylene, metallocene linear low-density, in this handbook may be closely related. Engage® POE, primarily based on butane-1 and octene-1 copolymer, is now commercially available in a wide variety of melt indexes and density ranges (over 18 grades) from Dow Chemical Company produced using Dow's proprietary single-site, constrained geometry catalyst (Insite) technology.[8,11] Engage® POEs are known to have small amounts of long chain branching (LCB) to improve processibilty.[17,25,26] Some butene-1 and hexene-1 copolymers of density lower than 0.880 g cm^{-3} made by the Exxon's metallocene catalyst are also known to exhibit moderate elastomeric properties.[12] Recently Exxon has also introduced commercial grade octane-1 based POE.[18] The moderate elastomeric properties in these ethylene copolymers have been attributed to the high fractional volume of amorphous phase anchored at multiple points to the minor crystalline domain on the same chains akin to a cross-linked system.[6,8-10,12,14,15] Properties listed below are intended to represent best published examples of the most commonly available commercial grades of POEs in the density range of 0.863–0.886 gcm^{-3}.

Major Applications POEs have been used in both plastic and rubber applications. These elastomeric polyethylenes have the ability to be cross-linked via peroxide, irradiation, and moisture (if silane grafted). Applications include blown film, cast film, extrusion coating application, hot-melt adhesives, tubing, impact modifiers, low-voltage cable insulation, elastic films, foams, shoe soles, belts, automotive hoses, medical applications, gasket seals, foams, roof membrane, fiber and other electrical applications.

Properties of Special Interest Modulus/fexibility, elasticity, toughness, processibility, excellent optics and electrical properties, superior heat resistance, and UV stability over cross-linked rubbers such as EPDM and EPM, low brittleness temperature, good chemical resistance to common solvents, and good heat seal.

Limitations Relatively poor wear resistance, exposure to hydrocarbon causes swelling, moderate heat resistance, moderate elastomeric recovery unless cross-linked.

Major Suppliers Dow Chemical Co., DuPont Dow Elastomers, Exxon Mobil Chemicals

Property	Units	Conditions	Value	Reference
Typical comonomer	–	Butane, hexene, octene	–	(10–13)
Degree of branching, commercial resins	mol%	NMR, total short chain branches	4.7–13.5	(8, 9, 25, 26)
NMR peaks	ppm	^{13}C NMR, octene-1	Characteristic δ peaks at 22–24, 24–25, 26.5–28.5, 25.2–33.2, 33.2–36.8, 35.8, 37–39.5, 39.5–40.5, 40.5–41.5	(27)
IR (characteristic absorption frequencies)	cm^{-1}	D2238	See entry on LLDPE in this handbook	(27)
Ratios of different unsaturated groups	–	FTIR Thickness, $t = 0.22$ mm	See table below	(27)

Ratios of different unsaturated groups in octene-1 LLDPE polymers

Mol %	Density g cm^{-3}	MI g (10 min^{-1})	Vinyl CH$_2$=CH— (908 cm^{-1})	Vinylidene CH$_2$=C< (889 cm^{-1})	trans-vinylene —CH=CH— (965 cm^{-1})
8.9	0.885	1.07	1	8	14
12.4	0.869	0.55	1	13	25

Melt index	g (10 min^{-1})	D 1238	0.5–30.0	(11, 25, 26)
Density range	g cm^{-3}	D 792 (commercial resins)	0.863–0.885	(11, 25, 26)
Mooney viscosity range	–	D 1646, ML 1 + 4 at 394 K		(11)
		Butene-1 copolymer:	7–54	
		Octene-1 copolymer	2–33	
Glass transition Temperature	K	DSC, density range = 0.862–0.886 g cm^{-3}	215–227	(11, 25, 26, 29)
		DSC, density = 0.8717 g cm^{-3} (12 mol% octene-1)	221	(25, 26, 29)
		DMA	224	(8)
		DMA, tanδ peak at 1 Hz, density	244–239	(12)
		Density range = 0.880–0.886 g cm^{-3} (butene-1)		

Polyethylene, Elastomeric (very highly branched)

Property	Units	Conditions	Value	Reference
Melting temperature	K	DSC (broad melting range from 253 to 363 K), peak endotherm values, density range = 0.862–0.886 g cm^{-3} Butene and octane copolymer	307–353	(8, 9, 11–13, 15, 20, 26)
Crystallinity	%	DSC, density range = 0.862–0.886 g cm^{-3}	9–25	(8, 9, 11–13, 15, 25)
Heat of fusion	kJ mol^{-1}	DSC	0.35–1.1	(8,9,12,13,15)
Lamella thickness	Å	SAXS	32–53	(14–16,19)
Avrami exponent	–	DSC and microscopy, octene-1: 46 CH$_3$/1000 C Isothermal crystallization temperature range:		(16,19)
		319–325 K	2.0	
		327–334 K	1.0	
Tensile modulus	MPa	D 1708, quenched		
		octene-1, density = 0.8702 g cm^{-3}	7.0	(8, 25, 26, 28)
		25 mm min^{-1} draw	12.5	(10)
		D 412, 5 mm min^{-1} draw, (octene-1, density range = 0.856–0.886 g cm^{-3})	1.5–12.5	(7)
Yield stress	MPa	D 1708, quenched (octene-1, density = 0.8702 g cm^{-3})	None detected	(8, 25, 26, 28)
1% Secant modulus	MPa	D 790 (butene-1, density = 0.886 g cm^{-3})	32–35	(12)
2% Secant modulus	MPa	D 790, butene-1 density = 0.862–0.886 g cm^{-3}	4–27	(11)
		Octene-1, density = 0.857–0.886 g cm^{-3}	4–33	(11)
		density = 0.8717 g cm^{-3} (12 mol% octene-1)	14.8	(11, 29)
Tensile strength at break	MPa	D 638M, 50 mm min^{-1}	9–30	(10, 11, 25, 26, 29)
Elongation at break	%	D 638M, 50 mm min^{-1}	750– >1000	(10, 11, 25, 26, 29)
Ultimate tensile strength	MPa	ASTM D 638, 508 mm min^{-1}		(11)
		Butene-1, density = 0.862–0:886 g cm^{-3}	2–11	
		Octene-1, density = 0.857–0:886 g cm^{-3}	3–18	
Tensile strain recovery	%	25.4 cm min^{-1}, butene-1		(12, 25, 26, 29)
		28% strain	100	
		70% strain	96	
		143% strain	89	
Permanent tension set	%	4th pull, 100% strain	5–35	(10)
Dynamic compression set	%	D 395B, at 20 Hz, 20% strain, 12 000 cycles at 296 K, density = 0.870–0.886 g cm^{-3}	4.5–5.75	(13)

Property	Units	Conditions	Value	Reference
Hysteresis loss	%	12 000 cycles at 296 K, density Range = 0.870–0.886 g cm^{-3}	8–11	(13)
		Peroxide cross-linked (4 pph) at 393 K	1–4	
Hardness	^{0}Shore	D 2240, Shore A	40–85	(11,29)
Hardness	^{0}Shore	D 2240, Shore D	11–28	(11)
Vicat softening temperature	°C	mol % octane = 8.9 density = 0.882 g cm^{-3}	85	(18)

Some typical characteristics of metallocene-based ethylene–octene elastomer Film (0.1524 mm thick)[30]

Density	MI	Melting point, K	Ultimate tensile strength, MPa	Ultimate elongation (%)	Stress at 200% elongation, MPa
0.855	0.5	306	8.3	1940	1.35
0.863	0.5	327	14.6	831	2.14
0.870	0.5	333	18.1	650	3.30
0.870	5.0	336	17.8	1030	2.14
0.885	1.0	354	34.7	665	5.86

References

1. Hoewing S, et al. In: Benedikt GM, Goodall BL, eds. *Metallocene-Catalyzed Polymers: Materials, Properties, Processing and Market.* New York: William Andrew Publication, Norwich; 1999:253.
2. Ho T, Mortin JM. In: Schiers J, Kaminsky W, eds. *Metallocene-Based Polyolefins.* John Wiley & Sons Ltd; 2. 2000:175.
3. Swogger KW, et al. *Society of Plastics Engineers Annual Technical conference Proceedings (SPE ANTEC), 64th,* 2006:1008.
4. Karande SV, et. al. *Society of Plastics Engineers Annual Technical conference Proceedings (SPE ANTEC), 64th,* 2006:1005.
5. Harrington BA, et al. *ACS Rubber Division Technical paper, 164th,* Oct. 14–17, 2003:853.
6. Hwang YC, et al. *Society of Plastics Engineers Annual Technical Conference Proceedings (SPE ANTEC), 52nd,* 1994:3414.
7. Sehanobish K, et al. *J. Appl. Polym. Sci.* 1994;51:887.
8. Bensason S, et al. *J. Polym. Sci., Polym. Phys. Ed.* 1996;34:1301.
9. Minick J, et al. *J. Appl. Polym. Sci.* 1995;58:1371.
10. Chum PS, Kao CI, Knight GW. *Plast. Eng.* (June 1995):21.
11. (a) Data supplied courtesy of Dr. D. Parekh, Dow Chemical Company, Freeport, Texas. (b) Technical Information – Product Selection Guide Literature from Dow Chemicals.
12. Woo L, Westphal SP, Ling MTK. *Society of Plastics Engineers Annual Technical Conference Proceedings (SPE ANTEC), 51st,* 1993:358.
13. Minick J, Sehanobish K. *Society of Plastics Engineers Annual Technical Conference Proceedings (SPE ANTEC), 54th,* 1996:1883.

14. Phillips PJ, Monar K. *Society of Plastics Engineers Annual Technical Conference Proceedings (SPE ANTEC), 54th*, 1996:1624.
15. Phillips PJ, Monar K. *Society of Plastics Engineers Annual Technical Conference Proceedings (SPE ANTEC), 55th*, 1997:1506.
16. Phillips PJ, Kim M, Monar K. *Society of Plastics Engineers Annual Technical Conference Proceedings (SPE ANTEC), 53rd*, 1995:1481.
17. Lai S, Knight GW. *Society of Plastics Engineers Annual Technical Conference Proceedings (SPE ANTEC), 51st*, 1993, p. 1188.
18. Dharmarajan N, et al. *Society of Plastics Engineers Annual Technical Conference Proceedings (SPE ANTEC)*, 59th 2001;3:1.
19. Phillips P, et al. In: Benedikt GM, Goodall BL, eds. *Metallocene-Catalyzed Polymers: Materials, Properties, Processing and Market.* New York: William Andrew Publication, Norwich; 1999:127141.
20. Phillips P, et al. In: Benedikt GM, Goodall BL, eds. *Metallocene-Catalyzed Polymers: Materials, Properties, Processing and Market.* New York: William Andrew Publication, Norwich; 1999:169.
21. Zhang M, et al., *Polym. Eng. Sci.* 2003;87:1878.
22. Razavi-Nouri, Mohammad, et al. *Iranian Polymer J.* 2007;16:105.
23. Young, Mu-Jen, et al. *J. Polym. Eng.* 2002;22:75.
24. Shanks RA, et al. *J. Thermal Analysis & Calorimetry.* 2000;59:471.
25. Bensason S, et al. *Macromolecules.* 1997;30:2436.
26. (a) Bensason S, et al. *Polymer.* 1997;38:3913.
    (b) Bensason S, et al. *Society of Plastics Engineers Annual Technical Conference Proceedings (SPE ANTEC), 54th*, 1996:1982.
27. Al-Malaika S, et al. *Polymer Degradation and Stability.* 2006;91:3131.
28. Haward RN. *Polymer.* 1999;40 9:5821.
29. Dibbern JA, Laughner MK, Silvis HC. *SPE Proceedings from the X International Conference on Polyolefins,* Texas: Houston; 1997:185.
30. Eckersley ST, et al. *J. Appl. Polym. Sci.* 2001;80:2545.

Poly(ethylene imine)

Sizhu Wu and Tarek M. Madkour

Acronym PEI

Class Polyamines

Structure $-[CH_2-CH_2-NH]_n$

Major Applications PEI offers potential cosmetic uses and new directions for clear antidandruff hair products and antiperspirants. Also used as a wet-strength agent in the paper-making process, a flocculating agent with silica sols, and in the coating of composite hollow-fiber membranes. PEI is obtained by cationic ring-opening polymerization of ethyleneimine.

Properties of Special Interest A special highly branched poly(ethylene amine). A cationic surfactant with natural affinity for hair and skin. A chelating agent with the ability to complex with heavy metal salts such as zinc and zirconium salts.

Property	Units	Conditions	Value	Reference
Molecular weight (of repeat unit)	g mol^{-1}	–	43.07	(1)
Degree of branching	%	Primary amine groups	30	(2)
		Secondary amine groups	40	
		Tertiary amine groups	30	
Molecular weight range	g mol^{-1}	Ring-opening polymerization	600–70 000	(1)
Typical polydispersity range (M_w/M_n)	–	Ring-opening polymerization	1.9–56.8	(3)
Heat of polymerization	kJ mol^{-1}	–	−83.7	(4)
Density	g cm^{-3}	Low mol. wt. PEI at 20°C		(1)
		Mol. wt. (g mol^{-1})		
		60.1	0.8994	
		103.1	0.9586	
		146.2	0.9839	
		189.2	0.9994	
Mark–Houwink parameters: K and a	K = ml g^{-1} a = None	25°C, 0.1 M Na	K a	(5,6)
		$1 \times 10^3 < M_w < 2 \times 10^4$	2.32 0.14	
		$2 \times 10^4 < M_w < 3 \times 10^6$	0.075 0.43	
pH	–	Commercial form	11–12	(4)
Water permeability	(m^{2}hkPa)$^{-1}$		3.9	(10)

Property	Units	Conditions	Value	Reference
Characteristic ratio	–	Calculated at 27°C for the isotactic polymer	6.21	(7)
		For the syndiotactic polymer	6.56	
Temperature coeffcient	$K^{-1}(\times 10^3)$	Calculated theoretically for the isotactic polymer	1.93	(7)
		For the syndiotactic polymer	2.85	

Unit cell dimensions[1]

Isomer	Lattice	Monomers per Unit Cell	Cell Dimension (Å)			Cell Angle (Degrees)	Density
			a	b	c	β	
Anhydrate	Ortho	40	29.8	17.2	4.79	Double	1.165
Hemihydrate	Mono	8	10.89	9.52	7.31	127.6	1.152
Sesquihydrate	Mono	8	11.55	9.93	7.36	104.5	1.139
Dihydrate	Mono	4	13.26	4.61	7.36	101.0	1.190

Property	Units	Conditions	Value	Reference
Melting point	K	Mol. wt. (g mol^{-1})		(1)
		60.1	284	
		146.2	285	
		404.7	331	
Boiling point	K	Mol. wt. (g mol^{-1})/mbar		(1)
		60.1/1013	389.5	
		103.1/1013	480.1	
		146.2/1013	550.9	
		189.2/1013	606	
		318.5/11	382–383	
		404.7/1	472–473	
Cationic charge density	meq g^{-1}	–	20	(4)
Refractive index n	–	25°C, mol. wt. = 404 gmol^{-1}	1.5161	(1)
Maximum adsorption on pulp fiber	mg g^{-1}	pH = 6, $M_w = 6 \times 10^5$	0.67	(8)
Adsorption equilibrium constant	g l^{-1}	–	3.5	(8)
Free energy of adhesion	kJ mol^{-1}	–	0.191	(8)
Maximum sorption	%	60 min on virgin hair	1.25	(4)
		Low mol. wt. on damaged hair	1.5	
		High mol. wt. on damaged hair	3.4	
Optimum flocculation dosage	eq l^{-1}	PEI	6.0×10^{-4}	(9)
		PEI : HCl (1:1)	3.0×10^{-4}	
		PEI : HCl (4:1)	4.0×10^{-4}	
Maximum use temperature	K	–	523	(2)
Toxicity (LD50)	g kg^{-1}	–	3	(4)

References

1. In: Brandrup J, Immergut EH, eds. *Polymer Handbook*. 3rd ed. New York: John Wiley and Sons; 1989.
2. In: Mark H, et al, eds. *Kirk-Othmer: Encyclopedia of Chemical Technology* 3rd ed. New York: Wiley-Interscience; 1984.
3. Dermer O, Ham G. *Ethylenimine and other Aziridines*. New York: Academic Press; 1969.
4. Feigenbaum H. *Cosmet. Toiletries*. 1993;108(8):73.
5. Hostetler R, Swanson J. *J. Polym. Sci. Polym. Chem.*. 1974;12:29.
6. Van der Berg J, Bloys von Treslong C, Polderman A. *Recl. Trav. Chim. Pays-Bas*. 1973;92:3.
7. Wang S, DeBolt L, Mark JE. *Polym. Prepr*. 1993;34(2):478.
8. Van de Ven T. *Adv. Colloid Interface Sci*. 1994;48:121.
9. Ishikawa M. *J. Colloid Interface Sci*. 1976;56:596.
10. Trimpert, Christiane, Boese, et al. *Macromol. Biosci*. 2006;6(4):274–284.

Polyethylene, Linear High Density

Rufina G. Alamo and Leo Mandelkern[a]

Acronyms, Trade Names PE, HDPE, LPE

Class Poly alkene

Structure $-CH_2-CH_2$-poly(ethylene) or $-CH_2$-poly(methylene)

Preparative Techniques Type of polymerization: Coordination polymerization

Typical Comonomers 1-alkenes, vinyl acetate, methacrylates, acrylates, methacrylic acid, acrylic acid

Property	Units	Conditions	Value	Reference
Molecular mass weight of repeat unit	$g\,mol^{-1}$	Ethylene	28	–
		Methylene	14	–
Typical Molecular mass range	$g\,mol^{-1}$	Very wide range available	1×10^3–8×10^6	–
Typical polydispersity (M_w/M_n)	–	Very wide range available	1.07–>10	–

IR (Characteristic absorption frequencies)[1]

Frequency (cm^{-1})	Phase	Transition Moment Orientation*	Assignment
720	Crystalline	∥ b-axis	Out-of-phase CH$_2$ rock of the two chains in the unit cell
	Amorphous	⊥ b-axis	CH$_2$ rock (tttt)$_n$ $n > 4$
731	Crystalline	∥ a-axis	In-phase CH$_2$ rock of the two chains in the unit cell
888	Amorphous	∥	CH$_2$ rock
1050	Crystalline	∥	CH$_2$ twist
1078	Amorphous	⊥	Skeletal C–C stretch (g and t conformation)
1176	Crystalline	∥	CH$_2$ wag
1303	Amorphous	∥	CH$_2$ wag (*gtg* conformation)
1353	Amorphous	∥	CH$_2$ wag (*gg* conformation)
1368	Amorphous	∥	CH$_2$ wag (*gtg* conformation)

[a] Deceased

Frequency (cm^{-1})	Phase	Transition Moment Orientation*	Assignment
1463	Crystalline	∥ b-axis	CH$_2$ bend
	Amorphous		CH$_2$ bend
1473	Crystalline	∥ a-axis	CH$_2$ bend
1820	Crystalline	∥	Combination of 1100 or 1130 + 720, 730 (weak)
1894	Crystalline	⊥	Combination of CH$_2$ rock. 1168 + 720, 730 (weak)
2016	Both	∥	Combination of 1294 + 720, 730 (weak)
2150	Both	⊥	Combination of CH$_2$ 1440 + 720, 730 (weak) or 1100 + 1050
2850			CH$_2$ symmetric stretch
2918			CH$_2$ asymmetric stretch

* With respect to uniaxial stretch.

Property	Units	Conditions	Value		Reference
			T_1	τ_c	
NMR(Solution) T_1 relaxation time. τ_c correlation time. ^{13}C	T_1 = s τ_c = ns	1,2,4 Trichlorobenzene, 383 K	2.52	0.019	(2)
		o-Dichlorobenzene, 373 K	2.70	0.018	(3)
		o-Dichlorobenzene, 303 K (extrapolated)	1.24	0.040	(3, 4)
NMR (melt) ^{1}H-Chemical shift-(CH$_2$)$_n$–	Ppm	Reference: TMSi, 220 MHZ	1.3		(5, 6)
NMR (solid, state) ^{13}C chemical shift	Ppm	Melt crystallized Reference: TMSi and solid adamantane 4.7 T	Crystalline component = 32.9		(7)
			Liquid-like component = 31.1		(7)
^{1}H Dipolar-decoupled ^{13}C Chemical shifts	Ppm	Uniaxially oriented films, $M_v = 3.8 \times 10^5$ Reference: TMSi and solid adamantane 50 MHz, 1.4 T	Crystalline component = 11.8		(8)
			Liquid-like component = 32.6		
Spin relaxation times	As indicated	Uniaxially oriented films, $M_v = 3.85 \times 10^5$, Reference: TMSi and solid adamantane 50 MHz, 1.4 T	Crystalline component (T_{1C}) = 1100, 60.5, 5 s Liquid-like component (T_{1C}) = 0.37 s Liquid-like component (T_{2C}) = 370, 32 µs		(8)

Property	Units	Conditions	Value	Reference
[1]H Dipolar-decoupled, MAS pulse, ^{13}C NMR, chemical shifts	Ppm	Melt crystallized reference peak: TMSi and solid adamatane 50 MHz, 4.7 T	Crystalline component, 33 Liquid-like component, 31 Interfacial component, 31.3	(9)
Spin relaxation times	s	Melt crystallized $M_v = 3 \times 10^6$ (unfractionated) $T_c = 403$ K, 4 weeks	Crystalline component $T_{1H} = 1.87$ $T_{1C} = 2560, 263, 1.7$	(9)
	As indicated	Melt crystallized $M_v = 3 \times 10^6$ M (unfractionated) $T_c = 403$ K, 4 weeks	Liquid-like component $T_{1H} = 0.39$ s $T_{1C} = 0.37$ s $T_{2C} = 2.4$ ms	(9)
	As indicated	Melt crystallized $M_v = 3 \times 10^6$ (unfractionated) $T_c = 403$ K, 4 weeks	Interfacial component $T_{1H} = 1.61$ s $T_{1C} = 0.37$ s $T_{2C} = 0.044$ ms	(9)
	s	Melt crystallized. $M_v = 2.48 \times 10^5$ (fraction) $T_c = 402$ K, 23 days	Crystalline component $T_{1H} = 2.20$ s $T_{1C} = 2750, 111, 1.3$ s	(9)
	s		Liquid-like component $T_{1H} = 0.50$ s $T_{1c} = 0.41$ s	(9)
	s		Interfacial component $T_{1H} = 2.04$ s $T = 0.41$ s	(9)
Spin relaxation times	s	Solution crystallized $M_v = 9.1 \times 10^4$ (fraction) $T_c = 358$ K, 0.08 w/v% in toluene	Crystalline component $T_{1H} = 1.90$ s $T_{1C} = 220, 21, 2$ s	(9)
	s		Interfacial component $T_{1H} = 1.9$ s $T_{1C} = 0.46$ s	(9)
Spin–spin relaxation times ([1]H T_2)	µs	Melt crystallized $M_w = 78 \times 10^3$ $M_w/M_n = 5.2$	Crystalline component $T_{2H} = 11$	(10)
	µs		Interfacial (semi-rigid) component $T_{2H} = 40$	(10)
	µs		Liquid-like (soft) component $T_{2H} = 100$	(10)
Thermal expansion coefficient	°C^{-1}	Liquid state (t in °C) 130–207°C 140°C	$\alpha_l = (0.727 \times 10^{-3})$ $-(0.030 \times 10^{-5})t$ $-(0.0120 \times 10^{-7})t^2$ $+(0.0021 \times 10^{-9})t^3$ $\alpha_l = 7.151 \times 10^{-4}$	(11)

Property	Units	Conditions	Value	Reference
	K^{-1}	Crystalline state (semicrystalline) (T in K) 293–383 K (orthorhombic unit cell)	$\alpha_c = (1.734 \times 10^{-3})$ $+ (6.523 \times 10^{-6}\, T)$ $\alpha_c = 3 \times 10^{-4}$	(12)
	K^{-1}	298 K	$\alpha_c = 2.10 \times 10^{-4}$	
Compressibility coefficient	Bar^{-1}	Liquid state 413–473 K (t in °C)	$(0.0894/1767)e^{-4.661 \times 10^{-3}t}$	(12)
	Bar^{-1}	Crystalline state 293–383 K (t in °C)	$0.0894/(4758 - 22.7t)$	(12)
Density	$g\, cm^{-3}$	Liquid state 403–480 K (t in °C)	$\rho_l = 0.8674 - (0.06313 \times 10^{-2})t$ $+ (0.00367 \times 10^{-4})t^2$ $- (0.00055 \times 10^{-6})t^3$	(11)
Reducing variables		(A) from Simha-Somcynsky (B) from Flory-Orwoll-Vrij (C) from Sanchez-Lacombe		(12–14) (11, 15) (16, 17)
Reducing temperatures	K	A: 413–473 K, 0–200 MPa 423–476 K, 0–200 MPa 415–473 K, 0–200 MPa B: 413 K, 0–3.55 MPa C: 408–471 K, 0–100 MPa	9250 10046 9772 7300 649	(12) (12) (12) (11) (17)
Reducing pressure	MPa	A: 413–473 K, 0–200 MPa 423–476 K, 0–200 MPa 415–474 K, 0–200 MPa B: 413 K,0–3.55 MPa C: 408–471 K, 0–100 MPa	897 716 748 460 358	(12) (12) (12) (11) (17)
Reducing volume	$cm^3\, g^{-1}$	A: 413–473 K, 0–200 MPa 423–476 K, 0–200 MPa 415–473 K, 0–200 MPa B: 408–471 K, 0–100 MPa	1.129 1.155 1.142 1.127	(12) (12) (12) (11)
Solvents		Solubility only above 353 K. Hydrocarbons, halogenated hydrocarbons and aromatics, higher aliphatic esters, ketones, di-n-amyl ether		
Nonsolvents		All common solvents below 353 K. Most polar organic solvents even at elevated temperatures, inorganic solvents		
Solubility parameters	$(MPa)^{1/2}$	Calculated Calculated Measured	16.0, 16.8 16.2 17.1	(18) (19) (20)
Theta temperature	K	Solvent Biphenyl Biphenyl Diphenylene oxide Diphenyl ether Dodecanol-1	398 401 ~391 434 437 411 417	(21) (22–25) (21) (24, 26) (25) (27, 24) (23)

Property	Units	Conditions	Value	Reference
Theta temperature	K	Solvent		
		n-Octyl alcohol	453	(22, 25)
			458	(24)
		n-Decyl alcohol	427	(22, 24, 25)
		n-Lauryl alcohol	411	(22, 25)
		p-Tertiary amyl alcohol	472	(22, 25)
		p-Octyl phenol	448	(22, 25)
		p-Nonyl phenol	436	(22, 25)
		2-Ethyl hexyl sebacate	423	(21)
		2-Ethyl hexyl adipate	443	(27)
		3,5,5 trimethyl hexyl acetate	394	(23)
		Anisole	427	(22, 25)
		Benzyl phenyl ether	465	(22, 25)
		Nitrobenzene	>473	(21)
		Di-butyl phthalate	>473	(21)
		n-Pentane	~358	(28)
		n-Hexane	407, 437	(28)
		n-Octane	504, 489	(29)
		Diphenyl methane	415	(22, 25)
Interaction parameter, χ (HDPE)	–	Solvent/Temperature (K)		
		cis-Decahydronapthalene,419	0.08	(30)
		cis-Decahydronapthalene, 426	0.06	(30)
		trans-Decahydronapthalene, 419	0.06	(30)
		trans-Decahydronapthalene, 426	0.05	(30)
		n-Decane, 419	0.32	(30)
		n-Decane, 426	0.31	(30)
		n-Decane, 418–463	0.18	(31)
		n-Decane, 458	0.12	(32)
		2,4 Dimethyl hexane, 419	0.39	(30)
		2,4 Dimethyl hexane, 426	0.36	(30)
		2,5 Dimethyl hexane, 419	0.43	(30)
		2,5 Dimethyl hexane, 426	0.38	(30)
		3,4 Dimethyl hexane, 419	0.32	(30)
		3,4 Dimethyl hexane, 426	0.30	(30)
		n-Dodecane, 419	0.29	(30)
		n-Dodecane, 426	0.28	(30)
		Ethyl benzene, 419	0.37	(30)
		Ethyl benzene, 426	0.37	(30)
		Mesitylene, 419	0.29	(30)
		Mesitylene, 426	0.27	(30)
		3-Methyl hexane, 419	0.42	(30)
		3-Methyl hexane, 426	0.39	(30)
		2-Methyl heptane, 419	0.39	(30)
		2-Methyl heptane, 426	0.39	(30)
		3-Methyl heptane, 419	0.37	(30)

Property	Units	Conditions	Value		Reference
		3-Methyl heptane,426	0.36		(30)
		n-Nonane, 491	0.35		(30)
		n-Nonane, 426	0.33		(30)
		n-Octane, 419	0.37		(30)
		n-Octane, 426	0.35		(30)
		1,2,3,4 Tetrahydronapthalene, 419	0.33		(30)
		1,2,3,4 Tetrahydronapthalene, 426	0.32		(30)
		1,2,3,4 Tetrahydronapthalene, 383	0.32		(33)
		Toluene, 419	0.39		(30)
		Toluene, 426	0.40		(30)
		2,2,4 Trimethyl hexane, 419	0.37		(30)
		2,2,4 Trimethyl hexane, 426	0.33		(30)
		2,2,4 Trimethyl pentane, 419	0.41		(30)
		2,2,4 Trimethyl pentane, 426	0.39		(30)
		p-Xylene, 419	0.32		(30)
		p-Xylene, 426	0.32		(30)
		m-Xylene, 419	0.34		(30)
		m-Xylene, 426	0.34		(30)
Interaction parameter, χ, PE1000 [$M_w = 1260$, $M_w/M_n = 1.11$]	–	Isododecane, 380	0.22 ± 0.05		(34)
		Octamethyl Cyclotetrasiloxane, 400	1.4 ± 0.1		(34)
Second virial coefficient	mol cm^3 g$^{-2} \times 10^4$	Solvent/Temperature (K), $M_w \times 10^{-5}$			
		1-Chloronaphthalene,398, 1.10–21.6	12.0–0.78		(35)
		1-Chloronaphthalene, 398, 1.75	10.0		(36)
		1-Chloronaphthalene, 398, 1.44	8.6		(36)
		1-Chloronaphthalene, 398, 0.5–5.6	12.4–2.7		(37)
		1-Chloronaphthalene, 408, 1.20	4.0		(38)
		1-Chloronaphthalene, 408, 0.14–1.20	15.9–10.3		(31)
		Diphenyl methane, 415, 0.82–0.89	$-0.25 - 0.93$		(38)
		n-Decane, 388, 1.44	5.9		(36)
		1,2,4-Trichlorobenzene, 408, 0.94	20.6		(38)
		1,2,4-Trichlorobenzene, 413, 0.11–0.30	45.2–41.1		(39)
		1,2,3,4 Tetrahydronaphthalene, 378,1.44	21.8		(36)
		1,2,3,4 Tetrahydronaphthalene, 378,1.25–4.65	23.1–15.9		(40)
		1,2,3,4 Tetrahydronaphthalene, 398,0.92–2.19	26.8–1.7		(35)
Mark-Houwink parameters k and a	As indicated	Solvent, Temperature (K), $M_w \times 10^{-4}$	k $\times 10^2$ (ml g^{-1})	a	
		1,2,4 Trichlorobenzene, 408, 0.8–123.0	5.1	0.71	(41)

Property	Units	Conditions	Value		Reference
		Solvent, Temperature (K), $M_w \times 10^{-4}$	$k \times 10^2$ $(ml\ g^{-1})$	a	
		1,2,4 Trichlorobenzene, 408, –	5.2	0.69	(27)
		1,2,4 Trichlorobenzene, 408, 0.6–20.0	5.6	0.70	(42)
		1,2,4 Trichlorobenzene, 408, 0.07–6.9	3.9	0.73	(43)
		Decalin, 408, 0.2–10.0	6.2	0.70	(26, 44)
		Decalin, 408, 0.3–10.0	6.8	0.67	(45)
		Decalin, 408, 0.3–6.4	4.6	0.73	(46)
		Decalin, 408, –	5.3	0.73	(47)
		Decalin, 408, 0.3–11.7	6.2	0.70	(41, 26)
		Diphenyl ether, 434.6, 0.2–10.0	29.5	0.50	(26)
		1-Chloronapthalene, 398, ___	14.0	0.58	(48)
		1-Chloronapthalene, 398, 0.5–5.6	4.3	0.67	(37)
		1-Chloronapthalene, 402, ___	2.7	0.71	(49)
		1-Chloronapthalene, 402, ___	9.1	0.69	(49)
		1-Chloronapthalene, 403, 0.6–20.	5.6	0.68	(43)
		Tetralin, 378, 1.3–5.7	1.6	0.83	(40)
		Tetralin, 393, 0.5–10.0	2.4	0.78	(50)
		Tetralin, 393, 0.03–5.5	3.3	0.77	(51)
		Tetralin, 403, 0.04–5.0	4.4	0.76	(52)
		Tetralin, 403, 0.08–2.0	3.8	0.72	(53)
		p-Xylene, 278, 1.3–5.0	1.7	0.83	(40)
		p-Xylene, 278, 0.1–1.2	1.8	0.83	(54)
		3,5,5 Trimethyl hexyl acetate, 394, 0.1–5.8	–	0.55	(23)
		Dodecanol-1, 401, 0.09–5.8	–	0.61	(23)
		Biphenyl, 401, 0.18–5.8	–	0.60	(23)
Huggins constant k'	–	Solvent, Temperature (K), $M_w \times 10^{-5}$			
		Decalin, 408, 0.1–10.0	0.70		(45)
		1-Chloronaphthalene, 403, 0.07–6.9	0.22–0.72		(43)
		1,2,4-Trichlorobenzene, 403, 0.07–6.9	0.36–0.79		(43)
Characteristic ratio $\langle r^2 \rangle_0 / nl^2$	–	Solvent, Temperature			
		Theoretical, 413 K	6.9		(55)
		Theoretical, 413 K	7.4, 7.6		(56)
		Dodecanol, 411 K	6.7		(24, 26, 55)
		Dodecanol, 401 K	7.1		(25)
		Diphenyl methane, 415 K,	6.8		(26)
		Diphenyl methane, 415 K	7.0		(25)
		1-Chloronapthalene, 413 K	6.8		(26)
		bis-2 Ethyl hexyl adipate, 418 K	10.3		(49)
		Biphenyl, 401 K	7.0		(24)
		Diphenyl ether, 434 K	6.4		(26)
		Diphenyl ether, 437 K	6.8		(25)
		Octanol, 453 K	6.4		(26)

Property	Units	Conditions	Value	Reference
Crystalline-state properties				
Lattice	–	Most stable, 1 atm	Orthorhombic	(57, 58)
Space group	–	Orthorhombic	Pnam	(57, 58)
Chain conformation	–	Orthorhombic	Planar zig-zag	(57, 58)
Unit cell dimensions	Å	Orthorhombic, oriented sheet	a = 7.40 b = 4.93 c = 2.53	(57)
	Å	Orthorhombic, fiber	a = 7.41 b = 4.95 c = 2.55	(58)
	Å	Orthorhombic, powder, melt crystallized	a = 7.40 b = 4.93 c = 2.53	(59)
	Å	Orthorhombic, powder, slow, melt crystallized	a = 7.42 b = 4.95 c = 2.55	(60)
	Å	Orthorhombic, solution, expitaxial	a = 7.48 b = 4.97 c = 2.55	(61)
Unit cell content	–	Orthorhombic	4 CH_2 units	(57, 58)
Lattice	–	Metastable, requires deformation	Monoclinic	(62)
Space group	–	Monoclinic	$P2_1$ m	(62)
Chain conformation		Monoclinic	Planar zig-zag	(62)
Unit cell dimensions	Å	Monoclinic	a = 8.09 b = 4.79 c = 2.53	(62)
Unit cell angle	Degrees	Monoclinic	β = 107.9	
Unit cell content	None	Monoclinic	4 CH_2 units.	(62)
Lattice	–	Requires >3 k bar, near melting point	Hexagonal	(63, 64)
Unit cell dimension	Å	Referred to orthohexagonal axis	a = 8.46 b = 4.88 c = 2.45	(63, 64)
		Referred to hexagonal axis	a = 4.88	(63, 64)
Unit cell content	–	Hexagonal	4 CH_2 units	(63, 64)
Degree of crystallinity	%	Depends on molecular weight, crystallization conditions, method of measurement	35–90	(65–67)

Property	Units	Conditions	Value	Reference
Heat of fusion	J mol^{-1} (of CH$_2$ units)	Macroscopic crystal, melting point depression by diluent	4140	(68–71)
		Actual finite crystal, depends on molecular weight, crystallization conditions, and method of measurement.	1450–3730	(65–67)
Entropy of fusion	J K^{-1} mol^{-1} (of CH$_2$ units)	Macroscopic ideal crystal, from heat of fusion and equilibrium melting temperature	9.9	(68–72)
		Actual finite crystal, depends on measured enthalpy of fusion	3.5–8.9	(65–67, 72)
Density (crystalline)	g cm^{-3}	Orthorhombic unit cell	0.996	(57, 58,
		Observed depends on molecular weight and crystallization conditions	0.92–0.99	65–67)
Polymorph		Stable at atmospheric pressure	Orthothombic	(57, 58)
		Metastable, involves deformation	Monoclinic	(62)
		Pressure >3 k	Hexagonal	(63, 64)
		bar, near melting temperature		
Avrami exponent		M (g mol^{-1}) = 4,800–5,800, T_c = (125–128°C)	4	(66)
		M (g mol^{-1}) = 7,800–11,500 T_c = 129–128°C	4	(66)
		T_c = 125–128°C	3	(66)
		M (g mol^{-1}) = 1.4 × 10^4–1.2 × 10^6 T_c = 125–132°C	3	(66)
		M (g mol^{-1}) = 3 × 10^6–8 × 10^6 T_c = 125–130°C	2	(66)

Transition temperatures

Property	Units	Conditions	Value	Reference
Glass transition temperature	K	Expansion Coefficient	153	(73)
		Expansion Coefficient	140	(74)
		Differential Scanning Calorimetry	150	(74)
		Adiabatic Calorimety	148	(75, 76)
		Dynamic Mechanical (5 Hz)	150	(74)
		Dynamic Mechanical (0.1–1.0 Hz)	146–155	(77)
		Dynamic Mechanical 0.67 Hz	140	(78)
		Dynamic Mechanical 4.8 Hz	149	(79)
		Dynamic Mechanical 10^2 Hz	160	(80)
		Small Angle x-ray, expansion coefficient	148	(81)
		Vibrational spectroscopy	<180	(82)
β -Transition	K	Dynamic Mechanical (3.5 Hz)	258 ± 5	(83, 84)
		Dynamic Mechanical (0.67 Hz)	253	(85)

Property	Units	Conditions	Value	Reference
		Dynamic Mechanical (1 Hz)	253	(86)
		Dynamic Mechanical (10^2 Hz)	283	(80)
		Expansion Coefficient	243	(87)
α -Transition	K	Dynamic Mechanical (3.5 Hz)	303–341	(84)
		Dynamic Mechanical (0.1 Hz)	323–383	(77)
		Value depends on crystallite thickness		
Equilibrium melting temperature	K	Theoretical	418 $\pm$ 1	(88)
		Dilatometry	419	(89)
		Extrapolated, T_m/T_c	419	(90)
		Extrapolated, Gibbs–Thomson	419	(91)
		Extrapolated, Gibbs–Thomson	419	(92–95)
Directly observed melting temperature	K	Depends on molecular weight, crystallization conditions, and method of measurement.	391–419	(66, 96)
Heat capacity	J mol^{-1}K^{-1}	Experimental 100 K, crystalline	9.45	(97)
		Experimental, liquid 608 K	4.39	(97)
		Extrapolated, liquid 300 K	3.09	(97)

Mechanical properties

Property	Units	Conditions	Value	Reference
tensile modulus	MPa	Initial modulus: Depends on molecular mass and morphological structure.	60–290	(98)
Bulk modulus	–		Reciprocal of compressibility.	–
Storage modulus	MPa	$T = 298$ K, slow cooled	800	(79)
		$T = 253$ K, $d = 0.936$ g cm^{-3}, 0.67 Hz	600	(85)
		$T = 253$ K, crystallinity 0.40,1 Hz	400	(77)
Loss modulus	MPa	$T = 298$ K, slow cooled, 0.67 Hz	6.2	(79)
		$T = 253$ K, d $= 0.936$ g cm^{-3}, 0.67 Hz,	7.6	(85)
		$T = 253$ K, crystallinity 0.40, 1 Hz	8.0	(77)
Tensile strength	MPa	Depends on molecular mass, based on original cross-section, strain rate 1 inch min^{-1} $T = 298$ K	10–60	(98)
Yield stress	MPa	Depends on crystallinity level, strain rate 1 inch min^{-1} $T = 298$ K.	18–32	(98)

Property	Units	Conditions	Value		Reference
Maximum extensibility $(L/L_\circ)$	–	Depends on molecular mass, strain rate 10^{-1} s^{-1}, $T = 298$ K.	18–4		(98)
Impact Strength	$J\,m^{-1}$	Izod (notched), $d = 0.94$–0.97 $g\,cm^{-3}$	30–200		(99)
Hardness	D1706	Shore D	55–70		(100)
Plateau Modulus	MPa	378 K	2.2		(101)
		413 K	2.6		(102)
Entanglement Mol. Wt.	$g\,mol^{-1}$	378 K	1100		(101)
		413 K	800		(102)
WLF parameters C_1, C_2	–	$M_v = 2 \times 10^6$ (unfractionated), calculated from ^{13}C nmr correlation times, $T_g = 173$ K $= T_{ref}$	$C_1 = 12.5$, $C_2 = 34.3$		(103) (77)
		$M_n = 6 \times 10^5$, $M_w = 4 \times 10^6$, Dynamic Mechanical, 1 Hz, $T_g = 155$ K $= T_{ref}$			
		degree of crystallinity $= 0.40$	$C_1 = 15.0$, $C_2 = 50.5$		
		degree of crystallinity $= 0.50$	$C_1 = 15.4$, $C_2 = 50.0$		
		degree of crystallinity $= 0.70$	$C_1 = 16.3$, $C_2 = 48.0$		
Abrasion resistance	$g\,MHz^{-1}$	Tabor	2–10		(100)

Electo-optical and magnetic properties

Property	Units	Conditions	Value		Reference
Index of refraction	–	Crystal, $\lambda = 5461$ Å, $T = 298$ K	$\alpha \sim \beta = 1.520$, γ 1.582		(104)
		Amorphous, $\lambda = 5461$Å			(105)
		T $= 403$ K	1.4327		
		T $= 412.9$ K	1.4297		
		T $= 423.6$ K	1.4261		

Property	Units	Solvent, Temperature (K)	$\lambda = 436$ nm	$\lambda = 546$ nm	Reference
Refractive index increment	$ml\,g^{-1}$	Biphenyl, 396	–	−0.174	(105)
		Biphenyl, 408	−0.195	−0.172	(106)
		Biphenyl, 400	−0.202	−0.176	(106)
		Bromobenzene, 408	−0.101	−0.089	(106)
		1-Chloronaphtalene, 363	–	−0.198	(44)
		1-Chloronaphtalene, 387–424	–	−0.196–0.194	(107)
		1-Chloronaphtalene, 398	–	−0.195	(44)
		1-Chloronaphtalene, 408	–	−0.190	(44)
		1-Chloronaphtalene, 400	–	−0.191	(108)
		1-Chloronaphtalene, 403	–	−0.193	(109)

Property	Units	Conditions	Value		Reference
		Solvent, Temperature (K)	$\lambda = 436$ nm	$\lambda = 546$ nm	
		1-Chloronaphtalene, 408	–	−0.193	(110)
		1-Chloronaphtalene, 418	–	−0.196	(109)
		1-Chloronaphtalene, 418	–	−0.188–0.193	(111)
		1-Chloronaphtalene, 418	−0.215	−0.192	(112)
		n-Decane, 384–422	–	0.087–0.099	(107)
		n-Decane, 408	0.117	0.114	(106)
		n-Decane, 379–408	0.116–0.132	0.113–0.126	(113)
		p-Dibromobenzene, 408	−0.179	−0.162	(106)
		o-Dichlorobenzene, 408	−0.091	−0.081	(106)
		o-Dichlorobenzene, 408	−0.095	−0.083	(106)
		Diphenyl methane, 415	−0.146	−0.129	(106)
		1-Dodecanol, 410	0.048	0.046	(106)
		1-Methyl napthalene, 408	−0.206	−0.177	(106)
		Tetrahydronapthalene, 408	−0.087	−0.077	(106)
		Tetrahydronapthalene, 368–417	–	−0.091–0.080	(107)
		1,2,4 Trichlorobenzene, 408	−0.125	−0.192–0.11	(106, 112)

Surface and interfacial properties

Property	Units	Conditions	Value	Reference
Surface tension	Nm^{-1} $\times 10^{-5}$	Method, Temperature (K)		
		Pendant drop, 413	28.8	(114, 115)
		Pendant drop, 453	26.5	(114, 115)
		Pendant drop, 298 (extrapolated)	35.7	(114, 115)
		Pendant drop, 423	28.1	(116)
		Pendant drop, 423	26.4	(117)
		Wilhelm plate, 485	24.5	(118)
		Wilhelm plate, 458	26.0	(118)
		Wilhelm plate, 293 (extrapolated)	36.0	(118)
		Maximum bubble pressure, 423	22.8	(119)
		Method, second component, T/K		
		Pendant drop, poly(styrene),		
		293 (extrapolated)	8.6	(120)
		413	5.9	(120)
		453	5.1	(120)
		Pendant drop, poly(n-butyl methacrylate)		
		293 (extrapolated)	7.1	(120)
		413	5.3	(120)
		453	4.7	(120)

Property	Units	Conditions	Value	Reference
Surface tension	N m^{-1} $\times 10^{-5}$	Pendant drop, poly(methyl methacrylate)		
		293 (extrapolated)	11.9	(120)
		413	9.7	(120)
		453	9.0	(120)
		Pendant drop, poly(ethylene oxide), 423	9.5	(117)
		Pendant drop, poly(dimethyl siloxane), 423	5.1	(117)
		Pendant drop, poly(tetrahydrofuran), 423	4.1	(117)
		Pendant drop, poly(ethylene-vinyl acetate), 423	1.3	(117)
		Pendant drop, poly(vinyl acetate), 453	10.2	(115)
		Pendant drop, poly(vinyl acetate), 423	9.8	(117)
		Pendant drop, poly(vinyl acetate), 413	11.3	(115)
		Pendant drop, poly(vinyl acetate), 293 (extrapolated)	14.5	(115)
		Spinning drop, poly(styrene), 473	4.4	(121)
		Spinning drop, poly(hexamethylene adipamide), 523	10.7	(121)
		Spinning drop, poly(methyl methacylate), 473	10.0	(121)
Transport properties				
permeability coefficient, P	cm^3(STP) cm^{-1} s^{-1} atm^{-1} ($\times 10^{-8}$)	Semi-crystalline, $d = 0.964$ g cm^{-3}, permeant		(122)
		He, 298 K	0.87	
		O_2, 298 K	0.31	
		Ar, 298 K	1.29	
		CO_2, 298 K	0.28	
		CO, 298 K	0.15	
		N_2, 298 K	0.11	
		CH_4, 298 K	0.30	
		C_2H_6, 298 K	0.45	
		C_3H_4, 298 K	3.06	
		C_3H_6, 298	0.88	
		C_3H_8, 298	0.41	
		SF_6, 298	0.0064	
		H_2S, 293	6.5	
Thermal conductivity	W m^{-1} K^{-1}	Unblended	0.52	(123)
		Nanocomposites HDPE,0.2 vol% Single-walled Carbon Nanotubes, SWCNT	3.50	(124)
Electrical conductivity	S m^{-1}	Unblended	1×10^{-9}	(125)
		Nanocomposites HDPE,10 wt% Single-walled Carbon Nanotubes (SWCNT)	100	(125)

Property	Units	Conditions	Value			Reference
Melt Viscosity	Pa s	Zero shear, fractions Temp. (K)	410 K	465 K	468 K	(126)
		M_w, M_w/M_n 13,600, 1.19	–	–	2.52	
		M_w, M_w/M_n 19,300, 1.12	25.7	10.1	–	
		M_w, M_w/M_n 32,100, 1.11	–	–	75.3	
		M_w, M_w/M_n 33,900, 1.11	157.0	64.5	–	
		M_w, M_w/M_n 58,400, 1.10	708.0	280	–	
		M_w, M_w/M_n 77,400, 1.10	1630.0	640	–	
		M_w, M_w/M_n 119,600, 1.19	–	–	8000.0	
		M_w, M_w/M_n 520,000, 1.18	–	–	$>10^7$	
Coefficient of sliding fraction		Sliding on steel				
		Polished	0.60			(127)
		Abraded	0.33			(127)
Speed of sound	m s^{-1}	273 K	1600			(128)

References

1. Noda I, Dowrey AE, Haynes JL, Marcott C. In: Mark JE, ed. : *Physical Properties of Polymers Handbook*. 2nd ed. Springer; 2007:395.
2. Bovey FA, In Sharma RH, ed. *Stereodynamics of Molecular Systems*. Oxford: Pergamon; 1979.
3. Inoue Y, Nishioka A, Chujo R.*Makromol. Chem.* 1973;168:163.
4. Heatley F. *Polymer.* 1975;16:493.
5. Ferguson RC. *ACS Polymer Preprints.* 1967;8(2):1026.
6. Bovey FA. *High Resolution NMR of Macromolecules*. New York: Academic Press; 1972.
7. VanderHart DL. *J. Chem. Phys.* 1986;84:1196.
8. Nakagawa M, Horii F, Kitamaru R. *Polymer.* 1990;31:323.
9. Kitamaru R, Horii F, Muruyama K. *Macromolecules.* 1986;19:636.
10. Hedesiu C, Demco DE, Kleppinger R, et al. *Polymer.* 2007;48:763.
11. Orwoll RA, Flory PJ. *J. Amer. Chem. Soc.* 1967;89:6814.
12. Olabisi O, Simha R. *Macromolecules.* 1975;8:206.
13. Simha S, Somcynsky T. *Macromolecules.* 1969;2:342.
14. Simha R. *Macromolecules.* 1977;10:1025.
15. Flory PJ, Orwoll RA, Vrij A. *J. Amer. Chem. Soc.* 1964;86:3507.
16. Sanchez IC, Lacombe RH. *J. Phys. Chem.* 1976;80:2352.
17. Sanchez IC, Lacombe RH. *J. Polym. Sci., Poly. Ltrs.* 1977;15B:71.
18. Hayes RA, *J. Applied Polym. Sci.* 1961;5:318.
19. Tobolsky AV. *Properties and Structure of Polymers*. Wiley, N. Y; 1960:64.
20. Allen G, Gee G, Mangaraj D, Sims D, Wilson GJ. *Polymer.* 1960;1:467.
21. Stacey CJ, Arnett RL. *J. Phys. Chem.* 1965;69:3109.
22. Nakajima A, Fujiwara H, Hamada F. *J. Polym. Sci., Pt. A-2.* 1960;4:507.
23. Wagner HL, Hoeve CA. *J. Polym. Sci.* 1976;54C:327.
24. Chiang R. *J. Phys. Chem.* 1966;70:2348.
25. Nakajima A, Hamada F, Hayashi S. *J. Polym. Sci.* 1966;15C:285.
26. Chiang R. *J. Phys. Chem.* 1965;69:1645.
27. Constantin D. *Europ. Polym. J.* 1977;13:907.
28. Nakajima A, Hamada F. *Report Polymer Phys. Japan.* 1966;9:41.

29. Hamada F, Fujisawa K, Nakajima A. *Polymer.* 1973;4:316.
30. Schreiber HP, Tewari YB, Patterson D. *J. Polym. Sci.: Phy. Ed.* 1973;11:15.
31. Patterson D, Tewari YB, Schreiber HP. *Macromolecules.* 1971;4:356.
32. Brockmeier NF, McCoy RW, Meyer JA. *Macromolecules.* 1972;5:130.
33. Tung LH. *J. Polym. Sci.* 1957;24:333.
34. Chatterjee J, Alamo RG. *J. Polym. Sci.: Phy. Ed.* 2002;40:878.
35. Tung LH. *J. Polym. Sci..* 1964;A2:4875.
36. Kokle V, Billmeyer FW Jr, Muus LT, Newitt EJ. *J. Polym. Sci.* 1962;62:251.
37. Atkins JT, Muus LT, Smith CW, Pieski ET. *J. Amer. Chem. Soc.* 1957;79:5089.
38. Stejskal J, Horska J, Kratocvichil P, *J. Appl. Polym. Sci.* 1982;27:3929.
39. Mirabella FM Jr. *J. Appl. Polym. Sci..* 1980;25:1775.
40. Trementozzi QA. *J. Polym. Sci..* 1959;36:113.
41. Otocka EP, Roe RJ, Hellman MY, Miglia PM. *Macromolecules.* 1971;4:507.
42. Wagner HL, Hoeve CAJ. *J. Polym. Sci.: Polym. Phys. Ed.* 1973;11:1189.
43. Hert M, Strazielle C. *Makromol. Chem.* 1983;184:135.
44. Chiang R. *J. Polym. Sci.* 1959;36:91.
45. Francis PS, Cooke R Jr, Elliot JH. *J. Polym. Sci.* 1957;31:453.
46. Henry PM. *J. Polym. Sci.* 1959;36:3.
47. Tung LH. *J. Polym. Sci.* 1959;36:287.
48. Wesslau H. *Makromol. Chem.* 1956;20:111.
49. Kotera A, Saito T, Takamisawa K, Miyazawa Y. *Report Prog. Polymer Soc. Japan.* 1960;3:58.
50. Duch E, üchler LK. *Z. Electrochem.* 1956;60: 218.
51. Wesslau H. *Makromol. Chem..* 1952;26:96.
52. Kaufman HS, Walsh EK. *J. Polym. Sci.* 1957;26:124.
53. Stacy CJ, Arnett RL. *J. Polym. Sci. A.* 1964;2:167.
54. Krigbaum WR, Trementozzi QA. *J. Polym. Sci.* 1958;28:295.
55. Flory PJ. *Statistical Mechanics of Chain Molecules,* revised Ed., Hanser Publishers (1988).
56. Abe A, Jernigan RL, Flory PJ. *J. Amer. Chem. Soc.* 1966;88:631.
57. Bunn CW. *Trans. Farad. Soc.* 1939;35:482.
58. Busing WR. *Macromolecules.* 1990;23:4608.
59. Kawaguchi A, Ohara M, Kobayashi K. *J. Macromol. Sci. Phys.* 1973;B16:193.
60. Zugenmaier P, Cantow. *Kolloid-Z. Z. Polymer* 1968;230:229.
61. Hu H, Dorset DL. *Acta. Cryst.* 1989;B45:283.
62. Seto T, Hara T, Tanaka T. *Japan J. Appl. Phys.* 1968;7:31. D.L. Dorset, Crystallography of the Polymethylene Chain. An Inquiry into the Structure of Waxes. IUCr Monographs on crystallography, No. 17. Oxford University Press (2005).
63. Bassett DC, Block S, Piermarina S. *J. Appl. Phys.* 1974;45:4146.
64. Yasuniwa F, Enoshito R, Takemura T. *Japan J. Appl. Phys.* 1970;15:142.
65. Fatou JG, Mandelkern L. *J. Phys. Chem.* 1965;69:417.
66. Ergoz E, Fatou JG, Mandelkern L. *Macromolecules.* 1972;5:147.
67. Mandelkern L. *Polym. J.* 1985;17:337.
68. Flory PJ, Vrij A. *J. Amer. Chem. Soc.* 1963;85:3548.
69. Quin FA Jr, Mandelkern L. *J. Amer. Chem. Soc..* 1958;80:31781.
70. Mandelkern L. *Rubber Chem. Tech.* 1959;32:1392.
71. Nakajima A, Hamada F. *Koll. Z. Z. Polymer.* 1965;205:55.
72. Sharma RK, Mandelkern L. *Macromolecules.* 1969;2:266.
73. Dannis ML. *J. Appl. Polym. Sci.* 1959;1:121.
74. Stehling FC, Mandelkern L. *Macromolecules.* 1970;3:242.
75. Beatty CL, Karasz FE. *J. Macromol. Sci. Rev. Macromal. Chem.* 1971;C17:37.

76. Simon J, Beatty CL, Karasz FE. *J. Thermal Anal.* 1975;7:187.

77. Alberola N, Cavaille JY, Perez J. *Euorpean Polym. J.* 1992;28:935.

78. Gray RW, McCrum NG. *J. Polym. Sci., Pt. A-2.* 1969;7:1329.

79. Willbourn AH. *Trans. Farad. Soc.* 1958;54:717.

80. Flocke H. *Kolloid Z. Z. Polymere.* 1962;180:118.

81. Fischer EW, Kloos F. *J. Polym. Sci. Polym. Ltrs.* 1970;8B:685.

82. Hendra PJ, Jobic H Holland-Moritz K. *J. Polym. Sci.* 1975;13B:365.

83. Popli R, Mandelkern L. *Polym. Bull.* 1983;9:260.

84. Popli R, Glotin M, Mandelkern L, Benson RS. *J. Polym. Sci., Polym. Phys. Ed.* 1984;22:407.

85. Cooper JW, McCrum NG. *J. Material Sci. Ltrs.* 1972;7:1221.

86. Moore RS, Matsuoka S. *J. Polym. Sci.* 1964;5C:163.

87. Magill JH, Pollack SS, Wyman DP. *J. Polym. Sci., Pt. A.* 1965;3:3781.

88. Flory PJ, Vrij A. *J. Amer. Chem. Soc.* 1963;85:3548.

89. Rijke AM, Mandelkern L. *J. Polym. Sci., A-2.* 1970;8:225.

90. Gopalan M, Mandelkern L. *J. Phys. Chem.* 1967;71:3833.

91. Chivers RA, Barham PJ, Martinez-Salazar I, Keller A. *J. Polym. Sci., Poly. Phys. Ed.* 1982;20:1717.

92. Brown RJ, Eby RK. *J. Appl. Phys.* 1964;35:1156.

93. Huseby TW, Bair HE. *J. Appl. Phys.* 1968;39:4969.

94. Hoffman JD, Davis GT, Lauritzen JI Jr. In: Hannay NB, ed. *Treatise in Solid State Chemistry.* Vol 3. Plenum Press; 1976:497.

95. Bair HE, Huseby TW, Salovey R. *ACS Polym. Preprints.* 1968;9:795.

96. Fatou JG, Mandelkern L. *J. Phys. Chem.* 1965;69:417.

97. Gaur U, Wunderlich B. *J. Phys. Chem. Ref. Data.* 1981;10:119.

98. Kennedy MA, Peacock AJ, Mandelkern L. *Macromolecules.* 1994;27:5279.

99. Brostow W, Kubát J, Kubát MM. In: Mark JE, ed. *Physical Properties of Polymers Handbook.* American Institute of Physics; 1996:313.

100. In: Brandrup J, Immergut E, ed. *Polymer Handbook.* 3rd ed. New York: V/23, John Wiley; (1989).

101. Graessley WW. In: Mark JE, ed. *Physical Properties of Polymers.* 2nd ed. American Chemical Society; 1992:97.

102. Fetters LJ, Lohse DJ, Colby RH. In: Mark JE, ed. *Physical Properties of Polymers Handbook.* American Institute of Physics; 1996:335.

103. Dekmezian A, Axelson DE, Dechter JJ, Borah B, Mandelkern L. *J. Polym. Sci.: Polym. Phys. Ed.* 1985;23:367.

104. Bryant WMD. *J. Polym. Sci.* 1947;2:547.

105. Th G, Scholte. *J. Polym. Sci. A-2.* 1968;6:91.

106. Horska J, Stejkal J, Kratocvichil P. *J. Appl. Polym. Sci.* 1979;24:1845.

107. Ehl J, Loucheux C, Reiss C, Benoit H. *Makromol. Chem.* 1964;75:35.

108. Casper R, Bishop U, Lange H, Pohl U. *Makromol. Chem.* 1976; 177:1111.

109. Peyrouset A, Prechner R, Panaris R,Bonoit H. *J. Appl. Polym. Sci.* 1975;19:1363.

110. Suzuki H, Muraoka Y, Inagoki H. *J. Polym. Sci. Polym. Phys. Ed.* 1981;19:189.

111. Wagner HL. *J. Res. Natl. Bur. Stnds.* 1972;76A:151.

112. Horska J, Stejkal J, Kratocvichil P. *J. Appl. Polym. Sci.* 1983;28:3873.

113. öhn LLB, Lanier U, Lechner MD. *Makromol. Chem.* 1983;184:585.

114. Wu S. *J. Polym. Sci.,***C34**. 1971;19.

115. Wu S, *J. Colloid and Interface Sci.* 1969;31:153.

116. Roe RJ. *J. Phys. Chem.* 1968;72:2013.

117. Roe RJ. *J. Colloid and Interface Sci..* 1969;31:228.

118. Dettre RH, Johnson RE Jr. *J Colloid and Interface Sci.* 1966;21:367.

119. Hybart FJ, White TR. *J. Appl. Polym. Sci.* 1960;3:118.

120. Wu S. *J. Phys. Chem.* 1970;74:632.

121. Elmendorp JJ, DeVos G. *Polym. Eng. Sci.* 1986;26:415.

122. Michaels AS, Bixler HJ. *J. Polym. Sci.* 1961;50:413.

123. Yang Y. In: Mark JE, ed. *Physical Properties of Polymer Handbook.* American Institute of Physics; 1996:111.

124. Hagenmueller R, Guthy C, Lukes JR, Fischer JE, Winey KI. *Macromolecules.* 2007;40:2417.

125. Jeon K, Lumata Ll, Tokumoto T, Steven E, Brooks J, Alamo RG. *Polymer.* 2007;48:4751.

126. Raju VR, Smith GG, Marid G, Knox JR, Graessley WW. *J. Polym. Sci. Polym. Phys.* 1979;17:1183.

127. Brandrup J, Immergut E, ed. *Polymer Handbook*, 3rd ed. New York: V/18, John Wiley; 1989.

128. Baccaredda M, Butta E, Frosiui V. *Makromol. Chem.*, **61**, 14 (1963).

Polyethylene, Linear Low-density

A. Prasad

Acronyms LLDPE, low-pressure PE, Poly(α-olefin) copolymer

Class Poly(α-olefin)

Structure $—(CH_2—CH_2—CHR—CH_2)_n—$ (R = α-olefin)

Introduction LLDPE is the common name for copolymers of ethylene with α-olefin comonomer. The comonomers most frequently used commercially are butene, hexene, and octene. Commercial grade LLDPE resins with 4-methyl-1-pentene (4-MP-1) as comonomer is also available. LLDPE prepared by the conventional Ziegler-Natta (Z-N) catalyst system always exhibit high heterogeneity in the intermolecular distribution of comonomer units along the polymer chains.[1–5] The branches are preferentially located in the lower molecular weight chains; thus the bulk of LLDPE behaves as if it were a blend of high molecular weight, linear molecules and low molecular weight, branched molecules. LLDPE differs from LDPE principally through a lack of long-chain branching (LCB) and a narrower molecular weight distribution (MWD). New types of LLDPEs based on the metallocene catalyst technology have been introduced recently in the market place. Such LLDPEs are characterized by narrower molecular weight and homogeneous short-chain branching distribution.[6–9] Some of the metallocene catalyst-based octene-1 LLDPE copolymers made by the Dow Chemical Company are known to have LCB.[9] For the properties of metallocene LLDPE see the entry Polyethylene, metallocene linear low density, in this handbook. LLDPE is commercially available in wide variety of melt indexes (MI) and density ranges. The properties of LLDPE are functions of molecular weight (MW), MWD, density, type, and amount of comonomer.[10–13] The comonomers are also referred to as short-chain branches (SCB). Consequently, physical and mechanical properties also vary accordingly. Mechanical properties such as tensile, tear, and impact are strongly dependent on the chemical nature of the comonomer type. Therefore, it is difficult to list all properties separately. The values of the properties shown in the following table are given in ranges because of their dependence on molecular structure and type of comonomer and are intended to represent the best published examples of the most commonly used commercial grades of LLDPE resins. The physical properties of extruded materials may vary substantially from those of the compression molded materials. For illustration purposes, a few of the physical properties that depend on the chemical nature of the comonomer are presented in Tables 3, 6, 7, and 8.

Major Applications Major applications include blown and cast films for bags, shrink-wrap, packaging, and injection molding. Such films exhibit exceptional toughness, dart impact, and puncture resistance when compared to blown films of LDPE. Other applications include blow molding, pipe and conduit, lamination, coextrusion, rotomolding, and wire and cable coatings. There is considerable use of blends of LLDPE with LDPE in a wide variety of applications.

Properties of Special Interest Low cost, flexibility, toughness, high impact strength, low brittleness temperature, good chemical resistance to acids and aqueous solvents, good dielectric properties, good heat seal properties, and much better thermal, stress-crack resistance, and moisture barrier properties when compared to LDPE. The limitations include poor resistance to oxidizing agents; aliphatic, aromatic and polar liquids; and chlorinated solvents. LLDPE is relatively difficult to process by extrusion due to narrower MWD and poor optical clarity when compared to LDPE.

Major Suppliers Lyondell Chemical Co. (Equistar Chemicals), Dow Chemical Co., Chevron. Westlake Polymers, Nova Chemicals, Huntsman Polymers, Borealis, Phillips Chemical Co., Du Pont Co., Exxon Mobil Chemical Co., Eastman Chemical Co., Solvay Polymers, Inc.

Catalyst for LLDPE[11,14,15]

Polymerization Process	Catalyst Specification	Polymerization Condition
Gas-phase fluidized bed polymerization, solution polymerization, slurry polymerization, and polymerization in melt under high ethylene pressure	LLDPEs are produced with two broad class of catalysts: (1) Ziegler catalyst (Z-N) : derivative of a transition metal (such as titanium) and organoaluminium compound (such as triethylaluminium) supported on inorganic and organic support (such as silica, magnesium dichloride etc.) (2) Chromium oxide-based catalysts (Cr) From Phillips Petroleum Co : these are mixed silica titania support containing 2–20 wt% of titania and a co-catalyst (i.e., trialkylaluminum compounds). These catalysts produce LLDPEs of very broad MWD (M_w/M_n in the range of 12–35) and MI in the 80–200 range	Typical heterogeneous Ziegler catalysts operate at temperature range of 343–373 K and low pressures of 0.1–2 MPa in inert liquid medium (e.g., hexane and isobutane) or in the gas phase

Property	Units	Conditions	Value	Reference
Typical comonomers	–	Butene, hexene, octene, and 4-MP-1	–	(1, 3, 4, 11, 12, 16–19)
Degree of branching, commercial grades	mol%	D 2238, NMR	2–4	(11, 17–19)
Typical molecular weight range (M_w)	g mol^{-1}	GPC, in 1,2,4-trichlorobenzene (TCB) at 408 K	5–$20 \times (10^4)$	(11, 17–19)
Typical polydispersity index (M_w/M_n)	–	GPC	4–35	(11, 17–19)
IR (characteristic absorption frequencies)	cm^{-1}	D2238	see Table below	(20–27)

Characteristic IR bands used to identify the type short-chain branching*

Comonomer Type	Methyl Deformation Band Position (cm^{-1})	Methyl Deformation Band[†] Position (cm^{-1})	Reference
Butene-1	1,379	908, 887, 771(vs)	(21–27)
Hexene-1	1,377.8	908, 894(vs), 837(s), 779(w)	(21–27)
Octene-1	1,377.6	908(vs), 889(s)	(21–27)
4-MP-1	1,383	908, 920(s)	(26)

* See also the entry on LDPE in this handbook.

[†] vs, s, w refer to the intensities of the absorbance bands: very strong, strong, and weak, respectively.

Ratios of different unsaturated groups in Octene-1 Z-N LLDPE polymers[28]

Mole %	Density (g cm^{-3})	MI (g (10 min^{-1}))	Vinyl CH$_2$=CH— (908 cm^{-1})	Vinylidene CH$_2$=C (889 cm^{-1})	trans-vinylene —CH=CH— (965 cm^{-1})
2.84	0.920	1.15	1	0.3	0.2

Property	Units	Conditions	Value	Reference
NMR	ppm	TCB/d6-benzene	See Table 1 Solution at 398 K	(29–32)
Linear thermal expansion coefficient	K^{-1}	D 696, 308–423 K	16–$20 \times (10^{-5})$	(33)
Solvents	–	368 K	Decalin, toluene	(34)
		369 K	Xylene	(34)
		371 K	Tetralin	(34)
		341 K	Cyclohexene	(35)
		374 K	n-Tetracosane	(35)
Nonsolvents	–	359 K	Methylene chloride	
		361 K	o-Dichloro benzene	
		366 K	1,2-Dichloropropane	
Mark-Houwink parameter: K and a	K = ml g^{-1} a = None	Decahydronaphthalene, 410 K	$K = 4.6 \times 10^{-4}$ $a = 0.73$	(36, 37)
		TCB, 408 K	$K = 3.63 \times 10^{-4}$, $a = 0.72$	(38)
Crystallographic data	Å	Unit cell dimensions depends on comonomer type and amount, and lamellae thickness	See Table 2	(13, 29, 39–41)
Degree of crystallinity	%	DSC (see also Table 3)	33–53	(3–6, 11, 29, 40, 42)
Heat of fusion	kJ mol^{-1}	DSC (see also Table 3)	1.37–2.18	(3–6, 11, 29, 40, 42)
Density, commercial resin	g cm^{-3}	D 1505–85 D 792	0.912–0.930	(10–12) (33)

Property	Units	Conditions	Value	Reference
Avarami exponent	–	Dependent on counit content and is independent of counit type; copolymer fractions of butene-1 4-MP-1 and octene-1 = 0.7–5.2 mol% range; isothermal crystallization range = 365–385 K	1.8–2.8	(4)
Long period spacing and lamellae thickness	Å	Raman longitudinal acoustic mode (LAM) and small-angle X-ray scattering (SAXS)	See Table 4	(5, 29, 40–42)
Surface free energy σ_e (chain- folding crystal face)	J m^{-2}	Dependent on counit content; counit content range = 0.70–7.6 mol%	0.067–0.225	(4, 43, 44)
Crystal phase structure	%	Raman LAM	See Table 3	(45)
Crystal orientation and Birefringence	–	Wide-angle X-ray (WAXD) and infrared diachroism	See Table5	(46)
Radius of gyration $R_G/M^{0.5}$	Amol$^{0.5}$g$^{0.5}$	Hydrogenated polybutadiene, 18 ethyl/1,000 C, small angle Neutron scattering (SANS)	0.440	(47)
Melting temperature	K	DSC peak endotherms (dual endotherm, peak range)	378–383 and 394–398	(3–6, 11, 29, 40, 42)

Equilibrium melting point T_m^o [4,42–44,48,49]

Copolymer	M_w	M_w/M_n	Co-unit (mol%)	Method	T_m^0(K)	Reference
Butene-1	–	–	2.2	Thompson-Gibbs	406	(42, 43)
Butene-1	–	–	7.3	Thompson-Gibbs	407, 411	(42, 43)
Octene-1 (metallocene)	98400	2.2	1.5	Thompson-Gibbs	412.5	(49)
Octene-1 (metallocene)	102,700	2.1	3.6	Thompson-Gibbs	407.3	(49)

Note: The equilibrium melting temperature (T_m^0) of copolymers depends on the molecular weight, sequence distribution and counit content. The T_m^0 value is determined by two commonly used techniques: the Hoffman-Weeks plot and the Thompson-Gibbs plot. The application of the Hoffman-Weeks method to determine the T_m^0 of a copolymer is unreliable (see reference 48). The more reliable method is to use the Thompson-Gibbs relationship of T_m as a function of lamellar thickness, provided a large range of lamella thickness can be obtained. Considerable disagreement exists between different authors on the exact value of transition that can be identified for the copolymers. Consequently, values tabulated in this table must be used cautiously. See references [44,48,49] for detailed discussions

Transition temperatures and activation energy*

Copolymer	Designation	Temperature Range (K)	Activation Energy (kJ mol^{-1})	Reference
Octene-1	α	333	62	(50)
(Dow 321)	β	253	319	(50)
tan δ peak at 10 Hz	γ	153	40	(50)
Octene-1	α	333	–	(51)
MI = 3.3, density = 0.912 g cm^{-3}	β	256	–	(51)
tan δ peak at 1 Hz	γ	150	–	(51)
Butene-1	α	304	–	(51, 52)
MI = 1, density = 0.890 g cm^{-3}	β	253	–	(51, 52)
tan δ peak at 1 Hz	γ	155	–	(51, 52)

* Conditions: DMA.

Note: The transitions and relaxation temperatures associated with amorphous regions are designated as α, β, γ, etc. in descending temperature order. The values of T_α depend only on crystallite thickness. The temperature of beta transition, T_β, does not depend on the crystallite thickness but rather on the comonomer type and content. The γ transition is associated with glass transition. All transition values depend on the frequency of the DMA test. See reference[18,53] for a detailed discussion.

Property	Units	Conditions	Value	Reference
Vicant softening point	K	D1525	353–367	(33, 38)
Tensile modulus	MPa	D 638	137–520	(10–12, 33, 38, 52, 54)
Tensile yield strength	MPa	D 638	9–20	(10–12, 33, 38, 52, 54)
Elongation at break	%	D 638	100–1,200	(10–12, 33, 38, 52, 54)
Yield stress	MPa	D 638	6.2–11.5	(10–12, 33, 38, 52, 54)
Flexural modulus	MPa	D 790, 298 K	235–800	(10–12, 33, 38)
Impact strength, notched Izod	J m^{-1}	D 256A	53.0–no break	(10, 33, 38)
Hardness	Shore D	D 676	47–58	(10, 33, 38)
Low temperature brittleness F50	K	D 746	<197	(10–12, 38)
Blown film properties	–	See Tables 6 and 7	–	(11, 12, 16, 55–58)
Refractive index n_D^{25}	–	D 542	1.52	(10)
Dielectric constant	–	D 150	2.3	(33)
Loss factor, tan δ	–	D 150, up to 100 MHz (at 1 MHz)	<0.0005	(33)
Melt index	g (10 min)$^{-1}$	D 1238	0.2–50	(11, 12)
Sonic velocity	m s^{-1}	MD and TD, blown films	See Table 5	(46)

Property	Units	Conditions	Value	Reference
Flow activation energy	kJ mol^{-1}	RMS, MI = 1.0, density = 0.918 g cm^{-3}, temp. range = 423–483 K	30–32	(59)
Water-vapor transmission Rate (WVTR)	g m^{-2} day^{-1} (normalized To 25.4 μm film Thickness)	30°C, 35% RH density = 0.9188 g cm^{-3} density = 0.921 g cm^{-3}	1.10 0.83	(60)

Table 1. Characteristic ^{13}C NMR bands by short-chain branching type[29,30]

Chemical Shift (ppm)		Chemical Shift (ppm)		Chemical Shift (ppm)		Chemical Shift (ppm)	
Butene	Sequence Assignment*	Hexene	Sequence Assignment*	Octene	Sequence Assignment*	4-MP-1	Carbon Assignment*
40.2	BBBB	41.4	HHHH	40.33	OOOO	44.5	3-C
39.56	BBBE + EBBB	40.86	HHHE + EHHH	38.24	EOE	35.7	—CH—
37.24	EBB + BBE	40.18	EHHE	34.62	EOEE + EEOE	34.6	alpha-C
35.0	BBB	38.13	EHE	32.22	EOE	26.8	beta-C
34.5	EBEB + BEBE	35.85	EHH + HHE	30.47	OEEE + EEEO	25.8	2-C
34.33	EBEE + EEBB	35.0	HHEH + HEHH	27.27	EOE	23.0	3-C
30.92	BEEB	34.9	EHH + HHE	27.09	OOEE + EEOO		
30.47	BEEE + EEEB	34.54	EHEE + EEHE	22.89	EOE + EOO + OOE + OOO		
27.7	BBB	34.13	EHE	14.17	EOE + EOO + OOE + OOO		
27.27	EBEE + EEBE	30.94	HEEH				
27.1	BBEE + EEBB	30.47	HEEE + EEEH				
26.68	EBE	29.51	EHE				
11.2	EBE	29.34	EHH + HHE				
11.0	EBB + BBE	27.28	EHEE + EEHE				
10.81	BBB	27.09	HHEE + EEHH				
		24.39	EHEHH + HHEHE				
		24.25	HHEHH				
		23.37	EHE + EHH + HHE + HHH				
		14.12	EHE + EHH + HHE + HHH				

* E, B, H, and O refers to ethylene, butene, hexene, and octene comonomers. ^{13}CNMR assignment at 50.3 MHz in 10% 1,2,4-trichlorobenzene solution at 125°C. Internal standard = tetramethylsilane.

Table 2. Unit cell information*

Lattice	Space Group	Comonomer		Unit Cell Dimensions (Å)			Reference
		Type	mol%	a	b	c	
Orthorhombic	Pnam	Butene	0.3	7.430	4.950	–	(39)
		Butene	0.6	7.450	4.950	–	(39)
		Butene	1.29	7.460	4.950	2.571	(41)
		Butene	3.85	7.480	4.970	2.571	(41)
		Butene	8.45	7.500	5.010	2.571	(41)
		Hexene	0.3	7.420	4.940	–	(39)
		Hexene	0.6	7.420	4.950	–	(39)
		Octene	0.065	7.429	4.950	–	(40)
		Octene	3.76	7.500	4.966	–	(40)
		Octene	5.0	7.571	4.987	–	(40)
		Octene	5.5	7.480	4.970	–	(39)

* Measured unit cell dimensions are dependent on lamellae thickness, which in turn depends on the crystallization temperature, comonomer type, and amount [see references[13,39–41] for details].

Table 3. Physical properties of commercial grade LLDPE as a function of counit content and type*[45,52]

Counit	$M_w(\times 10^{-3})$	Counit (mol%)	Crystallinity (%)	$\alpha_{c(\%)}$	$\alpha_{b(\%)}$	$\alpha_{a(\%)}$	Modulus (MPa)	Yield Stress (MPa)	Ultimate Tensile Stress	Draw Type Ratio at Break
Butene	104	0.6	39	46	36	18	152	20.3	296	9.9
	105	1.88	22	30	52	18	106	12.3	195	6.8
Hexene	65	0.34	63	63	27	10	361	32.1	264	10.6
Octene	98	0.62	41	48	35	16	160	18.4	224	7.4
	99	0.90	37	43	42	15	144	16.2	219	7.0
	70	0.94	39	40	42	18	163	16.4	191	6.9
	103	1.33	35	35	47	18	123	13.2	197	6.4
	81	1.77	49	26	52	22	75	9.5	183	8.1
	128	2.58	12	14	67	19	35	5.9	93	7.0
	228	5.38	4	7	76	17	3	1.8	–	–

* Samples were compression molded and slow cooled in air; tensile properties were determined at a draw rate of 25.4 mm min^{-1}; percent crystallinity was determined by DSC at a heating rate of 10°C min^{-1}; crystal phase structures were determined by Raman internal mode technique (LAM); α_c, α_b, and α_a refer to fraction of chain units in the perfect crystals, interfacial region, and amorphous region, respectively of a lamella.

Table 4. Long-period spacings and lamellae thickness of LLDPE fractions by short-chain branching type*

Comonomer Type	Mol% Branching	$M_w (\times 10^{-4})$	M_w/M_n	Conditions	Long-Period Spacing (Å)	Lamellae Thickness (Å)	Reference
Butene-1	1.29	13.0	2.96	SAXS	162	75	(41)
	2.29	19.0	5.55		118	52	(41)

(Continued)

Table 4. *(Continued)*

Comonomer Type	Mol% Branching	M_w ($\times 10^{-4}$)	M_w/M_n	Conditions	Long-Period Spacing (Å)	Lamellae Thickness (Å)	Reference
	3.17	9.93	3.09		105	55	(41)
	3.85	8.22	3.11		105	45	(41)
	5.10	4.61	2.90		92	50	(41)
	8.45	2.66	4.73		80	43	(41)
Butene-1	0.42	14.0	7.14	Raman LAM	–	91	(5)
	1.15	–	–		–	75	(5)
	3.43	–	–		–	72	(5)
	4.10	–	–		–	56	(5)
Hexene-1	1.57	259	11.5	Raman LAM	153	–	(29)
4-MP-1	1.76	94.5	5.0	Raman LAM	157	–	(29)
	2.74	91.2	4.0		130	–	(29)
	3.53	107.0	4.0		122	–	(29)
Octene-1	1.49	5.35	2.60	Raman LAM	–	80	(5)
	2.49	235.0	5.9		132	–	(29)
Octene-1	0.065	25.20	2.20	SAXS	369	163	(40)
	1.32	17.63	2.44		275	130	(40)
	2.22	14.62	2.75		187	70	(40)
	2.81	12.82	2.76		166	51	(40)
	3.76	9.40	2.86		140	43	(40)
	5.0	4.93	3.37		116	24	(40)

* Crystallization conditions are not known.

Table 5. Morphological properties and crystal orientation of LLDPE (hexene-1) blown tubular films[46,57]

Thickness (µm)	Density (g cm^{-3})	Birefringence ($\times 10^3$)	Sonic Velocity (km s^{-1})		Infrared Diachronic Ratio at 730 cm^{-1}	X-ray Angle (Degrees)			
						a-axis		*b*-axis	
			MD	TD	MD/TD	MD-TD	ND-MD	MD-TD	ND-MD
27.1	0.9116	−0.7478	828	1,009	1.26	90	90	0	0
24.7	–	+1.99	818	951	1.39	90	90	0	0
17.0	0.9172	0.00	843	717	1.24	90	90	0	0

Table 6. Melt rheology and molecular characteristics of some Z-N and Cr-based LLDPE[57]

Catalyst	MI g cm^{-3}	Density	$M_w \times 10^3$	$M_n \times 10^3$	M_w/M_n	Carreau-Yasuda Parameters	
						Zero-Shear Viscosity, η_0 (Pa s)	Relaxation Time, τ (s)
Z-N	1.0	0.920	105	27	4.0	1.04×10^4	0.026
Z-N	0.9	0.921	119	22	7.1	1.75×10^4	0.032
Z-N	0.8	0.923	144	20	7.1	3.17×10^4	0.046
Z-N	0.5	0.919	135	29	4.7	2.86×10^4	0.052
Cr	0.2	0.922	208	11	18.9	4.7×10^4	1.27

Table 7. Physical properties of commercial grade LLDPE film samples as a function of comonomer type*

Property	Units	Conditions	Butene	Hexene	Octene
Melt index	g (10 min)$^{-1}$	D 1238	1.0	1.04	1.04
Density	g cm^{-3}	D 1505	0.919	0.919	0.920
Total SCB	CH3/1,000C	NMR	21.7	17.9	13.7
Peak melting point	K	DSC, 10°C min^{-1}, cooling and heating rates	372 and 395	372 and 397.5	371 and 394
Heat of fusion	kJ mol^{-1}	DSC, 10°C min^{-1}, cooling and heating rates	1.50	1.68	1.65
M_w ($\times 10^{-5}$)	g mol^{-1}	GPC, 408 K in TCB	1.3	1.24	1.36
M_w/M_n	–	GPC	3.7	4.1	4.6
Total haze	%	D 1003	15	16	12
Gloss, 45°D	Units	D 2457	46	41	48
Narrow angle scattering	%	D 1746	7	20	34
Modulus, 1% secant	MPa	D 882, MD	185	206	200
		D 882, TD	227	250	230
Dart impact (per 25.4 μm)	g	D 1709, D 4272	74	187	201
Elmendorf tear	g	D 1922, MD	58	270	340
		D 1922, TD	520	710	750
Tensile strength at break	MPa	D 882, MD	41	39	58
		D 882, TD	18	20	31
Elongation at break	%	D 882, MD	430	430	440
		D 882, TD	560	550	600

* See references[11,55–58,61,62] for the effect of blowing conditions on film properties of LLDPEs. The result in this table was obtained for the following blown film conditions: blow-up ratio = 2.5: 1; die gap = 2.54 mm; output = 32 kg h^{-1}; film thickness = 25.4 μm; die size = 102 mm; frost line height = 230 mm; melt temperature = 483 K[58]

Table 8. Comparison of blown film properties of butene-1, 4-MP-1 copolymer, and butene/4-MP-1 terpolymer*[16]

Property	Units	Conditions†	Butene Copolymer	Butene/4-MP-1 Terpolymer	4-MP-1 Copolymer
Modulus, 1% secant	MPa	D 882, MD	154	205	277
		D 882, TD	205	234	354
Dart drop, F-50–66 cm (per mil)	g	D 1709 and D 4272	140	161	180
Elmendorf tear (per mil)	g	D 1922, MD	200	240	250
		D 1922, TD	470	540	720
Tensile strength at Break	MPa	D 882, MD	24.7	33.5	42
		D 882, TD	18.8	25.8	31.7

(Continued)

Table 8. *(Continued)*

Property	Units	Conditions[†]	Butene Copolymer	Butene/4-MP-1 Terpolymer	4-MP-1 Copolymer
Tensile strength at yield	MPa	D 882, MD	7.9	10.7	12.7
		D 882, TD	8.75	11.7	13.5
Elongation at break	%	D 882, MD	460	460	510
		D 882, TD	620	600	680

* Approximate melt index of 1 g $(10 \text{ min})^{-1}$ and density 0.920 g cm^{-3}. Extrusion conditions: blow-up ratio = 2: 1; die gap = 2.03 mm; output = 32 kg h^{-1}; film thickness = 31.75 μm; die size = 63.5 mm; frost line height = unknown; melt temperature = 466 K.

† MD and TD refer to machine and transverse directions film properties.

Note: The film properties of Tables 7 and 8 should not be compared due to different extrusion conditions.

References

1. Wild L, et al. *J. Polym. Sci., Poly. Phys. Ed.* 1982;20:441.
2. Mathot VBF, MF, Pijpers. *Polym. Bull.* 1984;11:297.
3. Hosoda S. *Polym. J.* 1986;20:383.
4. Zhou XQ, Hay JN. *Polym. J.* 1993;29:291.
5. Alamo R, Domszy R, Mandelkern L. *J. Phys. Chem.* 1984;88: 6587.
6. Speed CS, et al. In: *SPE RETEC, VII Polyolefins International Conference*, 1991:45.
7. Alamo RG, Viers BD, Mandelkern L. *Macromolecules.* 1993;26:5740.
8. Elston CTUS. *Patent.* 1972;3645992.
9. Lai S, Knight GW. In: *Society of Plastics Engineers Annual Technical Conference Proceedings (SPE ANTEC), Preprints.* 1993:1188.
10. Toensmeier PA, ed. In: *Modern Encyclopedia.* New York: McGraw-Hill; 1996:B–185.
11. Kissin YV. In: Graysen M, Eckroth D, eds. *Kirk-Othmer Encyclopedia of Chemical Technology.* 4th ed. Vol 17. New York: Wiley-Interscience; 1991:756.
12. James DE. In: Mark HF, et al, eds.*Encyclopedia of Polymer Science and Engineering*, 2nd ed. vol 6. John Wiley and Sons, 1985:429.
13. Schouterdem P, Groeninckx G, Reynaers H. In: Seymour RB, Cheng T, eds. *Advances in Polyolefins.* New York: Plenum Publishing; 1985, p. 373.
14. Kissin YV. In: *Isospecific Polymerization of Olefins with Heterogeneous Ziegler-Natta Catalyst.* New York: Springer-Verlag; 1985.
15. McDaniel MP, Benham EA. *U. S. Patent* 5,208,309 (4 May 1993); and U. S. Patent 5,274,056 (28 December 1993); (both to Phillips Petroleum Co.)
16. Leaversuch R. *Modern Plast.* (August 1996): 42.
17. Tso CC, et al. *Polymer.* 2004;45:2657.
18. Keating M, et al. *J. Macro. Sci., Phys.* 1999;B8:379.
19. Mirabella FM, et al. *J. Polym. Sci., Polym. Phys.* 2004;42:3416.
20. Ferhat-Hamid Z. et al. *J. Matr. Sci.* 200;42:3138.
21. ASTM Designation D 2238-92. "Standard Test Methods for Absorbance of Polyethylene Due to Methyl Groups at 1378 cm^{-1}. Annual Book of ASTM Standards, 1996.
22. Usami T, Takayama S. *Polym. J.* 1984;16:731.
23. McRae MA, Maddams W. *Makromol. Chem.* 1976;177:449.
24. Rugg FM, Smith JJ Wartman LH. *J. Polym. Sci.* 1953;11:1.

25. Willbour AHJ. *Polym. Sci.* 1959;34:569.

26. Blitz JP, McFaddin D. *J. Appl. Polym. Sci.* 1994;51:13.

27. Prasad A, Mowery D. In *Society of Plastics Engineers Annual Technical Conference Proceedings (SPE ANTEC)*, Preprints, 1997:2310.

28. Al-Malaika S, et al. *Polymer Degradation and Stability.* 2006;91:3131.

29. Bodor G, Dalcolmo HJ, Schroter O. *Coll. Polym. Sci.* 1989;267:480.

30. Randall JJ. *Macromol. Sci. Rev. Macromol. Chem. Phys.* 1989. C29:201.

31. Cheng H. *Macromolecules* 1991;24:4813.

32. De Pooter M, et al. *J. Appl. Polym. Sci.* 1991;42:399.

33. Pate TJ. In: Rubin II, ed. *Handbook of Plastics Material and Technology.* New York: John Wiley and Sons; 1990:327.

34. Coran AY, Anagnostopoulos CE. *J. Polym. Sci.* 1962;57:13.

35. Cernia EM, Mancini C, Saini A. *J. Appl. Poly. Sci.* 1968;12:789.

36. Wagner HLJ. *Phys. Chem. Ref. Data.* 1985;14(2):611.

37. Springer H, Hengse A, Hinrichsen G. *Colloid and Polym. Sci.* 1993;271:523.

38. Prasad, A. Unpublished data.

39. Bailey FE Jr, Walter ER. *Polym. Eng. Sci.* 1975;15:842.

40. Defoor F, et al. *Macromolecules* 1993;26:2575.

41. Heink M, Haberle KD, Wilke W. *Coloid and Polym. Sci.* 1991;269:675.

42. Martuscelli E, Pracella M. *Polymer.* 1974;15:306.

43. Darras O, Seguela R. *Polymer* 1993;34:2946.

44. Lambert WS, Phillips PJ. *Macromolecules.* 1994;27:3537.

45. Peacock AJ, Mandelkern L. *J. Polym. Sci., Polym. Phys. Ed.* 1990;28:1917.

46. Haber A, Kamal MR. *Plastics Engineering.* 1987;47(10):43.

47. Crist B, Graessley WW, Wignall GD. *Polymer.* 1982;23:1561.

48. Alamo RG, Chan EK, Mandelkern L. *Macromolecules.* 1992;25:6381.

49. Kim M, Phillips PJ. In: *Society of Plastics Engineers Annual Technical Conference Proceedings (SPE ANTEC), Preprints,* 1996:2205.

50. Jang YT, Parikh D, Phillips PJ. *J. Polym. Sci., Polym. Phys. Ed.* 1985;23:2483.

51. Woo L, Ling TK, Westphal S. *J. Plast. Film and Sheet.* 1994;10:116.

52. Konton E, et al. *Eur. Polym. J.* 2002;38:2477

53. Mandelkern L. In: Mark J, ed. *Physical Properties of Polymers.* 2nd ed. Washington, D.C: ACS Professional Reference Book, American Chemical Society; 1993:189.

54. Islam, Ashraful, et al. *J. Appl. Polym. Sci.* 2006;1009:5019

55. Dighton GL. In: Lappin GR, and Sauer JD, eds. *Alpha Olefins Applications Handbook.* New York: Marcel Dekker; 1989:63.

56. Patel RM, et al. *Polym. Eng. Sci.* 1994;34:1506.

57. Krishnaswamy RK, et al. (a) *Polymer.* 2000;41:9205.
    (b) *Polym. Eng. Sci.* 200040:2385.

58. Data supplied through the courtesy of Lyondell Chemical Co. (Equistar Chemicals), Cincinnati, Ohio.

59. Mavridis H, Shroff R. *Polym. Eng. Sci.* 1992;32:1778.

60. Shah K, et al. *J. Appl. Medical Polym.* 1999;3:34.

61. Sukhadia AM. *J. Plastic Film and Sheet.* 1994;10:213.

62. Halle RW, et al. *J. Plast. Film and Sheet.* 2005;21:13

Polyethylene, Low-density

A. Prasad

Acronyms LDPE, branched PE, high-pressure PE

Class Poly(α-olefin)

Structure $—(CH_2—CH_2)_n—$

Introduction LDPE is produced under high pressure (82–276 MPa) and high temperature (405–605 K) with a free radical initiator (such as peroxides and oxygen) and contains some long chain branches (LCB), which could be as long as chain backbones,[1,2] and short chain branches (SCB).[3] It is produced by either a tubular or a stirred autoclave reactor.[4] The autoclave process can produce LDPE resins having a wide range of molecular weight distribution (MWD) and less LCB in comparison with a tubular reactor.[4] Long chain branching has a strong influence on MWD, and hence on resin properties, such as processibility, melt strength, and film optical properties.[4] SCBs disrupt chain packing and are principally responsible for lowering the melting temperature and the crystal density for hydrocarbon polymers. LDPE is commercially available in a wide variety of molecular weight, MWD, SCB, and LCB contents, and density ranges.[4–6]

Thermal and mechanical properties of semicrystalline polymers are strongly dependent on MW, MWD, branching content, and density.[5,7,8] Controlled variations in these structural parameters result in a broad family of products with wide differences in thermal and mechanical properties. Shear modified LDPE samples are also available commercially.[9,10] The deliberate shearing of polymer melt results in a reduction in melt viscosity and elasticity without significant change in MW.[11] These reversible changes are advantageous for molding and extrusion and also result in film with better optical properties.[9,10,12] Shear modified LDPEs show vastly different crystallization rates.[12] The properties shown in the following table are given in ranges because of their dependence on molecular structure and are intended to represent best published examples of most commonly used commercial grades of LDPE for blown film, molding, and extrusion coating applications. Table 3 contains properties of tubular blown films.

Major Applications Major applications include blown film for bags and packaging; extrusion coatings for paper, metal, and glass; and injection molding for can lids, toys, and pails. Other applications include blow molding (squeeze bottles), rotomolding and wire and cable coatings, carpet backing, and foam for packaging material. There is considerable use of blends of LDPE with high-density polyethylene (HDPE) and linear low-density polyethylene (LLDPE) in a wide variety of applications. These blends are deliberately excluded in the data presented below.

Properties of Special Interest LDPE has a good balance of mechanical and optical properties with easy processibility and low cost. It can be fabricated by many different methods for a broad range of applications. Special properties of interest include: optical clarity, flexibility, toughness, high impact strength, good heat seal, low brittleness

662

temperature, good chemical resistance to aqueous solvents, and good electrical properties. LDPE may not be suitable for applications that require high stiffness and high tensile strength. Other limitations include: poor resistance to oxidizing agents, aliphatic solvents, aromatic solvents, polar liquids, chlorinated solvents, low softening point, poor scratch resistance, poor gas and moisture permeability, and relatively lower stress-crack resistance when compared to other types of polyethylene. LDPE undergoes thermal degradation at high temperatures and chain extension under shear conditions.

Major Suppliers Lyondell Chemical Co. (Equistar Chemicals), BP Amaco, Dow Chemical Co., Chevron Phillips Chemical Co., Huntsman Polymers, Westlake Polymers., Du Pont Co., ExxonMobil Chemical Co., Eastman Chemical Co., Rexene Corp., Nova Chemicals, Inc., Federal Plastics; Bamberger.

Property	Units	Conditions	Value	Reference
Type of branching	–	FTIR, NMR	Methyl, ethyl, butyl, amyl and longer branches	(13–15)
Type of unsaturation	%	FTIR		(16)
		Vinylidene	80	
		Vinyl	10	
		Trans	10	
Degree of SCB, commercial grades	Methy/1000 carbon	FTIR, NMR	10–33	(13–15)
Typical M_w range, commercial grades	g mol^{-1}	GPC	$3–40 \times 10^4$	(5, 16)
Typical polydispersity index (M_w/M_n)	–	GPC, strongly influenced by the amount of LCB	4–30	(17–19)
IR (characteristic absorption frequencies)	cm^{-1}	Thin film sample at room temperature	See table below	(14, 15, 20)

Characteristic frequencies of crystalline LDPE*

Wave Number	Intensity	Assignments
720	Very strong	CH_2 rocking
731	Very strong	CH_2 rocking
888	Very weak	Vinylidene groups
890	Very weak	CH_3 rocking
908	Medium	Terminal vinyl groups
964	Very weak	Trans double bond
990	Weak	Terminal vinyl groups
1,050	Very weak	CH_2 twisting
1,176	Very weak	CH_2 wagging
1,375	Weak-medium	CH_3 symmetrical bending
1,457	Very weak	CH_3 asymmetrical bending
1,463	Very strong	CH_2 bending

Wave Number	Intensity	Assignments
1,473	Very strong	CH_2 bending
2,850	Very strong	CH_2 symmetrical stretching
2,857	Very strong	CH_2 symmetrical stretching
2,874	Weak	CH_3 symmetrical stretching
2,899	Very strong	CH^2 asymmetrical stretching
2,924	Very strong	CH_2 asymmetrical stretching
2,960	Weak	CH_3 asymmetrical stretching

* Observed in the infrared spectra and band assignments.

Property	Units	Conditions	Value	Reference
Linear thermal expansion Coefficient	K^{-1} ($\times 10^{-5}$)	D 696 Temperature range = 238–423K	10.0–51.0 See table below	(21–23) (5, 16)

Temp. (K)	Coefficient of Expansion ($\times 10^{-5}$)	
	Linear	Cubical
238	10.0	30.0
273	18.3	55.0
293	23.7	71.0
313	29.0	87.0
333	33.7	101.0
353	40.3	121.0
373	46.6	140.0
383	51.0	153.0
388–423	25.0	75.0

Density (amorphous)	g cm^{-3}	At 298 K, extrapolated from melt temperature	0.855	(24, 25)
Solvents	–	368 K	Decalin, toluene	(26, 27)
		369 K	Xylene	
		371 K	Tetralin	
		341 K	Cyclohexene	
		374 K	n-Tetracosane	
Nonsolvents	–	359 K	Methylene chloride	(27)
		361 K	o-Dichloro benzene	
		366 K	1,2-Dichloropropane	
Solubility parameter	(MPa)$^{1/2}$	n-Tetracosane, 374 K	15.20	(27)
		Cyclohexene, 341 K	17.10	(27)
		n-Heptane, 359 K	14.57	(27)
		n-Octane, 361 K	14.50	(27)
		n-Hexane, 359 K	12.46	(27)
		n-Pentane, 361 K	12.52	(27)

Property	Units	Conditions	Value	Reference
		Chloroform, 350 K	16.00	(27)
		Carbon tetrachloride, 346 K	14.77	(27)
		Tetrachloroethylene, 346 K	15.69	(27)
		Chlorobenzene, 349 K	18.13	(27)
		o-Dichlorobenzene, 361 K	18.62	(27)
		Methylene chloride, 359 K	13.75	(27)
		See also the other references for more data	–	(27, 28–31)
Theta temperature Θ	K	Bis(2-ethylhexyl)adipate	418–443	(28)
		Bis(2-ethylhexyl)sebacate	423	
Flory interaction parameter χ	–	n-Tetracosane, 374 K	−0.09	(27)
		Cyclohexene, 341 K	−0.03	(27)
		n-Heptane, 359 K	0.13	(27)
		n-Octane, 361 K	0.18	(27)
		n-Octane, 393 K	0.31	(29)
		n-Octane, 418 K	0.30	(29)
		n-Hexane, 359 K	0.23	(27)
		n-Pentane, 361 K	0.35	(27)
		Toluene, 393 K	0.34	(29)
		Xylene, 350 K	0.51	(29)
		m-Xylene, 393 K	0.29	(29)
		p-Xylene, 393 K	0.27	(29)
		Chloroform, 350 K	0.25	(27)
		Tetrachloroethylene, 346 K	−0.05	(27)
		Chlorobenzene, 349 K	0.04	(27)
		o-Dichlorobenzene, 361 K	0.41	(27)
		Methylene chloride, 359 K	0.60	(27)
		Carbon tetrachloride, 346 K	0.01	(27)
		See also the other references for more data	–	(27–29)
Second virial coefficient, A_2	mol cm^3 g^{-2}	Tetralin, 334K		(32, 33)
		$M_w = 5.73 \times 10^5$	0.92×10^{-4}	
		$M_w = 1.98 \times 10^6$	0.84×10^{-4}	
Mark-Houwink parameters: K and a	$K = ml\,g^{-1}$ $a = None$	Should be used for approximate MW values only because LDPE contains long-chain branching	–	(32–36)
		Decalin, 342 K	$K = 3.87 \times 10^{-4}$, $a = 0.78$	(36)
		p-Xylene, 347 K	$K = 1.05 \times 10^{-3}$, $a = 0.63$	(33)
		Xylene, 353 K	$K = 1.35 \times 10^{-4}$, $a = 0.63$	(34)
Characteristics ratio $<r^2>/nl^2$	–	Tetralin, 334K	7.26, 7.58	(32)
Crystallographic data	Å	Wide-angle X-ray (WAXD)	See table below	(37–40)

Unit cell information

Lattice	Space Group	Unit Cell Dimensions (Å)			Monomers Per Unit Cell	Reference
		a	b	c		
Orthorhombic	Pnam	7.40	4.93	2.534	2	(37)
		7.36	4.92	–	–	(38)
		7.51	4.97	–	–	(39)
Monoclinic	$C2\,m^{-1}$	8.09	2.53	4.79	2	(40)

Property	Units	Conditions	Value	Reference
Crystallinity	%	DSC	33–53	(41–48)
Heat of fusion (per repeat unit)	$kJ\,mol^{-1}$	DSC	1.37–2.18	(41–48)
Density	$g\,cm^{-3}$	Commercial resins, ASTM D 1505	0.910–0.935	(4–6)
		Unit cell, 100% crystalline	1.00	(37)
		Unit cell, 100% crystalline	1.014	(38)
Lamellae thickness	Å	Raman longitudinal acoustic mode (LAM), various crystallization conditions used	See Table 1	(16, 48)
Crystal phase structure	%	LAM	See Table 1	(48)
Crystal orientation and birefringence	–	WAXD and infrared diachroism	See Table 2	(62)
Melting temperature	K	DSC, peak endotherms	378–388	(5, 6, 41, 44)
Glass transition Temperature	K	DMA, TMA	140–170 (See also transition and relaxation temperature table below)	(45–47)
Transition and relaxation Temperatures; and activation energy*	–	DMA	See table below	(16, 45–49)

* The transitions and relaxation temperatures associated with amorphous regions are designated as α, β, γ, etc. in descending temperature order. Several conflicting interpretations and values have been given regarding the origin and molecular nature of the transitions in LDPE (see references 45 to 49). It is believed that the values of T_α depends on crystallite thickness. The temperature of beta transition, T_β, does not depend on the crystallite thickness but rather the comonomer type and content. The transition is associated with glass transition. All transition values depend on the frequency of the DMA test. See references (46) and (47) for detailed discussions. The transition temperatures associated with peaks in dynamic loss listed in this table are given in ranges because of a wide range of values cited in the literature (for a specific examples see Table Below).

Transitions	Temperature Range (K)	Approximate Activation Energy ($kJ\,mol^{-1}$)
α	293–360	>420
β	233–280	160–200
γ	140–170	32–35

Property	Units	Conditions	Value	Reference
Heat capacity (of repeat units)	$kJ\,k^{-1}\,mol^{-1}$	At constant pressure and temperature of 298 K, (density $= 0.921\,g\,cm^{-3}$)	1.6536×10^{-4}	(50, 51)
		At other temperatures	–	(51)
Enthalpy (of repeat unit)	$kJ\,mol^{-1}$	Calorimeter		(51)
		Temperature (K)		
		80	0.287	
		140	0.883	
		200	1.722	
		260	2.887	
		320	4.545	
		380	7.619	
		415	9.384	
Entropy (of repeat unit)	$kJ\,k^{-1}\,mol^{-1}$	Calorimeter		(51)
		Temperature (K)		
		80	0.00586	
		140	0.0114	
		200	0.0164	
		260	0.0214	
		320	0.0273	
		380	0.0357	
		415	0.0406	
Defection temperature	K	D 648, at 273K and 0.45MPa	311–322	(23)
Heat distortion temperature	K	D 648, 455 KPa	313–323	(21)
Vicant softening point	K	D 1525	363–375	(5)
Tensile modulus	MPa	D 638	102–310	(5, 21, 52, 53)
Compressive strength	MPa	D 695	18–25	(23)
Tensile yield strength	MPa	D 638	9–15	(5, 21, 554)
Elongation at break	%	D 638	100–800	(5, 21, 52–54)
Yield stress	MPa	D 638	6.2–11.5	(5, 21, 52–54)
Flexural modulus	MPa	D 790 at 298K	240–330	(6, 21, 52)
Impact strength, notched Izod	$J\,m^{-1}$	D 256A	No break	(6, 21, 52)
Hardness	Shore D	D 676	40–60	(5, 6, 21, 52)
Low-temperature brittleness F_{50}	K	D 746	<197	(21, 52)
Refractive index	–	D 542, crystalline (value depends on density and chain branching)	1.5168–1.5260	(21, 55)
		Amorphous	1.49	(16, 56)
		Refractive index in melt	See table below	(56)
Specific volume	$cm^{3}\,g^{-1}$	Differential refractometer And dilatometer, Alathon 10	See table below	(56)
Specific refractivity	$cm^{3}\,g^{-1}$	Differential refractometer And dilatometer, Alathon 10	See table below	(56)

Temperature (K)	Specific Volume (cm^{3}g^{-1})	Refractive Index	Refractivity (cm^{3}g^{-1})
363.16	1.159	1.4801	0.3293
368.16	1.168	1.4736	0.3281
373.16	1.178	1.4693	0.3283
378.16	1.195	1.4630	0.3291
381.16	1.209	1.4575	0.3297
384.16	1.282	1.4510	0.3290
386.16	1.239	1.4432	0.3286
391.16	1.250	1.4392	0.3289
387.56	1.256	1.4368	0.3288

Property	Units	Conditions	Value	Reference
Dielectric constant ε'	–	D 150 at 1 kHz		(5, 21, 57, 58)
		Density (g cm^{-3})		
		0.920	2.28	
		0.930	2.30	
		0.935	2.31	
Loss factor, tan δ	–	Up to 100MHz	10^{-4}–10^{-3}	(5, 21)
Dielectric strength	M_V cm^{-1}	D 149, 283K	7.0	(21, 57–59)
		D 149, 373K	2.0	
Power factor	–	D 150 at 1 kHz	0.0003	(5)
Melt index	g (10 min)$^{-1}$	D 1238	0.2–50	(4, 6)
Flow activation energy:	kJ mol^{-1}	RMS, melt index = ~1 g (10 min)$^{-1}$	E_H = 61–67	(60)
E_H and E_V*		density = 0.919–0.931 g cm^{-3}, temp. range = 423–483 K	E_V = 8–10.5	
Coefficient of sliding friction μ	–	D 1894, dynamic c.o.f. to stainless steel, melt index = 2 g (10 min)$^{-1}$ density = 0.915 g cm^{-3}	0.7	(61)
Sonic velocity	m s^{-1}	Blown films	See Table 2	(62)
Water absorption	%	D 570, 24 h	<0.02	(21)

* The horizontal shift factor reflects the temperature dependence of relaxation time, and the vertical shift factor reflects the temperature dependence of modulus.

Table 1. Morphological and relaxation properties of LDPE as a function of crystallization condition*[(48)]

Condition	α-Relaxation (K, 3.5 Hz)	β-Relaxation (K, 3.5 Hz)	Crystallinity (%)	Crystallite Thickness (Å)	% Crystal Phase Structure		
					α_c	α_a	α_b
Slow cooled	348	258	36	105	37	49	14
Cooled in air	323	258	–	82	–	–	–
Quenched 80°C	323	258	40	80	36	53	11

Condition	α-Relaxation (K, 3.5 Hz)	β-Relaxation (K, 3.5 Hz)	Crystallinity (%)	Crystallite Thickness (Å)	% Crystal Phase Structure		
					α_c	α_a	α_b
Quenched 40°C	278	–	–	70	–	–	–
Quenched 0°C	273	–	32	–	–	–	–
Quenched −120°C	270	–	29	65	38	49	13

* $M_w = 3.46 \times 10^5$, $M_w = M_n = 18.5$, total SCB/1,000C = 10.6, LCB/1,000C = 2.2.

Note: Compression molded specimens were rapidly quenched to the specified quenching temperatures. Relaxation temperatures (tan δ peaks) were obtained on a DMA instrument at a heating scan rate of 1°C min^{-1}. Crystal phase structures were obtained by Raman LAM. The degree of crystallinity was determined from the heat of fusion data obtained on a DSC instrument.

Table 2. Morphological properties and crystal orientation of LDPE (density 0.920 g cm^{-3}) blown tubular films[62]

Thickness (µm)	Density (g cm^{-3})	Birefringence ($\times 10^3$)	Sonic Velocity (m s^{-1})		Infrared Diachronic Ratio (at 730 cm^{-1})	X-ray Angle (Degree)			
						a-axis		b-axis	
			MD*	TD*	MD/TD*	MD-TD*	ND-MD*	MD-TD*	ND-MD*
54	0.9182	−9.79	953	1,128	1.15	45–50	45–60	0	0
26	–	−8.23	870	1,003	0.88	45	45–60	0	0
23	0.9175	−3.07	852	998	1.13	70	60–70	0	0

* MD, TD, and ND refers to the machine, transverse, and normal direction of a blown film sample.

Table 3. Blown film properties of high clarity grade LDPE[64−70]*

Property	Units	Conditions	Value	Reference
Melt index	g (10 min)$^{-1}$	D 1238	2.0	(66)
Density	g cm^{-3}	D 1505	0.924	(66)
Peak melting point	K	DSC, 10°C min^{-1} cooling and heating rate	382.5	(66)
Heat of fusion	kJ mol^{-1}	DSC, 10°C min^{-1} cooling and heating rate	1.50	(66)
M_w	g mol^{-1}	GPC, 408K in 1,2,4-trichlorobenzene	79,200	(66)
M_w/M_n	–	GPC	3.8	(66)
Total haze	%	D 1003	5.2	(66)
Gloss, 45°D	Units	D 2457	71	(66)
Narrow angle scattering	%	D 1746	72	(66)
Modulus, 1% secant	MPa	D 882, MD	200	(66)
		D 882, TD	240	
Dart impact (per 25.4 µm)	g	D 1709, D 4272	74	(66)

Property	Units	Conditions	Value	Reference
Elmendorf tear	g	D 1922, MD	360	(65)
		D 1922, TD	220	
Tensile strength at break	MPa	D 882, MD	25	(66)
		D 882, TD	17	
Elongation at break	%	D 882, MD	290	(66)
		D 882, TD	500	

* Film and optics properties depend upon processing conditions (see references 9, 10, 58, 64–70). This table presents some representative values of typical high clarity grade LDPE films processed using the following conditions: blow-up ratio = 2:1; die gap = 0.635 mm; output = 48 kg h^{-1}; film thickness = 30.5 μm; die = 203 mm; frost line height = 330 mm; melt temperature = 463 K.

References

1. Flory PJ. *J. Am. Chem. Soc.* 1937;59:241.
2. Flory PJ. *J. Am. Chem. Soc.* 1947;69:2893.
3. Roedel MJ *J. Am. Chem. Soc.* 195375:6110.
4. Doak KW, Schrage A. In: Raff RAV, Doak KW, eds. *Crystalline Olefin Polymers Part 1.* New York: Interscience Publishers; 1965:chap. 8.
5. Bibee DV. In: Rubin II, ed. *Handbook of Plastics Materials and Technology.* John Wiley and Sons; 1990:317.
6. Toensmeier PA, ed. In: *Modern Plastics Encyclopedia.* New York: McGraw-Hill; 1996, p. B-185.
7. Popli R, Mandelkern L. *J. Polym. Sci., Polym. Phys. Ed.* 1987;25:441.
8. Peacock AJ, Mandelkern L. *J. Polym. Sci., Polym. Phys. Ed.* 1990;28:1917.
9. Meissner. *J. Pure Appl. Chem.* 1975;42:551.
10. Stehling FC, Speed CS, Westerman L. *Macromolecules* 1981;14:698.
11. Van Prooyen K, Bremner T, Rudin A. *Polym. Eng. Sci.* 1994;34:570.
12. Magill JH, Peddada SV, McManus GM. *Poly. Eng. Sci.* 198;21:1.
13. Mandelkern L, Maxfield J. *J. Polym. Sci., Polym. Phys. Ed.* 1979;17:1913.
14. Willbourn AH. *J. Polym. Sci.* 1959;34:569.
15. Blitz JP, McFaddin DC. *J. Appl. Polym. Sci.* 1994;51:13.
16. Aggarwal SL. In: Brandrup J, Immergut EH, eds. *Polymer Handbook.* 2nd ed. New York: John Wiley and Sons; 1975:13., chap. V, p.
17. Guillet JE, et al. *J. Appl. Polym. Sci.* 1965;8:757.
18. Lecacheux D, Lesec J, Quivoron C. *J. Appl. Polym. Sci.* 1982;27:4877.
19. Han CD, et al. *J. Appl. Polym. Sci.* 1983;28:3435
20. Groenewege MP, et al. In: Raff RAV, Doak KW, eds. *Crystalline Olefin Polymers Part 1.* New York: Interscience Publishers; 1965:chap. 14,
21. Boysen RL. In: Kroschwitz JI, ed. *Kirk-Othmer Encyclopedia of Chemical Technology.* 3rd ed. Vol 16. New York: Wiley-Interscience; 1981:402–420.
22. Hann FC, Macht ML, Fletcher DA. *Ind. Eng. Chem.* 1945;37:526.
23. Chanda M, Roy SK. *Plastics Technology Handbook.* New York: Marcel Dekker; 1987:519.
24. Allen G, Gee G, Wilson GJ. *Polymer* 1960;1:456.
25. Chiang R, Flory PJ. *J. Am. Chem. Soc.* 1961;83:2857.

26. Coran AY, Anagnostopoulos CE. *J. Poly. Sci.* 1962;57:13.

27. Cernia EM, Mancini C, Saini A. *J. Appl. Polym. Sci.* 1968;12:789.

28. Barton AFM. *CRC Handbook of Polymer-Liquid Interaction Parameters and Solubility Parameters.* Boca Raton, Fla: CRC Press; 1990:161–177.

29. Barton AFM. *CRC Handbook of Solubility Parameters and Other Cohesion Parameters.* Boca Raton, Fla: CRC Press; 1983:256.

30. Richards RB. *Trans. Faraday Soc.* 1946;42:10.

31. Muthana MS, Mark H. *J. Polym. Sci.* 1949;4:527.

32. Trementozzi QA. *J. Polym. Sci.* 1959;36:113.

33. Trementozzi QA. *J. Polym. Sci.* 1957;23:887.

34. Harris I. *J. Polym. Sci.* 1952;8:353.

35. Billmeyer JW. *J. Am. Chem. Soc.* 1953;75:6118.

36. (a) Tung LH. In: Raff RAV, Doak KW, eds. *Crystalline Olefin Polymers Part 1.* New York: Interscience Publishers; 1965:chap. 11;
    (b) Trementozzi QA, Newman S. In: Raff RAV, Doak KW, eds. *Crystalline Olefin Polymers Part 1.* New York: Interscience Publishers; 1965:chap. 9.

37. Bunn CW. *Trans. Faraday Soc.* 1939;35:482.

38. Walter ER, Reading PF. *J. Polym. Sci.* 1956;21:561.

39. Bailey FE, Walter ER. *Polym. Eng. Sci.* 1975;15:842.

40. Tanaka, K., T. Seto, and T. Hara. *J. Phys*

41. Prasad, A. Unpublished data.

42. Glotin M, Mandelkern L. *Colloid and Polym. Sci.* 1982;260:182.

43. Strobl GR, Hagedorn W. *J. Polym. Sci., Polym. Phys. Ed.* 1978;16:1181.

44. Doak KW. In: Mark HF, et al, eds. *Encyclopedia of Polymer Science and Engineering.* 2nd ed. Vol 6. New York: John Wiley and Sons; 1985:387.

45. Stehling FC, Mandelkern L. *Macromolecules* 1970;3:242.

46. Mandelkern L. In: Mark J, ed. *Physical Properties of Polymers.* 2nd ed. Washington, D.C: ACS Professional Reference Book, American Chemical Society; 1993:189.

47. Klein DE, Sauer JA, Woodward AE. *J. Polym. Sci.* 1956;22:455.

48. Popli R, et al. *J. Polym. Sci., Polym. Phys. Ed.* 1984;22:407.

49. Illers KH, Kolloid Z. *Z. Polym.* 1972;250:426.

50. Wilski H. In: Brandrup J, Immergut EH, eds. *Polymer Handbook.* 2nd ed. New York: John Wiley and Sons; 1975:215. chap. III.

51. Passaglia E, Kevorkian HK. *J. Appl. Polym. Sci.* 1963;7:119.

52. Prasad A. Unpublished data.

53. Brother MJ, et. al. *Int. Polym. Processing.* 2004;19:236

54. Basfer AA, et. al. *J. Appl. Polym. Sci.* 2003;88:459

55. Baccaredda M, Schiavinato G. *J. Polym. Sci.* 1954;12:155.

56. Bianchi JP. *J. Polym. Sci.* 1958;27:561.

57. Lanza VL, Herrmann DB. *J. Polym. Sci.* 1958;28:622.

58. Dodbin S, et. al. *J. Appl. Polym. Sci.* 2002;86:1959

59. Doepken HC, Kiss KD, Mangaraj D. In: *Preprints Organic Coatings and Plastics Chemistry.* Vol 38. Washington, D.C: American Chemical Society; 1978: 418.

60. Mavridis H, Shroff R. *Polym. Eng. Sci.* 1992;32:1778.

61. Lundberg RD. In: Walker BM, ed. *Handbook of Thermoplastic Elastomers.* New York: Van Nostrand Reinhold; 1979:250. chap. 6.

62. Haber A, Kamal MR. *Plastics Engineering.* 1987;43(10):43.

63. Silvestre C, et. al. *Macromol. Mat. Eng.* 2006;291:1477

64. Choi K, Spruiell JE, White JL. *J. Polym. Sci., Polym. Phys. Ed.* 1982;20:27.

65. Kwack TH, Han CD. *J. Appl. Polym. Sci.* 1983;28:3419.
66. Data supplied through the courtesy of Equistar Chemicals,Cincinnati, Ohio.
67. Lindenmeyer PH, Lustig S. *J. Appl. Polym. Sci.* 1965;9:227.
68. Gupta A, Simpson DM, Harrison IR. *J. Appl. Polym.Sci.* 1993;50:2085.
69. Simpson DM, Harrison IR. *J. Plast. Film and Sheet.* 1992;8:192.
70. Shang SW, Kamala RD. *J. Plast. Film and Sheet.* 1995;11:21.

Polyethylene, Metallocene Linear Low-density

A. Prasad

Acronyms, Alternative Names mLLDPE, metallocene PE, single site catalyzed LLDPE (SSC), polyolefin plastomers (POP), homogeneous ethylene copolymers

Class Poly (α-olefins)

Structure $-[CH_2-CH_2-CHR-CH_2]_n-$ (R = α-olefin)

Introduction New types of linear low-density polyethylenes (LLDPE) based on the metallocene catalyst technology was introduced in the market place in 1991. There are now a number of well-established metallocene PE families commercially available and new products are constantly being introduced into the marketplace. Current production based on metallocene catalyst technology exceeds 3 billion pounds. Metallocene share of global LLDPE consumption is projected to grow between 11% and 22% by 2012. Metallocene-based Ziegler-Natta catalysts utilize a new synthetic approach for the polymerization of poly (α-olefins).[1−8] Metallocene precatalysts are based primarily on group IV transition metals (primarily titanium and zirconium straddled by a pair of cyclic alkyl molecules) and require a coactivator, which is typically methylalumoxane but certain acids containing noncoordinating anions as bases also work well. This new family of polyolefin copolymers has a significantly different chain microstructure than conventional LLDPE.[9−18] The single site characteristics of metallocenes, with the catalyst site being identical, are known to produce materials having the most probable molecular weight distribution ($M_w/M_n = 2.0$), with essentially a random comonomer distribution and narrow composition distribution. The comonomers most frequently used commercially are butene, hexene, and octene. Copolymerization of ethylene with 4-methyl-1-penetene (4-MP-1), 1-decene, octadecene, and cyclic and bicyclic groups has been also reported in the literature.[19−22] Several terpolymers are also commercially available.[14] Exxon Chemical Company manufactures ethylene-butene copolymers, ethylene-hexene copolymers, and terpolymers of butene and hexene comonomers. Exxon markets these mLLDPEs under the trademark name of Exxpol™ Exact PE (density range of 0.910–0.865 g cm^{-3}) and Exxpol™ Exceed PE (density range of 0.925–0.910 g cm^{-3}). Dow Chemical Company manufactures ethylene-hexene and ethylene-octene copolymers using constrained geometry catalyst technology (CGCT). The Dow mLLDPE trade mark names are Engage™, Affinity™, Elite™ and Enhanced™ PE. The Affinity resins range in density from 0.902 to 0.935 g cm^{-3}, in weight percent comonomer from 2 to 12% octene comonomer, and melt index (MI) from 1.0 to 3.5 g (10 min)$^{-1}$. Elite™ resin range in density from 0.915 to 0.935 g cm^{-3} and melt index (MI) from 0.8 to 15 g (10 min)$^{-1}$. Metallocene LLDPEs are relatively difficult to process because of narrow molecular weight distribution (MWD) when compared to conventional Ziegler LLDPEs.[13,23−29] Metallocene catalyst based octene-1 LLDPE copolymers made by the Dow Chemical Company is known to process better as a result of their long-chain branched (LCB) Structure[25,26,30−40], referred to as Dow Rheology Index (DRI) numbers.[13,41] The LCB is also responsible for improved melt strength in mLLDPEs.[16] Exxon has also addressed the processibility issue with advanced performance terpolymers[14] and Univation process that uses super condensed mode gas phase technology.[28,29] LCB bimodal mLLDPE resins are commercially available from BP Chemicals.[42] Such mLLDPEs are produced by BP's proprietary gas phase fluid bed technology called Innovene™ technology.[42]

Besides molecular weight (MW) and molecular weight distribution (MWD), mechanical and thermal properties of LLDPE depend on the comonomer amount (density), composition distribution, and comonomer type.[8,9,11,16,18,43−45] The comonomer type is also referred to as short-chain branches (SCB). Consequently, mLLDPEs have quite different mechanical properties than conventional LLDPE made by Ziegler-Natta type catalyst. The mLLDPEs are commercially available in wide variety of MI and density ranges. The materials in the density range of 0.886 to 0.863 g cm^{-3} are called elastomeric PE and are presented in the entry on Polyethylene, elastomeric (very highly branched), in this handbook. Metallocene LLDPEs of density greater than 0.886 g cm^{-3} is called plastomers. This entry covers properties of mLLDPE plastomers in the density range of 0.887 to 0.935 g cm^{-3}. Due to wide range of MI and density, mLLDPE properties shown in the following table are given in ranges. Here, only those properties are listed that differ substantially from the conventional Ziegler-Natta type LLDPE and are intended to represent best published examples of commercially available grades of mLLDPE resins. Physical property data of mLLDPE copolymers of less common comonomer such as 1-decene of 1-octadecene can be found in references (21 and 22) The physical properties of extruded materials may vary substantially from those of the compression molded samples. For illustration purposes, some of the compression molded samples and blown film properties that depend on the chemical nature of the comonomer are listed in Tables 2, 3, and 4.

Major Applications Major applications include blown and cast packaging films, stretch film, injection molding goods, medical devices, automotive applications, wire and cable coatings, electrical cables, adhesives, and sealants, lamination, freezer-film, consumer bags. Other applications include blow molding, pipe and conduit, rotomolding, fibers and nonwoven fabrics, foams for sporting goods and houseware goods.

Properties of Special Interest Flexibility, low extractability, high shock resistance, high toughness, exceptionally high dart-impact strength and puncture resistance, balanced machine and transverse direction tear strength, better clarity, and low heat seal temperature, better electrical/abrasion properties, good organoleptic properties and better biaxial orientation than conventional LLDPEs. Other properties of interest include low brittleness temperature, good chemical resistance to acids and aqueous solvents, good heat seal, good stress-crack resistance properties, and good structural stability at high temperatures.

Limitations mLLDPEs without the long-chain branching are relatively difficult to process because of narrower MWD. Other limitations include: poor stretchability, no significant advantage in film tear properties, and higher resin cost when compared to conventional Ziegler LLDPE.

Major Suppliers Dow Chemical Co., Exxon Mobil Chemical Co., BP Amoco Chemical Co., BASF, Mitsubishi Chemical Corp, Borealis, Mitsui Petrochemicals, Chevron Phillips Chemical Co.

Property	Units	Conditions	Value	Reference
Typical comonomer	–	Butene, hexene, octene	–	(4, 8, 9–11)
Degree of branching, commercial resins	mol%	NMR, ethyl, butyl, and hexyl branches	0.5–8.5	(6, 18, 46–48)
NMR peaks	ppm	^{13}C NMR, Octene-1	characteristic δ Peaks @ 22–24, 24–25, 26.5–28.5, 25.2–33.2, 33.2–36.8, 35.8, 37–39.5, 39.5–40.5, 40.5–41.5	(47)
IR (characteristics Absorption frequencies)	cm^{-1}	D2238	see entry on LLDPE in this handbook	(47)
Ratios of different unsaturated groups	–	FTIR Thickness, $t = 0.22$ mm	See table below	(47)

Ratios of different unsaturated groups in Octene-1 mLLDPE polymers

Mole %	Density g cm^{-3}	MI g (10 min^{-1})	Vinyl CH$_2$=CH— (908 cm^{-1})	Vinylidene CH$_2$=C$\big\langle$ (889 cm^{-1})	trans-vinylene —CH=CH— (965 cm^{-1})
3.2	0.916	1.42	1	3	4
4.9	0.916	1.09	1	4	7

Property	Units	Conditions	Value	Reference
Extinction coefficient, ε'	l mol^{-1} cm^{-1}	FTIR, film, $t = 0.22$ mm	@ 965 cm^{-1} = 100 ± 7 @ 908 cm^{-1} = 122 ± 7 @ 889 cm^{-1} = 158 ± 7	(49)
Typical molecular weight range (M_w)	g mol^{-1}	GPC, in 1,2,4-trichlorobenzene	4–11 ($\times 10^4$)	(1, 2, 50)
Typical polydispersity Index (M_w/M_n)	–	GPC	2–2.5	(1, 2, 8, 41, 43)
Crystallographic data*	Å	Wide angle X-ray	See table below	(51–53)

* Crystalline unit cell parameters depend on crystallite thickness. The comonomer amount, comonomer type, and crystallization conditions determine the crystallite thickness in ethylene copolymers. The major cause of lattice expansion in ethylene copolymers is due to decrease in lamellae thickness by exclusion of branch points from the lamellar crystals coupled with surface stress on thin lamellae (see references 54 and 55 for details). The table below is for butene-1 mLLDPE ($M_w = 122{,}000$, $M_w/M_n = 2$), crystallized from the melt at a cooling rate of 7°C min^{-1}.

Lattice	Mol%	Unit Cell Dimensions (Å)			Unit Cell Volume (nm³)	Unit Cell Density (g cm⁻³)
		a	b	c		
Orthorhombic	3.0	7.53	5.00	2.54	0.0959	0.9724
	5.2	5.58	4.99	2.54	0.0963	0.9679

Property	Units	Conditions	Value	Reference
Crystallinity	%	DSC	33–53	(8–12, 43, 56, 57)
Heat of fusion	kJ mol⁻¹	DSC	1.37–2.18	(8–12, 43, 56, 57)
Density, commercial resins	g cm⁻³	D1505–85	0.887–0.940	(18, 58)
Avrami exponent*	–	Depends on counit content and crystallization temperature	See table below	(59–62)

* A caution should be exercised in using Avrami exponent values for the copolymers. In contrast to homopolymer crystallization, the isotherms of copolymers do not superpose one with the other; deviations from the Avrami relation occurs at low levels of crystallinity; and retardation in crystallization rate is pronounced with the extent of transformation due to continuous change in both composition and sequence distribution during crystallization. See reference (60) for the detailed discussion. Avrami exponent values for selected mLLDPE are shown below. See also reference (57, 61) for Avrami exponent value for non-isothermal crystallization conditions

Comonomer	Mole (%)	Crystallization condition	Value	Reference
Octene	<7.5	Not known	2–4	(59)
Octene	>7.5	Isotherm crystallization: <328K	2	(59)
Octene	>7.5	Isotherm crystallization: >328K	1	(59)
Hexene	1.21	Isotherm crystallization: 381–388K	3	(60)
Butene	3.3	Isotherm crystallization: 384.4–388K	2.8	(62)

Property	Units	Conditions	Value	Reference
Lamellae thickness and crystal phase structure	Å and %	Raman longitudinal acoustic mode (LAM), small-angle X-ray (SAXS), transmission electron microscopy (TEM)	See Table 1	(17, 43, 44, 51)
Surface free energy, σ_e (chain-folding crystal face)	J m⁻²	Thomson-Gibbs equation; value is dependent on counit content:		
		Octene-1 0.9 mol%	0.066	(44)
		Octene-1 3.9 mol%	0.096	(44)
		Octene-1 2.4 mol%	0.214-0.286	(63)
		Octene-1 4.7 mol%	0.268-0.357	(63)
		Butene-1 7.5 mol%	0.268-0.357	(63)

Property	Units	Conditions	Value	Reference
Melting temperature T_m	K	DSC peak endotherm, density range: 0.886–0.935 gm cm^{-3} (T_m depends on MW and SCB content but not on SCB type. Single and multiple endotherms have been observed)	363–398	(8–12, 56, 57)
Transition and relaxation Temperatures	K	DMA tan δ peaks at 1 Hz, heating rate = 3°C min^{-1} (values depend on mol% branching); value for 2.8–8.2 mol%, octene-1	$\alpha = 322$–373 $\beta = 245$–232 $\gamma = 153$–163	(11, 64, 65)
Vicant softening point	K	D 1525, density range = 0.920–0.912 gm cm^{-3}	382–368	(15, 67)
Tensile modulus	MPa	D 412, <1% strain, density range = 0.887–0.935 gm cm^{-3}	20–550	(18, 66)
		2–3% strain, independent of comonomer type, depends on crystallinity, value for crystallinity range of 7–50%	4–70	(43)
		D 1708, octene-1		(11)
		Density = 0.916 gm cm^{-3}	400	
		Density = 0.9014 gm cm^{-3}	120	
		ISO 527, octene-1		(67)
		298 K, density = 0.909 gm cm^{-3}	200	
		Density = 0.935 gm cm^{-3}	700	
Elongation at break	%	D 638	>700, no break	(66, 68)

Yield stress*[11,43,61,67,69]

Comonomer	Mole (%)	$M_w \times 10^{-4}$	Density (gm cm^{-3})	Crystallinity (%)	Conditions	Yield stress (MPa)	Reference
Butene	0.95	5.3	–	40.0	Specimen quenched to 195 K, draw rate = 2.54 cm min^{-1}	12.0	(43)
	1.26	9.0	–	37.0		10.4	(43)
	2.05	12.5	–	27.0		8.3	(43)
Hexene	0.6	5.6	–	40.0	Same as above	11.4	(43)
	1.1	5.8	–	36.0		10.3	(43)
	2.8	5.7	–	19.0		5.6	(43)
	3.5	6.3	–	15.0		5.5	(43)
4-MP-1	0.7	12.7	–	37.0	Same as above	10.0	(43)
	1.3	13.7	–	27.0		7.0	(43)
	2.0	10.6	–	26.0		7.0	(43)
	2.1	23.8	–	18.0		3.5	(43)

Comonomer	Mole (%)	$M_w \times 10^{-4}$	Density (gm cm^{-3})	Crystallinity (%)	Conditions	Yield stress (MPa)	Reference
Octene	0.7	11.7	–	40.0	Same as above	12.0	(43)
	1.4	7.9	–	28.0		7.5	(43)
	4.6	14.9	–	7.0		2.1	(43)
	0.005	8.25	0.935	–	ISO 527	20.0	(67)
	1.8	9.67	0.909	–	ISO 427	8.0	(67)
	2.8	–	0.9209	46.0	D 1708	13.0	(11)
	5.2	–	0.9029	33.0	D 1708	7.8	(11)

* Yield stress value is independent of comonomer type but depends on crystallinity value (see also ref. 70)

Property	Units	Conditions	Value	Reference
Flexular modulus	MPa	ISO 178, octane-1		(67)
		298 K, density = 0.909 gm cm^{-3}	125	
		Density = 0.935 gm cm^{-3}	635	
Ultimate tensile stress	MPa	Depends on MW and counit		(43, 66)
		Content, has maximum value		
		at $\sim M_w$ of 10^5 for all the copolymer;		
		for $M_w = 10^5$		
		mole% branch ≤ 1	48	(43)
		mole% branch = 1–3.5	38	(43)
		mole% branch ≥ 3.5	32	(43)
		mole% octene = 2.8	34.6	(11)
		mole% octene = 5.2	30.9	(11)
Impact Strength, notched Izod	J m^{-1}	ISO 180, octane-1		(67)
		298 K, density = 0.909 gm cm^{-3}	No break	
		Density = 0.935 gm cm^{-3}	2,500	
Dynatup impact	J	D 3763-86, several mLLDPE used		(71)
		Butene-1: density = 0.912 gm cm^{-3}	17	
		Butene-1: density = 0.921 gm cm^{-3}	13	
		Octene-1: density = 0.912 gm cm^{-3}	27	
		Octene-1: density = 0.921 gm cm^{-3}	19.5	
Hardness	Shore A	D 2240, octene-1, 6.2 mol%	75	(72)
Refractive Index n_D^{25}	–	D 542	see table below	(70)

Copolymer	Conditions	Density (g cm^{-3})	n_{ND}	n_{TD}	n_{MD}
Butene-1	Quenched	0.9192	1.5142	1.5148	1.5150
Butene-1	Slow cooled	0.9216	1.5156	1.5167	1.5162
Butene-1	Blown film	0.9169	1.5127	1.5132	1.5140
Hexane-1	Quenched	0.9186	1.5141	1.5141	1.5143

Copolymer	Conditions	Density (g cm^{-3})	n_{ND}	n_{TD}	n_{MD}
Hexane-1	Slow cooled	0.9210	1.5147	1.5168	1.5165
Hexane-1	Blown film	0.9167	1.5129	1.5128	1.5140
Octene-1	Quenched	0.9186	1.5135	1.5147	1.5152
Octene-1	Slow cooled	0.9197	1.5136	1.5161	1.5159
Octene-1	Blown film	0.9186	1.5122	1.5129	1.5137

n_{ND}, n_{TD}, n_{MD} refers to refrective index in normal, transverse and machine direction to the film

Property	Units	Conditions	Value	Reference
Volume resistivity	ohm cm ($\times 10^{14}$)	1mm thick sample Density = 0.915 gm cm^{-3}		(73)
		293K	10,800	
		313K	2,220	
		333K	54.0	
		363K	1	
		Density = 0.935 gm cm^{-3}		
		293 K	32,000	
		313 K	10,000	
		333 K	550	
		363 K	18.0	
Melt index	g (10 min)$^{-1}$	D 1238	0.8–30	(58)
Water vapor transmission rate (WVTR)	g m^{-2} day^{-1} (normalized to 25.4 μm thick film)	Test temperature = 311 K; relative humidity = 90%; butene-1: density range = 0.900–0.910 gm cm^{-3}	15.5–19.5	(74)
Oxygen transmission rate (OTR)	mmol cm/ cm^2 h kPa	Test temperature = 273–298 K; various types of mLLDPE	6–10($\times 10^{-8}$)	(74, 75)
Carbon dioxide transmission rate (CO$_2$TR)	mmol cm/ cm^2 h kPa	Test temperature = 273–298 K; various types of mLLDPE	8–50 ($\times 10^{-8}$)	(74, 75)

Table 1. Crystal phase structure and lamellae thickness of mLLDPE by short chain branching type

Comonomer	Mole % Branch	M_w ($\times 10^{-4}$)	Conditions[†]	Lamella Thickness (Å)	% Crystallinity (DSC)	α_a (%)	α_b (%)	α_c (%)	Reference
Butene	0.95	5.3	A	87	34	44	16	40	(43)
	1.26	9.0		93	31	46	17	37	(43)
	2.05	12.5		72	24	62	11	27	(43)
Butene	3.0	12.2	B	47	33	45	10	45	(51)
	5.2	11.9		29	25	47	15	38	(51)

(Continued)

Table 1. *(Continued)*

Comonomer	Mole % Branch	M_w ($\times 10^{-4}$)	Conditions[†]	Lamella Thickness (Å)	% Crystallinity (DSC)	α_a (%)	α_b (%)	α_c (%)	Reference
Hexene	0.6	5.6	A	95	41	48	12	40	(43)
	1.1	5.8		82	35	46	18	36	(43)
	1.2	10.4		77	–	58	10	32	(43)
	2.2	8.8		62	–	68	12	20	(43)
	3.5	6.27		57	12	67	18	15	(43)
Hexene	1.9	9.65	C	86	49	9	41	50	(17)
			D	71	–	13	42	45	(17)
4-MP-1	0.7	12.7	A	72	40	55	8	37	(43)
	1.3	13.7		67	38	61	12	27	(43)
	2.0	10.6		–	28	55	19	26	(43)
	3.6	6.3		50	19	67	22	11	(43)
Octene	0.7	11.7	A	87	34	47	13	40	(43)
	1.4	7.9		78	27	52	20	28	(43)
	4.6	14.9		45	8	82	11	7	(43)
Octene	0.9	7.7	E	140	60	–	–	–	(44)
	3.9	8.14		72	41	–	–	–	(44)

α_c, α_b, and α_a refers to fraction of chain units in the perfect crystals, interfacial region, and amorphous region, respectively of a lamella. See references (10) and (43) for more data and detailed discussions.

[†] A = LAM, quenched to 195 K. B = SAXS, sample cooled at 7°C min^{-1}. C = Lamella thickness by TEM, crystal phase structure by ^{13}C NMR; samples were crystallized isothermally at 383K for 18 h. D = quench cooled to 73 K. E = SAXS, sample cooled at 20°C min^{-1}. See also reference 17 and 76 for more data. See references 52 and 77 for lamellae thickness for mLLDPE samples crystallized at high pressure ~495 MPa.

Several publications now discuss the structure-property behavior and morphology development in films of mLLDPE as a function of comonomer content and type.[78–89]

Table 2. Blown film properties comparison of ethylene-octene mLLDPEs of different densities and conventional Ziegler ethylene-octene LLDPE*

Property	Units	Conditions	mLLDPE1 (Octene)	mLLDPE2 (Octene)	mLLDPE3 (Octene)	LLDPE (Octene)
Melt index	g (10 min)$^{-1}$	D 1238	0.85	1.5	1.6	1.0
Density	g cm^{-3}	D 1505	0.920	0.912	0.895	0.920
Total short-chain branch	Mol%	NMR	1.7	2.4	3.7	2.7
M_w/M_n	–	GPC	2.0	2.0	2.0	4.6
Total haze	%	D1003	12	11	1.1	11.3
Gloss, 45°D	Units	D2457	61	63	90	61
Modulus, 2% secant	MPa	D882, MD	206	152	53	190
		D882, TD	230	152	55	215

Table 2. (*Continued*)

Property	Units	Conditions	mLLDPE1 (Octene)	mLLDPE2 (Octene)	mLLDPE3 (Octene)	LLDPE (Octene)
Dart impact	g	D1709, D4272	>850 (no break)	650	>850 (no break)	266
Elmendorf tear	g	D1922, MD	740	1,190	550	980
		D1922, TD	990	1220	590	1,210

* The results in Table 2 were obtained for the following blown film conditions: blow-up ratio = 2:5: 1; die gap = 1.78 mm; output not mentioned; film thickness = 50.5 μm; die size = 152.4 mm; frost line height not mentioned; melt temperature = 508 K.[15;90]

Note: It is well-known that film properties depend on the chemical nature of the comonomer.[91−93] However, blown film properties also depend on the processing conditions.[91;93−95] Properties listed in Tables 2 and 3 were obtained at different extrusion conditions and, therefore, should not be compared.

Table 3. Blown film properties comparison of ethylene-hexene mLLDPEs of different densities and conventional Ziegler ethylene-hexene LLDPE*

Property	Units	Conditions	mLLDPE1 (Octene)	mLLDPE2 (Octene)	mLLDPE3 (Octene)	LLDPE (Octene)
Melt index	g (10 min)$^{-1}$	D 1238	0.80	1.0	1.0	1.04
Density	g cm^{-3}	D 1505	0.925	0.918	0.917	0.919
Total short-chain branch	Mol%	NMR	2.18	3.38	3.66	3.88
Peak melting point	K	DSC, at a cooling and heating rate of 5°C min^{-1}	385 & 394	379.5 & 392.5	379.5 & 391.5	372 & 397.5
Heat of fusion	kJ mol^{-1}	DSC, at a cooling and heating rate of 5°C min^{-1}	1.93	1.49	1.49	1.68
$M_{\mathrm{w}} \times 10^{-5}$	g mol^{-1}	GPC, 408 K in 1,2,4-trichlorobenzene	1.37	1.22	1.25	1.24
$M_{\mathrm{w}}/M_{\mathrm{n}}$	–	GPC	2.7	2.4	2.6	4.1
Total haze	%	D1003	19	12	10	16
Gloss, 45°D	Units	D2457	38	58	61	41
Modulus, 1% secant	MPa	D882, MD	305	187	175	206
		D882, TD	290	185	168	250
Dart impact	g	D1709, D4272	230	860	1040	215
Elmendorf tear	g	D1922, MD	300	230	260	360
		D1922, TD	450	480	420	710
Tensile yield	MPa	D 882, MD	11.8	9.0	8.9	9.6
		D 882, TD	14.0	8.7	11.1	11.2

(*Continued*)

Table 3. *(Continued)*

Property	Units	Conditions	mLLDPE1 (Octene)	mLLDPE2 (Octene)	mLLDPE3 (Octene)	LLDPE (Octene)
Elongation at yield	%	D 882, MD	6	8	8.5	8
		D 882, TD	16	12	20	14
Tensile strength break	MPa	D 882, MD	56	61	58	35
		D 882, TD	57	49	47	22
Elongation break	%	D 882, MD	530	580	560	420
		D 882, TD	650	600	580	540
Hexene extractables	%	312 K for 2 h	<1	<1	<1	>3

* The results in Table 3 were obtained for the following blown film conditions: blow-up ratio = 2.5 : 1; die gap = 2.54 mm; output = 30 kg h^{-1}; film thickness = 28 μm; die size = 102 mm; frost line height = 330 mm; melt temperature = 475 K.[96]
See also references (8, 14, 16, 71, 97) for more data.

Table 4. Blown film properties comparison of ethylene-octene and ethylene-butene mLLDPEs of different densities with the compression molded specimens*

Property	Units	Conditions	Butene-1	Octene-1	Butene-1	Octene-1
Melt index	g (10 min)$^{-1}$	D 1238	0.94	0.93	0.97	1.02
Density	g cm^{-3}	D 792	0.912	0.912	0.921	0.921
Total short-chain branch	Mol%	NMR	4.17	3.04	3.04	1.77
Peak melting point	K	DSC, at a cooling and heating rate of 10°C min^{-1}	374.2	379.4	383.3	386.5
$M_w \times (10^{-4})$	g mol^{-1}	GPC, 423K in 1,2,4-trichlorobenzene	7.54	9.0	7.28	8.3
M_w/M_n	–	GPC	2.22	2.12	2.25	2.19
Compression molded sample properties						
Intrinsic tear	g	Elmendorf A tear test using 25.4 μm sample (normalized to per mil)	86	345	63	300
Dynatup impact	J	D 3763-86	17.0	27.0	13.0	19.5
Tensile yield	MPa	D 638	8.68	8.96	12.00	11.64
Tensile break	MPa	D 638	15.73	25.61	14.63	24.14
Tensile strain at break	%	D 638	743	697	640	767
Blown film sample properties						
Density (film)	g cm^{-3}	D792	0.9086	0.9085	0.9177	0.9173
Film haze	%	D1003	5.01	3.92	6.86	5.33

Table 4. *(Continued)*

Property	Units	Conditions	Butene-1	Octene-1	Butene-1	Octene-1
Dart impact	g	D1709	184	>860	50	188
Elmendorf tear	g	D1922, MD	85	237	26	208
		D1922, TD	475	475	163	392
Tensile yield	MPa	D 882, MD	6.2	6.4	11.8	12.4
		D 882, TD	5.6	6.2	11.9	12.9
Tensile strength at break	MPa	D 882, MD	29.1	53.8	28.2	57.0
		D 882, TD	20	51.0	22.1	45.5
Elongation at break	%	D 882, MD	586	622	497	571
		D 882, TD	651	744	567	629

* The results in Table 4 were obtained for the following blown film conditions: blow-up ratio = 2.5 : 1; die gap = 1.78 mm; output = 14.5 kg h^{-1}; film thickness = 25.4 μm; die size = 76.2 mm; frost line height = 280 mm; melt temperature unknown.[71]

References

1. In: Benedikt GM, Goodall BL, eds. *Metallocene-Catalyzed Polymers: Materials, Properties, Processing and Market.* Norwich, NY: William Andrew Publication; 1999.

2. In: Schiers J, Kaminsky W, eds. *Matallocene-Based Polyolefins.* John Willey & Sons Ltd; 2000.

3. Benedikt GM. In: *Metallocene Technology in Commercial Application, Chem Tec Publication, Ont..* Canada: 1998.

4. Elston CT. *U.S. Patent.* 1972;3645992.

5. Kaminsky W, et al. *U.S. Patent.* 1985;4542199.

6. Spaleck W. *Organometallics.* 1994;13:954.

7. Soga K. *Macromol. Symp.* 1996;101:281.

8. Speed CS, et al. In: *SPE RETEC, VII Polyolefins International Conference.* 1991:45.

9. Alamo RG, Viers BD, Mandelkern L. *Macromolecules.* 1993;26:5740.

10. Alamo RG, Mandelkern L. *Macromolecules.* 1989;22:1273.

11. Bensason S, et al. *J. Polym. Sci., Polym. Phys. Ed.* 1996;34:1301.

12. de Garavilla JR. In: *Proceedings of Fourth International Business Forum on Specialty Polyolefins, Preprints.* Houston: Schotland Business Research; 1994:323.

13. Swogger KW. In *Proceedings of Second International Business Forum on Specialty Polyolefins, Preprints.* Schotland Business Research; 1992:155.

14. Michiels DJ. In: *Worldwide Metallocene Conference Proceedings, Proceedings, Metcon '94, Preprints.* Houston: 1994.

15. Whiteman NF, et al. In: *Society of Plastics Engineers IX International Polyolefins RETEC Conference Proceedings, Preprints,* Houston: 1995:575.

16. Todo A, Kashiwa N. *Macromol. Symp.* 1996:101: 310.

17. Kuwabara K, et al. *Macromolecules.* 1997;30:7516.

18. Sehanobish K, et al. *J. Appl. Polym. Sci.* 1994;51:887.

19. Marathe S, Mohandas TP, Sivaram S, Kashiwa. *Macromolecules.* 1995;28:7318.

20. Connor, Eric F, et al. *J. Polym. Sci., Part A: Polym. Chem.* 2002;40:2842.

21. Hong Han, Zhang, Zhicheng, Chung TCM, Lee RW. *J. Polym. Sci., Part A: Polym. Chem..* 2007;45:639.

22. Perez, Ernesto, et al. *J. Polymer Sci., Polym. Phys. Ed..* 2000;38:1440.

23. Ferhat-Hamid Z, et al. *J. Matr. Sci..* 2007;42:3138.

24. Chen J, et al. *J. Appl. Polym. Sci..* 2007;103:1927.

25. Hussein IA, et al. *J. Appl. Polym. Sci..* 2006;102:1717.

26. Kim K, et al. *International Polymer Processing.* 2006;21:81.

27. Lee SM, et al. In: *Society of Plastics Engineers Annual Technical conference Proceedings (SPE ANTEC),* 58th 2000;(3):2862.

28. Litteer DL, et al. *Popular Plastic & Packaging.* 1998;43:67.

29. Lue, Ching-Tai, et al. In: *Society of Plastics Engineers Annual Technical conference Proceedings (SPE ANTEC),* 56th 1998;(2):1816.

30. Leal V, et al. *Macromolecular materials and Eng..* 2006;291:670.

31. La Mantia Francesco P, et al., *Macromolecular Materials and Eng..* 2005;290:159.

32. Doerpinghaus, Phillip J. *J. Rheology.* 2003;47:717.

33. Malmberg, Anneli, et al. *Macromolecules.* 2002;35:1038.

34. Malmberg, Anneli, et al. *Annual Transactions of the Nordic Rheology Society.* 2001;8/9:163.

35. Wood-Adams PM, et al. *Macromolecules.* 2000;33:7481.

36. Bin Wadud SE, et al. *J. Rheology.* 2000;44:1151.

37. Gabriel, Claus, *Rheologica Acta.* 1999;38:393.

38. Harvard T, et al. *ACS Symp. Ser.* 1999;731:232.

39. Beigzadeh D, et al. *Polym. Reaction Eng.* 1997;5:141.

40. Soares JBP. In: Benedikt GM, Goodall BL, eds. *Metallocene-Catalyzed Polymers: Materials, Properties, Processing and Market.* Norwich, NY: William Andrew Publication; 1999:103.

41. Swogger KW, et al. *J. Plast. Film Sheet.* 1995;11:102; (Catalyst) *Macromolecules.* 1995;28:7318.

42. Howard P, et al. In *Proceedings of Fifth International Business Forum on Specialty Polyolefins, Preprints.* Schotland Business Research; 1995:313.

43. Kennedy MA, et al. *Macromolecules.* 1995;28:1407.

44. Miri VG, Elkoun S, Seguela R. *Polym. Eng. Sci.* 1997;37(10):1672.

45. Plumley TA, et al. *J. Plast. Film Sheet.* 1995;11:269.

46. Haward RN, *Polymer.* 1999;409:5821.

47. Al-Malaika S, et al. *Polymer Degradation and Stability.* 2006;91:3131.

48. De Wet-Roos D, et al. *J. Polymer Sci., Polym. Chem. Ed.* 2006;44:6847.

49. Arnold DR, et al. *Macromolecules.* 1984;17:332.

50. Trudell BC, Malpass CD. In: *Proceedings of Fifth International Business Forum on Specialty Polyolefins, Preprints.* Schotland Business Research. 1995:45.

51. Marigo A, Zannetti R, Milani F. *Eur. Polym. J.* 1997;33(5):595.

52. Mathot VBF, et. al *Polym. Mat. Sci. Eng.* 2001;84:409.

53. Cerrada ML, et al. *Polymer.* 2000;41:5957.

54. Defoor F, et al. *Macromolecules.* 1993;26:2575.

55. (a) Howard PR, Crist B. *J. Polym. Sci., Polym. Phys. Ed.* 1989;27:2269.
    (b) Bunn CW. In: Renfrew A, Morgan P, eds. *Polyethylene.* New York: Wiley-Interscience; 1957:chap. 7.

56. Mirabella FM, et al. *J. Polymer Sci., Polym. Phys.Ed..* 2004;42:3416.

57. Chiu F-C, et al. *J. Polymer Sci., Polym. Phys. Ed.* 2002;40:325.

58. (a) The Metallocene Monitor, June 1994, vol. II, p.7;
    (b) Childress BC. In: *Worldwide Metallocene Conference Proceedings, Metcon '94.* Houston: Preprints; 1994.

59. (a) Phillips PJ, M-H Kim, Monar K. *Society of Plastics Engineers Annual Technical conference Proceedings (SPE ANTEC)*, 53rd 1995:1481.
    (b) Phillips P. In: Benedikt GM, Goodall BL. *Metallocene-Catalyzed Polymers: Materials, Properties, Processing and Market.* Norwich, NY: William Andrew Publication; 1999:169.

60. Alamo R, Mandelkern L. *Macromolecules.* 1991;24:6480.

61. Islam MA, et al. *Eur. Polym. J.* 2007;43:599

62. Razavi-Nouri M, et al. *Polymer Testing.* 2006;25:1052.

63. Minick J, et al. *J. Appl. Polym. Sci.* 1995;58:1371.

64. Starck P. *Eur. Polym. J.* 1997;33:339.

65. Woo L, Ling M, Westphal S. *Thermochim. Acta.* 1996;272:171.

66. Kontou K, et al. *Eur. Polym. J.* 2002;38:2477.

67. Schellenberg J. *Adv. Polym. Tech.* 1997;16(2):135.

68. Chum PC, Kao CI, Knight GW. *Plast. Eng.* (July 1995): 21.

69. Graham JT, Alamo RG, Mandelkern L. *J. Polym. Polym. Sci., Polym. Phys. Ed..* 1997;35:213.

70. Gupta, Pankaj, et al. *Polymer.* 2006;46:8819.

71. Kale LT, et al. In: *Society of Plastics Engineers Annual Technical conference Proceedings (SPE ANTEC)*, 53rd 1995:2249.

72. Huang J-C H-L. *Huang. J. Poly. Eng.* 1997;17(3):213.

73. Wang S, et al. *J. Electrostatics.* 1997;42:219.

74. Michiels DJ. *Society of Plastics Engineers Annual Technical conference Proceedings (SPE ANTEC)*, 53rd, 1995:2239.

75. Young GL. *Society of Plastics Engineers Annual Technical Conference Proceedings (SPE ANTEC)*, 53rd, 1995:2234.

76. Leal, Virginia, et al. *Macromolecular Mat. and Eng..* 2006;291 (6):670.

77. Parker JA, et al. *J. Polymer Sci., Polym. Phys. Ed.* 2005;43:1986.

78. Silvestre, Clara, et al. *Macromolecular Mat. and Eng.* 2006;291 (12):1477.

79. Cran, Marlene J. *J. Plastic Film and Sheet.* 2006;22:121.

80. Johnston AD, et al. *Society of Plastics Engineers Annual Technical conference Proceedings (SPE ANTEC)*, 61st 2003;(3):3236–3256.

81. Johnston AD, et al. *Society of Plastics Engineers Annual Technical conference Proceedings (SPE ANTEC)*, 61st (3) 2003:3256.

82. Kulshreshtha AK. *Popular Plastic & Packaging.* 2001;46:75–81.

83. Krishnaswamy RK., et al. *Polym. Eng. Sci.* 2000;40:2385.

84. Raghu P, et al. *Popular Plastic & Packaging.* 2000;45:77.

85. Sukhadia AM. et al. *Society of Plastics Engineers Annual Technical conference Proceedings (SPE ANTEC)*, 58th (2) 2000:1578.

86. Seguela R, et al. *Poly. Mat. Sci. Eng.* 2000;82:82.

87. Sukhadia AM. *J. Plast Film Sheeting.* 1998;14:54.

88. Sukhadia AM. In: Benedikt GM, Goodal LBB, eds. *Metallocene-Catalyzed Polymers: Materials, Properties, Processing and Market.* Norwich, NY: William Andrew Publication; 1999:291 & 1999, p. 291 & 305.

89. Kim K, Kang H. *J. Intr. Polym. Processing.* 2006;21:81.

90. Jain P, Hazlitt LG, deGroot JA. In: *Society of Plastics Engineers X International Polyolefins RETEC Conference Proceedings, Preprints.* Houston, 1997:109.

91. Kissin YV. In: Graysen M, Eckroth D, eds. *Kirk-Othmer Encyclopedia of Chemical Technology.* 4th ed. Vol 17. New York: Wiley-Interscience; 1991:756.

92. James DE. In: Mark HF, et al. eds. *Encyclopedia of Polymer Science and Engineering..* 2nd ed. Vol 6. New York: John Wiley and Sons; 1985:429.

93. Dighton GL. In: Lappin GR, Sauer JD, eds. *Alpha Olefins Applications Handbook.* New York: Mercel Dekker; 1989:63.

94. Patel RM, et al. *Polym. Eng. Sci.* 1994;34:1506.

95. Sukhadia AM. *J. Plast Film Sheeting.* 1994;10:213.

96. Data supplied courtesy of Equistar Chemicals LP, Cincinnati, Ohio.

97. Sukhadia AM. *J. Plastic Film and Sheet.* 1998;14:54.

Poly(ethylene-2,6-naphthalate)

Jude O. Iroh

Acronyms, Trade Names PEN

Class Linear aromatic rigid polyester, thermoplastic

Structure

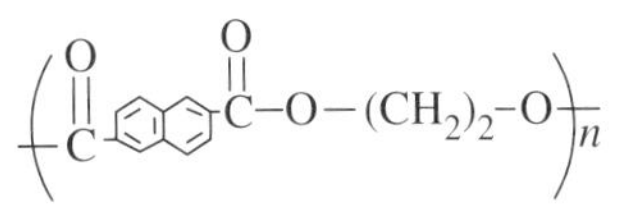

Synthesis

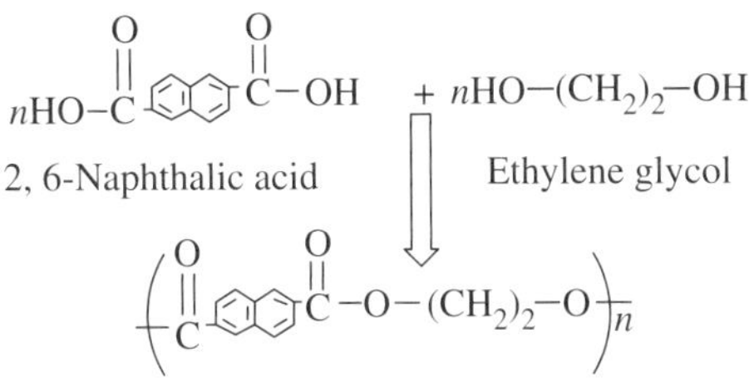

2, 6-Naphthalic acid   Ethylene glycol

Poly(ethylene-2,6-naphthalate)

Molecular Conformation Nearly planar

Major Applications Films, rigid thermoplastic polyesters

Properties of Special Interest Mostly synthesized as semi-crystalline thermoplastic. PEN is a clear and rigid polyester

Preparative Techniques Synthesized by step growth polymerization of ethylene glycol and naphthalene-2,6-dicarboxylate[1]

Property	Units	Conditions	Value	Reference
Thermal				
Glass transition	°C	DSC	117–121	(2–5)
temperature (T_g)		DMA	137.5	(6)
Melting temperature (T_m)	°C	DSC	265–266	(2–5)
Crystallization temperature	°C	DCS	198–203	(2, 3)
Heat of fusion (ΔH)	KJ mol^{-1}	DSC	9.2	(2, 3)

687

Property	Units	Conditions	Value	Reference
Heat of fusion (ΔH) for 100% crystallinity	KJ mol^{-1}	DSC	46	(2, 3)
Heat of cold crystallization (ΔH_{cc})	KJ mol^{-1}	DSC	7.3	(2, 3)
Coefficient of thermal expansion (α)	cm cm^{-1}°C^{-1}	–	4.4×10^{-5}	(2)
Molecular weight of repeat unit	g mol^{-1}	–	242	–
Inherent viscosity	dl g^{-1}	Dilute solution viscometry	0.51–0.53	(2, 4)
Intrinsic viscosity	dl	Dilute solution viscometry at 25°C	0.59	(5)
Solvent	phenol/ o-dichlorobenzene		–	(5, 7)
Physical state	–	Semi crystalline	–	–
Unit cell	–	–	Triclinic	
Lattice constants		X-ray diffraction	$a = 6.57$	(3, 8, 9)
			$b = 5.75$	(3, 8, 9)
			$c = 13.2$	(3, 8, 9)
			$\alpha = 81°20'$	(3, 8, 9)
			$\beta = 144°$	(3, 8, 9)
			$\gamma = 100°$	(3, 8, 9)
No of chains per unit cell	–	–	1	(8, 9)
		X-ray diffraction	–	
Number of chains	–	–	1	(8, 9)
No of monomers	–			(8, 9)
Mechanical properties		–	1	
Breaking strength, (σ_B)	MPa	Tensile	83	(7)
Tensile (Young's) modulus, E	GPa	–	2.0	(7)
Flexural strength ($\sigma*$)	MPa	3-Point Flexure	108	(7)
Flexural modulus (E)	GPa	3-Point Flexure	2.5	(7)
Elongation, ε_B	%	Tensile	48.53	(7)
Measured density	g cm^{-3}	Autodensimeter	1.3375–1.3502	(7, 10)

References

1. Fried JR. *Polymer Science and Technology*. New Jersey: Prentice Hall PTR; 1995:350–351.
2. Kim BS, Jang SH. *Polym. Eng. & Sci.* 1995;35(18):1421.
3. Cakmak M, Wang YD, Simhambhatla M.,*Polym. Eng. & Sci.* 1990;30(12):721.
4. Yoon KH, Lee SC. *Polym. Eng. & Sci.* 1995;35(22):1807.

5. Kit KM, Gohil RM. *Polym. Eng. & Sci.* 1995;35(8):680.

6. Aoki Y, et al. *Macromolecules.* 1999;32:1923.

7. Jang SH, Kim BS. *Polym. Eng. & Sci.* 1995;35(6):538.

8. Mencik Z. *Chem. Prim.* 1967;17(2):78.

9. Zachmann HG, Wiswe D, Gehrke R, Riekel C. Makromol. *Chem. Suppl.* 1985;12: 175.

10. Wu G, Li Q, Cuculo JA. *Polymer.* 2000;41:8139.

Poly(ethylene oxide)

Qingwen Wendy Yuan-Huffman

Acronyms PEO

Class Polyether

Structure $[-CH_2-CH_2-O-]$

Mol. Wt. of Repeat Unit 44

Property	Unit	Conditions			Values	Reference
Solvents	None				Benzene, Alcohols, Chloroform, Esters, Cyclohexanone, N,N-dimethyylacetamide, Acetonitrile, Water (cold), aqu. K_2SO_4 (0.45 M above 35°C), aqu. $MgSO_4$ (0.39 M above 45°C)	(1)
Nonsolvents	None				Ethers, Dioxane(sw), Water (hot), Aliphatic Hydrocarbons	(1)
Theta temperature	K	Solvent	Method			
		Acetonitrile/i-propyl ether (45/55)	CP	293.5		(1, 2)
		Benzene/isooctane (100/48)	CT	344.3		(1, 3)
			PE	344.5		(1, 3)
		$CaCl_2$/water (2 mol L^{-1})	CT	355.5		(1, 4)
			CT	359.5		(1, 5)
		Chloroform/n-hexane (54/46)	CT	293.5		(1, 6)
		(47.4/52.6)	CT, VM	293.5		(1, 7)
		CsCl/water (2 mol L^{-1})	CT	333.5		(1, 4)
		Diethylene glycol diethylether	VM	323.5		(1, 8)
		KCl/water (2 mol L^{-1})	CT	327.5		(1, 5)
			CT	330.5		(1, 4)
		KNO_3/water (2 mol L^{-1})	CT	338.5		(1, 5)
		K_2SO_4/water (0.45 mol L^{-1})	CT	307.5		(1, 9)
			CT	307.5		(1, 5)
			PE	308.5		(1, 10)
			VM	308.5		(1, 8)

Property	Unit	Conditions		Values	Reference
		Solvent	Method†		
		LiCl/water (2 mol L^{-1})	CT	363.5	(1, 5)
			CT	363.5	(1, 4)
		Methyl i-butyl ketone	VM	323.5	(1, 8)
		MgCl$_2$/water (2 mol L^{-1})	CT	353.5	(1, 4)
			CT	363.5	(1, 5)
		MgSO$_4$/water	CT	315.5	(1, 9)
		(0.39 mol L^{-1})	CP	315.5	(1, 5)
			PE	318.5	(1, 10)
		NaCl/water (2 mol L^{-1})	CT	333.5	(1, 4)
			CT	334.5	(1, 5)
		NH$_4$Cl/water (2 mol L^{-1})	CT	349.5	(1, 4)
			CT	350.5	(1, 5)
		Nitroethane/i-propyl ether (45/55)	CP	293.5	(1, 6)
		RbCl/water (2 mol L^{-1})	CT	329.5	(1, 4)
		SrCl$_2$/water (2 mol L^{-1})	CT	346.5	(1, 4)
			CT	355.5	(1, 5)
		Water	DM, VM	278.6	(1, 11)
			CT	369.5	(1, 5)
			CT	390.5	(1, 4)
Interaction parameter, χ	None	Method: vapor pressure Benzene, $T = 323.8$ K			
		$v_2 = 0.2$		0.18	(12)
		0.4		0.14	(12)
		0.6		0.10	(12)
		Benzene $T = 343.5$ K			
		$v_2 = 0.2$		0.19	(12)
		0.4		0.14	(12)
		0.6		0.12	(12)
		0.8		0.09	(12)

Property	Unit	Conditions			Values	Reference
Second virial coefficient	mol cm^3 g^{-2}	Solvent	Temp (°C)	Mol. wt (g mol^{-1})		
		Benzene,	25	7.70×10^3	27.4×10^{-4}	(1, 13)
			25	3.79×10^3	78×10^{-4}	(1, 13)
		Dimethylformamide	–	3.79×10^3	30×10^{-4}	(1, 13)
			(25–120)	$\sim 3.5 \times 10^3$	$(37–47) \times 10^{-4}$	(14)
		Methanol	25	7.70×10^3	66.0×10^{-4}	(1, 13)
				3.79×10^3	56.0×10^{-4}	(1, 13)
				$(0.316–6.75) \times 10^3$	$(18.0–16.4) \times 10^{-4}$	(1, 15)
					$(170–34.8) \times 10^{-4}$	(1, 16)
				$(0.062–37.3) \times 10^3$	$(1220–46) \times 10^{-4}$	(1, 16)
				$(1–31) \times 10^3$	$(102.5–39) \times 10^{-4}$	(1, 17)

Property	Unit	Conditions			Values		Reference
		Solvent	Temp (°C)	Mol. wt (g mol^{-1})			
				$(1–10) \times 10^3$	$(84.5–47.5) \times 10^{-4}$		(1, 18)
				$(4–23) \times 10^3$	$(87–46) \times 10^{-4}$		(1, 19)
				$(3–48) \times 10^3$	$(48–27.5)$v		(1, 20)
		Water,	25	$(10.9–800) \times 10^3$	$(116–30.4)$ $\times 10^{-4}$		(1, 21)
				10.1×10^3	62×10^{-4}		(1, 19)
Huggins coefficient	None	Solvent	Temp (°C)	$[\eta]$			
		Benzene	20	2	$\sim$3.7		(1)
				48	0.4		(1)
		Chloroform	20	3	$\sim$0.9		(1)
				82	0.4		(1)
		Dimethylformamide	20	3	$\sim$0.6		(1)
				45	0.4		(1)
		Toluene	30	3	$\sim$2.50		(1)
				39	0.4		(1)
			35	1.1	8.16		(1)
				1.9	3.41		(1)
				2.4	2.06		(1)
				5.5	2.3		(1)
		Water	20	3	1.1		(1)
				5	0.4		(1)
			35	1.7	4.95		(1)
				3.0	2.42		(1)
				5.0	0.93		(1)
				10.8	0.44		(1)
Mark-Houwink parameters (K and α)	ml g^{-1} and none	Solvent	Temp (°C)	Mol. wt (g mol^{-1})	K ($\times 10^{-3}$)	a	
		Acetone	25	$(7–100) \times 10^4$	32	0.67	(1, 9)
				$(0.02–0.3) \times 10^4$	156	0.50	(1, 22)
		Benzene	20	$(0.01–1.9) \times 10^4$	48	0.68	(1, 18)
			25	$(8–520) \times 10^4$	30.7	0.686	(1, 23)
			25	$(0.02–0.8) \times 10^4$	129	0.5	(1, 22)
		Carbon Tetrachloride	20	$(0.02–1.1) \times 10^4$	69	0.61	(1, 18)
			25	$(7–100) \times 10^4$	62	0.64	(1, 9)
		Chloroform	25	$(0.02–0.15) \times 10^4$	206	0.50	(1, 22)
		Cyclohexane	20	$(0.006–1.1) \times 10^4$	$[\eta] = 0.5 + 0.035\,M^{0.64}$		(1, 18)
		Diethylene Glycol Diethyl Ether	50	$(7–100) \times 10^4$	140	0.51	(1, 9)
		Dimethylformamide	25	$(0.1–3) \times 10^4$	$[\eta] = 2.0 + 0.024\,M^{0.73}$		(1, 17)
		Dioxane	20	$(0.006–1.1) \times 10^4$	$[\eta] = 0.75 + 0.035\,M^{0.71}$		(1, 18)
			25	$(0.02–0.15) \times 10^4$	138	0.5	(1, 22)

Poly(ethylene oxide)

Property	Unit	Conditions			Values		Reference
		Solvent	Temp (°C)	Mol. wt (g mol^{-1})	K ($\times 10^{-3}$)	a	
		Methanol	20	$(0.06-1.9) \times 10^4$	$[\eta] = 2.0 + 0.033$	$M^{0.72}$	(1, 18)
			25		82.5	0.57	(1, 24)
		4-Methylpenta-2-one	50	$(7-100) \times 10^4$	120	0.52	(1, 9)
		Toluene,	35	$(0.04-0.4) \times 10^4$	14.5	0.70	(1, 25)
		Water,	20	$(0.006-1.1) \times 10^4$	$[\eta] = 2.0 + 0.016$	$M^{0.76}$	(1, 24)
			25	$(0.019-0.1) \times 10^4$	156	0.50	(1)
			30	$(2-500) \times 10^4$	12.5	0.78	(1, 26)
			35	$(3-700) \times 10^4$	6.4	0.82	(1, 10)
			35	$(0.04-0.4) \times 10^4$	16.6	0.82	(1, 25)
			45	$(3-700) \times 10^4$	6.9	0.81	(1, 10)
		Aqueous K$_2$SO$_4$ (0.45M)	35	$(3-700) \times 10^4$	130	0.50	(1, 10)
			35	$(7-100) \times 10^4$	280	0.45	(1, 9)
		Aqueous MgSO$_4$ (0.39M)	45	$(3-700) \times 10^4$	100	0.50	(1, 10)
Solubility parameter	(Mpa)$^{1/2}$	Method: IPGC, 25°C		20.2 ± 2			(1, 27)
Heat of solution	J g^{-1}	Semicrystalline polymer					(1)
		Benzene, 30°C, 4.3×10^4 g mol^{-1}		170			(1)
		Chloroform, 30°C, 6×10^3 g mol^{-1}		52			(1)
		Water, 25°C, 2×10^4 g mol^{-1}		10			(1)
		Water, 30°C, 6×10^3 g mol^{-1}		24			(1)
		Water, 30°C, 4.3×10^4 g mol^{-1}		40			(1)
Heat of Fusion	kJ mol^{-1}			8.29			(1)
				9.5			(1)
				8.04			(1)
				11.7			(1)
				9.41			(1)
				7.86			(1)
				8.7			(1)
				7.33			(1)
Glass Transition Temperature	K	Conflicting data, value range from 158–233 K		232			(28–35, 53)
		Method: differential microcalorimeter					(36)
		R[((CH$_2$)$_2$O)]$_n$(CH$_2$)$_2$R					
		R = OH					
		$n = 1$		162.5			
		$n = 2$		170.5			
		$n = 3$		174.5			
		$n = 4$		176.5			
		$n = 5$		181.5			
		$n = 7$		183.5			
		R = Cl					
		$n = 1$		139.5			
		$n = 2$		157.5			

Property	Unit	Conditions	Values		Reference
		$n = 3$	168.5		
		$n = 4$	177.5		
		$n = 5$	185.5		
		$n = 7$	187.5		
		Electron Spin Resonance	213 K		(37)
		Highly crystalline	206.5		(38)
Melting temperature	K	method: differential microcalorimeter $R[(CH_2)_2O)]_n(CH_2)_2R$ $R = OH$			(36)
		$n = 1$	265.5		
		$n = 3$	262.5		
		$n = 4$	273.5		
		$n = 5$	282.5		
		$n = 7$	285.5		
		$R = Cl$			
		$n = 1$	209.5		
		$n = 3$	261.5		
		$n = 4$	254.5		
		$n = 5$	258.5		
		$n = 7$	249.5		
		Highly crystalline	339.5		(38)
Heat capacity C_p	$KJ K^{-1} mol^{-1}$ $\times 10^{-3}$	Temp. (K)	Solid	Melt	(1, 39, 40)
		10	0.51		
		20	3.18		
		30	7.24		
		40	11.16		
		50	14.62		
		60	17.60		
		70	20.13		
		80	22.33		
		90	24.90		
		100	26.93		
		110	28.78		
		120	30.44		
		130	32.10		
		140	33.57		
		150	35.05		
		160	36.53		
		170	37.80		
		180	39.11		
		190	40.40		
		200	41.88		
		210	43.17	81.88	
		220	44.64	82.55	
		230	45.91	83.21	
		240	47.26	83.88	

Property	Unit	Conditions			Values		Reference
		Temp. (K)			Solid	Melt	(1, 39, 40)
		250			48.61	84.55	
		260			49.96	85.22	
		270			51.31	85.89	
		280			52.66	86.55	
		290			54.01	87.22	
		300			55.36	87.89	
		310			56.71	88.56	
		320			58.06	89.23	
		330			59.41	89.89	
		340			60.76	90.56	
		350				91.23	
		360				91.90	
		370				92.57	
		380				93.23	
		390				93.90	
		400				94.57	
		410				95.24	
		420				95.91	
		430				96.57	
		440				97.24	
		450				97.91	
Index of refraction	None				1.4563		(1)
					1.51–1.54 (High Mol. Wt.)		

Property	Unit	Solvent	Temp (°C)	Mol. wt (g mol^{-1})	$\lambda_o = 436$ nm	$\lambda_o = 546$ nm	Reference
Specific refractive index increment, dn/dc	ml g^{-1}	Acetonitrile	25	62		0.0964	(1, 16)
				100		0.106	(1, 16)
				161		0.114	(1, 16)
				205		0.121	(1, 16)
				316		0.123	(1, 16)
				407		0.130	(1, 16)
				970		0.135	(1, 16)
				9400		0.135	(1, 16)
		Benzene		106	−0.086		(1, 41)
				194	−0.073		(1, 41)
				282	−0.066		(1, 41)
				810	−0.059		(1, 41)
			25	3510		−0.016	(1, 16)
			30	3510		−0.018	(1, 42)
			54	3510		−0.013	(1, 42)
		Benzene	25	1.5×10^3– 5.3×10^5	−0.017 to −0.010		(1, 43)

Property	Unit	Conditions		Values		Reference
		Solvent	Temp (°C)	Mol. wt (g mol⁻¹)	$\lambda_o = 436$ nm	$\lambda_o = 546$ nm
		Bromoform	23		−0.108	−0.090	(1, 44)
		n-Butanol				0.076	(1, 45)
		Carbon tetrachloride/ methanol	25	(75/25 vol.)		0.066	(1, 46)
				(50/50 vol.)		0.091	(1, 46)
				(20/80 vol.)		0.128	(1, 46)
		Chlorobenzene	23		−0.039	−0.030	(1, 44)
		Chloroform	23		0.054	0.053	(1, 44)
			30		0.054		(1, 42)
		Chloroform/ n-hexane (47/53 vol.)	20		0.091		1, 7
		1,2-Dibromoethane	23		−0.048	−0.044	1, 44
		Dioxane		10,000		0.045	(1, 41)
		Dioxane	45		0.061		1, 42
		Methyl ethyl ketone		810	0.092		1, 41
				10,000		0.094	(1, 41)
		Methanol			0.150		1, 45
			25	62		0.118	(1, 16)
				100		0.127	(1, 16)
				161		0.135	(1, 16)
				205		0.139	(1, 16)
				316		0.141	(1, 16)
				445		0.142	(1, 16)
				810		0.143	(1, 41)
				1020		0.144	(1, 16)
				3000		0.149	(1, 41)
				6000		0.150	(1, 41)
				9400		0.150	(1, 16)
				10,000		0.148	(1, 41)
				31,000		0.150	(1, 16)
			25			0.143	(1, 7)
			45			0.152	(1)
			30		0.145	0.142	(1)
			45		0.150		(1, 42)
		Methyl Acetate	25	6700		0.111	(1, 16)
		Pyridine	23	6700	−0.026	−0.018	(1, 44)
		1, 1, 2, 2-Tetrachloroethane	23	6700	0.006	0.007	(1, 44)
		Tetrahydrofuran		6700		0.068	(1, 45)
		Water	27	6700	0.134	0.132	(1, 47)
						0.139	(1, 45)
			20	6700	0.138	0.135	(1, 47)
			25		0.138		(1, 47)
			30		0.136		(1, 47)

Poly(ethylene oxide)

Property	Unit	Conditions			Values		Reference
		Solvent	Temp (°C)	Mol. wt (g mol^{-1})	$\lambda_o = 436$ nm	$\lambda_o = 546$ nm	
			40		0.134	0.132	(1, 47)
			50		0.133	0.131	(1, 47)
		Water	25	62	0.093		(1, 41)
				106	0.108		(1, 41)
				194	0.124		(1, 41)
				300	0.126	0.123	(1, 41)
				600	0.135	0.131	(1, 41)
				810	0.136	0.128	(1, 41)
				1200	0.139	0.134	(1, 41)
				3000	0.141		(1, 41)
				6000	0.145	0.139	(1, 41)
				9400		0.135	
				10,000	0.142		(1, 41)
				14,400		0.139	(1, 48)
			80	14,400		0.115	(1, 48)
				31,000		0.135	(1, 16)
				340,000		0.149	(1, 49)

Property	Unit	Conditions	Mol. wt (g mol^{-1})	20°C	150°C	200°C	Reference
Surface tension	mN m^{-1}	Solvent					(1, 50–52)
		Diol	86 to 17,000	42.9	30.1	25.2	
			6,000	42.9	33.0	29.2	
			6,000	42.5	30.1	25.4	
		Dimethylether	114	28.6	16.0	11.1	
			148	31.1	18.6	13.8	
			182	32.9	20.5	15.8	
			600	37.5	26.1	21.7	
			5000	44.1	32.7	28.3	
			100,000	44.2	32.8	28.4	

Property	Unit	Conditions	Temp. (°C)	Mol. wt (g mol^{-1})			Reference
Diffusion coefficient	$\times 10^{-7}$ cm^2 s^{-1}	Solvent					
		Acetone	25	4.3×10^3	37.8		(1)
		Formamide	25	4.3×10^3	5.76		(1)
		Methanol	25	4.3×10^3	23.5		(1)
				23.8×10^3	10.7		(1)
				19.2×10^3	10.7		(1)
		Water	20	0.29×10^3	37.0		(1)
				0.625×10^3	29.2		(1)
				1.25×10^3	24.0		(1)
				3.3×10^3	13.3		(1)
				5.8×10^3	11.6		(1)
				8.8×10^3	10.3		(1)
				10.6×10^3	7.2		(1)
				1.426×10^3	23.6		(1)

Property	Unit	Conditions			Values	Reference
		Solvent	Temp. (°C)	Mol. wt (g mol^{-1})		
		Water	25	1.470×10^3	22.6	(1)
				1.778×10^3	19.7	(1)
				1.822×10^3	20.1	(1)
				4.3×10^3	11.5	(1)
				238×10^3	4.85	(1)
				12×10^3	7.35	(1)
				23.8×10^3	4.85	(1)
				37.3×10^3	3.95	(1)
				17.7×10^3	1.33	(1)
				30.0×10^3	1.08	(1)
				67.9×10^3	0.79	(1)
				119×10^3	0.63	(1)
				1130×10^3	1.33	(1)
				1470×10^3	1.36	(1)
				1900×10^3	1.15	(1)
				2000×10^3	1.12	(1)
				2610×10^3	0.93	(1)
				2630×10^3	0.93	(1)
				2670×10^3	0.94	(1)
				320×10^3	1.0	(1)
Dipole moment Per momomer unit	D	Solvent	Temp. (°C)	P_n		
		Dioxane	25	1–7	1.68–1.29	
		Benzene	20	1.0–33.6	1.41–1.09	
		Benzene	20	2–227	1.46–1.07	
		Benzene	25	4.1–153.0	1.61–1.13	
		Benzene	25	4.0–176.2	1.68–1.13	
		End group: —OC$_2$H$_5$				
		Benzene	20	2 and 6	1.15 and 1.11	
		Benzene	25	1–6	1.14–1.07	
		Benzene	50	1–6	1.14–1.09	
Polymerization	None				Anionic ring-opening	(53)
Applications	None	Textile applications, Cosmetics, Antifoaming agents, Others (chemical intermediates, ink, and dye solvents, demulsifiers, plasticizers, etc.)				(53, 54)

Notes
1. Abbreviation of Method: CP—Cloud Point, CT—Cloud Temperature, VM—Intrinsic viscosity/Molar Mass, PE—Phase Equilibria, DM—Diffusion Coefficient/Molar Mass
2. Numbers in parenthesis are compositions in volume/volume

Poly(ethylene oxide)

Crystalline-state properties

Crystl. Syst. (Lattice)	Space Group	Unit Cell Parameters (Å)			Angles (Degree)	Monomer Per Unit Cell	Density (g cm^{-3})	Reference
		A	B	C				
Monoclinic		9.5	19.5	12.0	B = 101	36	1.207	(1)
Monoclinic	C2H-5	8.05	13.04	19.48	B = 125.4	28	1.229	(1)
Monoclinic	CS-2	8.03	13.09	19.52	B = 125.1	28	1.220	(1)
Monoclinic		7.95	13.11	19.39	B = 124.6	28	1.231	(1)
Monoclinic		8.02	13.4	19.25	B = 126.9	28	1.238	(1)
Monoclinic		8.16	12.99	19.30	B = 126.1	28	1.239	(1)
Monoclinic		7.51	13.35	19.90	B = 118.6	28	1.169	(1)
Triclinic	CI-1	4.71	4.44	7.12	63,93111	2	1.197	(1)

References

1. Brabdrup J, Immergut EH. *Polymer Handbook.* 3rd ed. New York: Wiley Interscience; 1989.
2. Hay JN, Afifi-Effat AM. *Brit. Polym. J.* 1977;9:1.
3. Afifi-Effat AM, Hay JN, Wiles M. *J. Polym. Sci.* B. 1973;11:87.
4. Napper DH. *J. Colloid Interface Sci.* 1970;33:384.
5. Boucher EA, Hines PM. *J. Polym. Sci. – Polym Phys. Ed.* 1978;16:501.
6. Elias H-G, Gruber U. *Makromol. Chem.* 1964;78:72.
7. Elias H-G, Gruber U. *Makromol. Chem.* 1961; 50:1.
8. Beech DR, Booth C. *J. Polym. Sci. [A-1].* 1969;7:575.
9. Napper DH. *J. Colloid Interface Sci.* 1970;32:106.
10. Bailey FE Jr, Callard RW. *J. Appl. Polym. Sci.* 1959;1:56.
11. Chew BA, Couper A. *J. Chem. Soc. Faraday Soc.* 1976;[1] 72:382.
12. Booth C, Devoy CJ. *Polymer.* 1971;12:309.
13. Elias H-G, Schlumpf H. *Makromol. Chem.* 1965;85:118.
14. Elias H-G, Maenner E. *Makromol. Chem.* 1960;40:207.
15. Kamide K, Sugamiya K, Nakayama C. *Makromol. Chem.* 1970;132:75.
16. Elias H-G, Lys HP. *Makromol. Chem.* 1966;92:1.
17. Ritscher TA, Elias H-G. *Makromol. Chem.* 1959;30:48.
18. Sadron C, Rempp P. *J. Polym. Sci.* 1958;29:127.
19. Elias H-G, *Z. Phys. Chem. (Frankfurt).* 1961;28:303.
20. Elias H-G. *Chem. Ing. – Tech.* 1961;33:359.
21. Elias H-G. *Angew. Chem.* 1961;73:209.
22. Rossi C, Cuniberti C. *J Polym. Sci.* B. 1964;2:681.
23. Allen G, Booth C, Hurst SJ, Jones MN, Price C. *Polymer.* 1967;8:391.
24. Elias H-G, *Kunststoffe-Plastics.* 1961;4:1.
25. Thomas DK, Charlesby A. *J. Polym. Sci.* 1960;42:195.
26. Bailey FE Jr, Kucera JL, Imhof LG. *J. Polym. Sci.* 1958;32:517.
27. DiPaola-Baranayi G. *Macromolecules.* 1982;15:622.
28. Faucher JA, Koleske JV, Santee ER, Stratta JJ, Wilson CW. *J. Appl. Phys.* 1966;37:3962.
29. Hellwege K-H, Hoffman R, Knappe W, Kolloid Z-Z. *Polymer.* 1968;226:109.
30. Faucher JA, Koleske JV. *Polymer.* 1968;9:44.
31. Swallow JC. *Proc. Roy. Soc., London.* 1957;A238:1.

32. Mabdlekern L, Jain NL, Kim H. *J. Polym. Sci. A-2.* 1968;6:165.

33. Ishida Y, Matsuo M, Takayanagi M, *J. Polym. Sci. B.* 1965;3:321.

34. Miller WG, Saunders JH. *J. Appl. Polym. Sci.* 1969;3:1277.

35. Vandenberg EJ, Ralston RH, Kocher BJ. *Rubber Age.* 1970;102:47.

36. Privalko VP, Lobodina AP. *Europ. Polym. J.* 1974;10(11):1033.

37. Tormala P. *Europ. Polym. J.* 1974;10(6):519.

38. Rodriguez F. *Principles of Polymer Systems.* 4th ed. Taylor & Francis Publishers; 1996.

39. Guar U, Wunderlich B. *J. Phys. Chem. Ref. Data.* 1981;10(4):1010.

40. Suzuki H, Wunderlich B. *J. Polym. Sci., Polym. Phys. Ed.* 1985;23:1671.

41. Rempp P. *J. Chem. Phys.* 1957;54:421.

42. Carpenter DK, Santiago G, Hung AH. *J. Polym. Sci., Polym. Symp.* 1974;44:75.

43. Candau F, Dufour C, Francois J. *Makromol. Chem.* 1976;177:3359.

44. Spatorico AL. *J. Appl. Polym. Sci.* 1974;18:1793.

45. Strazielle S. *Makromol. Chem.* 1968;119:50.

46. Hert M, Strazielle C. *Europ. Polym. J.* 1973;9:543.

47. Polik WF, Burchard W. *Macromolecules.* 1983;16:978.

48. Schnabel W, Borgwadt U. *Makromol. Chem.* 1969;123:73.

49. Teramoto A, Fujita H. *Makromol. Chem.* 1965;85:261

50. Wu S. *J. Macromol. Sci.* 1974;C10:1.

51. Bender GW, LeGrand DG, Gaines GL Jr. *Macromolecules.* 1969;2:681.

52. Rastogi AK, Pierre LE St. *J. Colloid Interface Sci.* 1971;35:16.

53. Odian G. *Principles of Polymerization.* 3rd ed. Wiley Interscience, 1991.

54. Mark HS, et al. *Encyclopedia of Polymer Science and Engineering.* Vol 6. Wiley Interscience; 1986.

Poly(ethylene sulfide)

Junzo Masamoto

Acronym PES

Class Polysulfides

Structure $-(S-CH_2CH_2)-$

Major Application Poly(ethylene sulfide) is a high melting plastics material. However, this polymer has not yet achieved commercial production, although its properties makes it warrant serious consideration as a plastic.[1]

Properties of Special Interest Poly(ethylene sulfide) is a high melting crystalline material. High-molecular-weight polymers prepared by ring-opening polymerization of ethylene sulfide have melting points generally above 478 K. The melting point of PES is much higher compared to poly(ethylene oxide) (melting point = 341 K). Only a few solvents are known that dissolve PES, but only at temperatures above 413 K. PES requires addition of stabilizers to permit processing in standard molding equipment, the best of which are polyamines with high boiling points and their derivatives. The polymer, properly stabilized, can best be molded using screw injection-molding equipment at temperature of 488–523K.[2]

Preparative Techniques There are two different pathways (polycondensation and ring-opening polymerization) for preparation of PES.[2] The first method leads to a polymer of relatively low molecular weight, whereas the ring-opening polymerization reaction can lead to high molecular weight polymers under particular circumstances.

Following is the polycondensation reaction:[3,4]

$$Br-CH_2 CH_2-Br+ nK_2 S \rightarrow (-C_2 H_4 S-)_n + 2nKBr$$

Condensation of an ethylene dihalide with an alkali metal sulfide[3] was studied in the latter part of the nineteenth century. The acid catalyzed polycondensation of certain mercapto alcohols represents a second method of synthesis.[4]

Following is the ring-opening polymerization reaction:[2,5]

$$n \underset{S}{\triangledown} \rightarrow (-CH_2CH_2S-)_n$$

It was found by Thiokol Chemical Corp. (Trenton, New Jersey, USA) that a catalyst formed by the reaction of diethyl zinc with water readily produces ethylene sulfide polymers that melted at 481–485 K.[5]

Property	Units	Conditions	Value	Reference
Molecular weight (of repeat unit)	g mol^{-1}	–	60	–
Typical molecular weight range of polymer	g mol^{-1}	Determined from zero-intercept melt viscosity	$1.4–8 \times 10^5$	(6, 7)
IR (characteristic absorption frequencies)	cm^{-1}	Rocking motion CH$_2$ Stretching C–S Twisting CH$_2$ Wagging CH$_2$ Symmetric deformation CH$_2$	672 724 1,183 1,259 1,427	(8)
Solvents	A few solvents are known that dissolve PES at temperature above 140°C		α-methylnaphthalene, nitorobenzene, o-dichlorobenzene, dithiolane, dimethyl sulfoxide	(2)
Nonsolvents	No ordinary solvent is known that dissolve PES at temperatures below 140°C			(2)
Mark-Houwink parameters: K and a	$K =$ ml g^{-1} $a =$ None	Solvent: dithiolane and stabilizer, at 160°C	$K = 2.2 \times 10^{-3}$ $a = 0.65$	(9)
Melt viscosity-molecular weight	–	$M =$ molecular weight; $G =$ melt index (g min^{-1})	$\log M = 5.14 – 0.4167(\log G)$	(7)
Characteristic ratio $\langle r^2 \rangle / nl^2$	–	–	4.2	(10, 11)
Lattice	–	–	Orthorhombic	(12–14)
Space group	–	–	Pbcn-D2h-6	(12–14)
Chain conformation	–	–	CH$_2$ — CH$_2$ trans	(8, 12, 15, 16)
			CH$_2$ — S Gauche (right-handed) S — CH$_2$ Gauche (right-handed) CH$_2$ — CH$_2$ trans CH$_2$ — S Gauche (left-handed) S — CH$_2$ Gauche (left-handed)	
Crystalline state conformation	–	–	(2/0) glide plane	(12)
Unit cell dimensions		X-ray photograph of oriented sampe	$a = 8.50$, $b = 4.95$, $c = 6.70$ (fiber axis)	(12)
		Electron diffraction of single crystal and fiber	$a = 8.508$, $b = 4.938$, $c = 6.686$ (fiber axis)	(13, 14)

Poly(ethylene sulfide)

Property	Units	Conditions	Value	Reference
Unit cell contents	–	–	4 monomeric units per unit cell (2 molecular chains)	(12)
Degree of crystallinity	%	X-ray diffraction and density	50–68	(9)
Heat of fusion	kJ mol^{-1}	100% crystallinity	$H_0 = 14.1$	(9)
	J g^{-1}	54% crystallinity sample	126	
Entropy of fusion	J K^{-1} mol^{-1}	–	$S_0 = 28.8$	(9)
Density	g cm^{-1}	Theoretical density for crystalline PES	1.41	(12)
		Observed density	1.33–1.34	(12)
		Amorphous density	1.295	(9)
Glass transition temperature	K	Extraporation from the T_g values of amorphous copolymers ethylene sulfide/isobutylene sulfide. DSC, heating rate $= 10°C$ min^{-1}	223	(17)
Melting point	K	–	481–485	(2)
			489	(11, 19)
Deflection temperature	K	1.8 MPa	432	(2)
Tensile modulus	MPa	–	2,200	(2)
		At 20°C	1,800	(9)
		At 70°C	1,060–1,150	(9)
		At 125°C	770–790	(9)
		Unaged	2,070	(1)
		Aged 7 days at 120°C	2,200	(1)
		Aged 7 days at 150°C	2,500	(1)
Tensile strength	MPa	–	72	(2)
		At 20°C	70–78	(9)
		At 60°C	56–62	(9)
		At 100°C	38–48	(9)
		At 125°C	30–38	(9)
		Unaged	68	(1)
		Aged 7 days at 120°C	39	(1)
		Aged 7 days at 150°C	32	(1)
Maximum extensibility (elongation)	%	–	10	(2)
		At 20°C	10–14	(9)
		At 60°C	12	(9)
		At 100°C	25–40	(9)
		At 125°C	40–50	(9)

Property	Units	Conditions	Value	Reference
		Unaged	15	(1)
		Aged 7 days at 120°C	4.2	(1)
		Aged 7 days at 150°C	3.7	(1)
Flexural modulus	Mpa	–	2,070	(1)
Flexural strength	MPa	–	72	(1)
Notched Izod impact strength	$J\,m^{-1}$	–	69	(2)
		Unaged	64	(1)
		Aged 7 days at 120°C	16	(1)
		Aged 7 days at 150°C	16	(1)
Melt viscosity	Pa s	Theoretical molecular weight		(9)
		80,000	4,950	
		9,500	570	
Melt index	$g\,min^{-1}$	235°C	0.01–0.2	(6)
Pyrolyzability, nature of product	Without stabilizer, above the melting point (225°C), the polymer viscosity falls rapidly with pronounced darkening and the liberation of volitiles		Ethylene, hydrogen sulfide, dithiane, methyldithiolane	(1, 6)
Pyrolyzability, amount of product	mol gas (unit $ES)^{-1}\,min^{-1}$	223°C, N_2	6.0×10^{-5}	–
		230°C, N_2	8.7×10^{-5}	
		240°C, N_2	39.4×10^{-5}	
		250°C, N_2	76.1×10^{-5}	
Pyrolyzability, amount of impurities	Stability is influenced by the nature of initiator. Polymers initiated by zinc or cadmium thiolate, or triethylene diamine are less stable than those prepared with the zinc ethyl/water catalyst. Acid and oxygen initiate degradation			(1)
Decomposition temperature	K	Without stabilizer, the polymer viscosity falls rapidly	Above the melting point (498)	–
		Polymers initiated by zinc ethyl/water system with the stabilizers (polyamines in conjunction with zinc oxide or zinc hydroxychloride)	523	(1, 6)
Water absorption	%	24 h	0.03	(2)
Weight change	%	After immersion for 30 days		
		50% sulfuric acid, 121°C	−0.7	(2)
		15% sodium hydroxide, 121°C	−0.3	
		Benzene, 93°C	+4.0	
		Perchloroethylene, 121°C	7.0	
Creep	%	Room temperature, 34 MPa, 500 h	1	(2)

Poly(ethylene sulfide)

Property	Units	Conditions	Value	Reference
Important patents	–	Polymerization	U.S. Patent 3,365,431	(19)
		Stabilizer	Canadian Patent 736,026	(20)
		Stabilizer	Canadian Patent 778,848	(21)
Cost	–	–	Expensive	
Availability			Not commercially available (no commercial availability of ethylene sulfide monomer)	

References

1. Cooper W. *Br. Polym. J.* 1971;3:28–35.
2. Gobran, RH. In: Mark HF, et al. ed. *Encyclopedia of Polymer Science and Technology.* Vol 10. New York: Interscience; 1969:324–36.
3. Meyer V. *Ber.* 1886;19:325.
4. Berenbaum MB, Broderick E, Christina RC. *U.S. Patent.* 3,317,486 (1967), assigned to Thiokol Chemical Corp.
5. Gobran RH, Larseen R. *J. Polym. Sci., Part C.* 1970;31:77.
6. Casiff EH, Gillis MN, Gobran RH. *J. Polym. Sci., Part A-1.* 1971;9:1271.
7. Casiff EH. *J. Appl. Polym. Sci.* 1971;15:1641–1648.
8. Angood AC, Koenig JL. *J. Macromol. Sci.: Phys.* 1969;B3:321–328.
9. Nicco A, et al. *E. Polym. J.* 1970;6:1427–1435.
10. Abe A. *Macromolecules.* 1980;13:546–549.
11. Bhaumik D, Mark JE. *Macromolecules.* 1981;14:162.
12. Takahashi Y, Tadokoro H, Chatani Y. *J. Macromol. Sci.: Phys.* 1968;B2:361–367.
13. Hasegawa H, Claffey W, Geil PH. *J. Macromol. Sci.: Phys.* 1977;B13:89–100.
14. Dorset P, McCourt MP. *J. Macromol. Sci.: Phys.* 1997;B36:301–313.
15. Yokoyama M, et al. *J. Macromol. Sci.: Phys.* 1973;B7:465–485.
16. Rinde E, Guzman J. *Macromolecules.* 1981;14:1234–1238.
17. Sorta E, De Chirico A. *Polymer.* 1976;17:348–349.
18. Chiro A, Zotteri L. *E. Polym. J.* 1975;11:487–490.
19. Gorban RH, Osborn SW. *U.S. Patent.* 3,365,431 (1968), assigned to Thiokol Chemical Corp.
20. Bulbenko GF, et al. *Canadian Patent.* 736,026 (1968), assigned to Thiokol Chemical Corp.
21. Larsen R. *Canadian Patent.* 778,848 (1968), assigned to Thiokol Chemical Corp.

Poly(ethylene terephthalate)

Jude O. Iroh

Acronyms, Trade Names PET, Dacron

Class Linear aromatic polyester, thermoplastic

Structure

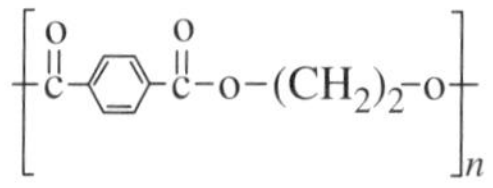

Synthesis

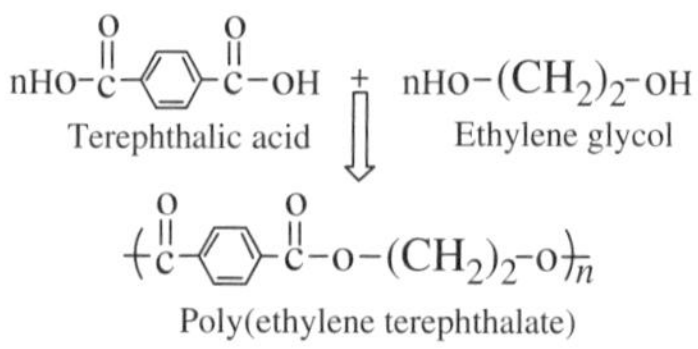

Poly(ethylene terephthalate)

Molecular Conformation Nearly planar

Major Applications Fibers, films, barrier film, soft drink bottle (amorphous PET), film for compression molding PE, PP, and for replacement of commodity metals such as steel and aluminum (1–3). Copolymers of PET are proposed for use as fire retardants (4) and biodegradable films (5, 6).

Preparative Techniques Synthesized by condensation/step growth polymerization between ethylene glycol and terephthalic acid. Low viscosity and easily spinnable PET are synthesized by ester interchange. Dimethyl terephthalate is reacted with ethylene glycol in a 1:1.7 ratio at 0.020 atm and 160–230°C. Final reaction occur at 260–300°C under vacuum at 0.001 atm. Synthesis of PET is done by using aromatic sulphonates as catalysts (7–10).

Properties of Special Interest Mostly synthesized as semi-crystalline thermoplastic, amorphous PET is clear, and is formed by quenching the polymer melt. Excellent film properties and easy to process. Impearmeable to air and hydrophobic.

Property	Units	Conditions	Value	Reference
Molecular weight of repeat unit	g mol^{-1}	–	192	–
Mark-Houwink constants	–	Solution viscometry (30°C)	$K = 3.72 \times 10^{-4}$ $a = 0.73$	(10, 11)
Solvent	–	–	o-chlorophenol	(10, 11, 12)

Poly(ethylene terephthalate)

Property	Units	Conditions	Value	Reference
Weight average molecular weight	g mol^{-1}	–	30,000–80,000	(10, 13, 14)
Unit cell	–	–	Triclinic:	
Lattice constants	degree	X-ray diffraction	$a = 4.56$	(10, 15–17)
			$b = 5.94$	(10, 15–17)
			$c = 10.75$	(10, 15–17)
			$\alpha = 98.5$	(10, 15–17)
			$\beta = 112$	(10, 15–17)
			$\gamma = 111.5$	(10, 15–17)
No of chains per unit cell	–	–	1	(10, 15–17)
Unit cell density	g cm^{-3}	X-ray diffraction	1.501	(10, 15)
Measured density	g cm^{-3}	–	1.41	(10, 15)
Number of chains	–	–	1	(10, 15)
No of monomers	–	–	1	(10, 15–17)
Glass transition temperature (T_g)	°C	DSC	69–115	(8, 10, 18–20)
Melting temperature (T_m)	°C	DSC	265°C	(8, 10, 18–20)
Heat of fussion (ΔH)	KJ mol^{-1}	DSC	24.1	(10, 18, 20)
Mechanical Properties				
Breaking strength (σ_B)	MPa	Tensile	50	(1, 2, 19, 20)
Tensile (Young's) modulus, E	GPa	–	1.7	(1, 2, 19, 20)
Flexural modulus (rigidity), E	GPa	3-point flexure	2.0	(1, 2, 19, 20)
Ultimate strain, ε_B	%	Tensile	180	(1, 2, 19, 20)
Yield strain, ε_Y	%	Tensile	4	(1, 2, 19, 20)
Impact strength	J m^{-1}	Notched Izod ASTM D256-86	90	(1, 2, 19, 20)
Hardness	–	Rockwell	R105	(1, 2, 19, 20)
Deflection, T	°C	HDT	T@ 264 psi = 63	(1, 2, 19, 20)
			T@ 66 psi = 71	(1, 2, 19, 20)
Thermal				
Thermal expansion coefficient, α	mm mm^{-1}°C	TMA	9.1×10^{-5}	(1, 2, 20, 21)
Water absorption	%	After 24 h	0.5	(1, 2, 20, 21)
Dielectric strength	KV mm^{-1}	–	1/8″ = 15.7	(1, 2, 20, 21)
			1/16″ = 22.1	(1, 2, 20, 21)
Dielectric constant	10^6 Hz	–	3.2	(1, 2, 20, 21)
Electrical				
Volume resistivity	Ω cm $\times 10^{16}$	ASTM D257	0.1	(1, 2, 20, 21)
Power factor (10^6 Hz)	10^6 Hz	D150	0.019	(1, 2, 20, 21)

Poly(ethylene terephthalate)

Property	Units	Conditions	Value	Reference
Dielectric strength	KV mm^{-1}	ASTM D149	26	(1, 2, 20, 21)
Dielectric constant	10^6 Hz	ASTM D150	3.3	(1, 2, 20, 21)

References

1. Jaquiss DBG, Borman WFH, Campbell RW. In: Grayson M, ed. *Encyclopedia of Chemical Technology.* Vol 18. New York: John Wiley & Sons, Inc; 1982:549.
2. Brozenick NJ. *Modern Plastics Encyclopedia.* New York: McGraw-Hill, Inc; 1986–1987:464.
3. In: Margolis JM ed. *Engineering Thermoplastics: Properties and Applications.* New York: Marcel Dekker, Inc. 1986–1987:42.
4. Asrar JPA, Berger J, Hurlbut J. *Polym. Sci A: Polym. Chem.* 1999;37:3119.
5. Deng L-M, Wang Y-Z, Yang K-K, Wang X-L, Zhou Q, Ding S-D. *Acta Materialia.* 2004;52:5671.
6. Kink D, Munoz-Guerra S. *Polym. Inter.* 1999;48:346.
7. *U.S. Pat.* 2,647,885 (August. 4, 1953), Billica, H. R. (to E. I. du Pont de Nemours & Co., Inc.)
8. Rodriguez F. *Principles of Polymer Systems.* Second ed. London: MacGraw Hill, International Student Ed; 1983:432, 435.
9. Wilfong RE. *J. Polym. Sci.* 1961;54:385.
10. In: Mark HF, Bikales NM, Overberger CG, Menges G, Kroschwitz JI, eds.*Encyclopedia of Polymer Science and Engineering.* Vol 12. New York: Wiley; 1985:226.
11. Hergenrother Wl, Nelson CJ. *J. Polym. Sci. Polym. Chem. Ed.* 1974;12:2905.
12. In: Mark HF, Bikales NM, Overberger CG Menges G, Kroschwitz JI, eds. *Encyclopedia of Polymer Science and Engineering.* Vol 12. New York: Wiley; 1985:50.
13. Kamiya T, Okamura I, Yamamoto Y. *Proceedings of the 29th SPI Annual Technical Conference.* New York: Society of the Plastics Industry; 1974, Sect. 24-D:1
14. Dixon ER, Jackson JB. *J. Mater. Sci.* 1968;3:464.
15. Hall IH. *Structure of Crystalline Polymers.* Barking, UK: Elsvier Applied Science Publishers, Ltd; 1984:39.
16. Sperling LH. *Introduction to Physical Polymer Science.* Second ed. New York: John Wiley and Sons, Inc; 1992:212.
17. Tadokoro H. *Structure of Crystalline Polymers.* New York: Wiley-Intersciences; 1979.
18. Sperling LH. *Introduction to Physical Polymer Science.* Second ed. New York: John Wiley and Sons, Inc. 1992:199.
19. Palys CH, Phillips PJ. *J. Polym. Sci. Polym. Phys. Edu.* 1980;18:829.
20. *Hand Book of Plastics Materials and Technology, Rubin II, Ed.* New York: John Wiley and Sons, Inc; 1990:644–645.
21. In: Mark HF, Bikales NM, Overberger CG, Menges G, Kroschwitz JI, eds. *Encyclopedia of Polymer Science and Engineering.* Vol 12. New York: Wiley; 1985:230–244.

Poly(ferrocenyldimethylsilane)

Ian Manners

Class Inorganic and Semi-Inorganic Polymers

Structure $[(C_5H_4)Fe(C_5H_4SiMe_2)]_n$

Major Applications Electroactive gels for photonic crystal displays, plasma etch resists, and catalyst precursors (2,4,5).

Properties of Special Interest Low cost. Ease of Synthesis. Interesting redox properties, high refractive index, photoconductor. Precursors to CSiFe solid state ceramic materials. Block copolymers known.

Property	Units	Condition	Value	Reference
UV-Vis Absorption ($\lambda_{max.}$)	Nm	THF Solution	430	(1)
UV-Vis Absorption Coefficient (ε)	$M^{-1}cm^{-1}$	THF Solution	190	(1)
Glass Transition Temperature	°C	DMA Experiment	33	(1)
Glass Transition Temperature	°C	DSC Experiment	25	(1)
Melting Temperature	°C	DSC Experiment	145	(2)
Refractive Index			1.68	(3)

For Monomer $(C_5H_4)FeSiMe_2$

Unit cell dimensions

Lattice	Monomers per Unit Cell	Cell Dimensions			Cell Angles		
		a	b	c	α	β	γ
Monoclinic	4	7.438(3)	10.322(4)	15.575(6)	90	99.04(3)	90

References

1. Foucher DA, Ziembinski R, Tang BZ, et al. *Macromolecules.* 1993;26:2878.
2. Manners I. *Chem. Comm.* 1999;857.
3. Paquet C, Cyr PW, Kumacheva E, Manners I. *Chem. Mater.* 2004;16:5205.
4. Arsenault AC, Puzzo DP, Manners I, Ozin GA. *Nature Photonics.* 2007;1:468.
5. Lastella S, Mallick G, Woo R, et al. *J. Appl. Phys.* 2006;99:024302.

Polygermanes

Robert West

Alternative Name Polygermylenes

Class Inorganic and semi-inorganic polymers

Structure $[-R_2Ge-]$

Properties of Special Interest Polygermane polymers, with their main chain consisting entirely of germanium atoms, resemble polysilanes in showing properties resulting from delocalization of the sigma-electrons along the polymer backbone. Thus the polygermanes have strong UV absorption bands, are thermochromic and photoactive, and become semiconducting when doped with SbF_5. Polysilane-polygermane copolymers have also been prepared, and are of interest as possible superlattice polymers. Listed below are known polygermanes and copolymers, with selected properties.

For an overview of properties of polygermanes, see reference 1.

Synthesis Methods (a) Reaction of $GeCl_2$ with dioxane with RLi. (b) R2GeCl2, Na, toluene, 110°C. (c) Electroreduction.

Polygermane	Synthesis*	Yield (%)	M_w	M_w/M_n	λ Max.[†]	Reference
$(Me_2Ge)_n$	a	25	31,000	—	—	(2)
$(Et_2Ge)_n$	b	—	3,400	1.2	303	(3)
$(n\text{-}Pr_2Ge)_n$	b	15	6,300	1.3	300	(3)
$(n\text{-}Bu_2Ge)_n$	b	25	423,000	1.74	333	(4)
			8,200	1.3		
	b	—	10,000	1.16	329	(5)
	b	52.5	6,800	1.6	320	(3)
	a	38	17,900	—	324	(2)
$(i\text{-}Bu_2Ge)$	b	2	3,900	1.12	330	(3)
$(n\text{-}Pent_2Ge)_n$	b	8.3	25,900	1.53	338	(6)
$(n\text{-}Hex_2Ge)_n$	b	20	15,100	1.5	325	(3)
	b	9.5	976,000	1.56	340	(6)
			6,500	1.12	—	(3)
	c	23	5,350	1.24	—	(6)

Polygermanes

Polygermane	Synthesis*	Yield (%)	M_w	M_w/M_n	λ Max.[†]	Reference
$(n\text{-Oct}_2\text{Ge})_n$	b	70	4,437	–	–	(6)
$(\text{PhGeMe})_n$	b	–	8,600	2.07	330	(5)
	b	–	5,000	1.4	327	(3)
$(n\text{-BuGePh})_n$	b	–	19,900	–	355	(7)

* See "Synthesis Methods" above.
[†] UV absorption maxima in solution.

Polygermane-Polysilane Copolymer	Synthesis*	Yield (%)	M_w	M_w/M_n	λ Max.[†]	Reference
$(n\text{-HexGePh})_n$	b	3.3	12,500	–	355	(3)
$[(n\text{-Bu}_2\text{Ge})(\text{Me}_2\text{Si})2]_n$	b	7.9	13,000	–	312	(8)
$[(n\text{-Bu}_2\text{Ge})(n\text{-HexSiMe})_{4.3}]_n$	b	10	637,000	1.91	314	(4)
$[(\text{Ph}_2\text{Ge})_{1.2}(\text{Cy-HexSiMe})]_n$	b	22	509,000	1.55	317	(4)
			5,600	1.5		
$[(\text{Ph}_2\text{Ge})(n\text{-HexSiMe})_{1.6}]_n$	b	5	33,000	1.51	354, 305	(4)
			3,100	1.1		
$[(n\text{-Hex}_2\text{Ge})(n\text{-Hex}_2\text{Si})]_n$	b	9	275,600	2.74	326	(6)
$[(n\text{-BuGePh})_{1.08}(\text{PhSiMe})]_n$	b	33	20,600	–	335	(8)

* See "Synthesis Methods" above.
[†] UV absorption maxima in solution.

Other Properties of Polygermanes	Reference
Photoresist properties of $(n\text{-Bu}_2\text{Ge})_n$	(9)
Luminescence of $(n\text{-Hex2Ge})_n$	(10)
Flash photolysis of polygermanes	(3)
Polygermane-polysilane superlattice	(8, 11)
Conductivity of $(\text{PhGeMe})_n$ and $(n\text{-Bu}_2\text{Ge})$, doped with SbF_5	(5)
Hole transport properties	(12)

References

1. Carraher CE Jr, Pittman CU Jr, Zeldin M, Abd-El-Aziz AS. In: *Macromolecules Containing Metal and Metal-Like Elements*. Chap. 4. Wiley; 2005.
2. Kobayashi S, Cao S. *Chem. Letters*. 1993;1385.

Polygermanes

3. Mochida K, et al. *Organometallics.* 1994;13:404. Mochida K, Chiba H. *J. Organomet Chem.* 1994;473:45.
4. Trefonas P, West R. *J. Polym. Sci., Polym. Chem. Ed.* 1985;23:2099.
5. Hayashi T, Uchimaru Y, Reddy P, Tanaka M. *Chem. Lett.* 1992;647.
6. Miller RD, Sooriyakumaran. *J. Polym. Sci., Polym. Chem. Ed.* 1985;25:111.
7. Shono T, Kushimura S, Murase H. *J. Chem. Soc., Chem. Commun.* 1992;896.
8. Isaka H, Funjiki M, Fujino M, Matsumoto N. *Macromolecules.* 1991;24:2647.
9. Ban H, Deguchi K, Tanaka A. *J. Appl. Polym. Sci.* 1989;37:1589.
10. Tachibana H, et al. *Phys. Rev. B.* 1992;45:8752.
11. Takeda K, Shiraishi K, Matsumoto N. *J. Am. Chem. Soc.* 1980;112:5043.
12. Abkowitz M, Baessler H, Stolka M. *Phil. Mag.* 1991;B63:210.

Polyglycine

Douglas G. Gold, Wilmer G. Miller,

Taner Z. Sen and Andrzej Kloczkowski

Trade Name Nylon 2

Class Polypeptides and proteins

Structure

$$+\text{NH}-\text{CH}_2-\overset{\overset{\displaystyle O}{\|}}{\text{C}}+$$

Major Applications Serves as a model for various proteins.

Properties of Special Interest Two crystalline forms of polyglycine, I and II, have been observed. Form I is thought to have a β structure where the individual chains exist in a helical conformation and form sheets stabilized by hydrogen bonds.[1,2] The individual chains in form II also have a helical conformation but are packed in a hexagonal lattice with a three-dimensional array of hydrogen bonds.[2]

Synthesis The synthesis is similar to that of poly(γ-benzyl-L-glutamate). (See also the entry on *Poly(γ-benzyl-L-glutamate)* in this handbook.) It involves the conversion of the amino acid to the N-carboxyanhydride (NCA) monomer by reaction with phosgene gas followed by polymerization of the NCA with an appropriate initiator (e.g., triethylamine). Typical comonomers include other amino acid NCAs.

New methods of polycondensation of α-amino acid esters in the presence of yttrium triflate as a Lewis acid[3] and controlled polymerization of activated glycine esters by copper(II) chelate[4] have been recently proposed.

Property	Units	Conditions	Value	Reference
Molecular weight (of repeat unit)	g mol^{-1}	–	57	–
Typical molecular weight range	g mol^{-1}	–	<20,000	–
IR (characteristic absorption frequencies)	cm^{-1}	Polyglycine I Polyglycine II	3,308; 1,685; 1,636; 1,517; 1,432; 708 3,303; 1,644; 1,554; 1,420; 740	(5)
Solvents	–	25°C	Dichloroacetic acid, trifluoroacetic acid, concentrated Li$^+$ and NH$_4^+$ halides, phosphoric acid	(1, 6)
Nonsolvents	–	–	Water	–
Optical activity $[\alpha]_D$	–	–	0	–
Surface tension	mN m^{-1}	0°c	50.1	(7)
Axial translation per residue	Å	Form I	3.5	(2)
		Form II	3.1	(1, 2)

Property	Units	Conditions	Value		Reference
Observed solid state ^{13}C chemical shifts	ppm	Conformation 3_1-helix β-sheet	C=O 172.9 169.0	Cα 42.6 44.3	(8)
The calculated carbon-proton distances	Å	anti-parallel β-sheet	Cα-Hα Cα-HN C=O-Hα C=O-HN	1.8 2.15 2.15 1.98	(8)
Mulliken Atomic Charges of the atoms of enol-trans and keto-trans tautomers	e		enol-trans C 0.290 H 0.403 O −0.300 N −0.193	keto-trans 0.370 0.324 −0.272 −0.268	(9)
Bond distances	Å	β pleated sheet	O_1-C_2 C_2-N_3 N_3-C_4 C_4-C_2' N_3-H_5 C_4-H_6 C_4-H_7	1.23 1.34 1.45 1.52 1.01 1.09 1.09	(10)
Bond angles	degrees	β pleated sheet	$O_1-C_2-N_3$ $C_2-N_3-C_4$ $N_3-C_4-C_2'$ $C_4-C_2'-N_3'$ $C_2-N_3-H_5$ $N_3-C_4-H_6$ $N_3-C_4-H_7$	122.9 121.9 110.4 116.6 119.5 109.47 108.85	(10)
Dihedral angles	degrees	β pleated sheet	$O_1-C_2-N_3-H_5$ $O_1-C_2-N_3-C_4$ $C_2-N_3-C_4-C_2' = \Phi$ $N_3-C_4-C_2'-N_3' = \Psi$ $H_6-C_4-N_3-C_2$ $H_7-C_4-N_3-C_2$ $C_4-C_2'-N_3'-C_4'$	180.0 0.0 −140.0 140.0 100.58 −19.42 180.0	(10)
Conduction and valence bands of PolyGly(8)[a]	eV eV	k-values in parantheses	CB_{max} CB_{min} CB_{width} VB_{max} VB_{min} VB_{width}	0.29(8)[b] −0.22(0) 0.51 −7.29(0) −7.43(4) 0.14	(10)
Fundamental gap value	eV			7.07	(10)

References

1. Bamford CH, Elliott A, Hanby WE. *Synthetic Polypeptides: Preparation, Structure, and Properties.* New York: Academic Press; 1956.
2. Fraser RDB, MacRae TP. *Conformation in Fibrous Proteins and Related Synthetic Polypeptides.* New York: Academic Press; 1973.
3. Nemoto T, Ando D, Naka K, Chujo Y. *J. Polym. Sci: Part A: Polym. Chem.* 2006;44:4731.
4. Naka K, Nemoto T, Chujo Y. *J. Polym. Sci: Part A: Polym. Chem.* 2006;41:1504.
5. Suzuki S, Iwashita Y, Shimanouchi T. *Biopolymers.* 1966;4:337.
6. Sober HA, ed. *Handbook of Biochemistry: Selected Data for Molecular Biology.* 2nd ed. Cleveland: CRC Company; 1970.
7. Brandrup J, Immergut EH, eds. *Polymer Handbook.* 3rd ed. New York: John Wiley and Sons; 1989.
8. Murata K, Kono H, Katoh E, Kuroki S, Ando I. *Polymer.* 2003;44:4021.
9. Kamiya K, Boero M, Shiraishi K, Oshiyama A. *J. Phys. Chem.* 2006;B110:4443.
10. Bogar F, Ladik J. *Int. J. Quant. Chem.* 2004;99:47.

Poly(glycolic acid)

Lichun Lu and Antonios G. Mikos

Acronyms, Trade Names PGA, Dexon (Davis and Geck)

Class Poly(α-hydroxy ester)

Structure

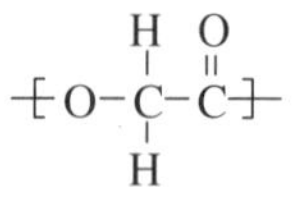

Major Applications Sutures, drug delivery devices, and scaffolds for use in cell culture, transplantation, and organ regeneration.

Properties of Special Interest Good biocompatibility; biodegradable mainly by simple hydrolysis; bioresorbable; good processability; a wide range of degradation rates, physical, mechanical, and other properties can be achieved by PGA of various molecular weights and its copolymers.

Preparative Techniques Practically useful high molecular weight PGA can be synthesized by a cationic ring opening polymerization of glycolide using organometallic compounds or Lewis acids as catalysts and alcohol as molecular weight and reaction rate control agent at high temperature and low pressure.

Property	Units	Conditions	Value	Reference
Degree of crystalline X_c	%	Dexon suture	46–52	(1)
Density ρ	g cm^{-3}	Typical range	1.5–1.64	(2)
		Complete amorphous	1.450	(3)
			1.50	(4)
		Complete crystalline	1.69	(5, 6)
			1.707	(4)
Heat of fusion ΔH_f	kJ mol^{-1}	Complete crystalline	8.1	(7)*
			11.1	(8)
Entropy of fusion ΔS_f	kJ K^{-1} mol^{-1}	–	0.022	(9)

* Data calculated from $X_c(0.52) = \Delta H_f$ (4.2 kJ mol^{-1})/ΔH_f(100% crystalline).

716

Poly(glycolic acid)

Unit cell dimensions

Lattice	Space Group	Monomers Per Unit Cell	Cell Dimension# (Å)			Packing Density k	Chain Conformation $N \times P/Q$	Reference
			a	b	c (fiber axis)			
Orthorhombic	$P2_12_12_1$	4	6.36	5.13	7.04	–	$3 \times 2/1$	(5)
Orthorhombic	Pcmn	4	5.22	6.19	7.02	0.81	$3 \times 2/1$	(5)
Orthorhombic	–	2	–	–	7	–	–	(6)

Cell angles $\alpha = \beta = \gamma = 90°$.

Property	Units	Conditions	Value	Reference
Refractive index	–	Highly oriented fiber	$n_{//} = 1.556$	(6)
			$n_\perp = 1.466$	
		Single crystal hedrite	$\alpha = 1.46$	(6)
			$\beta = 1.50$	
			$\gamma = 1.66$	
Glass transition temperature T_g	K	$M_w > 20,000$	318	(7)
		$M_w = 50,000$	309	(1)
Melting point T_m	K	$M_w > 20,000$	495	(7)
		$M_w = 50,000$	483	(10)
		$M_w = 50,000$	503	(1)
		–	500–503	(4)
		–	506	(5)
Heat capacity C_p	$J\,K^{-1}mol^{-1}$	Crystalline PGA in temperature range $= 0$–318 K	–	(3)*
			–	
		$T = 273.15$	115.0	(3)
		$T = 298.15$	121.4	(3)
		$T = 318.0\ (T_g)$	126.5	(3)
		Molten PGA, $T = 318.0$–550.0 K	226.5–243.4	(3)
		Predicted results in temperature range $= 0$–1000 K	–	(11)
Solvent	–	Glycolide at high temperature	–	(12)
		Hexafluoroisopropanol at	–	(13)
		room temperature	–	(12, 14)
		Phenol/trichlorophenol at		
		$T = 190°C$		
Inherent viscosity	$dl\,g^{-1}$	In hexafluoroisopropanol	0.5–1.6	(2)
		In phenol/trichlorophenol at $T = 30°C$	0.35	(14)
Water uptake	%	250 μm films in 0.2 M pH7 phosphate buffer	28	(1)

717

Poly(glycolic acid)

Property	Units	Conditions	Value	Reference
Apparent permeability, K	$m^4\,N^{-1}\,s^{-1}$	Prewetted nonwoven discs	4.29×10^{-10}	(15)
Degradation rate	–	*In vitro*	–	(13, 15–17)
		In vivo	–	(2, 17)
Decomposition temperature T_d	K	$M_w = 50{,}000$, $X_c = 0.52$ at a heating rate = 20°C min^{-1} under nitrogen	527	(10)
G factor	–	Under ^{60}Co irradiation at $T = 25$°C chain scission factor approximately equal to crosslinking factor	–	(14)
Tensile strength	MPa	Dexon plus (PGA multifilament)		(18)
		gauge 0	339	
		gauge 1	394	
Tenacity	MPa	Melt-spun fiber (diameter = 15–25 µm)	690–1,380	(2)
		Dexon suture	6,050	(16)
Knot pull strength	MPa	Size 3/0 suture	343	(2)
Straight pull strength	MPa	Size 3/0 suture	536	(2)
Knot/straight tenacity	%	Melt-spun fiber (diameter = 15–25 µm)	50–80	(2)
Elongation at break	%	Melt-spun fiber (diameter = 15–25 µm)	15–35	(2)
Confined compressive modulus	MPa	Prewetted nonwoven discs	2.86×10^{-3}	(15)
Aggregate modulus	MPa	Prewetted nonwoven discs	1.22×10^{-3}	(15)

* Data in Ref. (3) referred to $(CH_2-COO-CH_2-COO)$.

References

1. Gilding DK, Reed AM. *Polymer.* 1979;20:1459.
2. Frazza EJ, Schmitt EE. *J. Biomed. Mater. Res. Symp.* 1971;1:43.
3. Gaur U, Lau S-F, Wunderlich BB, Wunderlich B. *J. Phys. Chem. Ref. Data.* 1983;12:65.
4. Chujo K, Kobayashi H, Suzuki J, Tokuhara S. *Makromol. Chem.* 1967;100:267.
5. Chatani Y, et al. *Makromol. Chem.* 1968;113:215.
6. Grabar DG. *Microscope.* 1970;18:203.
7. Cohn D, Younes H, Marom G. *Polymer.* 1987;28:2018.
8. Chu CC, Browning A. *J. Biomed. Mater. Res.* 1988;22:699.
9. Wunderlich B. *Macromolecular Physics: Vol. 3, Crystal Melting.* New York: Academic Press; 1980.
10. Engelberg I, Kohn J. *Biomaterials.* 1991;12:292.
11. Lim S, Wunderlich B. *Polymer.* 1987;28:777.
12. Chujo K, et al. *Makromol. Chem.* 1967;100:262.
13. Suggs LJ, Mikos AG. In: Mark JE, ed. *Physical Properties of Polymers Handbook.* Woodbury, N.Y: American Institute of Physics Press; 1996:615–624.
14. Pittman CU Jr, Iqbal M, Chen CY, Helbert JN. *J. Polym. Sci. Polym. Chem. Ed.* 1978;16:2721.

15. Ma PX, Langer R. In: Mikos AG, et al. ed. *Polymers in Medicine and Pharmacy.* Pittsburgh: Materials Research Society; 1995:99–104.
16. Chu CC. *Polymer.* 1985;26:591.
17. Chu CC. In: Williams DF, ed. *Critical Reviews in Biocompatibility.* Boca Raton: CRC Press; 1985:261–322.
18. Singhal JP, Singh H, Ray AR. *J. Macromol. Sci.-Rev. Macromol. Chem. Phys.* 1988;C28:475.

Poly(hexene-1)

R. P. Patki, D. R. Panse and P. J. Phillips

Acronyms and Trade Names PHE, PHEX

Class α-Olefins

Structure of the Repeat Unit
$$-[-CH_2-CH-]-$$
$$CH_2CH_2CH_2CH_3$$

Major Applications Comonomer for ethylene and 4-methyl pentene-1, flow modifier, viscosity modifier in lubricants and petroleum oils.

Preparative techniques

Type of Polymerization	Conditions	Reference
Coordination	1. Bis(cyclopentadienyl) zirconium dichloride + methylaluminoxane catalyst	(1)
	2. Soluble magnesium-titanium catalyst in xylene + diethylaluminium chloride cocatalyst at 40°C for 420 min.	(2)
	3. MgCl$_2$ supported TiCl$_3$ catalyst	(3)
	4. TiCl$_4$ + Al(C$_2$H$_5$)$_3$, TiCl$_4$ + Al(C$_2$H$_5$)$_2$Cl, TiCl$_4$ + Al(iso-C$_4$H$_9$)$_3$ catalysts fixed on surface of carries such as pearlite and graphite	(4)
	5. Nickel(-diimine)/methylaluminoxane catalyst	(5)
	6. [2-(2,6-dialkylphenylamino)-1-phenylethoxy TiCl$_2$]/methylaluminoxane catalyst	(6)
	7. bis(phenoxyimine)Ti + i-Bu$_3$Al/Ph$_3$CB-(C$_6$F$_5$)$_4$	(7)
	8. Metallocene En(Ind)$_2$ZrCl$_2$ and iPr(Cp)(Flu)ZrCl$_2$ with methylaluminoxane	(8)

Typical comonomers used: ethylene, 4-methyl pentene-1, 1-pentene.

Structure and morphology

Property	Units	Condition	Value	Reference
Molecular weight of repeat unit	g mol^{-1}	–	84.16	
Stereoregularity		Soluble magnesium-titanium catalyst in xylene + diethylaluminium chloride cocatalyst	Mainly isotactic	(2)
		TiCl$_4$ + Al(C$_2$H$_5$)$_3$, TiCl$_4$ + Al(C$_2$H$_5$)$_2$Cl, TiCl$_4$ + Al(iso-C$_4$H$_9$)$_3$ catalysts fixed on surface of carries such as pearlite and graphite	98.5% isotactic	(4)

Poly(hexene-1)

Property	Units	Condition	Value	Reference
		Nickel(-diimine)/methylaluminoxane catalyst Polymerization temperature 70 to −7°C.	Butyl (C4), longer than butyl (>C4), methyl and regio-irregular methyl branches	(5)
		[2-(2,6-dialkylphenylamino)-1-phenylethoxy $TiCl_2$]/methylaluminoxane catalyst	80–90% isotactic	(6)
		bis(phenoxyimine)Ti + i-Bu_3Al/Ph_3CB-$(C_6F_5)_4$	mainly atactic with 50 mol% regio-irregular units	(7)
Typical molecular weight range	g mol^{-1}	$MgCl_2$ supported $TiCl_3$ catalyst (weight average mol. wt.)	100,000	(3)
		(weight avg.)	1–10×10^6	(4)
			400,000	(6)
			0.84–1.45×10^6	(7)
			29.5–48×10^3	(8)
		Stereorigid metallocene catalysts (number avg.)	<30,000	(9)
Typical polydispersity index		Soluble magnesium-titanium catalyst	2.0–2.7	(2)
		Heterogeneous $MgCl_2$ supported $TiCl_3$ catalyst	5–11	(3)
			4–10	(4)
			1.07–2.9	(6)
			1.4–1.9	(7)
			1.8–2.5	(8)
		Stereorigid metallocene catalysts	2–3	(9)

Solution properties

Solvents	Nonsolvents	Reference
For amorphous polymer: Saturated and aromatic hydrocarbon solvents at ambient temperature.		(10)
Good solvents: cyclohexane, tetrahydrofuran, toluene. Poor solvent: phenetole for crystalline isotactic polymer: Decalin at 135°C		(11)

Structure and morphology

Property	Units	Condition	Value	Reference
Theta temperature	K	Phenetole/VM	334.3	(12)
Characteristic ratio	–	Method—viscometry at 27°C	13.0	(13)

Crystalline state properties

Property	Units	Condition	Value	Reference
Lattice			Monoclinic	(14)
Unit cell dimensions	Å	Stretched fiber below −20°C for one week		(14)
A			22.2	
B			8.89	
C			13.7	
Unit cell angles	degrees	Isotactic polymer		(14)
α			90	
β			90	
γ			94.5	
Monomers per unit cell		Isotactic polymer	14	(14)
Helix conformation		Isotactic polymer	7.2	(14)
Crystalline density	g cm^{-3}	Isotactic polymer	0.83	(14)
Lattice			Orthorhombic	(14)
Unit cell dimensions	Å	Stretched fiber below −20°C for one week		(14)
A			11.7	
B			26.9	
C			13.7	
Unit cell angles			All 90	
Chains per unit cell			4	
Crystalline density	g cm^{-3}		0.91	

Transition temperatures

Property	Units	Condition	Value	Reference
Glass transition	K	calorimetry	215	(15)
			223	(16)
Melting temperature	K		<293	(10)

Thermodynamic and other properties

Property	Units	Condition	Value	Reference
Heat capacity	kJ (mol^{-1} K^{-1})	Temperature		(16)
		100 K	0.059 (a)*	
		200 K	0.112	
		250 K	0.160 (a)	
		290 K	0.1749 (a)	
Polymers with which compatible		Single T_g criteria	1-pentene	(17)
WLF constants				(18)
C1			17.4	
C2			51.6	
Refractive index increment	ml g^{-1}	cyclohexane at 25°C	−0.063	(11)
		toluene at 25°C	−0.042	

References

1. Sivaram S, Marathe S. *Unpublished work mentioned in reference 2.*
2. Satyanarayana G, Sivaram S. *Makromol. Rapid Comm.* 1994;15:601.
3. Chien JCW, Gong BM. *J. Polym. Sci., Pt. A: Polym. Chem.* 1993;31:1747.
4. Sukhova TA, Dyachkovskii FS. *Polym. Adv. Tech.* 1993;4:475.
5. Subramanyam U, Rajamohanan PR, Sivaram S. *Polymer.* 2004;45:4063.
6. Vijayakrishna K, Sundararajan G. *Polymer.* 2006;47:3363.
7. Saito J, Suzuki Y, Makio H, Tanaka H, Onda M, Fujita T. *Macromolecules.* 2006;39:4023.
8. Kawahara N, Saito J, Matsuo S, et al. *Polymer.* 2007;48:425.
9. Asanuma T, Mishimai Y, Ito M, Uchikawa N, Shimoura T. *Polym. Bull.* 1991;25:567.
10. Olefin Polymers (Higher Olefins). In: Othmer K, ed. *Encyclopedia of Chemical Technology.* John Wiley & Sons; 1996.
11. Lin FC, Stivala SS, Biesenberger JA. *J. Appl. Polym. Sci.* 1973;17:1073.
12. Lin FC, Stivala SS, Biesenberger JA. *J. Appl. Polym. Sci.* 1973;17:3465.
13. Wang JS, Porter RS, Knox JR. *Polym. J.* 1978;10:619.
14. Turner-Jones A. *Makromol. Chem.* 1964;71:1.
15. Bourdariat J, Isnard R, Odin J. *J. Polym. Sci: Polym. Phys. ed.* 1973;11:1817.
16. Gaur U, Lau SF, Wunderlich B, et al. *J. Phys. Chem.* 1983;12:29.
17. Dacroix JY, Piloz A. *Rev. Gen. Caoutsch. Plast.* 1977;54:91.
18. Kurath SF, Passaglia E, Pariser R. *J. Appl. Phys.* 1957;28:499.

Poly(*n*-hexyl isocyanate)

Jagath K. Premachandra, Chandima Kumudinie
and Junzo Masamoto

Acronym PHIC

Class Poly(isocyanates); N-substituted 1-nylons

Structure

$$\begin{array}{c} C_6H_{13} \\ | \\ -\!\!\!+\!N\!-\!\!\!\underset{\underset{O}{\|}}{C}\!\!\!+\!\!\!- \end{array}$$

Major Applications An ideal example of a polymer model for a semi-rigid macromolecular chain material amenable to physical studies.

Properties of Special Interest Polymer model for a semi-flexible macromolecular chain material. Stiff-chain solution characteristics due to helical configuration;[1,2] liquid crystalline properties[3] and molecular weight dependent chain dimensions in solution.[4]

Other Polymers Showing this Special Property Poly(*n*-butyl isocyanate), poly(γ-benzyl-L-glutamate)

Preparative techniques*

Polymerization Process	Conditions	Reference
Anionic	Temp.: $-58°C$; catalyst: NaCN in dimethylformamaide; solvent: benzene	(5)
Anionic	Temp.: -78 to $-100°C$; catalyst: NaCN in dimethylformamaide; solvent: toluene	(6a)
Anionic	Temp.: $-98°C$: catalyst: sodium naphthalenide in the presence of crown ether or sodium tetraphenyl borate ($NaBPh_4$)	(6b, 6c)
Anionic	Temp.: $-98°C$: catalyst: sodium benzanilide in THF	(6d)
Coordination	Catalysts: $TiCl_3(OCH_2CF_3)$, $TiCl_3(OCH_2CF_3)$-$(THF)_2$	(7)
Coordination	Catalysts: $C_pTiCl_2N(CH_3)_2$, $C_p = n^5$-cyclopentadienyl; 1 equivalent of Lewis base per monomer; no solvent	(8a)
Coordination	Catalsts: lanthanide phosphate/tri-i-butyl alminium	(8b)
Coordination	Catalsts: La(2,6-di-ter-butyl-4-methyl phenolate)$_3$	(8c)

* For synthesis of the monomer, *n*-hexyl isocyanate, see reference (2).

Poly(*n*-hexyl isocyanate)

Property	Units	Conditions	Value	Reference
Ceiling temperature	K	–	316.4	(7)
			251	(9)
Typical comonomers	–	–	Styrene, methyl methacrylate for block copolymer	(10)
Molecular weight (of repeat unit)	g mol^{-1}	–	127	–
Typical molecular weight range of polymer	g mol^{-1}	–	$2.4 \times 10^4 - 5.14 \times 10^5$	(11)
			$>10^6$	(12)
			$4 \times 10^3 - 7 \times 10^6$	(13)
Typical polydispersity index (M_w/M_n)	–	–	1.4	(14)
			1.05–1.2	(15)
			1.05–1.1	(16)
IR (characteristic absorption frequencies)	cm^{-1}	Solid state	C=O absorption at 1,700	(17)
		Solid state	C=O absorption at 1,709	(5)
		Dilute solution in tetrachloroethane	C=O absorption at 1,700	(17)
		Solid state	Disubstituted amide at 1,282 and 1,390	(5)
UV (absorption maxima at the high wavelength band, λ_{max})	nm	In *n*-hexane at room temperature	252	(18)
Extinction coefficient	L mol^{-1} cm^{-1}	In *n*-hexane at room temperature	4,572	(18)
NMR	^{13}C NMR, 300- and 500-MHz instrument, spin-lattice and spin-spin relaxation times and nuclear Overhauser enhancements of carbon atoms in or near the backbone of the extended-chain polymer			(19)
	NMR line-width measurement			(20)
	^{1}H NMR			(21)
Solvents	Aromatic and chlorinated hydrocarbons			(4, 5)
	Hexane, *n*-butyl benzene, toluene, many solvents in the presence of trifluroacetic acid			(4)
Nonsolvents	Methanol, cyclohexyl benzene, dodecyl benzene, higher paraffins (number of carbons >6)			(4)

Property	Units	Conditions	Value	Reference
Theta temperature θ	K	In toluene	289.4	(22)
		In methanol/toluene (19.5% v/v), turbidity point method	298	(4)
		In methanol/carbon tetrachloride (18.5% v/v), turbidity point method	298	(4)
			297	(23)
Second virial coefficient	mol cm^3 g^{-2}	In tetrahydrofuran at 25°C, $M_w = (3.8 \times 10^4$–$4.24 \times 10^5)$ g mol^{-1}, light scattering	$(9.7$–$16.6) \times 10^{-4}$	(4)
		In toluene at 34.4°C, $M_n = (3.9 \times 10^4$–$3.85 \times 10^5)$ g mol^{-1}, osmometry	$(3.66$–$8.34) \times 10^{-4}$	(4)
		In tetrachloroethane, $M_w = 9.6 \times 10^4$ g mol^{-1}, in anisotropic phase, light scattering	1.8×10^{-3}	(14, 24)
		In tetrachloroethane, $M_w = 5.6 \times 10^4$ g mol^{-1}, in isotropic phase, light scattering	1.8×10^{-3}	(14, 24)
		In tetrachloroethane, $M_w = 7 \times 10^4$ g mol^{-1}, in anisotropic phase, osmometry	1.9×10^{-3}	(14, 24)
		In tetrachloroethane, $M_n = 4 \times 10^4$ g mol^{-1}, in isotropic phase, osmometry	3.1×10^{-3}	(14, 24)
		In toluene at 34.4°C, $M_n = 7.9 \times 10^4$ g mol^{-1}, osmometry	7.8×10^{-4}	(23)
		In hexane at 25°C, $M_w = 6.8 \times 10^4$ g mol^{-1}, light scattering	9.8×10^{-4}	(23)
		In hexane at 25°C, $M_w = 4.63 \times 10^6$ g mol^{-1}, light scattering	7.5×10^{-4}	(23)
		In hexane at 25°C, $M_w = 7.24 \times 10^6$ g mol^{-1}, light scattering	5.4×10^{-4}	(23)
		$M_w = (1.1 \times 10^5$–1.06×10 g mol^{-1}, light scattering	$(5.97$–$6.49) \times 10^{-4}$	(16)
		In tetrahydrofuran, $M_w = (1.46$–$3.55) \times 10^5$ g mol^{-1}, light scattering	$(6.9$–$8.2) \times 10^{-4}$	(25)

Mark-Houwink parameters: K and a

Solvent	Temp. (°C)	Molecular Weight (g mol^{-1})	$K_m \times 10^5$ (dl g^{-1})	a	Reference
Toluene	25	$M_w = (0.038$–$4.3) \times 10^5$	2.48	1.05	(4)
Methanol/toluene	25	$M_w = (0.038$–$4.3) \times 10^5$	2.72	1.04	(4)
Carbon tetrachloride	25	$M_w = (0.038$–$4.3) \times 10^5$	1.52	1.10	(4)
Trifluro acetic acid/carbon tetrachloride	25	$M_w = (0.038$–$4.3) \times 10^5$	5.68	0.96	(4)
Tetrahydrofuran	25	$M_w = (0.038$–$4.3) \times 10^5$	2.20	1.06	(4)

Poly(*n*-hexyl isocyanate)

Solvent	Temp. (°C)	Molecular Weight (g mol^{-1})	$K_m \times 10^5$ (dl g^{-1})	a	Reference
Chloroform	25	$M_w = (0.038–4.3) \times 10^5$	4.99	0.97	(4)
Dichloromethane	20	$M_w = (0.24–5.14) \times 10^5$	6.6	0.923	(11)
Hexane	25	$(0.68–3.9) \times 10^5$	100	1.2	(23)
Hexane	25	$>10^6$	8,800	0.77	(23)
Hexane	25	$(0.091–2.3) \times 10^6$	540	0.97	(26)

Property	Units	Conditions	Value	Reference
Huggins constant	–	Toluene at 25°C	0.346	(4)
		Carbon tetrachloride at 25°C	0.341	(4)
		Chloroform at 25°C	0.319	(4)
		Tetrahydrofuran at 25°C	0.331	(4)
		1.5% acetic acid/carbon tetrachloride	0.328	(4)
		1.5% methanol/toluene	0.326	(4)
		Hexane at 25°C, $M_w = (0.68–3.34) \times 10^5$ g mol^{-1}	0.38	(23)
		Hexane at 25°C, $M_w = 7.24 \times 10^6$ g mol^{-1}	0.50	(23)
		Carbon tetrachloride, $M_w = 3.55 \times 10^5$ g mol^{-1}	0.56	(25)
		Benzene, $M_w = 3.55 \times 10^5$ g mol^{-1}	0.51	(25)
		Tetrahydrofuran, $M_w = 3.55 \times 10^5$ g mol^{-1}	0.43	(25)
		Tetrahydrofuran/dimethylformamide (4:1), $M_w = 3.55 \times 10^5$ g mol^{-1}	0.75	(25)
		Chloroform	1.10	(25)
Characteristic ratio $\langle r^2 \rangle_0 nl^2$	–	–	410	(27)
Radius of gyration	Å	In tetrahydrofuran at 25°C, $M_w = (0.38–4.3) \times 10^5$ g mol^{-1}	162–966	(4)
		In tetrachloroethane, $M_w = 9.6 \times 10^4$ g mol^{-1}, in anisotropic phase	315	(14)
		In tetrachloroethane, $M_w = 5.6 \times 10^4$ g mol^{-1}, in isotropic phase, light scattering	180	(14)
		In butyl chloride, light scattering, $M_w = (0.043–2.1) \times 10^6$ g mol^{-1}	300–2,190	(28)
Monomer projection length	Å	In carbon tetrachloride	~2.1–1.1	(4)
		In toluene	~2.1–1.0	
		In chloroform	~2.0–0.7	
Persistence length	Å	In 1-chloronaphthalene at 25°C; chain diameter $d = 10.3$ Å	230	(29)
		In 1-chloronaphthalene at 25°C, $d = 16.4$ Å	200	(29)
		In 1-chloronaphthalene at 45°C, $d = 10.3$ Å	190	(29)
		In 1-chloronaphthalene at 45°C, $d = 16.4$ Å	165	(29)
		In 1-chloronaphthalene at 90.2°C, $d = 10.3$ Å	110	(29)

Property	Units	Conditions	Value	Reference
		In 1-chloronaphthalene at 90.2°C, $d = 16.4$ Å	95	(29)
		In 1-chloronaphthalene at 110.8°C, $d = 10.3$ Å	90	(29)
		In 1-chloronaphthalene at 110.8°C, $d = 16.4$ Å	80	(29)
		In chloroform	200	(30)
		In dichloromethane	185	(11)
		In dichloromethane at 20°C	210	(31)
		In hexane	420	(23)
		In toluene	375	(11)
		In toluene at 10°C	410	(31, 37)
		In toluene at 25°C	370	(31, 37)
		In toluene at 40°C	340	(31, 37)
		In tetralin	400	(32)
Persistence length	Å	In tetrahydrofuran	425	(4)
		In tetrahydrofuran	400–500	(33)
		In butyl chloride	450	(28)
Lattice	–	Only a few d spacings were given	–	(34, 35)
Chain conformation		In the solid state by X-ray scattering: 12_5 helix		(34)
		A helical rodlike conformation at low molecular weights and a transition to random-coil conformation at high molecular weight		(1, 4)
Degree of crystallinity	%	X-ray diffraction	Low order of crystallinity	(5)
Density	g cm^{-3}	–	1.000	(14, 36)
Partial specific volume	cm^3 g^{-1}	In toluene at 25°C	0.987	(37)
		In toluene at 40°C	1.002	
		In toluene at 10°C	0.972	
		In dichloromethane at 20°C	0.992	
Glass transition temperature	K	By thermally stimulated discharge currents	223	(38)
		Hydrosilation cross-linked, DSC	258	(39)
		Copolymer with 10 mol% allyl composition, DSC	273	(39)
Melting point	K	DTA-DSC, Ar atmosphere, heating rate $= 10°C$ min^{-1}	$\sim$428	(17)
		$M_w = 9.6 \times 10^4$ g mol^{-1}	468	(40)
			468	(5)
			473	(41)
Mesomeric transition temperatures	K	Crystalline-nematic transition	438	(41)
		Crystalline-nematic transition, $M_w = 9.4 \times 10^4$	450	(40)
Sub-T_g transition temperatures	K	Chain skeleton rearrangement relaxations (or transition), $M_w = 9.4 \times 10^4$ g mol^{-1}	293 and 313	(42)
			393	(40)
Softening temperature	K	–	393	(5)

Poly(*n*-hexyl isocyanate)

Property	Units	Conditions	Value	Reference
Tensile modulus	MPa	At 23°C, strain rate $= 8.3 \times 10^{-4}$ s^{-1}	~420	(43)
		At -100°C	1,600	(44)
		At -40°C	1,000	(44)
		At 20°C	300	(44)
		Highly oriented fibers, orientation angle $= 4.6$°, room temperature, strain rate $= 10$ mm min^{-1}	~4,000	(45)
Theoretical axial modulus	MPa	Side chain reduced the axial modulus of rigid-chain polyamides by ~90%	~6,000	(45)
Tensile strength	MPa	At 23°C, strain rate $= 8.3 \times 10^{-4}$ s^{-1}	~10	(43)
		Highly oriented fibers, room temperature, strain rate $= 10$ mm min^{-1}	~200	(45)
Maximum extensibility	%	At 23°C, strain rate $= 8.3 \times 10^{-4}$ s^{-1}	~8	(43)
		Highly oriented fibers, room temperature, strain rate $= 10$ mm min^{-1}	~6.6	(45)

Mechanical properties of cross-linked PHIC[46, 47]

Conditions	Properties		
	Tensile Modulus (MPa)	Tensile Strength (MPa)	Elongation at Break (%)
Uncross-linked, 10% allyl concentration	140	10.6	11.0
Cross-linked, high-temp., hydrosilation	30	2.6	46.0
Oriented, high-temp., hydrosilation	260	15.6	12.5
Crosslinker: hydrido-oligo (dimethylsiloxane)			
Unoriented	30	4.55	166.6
Oriented, uniaxial //			
Extension ratio 35%	83	9.72	84.4
Extension ratio 70%	233	18.76	65.5
Extension ratio 100%	310	24.13	49.4
Oriented, uniaxial ⊥			
Extension ratio 20%	49	5.77	79.5
Extension ratio 50%	27	3.48	59.2
Oriented, biaxial			
Extension ratio 20%	62	13.71	157.5
Cross-linker: hexamethyl-trisiloxane			
Unoriented	16	7.05	172.7
Oriented, uniaxial //			
Extension ratio 80%	103	23.60	80.8
Extension ratio 120%	309	34.21	33.8
Extension ratio 140%	394	39.82	37.1

Property	Units	Conditions	Value	Reference
Entanglement molecular weight M_c	g mol^{-1}	–	6,350	(27, 33)
Refractive index increment dn/dc	ml g^{-1}	In *n*-butyl chloride at 25°C, $\lambda_0 = 546$ nm	0.092	(13)
		In *n*-hexane at 25°C, $\lambda_0 = 546$ nm	0.134	(23)
		In tetrahydrofuran, $\lambda_0 = 436$ nm	0.100	(48)
		In tetrahydrofuran, $\lambda_0 = 546$ nm	0.097	(48)
		In tetrahydrofuran, $\lambda_0 = 436$ nm	0.099	(33)
		In tetrahydrofuran at 25°C, $M_w = 3.8 \times 10^4$ g mol^{-1}	0.088	(4)
Refractive index increment dn/dc	ml g^{-1}	In tetrahydrofuran at 25°C, $M_w = 6.8 \times 10^4$ g mol^{-1}	0.0900	(4)
		In tetrahydrofuran, $M_w = 7.2 \times 10^4$ and 7.9×10^4 g mol^{-1}	0.0910	(4)
		In tetrahydrofuran, $M_w = (1.11, 1.35, 1.62, 2.04, 3.12,$ and $4.24) \times 10^5$ g mol^{-1}	0.0934	(4)
Dielectric constant ε'	–	In toluene at 292.2 K, 21.4% polymer (w/w), frequency range $\sim(2-1 \times 10^5)$ Hz	$\sim(10-120)$	(49)
Dielectric loss ε''	–	In toluene at 292.2 K, 21.4% polymer (w/w), frequency range $\sim(2-1 \times 10^5)$ Hz	$\sim(4-24)$	(49)
Dielectric critical frequency f_c	Hz	In toluene at 292.2 K, 21.4% polymer (w/w), frequency range $\sim(2-1 \times 10^5)$ Hz	~400	(49)
Optical activity: specific rotation $[\alpha]_D$	Degree	Poly((R)-1-deuterio-*n*-hexyl isocyanate) at the sodium *d*-line in dilute solution of chloroform at:		
		10°C	−450	(50)
		25°C	−367	(50)
		47°C	−258	(50)
		Poly((R)-1-deuterio-*n*-hexyl isocyanate), degree of polymerization = 6,800:		
		In chloroform and in hexane, (60 to −20°C)	$\sim$3 fold increase in $[\alpha]_D$	(51)
		In chloroform, (70 to −30°C)	$\sim(-550$ to $-150)$	(52)
		Poly((R)-1-deuterio-*n*-hexyl isocyanate) and poly((R)-2-deuterio-*n*-hexyl isocyanate), in dicloromethane, in 1-chlorobutane, in toluene and in hexane, (100 to −20°C)	$[\alpha]_D$ increases with decreasing temperature	(53)
		Copolymer of 99.5% *n*-hexyl isocyanate and 0.5% (R)-2, 6-dimethylheptyl isocyanate, (40 to −20°C) (for *n*-hexane ~-5°C, for *n*-octane ~10°C)	Sudden increase in $[\alpha]_D$	(54)
Decomposition temperature	K	TGA, N$_2$ atmosphere, heating rate = 20°C min^{-1}	463	(3)
		TGA, heating rate = 10°C min^{-1}	~453	(17)
		At polymer melting point	468	(5)

Poly(*n*-hexyl isocyanate)

Intrinsic viscosity

Molecular Weight (g mol^{-1})	Solvent	Temp. (°C)	$[\eta]$ (dl g^{-1})	Reference
$M_v = (1.01 - 2.78) \times 10^5$	Toluene	25	4.4–12.9	(29)
$M_v = (1.01 - 2.78) \times 10^5$	1-Chloronaphthalene	25	3.4–9.1	(29)
$M_v = (1.01 - 2.78) \times 10^5$	1-Chloronaphthalene	45	3.1–7.6	(29)
$M_v = (1.01 - 2.01) \times 10^5$	1-Chloronaphthalene	90.2	2.0–3.4	(29)
$M_v = (1.01 - 2.78) \times 10^5$	1-Chloronaphthalene	110.8	1.6–3.8	(29)
$M_w = 3.55 \times 10^4$	Carbon tetrachloride	–	12.1	(25)
$M_w = 3.55 \times 10^4$	Benzene	–	11.5	(25)
$M_w = 3.55 \times 10^4$	Tetrahydrofuran	–	10.0	(25)
$M_w = 3.55 \times 10^4$	Tetrahydrofuran-dimethylformamide (4:1)	–	7.8	(25)
$M_w = 3.55 \times 10^4$	Chloroform	–	7.3	(25)
$M_w = 6.5 \times 10^4$	Toluene	25	2.6	(17)
$M_w = 6.5 \times 10^4$	Chloroform	–	2.2	(17)
$M_w = 6.5 \times 10^4$	Methanol/toluene (19.5 v/v%)	–	2.38	(17)

Pyrolyzability

	Conditions	Observation	Reference
Nature of product	Direct pyrolysis mass spectrometry	Cyclic trimer of *n*-hexyl isocyanate as the principle decomposition product, and small amounts of hexyl isocyanate	(3)
	Thermal degradation tandem mass spectrometry	Principal pyrolysis product is the neutral trimer, minor amounts of monomer	(55)
Amount of product	At 138°C in xylene solution (conc. 0.1%)	Cyclic trimer of *n*-hexyl isocyanate	(5)
		Relative viscosity reduced quickly (within 30 min)	(5)
Impurities remaining	At room temperature in dimethylformamide	Depolymerization occurs in the presence of anionic initiator such as sodium cyanide	(5)
Electrical Functionality			
Displacement-field hysteris loop	Displacement-field hysteresis curve at 80°C	Coercive electric field: Ec = 6.8 MV/m Remanent polarization: Pr = 3.0 mC/m^2	(56)
Pyroelectric cefficient β	Heating rate of 3 °C/min. Poling treatment: external DC field 20 MV/m at 80 °C for 10 min.	9.5 μC/m^2 K up to 60 °C	(56)
Block Copolymer			
Diblock copolymer	Block copolymer of n-hexyl isocyanate with styrene or isoprene using the catalyst of benzyl sodium with the addition of sodium tetraphenyl borate (NaBPh$_4$) in THF at -98 °C.		(57)
	Block copolymer of 2-vinyl pyridine with n-hexyl isocyanate using the catalyst of potassium diphenyl methane with the addition of NaBPh4 in THF at -98 °C.		(58)

Conditions	Observation	Reference
	Block copolymer of n-hexyl isocyanate (HIC) with 2-vinyl pyrrolidone (NVP). Sodium 4-oxy-2,2,6,6-tetramethylpiperidinyloxy (TEMPO-ONa) as an intiator in THF at −98 °C for the polymerization HIC and AIBN at 90 °C for the polymerization of NVP.	(59)
Triblock copolymer	Triblock copolymer of Poly(n-hexyl isocyante)-b-poly(2-vinylpyridine)-b-poly(n-hexhylisocynate) obtained, first homopolymeization of 2-vinylpyridine at −78 °C in THF using sodium naphtalenide, and next polymerization of n-hexylisocynate at −98 °C in the presence of sodium tetraphenyl borate.	(60)

References

1. Bur AJ, Fetters LJ. *Chem. Rev.* 1976;76(6):727.
2. Ulrich H. In: Mark HF, et al. ed. *Encyclopedia of Polymer Science and Engineering.* Vol 8. New York: John Wiley and Sons; 1987:448–462.
3. Durairaj B, Dimock AW, Samulski ET, Shaw MT. *J. Polym. Sci., Part A, Polym. Chem.* 1989;27:3211.
4. Berger MN, Tidswell BM. *J. Polym. Sci.* 1973;42:1063.
5. Shashoua VE, Sweeny W, Tietz RF. *J. Am. Chem. Soc.* 1959;82:866.
6. (a) Okamoto Y, et al. *Macromolecules,* 1992;25:5536.
   (b) Lee J-S, Ryu SW. *Macromolecules,* 1999;32:2085.
   (c) Shin Y-D, Kim SY, Ahn J-H, Lee J-S. *Macromolecules.* 2001;34:2408.
   (d) Ahn J-H, Shin Y-D, Nath GY, et al. *J. Am. Chem. Soc.* 2005;127:4132.
7. Patten TE, Novak BM. *J. Am. Chem. Soc.* 1991;113:5065.
8. (a) Patten TE, Novak BM. *Macromolecules.* 1993;26:436.
   (b) Xu X, Ni X, Shen Z. *Polymer Bulletine.* 2005;53:81.
   (c) Xu X, Ni X, Hong H, Shen Z. *Chem. Res. Chinese U.* 2005;21:224.
9. Ivin KJ. *Angew. Chem., Int. Ed. Engl.* 1973;12:487.
10. Chen JT, Thomas EL. *J. Mater. Sci.* 1996;31:2531.
11. Conio G, Bianchi E, Ciferri A, Krigbaum WR. *Macromolecules.* 1984;17:856.
12. Fukuwatari N, Sugimoto H, Inoue S. *Macromol. Rapid Commun.* 1996;17:1.
13. Kuwata M, Murakami H, Norisue T, Fujita H. *Macromolecules.* 1984;17:2731.
14. Aharoni SM, Walsh EK. *Macromolecules.* 1979;12(2):271.
15. Patten TE, Novak BM. *J. Am. Chem. Soc.* 1996;118:1906.
16. Jinbo Y, Sato T, Teramoto A. *Macromolecules.* 1994;27:6080.
17. Aharoni SM. *Macromolecules.* 1979;12(1):94.
18. Munoz B, Zero K, Green MM. *Polym. Prepr.* 1992;33(2):294.
19. (a) DuPre DB, Wang H. *Macromolecules.* 1992;25:7155.
    (b) Nakatsuji N, Takeno Y, Yamakawa H. *Polymer J.* 2005;37:535.
20. Cook R, et al. *Macromolecules.* 1990;23:3454.
21. Aharoni SM. *Polymer.* 1980;21:21.
22. Kim YC, Lee DC. *Pollimo.* 1979;3:115.
23. Murakami H, Norisue T, Fujita H. *Macromolecules.* 1980;13:345.
24. Aharoni SM. *Polym. Prepr.* 1980;21(1):211.
25. Schneider NS, Furusaki S, Lenz RW. *J. Polym. Sci., Part A.* 1965;3:933.
26. Rubingh DN, Yu H. *Macromolecules.* 1976;9:681.

27. Aharoni SM. *Macromolecules*. 1983;16:1722.

28. Wang H, DuPre DB. *J. Chem. Phys*. 1992;96(2):1523.

29. Bianchi E, Ciferri A, Conio G, Krigbaum WR. *Polymer*. 1987;28:813.

30. Cantor AS, Pecora R. *Macromolecules*. 1994;27:6817.

31. Itou T, Chikiri H, Teramoto A, Aharoni S. *Polym. J*. 1988;20:143.

32. Nemoto N, Schrag JL, Ferry JD. *Polym. J*. 1975;7:195.

33. Fetters LJ, Yu H. *Macromolecules*. 1971;4:385.

34. Clough SB. In: Burke JJ, Weiss V, eds. *Characterization of Materials in Research, Ceramics and Polymers*. New York: Syracuse University Press; 1975:417–436.

35. Aharoni SM. *Macromolecules*. 1981;14:222.

36. Aharoni SM, Walsh EK. *J. Polym. Sci., Polym. Lett. Ed*. 1979;17:321.

37. Itou T, Teramoto A. *Macromolecules*. 1988;21:2225.

38. Mano JF, Correia NT, Ramos JJM. *J. Chem. Soc. Faraday Trans*. 1995;91(13):2003.

39. Zhao W, et al. *Macromolecules*. 1996;29:2796.

40. Aharoni SM. *J. Polym. Sci., Polym. Phys. Ed*. 1980;18:1303.

41. Aharoni SM. *Polym. Prepr*. 1980;21(1):209.

42. Pierre J, Marchal E. *J. Polym. Sci., Polym. Lett. Ed*. 1975;13:11.

43. Aharoni SM. *Polymer*. 1981;22:418.

44. Owadh AA, Parsons IW, Hay JN, Harward RN. *Polymer*. 1978;19:386.

45. Postema AR, Liou K, Wudl F, Smith P. *Macromolecules*. 1990;23:1842.

46. Zhao W, et al. *Macromolecules*. 1996;29:2805.

47. Zhao W. *Ph. D. Thesis*. University of Cincinnati; 1995.

48. Plummer H, Jennings BR. *Eur. Polym. J*. 1970;6:171.

49. Moscicki JK, Williams G, Aharoni SM. *Macromolecules*. 1982;15:642.

50. Green MM, Andreola C, Munoz B, Reidy MP. *J. Am. Chem. Soc*. 1988;110:4063.

51. Lifson S, Andreola C, Peterson NC, Green MM. *J. Am. Chem. Soc*. 1989;111:8850.

52. Andreola C, Green MM, Peterson NC, Lifson S. *Polym. Prepr*. 1991;32(3):643.

53. Okamoto N, et al. *Macromolecules*. 1996;29:2878.

54. Green MM, Khatri CA, Reidy MP, Levon K. *Macromolecules*. 1993;26:4723.

55. Majumdar TK, et al. *J. Am. Soc. Mass Spectrom*. 1991;2:130.

56. Sugita A, Yamashita Y, Tasaka S. *J. Polym. Sci., Part B: Polym. Phys*. 2005;43:3093.

57. Zorba G, Vazaios A, Pitsikalis M, Hadjichristidis N. *J. Polym. Sci., Part A: Polym. Chem*. 2005;43:3533.

58. Shin Y-D, Han S-H, Samal S, Lee J-S. *J. Polym. Sci., Part A: Polym. Chem*. 2005;43:607.

59. Bilalis P, Zorba G, Pitsikalis M, Hasjichristidis N. *J. Polym. Sci., Part A: Polym. Chem*. 2006;44:5719.

60. Rahman MS, Samal S, Lee J-S. Macromolecules, 2006;39:5009.

Poly(hydridosilsesquioxane)

Gui Lin and Ronald H. Baney

Acronyms, Alternative Names, Trade Name HSQ, LPHSQ, H-T, hydridosilsesquioxane, hydrogen silsesquixoane, polyhydrosilsesquioxane, FOx® (Dow Corning Corp.), hydridospherosiloxanes, Octahydro-POSS®

Class Polysiloxanes (siloxane ladder polymers)

Structure $[HSiO_{1.5}]_n$ ($n > 3$, such as 8, 10, 12, 14, 16,18). The structure of poly(hydridosilsesquioxanes) depends upon the preparation of methods, The structure is a function of the concentration of the initial monomer, the nature of the solvent, the nature of the hydrolysable substituents, the concentration of water, the temperature, the type of catalyst, and the nature of the non-hydrolysing substituent. The structure may include ladder structures, cage structures (oligomer), and partial cage structures (bridged)[1,2] and tube-like structure[3].

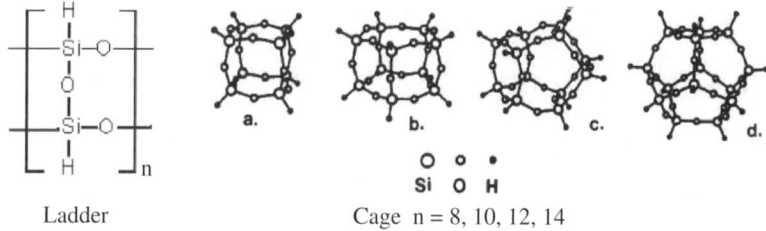

Ladder   Cage n = 8, 10, 12, 14

Major Applications Interlayer dielectrics, high-temperature resins, crosslinking aid, surface modification, glassification.

Properties of Special Interest Very high thermal stability (>500°C) and good dielectric properties.

Related Polymers Polyalkylsilsesquixoane and poly-*co*-silsesquioxanes. There are many references to these classes of materials[4], such as poly(methylsilsesquioxane-co-hydrosilsesquioxane)[5–8] and poly(phenylsilsesquioxane-co-hydridosilsesquioxane)[7]. But they are generally poorly characterized. They are not included in this handbook.

Poly(hydridosilsesquioxane)

Preparation

Acronym	Process	Reference
PHSQ $[HSiO_{1.5}]_n$ ($n = 10, 12, 14, 16$)	Hydrolysis and condensation of $HSiCl_3$ with sulfuric acid in aromatic solvents such as Benzene	(9–15)
PHSQ $[HSiO_{1.5}]_8$	Hydrolysis of $HSi(OR)_3$ in cyclohexane-acetic acid in the present of concentrated hydrochloric acid.	(9, 10)
LPHSQ	Pre-aminolysis of $HSiCl_3$ with 1,4-phenylene diamine (PDA) (2:1 mole ratio) and then hydrolysis in acetone-toluene solvent mixtures	(16)
$(HSiO_{3/2})_n$ $n = 8, 10$	Hydrolysis of $HSiCl_3$ with $FeCl_3$ in aromatic solvents and HCl	(17)
PHSQ	Slowing adding $HSiCl_3$ to MIBK containing 2 wt% H_2O, followed by bubbling N_2 carrying H_2O vapor	(18)
$(HSiO_{3/2})_8$	Adding a pentane solution of $HSiCl_3$ to a mixture of HCl, Sodium dodecyl sulfur, anhydrous $FeCl_3$ and MeOH.	(19)
LPPMHSQ	An oligomer of LPHSQ was reacted with an oligomer of $MeSiO_{3/2}$ prepared in the same way as LPHSQ	(16)

Characterization of hydridosilsesquioxanes

Material	Solubility	Solvent Insolubility	Resin Solubility	GPC M_w	Density (g cm^{-3})	Dielectric Constant (1 MHz)	Reference
PHSQ, 400°C cure	–			–	1.7	3.0	(20)
LPHSQ	Toluene			10^5–10^6	–	–	(16)
LPPMHSQ	Toluene			10^6	–	–	(16)
OctaSilane POSS®	THF, Chloroform, hexane	Water, methanol	Organic monomers				(21)
Octahydro POSS®	Slightly soluble in hexane and cyclohexane, chloroform, methylene chloride	Water, acetonitrile					(21, 22)

IR characteristic frequencies and NMR

Material	IR		^{1}H NMR		^{29}Si NMR	Reference
	Group	Peak (cm^{-1})	Bond	Peak		
LPHSQ	(Si-H)	2,257				(16)
	(Si-O-Si)	1,076, 1,132				
	(Si-OH)	853				

Material	IR		^{1}H NMR		^{29}Si NMR		Reference
	Group	Peak (cm^{-1})	Bond	Peak			
LPPMHSQ	(Si-H)	2,259					(16)
	(Si-O-Si)	1,113					
	(Si-OH)	835					
	(Si-Me)	769, 1,274					
$(HSiO_{3/2})_8$	$v(H\text{-}SiO_3)$	2,273	Si-H	$\delta = 4.203$ ppm	$^1J_{Si\text{-}H}$	$\sim$170 Hz	(17, 21, 23)
	$v_a(Si\text{-}O\text{-}Si)$	1,178					
	$\delta(H\text{-}SiO_3)$	859					
$(HSiO_{3/2})_{10}$			Si-H	$\delta = 4.424$ ppm			(17)

Solubility

Material	Solvent	Reference
$(HSiO_{3/2})_8$	C_6D_6, Hexane (partial soluble)	(17)
$(HSiO_{3/2})_{10}$	Hexane	(17)

Applications

Application	Reference
Glassification agent	(21, 23)
Cross-linker	(21, 24)
Low dielectric layers	(1, 25–30)
Coatings	(31)
Silica-supported catalyst	(32)

References

1. Jones RG, Ando W, Chojnowski J, Knovel F. *Silicon-containing polymers the science and technology of their synthesis and applications.* Dordrecht, Boston: Kluwer Academic Publishers; 2000.
2. Li GZ, Wang LC, Hi HL, Pittman CU Jr. *J. Inorgan Organ Polym.* 2001;11:123–154.
3. Schubert U, Jutzi P. *Silicon Chemistry: From the Atom to Extended Systems.* Wiley-VCH; 2003.
4. Baney RH, Itoh M, Sakakibara A, Suzuk T. *Chem. Rev.* 1995;95:1409.
5. Duan QH, Zhang TY, Deng KL, Xie P, Zhang RB. *Chinese J. Polym. Sci.* 2005;4:355–361.
6. Duan QH, Zhang Y, Jiang JQ, et al. *Polym. Intern.* 2004;53:113
7. Xie ZS, Jin SZ, Wan YZ, Zhang RB. *J. Polym. Sci.* 1992;10:361.
8. Cui L, Jin W, Liu JN, Tang YX, Xie P, Zhang RB. *Liqui. Cryst.* 1998;25:757–764.
9. Muller R, Kohne F, Sliwinski S. *J. Pract. Chem.* 1959;9:71–74.
10. Frye CL, Collins WT. *J. Am. Chem. Soc.* 1970;92:5586.
11. Cao M, Li Z, Zhang Y, et al. *React. Funct. Polym.* 2000;45:119–130.

12. Pauthe M, Phalippou J, Belot V, Corriu R, Leclercq D, Vioux A. *J. Non-Cryst. Solids.* 1990;125:187.

13. Belot V, Corriu RJP, Leclercq D, Mutin PH, Vioux A. *Chem. Mater.* 1991;3:127.

14. Campostrini R, D'Andrea G, Carturan G, Ceccato R, Soraru GD. *J. Mater. Chem.* 1996;6:585.

15. Agaskar PA, Day VW, Llemperer WG. *J. Am. Chem. Soc.* 1987;109:5554.

16. Xie Z, Jin S, Wang Y, Zhang R. *Chinese J Polym. Sci.* 1992;10(4):362.

17. Agaskar PA. *Inorg. Chem.* 1991;30:2707–2708.

18. Nishii K, Yoneda Y, Miyagawa M. *Jpn Patent* Kokai-S-60-86017. 1985.; *Chem. Abstr.* 1986;104:7840.

19. Nayman MD, Sesu SB, Peng CH. *Chem. Mater.* 1993;5:1636–1640.

20. Trade literature on FOx®. Dow Corning, Midland.

21. www.hybridplastics.com.

22. Greely JN, Meeuwenberg LM, Banaszak HMM. *J. Am. Chem. Soc.* 1998;120:7776–7782.

23. Said MA, Roesky HW, Rennekamp C, Andruh M, Schmidt HG, Noltemeyer M. *Angew. Chem. Int. Ed.* 1999;38:661–664.

24. Perry RJ, Karageorgis M, Hensler J. *Macromolecules.* 2007;40:3929–3928.

25. Schneider KS, Owens TM, Nicholson KT, et al. *Langmuir.* 2002;18:6233–6241.

26. Hu JC, Chen LJ. *US Patent.* 6,417,118 B1. 2002:Jul. 9.

27. Park CE, Kang JH. *US Patent.* 6,849, 926 B2. Feb. 1, 2005.

28. Zhang Z, Nguyen TD, Nguyen T. *US Patent.* 6,919,101 B2. July 19, 2005.

29. Kloster GM, Obrien KP, Brask JK, Goodner MD, Bruner D. *US Patent.* 7,034,399 B2. April 25, 2006.

30. Nguyen CV, Carter KR, Hawker CJ, et al. *Chem. Mater.* 1999;11:3080–3085.

31. Boisvert RP, Bujalski DR, Harkness BR, Li Z, Su K, Zhong B. *US Patent.* 6,737,117 B2 2004 May 18.

32. Ketelson HA, Brook MA, Pelton R, Heng YM. *Chem. Mater.* 1996;8:2195–2198.

Poly(4-hydroxy benzoic acid)

Sizhu Wu and Tarek M. Madkour

Acronym, Trade Name PHBA, Ekonol® (Norton)

Class Polyesters

Synthesis Polycondensation

Structure

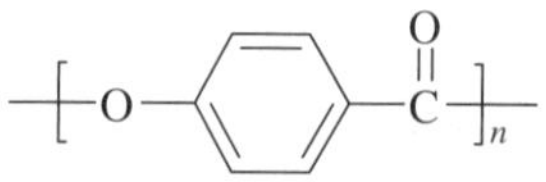

Major Applications A component in a family of random copolymers that show thermotropic liquid crystalline behavior and are marketed as structural materials.

Properties of Special Interest Intractable polymer with no melting behavior below temperatures of significant degradation. It polymerizes directly into the crystalline state, thus showing high tensile stiffness, low dielectric constant, and dimensional stability at high temperatures.

Property	Units	Conditions	Value		Reference
Molecular weight (of repeat unit)	g mol^{-1}		120.11		(1)
Typical molecular weight range	g mol^{-1}		1000–20,000		(2)
Infrared bands (frequency)	cm^{-1}		3440; 1735; 1600; 1540; 1420; 1256; 1155; 1048; 880		(3, 4)
^{13}C NMR bands		Solid state from CP/MAS data			(5)
		Phenoxy	155.0		
		Carboxyl	162.9		
^{13}C NMR shifts of substituent carbon atoms	ppm	substituent	Theoretical shift	Observed shift	(6)
		Carboxy carbon-non reacted	168.0	168.4	
		Carboxy carbon-reacted	166.8	165.9	
		Hydroxyl carbon-non reacted	159.4	161.6	
		Hydroxyl carbon-reacted	159.9	162.0	

Poly(4-hydroxy benzoic acid)

Property	Units	Conditions	Value	Reference
Thermal expansion coefficient	K^{-1}		5.04×10^5	(7)
Characteristic ratio $\langle r^2 \rangle_0 / nl^2$		Calculated at 300 K, for a chain with 30 monomeric units	20	(8)
Persistence length	Å	Calculated at 300 K, for a chain with 30 monomeric units	65	(8)
Radius of gyration	Å	Calculated at 300 K, for a chain with 30 monomeric units	42	(8)

Unit cell dimensions[1,9]

Isomer	Lattice	Space Group	Cell Dimension (Å)			Chain Confirmation	Density (g cm^{-3})
			a	b	c		
Phase I	Ortho	Pbc2$_1$	7.42	5.70	12.45	6*2/1	1.51
Phase II	Ortho	Pbc2$_1$	3.83	11.16	12.56	6*2/1	1.48
Phase III	Ortho		9.2	5.3	12.4	6*2/1	

Property	Units	Conditions	Value	Reference
Unit cell contents (number of repeat units)			4	(10)
Unit cell volume	Å^3		532	(10)
Glass transition temperature	K		434	(11)
Crystal-plastic crystal transition temperature	K		623	(9, 12)
Enthalpy of fusion	kJ mol^{-1}		4.204	(12)
Plastic crystal—nematic transition temperature	K		718	(9)
Heat capacity C_p	kJ K^{-1} mol^{-1}	Under constant pressure		(11)
		170 K	0.071	
		300 K	0.123	
		400 K	0.164	
		434 K	0.176	

Poly(4-hydroxy benzoic acid)

Property	Units	Conditions	Value	Reference
Specific heat increment	KJ K^{-1} mol^{-1}	At T_g	0.034	(11)
Elastic modulus	MPa		6896	(7)
Flexural strength	MPa		75.8	(7)
Dielectric strength	kV mm^{-1}		26	(7)
Dielectric loss temp.	K		211, 286	(13)
Volume resistivity	Ohm cm	Unfilled samples	1×10^{15}	(7)
Dissipation factor			2×10^{-4}	(7)
Thermal conductivity k	W m^{-1} K^{-1}		0.013	(7)
Intrinsic viscosity [η]		Polymerized for 3 h in water bath:		(3)

Solvent	Mesogenic character			
Xylen	Smectic	0.0321		
Toluene	Smectic-nematic	0.0330		
Benzene	Smectic	0.0355		
Nitrobenzene	Nonmesogenic	0.0384		

Property	Units	Conditions	Value	Reference
Maximum use temperature	K		723	(7)
TGA weight loss	%	Heating rate 10°C per minute		(3)
		250°C	2.97	
		300°C	4.8	
		350°C	5	
		400°C	11.45	
		500°C	22.9	
		600°C	98.43	

References

1. Brandrup J, Immergut EH, eds. *Polymer Handbook*. 3rd ed. New York: John Wiley and Sons, 1989.
2. Geiss R, et al. *J. Polym. Sci., Polym. Lett. Ed.* 1984;22:433.
3. Vora R, et al. *Mol. Cryst. Liq. Cryst.* 1984;108:187.
4. Wiberg G. et al. *Polym. Eng. & Sci.* 1998;38:1640.
5. Johnson R, et al. *Polym. Commun.* 1990;31:383.
6. Stephen GB, et al. *Progress in Organic Coatings.* 2000;39:137–143.
7. Economy J, Nowak B, Cottis S. *Polym. Prepr.* 1970;11(1):332.
8. Jung B, Schuèermann B. *Macromolecules.* 1989;22:477.
9. Iannelli P, Yoon D. *J. Polym. Sci., Polym. Phys. Ed.* 1995;33:977.
10. Yoon D, et al. *Macromolecules.* 1990;23:1793.
11. Cao M, Wunderlich B. *J. Polym. Sci., Polym. Phys. Ed.* 1985;23:521.
12. Hanna S, Windle A. *Polym. Commun.* 1988;29:236.
13. Han MS, et al. *Polym. Bull.* 2000;45:151.

Poly(hydroxybutyrate)

Isao Noda, Robert H. Marchessault and Mikio Terada

Acronyms, Alternative Names, Trade Names PHB, poly(3-hydroxybutyrate), P(3HB)*, poly(oxy-1-oxo-3-methyl-trimethylene), PHA for copolymers of hydroxyalkanoates

Class Chiral aliphatic polyesters

Structure [—O—CH(CH$_3$)—CH$_2$—CO—]

Major Applications In bacteria, PHB is a carbon reserve. The purified product is used as biodegradable packaging (bottles, containers, sheets, films, laminates, fibers, and coatings), especially as a copolymer of β-hydroxybutyrate and β-hydroxyvalerate. In biomedical applications, it is an excipient, a prosthetic material, etc. In organic syntheses, it provides chiral synthons.

Properties of Special Interest Biocompatibility and biodegradability. Biologically produced PHB is a semicrystalline isotactic stereoregular polymer of 100% R configuration that allows a high level of degradability. PHB is obtained by fermentation of bacteria capable of biosynthesizing polyesters as energy storage media. It can be completely biodegraded by numerous microorganisms. Synthetic racemic stereoblock structures degrade more slowly than the bacterial products. Copolymers, such as poly(3-hydroxybutyrate-co-3-hydroxyvalerate), are also available.

Preparative Techniques The 100% R configuration isotactic polymers are prepared by bacterial fermentation. Production in transgenic plants promises an agrotechnological production method similar to that for starch. Optically active synthetic polymer can also be prepared either by starting with optically active β-butyrolactone or by using a stereoselective catalyst with racemic β-butyrolactone. In vitro enzymatic synthesis using cloned synthase and (R)-β-hydroxybutyryl-CoA monomer.

Property	Units	Conditions	Value	Reference
Molecular weight (of repeat unit)	g mol^{-1}	–	86.09	–

Tacticity (stereoregularity)

Catalyst	Monomer	Isotactic Dyads (%)	Syndiotactic Dyads (%)	Reference
Alcaligenes eutrophus	(R)-β-hydroxybutyryl-CoA	100	0	(1)
ZnEt$_2$/H$_2$O	(S)-β-butyrolactone	100	0	(1)

* P(3HB-co-3HV) is a copolymer with valerate.

Tacticity (stereoregularity)

Catalyst	Monomer	Isotactic Dyads (%)	Syndiotactic Dyads (%)	Reference
1-Ethoxy-3-chlorotetrabutyldistannoxane	(R)-β-butyrolactone	94	6	(2)
Methylaluminoxane	(R, S)-β-butyrolactone	32 ± 5	68 ± 5	(3)
1-Ethoxy-3-chlorotetrabutyldistannoxane	(R, S)-β-butyrolactone	30	70	(2)

Polydispersity index (M_w/M_n)

Bacterial products in vivo depending on bacterial strain and carbon source[4]

Strain	Carbon Source	Method	M_w	M_w/M_n
*Alcaligenes eutrophus**	Fructose	GPC	7.37×10^5	1.9
*Alcaligenes eutrophus**	Butyric acid	GPC	4.32×10^5	2.1
Bacillus megaterium	Glucose	GPC	1.66×10^5	2.9
Zoogloea ramigera	Glucose	GPC	5.42×10^5	2.5

* Also known as *Ralstonia eutropha, Wautersia eutropha, Cupriavidus necator.*

Fractionated bacterial products in vivo*[5]

Sample Code	M_w	Method	M_n	Method	M_w/M_n
A-12	3.99×10^6	Light scattering	–	–	–
A-22	1.64×10^6	Light scattering	–	–	–
AB-12	1.35×10^6	Light scattering	–	–	–
AB-22	8.57×10^5	Light scattering	–	–	–
B-23	6.30×10^5	Light scattering	2.80×10^5	Osmotic pressure	2.25
B-32	5.33×10^5	Light scattering	2.77×10^5	Osmotic pressure	1.92
B-42	3.74×10^5	Light scattering	2.25×10^5	Osmotic pressure	1.66
B-5	2.36×10^5	Light scattering	1.91×10^5	Osmotic pressure	1.24
B-62	2.29×10^5	Light scattering	1.60×10^5	Osmotic pressure	1.43
B-72	1.15×10^5	Light scattering	8.55×10^4	Osmotic pressure	1.35

* Strain: *Azotobacter vinerand* ATCC 12837. Carbon source: sucrose.

Bacterial products in vitro

Enzyme	Monomer	Method	M_w	Reference
PHB synthase	(R)-β-hydroxybutyryl-CoA	GPC	1.3×10^7	(6)
Porcine pancreatic lipase	(R, S)-β-butyrolactone	MALDI-TOF MS*	$256 \sim 1,045$	(7)
Pseudomons cepacia lipase	(R, S)-β-butyrolactone	MALDI-TOF MS*	$643 \sim 681$	(7)

* MALDI-TOF MS: matrix-assisted laser desorption and ionization time of flight mass spectroscopy.

Poly(hydroxybutyrate)

Synthetic products

Catalyst	Monomer	M_w*	M_w/M_n	Isotactic Dyads (%)	Reference
ZnEt2/H20	(S)-β-butyrolactone	20,000	1.5	0	(1)
Methylaluminoxane	(R, S)-β-butyrolactone	130,000	8.1	79 ± 5	(3)
1-Ethoxy-3-chlorotetrabutyldistannoxane	(R)-β-butyrolactone	424,000	2.4	94	(2)
1-Ethoxy-3-chlorotetrabutyldistannoxane	(R, S)-β-butyrolactone	261,000	1.8	30	(2)

* Determined by GPC.

Property	Units	Conditions	Value	Reference
IR (characteristic absorption frequencies)	cm^{-1}	C=O stretching	1,754	(8)

IR (characteristic absorption frequencies)

Frequency (cm^{-1})	Relative Intensity	Polarization	Interpretation
2,990	Shoulder	$\perp$	CH$_3$
2,960	Medium	$\perp$	CH$_2$
2,930	Medium	$\perp$	CH$_3$
2,860	Shoulder	–	CH$_2$
1,730	Strong	$\parallel$	C=O
975	Medium	$\perp$	–

Assignment	Frequency (cm^{-1})	Intensity	Polarization
CH$_3$ rock and C—C skeletal	973	Medium	$\perp$
C—O stretch and others	1,053	Strong	–
C—O stretch and C—C skeletal	1,130	Medium-strong	–
C—O stretch and C—C skeletal	1,181	Strong	–
C—C—O stretch	1,276	Strong	–
CH$_3$ symmetric deformation (umbrella)	1,378	Medium	–
CH$_2$ deformation, CH3 asymmetric deformation	1,454	Medium	–
C=O stretch	1,726	Very strong	$\parallel$
CH$_2$ symmetric stretch	2,854	Medium	–
CH$_2$ asymmetric stretch	2,927	Medium	$\perp$
CH$_3$ asymmetric stretch	2,974	Weak	$\perp$

Property	Units	Conditions	Value	Reference
^{1}H chemical shift	ppm	Solution state: 2.0 (w/v)% polymer solution in CDCl$_3$, 500 MHz (27°C). Bacterial product (strain: B. *megaterium*)		(9)
		d, CH$_3$	1.27	
		m, CH	5.26	
		m, CH$_2$	2.45–2.65	

Property	Units	Conditions	Value	Reference
Spin-spin coupling constant $\begin{array}{ccc} H_A & CH_{X_3} & \\ | & | & \\ -C-C-\!\!-\!C-O- \\ \| & | & | \\ O & H_B & H_M \end{array}$	Hz	Solution state: 2.0 (w/v)% polymer solution in CDCl$_3$, 500 MHz (27°C). Bacterial product (strain: *B. megaterium*)		(9)
		J_{AB}	−15.5	
		J_{AM}	5.7	
		J_{BM}	7.3	
		J_{MX}	6.4	
^{13}C chemical shift	ppm	Bacterial product (strain: *A. eutrophus*, 100% isotactic). Solution state: 2.0 (w/v)% polymer solution in CDCl$_3$, 125.7 MHz (27°C).		(10)
		CH$_3$	19.8	
		CH$_2$	40.8	
		CH	67.6	
		C=O	169.2	
		Solid state: CP/DD/MAS 67.8 Hz		(10)
		CH$_3$	21.3	
		CH$_2$	42.8	
		CH	68.4	
		C=O	169.8	
		Solution state: 3–5 (w/v)% polymer solution in CDCl$_3$, 125 MHz (room temperature)		(11)
		Isotactic (88% dyad), C=O	169.10	
		Syndiotactic (66% dyad), C=O	169.09, 169.11, 169.20, 169.22	
		Solid state: CP/MAS, 75.3 MHz		(12)
		Isotactic		
		CH$_3$	21.3	
		CH$_2$	42.9	
		CH	68.5	
		C=O	170.2	
		Syndiotactic		
		CH$_3$	20.4	
		CH$_2$	40.7	
		CH	68.7	
		C=O	170.7	
Common solvents	–	Soluble in:		
		Chloroform, trichloroethylene, 2,2,2-trifluoroethanol, dimethylformamide, ethylacetoacetate, triolein, comphor, glacial acetic acid, 0.5 N aqueous phenol, N-NaOH, *N*-hyamine hydroxide (NH$_4$OH)		(13)
		Methylene chloride		(14)
		1,1,2,2-Tetrachloroethane		(15)
		Triacetin (glycerol triacetate) (110°C)		(16)
		Dichloroacetic acid		(17)
		Propylene carbonate, 1,2-dichloroethane		(4)

Poly(hydroxybutyrate)

Property	Units	Conditions	Value	Reference
		Partially soluble in:		
		Dioxane, toluene, octanol, pyridine		(13)
		Benzene, xylene, aniline, oleic acid dibutyphthalate		(18)
Nonsolvents	–	n-Hexane, carbon tetrachloride, acetone*, ethyl acetate, ether, methanol, ethanol, water, dilute mineral acid, alkaline hypochlorite		(13)
		Isopropanol, n-butanol, methylal, glycerides		(18)
Solubility parameter	$(MPa)^{1/2}$	Calculated using Hoy's group contributions of molar attraction constant	19.2	(19)
Solubility parameter of solvent	$(MPa)^{1/2}$	Soluble in:		(20)
		Chloroform	19.0	
		Trichloroethylene	18.8	
		2,2,2-Trifluoroethanol[†]	22.2	
		Dimethylformamide	24.8	
		Ethylacetoacetate	17.7	
		Triolein[†]	18.5	
		Camphor[†]	15.2	
		Glacial acetic acid	20.7	
		Methylene chloride	19.8	
		1,1,2,2-Tetrachloroethane	19.8	
		Triacetin[†]	19.4	
		Dichloroacetic acid	22.5	
		Propylene carbonate	27.2	
		1,2-Dichloroethane	20.1	
		Partially soluble in:		(20)
		Dioxane	20.5	
		Toluene	18.2	
		Octanol	21.1	
		Pyridine	21.9	
		Benzene	18.8	
		Xylene	18.0	
		Aniline	21.1	
		Oleic acid[†]	17.2	
		Dibutylphthalate	19.0	
Solubility parameter of nonsolvent	$(MPa)^{1/2}$	n-Hexane	14.9	(20)
		Carbon tetrachloride	17.6	
		Acetone	20.3	
		Ethyl Acetate	18.6	
		Diethyl ether	15.1	
		Methanol	29.7	
		Ethanol	26.0	
		Isopropanol	23.5	
		n-Butanol	23.3	
		Methylal[†]	17.4	

Property	Units	Conditions	Value	Reference
Interaction parameter χ	–	Chloroform, 30°C, $M_n = 127,000$	0.361	(21)

* Acetone is a solvent for amorphous PHB.
† Solubility parameters calculated from Klevelen and Hoftyzer's group contributions of solubility parameter components in reference (19).

Second virial coefficient A_2

$M_n \times 10^{-4}$	Light Scattering in Trifluoroethanol, 25°C	Osmotic Pressure in Chloroform, 35°C	Reference
	$A_2 \times 10^4$ (mol cm^3 g^{-2})	$A_2 \times 10^4$ (mol cm^3 g^{-2})	
910 ± 20	6.20 ± 0.2	–	(22)
761 ± 30	6.28 ± 0.2	–	(22)
667 ± 20	6.38 ± 0.2	–	(22)
590 ± 10	6.52 ± 0.2	–	(22)
380 ± 10	6.88 ± 0.2	–	(22)
335 ± 5	7.12 ± 0.2	–	(22)
252 ± 5	8.06 ± 0.2	–	(22)
183 ± 2	8.24 ± 0.2	–	(22)
120 ± 1	9.56 ± 0.1	–	(22)
77.9 ± 1	10.2 ± 0.1	–	(22)
63.0	10.6	6.07	(5)
53.3	10.6	8.40	(5)
37.4	12.4	8.72	(5)
23.6	13.5	9.34	(5)
22.9	12.9	10.3	(5)
11.5	16.4	12.5	(5)

Mark-Houwink-Sakurada parameters $[\eta] = K (M_w)^a$

Solvent	Temp. (°C)	$K \times 10^3$ (ml g^{-1})	a	Reference
Chloroform	30	7.7	0.82	(17)
	30	11.8	0.78	(5)
	30	16.6	0.76	(5)
Trifluoroethanol	30	25.1	0.74	(23)
	25	12.5	0.80	(5)
	25	22.2	0.76	(24)
	30	17.5	0.78	(24)
1,2-Dichloroethane	30	9.18	0.78	(5)
	30	16.8	0.74	(24)
n-Butyl chloride	13 (θ)	100	0.5	(24)
1-Chloronaphthalene	40	39.6	0.62	(24)

Poly(hydroxybutyrate)

Property	Units	Conditions			Value	Reference
Huggins constant k'	–	Solvent	Temp. (°C)	M_n		
		Chloroform	30	138,000	0.42	(17)
		Chloroform	30	127,000	0.35	(17)
		Chloroform	30	85,000	0.49	(17)
		Chloroform	30	66,000	0.65	(17)
		Chloroform	30	36,000	0.35	(17)
		Chloroform	30	31,700	0.44	(17)
		Chloroform	30	31,360	0.50	(17)
		Chloroform	30	20,400	0.63	(17)
		Chloroform-d	–	400,000	0.478	(15)
		Tetrachloro-ethane-d	–	400,000	1.72	(15)
Radius of gyration $\langle S^2 \rangle^{1/2}$	Å	Light scattering, trifluoroethanol, 25°C				(22)
		$M_w \times 10^{-4}$				
		910 ± 20			$2,560 \pm 50$	
		761 ± 30			$2,320 \pm 50$	
		667 ± 20			$2,140 \pm 50$	
		590 ± 10			$1,980 \pm 20$	
		434 ± 20			$1,710 \pm 50$	
		380 ± 10			$1,490 \pm 20$	
		335 ± 5			$1,387 \pm 20$	
		252 ± 5			$1,191 \pm 20$	
		183 ± 2			952 ± 15	
		120 ± 1			769 ± 5	
		77.9 ± 1			586 ± 10	
		51.0 ± 0.5			431 ± 5	
Flory viscosity constant Φ	10^{-21} mol^{-1}	Calculated data from M_w, $[\eta]$, and $\langle S^2 \rangle$				(5)
		$M_w \times 10^{-4}$				
		339			1.95	
		164			1.94	
		135			1.80	
		85.7			2.24	
		63.0			2.01	
		53.3			2.18	
		37.4			2.10	
		23.6			1.99	
		22.9			1.74	
		11.5			1.61	
Intrinsic sedimentation $S°$	10^{-13}s^{-1}	M_w				(17)
		780,000			−9.95	
		370,000			−8.16	
		156,000			−4.96	
		83,500			−3.50	
		21,100			−2.00	
Degree of crystallinity	%	Purification	Drying	Annealing		
		Enzyme	Spray-dry	None	70	(25)
		Enzyme	Spray-dry	120°C, 48 h	80	(25)

Property	Units	Conditions			Value	Reference
		Purification	Drying	Annealing		
		Hypochlorite	Freeze-dry	None	30	(25)
		Solution precipitation	Vacuum	160°C, 24 h	86	(26)
Lattice	–	Isotactic, α form			Orthorhombic	(8)
Space group	–	Isotactic, α form			$P2_12_12_1$	(8)

Form	Cell Dimension (A)			Packing Energy (Kcal mol^{-1})*			Torsional Angle Summary				Reference
	a	b	c	v. der W.	C.E.	Total	\multicolumn{4}{l}{$-[CH(CH_3)-CH_2-C(O)-O]_n-$}	
α	5.76	13.20	5.96	−28.7	−13.5	−42.2	142	−57	−31	180	(27)
α	5.76	13.20	5.96	−28.5	−14.7	−43.2	152	−52	−42	−175	(28)
α	5.73	13.14	5.93	−29.1	−47.4	−76.5	149	−59	−35	−173	(29)
β	–	–	4.6	–	–	–	112	−179	−110	174	(30)

* Packing energies calculated in a Dreiding II force field (Cerius^2 from Molecular Simulation, Inc.)

Property	Units	Conditions		Value	Reference
Heat of fusion ΔH	kJ mol^{-1}	86% crystallinity, DSC (differential scanning calorimetry)		12.5	(26)
Density	g cm^{-3}	Amorphous (extrapolation data from specific volumes)		1.177	(26)
		Crystalline (calculated data from crystal lattice parameters)		1.262	(8)
		Sample	Pretreatment		
		Film	None	1.232	(21)
		Film	Hot-stretched	1.250	(21)
Glass transition temperature T_g	K	Amorphous PHB			
		Dilatometry		269–274	(26)
		Dynamic mechanical measurement		268–278	(26)
		3 mol% 3HV		281	(31)
		9 mol% 3HV		279	(31)
		14 mol% 3HV		277	(31)
		20 mol% 3HV		272	(31)
		25 mol% 3HV		267	(31)
Melting temperature T_m (optical observation)		3 mol% 3HV		443	(31)
		9 mol% 3HV		435	
		14 mol% 3HV		423	
		20 mol% 3HV		418	
		25 mol% 3HV		410	

Poly(hydroxybutyrate)

Property	Units	Conditions		Value	Reference
		M_n	Method		(32)
		85,500	Osmotic pressure	453	
		52,400	Osmotic pressure	453	
		42,000	Osmotic pressure	447	
		31,360	Osmotic pressure	444	
		20,400	Osmotic pressure	443	
		15,100	Osmotic pressure	438	
		4,970	Osmotic pressure	419	
		1,870	Osmotic pressure	387	
		688	Chromatography	374	
		602	Chromatography	362	
		516	Chromatography	354	
		430	Chromatography	338	
		344	Chromatography	320	
Equilibrium melting point of infinite crystal $T_m°$	K	Calculated data from fitting of crystalline growth rate to Hoffman's theory		470 ± 2	(26)
Tensile modulus	MPa	–		1,400–2,200	(33)
Tensile strength	MPa	–		40	(31, 34, 35)
		Cold-rolling treatment		60	(33)
		10 mol% 3HV		25	(36)
		20 mol% 3HV		20	(36)
Flexural modulus	MPa	–		4,000	(35)
		–		3,500	(36)
		10 mol% 3HV		1,200	(36)
		20 mol% 3HV		800	(36)
Young's modulus	MPa	–		3,500	(31)
		Biaxially drawn film		4,000	
		3 mol% 3HV		2,900	
		9 mol% 3HV		1,900	
		14 mol% 3HV		1,500	
		20 mol% 3HV		1,200	
		25 mol% 3HV		700	
Extension at break	%	–		6	(31, 35)
		–		6 ∼ 8	(34)
		–		8	(36)
		Biaxially drawn film		75	(31)
		10 mol% 3HV		20	(36)
		20 mol% 3HV		50	(36)
Stress at break	MPa	Biaxially drawn film		100	(31)
Breaking strain	%	Cold-rolling treatment		130	(33)
Notched Izod impact strength	Jm^{-1}	–		35	(37)
		–		50	(31)
		3 mol% 3HV		60	(31)
		9 mol% 3HV		95	(31)

Property	Units	Conditions		Value	Reference
Notched Izod impact strength	Jm^{-1}	–		35	(37)
		14 mol% 3HV		120	(31)
		20 mol% 3HV		200	(31)
		25 mol% 3HV		400	(31)
Solvent resistance	–	–		Poor	(34, 35)
UV resistance	–	–		Good	(34, 35)
Fold surface free energy σ_e	$mJ\ m^{-2}$	Calculated data from plot of T_m as function of inverse lamellar thickness		38 ± 6	(26)
Specific rotation $[\alpha]^{30}$	Degree	Solvent	λ (nm)		
		Chloroform (room temp.)	350	11	(38)
			589	−2	(38)
		Ethylene dichloride (at 25°C)	350	21	(5)
			589	0	(5)
		Dimethyl formamide (at 60°C)	350	16	(5)
			589	−3	(5)
		Trifluoroethanol	350	28	(5)
			589	0	(5)
		Dichloroacetic acid	350	80	(5)
			589	14	(5)
		2-Chloroethanol	350	33	(5)
			589	2	(5)
Oxygen permeability coefficient	m^3 (STP) $ms^{-1}m^{-2}Pa^{-1}$			2.1	(31, 34)

Pyrolysis[39]

Degradation Temperature (°C)	$M_n \times 10^{-3} (M_w/M_n)$						Rate Constant (k_d)
	0 min.	1 min.	2 min.	5 min.	10 min.	20 min.	
175	546 (2.1)	505 (2.2)	417 (2.2)	–	–	146 (1.9)	$2.2 \pm 0.5 \times 10^{-5}$
180	477 (2.2)	432 (2.0)	434 (1.8)	312 (2.0)	169 (2.0)	93 (2.1)	$3.8 \pm 0.1 \times 10^{-5}$
190	402 (2.1)	192 (2.9)	136 (3.3)	–	52 (2.4)	–	$1.4 \pm 0.1 \times 10^{-4}$
200	282 (2.0)	121 (2.2)	121 (2.2)	37 (2.4)	–	7 (3.9)	$6.2 \pm 0.5 \times 10^{-4}$

Property	Units	Conditions	Value	Reference
Activation energy of random chain session	$kJ\ mol^{-1}$	Calculated from Arrhenius plot for rate constant k_d. Temperature range: 170–200°C	212 ± 10	(39)
Biodegradability of PHB single crystals		Sample films incubated with *P. lemoignei* and *A. fumigatus* extracellular depolymerase		(40)

Poly(hydroxybutyrate)

Sample (% Isotactic Dyads)	M_w (g ml^{-1})	Heat of Fusion (J g^{-1})	Degradation Time (h)	Weight Loss (%)
100	500,000	85	50	100
79	130,000	51	890	64
55	9,500	14	730	84
34	2,700	30	890	45

Biodegradation*[(41)]

Environment	Temperature (°C)	Degradation Time (Weeks)
Sea water	15	350
Soil	25	75
Aerobic sewage	–	60
Anaerobic sewage	–	6

* All 1-mm thick samples.

General reference and review (42, 43)

Refractive index of PHB thin films[44]

Wavelength (nm)	n
1000	1.475
600	1.485
500	1.492
400	1.550

References

1. Zang Y, Gross RA, Lenz RW. *Macromolecules.* 1990;23:3206.
2. Hori Y, Takahashi Y, Yamaguchi A, Hagiwara T. *Can. J. Microbiol.* 1995;41:282.
3. Hocking PJ, Marchessault RH. *Polym. Bull.* 1994;30:163.
4. Doi Y. *Microbial Polyesters.* New York: VCH; 1990.
5. Akita S, Einaga Y, Miyaki Y, Fujita H. *Macromolecules.* 1976;9:774.
6. Gerngross TU, Martin DP. *Proc. Natl. Acad. Sci. USA.* 1995;92:6279.
7. Nobes GAR, Kazlauskas PJ, Marchessault RH. *Macromolecules.* 1996;29:4829.
8. Okamura K, Marchessault RH. In: Ramachandran GN, ed. *Conformation of Biopolymers.* Vol 2. New York: Academic Press, 1967:709.
9. Doi Y, Kunioka M, Nakamura Y, Koga K. *Macromolecules.* 1986;19:1274.
10. Doi Y, Kunioka M, Nakamura Y, Koga K. *Macromol. Chem, Rapid Commun.* 1986;7:661.
11. Hocking PJ, Marchessault RH. *Macromolecules.* 1995;28:6401.
12. Hocking PJ. *Characterization and Enzymatic Degradation of Poly [(R,S)-β-hydrobutyrate] of Varied Tacticities.* Montreal, Quebec, Canada: Doctoral thesis, McGill University; 1995.
13. Dawes EA, Senior PJ. *Adv. Microb. Physiol.* 1973;10:203.
14. Schlegel HG, Gottschalk G, von Bartha R. *Nature.* 1961;191:463.
15. Nedea ME, Morin FG, Marchessault RH. *Macromolecules.* 1989;22:4208.
16. Lauzier C, Marchessault RH. *Polymer.* 1992;33:823.
17. Marchessault RH, Okamura K, Su CJ. *Macromolecules.* 1970;3:735.
18. Kepes A, Peaud Lenoel C. *Bull Soc. Chim. Biol.* 1952;34:563.

19. van Krevelen DW, Hoftyzer PJ. *Properties of Polymers: Their Estimation and Correlation with Chemical Structure.* Amsterdam: Elsevier; 1976.

20. Gruke EA. In: Brandrup J, Immergut EH, eds. *Polymer Handbook.* 3rd ed. New York: Wiley-Interscience; 1989.

21. Okamura K. *X-Ray Structure and Morphology of Poly (β-Hydroxybutyrate).* Syracuse, New York: Master of Science thesis, State University of New York, College of Forestry; 1965.

22. Akita S, Einaga Y, Miyaki Y, Fujita H. *Macromolecules.* 1977;10:1356.

23. Cornibert J, Marchessault RH, Benoit H, Weill G. *Macromolecules.* 1970;3:741.

24. Hirose T, Einaga Y, Fujita H. *Polym. J.* 1979;11:819.

25. Lauzier C, Revol JF, Debzi EM, Marchessault RH. *Polymer.* 1994;35:4156.

26. Barham PJ, Keller A, Otun EL, Holmes PA. *J. Mater. Sci.* 1984;19:2781.

27. Cornibert J, Marchessault RH. *J. Mol. Biol.* 1972;71:735.

28. Yokouchi M, et al. *Polymer.* 1973;14:267.

29. Bruckner S, et al. *Macromolecules.* 1988;21:967.

30. Orts WJ, Marchessault RH, Bluhm TL, Hamer GK. *Macromolecules.* 1990;23:5368.

31. Holmes FA. In: Basset DC, ed. *Developments in Crystalline Polymers.* Vol 2. London: Elsevier; 1988:1–65.

32. Marchessault RH, et al. *Can. J. Chem.* 1981;59:38.

33. Barham PJ, Keller A. *J. Polym. Sci. Polm. Phys. Ed.* 1986;24:69.

34. Brandle H, Gross RA, Lenz RW, Fuller RC. *Adv. Biochem. Eng. Biotechnol.* 1990;41:77.

35. Howells ER. *Chem. Ind.* 1982;15:508.

36. Liddell JM. *Spec. Publ. R. Soc. Chem. (Chem. Ind. Friend. Environ.).* 1992;103:10.

37. Byrom D. In: Mobley DP, ed. *Plast. Microbes.* Munich: Hanser; 1994:5.

38. Alper R, Lundgren DG, Marchessault RH, Core W. *Biopolymers.* 1963;1:545.

39. Kunioka M, Doi Y. *Macromolecules.* 1990;23:1933.

40. Hocking PJ, et al. *J. Macro. Sci.* 1995;A32:889.

41. Winton JM. *Chem. Week.* 1985;28:55.

42. Doi Y, Steinbüchel A, Eds. *Biopolymers: Polyesters II.* Weinheim: Wiley-VCH; 2002.

43. Lenz RW, Marchessault RH. *Biomacromolecules.* 2005;6:1.

44. Kesckemeti G, et al. *Appl. Surf. Sci.* 2006;253: 1185.

Poly(2-hydroxyethyl methacrylate)

N.A. Peppas

Acronyms, Trade Names PHEMA

Class Vinylidene Polymers

Structure

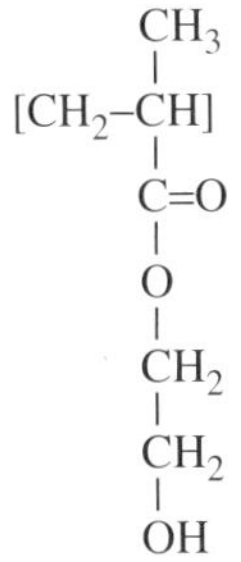

$$
\begin{array}{c}
CH_3 \\
| \\
[CH_2-CH] \\
| \\
C=O \\
| \\
O \\
| \\
CH_2 \\
| \\
CH_2 \\
| \\
OH
\end{array}
$$

Major Applications Contact lenses, intraocular lenses, drug delivery systems, biomedical applications, chromatographic columns, hemodialysis membranes, flocculating agents.

Properties of Special Interest Hydrophilicity. Good swelling in water and electrolytic solutions.

Property	Units	Conditions	Value	Reference
Glass transition temperature	°C	isotactic	35	(1)
		atactic	55	(2)
		atactic	86	(3)
		atactic	87	(4)
		atactic	90	(5)
		atactic	98	(6)
		atactic	100	(7)
		syndiotactic	109	(1)
		as a function of crosslinking ratio	115–126	(8)
		as a function of crosslinking ratio	95.9–99.9	(9)
Polymer-water interaction parameter, χ_1	–	–	$\chi_1 = 0.32 + 0.904\upsilon_2$	(10)
			0.77–0.83	(11)

Property	Units	Conditions	Value	Reference
Water equilibrium volume fraction	–	swelling	0.40	(2, 12)
			0.421	(13)
			0.395–0.431	(14, 15)
Water diffusion coefficient	$cm^2 s^{-1}$	diffusion at 7°C	2.17×10^{-6}	(16)
		diffusion at 23°C	3.46×10^{-6}	(16)
		diffusion at 34°C	4.78×10^{-6}	(16)
Water self diffusion coefficient	$cm^2 s^{-1}$	–	$0.59–5.37 \times 10^{-6}$	(17)
Ion diffusion coefficient	$cm^2 s^{-1}$	KF at 37°C	1.04×10^{-6}	(18)
		KCl at 37°C	1.34×10^{-6}	(18)
		KBr at 37°C	1.42×10^{-6}	(18)
		KI at 37°C	1.56×10^{-6}	(18)
		$KHCO_3$ at 37°C	8.1×10^{-7}	(18)
		K_2SO_4 at 37°C	4.1×10^{-7}	(18)
		KNO_3 at 37°C	1.5×10^{-6}	(18)
		K_2CO_3 at 37°C	8×10^{-7}	(18)
Linear expansion coefficient, α_g	K^{-1}	Solid	3.7×10^{-4}	(19)
Storage modulus, G' (sheer)	MPa	from −20°C to −160°C	1.03–2.01	(20)
Swelling rate, $\dot{v}$	h^{-1}	as a function of 0–1 wt% crosslinking agent	7–13.3	(21)

References

1. Sung YK, Gregonis DE, Russell GA, Andrade JD. *Polymer.* 1996;19:1362.
2. Franson NM, Peppas NA. *J. Appl. Polym. Sci.* 1983;28:1299.
3. Shen MC, Strong JD, Matusik FJ. *J. Macromol. Sci.* 1967;B1:15.
4. Caykara T, cengiz O, Ömer K, Baki E. *Polym. Degrad. Stab.* 2003:80:339.
5. Ilavsky M, Prins W. *Macromolecules.* 1970;3:415.
6. Ilavsky M, Hasa J. *Coll. Czech. Chem. Commun.* 1968;33:2142.
7. Kolarik J, Janacek J. *J. Polym. Sci., A2.* 1972;10:11.
8. Roorda WE, Bouwstra JA, de Vries MA, Junginger HE. *Pharm. Res.* 1988;5:722.
9. Seidel M, Malmonge SM. *Mater. Res.* 2000;3:3.
10. Janacek J, Hasa J. *Coll. Czech. Chem. Commun.* 1966;31:2186.
11. Shen MC, Tobolsky AV. *J. Polym. Sci., A2.* 1964;2:2513.
12. Rosenberg M, Bartl P, Lesko J. *J. Ultrastruct. Res.* 1960;4:298.
13. Allen L. *Polym. Prepr.* 1974;15:395.
14. Ratner BD, Hoffman AS. In: Andrade JD, ed. *Hydrogels for Medical and Related Applications. ACS Symposium Series.* Vol 31. 1, Washington, DC: ACS; 1976.
15. Roorda WE, Bouwstra JA, de Vries MA, Junginger HE. *Biomaterials.* 1988;2:494.
16. Wisniewski S, Kim SW. *J. Membr. Sci.* 1980;6:309.

17. Peschier LJC, Bouwstra JA, de Bleyser J, Junginger HE, Leyte JC. *Biomaterials*. 1993;14:945.
18. Hamilton CJ, Murphy SM, Atherton ND, Tighe BJ. *Polymer*. 1988;29:1879.
19. Moynihan HJ, Honey MS, Peppas HA. *Polym. Eng. Sci*. 1986;26:1180.
20. Wilson TW, Turner DT. *Macromolecules*. 1988;21:1184.
21. Wood JM, Attwood D, Collett JH. *Intern. J. Pharmac*. 1981;7:189.

Poly(4-hydroxystyrene)

Ruzhi Zhang

Acronyms, Alternative Names, Trade Names Poly(4-hydroxystyrene), Poly(p-hydroxystyrene), PHS, PHOST, Poly(4-hydroxyvinylbenzene), Poly(p-hydroxyvinylbenzene), Poly(4-vinylphenol), Poly(p-vinylphenol), PVP, Resin M, Maruka Lyncur

Class Vinyl polymers

Structure

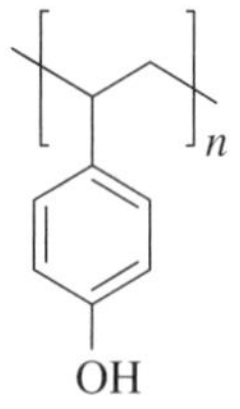

Properties of Special Interest and Major Applications Poly(4-hydroxystyrene) is an analogue of Novolac that has comparable solubility in organic solvents and in aqueous bases. It also has comparable resistance to dry etching environments. It is transparent at DUV ($\sim$250 nm) wavelength and is the polymer backbone of a number of positive resists used for KrF lithography.[1,2] Its high glass transition temperature and hydrogen bonding capability, in comparison to polystyrene, are also of special interests to industrial applications.

Preparative Methods Poly(4-hydroxystyrene) is commonly made by hydrolysis of poly(4-acetoxystyrene) which is synthesized by free radical polymerization of 4-acetoxystyrene.[3,4]

Physical constants of poly(4-hydroxystyrene) (CAS number [24979-70-2])

Property	Units	Conditions	Value	Reference
Physical Apperance	–	–	White to off-white, free flowing powder	(3)
Density	$G\ cm^{-3}$	–	1.16	(3)
		by pycnometry at 298.15 K	1.15	(5)
		at $(T_g + 15)$; $T_g = 177°C$	1.079	(6)
Glass transition temperature	°C	$M_w = 92,000$	182	(4)
		$M_w = 5,000–40,000$	165–180	(3)
Decomposition temperature	°C	–	>360	(3)

Poly(4-hydroxystyrene)

Property	Units	Conditions	Value	Reference
Change in heat capacity upon glass transition ΔC_p	$J\,g^{-1}\,{}^\circ C^{-1}$	$M_w = 8,000$	0.386	(7)
dn/dc	$cm^3\,g^{-1}$	SEC-MALLS with DMF eluent; $M_w = 153,000$	0.154	(8)
z-Average RMS radius $R_{g,z}$	Nm	SEC-MALLS with DMF eluent; $M_w = 153,000$	15.1	(8)
Persistence length ξ	Nm	SEC-MALLS with DMF eluent; $M_w = 153,000$	2.1	(8)
Molecular weight between entanglements	$G\,mol^{-1}$	at 192°C	$29,300 \pm 1,200$	(6)
Zero-shear viscosity	Pa s	at 192°C	6.33×10^6	(6)

Chemically dissimilar polymer pairs miscible in the amorphous state at room temperature [poly(4-hydroxystyrene) is Polymer I]

Polymer II of	Method	Comments	Reference
Amide	Single Tg	Polymer II was nylon 6, nylon 6,6 or nylon 11 (three miscible systems); semicrystalline	(9)
n-Butyl acrylate-co-t-butyl acrylate	Single Tg	II had 64 mol% or higher of n-butyl acrylate	(10)
Butylene terephthalate	Single Tg; FTIR	Semicrystalline	(9, 11)
Ethyl methacrylate-co-methyl methacrylate	Single Tg; FTIR	II had 30 or 60 wt.% MMA	(12, 13)
Ethylene terephthalate	Single Tg; FTIR	Semicrystalline	(11)
Methyl methacrylate	Single Tg; FTIR; NMR	–	(13–17)
Vinyl acetate	Single Tg; FTIR	–	(18, 19)
Vinyl methyl ether	Single Tg; FTIR	–	(20)
N-vinylpyrrolidone	Single Tg	Formed complexes from methanol	(21)

Infrared absorption[22]

Frequency (cm^{-1})	Assignment
3,200–3,500	O—H stretch
3,012	C—H stretch of aromatic ring
2,929	CH$_2$ asymmetric stretch

Frequency (cm^{-1})	Assignment
1,600	Aromatic ring modes
1,515	Breathing mode of aromatic ring
1,450	CH$_2$ symmetric stretch
1,240	C—C—C stretch
1,170	C—H in-plane bend of aromatic ring
833	C—H out-of-plane bend of aromatic ring

^{1}H NMR (DMSO-d_6, 300 MHz)[8]

Chemical Shift (ppm)
8.9–9.1 (br, 1H), 6.1–6.9 (br, 4H), 1.1–2.2 (br, 3H)

Solvents and non-solvents for poly(4-hydroxystyrene)[23]

Solvents	Non-solvents
Methanol and higher alcohols, methyl acetate, THF, THF/chloroform (1:1)	Chloroform, toluene

Fractionation of poly(4-hydroxystyrene)[24]

Method of Fractionation	Solvent or Solvent/Non-Solvent Mixture	Remarks
Chromatography	Tetrahydrofuran	SEC, styragel

Solubility parameter of poly(4-hydroxystyrene)[25,26]

δ_d [(MPa)$^{1/2}$]	δ_p [(MPa)$^{1/2}$]	δ_h [(MPa)$^{1/2}$]	δ [(MPa)$^{1/2}$]	Reference
			22.5	(27)
17.6	10	13.7	24.55	(28)

Unperturbed dimensions of poly(4-hydroxystyrene)[27]

Property	Unit	Value	Temp. (°C)	Method	Solvent
K$_0$ × 10^3	ml g^{-1}	82.7	25	VG	Dioxane
		90.0	25	VG	Ethyl propionate
		90.0	25	VG	Isobutyl acetate
		83.7	25	VG	Tetrahydrofuran

Poly(4-hydroxystyrene)

Property	Unit	Value	Temp. (°C)	Method	Solvent
$\sigma = r_0/r_{0f}$	–	2.38	25	VG	Dioxane
		2.45	25	VG	Ethyl propionate
		2.45	25	VG	Isobutyl acetate
		2.39	25	VG	Tetrahydrofuran
$C_\infty = r_0^2/nl^2$	–	11.3	25	VG	Dioxane
		12.0	25	VG	Ethyl propionate
		12.0	25	VG	Isobutyl acetate
		11.4	25	VG	Tetrahydrofuran

Mark-Houwink parameters (K and a) of poly(4-hydroxystyrene) (viscosity–molecular weight relationships, $[\eta] = KM^a$)

Solvent	T (°C)	$K \times 10^2$ (ml g^{-1})	a	Molecular Weight Range, M $\times 10^{-4}$	Method(s)	Reference
Dioxane	25	20.3	0.66	3.8–583	LS	(27)
Ethyl propionate	25	63.7	0.52	3.8–583	LS	(27)
Isobutyl acetate	25	59.4	0.53	3.8–583	LS	(27)
Tetrahydrofuran	25	37.2	0.60	3.8–583	LS	(27)
1-Chloro-n-hexane	5.7	91.6	0.468	5.9–90.3	LS, SEC, OS	(28)

Flory-Huggins polymer-solvent interaction parameter χ for poly(4-hydroxystyrene)[5]

Solvent	$\chi = \chi(T, \phi)$	Value
Acetone	$\chi = \chi_0 + \chi_1\phi_2 + \chi_2\phi_2^2$ $\chi_0 = \chi_{01} + \chi_{02}T^{-1}$	$\chi_{01} = 0.290$ $\chi_{02} = 25.654$ K $\chi_1 = -0.369$ $\chi_2 = -1.983$ $\phi_2 = $ volume fraction of polymer

Dissolution rate at varying base concentration for poly(4-hydroxystyrene)[29]

M_n	Rate (Å s^{-1})				
	0.109 N	0.122 N	0.134 N	0.161 N	Critical Base Concentration c_0 (N)
3,600	112	318	572	–	0.037
5,160	37	107	212	–	0.056
8,770	11	31	74	435	0.075
14,500	5	15	24	119	0.085
29,700	1	5	12	50	0.101

References

1. Levinson HJ. *Principles of Lithography.* 2nd ed. Bellingham, Washington: SPIE Press; 2004.
2. Willson CG. In: Thompson LF, Willson CG, Bowden MJ, eds. *Organic Resist Materials* in *Introduction to Microlithography.* 2nd ed. Washington, DC: American Chemical Society; 1994.
3. MSDS and Data Sheet from DuPont Electronic Polymers. http://www.dupont.com/et.
4. Nakamura K, Hatakeyama T; Hatakeyama H. *Polymer.* 1981;22:473.
5. Luengo G, Rojo G, Rubio RG, Prolongo MG, Masegosa R. *Macromolecules.* 1991;24:1315.
6. Cai H, Ait-Kadi A, Brisson J, *J. Appl. Polym. Sci.* 2004;93(4):1623.
7. Lin Q, Simons J, Angelopoulos M, Sooriyahumaran R. *Proc. SPIE – Int. Soc. Optical Eng.* 2002;4690:410.
8. Yoshida M, Fresco ZM, Ohnishi S, Frechet JMJ. *Macromolecules.* 2005;38:334.
9. Landry MR, Massa DJ, Landry CJT, Teegarden DM, Colby RH, Long TE, Henrichs PM. *J. Appl. Polym. Sci.* 1994;54:991.
10. Zhu KJ, Chen SF, Ho T, Pearce EM, Kwei TK. *Macromolecules.* 1990;23:150.
11. Landry CJT, Massa DJ, Teegarden DM, Landry MR, Henrichs PM, Colby RH, Long TE. *Macromolecules.* 1993;26:6299.
12. Pomposo JA, Cortazar M, Calahorra E. *Macromolecules.* 1994;27:245.
13. Pomposo JA, Eguiazabal I, Calahorra E, Cortazar M. *Polymer.* 1993;34:95.
14. Goh SH, Siow KS. *Polym. Bull.* 1988;17:453.
15. Landry CJT, Teegarden DM. *Macromolecules.* 1991;24:4310.
16. Serman CJ, Painter PC, Coleman MM. *Polymer.* 1991;32:1049.
17. White JL, Mirau PA. *Macromolecules.* 1994;27:1648.
18. Coleman MM, Lichkus AM, Painter PC. *Macromolecules.* 1989;22:586.
19. Coleman MM, Serman CJ, Painter PC. *Macromolecules.* 1987;20:226.
20. Pedrosa P, Pomposo JA, Calahorra E, Cortazar M. *Macromolecules.* 1994;27:102.
21. Wang LF, Pearce EM, Kwei TK. *J. Polym. Sci., Polym. Phys. Ed.* 1991;29:619.
22. Tan TL, Kudryashov VA, Tan BL. *Applied Spectroscopy.* 2003;57(7):842.
23. Xing P, Dong L, An Y, Feng Z, Avella M, Martuscelli E. *Macromolecules.* 1997;30:2726.
24. Yoshida H, Nakamura K. *Polym. J.* 1982;14:855.
25. Coleman MM, Serman CJ, Bhagwagar DE, Painter PC. *Polymer.* 1990;31:1187.
26. Arichi S, Himuro S. *Polymer.* 1989;30:686.
27. Arichi S, Sakamoto N, Yoshida M, Himuro S. *Polymer.* 1986;27:1761.
28. Mays JW, Hadjichristidis N, Graessley WW, Fetters LJ. *J. Polym. Sci., Polym. Phys. Ed.* 1986;24:2553.
29. Tsiartas PC, Flanagin LW, Henderson CL, Hinsberg WD, Sanchez IC, Bonnecaze RT, Willson CG, *Macromolecules.* 1997;30:4656.

Poly(isobutylene), Butyl Rubber, Halobutyl Rubber

Suresh Murugesan, Gary W. Ver Strate and David J. Lohse

Acronyms PIB, IIR (isobutylene isoprene rubber), CIIR (chlorinated IIR), Br IIR (brominated IIR) BIMSM (brominated isobutylene paramethyl styrene copolymer)[1,2]

Class Vinylidine polymers (IIR and halogenated derivatives are olefin-containing elastomers)

Structure[2]

PIB | IIR comonomer ~ 1% | Cl or Br IIR comonomer ~ 1% (X = Cl or Br)

Properties of Special Interest High bulk density (0.917 g cm^{-3} at 20°C for PIB and IIR) for an amorphous elastomer, which leads to low gas permeability and high hysteresis at a given temperature. The introduction of a mole percent of olefin or halogen dramatically changes chemical reactivity but not physical properties.[2,3]

Preparative Techiniques Type of polymerization: cationic Lewis acids (e.g., AlCl$_3$/H$_2$O), −80°C[2,4]

Property	Units	Conditions	Value	Reference
Ceiling temperature	K	CO$_2$ at 139 bar	361	(5)
Typical comonomers		Isoprene, paramethyl styrene, halogen, are introduced in a post polymerization process		(2)
Molecular weight of repeat units	g mol^{-1}	Isobutene	56	(2)
Tacticity (stereoregularity)		Not applicable, backbone symmetrically disubstituted. PIB will crystallize below 20°C or under stress (see below)		
Head-to-head contents	–	Negligible	–	(2, 6)
Degree of branching		Long chain branching is negligible except for intentionally branched commercially made products (e.g., Star branched butyl, ExxonMobil Chemical)		(2)

Property	Units	Conditions	Value	Reference
Typical molecular weight range of polymer	g mol^{-1}	As dispersants, As elastomers In blends, viscosity modifiers, chewing gums	500–5,000 1–6 ($\times 10^5$) 5×10^3–6×10^6	(2)
Typical poly dispersity index (M_w/M_n)	–	–	2.0–4.0	(2, 4)
Mooney viscosity	ML 1 + 8	125°C, Butyl Halobutyl	33–51 ± 3–5 32–50 ± 4	(7)
Antioxidant	wt%	Non-staining	≥ 0.3	
Stabilizer	wt%	–	1.3 ± 0.3	
IR (characteristic absorption frequencies)	cm^{-1}	Ambient, film, doublet	922; 948; also 1,225;1,365; 1,385;1,470	(8)
UV (characteristic absorption frequencies)	nm	Ambient, THF, hexane, broad centered at:	< 200	(9)
NMR (compositional analysis)	^{13}C and ^{1}H NMR (see reference for detailed peak assignments)			(10–12)
Thermal expansion coefficients	K^{-1}	1 atm, 27°C $(1/V)(dV/dT)_p$	5.5×10^{-4}	(13)
Compressibility coefficients	bar^{-1}	$(1/V)(dV/dP)_T$	4.8×10^9	(13)
Reducing temperature T^*	K	150–250°C, 10–200 MPa	7,693	(13)
Reducing pressure P^*	J cm^3	150–250°C, 10–200 MPa	469	(13)
Reducing volume V^*	cm^3 g^{-1}	150–250°C, 10–200 MPa	0.959	(13)
Density (amorphous)	g cm^{-3}	1 atm no halogen Halobutyl 20°C	0.917(1–30/M_n) 0.92–0.93	(14) (7)
(crystalline)	g cm^{-3}	1 atm	0.964	(23)
Solvents	Aromatic and aliphatic hydrocarbons, lubricating oils, nonpolar oxygen containing liquids			(2, 15)
Nonsolvents	Polar compounds, organic acids, ketones, alcohols with low carbon number, methyl chloride			(2, 15)
Solubility parameter	(MPa)$^{1/2}$	1 atm, 20°C	16.5	(16)
Theta temperature Θ	K	Toluene Ethyl hexanoate Benzene	260 330 296	(15, 17) (15, 18) (15, 18–20)
Interaction parameter χ	–	Cyclohexane, 25°C	0.44	(21)
Second virial coefficient A_2	mol cm^3 g^{-2}	Cyclohexane,25°C	$(6.9 \times 10^3)M_w^{0.2}$	(20)
Mark-Houwink parameter, K and a	K = ml g^{-1} a = None	Cyclohexane, 25°C Benzene, 25°C	K = 0.0135 a = 0.74 K = 0.10 a = 0.504	(20)
Huggins constant k′	–	Cyclohexane, 25°C	0.233 $M_w^{0.095}$	(20)
Characteristic ratio $\langle r^2 \rangle_0/nl^2$	–	SANS, bulk polymer Benzene, 24°C	6.9 6.6	(22) (15)

Property	Units	Conditions	Value	Reference
Space group	–	Orthorhombic	$P2_12_12_1$	(23)
Chain conformation (ρ_n of helix)	–	–	2*8/3	(23)
Unit cell dimensions	Å	20°C, 1 atm	$a = 6.94$ $b = 11.96$ $c = 18.63$	(23)
Number of repeat units	–	–	16	(23)
Heat of fusion	cal g^{-1}	At the melting temperature, 1 atm	52	(23)
Glass transition temperature	K	1 atm, DSC	202, 208	(2, 15, 24)
Melting point	K	1 atm, depends on annealing conditions	275 317	(2, 15) (23)
Oxidation rate	u.r. g^{-1} min^{-1}	Butyl at 175°C Chlorobutyl at 175°C Bromobutyl at 175°C	438 566 196	(26)
Activation energy	kJ mol^{-1}	Butyl at 175°C Chlorobutyl at 175°C Bromobutyl at 175°C	102.6 117.2 118.2	(26)
Deflection temperature	K	–	< 210	–
Heat capacity, $C_p = (dH/dT)_P$	kJ K^{-1} mol^{-1}	1 atm, 27°C	0.110	(15, 27)
Polymers with which miscible		Ethylene-butene copolymers from 52 to 78 wt% Butene, LCST from 25 to 120°C Head to head polypropylene, LCST = 180°C		(13)
Sub-T_g transition temperatures		None of significance, but see reference; T_g behavior is complex		(25)
Tensile modulus	MPa	Depends on compounding ingredients, temperature, strain rate	0.5–50	(28)
Bulk modulus	MPa	Depends on compounding ingredients, temperature, strain rate	2,000	(15)
Shear modulus	MPa	Depends on compounding ingredients, temperature, strain rate	0.3–20	(15, 28)
Storage modulus	MPa	Depends on compounding ingredients, temperature, strain rate	0.3–20	(15, 24, 29, 30)
Loss modulus, tan δ	–	1 atm, 20°C, 1 Hz, uncross-linked, high molecular weight	0.3	(15, 24)
Tensile strength	MPa	Depends on compounding ingredients, temperature, strain rate	0.5–50	(28)

Property	Units	Conditions	Value	Reference
Maximum extensibility $(L/L_o)_r$	–	Depends on compounding ingredients, temperature, strain rate	8	(28)
Hardness	Shore A	Depends on compounding ingredients, temperature, strain rate	5–100	(29, 30)
Poisson's ratio	–	20°C	0.49	(15)
Plateau modulus	MPa	75 and 250°C, high M	0.032	(24)
WLF parameters: C_1 and C_2	$C_1 =$ None $C_2 = K$	Uncross-linked PIB	$C_1 = 7.5$ $C_2 = 190$	(24)
Index of refraction n	None	1 atm, 25°C, n_D	$1.5092-13.9/M_n$	(14)
Refractive index increment, dn/dc	ml g^{-1}	1 atm, THF, 27°C	$0.115\ (1-22/M_n)$	(14)
Dielectric constant ε'	–	1 atm, 20°C, 1 KHz	2.4	(15, 31)
Dielectric constant ε''	–	1 atm, 20°C, 1 KHz	0.003	(15, 31)
Segment anisotropy	10^{25} cm^3	Benzene, xylene	45–59	(15, 32)
Electronic conductivity	(ohm cm)$^{-1}$	20°C, gum vulcanizate	10^{-14}	(26, 33)
Resistivity	log R, ohms	20°C, depends on carbon black, all 0.2 volume fraction	1.6–11.0	(33, 34)
Stress-optical coefficient	–	Data for a range of conditions	–	(25, 29)
Surface tension	mN m^{-1}	1 atm, 20°C	33.6	(15)
		150°C	25.3	
Interfacial tension	nM m^{-1}	With PDMS, 20°C	4.0	(35)
		With PVA		
		20°C	9.9	(36)
		100°C	8.3	
Permeability coefficient	$\dfrac{[m^3][cm]}{[s][m^2][Pa]}$	H$_2$, 250°C	5.43×10^{-17}	(37, 38)
		He, 25°C	6.37×10^{-17}	
		N$_2$		
		25°C	0.243×10^{-17}	
	cm^2 s^{-1} MPa^{-1}	60°C	15×10^{-8}	(7)
		80°C	35×10^{-8}	
		Air		
		60°C	20×10^{-8}	
		80°C	50×10^{-8}	
		CO$_2$		
		60°C	130×10^{-8}	
		80°C	290×10^{-8}	(39, 40)
		Miscellaneous organic solvents	–	
	–			
Melt viscosity	Poise	Newtonian, 25°C	$(4.3 \times 10^7)\,[\eta]^{4.66}$	(24)
			$(4.7 \times 10^{-11})M^{3.43}$	
		Newtonian, 175°C	$(2.14 \times 10^4)[\eta]^{4.74}$	

Property	Units	Conditions	Value	Reference
Thermal conductivity	$W m^{-1} K^{-1}$	1 atm, 20°C, gum vulcanizate	0.13	(27, 41)
		50 phr carbon black	0.23	
Coefficient of sliding friction μ	–	20°C, compounded vulcanizate, sliding on emery paper at 10^{-1} to 10^3 cm s^{-1}	1.6	(42)
Pyrolyzability, nature of product	Pure PIB is completely combustible			(43)
Biodegradability, effective microorganisms	Inert			
Maximum use temperature	Up to 150°C continuous service (see manufacturers for compounding information)			(43)
Decomposition temperature	50% volatile for 30 min at 320°C			(43)
Cross-linking, quantum yield	PIB or IIR cross-link very poorly with chemically generated radicals or with radiation, these processes are not used commercially for those rubbers but see Cl, Br, IIR below			

^{60}Co Irradiation at 77 K	Radicals G(R)	Cross-Linking G(x)	Scission G(s)	Gas G(isobutene)	Reference
PIB	2.3	0.0	3.7	0.62	(11)
ClIIR	4.3	3.6	1.7	0.03	(11)
BrIIR	3.7	3.7	0.44	0.03	(11)

Property	Units	Conditions	Value	Reference
Scission		25°C, γ radiation, PIB		
G(s)	mol J^{-1}		6.2×10^{-5}	(44)
G(x)	events/100 eV adsorbed		4	(15)
Water absorption	ppm	Pure polymer, 20°C, total immersion, compounding increases	<200	(38)
Storage life	–	Butyl rubber	~5 years	(7)
		Halobutyl rubber	~2 years	
Cost	US$ kg^{-1}	–	2.8–3.2	–
Suppliers	–	–	ExxonMobil Chemical, Lanxess	–

References

1. D1418-94 *Rubber and Rubber Lattices: Nomenclature*. American Society for Testing and Materials, Philadelphia.
2. Kresge E, Schatz R, Wang H-C. In: Mark HF, et al, eds. *Encyclopedia of Polymer Science and Engineering*. 2d ed. Vol 8. New York: John Wiley and Sons; 1987.
3. Boyd RH, Pant PVK. *Macromolecules*. 1991;24:6325.

4. Odian G. *Principles of Polymerization.* New York: John Wiley and Sons; 1991.

5. Deak G, Pernecker T, Kennedy JP. *Polym. Bull.* 1994;33:259.

6. Malanga M, Vogl O. *Polym. Bull.* 1983;7:236.

7. Lanxess Manual for Rubber Industry.

8. Hummel/Scholl D. *Atlas of Polymer and Plastics Analysis.* 2d ed. Vol 1. Germany: C. Hauser Verlag; 1978:25.

9. DePierri W. Exxon Chemical Co., Baton Rouge. Personal communication.

10. Chu CY, Watson KN, Vokov R. *Rubber Chem. and Tech.* 1987;60:636.

11. Hill DJT, et al. *Polymer.* 1995;36:4185.

12. Cheng DM, et al. *Rubber Chem. and Tech.* 1990;63:265.

13. Krishnamoorti R, et al. *Macromolecules.* 1995;28:1252–1259.

14. Chance R, et al. *Int. J. Polymer Analysis and Characterization.* 1995;1:3–34.

15. Brandrup J, Immergut E, eds. *Polymer Handbook.* 3rd. ed. New York: John Wiley and Sons; VI, 1989:413, #26.

16. Olabisi O, Robeson I, haw M. *Polymer–Polymer Miscibility.* New York: Academic Press; 1979:54.

17. Fox T, Flory PJ. *J. Am. Chem. Soc.* 1951;73:1909.

18. Carpenter D. *ACS Preprints.* 1972;13:981.

19. Tsuji T, Fujita H. *Polymer J.* 1973:4:409.

20. Fetters LJ, et al. *Macromolecules.* 1991;24:3127.

21. Eichinger BE, Flory PJ. *Trans. Faraday Soc.* 1968;64:2061.

22. Krishnamoorti R. *PhD Thesis.* Princeton University, New Jersey.

23. Wunderlich B. *Macromolecular Physics.* Vol 1. New York: Academic Press; 1973.

24. Fetters LJ, Graessley WW, Kiss AD. *Macromolecules.* 1991;24:3136.

25. Inoue T, Osaki K. *Macromolecules.* 1996;29:1595.

26. Jipa S, Giurginca M, Setnescu T, Setnescu R, Ivan G, Mihalcea I. *Polym. Degradation Stab.* 1996;54:1–6.

27. Nasr GM, et al. *Polym. Degradation Stab.* 1995;48:237.

28. Smith TL. In: Eirich F, ed. *Rheology.* Vol 5. New York: Academic Press; 1969.

29. Trexler HE, Lee MCH. *J Appl. Polym. Sci.* 1986;32:3899.

30. Dutta NK, Tripathy DK. *J Appl. Polym. Sci.* 1992;44:1635.

31. Caps RN, Bums J. *J. Non-Cryst. Solids (Part 2).* 1991;131:877.

32. Frisman EV, Dadivananyan A. *J. Polym. Sci.* 1967;C16:1001.

33. Medalia AI. *Rubber Chem. and Tech., Rubber Reviews.* 1986;59:432.

34. Aminabhavi TM, Cassidy PE, Thompson CM. *Rubber Chem. and Tech., Rubber Reviews.* 1990;63:451.

35. Wagner M, Wolf BA. *Macromolecules.* 1993;26:6498.

36. Wu S. *Polymer Interface and Adhesion.* New York: Marcel Dekker; 1982.

37. Van Amerongen GJJ. *Polym. Sci.* 1950;5:307.

38. Crank J, Park G. *Diffusion in Polymers.* New York: Academic Press; 1968.

39. Guo CJ, deKee D, Harrison A. *J Appl. Polym. Sci.* 1995;56:823.

40. Aminabhavi TM, Khinnavar, Rajashekhar S. *Polymer.* 1993;34:4280.

41. Goldsmith TE, Waterman, Hirschbom J, eds. *Handbook of Thermoproperties and Solid Materials.* Vol IV. New York: Macmillan; 1961.

42. Grosch KA, Schallamach A. *Rubber Chem. and Tech., Rubber Reviews.* 1976;49:862.

43. Fabris HJ, Sommer JG. *Rubber Chem., and Tech.* 1977;50:523.

44. Charlesby A, Bridges BJ. *Radiation Phys. Chem.* 1982;20:359.

45. Veith AG. *Rubber Chem. and Tech., Rubber Reviews.* 1992;65:601.

46. Bohm GGA, Tveekrem JO. *Rubber Chem. and Tech., Rubber Reviews.* 1982;55:675.

47. Medalia AI. *Rubber Rev.* 1991;64:481.

48. Mark HF, et al, eds. *Encyclopedia of Polymer Science and Engineering.* 2nd ed. Vol 1. New York: John Wiley and Sons; 1985:147 (acoustic properties).

49. Schuster RH, Issei HM, Peterseim V. *Rubber Chem. and Tech.* 1996;69:769.

50. Thompson CM, Allen JS. *Rubber Chem. and Tech.* 1994;62:107.

51. Cassidy PE, Aminabhavi TM, Thompson CM. *Rubber Chem. and Tech., Rubber Rev.* 1983;56:594.

52. Hess WM, Herd CR, Vegvani PC. *Rubber Chem. and Tech., Rubber Reviews.* 1993;66:329.

53. D3958 *Rubber-Evaluation of BIlR, CIlR.* D3188-91 *Rubber-Evaluation of IIR.* American Society for Testing and Materials.

54. ExxonMobil Chemical Co., Houston.

55. See *Rubber World*, rubber *Blue and Red Books.* issued on a regular basis.

56. Gursky LJ, et al. *Kautchuk and Gummi. Kunststoffe.* 1990;43:692.

57. Gursky LJ, et al. *Rubber World.* 1990;41:202.

58. Caporal JY, Boucher JN. *Revue Generale des Caoutchoucs & Plastiques* 607–608. 1981:61–66.

Cis-1,4-Polyisoprene

Ruzhi Zhang and James E. Mark

Acronyms, Alternative Names, Trade Names *cis*-PIP, CPI, IR, Natural Rubber (NR, NK), Hevea, *cis*-1,4-Poly(2-methylbutadiene) (PMBD), Natsyn, Cariflex, Ebonite[1–3]

Class Diene elastomers

Structure

Major Applications *cis*-1,4-Polyisoprene is used in tires and tire products, foam rubber, rubber sheeting, rubber bands, hoses, gaskets, belts, molded and mechanical goods, footwear and sporting goods, gloves, sealants, adhesives, bottle nipple, caulking, and other typical elastomer applications.

Properties of Special Interest High degree of stereoregularity in structure, presence of the reactive double bonds (unsaturation), strain-induced crystallization, high gum tensile strength, superior building tack, green stock strength, better processing, high strength in non-black formulations, hot tear resistance, retention of strength at elevated temperatures, high resilience, low hysteresis (heat build-up), excellent dynamic properties, and general fatigue resistance.

Natural Sources Natural rubber occurs in over 200 species of plants. However, only one tree source, Hevea Brasiliensis, is of commercial importance, and it accounts for over 99% of the world's natural rubber production.[4, 5]

Preparative Methods *cis*-1,4-Polyisoprene is made by coordination, anionic, or cationic polymerization of isoprene through the use of coordination catalysts, alkali metal catalysts, Alfin catalysts, organoalkali catalysts, or conventional Lewis acids.[6]

Chemical Modification The following chemical modifications of *cis*-1,4-polyisoprene are employed as a convenient way of altering physical and mechanical properties: hydrohalogenation, halogenation, oxidation, ozonolysis, epoxidation, hydrogenation, carbene addition, or cyclization.[6]

Typical composition (%) of natural rubber latex[7, 8]

Total solid content	36
Dry rubber content	33
Proteineous substances	1–1.5
Resinous substances	1–2.5
Ash	<1
Sugars	1
Water	60

Physical constants of *cis*-1,4-polyisoprene (unvulcanized, CAS number [9003-31-0])

Property	Units	Value	Conditions	Reference
Density	g cm^{-3}	0.913	–	(9)
		0.9283	0°C	(10)
		0.9162	20°C	(10)
		0.9283–6.10 $\times 10^{-4}$ T	Temp. range: 0–25°C, T in °C, densities as a function of temperature (measured above T_g)	(10)
Thermal expansion coefficients	K^{-1}	6.6×10^{-4}	0°C	(10)
		6.6×10^{-4}	20°C	(10)
Tait equation parameters: C, b_o, and b_1	C = None	$C = 0.0894$	0–25°C, 0–500 bar, densities as a function of pressure	(10)
	b_o = bar	$b_o = 1,937$	0–25°C, 0–500 bar	
	$b_1 = °C^{-1}$	$b_1 = 0.00517$	0–25°C, 0–500 bar	
Isothemal compressibility	bar^{-1}	4.6×10^{-5}	0°C, atmospheric pressure	(10)
		5.0×10^{-5}	20°C, atmospheric pressure	
Thermal conductivity k	W m^{-1} K^{-1}	0.13	–	(9, 11, 12)
Specific heat C_p	J kg^{-1} K^{-1}	1.905×10^3	–	(13, 14)
$\partial C_p / \partial T$	J kg^{-1} K^{-2}	3.54	–	(13, 14)
Glass transition temperature	K	201	–	(15)
		199–204	–	(15, 16)
Melting point	K	308.6	–	(17, 18)
Heat of fusion ΔH_u	kJ mol^{-1}	4.393	Determined by use of diluent equation	(17, 18)
$\Delta H_u / M_o$	J g^{-1}	64.6	Determined by use of diluent equation	(17, 18)
Entropy of fusion ΔS_u	J K^{-1} mol^{-1}	14.2	Determined by use of diluent equation	(17, 18)
Heat of combustion	kJ g^{-1}	45.2	–	(9)
Temperature of most rapid crystallization	K	248	–	(19)
Refractive index n_D	–	1.5191	–	(20)

Property	Units	Values	Conditions	Reference
$\partial n_D/\partial T$	K^{-1}	−0.0037	–	(20)
Dielectric constant	–	2.37–2.45	1 kHz	(9, 21)
Dissipation factor	–	0.001–0.003	1 kHz	(21)
Conductivity	$S\,m^{-1}$	2.57×10^{-15}	60 s	(9, 21)
Bulk modulus	Pa	$1,940 \times 10^{6}$	Isothermal K	(10)
		$2,270 \times 10^{6}$	Adiabatic K_a	(10)
Bulk wave velocity V_b	$m\,s^{-1}$	1,580	Longitudinal wave	(10)
$\partial V_b/\partial T$	$m\,s^{-1}\,K^{-1}$	−3	–	(10)
Storage modulus G'	log Pa	5.61 (5.53–5.75)	Values of log G′	(22)
Loss modulus G''	log Pa	4.46 (4.43–4.65)	Values of log G″	(22)
Loss tangent G''/G'	–	0.09 (0.07–0.13)	–	(22)
Resilience	%	75–77	Rebound	(23, 24)
Unperturbed dimension $r_{0f}/M^{1/2}$	$nm\,mol^{1/2}$ $g^{-1/2}$	$0.402/M_u^{1/2}$	Calculated unperturbed dimensions of freely rotating chains	(25)
		$0.201/m^{1/2}$		
Surface tension	$mN\,m^{-1}$	32	Contact angle	(26)

Physical constants of *cis*-1,4-polyisoprene (pure-gum vulcanizate)

Property	Units	Value	Conditions	Reference
Density	$g\,cm^{-3}$	0.970	–	(27, 28)
		0.9211	0°C	(10)
		0.9093	20°C	(10)
		0.9210–$5.86 \times 10^{-4}\,T$	Temp. range: 0–25°C, T in °C, densities as a function of temperature (measured above T_g)	(10)
Thermal expansion coefficients	K^{-1}	6.5×10^{-4}	0°C	(10)
		6.4×10^{-4}	20°C	(10)
		6.7×10^{-4}	20°C	(29)
Tait equation parameters: C, b_o, and b_1	C = None	0.0894	0–25°C, 0–500 bar, densities as a function of pressure	(10)
	b_o = bar	1,916	0–25°C, 0–500 bar	
	$b_1 = °C^{-1}$	0.00425	0–25°C, 0–500 bar	
Isothemal compressibility	bar^{-1}	4.6×10^{-5}	0°C, atmospheric pressure	(10)
		5.0×10^{-5}	20°C, atmospheric pressure	(10)
		5.3×10^{-5}	20°C, atmospheric pressure	(29)
Thermal conductivity k	$W\,m^{-1}\,K^{-1}$	0.153	–	(30, 31)
$(\partial k/k)/\partial T$	$\%\,K^{-1}$	0	–	(32)
		−0.1	–	(30)

Property	Units	Values	Conditions	Reference
Specific heat C_p	J kg^{-1} K^{-1}	1.828×10^3	–	(33)
Glass transition temperature	K	210	–	(34)
		201–212	–	
Melting point	K	313	–	(35)
Heat of combustion	kJ g^{-1}	44.4	–	(9)
Refractive index n_D	–	1.5264	–	(9)
$\partial n_D/\partial T$	K^{-1}	−0.0037	–	(9)
Dielectric constant	–	2.68	1 kHz	(21)
		2.5–3.0		(9, 21)
Dissipation factor	–	0.002–0.04	1 kHz	(21)
Conductivity	S m^{-1}	$2–100 \times 10^{-15}$	60 s	(9, 21)
Bulk modulus	Pa	$1{,}950 \times 10^6$	Isothermal K	(10)
		$2{,}260 \times 10^6$	Adiabatic K_a	(10)
Bulk wave velocity V_b	m s^{-1}	1,580	Longitudinal wave	(10)
		1,500–1,580		(10, 36, 37)
$\partial V_b/\partial T$	m s^{-1} K^{-1}	−3	–	(10)
Strip velocity v_l	m s^{-1}	45	Longitudinal wave, 1 kHz	(36)
		35–51		(9, 36, 38)
$\partial v_l/\partial T$	m s^{-1} K^{-1}	−0.2	–	(9)
Ultimate Elongation	%	750–850	–	(23, 39)
Tensile strength	MPa	17–25	–	(23, 39)
Initial slope of stress-strain curve, Young's modulus, E	MPa	1.3	60 s	(40, 41)
		1.0–2.0		(23, 34, 40, 41)
Shear modulus G	MPa	0.43	60 s	(34, 41)
		0.3–0.7		(34, 42)
Shear compliance J	(MPa)$^{-1}$	2.3	60 s	(34, 41)
		1.5–3.5		(34, 42)
Creep rate *(1/J)* $(\partial J/\partial \log t)$	% (unit $\log t)^{-1}$	2	–	(34, 40)
		1–3		(34, 40, 41, 43–45)
Poisson's ratio μ	–	0.49989	Calculated as ½−(1/6) (E/K)	(10, 46, 47)
E/G	–	2.9978	Calculated as 3−(1/3) (E/K)	(10, 46, 47)
Storage modulus G'	log Pa	5.61 (5.49–5.78)	Values of log G'	(48)
Loss modulus G''	log Pa	3.80 (3.72–4.48)	Values of log G''	(48)
Loss tangent G''/G'	–	0.016 (0.01–0.05)	–	(48)
Resilience	%	75–84	Rebound	(24, 49)

Unit cell dimensions

Lattice	Space Group	Monomers per Unit Cell	Cell Dimensions (Å)			Cell Angles	Reference
			a	b	c	γ	
Mono	$C_{2h}{}^5$	8	12.46	8.89	8.10	92°	(50)
Ortho	–	16	25.2	8.97	8.20	–	(51)
Mono	–		26.3	8.15	8.9	109.5°	(52)
Ortho	–		12.4	8.15	8.9	–	(52, 53)

Permeability and diffusion data of *cis*-1,4-polyisoprene

Permeant	T (°C)	$P \times 10^{13}$	$D \times 10^6$	$S \times 10^6$	Temp. Range (°C)	$P_o \times 10^7$	E_P	E_D	E_S	Reference
He	25	–	21.6	–	20–50	–	–	19.7	–	(54)
O_2	25	17.6	1.73	1.02	20–50	2.38	29.3	33.5	−4.2	(54, 55)
Ar	25	17.2	1.36	1.26	20–50	9.15	32.7	33.1	−0.4	(54)
CO_2	25	115	1.25	9.20	20–50	0.755	21.8	34.3	−12.5	(54, 55)
CO	25	11.8	1.35	0.874	20–50	3.16	31.0	31.0	0	(54)
N_2	25	7.11	1.17	0.608	20–50	12.2	35.6	33.5	2.1	(54, 55)
CH_4	25	22.7	0.89	2.55	20–50	6.08	31.0	36.4	−5.4	(54)
C_2H_6	25	–	0.40	–	20–50	–	–	42.7	–	(54)
C_3H_6	25	154	0.31	49.7	20–50	17.7	28.9	42.7	−13.8	(54)
C_3H_8	25	126	0.21	60.0	20–50	1.34	23.0	46.5	−23.5	(54, 55)
SF_6	25	2.70	0.115	2.35	20–50	4.62	35.6	50.2	−14.6	(54)
C_2H_2	25	74.5	0.467	16.0	25–50	17.0	30.6	39.8	−9.2	(55)
H_2O	25	1,720	–	–	–	–	–	–	–	(56)

Polymer pairs compatible in the amorphous state at room temperature (*cis*-1,4-polyisoprene is polymer I)[57]

Polymer II of	Method	Comments
Styrene	Single dynamic mechanical loss peak	I was natural rubber; II had $M_w \leq 350$; two peaks when II had $M_w \geq 600$
Vinyl cyclohexane	Single dynamic mechanical loss peak	I was natural rubber; II had $M_w \leq 375$; two peaks when II had $M_w \geq 650$

Infrared absorption[58–62]

Frequency (cm^{-1})	Assignment
836	Trisubstituted olefin out-of-plane CH wag[a]
1,129	CH_3 rock
1,300	CH_2 wag
1,376	CH_3 symmetric (umbrella) deformation
1,450	CH_2 symmetric (scissors) and CH_3 asymmetric deformation
1,664	C=C stretch
2,720	Overtone of CH_2 umbrella

Cis-1,4-Polyisoprene

Frequency (cm^{-1})	Assignment
~2,850	CH_2 and CH_3 symmetric stretch
2,920	CH_2 asymmetric stretch
2,962	CH_3 asymmetric stretch
3,030	Olefin CH stretch

(a): Intensity increases with crystallinity.

^{1}H NMR (CDCl$_3$, 500 MHz)[63]

Chemical Shift (ppm)	Assignment
5.1 (b)	$-CH_2-CH=C(CH_3)-CH_2-$
4.7 (m)	$-CH_2-C(H)-C(CH_3)=CH_2$
2.1 (b)	$-CH_2-CH=C(CH_3)-CH_2-$ and $-CH_2-C(H)-C(CH_3)=CH_2$
1.7 (m)	$-CH_2-CH=C(CH_3)-CH_2-$ and $-CH_2-C(H)-C(CH_3)=CH_2$
1.4 (m)	$-CH_2-C(H)-C(CH_3)=CH_2$

^{13}C NMR (CDCl$_3$, 75 MHz)[63]

Chemical Shift (ppm)	Assignment
148	$-CH_2-C(H)-C(CH_3)=CH_2$
135	$-CH_2-CH=C(CH_3)-CH_2-$
125	$cis-CH_2-CH=C(CH_3)-CH_2-$
124	$trans-CH_2-CH=C(CH_3)-CH_2-$
112	$-CH_2-C(H)-C(CH_3)=CH_2$
48	$-CH_2-C(H)-C(CH_3)=CH_2$
40	$trans-CH_2-CH=C(CH_3)-CH_2-$
32	$cis-CH_2-CH=C(CH_3)-CH_2-$
31	$-CH_2-C(H)-C(CH_3)=CH_2$
27	$-CH_2-CH=C(CH_3)-CH_2-$
23	$cis-CH_2-CH=C(CH_3)-CH_2-$
19	$-CH_2-C(H)-C(CH_3)=CH_2$
16	$trans-CH_2-CH=C(CH_3)-CH_2-$

Radiation resistance

Property	Half-Value Dose (MGy) in Air at Different Dose Rates (Gy h^{-1})					
	$\geq 10^5$	Reference	10^4	Reference	5	Reference
σ_R	3	(64)	>1	(65)	0.1	(65)
			1	(66)		
ε_R	1–1.5	(64)	>1	(65)	0.07	(65)
			1.5	(66)		

σ_R Tensile strength at break (ultimate strength).

ε_R Elongation at break (ultimate elongation).

Solvents and non-solvents for *cis*-1,4-polyisoprene[67-69]

Solvents	Non-solvents
Hydrocarbons, THF, higher ketones, higher aliphatic esters	Alcohols, lower ketones and esters, nitromethane, propionitrile, water, diluted acids, diluted alkalies, hypochlorite solutions

Fractionation of *cis*-1,4-polyisoprene[70]

Method of Fractionation	Solvent or Solvent/Non-Solvent Mixture	Remarks
Fractional precipitation	Benzene/acetone	Hevea
	Benzene/*n*-butanol	–
	Benzene/isopropanol	Low temperature
	Benzene/methanol	Low temperature
	Chloroform/acetone	Pale crepe
	Dichloroethane/2-butanone	–
	Toluene/*n*-butanol	30°C
	Toluene/boiling methanol	Chlorinated natural rubber
	Toluene/methanol	–
Fractional solution	Acetone	Hevea, extraction
	Acetone, *n*-hexane	Guayule, extraction
	Benzene/methanol	25°C, column extraction, natural rubber
Chromatography	Benzene/methanol	Precipitation chromatography
	Chloroform	30°C, GPC, styragel
	Cyclohexane	GPC
	Cyclohexanone	Partition on paper
	o-Dichlorobenzene	GPC, 135°C
	Dichloromethane	GPC, μ-styragel
	Toluene	Preparative GPC, styragel
	Toluene/isopropyl alcohol	Precipitation chromatography

Solubility parameter

Polyisoprene	$\delta[(\text{MPa})^{1/2}]$	$\delta[(\text{cal cm}^{-3})^{1/2}]$	Method	T [°C]	Reference
1,4-*cis*	15.18	7.42	Calculated	25	(71)
	20.46	10.0	Swelling	35	(72)
	16.57	8.10	Average	35	(72)
	16.47	8.05	Swelling	35	(72)
	16.68	8.15	Calculated	35	(72)
	16.68	8.15	Calculated	25	(73)
	16.2	7.9	Observed	25	(73)
	17.09	8.35	Observed	25	(73)
Natural rubber	17.0	8.3	Observed	25	(74)
	16.6	8.1	Observed	25	(75)
	17.09	8.35	Observed	25	(76)
	16.33	7.98	Observed	25	(77)
	16.49–16.42	8.06–8.12	Swelling	25	(78)

Anisotropy of segments and monomer units of *cis*-1,4-polyisoprene

Solvent	$(\alpha_1-\alpha_2) \times 10^{25}$ cm^3	$(\alpha_\parallel -\alpha_\perp) \times 10^{25}$ cm^3	Reference
Benzene	+48	+30.5	(79)
Squalene sw. p.	+56.5	–	(80)

Unperturbed dimensions of linear *cis*-1,4-polyisoprene

Property	Unit	Value	Temp. (°C)	Remarks	Reference
$S_{0z}/M_{\rm w}^{1/2} \times 10^4$	Nm	0.76	22	Diisopropyl ether; 100% *cis*	(81)
$K_0 \times 10^3$	Ml g^{-1}	130 ± 20	20	Benzene; 2-pentanone; 100% *cis*	(82, 83)
		119	14.5	2-Pentanone; 100% *cis*	(84)
$r_0/M^{1/2} \times 10^4$	Nm	810 ± 45	20	Benzene; 2-pentanone; 100% *cis*	(82, 83)
		847	22	Diisopropyl ether; 100% *cis*	(81)
$r_{0f}/M^{1/2} \times 10^4$	Nm	485	20	Benzene; 2-pentanone; 100% *cis*	(82,83)
		485	22	Diisopropyl ether; 100% *cis*	(81)
$\sigma = r_0/r_{0f}$	–	1.67 ± 0.09	20	Benzene; 2-pentanone; 100% *cis*	(82, 83)
		1.74	22	Diisopropyl ether; 100% *cis*	(81)
$C_\infty = r_0^2/nl^2$	–	5.0	20	Benzene; 2-pentanone; 100% *cis*	(82, 83)
		5.5	22	Diisopropyl ether; 100% *cis*	(81)
		4.7	14.5	2-Pentanone; 100% *cis*	(84)
$d\ln r_0^2/dT$	Deg^{-1}	0.41×10^{-3}	$-10-70$	Undiluted; 100% *cis*	(84)
		0.56×10^{-3}	30–70	Undiluted; 100% *cis*	(84)

Mark-Houwink parameters (K and a) (viscosity–molecular weight relationships, $[\eta] = K\,M^a$)

Polyisoprene	Solvent	T(°C)	$K \times 10^2$ (ml g^{-1})	a	Molecular Weight Range, M $\times 10^{-5}$	Method of Calibration	Reference
Natural rubber	Benzene	30	1.85	0.74	0.8–2.8	OS	(83)
	Cyclohexane	27	3.0	0.70	18.5	LS, SD	(85)
	4-Methyl-2-pentanone	35	6.07	0.57	0.5–10	LS	(86)
	2-Pentanone	14.5	11.9	0.50	0.8–2.8	OS	(83)
	Toluene	25	5.02	0.667	0.7–10.0	OS	(87)
Synthetic *cis*	Hexane	20	6.84	0.58	0.5–8.0	SD	(88)
	Toluene	30	0.851	0.77	2.0–10.0	LS	(89)

Huggins coefficients for natural rubber[90]

Solvent	T (°C)	$[\eta]$	k'
Benzene	30	354	0.32
n-Hexane	30	170	0.359

Dipole moment of *cis*-1,4-polyisoprene in solution[91]

Solvent	T (°C)	P_n	$(\mu^2/N)^{1/2}$ (D)	φ	Remarks
Benzene	25.0	13,762	0.28	0.70	$u_0 = 0.34$ D (2-methyl-2-butene in benzene)

Heat of solution[92]

Solvent	Heat of Solution (J g^{-1} Polymer)	Remarks
Benzene	12	$16°C, 4 \times 10^3$ g mol^{-1}

Second virial coefficient (A_2)

Polyisoprene	Solvent	Temp (°C)	M × 10^{-6} (g mol^{-1})	A_2 × 10^4 (mol cm^3 g^{-2})	Reference
cis	Cyclohexane	20	1.6	6.5	(93)
		25	0.62	5.0	(94)
Natural rubber	Cyclohexane	7	1.7	14.2	(85)
		27	1.7	14.3	(85)
		7	1.3	11.7	(95)
		27	1.3	12.7	(95)
		25	0.3	6.2	(94)

Sedimentation coefficients, diffusion coefficients, and frictional ratios for polyisoprene in solution

Polyisoprene	Solvent	Temp (°C)	M × 10^{-3} (g mol^{-1})	s_0 × 10^{13} (s)	D_0 × 10^7 (cm^2 s)	f_0/f_{sp}	Reference
Linear	Carbon tetrachloride	50	<5	–	$D_0 = 10^{-7.73\pm0.08} \times M^{-0.54\pm0.04}$	–	(96)
			>5	–	$D_0 = 10^{-7.47\pm0.05} \times M^{-0.61\pm0.01}$	–	
Natural rubber, crepe	Chloroform	20	270	15.5	2.24	3.32	(97)
			485	15.5	1.26	5.10	
			930	27.5	1.16	5.26	
Natural rubber	Hexane	20	270	9	3	–	(97)
			1,660	21	1.01	–	

Polymer-solvent interaction parameter χ

Solvent	Temp. (°C)	Volume fraction ϕ_2	χ	Reference
Acetone	0	1	2.1	(98)
	25	0.8–1	1.27–1.8	(98)
Benzene	10	0.6–0.8	0.48–0.51	(99)
	25	0–1	0.41–0.51	(99–101)
	25–55	1	0.46–0.43	(102)
	40	0.8–1	0.49–0.50	(99)
2-Butanone	25	0.6–1	0.86–1.43	(98)
	45	0.6–1	0.83–1.2	(98)
Ethyl acetate	25	0.4–1	0.69–1.24	(98, 103)
	50	0.4–1	0.68–1.0	(98, 103)
Ethylbenzene	25–55	1	0.34–0.30	(102)
n-Heptane	25–55	1	0.51–0.49	(102)
n-Hexane	25–55	1	0.54–0.50	(102)
2-Methylheptane	25–55	1	0.50–0.47	(102)
2-Methylhexane	25–55	1	0.52–0.50	(102)
2-Methylpentane	25–55	1	0.56–0.52	(102)
n-Octane	25–55	1	0.49–0.46	(102)
n-Pentane	25–55	1	0.61–0.53	(102)
Toluene	25–55	1	0.36–0.32	(102)
2,2,4-Trimethylpentane	25–55	1	0.49–0.46	(102)
p-Xylene	25–55	1	0.28–0.26	(102)

Theta temperature

Polyisoprene	$M_w \times 10^{-4}$	Solvent	Theta Temp. (°C)	$K_\theta \times 10^4$ [dl g^{-1}(g mol wt)$^{-1/2}$]	Reference
cis	5–100	*n*-Hexane/isopropanol (50/50)	21.0	16.6	(104)
cis (96%)	6.9–75	Dioxane	31.2	13.4	(105)
cis (94%) linear	9.4	Methyl isobutyl ketone	16.5	–	(106)
		Methyl propyl ketone	33.0	–	(106)
3 branches	5.7 (Br:1.75)	Methyl propyl ketone	33.0	–	(106)
11 branches	18 (Br:1.6)	Methyl propyl ketone	27.8	–	(106)
22 branches	34.2 (Br:1.6)	Methyl propyl ketone	23.5	–	(106)
		Methyl isobutyl ketone	15.0	–	(106)

Specific refractive index increment in dilute solution, dn/dc (ml g^{-1})

Polyisoprene	Solvent	$\lambda_0 = 436$ nm	$\lambda_0 = 546$ nm	T(°C)	Reference
Synthetic, high *cis*	Chloroform	0.104	0.100	25	(109)
Natural Hevea	Chloroform	0.107	0.104	25	(109)
Synthetic, high *cis*	*n*-Hexane	0.192	0.191	25	(109)
Natural Hevea	*n*-Hexane	0.200	0.198	25	(109)

Polyisoprene	Solvent	$\lambda_0 = 436$ nm	$\lambda_0 = 546$ nm	$T(°C)$	Reference
Synthetic, high *cis*	Tetrahydrofuran	–	0.128	20	(107)
		–	0.160	19–21	(108)
		0.153	0.149	25	(109)
Natural Hevea	Tetrahydrofuran	0.160	0.156	25	(109)
Synthetic	Toluene	0.030	0.028	25	(109)
Natural Hevea	Toluene	0.034	0.032	25	(109)

References

1. Brandrup J, Immergut EH, Grulke EA, eds. *Polymer Handbook*. 4th ed. New York: John Wiley and Sons; 1999.

2. Mark HF, Bikales NM, Overberger CG, Menges G, eds. *Encyclopedia of Polymer Science and Engineering*. 2nd ed. New York: John Wiley and Sons; 1989.

3. Salamone JC, ed. *Polymeric Materials Encyclopedia*. New York: CRC Press; 1996.

4. Brown H. *Rubber: Its Source, Cultivation, and Preparation*. London: John Murray; 1918.

5. Martin G. *IRI Trans*. 1943;19:38.

6. Senyek ML. In: Mark HF, Bikales NM, Overberger CG, Menges G, eds. *Isoprene Polymers* in *Encyclopedia of Polymer Science and Engineering*. 2nd ed. Vol 8. New York: John Wiley and Sons; 1989.

7. St. Cyr DR. In: Mark HF, Bikales NM, Overberger CG, Menges G, eds. *Encyclopedia of Polymer Science and Engineering*. 2nd ed. Vol 8. New York: John Wiley and Sons; 1989.

8. Hourston DJ, Tabe JO. In: Salamone JC, ed. *Polymeric Materials Encyclopedia*. Vol 2. New York: CRC Press; 1996.

9. Wood LA. *Values of the Physical Constants of Rubber*, in *Proceedings of the Rubber Technology Conf.* (Institution of the Rubber Industry, London), 1938: 933; *Rubber Chem. Technol.* 1939;12:130.

10. Wood LA, Martin GM. *J. Res. Nat. Bur. Stand.* 1964;68A:259; *Rubber Chem. Technol.* 1964;37:850.

11. Thompson EV. In: Mark HF, Bikales NM, Overberger CG, Menges G, Kroschwitz JI, eds. *Encyclopedia of Polymer Science and Engineering*. Vol 16. New York: Wiley-Interscience; 1985:711–737.

12. In: Mark HF, Othmer DF, Overberger CG, Seaborg GT, eds. *Encyclopedia of Chemical Technology*. 3rd ed. New York: Wiley-Interscience; 1978.

13. Chang SS, Bestul AB. *J. Res. Nat. Bur. Stds.* 1971;75A:113.

14. Wood LA, Bekkedahl N. *J. Polym. Sci., Part B, Polym. Lett.* 1967;5:169.

15. Wood LA. *Synthetic Rubbers: A Review of Their Compositions, Properties, and Uses*, in *Natl. Bur. Std. Circ., C 427* (1940); *Rubber Chem. Tech.* 1940;13:861; *India Rubber World*. 1940;102:433.

16. Dannis ML. *J. Appl. Polym. Sci.* 1959;1:121.

17. Dalai EN, Taylor KD, Phillips PJ. *Polymer*. 1983;24:1623.

18. Roberts DE, Mandelkern L. *J. Am. Chem. Soc.* 1955;77:781.

19. Wood LA, Bekkedahl N. *J. Res. Natl. Bur. Stds.* 1946;36:489; RP 1718; *J. Appl. Phys.* 1946;17:362; *Rubber Chem. Technol.* 1946;19:1145.

20. Wood LA, Tilton LW. *Proc. 2nd Rubber Technology Conference*, London, 1948, p. 142, Institution of the Rubber Industry, London; *J. Res. Natl. Bur. Stds.* 1949;43:57, RP 2, 004.

21. McPherson AT. *Rubber Chem. Technol. (Rubber Rev.)*. 1963;36:1230.

22. Zapas LJ, Shufler SL, deWitt TW. *J. Polym. Sci.* 1955;18:245, *Rubber Chem. Technol.* 1956;29:725.

23. Boonstra BBST. In: Houwink R, ed. *Properties of Elastomers*, Chapter 4 of Vol III. *Elastomers: Their Chemistry, Physics and Technology*. New York,: Elsevier; 1948.

24. Boonstra BBST. *Rev. Gen. Caoutchouc.* 1950;27:409. Translated in *Rubber Chem. Technol.* 1951;24:199.

25. Benoit H. *J. Polym. Sci.* 1948;3:376.

26. Lee LH. *J. Polym. Sci., Part A-2.* 1967;5:1103.

27. Wood LA. In: Brandrup J, Immergut EH, eds. *Polymer Handbook*. 3rd ed. New York: John Wiley and Sons; 1989.

28. Wildschut J. *Technological and Physical Investigations on Natural and Synthetic Rubbers*. New York: Elsevier; 1946.

29. Allen G, Gee G, Mangaraj D, et al. *Polymer.* 1960;1:467.

30. Carwile LC, Hoge HJ. *Thermal Conductivity of Soft Vulcanized Natural Rubber, Selected Values* in *Advances in Thermophysical Properties at Extreme Temperatures and Pressures*. New York: American Society of Mechanical Engineers; 1965; *Rubber Chem. Technol.* 1966;39:126.

31. Pillsworth MN Jr, Hoge HJ, Robinson HE. *J. Mater.* 1972;7(4):550.

32. Hands D. *Rubber Chem. Technol. (Rubber Rev.).* 1977;50:480.

33. Hamill WH, Mrowca BA, Anthony RL. *Ind. Eng. Chem.* 1946;38:106; *Rubber Chem. Technol.* 1946;19:622.

34. Wood LA, Roth FL, *Proc. 4th Rubber Technology Conference*, London, 1962:328, London: Institution of the Rubber Industry; 1963; *Rubber Chem. Technol.* 1963;36:611.

35. Furukawa GT, Reilly ML. *J. Res. Nat. Bur. Stds.* 1956;56:285; RP 2676.

36. Cramer WS, Silver I. NSVORD Report 1778, Feb. 1951, U. S. Naval Ordnance Lab., White Oak, MD.

37. Ivey DG, Mrowca BA, Guth E. *J. Appl. Phys.* 1946;20:486; *Rubber Chem. Technol.* 1950;23:172.

38. Payne AR, Scott JR. *Engineering Design with Rubber*. New York: Interscience; 1960.

39. Ball JM. Maassen GC. American Society for Testing Materials, *Symp. on the Applications of Synthetic Rubbers*, March 2, 1944:27.

40. Martin GM, Roth FL, Stiehler RD. *Trans. Inst. Rubber Ind.* 1956;32:189; *Rubber Chem. Technol.* 1957;30:876.

41. Roth FL, Bullman GW, Wood LA. *J. Res. Natl. Bur. Stds.* 1965;69A:347; *Rubber Chem. Technol.* 1966;39:397.

42. Philipoff W. *J. Appl. Phys.* 1953;24:685.

43. Chasset R, Thirion P. In: Prins JA, ed. *Proc. Int. Conf. Non-Crystalline Solids*, Delft, 1964. p. 345, New York: North Holland, Amsterdam, Interscience; *Rubber Chem. Technol.* 1966;39:870; *Rev. Gen. Caoutchouc.* 1967;44:1041.

44. Wood LA. *J. Rubber Res. Inst. Malaysia.* 1969;23 (3):309; *Rubber Chem. Technol.* 1970;43:1482.

45. Wood LA, Bullman GW. *J. Polym. Sci.* 1972;A-2:1043.

46. Holownia BP. *J. Inst. Rubber Ind.* 1974;8:157; *Rubber Chem. Technol.* 1975;48:246.

47. Glenn K, Rightmire. *Am. Soc. Mech. End. Trans. Series F, J. Lubrication Technol.* 381, July 1970.

48. Perry JD, Mancke RG, Maekawa E, Oyanagi Y, Dickie RA. *J. Phys. Chem.* 1964;68:3414.

49. Dillon JH, Prettyman IB, Hall GL. *J. Appl. Phys.* 1944;15:309; *Rubber Chem. Technol.* 1944;17:597.

50. Bunn CW. *Proc. R. Soc. London, Ser. A.* 1942;180:40.

51. Meyer KH. *Natural and Synthetic High Polymers*. New York: Interscience; 1950.

52. Morss HA Jr. *J. Am. Chem. Soc.* 1938;60:237.

53. Natta G, Corradin P. *Angew. Chem.* 1956;68:615; *Nuovo Cimento, Suppl.* 1960;15:111.

54. Michaels AS, Bixler HJ. *J. Polym. Sci.* 1961;50:413.

55. Amerongen GJ. *J. Polym. Sci.* 1950;5:307.

56. Taylor RL, Hermann DB, Kemp AR. *Ind. Eng. Chem. Int. Ed.* 1936;28:1255.

57. Class JB, Chu SG. *J. Appl. Polym. Sci.* 1985;30:815.

58. Colthup NB, Daly LH, Wiberley SE. *Introduction to Infrared and Raman Spectroscopy.* 2nd ed. New York: Academic; 1975.

59. *The Infrared Spectra Atlas of Monomers and Polymers.* Philadelphia: Sadtler Res. Labs; 1980.

60. Jasse B, Koenig JL. *J. Makromol. Sci.-Rev. Macromol. Chem.* 1979;C17:61.

61. Shindo Y, Read BE, Stein RS. *Makromol. Chem.* 1968;118:272.

62. Gotoh R, Takenaka T, Hayama N. *Kolloid Z.* 1965;205:18.

63. Switek KA. *et al. Macromolecules.* 2004;37:6355.

64. Collins GC, Calkins VP. APEX-261, 1956.

65. Wuckel L, Koch W. *Isotopenpraxis.* 1972;8:1.

66. Wundrich K. In: Brandrup J, Immergut EH, eds. *Polymer Handbook.* 3rd ed. New York: John Wiley and Sons; 1989.

67. Dexheimer H, Fuchs O, In: Nitsche R, Wolf KA, eds. *Struktur and Physikalisches Verhalten der Kunststoffe.* Vol 1. Berlin-Goettingen-Heidelberg: Springer Verlag; 1961.

68. Kurata M, Stockmayer WH. *Adv. Polym. Sci.* Vol 3. Berlin-Goettingen-Heidelberg: Springer Verlag; 1963:196.

69. Roff WJ. *Fibers, Plastics, and Rubbers,* New York: Academic Press; 1956.

70. Bello A, Barrales-Rienda JM, Guzman GM. In: Brandrup J, Immergut EH, eds. *Polymer Handbook.* 3rd ed. New York: John Wiley and Sons; 1989.

71. DiBenedetto AT. *J. Polym. Sci.* 1963;A1:3459.

72. Mangaraj D, Bhatnagar SK, Rath SB. *Makromol. Chem.* 1963;67:75.

73. Small PA. *J. Appl. Chem.* 1953;3:71.

74. Vocks F. *J. Polym. Sci.* 1964;A2:5319.

75. Tobolsky AV. *Properties and Structure of Polymers.* New York: Wiley; 1960:64–66.

76. Mark H, Tobolsky AV. *Physical Chemistry of High Polymers,* New York: Interscience; 1950:263.

77. Gee G. *Trans. Inst. Rubber Ind.* 1943;18:266.

78. Bristow GM, Watson WF. *Trans. Faraday Soc.* 1958;54:1731.

79. Poddubnyi IY, Erenburg EG, Eryomina MA. *Vysokomol. Soedin.* 1968;10A:1381.

80. Treloar LRG. *Trans. Faraday Soc.* 1947;43:284.

81. Kratky O, Sand H. *Kolloid-Z.* 1960;172:18.

82. Kurata M, Stockmayer WH. *Fortschr. Hochpolymer. Forsch.* 1963;3:196.

83. Wagner HL, Flory PJ. *J. Am. Chem. Soc.* 1952;74:195.

84. Mark JE. *J. Am. Chem. Soc.* 1966;88:4354.

85. Altgelt K, Schulz GV. *Makromol. Chem.* 1960;36:209.

86. Corbin N, Prudhomme J. *J. Polym. Sci., Polym. Phys. Ed.* 1977;15:1937.

87. Carter WC, Scott RL, Magat M. *J. Am. Chem. Soc.* 1946;68:1480.

88. Poddubnyi IY, Grechanovskii VA, Podalinskii AV. *Vysokomol. Soedin.* 1964;5:1588.

89. Abe M, Iwama M, Homma T. *Kogyo Kagaku Zasshi (J. Chem. Soc. Jpn. Ind. Chem. Sec.).* 1969;72:2313.

90. Kapur SL, Gundiah S. *Makromol. Chem.* 1958;26:119.

91. Le Fèvre RJW, Sundaram KMS. *J. Chem. Soc.* 1963:3547.

92. Gee G, Orr WJC. *Trans. Faraday Soc.* 1946;42:507.

93. Altgelt K, Schulz GV. *Makromol. Chem.* 1959;32:66.

94. Ng TS, Schulz GV. *Makromol. Chem.* 1969;127:165.

95. Schulz GV, Altgelt K, Cantow H-J. *Makromol. Chem.* 1956;21:13.

96. Xuexin C, Zhongde X, von Meerwall E, Seunge N, Hadjichristidis N, Fetters LJ. *Macromolecules.* 1984;17:1343.

97. Bywater S, Johnson P. *Trans. Faraday Soc.* 1951;47:195.

98. Booth C, Gee G, Holden G. *et al. Polymer.* 1964;5:343.

99. Eichinger BE, Flory PJ. *Trans. Faraday Soc.* 1968;64:2035.

100. Gee G. *J. Chem. Soc.* 1947;280.

101. Gee G, Herbert JBM, Roberts RC. *Polymer.* 1965;6:541.

102. Tewari YB, Schreiber HP. *Macromolecules.* 1972;5:329.

103. Booth C, Gee G, Williamson GR. *J. Polym. Sci.* 1957;23:3.

104. Poddubnyi IY, Grechanovskii VA, Podalinskii AV. *J. Polym. Sci., Part C.* 1968;16:3109.

105. Ansorena FJ, Revuelta LM, Guzmán GM. *et. al. Eur. Polym. J.* 1982;18:19.

106. Candau F, Strazielle C, Benoit H. *Makromol. Chem.* 1973;170:165.

107. Vavra J. *J. Polym. Sci., C.* 1967;16:1103.

108. Bristow GM, Westall B. *Polymer.* 1967;8:609.

109. Angulo-Sanchez JL, Gallegos A, Ponce-Vélez MA, Campos-López E. *Polymer.* 1977;18:922.

trans-1,4-Polyisoprene

Guru Sankar Rajan

Acronym, Alternative Names, Trade Names[1-7] *trans*-1,4-polyisoprene (TPIP), gutta percha, balata, TP 301

Class Diene elastomers

Structure

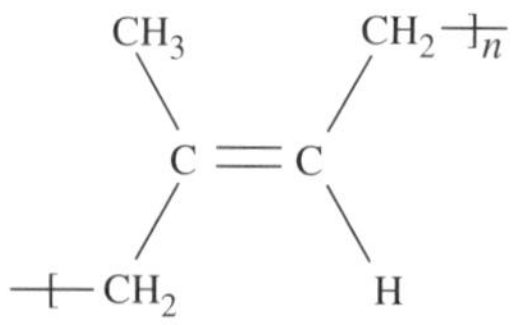

General Information[1-8] Gutta percha from Malaysia (Palaquim gutta and Dichopsis gutta), balata from Brazil (Bolle tree); hard, crystalline thermoplastic material; synthetic *trans*-1,4-polyisoprene (TP 301: Kuraray Co., Ltd., Japan).

Major Applications[1-9] Used in high-quality golf ball covers; in transmission belts, cable coverings, and adhesives; in prosthetics, braces, casts, and attachments for artificial limbs; and in dental applications.

Properties of Special Interest TPIP resists abrasion, scuffing, and cutting; it is a tough, rigid, durable, and lightweight polymer at room temperature; it can be extruded, calendered, injection molded, and compression molded; it can be compounded with fillers, and used in blends with other polymers; resistant to ozone, alkalies, fats, oils, and some concentrated acids except nitric acid and sulfuric acid.

Catalysts used for synthesis of TPIP[5,7,10-12]
AlR_3 or $AlR_2Cl + VCl_3$
AlR_3 + supported VCl_3
$AlR_3-VCl_3-Ti(OR)_4$
Allylsodium–sodium isoperoxide–sodium chloride
Sodium or potassium metals in *n*-heptane
Borohydridoneodymium catalysts—$Cp^{*\prime}Nd(BH_4)_2(THF)_2/Mg(nBu)_2$
$\{(Cp^{*\prime} = C_5Me_4nPr)\}$; $Nd(BH_4)_3(THF)_3/Mg(nBu)_2$

trans-1,4-Polyisoprene

Mark–Houwink parameters[13,14]

Polymer Type	Conditions	Molecular Weight Range ($M \times 10^{-4}$)	$K \times 10^3$ ml g^{-1}	a
Synthetic *trans*	Benzene, 32°C	8–140	43.7	0.65
Synthetic *trans* (98%)	Benzene, 30°C	14–77	18.1	0.72
Synthetic *trans* (98%)	Cyclohexane, 30°C	14–77	16.2	0.74
Synthetic *trans* (98%)	Hexane, 30°C	14–77	13.8	0.71
Synthetic *trans* (98%)	Toluene, 30°C	14–77	17.6	0.73
Gutta percha	Benzene, 25°C	0.2–5	35.5	0.71
Gutta percha	Dioxane, 47.7°C	0.2–5	191	0.50
Gutta percha	Propyl acetate, 60°C	10–20	232	0.50

Mooney viscosity[5,10]

Polymer Type	ML 1 + 4, 100°C	After 4-min Milling at 150°C	Reference
Natural balata	20–30	–	(10)
Synthetic balata	>150	24–36	(10)
TP301	30	–	(5)

Number average molecular weights and temperature coefficients of intrinsic viscosity of 97% TPIP (Polysar, Canada) in cyclohexane[15]

Fraction	10^{-5} M, g mol^{-1}	10^3 $d$ln $[\eta]$/d T, K^{-1}
1	2.650	−0.75
2	2.119	−0.41
3	1.370	−1.16
4	0.911	−0.72
5	0.390	−0.84

Effect of degree of crosslinking ($\nu/2V$, moles of crosslinks per cm^3 of unswollen network) on force temperature quantities (f_e/f) of TPIP[16,17]

($10^5 \nu/2V$)	f_e/f
5.41	−0.04 (±0.05)
9.47	−0.09 (±0.02)
10.5	−0.10 (±0.03)
13.3	−0.08 (±0.03)
16.5	−0.12 (±0.04)

Effect of type of deformation on f_e/f of TPIP[17]

Type of Deformation	f_e/f
Elongation	−0.11 (±0.04)
Torsion	−0.14 (±0.04)

Effect of dilution on f_e/f of TPIP[17]

Diluent	Volume Fraction of Polymer in Network v_2	f_e/f
None	1.00	$-0.10\ (\pm 0.05)$
Paraffin oil	0.40	$-0.13\ (\pm 0.02)$
Decalin (swelling equilibrium)	0.18	$-0.20\ (\pm 0.04)$

Unperturbed dimensions of *trans*-polyisoprene*[18,19]

Solvent	$K \times 10^3$ ml g^{-1}	$r_0/M^{1/2} \times 10^4$, nm	$r_{0f}/M^{1/2} \times 10^4$, nm	$S = r_0/r_{0f}$	$C_\infty = r_0^2/nl^2$
Propyl acetate (60°C)	232	970	703	1.38	7.20
Propyl acetate (60°C)	232	1030	703	1.47	
Dioxane (47.7°C)	191	910	703	1.30	6.35

* Calculated values of $r_{0f}/M^{1/2} \times 10^4$, nm mol$^{1/2}$ g$^{-1/2}$: $0{:}580/M_u^{1/2}$ and $0{:}290/m_{1/2}$, where r_0 is the unperturbed root mean square end-to-end distance, r_{0f} is the unperturbed root mean square end-to-end distance of the freely-rotating chain, s is the effect of steric hindrance on the average chain dimension, C_∞ is the characteristic ratio, M_u is the molecular weight of the repeating unit, and m is the average molecular weight per skeletal link.

Observed infrared frequencies (cm^{-1}) of α-, β-, and amorphous TPIP, and band assignments[20,21]

α-TPIP	β-TPIP	Amorphous TPIP	Type of Vibration	α-TPIP	β-TPIP	Amorphous TPIP	Type of Vibration
3018 sh	3022 sh	3015 sh	$\nu_s(=CH)$	1280 w	1280 w	1281 w	$\gamma(CH_2)$
2970 s	2979 sh	2975 sh	$\nu_{as}(CH_3)$	1260 w	1260 m	1253 w	$\gamma(CH_2)$
				1250 w			
2941 vs	2965 vs	2961 s	$\nu_{as}(CH_3)$	1205 m	1212 m	1220 w 1205 w	$\delta(=C-H)_{ip}$
2918 vs	2943 vs	2925 vs	$\nu_{as}(CH_2)$	1150 m–w	1150 m	1150 m	$\gamma(CH_2)$
2879 vs	2914 vs	2920 vs	$\nu_{as}(CH_2)$	1105 sh	1107 m	1099 m–s	$\nu(C-C)$
				1099 m			
2872 vs	2906 vs	2912 vs	$\nu_s(CH_3)$	1050 m	1058 vw	1033 m	$\gamma_r(CH_3) + \nu(C-C)$
2851 s	2855 s	2848 s	$\nu_s(CH_2)$	1030 m	997 m	987 m	$\nu(C-C) + \gamma_r(CH_3)$
2830 s	2846 s						
1672 m	1664 m	1665 m	$\nu(C=C)$	992 m	978 w	976 m	$\gamma_r(CH_3)$
1450 s	1450 s	1450 s	$\delta(CH_2); \delta_{as}(CH_3)$	882 s 862 s	877 s	884 sh 860 sh	$\delta(=C-H)_{op}$
						842 s	
1430 sh	1430 sh	1430 sh	$\delta(CH_2)$	800 s 780 sh	800 s	800 sh	$\nu(C-CH_3)$
1383 s	1384 s	1383 s	$\delta_s(CH_3)$	750 w 730 w	750 m–s	764 m 743 m	$\gamma_r(CH_2)$
		1360 m					
1340 m	1348 m	1329 m	$\alpha : \delta_s(CH_3) + \gamma(CH_2)$ $\beta : \delta_s(CH_3) + \nu(C-C)$	618 m 555 m	600 ms	595 m 570 m	$\delta(C=C-C)$
1320 w	1324 w	1307 sh	$\gamma(CH_2)$	490 m 470 m	474 s	520 m <505 m	$\delta(=C-CH_3)_{op}$

s: strong; m: medium; w: weak; sh: shoulder; ν: stretch; δ: bend; γ: twist, wag; γ_r: rock; as: asymmetrical; s: symmetrical; ip: in plane; op: out of plane.

Absorptivity values of gutta percha (98.7% *trans*) using near infrared spectra[22]

Polymer	Concentration, g per 50 ml	$A_{2.46\mu}$	$A_{2.315\mu}$	$A_{2.46\mu}/A_{2.315\mu}$
Gutta percha	0.5099	0.060	0.615	0.0975

Raman spectra: frequencies and band assignments[23]

Wave Number, cm^{-1}	Assignment	Wave Number, cm^{-1}	Assignment
3010	$\nu(=CH)$	1380	$\delta_s(CH_3)$
2969	$\nu_{as}(CH_3)$	1354	$\gamma_w(CH_2)$
2879	$\nu_s(CH_3)$	1323	$\delta(=CH)_{ip}$
2839	$\nu_{op}(CH_2)$	1148	$\nu(C-C)$
1662	$\nu(C=C)$	1032	$\gamma_r(CH_3)$
1434	$\delta(CH_2)$		

Film cast from *trans*-1,4-polyisoprene (Polymer Corp)/hexane solvent.

ν: stretch; δ: bend; γ_w: wag; γ_r: rock; as: asymmetrical; s: symmetrical; ip: in plane; op: out of plane.

Carbon-13 spin relaxation times, Overhauser enhancement factors, and chemical shift for TPIP-balata (α-form) at 40°C[24]

Carbon[a]	T_1, sec	$1 + ((C_Z-C_0)/C_0)$	Chemical Shift, ppm
1	0.66	2.79	39.85
2	7.79	2.37	134.96
2'	3.80	2.58	16.10
3	1.31	2.79	124.43
4	0.68	2.77	26.88

[a]: $-CH_2CCH_3=CHCH_2-$: 122'34; in CDCl$_3$

$M_n = 290,000$; $M_w/M_n = 1.4$; density $= 0.975$ g cm^{-3}.

X-ray diffraction:[25-31] *d*-spacing and intensity (Cu-K radiation)

α-Gutta Percha[25]		β-Gutta Percha[25]	
d Spacing, Å	Intensity	*d* Spacing, Å	Intensity
12.2	Very weak	4.75β	Medium
4.97β	Medium	4.73	Very strong
4.96	very strong	3.90β	Medium
4.56	Strong	3.89	Strong
3.94	Strong	3.29	Very weak
3.32	Strong	2.95	Medium
2.98	Weak	2.77	Weak
2.74	Weak		

Miller indices and calculated intensity of β-gutta percha[29,30]

hkl	I_{calc} (29)	I_{calc} (30)	*hkl*	I_{calc} (29)	I_{calc} (30)	*hkl*	I_{calc} (29)	I_{calc} (30)
120	325	538	240	12.2	11	211	32.8	69
200	207	295	160	28.6	19	411	13.6	14
040	24.8	39	011	74.0	93	112	26.0	1
140	38.1	59	121	10.7	10	022	52.0	10
320	9.8	17	031	15.9	24	202	11.2	8
						222	29.1	12

Unit cell dimensions[32]

Polymer Type	Crystal System and Space Group	Cell Dimension, Å		
		a	*b*	*c*
Gutta percha-α	Monoclinic, P2$_1$/c	7.98	6.29	8.77
Gutta percha-β	Orthorhombic, P2$_1$2$_1$2$_1$	7.78	11.78	4.72

Glass transition temperatures (T_g) by differential scanning calorimeter[33]

Sample Type	Glass Transition Temperature, K		ΔC_p, cal g^{-1} K^{-1}	ΔH_{fus}, cal g^{-1}
	Onset	Midpoint		
Amorphous				
Gutta percha	203.0 ± 0.1	205.0 ± 0.1	0.107 ± 0.003	
Balata	201.8	203.7	0.101	
TPIP	201.6	203.4	0.103	
Crystalline				
Gutta percha	204.4 ± 0.5	207.9 ± 0.5	0.027 ± 0.005	12.1
Balata	202.3 ± 0.3	205.4 ± 0.5	0.023 ± 0.004	10.8
TPIP	202.8	206.8	0.042	10.8

Some more T_g values (references/methods to obtain T_g are given in parentheses)[7,33−35]

Sample Type	Glass Transition Temperature, K	
Gutta percha	205 (34)	
Balata	204 (34)	213 (33, dilatometry)
TPIP (synthetic or natural)	203 (34), 213 (35)	203, 207, 213 (33, DSC)
Synthetic TPIP	202 (7, DSC)	200 (33, DMA)

trans-1,4-Polyisoprene

Thermodynamic quantities determined using diluent equation[36]

Polymer	Equilibrium Melting Point T_m^0, K	Enthalpy of Fusion per Repeating Unit ΔH_u, J mol^{-1}	$\Delta H_u/M_0$, J g^{-1}	Entropy of Fusion ΔS_u, J (K mol)$^{-1}$
Gutta percha-α	360.2	12719	187.0	35.3
Gutta percha-β	354.2	10544	155.1	29.8

Melting temperatures*[5,7,10,35]

Polymer Type	Melting Temperature, K
Natural balata	329, 336 (10, dilatometry)
Natural balata	340 (5)
Synthetic balata	329, 335 (10, prepared from benzene, dilatometry)
Synthetic balata	324, 334 (10, prepared from heptane, dilatometry)
Synthetic TPIP	333 (35)
TP301	340 (5)
Synthetic TPIP	320 (7)

* References/methods to obtain melting temperatures are given in parentheses.

Theta temperatures[37]

Solvent	Theta Temperature, K	$K_\theta \times 10^4$, dl g^{-1} (g mol. wt.)$^{-1/2}$
n-Propyl acetate*	333	
Toluene/*n*-propanol		
(68.4/31.6)	298	22.2
(67.6/32.4)	303	21.9
(66.5/33.5)	308	21.7
(65.8/34.2)	313	21.4
(64.5/35.5)	318	21.3
(63.8/36.2)	323	21.1
Dioxane	320.7	19.1

* For gutta percha; rest of data for other 96% *trans*-polymer.

Property	Units	Conditions	Value	Reference
Surface tension γ	mN m^{-1}	20°C, contact angle	31	(38)
Solubility parameter δ	(MPa)$^{1/2}$	calculated	16.6	(38)
Dipole moment per monomer unit $(\mu^2 N^{-1})^{1/2}$	D	benzene, 25°C, number average degree of polymerization = 3125	0.31	(39)
Solvents		gutta percha	Hot petroleum ether, benzene, chloroform	(1–12, 34, 35)
Nonsolvents		gutta percha	Alcohol, water	(1–12, 34, 35)

Crystallization constants of gutta percha[40]

Isothermal Crystallization Temperature T_C, K	Half Time of Crystallization $t_{1/2}$, s
308	768
313	1260
318	6780
324	31800
330	291000

Lamellar thickness (L), crystalline stem length (L_C), and average number of monomer units per fold and interlamellar traverse (U) for TPIP (α-form) (1% amyl acetate solution)[41]

Crystallization Temperature T_C, °C	$M_n \times 10^{-5}$	L, nm (Electron Microscopy)	L_C, nm (From L, Calculated)	U
10	2.5	7.8	3.2	8.9
20	1.0	9.0	4.3	9.0
	2.5	9.0	4.1	9.7
25	1.0	11.5	5.8	11.0
	2.5	11.5	5.2	12.0
30	2.5	12.5	6.0	13.0

Density (g cm^{-3}) of TPIP (balata) crystals grown from dilute amyl acetate solution at different crystallization temperatures (T_C) and molecular weights[42]

T_C, °C	$M_n \times 10^{-5}$*								
	3.8	2.9	2.8	2.3	2.1	1.1	0.69	0.26	Av.
10	0.970	0.970 (2)	0.968 (2)			0.963			0.969
20	0.974 (5)	0.975 (5)	0.970 (2)	0.972 (2)	0.973	0.974	0.973	0.978 (2)	0.974
30	0.982	0.978 (4)	0.972 (2)		0.970		0.976	0.981	0.977

* Number of samples used given in parentheses if more than one.
Density (g cm^{-3}):[41] ρ (amorphous) = 0.905; ρ (crystalline, α form) = 1.05; ρ (crystalline, β form) = 1.02.

Optical properties[43]

Polymer Type	Refractive Index[a]	Optical Configuration Parameter $\Delta\alpha$, Å^3	Stress Optical Coefficient C, 10^9 Pa^{-1}
Gutta percha-α	1.514		
Gutta percha-β	1.509		
TPIP		49 (diluent: benzene)[b]	
Gutta percha crosslinked			3.0 (temperature 85–250°C)[b]

[a] Wavelength—5893 Å.
[b] Wavelength—6328 Å.

trans-1,4-Polyisoprene

Mechanical properties of balata[2,10]

Properties	Natural Balata	Synthetic Balata
Tensile strength, psi at 23°C	3860	4400
Elongation at break, % at 23°C	440	480
Tear strength, kg cm^{-1}	21.0	20.5
Torsional modulus, psi		
50°C	666	384
25°C	37000	33700
0°C	49800	44500
–30°C	79500	61200

Hot strength of cured and uncured TPIP[2]

Properties	Cured	Uncured
Tensile strength, psi	64	9
Elongation at break, %	1600	1000

Trans-1,4-Polyisoprene	Reference
Synthesis	(5, 7, 10–12, 44, 51)
XRD	(2, 25–32, 41, 42, 52)
IR	(2, 20–22, 44–46)
NMR	(7, 24, 47–53)
Morphology	(41, 42, 52, 54)
Microscopy	(41, 42, 52, 54, 55)
Viscosity, molecular weight	(2, 7, 10, 13–15)
Moments of end-to-end vectors, molecular parameters/entanglements, force/thermodynamic constants, rotational isomeric state, conformation analysis	(15–19, 56, 57, 59)
Melting entropy—PVT data	(58)
Mechanical properties	(2, 10)
Optical properties	(43)
Thermal transition, thermal behavior	(7, 18, 33–37, 41, 42, 52)
Solution properties	(1–19, 33–39)
Crystallization	(29, 31, 33, 40–42, 52, 54)
Simulation	(59)
Degradation with enzyme mediator systems	(60)

References

1. Dean JN. In: Davis CC, ed. *Chemistry and Technology of Rubber.* New York: Reinhold Publishing; 1937:705–719.
2. Kent EG, Swinney FB. *Ind. Eng. Chem. Prod. Res. Dev.* 1966;5:134.

3. Miles DC, Briston JH. *Polymer Technology*. New York: Chemical Publishing; 1979:447.

4. Senyek ML. In: Mark HF, et al, eds. *Encyclopedia of Polymer Science and Engineering*. 2nd ed. Vol 8. New York: John Wiley and Sons; 1989:499.

5. Senyek ML. In: Howe-Grant M, ed. *Kirk-Othmer Encyclopedia of Chemical Technology*. 4th ed. Vol 9. New York: John Wiley and Sons; 1994:1–14.

6. Brydson JA. *Plastics Materials*. 6th ed. Oxford: Butterworth-Heinemann; 1995:844.

7. Bonnet F, et al. *Macromolecules*. 2005;38:3162.

8. Sperling LH, Carraher CE. In: Mark HF, et al, eds. *Encyclopedia of Polymer Science and Engineering*. Vol 12. 2nd ed. New York: John Wiley and Sons; 1989:668.

9. Rootare HM, Powers JM. *J. Dental Res*. 1977;56:1453.

10. Lasky JS, *et al. Ind. Eng. Chem. Prod. Res. Dev*. 1962;1:82.

11. Pasquon I, Giannini U. In: Mark HF, et al, eds. *Encyclopedia of Polymer Science and Engineering*. 2nd ed. Vol 15. New York: John Wiley and Sons; 1989:674.

12. Senyek ML. In: Mark HF, et al, eds. *Encyclopedia of Polymer Science and Engineering*. 2nd ed. Vol 8. New York: John Wiley and Sons; 1989:516.

13. Chaturvedi PN, Patel CK. *J. Polym. Sci., Polym. Phys*. 1985;23:1255.

14. Cooper W, et al. *J. Polym. Sci*. 1962;59:241.

15. Fernandez-Berridi MJ, et al. *Macromolecules*. 1980;13:190.

16. Mark JE. *J. Am. Chem. Soc*. 1967;89:6829.

17. Erman B, Mark JE. *Structures and Properties of Rubberlike Networks*. New York: Oxford University Press; 1997:116–121.

18. Wagner HL, Flory PJ. *J. Am. Chem. Soc*. 1952;74:195.

19. Kurata M, Tsunashima Y. In: Brandrup J, Immergut EH, eds. *Polymer Handbook*. 3rd ed. New York: John Wiley and Sons; 1989:VII 32–33.

20. Gavish M, et al. *Macromolecules*. 1988;21:2075.

21. Gavish M, et al. *Macromolecules* 1988;21:2079.

22. Corish PJ. *Rubber Chem. Technol*. 1960;33:975.

23. Cornell SW, Koenig JL. *Macromolecules*. 1969;2:546.

24. Schilling FC, *et al. Macromolecules*. 1985;18:2688.

25. Hopff H, Susich GV. *Rubber Chem. Technol*. 1931;4:75.

26. Stillwell CW, Clark GL. *Ind. Eng. Chem*. 1931;23:706.

27. Bruni G, Natta G. *Rubber Chem. Technol*. 1934;7:603.

28. Fuller CS. *Ind. Eng. Chem*. 1936;28:907.

29. Bunn CW. *Proc. Royal Soc. Series A*. 1942;180:40.

30. Jeffrey GA. *Trans. Faraday Soc*. 1944;40:517.

31. Mandelkern L, et al. *J. Am. Chem. Soc*. 1956;78:926.

32. Patterson DJ, Koenig JL. *Polymer*. 1988;29:240.

33. Burfield DR, Lim K-L. *Macromolecules*. 1983;16:1170.

34. Senyek ML. In: Mark HF, et al. *Encyclopedia of Polymer Science and Engineering*. 2nd ed. Vol 8. New York: John Wiley and Sons; 1989:506.

35. Senyek ML. In: Mark HF, et al, eds. *Encyclopedia of Polymer Science and Engineering*. 2nd ed. Vol 8. New York: John Wiley and Sons; 1989:550.

36. Mandelkern L, Alamo RG. In: Mark JE, ed. *Physical Properties of Polymers Handbook*. Woodbury, NY: AIP Press; 1996:119.

37. Ansorena FJ, et al. *Euro. Polym. J*. 1982;18:19.

38. Lee L-H. *J. Poly. Sci., Part A-2*. 1967;5:1103.

39. Fevre RJW, Sundaram KMS. *J. Chem. Soc*.1963:3547.

40. Mezghani K, Phillips PJ. In: Mark JE, ed. *Physical Properties of Polymers Handbook*. Woodbury, NY: AIP Press; 1996:417.

41. Kuo C-C, Woodward AE. *Macromolecules*. 1984;17:1034.

42. Anandakumaran K, et al. *Macromolecules.* 1983;16:563.

43. Galiatsatos V, et al. In: Mark JE, ed. *Physical Properties of Polymers Handbook.* Woodbury, NY: AIP Press; 1996:535.

44. Richardson WS, Sacher A. *J. Polym. Sci.* 1953;10:353.

45. Binder JL, Ransaw HC. *Anal. Chem.* 1957;29:503.

46. Golub MA. *J. Polym. Sci.* 1959;36:523.

47. Chen HY. *Anal. Chem.* 1962;34:1793.

48. Golub MA, et al. *J. Am. Chem. Soc.* 1962;84:4981.

49. Duch MW, Grant DM. *Macromolecules.* 1970;3:165.

50. Schaefer J. *Macromolecules.* 1972;5:427.

51. Sato H, Tanaka Y. *J. Polym. Sci. Polym. Chem.* 1979;17:3551.

52. Xu J-R, Woodward AE. *Macromolecules.* 1988;21:83.

53. Oldfield E, et al. *Macromolecules.* 1992;25:3027.

54. Xu J-R, Woodward AE. *Macromolecules.* 1986;19:1114.

55. Winesett DA, et al. *Rubber Chem. Technol.* 2003;76:803.

56. Tonelli AE. *Macromolecules.* 1990;23:3129.

57. Herman MF. *Macromolecules.* 2001;34:4580.

58. Naoki M, Tomomatsu T. *Macromolecules.* 1980;13:322.

59. Faller R, et al. *Macromolecules.* 2001;34:1436.

60. Enoki M, et al. *Biomacromolecules.* 2003;4:314.

Poly(N-isopropyl acrylamide)

N.A. Peppas

Acronyms, Trade Names PNIPA, PNIPAAm

Class Vinyl polymer

Structure $[CH_2\!-\!CH]_n$
$$| $$
$$C=O$$
$$| $$
$$NH$$
$$| $$
$$CH$$

$$CH_3 \qquad CH_3$$

Major Applications Membranes, chromatographic resins, size exclusion particles, drug delivery systems, actuators, intelligent systems, separation units.

Properties of Special Interest Exhibition of low critical solubility temperature in water at 32°C provides for interesting applications in separation science. This temperature maybe modified with copolymerization.

Property	Units	Conditions	Value	Reference
Density, ρ	g/cm^3	dry state	1.386	(1)
Polymer-water interaction parameter, χ_1	–	Varying pressures, P (in MPa)	$\chi_1 = 0.505 - 1.39 \times 10^{-3}\,P$	(2)
		20°C	0.51	(1)
		40°C	0.95	(1)
		25°C	0.518	(3)
Polymer-methanol interaction parameter, χ_1	–	25°C	0.45	(4)
Lower critical solution temperature, T_c	°C	hydrogels	30–35	(5)
		aqueous solution	31	(6, 7)
		copolymers	30–33	(8)
Intrinsic viscosity, $[\eta]$	dL/g	aqueous solution at		
		$T = 15$°C	2.74	(6)
		$T = 25$°C	1.70	(6)
		$T = 33$°C	1.44	(6)

Poly(N-isopropyl acrylamide)

Property	Units	Conditions	Value	Reference
Mark-Houwink coefficient, a	–	solution in water at		
		$T = 15°C$	0.93	(9)
		$T = 25°C$	0.97	(9)
		solution in methanol		
		at $T = 25°C$	0.64	(9)
Refractive index, n	–	dry polymer	1.5	(10)
		swollen polymer	1.36	(10)
Storage modulus, E' (compressive)	MPa	polymer gel at		
		$T = 30°C$	0.1	(11)
		$T = 60°C$	1.3	(11)
Loss modulus, E'' (compressive)	MPa	polymer gel at		
		$T = 30°C$	0.005	(11)
		$T = 60°C$	1.05	(11)

References

1. Bae YH, Okano T, Kim SW. *J. Polym. Sci., Polym. Phys.* 1990:28:923.
2. Nakamoto C, Kitada T, Kato E. *Polym. Gels & Networks.* 1996;4:17.
3. Hirotsu S. *J. Chem. Phys.* 1991;94:3949.
4. Tao CT, Young TH. *Polymer.* 2005;46:10077.
5. Shibayama M, Mizutani S, Nomura S. *Macromolecules.* 1996;29:2019.
6. Heskins M, Guillet JE. *J. Macromol. Sci., Chem.* 1968;A2:1441.
7. Pelton RH, Pelton HM, Morphesis A, Rowell RL. *Langmuir.* 1989;5:816.
8. Zhang J, Peppas NA. *Macromolecules.* 2000;33:102.
9. Chiantore O, Guaita M, Trossarelli L. *Makromol. Chem.* 1979;180:969.
10. Zhou S, Wu C. *Macromolecules.* 1996;29:4998.
11. Shibayama M, Morimoto M, Nomura S. *Macromolecules.* 1994;27:5060.

Poly(lactic acid)

Lichun Lu and Antonios G. Mikos

Acronyms, Trade Names PLA

Class Poly(α-hydroxy ester)

Structure

$$\left[\!\!\left[O-\underset{\underset{CH_3}{|}}{\overset{\overset{H}{|}}{C}}-\overset{\overset{O}{\|}}{C} \right]\!\!\right]$$

Major Applications L-PLA is used as sutures and dental, orthopaedic, and drug delivery devices. D,L-PLA is used mainly for drug delivery. Both are of interest in the area of tissue engineering.

Properties of Special Interest Good biocompatibility; biodegradable mainly by simple hydrolysis; bioresorbable; very good processability; a wide range of degradation rates, physical, mechanical, and other properties can be achieved by PLA of various molecular weights and its copolymers.

Preparative Techniques Practically useful high molecular weight PLA can be synthesized by a cationic ring opening polymerization of lactide using antimony, zinc, lead, or tin as catalyst and alcohol as molecular weight and reaction rate control agent at high temperature and low pressure.

IR of 1% w/v D,L-PLA in chloroform[1]

Structure	Absorption Frequency (cm^{-1})
OH, alcohol and carboxylic acid	3,700–3,450
C=O	1,750–1,735
COO	1,600–1,580
C—O	1,200–1,000
CH	950–700

^{1}H NMR of 10% w/v D,L-PLA in deuteriochloroform[1]

Structure	Chemical Shift (ppm) and Peak Multiplicity
OH	7.30, s
CH—CH$_3$	5.20, m
CH—CH$_3$	1.55, d

Poly(lactic acid)

^{13}C NMR of 10% w/v D,L-PLA in deuteriochloroform[1]

Structure	Chemical Shift (ppm)
C=O	169.3
C—O	69.0
CH$_3$	16.7

Unit cell dimensions of L-PLA

Lattice	Monomers per Unit Cell	Cell Dimension (Å)			Cell Angles* γ (Degree)	Chain Conformation ρ_n of Helix	Reference
		a	b	c (Fiber Axis)			
Hexagonal	–	5.9	5.9	–	120	–	(2)
Orthorhombic	–	10.31	18.21	9.00	90	3$_1$	(3)
Pseudo-orthorhombic	20	10.34	5.97	–	90	10$_3$	(2)
Pseudo-orthorhombic	20	10.6	6.1	28.8	90	10$_3$	(3)
Pseudo-orthorhombic	20	10.7	6.45	27.8	90	10$_3$	(4)

* Cell angles $\alpha = \beta = 90°$

Property	Units	Conditions	Value	Reference
Degree of crystalline X_c	%	D-PLA	semicrystalline	(5)
		L-PLA	0–37	(6)
		D,L-PLA	amorphous	(6)
Density ρ	g cm^{-3}	P(L-co-DL)LA		(7)
		amorphous	1.248	
		single crystal	1.290	
Heat of fusion ΔH_f	kJ mol^{-1}	L-PLA complete crystalline	146	(8)
		L-PLA fiber		(9)
		as-extruded	2.5	
		after hot-drawing	6.4	
Heat capacity C_p	J K^{-1} g^{-1}	L-PLA of		(10)
		$M_v = 5,300$	0.60	
		$M_v = 20,000–691,000$	0.54	
Glass transition temperature T_g	K	L-PLA of various molecular weights with dichloromethane	326–337	(5, 6, 8, 10, 11)
			–	(12)
		D,L-PLA of various molecular weights with dichloromethane	323–330	(1, 5, 8)
			–	(12)
Melting point T_m	K	D-PLA injection-molded $M_v = 21,000$	444.4	(13)
		L-PLA of various molecular weights	418–459	(5, 6, 8, 10, 11)
Equilibrium melting point	K	L-PLA of $M_v = 550,000$	488	(2)
		L-PLA	480	(14)

Property	Units	Conditions	Value	Reference
Secondary relaxation temperature	K	β-relaxation at 1 Hz (95% L-isomer)	228	(15)
Decomposition temperature T_d	K	L-PLA of $M_w = 50,000–300,000$	508–528	(5)
		D,L-PLA of $M_w = 21,000–550,000$	528	(5)

Mark-Houwink parameters: K and a

Isoform	Solvent	Temperature (°C)	Conditions	$K \times 10^3$ (ml g^{-1})	a	Reference
Atactic	Benzene	30	For M_v	2.27	0.75	(16)
Atactic	Chloroform	25	For M_n	6.60	0.67	(1)
Atactic	Chloroform	25	For M_w	6.06	0.64	(1)
Atactic	Chloroform	25	For M_v	1.33	0.79	(12)
Atactic	Chloroform	30	For M_v	2.21	0.77	(16)
Atactic	Ethyl acetate	25	For M_n	1.58	0.78	(1)
Atactic	Ethyl acetate	25	For M_w	1.63	0.73	(1)
Isotactic	Benzene	30	For M_v	5.72	0.72	(16)
Isotactic	Chloroform	25	For M_v	2.48	0.77	(12)
Isotactic	Chloroform	30	For M_v	5.45	0.73	(16)

Property	Units	Conditions	Value	Reference
Solvent	–	Acetone at room temperature	–	(17)
		Benzene at $T = 30°C$	–	(16)
		Bromobenzene at $T = 85°C$	–	(18)
		Chloroform at $T = 25°C$	–	(12)
		m-cresol at room temperature	–	(18)
		Dichloroacetic acid	–	(18)
		Dichloromethane at $T = 25°C$	–	(9)
		Dioxane at $T = 25°C$	–	(12)
		Dimethylformamide	–	(17)
		Ethyl acetate at $T = 25°C$	–	(1)
		Isoamyl alcohol	–	(2)
		N-methyl pyrrolidone	–	(12)
		Toluene	–	(19)
		Tetrahydrofuran	–	(1)
		Trichloromethane at $T = 25°C$	–	(9)
		p-xylene	–	(2)
Equilibrium dissolution temperature	K	L-PLA in p-xylene	399.5	(2)
Cloud point temperature	K	D,L-PLA in dioxane/water mixture	–	(12)
Interaction parameter χ	–	L- or D,L-PLA in dioxane or chloroform at $T = 25°C$	0.1–0.3	(12)

Poly(lactic acid)

Property	Units	Conditions	Value	Reference
Swelling	%	L- or D,L-PLA in methanol or water	–	(12)
		L-PLA film in 0.2 M pH7 buffer	2	(6)
Second virial coefficient A_2	mol cm^3 g^{-2}	L-PLA in bromobenzene at 85°C $M_w = 80{,}000$–490,000	3.5–3.2×10^{-4}	(18)
Huggins constant	–	atactic or isotactic in benzene	0.33–0.41	(16)
Steric hindrance parameter	–	atactic in benzene or chloroform	1.98	(16)
		isotactic in benzene or chloroform	2.69	(16)
Characteristic ratio $\langle r^2 \rangle_0 / nl^2$	–	L-PLA in bromobenzene at 85°C	2.0	(18)
Intrinsic viscosity $[\eta]$	dl g^{-1}	L-PLA		
		in chloroform at $T = 25$°C	3.8–8.2	(19)
		in chloroform at $T = 30$°C	2.63	(18)
		in bromobenzene at $T = 85$°C	1.38	(18)
		in chloroform	4.2	(12)
		in dioxane	3.2	(12)
		in N-methyl pyrrolidone	2.3	(12)
		L- or D,L-PLA in mixture of		
		chloroform/methanol	–	(12)
		dioxane/water	–	(12)
		D,L-PLA in chloroform at $T = 25$°C	0.1–1.5	(1)
Nucleation constant K_g	–	L-PLA crystallized from melt		
		isothermal	2.44×10^5	(14)
		non-isothermal	2.69×10^5	(20)
Fold surface energy σ_e	J m^{-2}	L-PLA crystallized from melt	60.89×10^{-3}	(14)
		L-PLA single crystals in p-xylene	75×10^{-3}	(2)
Lateral surface energy σ	J m^{-2}	L-PLA crystallized from melt		
		isothermal	12.03×10^{-3}	(14)
		non-isothermal	13.6×10^{-3}	(20)
Refractive index increment dn/dc	ml g^{-1}	L-PLA		
		in bromobenzene at $T = 85$°C	−0.06	(18)
		in tetrahydrofuran	0.0558	(21)
Optical rotation $[\alpha]_D$	Degrees	L-PLA at T $= 25$°C		
		in chloroform	−151	(18)
		in chloroform at $\lambda = 589$ nm	−161	(19)
		in dichloromethane	−162.8	(16)
		in chlorofor/methanol mixture	−158 to −173	(12)
		in dioxane/water mixture	−165	(12)
		L-PLA in p-dioxane at $\lambda = 365$ nm	−443	(7)
Degradation rate	–	in vitro L-PLA	–	(17, 22–24)
		D,L-PLA	–	(17)
		in vivo D-PLA	–	(13)
		L-PLA	–	(13,17)
		D,L-PLA	–	(13,17)
		Mechanism	–	(25)

L-PLA ($[\eta] = 16$ cm^3 g^{-1} in benzene at $T = 30°$C) under ^{60}Co radiation[26]

Atmosphere	Dose (M Gy)	Chain scission G factor	Crosslinking G factor
N$_2$	Low (<2.5)	26.5	4.5
N$_2$	Pregel region (>2.5)	40.5	11.0
air	Low (<2.5)	14.5	0.4
air	Pregel region (>2.5)	23.0	6.5

Property	Units	Conditions	Value	Reference
Tensile strength	MPa	L-PLA film or disk	28–50	(5)
		$M_w = 50,000–3000,000$		
		L-PLA melt-spun fiber	up to 870	(27)
		L-PLA solution-spun fiber from		
		toluene	up to 1000	(19)
		trichloromethane	up to 1200	(9)
		chloroform/toluene mixture	up to 2300	(3)
		D,L-PLA film or disk	29–35	(5)
		$M_w = 107,000–550,000$		
Tensile modulus	MPa	L-PLA film or disk	1200–3000	(5)
		$M_w = 50,000–300,000$		
		L-PLA melt-spun fiber	up to 9200	(27)
		L-PLA solution-spun fiber from		
		toluene	up to 10,000	(19)
		trichloromethane	12,000–15,000	(9)
		chloroform/toluene mixture	up to 16,000	(3)
		D,L-PLA film or disk	1900–2400	(5)
		$M_w = 107,000–550,000$		
Tensile storage modulus E'	MPa	L-PLA varies with temperature		
		melt-spun monofilament at 1 Hz	–	(28)
		injection-molded bar at 3 Hz	–	(10)
		D,L-PLA varies with temperature		
		film at 110 Hz	–	(8)
		film at 11 Hz	–	(8)
Tensile loss modulus E''	MPa	L-PLA varies with temperature		
		melt-spun monofilament at 1 Hz	–	(28)
		film or injection-molded bar at various frequencies	–	(10)
		D,L-PLA varies with temperature		
		film at 110 Hz	–	(8)
		film at 11 Hz	–	(8)
Flexural storage modulus	MPa	L-PLA film or disk	1400–3250	(5)
		$M_w = 50,000–300,000$		
		D,L-PLA film or disk	1950–2350	(5)
		$M_w = 107,000–550,000$		

Poly(lactic acid)

Property	Units	Conditions	Value	Reference
Shear strength	MPa	L-PLA pin	54.5	(29)
Shear modulus	MPa	L-PLA melt-spun monofilament	1210–1430	(28)
Bending strength	MPa	L-PLA pin	132	(29)
Bending modulus	MPa	L-PLA pin	2800	(29)
Elongation at yield	%	L-PLA film or disk $M_w = 50,000–300,000$	3.7–1.8	(5)
		D,L-PLA film or disk $M_w = 107,000–550,000$	4.0–3.5	(5)
Elongation at break	%	L-PLA film or disk $M_w = 50,000–300,000$	6.0–2.0	(5)
		L-PLA fiber spun from toluene	12–26	(19)
		L-PLA melt-spun fiber $M_v = 180,000$	25	(19)
		D,L-PLA film or disk $M_w = 107,000–550,000$	6.0–5.0	(5)

References

1. Rak J, Ford JL, Rostron C, Walters V. *Pharm. Acta Helv.* 1985;60:162.
2. Kalb B, Pennings AJ. *Polymer.* 1980;21:607.
3. Hoogsteen W, et al. *Macromolecules.* 1990;23:634.
4. De Santis P, Kovacs AJ. *Biopolymers.* 1968;6:299.
5. Engelberg I, Kohn J. *Biomaterials.* 1991;12:292.
6. Gilding DK, Reed AM. *Polymer.* 1979;20:1459.
7. Fischer EW, Sterzel HJ, Wegner G. *Kolloid-Z. U. Z. Polymere.* 1973;251:980.
8. Jamshidi K, Hyon S-H, Ikada Y. *Polymer.* 1988;29:2229.
9. Gogolewski S, Pennings AJ. *J. Appl. Polym. Sci.* 1983;28:1045.
10. Celli A, Scandola M. *Polymer.* 1992;33:2699.
11. Cohn D, Younes H, Marom G. *Polymer.* 1987;28:2018.
12. Van De Witte P, Dijkstra PJ, Van Den Berg JWA, Feijen J. *J. Polym. Sci., Polym. Phys.* 1996;B34:2553.
13. Gogolewski S, et al. *J. Biomed. Mater. Res.* 1993;27:1135.
14. Vasanthakumari R, Pennings AJ. *Polymer.* 1983;24:175.
15. Starkweather HW Jr, Avakian P, Fontanella JJ, Wintersgill MC. *Macromolecules.* 1993;26:5084.
16. Schindler A, Harper D. *J. Polym. Sci. Polym. Chem. Ed.* 1979;17:2593.
17. Suggs LJ, Mikos AG. In: Mark JE, ed. *Physical Properties of Polymers Handbook.* Woodbury, N.Y: American Institute of Physics Press; 1996:615–624.
18. Tonelli AE, Flory PJ. *Macromolecules.* 1969;2:225.
19. Eling B, Gogolewski S, Pennings AJ. *Polymer.* 1982;23:1587.
20. Kishore K, Vasanthakumari R. *Colloid Polym. Sci.* 1988;266:999.
21. Sosnowski S, Gadzinowski M, Slomkowski S. *Macromolecules.* 1996;29:4556.
22. Cam D, Hyon S-H, Ikada Y. *Biomaterials.* 1995;16:833.
23. Vert M, Li SM, Garreau H. *J. Biomater. Sci. Polym. Ed.* 1994;6:639.
24. Lu L, et al. *Biomaterials.* 2000;21:1595.
25. Göpferich A. *Biomaterials.* 1996;17:103.

26. Gupta MC, Deshmukh VG. *Polymer.* 1983;24:827.
27. Fambri L, et al. *Polymer.* 1997;38:79.
28. Agrawal CM, Haas KF, Leopold DA, Clark HG. *Biomaterials.* 1992;13:176.
29. Gogolewski S, Mainil-Varlet P. *Biomaterials.* 1996;17:523.

Polymeric Selenium

Stephen J. Clarson

Class Inorganic and semi-inorganic polymers

Selenium and its Applications Selenium derives its name from the Greek name for the moon (Selene)—in part due to its silvery appearance (Atkins). The most stable form of this element is grey selenium (metallic) which has a melting point of 494K. This trigonal grey selenium consists of parallel spiral chains of Se atoms which repeat after three atoms. Red crystalline selenium Se_8 rings (Se_{alpha} and Se_{beta}) can be obtained by crystallization from a carbon disulphide solution of black selenium. Black selenium is formed by pouring molten selenium in water. Detailed calculations on the conformations of polymeric selenium chains and on their ring-chain equilibrium distributions have been described by Semlyen. In the gas phase above the boiling point the dominant species is Se_2. Grey selenium has an electrical conductivity which increases with temperature and thus exhibits semiconducting behavior.

Major Applications Selenium has also been demonstrated to be a photoconductor and hence finds applications as selenium photocells used for the measurement of the intensity of light.

Selected properties

Properties	Values
Atomic number	34
Atomic Mass	78.96 g mol^{-1}
Bond Length l_{Se-Se}	2.34 Å
Bond Energy Se—Se	172 kJ mol^{-1}
Bond Energy Se=Se	290 kJ mol^{-1}
Bond Angle Se—Se—Se in Se_8 ring	106°
van der Walls radius	1.90 Å
Glass Transition Temperature α-Se	310 K
Enthalpy Relaxation Time (at T_g)	200 s
Activation Energy Stress Relaxation E_s	485 kJ mol^{-1}
Kauzmann temperature α-Se	240 K
Melting Point Grey Se	494 K
Young's Modulus α-Se	9.8 GPa
Density (220°C)	4.06 g cm^{-3}
Crystal Structure	Hexagonal $a = 4.3640$ Å, $c = 4.9594$ Å
Standard Entropy of Cyclization (polymeric selenium to cyclooctaselenium)	-5.5 cal deg^{-1} mol^{-1}
Enthalpy Change for the Formation of Cyclooctaselenium	-2.3 kcal mol^{-1}
Boiling point Se	680°C

Selected Bibliography

1. Brieglieb GZ. *Phys. Chem.* 1929;A 144:321.
2. Astakhov KV, Penin NA, Dobkina EI. *Chem, Abstr.* 1948;42:11.
3. Semlyen JA. *Trans. Faraday Soc.* 1967;63:743.
4. Semlyen JA. *Trans. Faraday Soc.* 1967;63:2342.
5. Massey AJ. *The Typical Elements* Penguin Books, 1972:220–223.
6. Powell P, Timms PL. *The Chemistry of the Non-Metals.* London: Chapman and Hall; 1974:148–150 and 182.
7. Misawa M, Suzuki K. *Trans. Jpn. Inst. Met.* 1977;18:427.
8. Misawa M, Suzuki K. *J. Phys. Soc. Jpn.* 1978;44:1612.
9. Stephens RBJ. *Non-Cryst. Solids.* 1976;20:75.
10. Gerlach E, Grosse P. *The Physics of Selenium and Tellurium.* Berlin: Springer; 1979.
11. Steudel R, Strauss E-M. *Adv. Inorg. Chem.Radiochem.* 1984;28:135.
12. Flory PJ. *Statistical Mechanics of Chain Molecules.* New York: Hanser/Oxford University Press; 1988;157–159.
13. Cotton FA, Wilkinson G. *Advanced Inorganic Chemistry.* 5th ed. New York: Wiley; 1988:496–497.
14. Bohmer B, Angell CA. *Physical Rev B.* 1993;48:5857.
15. Atkins PW. *The Periodic Kingdom.* New York: Basic Books; 1995.

Polymeric Sulfur

Stephen J. Clarson

Class Inorganic and semi-inorganic polymers

Introduction The interesting color changes (yellow to orange to red to black) that are seen upon heating sulfur from its melting point up to its boiling point are fascinating. Elemental sulfur has been studied by mankind for many, many years—for example, the self-depression of the freezing point of sulfur was reported by Germez in 1876. According to Kelly, there are some 15 references to the element in the Bible. Sulfur is abundant in nature both in its elemental form and also as compounds in the form of H_2S, SO_2, sulfide minerals and various sulfates. Sulfur can exist as a large variety of allotropes which depend upon the temperature and pressure of the system. These consist of linear and cyclic $-[S]_n-$ species commonly from $n = 2$–20 for the rings and even much higher for the chains (species as long as 8×10^5 have been reported). The most stable form of sulfur at 25°C is the cyclooctasulfur S_8 ring. When sulfur is poured onto ice water in the molten state plastic sulfur or catenasulfur is produced. Unlike the other sulfur allotropes, catenasulfur is insoluble in carbon disulfide (CS_2). Sulfur fibers can be produced which have helical conformations with approximately 3.5 S atoms per turn. Detailed calculations on the conformations of polymeric sulfur chains and on their ring-chain equilibration distributions have been described in a series of articles by Semlyen.

Major Applications Although sulfur has many industrial applications in the area of organic and inorganic synthesis, in the field of polymer science and engineering it is probably best know for it use in the vulcanization of natural rubber and related unsaturated polymer chains (see Coran and pertinent references cited therein). In this application sulfur forms shorts chains which link the polymer network precursor chains together by covalent bonding between the carbon containing chains at the sites of unsaturation. Another important area worth mentioning is the chemical reaction seen in proteins between two cysteine ($S-H$) units either on the same or on different chains to give covalent disulfide linkages (cystine units) (see Creigton and pertinent references cited therein).

Selected properties

Properties	Values
Atomic Number	16
Atomic Mass	32.06 g mol^{-1}
Bond Length l_{S-S}	2.06 ± 0.02 Å
Bond Energy $S-S$	265 kJ mol^{-1}
Bond Angle $S-S-S$ in S_8 ring	108°
van der Walls radius	1.80 Å

Properties	Values
Standard Entropy of Cyclization (polymeric sulphur to cyclooctasulfur)	-4.63 cal deg^{-1} mol^{-1}
S$_{alpha}$ (orthorhombic) to S$_{beta}$ (monoclinic) transition temperature	95.5°C
Melting point S$_{alpha}$	112.8°C
Melting point S$_{beta}$	119°C
Boiling point S	444.6°C
S$_8$ –> polymer critical polymerization temperature	159°C
Enthalpy S$_8$ –> polymer	13.4 kJ mol^{-1} at 159°C
Temperature of maximum melt viscosity	~ 200°C
Entanglement Molecular Weight M$_e$	178 g mol^{-1} (575 K)
Plateau Modulus	0.03 MPa (575 K)

Selected Bibliography

1. Gernez MD. *Comp. Rend.* 1876;82:115.
2. Abrahams SC. *Quart. Rev.* 1956;10:407.
3. Tobolsky AV, Eisenberg A, *J. Am. Chem. Soc.* 1959;81:780.
4. Tobolsky AV, Eisenberg A, *J. Am. Chem. Soc.* 1959;81:2803.
5. Tobolsky AV, Eisenberg A, *J. Am. Chem. Soc.* 1960;82:289.
6. Pauling L. *The Nature of the Chemical Bond.* 3rd ed. Ithaca, NY: Cornell University Press; 1960:134–136.
7. Tobolsky AV, MacKnight WJ. In: Mark HF, Immergut EH, eds. *Polymeric Sulfur and Related Polymers.* New York: Wiley-Interscience; 1965.
8. Semlyen JA. *Trans. Faraday Soc.* 1967;63:743.
9. Semlyen JA. *Trans. Faraday Soc.* 1967;63:2342.
10. Semlyen JA. *Trans. Faraday Soc.* 1968;64:1396.
11. Semlyen JA. *Polymer.* 1971;12:383.
12. Puddephatt RJ. *The Periodic Table of the Elements.* Oxford: Oxford University Press; 1972.
13. Semlyen JA. *Chemistry in Britain.* 1982;18:704.
14. Steudel R, Mausle HJ, Rosenbauer D, Mockel H, Freyholdt T. *Angew. Chem. Int. Ed. Engl.* 1981;20:394 .
15. Creighton TE. *Proteins : Structure and Molecular Properties.* New York: Freeman; 1984.
16. Meyer B. *Elemental Sulphur – Chemistry and Physics.* New York: Interscience; 1985.
17. Semlyen JA. In: Semlyen JA, ed. *Cyclic Polymers.* Elsevier; 1986:33–37.
18. Flory PJ. *Statistical Mechanics of Chain Molecules.* New York: Hanser/Oxford University Press; 1988:157–159.
19. Cotton FA, Wilkinson G. *Advanced Inorganic Chemistry.* 5th ed. New York: Wiley; 491–543 (1988).
20. Coran AY. In: Mark JE, Erman BE, Eirich FR, ed. *Science and Technology of Rubber.* 2nd ed. San Diego, CA: Academic Press; 1994:339–366.

21. Atkins PW. *The Periodic Kingdom.* New York: Basic Books; 1995.

22. Kelly P. *Chemistry in Britain.* 1997;33:25–27.

23. Haiduc I, King RB. In: Semlyen JA, ed. *Large Ring Molecules.* New York: Wiley; 1997:350–352.

24. Fetters LJ, Lohse DJ, Graessley WW. *J. Polym. Sci.: Part B: Polym. Phys.* 1999;37:1023.

25. Dodgson K, Heath RE, Semlyen. *J. A. Polymer.* 1999;40:3995.

Poly(methacrylic acid)

Jianye Wen

Acronyms, Trade Names PMAA or PMA

CAS number 25087-26-7

Class Vinylidene polymers

Structure

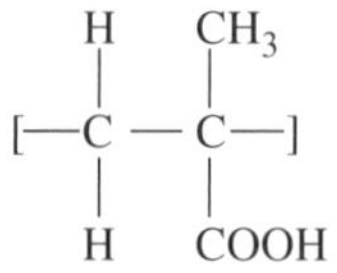

$$[-\underset{\underset{\text{H}}{|}}{\overset{\overset{\text{H}}{|}}{\text{C}}} - \underset{\underset{\text{COOH}}{|}}{\overset{\overset{\text{CH}_3}{|}}{\text{C}}}-]$$

Major Applications Various applications in the fields of mining, textile manufacture, cosmetics, oil recovery, agriculture, and water clarification as thickening agent for latices and adhesives, ion-exchange resins, adhesives, binders, dispersants, and floculating agents

Properties of Special Interest Weak acid, brittle solid that cannot be molded, cross-link on heating, decompose without softening at high temperature, too water sensitive to be plastics, generation of viscosity and thixotropy at low concentrations, interaction with counterions or charged particulate matter, inverse solubility-temperature behavior.

Property	Units	Conditions	Value	References
^{13}C-NMR spectra	ppm	Selected group assignments Carbonyl C=O	179.2	(2)
Density	g cm^{-3}	at 25°C	1.285	(3)
FTIR spectra	cm^{-1}	Selected group assignments C=O: free	1742	(4)
		C=O: hydrogen-bonded dimmer	1700	
Glass-transition temp. (T_g)	°C		130	(5)
			185	(6)
			186	(7)
			150.8	(8)
Heat capacity	J mol^{-1} K^{-1}	100 K	45.18	(9)
		200 K	81.41	
		300 K	112.50	
Melting temperature (T_m)	°C		205	(8)

Poly(methacrylic acid)

Mark-Houwink constants

Solvent	Temp. (°C)	Mol. Wt. Range ($M \times 10^{-4}$)	$K \times 10^3$ (ml g^{-1})	a	References
Methanol	26	−20	242	0.51	(10)
Aqua. HCl (0.002 M)	30	−90	66	0.50	(11)
Aqua. NaNO$_3$(2 M)	25	−70	44.9	0.65	(12)

Property	Units	Conditions	Value	References
Molecular weight of repeat unit	g mol^{-1}	–	86.09	
Properties of monomer	–	–	–	(13)
Melting point	°C	–	14	
Boiling point	°C	–	159–163	
Refractive index n^D_{25}	–	–	1.4288	
Specific gravity	–	–	1.015	
Heat capacity	J g^{-1} K^{-1}	–	2.1–2.3	
Dissociation constant pK	–	–	4.66	
Heat of polymerization	kJ mol^{-1}	–	56.5	
Solubility parameter	(MPa)$^{1/2}$	Isobutyl ester, 140°C	14.7	(14)
		Ethyl ester	18.31	(15)
		Methyl ester, 25°C	18.58	(16)
		Poor solvent hydrogen bonding	0	(17
		Moderate	20.3	(17)
		Strong	26.0–29.7	(17)
Solvents		Ethanol, methanol, water, dioxane, dimethylformamide		(18)
		Alcohols, aqua. Hydrogen chloride (0.002 M, above 30°C), dil. Aqua. Sodium hydroxide		(19, 20)
Non-solvents		Acetone, diethyl ether, benzene, aliphatic hydrocarbons		(18)
		Ketones, carboxylic acids, esters		(19, 20)
Sound speed	m s^{-1}	Longitudinal	3350	(3)

Tacticity[21,22]

Polymerization Condition	Product
Free-radical polymerization in methyl ethyl ketone at 60°C	57% syndiotactic triads
Hydrolysis of poly(methacrylic anhydride) at 40°C	Atactic
Hydrolysis of esters having appropriate configurations	Syndiotactic

References

1. General references: (a) Brandrup J, Immergut EH. *Polymer Handbook.* Third ed., New York: Wiley-Interscience; 1989; (b) Daniels W. In: Bikales NM, eds. *Encyclopedia of Polymer Science & Technology.* Vol 17. New York: Wiley-Interscience; 1987:402; (c) Mark JE, ed. *Physical Properties of Polymers Handbook.* Woodbury, New York: AIP Press; 1996.

2. Yi JZ, Goh SH. *Polymer.* 2001;42:9313.
3. Brandrup J, Immergut EH. *Polymer Handbook.* Third ed. Vol 1. New York: Wiley-Interscience; 1989:147.
4. Motzer HR, Painter PC, Coleman MM. *Macromolecules.* 2001;34:8390.
5. Odajima A, Woodward AE, Sauer JA. *J. Polym. Sci.* 1961;55:181.
6. Greenwald HL, Luskin LS. In: Davison RL, ed. *Handbook of Water Soluble Gums and Resins.* Chap. 17. New York: McGraw-Hill; 1980:1–19.
7. Oliverira HPM, Gehlen MH. *J. Luminescence.* 2006;121:544.
8. Mansur C, Tavares M, Monteiro E. *J. Appl. Polym. Sci.* 2000;75:495.
9. Gaur U, Lau SF, Wunderlich BB, et al. *J. Phys. Chem. Ref. Data.* 1982;11:1065.
10. Weiderhorn NM, Brown AR. *J. Polym. Sci.* 1952;8:651.
11. Katchalsky A, Eisenberg H. *J. Polym. Sci.* 1951;6:145.
12. Arnold R, Caplan SR. *Trans. Faraday Soc.* 1955;51:857.
13. Kine BB, Novak RW. In: Bikales NM, ed. *Encyclopedia of Polymer Science & Technology.* Vol. 1. New York: Wiley-Interscience; 1987:241.
14. DiPaola-Baranayi G. *Macromolecules.* 1982;15:622.
15. Mangaraj D, Patra S, Rashid S. *Makromol. Chem.* 1963;65:39.
16. Bristow GM, Watson WF. *Trans. Faraday Soc.* 1958;54:1731.; 1958;54:1742.
17. Grulke EA. In: Brandrup J, Immergut EH, ed. *Polymer Handbook.* Third ed. New York: Wiley-Interscience; 1989:VII-519.
18. Hughes LJ, Britt GE. *J. Appl. Polym. Sci.* 1961;5:337.
19. Dexheimer H, Fuchs O. In: Nitsche R, Wolf KA, ed. *Strukur und Physikalisches Verhalten der Kunststoffe, V1.* Springer Verlag, Berlin-Goettingen-Heidelberg; 1961.
20. Kurata M, Stockmayer WH. *Adv. Polym. Sci, V3.* Springer Verlag, Berlin-Goettingen-Heidelberg; 1963:196.
21. Loebl EM, O'Neill JJ. *J. Polym. Sci., Polym. Lett. Ed.* 1963;1:27.
22. Greber G, Egle G. *Makromol. Chem.* 1960;40:1.

Poly(methyl acrylate)

Jianye Wen

Acronyms, Trade Names PMA

CAS number 9003-21-8

Class Vinyl polymers

Structure

$$[-CH_2-CH-]$$
$$|$$
$$COOCH_3$$

Major Applications Coatings, textile finishing, paper saturants, leather finishing

Properties of Special Interest A tough, rubbery, and moderately hard polymer, with little or no tack at room temperature; superior resistance to degradation and remarkable retention of their original properties under use conditions.

Property	Units	Conditions	Value	Reference
^{13}C-NMR spectra	Ppm	Selected group assignments		(2)
		Carbonyl C=O	174.65	
		Methine		
		CH_2–CHCOOCH$_3$	34.74	
		Methylene		
		CH_2–CHCOOCH$_3$	34.74	
		Methyl, CH$_3$	51.56	
H-NMR spectra	Ppm	Selected group assignments		(3)
		Methine		
		CH_2–CHCOOCH$_3$	2.25	
		Methylene		
		CH_2–CHCOOCH$_3$	1.85	
			1.60	
			1.45	
		Methyl, CH$_3$	3.55	
Cohesive energy E_{coh}	J mol^{-1}		31678	(4)
Density	g cm^{-3}	25°C	1.22	(5–7)

Property	Units	Conditions	Value	Reference
FTIR spectra	cm^{-1}	Selected group assignments		(8)
		C=O	1736	
		C—CH$_3$ bending	1380, 1437	
		C—O—C stretch	263	
		Aliphatic C—H stretches	2954	
Glass transition temperature	°C		6	(9, 10)
		conventional	10, 11	(11–15)
		head to tail	6	(16–20)
		head to head	31	(21–26)
Heat capacity	J (g K)$^{-1}$	–173°C	0.6154	(27)
		–73°C	0.9816	(27)
		27°C	1.765	(27)
		227°C	2.143	(27)
		ΔC_p	42.30/86.09	(27)
Interaction parameter χ, with;				
		butane 70–90°C	2.392–1.753	(28)
		hexane 70–110°C	2.731–1.885	(28)
		heptane 70–110°C	2.808–1.983	(28)
		decane 70–110°C	3.107–2.434	(28)
		cyclohexane 70–110°C	2.316–1.460	(28)
		benzene 70–110°C	0.471–0.359	(28)
		toluene 70–100°C	0.624–0.511	(28)
		chloroform 70–110°C	–0.222–0.075	(28)
		carbon tetrachloride 70–110°C	0.986–0.658	(28)
		acetone 70–110°C	0.482–0.384	(28)
		methyl ethyl ketone 70–110°C	0.459–0.388	(28)
		tetrahydrofuran 70–100°C	0.425–0.316	(28)
		dioxane 70–100°C	0.205–0.198	(28)
		methyl acetate 70–110°C	0.396–0.375	(28)
		ethyl acetate 70–110°C	0.471–0.428	(28)
		butanone 100°C	0.40	(29)
		ethanol 100°C	1.01	(29)
		n-octane 90–100°C	2.4–2.2	(29, 30)
		1-propanol 100°C	0.82	(29)
Interfacial tension	mN m^{-1}	with poly(n-butyl acrylate)20°C	4.0	(31)
		with PE	10.6	(31)

Mark-Houwink coefficients

Solvent	Temperature (°C)	Molecular Weight Range ($M \times 10^4$)	$K \times 10^3$ (ml g^{-1})	a	Reference
acetone	25	–160	5.5	0.77	(32)
	30	–45	28.2	0.52	(32)
benzene	25	–130	2.58	0.85	(32)
	30	–160	4.5	0.78	(32)

Poly(methyl acrylate)

Solvent	Temperature (°C)	Molecular Weight Range ($M \times 10^4$)	$K \times 10^3$ (ml g^{-1})	a	Reference
butanone	20	−240	3.5	0.81	(32)
	25	−68	14.1	0.67	(33)
	30	−190	3.97	0.772	(34)
diethyl malonate	30	−190	3.51	0.793	(34)
ethyl acetate	35	−148	11	0.69	(35)
toluene	30	−190	7.79	0.697	(34)
	35	−69	21	0.60	(35)

Property	Units	Conditions	Value	Reference
Mechanical properties				
Tensile strength	MPa		6.9	(36, 37)
Elongation at break	%		750	(36, 37)
Refractive index, n_D^{25}			1.479	1(b)
			1.479	(38)
2nd virial coefficient A$_2$	10^4 (mol cm^3 g^{-2})	acetone 20°C $M \times 10^{-3} = 77-880$	4.5–2.8	(39)
		25°C $M \times 10^{-3} = 280-2500$	4.2–2.4	(40, 41)
		ethyl acetate 35°C $M \times 10^{-3} = 362-1480$	1.92	(35)
		butanone/isopropanol 58/42 20°C $M \times 10^{-3} = 290-1720$	0.1–0.06	(40)
Solvents	aromatic hydrocarbons, chlorinated hydrocarbons, tetrahydrofuran, esters, ketones, glycolic ester ethers, phosphorus trichloride			(42)
Non-solvents	aliphatic hydrocarbons, hydrogenated naphthalenes, diethyl ether, alcohols, carbon tetrachloride			(42)
Solubility parameter	(J cm^{-3})$^{1/2}$		20.7	(43)
			21.4	(44)
Surface tension	mN m^{-1}	$M_w = 25\,000$		(45)
		at 20°C	41.0	
		at 150°C	31.0	
		at 200°C	27.2	
	mN (m K)$^{-1}$	$-d\gamma/dT$	0.070	
	χ^p	polarity	0.248	

Poly(methyl acrylate)

Unperturbed dimension[a]

Conditions	$r_0/M^{1/2} \times 10^4$ (nm)	$r_{of}/M^{1/2} \times 10^4$ (nm)	$\sigma = r_0/r_{of}$	$C_\infty = r_0^2/nl^2$	Reference
Various solvents 30°C	680 ± 30	332	2.05 ± 0.10	8.4	(46)
Butanone/2-propanol 42/58 vol, 20°C	680	332	2.05	8.4	(40, 41)
50/50 vol, 30°C	665	332	2.00	8.0	(34)
undiluted, 60°C	$\mathrm{d}lnr_0^2/\mathrm{d}T = -0.2 \times 10^{-3}$ $[\deg^{-1}]$				(47)

[a] See reference 1(b) for details.

References

1. General references: (a) Brandrup J, Immergut EH, eds. *Polymer Handbook*. Third ed. New York: Wiley-Interscience; 1989; (b) Kine BB, Novak RW. In Bikales NM, eds. *Encyclopedia of Polymer Science & Technology*. Vol 1. New York: Wiley-Interscience; 1987:234; (c) Mark JE, ed. *Physical Properties of Polymers Handbook*. Woodbury, New York: AIP Press; 1996.
2. Mathakiya I, Rakshit AK. *Int. J. Polym. Anal. Charact.* 2003;8:339.
3. Cunningham ID, Fassihi K. *Polym. Bull.* 2005;53:359.
4. Yu X, Wang X, Li X, Gao J, Wang H. *J. Polym. Sci. Part B*. 2006;44:409.
5. Van Krevelen DW. *Properties of Polymers*. Amsterdam: Elsevire Publishing Co; 1976.
6. Shetter JL. *J. Polym. Sci., Part B*. 1963;1:209.
7. Kine BB, Novak RW. In: Bikales NM, ed. *Encyclopedia of Polymer Science and Technology*. Vol 1. New York: Wiley-Interscience; 198?:257.
8. Charles CJ, ed. *The Aldrich library of FT-IR spectra*. 1 Milwaukee, Wis. Aldrich Chemical Co, V2, P1186; 1985.
9. Van Krevelen DW. *Properties of Polymers*. Amsterdam: Elsevire Publishing Co; 1976.
10. Crawford JWC. *J. Soc. Chem. Ind. London*. 1949;68:201. Boyer RF, Spercer RS. *Advance in Colloid Science*, Vol II. New York: Wiley-Interscience; 1946:1.
11. Wiley RH, Brauer GM. *J. Polym. Sci.* 1948;3:455.
12. Wiley RH, Brauer GM. *J. Polym. Sci.* 1948;3:647.
13. Riddle EH. *Monomeric Acrylic Esters*. New York: Reinhold; 1954:59.
14. Aida H, Senda H. *Fukui Daigaku Kogakubu Kenkyu Hokoku*. 1980;28:95.
15. Gerke RH. *J. Polym. Sci.* 1954;13:295.
16. Utracki LA, Simha R. *Makromol. Chem.* 1968;117:94.
17. Hughes LJ, Brown GL. *J. Appl. Polym. Sci.* 1961;5:580.
18. Jenckel E, Ueberriter K. *Z. Physik. Chem. (Leipzig)*. 1938;A182:361.
19. Wuerstlin F, Thurn H. In: Stuart HA, ed. *Die Physik der Hochpolymeren*. Springer-Verlag, Berlin; 1956.
20. Rehberg CE, Fisher CH. *Ind. Eng. Chem*, 1948;40:1429.
21. Shetter JA, *J. Polym. Sci., A1*. 1966;4:2381.
23. McCurdy RM, Prager JH. *J. Polym. Sci. A*, 1964;2:1885.
24. Wuerstlin F. In: Stuart HA, ed. *Die Physik der Hochpolymeren*. Chap. 11. Springer-Verlag, Berlin; 1955.
25. Haldon RA, Simha R. *J. Appl. Phys.* 1968;39:1890.
26. Otsu T, Aoki S, Nakatani R. *Makromol. Chem.* 1970;134:331.

27. Gaur U, Lau SF, Wunderlich B, et al. *J. Phys. Chem. Ref. Data.* 1982;11:1065.

28. Tian M, Munk P. *J. Chem. Eng. Data.* 1994;39:742.

29. Munk P, Hattam P, Du Q, et al. *J. Appl. Polym. Sci.: Appl. Polym. Symp.* 1990;45:289.

30. DiPaola-Baranyi G, Guillet JE. *Macromolecules.* 1978;11:228.

31. Wu S. In: Brandrup J, Immergut EH, eds. *Polymer Handbook.* Third ed. New York: Wiley-Interscience; 1989:VI–411.

32. Krause S. *Dilute Solution Properties of Acrylic and Methacrylic Polymers.* Part 1, Revison 1. Philadelphia, Pennsylvania: Rohm & Haas Co; 1961.

33. Kotera A, Saito T, Watanabe Y, Ohama M. *Makromol. Chem,* 1965;87:195.

34. Matsuda H, Yamano K, Inagaki H. *J. Polym. Sci. A2.* 1969;7:609.

35. Karunakaran K, Santappa M. *J. Polym. Sci. A2.* 1968;6:713.

36. Brendley WH Jr. *Paint Varn. Prod.* 1973;63:19.

37. Craemer AS. *Kunststoffe.* 1940;30:337.

38. Holder AJ. *QSAR & Combinatorial Science.* 2006;25(10):905.

39. Wunderlich W. *Angew. Makromol. Chem.* 1970;11:189.

40. Trossarelli L, Saini G. *Atti Accad. Sci. Torino: Classe Sci. Fix. Mat. Nat.* 1955–56;90:419.

41. Saini G, Trossarelli L. *Atti Accad. Sci. Torino: Classe Sci. Fix. Mat. Nat.* 1955–56;90:431.

42. Fuchs O. In: Brandrup J, Immergut EH, eds. *Polymer Handbook,* Third ed. New York: Wiley-Interscience; 1989:VII–379.

43. Yu X, Wang X, Li X, Gao J. *QSAR Comb. Sci.* 2006;25:156.

44. Gardon JL. In: Bikales NM, ed. *Encyclopedia of Polymer Science and Technology.* Vol 3. New York: Wiley-Interscience; 1965:833.

45. Partington JR. *An Advanced Treatise of Physical Chemistry: Physico-Chemical Optics.* Vol IV. London: Longmans, Green and Co; 1960.

46. Kurata M, Stockmayer WH. *Fortschr. Hochpolymer. Forsch.* 1963;3:196.

47. Tobolsky AV, Carlson D, Indictor N. *J. Polym. Sci.* 1961;54:175.

Poly(methacrylonitrile)

J. R. Fried

Acronym PMAN

Class Polynitriles

Structure

$$-CH_2-\underset{\underset{C\equiv N}{|}}{\overset{\overset{CH_3}{|}}{C}}-$$

Major Applications Films, coatings, elastomers, packaging, photoresists.

Properties of Special Interest Good resistance to many solvents, acids, and water, but attacked by polar solvents and decomposed by concentrated alkali and hot dilute alkali.

Types of Polymerization Free-radical or ionic polymerization of methacrylonitrile (2-cyanopropylene) in bulk, emulsion, or solution; group-transfer polymerization also has been used. Ionic polymerization in inert solvents can produce either amorphous poly(methacrylonitrile) (by use of anionic catalysts such as *n*-butyllithium) or primarily isotactic poly(methacrylonitrile) (by use of coordination catalysts such as ethylberyllium or diethylmagnesium).

Property	Units	Conditions	Value	Reference
Molecular weight (of repeat unit)	g mol^{-1}	—	67.09	—
Typical comonomers	Butadiene, styrene, α-methylstyrene, methacrylic acid			
Solvents	Trifluoroacetic acid, acetone, acetonitrile, acrylonitrile, aniline, benzaldehyde, *m*-cresol, cyclohexanone, *N*,*N*-dimethyl acetamide, *N*,*N*-dimethyl formamide, dimethyl sulfoxide, ethanol amine, formic acid, *N*-methyl-2-pyrolidone, nitrobenzene, propylene carbonate, pyridine, triethyl phosphate			
Nonsolvents	Acetic acid, benzene, 1-butanone, *n*-butyl acetate, chlorobenzene, cyclohexane, diethyl ether, diethylene glycol, diisobutyl ketone, ethyl acetate, 2-ethyl hexanol, *n*-heptane, isoamyl alcohol, isopropyl alcohol, *n*-octyl alcohol, 1-propanol propylene glycol, styrene, tetralin, 1,1,1-trichloroethane, toluene			

Poly(methacrylonitrile)

Property	Units	Conditions	Value	Reference
Ceiling temperature	K	In benzonitrile	418	(1)
Density	$g\ cm^{-3}$	Amorphous	1.13	(2)
Dielectric constant ε' (D150)	—	60 Hz 1 kHz 1 MHz	4.14 3.83 3.30	(2)
Dissipation factor (D150)	—	60 Hz 1 kHz 1 MHz	0.046 0.038 0.025	(2)
Maximum extensibility $(L/L_0)_r$	%	D638	2–3	(2)
Flexural modulus	GPa	D790	3.86–4.48	(2)
Flexural strength	MPa	D790	83–97	(2)
Glass transition temperature	K	Amorphous (free-radical polymerization)	285	(3)
Hardness	M scale	Rockwell (D785)	95	(2)
Heat deflection temperature	K	D648 (1.8 MPa)	370–373	(2)
Impact strength	$J\ m^{-1}$	Notched Izod (D256)	21	(2)
Infrared spectrum (principal absorptions)	cm^{-1}	Assignment CH_3 stretching $C{\equiv}N$ stretching	Wavenumber 2,990 2,234.5	(4)
Index of refraction n		At 20°C	1.5932	(2)
Permeability coefficient	m^3 (STP) m s^{-1} $m^{-2}\ Pa^{-1}s$	At 25°C O_2 O_2 H_2O	 9×10^{-21} 2.4×10^{-20} 3.10×10^{-15}	(5)
Solubility parameter	$(MPa)^{1/2}$	—	21.9, 25.4	(6, 7)
Tensile strength	MPa	D638	55–69	(2)
Volume resistivity	ohm cm	—	1.14×10^{16}	(2)
Water absorption	%	144 h at ambient temperature	0.24	(2)

Unit cell dimensions[(2)]

Lattice	Monomers per Unit Cell	Cell Dimension (Å)			Cell Angles (deg)		
		a	b	c^*	α	β	γ
Pseudohexagonal (modification I)	–	9.03	–	6.87	–	–	–
Monoclinic (modification II)	8	13.5	7.71	7.62	–	97°49′	–

* Fiber identity period.

References

1. Brandrup J, Immergut EH, eds. *Polymer Handbook.* 2d ed. New York: John Wiley and Sons; 1989:II–439.
2. Ball LE, Curatolo BS. In: Kroschwitz. ed. *Encyclopedia of Polymer Science and Engineering.* Vol 9. New York: John Wiley and Sons; 1990:669–705.
3. Nielson LE. *Mechanical Properties of Polymers.* Stamford, Conn: Reinhold Publishing Company; 1962:19.
4. Nagata A, Ohta K, Iwamoto R. *Macromol. Chem. Phys.* 1996;197:1959.
5. Salame MJ. *Polym. Sci. Symp.* 1973;41:1.
6. Small PA. *J. Appl. Chem.* 1953;3:71.
7. Ho B-C, Chin W-K, Lee Y-D. *J. Appl. Polym. Sci.* 1991;42:99.

Poly(N-methylcyclodisilazane)

Donna M. Narsavage-Heald

Alternative Names Poly(1,3-dimethyl-2,2,4,4-tetramethylcyclodisilazane);
poly(hexamethylcyclodisilazane)

Class Polysilazanes

Repeat Unit —[Me$_2$Si—NMe]—

Property	Units	Conditions	Value	Reference
Preparative techniques	–	Anionic ring opening		
		MeLi/THF	–	(1, 2)
		n-BuLi/THF		(1)
		n-BuLi/hexane		(2)
		t-BuLi/THF		(2)
		PhLi/THF		(1)
		MeLi/Me$_3$CONa/THF		(1)
		n-BuLi/Me$_3$CONa/THF		(1)
		Naphthalene-Na		(2)
		α-MeStyNa		(2)
		Cationic ring opening		
		CF$_3$SO$_3$Me		(2)
		CF$_3$SO$_3$SiMe$_3$		(2)
Molecular weight (of repeat unit)	g mol^{-1}	–	87	–
Molecular weight (M_n)	g mol^{-1}	MeLi/THF	3,000	(1)
		MeLi/THF	4,400	(2)
		n-BuLi/THF	1,500	(1)
		n-BuLi/hexane	1,020	(1)
		t-BuLi/THF	4,200	(2)
		PhLi/THF	4,200	(2)
		MeLi/Me$_3$CONa/THF	1,500	(1)
		Naphthalene-Na	16,000	(2)
		α-MeStyNa	4,000	(2)
		CF$_3$SO$_3$Me	16,800	(2)
		CF$_3$SO$_3$SiMe$_3$	18,000	(2)
Typical polydispersity index (M_w/M_n)	–	–	1.2	(2)

Poly(N-methylcyclodisilazane)

Property	Units	Conditions	Value	Reference
NMR	ppm	^{1}H	0.12 (SiMe), 2.39 (NMe)	(1, 2)
		^{13}C	1.94 (SiMe), 30.25 (NMe)	(1, 2)
		^{29}Si	−2.2	(1, 2)
		^{15}N	−361	(2)
Glass transition temperature	K	DSC	235	(2)
Melting temperature	K	DSC	500	(2)
Phase transition	K	DSC	428	(2)
Pyrolyzability, amount of product	–	TGA, argon flow, $10°C\ min^{-1}$	2.85%	(1)

References

1. Seyferth D, Schwark JM, Stewart RM. *Organometallics.* 1989;8:1980–1986.
2. Duguet E, Schappacher M, Soum A. *Macromolecules.* 1992;25(19):4835–4839.

Poly(methylene oxide)

Allan S. Hay and Yong Ding

Alternative Names, Trade Names Polyacetal, polyoxymethylene, acetal, Delrin®, Celcon® (copolymer), Ultraform® (copolymer)

Class Polyether engineering thermoplastics

Structure $[-CH_2-O-]$

Major Applications Poly (methylene oxide) resin has been widely used in mechanical, automotive, plumbing, appliance, industrial, and electrical conponents along with its copolymer resins. It is continuing to replace die-cast zinc, brass, aluminum, steel, and other metals in the various end-use industries.

Properties of Special Interest Poly (methylene oxide) resin, like other polyacetal resins, is a highly crystalline polymer characterized by its metallic qualities of hardness, strength, and stiffness. It also has good lubricity properties under a wide variety of enviromental conditions of moisture and heat, good fatigue resistance, a low coefficient of friction, and springiness. In addition, it has good chemical resistance to most solvents. It cannot, however, be flameproofed.

Preparative Techniques The homopolymer is prepared by anionic polymerization of purified formaldehyde with the addition of an initiator such as an amine, phosphine, or metal alcohol. The copolymers are manufactured commercially by copolymerization of trioxane, the cyclic trimer of formaldehyde, with small amounts of a comonomer. Typically, acetal copolymer resins have 95% or more oxymethylene units.

Property	Units	Conditions	Value	Reference
Molecular weight of repeat unit	$g\,mol^{-1}$	–	30.03	–
Typical molecular weight range of polymer	$g\,mol^{-1}$	–	$2-9 \times 10^4$	–
IR (characteristic absorption frequencies)				(1, 2)
NMR				(3, 4)
Thermal expansion coefficients	K^{-1}	233–303 K	7.5×10^{-5}	(5)
Density (amorphous)	$g\,cm^{-3}$	D792		(6)
		Homopolymer	1.42	
		Copolymer	1.41	

Solvents[5]

Solvent	Gel Temp. (K)	Dissolving Temp. (K)
m-Chlorophenol	328	362
Phenol	331	382
p-Chlorophenol	333	371
3,4-Xylenol	361	401
Aniline	375	403
γ - Butyrolactone	385	407
N,N-Dimethylformamide	388	408
Pentachloroethane	390	413
Ethylene carbonate	390	418
Benzyl alcohol	392	405
Styrene oxide	398	419
Formamide	403	423
Nitrobenzene	407	421
Cyclohexanol	413	423
Propionic anhydride	417	428

Mark-Houwink parameters: K and a

Solvent	Temp. (K)	$M_w \times 10^{-3}$ (g mol)$^{-1}$	$K \times 102$ (ml g^{-1})	a	Reference
p-Chlorophenol	403	–	5.43	0.66	(7)
p-Chlorophenol, 2% α-pinene	333	62–129	4.13	0.724	(8)
1H,1H,5H-octafluoropentanol-1	383	62–129	1.33	0.81	(8)
Phenol-tetrachloroethane (25–75 wt.)/2% α-pinene	363	1.1–92	1.216	0.64	(9)
Phenol	363	–	1.13	0.76	(10)
Dimethylformamide	423	89–285	4.4	0.66	(11)
Dimethylformamide	403	1.5–15	2.24	0.71	(12)
Hexafluoroacetone-sesquihydrate + triethylamine	298	1.5–15	4.60	0.74	(12)

Property	Units	Conditions	Value		Reference
Lattice	–	–	TRIG	ORTH	(13)
Space group	–	–	P3$_1$ or P3$_2$	P2$_1$2$_1$2$_1$	(13)
Chain conformation	–	–	2*9/5	2*2/1	(13)
Unit cell dimensions	Å	–	a = 4.471	a = 4.767	(13)
			b = 4.471	b = 7.660	
			c = 17.39	c = 3.563	
Unit cell contents (number of repeat units)			9	4	(13)
Density (crystalline)	g cm^{-3}	–	1.491	1.533	(13)
Heat of fusion (of repeat units)	kJ mol^{-1}	–	9.79	–	(14)
Entropy of fusion (of repeat units)	kJ K^{-1}mol^{-1}	Constant pressure	8.21 × 10^{-3}		(15)
		Constant volume	4.98 × 10^{-3}		(15)
		Equilibrium value	10.70 × 10^{-3}		(16)

Poly(methylene oxide)

Property	Units	Conditions	Value	Reference
Degree of crystallinity	%	Homopolymer		
		293 K (density)	64–69	(17)
		298 K (x-ray)	77	(18)
		408 K (x-ray)	75	(18)
		413 K (x-ray)	73.0	(19)
		418 K (x-ray)	73.1	(19)
		423 K (x-ray)	75.4	(19)
		428 K (x-ray)	76.9	(19)
		430 K (x-ray)	67	(18)
		433 K (x-ray)	80.0	(19)
		440 K (x-ray)	95.5	(19)
		Copolymer, Hostaform C 2520, $M_w = 80,000$, 298 K (density)	56–59	(17)
		Copolymer, Hostaform C9020, $M_w = 58,000$, 298 K (density)	56.6	(17)
Glass transition temperature	K	–	198	(18)
Melting point	K	Delrin 500, ASTM D2133	448	(6)
		Celcon M90, ASTM D2133	438	
Heat capacity (of repeat units)	$JK^{-1} mol^{-1}$	Homopolymer		(20)
		0 K	0	
		50 K	9.94	
		100 K	16.69	
		200 K	28.82	
		300 K	42.79	
		Copolymer		(20)
		0 K	0	
		50 K	9.97	
		100 K	16.40	
		200 K	26.56	
		300 K	41.11	
		Crystalline polymer		(21)
		0 K	0	
		50 K	10.10	
		100 K	16.68	
		200 K	27.15	
		300 K	38.52	
Deflection temperature	K	Delrin 500		(6)
		ASTM D648, 1.82 Mpa	409	
		ASTM D648, 0.45 Mpa	445	
		Celcon M90		(6)
		ASTM D648, 1.82 Mpa	383	
		ASTM D648, 0.45 MPa	431	
Tensile modulus	MPa	296 K, ASTM D638		(5)
		Homopolymer	3,100	
		Copolymer	2,825	

Property	Units	Conditions	Value	Reference
Tensile strength	MPa	296 K, ASTM D638		(5)
		Homopolymer	68.9	
		Copolymer	60.6	
Maximum extensibility $(L/L_0)_r$	%	296 K, ASTM D638		(5)
		Homopolymer	23–75	
		Copolymer	40–75	
Flexural modulus	MPa	296 K, ASTM D790		(5)
		Homopolymer	2,830	
		Copolymer	2,584	
Flexural strength	MPa	296 K, ASTM D790		(5)
		Homopolymer	97.1	
		Copolymer	89.6	
Impact strength	$J\,m^{-1}$	296 K, notched, 3.175 mm, ASTM D256		(5)
		Homopolymer	69–122	
		Copolymer	53–80	
		233 K, notched, 3.175 mm, ASTM D256		
		Homopolymer	53–95	
		Copolymer	43–64	
Hardness	–	Rockwell hardness, ASTM D785		(5)
		Homopolymer	94	
		Copolymer	80	
Shear stress	MPa	296 K, ASTM D732		(5)
		Homopolymer	65	
		Copolymer	53	
Dielectric constant ε'	–	102–106 Hz, ASTM D150		(5)
		Homopolymer	3.7	
		Copolymer	3.7	
Dielectric loss ε''	–	Copolymer, ASTM D150		(5)
		102 Hz	0.0010	
		103 Hz	0.0010	
		104 Hz	0.0015	
		106 Hz	0.006	

References

1. Novak A, Whalley E. *Trans. Faraday Soc.* 1959;55:1484.
2. Mucha M. *Colloid Polym. Sci.* 1972;162:103.
3. Fleischer D, Schulz RC. *Makromol. Chem.* 1972;162:103.
4. Yamashita Y, Asakura T, Okada M, Ito K. *Makromol. Chem.* 1969;129:1.

5. Dolce TJ, Grates JA. In: Mark HF, Bikales NM, Overberger CG, Menges G, eds. *Encyclopedia of Polymer Science Engineering.* Vol 1. New York: John Wiley and Sons; 42.

6. Serle AG. In: Margolis JM, ed. *Engineering Thermoplastics: Properties and Application.* New York: Marcel Dekker; 1985:151.

7. Tanaka A, Uemura S, Ishida Y. *J. Polym. Sci. Part A-2.* 1970;8:1585.

8. Wagner HL, Wissbrun KF. *Makromol. Chem.* 1965;81:14.

9. Doerffel K, Friedrich H, Grohn H, Wimmers D. *Plaste-Kautschuk.* 1965;12:524.

10. Thuemmler W. *Plaste-Kautschuk.* 1965;12:582.

11. Bel'govskii IM, Enikolopyan NS, Sakhonenka LS. *Polym. Sci. (USSR).* 1963;4:367.

12. Hoehr L, et al. *Makromol. Chem.* 1967;103:279.

13. Wunderlich B. *Macromolecular Physics: Vol. 1. Crystal Structure, Morphology, Defects.* New York: Academic Press; 1973:118.

14. Uchida T, Tadokoro H. *J. Polym. Sci. Part A-2.* 1967;5:63.

15. Starkweather HW, Boyd RH. *J. Phys. Chem.* 1960;64:410.

16. Wunderlich B. *Macromolecular Physics: Vol. 3, Crystal Melting.* New York: Academic Press; 1980:64.

17. Wilski H. *Makromol. Chem.* 1971;150:209.

18. Aoki Y, Nobuta A, Chiba A, Kaneko M. *Polymer J.* 1971;2:502.

19. Salaris F, Turturro A, Bianchi U, Matruscelli E. *Polymer.* 1978;19:1163.

20. Dainton FS, Evans DM, Hoare FE, Melia TP. *Polymer.* 1962;3:263.

21. Gaur U, Wunderlich B. *J. Phys. Chem. Ref. Data.* 1981;10:1005.

Poly(methylene terephthalate)

Xiujuan Zhang and Gui Lin

Acronym PMT, 1GT, Poly(oxymethyleneoxyterephthaloyl)

Class Polyesters; linear and flexible aromatic polyester; thermoplastics

Structure

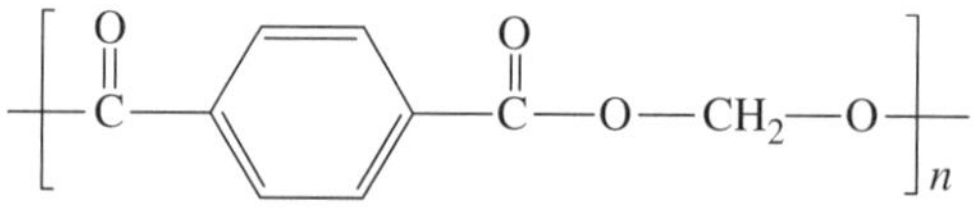

$$\left[\!\!\begin{array}{c} \underset{\underset{C}{\parallel}}{O} \end{array}\!\!-\!\!\left\langle\!\!\bigcirc\!\!\right\rangle\!\!-\!\!\underset{\underset{C}{\parallel}}{O}\!\!-\!\!O\!\!-\!\!CH_2\!\!-\!\!O\!\!\right]_n$$

Major Applications Fiber, filament, yarn, film and the like[1]

Properties of Special Interest PMT can be viewed as a candidate of higher strength and modulus material.

Related Polymers Poly(methylene/ethylene terephthalate) copolymers[2,3]

Preparation/Polymerization Conditions

(1) Synthesized by triethylamine-mediated reactions of terephthalic acid and dihalomethanes.[3]
(2) Synthesized as white powders via the reaction of cesium or potassium terephthalates with dibromomethane or bromochloromethane in N-methylpyrrolidone at temperatures of 80–125°C.[4–6]

Properties

Solubility Certain mixtures oftetrachloroethane/trifluoroacetic acid (18/82-70/30 v/v) dissolved the polymer in the temperature range of 45–65°C.[5]

IR results[5]

Group	Unit: cm^{-1}
Strong C=O absorption	1740
Overlapping C—O stretching	1000–1300
C—H bending vibrations	1000–1300
C—H deformation in p-substituted aromatic rings	725
—C=C— ring stretching	1580, 1410
C—H stretching vibrations	3000–3100

Poly(methylene terephthalate)

WAXD result

2θ (°)	hkl	d_{hkl} (Å)	a (Å)	b (Å)	C (Å)	α (°)	β (°)	γ (°)
16.74	010	5.29	4.43	5.94	9.42	97.3	126	106.1
24.06	110	3.70						
27.90	100	3.20						
32.46	021	2.76						
41.36	105	2.18						

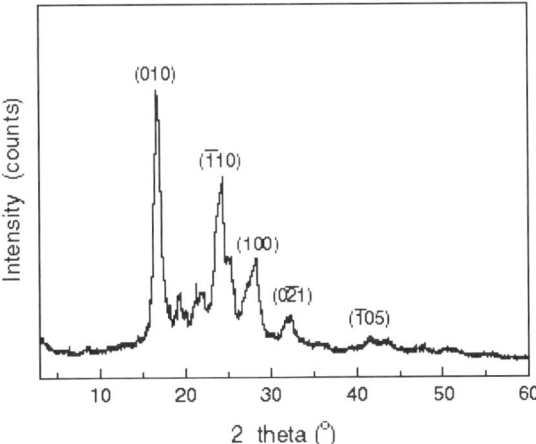

WAXD pattern of crystalline PMT[3,6]

Property	Units	Conditions	Value	Reference
Molecular conformation	–	–	Nearly planar	–
Molecular weight (of repeat unit)	g mol^{-1}	–	178	–
Solvent	–	In the range of 45–65°C	tetrachloroethane/ trifluoroacetic acid (18/82–70/30 v/v)	(6)
Unit cell	–	–	Triclinic crystal systems	(6)
Lattice constant		X-ray diffraction	a = 4.43 Å b = 5.94 Å c = 9.42 Å α = 97.3° β = 126° γ = 106.1°	(6)
Glass transition temperature T_g	°C	DSC, 20°C min^{-1}, N$_2$	95–104	(5)
Melting temperature T_m	°C	DSC, 20°C min^{-1}, N$_2$	245.1–269	(5, 6)

Poly(methylene terephthalate)

Property	Units	Conditions	Value	Reference
Crystallization Temperature, T_c	°C	DSC, 20°C min^{-1}, N_2	135–192.8	(5, 6)
The Avrami exponents n		Isothermal crystallization; non-isothermal crystallization	2–3; 3–4	(6)
The Ozawa exponents m		non-isothermal crystallization in the range of 135–155°C	1–3	(6)
Heat of fusion ΔH	kJ mol^{-1}	Isothermal crystallization; non-isothermal crystallization	78.8; 94.5	(6)
Degradation Temperature	°C	TGA, 20°C min^{-1}, in N_2	Step I: 275–400 Step II: 420–500	(5)

References

1. Kuwayama S, Sato M. Textiles for processing of magnetic disk texture. *Jpn. Kokai Tokkyo Koho.* 2005:10 pp. JP 2005256241 A 20050922.
2. Pinkus AG, Hariharan R. Preparation of poly(methylene/ethylene terephthalate) copolymers. *U.S. Pat. Appl. Publ.* 2003.
3. Pinkus AG, Hariharan R, Thrasher LP, Kesse AP. Synthesis of poly(methylene terephthalate) and copolymers with poly(ethylene terephthalate). *J. Macromolecular Sci., Pure Appl. Chem.* 2000;A37(9):1037–1051.
4. East GC, Morshed M. The preparation of poly(methylene esters). *Polymer.* 1982;23:1555.
5. Cimecioglu AL, East GC, Morshed M. The synthesis and characterization of poly(methylene terephthalate). *J. Polym. Sci., Part A: Polym. Chem.* 1988;26:2129–2139.
6. Run MT, Yao CG, Wang YJ. Morphology, isothermal and non-isothermal crystallization kinetics of poly(methylene terephthalate). *Euro. Polym. J.* 2006;42(3):655–662.

Poly(methyl methacrylate)

Shaw Ling Hsu

Acronyms, Trade Names PMMA, Plexiglas, Lucite, Elvacite, Plex, Diakon

Class Acrylics

Structure $-[CH_2-C(CH_3)(OCOCH_3)]-$

Chemical Registry Number, 9011-14-7

Commercial grade materials generally has 50–70% syndiotactic, ∼30% atactic, and <10% isotactic.[1]

Special Properties Optically clear (92% transmission, theoretical limit for normal incidence, in the visible region) through the visible wavelength range; very little ultra-violet absorption until 260 nm.

Major Applications Replacement for glass. Good mechanical properties. Can be used in various biomedical applications. Extremely high weatherability. Commercial materials are usually atactic polymers (∼75% syndiotactic), although isotactic and syndiotactic polymers have been synthesized. High sensitivity to electron radiation. Can be used as one component deep UV, electron beam, or ion beam resists in microelectronics chips manufacturing.[2,3]

Infrared Absorption[4] Methylene stretching vibrations, 2958 and 2933 cm^{-1} assigned to asymmetric and symmetric CH_2 stretching vibrations.
  Ester methyl stretching vibrations, 2995, 2948, and 3025 cm^{-1}.
  Carbonyl vibration, 1733 cm^{-1}.

NMR Transitions Atactic, 1.9 ppm vs. TMS.[5,6]
  Syndiotactic.[5]
  Isotactic, pair of doublets between 1.5–2.5 ppm vs. TMS.[5]

Effects of Radiation Sensitivity to electron beam is 10^{-5} C cm^{-2} at an e-beam energy of 25 KeV;[7] main chain scission, 0.46 (0.5 J cm^{-2} at 26 KeV).[7]
  UV, 50% (2000 Å), 57% (2200 Å), 78% (2400 Å), 78% (2600 Å), 78% (3000 Å);[8] main chain scission is 0.22 (4–6 eV, 0.6 J cm^{-2}).[7]
  ion beam, 0.75 (0.48 J cm^{-2}; 300 KeV).[7]
  visible range, effectively 92% transmission, 2% haze.

Poly(methyl methacrylate)

Density 1.17–1.20 gm cm$^{-3(9)}$

Glass Transition Temperature Atactic polymer 106°C$^{(9)}$; 114.3°C$^{(10)}$; 113°C$^{(11)}$
Isotactic polymer 45°C$^{(1,11)}$; 51°C$^{(12)}$
Syndiotactic polymer $120 \geq 140$°C$^{(1,11,12)}$

Heat Deflection Temperature 155–210°F, 1.82 MPa$^{(9)}$

Dielectric Constants 3.5–3.7 (50 Hz, 25°C);$^{(13-15)}$ 3.3 (1000 Hz);$^{(15)}$ 2.2–2.5 $(1.0 \times 10^6$ Hz)$^{(15)}$

Water Absorption 0.3–0.4% (1/8 inch bar, 24 h)$^{(9)}$; 2%$^{(14)}$; 0.1–0.3%$^{(15)}$

Thermal Expansion Coefficient $6 \times 10^{-4} > T_g$; $2-3 \times 10^{-4} < T_g$$^{(13)}$

Crystalline Structures for PMMA Isotactic PMMA
Unit cell parameters, isotactic isomer.$^{(16)}$
$a = 20.98$ Å, $b = 12.06$ Å, c (fiber axis) = 10.40 Å
Syndiotactic PMMA
Syndiotactic PMMA can only crystalline phase when complexed with various solvents.$^{(17,18)}$
$a = 25.8$ Å, $b = 35.1$ Å (with chloroacetone); fiber repeat $c = 35.4$ Å irrespective of the type of solvent.$^{(18)}$

Index of Refraction 1.49.$^{(9,13)}$

Tensile Strength 48–76 MPa.$^{(1,9)}$

Fracture Toughness 1.21 MPa m$^{1/2}$ (23°C, air); 1.76 MPa m$^{1/2}$ (37°C, water).$^{(10)}$

Elongation 2–10%.$^{(9)}$

Tensile Modulus 3.1 GPa$^{(9)}$; 3.18 GPa (23°C, air), 2.70 GPa (37°C, water).$^{(10,15,19)}$

Poisson's Ratio 0.35.$^{(20)}$

Flexural Modulus 2.9–3.1 GPa.$^{(1)}$

Melt Flow Rate 20–30 (low heat resistance material); 2–4 (high heat resistance material).$^{(1)}$

Impact Strength 0.3–0.5 (ft-lb/in of notch).$^{(9)}$

Continuous Use Temperature 91–109°C$^{(1)}$

Typical Solvents Ethanol, isopropanol, methyl ethyl ketone, formic acid, nitroethane; any alcohol solutions containing 10% alcohol may attack PMMA.$^{(15)}$

Typical Non-solvent Turpentine, carbon tetrachloride, butylene glycol, diethyl ether, isopropanol ether, m-cresol.

Suppliers Rohm and Haas and a large number of Asian suppliers.

References

1. Salamone JC, Ed. *Polymeric Materials Encyclopedia.* New York: CRC Press; 1996.
2. Htoo MS, Ed. *Microelectronic Polymers.* New York, NY: Marcel Dekker; 1989.
3. Thompson LF, Willson CG, Frechet JMJ, Eds. *Materials for Microlithography: Radiation-Sensitive Polymers.* Vol 266. Washington D. C: American Chemical Society; 1984.
4. Lipschitz I. *Polym-Plast. Technol. Eng.* 1982;19:53.
5. Schilling FC, Bovey FA, Bruch MD, Kozlowski SA. *Macromolecules.* 1985;18:1418.

6. Wu J, Beshah K. *NMR Spectroscopy of Polymers in Solution and in the Solid State ACS Symposium Series.* Vol 834. Washington DC: Amer Chemical Soc; 2003:345–357.

7. Clough RL, Shalaby SW, Eds. *Radiation effects on polymers.* Vol 475. Washington D. C: American Chemical Society; 1991.

8. Lin BJ. *J. Vac. Sci. Technol.* 1975;12:1317.

9. Billmeyer FW. *Textbook of Polymer Science.* New York: John Wiley & Sons; 1984.

10. Johnson JA, Jones DW. *J. Mat. Sci.* 1994;29:870.

11. John E, Ree T. *J. Polym. Sci. Part A.* 1990;28:385–398.

12. Kitayama T, Fujimoto N, Terawaki Y, Hatada K. *Polymer Bulletin.* 1990;23:279–286.

13. Wunderlich W, Ed. *Physical Constants of Poly(methyl methacrylate).* Second Ed. New York: John Wiley & Sons; 1975.

14. Mazur K. *Journal of Physics D: Applied Physics.* 1997;30:1383–1398.

16. Tadokoro H. *Structures of Crystalline Polymers.* New York: John Wiley & Sons; 1979.

17. Fox TG, Garret BS, Goode WE, Gratch S, Kincaid JF, Spell A, Stroupe, JD. *J. Am. Chem. Soc.* 1958;80:1768.

18. Kusuyama H, Miyamoto N, Chatani Y, Tadokoro H. *Polymer Communications.* 1983;24:119–122.

19. De Santis R, Mollica F, Ambrosio L, Nicolais L, Ronca D. *Journal of Materials Science-Materials in Medicine.* 2003;14:583–594.

20. Bailey FE, Koleske JV. *Alkylene Oxides and Their Polymers.* Vol 35. New York: Marcel Dekker; 1991.

Poly(4-methyl pentene-1)

R. P. Patki, D. R. Panse and P. J. Phillips

Acronyms and Trade Names Polymethylpentene, PMP, P4MPE, TPX, Crystalor

Class α-Olefins

Structure of the Repeat Unit

$$-[-CH_2-CH-]-$$
$$|$$
$$CH_2CH(CH_3)_2$$

Major Applications Hypodermic syringes, needle hubs, blood collection and transfusion equipment, pacemaker parts, cells for spectroscopic and optical analysis, laboratory ware, light covers, automotive and electrical components, microwave components, cookware.

Properties of Special Interest High optical transparency, excellent dielectric properties, high thermal stability, chemical resistance, crystalline density lower than amorphous density.

Preparative techniques

Type of Polymerization	Conditions	Reference
Coordination	Catalytic systems used: α and δ-TiCl$_3$ in combination with Al(C$_2$H$_5$)$_2$Cl, VCl$_3$-Al(i-C$_4$H$_9$)$_3$	(1)
	Supported catalysts such as TiCl$_4$/MgCl$_2$-Al(C$_2$H$_5$)$_3$ modified by aromatic acid esters, diesters in temperature range of 30–70°C	(2)
	Dibenzyl titanium and zirconium complexes of three amine bis(phenol-late) ligands.	(3)
	Homogeneous Ziegler-Natta catalyst systems: M(acac)$_3$–AlEt$_2$Br; (M = Cr, Mn, Fe and Co) at 40°C in benzene	(4)
Metallocene	Me$_2$Si(η1-C$_{29}$H$_{36}$)-(η^1-N-^{t}Bu)ZrCl$_2$.OEt$_2$/MAO cocatalyst, (MAO = methylaluminoxane)	(5)
	Monocyclopentadienylamido (CpA) catalyst complex [η^5: η^1-(2,3-Me$_2$Benz[e]Ind)SiMe$_2$N^{t}Bu]TiCl$_2$	(6)
	En(Ind)$_2$ZrCl$_2$ and iPr(Cp)(Flu)ZrCl$_2$, with MAO	(7)
Cationic	Catalysts AlCl$_3$, AlBr$_3$, AlC$_2$H$_5$Cl$_2$ and cocatalysts RCl with R=CH$_3$, C$_2$H$_5$, C$_6$H$_5$	(1)

Typical comonomers used: 1-hexene, 1-pentene, 1-octene, 1-decene, 1-octadecene.

Poly(4-methyl pentene-1)

Structure and morphology

Property	Units	Condition	Value	Reference
Molecular weight of repeat unit	g mol^{-1}	–	84.16	
Stereoregularity	% isotactic	Catalyst system: $\delta-$TiCl$_3$-Al(i-C$_4$H$_9$)$_3$ $\delta-$TiCl$_3$-Al(C$_2$H$_5$)$_2$Cl rac-Me$_2$Si(2-Me-Benz[e]Ind)$_2$ZrCl$_2$	60 90 94	(8) (9) (6)
	% syndiotactic	Me$_2$Si(η^1-N-^{t}Bu)(η^1-C$_{29}$H$_{36}$) ZrCl$_2$.OEt$_2$/MAO	97	(10)
Typical molecular weight range	g mol^{-1}	Cationic polymerization Coordination polymerization Metallocene	2000–250 000 6000–115 000 75 000–460 000	(1) (3, 4) (5–7)
Typical polydispersity index	–	Cationic polymerization at −78°C −50°C 5°C	2.76 2.85 4.11	(1)
		Catalyst system: Dibenzyl titanium and zirconium complexes of three amine bis(phenol-late) ligands	1.26–1.57	(3)
		Monocyclopentadienylamido (CpA) catalyst complex [η^5: η^1-(2,3-Me$_2$Benz[e]Ind) SiMe$_2$N^{t}Bu]TiCl$_2$	1.95–2.51	(6)
		En(Ind)$_2$ZrCl$_2$ and iPr(Cp) (Flu)ZrCl$_2$, with MAO	1.81–1.85	(7)

Equation of State properties

Property	Units	Condition	Value	Reference
Thermal expansion coefficient	K^{-1}	ASTM D696	1.17×10^{-4}	(1, 11)
Reducing temperature	K	235–320°C	11481	(12)
Reducing pressure	Pa	0–200 MPa	453×10^6	
Reducing volume	cm^3 g^{-1}		1.2303	
Amorphous density	g cm^{-3}		0.838	(13)

Solution properties

Solvents	Nonsolvents	Reference
Above 100°C: cyclohexane, tetralin, decalin, xylene, chlorobenzene	Any organic solvents at room temperature	(13)

Property	Units	Condition	Value	Reference
Solubility parameter	$(MPa)^{0.5}$		15.14–16.36	(14)
Theta temperature	K	90 for 94% isotactic polymer solvent/method		(15)
		diphenyl/PE, VM	467.6	
		diphenyl ether/PE, VM	483	
		diphenyl methane/PE, VM	449.6	
Characteristic ratio C∞		Temperature(°C)/method		
		205/VM*	6.7	(16)
		210/VM*	13.2	(15)

* PE-Phase equilibria, VM-intrinsic viscosity/molar mass.

Mark-Houwink parameters

Solvent/method	Temperature (°C)	Mol. wt. $\times 10^{-4}$ (g mol^{-1})	K $\times 10^3$ (ml g^{-1})	A	Reference
Biphenyl/OS	$\Theta = 194.6°C$	30	152	0.5	(15)
Decalin/OS	130°C	30	19.5	0.75	
Diphenyl ether/OS	$\Theta = 210°C$	30	158	0.5	
Diphenyl methane/OS	$\Theta = 176.6°C$	30	160	0.5	

* OS-Osmotic pressure, Θ-Theta temperature.

Crystalline state properties
Isotactic polymorphs

Property	I	II	III	IV	Reference
Lattice	Tetragonal	Tetragonal	Tetragonal	Hexagonal	(17–25)
Unit cell dimensions (Å)					
A	18.6–18.7	19.16	19.46	22.17	
B	18.6–18.7	19.16	19.46		
C	13.8	7.12	7.02	6.5	
Unit cell angles (°)	All 90	all 90	all 90		
Monomers per unit cell	28				
Helix conformation	7_2	4_1	4_1	3_1	
Space group	S4–1, P⁻–4		$I4_1$		
Crystalline density@ 23°C	0.814				

Poly(4-methyl pentene-1)

Syndiotactic polymorphs

Property	I	II	Reference
Lattice		Tetragonal	(17, 26)
Unit cell dimensions (Å)			
A		18.03	
B		18.03	
C		46.91	
Unit cell angles (°)		all 90	
Helix conformation	24_7	$(12/7)_2$	
Space group			

Property	Units	Condition	Value	Reference
Degree of crystallinity	%	annealed	70	(27)
		strongly oriented fiber	85	
		moldings	55–60	
Heat of fusion	kJ mol^{-1}	Clapeyron equation	5.297	(28)
			5.205	(27)
Entropy of fusion	kJ K^{-1} mol^{-1}	Clapeyron equation	10.1×10^{-3}	(28)
			10.3×10^{-3}	(27)
Avrami exponent n	–	DSC, crystallization temperature	1.64–1.92	(29)
Avrami rate constant k	min^{-n}	range: 221–216°C	1.9×10^{-2}–5.7×10^{-1}	

Transition temperatures

Property	Units	Condition	Value	Reference
Glass transition	K	DSC	323	(30)
			303	(31)
Melting temperature	K	isotactic polymer syndiotactic polymer	518	(1)
			483	(26)
Sub T_g transition	K		153–123	(1, 13)
			23	(1, 13)
Crystalline phase disordering	K		403–453	(13)

Thermodynamic and other properties

Property	Units	Condition	Value	Reference
Heat capacity	kJ mol^{-1} K^{-1}	Temperature (K)		(31)
		80	0.0472	
		180	0.0917	
		250	0.121	
		300	0.145	
Deflection temperature	K	Under flexural load		(1)
		0.46 MPa	353–363	
		1.82 MPa	321–323	

Mechanical properties

Property	ASTM Method	Units	Value	Reference
Tensile modulus	D638	MPa	1500–2000	(1)
Bulk modulus	–	MPa	2670	
Tensile strength at yield	D638	MPa	23–28	
Tensile strength at break	D638	MPa	17–20	
Elongation at break	–	%	10–25	
Flexural strength	D790	MPa	25–35	
Flexural modulus	D790	MPa	1300–1800	
Notched Izod impact strength	D256	kJ m^{-1}	100–200	(11)
			L80–90	(32)
Rockwell hardness	–	–	0.34	(33)
Poisson ratio	@ RT and ambient pressure	–	0.43	(1)
Shear modulus	@ RT and ambient pressure	MPa	970	

Electro-Optical properties

Property	Units	Condition	Value	Reference
Index of refraction n	–	Isotactic polymer	1.463	(13)
Haze	%	ASTM D1003	1.2–1.5	
Optical transparency	%	ASTM D1003	90–92	
Dielectric constant	–	25°C, 102–106 Hz	2.12	(1)
Dielectric loss factor		20°C, frequency range:		
		50 Hz	60×10^{-6}	(11)
		1 kHz	35–140×10^{-6}	(11)
		1 MHz	25–50×10^{-6}	
Dielectric breakdown voltage	kV mm^{-1}	–	42–65	
Volume resistivity	Ω cm	–	>1016	

Poly(4-methyl pentene-1)

Surface and interfacial properties

Property	Units	Condition	Value	Reference
Surface tension	mN m^{-1}	@ 20°C, contact angle method	25	(11)

Transport properties

Property	Units	Condition	Value	Reference
Thermal conductivity	W m^{-1} K^{-1} m^3 (STP)	ASTM C177	0.167	(11)
Permeability coefficient	m s^{-1} m^{-2} Pa^{-1} × 10^{-16}	film thickness = 78 μm permeant:		(34)
		O$_2$	317.2	
		N$_2$	74	
		He	1020	
	–	H$_2$	1342	
Gas separation factor		CO$_2$	960	(35)
		gas 1/gas 2		
		O$_2$/N$_2$	4.1	
		H$_2$/N$_2$	16.5	
		CO$_2$/N$_2$	8.6	
		CO$_2$/O$_2$	2.1	
		H$_2$/O$_2$	4.1	
		H$_2$/CO$_2$	1.9	(11)
Melt index	g per 10 min	@ 260°C, 5 kg load	20	
Speed of sound, longitudinal	ms^{-1}	–	2180	(32)
Speed of sound, shear	ms^{-1}		1080	(32)

Stabilities

Property	Units	Condition	Value	Reference
Pyrolyzability, amount of product	%	name of product		(36)
		propene	0.8	
		propane	33.9	
		2-methylpropene	55.6	
		2-methylpropane	3.5	
		2-methylbutene	2.0	
		pentane	0.3	
		4-methyl 1-pentene	2.2	
		2,3-dimethylbutane	1.0	
		others	0.7	
Vicat softening point	K	ASTM D1525	446	(11)
Degradation temperature	K	–	553	(1)
Radiation G (product)	–	per 100 eV of absorbed radiation	0.3	(37)

Property	Units	Condition	Value	Reference
G(S)/G(X)	–	irradiated in air	0.6	(37)
Water absorption	%	saturation	0.01	(1)
Flammability, flame propagation rate	cm min^{-1}	ASTM D635	2.5	(11)

Practical matters

Manufacturing/Trading Company	Trade Name
Mitsui Petrochemical Industries	TPX
Phillips 66	Crystalor
British Petroleum Company	–

References

1. Kissin YV. In: Kroschwitz JI, ed. "Encyclopedia of Polymer Science and Engineering," Vol 2. 2nd ed. NY: John Wiley & Sons; 1985.
2. Gaylord NG, Mark HF. "Linear and Stereoregular Addition Polymers". Interscience Publishers Inc; 1959.
3. Gendler S, Groysman S, Goldschmidt Z, et al. *J. Polym. Sci., Polym. Chem.* 2005;44:1136.
4. Balaji R, Kothandaraman H, Rajarathnam D. *Eur. Polym. J.* 2002;38:1055.
5. Irvin LJ, Reibenspies JH, Miller SA. *J. Am. Chem. Soc.* 2004;126:16716.
6. Xu G, Cheng D. *Macromolecules.* 2001;34:2040.
7. Kawahara N, Saioto J, et al. *Polymer.* 2007;48:425.
8. Kissin YV. "Isospecific Polymerization of Olefins with Heterogeneous Ziegler-Natta Catalysts". Springer-Verlag, NY: 1985.
9. Tait PJT. In Chien JCW, ed. "Coordination Polymerization". NY: Academic Press Inc; 1975.
10. Irvin LJ, Miller SA. *J. Am. Chem. Soc.* 2005;127:9972.
11. Geoffrey Heggs T. "Ullmann's Encyclopedia of Industrial Chemistry". Vol A21. VCH Publishers Inc; 1992.
12. Zoller P. *J. Polym. Sci., Polym. Phys.* 1978;16:1491.
13. Othmer K, ed. "Olefin Polymers (Higher Olefins)," in "Encyclopedia of Chemical Technology". NY: John Wiley & Sons; 1996.
14. Fedors RF. *Polym. Eng. Sci.* 1974;14:147.
15. Tani S, Hamada F, Nakajima A. *Polym. J.* 1973;5:86.
16. Neuenschwander P, Pino P, *Eur. Polym. J.* 1983;19:1075.
17. Frank FC, Keller A, O'Connor A. *Phil. Mag.* 1959;4:200.
18. De Rosa C. *Macromolecules.* 2003;36:6087.
19. Natta G, Corradini P, Bassi W. *Rend. Fis. Acc. Lincei.* 1955;19:404.
20. Charlet G, Delmas G. *Polymer.* 1984;25:1619.
21. Takayanagi M, Kawasaki N. *J. Macromol. Sci. Phys.* 1967;B1:741.
22. De Rosa C, Borriello A, Venditto V, Corradini P. *Macromolecules.* 1994;27:3864.
23. De Rosa C, Auriemma F, Borriello A, Corradini P. *Polymer.* 1995;36:4723.
24. Charlet G, Delmas G. *Polym. Bull.* 1982;6:367.
25. De Rosa C. *Macromolecules.* 1999;32:935.

26. De Rosa C, Grassi A, Capitani D. *Macromolecules.* 1998;31:3163.

27. Zoller P, Starkweather HW, Jones GA. *J. Polym. Sci., Polym. Phys.* 1986;24:1451.

28. Charlet G, Delmas G. *J. Polym. Sci., Polym. Phys.* 1988;26:1111.

29. Suh J, White JL. *J. Appl. Polym. Sci.* 2007;106:276.

30. Brydson JA. "Plastic Material". 4th ed. Kent, UK: Butterworth & Co. Ltd; 1982.

31. Gaur U, Lau SF, Wunderlich BB, et al. *J. Phys. Chem., Ref. Data.* 1983;12:29.

32. Hartmann B. *J. Appl. Phys.* 1980;51:310.

33. Warfield RW, Barnet FR. *Die. Angew. Makromol. Chem.* 1972;27:215.

34. Yasuda H, Rosengren KJ. *J. Appl. Polym. Sci.* 1970;14:2839.

35. Levasalmi JM, McCarthy TJ. *Macromolecules.* 1995;28:1733.

36. Regianto L. *Makromol. Chem.* 1970;132:113.

37. Soboleva NS, Leshchenko SS, Karpov VL. *Polym. Sci. USSR.* 1983;25:446.

Polymethylphenylsiloxane

Alex C. M. Kuo

Acronym, Alternate Names, Trade Names PMPS; phenylmethyl siloxane oil: poly[oxy(methylphenylsilylene)]; Dow Corning® 710 Fluid.

Class Polysiloxanes

Structure $-[(CH_3)(C_6H_5)Si-O-]_n$; $MD_n^{ph}M$

Major Applications Heat transfer fluids; high temperature lubricating oil for instruments and bearings; glass sizing agents; greases; chromographic stationary phases. hydraulic fluids and elastomer and sealant using copolymer with dimethylpolysiloxane for low temperature application.

Properties of Special Interest Thermal stability, oxidative stability, wide serviceable temperature (-70 to $260°C$) and minimal temperature effect, good shear resistance and resistance to UV radiation, excellent antifriction, lubricity and good dielectric strength.

CAS Registry Number [9005-12-3] [63148-58-3]

Preparative Techniques[1,2,3] Hydrolysis of dichloromethylphenylsilane, polycondensation of methylphenylsiloxane diol: ring-opening polymerization of methylphenylsiloxane oligomeric cyclics.

Property	Units	Conditions	Value	Reference
Infrared absorption	cm^{-1}	$Si-O-Si$	1,000–1,130	(4, 5)
		$Si-(C_6H_5)$	3,020–3,080; 1,590; 1,430; 1,120; 700;730	(4, 5)
		$Si-(CH_3)$	760–845; 1,245–1,275	(4, 5)
		$Si-H$	2,100–2,300; 760–910	(4, 5)
		$Si-OH$	3,200–3,695; 810–960	(4,5)
		$Si-CH=CH_2$	1,590–1,610; 990–1,020; 980–940	(4, 5)
Ultraviolet (UV) absorptiony	nm	$Si-(C_6H_5)$	270; 264; 259	(6)
^{29}Si Nuclear Magnetic Resonance Spectroscopy	ppm	$-[O-Si(CH_3)_2(C_6H_5)]$	-1	(7, 8)
		$-[O-Si(CH_3)(C_6H_5)]-$	-31 to -35	(7, 8)
		$[O-Si(CH_3)(C_6H_5)]_3$	-21	(7, 8)
		$[O-Si(CH_3)(C_6H_5)]_4$	-30.5	(7, 8)
		$(-O_{0.5}-)_3Si-C_6H_5$	-77 to -82	(7, 8)
		$(-O_{0.5}-)_4Si$	-105 to -115	(7, 8)
TOF-SIMS spectra	Mass/Da	Relative high intensity of silver cationized oligomer	1,357; 1,493; 1,629	(9)

Property	Units	Condition	Value				Reference
Specific gravity	g cm^{-3}	PMPS (3.5 cs) at 20°C	0.9809				(10)
		PMPS (27 cs) at 20°C	1.0735				(10)
		PMPS (102 cs) at 20°C	1.0787				(10)
		PMPS (500 cs) at 25°C	1.110				(11)
		PMPS ($M = 3.27 \times 10^5$) at 25°C	1.115				(12)
		Phenyl terminated PMPS (600 cs) at 25°C	1.115				(13)
Density-temperature-molecular weight relationship	g cm^{-3}	Trimethylsiloxy-ended PMPS at 0–60°C	$1/\rho = 0.7303 + (4.4893 \times 10^{-4})T + (0.1814T + 16.3684)/M$				(14)
Solubility parameter δ	(MPa)$^{1/2}$	Silica filled PMPS elastomer measured by swelling	18.4				(15)
Gas solubility coefficient S	cm^{-3} (STP)/cm^3 polym. atm			10°C	35°C	55°C	
			CO_2	1.19	0.81	0.76	(16)
			CH_4	0.3	0.25	0.20	(16)
			C_3H_8	8.57	3.79	2.65	(16)
Solvents		toluene, chloroform, butanol, diethyl ether, ethyl acetate, acetone (hot), benzene, cyclohexane, ethylbenzene, methyl cyclohexane, THF.					(15, 17, 32)
Non-solvents		methanol, ethanol, ethylene glycol, acetonitrile					(15, 17)
Theta temperature Θ	K	Diisobutylamine	303.4				(18)
Second virial coefficients A_2	mol cm^3 g^{-2}	PMPS ($M_n = 4.06 \times 10^5$) in cyclohexane at 25°C	1.52×10^{-4}				(18)
Characteristic ratio of unperturbed dimension $(\langle r_o^2 \rangle/\langle r_{of}^2 \rangle)^{1/2}$	–	PMPS ($5 \times 10^4 < M < 1.5 \times 10^6$) diluted in toluene and cyclohexane at 25°C	1.56				(18)
Characteristic ratio $C_\infty = \langle r^2 \rangle / nl^2$	–	Undiluted PMPS with 100 bonds equilibrated at 383 K	10.7				(19)
Mean square end-to-end chain length $(\langle r^2 \rangle / M)^{1/2}$	Nm mol$^{1/2}$ g$^{-1/2}$	PMPS ($5 \times 10^4 < M < 1.5 \times 10^6$) at 25°C	5.65×10^{-2}				(18)
		Value calculated for $l = 1.65$Å, $\theta_1 = 110°$, $\theta_2 = 143°$	3.63×10^{-2}				(18)
Z-average radii of gyration $\langle s^2 \rangle_z$	–	PMPS in benzene-d$_6$ at 293K ($M_z = 3,890$)	11.9				(20)
		PMPS in benzene-d$_6$ at 293K ($M_z = 8,500$)	18.6				(20)
		PMPS in benzene-d$_6$ at 293K ($M_z = 21,130$)	26.7				(20)

Property	Units	Condition	Value	Reference
Enthalpy of fusion ΔH_u	$J\,g^{-1}$	Semicrystalline PMPS	4.5	(2)
Viscosity temperature coefficient VTC	–	$MD_3{}^{ph}M$ PMPS polymer (500 cs)	0.692	(14)
			0.79	(11)
		PMPS polymer (482 cs)	0.88	(21)
		50% Phenylmethyl 50% dimethyl polysiloxane (115 cs)	0.78	(21)
		10% Phenylmethyl 90% dimethyl polysiloxane(42 cs)	0.61	(21)
Activation energy for viscous flow ΔE_{visc}	$kJ\,mol^{-1}$	PMPS polymer	50.2	(22)
		PMPS polymer	49.8	(23)
Refractive index ($n_D{}^{25}$)	–	$MD^{ph}M$ at 25°C	1.4442	(12)
		$MD_2{}^{ph}M$ at 25°C	1.4744	(12)
		$MD_3{}^{ph}M$ at 25°C	1.4889	(12)
		PMPS (500 cs) at 25°C	1.533	(11)
		PMPS ($M = 40,000$)	1.550	(24)
Refraction Index n vs. T (°C)	–	Cured elastomer ($M_c = 40,000$) at 10 to 50°C	$n = 1.5626-$ $3.54 \times 10^{-4}\,T$	(12)
Coefficients of cubical expansion α	K^{-1}	3.5 cs PMPS at 20°C	8.7×10^{-4}	(10)
		27 cs PMPS at 20°C	7.2×10^{-4}	(10)
		102 cs PMPS at 20°C	7.1×10^{-4}	(10)
		500 cs PMPS (273–427 K)	4.3×10^{-4}	(11)
		PMPS cured elastomer at –20 to 25°C	4.69×10^{-4}	(25)
		PMPS elastomer elastomer at 30 to 90°C	8.52×10^{-4}	(26)
Glass transition temperature T_g	K	PMPS ($M_n = 300$)	155.8	(27)
		PMPS ($M_w = 8,000$)	242	(2)
		PMPS ($M_n = 27,300$)	247	(28)
		PMPS ($M_n = 93,000$)	240.5	(27)
		PMPS ($M \rightarrow \infty$)	251.3	(2)
		PMPS cured elastomer from Volume–temperature data	187	(25)
Stiffening temperature T_s	K	PMPS cured elastomer measured by Gehman cold flex testing	246.3	(22)
Pseudo-equilibrium crystallization temperature T_{pc}	K	PMPS cured elastomer	237.6	(29)
Melting point T_m	K	Semicrystalline PMPS	308	(2)
Compressibility β	%	710 fluid at 34.5 MPa	1.7	(11)
		710 fluid at 69 MPa	3.15	(11)
		710 fluid at 138 MPa	5.5	(11)
Water contact angle θ	Degrees	Film of PMPS (610 cs) on soda-lime glass, after 15 min treatment at 100°C	77	(30)
		200°C	81	(30)
		300°C	83	(30)

Polymethylphenylsiloxane

Property	Units	Condition	Value	Reference
		400°C	81	(30)
		450°C	60	(30)
		475°C	0	(30)
Surface tension γ	mN m^{-1}	3.5 cs PMPS at 20°C	29.6	(10)
		27 cs PMPS at 20°C	28.6	(10)
		50 cs PMPS at 20°C	27.2	(10)
		102 cs PMPS at 20°C	26.1	(10)
		500 cs PMPS at 25 °C	28.5	(11)
Temperature coefficient of surface tension $-d\gamma_{12}/dT$	mN m^{-1} K^{-1}	3.5 cs PMPS at 20°C	0.080	(10)
		27 cs PMPS at 20°C	0.080	(10)
		50 cs PMPS at 20°C	0.11	(10)
		102 cs PMPS at 20 °C	0.11	(10)
Flash Point	K	PMPS polymer (500 cs)	575	(11)
		Phenyl terminated PMPS (600 cs)	588.6	(13)
Freeze point (pour point)	K	PMPS polymer (500 cs)	251	(11)
		Phenyl terminated PMPS (600 cs)	271	(13)
Thermal conductivity	W m^{-1} K^{-1}	PMPS polymer (500 cs) at 50°C	0.14	(11)
Specific heat	kJ kg^{-1} K^{-1}	PMPS polymer (500 cs) at 40°C	1.52	(11)
		PMPS polymer (500 cs) at 100°C	1.64	(11)
		PMPS polymer (500 cs) at 200°C	1.84	(11)
Radiation resistance	rads	PMPS polymer (500 cs)	1.7×10^8	(11)
		600 cs PMPS phenyl terminated	4.5×10^8	(13)
Diamagnetic susceptibility X_m	cm^3 g^{-1}	PMPS fluid	5.97×10^{-7}	(31)
Sound velocity	m s^{-1}	PMPS polymer (500 cs) at 278 K	1,370	(11)
Load bearing capacity	kg	High phenyl content PMPS film	200–350	(3)
		Low phenyl content PMPS film	100–200	(3)
Lubricity, shell four ball test (wear scare)	mm	Steel on steel 40 mol% phenyl PMPS-co-PDMS at 1 h/600 rpm/50 kg load/ambient temp	4.13	(1)
		Steel on bronze 40 mol% phenyl PMPS-co-PDMS at 1 h/600 rpm/50 kg load/ambient temp	0.42	(1)
		Steel on steel 25 mol% phenyl PMPS-co-PDMS at 1 h/600 rpm/50 kg load/ambient temp	4.18	(1)
		Steel on bronze 25 mol% phenyl PMPS-co-PDMS at 1 h/600 rpm/50 kg load/ambient temp	2.53	(1)
X-ray diffraction pattern	Å	Semicrystalline PMPS	8.33, 7.69, 4.83, 4.40, 3.8	(2)

Mark-Houwink parameters, *K and a*

Solvents	Temperature (°C)	$K \times 10^3$, $(ml\, g^{-1})$	a	Reference
Toluene	25°C	3.90	0.78	(18)
Diisobutylamine	30.4°C	51.5	0.5	(18)
Cyclohexane	25°C	5.52	0.72	(18)
Cyclohexane	25°C	27.3	0.60	(32)
Cyclohexane	50°C	15.6	0.65	(32)
Methylcyclohexane	20°C	30.6	0.58	(32)
THF	25°C	16.5	0.69	(32)
Toluene	25°C	12.3	0.684	(2)
Toluene	25°C	6.7	0.78	(33)
Benzene	20°C	110.6	0.57	(33)

Interaction parameter χ_{12}

Compounds	Conditions	χ_{12}	Method	Reference
PMPS network/Toluene	at 298 K	0.485	Swelling	(6)
PMPS network/Benzene	at 298 K	0.489	Swelling	(6)
PMPS network/Chloroform	at 298 K	0.496	Swelling	(6)
PMPS network/Cyclohexane	at 298 K	0.632	Swelling	(6)
PMPS network/Hexane	at 298 K	0.891	Swelling	(6)
$MD_{28}^{ph}M/MD_{13}M$	at critical point, $T_c = 458$ K	0.112	Light Scattering	(34)
$MD_{23}^{ph}M/MD_{13}M$	at critical point, $T_c = 518$ K	0.122	Light Scattering	(34)
$MD_{23}^{ph}M/M^{OH}D_{15}M^{OH}$	at critical point, $T_c = 446$ K	0.111	Light Scattering	(35)
PMPS/PDMS (cyclic)	at critical point, $T_c = 442$ K	0.095	Light Scattering	(36)
$MD_3^{ph}M/PDMS$ network	at 298 K	0.345	Swelling	(37)
$MD_2^{ph}M/PDMS$ network	at 298 K	0.438	Swelling	(37)
$MD^{ph}M/PDMS$ network	at 298 K	0.356	Swelling	(37)

Gas permeability coefficient (Pr) of silica filled PMPS membrane at 35°C[38,39]

Gas	$Pr \times 10^8$ $(cm^3(STP)\, cm\, s^{-1}\, cm^{-2}\, cm^{-1}\, Hg)$	Gas	$Pr \times 10^8$ $(cm^3(STP)\, cm\, s^{-1}\, cm^{-2}\, cm^{-1}\, Hg)$
NH_3	10.97	CH_4	0.36
H_2S	8.73	O_2	0.32
C_3H_8	1.39	N_2	0.103
C_2H_6	0.91	H_2	1.15
CO_2	2.26	He	0.35
C_2H_4	0.93		

Polymethylphenylsiloxane

WLF parameters for PMPS

M_n	T_0 (K)	C_1	C_2/K	T_g (K)	$a_{T,\alpha}$ method	Reference
5,000	181.2	20.4	56.76	223.3	Photon correlation spectroscopy	(40,41)
12,000	237.4	23.96	48.8	237.4	Dynamic mechanical measurement	(41,42)
12,000	258.4	7.32	32.5	237.4	Data from dielectric relaxation	(40,42)
27,300	273.2	14.8	66.4	247.2	Photon correlation spectroscopy	(28)
27,300	248.2	14.8	55.9	248.2	Photon correlation spectroscopy	(28)
27,300	273.2	11.8	67.9	247.2	Data from dielectric relaxation	(28)
130,000	243.2	17.69	34.71	243.2	Dynamic mechanical measurement	(40,42)
130,000	261.8	7.47	36.1	243.2	Data from dielectric relaxation	(40,42)

Property	Units	Condition		Value	Reference
Electric properties of PMPS (500 cs)		Frequency, Hz	Dielectric constant	dissipation factor ($\tan \delta \times 10^4$)	(43)
at 25°C		1×10^2	2.98	13	
		1×10^3	2.98	1.6	
		1×10^4	2.98	0.7	
		1×10^5	2.98	3	
		1×10^6	2.98	10	
		3×10^7	2.93	50	
		3×10^8	2.93	200	
		3×10^9	2.79	140	
		1×10^{10}	2.60	170	
Dielectric strength	volts mil^{-1}	PMPS polymer (500 cs) at 100 mil gap, rapid rise		350	(11)
Volume resistivity	Ohm cm^{-1}	PMPS polymer (500 cs)		1.0×10^{13}	(11)
Optical configuration parameter Δa	cm^3	PMPS ($M = 4 \times 10^4$) in benzene		-8.55×10^{-24}	(24)
		Peroxide cure PMPS elastomer at 25°C		-1.21×10^{-25}	(12)
		Peroxide cure PMPS elastomer at 50°C		-1.27×10^{-25}	(12)
		Peroxide cure PMPS swelled in decalin at 25°C		-8.5×10^{-24}	(12)
		Theoretical value for PMPS at energy parameter $E_\delta = -0.8$		-1.16×10^{-25}	(12)
Stress-optical coefficient C	m^2 N^{-1}	PMPS elastomer with $M_c = 4.4 \times 10^4$ at 25°C		5.73×10^{-9}	(12)
Dipole moment ratio per repeat unit $\langle\mu_0\rangle/nm^2$	D	PMPS with $M_w = 1.2 \times 10^5$ in cyclohexane at 25°C		0.31	(44)
		Theoretical value of polymer with 0.5% meso diads		0.2679	(44)
		PMPS elastomer at 25°C		0.26	(45)
Fragility index $F_{1/2}$	–	Crosslinked PMPS elastomer measured by dielectric relaxation spectroscopy		0.77	(45)
Steepness index m_T	–	PMPS polymer measured by dielectric relaxation spectroscopy at pressure of 0.1 MPa		86	(46)

Property	Unit	Condition	Value	Reference
Thermal decomposition point	K	PMPS polymer (500 cs)	643	(11)
Gel time at 550 °F in an air circulating oven	Hours	PMPS polymer (500 cs)	350–400	(13)
		Phenyl terminated PMPS (600 cs)	500–600	(13)
Spontaneous ignition temperature	K	PMPS polymer (500 cs)	761	(11)
Activation energy of depolymerization	kJ mol^{-1}	Trimethylsiloxy terminated PMPS	180	(47)
Decomposition products (condition: random scission at $T > 300°C$)		Mixture of stereoisomeric cyclic trimers and tetramers with small amount of pentamer, benzene, and two more complex oligomers		(48)
Fire parameters		Cone calorimeter test		
Peak rate of heat release	kW m^{-2}	External heat flux 60 kW m^{-2}	90	(49)
Yield of carbon monoxide	kg kg^{-1}	External heat flux 60 kW m^{-2}	0.016	(49)
Specific extinction area	m^2 kg^{-1}	External heat flux 60 kW m^{-2}	1,800	(49)

References

1. Meals RN, Lewis FM. *Silicones.* Chap. 2. New York: Reinhold Publishing; 1959.
2. Momper B, et al. *Polymer Commun.* 1990;31:186.
3. Noll W. *Chemistry & Technology of Silicone.* Chap. 6. New York: Academic Press; 1968.
4. Anderson DR. In: Smith AL, ed. *Analysis of Silicone.* Chap. 10. New York: John Wiley and Sons; 1974.
5. Mayhan KG, Thompson LF, Magdalin CF. *J. Paint Tech.* 1972;44:85.
6. Kuo CM Ph.D. *Dissertation.* University of Cincinnati; 1991.
7. Taylor RB, Parbhooand B, Fillmore DM. In: Smith AL, ed. *Analysis of Silicone.* 2d ed. Chap. 12. New York: John Wiley and Sons; 1991.
8. Williams EA. In: Webb GA, ed. *Annual Reports on NMR Spectroscopy.* Vol 15. London: Academic Press; 1983:235.
9. Dong, Xia Hercules DM. *J Phys. Chem B.* 2001;105:3942.
10. Fox HW, Taylor PW, Zisman WA. *Ind. Eng. Chem.* 1947;39:1401.
11. Dow Corning® 710 Fluid, Product information about Dow Corning Silicone Fluid, Dow Corning Corp., Midland, Michigan, Form No. 22-0281G-01 and 22-281A-76.
12. Llorente MA, de Pierola IF, Saiz E. *Macromolecules.* 1985;18:2663.
13. Schiefer HM, Awe RW, Whipple CL. *J. Chem. & Eng. Data.* 1961;1:155.
14. Nagy J, Gabor T, Becker-Palossy K. *J. Orgamometal. Chem.* 1966;6:603.
15. Yerrick KB, Beck HN. *Rubber Chem. Technol.* 1964;37:261.
16. Shah VM, Hardy BJ, Stern SA. *J. Polym. Sci.: Part B, Polym. Phys.* 1986;24:2033.
17. Kiselov BA, Stepina IA, Ablekova ZP. *Soviet Plastics.* 1970;13.
18. Buch RR, Klimisch HM, Johnanson OK. *J. Polym. Sci.: Part A-2.* 1970;8:541.
19. Beevers MS, Semlyen JA. *Polymer.* 1971;12:373.
20. Clarson SJ, Dodgson K, Semlyen JA. *Polymer.* 1987;28:189.
21. Barry AJ, Beck HN. In: Stone FGA, Graham WAG, eds. *Silicone Polymer.* New York: Academic Press; 1962.
22. Polmanteer KE. *J. Elastoplas.* 1970;2:165.

23. Polmanteer KE. *Rubber Chem. & Technol.* 1987;61:470.

24. Tsvetkov VN, et al. *Vysokomol. Soyed.* 1967;9A;3.

25. Polmanteer KE, Hunter MJ.*J. Appl. Polym. Sci.* 1959;1:3.

26. de Candia F, Turturro A. *J. Macromol. Sci. Chem.* 1972;A6:1417.

27. Clarson SJ, Semlyen JA, Dodgson K. *Polymer.* 1991;32:2823.

28. Boese D, et al. *Macromolecules.* 1989;22:4416.

29. Fischer DJ. *J. Appl. Polymer Sci.* 1961;16:436.

30. Hunter MJ, et al. *Ind. Eng. Chem.* 1947;39:1389.

31. Bondi A. *J. Phys. Coll. Chem.* 1951;55:1355.

32. Salom CJ, Freire J, Hernandez-Fuentes I. *Polymer.* 1989;30:615.

33. Andrianov KA, et al. *Vysokomol. Soedin.* 1972;A14:1816.

34. Kuo CM, Clarson SJ. *Macromolecules* 1992;25:2192.

35. Kuo CM, Clarson SJ. *Eur. Polym, J.* 1993;29:661.

36. Kuo CM, Clarson SJ, Semlyen JA. *Polymer.* 1994;35:4623.

37. Clarson SJ, Galiatsatos V, Mark JE. *Macromolecules.* 1990;23:1504.

38. Stern SA, et al. *J. Polym. Sci.: Part B: Polym. Phys.* 1987;25:1263.

39. Bhide BD, Stern SA. *J. Appl. Polym. Sci.* 1991;42:2397.

40. Ngai KL, Plazek DJ. In: Mark JE, ed. *Physical Properties of Polymers Handbook.* Chap. 25. Woodbury, N.Y: AIP Press; 1996.

41. Plazek DJ, et al. *Colloid Polym. Sci.* 1994;272:1430.

42. Santangelo PG, et al. *J. Non-cryst. Solids.* 1994;172–174:1084.

43. *Table of Dielectric Materials, O.N.R. Contracts N5ori-07801 and N5ori-07858.* Vol 4. Cambridge, Massachusetts: Laboratory for Insulation Research, MIT; 1953:67.

44. Salom C, Freire JJ, Hernanez-Fuentes I. *Polymer J.* 1988;20:1109.

45. Fitz BD, Mijovic J. *Macromolecules.* 1999;32:3518.

46. Paluch M, Roland CM, Pawlus SJ. *J. Chem Phys.* 2002;116:10932.

47. Thomas TH, Kendrick TC. *J. Polym. Sci.: Part A-2.* 1970;8:1823.

48. Grassie N, Macfarlane IG, Francey KF. *Eur. Polym. J.* 1979;15:415.

49. Buch RR. *Fire Safety Journal.* 1991;17:1.

Poly(methylphenylsilylene)

Robert West

Acronym, Alternative Name PMPS, polymethylphenylsilane

Class Polysilanes

Structure $[-CH_3SiC_6H_5-]$

Major Applications Hole transport agent in electrophotography, light-emitting diodes, display devices, and printing processes.

Properties of Special Interest Good film-forming characteristics and efficient hole conductor.

General Information Polysilanes, or poly(silylene)s, are polymers in which the entire main chain is made up of silicon atoms. This structure permits delocalization of the σ-electrons, giving the polysilanes unique electronic properties. Polysilanes have strong UV absorption bands in the near UV region ($\sim$300–400 nm). The excitation energy depends on the polymer chain conformation, which may change with temperature, so many polysilanes are thermochromic. Polysilanes undergo photodegradation with UV light; they can be patterned in photolithographic processes and used as free-radical photoinitiators. They are excellent hole conductors, and display nonlinear optical behavior.

For an overview of polysilanes, see references (1, 2, 3).

Preparative techniques

Reactants	Temp. (°C)	Yield (%)	$M_w \times 10^{-3}$	Reference
PhMeSiCl$_2$, Na, toluene	110	41	200, 6	(4)
PhMeSiCl$_2$, Na, Et$_2$0, 15-crown-5	35	88	66	(5)
PhMeSiCl$_2$, Na, toluene (15% heptane), 15-crown-5	65	40	10.2	(6)
PhMeSiCl$_2$, Na, toluene, ultrasound	110	55	107, 3.3	(7)
PhMeSiCl$_2$, Na, toluene, 2% EtOAc	110	16	431, 11.6	(8)

Property	Units	Conditions	Value	Reference
Repeat unit	g mol^{-1}	C$_6$H$_5$SiCH$_3$	120	–
Molecular weight	Varies greatly depending on polymerization conditions			
Polydispersity	Varies greatly depending on polymerization conditions			
Glass transition temperature T_g	K	Polymer is ordinarily atactic and amorphous	$\sim$393	–

Poly(methylphenylsilylene)

Property	Units	Conditions		Value	Reference
Melting temperature, T_m	K	Polymer is ordinarily atactic and amorphous		~493	
Infrared spectrum	cm^{-1}	–		3,030, 2,960, 2,870, 2,000–1,660, 1,600, 1,530, 1,430, 1,100, 1,265, 830–650, 430	(7)
UV absorption	nm	$M_w = 10^6$		342	(8)
		$M_w = 10^4$; 9,300 (ε = repeat)		341	(5, 9)
		$M_w = 10^3$		332	(9)
Emission spectrum	nm	2-MeTHF solution, $\phi = 0.75, \tau = 0.025$ ps		353	(1)
		Solid, 77 K		350, 480	(10)
		Solid, 298 K		365, 530	(11)
NMR spectra	δ (ppm)	Nucleus	Condition		
		^{29}Si	C$_6$D$_6$	−39.2, −39.9, −41.2	(4, 7)
		^{13}C	C$_6$D$_6$	−6.7 to −5.4	(7)
		^{13}C	C$_6$D$_6$	127.6–129.3	(7)
		^{13}C	C$_6$D$_6$	135.0–136.3	(7)
		^{1}H	C$_6$D$_6$	0.5–1.0, b, CH$_3$	(7)
		^{1}H	C$_6$D$_6$	6.0–7.5, b, C$_6$H$_5$	(7)
Solvents	THF, toluene, CH$_2$Cl$_2$, hexane, 25°C				
Nonsolvents	Ethanol, 2-propanol				
Properties from light scattering study					(12)
M_w	g mol^{-1}	THF solution		46,000	
M_w/M_w	–			4.2	
$10^4 A_2$	mol cm^3 g^{-2}	–		3.6 ± 0.5	
R_g	nm			21	
$R_{g'}^o$ w	nm	–		15	
C_∞	–	–		64 ± 20	
Electrical conductivity	S cm^{-1}	Doped with SbF$_5$		2×10^{-4}	(13)
Hole drift mobility	cm^2 V^{-1} s^{-1}	$M_w = 69,000$, field = 2×10^5 V cm^{-1}, 298 K		2×10^{-4}	(14)
		$M_w = 11,000$		7×10^{-5}	
Surface tension	mN m^{-1}	–		43.3, 44.1	(15)
Scission, quantum yield, ϕ_s	mol Einstein^{-1}	THF solution, $\lambda = 313$ nm		0.97	(1)
		Solid, $\lambda = 313$ nm		0.015	
Cross-linking, quantum yield, ϕ_x	mol Einstein^{-1}	THF solution, $\lambda = 313$ nm		0.12	(1)
		Solid, $\lambda = 313$ nm		0.002	
Suppliers	Nippon Soda Co. Ltd., 2–1, Ohtemachi 2-chome, Chiyoda-ku, Tokyo 100, Japan Gelest Inc., 612 William Leigh Drive, Tullytown, PA 19007-6308, USA				

Poly(methylphenylsilylene)

Nonlinear optical properties[17]

M_w (g mol^{-1})	Temp. (°C)	λ (nm)	Lp (nm)	χ^{131} (esu)
>300,000	23	1,064	120	7.2×10^{-12}
–	23	1,907	120	4.2×10^{-12}
–	23	1,907	1,200	1.9×10^{-12}

References

1. West R. In: Mark JE, Allcock HR, West R, eds. *Inorganic Polymers.* 2nd Ed. Chap. 5. Oxford University Press; 2005.
2. West R. In: Rappoport Z, Apeloig Y, eds. *The Chemistry of Organic Silicon Compounds.* Vol 3. Chap. 9. Wiley; 2001.
3. Koe JR. In: Housecroft CE, ed. *Comprehensive Organometallic Chemistry III.* Vol 3. Chap. 11. Elsevier; 2006.
4. West R, Trefonas P. *Inorg. Synth.* 1988;25:58.
5. Cragg RH, Jones RG, Swain AC, Webb SJ. *J. Chem. Soc., Chem. Commun.* 1990;1147.
6. Miller RD, Thompson D, Sooriyakumaran R, Fickes GN. *J. Polym. Sci., Polym. Chem. Ed.* 1991;29:813.
7. Matyjaszewski K, Greszka D, Hrkach JS, Kim HK. *Macromolecules.* 1995;28:59.
8. Miller RD, Jenkner PK. *Macromolecules.* 1994;27:5921.
9. DeMahiu AF, Daoust D, Devaux J, de Valete M. *Eur. Polym. J.* 1992;28:685.
10. Kagawa T, Fujino M, Takeda K, Matsumoto N. *Solid State Commun.* 1986;57:635.
11. Miller RD, Michi J. *J. Chem. Rev.* 1989;89:1359.
12. Nakayama Y, et al. *J. Non-Cryst. Solids.* 1992;198.
13. Cotts PM, et al. *Macromolecules.* 1987;20:1046.
14. Hayashi T, Uchimaru Y, Reddy P, Tanaka M. *Chem. Letters.* 1992;647.
15. Dohmaru T, et al. *Phil. Mag..* 1995;B 71:1069.
16. Fujisaka T, West R, Murray C. *J. Organometal. Chem.* 1993;449:105.
17. Baumert JC, et al. *Appl. Phys. Lett.* 1988;53;1147.

Poly(methylsilmethylene)

Q. H. Shen and L. V. Interrante

Acronyms PC, PCS

Class Polycarbosilanes

Structure Si(Me)HCH$_2$ (branched, partially x-linked)

Preparative Techniques The polycarbosilane* employed to make commercial Nicalon SiC ceramic fiber is prepared via thermally induced rearrangement reaction of poly(dimethylsilane) or dodecamethylcyclohexasilane.

Major Application Precursor for the commercial Nicalon™ fiber, SiC composites. The polymer itself is no longer available for sale in the United States and Canada.

Properties of Special Interest Relatively low cost. High yield for SiC ceramic. Fuseable solid, soluble in hydrocarbons. Poor resistance to base and oxidation by air.

Property	Units	Conditions			Value	Reference
Molecular weight, M_n	g mol^{-1}	Polymer	Starting materials	Reaction temp. (°C)		
		PC-450	Polydimethylsilane	450	1,250	(1)
		PC-460	Polydimethylsilane	460	1,450	(1)
		PC-470	Polydimethylsilane	470	1,750	(1)
		PC-B5.5	Polydimethylsilane	320	1,312	(2)
			Borodiphenylsiloxane	–	–	–
		PC-B3.2	Polydimethylsilane	280	1,730	(2)
			Borodiphenylsiloxane	–	–	–
IR (characteristic absorption frequencies)	cm^{-1}	For SiCH$_2$Si For Si-H			1,050, 1,350, 2,100	(1)
NMR spectra	ppm	^{1}H NMR, solution			4.4, 0.2, −0.3	(1)
		^{13}C NMR, solution			3	(3)
		^{29}Si NMR, solution			−0.75 to 0.5; −17.5 to −16.01	(2)
		^{29}Si NMR, solid state			–	(3, 4, 5)

* Polycarbosilanes with the [SiMeHCH$_2$]$_n$ formula can also be prepared via the Grignard coupling reaction of Cl$_2$(Me)SiCH$_2$Cl, followed by reduction with LiAlH$_4$, or via ROP of 1,3-dichloro-1,3-dimethyl-1,3-disilacyclobutane, followed by LiAlH$_4$ reduction, or via chlorination of poly(dimethylsilylenemethylene), followed by reduction with LiAlH$_4$. The products of these latter reactions differ considerably in structure and properties from the "PCS" obtained from [Me$_2$Si]$_n$, have lower yields as SiC precursors, and are not widely used for this purpose.

Poly(methylsilmethylene)

Property	Units	Conditions	Value	Reference
Density	$g\ ml^{-1}$	25°C	1.116	(6)
Decomposition temperatures	K	For cured PC fibers in N_2		(7)
		Starting decomp. temp.	~673	
		Ending decomp. temp.	~1,573	

Pyrolyzability

Conditions	Pyrolysis Temp. (K)	Value	Reference
Nature of the product (under N_2);		Empirical formula for pyrolyzed SiC	(2)
PC precursors		fibers (amorphous)	
PC-TMS	1,573	$SiC_{1.79}H0._{037}O_{0.191}$	
PC-470	1,573	$SiC_{1.40}H_{0.046}O_{0.038}$	
PC-B3.2	1,573	$SiC_{1.48}H_{0.139}O_{0.145}$	
PC-B5.5	1,573	$SiC_{1.57}H_{0.051}O_{0.145}B_{0.006}$	
Amount of product (under N_2);		Ceramic yield (%)	(2)
PC precursors			
PC-470	1,573	54	
PC-TMS	1,573	76	
PC-B-5.5	1,573	61	
PC-B3.2	1,573	64	
Impurities remaining (under N_2)	1,573	Solid impurities Free C, SiO_2	(8, 9)
Gaseous products	673–873	H_2, C_nH_{2n+2}	(2)
(under vacuum or N_2)*	873–1,273	H_2, CH_4	
	1,273–1,573	H_2	
	>1,773	CO	
Gaseous products (under He)	873	CH_4	(10)
from PCS precursors	973	CH_4, C_2H_6, Me_2SiH_2, Me_3SiH, Me_4Si	
	1,073	CH_4, C_2H_6, Me_3SiH, Me_4Si	
	1,273	CH_4, C_2H_6, CO, C_2H_4 Me_3SiH, Me_4Si	

* From PC-470 and PC-B precursors.

References

1. Yajima S, Hasegawa Y, Hayashi J, Imura M. *J. Mater. Sci.* 1978;13:2569.
2. Hasegawa Y, Okamura K. *J. Mater. Sci.* 1983;18:3633.
3. Soraru GD, Babonneau F, Mackenzie JD. *J. Mater. Sci.* 1990;25:3886.
4. Taki T, et al. *J. Mater. Sci. Lett.* 1987;6:826.
5. Taki T, Okamura K, Sato M. *J. Mater. Sci.* 1989;24:1263.
6. Ichikawa H, Machino F, Teranishi H, Ishikawa T. *Silicon-based Polymer Science, Advances in Chemistry Series.* 1990;224:619.
7. Hasegawa Y, Iimura M, Yajima S. *J. Mater. Sci.* 1980;15:720.
8. Yajima S, et al. *Nature.* 1979;279:706.
9. Okamura K, Sato M, Hasegawa Y. *J. Mater. Sci. lett.* 1983;2:769.
10. Bouillon E, et al. *J. Mater. Sci.* 1991;26:1333.

Poly(methylsilsesquioxane)

Gui Lin and Ronald H. Baney

Acronyms, Alternative Names, Trade Name Methyl-T, PMSQ, MH-T, Glass Resin®
(Owens Illinois/Showa Denko), OctaMethyl-POSS®

Class Polysiloxanes (siloxane ladder polymers)

Structure $[CH_3SiO_{3/2}]_n$ ($n > 3$). The structure of poly(phenylsilsesquioxane) depends
upon the method of preparation. The structure is a function of the concentration of the
initial monomer, the nature of the solvent, the nature of the hydrolysable substituents, the
concentration of water, the temperature, the type of catalyst, and the nature of the
non-hydrolysing substituent. The methylsilsesquioxanes include random structures, ladder
structures, cage structures, and partial cage structures.[1] Structural studies on
methylsilsesquioxane are virtually nonexistent though the term ladder structure is
frequently used.[2]

$N = 8$: $C_8H_{24}O_{12}Si_8$ FW 536.96

Major Applications Interlayer dielectrics, high-temperature resins, and organic
antireflective coatings. Improves hydrophobicity, printability, processing etc.

Properties of Special Interest Very high thermal stability ($>500°C$) and good dielectric
properties.

Related Polymers There are many references to poly(alkylsilsesquioxane) and
poly(methyl-co-silsesquioxane)s such as Poly(methylsilsesquioxane)-block-
poly(N-isopropyl-acrylamide) (PMSSQ-b-PNIPAM),[3] poly(methylsilsesquioxane)-
block-polyimide, poly(methylsilsesquioxane-co-hydrosilsesquixoane),[4–7]
poly(methyl-co-aminopropylsilsesquioxane),[4] poly(methyl-co-epoxysilsesquioxane),[8]
methylsilsesquioxane-co-phenylsilsesquioxane,[9] polymer brush (PSQ-PAA) with a PSQ
core and PAA hair chains.[10,11] But they are generally poorly characterized. Thus, they are
not included in this handbook.[12]

Poly(methylsilsesquioxane)

Preparation

Acronym	Process	Molecular Weight $(g\,mol^{-1})$	Reference
$[CH_3SiO_{1.5}]_6$	Condensation of $CH_3Si(OC_2H_5)_3$ in the present of solvent benzene and catalyst HCl		(13)
$[CH_3SiO_{1.5}]_8$	Condensation of CH3SiCl3 in the present of solvent methanol and catalyst HCl		(13–16)
PMSQ-1	Add H_2O to $MeSiCl_3$ in THF and/or MIBK + Et_3N at 0°C then heat to 110°C	$M_w = 10^5$	(17, 18)
PMSQ-2	Add H_2O to $MeSiCl_3$ in THF and/or MIBK + Et_3N at 0°C then heat to 110°C at 3,000 Pa N_2	$M_w = 10^6$	(19)
PMSQ-3	Two layer system of sodium acetate in H_2O and toluene with 2-propanol	$M_w = 5 \times 10^3$	(20)
PMSQ-4	$MeSiCl_3$ + ethylenediamine (2:1) then hydrolysis in acetone–water–HCl, dried solid heated in xylene at 35°C	$M_w = 10^5$–10^6	(21)
PMSQ-5	$MeSi(OMe)_3$ at interface of aqueous ammonia	Insoluble spheres	(22, 23)
PMSQ-6	Partial hydrolysis and condensation of $MeSi(OMe)_3$	–	(24)
PMSQ-7	$MeSiOAc(OMe)_2$ reacted with $NaHCO_3$ suspended in MIBK at 100°C gave prepolymer which was then heated with 1 wt% KOH	$M_w = 1.4 \times 10^5$	(24)
PMSQ-Insoluble	Direct hydrolysis of $MeSiCl_3$ with no solvent	Insoluble gel	(21)
Ladder-PMSQ	Pre-aminolysis of $MeSiCl_3$ with 1,4-phenylenediamine (PDA) by dropwise and low temperature, which was then hydrolysis and polycondensation based on the H-bonding template effect.	18×10^3, 2,650, 2,800 (by VPO method)	(25–27)

* See reference (1).

Characteristic IR bands (Si—O—Si stretch) for "ladder" structure*

PMSQ-	Characteristic IR (cm^{-1})	d Spacing (Å)	^{29}Si NMR (ppm)	Reference
1	1,180, 1,020			(17)
2	1,130, 1,035			(19)
3	1,125, 1,040			(20)
4	1,120, 1,030	8.7, 3.6	−55.3, −64.8	(21)
7	1,125, 1,040			(28)
Ladder-PMSQ	1030–1114.78 1030–1120.50			(25–27)

* Not definitive.

Poly(methylsilsesquioxane)

Thermal stability

Material	Conditions	Temp. (°C)		Reference
		Air	N₂	
MeSiCl₃ hydrolyzed with "organic solvent" and condensed with Et₃N catalyst	Onset, decomposition	460	–	(29)
PMSQ-3	Onset, decomposition	400	660	(20)
PMSQ-4	5% N₂, 9% air	400	400	(21)

Properties

Material	Value	Reference
Dielectric constant	2.7, 2.6–2.8	(30, 31)
Low moisture absorption		(30)
Good mechanical hardness		(32–34)
Thermal expansion coefficient	80 ppm per °C	
Solvent solubility	Very slightly soluble, tetrahydrofuran, chloroform	(35)
Solvent insolubility	benzene, acetone, acetonitrile, methanol	(35)
Resin solubility	Most thermoplastic (PP, PE, PA, PU)	(35)
Appearance	White powder	(35)

Applications

Applications	Reference
Resists	(36)
Electrical insulation, interlayer dielectrics	(17–20, 37, 38)
Additives for cosmetics	(39)
Additives for toughening plastics	(40, 41)
Cladding for glass fiber	(42)
Ceramic binder	(43–46)
Si—C—O ceramic precursor	(47, 48)

References

1. Li GZ, Wang LC, Hi HL, Pittman CU Jr. *J. Inorgan Organ Polym.* 2001;11:123–154.
2. Baney RH, Itoh M, Sakakibara A, Suzuki T. *Chem. Rev.* 1995;95(5):1409.
3. Daniel K, Patrick T. *Macromol. Symp.* 2007;249–250:424–430
4. Duan QH, Zhang TY, Deng KL, Xie P, Zhang RB. *Chin. J. Polym. Sci.* 2005;4:355–361.
5. Duan QH, Zhang Y, Jiang JQ, Deng KL, Zhang TY, Xie P, Zhang RB, Fu PF, *Polym. Intern.* 2004;53:113.
6. Xie ZS, Jin SZ, Wan YZ, Zhang RB. *J. Polym. Sci.* 1992;10:361.

7. Cui L, Jin W, Liu JN, Tang YX, Xie P, Zhang RB. *Liqui. Cryst.* 1998;25:757–764.

8. Cao M, Li Z, Zhang Y, Xie P, Dai DR, Zhang RB, Lin YH, Chung NT. *React. Funct. Polym.* 2000;45:119–130.

9. Sosa JM. *Macromolecules.* 1980;13:1260–1264.

10. Li C, Schmidt M, Chen Y. *Abstr. Pap. Am. Chem. Soc.* 2003;225. U623 511-POLY (Part 2).

11. Yang SG, Zhang YJ, Wang L, Hong S, Xu J, Chen YM. *Langmuir.* 2006;22:338–343.

12. Franco R, Kandalam AK, Pandey R. *J. Phys. Chem. B.* 2002;106:1709–1713.

13. Sprung MM, Guenther FO. *J. Am. Chem. Soc.* 1955;77:3990.

14. Barry AJ. *J. Am. Chem. Soc.* 1955;77:4248.

15. Sprung MM, Guenther FO. *J. Am. Chem. Soc.* 1955;77:6045.

16. Vogt LH, Brown JF. *Inorg. Chem.* 1963;2:189.

17. Suminoe T, Matsumura Y, Tomomitsu O. *Japanese Patent Kokoku*-S-60-17214 (1985) [Kokai-S-53-88099 (1978)]; *Chem Abstr.* 1978;89:180–824.

18. Matsumura Y, et al. *U.S. Patent.* 1983;4399266.; *Chem. Abstr.* 1983;99:159059.

19. Fukuyama S, et al. *European Patent* 0 406 911A 1. 1985; *Chem. Abstr.* 1986;105:115551.

20. Nakashima H. *Japanese Patent Kokai*-H-3-227321 (1991); *Chem. Abstr.* 1992;116:60775.

21. Xie Z, He Z, Dai D, Zhang R. *Chinese J. Polym. Sci.* 1989;7(2):183.

22. Nishida M, Takahashi T, Kimura H. *Japanese Patent Kokai*-H-1-242625 (1989); *Chem. Abstr.* 1990;112:99962.

23. Terae N, Iguchi Y, Okamoto T, Sudo M. *Japanese Patent Kokai*-H-2-209927 (1990); *Chem. Abstr.* 1991;114:43819.

24. Abe Y, et al. *J. Polym. Sci., Part A, Polym. Chem.* 1996;33:751.

25. Duan QH, Zhang Y, Jiang JQ, Deng KL, Zhang TY, Xie P, Zhang RB, Fu PF, *Polym. Inter.* 2004;53m:113–120.

26. Xie P, Zhang RB. *Polym. Adv. Tech.* 1997;8:649–656.

27. Zhang RB, Dai DR, Cui L, Xu H, Liu CQ, Xie P. *Mater. Sci. Eng. C.* 1999;10:13–18.

28. Morimoto N, Yoshioka H. *Japanese Patent Kokai*-H-3-20331 (1991); *Chem. Abstr.* 1991;115:30554.

29. Adachi H, Adachi E, Hayashi O, Okahashi K. *Rep. Prog. Polym. Phys. Japan.* 1986;29:257.

30. Lee J-K, Char K, Rhee H-W, Ro HW, Yoo DY, Yoon DY. *Polymer.* 2001;42(21):9085–9089.

31. Yang S, MIrau PA, Pai CS, Nalamasu O, Reichmanis E, Lin EK, Lee HJ, Gidley DW, Sun JN. *Chem. Mater.* 2001;13:762–2764.

32. Baney RH, Itoh M, Sakakibara A, Suzuki T. *Chem Rev.* 1995;95:1409.

33. Loy DA, Shea KJ. *Chem Rev.* 1995;95:1431.

34. Kim SM, Yong DY, Nguye CV, Han J, Jaffe RL. *MRS Symp Proc.* 1999;511:39.

35. www.hybridplastics.com.

36. Gozdz AS. *Polym. Adv. Technol.* 1994;5:70.

37. Lee HJ, Soles CL, Liu DW, Bauer BJ, Lin EK, Wu WL. *J. Appl. Phys.* 2006;100:064104.

38. Mikoshiba S, Hayase S. *J Mater. Chem.* 1999;9:591–598.

39. Hase N, Tokunaga T. *Japanese Patent Kokai*-H-5-43420 (1993); *Chem. Abstr.* 1993;119:34107.

40. Kugimiya Y, Ishibashi T. *Japanese Patent Kokai*-H-1-135840 (1989); *Chem. Abstr.* 1989;111:215766.

41. Dote T, Ishiguro K, Ohtaki M, Shinbo Y. *Japanese Patent Kokai*-H-2-194058 (1990); *Chem. Abstr.* 1990;113:213397.

42. Honjo M, Yamanishi T. *Japanese Patent Kokai*-H-3-240002 (1991); *Chem. Abstr.* 1992;116:107865.

Poly(methylsilsesquioxane)

43. Mine T, Komasaki S. *Japanese Patent Kokai*-S-60-210569 (1985); *Chem. Abstr.* 1986;104:154451.
44. Park CE, Kang JH. *US Patent.* 6,849, 926 B2. Feb. 1, 2005.
45. Zhang Z. Nguyen TD, Nguyen T. *US Patent.* 6,919,101 B2. Jul. 19, 2005.
46. Kloster GM, Obrien KP, Brask JK, Goodner MD, Bruner D. *US Patent.* 7,034,399 B2. Apr. 25, 2006.
47. Laine RM, et al. *Chem. Mater.* 1990;2:464.
48. Dong HJ, Brennan JD. *Chem. Mater.* 2006;18:541–546.

Poly(α-methylstyrene)

Michelle K. Gaines, Lisaleigh Kane and Richard J. Spontak

Acronyms PαMS, PAMS

Class Vinyl polymers

Cas Number 25014-31-7

Structure

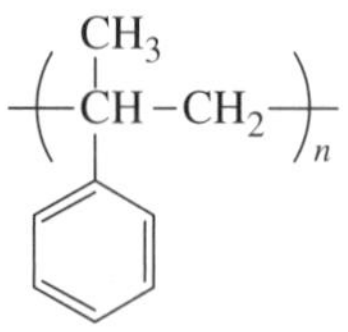

Major Application Copolymerized with styrene for improved heat resistance

Property	Units	Conditions	Value	Reference
Density	g cm^{-3}	–	1.07	(1)
		$\overline{M}_w = 135,000$	1.065	(2)
		$\overline{M}_w = 3.5k–55k$	1.054–1.062	(3)
Glass transition temperature T_g	K	$\overline{M}_w = 700,000$	435	(4)
		400,000	444	(4)
		113,000	441	(4)
		76,500	447	(5)
		61,000	443	(6)
		55,000	442, 453	(3, 7, 8)
		50,000	453	(1)
		25,000	440	(5)
		19,500	442	(3)
		6,700	433	(3)
		3,500	414	(3)
		2,510	366	(5)
Heat capacity C_p	J K^{-1} mol^{-1}	300 K to T_g	$29.42 + 0.4498T - (1.280 \times 10^6)T^{-2}$	(9)
		T_g to 490 K	$-6.43 + 0.5758T$	(9)
	J g^{-1} K^{-1}	PαMS/Pentamer		(10)
		$T = 453$ K	2.22	
		$T = 433$ K	2.15	
		$T = 413$ K	2.08	
		$T = 395$ K	2.01	
		$T = 363$ K	1.90	
		$T = 343$ K	1.84	
		$T = 323$ K	1.75	

Poly(α-methylstyrene)

Property	Units	Conditions	Value	Reference
Ceiling temperature	K	–	334	(7, 11)
Polymerization conversion		Living polymerization with Sn initiators		(12)
Depolymerization temperature	K	–	510	(13, 14)
Activation energy for pyrolysis	kJ (per repeat unit)	–	188–243	(15)
Dielectric constant	–	–	2.56, 2.60	(16)
Specific free volume	$cm^3 \ g^{-1}$	$\overline{M}_w = 135{,}000$	0.135	(2)
Thermal expansion coefficient α	K^{-1}		5.72×10^{-4}	(17)

Optical Property	Units	Conditions		Value	Reference
Blue transmitted λ	Nm	12.7/1.3/86 PαMS/polyisoprene/poly(methyl methacrylate) blend		180 (λ_{max}: 440 nm) from reflected λ_{max}	(18)
Green transmitted λ	Nm	6.1/0.9/93 PαMS/polyisoprene/poly(methyi methacrylate) blend		216 (λ_{max}: 530 nm) from reflected λ_{max}	
Refractive index	–			1.618	(14)
	–			1.61	(19)
Refractive index increment d_n/d_c	$mL \ g^{-1}$	633 nm in tetrahydrofuran; $T = 25°C$			(20)
		$\overline{M}_w = 19{,}700$		0.2044 ± 0.0023	
		$\overline{M}_w = 72{,}000$		0.2056 ± 0.0015	
	$mL \ g^{-1}$	Solvent	λ (nm)		(20)
		Toluene	546	0.129, 0.126	
			436	0.131, 0.130	
			366	0.135	
		Cyclohexane	546	0.192	
			436	0.200, 0.206, 0.197, 0.204	
			366	0.219	
		Benzene	546	0.138	
			436	0.134, 0.124	
		Ethylene dichloride	436	0.176	
		CCl_4	546	0.168	
			436	0.178	
			366	0.193	

Spectral Property	Conditions	Reference
NMR spectroscopy	Solvent = d-chloroform, 25 MHz, $T = 30°C$, conc. = 10% (w/v)	(21)
	Solvent = chlorobenzene, $T = 120°C$, conc. = 20% (w/v)	(22)
	Solvent = o-dichlorobenzene, $T = 100°C$, conc. = 10 wt%	(23)
	Solvent = cholorobenzene-d$_5$, $T = 30, 70°C$, conc. = 7.5% (w/v)	(24)
	Solvent = methylene chloride, $T = -78°C$	(25, 26)
	Evaluated at 25°C	(27)
FTIR spectroscopy	Includes a p-perfluoro{1-2-(2-fluorosulfonyl-ethoxy) propoxy]}ethylated derivative	(28)

Mechanical Property	Units	Conditions	Value	Reference
Elastic modulus	GPa	$T = 25°C$	2.5	(29)
Rubbery modulus	MPa		16	(30)
Glassy modulus	GPa		2.3	(30)
Rockwell hardness test		$^1/_4$ Rockwell superficial hardness ball	Indentation/Recovery	(31)
	–	15 kg load	160/197	
		30 kg load	107/193	
		45 kg load	85/190	
Bending modulus	MPa		417	(31)
Tensile modulus	MPa		2.07	(31)
Flexural strength	MPa		51.7	(31)
Rubbery stress-optical coefficient	Pa^{-1}		-3400×10^{-12}	(30)
Glassy stress optical coefficient	Pa^{-1}		20×10^{-12}	(30)
Photoelastic stress-optical coefficient	Pa^{-1}		-3.6×10^{-12}	(30)

Thermodynamic Property	Units	Conditions	Value	Reference
Flory-Huggins interaction parameter χ	–	d-Polystyrene ($\phi = d$-PS volume fraction)	$\chi(\phi,T) = 59.2/T (0.0626 - 0.0018\phi - 5.6 \times 10^{-5}T)$	(32)
		Polystyrene, $T = 25°C$	0.032	(33)
		Tetrahydrofuran, $T = 30°C$	0.462	(34)
		α-Chloronaphthalene		(34)
		$T = 45.5°C$	0.428	
		$T = 30°C$	0.440	
		Toluene		(34)
		$T = 30°C$	0.452, 0.463–0.465	
		$T = 25°C$	0.425, 0.466	
		$trans$-Decalin		(34)
		$T = 30°C$	0.473	
		$T = 10°C$	0.500	

Thermodynamic Property	Units	Conditions	Value	Reference
		1-Chlorobutane		(34)
		$T = 50°C$	0.489, 0.492	
		$T = 25°C$	0.490	
		$T = 5°C$	0.492	
		Cyclohexane		(34)
		$T = 46°C$	0.496, 0.498	
		$T = 39°C$	0.499	
		$T = 38.6°C$	0.500	
		$T = 36°C$	0.500	
		$T = 35°C$	0.500	
		$T = 32°C$	0.503	
		$T = 28°C$	0.505, 0.506	
		$T = 24°C$	0.508	
		$T = 20°C$	0.509	
		p-Xylene, $T = 30°C$	0.459	(34)
		Nitrobenzene, $T = 30°C$	0.481	(34)
		Chlorobenzene, $T = 30°C$	0.455	(34)
		Tetralin, $T = 50°C$	0.427	(34)
		p-Dioxane, $T = 30°C$	0.463	(34)
		2-Hexanone, $T = 30°C$	0.532	(34)
		n-Butyl acetate, $T = 30°C$	0.526	(34)
		Dimethylformamide, $T = 30°C$	0.525	(34)
Solubility parameter	$(MPa)^{1/2}$	–	18.6	(8)
Interaction energy $(P^*)^{1/2}$	$(MPa)^{1/2}$	–	20.6	(8)
Sanchez-Lacombe equation of state parameters		Temperature range: 220–270°C		(8, 14)
	K	Characteristic temperature T^*	827	
	bar	Characteristic pressure P^*	4258	
	g cm^{-3}	Characteristic density ρ^*	1.13	
Flory-Huggins interaction energy density	cal cm^{-3}	Various PαMS/polycarbonate blends		(14)
		Polycarbonate Dimethyl bisphenol-A	0.39–0.44	
		Polycarbonate Tetramethyl bisphenol-A	>0.18	
		Polycarbonate Trichloro bisphenol-A	0.26	
		Polycarbonate Bisphenol-Z	>0.31	
		Polycarbonate Hexafluoro bisphenol-A	>0.31	
		Polycarbonate Bisphenol chloral	0.22–0.72	
		Polycarbonate Tetramethyl bisphenol-P	>0.33	
		Polycarbonate	0.21–0.29	
	cal cm^{-3}	Various PαMS/polysulfone blends		(35)
		Bisphenol-A polysulfone	>0.43, 0.32	
		Dimethyl bisphenol-A polysulfone	>0.30	
		Tetramethyl bisphenol-A polysulfone	>0.35, 0.24–0.27	
		Hexamethyl bisphenol-A polysulfone	>0.29	
		Polyethersulfone	>0.31	
		Hexafluoropolysulfone	>0.20	
		Tetramethylhexafluoro polysulfone	>0.30	
		Tetramethyl bisphenol-P polysulfone	>0.29	

Thermodynamic Property	Units	Conditions	Value	Reference
	cal cm^{-3}	Polyacrylonitrile (50/50 w/w) blend	7.30	(36)
	cal cm^{-3}	Polystyrene (50/50 w/w) blend	0.02	(36)
	J cm^{-3}	Polystyrene (50/50 w/w) blend	4.3, 0.15	(20, 33)
	J cm^{-3}	Polystyrene (50/50 w/w) blend (group contribution method)	0.7	(33, 37)
	J cm^{-3}	Polystyrene (50/50 w/w) blend in toluene	-0.223	(37)
Interaction pair $(\delta_i - \delta_j)^2$	MPa	Polyacrylonitrile	92.42	(8)
		Poly(methyl methacrylate)	0.00	(8)
		Tetramethylbisphenol A polycarbonate	0.67	(8)
		Poly(vinyl chloride)	2.68	(8)
		Poly(2,6-dimethyl-1,4-phenylene oxide)	2.05	(8)
		Poly(ε-caprolactone)	0.38	(8)
Second virial coefficient A$_2$	mol cm^3 g^{-2}	n-Butyl chloride, $\overline{M}_w = 6,900–3,540,000$ g mol^{-1}, $T = 25°C$	$(3.1 \times 10^{-3}) \overline{M}_w^{-0.255}$	(38)
		Cyclohexane, $\overline{M}_w = 5,900–341,000$ g mol^{-1}		(39)
		$T = 30°C$	$-(5.5 \times 10^{-10}) \overline{M}_w^{0.84}$	
		$T = 24°C$	$-(6.0 \times 10^{-9}) \overline{M}_w^{0.72}$	
		$T = 20°C$	$-(2.4 \times 10^{-7}) \overline{M}_w^{0.50}$	
		Toluene, $\overline{M}_w = 3,000–804,000$ g mol^{-1}, $T = 25°C$	$(2.5 \times 10^{-2}) \overline{M}_w^{0.32}$	(40)
		Cyclohexane, $\overline{M}_w = 0.31 \times 10^6$		(41)
		$T = 50°C$	5.75×10^{-5}	
		$T = 45°C$	4.04×10^{-5}	
		$T = 40°C$	$2.23 \times 10^{--5}$	
		$T = 38°C$	1.34×10^{-5}	
		$T = 36°C$	0.41×10^{-5}	
		$T = 34°C$	-0.37×10^{-5}	
		$T = 32°C$	-1.42×10^{-5}	
		Cyclohexane, $\overline{M}_w = 1.73 \times 10^6$		(41)
		$T = 50°C$	4.61×10^5	
		$T = 45°C$	3.50×10^{-5}	
		$T = 40°C$	2.06×10^{-5}	
		$T = 38°C$	1.31×10^{-5}	
		$T = 36°C$	0.50×10^{-5}	
		$T = 34°C$	-0.25×10^{-5}	
		$T = 32°C$	-1.16×10^{-5}	

Poly(α-methylstyrene)

Thermodynamic Property	Units	Conditions	Value	Reference
		Cyclohexane, $\overline{M}_w = 3 \times 10^6$		(41)
		$T = 50°C$	4.06×10^{-5}	
		$T = 45°C$	3.13×10^{-5}	
		$T = 40°C$	1.95×10^{-5}	
		$T = 38°C$	1.18×10^{-5}	
		$T = 36°C$	0.50×10^{-5}	
		$T = 34°C$	-0.19×10^{-5}	
		$T = 32°C$	-1.09×10^{-5}	
		Cyclohexane, $\overline{M}_w = 1.22 \times 10^6$		(41)
		$T = 50°C$	5.09×10^{-5}	
		$T = 40°C$	2.18×10^{-5}	

Solution Property	Units	Conditions	Value		Reference
Radius of gyration R_g	nm	n-Butyl chloride, $\overline{M}_w = 66{,}600–3{,}540{,}000$ g mol^{-1}, $T = 25°C$	$(2.10 \times 10^{-2})\,\overline{M}_w^{0.526}$		(38, 42)
		n-Butyl chloride, $T = 35°C$	$(2.89 \times 10^{-2})\overline{M}_w^{0.50}$		(42)
		RIS prediction, $w_m = 0.4$ (fraction of meso dyads)	$(2.63 \times 10^{-2})\,\overline{M}_w^{0.50}$		(42)
		Cyclohexane, $\overline{M}_w = 59{,}000–3{,}410{,}000$ g mol^{-1}			(39)
		$T = 36°C$	$(2.82 \times 10^{-2})\,\overline{M}_w^{0.499}$		
		$T = 28°C$	$(4.08 \times 10^{-2})\overline{M}_w^{0.463}$		
		$T = 24°C$	$(4.65 \times 10^{-2})\overline{M}_w^{0.450}$		
		$T = 20°C$	$(6.54 \times 10^{-2})\overline{M}_w^{0.414}$		
Intrinsic viscosity $[\eta]$	mL g^{-1}	RIS prediction, $w_m = 0.4$ (fraction of meso dyads)	$(7.36 \times 10^{-2})\,\overline{M}_w^{0.497}$		(42)
		n-Butyl chloride, $T = 25°C$	$(3.36 \times 10^{-2})\,\overline{M}_w^{0.87}$		(38, 42)
		n-Butyl chloride, $T = 35°C$	$(7.30 \times 10^{-2})\,\overline{M}_w^{0.50}$		(42)
Mark-Houwink parameters: K and a	$K = $ mL g^{-1} $a = $ none		K	a	
		n-Butyl chloride, $\overline{M}_w = 53{,}800–3{,}540{,}000$ g mol^{-1}, $T = 25°C$	2.70×10^{-2}	0.590	(38)
		Toluene, $\overline{M}_w = 26{,}000–603{,}000$ g mol^{-1}, $T = 25°C$	7.81×10^{-5}	0.73	(43)
		Toluene, $\overline{M}_w = 3{,}400–840{,}000$ g mol^{-1}, $T = 25°C$	1.10×10^{-4}	0.71	(40)

Solution Property	Units	Conditions	Value		Reference
	$K = \mathrm{dL}$ $\mathrm{g}^{-3/2}\,\mathrm{mol}^{1/2}$	Cyclohexane, $\overline{M}_\mathrm{w} = 59{,}000\text{--}3{,}410{,}000\ \mathrm{g\ mol}^{-1}$			(44)
		$T = 46°\mathrm{C}$	4.87×10^{-4}	0.539	
		$T = 39°\mathrm{C}$	6.66×10^{-4}	0.508	
		$T = 36°\mathrm{C}$	7.47×10^{-4}	0.498	
		$T = 32°\mathrm{C}$	10.5×10^{-4}	0.466	
		$T = 28°\mathrm{C}$	14.7×10^{-4}	0.434	
		$T = 24°\mathrm{C}$	18.9×10^{-4}	0.407	
		$T = 20°\mathrm{C}$	22.7×10^{-4}	0.386	
Mutual diffusion coefficient D_0	$\mathrm{cm}^2\,\mathrm{s}^{-1}$	n-Butyl chloride, $\overline{M}_\mathrm{w} = 53{,}800\text{--}3{,}540{,}000\ \mathrm{g\ mol}^{-1}$, $T = 25°\mathrm{C}$	$(3.81 \times 10^{-4})\,\overline{M}_\mathrm{w}^{\,0.531}$		(38)
		Benzene, $T = 30°\mathrm{C}$	$(3.26 \times 10^{-4})\,\overline{M}_\mathrm{w}^{\,-0.561}$		(45)
1st Diffusion coefficient k_d	$\mathrm{cm}^3\,\mathrm{g}^{-1}$	Benzene, $\overline{M}_\mathrm{w} = 3.76 \times 10^5 - 6.85 \times 10^6\ \mathrm{g\ mol}^{-1}$, $T = 25°\mathrm{C}$	$(3.28 \times 10^{-3})\,\overline{M}_\mathrm{w}^{\,0.769}$		(45)
2nd Diffusion coefficient k_{d2}	$\mathrm{cm}^6\,\mathrm{g}^{-2}$		$(-1.40 \times 10^{-6})\,\overline{M}_\mathrm{w}^{\,1.54}$		
Hydrodynamic radius R_H	cm	Benzene, $T = 30°\mathrm{C}$	$(1.22 \times 10^{-9})\,\overline{M}_\mathrm{w}^{\,0.561}$		(45)
Sedimentation constant S_0	s	Toluene, $\overline{M}_\mathrm{w} = 26{,}000\text{--}603{,}000\ \mathrm{g\ mol}^{-1}$, $T = 25°\mathrm{C}$	$(1.72 \times 10^{-2})\,\overline{M}_\mathrm{w}^{\,0.49}$		(43)
		Benzene, $T = 30°\mathrm{C}$	$(5.46 \times 10^{-15})\,\overline{M}_\mathrm{w}^{\,0.391}$		(45)
1st Sedimentation coefficient k_s	$\mathrm{cm}^3\,\mathrm{g}^{-1}$	Benzene, $\overline{M}_\mathrm{w} > 3 \times 10^5\ \mathrm{g\ mol}^{-1}$, $T = 25°\mathrm{C}$	$(3.88 \times 10^{-2})\,\overline{M}_\mathrm{w}^{\,0.663}$		(45)
2nd Sedimentation coefficient k_{s2}	$\mathrm{cm}^6\,\mathrm{g}^{-2}$		$(-6.93 \times 10^{-6})\,\overline{M}_\mathrm{w}^{\,1.48}$		(45)

Solvents	Reference
α-Chlorophthalene	(34)
α-Methyl nathpthalene	(13)
Benzene	(34)
Benzyl chloride	(46)
1-Chlorobutane	(34)
Chlorobenzene	(22, 34)
Chloroform	(47, 48)
Cyclohexane	(34, 43)
Decalin	(34, 49)
Dichloromethane	(46)
Dimethylformamide	(34)
9,10-Dithdroanthracene	(49)
Diphenylamine	(49)
Diphenyl ether	(13)
2-Hexanone chloride	(34)
n-Hexane	(47, 48)

Poly(α-methylstyrene)

Solvents	Reference
1-Methylnaphthalene	(49)
Methylene chloride	(48)
2-Naphthol	(49)
Nitrobenzene	(34)
n-Butyl acetate	(34)
n-Butyl chloride	(38)
Phenol	(49)
p-Xylene	(34)
p-Dioxane	(34)
Sulfur dioxide (liq)	(46)
Tetralin	(34, 49)
Tetrahydrofuran	(34, 43, 50, 51)
Toluene	(4, 34, 43, 47, 48, 52)
trans-Decalin	(34)
Triphenylmethane	(49)
1,2,4-Trichlorobenzene	(13)
Nonsolvents	
Methanol	(4, 50, 53)
n-Hexane	(54)
Ethanol	(55)

Phase Property	Units	Conditions	Value		Reference
Critical temperatures	K	For $\overline{M}_w = 114{,}000$ g mol^{-1}			(56)
		Solvent =	T_{UCST}	T_{LCST}	
		Cyclopentane	299	418	
		Cyclohexane	287	481	
		Trans-decalin	267	–	
		n-Butyl acetate	–	447	
		n-Pentyl acetate	312	476	
		n-Hexyl acetate	303	501	
		For $\overline{M}_w = 5{,}700 - 1.40 \times 10^6$ g mol^{-1}			(17)
		Solvent			
		Cyclohexane	308	456	
		Methylcyclohexane	366	431	
		n-Butyl chloride	263	412	

Heats of solution for PαMS/PS mixtures and blends in toluene at 60°C.[52]

PαMS/PS (w/w)	$\Delta H_{\mathrm{mixture}}$ (J g^{-1})	$\Delta H_{\mathrm{blend}}$ (J g^{-1})
100/0	–	-15.5 ± 0.3
80/20	-16.5 ± 0.6	-7.4 ± 0.3
50/50	-9.2 ± 0.2	-8.0 ± 0.2
20/80	-8.5 ± 0.5	-7.4 ± 0.5
0/100	–	-6.8 ± 0.3

Thermal degradation evaluated at 275°C.[49]

Solvent	Boiling Point (°C)	% Conversion
2-Naphthol	286	33.1
Phenol	182	41.9
1-Methylnaphthalene	242	35.7
Decalin	187	23.9
Diphenylamine	302	30.2
Tetralin	207	33.8
Triphenylmethane	360	–
9,10-Dihydroanthracene	312	30.3

Polymerization Property	Units	Conditions	Value	Reference
Heat of polymerization $\Delta H°$	kcal mol^{-1}	Anionic polymerization, sodium naphthalene complex initiator, THF solution	-6.96	(11)
Entropy of polymerization $\Delta S°$	kcal mol^{-1} K^{-1}	Anionic polymerization, sodium naphthalene complex initiator, THF solution	-24.8	(11)
Rate of polymerization	mol^{-1} h^{-1}	PαMS/polystyrene copolymer (48 mol% PαMS)	$k_i = 5.37 \times 10^{17}$ $e^{-(38731 \text{ cal/mol})/RT}$	(7)
Rate of depolymerization		$T = 236.5°C$		
		α-Methyl naphthalene	$k_i = 0.19 \times 10^{-4}$	(13)
		Diphenyl ether	$k_i = 0.24 \times 10^{-4}$	(13)
		Trichlorobenzene	$k_i = 0.66 \times 10^{-4}$	(13)
		PαMS/polystyrene copolymer (48 mol% PαMS)	$k_i = -1.43 \times 10^{13}$ $e^{-(35330 \text{ cal/mol})/RT}$	(7)
Thermal degradation temperature	°C		239–395	(57, 58)

Polymerization

Initiator	Solvent/Conditions	T (°C)	$\bar{M}_w/\bar{M}_w$	Reference
Sodium naphthalide	Tetrahydrofuran	−78	< 1.05	(59, 60)
n-C$_4$H$_9$Li	Tetrahydrofuran	−78	< 1.1	(50)
	Methylcyclohexane	–	–	(51)
α, α–Azobisisobutyronitrile (AIBN)	Tetrahydrofuran	60	–	(61)
AIBN	100 psi O$_2$–100 psi	45–65	1.45, 1.11, 1.32, 1.56	(62)
AIBN	n-Butyl acetate as solvent; bis(difluoroboryl) dimethylglyoximato cobalt II (COBF) as chain transfer agent	60	–	(55)

Poly(α-methylstyrene)

Polymerization

Initiator	Solvent/Conditions	T (°C)	$\bar{M}_w/\bar{M}_w$	Reference
Sodium naphthalene	Tetrahydrofuran	−78	1.00–1.03	(11, 43)
Sec-Butyl lithium	Tetrahydrofuran	−78	1.15	(39)
4-t-Butyl catechol	Ethyl benzene	100	–	(7)
n-Butyl lithium	Tetrahydrofuran	−78	<1.05	(45)
TiCl$_4$	Methylene chloride	−78	3.54	(48)
BF$_3$·OEt$_2$	Hexane/chloroform	−78	–	(48)
BF$_3$	Hexane/chloroform	−78	–	(48)
BF$_3$	Toluene	−78	–	(48)
BF$_3$	Hexane	−78	–	(48)
p-Methoxybenzyl chloride	Sulfur dioxide (liq)	−20, −40, −60	1.02–2.33	(46)
Iodine	Sulfur dioxide (liq)	−60	–	(63)
SnBr$_4$	Methylene chloride	−78	∼1.15	(26)
SnCl$_4$	Methylene chloride	−78	1.14	(25)
SnBr$_4$	Methylene chloride	−78	1.12–1.93	(25)
	Hexane/methylene chloride (60/40, 50/50, 40/60)	−80, −60	<1.1	(12)
SnBr$_4$ with chloride-vinyl ether adduct	Methylene chloride	−78	–	(64)
AlCl$_3$	Chloroform/hexane	−78, −130	–	(47)
	Toluene/hexane	−78, −130	–	(47)
	Ethyl chloride	−130, −50	–	(65)
	Toluene	−78	–	(22)
	Carbon disulfide	−50	–	(49)
B(C$_6$F$_5$)$_3$	Toluene/methylene chloride	−78, 20	2.7	(66)

Syndiospecific polymerization

Catalyst (cationic)	Solvent	T (° C)	Reference
BF$_3$ · O(C$_2$H$_5$)$_2$	Toluene	−78	(22, 48)
	Toluene/methylcyclohexane	−78	(22)
	Methylcyclohexane	−78	(22)
	Hexane/chloroform	−78	(48)
	Hexane	−78	(48)
BF$_3$	Hexane/chloroform	−78	(48)
	Toluene	−78	(48)
	Hexane	−78	(48)
	Methylene chloride	−78	(48)
AlCl$_3$	Toluene	−75, −78	(22)
	Carbon disulfide	−50	(49)
	Toluene/methylcyclohexane	−78	(22)
	Methylene chloride	−78	(48)
AlBr$_3$ · trichloroacetic acid	Toluene	−78	(54)
	Methylene chloride	−78	(54)

Poly(α-methylstyrene)

Catalyst (cationic)	Solvent	T (° C)	Reference
TiCl$_4$	Toluene	$-75, -78$	(22, 48, 54)
	Toluene/methylcyclohexane	-78	(22)
	Methylcyclohexane	-78	(22)
	Methylene chloride	-78	(48, 54)
	Methylene chloride/toluene	-78	(48)
	Methylene chloride/hexane	-78	(48)
	Hexane	-78	(48)
SnCl$_4$	Toluene	-75	(22)
	Toluene/methylcyclohexane	-78	(22)
	Methylcyclohexane	-78	(22)
	Methylene chloride	-78	(48)
n-Butyl lithium	Tetrahydrofuran	-78	(45)

Gas Transport Property	Units	Conditions	Value	Reference
Permeability coefficient P	cm^3 (STP) cm (cm^2 s cm Hg)$^{-1}$	35°C, 1 atm		(2)
		He	14.5×10^{-10}	
		CH$_4$	0.14×10^{-11}	
		O$_2$	0.82×10^{-11}	
		N$_2$	0.15×10^{-11}	
		CO$_2$	3.0×10^{-10}	
Ideal selectivity P_1/P_2	–	35°C, 1 atm		(2)
		He/CH$_4$	100	
		O$_2$/N$_2$	5.4	
		CO$_2$/CH$_4$	20.8	
Diffusion coefficient D	cm^2 s^{-1}	35°C, 1 atm		(2)
		CH$_4$	1.6×10^{-9}	
		O$_2$	2.1×10^{-8}	
		N$_2$	4.0×10^{-9}	
		CO$_2$	7.4×10^{-9}	
Diffusion selectivity D_1/D_2	–	35°C, 1 atm		(2)
		O$_2$/N$_2$	4.6	
		CO$_2$/CH$_4$	4.9	
Solubility coefficient	cm^3 (STP) cm (cm^3 atm)$^{-1}$	35°C, 1 atm		(2)
		CH$_4$	0.72	
		O$_2$	0.30	
		N$_2$	0.26	
		CO$_2$	3.0	
Solubility selectivity S_1/S_2	–	35°C, 1 atm		(2)
		O$_2$/N$_2$	1.2	
		CO$_2$/CH$_4$	4.2	

References

1. Yang H, Ricci S, Collin M. *Macromolecules*. 1991;24:5218.
2. Puleo AC, Muruganandam N, Paul DR. *J. Polym. Sci. B: Polym. Phys*. 1989;27:2385.
3. Callaghan TA, Paul DR. *Macromolecules*. 1993;26:2439.
4. Cowie JMG, Fernandez MD, Fernandez MJ, McEwen IJ. *Polymer*. 1992;33:2744.
5. Schneider HA, Dilger P. *Polym. Bull*. 1989;21:265.
6. Maier R-D, Kressler J, Rudolf B, Reichert P, Koopmann F, Frey H, Mülhaupt R. *Macromolecules*. 1996;29:1490.
7. Priddy DB, Traugott TD, Seiss RH. *J. Appl. Polym. Sci*. 1990;41:383.
8. Gan PP, Paul DR, Padwa AR. *Polymer*. 1994;35:1487.
9. Judovits LH, Bopp RC, Gaur U, Wunderlich B. *J. Polym. Sci. B: Polym. Phys*. 1986;24:2725.
10. Huang D, Simon SL, McKenna GB. *J. Chem. Phys*. 2003;119:3590.
11. McCormick HW. *J. Polym. Sci*. 1957;25:488.
12. Li D, Hadjikyriacou S, Faust R. *Marcomolecules*. 1996;29:6061.
13. Bywater S, Black PE. *J. Phys. Chem*. 1965;69:2967.
14. Callaghan TA, Paul DR. *J. Polym. Sci. B.: Polym. Phys*. 1994;32:1813.
15. Brown DW, Wall LA. *J. Chem. Phys*. 1958;62:848.
16. Gray DN, Grier JD. *U.S. Patent*. 3,661,615, May 9, 1972.
17. Cowie JMG, McEwen IJ. *Polymer*. 1975;16:244.
18. Ishizu K, Yasuda M, Sato Y, Tamura T. *Polym. Adv. Technol*. 2005;16:628.
19. Hanes MD. *U.S. Patent*. 6,040,382, March 21, 2000.
20. McManus NT, Penlidis A. *J. Appl. Polym. Sci*. 1998;70:1253.
21. Laurêtre Noël C, Monnerie L. *J. Polym. Sci.: Polym. Phys. Ed*. 1977;15:2143.
22. Kunitake T, Aso C. *J. Polym. Sci.: Part A-1*. 1970;8:665.
23. Malhotra SL, Baillet C, Minh L, Blanchard LP. *J. Macromol. Sci. – Chem*. 1978;A12:129.
24. Berger PA, Kotyk JJ, Remsen EE. *Macromolecules*. 1992;25:7227.
25. Higashimura T, Kamigaito M, Kato M, Hasebe T, Sawamoto M. *Macromolecules*. 1993;26:2670.
26. Sawamoto M, Hasebe T, Kamigaito M, Higashimura T. *J. Mater. Sci. – Pure Appl. Chem*. 1994;A31:937.
27. Kishore K, Paramasivam S, Sandhya TE. *Macromolecules*. 1996;29:6973.
28. Ni H-B, Yang J-C, Zhao C-X. *J. Appl. Polym. Sci*. 2006;100:3615.
29. Chun BC, Gibala R. *Polym. Eng. Sci*. 1996;36:744.
30. Osaki K, Inoue T, Hwang E-J, Okamoto H, Takiguchi O. *J. Non-Cryst. Solids*. 1994;172:838.
31. Jones GD, Friedrich RE, Werkema TE, Zimmerman RL. *Ind. Eng. Chem*. 1956;48:2123.
32. Geoghegan M, Jones RAL, Clough AS. *J. Chem. Phys*. 1995;193:2719.
33. Cowie JMG, McEwen IJ. *Polymer*. 1985;26:1662.
34. Chee KK, Ng SC. *J. Appl. Polym. Sci*. 1993;50:1115.
35. Callaghan TA, Paul DR. *J. Polym. Sci. B.: Polym. Phys*. 1994;32:1847.
36. Gan PP, Paul DR, Padwa AR. *Polymer*. 1994;35:3351.
37. Lanzavecchia L, Pedemonte E. *Thermochimica Acta*. 1988;137:123.
38. Mays JW, Nan S, Lewis ME. *Macromolecules*. 1991;24:4857.
39. Li J, Harville S, Mays JW. *Macromolecules*. 1997;30:466.
40. Burge DE, Bruss DB. *J. Polym. Sci. A*. 1963;1:1927.
41. Kato T, Miyaso K, Nagasawa M. *J. Phys. Chem*. 1968;72:2161.
42. Ma H-Z, Ye G-X. *Chin. J. Polym. Sci*. 2006;24:87.
43. McCormick HW. *J. Polym. Sci*. 1959;41:327.

44. Hadjichristidis N, Lindner JS, Mays JW, Wilson WW. *Macromolecules*. 1991;24:6725.
45. Tsunashima Y, Hashimoto T, Nakano T. *Macromolecules*. 1996;29:3475.
46. Rueda JC, Gomes AS, Soares BG. *Polym. Bull.* 1994;33:405.
47. Okamura S, Higashimura T, Imanishi Y. *J. Polym. Sci.* 1958;33:491.
48. Lenz RW. *J. Macromol. Sci – Chem.* 1975;A9:945.
49. Murakata T, Saito Y, Yosikawa T, Suzuki T, Sato S. *Polymer*. 1993;34:1436.
50. Roestamsjah, Wall LA, Florin RE, Aldridge MH, Fetters LJ. *J. Polym. Sci.: Polym. Phys. Ed.* 1975;13:1783.
51. Zheng KM, Greer SC, Corrales LR, Ruiz-Garcia J. *J. Chem. Phys.* 1993;98:9873.
52. Brunacci A, Pedemonte E, Cowie JMG, McEwen IJ. *Polymer*. 1994;35:2893.
53. Yagci Y, Acar MH, Ledwith A. *Eur. Polym. J.* 1992;28:717.
54. Oh Sumi Y, Higashimura T, Okamura S. *J. Polym. Sci. A-1.* 1966;4:923.
55. Chiu TYJ, Heuts JPA, Davis TP, Stenzel MH, Barner-Kowollik C. *Macromol. Chem. Phys.* 2004;205:752.
56. Pfohl O, Hino T, Prausnitz JM. *Polymer*. 1995;36:2065.
57. Richards DH, Salter DA. *Polymer*. 1967;8:127.
58. Denq B-L, Chiu W-Y, Chen L-W, Lee C-Y. *Polym. Degrad. Stab.* 1997;57:261.
59. Grant DH, Vance E, Bywater S. *Trans. Faraday Soc.* 1960;56:1697.
60. Andrews AP, Andrews KP, Greer SC, Boué F, Pfeuty P. *Macromolecules*. 1994;27:3902.
61. Ahmad S, Zulfiqar S. *Polym. Degrad. Stab.* 2002;76:173.
62. De P, Sathyanarayana DN. *Macromol. Chem. Phys.* 2002;203:2218.
63. Da Silva A, Gomes AS, Soares BG. *Polym. Bull.* 1993;30:133.
64. Fukui H, Deguchi T, Sawamoto M, Higashimura T. *Macromolecules*. 1996;29:1131.
65. Jordan DO, Mathieson AR. *J. Chem. Soc.* 1952;442:2363.
66. Wang Q, Quyoum R, Gillis DJ, Tudoret M-J, Jeremic D, Hunter BK, Baird MC. *Organometallics*. 1996;15:693.

Poly(*p*-methylstyrene)

Ali E. Ozcam, Archie P. Smith and Richard J. Spontak

Acronyms PpMS, PMS, P4MS, MST, *p*MS, P-*p*MS, PPMS, 4MS, P*p*MeS

Class Vinyl polymers

Cas Number 24936-41-2

Cas Name Benzene, 1-ethenyl-4-methyl homopolymer

Major Applications Similar to polystyrene (PS) (for example, molded objects, foam drinking cups, and packaging materials). Has a higher deformation temperature and flame resistance than PS. Functionalization of the methyl group of homopolymers/copolymers yields graft polymers and copolymers that are used to prepare new variants of butyl rubber. Specific syndiotactic forms of PpMS can be used as polymeric molecular sieves.

Polymerization Method	Conditions	Reference
Cationic photopolymerization	Cationic polymerization using phosphonium and arsonium salts as initiators. Illuminated with Xe arc lamp at 25°C	(1)
Living carbocationic	Polymerization in methyl chloride/methyl cyclohexane solution at −80°C without initiator and with $TiCl_4$ as initiator	(2)
Living carbocationic	Polymerization induced by cumyl acetate, cumyl propionate, 1(2,4,6-trimethylphenyl)ethyl acetate and 1(4-methylphenyl)ethyl acetate in the presence of BCl_3 in methyl chloride and ethyl chloride at −30 and −50°C	(3)
Living carbocationic	Polymerization in methyl chloride/methyl cyclohexane in the presence of *n*-Bu_4NCl at −30°C	(4)
Radical polymerization	Bulk polymerization at 50°C using azobisisobutyronitrile (AIBN) as initiator	(5)
Radical polymerization	Polymerization initiated by AIBN in cyclohexane over the temperature range of 50–70°C	(6)
Radical polymerization	Polymerization initiated by AIBN in benzene at 50°C	(7)
Cationic polymerization by radiation	Polymerization of both wet and dry monomers by exposure to ^{60}Co γ-ray radiation	(8)
Cationic polymerization	Polymerization conducted in dichloromethane, dichloromethane/nitrobenzene, benzene, carbon tetrachloride (CCl_4) and dichlorobenzene/CCl_4 solvents with acetyl perchlorate or iodine at 0°C	(9)

Polymerization Method	Conditions	Reference
Cationic polymerization	Polymerization performed in dichloromethane, chloroform, and methylcyclohexane/methyl chloride using various initiators at different temperatures	(10)
Cationic polymerization	Polymerization in chloroform initiated by 1-phenylethyl bromide and tin tetrachloride ($SnCl_4$) at −27 and 70°C	(11)
Living cationic polymerization	Polymerization performed with hydrogen iodide/zinc chloride or zinc iodide initiators in toluene or dichloromethane over a temperature range of −15 to 25°C	(12)
Living cationic polymerization	Polymerization in dichloromethane with $SnCl_4$	(10)
Anionic polymerization	Polymerization in cyclohexane initiated by n-BuLi	(13)
Anionic polymerization	Polymerization in THF/benzene with sodium naphthalene at −70°C	(14)
Anionic polymerization	Polymerization in toluene at 60°C initiated by n-BuLi	(15)

Propagation and termination constants

k_p	$2k_t \times 10^{-6}$	$k_p/2k_t \times 10^6$	$T(°C)$	Method*	Reference
84	66	1.28	30.0	A	(16)
102.7	–	–	40.2	B	(17)
134.7	–	–	50.5	B	
187.5	–	–	60.1	B	
265.4	–	–	69.8	B	
Temperature dependence of propagation rate constants			$2.27 \times 10^7 \exp(-7663/RT)$ $(7663 \text{ cal mol}^{-1})$		(17)

* A = rotating sector and inhibitor method; B = Smith-Ewart kinetic theory.

Chemical Property	Units	Conditions	Value	Reference
Chain transfer constants C_s	–	Solvent ($T = 60°C$)		(18)
		p-Isopropyl anisole	3.27×10^{-4}	
		p-Diisopropyl benzene	7.34×10^{-4}	
		p-Isopropyl benzonitrile	2.60×10^{-3}	
		Cumene	4.12×10^{-4}	
		p-Bromocumene	9.23×10^{-4}	
		p-Chlorocumene	7.67×10^{-4}	
Chain transfer rate constants k'	mol l^{-1}s^{-1}	p-Isopropyl anisole	2.75×10^{-2}	(19)
		p-Diisopropyl benzene	3.08×10^{-2}	
		p-Isopropyl benzonitrile	2.18×10^{-1}	
		Cumene	3.46×10^{-2}	
		p-Bromocumene	7.75×10^{-2}	
		p-Chlorocumene	6.44×10^{-2}	

Chemical Property	Units	Conditions	Value		Reference
Copolymerization reactivity ratios: r_1 and r_2 (depends on polymerization route and conditions)	–	Comonomer	r_1	r_2	
		Styrene	1.47	0.40	(20)
			2.40	0.55	(21)
			1.50	0.49	(22)
			2.49	0.42	(23)
		Methyl acrylate	1.54	0.17	(24)
		Methyl methacrylate	0.44	0.41	(25)
		N,N-Divinylaniline	11.8	0.05	(26)
		p-Chlorostyrene	0.61	1.15	(25)
			6.64	0.16	(23)
		Vinyl methyl sulfoxide	2.73	0.01	(27)

Comonomers used in copolymerizations.

Comonomer	Reference
p-Bromostyrene	(28)
p-Chlorostyrene	(23, 29, 30, 31)
Isobutylene	(2, 32–41)
Methacrylonitrile	(42)
Methyl acrylate	(24)
Methyl methacrylate	(25, 43)
Cyclohexyl methacrylate	(43)
Styrene	(21, 23, 25, 31, 36, 44–48)
Indene	(37, 49)
Ethylene	(50–54)
Propylene	(51, 55–56)
1-Octene	(51)
Acrylonitrile	(57)
Carbon monoxide	(58, 59)

Spectral Property	Units	Conditions	Value	Reference
Infrared spectroscopy (peak positions)	cm^{-1}	Peak assignments		
		Aromatic ν_{CH}	3100, 3080, 3040, 3910	(60)
		Aromatic ν_{CH_3}	2910	
		Aromatic ν_{CH_2}	2840	
		Typical of *p*-substitution	2000–1600	
		Phenyl ring	1495	
		δ_{CH_3}, δ_{CH_2}	1430, 1390	
		Helical chain structure	1363, 1314, 1290	(61)
		δ_{CH_3}	1360	(60)
		Stereoregular chain structure	1346, 1093	(61)
		Chain conformational regularity	1334, 1304, 1224, 1191	(61)

Spectral Property	Units	Conditions	Value	Reference
		In-plane CH bending of phenyl ring	1160, 1090, 1000	(60)
		Chain conformational regularity	977, 861, 749, 738	(61)
		Out-of-plane CH bending of phenyl ring	790	(60)
		Peaks assigned to syndiotactic Form I polymorph for the conformational regularity of s(2/1)2 helices	1314, 1307, 1273, 1162, 735, 608, 566, 540, 535, 519, 503	(62)
		Peaks assigned to conformational regularity of trans-planar	1334, 1263, 1222, 1191, 976, 858, 749, 738, 718, 564, 540	(62)
		Peaks assigned to syndiotactic Form III polymorph for the trans-planar conformation	1335, 1223, 861, 719, 540	(62)
NMR spectroscopy		Cross-polarization/magic angle spinning		(63, 64)
		15–20% (w/v) solution in CDCl$_3$ at 30°C		(42)
		Solution in CCl$_4$ at 75°C		(60)
		10% solution in CDCl$_3$ at 24°C		(65)
		^{1}H spin echo at 60 MHz of 10% (w/w) hexachlorobutadiene solution		(66)
		^{13}C at 25 MHz of 10% (w/v) solution in CDCl$_2$		(66)
		^{1}H and ^{13}C NMR in *o*-dichlorobenzene at 130°C		(61)
		^{1}H and ^{13}C NMR of poly(isobutylene-*co-p*-methylstyrene)		(67)

Physical Property	Units	Conditions	Value	Reference
Molecular weight	g mol^{-1}	Monomer	118.18	–
		Polymer range ($\overline{M}_\mathrm{w}$)	0.15–10 $\times 10^5$	–
Polydispersity	–	Depends on polymerization method	1.03–5.0	–
Melting temperature	K	Syndiotactic	446	(68)
Thermal expansion coefficient α	K^{-1}	Below T_g	2.1×10^{-4}	(69)
		150 K $< T < T_\mathrm{g}$	7.1×10^{-5}	(70)
		Above T_g	6.3×10^{-4}	(69)
		$T_\mathrm{g} < T <$ 440 K	1.6×10^{-4}	(70)
Density	g cm^{-3}	20°C	1.022	(69)
		21°C	1.012	(71)
		23°C	1.016	(65, 72)
		25°C	1.011	(70)
Solvents	–	Solvency		
		Good	Benzene	(73)
		Good	Butyl acetate	(73)
		–	Carbon tetrachloride	(60)
		Intermediate	Cyclohexane	(73,74)
		Intermediate	Dichloroethane	(74)
		Theta	Diethyl succinate	(73, 74)

Physical Property	Units	Conditions	Value	Reference
		Intermediate	Methyl ethyl ketone	(73, 74)
		–	Tetrahydrofuran	(60)
		Good	Toluene	(73–75)
		Poor	Ethanol/benzene (6:1 v/v)	(75)
Theta temperature	K	Diethyl succinate	289.6	(74)

Property	Units	PpMS Volume Fraction	Value	Reference
Flory-Huggins	–	0.20	0.378	(28)
interaction		0.24	0.401	
parameter χ		0.28	0.413	
in toluene at 22°C		0.32	0.412	
		0.36	0.404	
		0.40	0.382	
		0.44	0.364	
		0.48	0.356	
		0.52	0.355	
		0.56	0.352	
		0.60	0.341	
		0.64	0.322	

Flory-Huggins interaction parameter χ in polymer blends: $\chi = a \times 10^{-3} + b/T$

Property	Conditions			Value		Reference
	dPS Molecular Weight (g mol^{-1})	PpMS Molecular Weight (g mol^{-1})	PpMS Volume Fraction	a	b(K)	
PS/PpMS blends (small-angle neutron scattering)	32,700	58,800	0.75	-11 ± 1.0	7.0 ± 0.5	(76)
	32,700	58,800	0.50	-8.1 ± 0.9	5.7 ± 0.4	
	32,700	58,800	0.25	-8.3 ± 0.9	6.1 ± 0.4	
	32,700	131,000	0.75	-12 ± 0.6	7.2 ± 0.3	
	PS Molecular Weight (g mol^{-1})	d-PpMS Molecular Weight (g mol^{-1})	PpMS Volume Fraction	a	b(K)	
PS/PpMS blends (small-angle neutron scattering)	32,000	63,500	0.50	-6.6 ± 0.5	5.7 ± 0.2	(77)
	dPS Molecular Weight (g mol^{-1})	PpMS Molecular Weight (g mol^{-1})	PpMS Volume Fraction	a	b(K)	
PS/PpMS blends (neutron reflectivity; double-layer structures)	105,000	131,000	–	-11 ± 2.0	6.8 ± 1.0	(78)
	714,000	131,000	–			
	714,000	613,000	–			

Property	Conditions			Value				Reference
	PM2PO Molecular Weight (g mol^{-1})	d-PpMS Molecular Weight (g mol^{-1})	PpMS Volume Fraction	$\chi \times 10^3 (\pm 16\%)$				
				493 K	513 K	533 K	553 K	
Poly(2,6-dimethyl-1,	35,400	88,500	0.101	-10.5	–	-7.8	–	(79)
4-phenylene	35,400	88,500	0.198	–	-7.0	-6.8	-6.0	
oxide	35,400	88,500	0.301	-11.7	–	-12.5	–	
(PM2PO)/PpMS	35,400	88,500	0.399	–	–	-15.4	–	
blends	35,400	88,500	0.600	–	–	-20.4	–	

Solution characteristics

Property	Units	Conditions	Value		Reference
Mark-Houwink parameters K and a	$K = \text{mol g}^{-1}$ $a = \text{none}$	$\overline{M}_\text{w}$ range $= (20-155) \times 10^4$	$K \times 10^5$	a	(80)
		Cyclohexane at 30°C	8.07	0.72	
		Methyl ethyl ketone at 30°C	10.3	0.68	
		Toluene at 30°C	6.88	0.76	
Second virial coefficient A_2	mol cm^3 g^{-2}	–	See table below		(74)
Mean square radius $\langle r^2 \rangle$	cm^2	–	See table below		
Intrinsic viscosity $[\eta]$	dl g^{-1}	–	See table below		

Solvent	$\overline{M}_\text{w} \times 10^{-4}$	T(°C)	$A^2 \times 10^4$	$\langle r^2 \rangle \times 10^{12}$	$[\eta]$
Cyclohexane	84.7	30	1.56	12.4	1.25
Dichloroethane	118	30	1.70	18.9	1.83
Diethyl succinate	197	60	0.58	23.9	1.34
		40	0.41	20.8	1.16
		20	0.09	17.0	0.89
		18	0.03	16.1	0.85
		16	-0.02	15.5	0.80
	89.6	60	0.81	11.5	1.05
		40	0.60	10.4	0.94
		20	0.20	8.85	0.75
		18	0.11	8.54	0.72
		16	0.00	8.00	0.69
	76.7	60	0.90	9.25	0.95
		40	0.70	8.55	0.85
		20	0.17	7.37	0.70
		18	0.12	6.94	0.67
		16	-0.01	6.50	0.64
	68.4	60	0.83	8.18	0.90
		40	0.61	7.55	0.79
		20	0.12	6.37	0.65
		18	0.06	6.15	0.62
		16	-0.04	5.85	0.60

Poly(*p*-methylstyrene)

Solvent	$\overline{M}_w \times 10^{-4}$	T(°C)	$A^2 \times 10^4$	$\langle r^2 \rangle \times 10^{12}$	$[\eta]$
Methyl ethyl ketone	121	30	1.00	17.3	1.40
Toluene	180	30	2.18	36.9	3.72
	81.3	30	3.08	15.4	2.20
	47.6	30	3.41	8.70	1.36
	19.2	30	4.37	3.00	0.73

Crystal structure.

Polymorph	Description							Reference
Syndiotactic Form I	Chains have helical s(2/1)2 conformation, repeat distance of 7.8 Å, $T_m = 178$°C							(81,82)

Space Group	Monomers Per Unit Cell	Cell Dimensions (Å)			Cell Angles			
		a	*b*	*c*	α	β	γ	
P2$_1$/a	–	24.5	12.4	8.1	90	90	143.5	(83)

Calculated crystalline density   1.06 g cm^{-3}

Polymorph	Description	Reference
Syndiotactic form II	Chains have helical s(2/1)2 conformation, repeat distance of 7.8 Å, $T_m = 201$°C	(81,82)
Syndiotactic form III	Chains have trans planar conformation, repeat distance of 5.1 Å, $T_m = 224$°C	(81,82)

Space Group	Monomers Per Unit Cell	Cell Dimensions (Å)			Cell Angles			
		a	*b*	*c*	α	β	γ	
Pnam	8	13.36	23.21	5.12	90	90	90	(84)

Experimental density   1.00 g cm^{-3}
Calculated density   0.99 g cm^{-3}

Polymorph	Description	Reference
Syndiotactic form IV	Chains have trans planar conformation, repeat distance of 5.1 Å, $T_m = 194$°C	(81, 82)
Syndiotactic form V	Chains have trans planar conformation, repeat distance of 5.1 Å	(82)
Syndiotactic clathrates	Chains have helical s(2/1)2 conformation, repeat distance of 7.8 Å	(81, 82)

Class	Guest Molecule	Space Group	Cell Dimensions (Å)			Cell Angles (°)			
			a	*b*	*c*	α	β	γ	
β	Tetrahydrofuran	P2$_1$/a	18.8	12.7	7.7	90	90	100	(85, 86)

Experimental density   0.95 g cm^{-3}
Calculated crystalline density   1.13 g cm^{-3}

Class	Guest Molecule	Space Group	Cell Dimensions (Å)			Cell Angles (°)			
			a	*b*	*c*	α	β	γ	
β	Benzene	C222$_1$	19.5	13.3	7.7	90	90	90	(87)

Crystalline density   1.02 g cm^{-3}
Calculated crystalline density   1.05 g cm^{-3}

Polymorph	Description								Reference
Class	**Guest Molecule**	**Space Group**	**Cell Dimensions (Å)**			**Cell Angles (°)**			
			a	*b*	*c*	*α*	*β*	*γ*	
β	Carbon disulfide	C222$_1$	20.0	12.5	7.7	90	90	90	(88)
		Calculated crystalline density	1.08 g cm^{-3}						
			Cell Dimensions (Å)			**Cell Angles (°)**			
			a	*b*	*c*	*α*	*β*	**Γ**	
A	*o*-dichlorobenzene	P2$_1$/a	23.4	11.8	7.7	90	90	115	(86, 89)
		Density 1.10 g cm^{-3}							
		Calculated crystalline density 1.07 g cm^{-3}							
Γ	Cyclohexanone or cyclohexane	T2G2T2G2	Periodicity: 11.7 ± 0.1 Å						(90)
Syndiotactic clathrates of styrene-*co*-*p*-methylstyrene	*p*-methylstyrene concentration >35%	*s*-poly(*p*-methylstyrene)-like clathrates							(91,92)
	p-methylstyrene concentration <35%	*s*-polystyrene-like clathrates							

Property	Units	Conditions	Value	Reference
Glass transition temperature T_g	K	–	356	(69)
		Syndiotactic	379	(68)
		Isotactic	374	(68)
		Creep test	361	(93)
		Differential thermal analysis (DTA)	366	(94)
		Stress relaxation	366	(93)
		Differential scanning calorimetry (DSC)	374	(93)
		DSC	380	(95)
		DSC	383	(65)
		DSC	384	(96)
		DSC	379	(48)
		DSC	385	(48)
		DSC, extrapolated to infinite molecular weight	384	(60)
		Dynamic mechanical analysis (DMA) (1 Hz, 2°C min^{-1})	385	(93)
		Dielectric analysis (DEA) (1 Hz, 3°C min^{-1})	391	(93)
		T_g dependence on $\overline{M}_n$	$384 - (2.56 \times 10^5)/\overline{M}_n$	(60)
Sub-T_g transitions	K	β transition, DMA at 0.1 Hz, $E_a = 71$ kJ mol^{-1}	273	(93)
		γ transition, DMA at 0.1 Hz, $E_a = 29$ kJ mol^{-1}	212	(93)
		δ transition, resonance electrostatic method at 8 kHz	92	(97)

Poly(p-methylstyrene)

Property	Units	Conditions	Value	Reference
Heat capacity C_p	$J\,mol^{-1}\,K^{-1}$	300 K to T_g T_g to 500 K	$-3.54 + 0.5138T$ $90.85 + 0.3564T$	(95)
Deflection temperature	K	ASTM Test D-648 under 1.8-MPa load on injection-molded and annealed sample	365	(98)
Tensile modulus	MPa	ASTM Test D-638	2206	(98)
Dynamic storage modulus	MPa	DMA, 1 Hz, 20°C	3400	(93)
Dynamic loss modulus	MPa	DMA, 1 Hz, 20°C	87	(93)
Tensile strength at break	MPa	ASTM Test D-638	49.6	(98)
Tensile elongation at break	%	ASTM Test D-638	3.0	(98)
Flexural modulus	MPa	ASTM Test D-790	2992	(98)
Flexural strength at break	MPa	ASTM Test D-790	79.3	(98)
Impact strength	$J\,m^{-1}$	ASTM Test D-256, 73°F, notched 3.175 mm thick specimen	16	(98)
Hardness	–	ASTM Test D-785, Rockwell M scale	80	(98)
	$MPa\,mm^{-2}$	–	12.1	(72)
Nominal melt flow rate (200°C, 5000 g)	$g\,(10\,min)^{-1}$	D-1238	4.1	(98)
Vicat softening temperature	°C	D-1525 –	116 113	(98) (72)
Resonance frequency	kHz	Mechanical damping measurements of polymer discs	9.7	(97)
Index of refraction	–	20°C	1.577 1.58	(69) (72)
Dielectric constant	–	Dielectric spectroscopy, 1 kHz and 23°C	2.86	(96)
		Dielectric spectroscopy, 1 kHz and 25°C	2.47	(70)
		Dielectric spectroscopy, 10 kHz, varies linearly with temperature		
		-181°C	2.62	(99)
		52°C	2.54	
Specific free volume	$cm^3\,g^{-1}$	–	0.176	(100)
Permeability coefficient P	$10^{-10}\dfrac{cm^3(STP)\cdot cm}{cm^2\cdot s\cdot cmHg}$	CH_4 at 1 atm and 35°C	2.20	(100)
		CO_2 at 1 atm and 35°C	29.8	(100)
		CO_2 at 200 mm Hg pressure and 25°C	9.0	(101)

Property	Units	Conditions	Value	Reference
		He at 1 atm and 35°C	37.1	(100)
		N$_2$ at 1 atm and 35°C	1.50	(100)
		O$_2$ at 1 atm and 35°C	7.2	(100)
		O$_2$ at 200 mm Hg pressure and 25°C	1.04	(101)
Diffusion coefficient D	$10^{-8} \dfrac{cm^2}{s}$	CH$_4$ at 1 atm and 35°C	4.0	(100)
		CO$_2$ at 1 atm and 35°C	13.7	(100)
		CO$_2$ at 200 mm Hg pressure and 25°C	5.8	(101)
		N$_2$ at 1 atm and 35°C	10.4	(100)
		O$_2$ at 1 atm and 35°C	28.1	(100)
		O$_2$ at 200 mm Hg pressure and 25°C	10.2	(101)
Temperature dependence of D	$\dfrac{cm^2}{s}$	CO$_2$ at 200 mm Hg pressure and 25°C	$0.090 \exp\left(\dfrac{-9.7\text{kJ mol}^{-1}}{RT}\right)$	(101)
		O$_2$ at 200 mm Hg pressure and 25°C	$0.082 \exp\left(\dfrac{-9.4\text{kJ mol}^{-1}}{RT}\right)$	

Degradation Properties	Experimental Conditions	Degradation	Reference
	Irradiation with 253.7 nm UV light under vacuum at 25°C	Hydrogen was the major product and small amounts of methane, ethane, styrene, p-methylstyrene were obtained as degradation products	(102)
	Irradiation with 284 nm UV photons	C—H cleavage, polymer degradation	(75)
	Isothermal treatments between 250 and 365°C	Weight loss between 1 and 75% due to random chain scission and depolymerization; above 330°C cross-linking occurs	(60)

Activation energy for thermal decomposition with the assumed reaction order, n (kJ mol^{-1})

$n = 0$	$n = 1$
191.3	260.4

Poly(*p*-methylstyrene)

Degradation Properties	Experimental Conditions		Degradation		Reference
	Units		Conditions	Value	
Maximum thermal decomposition temperature	K		–	490	(103)
			–	603	(72)

Degradation Properties	Experimental Conditions		Degradation		Reference
	Units		Conditions	Value	
G value of scission	–		γ radiation at 130°C	0.043	(104)
G value of cross-linking	–		γ radiation at 65°C	0.061	
			γ radiation at 98°C	0.022	

References

1. Abu-Abdoun I, Ali A. *Eur. Polym. J.* 1993;29:1439.
2. Fodor Z, Faust R. *J. Macromol. Sci., Part A: Pure Appl. Chem.* 1994;A31:1985.
3. Faust R, Kennedy JP. *Polym. Bull.* 1988;19:29.
4. Nagy A, Majoros I, Kennedy JP. *J. Polym. Sci., Part A: Polym. Chem.* 1997;35:3341.
5. Gyöngyhalmi I, Földes-Berezsnich T, Tüdós F. *Eur. Polym. J.* 1993;29:219.
6. Mutschler H, Schröder U, Fahner E, Ebert KH, Hamielec AE. *Polymer.* 1985;26:935.
7. Gyöngyhalmi I, Nagy A, Földes-Berezsnich T, Tüdós F. *Makromol. Chem.* 1993;194:3357.
8. Hayashi K, Pepper DC. *Polym. J.* 1976;8:1.
9. Higashimura T, Kishiro O, Takeda T. *J. Polym. Sci.: Polym. Chem. Ed.* 1976;14:1089.
10. De P, Faust R. *Macromolecules.* 2005;38:5498.
11. Yang M, Li K. Stöver HD H. *Macromol. Rapid Commun.* 1994;15:425.
12. Kojima K, Sawamoto M, Higashimura T. *J. Polym. Sci., Part A: Polym. Chem.* 1990;28:3007.
13. Huang H, Chang C, Liu I, Tsai H, Lai M, Tsiang RC. *J. Polym. Sci., Part A: Polym. Chem.* 2005;43:4710.
14. Shen J, Chen Y, Cai R, Huang Z. *Polymer.* 2000;41:9291.
15. Stroeks A, Paquaij R, Nies E. *Polymer.* 1991;32:2653.
16. Imoto M, Kinoshita M, Nishigaki M. *Makromol. Chem.* 1965;86:217.
17. Paoletti KP, Billmeyer FW. *J. Polym. Sci., Part: A.* 1964;2:2049.
18. Yamamoto T, Otsu T. *Polym. Lett.* 1966;4:1039.
19. Yamamoto T, Otsu T. *J. Polym. Sci., Part A-1.* 1969;7:1279.
20. Soga K, Nakatani H, Monoi T. *Macromolecules.* 1990;23:953.
21. Grassi A, Longo P, Proto A, Zambelli A. *Macromolecules.* 1989;22:104.
22. Zambelli A, Pellecchia C, Oliva L, Longo P, Grassi A. *Makromol. Chem.* 1991;192:223.

23. Wood KB, Stannett VT, Sigwalt P. *Makromol. Chem. Suppl.* 1989;15:71.
24. Faber JWH, Fowler WF. *J. Polym. Sci., Part A-1.* 1970;8:1777.
25. Walling C, Briggs ER, Wolfstirn KB, Mayo FR. *J. Am. Chem. Soc.* 1948;70:1537.
26. Chang EYC, Price CC. *J. Am. Chem. Soc.* 1961;83:4650.
27. Fujihara H, Shindo T, Yoshihara M, Maeshima T. *J. Macromol. Sci., Part A: Pure Appl. Chem.* 1980;A14:1029.
28. Corneliussen R, Rice SA, Yamakawa H. *J. Chem. Phys.* 1963;38:1768.
29. Mashimo S, Nozaki R. *J. Non-Cryst. Solids.* 1991;131–133:1158.
30. Overberger CG, Arond LH, Tanner D, Taylor JJ, Alfrey T. *J. Am. Chem. Soc.* 1952;74:4848.
31. Wood KB, Stannett VT, Sigwalt P. *J. Polym. Sci., Part A: Polym. Chem.* 1995;33:2909.
32. Lubnin AV, Országh I, Kennedy JP. *J. Macromol. Sci., Part A: Pure Appl. Chem.* 1995;A32:1809.
33. Kuwamoto K. *Int. Polym. Process.* 1994;9:319.
34. Fodor Z, Faust R. *J. Macromol. Sci., Part A: Pure Appl. Chem.* 1995;A32:575.
35. Steinke JHG, Haque SA, Fréchet JMJ, Wang HC. *Macromolecules.* 1996;29:6081.
36. Taylor SJ, Storey RF, Kopchick JG, Mauritz KA. *Polymer.* 2004;45:4719.
37. Tsunogae Y, Kennedy JP. *J Macromol. Sci., Part A: Pure Appl. Chem.* 1993;A30:269.
38. Fodor Z, Faust R. *J. Macromol. Sci., Part A: Pure Appl. Chem.* 1995;A32:575.
39. Wang HC, Powers KW. *Elastomerics.* 1992;124:25.
40. Ma Y, Wu G, Yang W. *J. Polym. Sci., Part A: Polym. Chem.* 2003;41:408.
41. Everland H, Kops J, Nielsen A, Iván B. *Polym. Bull.* 1993;31:159.
42. Chen J, Goh SH, Lee SY, Siow KS. *J. Polym. Sci., Part A: Polym. Chem.* 1994;32:1263.
43. Múgica A, Remiro PM, Cortázar M. *Polymer.* 2000;41:5257.
44. Oh J, Kang S, Kwon O, Choi S. *Macromolecules.* 1995;28:3015.
45. Nyquist RA, Malanga M. *Appl. Spectrosc.* 1989;43:442.
46. Cardi N, Po R, Abbondanza L, Abis L, Conti G. *Macromol. Symp.* 1996;102:123.
47. Visse F, Marechal E. *Polymer.* 1974;15:485.
48. Nakatani H, Nitta K, Soga K, Takata T. *Polymer.* 1997;38:4751.
49. Tsunogae Y, Majoros I, Kennedy JP. *J Macromol. Sci., Part A: Pure Appl. Chem.* 1993;A30:253.
50. Hwu J, Chang M, Lin J, Cheng H, Jiang G. *J. Organomet. Chem.* 2005;690:6300.
51. Lu HL, Hong S, Chung TC. *Macromolecules.* 1998;31:2028.; 225b. Lu B, Chung TC. *J. Polym. Sci., Part A: Polym. Chem.* 2000;38:1337. (koysammmi??).
52. Chung TC, Lu HL, Ding RD. *Macromolecules.* 1997;30:1272.
53. Chung TC, Lu HL. *J. Polym. Sci., Part A: Polym. Chem.* 1997;35:575.
54. Chung TC, Lu HL. *J. Polym. Sci., Part A: Polym. Chem.* 1998;36:1017.
55. Lu HL, Hong S, Chung TC. *J. Polym. Sci., Part A: Polym. Chem.* 1999;37:2795.
56. Chung TC, Dong JY. *J. Am. Chem. Soc.* 2001;123:4871.
57. Chong YF, Goh SH. *Polymer.* 1992;33:127.
58. Binotti B, Carfagna C, Zuccaccia C, Macchioni A. *Chem. Commun.* 2005;1:92.
59. Carfagna C, Gatti G, Martini D, Pettinari C. *Organometallics.* 2001;20:2175.
60. Malhotra SL, Lessard P, Minh L, Blanchard LP. *J Macromol. Sci., Part A: Pure Appl. Chem.* 1980;A14:517.
61. Abis L, Albizzati E, Conti G, Giannini U, Resconi L, Spera S. *Makromol. Chem. Rapid Commun.* 1988;9:209.
62. Guerra G, Poggetto FD, Iuliano M, Manfredi C. *Makromol. Chem.* 1992;193:2413.

63. Guerra G, Iuliano M, Grassi A, Rice DM, Karasz FE, Macknight WJ. *Polym. Commun.* 1991;32:430.

64. Dickinson LC, Yang H, Chu CW, Stein RS, Chien JC W. *Macromolecules.* 1987;20:1757.

65. Gehlsen MD, Weimann PA, Bates FS, Harville S, Mays JW, Wignall GD. *J. Polym. Sci., Part B: Polym. Phys.* 1995;33:1527.

66. Lauprêtre F, Noël C, Monnerie L. *J. Polym. Sci.: Polym. Phys. Ed.* 1977;15:2143.

67. Ashbaugh JR, Ruff RR, Shaffer TD. *J. Polym. Sci., Part A: Polym. Chem.* 2000;38:1680.

68. Tomotsu N, Ishihara N, Newman TH, Malanga MT. *J. Mol. Catal. A: Chem.* 1998;128:167.

69. Kennedy GT, Morton F. *J. Chem. Soc.* 1949;2383.

70. Corrado LC. *J. Chem. Phys.* 1969;50:2260.

71. Fried JR, Lorenz T, Ramdas A. *Polym. Eng. Sci.* 1985;25:1048.

72. Kozorezov Y, Shilyaeva IY. *Int. Polym. Sci. Tech.* 1995;22:T58.

73. Ono K, Okada Y, Yokotsuka S, Sasaki T, Yamamoto M. *Macromolecules.* 1994;27:6482.

74. Tanaka G, Imai S, Yamakawa H. *J. Chem. Phys.* 1970;52:2639.

75. Tamai T, Hashida I, Ichinose N, Kawanishi S, Inoue H, Mizuno K. *Polymer.* 1996;37:5525.

76. Londono JD, Wignall GD. *Macromolecules.* 1997;30:3821.

77. Zirkel A, Gruner SM, Urban V, Thiyagarajan P. *Macromolecules.* 2002;35:7375.

78. Schnell R, Stamm M, Creton C. *Macromolecues.* 1998;31:2284.; 77b. Schnell R, Stamm M. *Physica B.* 1997;234–236:247.

79. Maconnachie A, Fried JR, Tomlins PE. *Macromolecules.* 1989;22:4606.

80. Kuwahara N, Ogino K, Kasai A, Ueno S, Kaneko M. *J. Polym. Sci.: Part A.* 1965;3:985.

81. Iuliano M, Guerra G, Petraccone V, Corradini P, Pellecchia C. *New Polymeric Mater.* 1992;3:133.

82. Rosa CD, Petraccone V, Guerra G, Manfredi C. *Polymer.* 1996;37:5247.

83. Esposito G, Tarallo O, Petraccone V. *Macromolecules.* 2006;39:5037.

84. Rosa CD, Petraccone V, Poggetto FD, Guerra G, Pirozzi B, Lorenzo MLD, Corradini P. *Macromolecules.* 1995;28:5507.

85. Petraccone V, Camera DL, Pirozzi B, Rizzo P, Rosa CD. *Macromolecules.* 1998;31:5830.

86. Rizzo P, de Ballesteros OR, Rosa CD, Auriemma F, Camera DL, Petraccone VG. *Polymer.* 2000;41:3745.

87. Camera DL, Petraccone V, Artimagnella S, de Ballesteros OR. *Macromolecules.* 2001;34:7762.

88. Petraccone V, Tarallo O. *Macromol. Symp.* 2004;213:385.

89. Petraccone V, Camera DL, Caporaso L, Rosa CD. *Macromolecules.* 2000;33:2610.

90. Petraccone V, Espesito G, Tarallo O, Caporaso L. *Macromolecules.* 2005;38:5668.

91. Loffredo F, Pranzo A, Guerra G, Venditto V, Longo P. *Macromol. Symp.* 2001;166:165.

92. Loffredo F, Pranzo A, Venditto V, Longo P, Guerra G. *Macromol. Chem. Phys.* 2003;204:859.

93. Gao H, Harmon JP. *Thermochim. Acta.* 1996;284:85.

94. Dunham KR, Faber JWH, Vandenberghe J, Fowler WF. *J. Appl. Polym. Sci.* 1963;7:897.

95. Judowits LH, Bopp RC, Gaur U, Wunderlich B. *J. Polym. Sci.: Part B Polym. Phys.* 1986;24:2725.

96. Gustafsson A, Wiberg G, Gedde UW. *Polym. Eng. Sci.* 1993;33:549.

97. Baccaredda M, Butta E, Frosini V, Petris DS. *Mater. Sci. Eng.* 1968;3:157.

98. Kaeding WW, Barile GC. In: Culbertson BM, Pittman CU, eds. *New Monomers Polymers.* New York: Plenum Press; 1984:223.

99. Nozaki M, Shimada K, Okamoto S. *Jpn. J. Appl. Phys.* 1971;10:179.
100. Puleo AC, Muruganandam N, Paul DR. *J. Polym. Sci.: Part B Polym. Phys.* 1989;27:2385.
101. Greenwood R, Weir N. *Makromol. Chem.* 1975;176:2041.
102. Weir NA. *J. Appl. Polym. Sci.* 1973;17:401.
103. Fares MM, Yalcin T, Hacaloglu J, Gungor A, Suzer S. *Analyst.* 1994;119:693.
104. Burlant W, Neerman J, Serment V. *J. Polym. Sci.* 1962;58:491.

Poly(norbornene)

Vassilios Galiatsatos

Alternative Names Polybicyclo[2.2.1]hept-2-ene (homopolymer), Polybicyclo[2.2.1]hept-5-ene (homopolymer), Poly(2-norbornene), Polynorbornenylene, Poly(l,3-cyclopentylenevinylene)

Trade Names Norsorex, Telene (copolymer of dicyclopentadiene)

Class Diene elastomers; Telene is a thermoset.

Major Applications The rubbery polymers are useful as vibration and noise dampening materials. Also for oil spill recovery, sound barrier materials, and for soft seals and gaskets. More recent applications for polynorborne derivatives include interlayer and intermetal dielectrics in microelectronics.

Structure

Preparation The polymer obtained by ring-opening polymerization of norbornene. Both *cis* and *trans* structures may result depending on the catalyst system used. The repeat unit contains both in-chain ring and a double bond. Polymer is typically free of oligomers and macrocycles. Ethylene copolymers are of interest as thermoplastics. Copolymers with a norbornene content of around 30 wt-% have a T_g of around 0°C.

Crosslinking Cross-linking can occur by conventional accelerated sulfur vulcanization.[1,2] Typically a higher than usual accelerator sulfur ratio is used. Crosslinking produces very soft elastomers that retain a useable tensile strength. For example a combination of hardness of 18 Shore-A with a tensile strength of 10 MPa.

Property	Units	Conditions	Value	Reference
Typical molecular weight of polymer	g mol^{-1}	–	2–3×10^6	–
Typical appearance	–	–	White powder	–
Glass transition temperature T_g	K	Commercial product	308–318	(3)
		Incorporation of a mineral oil extender, which gives useful rubbery properties, including very soft compositions	228–213	
		20% *cis* content polymer, which is totally amorphous	308	

Property	Units	Conditions	Value	Reference
Crystalline melting temperature	K	Hydrogenated polynorbornene	413.8	(3)
Heat of fusion	kJ g^{-1}	Hydrogenated polynorbornene	58.7×10^{-3}	(3)
Decomposition temperature	K	–	>673	(3)
Density	g cm^{-3}	–	0.30	(3)
Index of refraction	–	–	1.534	(3)
Hardness	Shore A	Cured for 10 min at 320°F	40	(3)
100% modulus	MPa	Cured for 10 min at 320°F	0.552	(3)
300% modulus	MPa	Cured for 10 min at 320°F	2.24	(3)
Tensile strength	MPa	Cured for 10 min at 320°F	15.1	(3)
Elongation	%	Cured for 10 min at 320°F	560	(3)
FTIR spectrum	cm^{-1}	*Cis* absorption	740	(4)
		Trans out of plane =C—H bending	960	
		Cis in plane =C—H bending	1,404	

Supercritical fluid behavior Polynorbornene, molecular weight = 2×10^6, 25°C, pressure = 19.0 MPa

Force field parameters for bond stretching[5]

Bond
C$_2$—C$_3$ Cl—C$_2$ Cl—C &
CH (averaged)
Bond length (Å) 1.551
1.560
1.545
1.086
Force constant (kJ Å^1)
2,358 2,975 3,050 3,248
Force field for angle bending [5]

Angle	Angle (degrees)	Force Constant (kJ Å^2)
(C7)H2	109.4	565
(C1-6)H2	107.8	573
C1-C7-C4	96.1	688
C2-C1-C6	108.3	1122
C2-C1-C7	101.6	426
C1-C2-C3	103.2	506

Characteristic ratio	–	Calculated	12.1	(5)
		–	11.4	
Entanglement molecular weight	g mol^{-1}	–	41,000	(5)

Property	Units	Conditions	Values	Reference
Van der Waals volume	$cm^3\,mol^{-1}$	Calculated	108	(5)
		Experimental	149.9	
Intrinsic viscosity	$dl\,g^{-1}$	In benzene at 30°C (at a strain rate = 100% min^{-1} at 25°C)	3.4, 4.3, 5.0, 9.0	(6)
Trans/cis	–	Deduced from the ratio of optical ratios at 10.35 and 13.8 μm (at a strain rate = 100% min^{-1} at 25°C)	3, 4, 4.2, 4.3	(6)
Tensile strength	psi ($\times 10^3$)	*At a strain rate = 100% min^{-1} at 25° C*	*3, 4.2, 4.8, 6.5*	(6)
Ultimate elongation	%	*At a strain rate = 100% min^{-1} at 25° C*	*16, 80, 85, 300*	(6)
Young's modulus	*MPa*	*At a strain rate = 100% min^{-1} at 25° C*	*90, 70, 50, 20*	(6)
Crystallographic identity period		*2 repeat units per unit cell, 1.18 nm*		(7)

Environmental Properties Polynorbornene has an operational range of –40 to +80°C. Formulation plays a critical role here. Polynorbornene displays good tolerance to room temperature water and poor tolerance to ozone. Clearly mineral oil contact should also be avoided.[8]

Control of Properties by Blending The presence of the ring in the main chain results in a high T_g as high as 35°C thus rendering the polymer unsuitable as a rubber at room temperature. Blending with aromatic oil or a choice of ester plasticizers will result in a rubbery material. The polymer can take up large quantities of oil resulting in T_g's as low as –60°C. Fillers may also be employed is those blends.

Suppliers Norsorex,[9] Zeon Chemicals www.zeonchemicals.com. Telene (copolymer of dicyclopentadiene), Telene S.A.S. (France) www.telene.com.

References

1. Makovetskii KL. *Polymer Sci. Ser. A.* 1994;36(10):1433.
2. Ivin KJ. *Olefin Methathesis.* London: Academic Press; 1983:249.
3. Ohm RF. *Chem. Tech.* 1980;10:183.
4. Cataldo F. *Polymer International.* 1994;34:49.
5. Haselwander TFA, et al. *Macromol. Chem. Phys.* 1996;197:3435.
6. Galperin I, Carter JH, Hein PR. *J. Appl. Polym. Sci.* 1968;12:1751.
7. Truett WL, et al. *J. Am. Chem. Soc.* 1960;82:2337.
8. Smith EH, Ed. *Mechanical's Engineers Reference Handbook.* 12th Ed. Elsevier; 1998:1248.
9. Bhowmick AK, Stein C, Stephens HL. *Plast. Eng.* 2001;61:775.

Polyoctenamer

Vassilios Galiatsatos

Acronym Alternative Name Trade Name TOR (for trans), poly(l-octenylene), cyclooctene homopolymer, poly(1,8-octenamer), poly(1-octenylene), poly(hexamethylenevinylene), poly(1-octene-1,8-diyl), Vestenamer (Degussa AG - www.degussa-hpp.com)

Class Diene elastomers

Structure $(CH{=}CH(CH_2)_6)_n$

Synthesis Ring-opening polymerization of cyclooctene in the presence of Ziegler Natta catalysts. Cyclooctene is polymerized to polyoctenamer (TOR) in a metathesis reaction that produces both linear and cyclic macromolecules. The cis/trans ratio, which determines the degree of crystallinity of TOR, is controlled by the polymerization conditions.

Fractionation Methods Gel permeation chromatography employing THF as a solvent.[1]

Property	Units	Conditions	Value	Reference
Molecular weight	$g\,mol^{-1}$	Commercial polymer	$\sim$100,000 (M_w by GPC)	www.degussa-hpp.com
Molar absorptivities of IR bands attributed to trans and cis units	$(mol\,cm)^{-1}$	ε_{trans} (10.35 μm)	135	(2)
		ε_{cis} (7.12 μm)	8.7	(2)
Mark-Houwink parameters: K and α	K = ml g^{-1} α = None	40–50% trans content at 30°C in toluene	$K = 8.0 \times 10^4$ $\alpha = 0.63$	(3)
Glass transition temperature, T_g	K	Cis-polyoctenamer	165 by DSC	(4)
Crystalline melting temperature, T_m	K	37.6% trans	290 by DSC	(5)
		75–85% trans	335,340 by X-ray	(6)
		100 (extrapolated)	350	(7)
		100 (extrapolated)	346 by dilatometry	(8)
		100 (extrapolated)	333	(8)
Heat of fusion, ΔH	$J\,g^{-1}$	37.6% trans	290	(5)
		100 (extrapolated)	220.1	(7)
		100 (extrapolated)	136.4	(8)
		100 (extrapolated)	185.8	(8)

Polyoctenamer

Property	Units	Conditions	Value	Reference
Crystallographic information	Unit cell parameters a, b and c	Monoclinic, 1 repeat unit in unit cell, 0.99 nm identity period	$a = 7.43$, $b = 5.00$, $c = 9.90$ (for C2h-5 space group – trans); $a = 4.58$, $b = 9.50$, $c = 17.11$ (for C2h-6 space group – cis);	(9, 11)
		Triclinic, 1 repeat unit in unit cell, 0.97 nm identity period	$a = 4.34$, $b = 5.41$, $c = 9.78$ (for Ci-1 space group – trans)	(10, 11)
Density	g cm^{-3}	Amorphous	0.867	

References

1. Arlie JP, et al. *Makromol. Chem.* 1974;175:861.
2. Tosi C, Ciampelli F, Dall' Asta G. *J. Polym. Sci., Polym. Phys. Ed.* 1973;11:529.
3. Glenz VW, et al. *Angew. Makromol. Chern.* 1974;37:97.
4. Dall' Asta G. *Pure Appl. Chem. (additional publ.).* 1974;1:133.
5. Dall' Asta G. *Pure Appl. Chem.* 1974;1:133.
6. Natta G, et al. *Makromol. Chem.* 1966;91:87.
7. Gianotti G, Capizzi A. *Eur. Polym. J.* 1970;6:743.
8. Calderon N, Morris MC. *J. Polym. Sci., Part A-2.* 1967;5:1283.
9. Natta G, Bassi IW, Fagherazzi C. *Eur. Polym. J.* 1967;3:339.
10. Bassi IW, Fagherazzi G. *Eur. Polym. J.* 1968;4:123.
11. Brandrup J, Immergut EH, Edmund H, Grulke EA, Abe A, Bloch DR, eds. In. *Polymer Handbook.* 4th ed. New York: John Wiley and Sons; 1999;2005.

Polypentenamer

Vassilios Galiatsatos

Alternative Name Poly(l-pentenylene), Poly(1-pentene-1,5-diyl), cyclopentene homopolymer, polycyclopentene

Class Diene elastomers

Structure $(CH{=}CH(CH_2)_3)_n$

Synthesis Ring-opening polymerization of cyclopentene. *Trans*-polypentenamer is produced by Ziegler-Natta polymerization employing a catalyst based on aluminum triethyl/tungsten hexachloride compound. Aluminum diethylchloride/molybdenum pentachloride compounds may be employed to produce the *cis* isomer. Both macrocycles and linear chains are produced during polymerization.

Fractionation Methods Fractional precipitation in toluene/methanol (solvent/nonsolvent) mixtures at 40/20°C.[1,2]

Property	Units	Conditions	Value		Reference
Gel permeation chromatography	–	Using THF as the solvent	–		(3)
Molar absorptivities of IR bands attributed to *trans* and *cis* units	$(mol\ cm)^{-1}$	ε_{trans} (10.35 µm) ε_{cis} (7.12 µm)	152 5.0		(4)
Mark-Houwnink parameters: K and a	$K = ml\ g^{-1}$ $a =$ None	*Trans*-polypentenamer (~85% *trans* content)	$K \times 10^4$	a	(5)
		Toluene, 30°C	5.21	0.69	
		Cyclohexane, 30°C	5.69	0.68	
		i-Amyl acetate (θ solvent), 38°C	23.4	0.63	
Specific refractive index increment	–	*n*-Hexane (dilute solution at 25°C)			(6)
		436 nm	0.175		
		546 nm	0.171		
Glass transition temperature T_g	K	*Cis*-polypentenamer			
		DTA	159		(7)
		TBA	163		(8)
		Trans-polypentenamer			
		DTA	176		(9)
		DTA	183		(10)
		DSC	178		(11)
		TBA	180		(8)
		DSC	182		(12)

Polypentenamer

Property	Units	Conditions			Value	Reference
Crystalline melting	K	*Trans* (%)	ΔH (J g^{-1})	Technique		
temperature T_m		1	–	DTA	232	(7)
		85	–	DTA	291	(9)
		100 (extrapolated)	Diluent	176.6	–	(13)
		100 (extrapolated)	–	DSC	317	(12)
Effect of microstructure	hours	*Trans* (%) at 0°C				(14)
on crystallization		93 (85 based on IR analysis)			0.3	
rate of *trans-*		90 (82 based on IR analysis)			0.8	
polypentenamer		89 (81 based on IR analysis)			13	
($T_{1/2}$)		87 (79 based on IR analysis)			45	
Crystallographic information		Orthorombic, 2 repeat units in unit cell, 1.19 nm identity period				(15)
Unit cell parameters		$a = 7.28$, $b = 4.97$, $c = 11.90$				
Unperturbed dimensions ($r_o/M^{1/2}$)	nm	At 38°C, utilizing the Flory-Fox theory of viscosity vs. molecular weight in a θ solvent			9.91×10^6	–
Theta temperature	°C	For 85% *trans* content in i-amyl acetate			38	
Relaxation behavior	K	By DMA, for 82% *trans* content $M_n = 94{,}400$ g mol^{-1} $M_w = 172{,}300$ g mol^{-1}) at 110Hz				–
		α relaxation			353	
		β relaxation			273	
		γ relaxation			158, 153	
Density	g cm^{-3}	Fully crystalline	1.051			
Solvents		Chlorinated hydrocarbons, hydrocarbons				
Non-solvents		Alcohols, aliphatic ketones, ethers				
Heat capacity, C_p	J (mol K)$^{-1}$	298.15 K	132.5			

References

1. Gianotti G, Bonicelli U, Borghi D. *Makromol. Chem.* 1973;166:235.
2. Witte J, Hoffman M. *Makromol. Chem.* 1978;179:641.
3. Arlie JP, et al. *Makromol. Chem.* 1974;175:861.
4. Tosi C, Ciampelli F, Dall' Asta G. *J. Polym. Sci., Polym. Phys. Ed.* 1973;11:529.
5. Gianotti G, Bonicelli U, Borghi D. *Makromol. Chem.* 1973;166:235.
6. Izyumnikov AL, Polyakova GR, Gantmakher AR. *Polym. Sci. USSR.* 1983;25:2721.
7. Dall' Asta G, Scaglione P. *Rubber Chem. Technol.* 1969;42:1235.
8. Gillam JK, Bencid JA. *J. Appl. Polym. Sci.* 1974;18:3775.
9. Dall' Asta G, Motroni G. *Angew. Makromol. Chern.* 1971;16–17:51.
10. Gunther G, et al. *Angew. Makromol. Chern.* 1970;14:82.
11. Minchak J, Tucker H. *ACS Symp. Ser.* 1982;193:155.

12. Wilkes GE, Pelko MJ, Minchak RJ. *J. Polym. Sci., Polym. Symp.* 1973;43:97.
13. Capizzi A, Gianotti G. *Makromol. Chem.* 1972;157:123.
14. Haas F, Theisen D. *Kaut. Gummi Kunstst.* 1970;23:502.
15. Natta G, Bassi I. *J. Polym. Sci., Part C.* 1967;16:2551.

Poly(1,4-phenylene)

Jacek Swiatkiewicz and Paras N. Prasad

Acronym, Alternative Name PPP, poly (*p*-phenylene)

Class Polyaromatics

Structure $[-C_6H_4-]$

Properties of Special Interest Electroactive and electroluminescent material. Electrical properties can be tuned by choice of doping and preparation procedure. Insoluble and infusible material, sustains high-temperature treatment.

Preparative Techniques Various aryl coupling reactions, pyrolysis of the polymer precursors, anodic polymerization.[1−4]

Property	Units	Conditions	Value	Reference
Density	g cm^{-3}	Amorphous	1.11 ± 0.02	(3)
		Semi-crystalline	1.228	(3)
		Highly crystalline, annealed	1.39	(2)

Unit cell dimensions

Lattice	Monomers Per Unit Cell	Cell Dimensions (nm)			Cell Angles			Reference
		a	*b*	*c*	α	β	γ	
Monoclinic	2	0.779	0.562	0.426	−	79°	−	(2)
Monoclinic	2	0.806	0.555	0.430	−	100°	−	(5)
Orthorhombic	2	0.781	0.553	0.420	−	−	−	(5)
Orthorhombic	2	0.780	0.556	0.420	−	−	−	(5)

Property	Units	Conditions	Value	Reference
IR (characteristic absorption frequencies)	cm^{-1}	−	3,027	(3)
			3,030	(4, 6)
			1,603	(3)
			1,600	(4, 6)
			1,482	(3)
			1,460	(4, 6)

Property	Units	Conditions	Value	Reference
			1,003	(3)
			1,000	(4, 6)
			808	(3)
			803	(4, 6)
			765	(3)
			760	(4, 6)
			509	(3)
			500	(4, 6)
Raman (characteristic frequencies)	cm^{-1}	–	1,600	(7)
			1,598	(8)
			1,280	(7)
			1,276	(8)
			1,220	(7, 8)
Wavelength at maximum of the band	nm	UV-V is absorption	362	(2)
			333–338	(3)
			350	(9)
		Photo-excitation	400	(9)
Emission band	nm	Photo-luminescence	500	(9)
			460	(10)
Electronic conductivity	S cm^{-1}	$T = 298$ K	1.6×10^{-13}	(11)
			3.3×10^{-13}	(9)
Energy gap	eV	–	2.7	(10)
			2.8	(12)
Electroluminescence emission peak	nm	–	460	(10)

References

1. Feast WJ. In: Skotheim TA, ed. *Handbook of Conducting Polymers.* New York: Marcel Dekker; 1986:1.
2. Elsenbaumer RL, Shacklette LW. In: Skotheim TA, ed. *Handbook of Conducting Polymers.* New York: Marcel Dekker; 1986:213.
3. Gin DL, Avlyanov JK, MacDiarmid AG. *Synth. Met.* 1994;66:169.
4. Goldenberg LM, Lacaze PC. *Synth. Met.* 1993;58:271.
5. In: Brandrup J, Immergut EH, eds. *Polymer Handbook.* 3rd ed. New York: Wiley-Interscience; 1989.
6. Goldenberg LM, et al. *Synth. Met.* 1990;36:217.
7. Krichene S, Buisson JP, Lefrant S. *Synth. Met.* 1987;17:589.
8. Buisson JP, Krichene S, Lefrant S. *Synth. Met.* 1989;29:E13.
9. Miyashita K, Kaneko M. *Synth. Met.* 1995;68:161.
10. Grem G, Leising G. *Synth. Met.* 1993;55–57:4105.
11. Edwards G, Goldfinger G. *J. Polym. Sci.* 1955;16:589.

12. Froyer G, Pelous Y, Olivier G. *Springer Ser. Solid State Sci.* 1987;76:303.

13. Mulazzi E, Ripamonti A, Athouel L, Wery J, Lefrant S. Theoretical and Experimental Investigation of the Optical Properties of Poly(paraphenylene): Evidence of Chain-length Distribution. *Physical Review B: Condensed Matter and Materials Physics.* 2002;65(8):085204/1–08524/9.

14. Geetha S, Trivedi DC. Electrochemical Synthesis and Characterization of Conducting Polyparaphenylene Using Room-temperature Melt as the Electrolyte. *Synth. Met.* 2005;155(2):306–310.

15. Mohammand F, Calvert PD, Billingham NC. Electrial and Eelctronics Properties of Polyparaphenylene. *Journal of Physics D: Applied Physics.* 1996;29(1):195.

16. Lynge Thomas Bastholm, Pedersen Thomas Garm. Analytic and Numerical Electro-optic Models of Poly(para-phenylene). *Synth. Met.* 2003;138(1–2):329.

Poly(*m*-phenylene isophthalamide)

Zhengcai Pu

Trade Names Nomex, Teijinconex, Fenilin

Class Aromatic polyamides

Structure

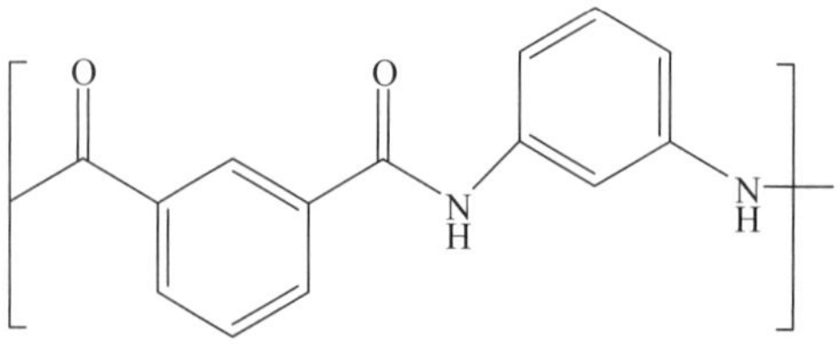

Major Applications Heat-resistant and flame-retardant apparel; (high-voltage) electrical insulation; low-, medium-, and high-density pressboard; honeycomb structure composite

Properties of Special Interest High extensibility relative to other aromatic polyamide, high degradation and glass transition temperature, excellent dielectric property, and good spinnability (1)

Producers and/or Suppliers Du Pont (Nomex); Teijin Ltd., Japan (Teijinconex); Russia (Fenilin)

Property	Units	Conditions	Value	Reference
Anistropy of segment	cm^{-3}	Sulfuric acid		(2)
		$\alpha_1 - \alpha_2$	3.6×10^{23}	
		$\alpha_{\parallel} - \alpha_{\perp}$	1.0×10^{23}	
Coefficient of linear thermal expansion	K^{-1}	294–477 K	6.2×10^6	(3)
Density ρ	g cm^{-3}	–	1.38	(3, 4)
Dielectric constant	–	60 Hz	1.6–2.9	(3)
Dielectric loss	–	60 Hz, 50% relative humidity	0.006	(3)
Dielectric strength	kV m^{-1}	23°C, 50% relative humidity	2.0–3.9 ($\times 10^4$)	(3)
Diffusion coefficient	m^2 s^{-1}	$M_w = 4.3$–112 kg mol^{-1}, 3% LiCl in DMF, 298 K	6.19–0.82 ($\times 10^{11}$)	(2)
Friction coefficient		Sliding over a period of 120 min at a sliding speed of 0.42 m s^{-1} and a load of 196 N at room temperature	~0.38	(14)

Poly(*m*-phenylene isophthalamide)

Property	Units	Conditions	Value	Reference
Glass transition temperature	K	Heating rate $= 2$ K min^{-1}	553	(2, 3, 5)
Heat capacity	kJ K^{-1} mol^{-1}	–	0.29	(3)
Inherent viscosity η_{inh}	dl g^{-1}	30°C, in 0.5 g fiber/100 ml sulfuric acid solution	1.86–2.11	(6)
Limiting oxygen index (LOI)	%	–	28	(3, 4, 7)
Mark-Houwink parameters: K and a	$K = $ ml g^{-1} $a = $ None	–	$K = 3.7 \times 10^{-4}$ $a = 0.73$	(3)
Melting point	K	DTA transition	708	(3, 5)
Modulus				
Dynamic storage	MPa	10% fiber in DMAc/LiCl, $\omega = 1$ s^{-1}	2×10^5	(3)
Flexure	MPa	3.2 mm thick pressboard	2.55–3.60	(3)
Initial tension	GPa	–	13.7	(8)
			9.6	(13)
Nonsolvents	Hexamethyl phosphoramide, *m*-cresol, formic acid			(3)
Refractive index increment *dn/dc*	ml g^{-1}	DMA	0.245	(2)
		DMA + LiCl, room temperature, $\lambda_0 = 546$ nm	0.219–0.200	

Resistance to chemicals[3]

Chemical	Effect On Breaking Strength					
	None			Appreciable		
	Conc. (%)	Temp. (K)	Time (h)	Conc. (%)	Temp. (K)	Time (h)
Hydrochloric acid	35	294	10	10	368	8
Nitric acid	10	294	100	70	294	100
Sulfuric acid	10	294	100	70	368	8
Acetic acid	100	294–366	10–1,000	–	–	–
Benzenesulfonic acid	–	–	–	100	366	10
Formic acid	91	294	1,000	–	–	–
Ammonium hydroxide	28	294	100	–	–	–
Sodium hydroxide	10	294	100	50	333	100
Acetone	100	294	1,000	–	–	–
Benzene	100	294	1,000	–	–	–
m-Cresol	100	294	1,000	–	–	–
Ethyl alcohol	100	294	1,000	–	–	–
Gasoline (leaded)	100	294	1,000	–	–	–
Nitrobenzene	100	294	1,000	–	–	–
m-Xylene	100	343	168	–	–	–

Resistance to radiation (β-ray)*[3]

Dose (Mgrads)	Retained Tensile Strength (%)	Retained Elongation (%)	Dielectric Strength (kV m^{-1})	Dielectric Constant[†]	Dissipation Factor[‡]
0	100	100	3.4×10^4	3.1–2.9	0.0083–0.0183
100	100	92	3.4×10^4	3.0–2.9	0.0135–0.0205
200	99	91	3.3×10^4	3.0–2.9	0.0104–0.0198
400	99	88	3.3×10^4	3.0–2.9	0.0120–0.0199
800	97	82	3.3×10^4	3.0–2.8	0.0089–0.0185
1,600	86	47	3.4×10^4	3.1–3.0	0.0137–0.0195
3,200	81	27	3.5×10^4	2.3–2.2	0.0071–0.0148
6,400	69	16	3.1×10^4	2.5–2.4	0.0095–0.0174

* 0.25 mm Nomex Type 410 paper, cross direction.
† 2 MeV electrons.
‡ 60 Hz to 10 kHz.

Resistance to radiation (X-ray)[3]

X-ray (kV)	Irradiation Time (h)	Breaking Strength Retained (%)
50	50	85
50	100	73
50	250	49

Resistance to temperature [3]

Temperature (K)	Breaking Tenacity (MPa)	Initial Modulus (MPa)	Breaking Elongation (%)
223	738	1.76×10^4	19.4
311	614	1.46×10^4	21.3
422	521	1.15×10^4	23.7
533	346	0.80×10^4	26.0

Property	Units	Conditions	Value	Reference
Resistivity	ohm cm	50% relative humidity	1,016	(3)
Secondary-relaxation	K	Torsion pendulum, 1 Hz		(7)
		T_b	550	
		T_g	352	
Solvents		Concentrated sulfuric acid, methanesulfonic acid, dimethyl acetamide, dimethylsulfoxide, DMF, N-methylpyrrolidone		(3)
Strength				
Bending	MPa		~195	(14)
Elongation at break	%	–	20–30	(3, 6, 9)
			11	(13)
Flexure	MPa	3.2 mm thick pressboard	0.08–0.09	(3)

Poly(*m*-phenylene isophthalamide)

Property	Units	Conditions	Value	Reference
Shear	N	–	31,000	(3)
Tensile at break	MPa	–	54–68	(6)
			70	(13)
Temperature	K	Begin to degrade	573	(3)
		10% weight loss	731	
Tenacity at break	N/tex	–	0.39–0.49	(6)
Thermal conductivity	$W\ m^{-1}\ K^{-1}$	–	0.13	(3)
Upper use temperature	K	In air	643	(10)
Upper use voltage	$kV\ m^{-1}$	23°C, 50% relative humidity	1.6×10^3	(3)
Water uptake	% (w/w)	20°C, 65% relative humidity	6.5–9.3	(6, 11)
Wear volume	mm^3	Sliding over a period of 120 min at a sliding speed of 0.42 m s^{-1} and a load of 196 N at room temperature	$\sim 1.61 \times 10^{-8}$	(14)
Zero-strength temperature	K	–	713	(12)

Sedimentation coefficient at zero concentration$^{(2)}$

Solvent	Temperature (K)	M_w (kg mol^{-1})	S_0 (s)
DMF	298	30.2–156	$(1.9 \times 10^{15})M^{0.44}$
LiCl (2.5 g l^{-1} + 96% H$_2$SO$_4$) in DMF	298	20.7–142	$(2.8 \times 10^{15})M^{0.39}$
3% LiCl in DMF	298	4.3–112	0.33–1.15 ($\times 10^{13}$)

Unit cell data

Crystallographic System	Triclinic$^{(3)}$	Ortho$^{(2)}$	Ortho$^{(2)}$
Space group	$P1 - C_1^1$	–	–
Cell dimension			
a_0 (Å)	5.27	6.7	5.1
b_0 (Å)	5.25	4.71	5.0
c_0 (Å)	11.3	11.0	23.2
α (°)	111.5	–	–
β (°)	111.4	–	–
γ (°)	88	–	–
Repeat unit per unit cell	1	1	2

References

1. Ulrich H. *Introduction to Industrial Polymers.* 2nd ed. Munich: Hanser Publishers; 1993.
2. Brandrup J, Immergut EH. *Polymer Handbook.* 3rd ed. New York: Wiley-Interscience; 1989.

3. Lewin M, Preston J, eds. *Handbook of Fiber Science and Technology.* Vol 3. New York: Marcel Dekker; 1983.
4. Elias H-G, Vohwinkel F. *New Commercial Polymers.* Vol 2. New York: Gordon and Breach Science Publishers; 1986.
5. Yang HH. *Aromatic High-Strength Fibers.* New York: John Wiley and Sons; 1989.
6. Mark HF, et al. *Encyclopedia of Polymer Science and Engineering.* Vol 6. New York: John Wiley and Sons; 1996.
7. Mark JE, ed. *Physical Properties of Polymers Handbook.* New York: AIP Press; 1996.
8. Wortmann F. *J. Polymer.* 1994;35:2108.
9. Dyson RW, ed. *Specialty Polymers.* London: Blackie and Son Limited; 1987.
10. Warner SB. *Fiber Science.* Prentice-Hall, Englewood Cliffs, N.J: 1995.
11. Salamone JC. *Polymer Materials Encyclopedia.* Vol 8. Boca Raton, Fla: CRC Press; 1996.
12. Mark HF, Atlas SM, Cernia E, eds. *Man-Made Fibers Science and Technology.* Vol 2. New York: Interscience Publishers; 1968.
13. Wang HH, Lin MF. *J. Appl. Polym. Sci.* 1998;68:1031, 1043.
14. Liu XJ, Li TS, Tian N, Liu WM. *J. Appl. Polym. Sci.* 2001;80: 2790, 2794.

Poly(*p*-phenylene oxide)

Allan S. Hay and Yong Ding

Acronyms PPO, PPE

Class Polyether thermoplastics

Structure

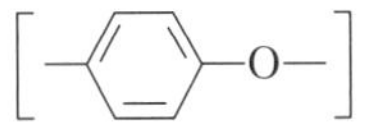

Properties Of Special Interest Highly crystalline polymer, excellent chemical and solvent resistance. Not commercially available.

Preparative Techniques Poly(*p*-phenylene oxide) is prepared from mono *p*-bromo- or *p*-chloro-phenolate at 170–200°C in the presence of cuprous salt as catalyst.[1-3]

Property	Units	Conditions	Value	Reference
Molecular weight of repeat unit	g mol^{-1}	–	92.03	–
IR (characteristic absorption frequencies)				(3)
Thermal expansion coefficients	K^{-1}	Amorphous sample, DSC		(4)
		Above T_g	249×10^{-6}	
		Below T_g	62×10^{-6}	
		Crystalline sample, DSC		(4)
		$0.7\,T_m < T < 0.95\,T_m$	93×10^{-6}	
Density (amorphous)	g cm^{-3}	–	1.27	(5)
Solvents		Boiling nitrobenzene, benzophenone, diphenyl ether, N-methylpyrrolidinone, tetralin, naphthalene, and hexamethylphosphoric acid triamide		(3)
Nonsolvents		Room temperature: acetone, alcohols, tetrahydrofuran, halogenated solvents		(3)
Lattice	–	–	ORTH	(5)
Space group	–	–	*Pbcn*	(5)
Chain conformation	–	–	7 * 2/1	(5)
Unit cell dimensions	Å	Compression-molded or uniaxially oriented	$a = 8.07$ $b = 5.54$ $c = 9.72$	(5)

Property	Units	Conditions	Value	Reference
Unit cell contents (number of repeat units)			4	(5)
Degree of crystallinity*	%	Hold at 230°C for 1 h, cooling rate > 1,000°C min^{-1}, x-ray	0	(4)
		Hold at 230°C for 1 h, cooling rate 100°C min^{-1}, x-ray	42	
		Hold at 230°C for 1 h, cooling rate 1°C min^{-1}, x-ray	45	
		Hold at 230°C for 1 h, cooling rate 0.1°C min^{-1}, x-ray	70	
		Hold at 112°C for 1 h, cooling rate 0.1°C min^{-1}, x-ray	58	
		25°C, 0.2% nitrobenzene solution quenched with alcohol, x-ray	15	
Heat of fusion (of repeat units)	kJ mol^{-1}	DSC	7.835 ± 0.419	(4)
Entropy of fusion (of repeat units)	kJ K^{-1} mol^{-1}	DSC	0.015 ± 0.003	(4)
Density (crystalline)	g cm^{-3}	–	1.407 ± 0.01	(5)
Glass transition temperature	K	DSC	363	(4)
Melting point	K	DSC	535 ± 10	(4)
Heat capacity (of repeat units)	kJ K^{-1} mol^{-1}	300–358 K	$C_p = (0.337T + 7.95) \times 10^{-3}$	
		358–620 K	$C_p = (0.1425T + 99.01) \times 10^{-3}$	(6)
Dielectric constant ε'	–	100 Hz, 296 K	4.76	(2)
		100 Hz, 348 K	4.72	
		100 Hz, 398 K	4.73	
		100 Hz, 448 K	4.76	
		100 Hz, 498 K	4.60	
		100 Hz, 523 K	4.59	
		100 Hz, 548 K	4.78	
		100 Hz, 573 K	7.01	
		1000 Hz, 296 K	4.76	
		1000 Hz, 348 K	4.71	
		1000 Hz, 398 K	4.71	

Poly(*p*-phenylene oxide)

Property	Units	Conditions	Value	Reference
		1000 Hz, 448 K	4.75	
		1000 Hz, 498 K	4.58	
		1000 Hz, 523 K	4.53	
		1000 Hz, 548 K	4.50	
		1000 Hz, 573 K	4.51	
		1×10^5 Hz, 296 K	4.76	
		1×10^5 Hz, 348 K	4.71	
		1×10^5 Hz, 398 K	4.68	
		1×10^5 Hz, 448 K	4.71	
		1×10^5 Hz, 498 K	4.54	
		1×10^5 Hz, 523 K	4.50	
		1×10^5 Hz, 548 K	4.47	
		1×10^5 Hz, 573 K	4.42	
Dielectric loss ε''	–	100 Hz, 296 K	0.0005	(2)
		100 Hz, 348 K	0.0005	
		100 Hz, 398 K	0.0047	
		100 Hz, 448 K	0.0079	
		100 Hz, 498 K	0.0311	
		100 Hz, 523 K	0.1745	
		100 Hz, 548 K	0.4417	
		100 Hz, 573 K	1.2085	
		1000 Hz, 296 K	0.0005	
		1000 Hz, 348 K	0.0007	
		1000 Hz, 398 K	0.0024	
		1000 Hz, 448 K	0.0027	
		1000 Hz, 498 K	0.0051	
		1000 Hz, 523 K	0.0180	
		1000 Hz, 548 K	0.0462	
		1000 Hz, 573 K	0.1876	
		1×10^5 Hz, 296 K	0.0013	
		1×10^5 Hz, 348 K	0.0006	
		1×10^5 Hz, 398 K	0.0016	
		1×10^5 Hz, 448 K	0.0027	
		1×10^5 Hz, 498 K	0.0092	
		1×10^5 Hz, 523 K	0.0023	
		1×10^5 Hz, 548 K	0.0023	
		1×10^5 Hz, 573 K	0.0026	

* Sample thickness: ca. 10 μm.

References

1. Stamatoff GS. *U.S. Patent*. 3,228,910 (to E. I. du Pont), 1966.
2. Taylor CW, Park SP, Davis SP. *U.S. Patent*. 3,491,085 (to 3M), 1970.
3. van Dort HM, et al. *Europ. Polym. J.* 1968;4:275.
4. Wrasidlo W. *J. Polym. Sci. Part A-2.* 1972;10:1719.
5. Boon J, Magré EP. *Makromol. Chem.* 1969;126:130.
6. Gaur U, Wunderlich B. *J. Phys. Chem. Ref. Data.* 1981;10:1005.

Poly(*p*-phenylene sulfide)

Junzo Masamoto

Acronym, Trade Names PPS, Ryton, Fortron, Torelina, Tohprene, DIC-PPS

Class Polysulfides

Structure

Major Applications Poly(*p*-phenylene sulfide) (PPS) is mainly used in the reinforced form with glass fiber or mineral fillers as a high-performance thermoplastic. It is used for electrical and electronic parts (e.g., plugs and multipoint connectors, bobbins, relays, switches, encapsulation of electronic component, etc.), automobile parts (air intake systems, pumps, valves, gaskets, components for exhaust gas recirculation systems, etc.), and as components for mechanical and precision engineering. Nonfiller PPS is used for fiber, film, sheet, nonwoven fabric, etc.

Properties of Special Interest PPS is a semicrystalline thermoplastic. PPS reinforced with glass fiber or mineral fillers shows excellent mechanical properties, high thermal stability, excellent chemical resistance, excellent flame retardance, good electrical and electronic properties, and good mold precision. Recently developed linear type PPS additionally shows improved properties of elongation and toughness and opens the new route for the use of a neat polymer.

Preparative Technique Condensation polymerization: Reaction between *p*-dichlorobenzene and sodium sulfide is accomplished in the presence of a polar solvent (e.g., N-methyl pyroridone). Polymer formation is accompanied by the production of sodium chloride as a byproduct. Medium-low molecular weight solid PPS powder is heated to below its melting point (448–553 K) in the presence of air. Several important properties of PPS change when the polymer is cured: (1) molecular weight increased; (2) toughness increased; (3) melt viscosity increased; (4) the color of the polymer changes from off-white to tan/brown. Modified high molecular weight linear polymer is directly obtained during polymerization by Phillips Petroleum using alkali metal carboxylate as a polymerization modifier. Kureha Chemical developed a modified process for obtaining linear type PPS, adding water during the last stage of polymerization.[1, 2]

Property	Units	Conditions	Value	Reference
Molecular weight (of repeat unit)	g mol^{-1}	–	109	–
Typical molecular weight range of polymer	g mol^{-1}	Dilute solution light scattering and gel permeation chromatographic studies (performed in 1-chloronaphthalene at 220°C), and the inherent viscosity (performed in 1-chloronaphthalene at 206°C) is 0.16. The polymer is as polymerized, just before the curing step	18,000	(6, 7)
		The linear type of modified high molecular weight PPS by the Phillips modified process	35,000	(6)
Typical polydispersity index ($M_\mathrm{w}/M_\mathrm{n}$)	–	–	1.7	(8–10)
IR (characteristic absorption frequencies)	cm^{-1}	Skeletal benzene	480	(11, 12)
		Skeletal benzene	556	
		Skeletal benzene	724	
		Out-of-plane C—H bending	818	
		Out-of-plane C—H bending	960	
		Skeletal benzene	1,011	
		Phenylene sulfur stretching	1,096	
		In-plane C—H bending	1,178	
		In-plane C—H bending	1,235	
		Skeletal benzene	1,390	
		Skeletal benzene	1,471	
		Skeletal benzene	1,571	
		Skeletal benzene	1,652	
		Skeletal benzene	1,906	
		Skeletal benzene	2,299	
		C—H stretching	3,065	
Thermal expansion coefficients	K^{-1}	Unfilled	4.9×10^5	(10)
		40 wt% glass fiber-filled	4×10^5	
		Glass fiber and mineral-filled	2.8×10^5	
Solvents	–	>200°C	1-Chloronaphthalene	(1)
		>200°C	Biphenyl, 3-chlorobiphenyl, *o*-terphenyl	(13)
Nonsolvents	–	<200°C	Almost insoluble in organic and inorganic solvents	(10)
Mark-Houwink parameter: K and a	K = ml g^{-1} a = None	1-Chloronaphthalene, at 208°C	$K = 8.91 \times 10^{-5}$ $a = 0.747$	(14)
Lattice	–	–	Orthorhombic	(15, 16)

Property	Units	Conditions	Value	Reference
Space group	–	–	D2H-14	(15, 16)
Chain conformation		All *trans* conformation defined by the plane of the C—S—C linkages (C—S—C bond angle 110°), while the phenyl rings are successively inclined at + and −45° to the plane		–
		All *trans* conformation defined by the plane of the C—S—C linkages (C—S—C bond angle 103–107°), while alternate phenyl rings are nearly coplanar with the C—S—C plane, and while the remaining ones are inclined to 60°C		(17)
Unit cell dimensions	Å	–	$a = 8.67$, $b = 5.61$, $c = 10.26$ (fiber identity period)	(15)
			$a = 8.68$, $b = 5.61$, $c = 10.26$ (fiber identity period)	(16)
Unit cell contents	Monomeric units	–	4	(15, 16)
Degree of crystallinity	%	X-ray diffraction method for fully crystalline PPS	65	(18)
Heat of fusion	kJ mol^{-1}	Typical heat of fusion of crystalline PPS	4.6–5.5	(19)
		100% crystalline material, by extrapolation	8.7	(16, 18)
Density	g cm^{-3}	Theoretical density for PPS	1.440	(15)
		crystalline	1.425	(16)
		Observed density		(20)
		For PPS neat	1.35	
		40% glass reinforced	1.6	
		Glass and mineral filled	1.6–1.8	
Glass transition temperature	K	DSC	358	(2)
		$M_w = 51,000$, DSC, heating rate $= 20°C$ min^{-1}	357	(16)
Melting point	K	DSC	558	(2)
			568	(15)
		Equilibrium melting temperature, DSC determined from the relationships between T_m and $T_{c'}$ 5°C min^{-1}		(21)
		$M_w = 15,000$	576	
		$M_w = 51,000$	588	
Heat capacity	kJ K^{-1}mol^{-1}	Unfilled, cured feed stock	0.112	(10)
Deflection temperature	K	Unfilled, cured feed stock, sample annealed at 260°C for 4 h, ASTM D648	408	(20)
		40% glass fiber reinforced PPS	>533	(20)
		Glass and mineral filled PPS	>533	(20)
		Unfilled, linear type PPS, ASTM D648 at 1.82 MPa	388	(22)

Property	Units	Conditions	Value	Reference
		40% glass fiber reinforced linear type PPS	538	(22)
		Glass and mineral filled linear type PPS	538	(22)
Tensile modulus	MPa	Biaxally oriented PPS film	2,600–3,900	(6)
		PPS fiber, draw ratio 3.8, 25.5 tex	3,500–4,700	
Tensile strength	MPa	Unfilled, cured feed stock, ASTM D638	65	(20)
		40% glass fiber reinforced	120	(20)
		Glass and mineral filled PPS	74	(20)
		Unfilled, linear type, ASTM D638	86	(22)
		40% glass fiber reinforced linear type	172	(22)
		Glass and mineral filled linear type	113	(22)
		Biaxially oriented PPS film	125–190	(6)
		PPS fiber, draw ratio 3.8, 25.5 tex	300	(6)
		PPS fiber	480	(22)
Yield stress	MPa	Unfilled, linear type	80	(23)
Yield strain $(L/L_0)_y$	%	Unfilled, linear type	5	(23)
Maximum extensibility	%	Unfilled, cured feed stock, ASTM D638	1.6	(20)
		40% glass fiber reinforced	1.2	(20)
		Glass and mineral filled	0.54	(20)
		Unfilled, linear type	12	(23)
		Unfilled, cured feed stock	2	(23)
		Unfilled, linear type, ASTM D638	3–6	(19)
		40% glass fiber reinforced linear type	1.7	(19)
		Glass and mineral filled linear type	1.0	(19)
		Unfilled, cured PPS	1.1	(6)
		Unfilled, linear type	21	(6)
		40% glass fiber reinforced cured PPS	0.5	(6)
		40% glass fiber reinforced linear type PPS	0.8	(6)
		Biaxially oriented PPS film	40–70	(6)
		PPS fiber, draw ratio 3.8, 25.5 tex	25–35	(6)
		PPS fiber	25–40	(22)
Flexural modulus	MPa	Unfilled, cured feed stock	3,860	(20)
		40% glass fiber reinforced	11,700	(20)
		Glass and mineral filled	15,200	(20)
		Unfilled, linear type	3,400	(23)
		Unfilled, linear type, ASTM D790	4,130	(22)
		40% glass fiber reinforced linear type	13,100	(22)
		Glass and mineral filled linear type	16,500	(22)
		Unfilled, cured PPS	3,845	(6)
		Unfilled, linear type	3,4041	(6)
		40% glass fiber reinforced cured PPS	1,5001	(6)
		40% glass fiber reinforced linear type PPS	1,800	(6)
Flexural strength	MPa	Unfilled, cured feed stock	96	(20)
		40% glass fiber reinforced	180	(20)
		Glass and mineral filled	100	(20)

Property	Units	Conditions	Value	Reference
		Unfilled, linear type	110	(23)
		Unfilled, linear type, ASTM D790	145	(22)
		40% glass fiber reinforced linear type	241	(22)
		Glass and mineral filled linear type	182	(22)
		Unfilled, cured PPS	104	(6)
		Unfilled, linear type	147	(6)
		40% glass fiber reinforced cured PPS	153	(6)
		40% glass fiber reinforced linear type	180	(6)
Impact strength, notched	$J\,m^{-1}$	ASTM D256		
		Unfilled, cured feed stock	16	(20)
		40% glass fiber reinforced PPS	69	(20)
		Glass and mineral filled	32	(20)
		Unfilled, linear type, ASTM D256	26	(20)
		40% glass fiber reinforced linear type	85	(22)
		Glass and mineral filled linear type	64	(22)
		Unfilled, cured PPS	10.7	(6)
		Unfilled, linear type	16.7	(6)
		40% glass fiber reinforced cured PPS	48.2	(6)
		40% glass fiber reinforced linear type	58.9	(6)
		Elastomer toughened PPS	500	(24)
		40% glass fiber reinforced elastomer toughened PPS	220	(24)
Impact strength, unnotched	$J\,m^{-1}$	ASTM D256		
		Unfilled, cured feed stock	101	(20)
		40% glass fiber reinforced PPS	240	(20)
		Glass and mineral filled	101	(20)
		Unfilled, linear type	900	(23)
		Unfilled, cured feed stock	60	(23)
		Unfilled, linear type	320–640	(23)
		40% glass fiber reinforced linear type	590	(23)
		Glass and mineral filled linear type	250	(23)
		Unfilled, cured PPS	80.3	(6)
		Unfilled, linear type	578	(6)
		40% glass fiber reinforced cured PPS	139	(6)
		40% glass fiber reinforced linear type	241	(6)
Compressive strength	MPa	Unfilled, cured feed stock	110	(20)
		40% glass fiber reinforced PPS	145	
		Glass and mineral filled	110	
Rockwell hardness	–	Unfilled, cured feedstock	R-120	(20)
		40% glass fiber reinforced PPS	R-123	
		Glass and mineral filled	R-121	
Entanglement molecular weight	$g\,mol^{-1}$	–	20,000	(8)

Property	Units	Conditions	Value	Reference
Dielectric strength	kV mm^{-1}	40% glass fiber filled, ASTM D149, transformer oil, rate of increase = 500 V s^{-1}, 1.6–3.2 mm thickness	17.7	(10)
		Glass fiber and mineral filled	13.4–15.7	
Dielectric constant	–	40% glass fiber filled, 1 MHz, ASTM D150	3.8	(20)
		Glass fiber and mineral filled	4.6	
Dissipation factor	–	40% glass fiber filled, 1 MHz, ASTM D150	0.0013	(20)
		Glass fiber and mineral filled	0.016	
Volume resitivity	ohm cm	40% glass fiber filled, 2 min, ASTM D257	4.5×10^{16}	(20)
		Glass fiber and mineral filled	2.0×10^{16}	(20)
		Biaxially oriented PPS film	10^{17}	(6)
Arc resistance	s	40% glass fiber filled, ASTM D 495	35	(20)
		Glass fiber and mineral filled	200	
Comparative tracking index	V	40% glass fiber filled, UL 746 A	180	(20)
		Glass fiber and mineral filled	235	
Insulation resistance	ohm	40% glass fiber filled	10^{11}	(20)
		Glass fiber and mineral filled	10^{9}	
Thermal conductivity	W m^{-1} K^{-1}	At 20°C	0.29	(25)
Melt index (melt flow values)	g (10 min)$^{-1}$	Uncured PPS (before curing steps)	3,000–8,000	(19)
		Powder coating PPS	1,000	
		PPS for mineral glass filled compounds	600	
		PPS for glass fiber filled compounds	60	
		Compression molding	0	
Maximum use temperature	K	UL temperature index for long-term use, for PPS resin	493	(26)
		PPS fiber for long-term use	505	(27)
			463	(28)
			>473	(29)
Decomposition temperature	K	Start of decomposition	698	(10)
		20% loss, thermogravimetric analyses of polymer, 10°C min^{-1}	823	
Water absorption	%	40% glass fiber reinforced PPS, 24 h immersion in water	0.03	(22)
		Glass and mineral filled PPS	0.03	
Oxygen index	–	Unfilled PPS, ASTM D2863	44	(10)
		40% glass fiber reinforced PPS	46.5	(10)
		Glass and mineral filled	53	(10)
		PPS fiber	34	(28)
			49	(29)
Flammability	–	Unfilled PPS, UL 94	V-0	(10)
		40% glass fiber reinforced PPS	V-0/5V	
		Glass and mineral filled	V-0/5V	

Poly(*p*-phenylene sulfide)

Property	Units	Conditions	Value	Reference
Flame spread index	mm	ASTM E 162	50.8	(20)
Autoignition temperature	K	–	813	(19)
Smoke density	min	Obscuration time, smoldering	15.5	(30)
		Obscuration time	3.2	
Important patents	U.S. Patent 3,354,129			(1)
	U.S. Patent 3,524,835			(31)
	U.S. Patent 3,717,620			(3)
	U.S. Patent 3,919,177			(4)
	U.S. Patent 4,645,826			(5)
Availability	kg yr^{-1}		41,400,000	(32)
Suppliers	Chevron Phillips Chemical, Woodlands, Texas, USA			
	Fortron, Wilmington, North Carolina, USA			
	Kureha Chemical, Tokyo, Japan			
	Toray, Tokyo, Japan			
	Idemitsu Kosan, Tokyo, Japan			
	Dainippon Ink and Chemicals (DIC), Tokyo, Japan			
	Tosoh, Tokyo, Japan			

Properties of special interest

Heat deflection temperature for glass fiber reinforced engineering plastics over 500 K: Poly(ether ether ketone) (PEEK), Nylon 6,6, poly(ethylene terephthalate), poly(butylene terephthalate)

UL temperature indices for long-term use over 450 K: Poly(ether ether ketone) (PEEK), poly(etherimide), poly(ether sulfone)

Flame resistance UL 94 V-O: Poly(ether ether ketone) (PEEK), poly(etherimide), poly(ether sulfone), polysulfone

Electrical conducting by the addition of dopants: Polyacetylene, poly(*p*-phenylene), polypyrrole[33]

References

1. Edmonds J, Hill HW Jr. *U.S. Patent* 3,354,129 (1967), assigned to Phillips Petroleum.
2. Brady DG. *J. Appl. Polym. Sci., Appl. Polym. Symp.* 1981;36:231.
3. Rohlfing RG. *U.S. Patent.* 3,717,620 (1973), assigned to Phillips Petroleum.
4. Campbell RW. *U.S. Patent.* 3,919,177 (1975), assigned to Phillips Petroleum.
5. Iizuka Y, et al. *U.S. Patent.* 4,645,826 (1987), assigned to Kureha Chemical.
6. Hill HW Jr. *Ind. Eng. Chem. Prod. Res. Dev.* 1979;18:252.
7. Stacy CJ. *Polym. Prepr.* 1985;26(1):180.
8. Kraus G, White WM. *IUPAC 28th Macromolecular Symposium*, Amherst, Mass., 12 July 1982 *Chem. Abstr.* 1983;99:123 454c.
9. Kinugawa A. *Jpn. J. Polym. Sci. Technol.* 1987;44:139.
10. Hill HW Jr, Brady DG. In: Mark HF, ed. *Encyclopedia of Polymer Science and Technology.* 2d ed. Vol 11. New York: Wiley-Interscience; 1988:531.
11. Piaggio P, et al. *Spectrochim. Acta.* 1989;45A:347.

12. Zhang G, Wang Q. *Spectrochim. Acta.* 1991;47A:737.

13. Frey DA. *U.S. Patent.* 3,380,951 (1968), assigned to Phillips Petroleum.

14. Stacy CJ. *J. Appl. Polym. Sci.* 1986;32:3959.

15. Tabor BJ, Magre EP, Boon J. *Eur. Polym. J.* 1971;7:1127.

16. Lovinger AJ, Padden FJ Jr, Davis DD. *Polymer.* 1988;29:229.

17. Garbarczk J. *Polymer Commun.* 1986;27:335.

18. Brady DJ. *J. Appl. Polym. Sci.* 1976;20:2541.

19. Hill HW Jr, Brady DJ. In: Kroschwitz JI, ed. *Kirk-Othmer Encyclopedia of Chemical Technology.* 3d ed. Vol 18. New York: John Wiley and Sons; 1982:793.

20. Geibel JF, Campbell RW. In: Allen SG, ed. *Comprehensive Polymer Science.* Vol 5. London: Pergoman Press; 1989:543.

21. Lovinger AJ, Davis DD, Padden FJ Jr. *Bull. Am. Phys. Soc.* 1985;30:433.

22. *Fortron Polyphenylene Sulfide (PPS).* Catalogue from Hoechst Celanese.

23. Yamada J, Hashimoto O. *Plastics.* 1987;38(4):109.

24. Masamoto J, Kubo K. *Polym. Eng. Sci.* 1996;36:265.

25. Thompson EV. In: Mark HF, ed. *Encyclopedia of Polymer Science and Technologies.* 2d ed. Vol 16. New York: Wiley-Interscience; 1988:711.

26. Shue RS. *Dev. Plast. Technol.* 1985;2:259.

27. Rebenteld L. In: Mark HF, ed. *Encyclopedia of Polymer Science and Technologies.* 2d ed. Vol 6. New York: Wiley-Interscience; 1988:647.

28. Catalogue in *PPS fiber.* Toray, Tokyo, Japan.

29. Catalogue in *Fortron KPS.* Kureha Chemical, Tokyo, Japan.

30. Hiado CJ. *Flammability Handbook for Plastics.* 2d ed. Westport, Conn: Technomic Publishing; 1974:60.

31. Edmonds J, Hill HW Jr. *U.S. Patent.* 3,524,835 (1970), assigned to Phillips Petroleum.

32. Yamagata K, Takano T. *Plastics.* 2006;57(1):90.

33. Rabolt JF, et al. *J. Chem. Commun.* 1980:347.

Poly(1,4-phenylene vinylene)

Jacek Swiatkiewicz and Paras N. Prasad

Acronym, Alternative Name PPV, poly (p-phenylene vinylene)[1]

Class Polyaromatics

Structure $[-C_6H_4-CH=CH-]$

Properties of Special Interest Electroactive and electroluminescent material. Electrical and electrooptical properties can be tuned by choice of doping and preparation procedure. Large third-order nonlinear optical susceptibility. Insoluble and infusible material, sustains high temperature treatment.

Preparative Techniques Thermal conversion of a soluble precursor polymer in oxygen free atmosphere.[2] Uniaxial stretch during thermal process yields highly anisotropic PPV films.[3]

Property	Units	Conditions	Value	Reference
Density	g cm^{-3}	Flotation method	1.24	(4)
		Unit cell dimensions	1.283	

Unit cell dimensions

Lattice	Monomers per Unit Cell	Cell Dimensions (nm)			Cell Angles			Setting Angle* ϕ_s	Reference
		a	b	c	α	β	γ		
Monoclinic	2	0.790	0.605	0.658	123°	–	–	56–68°	(4)
Monoclinic	2	0.815	0.607	0.66	123°	–	–	–	(5)
Monoclinic	2	0.805	0.591	0.66	122°	–	–	56–58°	(6)
Monoclinic	2	0.80	0.60	0.66	123°	–	–	50° ± 2°	(7)

* Position of projected molecular major axis with respect to the a-axis direction.

Property	Units	Conditions	Value	Reference
Characteristic frequencies	meV (cm^{-1})	Inelastic incoherent neutron scattering (IINS)	2.5 (20)	(8)
			7 (57)	
			15 (121)	
			25 (202)	
			37 (2990)	
			40 (3230)	

Poly(1,4-phenylene vinylene)

Property	Units	Conditions	Value	Reference
			51 (4120)	
			60 (4850)	
			68 (5500)	
			80 (6470)	
IR (characteristic absorption frequencies)	cm^{-1}	–	3,024	(9)
			1,594	
			1,519	
			1,423	
			965	
			837	
			784	
Raman (characteristic absorption frequencies)	cm^{-1}	–	1,628	(10)
			1,586	
			1,550	
			1,330	
			1,304	
			1,174	
			966	
Onset of the optical absorption band	eV	–	2.49	(11)
			2.4	(12)
			2.34	(13)
Wavelength at maximum of the band	nm	UV-Vis absorption	200	(11)
			244.8	(11)
			402	(11)
		80 K	511.9	(14)
Lowest even parity excited singlet state	eV	Two-photon fluorescence	2.95	(15)
		Two-photon absorption	3.58	(16)
Emission band	nm	Photo-luminescence	550	(17)
		80 K	522	(12)
		80 K	529	(14)
		77 K	531.5	(13)
		77 K	570.4	(13)
		77 K	615.3	(13)
		25 K	522	(18)
		25 K	562	(18)
		6 K	529	(19)
Tensile strength	MPa	Unoriented	41.2	(20)
		Oriented (draw ratio 6), in the machine direction	500	
		Oriented (draw ratio 5), transverse to the machine direction	31.7	
Young's modulus	MPa	Unstretched	3,200	–
		Oriented	37,000	

Poly(1,4-phenylene vinylene)

Property	Units	Conditions	Value	Reference
Elastic constants	MPa	Oriented (draw ratio 10) along 3 axis (draw direction)		(21)
		c_{11}	8,440	
		c_{13}	3,620	
		c_{33}	46,600	
		c_{44}	2,540	
Dielectric constant ε'	–	0.5 MHz	3.2	(22)
Index of refraction	–	3–25 μm, parallel*	2.1 ± 0.2	(9)
		3–25 μm, perpendicular* (oriented film)	1.5 ± 0.2	(9)
		1.064 μm, parallel	1.968	(23)
		1.064 μm, perpendicular (unoriented)	1.584	(23)
		0.633 μm, parallel	2.085	(23)
		0.633 μm, perpendicular	1.610	(23)
		0.633 μm, parallel	2.20	(24)
		0.602 μm, parallel	2.89(1)	(25)
		0.602 μm, perpendicular (oriented film)	1.63(1)	(25)
Nonlinear refraction coefficient (DFWM)	$cm^2\ W^{-1}$	0.800 μm, parallel (unoriented)	10^{-11}	(26)
Nonlinear absorption coefficient	$cm\ W^{-1}$	–	8.0×10^{-8}	–
		0.700 (probe), 0.620 (pump)	5.0×10^{-9}	(27)
		0.531 (probe), 1.064 (pump)	5.0×10^{-8}	(16)
$\chi^{(3)}$, DFWM	esu	0.580 μm	1.6×10^{-10}	(28)
		0.620 μm (unoriented)	1×10^{-10}	(28)
		0.602 μm, parallel	1.1×10^{-9}	(25)
		0.602 μm, perpendicular (oriented film)	5.8×10^{-11}	(25)
$\chi^{(3)}$, THG	esu	1.064/0.355 μm, parallel (oriented fim)	2×10^{-11}	(24)
		1.064/0.355 μm, parallel (unoriented film)	7.5×10^{-11}	(29)
Electronic conductivity	$S\ cm^{-1}$	$T = 298$ K	10^{-11}	(30)
			2.2×10^{-14}	(31)
Electroluminescence emission peak	nm	ITO/PPV/AuAl/PPV/Au	562	(32)
			550	(33)
Quantum efficiency	%	ITO/PPV/Au	0.01	(32)
		Al/PPV/Au	0.01–0.1	(33)

* Light polarization orientation vs. polymer chain direction.

References

1. Poly(1,4-phenylene-1,2-ethenediyl), CAS.
2. Gangon DR, et al. *Polymer*. 1987;28:567; Bradley DDC. *J. Phys D: Appl. Phys.* 1987;20:1389; Holiday DA, et al. *Synth. Met.* 1993;55–57:954.
3. Machado JM, et al. *New Polym. Mater.* 1989;1:189.
4. Granier T, et al. *J. Polym. Sci. Phys.* 1986;B24:2793
5. Moon YB, et al. *Synth.Met.* 1989;29:E79.
6. Martens JHF, et al. *Synth. Met.* 1991;41:301.
7. Chen D, Winokur MJ, Masse MA, Karasz FE. *Polymer*. 1992;33:3116.
8. Papanek P, et al. *Phys. Rev.* 1994;B50:15668.
9. Bradley DDC, Friend RH, Lindenberger H, Roth S. Polymer. 1986;27:1709.
10. Lefrant S, et al. *Synth. Met.* 1989;29:E91.
11. Obrzut J, Karasz FE. *J. Chem. Phys.* 1987;87:2349.
12. Colaneri NF, et al. Phys. Rev. 1990;B42:11670.
13. Bullot J, Dulieu BV, Lefrant S. *Synth. Met.* 1993;61:211.
14. Pichler K, et al. *Synth. Met.* 1993;55–57:230.
15. Baker CJ, Gelsen OM, Bradley DDC. *Chem. Phys. Lett.* 1993;201:127.
16. Yang J-P. *Chem. Phys. Lett.* 1995;243:129.
17. Hayes GR, Samuel IDW, Phillips RT. *Phys. Rev.* 1995;B52:R-11569.
18. Lee GJ, et al. *Synth. Met.* 1995;69:431.
19. Ramscher U, Bassler H, Bradley DDC, Hennecke M. *Phys. Rev.* 1990;B42:9830.
20. Machado JM, Masse MJA, Karasz FE. *Polymer*. 1989;30:1992.
21. Cui Y, Rao DN, Prasad PN. *J. Phys. Chem.* 1992;96:5617.
22. Nguyen TP, Tran VH, Lefrant S. *Synth. Met.* 1995;69:443.
23. Burzynski R, Prasad PN, Karasz FE. *Polymer*. 1990;31:627.
24. McBranch D, et al. *Synth. Met.* 1989;29:E90.
25. Swiatkiewicz J, Prasad PN, Karasz FE. *J. Appl. Phys.* 1993;74:525.
26. Samoc A, Samoc M, Woodruff M, Luther-Davies B. *Opt. Lett.* 1995;20:1241.
27. Lemmer U, et al. *Chem. Phys. Lett.* 1993;203:29.
28. Bubeck C, Kaltbeitzel A, Gramd A, LeClerc M. *Chem. Phys.* 1991;154:343.
29. Bradley DDC, Mori Y. *Jpn. J. Appl. Phys. Part I.* 1989;28:174.
30. Ueno H, Yoshino K. *Phys. Rev.* 1986;B34:7158.
31. Kossmehl GA. In: Skotheim TA, ed. Handbook of Conducting Polymers. New York: Marcel Dekker, 1986:351.
32. Burroughed JH, et al. *Nature*. 1990;347:539.
33. Cimrova V, Neher D. *Synth. Met.* 1996;76:125.
34. Aarab H, et al. Electrical and Optical Properties of PPV and Single-walled Carbon Nanotubes Composite Films. *Synthetic Metals*. 2005;155(1):63.
35. Mulazzi E, Botta C, Facchinetti D, Bolognesi A. Evidence of Conjugation Length Distributions in Electroluminescent Segmented PPV: Absorption, Photoluminescence and Raman Scattering. *Synthetic Metals*. 2004;142(1–3):85.
36. Mulazzi E, Ripamonti A, Wery J, Dulieu B, Lefrant S. Theoretical and Experimental Investigation of Absorption and Raman Spectra of Poly(Paraphenylene Vinylene). Physical Review B: Condensed Matter and Materials Physics. 1999;60(24):16519.

37. Chandross M, Mazumbar S. Coulomb, Interactions and Linear, Nonlinear, and Triplet Absorption in Poly(Para-Phenylenevinylene). Physical Review B: Condensed Matter. 1997;55(3):1497.

38. Shim Hong-Ku, Hwang Do-Hoon Lee, Jeong-Ik Kang, In-Nam Jin, Jung-Il. Optical Third-Harmonic Generation of Poly(1,4-phenylene Vinylene) Derivatives Containing Stilbene Moiety. Section A: Molecular Crystals and Liquid Crystals. 1996;28047.

Poly(α-phenylethyl isocyanide)

Chandima Kumudinie Jayasuriya and Jagath K. Premachandra

Class poly(isocyanide); poly(iminoethylene); poly(isonitrile)

Structure

Major Applications Potential applications in mimicking biological macromolecules and applications in the areas of liquid crystals, coatings, column chromatographic supports, and polymer supported chiral catalysts.[1,2]

Properties of Special Interest Chiral-helical rigid-rod structure and yields liquid crystals in solution.[1] Potentially useful as models for understanding of the structure and properties of biological molecules.[3] Unreactive toward hydrogenation at ambient temperature and pressure and resistant toward acid hydrolysis.[4] One of the few soluble polyisocyanides of high molecular weight.[1]

Others Polymers Showing This Special Property Chiral helical structure: poly (t-butyl isocyanide) and poly (σ-tolyl isocyanide). Rigid-rod molecule: poly (*n*-hexyl isocyanate) and poly(*n*-butyl isocyanate).

Preparative techniques*

Conditions	Yield (%)	Reference
No initiator or solvent; temp.:25°C	Small yield	(3, 5, 6)
Initiator: Ni(acetylacetonate)$_2$; solvent:ethanol; temp.:25°C	80	(3)
Initiator: NaHSO$_4$,O$_2$, glass dibenzoyl peroxide; solvent: n-heptane; temp.:50°C	60	(7)
Poly(1-α-phenylethyl isocyanide); initiator:H$_2$SO$_4$,O$_2$, glass dibenzoyl peroxide; solvent: n-heptane; temp.:27°C	32	(7)
Poly(d-α-phenylethyl isocyanide); initiator: H$_2$SO$_4$, O$_2$, glass dibenzoyl peroxide; solvent: n-heptane; temp.: 27°C	23	(7)
Catalyst: NiCl$_2$ 6 H$_2$O, (R)-(+)-α-phenylethyl isocyanide	–	(8)
Concentrated H$_2$SO$_4$ at 40°C in air for 43 h	24	(9)
H$_2$SO$_4$ as a fine droplet dispersion in heptane, 25–100°C	–	(3)
H$_2$SO$_4$ acid, coated on powdered glass	–	(6, 9)
At room temperature, 0.1–5 mol% NiCl$_2$ 6 H$_2$O, in methanol and with no solvent	60–95	(10, 11)

* For preparation of monomer.[9,10,12]

Poly(α-phenylethyl isocyanide)

Property	Units	Conditions	Value	Reference
Typical comonomers	Sec-butyl isocyanide, methyl α-isocyanopropionate			(3)
Molecular weight (of repeat unit)	g mol^{-1}	–	131	–
Typical molecular weight range of polymer	g mol^{-1}	Osmometry	$M_n = (0.3–1.3) \times 10^5$	(3, 5)
			$M_n = (0.25–2.7) \times 10^5$	(12)
			$M_w = (0.5–2) \times 10^5$	(9)
		(RS)-poly(α-phenylethyl isocyanide), light scattering	$M_w = 3.4 \times 10^4$	(8)
		(R)-poly(α-phenylethyl isocyanide), light scattering	$M_w = 1.07 \times 10^5$	(8)
		–	Strongly depends on amount of catalyst	(10)
		Light scattering in toluene at 35°C	$M_w = 1.2$ and 1.5 ($\times 10^5$)	(6)
		Osmometry in toluene at 37°C	$M_n = 5.49$ and 7.55 ($\times 10^4$)	(6)
Degree of polymerization	–	(R)-poly(α-phenylethyl isocyanide), light scattering	817	(8)
		(RS)-poly(α-phenylethyl isocyanide), light scattering	260	
Typical polydispersity index	–	Fractionated samples	1.6–2.8	(3, 5)
		–	1.1–1.3	(3)
		Polymerization: ground- glass – sulfuric acid catalyst system	1.7–2.0	(6, 9)
		–	1.6–3.1	
IR (Characteristic absorption frequencies)	cm^{-1}	N=C stretching	1,620–1,650	(10)
		Conjugated amine	1,625	(4)
		Nonconjugated amine	1,660	(4)
NMR		^{1}H NMR, in CDCl$_3$ and CCl$_4$		(13)
		^{13}C NMR, (R)-(+)-poly(α-phenylethyl isocyanide) at 23°C, in CDCl$_3$, 125.7 MHz		(8)
		d-poly(α-phenylethyl isocyanide)		(3, 7)
		^{1}H NMR, in tetracholoroethylene, at 25°C and solid-state NMR		(7, 9)
Solvents		Soluble in more than 40 solvents		(3, 9)
		Soluble in apolar solvents (chloroform, benzene, petroleum ether)		(10)
		Copolymers with sec-butyl isocyanide is sparingly soluble in common solvents		(3)
		Copolymers with methyl α-isocyanopropionate have solubilities suitable for conventional solution characterization methods		(3)
Nonsolvents		Insoluble in polar solvents (alcohols, water)		(10)
Second viral coefficient	Mol cm^3 g^{-2}	In toluene, Mn=20,000–123,000	Nearly invariant	(8)
		In toluene at 22°C, light scattering	0.2×10^{-4}	(14)
		In benzene at 22°C, light scattering	$10^{-5}–10^{-6}$	(14)
		–	2.86×10^{-4}	(15)
		–	5.87×10^{-4}	(15)

Poly(α-phenylethyl isocyanide)

Property	Units	Conditions	Value	Reference
Solubility parameters	$(MPa)^{1/2}$	$\delta_d =$ due to dispersion forces, $\delta_p =$ due to permanent dipole-dipole forces, $\delta_h =$ due to hydrogen-bonding forces	$\delta_d = 19.68$, $\delta_p = 2.41$, $\delta_h = 5.15$	(9)
Cohesive energy density	$(MPa)^{1/2}$	–	9.56	(9)
Mark-Houwink parameters: K and a	$K = ml\,g^{-1}$ $a =$ None	Unfractionated poly(d, 1-α-phenylethyl isocyanide), in toluene at 30°C	$K = 1.1 \times 10^{-2}$, $a = 0.8$	(3, 16)
		Fractionated poly(d, 1-α-phenylethyl isocyanide), in toluene at 30°C	$K = 3.8 \times 10^{-5}$, $a = 1.30$	(3, 16, 17)
		In toluene at 30°C	$K = 1.9 \times 10^{-5}$, $a = 1.36$	(9)
		In tetrahydrofuran at 30°C	$K = 2.769 \times 10^{-5}$, $a = 1.35$	(16)
Huggins constant	–	For some fractions of $M_n > 38,000$ and for the unfractionated sample	0.59	(9)
		For some fractions of $M_n < 32,000$	1.24	
Radius of gyration	Å	X-ray scattering, in toluene	Not proportional to the mol. wt.	(3, 14)
		$M_w = 13,000$	28	(3)
		$M_w = 45,800$	55	(3)
		$M_w = 91,500$	80	(3)
Hydrodynamic radius	Å	(R)-poly(α-phenylethyl isocyanide), light scattering	51	(8)
		(RS)-poly(α-phenylethyl isocyanide), light scattering	23	
Monomer projection length	Å	Calculated using density $= 1.12\ g\,cm^{-3}$	1.0	(1, 4)
		Using second virial coefficient of osmotic pressure data	1.02–1.04	(1)
Chain diameter	Å	X-ray scattering	15	(9)
			18	(3)
			15.1	(4)
Persistence length	Å	(R)- Poly(α-phenylethyl isocyanide), in tetrahydrofuran, ~room temperature	32	(8)
		(RS))- Poly(α-phenylethyl isocyanide), in toluene, ~room temperature, $M_w = 18,000\ g\,mol^{-1}$, $R_g = 28$ Å	27	
		(RS))- Poly(α-phenylethyl isocyanide), in toluene, ~room temperature, $M_w = 15,800\ g\,mol^{-1}$, $R_g = 28$ Å	32	
		(RS))- Poly(α-phenylethyl isocyanide), in toluene, ~room temperature, $M_w = 91,500\ g\,mol^{-1}$, $R_g = 80$ Å	30	
		(RS))- Poly(α-phenylethyl isocyanide), $M_w = 91,500\ gmol^{-1}$, by Ni^{II} initiation	21	

Poly(α-phenylethyl isocyanide)

Property	Units	Conditions	Value	Reference
Chain conformation		Nearly rigid rod like helix, by circular dichorism and optical rotatory studies		(3)
		Tightly wound helix with an overall shape of a cylindrical rod of about 15 Å diameter, 4_1 helix, by X-ray data		
Unit cell dimensions Lattice	–	–	Pseudo-hexagonal triclinic	(1, 3)
Cell dimensions	Å	–	$a = b = 14.92$ $c = 10.33$	
Cell angles	Degrees	–	$\alpha = 93.4\ \beta = 90.5$ $\gamma = 118.2$	
Density	g cm^{-3}	–	1.12	(1)
Optical activity, molar specific rotation, $[M]_d$	deg cm^2 g^{-1}	d- and 1-Poly(α-phenylethyl isocyanide) at 27°C in toluene	~ 500	(1, 7, 9)
		In choloroform; poly(d-α-phenylethyl isocyanide)	-458	(10, 11)
Electrical conductivity	ohm m	At 1,000 psi pressure	10^{10}	(1)
Intrinsic viscosity	dl g^{-1}	$M_w = 107{,}000$, in choloroform at 25°C	0.57	(8)
		In toluene at 30°C	0.94	(9)
		In benzene at 25°C	0.760	(3)
		In toluene at 50°C	0.204	(3)
		In toluene at 27°C	1.94, 1.26	(3)
Decomposition temperature	K	Heating rate = $10°$ min^{-1}		
		In N_2 or Ar atmosphere	543	(9)
		In Ar atmosphere	513	(3)

Circular dichoric measurements[1]

λ (nm)	Film Thickness (mm)	Solvent	Molar CD Ellipticity (deg cm^2 dmol^{-1})
550–700	5.0	Methylenechloride	-560
480–500	5.0	Methylenechloride	$+43{,}750$
280–320	5.0	Methylenechloride	$+257{,}320$
550–700	3.0	Choloroform	$-1{,}580$
480–500	3.0	Choloroform	$+23{,}830$
280–320	3.5.00	Choloroform	$+79{,}420$
550–700	5.0	Dioxane	$-13{,}230$
480–500	5.0	Dioxane	$-20{,}840$
280–320	5.0	Dioxane	$-1{,}620$
550–700	10.0	Benzene	$-39{,}000$
480–500	10.0	Benzene	$-50{,}180$
280–320	10.0	Benzene	$-14{,}280$

Poly(α-phenylethyl isocyanide)

Pyrolyzability[3]

	Conditions	Observation
Nature of product	IR spectroscopy	Pyrolysis at 500°C produces an intense broad infrared absorption band ~3,300 cm^{-1}, associated with N–H bonds Pyrolysates at 700°C reveal nitrile absorption at 2,270 cm^{-1} Nitrile absorption at 2,270 cm^{-1} becomes more intense in pyrolysates produced up to 1,300°C

References

1. Millich FJ. *Polym. Sci. Macromol. Rev.* 1980;15:207.
2. King RB. *Polym. News.* 1987;12:166.
3. Millich F. *Adv. Polym. Sci.* 1975;19:141.
4. Millich F, Sinclair RG. *J. Polym. Sci. Part C.* 1968;22:33.
5. Millich F, Sinclair RG. *Polym. Prepr, Am. Chem. Soc., Div. Polym.Chem.* 1965;6:736.
6. Millich F, Sinclair RG. *J. Polym. Sci., Part A-1.* 1968;6:1417.
7. Millich F, Baker GK. *Macromolecules.* 1969;2:122.
8. Green MM, et al. *Macromolecules.* 1988;21:1839.
9. Millich F. *Chem. Rev.* 1972;72:101.
10. Van Beijnen AJM, et al. *Macromolecules.* 1983;16:1679.
11. Nolte RJM. *Chem. Soc. Rev.* 1994;23(1):11.
12. Millich F. In: Mark HF, et al. ed. *Encyclopedia of polymer Science and Engineering.* Vol 12. New York: John Wiley and Sons; 1987:383–399.
13. Kamer PCJ, Drenth W, Nolte RJM. *Polym. Prepr., Polym. Chem.* 1989;30(2):418.
14. Huang SY, Hellmuth EW. *Polym. Prepr., Am. Chem. Soc., Div. Polym. Chem.* 1974;15:499.
15. Huang SY, Hellmuth EW. *Polym. Prepr., Am. Chem. Soc., Div. Polym. Chem.* 1974;15:505.
16. Millich F. In: Mark HF, Gaylord NG, Bikales NM, eds. *Encyclopedia of polymer Science and Technology.* Vol 15. New York: Wiley-Interscience; 1971:395.
17. Millich F, Hellmuth EW, Huang SY. *J. Polym. Sci., Polym. Chem.* 1975;13:2143.

Poly(phenylsilsesquioxane)

Gui Lin and Ronald H. Baney

Acronyms, Alternative Names, Trade Names Phenyl-T, PPSQ, PLOS, CLPHS, phenyl silicobenzoic anhydride, cyclolinear poly(phenylsiloxane), phenyl siliconic anhydride, Ladder Coat® (Mitsubishi Electric), Glass Resin® (Owens Illinois/Showa Denko), OctaPhenyl-POSS®

Class Polysiloxanes (siloxane ladder polymers)

Structures $[C_6H_5SiO_{3/2}]_n$, $n > 3$ such as 8, 10, 12, 14, 16 and so on. The structure of poly(phenylsilsesquioxane) depends upon the method of preparation. The structure is a function of the concentration of the initial monomer, the nature of the solvent, the nature of the hydrolysable substituents, the concentration of water, the temperature, the type of catalyst, and the nature of the non-hydrolysing substituent.[1,2] All of the structural types such as random structure, ladder structure (double chains polymers), partial cage structure and cage structure, or combination of the types shown may exist. The first table below summarizes the proposed structures and the evidence for such structures.

Ladder structure

Cage structures

(T_8)

Partial cage structures

Random structure

Major Application Interlayer dielectrics, high-temperature resins, and organic antireflective coatings, lithographic materials, gas-separation membrane.

Properties of Interest Very high thermal stability ($>500°C$) and good dielectric properties.

Related Polymers Poly(alkylsilsesquioxane) and poly(co-silsesquixoanes): There are many references to these classes of materials, such as the following copolymers, poly(phenylsilsesquioxane)-polyimide,[3] poly(phenylsilsesquioxane)-titania hybrid,[4] poly(phenylsilsesquioxane-co-dimethylsiloxane),[5] poly(phenylsilsesquioxane)-co-poly[(ethynediyl)(arylene)(ethynediyl)silylene],[6] poly(phenylsilsesquioxane-co-pyromellitic dianhydride),[7] poly(phenylsilsesquioxane-co-diphenylsiloxane),[8,9] poly(phenylsilsesquioxane-co-benzylsilsquioxane),[10] methylsilsesquioxane-co-phenylsilsesquioxane,[11] poly(phenylsilsesqioxane-co-p-chlorophenylsilsesquioxanes)[12] and poly(phenylsilsesquioxane-co-hydridosilsesquioxane).[13] But they are generally poorly characterized. Thus, they are not included in this handbook.

Poly(phenylsilsesquioxane)

Preparation

Structure	PPSQ	Monomer	Solvent + Condition	Catalyst	M_W (g mol^{-1})	Reference
Cage	$[C_6H_5SiO_{1.5}]_8$	$C_6H_5Si(OCH_3)_3$	Benzene	$PhCH_2(CH_3)_3NOH$	992	(14–16)
Cage	$[C_6H_5SiO_{1.5}]_8$	$C_6H_5Si(OCH_3)_3$	EtOH, H_2O	KOH (7.5 wt%)	705	(2)
Cage	$[C_6H_5SiO_{1.5}]_8$	$PhSiCl_3$	ether benzene	KOH	992	(17, 18)
Cage	$[C_6H_5SiO_{1.5}]_{10}$	$C_6H_5Si(OC_2H_5)_3$	THF	Me_4NOH		(16)
Cage	$[C_6H_5SiO_{1.5}]_n$ (n = 12, 22, 24)	$C_6H_5Si(OC_2H_5)_3$	THF	Me_4NOH		(14–16)
	LMW PPSQ		THF, Toluene, MeOH, EtOH	KOH (2.5 wt%)	1,300	(19)
	HMW PPSQ	LMW PPSQ	THF, Toluene, MeOH, EtOH	KOH (2.5 wt%)	1,300 ~61,000	(2)
	PPSQ	$PhSiCl_3$	Toluene, 25°C, 4 h	K_2CO_3/H_2O (1/1)	White cross-linked insoluble solid	(9)
	PPSQ	$PhSiCl_3$	Toluene, 20°C, 4 h	K_2CO_3/H_2O (1/1)	2,420 White soluble solid	(9)
	PPSQ White soluble solid	$PhSiCl_3$	Toluene, 50°C, 4 h	$Na_2B_4O_7.10H_2O$	1,490	(9)
	PPSQ White soluble solid	$PhSiCl_3$	Toluene, 50°C, 4 h	$Na_2B_4O_7.10H_2O$ + Na_2CO_3	1,390	(9)
	PPSQ	$PhSiCl_3$	Toluene, 80°C, 4 h	$Na_3PO_4.12H_2O$ (0.167)	White cross-linked insoluble solid	(9)
	PPSQ	$PhSiCl_3$	Toluene, 50°C, 4 h	$Na_3PO_4.12H_2O$ (0.167)	3,210 White fractionally soluble solid	(9)

Structure, process, and molecular weight

Proposed Structure	Process Conditions	Acronym*	Structural Evidence	Polymer $M_W \times 10^{-3}$ (g mol^{-1})	Reference
Cis-syndiotactic double chain	Equilibration method $PhSiCl_3$ + H_2O at 50% toluene to hydrolysate 0.1% KOH + 30% toluene at ~100°C to give "prepolymer" (I) or T-12 cage at ~250°C/90% solids in high boiling solvents	PPSQ-1	XRD, IR, UV Hypochroism, bond angle calculations, Mark-Houwink equation	4,100	(20–22)

Proposed Structure	Process Conditions	Acronym*	Structural Evidence	Polymer $M_W \times 10^{-3}$ (g mol^{-1})	Reference
Rigid chain polymers	Same as PPSQ-1 except final equilibration at 100% solids	PPSQ-2	High Kuhm segment Dynamo-optical (high negative segmental anisotropy)	–	(23–26)
Linked partial cages	–	PPSQ-1	Curvature in the Mark-Houwink equation Gelation at various temperatures, solvent types and concentrations	1,000	(27)
Cis-syndiotactic double chain	(1) PhSiCl$_3$ + H$_2$O in MIBK ~10°C to hydrolysate (2) 0.1% KOH + 50 wt% solids in xylene reflux	PPSQ-3	IR	165	(28) (29)
Cis-syndiotactic double chain	Fluoride ion catalyzed equilibration of hydrolyzate	PPSQ-4	–	1,200	(30)
"Branched" ladder	PhSiCl$_3$ + H$_2$O in ether or toluene to hydrolyzate to give "prepolymer" (I) with 30% dicyclohexyl-carboimide in xylene, 44% solids, 13 h, reflux	PPSQ-5a	FTIR, ^{1}H-NMR, ^{29}Si-NMR	12	(31, 32)
"Branched" ladder	(I) with 0.5% KOH in toluene, 44% solids 13 h, reflux	PPSQ-5b	FTIR, ^{1}H-NMR, ^{29}Si-NMR	12	(31, 32)
Gel	(I) in toluene with 5% KOH, 44% solids, 13 h reflux	PPSQ-5c	FTIR, ^{1}H-NMR, ^{29}Si-NMR	Gel	(31, 32)
Ladder	(I) in toluene and 8% diphenyl ether with 5% KOH, 40% solids, 13 h, 260°C	PPSQ-5d	FTIR, ^{1}H-NMR, ^{29}Si-NMR	26	(31, 32)
Ladder	(I) in 1:1 toluene and diphenyl ether with 0.005% KOH, 230°C, 5 h	PPSQ-5e	FTIR, ^{1}H-NMR, ^{29}Si-NMR	550	(31, 32)
Cis-isotactic double chain	(I) in 2:1:1:2 benzene-toluene-xylene-diphenyl ether with 10–4% KOH, 7 h	PPSQ-5f	Eximer fluorescene	340	(33, 34)

Poly(phenylsilsesquioxane)

Proposed Structure	Process Conditions	Acronym*	Structural Evidence	Polymer $M_W \times 10^{-3}$ (g mol^{-1})	Reference
"Ladder like"	PhSi(OEt)$_3$ in MIBK 20% solids with Et$_4$NOH, reflux, 12 h	PPSQ-6	Elemental analysis and molecular weight	5	(35)
Linked partial cages	Condensation of (PhOHSiO)$_4$	PPSQ-7	Insoluble amorphous gels	90	(36)
Cis-syndiotactic double chain	Condensation of PhSi(OK)$_3$	PPSQ-8 $C_{72}H_{60}O_{18}Si_{12}$	IR, XRD	72 155,026	(37) (38)
[C$_6$H$_5$SiO$_{1.5}$]$_{10}$	Hydrolysis of PhSiCl$_3$, followed by condensation of the resulting hydrolyzates in the presence of a catalytic amount of HCl in methyl isobutyl ketone (MIBK).			1.4	(39)

* See ref. (1).

Mark-Houwink parameter, *a*, for selected poly(phenylsilsesquixoanes)

PPSQ-	*a*	Molecular Weight	Reference
1	0.92	1.4×10^4 (M_n)	(20)
2	1.10	2×10^5	(24)
2	0.90	0.6×10^3	(24)
2	0.90	$(2.5–3) \times 10^5$	(25, 26)
1	0.898	$(0.26–4.88) \times 10^5$ (M_n)	(40, 41)
2	0.70	3×10^5	(24)
1	0.54	2×10^5	(22)

IR and NMR characterization

PPSQ	IR(cm^{-1})		NMR (ppm)			Reference
	Group	Characteristic Frequencies	^{1}H	^{13}C	^{29}Si	
1	Vs Si–Ar	1,130,				(31)
	Vas Si–O–Si	1,045,				
1 with "defects"		1,137				(31)
Cage, T-8	ν C–H	3,127–2,890			−75.9	(2)
	ν C–C	1,596				
	δ C–H	1,432				
	ν Si–O–Si	1,304–990				
	γ C–H	746, 698				

Poly(phenylsilsesquioxane)

IR and NMR characterization

PPSQ	IR(cm^{-1})		NMR (ppm)			Reference
	Group	Characteristic Frequencies	^{1}H	^{13}C	^{29}Si	
Cage, T-10	ν C–H	3,122–2,875			–74.9, –77.9	(2)
	ν C–C	1,596				
	δ C–H	1,433				
	ν Si–O–Si	1,311–975				
	γ C–H	743, 698				
LMW PPSQ	Si–OH	3,725–3,290	Ar(8.2–6.4)	135.3, 131.9, 129.1	–67.4, –76.0	(2)
	ν C–H	3,124–2,864				
	ν C–C	1,597				
	δ C–H	1,430				
	ν Si–O–Si	1,314–978				
	γ C–H	743, 699				
HMW PPSQ	ν C–H	3120–2835	Ar(8.2–6.4)	135.5, 132.0, 129.3	–77.7	(2)
	ν C–C	1597				
	δ C–H	1430				
	ν Si–O–Si	985				
	γ C–H	743, 698				

Solution properties

PPSQ-	Soluble at Room Temperature	Insoluble at Room Temperature	Theta Solvent	Resin Solubility	Reference
Oligomers	Benzene, chloroform, THF	Acetone, hexane, cyclohexane, ether, carbon tetrachloride, MIBK, isobutyl ether			(18)
1	Benzene, THF, methylene	–			(20, 21)
2	Benzene, bromoform	–	Benzene/ butylacetate (60:40)		(42)
5a, b, d, e, f	Benzene, toluene, THF	–	–		(31, 32)
8	Benzene, chloroform, ether, toluene, THF, methyl ethyl ketone, carbon tetrachloride, MIBK	Acetone, methanol, ethanol	–		(37)

Poly(phenylsilsesquioxane)

PPSQ-	Soluble at Room Temperature	Insoluble at Room Temperature	Theta Solvent	Resin Solubility	Reference
$[C_6H_5SiO_{1.5}]_{10}$	Slightly soluble in THF, toluene, chloroform	Acetone, acetonitrile, methanol		Thermoplastic resins including fluoropolymers	(43)
$[C_6H_5SiO_{1.5}]_{12}$	Slightly soluble in THF, toluene, chloroform	Acetone, acetonitrile, methanol		Thermoplastic resins including fluoropolymers	(44)

XRD

PPSQ	d spacing (Å)	Reference
1	5.0, 12.5	(20)
1	4.6, 12.3	(26)

Persistence length

PPSQ-	Persistence Length (Å)	Method	Reference
1	80	Yamakawa, Fujii method*	(45)
5f	64	Yamakawa, Fujii method*	(46)
2	100	Diffusion in butyl acetate	(24)
2	89	$M[\eta]$ in bromoform	(24)
2	68	$M[\eta]$ in benzene	(24)

* See reference (46)

Thermal stability

PPSQ-	Thermolysis		Thermal Expansion Coefficient (ppm)	Soften Temperature°C	Reference
	Condition	Temp. (°C)			
1	Thermal balance in air-onset	525			(47)
3	TGA air, 10°C, min-onset	500	110–140 (below 250°C)		(29)
			90 (above 220°C)		(48)
4	TGA air	505		121°C	(49)
PPSQ					(8, 50)

Mechanical properties

PPSQ-	Temp. (°C)	Tensile Strength (MPa)	Elongation (%)	Reference
1	Room temp.	27.6–41.5	3–10	(51)
2	100	39	25	(26)

Poly(phenylsilsesquioxane)

Mechanical properties

PPSQ-	Temp. (°C)	Tensile Strength (MPa)	Elongation (%)	Reference
4	Room temp.	18–30	–	(30)
3	Room temp.	800	0.4	(29)
3	250	400	2.7	(29)
3	250	559	2.6	(1)

Other physical properties

PPSQ-	Specific Dielectric Constant	Pencil Hardness	Reference
3	–	–	(29)
3	–	–	
3	3.2 (KHz)	–	(52)
3	–	5H	(53)
PPSQ	2.8 (1 MHz at 100°C)		(7)

Patented uses

Applications	Reference
Photoresists, protective coatings, and reflective coating	(54–60)
Interlayer dielectric	(61–66)
Liquid crystal display elements	(67, 68)
Semiconductor devices, Magnetic recording media	(69–72)
Optical fiber coatings	(73, 74)
Gas separation membranes	(75–77)
Binder for ceramics	(78)
Carsinostatic drugs	(79)
Porosity control, printability	(38)
Pigments	(80)
lithographic materials	(81, 82)

References

1. Baney RH, Itoh M, Sakakibara A, Suzuki T. *Chem. Rev.* 1995;95(5):1409.
2. Kim SG, Choi J, Tamaki R, Laine RM, *Polymer.* 2005;46:4514–4524.
3. Iyoku Yoshitake, Kakimoto Masa-aki, Imai Yoshio. *High Perfor. Polym.* 1994;6(1):53–62.
4. Takahashi K, Tadanaga K, Tatsumisago M, Matsuda A. *J Amer. Cera. Soc.* 2006;89(10):3107–3111.
5. Liang G, Liu Si, Ye H, Zhao J, Zhang L, *Hecheng Xiangjiao Gongye*, 2005;28(4):306.

6. Douglas WE, Klapshina LG, Kuzhelev AS, Peng W, Semenov VV. *J. Mater. Chem.* 2003;13(11):2809–2813.

7. Hedrick JL, Cha HJ, Miller RD, Yoon DY, Brown HR, Srinivasan S, Pietro RD. eds. *Macromolecules.* 1997;30:8512–8515.

8. Menaa B, Takahashi M, Tokuda Y, Yoko T. *J Sol-Gel Sci. Technol.* 2006;39:185–194.

9. Lesniak E, Michalska ZM, Chojnowski J. *J Inorg Organ Polym.* 1998;8:1–21.

10. Matsuda A, Sasaki T, Tanaka T, Tatsumisago M, Minami T. *J Sol-Gel Sci Tech.* 2002;23:247–252.

11. Sosa JM. *Macromolecules.* 1980;13:1260–1264.

12. Affrossman S, Angadji H, Bakhshaee M, Coffey K, Chow FL, Hayward D, McLeod GG, Pethrick RA, Wittaker P. *Polymer.* 1989;30:1022.

13. Duan QH, Zhang TY, Deng KL, Xie P, Zhang RB. *Chin. J. Polym. Sci.* 2005;4:355–361.

14. Brown JF Jr. *J. Am. Chem. Soc.* 1965;87:4317.

15. Sprung MM, Guenther FO. *J. Am. Chem. Soc.* 1955;77:6045.

16. Brown JF, Vogt LH. *J. Am. Chem. Soc.* 1964;86:1120.

17. Barry AJ, Daudt WH, Domicone JJ, Gilkey JW. *J. Am. Chem. Soc.* 1955;77:4248.

18. Sprung MM, Guenther FO. *J. Poly. Sci.* 1958;28:17.

19. www.gelest.com.

20. Brown JF, et al. *J. Am. Chem. Soc.* 1960;82(23):6194.

21. Brown JF Jr. *J. Poly. Sci.* 1964;1:83.

22. Brown JF Jr, Prescott PL. *J. Am. Chem. Soc.* 1964;86:1402.

23. Andrianov KA, Kurakov GA, Suschentsova FF, Miagkov VA. *Vysokomolek.Soedin.* 1965;7:1477.

24. Tsvetkov VN, Andrianov KA, Okhrimenko GI, Vitovskaya MG. *Eur. Polym. J.* 1971;7:1215.

25. Tsvetkov VN, et al. *Eur. Polym. J.* 1973;9:27.

26. Andrianov KA, Zhdanov AA, Levin, Yu V. *Ann. Rev. Mater. Sci.* 1978;8:313. (and references therein).

27. Frye CL, Klosowski JM. *J. Am. Chem. Soc.* 1971;93:4599.

28. Adachi H, Adachi E, Hayashi O, Okahashi K. *Rep. Prog. Polym. Phys. Japan.* 1985;28:261.

29. Adachi H, Adachi E, Yamamoto S, Kanegae H. *Mat. Res. Soc. Symp. Proc.* 1991;227:95.

30. Hata H, Komasaki S. *Jpn Patent Kokai*-S-59-108033 1984; *Chem. Abstr.* 1984;101:172654.

31. Zhang X, Chen S, Shi L. *Chin. J. Polym. Sci.* 1987;5:162.

32. Zhang X, Shi L. *Chin J. Polym. Sci.* 1987;5:197.

33. Huang C, Xu G, Zhang X, Shi L. *Chin. J. Polym. Sci.* 1987;5:347.

34. Zhang X, Shi L, Huang C. *Chin. J. Polym. Sci.* 1987;5:353.

35. Sprung MM, Guenther FO. *J. Polym. Sci.* 1958;28:17.

36. Brown JF Jr. *J. Am. Chem. Soc.* 1965;87:4317.

37. Takiguchi T, Fujikawa E, Yamamoto Y, Ueda M. *Nihon Kagakukaishi.* 1974;108.

38. Hybrid Plastics™. Superior Technology for superior Products 2006.

39. Yamamoto S, Yasuda N, Ueyama A, Adachi H. Ishikawa M. *Macromolecules.* 2004;37(8):2775–2778.

40. Heminiak TE, Benner CL, Gibbs WE. *ACS Polym. Prepr.* 1967;8:284.

41. Helminiak TE, Berry GC. *J. Polmy. Sci.* 1978;65:107.

42. Tsvetkov VN, et al. *J. Polym. Sci, Part C.* 1968;23:385.

43. Li GZ, Matsuda T, Nishioka A, Liang KW, Masubuchi Y, Koyama K, Pittman CU Jr. *J Appl Polym Sci.* 2007;104(1):352–359.

44. www.hybridplastics.com.

45. Shi L, et al. *Chin J. Polym. Sci.* 1987;5:359.

46. Yamakawa H, Fujii M. *Macromolecules.* 1974;7:128.

47. Brown JF Jr. *J. Polym. Sci., Part C,* 1963;1:83.
48. Adachi H, Adachi E, Hayashi O, Okahashi K. *Rep. Prog. Polym. Phys. Jpn.* 29 1986:257.
49. Zhang X, Shi L, Li S, Lin Y. *Polym. Degrad. Stab.* 1988;20:157.
50. Matsuda A, Sasaki T, Hasegawa K, Tatsumisago M, Minami T. *J. Ceram. Soc. Jpn.* 2000;108:830–835.
51. Brown JF Jr. *J. Polym. Sci., Part C.* 1963;1:83.
52. *Trade literature on "Ladder Coat".* Sanda City, Japan: Ryoden Kasei Co. Ltd.
53. Matsui F. *Kobunshi Kako.* 1990;39:299.
54. Yoneda Y, Kitamura T, Naito J, Kitakohji T. *Jpn Patent Kokai*-S-57-168246 1982; *Chem. Abstr.* 1984;100:43074.
55. Uchimura S, Sato M, Makino D. *Jpn Patent Kokai*-S-58-96654 (1983); *Chem. Abstr.* 1984;100:35302.
56. Yoneda Y, et al. *JpnPatent Kokai*-S-57-168247 (1982); *Chem. Abstr.* 1984;100:43075.
57. Uchimura S, Sato M, Makino D. *Jpne Patent Kokai*-S-58-96654 (1983); *Chem. Abstr.* 1984;100:35302.
58. Adachi H, Hayashi O, Okahashi K. *Jpn Patent Kokoku*-H-2-15863 (1990) [Kokai-S-60-108839 1985]; *Chem. Abstr.* 1986;104:120003.
59. Adachi H, Hayashi O, Okahashi K. *Jpn Patent Kokai*-S-60-108841 1985; *Chem. Abstr.* 1986;104:43184.
60. Adachi H, Adachi E, Hayashi O, Okahashi K. *Japanese Patent Kokoku*-H-4-56975 1992 [Kokai-S-61-279852 (1986)]; *Chem. Abstr.* 1987;106:224512.
61. Shoji F, Takemoto K, Sudo R, Watanabe T. *Japanese Patent Kokai*-S-55-111148 1980.
62. Adachi E, Aiba Y, Adachi H. *Jpn Patent Kokai*-H-2-277255 (1990); *Chem. Abstr.* 1991;114:124250.
63. Aiba Y, Adachi E, Adachi H. *Jpn Patent Kokai*-H-3-6845 (1991); *Chem. Abstr.* 1991;114:155372.
64. Adachi E, Adachi H, Hayashi O, Okahashi K. *JpnPatent Kokai*-H-1-185924 1989; *Chem. Abstr.* 1990;112:170346.
65. Hayashide Y, Ishii A, Adachi H, Adachi E. *Jpn Patent Kokai*-H-5-102315 (1993); *Chem. Abstr.* 1994;120:180306.
66. Adachi E, Adachi H, Kanegae H, Mochizuki H. *German Patent.* 4202 290 (1992); *Chem. Abstr.* 1992;117:193364.
67. Shoji FK, Sudo R, Watanabe T. *Jpn Patent Kokai*-S-56-146120 (1981); *Chem. Abstr.* 1982;96:208471.
68. Azuma K, Shindo Y, Ishimura S. *Jpn Patent Kokai*-S-57-56820 (1982); *Chem. Abstr.* 1982;97:227612.
69. Imai E, Takeno H. *Jpn Patent Kokai*-S-59-129939(1984); *Chem. Abstr.* 1984;101:221241.
70. Yanagisawa M. *Jpn Patent Kokai*-S-62-89228 1987.
71. Xie P, Rongben Z. *Polym Adv Technol.* 1997;8:649.
72. Tamaki R, Tanaka Y, Asuncion MZ, Choi J, Laine RM. *J Am Chem Soc.* 2001;123:12416–12417.
73. Mishima T, Nishimoto H. *Jpn. Patent Kokai*-H-4-247406 1992; *Chem. Abstr.* 1993;118:256243.
74. Mishima T, Nishimoto H. *Jpn Patent Kokai*-H-4-271306 1992; *Chem. Abstr.* 1993;118:256251.
75. Saito Y, Tsuchiya M, Itoh Y. *Jpn. Patent Kokai*-S-58-14928 1983; *Chem. Abstr.* 1983;98:180758.
76. Prado LAS, Radovanovic E, Pastore H, Yoshida P IV. Trriani IL. *J Polym Sci A: Poly Chem.* 2000;38:1580–1589.
77. Mi Y, Stern SA. *J Polym Sci Part B: Polym Phys.* 1991;29:389.

Poly(phenylsilsesquioxane)

78. Mine T, Komasaki S. *Jpn Patent Kokai-S*-60-210570 1985; *Chem. Abstr.* 1986;104:154450.
79. Tsutsui M, Kato S. *Jpn Patent Kokoku-S*-63-20210 1988 [Kokai-S-56-97230 1981]; *Chem. Abstr.* 1981;95:192394.
80. Sato H, Watanabe K, Otsuka N. *Jpn. Kokai Tokkyo Koho.* 2003;6:JP2003277640 A 20031002.
81. Ban H, Tanaka A, Kawai Y, Imamura M. *Polymer.* 1990;31:564.
82. Clarson SJ, Semlyen JA. *Siloxane Polymers.* PTR Prentice Hall, Englewood Cliffs, New Jersey: 1993.

Poly(phenyl/tolylsiloxane)

Dale J. Meier

Acronym PP/TS

Class Polysiloxanes

Repeat Triad Structures –PPP–, –PPP′–, –PP′P′–, –P′P′P′–, –PPP″–, –PP″P″–, –P″P″P″–, –PPM′–, PPM″–, –PM″M″–, –M″M″M″–.

where   P = –Si(Ph)$_2$–O–
P′ = –Si(Ph/p-T)–O–
P″ = –Si(p-T)$_2$–O–
M′ = –Si(Ph/m-T)–O–
M″ = –Si(m-T)$_2$–O–
Ph = phenyl
p-T = p-tolyl
m-T = m-tolyl.

Major Applications The various PP/TS polymers are not commercial.

Properties of Special Interest Highly crystalline, high melting point, excellent thermal stability, mesomophic state at high temperatures.

Preparative Techniques	Conditions	Reference
Anionic	Initiators for cyclic trimers	
	Li alkyl, solution	(1, 6, 7)
	KO–[Si(Ph/Tol)–O]$_n$–K, solution, bulk	(2–5)

Property	Units	Polymer	Conditions	Value	Reference
Solvents	K	–PPP– –P″P″P″–	Diphenyl ether c1-Chloronapthalene 1,2,4-Trichlorobenzene	>420	(1–3, 8)
		–PPP –P″P″P″–	Quenched from solution	315	(7)
		–PPP′– –PP′P′– –PPP″– –P″P″–	Toluene Chloroform	300	(1–4)

Poly(phenyl/tolylsiloxane)

Property	Units	Polymer	Conditions	Value		Reference
		–PPM″–				
		–PM″M″–				
		–M″M″M″–				
Mark-Houwink parameters: K and a	$K = \mathrm{ml\,g^{-1}}$ $a = $ None		Chloroform, 40°C	$K \times 10^{-3}$	a	(10)
		–PPP′–		2.1	0.83	
		–PPP″–		2.6	0.83	
		–P′P′P′–		2.4	0.83	
NMR chemical shifts	ppm	–PPP′–	^{29}Si	−46.16, −45.83		(1)
		–PPP″–	^{29}Si	−45.66, −56.99		(1)
		–P′P′P′–	^{29}Si	−46.49		(1)
		–M″M″M″–	^{13}C	20.87 (CH$_3$)		(5)
Tensile strength	MPa	–PPP′–	Films from toluene or chloroform	<0.2		(4)
		–PPP″–		<0.2		
		–PPM′–		<0.2		
		–PPM″–		<0.2		
		–PM″M″–		2.5		
		–M″M″M″–		3.5		
Elongation at break	%	–PM″M″–	Films from toluene or chloroform	130		(4)
		–M″M″M″–		13		

Crystalline state properties[3]

Polymer	Lattice	Cell Dimensions (nm)			Monomer per Cell
		a	b	c	
–PPP′–	Rhombic	2.106	1.053	1.036	2

d-spacings (nm)	Layer Line Number	Electron Diffraction	X-ray Diffraction
	0	1.053	1.052
		0.940	0.940
		0526	–
		0.476	0.467
		0.421	0.429
		0.391	0.393
	2	0.492	0.498
		0.464	0.467
		0.450	0.448
		0.425	–
		0.367	0.363
	3	0.326	–
		0.324	0.325
	4	0.259	–

Polymer	Lattice	Cell Dimensions (nm)			Monomer per Cell
		a	*b*	*c*	
–P′P′P′–	Rhombic	2.104	1.086	0.997	2

d-spacings (nm)	Layer Line Number	Electron Diffraction	X-ray Diffraction
	0	1.052	1.053
		0.960	0.960
		0.520	–
		–	0.526
		0.455	0.468
	1	–	0.757
		–	0.468
	2	0.483	–
		0.455	–
		0.445	0.443
		0.420	0.423
	3	0.335	–
		0.317	–
		0.249	–

Property	Units	Polymer	Conditions	Value	Reference
Density	g cm^{-3}	–PPP′–	From experimental	1.12	(3)
		–P′P′P′–	From unit cell	1.17	
			From experimental	1.13	
			From unit cell	1.24	
Melting temperature	K	–PPP–	To mesomorphic state	538, 545	(1, 8)
			To isotropic state	813	(1)
		–PPP′–	To mesomorphic state	458	(1)
			To isotropic state	733	(1)
		–PPP″-	To mesomorphic state	413	(1)
			To isotropic state	703	(1)
		–P′P′P′–	To mesomorphic state	433	(1)
			To isotropic state	723	(1)
		–P″P″P″–	To mesomorphic state	573	(1)
			To isotropic state	>753 (decomp.)	(1)
		–M″M″M″–	To mesomorphic state	404	(6)
			To isotropic state	803	(6)
Glass transition temperature	K	–PPP–	From DSC	313, 322	(1, 9)
		–PPP′		313	(1)
		–PPP″–		313	(1)
		–P′P′P′–		323	(1)
		–P″P″P″–		323	(1)
		–M″M″M″–		268	(6)

Poly(phenyl/tolylsiloxane)

Property	Units	Polymer	Conditions	Value	Reference
Thermal stability	K	–PPP′–	TGA, 10% weight loss,	756	(1)
		–PPP″–	$10°$ min^{-1} under N_2	727	(1)
		–P′P′P–		742	(1)
		–P″P″P″–		789	(1)
		–M″M″M″–		731	(6)

References

1. Lee MK, Meier DJ. *Polymer.* 1993;34:4882.
2. Korshak VV, et al. *Vysokomol. Soyed.* 1985;B27:300.
3. Babchinitser TM, et al. *Polymer.* 1985;26: 1527.
4. Vasilenko NG, et al. *Vysokomol. Soyed.* 1989;A31:1585; *Poly. Sci. USSR* (English translation). 1989;31:1737.
5. Vasilenko NG. et al. *Vysokomol. Soyed.* 1989;A31:2026. *Poly. Sci. USSR* (English translation). 1989;31:2225.
6. Lee MK, Meier DJ. *Polymer.* 1994;35:4197.
7. Ibemesi J, et al. In: Mark JE, Schaefer DW, eds. *Polymer Based Molecular Composites.* Pittsburgh: Materials Research Society; 1989.
8. Govodsky YK, Papkov VS. *Adv. Poly. Sci.* 1989;88:129.
9. Buzin MI, et al. *Vysokomol. Soedin.* 1992;34 Series B:66.
10. Lee MK, Meier DJ. *Polymer.* 1994;35:3282.

Polyphosphates

Leiyuchuan Ding and Bruce M. Foxman

Class Inorganic and semi-inorganic polymers

Structure $-[O-P(O)_2]_n-O-$ or $-[P(O)(OR')-O-R-O]_n-$

Major Applications Acids: Intermediate in fertilizer production. Catalysts for alkylation, dehydrogenation, polymerization, and isomerization. Dehydrating agent in dye and pigment production. Salts: Builders in detergent and cleaning formulations. Consistency control agents in foods. Deflocculants in clays, dyes, and ink. Anticalculus agents in toothpaste and mouthwash. Dispersants for solids in clay processing, drilling mud, and pigments. Flame retardation. Models for natural biopolymers.[1-3]

Preparative Techniques Principal synthetic routes: condensation and addition reactions. A useful survey is available.[1]

Property	Units	Conditions	Value	Reference
Theta temperatures Θ	K	Lithium polyphosphate in LiCl (0.4 M)/H_2O	293.2	(4)*
		Lithium polyphosphate in LiBr (1.80 M)/H_2O	298.2	(5)
		Sodium polyphosphate in NaBr (0.415 M)/H_2O	298.2	(6)
Glass transition temperature	K	Hydrogen polyphosphate	263	(7, 8†)
		Lithium polyphosphate	608	
		Sodium polyphosphate	553	
		Calcium polyphosphate	793	
		Strontium polyphosphate	758	
		Barium polyphosphate	743	
		Zinc polyphosphate	793	
		Cadmium polyphosphate	723	
Characteristic ratio $\langle r_2 \rangle_0 / nl_2$	–	Sodium polyphosphate (aqueous NaBr, 0.35–0.415 M) 25°C	6.6	(6)
		Cesium polyphosphate (aqueous CsCl, 0.96 M) 30°C	7.1	(9, 10)

* Strauss and Anders (1962) suggested that the results obtained for the theta temperature of lithium polyphosphate in 0.4 M LiCl should be "regarded with caution."

† In ref. 8, glass transition temperatures were measured by using an automatic device which measured the length of the polymer sample as a function of temperature.

Percentage composition of the strong phosphoric acids* [11]

P_2O_5 (wt. %)	$\frac{[P_2O_2]}{[H_2O]}$	1	2	3	4	5	6	7	8	9	10	11	12	13	14	HIGH-POLY
67.4	0.263	100.0														
68.7	0.279	99.7	0.33													
70.4	0.302	96.2	3.85													
71.7	0.321	91.0	8.86													
73.5	0.352	77.1	22.1	0.79												
73.9	0.360	73.6	25.1	1.34												
75.7	0.394	53.9	40.7	4.86	0.46											
77.5	0.438	33.5	50.6	11.5	2.68	0.74										
79.1	0.481	22.1	46.3	20.3	7.82	2.26	1.02	0.34								
80.5	0.523	13.8	38.2	21.0	13.0	6.86	3.38	1.67	1.03	0.22						
81.0	0.542	12.2	34.0	22.7	14.6	8.42	4.36	22.7	1.41	0.56						
81.2	0.549	10.9	32.9	22.3	15.0	9.36	5.41	2.85	1.75	0.97	0.36	0.05				
82.4	0.594	7.32	23.0	19.3	15.9	12.3	8.21	5.73	3.89	2.52	1.36	0.91	0.14			
84.0	0.667	3.92	11.8	12.7	12.0	10.5	8.97	7.99	6.62	5.63	4.54	3.72	3.03	2.46	1.68	6.63
85.0	0.717	2.28	6.36	7.32	8.01	8.17	7.67	7.22	6.93	6.42	5.89	5.27	4.69	3.99	3.83	16.9
85.3	0.736	1.87	4.73	6.33	6.58	6.66	6.71	6.36	6.11	5.88	5.46	5.07	4.90	4.64	4.38	25.6
86.1	0.787	1.46	2.81	3.74	4.43	4.52	4.77	4.79	4.93	4.67	4.54	4.67	4.63	4.38	4.17	43.5
87.1	0.860	0.83	1.81	2.17	2.53	3.09	3.39	3.46	3.33	3.55	3.47	3.45	3.52	3.26	3.24	61.1
87.9	0.920	0.50	0.82	1.56	1.76	1.72	2.03	2.13	2.26	2.07	2.26	2.06	2.20	1.99	2.30	76.4
89.4	1.066	1.88	1.52	0.77	0.61	0.62	0.68	0.54	0.71	0.86	1.03	0.98	1.16	1.23	1.37	86.8

* For total % $P_2O_5 \geq 86.1$, small amounts of trimeta- and tetrametaphosphoric acid were also detected. 1 = ortho-, 2 = pyro-, 3 = tri-, 4 = tetra-phosphate, etc. Highpoly = higher-molecular-weight material including 15-phosphoric acid. (Source: Jameson 1959. Reprinted with permission from the Royal Society of Chemistry.)

Composition of supernatant, syneresis liquid and gel obtained by admixture of iron (III) nitrate and sodium polyphosphate solutions[12]

Sample	Weight (g)	$n^*_{NA}(\times 10^3)$	$n_{Fe}(\times 10^3)$	$n_{PO3}(\times 10^3)$	$n^{\dagger}_{H2O}$	$n^{\ddagger}_{NO3}(\times 10^3)$
Supernatant						
Syneresis liquid	5.41	5.1 ± 0.3	0.7 ± 0.2	0.6 ± 0.2	0.25	
Gel	62.74	54.9 ± 0.4	29.4 ± 0.2	59.4 ± 0.2	2.80	21.8

* Deviations are calculated from triplicate measurements, i.e. readings on three different sample spot. *ns* are the respective mole numbers.

† Loosely bound water, calculated using loss at 120°C.

‡ Calculated from mass balance.

Composition of supernatant, syneresis liquid and gel obtained by admixture of calcium and iron (III) nitrate and sodium polyphosphate solutions[12]

z_{Ca}	Sample	Weight (g)	$n^*_{Na}(\times 10^3)$	$n_{Ca}(\times 10^3)$	$n_{Fe}(\times 10^3)$	$n_{PO3}(\times 10^3)$	$n^\dagger_{H2O}$	$n^\ddagger_{NO3}(\times 10^3)$
[P]/[Fe] + [Ca] = 2								
1.00	Supernatant	59.58	50.5 ± 5.3	23.3 ± 3.3		53.1 ± 3.1	2.89	
	Syneresis liquid	0.92	1.9 ± 1.0	1.3 ± 0.6		1.4 ± 0.6	0.04	
	Gel	7.04	7.7 ± 4.3	5.4 ± 2.6		5.6 ± 2.6	0.16	40.3
0.85	Supernatant							
	Syneresis liquid	16.39	14.4 ± 2.4	7.1 ± 1.7	0.5 ± 0.3	5.8 ± 2.9	0.78	
	Gel	51.84	45.7 ± 2.4	18.4 ± 1.7	4.0 ± 0.3	54.3 ± 2.9	2.26	29.9
0.70	Supernatant							
	Syneresis liquid	18.61	15.7 ± 0.8	4.2 ± 2.1	0.2 ± 0.1	2.6 ± 1.4	0.90	
	Gel	48.98	44.4 ± 0.8	16.7 ± 2.1	8.8 ± 0.1	57.5 ± 1.4	2.12	15.7
0.50	Supernatant	5.23	6.1 ± 1.7	0.5 ± 0.2	0.5 ± 0.2	3.2 ± 1.5	0.24	
	Syneresis liquid	42.20	24.2 ± 1.7	9.3 ± 0.2	1.7 ± 0.3	19.0 ± 1.6	2.05	
	Gel	20.96	29.7 ± 3.4	5.2 ± 0.3	12.9 ± 0.8	55.3 ± 1.4	2.04	18.2
0.30	Supernatant							
	Syneresis liquid	20.58	19.4 ± 6.6	1.4 ± 0.3	2.0 ± 0.8	4.7 ± 1.4	1.02	
	Gel	47.47	40.7 ± 6.6	7.6 ± 0.3	19.0 ± 0.8	55.3 ± 1.4	2.04	18.2
0.15	Supernatant							
	Syneresis liquid	15.24	1.9 ± 0.3	1.1 ± 0.2	0.4 ± 0.3	1.2 ± 0.4	0.75	
	Gel	52.17	58.2 ± 0.4	3.4 ± 0.2	25.0 ± 0.4	58.9 ± 0.4	2.29	17.6
[P]/[Fe] + [Ca] = 1								
1.00	Supernatant	61.58	26.1 ± 5.5	46.6 ± 2.0		24.7 ± 3.6	2.646	
	Syneresis liquid	2.39	12.3 ± 3.1	1.1 ± 0.3		1.9 ± 0.4	0.106	
	Gel	7.34	21.7 ± 7.7	12.3 ± 1.8		33.4 ± 3.3	0.162	
0.85	Supernatant	1.4	4.0 ± 1.0	0.4 ± 0.3	0.1 ± 0.04	0.3 ± 0.1	0.06	
	Syneresis liquid	58.72	37.5 ± 6.2	37.4 ± 3.3	0.2 ± 0.1	24.6 ± 6.3	2.56	
	Gel	9.65	18.5 ± 5.3	13.1 ± 3.1	8.7 ± 0.2	35.1 ± 6.3	0.16	19.8
0.70	Supernatant							
	Syneresis liquid	23.08	21.3 ± 4.4	7.9 ± 0.9	0.5 ± 0.1	3.8 ± 0.5	1.02	
	Gel	47.75	38.7 ± 4.4	34.0 ± 0.9	17.6 ± 0.2	56.2 ± 0.5	1.98	18.2
0.50	Supernatant							
	Syneresis liquid	6.77	11.8 ± 0.7	1.0 ± 0.3	0.6 ± 0.2	0.6 ± 0.3	0.29	
	Gel	64.47	48.2 ± 0.7	28.9 ± 0.3	29.5 ± 0.2	59.5 ± 0.3	2.71	21.0
0.30	Supernatant							
	Syneresis liquid	2.04	8.7 ± 2.3	0.4 ± 0.03	1.1 ± 0.04	0.5 ± 0.04	0.09	
	Gel	70.42	51.4 ± 2.2	17.6 ± 0.03	41.0 ± 0.04	59.6 ± 0.04	2.98	30.7
0.15	Supernatant							
	Syneresis liquid	1.23	24 ± 1.1	0.2 ± 0.1	1.0 ± 1.3	0.2 ± 0.1	0.05	
	Gel	71.49	57.7 ± 1.1	8.8 ± 0.1	50.0 ± 1.3	59.8 ± 0.1	3.03	17.6

* Deviations are calculated from triplicate measurements, i.e. readings on three different sample spot. *ns* are the respective mole numbers.

† Loosely bound water, calculated using loss at 120°C.

‡ Calculated from mass balance.

References

1. Kroschwitz J, ed. *Encyclopedia of Polymer Science and Engineering.* Vol 11. New York: Wiley-Interscience; 1988:96–126.
2. Kroschwitz J, Howe-Grant M, eds. *Encyclopedia of Chemical Technology.* Vol 10. New York: Wiley-Interscience; 1993:976–998.
3. Kroschwitz J, Howe-Grant M, eds. *Encyclopedia of Chemical Technology.* Vol 18. New York: Wiley-Interscience; 1996:669–718.
4. Saini G, Trossarelli L. *J. Polym. Sci.* 1957;23:563.
5. Strauss UP, Ander P. *J. Phys. Chem.* 1962;66:2235.
6. Strauss UP, Wineman PL. *J. Am. Chem. Soc.* 1958;80:2366.
7. Eisenberg A, Farb H, Cool LG. *J. Polym. Sci.* 1966;A-2,4:855.
8. Eisenberg A, Sasada T. In: J. Prins A, ed. *Physics of Non-crystalline Solids.* North-Holland; 1965:99–116.
9. Brandrup J, Immergut EH, eds. *Polymer Handbook.* New York: Wiley; 1989:VII/27, 43.
10. Peterson JK. *Thesis.* Columbus: Ohio State University; 1961.
11. Jameson RF. *J. Chem. Soc.* 1959:752–759.
12. Masson. *Colloids and Surfaces A Physicochemical and Engineering Aspects*, v. 121 issue 2–3, 1997, p. 247.

Polyphosphazenes

Harry R. Allcock

There are more that 700 different polyphosphazenes known, all of which have different engineering properties. Applications have been developed as hydrophobic and superhydrophobic materials, hydrogels, optical and photonic materials, solid and gel polymer electrolytes, high performance elastomers, membranes, and as fire resistant materials. Most are stable in the atmosphere and to aqueous media, and a few are bioerodible via hydrolysis mechanisms and are the basis of tissue engineering and drug delivery procedures.

In addition to the linear polymers with two organic or organometallic side groups attached to each phosphorus (derived by chlorine replacement from polymer 1), other architectures are known including block copolymers with organic polymer or organosiloxane blocks, star and dendritic structures, comb homo- and co-polymers, and polymers with cyclic phosphazene rings joined together through organic linkages as either cyclolinear or three-dimensional structures. In addition, related polymers are known that have carbon or sulfur in the backbone as well as phosphorus and nitrogen.

With this complexity it is not possible to describe here even a representative fraction of known polyphosphazenes in terms of their engineering properties. (Several classes of representative systems described in this Handbook, however, are "Poly(aryloxy)thionylphosphazenes", "Poly(phosphazene), bioerodible", "Poly(phosphazene) elastomer", and "Poly(phosphazene), semicrystalline".) Instead, it is recommended that the interested reader should consult two recent books that contain summaries of properties and structure-property relationships. These are "Inorganic Polymers", 2nd ed., by J. E. Mark, H. R. Allcock, and R. West, Oxford University Press, 2005, 338 pp.; and "Chemistry and Applications of Polyphosphazenes" by H. R. Allcock, Wiley-Interscience, 2003, 725 pp. This last book contains numerous tables and discussion sections about properties related to applications.

Poly(phosphazene), Bioerodible*

Joseph H. Magill

Acronym PPHOS

Class Polyphosphazene

Structures Poly[(*p*-methylphenoxy)-co-(ethylglycinto)phosphazene] (50/50: mole ratio)

Major Applications Polymers have shown promise as bioerodible materials capable of (controlled degradation and sustained drug delivery for therapeutic and other related uses.[1−14] Polyphosphazenes have been evaluated for approximately two decades, but research has become more focused in recent years.

Properties of Special Interest In general, tailored side groups (see the section on polyphosphazene synthesis below) enable a wide variety of hydrolytic properties to be designed into selected polymers for applicatons in biological environments for sustained drug administration without the release of harmful degradation products at physiological concentrations.[1,3,6−9] Limited modeling studies have been conducted.[14]

Synthesis Techniques and Types of Polymers Polymer with specific poly(phosphazene) structures that are susceptible to hydrolytic degradation under physiological conditions. Examples with glucosyl, amino acid ester, imidazolyl, glycerol side groups have been synthesized.[9] Besides this, side groups have also been grafted (through direct γ irradiation) onto PBFP polymers particularly for bicompatibility enhancement.[11,12]

In vitro evaluations have been made. Bioerodible poly(phosphazenes) have the advantage that the degradation products are biocompatible. The majority of bioerodible poly(phosphazenes) have been synthesized by the classical thermal procedure of Allcock et al. (1965) – reference (15). The copolymer in question is described.[1,3] In vivo performances in clinically relevant conditions are planned for PPHOS matrices.

---

* There is a paucity of tabulated release data on bioerodible polymers. Thus, the author has presented some results interpolated from graphical plots of controlled release for matrices at different Inulin, that is, $(C_6H_{10}O_5)_x$ loadings.[1] Different drug loadings and release rates were monitored in vitro and modulated through changes in pH. Surface inspections of the matrices were conducted by using surface-scanning electron microscopy techniques to characterize changes in texture.[1]

Poly(phosphazene), Bioerodible

Chemical structure and properties

Property	Units	Conditions	Value	Reference
Molecular mass (of repeat units)	$g\,mol^{-1}$	For the basic unit illustrated above	254.2	–
Typical molecular weight	$g\,mol^{-1}$	Variable for the same reasons; $M_w \sim 1 \times 10^6$ by GPC	–	–
Typical polydispersity M_w/M_n	–	Variable but usually broad for kinetic evaluations made to date on account of synthesis procedures that were employed		(2)

Release rates for 50/50 polyphosphazene copolymer[1]

Conditions	Time (days)	Value (% mass loss day^{-1})	Time (days)	Value (% mass loss day^{-1})
Inulin loadings at 40% in copolymer at pH = 2.0; S.D. ± 3; error <10%	0.0	0.0	5.1	96.4
	0.07	9.0	7.1	98.8
	0.1	25.4	9.0	98.9
	0.3	49.0	12.0	99.5
	1.0	68.6	15.0	99.6
	3.0	85.5	25.0	100.0
	4.0	89.4		
Ditto (at 10% loading)	0.0	0.0	7.2	37.8
	0.1	0.6	10.2	43.5
	0.5	4.8	13.1	47.3
	1.1	10.4	17.1	51.4
	2.2	18.3	21.1	54.5
	4.0	27.0	24.0	57.4
Ditto (at 1% loading)	0.2	0.84	10.2	29.3
	0.5	8.2	14.1	34.3
	1.1	11.9	18.0	38.8
	3.1	17.7	22.0	40.7
	6.2	23.9	24.0	42.3
Inulin loadings at 40% in copolymer at pH = 7.4; S.D. ± 3; error <10%	0.0	0.01	9.1	77.8
	0.1	0.5	12.0	78.6
	0.2	25.5	15.0	79.4
	0.5	42.1	18.0	79.9
	2.1	62.7	22.0	79.7
	4.2	70.5	24.0	79.5
	6.2	77.3		
Ditto (at 10% loading)	0.0	0.0	9.0	26.7
	0.2	4.4	12.0	28.7
	0.5	10.1	14.0	30.0
	1.0	14.4	17.0	32.0
	2.0	18.3	20.0	33.6
	4.2	21.5	24.0	35.0
	6.2	24.8		

Poly(phosphazene), Bioerodible

Conditions	Time (days)	Value (% mass loss day^{-1})	Time (days)	Value (% mass loss day^{-1})
Ditto (at 1% loading)	0.0	0.0	10.1	22.5
	0.6	7.9	13.1	23.9
	2.1	13.0	20.1	27.9
	4.2	16.0	22.1	28.7
	6.2	18.5	24.0	30.0
	8.1	20.7		
Inulin loadings at 40% in copolymer at pH = 10; S.D. ± 3; error <10%	0.0	0.0	7.1	88.1
	0.1	13.5	9.1	88.4
	0.4	43.0	13.1	88.7
	1.1	60.6	17.1	88.8
	3.0	77.8	24.1	89.2
	5.0	83.6		
Ditto (at 10% loading)	0.0	0.0	7.0	36.4
	0.5	8.9	10.1	38.4
	2.1	21.3	13.0	40.5
	2.1	29.3	16.0	42.5
	4.1	34.0	23.0	43.3
Ditto (at 1% loading)	0.0	0.0	11.0	33.6
	0.5	13.7	14.0	34.8
	2.2	18.0	17.1	35.7
	5.1	24.5	22.0	36.2
	8.1	29.4	24.0	36.2

References

1. Ibim SM, et al. *J. of Controlled Release.* 1996;40:31.
2. Davies BK. *Experiments.* 1972;28: 348; Langer R, Folkman J. *Nature.* 1976;261:797.
3. Allcock HR, et al. *Macromolecules.* 1977;10:824.
4. Laurencin CT, et al. *J. Biomed. Mater. Res.* 1987;21:1231.
5. In: Heller J. In: Salamone JH, ed. *Polymeric Materials Encyclopedia.* Vol 1. Boca Raton, Fla: CRC Press; 1996:600.
6. Heller J. *J. Ad. Drug Deliv. Revs.* 1993;10:163.
7. Chasin M, Langer R. *Biodegradable Polymers as Drug Delivery Systems.* New York: Marcel Dekker; 1990.
8. Cohen S, et al. *J. Amer. Chem. Soc.* 1990; 112:7832.
9. See for example: Allcock HR. In: Hatada K, Katayama T, Vogel O, eds. *Macromolecular Design of Polymeric Materials.* New York: Marcel Dekker; 1997.
10. Calciti P, et al. *Il Farmaco.* 1994:49: 69.
11. Lora S, et al. *Biomaterials.* 1994;15:937.
12. Carenja M, et al. *Radiation, Phys. Chem.* 1996;48:231.
13. Grommen JHL, Schacht EH, Mense EHG.*Biomaterials.* 1992;13:601.
14. Grolleman CWJ, et al. *J. of Controlled Release.* 1986;4:119.
15. Allcock HR, Kugel RL. *J. Amer. Chem. Soc.* 1965;87:4216.

Poly(phosphazene), Elastomer

Joseph H. Magill

Acronym, Trade Name PNF elastomer, EYPEL-F

Class Polyphosphazenes

Structure $-[-N=P-(OCH_2CF_3)(OCH_2(CF_2)CH_yF_x)-]_n-$
$(y = 0, F = 3, \text{ or } y = 1, F = 2)$

Major Applications Developmental quantities of PNF and EYPEL-F and other elastomers were manufactured in quantity for high-performance seals, collapsible storage tanks, O-rings, and vibration shock absorption mounts in military and other devices. The service life of these items is claimed to be relatively long and reliable.

Properties of Special Interest High-cost items for commercialization, but this is less critical where they have potential applications as biomaterials such as soft denture liners, blood-compatible parts (prostheses), drug-related release agents and the like. Other more mundane uses encompass fire-resistant paint additives, agrichemicals, and herbicides, proofing of textiles of diverse kinds, lubricants, and fire-resistant fluids (as low molecular weight and cyclic compounds), and many more possibilities.

Synthetic Techniques and Types of Synthesis (a) Thermal two-stage polymerization (ring-opening of hexachlorocyclotriphosphazene followed by nucleophilic substitution).[1] (b) Mixed nucleophiles have also produced useful elastomers[2-8] using the same two-step procedure. (c) Now better defined block and random polymers with elastomeric properties have been developed and characterized.[9,10]

Property	Value	Reference
Molecular mass of repeat unit	Variable, depending upon the side groups copolymer type and composition	–
Typical molecular weight	Variable, depending upon the side groups, copolymer type, and composition	–
Typical polydispersity index M_w/M_n	Variable (usually high and broad in thermal synthesis)	–
Solvents	Methyl isobutyl ketone, methyl ethyl ketone, acetone, dimethylformamide, tetrahydrofuran, 1-methyl-2-pyrrolidine, acetonitrile and related polar solvents	(6, 8, 11)
	Freon and freon ether type solvents are best for the more heavily fluorinated polymers	(8)
Nonsolvents	Hydrocarbons (aliphatic and aromatic) petroleum products, hydraulic fluids, water-glycol, aqueous ammonia, acetic acid, and the like	(6)

Poly(phosphazene), Elastomer

Mechanical properties

Property	Units	Conditions*	Value		Reference
Tensile modulus (100%)	MPa	–	1.4–10.5		(4, 6)
			Below T_g	Above T_g	
Dynamic modulus	–	PNF elastomer, radiation	(193 K)	(353 K)	(12)
Storage modulus	MPa	vulcanized, unfilled;	1,590	0.401	
Loss modulus	MPa	110 Hz	1,580	0.396	
Dynamic modulus	–	PNF elastomer, peroxide	–	0.0515	(12)
Storage modulus	MPa	vulcanized, (30 pph of FEP	2,520	4.56	
Loss modulus	–	carbon black)	2,520	62.2	
			4.48	86.8	
Dynamic modulus	MPa	PNF elastomer, peroxide	2,080	3.46	(12)
Storage modulus	–	vulcanized (30 pph Silanox	2,080	3.40	
Loss modulus	–	101 silica)	–	0.620	
Yield strain L/L_o	%	–	100–350		(4, 6)
Hardness	Shore D	–	40–90		(4, 6)
Tear strength	kN m^{-1}	–	7.0–17.5		(4)
Compression set	%	70 h at 423 K, in air	20–50		(4)
Flexible modulus	MPa	At 273 K	17.2		(4)
		At 233 K	44.8		
		At 200 K	222		
Flexural Gehman freeze point	K	ASTM D-1053	205		(4)

* Aging changes in mechanical behavior of PNF elastomers were reported with time, temperature, degree of cross-linking (radiation and chemical), and fluid and other environmental conditions for in-service evaluations ASTM and other practical tests were employed. Property changes and conditions are detailed in several references.[4,5,11]

Rheological measurements are expressed graphically as loss moduli G'' and storage moduli G' versus shear rate respectively.[4,6] Dynamic torsional braid analysis (TBA) spectra over a wide range of temperatures and several frequencies depicted significant transitional behavior in the region of T_g, T,[1] and beyond.[13]

Solution properties

Property	Units	Conditions	Value	Reference
Theta temperature Θ	K	Methyl isobutyl ketone	298	(8)
Interaction parameter χ	–	MIBK, 298 K	0.49	(8)
Second virial coefficient	mol cm^3 g^{-2}	DMF at 298 K; $M_w = 23$–126 ($\times 10^5$); $M_n = 3.2$–6.7 ($\times 10^5$)	2.8–11.0 [A2 = $(1.1 \times 10^{-3})M_w^{-0.35}$]	(8)
Mark-Houwink parameters: K and a	K = ml g^{-1} a = None	–	$K = 2.62 \times 10^{-3}$ $a = 0.52$	(8)

Property	Units	Conditions	Value		Reference
Huggins constants: k' and k''	–	MIBK; 298 K	$k' \times 10^2$	$k'' \times 10^2$	(8)
		THF	1.09	−2.08	
		MEK	1.47	−4.06	
		Acetone	0.89	−0.91	
		Acetonitrile	0.77	−0.98	
		DMF	0.46	−0.16	
			0.87	−0.28	
Characteristic ratio $\langle r^2 \rangle_0 / nl^2$	–	MIKB; 298 K	25–35		(8)
Persistence length	Å	MIKB; 298 K	42–64		(8)
Radius of gyration $\langle S^2 \rangle_z^{1/2}$	Å	DMF; 298 K; $M_w = 23\text{–}126\,(\times 10^5)$; $M_n = 3.2\text{–}6.7\,(\times 10^5)$	340–890		(3, 8)
		Solvent E2* (9.09% acetone) at 295 K; $M_n = 15.5 \times 10^4$; MW $= 6.8 \times 10^6$	870		(3)
		Solvent E2 at 295 K; MW $= 42.8 \times 10^4$; MW $= 10.0 \times 10^6$	930		(3)

* Solvent E2 is F–(CFCF$_3$CF$_2$O)$_2$CHFCF$_3$, manufactured by DuPont Freon Products Division, Wilmington, Delaware, USA.

Anomalous changes are frequently noted for fractions across the broad molecular weight distribution(s). For example, $k' + k'' \neq 0.5$ – see references (3) and (4). Intrinsic viscosity parameters as a function of percent acetone in E2 solvent are plotted in figure 3 of reference (3) for several PNF fractions. Many other values are tabulated by Hagnauer and Schneider in this reference along with many other solution parameters. Recently, the solution properties of polyphosphazenes have been critically reviewed.[14] Besides polymer quality, there have been problems with "tailing" in the fractionation of fluorinated polyphosphazenes as pointed out in reference (15). The quality polymers synthesized since the 1990s[9,10,16] should circumvent these problems that have been encountered with dilute solution and other kinds of characterization.

Stabilities: Flammability properties

Property	Units	Conditions	Value			Reference
Oxygen index:* LOI =	%	PNF sheet, mole %: Percent trifluoroethoxy/ fluoroalkoxy (65/35)	UF (Radiation vulcanized)	30 phr C (peroxide vulcanized)	30 phr silica (Peroxide vulcanized)	
$100[O_2]/[O_2] + [N_2]$	%		48	65	47	(17)
			48	–	–	(18)
Burn velocity†	mm s⁻¹	In 75% oxygen	1.65	0.05	0.7	(17)
Smoke density	Relative	Optical cell	~1	~2.5	~0.7	(17)

Poly(phosphazene), Elastomer

Property	Units	Conditions	Value			Reference
Average residue	Relative	At 773 K				(17)
		Air	4	26	26	
		N_2	6.5	28.5	29	
Residue	K	Temperature for 10% loss At 10 K min^{-1}				(17)
		Air	650	669	693	
		N_2	659	659	690	
		For 50% loss				
		Air	701	717	735	
		N_2	710	712	735	
Activation energy (degradation)	kcal mol^{-1}	Air	30.5	35.4	23.3	(17)
		Nitrogen	31.5	33.9	14.5	
Glass transition temperature	K	DSC	484	482	483	(17)

* Test = ASTM D2863.
† During burning, dripping may distort the result. These values fall sharply with increasing incident radiation (heat flux) on the specimen.

References

1. Allcock HR, Kugel RL. *J. Amer. Chem. Soc.* 1965;87:4216; Allcock HR, Kugel RL, Valan K. *J. Inorg. Chem.* 1966;5:1709.
2. Rose SH. *J. Polym. Sci., Polymer Letters.* 1968;6:837.
3. Hagnauer GL, Schneider NS. *J. Polym. Sci., Part A2,* 1972;10:699.
4. Kyker GS, Antkowiak TA. *Rubber Chem. Technol.* 1974;47:32.
5. Singler RE, Hagnauer GL, Sicka RW. *ACS Symp. Series.* 1982;193:229.
6. Tate DP. *J. Polym. Sci., Symp.* 1974;48:33. Tate DP, Antkowiak TA. In: Kroschwitz JI, ed. *In Kirk-Othmer Encyclopedia of Chemical Technology.* 3rd ed. Vol 10. New York: John Wiley and Sons; 1980:936.
7. Vicic JC, Reynard KA. *J. Appl. Polym. Sci.* 1977;21:3185.
8. Carlson DW, et al. *J. Polym. Sci, Polym. Chem. Edn.* 1976;14:1379.
9. White ML, Matyjaszewski K. *Macromol. Chem. Phys.* 1997;198:665.
10. Matyjaszewski K, White ML. In: Salamone JC, ed. *Polym. Mat. Encyl.* Vol 9. New York: CRC Press; 1996:6556.
11. Singler RE, Schneider NS, Hagnauer GL. *Polym. Eng. Sci.* 1975;15:321.
12. Choy IC, Magill JH. *J. Polym. Sci., Polym.Chem. Ed.* 1981;19:2495.
13. Connolly TM Jr, Gillham JK. *J. Appl. Polym.Sci.* 1975;19:2461.
14. Tarazona MP. *Polymer.* 1994;35:819.
15. Neilson RH, et al. *Macromolecules.* 1987;20:910.
16. Allcock HR, et al. *Macromolecules.* 1997;30:50.
17. Peddada SV, Magill JH. *J. Fire and Flamm.* 1980;11:63.
18. Lawson DF, Cheng TC. *Fire Research.* 1977–1978;1:223.

Poly(phosphazene), Semicrystalline

Joseph H. Magill

Acronyms, Alternative Name PBFP, PTFP, PBTFP, PFPN, PF, poly[(2,2,2,-trifluoroethoxy)phosphazene]

Class Polyphosphazenes

Structure $-[N{=}P(OCH_2CF_3)_2]_x-$

Major Applications Produced for many years in developmental quantities for evaluation in research and limited use in commercial tests and military applications; nonflammable fibers and films. Under evaluation for controlled drug delivery systems, hydrogels, implants, and membranes.

Properties of Special Interest Low-temperature flexibility and high-temperature stability, high oxygen index and low flame spread rate, hydrophobic (low surface tension), good carbon solvent resistance, biocompatibility, mesophase formation, and polymorphism.

Synthesis Techniques (a) Stokes[1] thermally polymerized hexachlorocyclotriphosphazene via a ring-opening process to provide a cross-linked elastomer, but it was not until 1965 that a high molecular poly(dichlorophosphazene) was isolated and subsequently transformed, via nucleophilic substitution, into thermally stable semicrystalline homopolymers.[2–4] This procedure was used widely to synthesize a variety until a few years ago, but it suffered from relatively low conversions (<70% so as to avoid cross-linking), unknown chain-end groups and lack of molecular weight control of the product. These difficulties obstructed its commercialization.

(b) Other thermally induced polymerization techniques have been developed employing Lewis acid catalysed solution polymerization[5–7] of the hexachlorocyclic monomer as well as by polycondensation of $Cl_3P{=}N-P(O)Cl_2^{[8]}$ and the thermal polymerization of phosphoranimines[9,10] to provide many alkyl and aryl substituted phosphazenes, but this procedure also has processing disadvantages.

(c) Still, well-defined poly(phosphazenes) with high conversions, known end-groups, and molecular weight control were first prepared less than ten years ago by employing the anionically initiated polymerization of phosphoranimines to produce well-defined homo, block, and random copolymers.[11–14]

(d) Recently, a living cationic polymerization of phosphoranimines with molecular weight control has been developed to produce polyphosphazenes of of similar quality to (c).[15–17]

Now that polymerization control has been established, these techniques may lead to cost-effective and new developments/applications in this interesting class of polymers. Some physical properties that are sensitive to structure and chain conformations may require further investigation. Some of these polyphosphazenes are to be found among the polymers that follow.

Poly(phosphazene), Semicrystalline

Property	Units	Conditions	Value
Molecular mass of repeat unit	g mol^{-1}	–	243.04
Typical molecular weight	g mol^{-1}	Daltons	$<2 \times 10^3$ to 3×10^7
Typical polydispersity index (M_w/M_n)*	–	–	$\sim$1.2–20

* Low polydispersity polyphosphazenes (P.I. between 1 and 2) are novel and were first synthesized less than a decade ago.

Morphology[18]

Property	Units	Conditions	Value
Birefringence (spherulites and mesophase moieties)*	–	Relative values measured on stepwise (a) heating/(b) cooling of thin solution-crystallized negative spherulites:	
		(a) Heating; orthorhombic (folded chains)	
		293 K	1.064
		318 K	1.121
		325 K	1.179
		328 K	1.250
		333 K	1.267
		338 K	1.297
		Mesophase (2-dimensional chain, extended)	
		393 K	3.298
		413 K	3.078
		433 K	3.500
		(b) Cooling; (2-dimensional chain, extended)	
		413 K	3.540
		402 K	3.345
		393 K	3.148
		373 K	3.308
		353 K	3.314
		(b) Orthorhombic (3-dimensional chain, extended)	
		332 K	2.783
		322 K	2.640
		313 K	2.756
		293 K	2.540

* The pattern observed here is analogous to that encountered in dilatometry measurements (see phase transitions in the tables on *Transition temperatures* below). The birefringence is always negative in sign but increases in magnitude as the morphology changes from the solution cast chain-folded to the columnar chain-extended pseudohexagonal form (see the table on *Crystalline-state properties* below). The birefringence stabilizes upon cycling (heating/cooling) and subsequently follows a set pattern after a few cycles. Hereafter, the pathway is reversible between orthorhombic to/from the mesophase during heating and cooling. Crystallization directly from the melt produces smectic and batonnet morphologies of high crystallinity, not classical spherulites (of 35–50% crystallinity) that only grow from moderately concentrated polymer solutions.

Properties of unique block and random methoxyethoxy (MEO)/(trifluoroethoxy) (TFO) phosphazene copolymers*[19]

Random	m	$M_w \times 10^{-3}$	M_w/M_n	T_g (K)[a]	$T_{(1)}$ (K)[b]	$\Delta H_{T_{(1)}}$ (J g^{-1})[c]	T_m (K)[d]	Density (g cm^{-3})[e]
Mole fraction	0.000	99.0	1.46	–	356	47.0	495	–
"m" of MEO	0.028	94.7	1.78	215	344	40.5	480	1.690
	0.053	77.0	1,92	212	335	36.0	464	1.612
Block								
Mole fraction	0.126	65.8	1.60	–	315	21.9	446	1.431
"m" of MEO	0.047	173.7	1.36	220	353	46.2	493	1.690
	0.080	178.2	1.56	211	341	38.8	478	1.664
	0.126	77.1	1.57	–	339	27.0	455	1.591
	0.134	62.2	1.34	211	321	21.7	421	1.583

* Side group placement and composition of methoxyethoxy produces a wide variety of properties that range from crystalline low "m" (<0.126) with morphologies akin to the PBFP homopolymer, to higher "m" (>0.134) where these crystalline transformations cease to exist. All polymer exhibit thermotropic behavior. Anionically initiated polymerization of $(CH_3OCH_2CH_2O)(CF_3CH_2O)_2P{=}NSi(CH_3)_3$ followed by the addition of $(CF_3CH_2O)_3P{=}NSi(CH_3)_3$, except for random copolymers where each of two polymerizations were conducted concomitantly.

[a] T_g = glass transition temperature.
[b] $T_{(1)}$ = mesophase transition temperature.
[c] $\Delta H_{T_{(1)}}$ = enthalpy of $T_{(1)}$ transition.
[d] T_m = melting temperature; values were determined optically.
[e] At 25°C via flotation in CsCl solution.

Spectroscopic properties

Property	Units	Conditions	Value	Reference
UV (characteristic absorption frequencies)	µm	Electronic spectra absorption	220; 400 weak 270–280 diffuse	(20, 21)
IR (characteristic frequencies)	cm^{-1}	FT-IR	1,420; 965; 880	(22, 23)
		P—O—C		
		P—N	1,280	
		P—O—P	870–1,000	
		Stretching and bending vibrations (for cross-linking) —P=N—stretching	1,250–1,330	
FT-IR (Nicolet 5DXB)	cm^{-1}	Solid (well-defined) PBFP polymer by KBr		(24)
		Aliphatic CH	2,981	
		CH	1,462	
		P—O—C	1,427	
		P=N (br)	1,271–1,308	
		P=O	1,173; 963	
		C—O	1,089	
NMR	ppm	^{31}P	7.5	(25)
(Solution)		Dipolar-^1H	−6.9	
(Brüker AM 500)		Decoupled ^{13}C	120.7–127.3	
202 MHz for ^{31}P and		^{19}F (3 atoms)		

Property	Units	Conditions	Value	Reference
125 MHz for ^{13}C measurements		^{1}H dipolar-coupled ^{13}C (reference peak) TMSi (reference 0 ppm)	64.3 -6.9	

Property	Conditions	Value	Reference
NMR (solid state); ^{1}H dipolar decoupled; ^{13}C (MAS at 2–4 kHz); no decoupling for ^{31}P spectra; reference H_3PO_4 (0 ppm); Brüker MSL 300; analysis made at 121.5 MHz for ^{31}P and 75.5 MHz for ^{13}C	Solution crystallized sample: $M_w = 300,000$; $M_w/M_n = 2.3$; $T_m = 515$ K, $T_g = 207$ K; heating/cooling spectra recorded/10 K stepwise from 293–373 K (i.e., through $T_{(1)}$ mesomorphic transition)	Two peaks (mobile/immobile) in the ^{31}P spectra below $T_{(1)}$ and one mobile narrower peak about this transition, where side groups and chains are all mobile; above $T_{(1)}$ the phase is 2-D pseudohexagonal; below $T_{(1)}$ a 3-D highly crystalline form exists	(26)
NMR (wide line)	^{1}H, ^{13}C, and ^{19}F nuclei studied under stepwise heating/cooling; $T_{(1)} = 353$ K; $T_m = 513$ K; intrinsic viscosity $= 1.06$ dl g^{-1} in THF, 298 K	Rotating backbone in a hexagonal lattice above $T_{(1)}$ and enhanced side group motions; single narrow line $\sim$0.4 Oe for ^{19}F; $\sim$0.6 Oe for ^{1}H; ^{31}P narrows to 1.1 Oe as compared to 2.4 Oe at 20°C indicating rigidity	(27)
Spin-lattice ^{1}H NMR relaxation times	Semicrystalline sample; $M_w = 2 \times 10^5$; $M_w/M_n = 1.75$; $T_{(1)}$ of cast film from THF = 348 K; measurement range = 303–443 K	The ^{13}C measurements made through $T_{(1)}$ for CH_2 and CF_3 side groups are 1.75 and 3.55 respectively with activation energies of $\sim$17.3 and 13.7 kJ mol^{-1} obtained from ln τ versus T (K^{-1})	(25)
NMR (solid echo), 90 MHz	Semicrystalline sample; $M_w/M_n = 1.75$; film cast from THF at 348 K; range of measurement = 303–443 K	Molecular motions above and below $T_{(1)}$ fitted with Weibull functions using τ_2 relaxation values below and above the $T_{(1)}$ transition	(28)

Equations of state

Property	Units	Conditions	Value	Reference
Thermal expansion coefficient	K^{-1} ($\times 10^4$)	X-ray		(29–31)
		(i) Solution cast film	1.74	
		(ii) Melt cast film	2.7	
		a-axis expansion (linear)	1.8	

Property	Units	Conditions	Value	Reference
		Dilatometry: semicrystalline crystalline orthorhombic phase (volume)	$\alpha_c = 2.48$	(31)
		Mesophase ($T_{(1)}$ to T_m)	$\alpha_t = 6.99$	(32)
		(volume)	$\alpha_t = 6.24$	(32)
		Isotropic ($\geq T_m$)	$\alpha_1 = 9.25$	(31)
		(volume)	$\alpha_1 = 8.67$	(32)
		Monoclinic (initially below $T_{(1)}$) (volume)	$\alpha_m = 6.83$	(25)
Volume change in transition regions	$\Delta V\%$	Dilatometry		(25, 31)
		$T_{(1)}$ transition (orthorhombic to mesophase)	~6	
		T_m transition (mesophase to isotropic)	~5–6	
		$T_{(1)}$ transition (monoclinic to mesophase)	~3	
Thermo-mechanical analysis (TMA)	$\Delta V\%$	$T_{(1)}$ transition	~5	(31–33)
		T_m transition	~6	
Thermal (volume) expansion coefficient	K^{-1} ($\times 10^{-4}$)	Solution cast α-form,* (monoclinic) below $T_{(1)}$	7.5	(25)
Density*	g cm^{-3}	α-form		(25)
		303 K	1.665	
		311 K	1.655	
		321 K	1.643	
		331 K	1.632	
		334 K	1.629	
		338 K	1.621	

* Densities are also presented graphically for the δ-hexagonal (columnar) phase and the chain-extended γ-orthorhombic form of high crystallinity and the isotropic phase above T_m under conditions of heating and cooling. See the corresponding birefringence and transitional data under corresponding temperatures in the table on *Morphology* above.

Pressure properties

Property	Units	Conditions	Value		Reference
Compressibility coefficient β	Bar^{-1} ($\times 10^{-4}$)	From P—V—T data* Pressure (MPa)	$< T_{(1)}$ (at 298 K)	$> T_{(1)}$ (at 498 K)	(34)
		0.1	0.46	1.12	
		50	0.42	0.95	
		100	0.37	0.71	
		200	0.33	0.59	
		300	0.25	0.37	
		400	0.21	0.28	
		650	0.19	0.22	
Thermal expansion coefficient α	K^{-1} ($\times 10^{-4}$)	Graphs for pressures = 0.1–700 MPa; 298–460 K (approximate)	$< T_{(1)} = 2.4$ At $T_{(1)} = 7.0$ $> T_{(1)} = 9.2$		(34)
Gruneisen parameter	–	At 303 and 220 K	4.6		(34)

Property	Units	Conditions	Value	Reference
Density	g cm^{-3}	As a function of pressure and temperature	Data represented graphically	(34)
Compression modulus K_v	bar or (GPa)	Pressure dependence of compression modulus	Plotted as function of pressure up to 600 MPa through $T_{(1)}$ transitions	(34)

* $M_w \sim 1 \times 10^6$; $M_w/M_n \sim 1.2$; solid, mesophase, and liquid states at various pressures and temperatures (interpolated).

Solution properties*

Property	Units	Conditions	Value	Reference
Theta Θ temperature	K	Tetrahydrofuran	298	(20)
Interaction parameter χ_1	–	In graphical form	Estimates	(36)
Second virial coefficient	mol cm^3 g^{-2}	THF, 298 K, $M_w = 1.48 \times 10^6$	1.0×10^{-3}	(23)
		Acetone, 298 K, $M_w = 1.54 \times 10^6$	5.1×10^{-4}	(23)
		Cyclohexanone, 298 K, $M_w = 1.42 \times 10^6$	6.7×10^{-5}	(23)
		Cyclohexanone, 298 K, $M_w = 2.92 \times 10^{-5}$, $M_n = 1.3 \times 10^5$	4.7×10^{-5}	(37)
Mark-Houwink parameters: K and a	K = ml g^{-1} a = None	THF (details not given since coefficients change across the MW distribution)	$K = 620$ $a = 0.85$	(37)
Huggins constant k_H	–	THF, 298 K, $M_w = 1.48 \times 10^6$	~ 0.02	(23)
		Acetone, 298 K, $M_w = 1.54 \times 10^6$	0.30	
		Cyclohexanone, 298 K, $M_w = 1.42 \times 10^6$	0.42	
Intrinsic viscosity $[\eta]$	–	Fractions in methyl isobutyl ketone (MIBK)	(4.89×10^{-3})	(37)

* Anomalous behavior is noted among some of the solution properties since polymers are not always well-defined.

Crystalline-state properties*[29,30,38]

Comments	Lattice	Monomers Per Unit Cell	Unit Cell Dimensions			Cell Angles			Crystal Density (g cm^{-3})
†			a	b	c	α	β	γ	
Form α	Orthorhombic	2	10.14	9.35	4.86	–	–	–	1.748
Form β	Monoclinic	2	10.03	9.37	4.86	–	91°	–	1.767
Form γ	Orthorhombic	4	20.60	9.40	4.86	–	–	–	1.715
Form δ	Hexagonal	(?)	d(100)γ	10.3(200°C)		–	–	–	1.354 (estimate)

* Crystalline modifications (semicrystalline polymorphic states and mesophase). See the expansion coefficient as a function of temperature in the table on *Equations of state* above.

† Form α = chain-folded from THF solution; Form β = low molecular weight from pseudohexagonal (columnar) mesophase; Form γ = melt quench from isotropic melt 250°C to room temperature as chain-extended orthorhombic form; Form δ = low molecular weight from pseudohexagonal (columnar) mesophase.

Property	Units	Conditions	Value	Reference
Crystallization kinetics	–	Isothermal growth rate and form depend upon undercooling measured by DSC and polarized light transmission method (also known as the DLI technique)	See reference for details	(39)
Avrami exponent	–	Transformations kinetics for:		(39)
		(1) Isotropic to (2-D) mesophase; that is, sub-T_m (K)	2	
		(2) 2-D mesophase to orthorhombic (3-D); that is, sub-$T_{(1)}$ (K)	2	

Transition temperatures*

Property	Units	Conditions	Value	Reference
Glass transition temperature T_g	K	Differential scanning calorimetry	207	(3, 25, 33, 36, 40–42)
		Dynamical mechanical	–	(43)
		Torsional braid analysis	220	(44)
Mesophase phase transition $T_{(1)}$	K	Dilatometry	338–365	(24, 30, 47)
		DSC and TMA	365	(33, 36, 40)
		DSC	339–363	(44–46)
Mesophase phase transition $T_{(1)}$	K	Dielectric analysis	(See graphed data in appropriate section below)	
		Torsional braid analysis	~331	(44)
		Creep compliance	338–343	(30)
		DLI (transmitted light)	–	(39, 46)
$T_{(1)}$ relationship equilibrium $T_{(1)}^0$	K	Best fit to data (estimated)	$T_{(1)} = 371 - (1{,}288)M_w^{-0.37}$ $T_{(1)}^0 = 371$	(41)
Melting (isotropization) transition (T_m)	K	DSC	515 513–515 519, 522	(All of the above and more)
Melting temperature ($T_m \equiv T_i$)	K	"Fit" to relevant experimental data	$T_m = 539 - 1{,}904 M_w^{-0.39}$	(41)
Equilibrium melting temperature	K	–	$T_m^0 = 539$	(44)
T_g, $T_{(1)}$, and T_m interrelationship	K	"Fitted" to oxy-type polymers from a linear plot of $(T_m - T_g)/(T_m - T_{(1)})$ vs. $T_{(1)}/T_m$	$3.2(T_m)^2 - [T_g + 8.2T_{(1)}]$ $T_m + 6[T_{(1)}]^2 = 0$	(25, 40)

Property	Units	Conditions	Value	Reference
		For all data including trifluoroethoxy/alkoxy copolymers	$2.4(T_m)^2 - [T_g + 6.2T_{(1)}]$ $T_m + 4.7[T_{(1)}]^2 = 0$	(41)

* All transition temperatures depend on factors such as sample MW and conditions of measurement. Note that α-, β-, and mesophase transition values depend upon the measurement method, molecular weight, and specimen history. Consult references for the techniques employed. For example, some authors claim (with good reason) that dynamical techniques are only related to classical dilatometry (1^0 min^{-1}) results. Logically, all comparisons should be on similar time scales.

Mechanical properties

Property	Units	Conditions	Value			Reference
Tensile modulus	MPa	$M_\eta = 2.97 \times 10^6$; $M_w/M_n < 1.4$	Unlisted			(48)
Tensile strength	MPa	Solution cast film	196			(48)
Elongation at break	%	Solution cast film	700			(48)
Some typical values in the transition regions†			$< T_g$ (165 K)	$< T_{(1)}$ (312 K)	$> T_{(1)}$ (380 K)	(43)
Dynamic modulus $(E)^*$	MPa	Unoriented cast film; $d = 1.695$ g cm^{-3}; $M_w > 10^6$; 110 Hz	1,170	130	11.8	
Storage modulus E'	MPa	Unoriented cast film; $d = 1.695$ g cm^{-3}; $M_w > 10^6$; 110 Hz	1,170	130	11.8	
Some typical values in the transition regions†			$< T_g$ (165 K)	$< T_{(1)}$ (312 K)	$> T_{(1)}$ (380 K)	(43)
Loss modulus E''	MPa	Unoriented cast film; $d = 1.695$ g cm^{-3}; $M_w > 10^6$; 110 Hz	29.0	19.1	0.89	
Dynamic modulus E^*	MPa	Same film oriented $\times 9$; $d = 1.692$; 110 Hz	8,670	455	76.8	
Storage modulus E'	MPa	Same film oriented $\times 9$; $d = 1.692$; 110 Hz	8,670	455	57.7	
Loss modulus E''	MPa	Same film oriented $\times 9$; $d = 1.692$; 110 Hz	971	12.9	9.28	

† Thermo-mechanical spectra have been measured from 153 to 413 K at 3.5 and 110 Hz respectively; only selected values are presented here.

Electrooptical and magnetic properties

Property	Units	Conditions	Value	Reference
Index of refraction n	–	PBFP in ethyl acetate solution; $M_w = 18 \times 10^6$; $[\eta]_{THF} = 410$	1.37	(49)
Refractive index increment dn/dc	g mol^{-1}	PBFP in ethyl acetate solution; $M_w = 18 \times 10^6$; $[\eta]_{THF} = 410$	0.004	(49)

Property	Units	Conditions	Value	Reference
Dielectric constant ε'	–	ε' plots from 100 Hz to 100 kHz in the range 78–393 K	See graphs	(50, 51)
Dielectric loss ε''	–	ε'' plots from 100 Hz to 100 kHz in the range ~78–393 K	See graphs	–
Dielectric strength	V mil^{-1}	–	360	(52)
Dipole moment of monomer unit μ_{110}	Debye	–	9.0	(49)
Optical anisotropy (segmental) $(\alpha_1 - \alpha_2)$	cm^3	*Cis-trans* conformation (assumed)	160×10^{-2}	(49)
Kerr constant	cm^{5}g^{-1} (300 V)$^{-1}$	Electric birefringence in EtOAc solution	7.0	(49)
Shear optical coefficient $[n]/[\eta]$	cm^{3}g^{-1}	Dynamic birefringence in solution	12	(49)
Relaxation time τ	s	–	2–9 ($\times 10^{-4}$)	(49)

Surface and interfacial properties

Property	Units	Conditions	Value	Reference
Surface tension	mNm^{-1}	Contact angle (microscopy)	16	(53)
		Zisman plots	16.5	(53)
		Contact angle (degradation after prolonged UV irradiation)	16.5–14.4	(54)
Interfacial free energy	erg^2 cm^{-4}	From isothermal crystallization studies	30	(39)
	erg cm^{-2}	At (2-D to 3-D) interface	~10 (estimate)	

Optical properties

Property	Units	Conditions	Value		Reference
Refractive index increment dn/dc	ml g^{-1}	Tetrahydrofuran at 298 K	0.023		(55)
			0.0233		(41)
			0.0232		(23)
		Cyclohexanone at 298 K	0.053		(23)
		Acetone at 298 K	0.019*		(23)
		Freon* E-2/acetone: 10/1 (v/v)	0.048		(56)
Refractive index n	–	Tetrahydrofuran at 298 K	1.405		(23)
		Acetone at 298 K	1.360		
		Cyclohexanone at 313 K	1.448		
Intrinsic viscosity $[\eta]$	(dl g^{-1})	Acetone with TBAN† at 298 K TBAN (mol)	$[\eta]$	k'	(23)
		0.0	3.70	0.03	
		0.01	2.02	0.29	
		0.02	2.04	0.19	
		0.05	2.07	0.18	

Property	Units	Conditions	Value	Reference
Radius of gyration	Å	$MW = 3.0 \times 10^6$	~660	–
Power factor	–	Frequency = 10^2–10^6 Hz	10^{-3} to 40×10^{-3}	(42, 52)

* Freon E2 is — (CFCF$_2$O)2 — CHFCF$_2$, from DuPont Freon Products Division, Wilmington, Deleware, USA.

† Tetrabutyl ammonium nitrate used as an "aggregate breaker." Other salts have been employed to prevent "tailing" in GPC analysis.[55]

Degradation stabilities

Property	Conditions	Value	Reference
Thermal degradation:* poly(tri-fluoroethoxy-phosphazene) (homopolymers)	Polymer made by ring-opening thermal synthesis: polymerization (MW uncontrolled, chain-ends unknown)	Depolymerization to cyclics, followed by chain scission at weak points (i.e., defects in the backbone) followed by rapid depolymerization to cyclic oligomers	(57)
		Random chain scission followed by partial unzipping of fragments to cyclic oligomers	(58)
		Depolymerization by chain scission and subsequently partial unzipping with some chain end initiation; reaction order of 0.8 proposed	(59)
		Initiation occurs at chain ends with subsequent depolymerization and chain transfer; some chain scission occurs at weak points in the backbone	(60)
		Two stage initiation, followed by backbone rearrangement and subsequently chain scission at resultant weak links within the backbone	(61)
Homo- and copolymers†	Polymer made by anionically initiated polymerization with MW, chain-end and chain-sequence control with defect-free chains	Depolymerization with chain end initiation followed by complete unzipping to cyclic trimer by a cationic mechanism; stability of the copolymers decreases by incorporating and increasing alkoxyalkoxy side group	(62)

* Phosphazene polymers with halogenated side chains give rise to toxic gaseous products, based upon "overall hazard rating," ALH, involving thermal stability, flammability, and toxicity parameters (RD_{50} and LC_{50}). Halogen-free polyphosphazenes are preferred over halogen-containing polymers for high-temperature applications. For example, see reference (64) for graphical details and analysis.

† The results represent the first thermal degradation study that has been conducted on well-defined polyphosphazenes. They also provide an unambiguous answer to the actual mechanism of degradation in these polymers.

Transport properties

Property	Units	Conditions	Value	Reference
Permeability P*: (sorption and time-lag techniques)	$cm^3(STP)cm\ cm^{-1}\ s^{-1}$ $Pa^{-1}\ (\times 10^{12})$	Semicrystalline film: $T_g = 198$ K; $T_m = 491$ K; $T_{(1)} = 343$ K; $\alpha = 6.24 \times 10^{-4}\ K^{-1}$.		(63)
		Permeant gas (298 K)		
		He	7.10	
		Ne	3.15	
		Ar	2.05	
		Kr	1.93	
		Xe	1.75	
		H_2	4.74	
		O_2	2.66	
		N_2	1.10	
		CO_2	1.47	
		N_2O	1.62	
		CH_4	1.43	
		C_2H_6	1.47	
		C_2H_4	2.69	
		C_3H_8	1.25	
		Permeant gas (300 K)		
		O_2	1.50	
		N_2	1.21	
	$cm^3(STP)cm\ cm^{-1}\ s^{-1}$ $Pa^{-1}\ (\times 10^{11})$	Mesophase, above $T_{(1)}$; $\alpha = 8.67 \times 10^{-4}$ (K).		(32)
		Permeant gas (348 K)		
		He	3.93	
		Ne	2.09	
		Ar	1.97	
		Kr	2.04	
		Xe	2.11	
		H_2	3.48	
		O_2	2.26	
		N_2	1.25	
		CO_2	8.78	
		CH_4	1.64	
		C_2H_6	1.68	
		C_2H_4	2.68	
		C_3H_8	1.44	
Permeability coefficient P	$cm^3\ cm\ cm^{-3}\ s^{-1}$ $cm^{-1}Pa^{-1}(\times 10^6)$	Semicrystalline solution cast film; $M_n \sim 3 \times 10^7$; $M_w/M_n < 1.4$.		(48)
		Permeant gas — Activation energy $(kJ\ mol^{-1})$		
		$He^\dagger$ — 15.8	17.2	
		$Xe^\dagger$ — 24.8	3.45	
		$O_2^\dagger$ — 21.5	5.40	
		$N_2^\dagger$ — 24.3	2.33	

Property	Units	Conditions		Value	Reference
		Permeant gas	$(kJ\ mol^{-1})$		
		$CO_2^{\dagger}$	13.1	27.7	
		$CH_4^{\dagger}$	25.4	2.7	
		$He^{\ddagger}$	15.7	68.6	
		$Xe^{\ddagger}$	15.2	31.6	
		$O_2^{\ddagger}$	16.3	37.5	
		$N_2^{\ddagger}$	19.3	21.3	
		$CO_2^{\ddagger}$	11.2	108.6	
		$CH_4^{\ddagger}$	16.3	28.1	
Diffusivity coefficient D: (sorption and time-lag techniques)	$cm^2\ s^{-1}\ (\times 10^7)$	Semicrystalline film; $d = 1.707\ g\ cm^{-3}$; $T_g = 198\ K$; $T_m = 491\ K$; $T_{(1)} = 343\ K$; $\alpha = 6.24 \times 10^{-4}\ (K)$.			(63)
		Diffusant gas (298 K)			
		He		343.0	
		Ne		438.5	
		Ar		21.61	
		Kr		10.52	
		Xe		4.46	
		H_2		161.8	
		O_2		27.83	
		N_2		17.15	
		CO_2		12.66	
		N_2O		13.42	
		CH_4		11.30	
		C_2H_6		3.61	
		C_2H_4		5.91	
		C_3H_8		1.29	
		Mesophase above $T_{(1)}$ transition; $\alpha = 8.67 \times 10^{-4}\ (K)$.			(32)
		Diffusant gas (348 K)			
		He		771.0	
		Ne		338.0	
		Ar		130.0	
		Kr		54.6	
		Xe		50.8	
		H_2		540.0	
		O_2		154.0	
		N_2		122.0	
		CO_2		97.3	
		CH_4		97.9	
		C_2H_6		47.4	

Property	Units	Conditions	Value	Reference
Diffusivity coefficient D: (sorption and time-lag techniques)	cm^2 s^{-1} ($\times 10^7$)	Mesophase above $T_{(1)}$ transition; $\alpha = 8.67 \times 10^{-4}$ (K). Diffusant gas (348 K)		
		C_2H_4	66.5	
		C_3H_8	25.1	
Solubility coefficient S	cm^3(STP) cm^{-3} Pa^{-1} ($\times 10^7$)	Semicrystalline film; $d = 1.707$ g cm^{-3}; $T_g = 198$ K; $T_m = 491$ K; $T_{(1)} = 343$ K; $\alpha = 6.24 \times 10^{-4}$ (K). Solubilant gas (298 K)		(63)
		He	2.07	
		Ne	2.72	
		Ar	9.47	
		Kr	18.36	
		Xe	39.2	
		H_2	2.93	
		O_2	9.55	
		N_2	6.44	
		CO_2	11.63	
		N_2	12.07	
		CH_4	12.62	
		C_2H_5	40.85	
		C_2H_4	45.4	
		C_3H_8	97.0	
		Mesophase above $T_{(1)}$ transition. Solubilant gas (348 K)		(32)
		He	5.10	
		Ne	6.20	
		Ar	15.15	
		Kr	37.35	
		Xe	41.4	
		H_2	6.44	
		O_2	14.7	
		N_2	10.2	
		CO_2	90.0	
		CH_4	16.65	
		C_2H_6	35.40	
		C_2H_4	40.20	
		C_3H_8	57	

* Measurements have also been reported – reference (48) – for some of these same gaseous permeants in the range 293–403 K for an HMW polymer ($\sim 10^7$ Daltons).

† Permeant gas, above $T_{(1)}$, $K > 350$.

‡ Permeant gas, below $T_{(1)}$, $K < 350$.

References

1. Stokes HN. *Amer. Chem. J.* 1897;19:782.

2. Allcock HR, Kugel RL. *J. Amer. Chem. Soc.* 1965;87:4216.

3. Allcock HR, Kugel RL, Valan K. *J. Inorg. Chem.* 1966;5:1709.

4. For an overview of the field, see for example: Allcock HR. In: Mark JE, Allcock HR, West R, eds. *Inorganic Polymers.* Englewood Cliffs, N.J: Prentice Hall; 1992:chap. 3.

5. Mujumdar AN, et al. *Macromolecules.* 1990;23:14.

6. Lee DC, et al. *Macromolecules.* 1986;19:1856.

7. Sennett S, et al. *Macromolecules.* 1986;19:959.

8. D'Hallium G, et al. *Macromolecules.* 1992;25:1254.

9. Neilson RH, Wisian-Neilson P. *J. Macromol. Sci.-Chem.* 1981;A16:425.

10. Neilson RH, Wisian-Neilson P. *Chem. Rev.* 1988;88:541.

11. Montague RA, Burkus II F, Matyjaszewski K. *ACS Polym. Prepr.* 1993;34(1):316.

12. White ML, Matyjaszewski K. *J. Polym. Sci., Part A: Chemistry.* 1996;34:277; *Makromol. Chem. Phys.* 1977;190:665.

13. White ML, Matyjaszewski K. *J.M.S.: Pure Appl. Chem.* 1995;A32(6):1115.

14. Matyjaszewski K. *J. Amer. Chem. Soc.* 1990;112:6721.

15. Honeyman CH, et al. *J. Amer. Chem. Soc.* 1995;117:7035.

16. Allcock HR, et al. *Macromolecules.* 1996;29:7740.

17. Allcock HR, et al. *Macromolecules.* 1997;30:50.

18. Magill JH, Petermann J, Reick U. *Colloid and Polym. Sci.* 1986;264:570.

19. Kojima M, et al. *Macromol. Chem. Phys.* 1994;195:1823.

20. Allcock HR. *Phosphorus-Nitrogen Compounds.* New York: Academic Press; 1972:21.

21. Hiraoka H, et al. *Macromolecules.* 1979;12:753.

22. Ferrar WT, Marshall AS, Whitefeld T. *Macromolecules.* 1987;20:357.

23. Mourey TH, et al. *Macromolecules.* 1989;22: 4286.

24. Montague RA, Green JB, Matyjaszewski K. *J.M.S.: Pure Applied Chem.* 1997;A32:497.

25. Young SG, et al. *Macromolecules, Polymer.* 1992;15:3215.

26. Young SG, Magill JH. *Macromolecules.* 1989;22:2549.

27. Alexander M, et al. *Macromolecules.* 1977;10:721.

28. Saito K, Masuko T. *Polym. Comm.* 1986;27:299.

29. Desper CR, Singler RE, Schneider NS. *IUPAC Symposium,* Amherst, Mass., 12–16 July 1982, p. 682.

30. Kojima M, Magill JH. *Makromol. Chem.* 1985;186:649.

31. Masuko T, et al. *Macromolecules.* 1984;17:2857.

32. Mizoguchi K, Kamiya Y, Hirose T. *J. Polym. Sci., Part B, Polym. Phys.* 1991;29:695.

33. Desper NS, et al. In: Carraher CE Jr, Sheats JE, Pittman CU Jr, eds. *Organometallic Materials.* New York: Academic Press; 1978.

34. Dreval VE, et al. *Polym. Sci. Ser. A.* 1995;37:179.

35. Aharoni SM. *Polym. Preprints, Amer. Chem. Soc.* 1981;22(1):116.

36. Allen G, Lewis CJ, Todd M. *Polymer.* 1970;11:44.

37. Tate DP. *J. Polym. Sci. Symp.* 1974;48:33.

38. Kuptsov SA, et al. *J. Polym. Sci.* 1993;35(5):635.

39. Ciora RJ Jr, Magill JH. *Macromolecules.* 1990;23:2350.

40. Zadorin AN, et al. *Polym. Sci., Series A.* 1994;75.

41. White ML. *Ph.D. Thesis in Chemistry.* Carnegie Mellon University, August 1994.

42. Singler RE, Schneider NS, Hagnauer GL. *Polym. Eng. Sci.* 1975;15:321.

43. Choy IC, Magill JH. *J. Polym. Sci., Polym. Chem. Ed.* 1981;19:2495.

44. Connelly TM Jr, Gillham JK. *J. Applied Polym. Sci.* 1976;20:473.

45. Sun DC, Magill JH. *Polymer.* 1987;28:1243.

46. Schneider NS, Desper CR, Beres JJ. In: *Liquid Crystalline Order in Polymers.* New York: Academic Press; 1978:chap. 9, p. 299.

47. Masuko T, et al. *Macromolecules.* 1989;22:4636.

48. Starannikova LE, et al.*Vysokomolekulyarnye-Soedineniya, Ser. A & B.* 1994;36(11):1906.

49. Rjumtsev EI, et al. *Eur. Polym. J.* 1992;28:1031.

50. Uzaki S, Adachi K, Kotaki T. *Polym. Joural.* 1988;20:221.

51. Murakami I, et al. *J. Inorg. and Organometallic Polym.* 1992;2:255.

52. Reynard KA, Gerber AH, Rose SH. *Synthesis of Polynitrilic Elastomers for Marine Applications.* Horizons Inc., Cleveland, Naval Ship Engineering Center, AMMRC CTR, 72-29, December 1972, (AD 755188).

53. Allcock HR, Smith DE. *Chem. Mat.* 1995;7:1469.

54. Reichert WM, Filisko FE, Barenberg SA. *J. Biomed. Mat. Sci.* 1982;16:301.

55. Neilson RH, et al. *Macromolecules.* 1989;22:4286.

56. Hagnauer GL, Schneider NS. *J. Polym. Sci., Part A2.* 1972;10:699.

57. Allcock HR, Cook WJ. *Macromolecules.* 1974;7:284.

58. MacCallum JR, Tanner JR. *J. Macromol. Sci. Chem.* 1970;A4(2):481.

59. Zeldin M, Jo WH, Pearce EM. *Macromolecules.* 1980;13:1163.

60. Peddada SV, Magill JH. *Macromolecules.* 1983;16: 1258.

61. Papkov VS, et al. *J. Polym. Sci. USSR.* 1989;31(11):2509.

62. White ML, Matyjaszewski K. *J.M.S.: Pure Appl. Chem.* 1995;A32(6):1115.

63. Hirose T, Kamiya Y, Mizouchi K. *J. Applied Polym.* 1989;38:809.

64. Lieu PJ, Magill JH, Alarie YC. *J. Fire and Flammability.*1980;11:167.

Poly(phosphonates)

Leiyuchuan Ding and Bruce M. Foxman

Class Inorganic and semi-inorganic polymers

Structure $-[P(O)(R')-O-R-O]_n-$

Major Applications Flame retardation. Corrosion-resistant and improved adhesion coatings. Prevention of gingivitis and dental caries. Adjuvants and thickeners in textile dyeing. Scale inhibitors. Molding resins.[1,2]

Preparative Techniques Principal synthetic routes: condensation and addition reactions. A useful survey is available.[1]

Property	Units	Conditions	Value	Reference
Glass transition temperatures	K	R = 4,4′-biphenol; R′ = phenyl	393	(3)
		R = 3-(4-hydroxyphenyl)-1,1,3-trimethyl-5-indanol; R′ = phenyl	401	(3)*
		R = 4,4′-biphenol; R′ = phenyl	438	(3)[†]
		R = 4,4′-sulfonyldiphenyl; R′ = phenyl	419	(4)
		R = 4,4′-thiodiphenyl; R′ = phenyl	362	(4)
Decomposition temperatures	K	R = 4,4′-biphenol; R′ = phenyl	668	(3)
		R = 3-(4-hydroxyphenyl)-1,1,3-trimethyl-5-indanol; R′ = phenyl	633	(3)
		R = 4,4′-biphenol; R′ = phenyl	683	(3)
		R = 4,4′-sulfonyldiphenyl; R′ = phenyl	738	(4)
		R = 4,4′-thiodiphenyl; R′ = phenyl	738	(4)

* In ref. 3, glass transition temperatures were determined from DTA and TMA curves. Also, decomposition temperatures were determined as the temperature at which 10% weight loss occurred, as determined by TGA.

† In ref. 4, glass transition temperatures were determined by using DSC results; the midpoint in the baseline shift was taken as the glass transition temperature. Also, decomposition temperatures are quoted as the temperature at which 10% weight loss occurred, as determined by TGA.

Poly(phosphonates)

Thermal degradation of diethyl vinylphosphonate/vinyl alcohol copolymers[5]

Vinylphosphontate Unit (mol%)	Sequence Distribution			Threshold Temperature	Residue at 450°C
	A–A	A–B	B–B		
71.2	1	47	52	245	48.1
42.2	34	42	24	120	45.5
21.9	59	33	8	135	30.2
7.16				140	22.4

Transesterification polymerization of methyl poly(phosphonate) oligomers[6]

Time (h)	Degree of Polymerization*	Number of Methyl Phosphonate End-Groups	Number of Phosphonic Acid End-Groups	% of Phosphonic Acid End-Groups
22	20.35	0.99	1.01	50.5
46	27.40	0.64	1.36	68.0
70	31.36	0.30	1.70	85.0
166	31.76	0.17	1.83	91.5
190	34.53	0.00	2.00	100
213	36.02	0.00	2.00	100
237	40.43	0.00	2.00	100

* The degree of polymerization is the number of repeating units of the polymer (n). This is calculated by intergrating the area of all three ^{31}P resonances and setting the combined areas of the end-groups equal to 2.

References

1. Kroschwitz J, ed. *Encyclopedia of Polymer Science and Engineering.* Vol 11. New York: Wiley-Interscience; 1988:96–126.
2. Kroschwitz J, Howe-Grant M, eds. *Encyclopedia of Chemical Technology.* Vol 10. New York: Wiley-Interscience; 1993:976–998.
3. Imai Y, Kamata HM-A. *Kakimoto. J. Polym. Sci.* 1983; 22:1259.
4. Kim K-S. *J. Appl. Polym. Sci.* 1983;28:1119.
5. Inagaki Norihiro, Goto, Kiyoshi, Katsuura, Kakui, *Polymer.* 1975;16(9):641–644.
6. Dustan Myrex, et al., *European Poltmer Journal.* 2003;39:1105–1115.

Polypropylene, atactic

Charles L. Myers and Anita Dimeska

Acronym, Trade Names a-PP, AFAX®, REXTAC®, EASTOFLEX®

Class Poly(α-olefins)

Structure $-[CH_2CH(CH_3)]-$

Major Applications Low molecular weight atactic polypropylene is used as a component of hot melt adhesive and sealant formulations,[1] laminations, bitumen modification, as well as wire and cable filling and flooding. "Atactic" polypropylene which is produced as a by-product of isotactic PP production is not ideally atactic or completely amorphous.[2,3] Ideally atactic polypropylene has been prepared by hydrogenation of poly(2-methyl-1,3-pentadiene), that is, poly(1,3-dimethyl-1-butenylene) or PDMB.[4] Directly synthesized atactic polypropylene and other amorphous poly(α-olephins) (APAO or APO) have been developed.[1,2,3,5,6] Lower molecular weight versions are commercial products.[1,6,7] Atactic polypropylenes with higher molecular weight have also been developed using metallocene catalysts.[8,9] High molecular weight versions are being evaluated as elastomers and as blend components for modification of isotactic polypropylene.[2,3,5,6]

Properties of Special Interest Tensile strength, extensibility, recovery, softening temperature, hardness, melt viscosity, and compatibility with other polyolefins and adhesive formula components.

Property	Units	Conditions	Value	Reference
Molecular weight	g mol^{-1}	metallocene catalyst liquid monomer	100,000–1,000,000	(10)
		metallocene catalyst, $M_w/M_n = 1.9$	3,670,000	(11)
Density	g cm^{-3}	(a) Nonmetallocene a-PP		(2)
		(b, c) Metallocene a-PP		
		(a) $M_w = 29,000$, $M_w/M_n = 6$	0.8626	
		(b) $M_w = 200,000$, $M_w/M_n = 3.3$	0.8606	
		(c) $M_w = 490,000$, $M_w/M_n = 2.3$	0.8550	
		Hydrogenated PDMB,	0.8542	(12)
		$M_w = 23,300$, $M_w/M_n = 1.03$		
		Temperature dependence, 80–120°C,	0.848–x	(13, 14)
		$x = (-0.19 \times 10^{-4})t - (3.05 \times 10^{-6})t^2$		
Thermal expansion coefficient	K^{-1}	80–120°C	$(6.1–9.3) \times 10^{-4}$	(13, 14)
Crystallinity	%	DSC, XRD		(2)
		(a) $M_w = 29,000$, $M_w/M_n = 6$	Some	
		(c) $M_w = 490,000$, $M_w/M_n = 2.3$	None detected	

Property	Units	Conditions	Value	Reference
Refractive index increment dn/dc	ml g^{-1}	Hydrogenated PMBD, cyclohexane		
		30°C	0.0989	(4)
		20°C	0.0844	(12)
Head-to-head content	%	NMR (metallocene a-PP) (Bernoullian statistics)	None detected	(3)
		NMR (hydrogenated PDMB) (Bernoullian statistics)	None detected	(4)
Glass transition temperature	K	DSC, hydrogenated PDMB		
		$M_w = 23,300$	268	(12)
		$M_w = 40,800$	266.9	(4)
		$M_w = 33,400$	270.6	(4)
		Hercules AFAX™ 600 HL-5	255	(4)
		Fractionated a-PP	265.5	(4)
		Average a-PP	260	(4, 15)
		Commercial APAO homopolymer, DSC		
		Rexene Rextac™ 2115	252	(1)
		Eastoflex™ P1010 and P1023	263	(7)
Radius of gyration, $R_G/M^{1/2}$	Å mol$^{0.5}$g$^{-0.5}$	Hydrogenated PDMB, 298 K, SANS	0.336	(12)
		Several a-PP citations, Theta, IV	0.333	
Chain dimension temperature coefficient $d \ln \langle R^2 \rangle_0 / dT$	K^{-1}	Hydrogenated PDMB, melt, SANS	-0.1×10^3	(12)
		Theta, IV several a-PP citations	$(-1.0$ to $-3.0) \times 10^{-3}$	
Characteristic ratio, $6R_G^2/N_w nl^2$	–	Hydrogenated PDMB, 298 K, SANS	6.1	(12)
		Several a-PP citations, 298 K, Theta, IV	6.2	(12)
		311K	5.8–5.9	(4)
Mark-Houwink parameters: K and a	K = ml g^{-1} a = None	Decalin 135°C	$K = 1.066 \times 10^{-4}$ $a = 0.804$	(12)
Theta temperature	K	2-Octanol, hydrogenated PDMB	310.6	(4, 16)
		1-Octanol	350	(16, 17)
		Biphenyl	402	(16, 18)
Entanglement molecular weight	g mol^{-1}	413 K, measured	4,600	(19)
		413 K, calculated	5,400	
		298 K, measured	3,500	
		298 K, calculated	4,100	
Tensile strength	MPa	Compression molded, ASTM D412		(2)
		(a) $M_w = 29,000$, $M_w/M_n = 6$	1	
		(b) $M_w = 200,000$, $M_w/M_n = 3.3$	1	
		(c) $M_w = 490,000$, $M_w/M_n = 2.3$	2	
		APAO, Eastoflex™ P1010 and P1023	1.38	(7)
		$M_w = 425,000$	5	(20)

Property	Units	Conditions	Value	Reference
Maximum extensibility	%	Compression molded, ASTM D412		(2)
		(a) $M_w = 29,000$, $M_w/M_n = 6$	110	
		(b) $M_w = 200,000$, $M_w/M_n = 3.3$	1,400	
		(c) $M_w = 490,000$, $M_w/M_n = 2.3$	2,000	
		APAO, Eastoflex™ P1010	200	(7)
		APAO, Eastoflex™ P1023	100	(7)
Tensile set	%	300% elongation, 20 cm min^{-1}, 10 min hold under stress, 10 min relax		(2)
		(a) $M_w = 29,000$, $M_w/M_n = 6$	Break	
		(c) $M_w = 490,000$, $M_w/M_n = 2.3$	76	
		(a) 100% elongation	45	
		(c) 100% elongation	14	
Flexural modulus	MPa	Compression molded, ASTM D5023		(2)
		(a) $M_w = 29,000$, $M_w/M_n = 6$	10	
		(b) $M_w = 200,000$, $M_w/M_n = 3.3$	8	
		(c) $M_w = 490,000$, $M_w/M_n = 2.3$	5	
Hardness	°Shore	Shore A, compression molded		(2)
		(a) $M_w = 29,000$, $M_w/M_n = 6$	67	
		(b) $M_w = 200,000$, $M_w/M_n = 3.3$	50	
		(c) $M_w = 490,000$, $M_w/M_n = 2.3$	55	
Hardness	dmm	Penetration, ASTM D-5, APAO		
		Homopolymers: Rextac™ 2115	5	(1)
		Eastoflex™ P1010 and P1023	20	(7)
Softening point	K	Ring and ball, ASTM E-28, APAO		
		Homopolymers. Rextac™ 2115	426	(1)
		Eastoflex™ P1010 and P1023	423, 428	(7)
Melt viscosity	Pa s	Brookfield, 190°C, ASTM D-3236, APAO		
		Homopolymers. Rextac™ 2115	1.425	(1)
		Eastoflex™ P1010 and P1023	1.0, 2.3	(7)

References

1. Sustic A, Pellon B. *Adhesives Age* Nov. 1991: 17.
2. Silvestri R, Resconi L, Pelliconi A. In: *Metallocenes '95 (Brussels) Conference Proceedings*. Skillman, NJ. Schotland Business Research; 1995:207.
3. Resconi L, Jones RL, Rheingold AL, Yap GPA. *Organometallics*. 1996;15:998.
4. Zhongde X, et al. *Macromolecules*. 1985;18:2560.
5. Canich JM, Yang HW, Licciardi GF. *U.S. Patent*. 1996;5516848.
6. Robe GR. *Adhesives Age* Feb. 1993: 26.
7. Eastman Chemical Company Publication WA-4D. *Eastoflex™ Amorphous Polyolefins*. Nov., 1995.
8. McKnight AL, Masood MA, Waymouth RM. *Organometallics*. 1997;16:2879.
9. Resconi L, et al. *J Organomet. Chem* 2002; 5: 664.

10. Pasquini N. *Polypropylene Handbook*. Cincinnati: Hanser Gardner Publications; 2005:126.
11. Eckstein A, et al. *Macromolecules*. 1998;31:1335.
12. Zirkel A, et al. *Macromolecules*. 1992;25:6148.
13. Orwoll RA. In: Mark JE. *Physical Properties of Polymers Handbook*. Woodbury, N. Y: AIP Press; 1996, Ch. 7, p. 82.
14. Rogers PA. *J. Appl. Polym. Sci.* 1993;48:1061.
15. Gaur U, Wunderlich B. *J. Phys. Chem. Ref. Data*. 1981;10:1051.
16. Sundararajan PR. In: Mark JE, ed. *Physical Properties of Polymers Handbook*. Woodbury, N.Y: AIP Press; 1996:Ch. 15, 203.
17. Mays JW, Fetters LJ. *Macromolecules*. 1989; 22:921.
18. Moraglio G, Gianotti G, Bonicelli U. *Eur. Polym. J.* 1973; 9:623.
19. Fetters LJ, Lohse DJ, Colby RH. In: Mark JE, ed. *Physical Properties of Polymers Handbook*. Woodbury, N. Y: AIP Press; 1996:335.
20. Pasquini N. *Polypropylene Handbook*. Cincinnati: Hanser Gardner Publications; 2005:129.

Polypropylene, elastomeric (stereoblock)

Charles L. Myers and Anita Dimeska

Acronyms, Trade Names ELPP, elPP, REXFLEX®, SUPERSOFTPP®

Class Poly(α-olefins)

Structure —[$CH_2CH(CH_3)$]—

Major Applications The polymers referred to in this chapter include those families of homopolymers of propylene which are known to have elastomeric recovery properties at reasonable molecular weight and for which properties have been attributed to a crystallizable-noncrystallizable (e.g., isotactic-atactic) stereoblock structure, or to a major component with a stereoblock structure, whether or not the compositions are homogeneous by solvent fractionation tests. Copolymers and blends are not deliberately included in the data presented, but are described in some of the references. (See also some of the closely related elastomeric polymers presented in the entry on *Polypropylene, atactic* in this handbook.) The criterion of multiple crystallizable blocks per polymer chain may be met in significant fractions of low-tacticity, low-stereoregularity polymers of very high molecular weight.

Elastomeric polypropylenes are being actively studied in academic and industrial laboratories. Some materials are in the pilot developmental stage. Some flexible polyolefins, with moderate elastomeric recovery, are currently being evaluated on a larger scale.[1] Potential applications include fiber, film, and extruded goods.

Properties of Special Interest Strength, modulus/flexibility, mechanical recovery, high transparency, degree of thermal resistance, and solubility/extractability. Mechanical properties in the following table are intended to represent best published examples of the respective types.

Property	Units	Conditions	Value	Reference
Density	$g\ cm^{-3}$	Buoyancy method, calculated from data in reference (2) for 19–31% crystallinity	0.8683–0.8787	(2)
		–	<0.9	(3)
		–	0.88–0.89	(1)
Intrinsic viscosity	$dl\ g^{-1}$	Ubbelohde, 135°C in 1,2,3,4-tetrahydronaphthalene	1.17	(4)
Glass transition temperature	K	DSC	262.9–261.5	(2)
			265	(5)
Melting temperature	K	DSC peak endotherms (broad temperature range)		(6)
		16% mmmm	325, 352	
		28% mmmm, peak range	398–418	

Property	Units	Conditions	Value	Reference
		DSC several examples, broad range, peak endotherm	417–418	(2)
		DSC, 45–54% mmmm, dual endotherm peaks ranges	323–327 352–346	(7)
		DSC, broad range, dual endotherm peaks	316.5, 338	(5)
		DSC, dual endotherm peaks		(8)
		35% mmmm	324, 339	
		40% mmmm	326, 337	
		DSC endotherm peak range	427–433	(1)
		DSC, peak, 35% mmmm	318	(4)
Heat of fusion (experimental)	kJ mol^{-1}	DSC, >20 polymers (10–70 J g^{-1})	0.4–2.9	(2)
		DSC, 45–54% mmmm (31–40 J g^{-1})	1.3–1.7	(7)
		DSC, 40% mmmm (14 J g^{-1})	0.59	(8)
		DSC, 40% mmmm (26 J g^{-1})	1.10	(9)
Crystallinity	%	DSC		(7)
		54% mmmm	19.1	
		45% mmmm	14.8	
		52% mmmm	16.7	
		DSC (ELPP type of reference (10))	13	(5)
		XRD	16	(5)
		Annealed, XRD, 30°C, 35–40% mmmm	26–27	(11)
		XRD, density methods, fractionated ELPP		(12)
		Whole ELPP, IV = 2.7 dl g^{-1}, XRD (density)	21 (19)	
		Ether soluble, 0.73 dl g^{-1}	8 (0)	
		Hexane soluble, 2.56 dl g^{-1}	14 (17)	
		Hexane insoluble, 4.16 dl g^{-1}	29 (44)	
		Whole ELPP, IV = 12.1 dl g^{-1}	17 (24)	
		Ether soluble, 3.42 dl g^{-1}	9 (9)	
		Hexane soluble, 7.80 dl g^{-1}	15 (25)	
		Hexane insoluble NA	22 (29)	
Equilibrium modulus	MPa	50°C, 0.5% strain, stress relaxed 10^4 s (2 examples)	0.56 1.47	(8)
Segment length between virtual cross-links	Daltons	M_n of amorphous segments between physical cross-links, estimated from $M_{n,a}$ = density × RT/G_{eq} (2 examples)	2,100 4,400	(8, 11)
		Mechanical rheometry, 25°C	900	(9)
		Mechanical rheometry, 25°C	940	(13)
Tensile strength	MPa	51 cm min^{-1}, ASTM D412	5–8	(2)
		51 cm min^{-1}	3.2	(6)
		25.5 cm min^{-1}	16–39	(7)
		20 cm min^{-1}	4–12	(8, 11, 14)
		10 mm min^{-1}, some necking	22	(15)
		51 cm min^{-1}	11.7–14.8	(16)
		51 cm min^{-1}, ASTM D1708	6.95–9.31	(17)
		51 cm min^{-1}, syndiotactic ELPP	11	(18)

Polypropylene, elastomeric (stereoblock)

Property	Units	Conditions	Value	Reference
Maximum extensibility	%	51 cm min^{-1}	>1,000	(2)
		51 cm min^{-1}	1,200	(6)
		25.5 cm min^{-1}	800	(7)
		20 cm min^{-1}	525–1,260	(8, 11, 14)
		10 mm min^{-1}, some necking	700	(15)
		51 cm min^{-1}	814–863	(16)
		51 cm min^{-1}, syndiotactic ELPP	750–908	(17)
		Not specified	>1,000	(1)
Tensile modulus	MPa	DIN 53457, 23°C	23–28	(3)
		Not specified	69–359	(1)
		51 in min^{-1}	1.7	(6)
		51 cm min^{-1}, ASTM D1708	13.45–15.98	(17)
Impact strength	kJ m^{-2}	Tensile impact, ISO 8256, 23°C	270–300	(3)
		Flexural impact, ISO 179 1 eu, −20°C	14–22	
Hardness	°Shore	Shore A	77–83	(3)
			81–96	(1)
Tensile set	%	300% extension, 51 cm min^{-1}, ASTM D412, 23°C, no hold at extension	80	(3)
			93	(2)
			60–130	(19)
			50	(6)
			82–93	(16)
			22–28	(17)
		300% extension, 20 cm min^{-1}, no hold at extension	24	(8, 11, 14)
		300%, conditions not specified	100–200	(1)
		400% extension, 51 cm min^{-1}, no hold at extension	65–110	(17)
Tensile recovery	%	100% extension, 25.5 cm min^{-1}	92–97	(7)
		No hold at extension, 2 min recovery after extension	97	(8, 11, 14)
		200% extension, 25.5 cm min^{-1}	90–97	(7)
		No hold at extension, 2 min recovery after extension	96	(8, 11, 14)

References

1. Pellon BJ. In: *SPO '93 (Houston, Texas) Conference Proceedings*. Skillman, NJ. Schotland Business Research; 1993, p. 399.
2. Collette JW, et al. *Macromolecules* 1989;22:3851.
3. Gahleitner M, et al. In: *SPO 1996 (Houston, Texas) Conference Proceedings*. Skillman, NJ. Schotland Business Research; 1996:281.
4. De Rosa C, et al. *Macromolecules*. 2004;37:6843.
5. Canevarolo S, DeCandia F. *J. Appl. Polym. Sci.* 1994;54:2013.
6. Coates GW, Waymouth RM. *Science*. 1995;267:217.

7. Gauthier WJ, Corrigan JF, Taylor NJ, Collins S. *Macromolecules.* 1995;28:3771.

8. Mallin DT, et al. *J. Am. Chem. Soc.* 1990;112:2030.

9. Carlson ED, et al. *Macromolecules.* 1998;31:5343

10. Ewen JA. *J. Am. Chem. Soc.* 1984;106:6355.

11. Llinas HL, et al. *Macromolecules.* 1992;25:1242.

12. Collette JW, Ovenall DW, Buck WH, Ferguson RC. *Macromolecules.* 1989;22:3858.

13. Carlson ED, et al. In: *68th Annual Society of Rheology Meeting.* Galveston, Tex: Society of Rheology; February 1997.

14. Chien JCW, et al. *J. Am. Chem. Soc.* 1991;113:8569.

15. Canevarolo SV, DeCandia F, Russo R. *J. Appl. Polym. Sci.* 1995;55:387.

16. Wilson SE, Job RC. *U.S. Patent.* 4, 971, 936(1990).

17. Job RC. *U.S. Patent.* 5, 270, 276(1993).

18. Hu Y, et al. *Macromolecules.* 1998;31:6908.

19. Tullock CW, et al. *J. Poly. Sci.: Part A: Polym. Chem.* 1989;27:3063.

20. Gauthier WJ, Collins WJ. *Macromolecules.* 1995;28:3779.

Polypropylene, Isotactic

David V. Howe and Anita Dimeska

Acronym PP

Class Poly(α-olefins)

Structure —[CH$_2$CH(CH$_3$)]—

Major Applications Fiber, slit tape, nonwoven fabrics, cast and biaxially oriented film, containers and closures, automotive interior and exterior parts, appliance housing and components, medical applications, component in elastomeric blends with polyethylene and olefinic rubbers.

Properties of Special Interest Low cost; easily processed by injection molding, extrusion, and spinning; can be oriented; excellent resistance to chemicals; low color; can be stabilized to provide good thermal aging stability; moderate strength and stiffness; good toughness when impact modified either in the reactor or by compounding; excellent flexural fatigue resistance; modest clarity.

Preparative Techniques Ziegler-Natta polymerization with titanium halide/aluminum alkyl catalyst, and optionally ether, ester, or silane activator. Catalyst may be deposited on magnesium chloride support. Slurry and gas phase processes are used. Polymerization with metallocene catalyst/aluminoxane cocatalyst systems is now commercially available. Typical comonomers are ethylene and 1-butene.

Isotacticity

Polymerization Conditions	Isotacticity				Reference
	Isotactic Index (% heptane insolubles)	Xylene Insolubles	% mmmm	% mm	
Ziegler-Natta catalyst systems					
MgCl$_2$/TiCl$_4$/DIBP* catalyst modified with TMPIP* and AlEt$_3$ prepared at 140 ^{o}C	–	94	89.3	–	(1)
MgCl$_2$/TiCl$_4$/DIBP* catalyst modified with (i-Bu)$_2$Si(OMe)$_2$ and AlEt$_3$	97	–	–	–	(2)
MgCl$_2$/TiCl$_4$/DE* catalyst modified with AlEt$_3$	95–99	–	–	–	(3)
Various MgCl$_2$ or TiCl$_3$ supported Ziegler-Natta catalysts	–	–	–	92.2–94.9	(4)
Metallocene catalyst systems					
rac-[Me$_2$Si(2-Me-4,5-BenzoInd)$_2$]ZrCl$_2$	98	–	–	–	(5)
rac-[Me$_2$Si(2-Me-4-Ph-Ind)$_2$]ZrCl$_2$	–	–	95.2	–	(6)

* DIBP = Diisobutyl phthalate; TMPIP = 2,2,6,6-tetramethylpiperidine; DE = 1,3-diether.

Molecular weight (M_w) and polydispersity index (M_w/M_n)

Polymerization Conditions	M_w (g mol^{-1})	M_w/M_n	Reference
MgCl$_2$/TiCl$_4$/DIBP catalyst modified with (i-Bu)$_2$Si(OMe)$_2$ and AlEt$_3$			(2)
H$_2$ concentration = 0 mol l^{-1}	560,000	3.8	
H$_2$ concentration = 6.9×10^{-3} mol l^{-1}	382,000	6.1	
Typical range (extrapolated from melt flow rates of commercial products)	<100,000–>600,000	5–12	(7, 8)
Borealis VC20 82C (MFR: 20 g per 10 min)	265,000	4.3	(9)
Typical for controlled rheology (chemically cracked products)	–	< 5	(10, 11)
Single site catalyst	–	~ 2	(12)
rac-[Me$_2$Si(2-Me-4-Ph-Ind)$_2$]ZrCl$_2$/MAO H$_2$ concentration = 0 mol l^{-1}	729,000	< 3	(6)
Melt flow index			(7)
0.63	646,000	–	
2.9	412,000	–	
11.9	297,000	–	

Property	Units	Conditions	Value	Reference
Molecular weight of repeat unit	g mol^{-1}	—CH$_2$—CH(CH$_3$)—	42.07	
Entanglement molecular weight	g mol^{-1}	463 K, calculated, metallocene	6,900	(13)
Morphology (blends, 'impact copolymer')	–	Elastomer content <~60%	Dispersed phase	(14)
		Elastomer content >~60% (depends upon processing conditions)	Dispersed or co-continuous phase	
IR (characteristic absorption frequencies)	cm^{-1}	CH$_3$, CH$_2$, CH stretching	2956 (s), 2951 (s), 2925 (sh), 2907 (sh), 2880 (s), 2868(s), 2843 (s)	(15, 16)
		CH$_3$ antisymmetric bending, CH$_2$ bending	1459 (sh), 1454 (s)	
		Various CH$_3$, CH$_2$, and CH bending, wagging, twisting, C–C stretching	1377 (s), 1359 (m), 1329 (w), 1305 (w), 1297 (w), 1257 (w), 1219 (w)	
		Various CH$_3$, CH$_2$, and CH bending, wagging, twisting, and rocking, C–C stretching	1167 (s), 1153 (sh), 997 (s), 973 (s), 841 (s), 809 (m)	
NMR		^{1}H NMR		(17–20)
		^{13}C NMR		(21, 22)
Coefficient of linear thermal expansion	K^{-1}	ASTM Method D696		(23)
		From 243 to 273K	6.5×10^{-5}	
		From 273 to 303K	1.05×10^{-4}	
		From 303 to 330K	1.40×10^{-4}	

Property	Units	Conditions	Value	Reference
Coefficient of thermal expansion (volume, melt)	K^{-1}	From 448 to 573K	6.6×10^{-4}	(24)
		From 453 to 503K	6.7×10^{-4}	(25)
Isothermal compressibility	bar^{-1}	453K	1.27×10^{-4}	(25, 26)
		493K	1.50×10^{-4}	
		533K	1.78×10^{-4}	
Density	$g\,cm^{-3}$	298 K, α-crystalline phase	0.936–0.946	(27–29)
		298 K, amorphous phase	0.850–0.855	
		298 K, typical commercial material	0.90–0.91	
Solvents		Room temperature	No common solvents	(30)
Solubility parameter	$(MPa)^{1/2}$	Inverse phase gas chromatography	18.8	(31)
		Montell Profax 6701	17.3	(32)
Theta temperature	K	$M_w = 28,000–564,000$		(33, 34)
		p-tert-amylphenol	414	
		dibenzyl ether	456	
		biphenyl	398	
		n-butanol	420	
Lattice	–	α_1, α_2-forms	Monoclinic	(27, 35)
		β-form	Hexagonal	(35)
		γ-form	Orthorhombic	(35–37)

Form	Bravais Lattice	Space Group	Chain Conformation	Unit Cell Dimension (Å)			Unit Cell Angle (Degrees)	Reference
				a	b	c		
α_1	Monoclinic	C_2/c	Helix (3/1)	6.67	20.8	6.50	98.67	(38)
α_2	Monoclinic	$P2_1/c$	Helix (3/1)	6.65	20.73	6.50	98.67	(38)
β	Hexagonal	$P3_121$	Helix (3/1)	11.03	11.03	6.49	–	(35)
γ	Orthorhombic	$Fddd$	Helix (3/1)	8.54	9.93	42.41	–	(35–37)

Property	Units	Conditions	Value	Reference
Heat of fusion	$J\,g^{-1}$	DSC, α-crystalline material (100%)	165	(28)
Degree of crystallinity	%	DSC, density depends upon tacticity and crystallization conditions	50–70	(4, 9, 29)
Glass transition temperature	K	DMA		(9)
		30Hz	283.7	
		1Hz	275.5	
Melting point	K	100% crystalline	~ 459	(39)

Commonly reported mechanical properties*

Property	Units	Conditions	Polymer Type			
			IPP[a]	RCP[b]	ICP-L[c]	ICP-H[c]
Yield stress	MPa	ASTM D638	34.5	27.6	26.2	22.0
Yield strain $(L/L_0)_y$	%	ASTM D638	10	14	12	14
Flexural modulus	MPa	ASTM D790	1,389	1,035	1,210	1,000
Izod impact strength[d]	J m^{-1}	ASTM D256	27	55	130	No break
Hardness	Rockwell	ASTM D785	R90	R80	R80	R60
Deflection temperature	K	ASTM D648, 0.45MPa outer fiber stress	380	360	360	345

* These are typical properties for the classes of materials based on the range of properties reported in references (40) and (41).

[a] IPP = isotactic propylene homopolymer.

[b] RCP = ethylene-propylene random copolymer with an ethylene content of about 3%.

[c] ICP = blends of isotactic propylene homopolymer with ethylene-propylene rubber. These materials are commonly called "impact copolymers," "heterophasic copolymers," or, incorrectly, "block copolymers." These are typically prepared during the polymerization process using a series of reactors. L = low rubber (less than about 15% rubber by weight; typically with an ethylene content of less than about 10%). H = high rubber content blends (greater than about 15% rubber by weight; typically with an ethylene content of at least 7%).

[d] Impact strength is very dependent upon the molecular weight of the polymer as well as the rubber content of the material. These data are for typical injection molding grade materials.

Property	Units	Conditions	Value	Reference
Storage modulus	MPa	293 K, Homopolymer, 30 Hz	1,400	(9)
Tan δ	–	293 K, Homopolymer, 30 Hz	0.086	(9)
Poisson ratio	–	296 K	0.38	–
Index of refraction n_D	–	293 K, density 0.9075 g cm^{-3}	1.5030	(42)
Refractive index increment dn/dc	–	1-chloronaphthalene and 1,2,4-trichlorobenzene solvents	(see reference)	(43)
Dielectric constant ε'	–	At 1KHz (D150)	2.2–2.3	(29)
		At 1MHz	2.1–2.3	(44)
Dielectric strength	V cm^{-1}	298K (D149)	240,000	(29)
		298K	217,000–300,000	(44)
		393K	170,000	(29)
Dissipation factor	–	60 Hz–100 MHz (D510)	0.3–1 ($\times 10^{-3}$)	(29)
		1 MHz	1–3 ($\times 10^{-4}$)	(44)
Volume resistivity	ohms cm	ASTM D257	10^{16}–10^{17}	(29, 44)
Surface tension γ	mN m^{-1}	438K	22.5	(45)
		473K	21.2	(46)
		495K	20.2	(45)
Surface free energy	mJ m^{-2}	298K (calculated from advancing contact angles)	29.0	(46)
Contact angle	degrees	H_2O; advancing, 298K	116	(46)
		CH_2I_2; advancing, 298K	64	

Property	Units	Conditions	Value	Reference
Permeability coefficient	m^3(STP) m $s^{-1}m^{-2}$ Pa^{-1}	H_2O, 298K	3.83×10^{-16}	(47)
		O_2, 298K (isotropic, all pressures)	7.73×10^{-18}	(48)
		O_2, 298K (12.5 :1 draw ratio)	2.12×10^{-18}	(48)
		CO_2, 298K (<1 atm)	2.37×10^{-17}	(49)
		CO_2, 298K (50 atm)	7.50×10^{-17}	(49)
Thermal conductivity	$W\,m^{-1}\,K^{-1}$	293K	0.12	(50)
			0.22	
Melt flow rate	$g\,(10\,min)^{-1}$	ASTM D1238, 503 K, 2.16 kg	$0.2->500$	–
Speed of sound	$m\,s^{-1}$	Unoriented		(51)
		298K	2.5×10^3	
		398K	125×10^3	
		Oriented		(52)
		Long. dir. 298K	3.3×10^3	
		Trans. dir. 298K	2.1×10^3	
Decomposition temperature	K	TGA in helium, Montell Profax 6501	623	(53)
Ignition temperature	K	Calculated from critical heat flux data	736	(54)
Oxygen index	%	ASTM D2863, Montell Profax 6505	17.4	(55)
Scission, G factor	–	γ irradiation		(56, 57)
		Initial	1.2	
		At doses above gel point	0.27	
Cross-linking, G factor	–	γ irradiation	0.07–0.30	(56, 57)
Producers	–	Worldwide in 2005 (see table below for examples)		
Technology providers and process licensors		(see table below for examples)		
Capacity	ktons	Worldwide in 2005 (see table below for examples)	43,942	(58)

Some producers and capacities (from 2005)[58]

Producer	Capacity (ktons)
LyondellBasell	6,587
Sinopec	3,263
Innovene/BP	2,648
Total PC	2,187
ExxonMobil	2,090

Technology providers and process licensors[59]

Company	Technology
LyondellBasell	Spheripol, Spherizone
DOW Chemical	Unipol
Novolen Technology	Novolen
BP	Innovene
Borealis	Borstar

References

1. Chadwick JC, et al. *Makromol. Chem.* 1992;193:1463–1468.
2. Proto A, et al. *Macromolecules.* 1990;23:2904–2907.
3. Moore EP Jr. *Polypropylene Handbook.* Cincinnati: Hanser/Gardner Publications; 1996:37.
4. Paukkeri R, Lehtinen A. *Polymer.* 1993;34:4075–4082.
5. Kaminsky W, Laban A. *Applied Catalysis A: General.* 2001;222:47–61.
6. Spaleck W, et al. *Organometallics.* 1994;13:954–963.
7. Bremner T, Rudin A, Cook DG. *J. Appl. Poly. Sci.* 1990;41: 1617–1627.
8. Aggarwal SL. In: Brandrup J, Immergut EH, eds. *Polymer Handbook.* 2nd ed. New York: John Wiley and Sons; 1975:V-23–V–28.
9. Järvelä P, Shucai L, Järvelä P. *J. Applied Polymer Sci.* 1996;62:813–826.
10. Moore EP Jr. *Polypropylene Handbook.* Cincinnati: Hanser/Gardner Publications; 1996:192–193.
11. Tzoganakis C, Vlachopoulos J, Hamielec AE. *Polym. Eng. and Sci.* 1988;28:170–179.
12. Moore EP Jr. *Polypropylene Handbook.* Cincinnati: Hanser/Gardner Publications; 1996:52.
13. Eckstein A, et al. *Macromolecules.* 1998;31:1335.
14. Moore EP Jr. *Polypropylene Handbook.* Cincinnati: Hanser/Gardner Publications; 1996:150ff.
15. Painter PC, Coleman MM, Koenig JL. *TheTheory of Vibrational Spectroscopy and Its Application to Polymeric Materials.* New York: John Wiley and Sons; 1982:379–389.
16. McDonald MP, Ward IM. *Polymer.* 1961;2:341–355.
17. Ferguson RC. *Macromolecules.* 1971;4:324–329.
18. Ferguson RC. *Trans. N.Y. Acad. Sci.* 1967;29:495–501.
19. Heatley F, Zambelli A. *Macromolecules.* 1969;2:618–619.
20. Heatley F, Salovey R, Bovey FA. *Macromolecules.* 1969;2:619–623.
21. Tonelli AE, Schilling FC. *Accts. Chem. Res.* 1981;14:233–238.
22. Wehrli FW, Wirthlin T. *Interpretation of Carbon-13 NMR Spectra.* London: Heyden and Son Ltd; 1980:218.
23. Crespi G, Luciana L. In: Kroschwitz JI, ed. *Kirk-Othmer Encyclopedia of Chemical Technology.* 3rd ed. Vol 16. New York: John Wiley and Sons; 1981:453–469.
24. Wang YZ., et al. *J. Appl. Polym. Sci.* 1992;44:1731–1736.
25. Zoller P. *J. Appl. Polym. Sci.* 1979;23:1057–1061.

26. Orwoll RA. In: Mark JE, ed. *Physical Properties of Polymers Handbook*. Woodbury, N.Y: AIP Press; 1996:87.

27. Natta G, Corradini P. *del Nuovo Cimento*. 1960;XV:40–51.

28. Wunderlich B. *Macromolecular Physics*. Vol 3. New York: Academic Press; 1980:61–64.

29. Brandrup J, Immergut EH, eds. *Polymer Handbook*. 3rd ed. New York: John Wiley and Sons; 1989:V-27ff.

30. Brandrup J, Immergut EH, eds. *Polymer Handbook*. 2nd ed. New York: John Wiley and Sons; 1975:IV-243.

31. Abe M, Iwama M, Homma T. *Kogyo Kagaku Zasshi (J. Chem. Soc. Jpn. Ind. Chem. Sec.)* 1969;72:2313–2318.

32. Barton AFM. *CRC Handbook of Solubility Parameters and Other Cohesive Parameters*. 2nd ed. Boca Raton, Fla: CRC Press; 1991:445.

33. Nakajima A, Saijyo A. *J. Polym. Sci., Part A-2*. 1968; 6:723–733.

34. Nakajima A, Saijyo A. *J. Polym. Sci., Part A-2*. 1968;6:735–744.

35. Phillips PJ, Mezghani K. In: Salamone JC, ed. *Polymeric Materials Encyclopedia*. Vol 9. Boca Raton, Fla: CRC Press; 1996:6637–6649.

36. Brückner S, Meille SV. *Nature*. 1989;340:455–457.

37. Campbell RA, Phillips PJ, Lin JS. *Polymer*. 1993;34:4809–4816.

38. Hikosaka M, Seto T. *Polym. J*. 1973;5:111–127.

39. Mezghani K, Campbell RA, Phillips PJ. *Macromolecules*. 1994;27:997–1002.

40. *Manufacturing Handbook and Buyers' Guide: Plastics Technology*. New York: Bill Communications; 1997–1998:601–621.

41. Moore EP Jr. *Polypropylene Handbook*. Cincinnati: Hanser/Gardner Publications; 1996:238.

42. In: Brandrup J, Immergut EH, eds. *Polymer Handbook*. 3d ed. John Wiley and Sons; 1989:VI-455.

43. HorskaÂ J, Stejskal J, Kratochvil P. *J. Appl. Polym. Sci*. 1983;28:3873–3874.

44. In: Johnson LR, ed. *International Plastics Selector*. Vol 2. Englewood, Colo: 17. D.A.T.A Business Publishing; 1996:1193–1453.

45. Schonhorn H, Sharpe LH. *J. Polymer Sci. B*. 1965;3:235–237.

46. Sauer BB, Diapaolo NV. *J. Colloid Interface Sci*. 1991;144:527–537.

47. Myers AW, Stannett V, Szwarc M. *J. Polymer Sci*. 1959;35:185–288.

48. Taraiya AK, Orchard GAJ, Ward IM. *J. Appl. Polym. Sci*. 1990;41:1659–1671.

49. Naito Y, et al. *J. Polym. Sci.: B, Polymer Physics*. 1991;29:457–462.

50. Thompson EV. In: Mark HF, et al, eds. *Encyclopedia of Polymer Science and Engineering*. Vol 16. New York: Wiley-Interscience; 1985:711–747.

51. Bikales NM, ed. *Encyclopedia of Polymer Science and Technology*. Vol 12. New York: John Wiley and Sons; 1970:702.

52. Price HL. *SPE Journal*. 1968;24(2):54–59.

53. Chien JW, Kiang J. In: Hawkins WL and DLA, eds. *Stabilization and Degradation of Polymers*. Advances in Chemistry Series, Vol 169. Washington, D.C: American Chemical Society; 1978:175–197.

54. Tewarson A. In: Mark JE, ed. *Physical Properties of Polymers Handbook*. Woodbury, N.Y: AIP Press; 1996:584.

55. Cullis CF, Hirschler MM. *The Combustion of Organic Polymers*. Oxford: Clarendon Press; 1981:53.

56. Schnabel W, Dole M. *J. Phys. Chem.* 1963;67:295–300.
57. Keyser RW, Clegg B, Dole M. *J. Phys. Chem.* 1963;67:300–303.
58. *2006 World Polyolefins Analysis*, Chemical Market Associates, Inc. 2005, 451.
59. Pasquini N. *Polypropylene Handbook*. Cincinnati: Hanser Gardner Publications; 2005:368–375.

Poly(propylene imine) Dendrimers

Donald A. Tomalia and Margaret Rookmaker

Alternative Name, Acronym, Trade Name Polypropylenimine (PPI) dendrimers, Astramol® dendrimers, sometimes POPAM-dendrimers

Class Dendritic polymers; dendrimers

Structure Dendrimers are three-dimensional macromolecules consisting of three major architectural components: a core, branch cells, and terminal groups. These products are constructed from repeat units called *branch cells* [e.g., $-CH_2-CH_2-CH_2-N(CH_2-CH_2-CH_2)_2$] in concentric generations (G) surrounding various cores according to dendritic rules and principles, where N_c = multiplicity of core; N_b = multiplicity of branch cell; and Z = terminal groups (i.e., $-CN$ or $-CH_2-NH_2$).

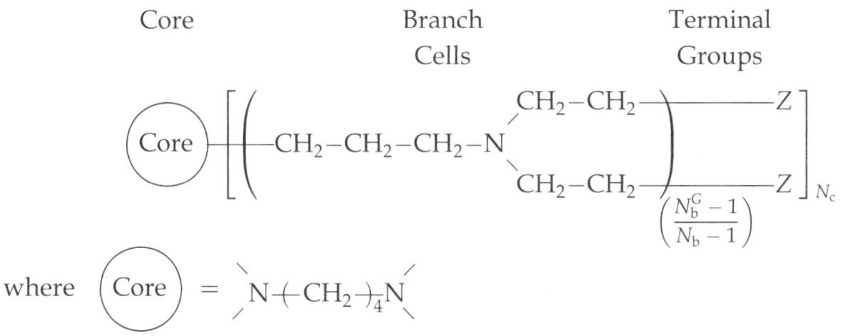

DSM uses its own designations to describe these dendritic products, wherein the three architectural components are noted as follows:

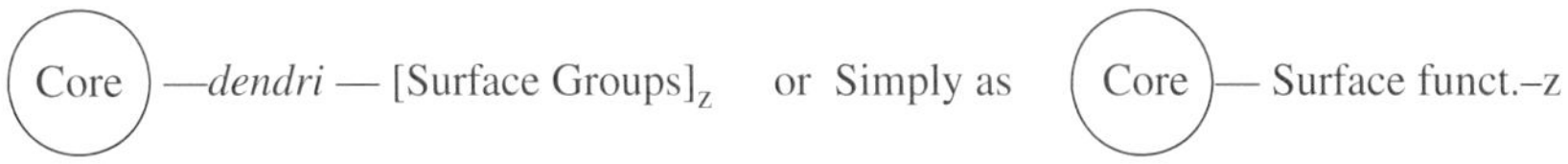

Examples: DAB–dendr–$(NH_2)_{16}$ or simply DAB–Am–16, where the core is diaminobutane (DAB) (i.e., 1,4-diaminobutane); *dendr* indicates the interior dendritic branch cell; and the last component defines the type and number of surface groups, Z.

The DSM generation (G') designation counts the number of iteration steps rather than the branch cell formation stages. Therefore, compared to the literature notation used here, $G' = G - 1$.

Preparative Techniques PPI dendrimers are synthesized by the divergent method starting from 1,4–diaminobutane (DAB) ($N_c = 4$). They are amplified by progressing through an iterative sequence consisting of (a) an double Michael addition of acrylonitrile to the primary amino groups followed by (b) hydrogenation under pressure in the presence of Raney cobalt.

Products are produced up to generation $= 4$ (literature); generation $= 5$ (DSM) nomenclature (Z $= 64$).[1]

Properties of Special Interest Unique dendrimer properties not found in traditional macromolecular architecture include: (1) a distinct parabolic intrinsic viscosity curve with a maximum as a function of molecular weight; (2) very monodispersed sizes and shapes (i.e., $\overline{M}_w/\overline{M}_n$ routinely below 1.1 even at high molecular weights);[2] (3) *exo* presentation of exponentially larger numbers of surface functional groups as a function of generation (i.e., up to several thousand); and (4) typical Newtonian-type rheology. This dendrimer family exhibits excellent hydrolytic and thermal stability.

Major Applications Used as templates for initiation of caprolactam polymerization to produce injection moldable star-like-Nylon-6 products.[3] This dendrimer family has been used in a variety of metal chelation, coatings, and lubrication-type applictions.[4]

Suppliers See general reviews[5,6] for more information. DSM, Het Overloon 1, Heerlen, P.O. Box 6500, 6401 JH Heelen, The Netherlands. SYMO-chem, Den Dolech 2, Eindhoven (www.symo-chem.nl) the Netherlands.

Data for amine terminated polypropylenimine dendrimer DAB-*dendr*-(NH$_2$)$_x$

Generation		DSM Designation	Molecular Weight (g mol^{-1})*	Number of Surface Groups	$[\eta]25°C$	$V_\eta(\text{Å}^3)$	$R_\eta(\text{Å})$	$R_g(\text{SANS})$ (D$_2$0) (Å)	Modeling (Å)	
Literature	DSM								R_g (cvff)	R_g (cvffrep)
0	1	DAB-*dendr*-(NH$_2$)$_4$	317	4	0.045	948	6.1	4.4	4.9	5.0
1	2	DAB-*dendr*-(NH$_2$)$_8$	773	8	0.055	2,824	8.8	6.9	6.0	7.6
2	3	DAB-*dendr*-(NH$_2$)$_{16}$	1,687	16	0.062	6,947	11.8	9.3	7.4	10.1
3	4	DAB-*dendr*-(NH$_2$)$_{32}$	3,514	32	0.068	15,872	15.6	11.6	10.0	12.9
4	5	DAB-*dendr*-(NH$_2$)$_{64}$	7,168	64	0.068	32,367	19.8	13.9	12.5	15.9

* Theoretical values.

Data for nitrile terminated polypropylenimine dendrimer DAB-*dendr*-(CN)x

Generation		DSM Designation	Molecular Weight (g mol^{-1})*	Number of Surface Groups	$[\eta]25°C$ (acetone) (dl g^{-1})	$V_\eta(\text{Å}^3)$	$R_\eta(\text{Å})$	$R_g(\text{SANS})$ (acetone-d_6) (Å)
Literature	DSM							
0	1	DAB-*dendr*-(CN)$_4$	300	4	0.024	478	4.9	–
1	2	DAB-*dendr*-(CN)$_8$	741	8	0.030	1,477	7.1	6.0
2	3	DAB-*dendr*-(CN)$_{16}$	1,622	16	0.034	3,663	9.6	8.0
3	4	DAB-*dendr*-(CN)$_{32}$	3,385	32	0.035	7,869	12.3	10.1
4	5	DAB-*dendr*-(CN)$_{64}$	6,910	64	0.036	16,523	15.8	12.2

* Theoretical values.

DSM Designation	Units	DAB/Acn4	DAB/ACN8	DAB/ACN16	DAB/ACN32	DAB/ACN64
Generation						
DSM		1	2	3	4	5
Literature		0	1	2	3	4

Poly(propylene imine) Dendrimers

DSM Designation	Units	DAB/Acn4	DAB/ACN8	DAB/ACN16	DAB/ACN32	DAB/ACN64
End groups	–	4*CN	8*CN	16*CN	32*CN	64*CN
Molecular weight	g mol^{-1}	300	741	1,622	3,385	6,910
Diameter	nm	1.4	1.9	2.6	3.3	4.3
Radius of gyration (acetone-d_4)	Å	–	6	8	10.1	12.2
Density	g cm^{-3}	–	1.0600	1.0582	–	–
Appearance	–	White powder	Sl. yellow vis. liq.	–	–	–
Melting point	K	326	–	–	–	–
Viscosity, bij 50°C	Pa s	–	2.6	10.3	15	50
Intrinsic viscosity	d g^{-1}					
25°C/acetone		0.026	0.031	0.035	0.038	0.038
25°C/THF		−0.028	−0.034	−0.042	−0.045	−0.045
T_g onset	K	213.5	218.8	225.8	227.8	232.9
T_gA maximum, 20°C min^{-1}	K	603	603	603	603	603
Thermal stability	K	483	483	483	483	483
Vapor pressure	–	<0.0001	<0.0001	<0.0001	<0.0001	<0.0001
Safety data						
Flash point	K	375	442	–	–	–
Autoignition	K	692	658	–	–	–
Ames test	–	Not carcinogen	Not carcinogen	Not carcinogen	Not carcinogen	Not carcinogen
Irritation	–	Mild irritating	–	Not irritating	–	–
Labeling by irritation	–	None	–	None	–	–
LD50 test	mg kg^{-1}	>5,000	–	–	4,000	–
Labeling by LD50 test	–	None	–	–	None	–

DSM Designation	Units	DAB/PA4	DAB/PA8	DAB/PA16	DAB/PA32	DAB/PA64
Generation	–					
DSM		1	2	3	4	5
Literature		0	1	2	3	4
End groups	–	4*NH$_2$	8*NH$_2$	16*NH$_2$	32*NH$_2$	64*NH$_2$
Molecular weight	g mol^{-1}	317	773	1,687	3,514	7,166
Diameter	nm	1.5	1.9	2.7	3.4	4.4
Radius of gyration (D_2O)	Å	4.4	6.9	9.3	11.6	13.9
Density	g cm^{-3}	0.9578	0.9785	0.989	1.0097	
Appearance	–	Light yellow oil	Light yellow oil	Light yellow oil	Light yellow oil	Light yellow oil

Poly(propylene imine) Dendrimers

DSM Designation	Units	DAB/PA4	DAB/PA8	DAB/PA16	DAB/PA32	DAB/PA64
Melting point	K	~298	–	–	–	–
Viscosity, bij 50°C	Pa s	0.028	0.28	1.1	2.5	6.7
Intrinsic viscosity						
25°C/D2O	d g^{-1}	0.045	0.055	0.062	0.068	0.068
25°C/THF		0.026	0.036	0.04	–	–
25°C/MeOH		0.046	0.055	0.061	0.064	0.059
T_g onset	K	166	176	183	186	189
T_gA maximum, 20°C min^{-1}	K	~623	~718	~718	~718	~718
Thermal stability	K	tot 573	tot 573	>573	>573	>573
Vapor pressure	–	<0.0001	<0.0001	<0.0001	<0.0001	<0.0001
Safety Data						
Flash point	K	406	–	–	–	–
Autoignition	K	597	–	–	–	–
Ames test	–	Not carcinogen	Not carcinogen	Not carcinogen	Not carcinogen	Not carcinogen
Irritation	–	Strong	–	Strong	Strong	–
Labeling by irritation	–	Corrosion, R41	–	Corrosion, R41	Corrosion, R41	–
LD50 test	mg kg^{-1}	977	–	1,373	–	–
Labeling by LD50 test	–	Harmful	–	Harmful	–	–

References

1. de Brabander-van den Berg EW, Meijer. *Angew. Chem. Int. Ed. Engl.* 1993;32(9):1308.
2. Hummelen JC, van Dongen JLJ, Meijer EW. *Chem. Eur. J.* 1997;3(9):1489.
3. Grinthal W. *Chemical Engineering.* 1993;51.
4. Dendrimer Breakthrough at DSM Could Pave the Way for Commercialisation. *Process Eng.* 74 (November 1993): 22.
5. Bosmau AW, Jausseu HM, Meijer EW. *Chem. Rev.* 1999;99:1665–1688.
6. *Materials and Technology Sekies.* Eds. R. W. Cahn, P. Haasen, E. J. Kramer, Volume: "Synthesis of Polymers", Ed. A. D. Schlüter, Chapter 12: "The Synthesis and Characterization of dendritic molecules", Jausseu, H. M., and Meijer, E. W. 403–458, Wiley-VCH Weinheim, 1999.

Poly(propylene oxide)

Qingwen Wendy Yuan-Huffman

Acronyms PPO

Class Polyether

Structure $[-CH_2-CH(CH_3)-O-]59$

Mol. Wt. of Repeat Unit

Property	Unit	Conditions			Values	References
Polymerization	None	Ring-opening polymerization				1
Typical copolymers	None	Ethylene oxide–propylene oxide copolymer				2
Solvents	None	Benzene, ethanol, dioxane, N,N-dimethylacetamide, chloroform, tetrahydrofuran, methanol (hot), acetone				3
Nonsolvent	None	Diethyl ether (sw), 2-Aminoethanol, ethyl acetate (sw), N,N-dimethylacetamide				3
Theta temp.	K	Iso-octane, virial coefficients			323.5	3, 4
Second virial coefficient	$mol\,cm^3 g^{-2}$	Solvent	Temp. (°C)	Mol. wt. $(g\,mol^{-1})$		
		Acetone	25	0.067×10^{-3}	-90×10^{-4}	3, 5
				0.125×10^{-3}	0	
				$(0.45–3.85) \times 10^{-3}$	15.2×10^{-4} (app.)	
		Hexane	46	$(783–901) \times 10^{-3}$	$(0.46–4.50) \times 10^{-4}$	3, 4
				$(34.2–4410) \times 10^{-3}$	$(3.16–0.523) \times 10^{-4}$	3, 6
		Methanol	20	$(0.535–3.31) \times 10^{-3}$	$(10.75–0.95) \times 10^{-7}$	3, 7
		Iso-octane	(48–85)	901×10^{-3}	$(0–1.58) \times 10^{-4}$	3, 4
			(50–89)	783×10^{-3}	$(-0.25–1.72) \times 10^{-4}$	3, 4

Property	Unit	Conditions			Values			References
Mark-Houwink parameter K and α	$ml\ g^{-1}$ and none	Solvent	Temp. (°C)	Mol. wt. (g mol^{-1})	$K\ (\times 10^{-3})$	α		
		Acetone	25	$(0.1–0.4) \times 10^4$	75.5	0.56	3, 8	
		Benzene	20	$(0.07–0.33) \times 10^4$	11.1	0.79	3, 9	
			25	$(3–70) \times 10^4$	11.2	0.77	3, 10	
			25	–	14	0.8	3	
		Isotactic						
		Benzene	25	$(0.5–92) \times 10^4$	38.5	0.73	3, 11	
			25	$(1–8) \times 10^4$	41.3	0.64	3, 12	
			25	$(0.05–0.4) \times 10^4$	41.5	0.65	3, 8	
		Hexane	46	$(3.4–367) \times 10^4$	19.7	0.67	3, 10	
		Methanol	20	$(0.05–0.33) \times 10^4$	40.6	0.64	3, 9	
			25	$(1–7) \times 10^4$	76.9	0.55	3, 12	
		Tetrahydrofuran	20	$(0.05–0.33) \times 10^4$	55.0	0.62	3, 9	
			25	$(3–70) \times 10^4$	12.9	0.75	3, 9	
		Toluene/2,2,4-trimethylpentane (5/7 vol)	39.5	$(1–7) \times 10^4$	107.5	0.50	3, 10	
		Oligomer						
		Acetone	20	$(0.1–0.4) \times 10^4$	75.5	0.56	3, 13	
		Benzene	20	$(0.04–0.4) \times 10^4$	41.5	0.65	3, 13	
Heat of solution	$J\ g^{-1}$	Above glass transition						
		Carbon tetrachloride, 30°C, 6×10^4 gmol^{-1}			−20		3	
		Chloroform, 30°C, 6×10^4 g mol^{-1}			−100		3	
		Methyl alcohol, 27°C, 10^3 g mol^{-1}			−7		3	
Glass transition temperature	K	(conflicting data)			198		3	
		Amorphous, atactic			201.5		2	
		Method: dynamic mechanical spectrum					14	
		PPO crosslinked with stoichiometric quantities of tris(p-isocyanatophenyl-thiophosphate)						
		$M_c = 452$ g mol^{-1}			328.3			
		725			281.3			
		1025			262.6			
		2000			241.0			
		3000			235.6			
		PPO crosslinked with stoichiometric quantities of an aromatic triisocyanate						
		$M_c = 425$ g mol^{-1}			321.1			
		725			277.6			
		1025			265.8			
Melting temperature	K	Isotactic			348.5		2	

Poly(propylene oxide)

Property	Unit	Conditions		Values		References
Heat capacity	$KJ\ K^{-1}\ mol^{-1}$ $\times 10^{-3}$	Temp. (K)		Solid	Melt	3, 14
		80		31.21		
		90		34.33		
		100		37.37		
		110		40.34		
		120		43.22		
		130		46.03		
		140		48.76		
		150		51.41		
		160		53.98		
		170		56.48		
		180		58.89		
		190		61.23		
		200			95.46	
		210			97.04	
		220			98.61	
		230			100.19	
		240			101.77	
		250			103.35	
		260			104.92	
		270			106.50	
		280			108.08	
		290			109.65	
		300			111.23	
		310			112.81	
		320			114.38	
		330			115.96	
		340			117.54	
		350			119.12	
		360			120.69	
		370			122.27	
Index of refraction	None			1.4495		3

Specific refractive index increment, dn/dc	$ml\ g^{-1}$	Solvent	Temp. (°C)	Mol. wt. $(g\ mol^{-1})$	$\lambda_0 = 436$ nm	$\lambda_0 = 546$ nm	
		Acetone	25	67	0.085		5
				125	0.0915		5
				450	0.096		5
				1,100	0.099		5
				1,200	0.099		5
				2,100	0.100		5
				3,270	0.100		5
				3,850	0.101		5
		Benzene	25		−0.0530	−0.0448	6
		Chlorobenzene	25		−0.0658	0.0638	6

Property	Unit	Conditions			Values		References
Specific refractive index increment, dn/dc	ml g^{-1}	Solvent	Temp. (°C)	Mol. wt. (g mol^{-1})	$\lambda_0 = 436$ nm	$\lambda_0 = 546$ nm	
		n-Hexane	25		0.0775	0.0775	6
			40		0.0460	0.0460	6
			46	9.6×10^5	0.0887	0.0887	6
				2.0×10^5	0.0895	0.0895	6
			57	9.6×10^5	0.101	0.101	6
			57	2.0×10^5	0.0104	0.0104	6
		Feron 113	25		0.118	0.115	6
		Iso-Octane	25		0.0655	0.0655	6
		Methanol	24		0.137	0.135	3
			25	12.2×10^5	0.118	0.118	6
			25	12.5×10^5	0.115	0.115	6
Dipole moment	D	Benzene, $T = 25°C$, $P_n = 6.6$–69.0			1.40–1.02		3, 15

Property	Unit	Conditions	Values			References
Surface tension	mN m^{-1}		20°C	150°C	200°C	3
		Diol $M = 2025$ g mol^{-1}	31.5	21.1	17.1	
		Diol $M = ?$	31.7	20.6	16.4	
		Diol $M = 3000$ g mol^{-1}	31.2	20.9	17.0	
		Diol $M = 400$ to 4100 g mol^{-1}	31.1	21.6	17.9	
		Poly(oxypropylene)-dimethylether $M = 3000$ g mol^{-1}	0.7	18.3	13.6	

Property	Unit	Conditions			Values	References
Diffusion coefficient	cm^2 s^{-1} $\times 10^{-7}$	Solvent	Temp. (°C)	Mol. wt. (g mol^{-1})		3
		Acetone	20	$(0.074$–$3.375)$ $\times 10^3$	$D_0 = K_s \times M^{-0.52}$	
		Benzene	20	$(0.134$–$3.375)$ $\times 10^3$	$D_0 = K_s \times M^{-0.55}$	
		Water	25	40×10^3	3.73	
				73×10^3	2.09	
				148×10^3	1.66	
				278×10^3	1.72	
				661×10^3	1.07	
			15	148×10^3	1.73	
			34		2.47	
			43		2.88	
			30.3		2.72	
			35.4		3.08	
			40.5		3.25	
			45.5		3.49	

Poly(propylene oxide)

Property	Unit	Conditions	Values		References
Permeability coefficient	10^{-19} m^3 m s^{-1} m^{-2} Pa^{-1}	PPO crosslinked with stoichiometric quantities of tris(p-isocyanatophenyl-thiophosphate)	H$_2$	CO	3, 16, 17
		$M_c = 425$	2.86	0.0608	
		725	7.26	0.65	
		1025	18.80	2.93	
		2000	28.73	7.65	
		3000	44.10	12.60	

Crystalline-state properties

Crystl. Syst. (Lattice)	Space Group	Unit Cell Parameters			Monomer per Unit Cell	Density (g cm^{-3})	Heat of Fusion (kJ mol^{-1})	References
		A	B	C				
		(A)	(A)	(A)				
Orthorhombic	C2V-9 or D2-4	10.52	4.67	7.16	4	1.097	8.4	3
Orthorhombic	D2-4	10.51	4.69	7.09	4	1.104		3
Orthorhombic	D2-4	10.52	4.68	7.10	4	1.104		3
Orthorhombic	D2-4	10.40	4.64	6.92	4	1.155		3
Orthorhombic	D2-4	10.46	4.66	7.03	4	1.126		3

References

1. Rodriguez F. *Principles of Polymer Systems.* 4th ed. Taylor & Francis Publishers; 1996.
2. Mark HS, et. al., *Encyclopedia of Polymer Science and Engineering,* Vol 6. Wiley Interscience; 1986.
3. Brabdrup J, Immergut EH. *Polymer Handbook.* 3rd ed. New York: Wiley Interscience; 1989.
4. Allen G, Booth C, Orice C. *Polymer.* 1966;7:167.
5. Meyerhoff G, Moritz U. *Makromol. Chem.* 1968;109:143.
6. Allen G, Booth C, Jones MN. *Polymer* (London). 1964;5:195.
7. Scholtan W, Lie SY. *Makromol. Chem.* 1967;108:315.
8. Meyerhoff G, Moritz U. *Makromol. Chem.* 1967;109:143.
9. Scholtan W, Lie SY. *Makromol. Chem.* 1967;108:104.
10. Alen G, Booth AC, Jones MM. *Polymer.* 1964;5:195.
11. Valles RJ. *Makromol. Chem.* 1968;113:147.
12. Moacanin J. *J. Appl. Polym. Sci.* 1959;1:272.
13. Meyerhoff G. *Makromol. Chem.* 1971;145:189.
14. Guar U, Wunderlich B. *J. Phys. Chem. Ref. Data.* 1981;10 (4):1033.
15. Loveluck GD. *J. Chem. Soc.* 1961:4729.
16. Andrady AL, Sefcik MD. *J. Polym. Sci.* 1983;21:2453.
17. Andrady AL, Sefcik MD. *J. Polym. Sci.* 1984; 22:237.

Poly(propylene sulfide)

Junzo Masamoto

Acronym PPS

Class Polysulfides

Structure $+\text{CH(CH}_3)\text{S}+_n$

Major Applications Poly(propylene sulfide) is an elastic material that compares with styrene-butadiene rubbers. However, this polymer has not yet achieved commercial production, although the PPS elastomer offers a combination of good solvent and weather resistance. Low molecular weight functional PPS is suitable for use in sealants, adhesive, etc.[1]

Properties of Special Interest Poly(propylene sulfide) is an elastic material with an excellent combination of good solvent- and weather-resistance with an acceptable level of physical and dynamic properties. It also gives both types of crystalline stereoregular polymer and amorphous atactic polymer depending on the initiator.[2] By using an optically active coordination initiator, an isotactic optically active polymer can be obtained.[3–7]

Other Polymers Showing These Special Properties Solvent resistance: polysulfide rubbers; second order transition temperature: styrene-butadiene rubbers; weather and ozone resistance: polychloroprene rubbers.

Preparative Technique Poly(propylene sulfide) can be prepared by ring-opening polymerization, using anionic, cationic, and coordinate catalyst. Anionic and cationic systems give an amorphous atactic polymer, while coordinate catalytic system, such as cadmium salts, give an isotactic or crystalline polymer.[2,8]

The monomer undergoes polymerization by an anionic mechanism with basic initiators:[1]

$$\text{(Anionic)} \quad -\text{CH}_2\overset{\overset{\displaystyle \text{Me}}{|}}{\text{C}}\text{HS}^-$$

$$\text{(Cationic)} \quad -\overset{\overset{\displaystyle \text{Me}}{|}}{\text{S}}\text{CHCH}_2\text{S}^+\!\!\!\stackrel{\displaystyle \overset{\text{Me}}{|}\ \ \ C}{\diagdown \text{CH}_2}\quad \text{B}^- \quad (\text{B}^-=\text{BF}_3\text{OH}^-)$$

$$\text{(Coordinate)} \quad -\text{Zn}^+\ \ ^-\!\!(\overset{\overset{\displaystyle \text{Me}}{|}}{\text{S}}\text{CHCH}_2)_n- \quad \longrightarrow \quad -\text{Zn}^+\ \ ^-\!\!(\overset{\overset{\displaystyle \text{Me}}{|}}{\text{S}}\text{CHCH}_2)_{n+1}{}^-$$

$$\overset{|}{\underset{\underset{\underset{\text{Me}}{|}}{\text{CH}_2-\text{CH}}}{\text{S}}}$$

Poly(propylene sulfide)

The cationic polymerization by initiators such as boronfluoride, etherate, probably involves the intermediary of sulphonium ions:[1] In polymerization initiated by zinc or cadmium compounds, the metal-sulfur bond will be predominantly covalent, and it is possible that the monomer is coordinated to the metal atom before insertion into the growing chain.

Property	Units	Conditions	Value	Reference
Molecular weight (of repeat unit)	$g\,mol^{-1}$	–	74	–
Tacticity (stereoregularity)	%	Coordination polymerization, cadmium thiolate catalyst	Isotactic: 90–100 meso dyads	(6, 9–11)
		Anionic polymerization, sodium thiolate active center	Atactic	(10)
		Zinc N-substituted porphyrins	Atactic	(12)
Typical molecular weight range of polymer	$g\,mol^{-1}$	Anionic polymerization, active center: sodium thiolates; determined by osmotic pressure method	M_n: $1-6 \times 10^5$	(13)
		For amorphous PPS, in toluene solution at 35°C	Intrinsic viscosity 0.5–3.0	(8)
		In benzene solution at 25°C, $ZnEt_2/H_2O$, cadmium tartrate initiator	Intrinsic viscosity 2.0–4.0	(14)
		For crystalline PPS, in toluene solution at 35°C	Intrinsic viscosity 0.5–2.5	(8)
		KSCN initiator with a cryptate	M_w: 1.7×10^7	(15)
		Rare earth coordination catalyst	M_w: $1-5 \times 10^6$	(16)
		Initiator: zinc N-substituted porphyrins	M_n: $1 - 27 \times 10^3$	(12)
		Initiator: cadmium thiolate	M_n: $3-15 \times 10^4$	(7)
Typical polydispersity index (M_w/M_n)	–	Anionic polymerization; active center: sodium thiolates, tetrahydrofuran solvent	1.1–1.2	(13)
		Anionic polymerization; initiator: sodium naphthalene, tetrahydrofuran solvent	<1.1	(17)
		Initiator: zinc N-substituted porphyrins	1.05	(11)
		Initiator: cadmium crotyl mercaptide, M_w: $(1.1 \sim 15.8) \times 10^4$	$1.7 \sim 2.3$	(18)
IR (characteristic absorption frequencies)	cm	$-CH_2-$ deformation	1,449	(8)
		Symmetrical $-CH_3$ deformation	1,379	
		Asymmetrical $-C-S-$ stretching vibration	735	
		Symmetrical $-C-S-$ stretching vibration	684	
NMR		^1H NMR, 300-MHz NMR at 17°C in deutrated chloroform or carbon tetrachloride		(19)
		^{13}C NMR, operating at 25 MHz in $CCl_4-C_6D_6$		(9)
		(90/10) at 160°C or at 60°C		(12)

Property	Units	Conditions	Value	Reference
Thermal expansion coefficients	K^{-1}	Atactic PPS, $M_w = 5 \times 10^5$	0.59×10^{-3}	(20)
Density (amorphous)	g cm^{-3}	At 25°C, by pyconometry measurement	1.0340	(20)
			1.130	(8)
Solvents		(Cyclic) propylene sulfide		(20)
		Benzene, tetrahydrofuran, toluene, carbon tetrachloride, o-dichlorobenzene		(8)
Nonsolvents	–	For atactic PPS	Methyl ethyl ketone	(21)
		For crystalline PPS	Heptane, cyclohexane, dibutylphthalate, aqueous hydrochloric acid, aqueous sodium hydroxide	(8)
Solubility parameter	MPa	–	17.9	(21)
Mark-Houwink parameters: K and a	$K = $ ml g^{-1} $a = $ None	Benzene, 20°C Benzene, 31°C	$K = 3.3 \times 10^{-5}$; $a = 0.86$	(23)
			$K = 5.036 \times 10^{-5}$; $a = 0.78$	(18)
		Tetrahydrofuran, 25°C	$K = 2.58 \times 10^{-4}$; $a = 0.656$	(24)
Characteristic ratio $\langle r^2 \rangle / nl^2$	–	Atactic PPS	4.0	(25)
Lattice	–	–	Orthorhombic	(21)
Space group	–	–	$P2_12_1$-D_24	(21)
Chain conformation	–	–	Planar zigzag (but not fully extended)	(21, 24)
Unit cell dimensions	Å	X-ray photograph of oriented samples of both types of optical active and racemic PPS; both: isotactic	$a = 9.95$, $b = 4.89$, $c = 8.20$ (fiber axis)	(21)
Unit cell contents	–	–	4 monomeric units per unit cell (2 molecular chains)	–
Degree of crystallinity	%	Initiator: Zn/H$_2$O, DTA and X-ray diffraction	60	(14)
		Initiator: cadmium tartrate, DTA and X-ray diffraction	85	
Density	g cm^{-3}	Theoretical density for crystalline PPS	1.24	(21)
		Observed density for crystalline PPS	1.152	(21)
		–	1.16	(8)
Glass transition temperature	K	–	220.5	(27, 28)
		Amorphous and crystalline PPS	225	(8)
			233	(29)

Poly(propylene sulfide)

Property	Units	Conditions	Value	Reference
		Sulfur-vulcanized carbon black filled propylene sulfide-allyl thioglycidil ether copolymer prepared by coordination catalyst	233	(2)
		Viscoelastic measurement	234.6	(29)
		Calorimetric	236	(31)
Melting point	K	Isotactic PPS	313–314	(8)
			325	(9)
		Calorimetric, Et$_2$Zn-S catalyst	326	(30)
		Isotacticity: >90%	331	(7, 14)
Tensile strength	MPa	Sulfur-vulcanized PPS-allyl thioglycidyl ether copolymer filled with carbon black	11	(2)
		Sulfur-vulcanized ethylene sulfide (28 mol%), propylene sulfide (69 mol%), allyloxymethyl thiarne (3 mol%), terpolymer filled with carbon black	15.9	(8)
		PPS homopolymer filled with carbon black	13	(1)
		PPS homopolymer, cured without carbon black	1.2	(1)
Maximum extensibility (elongation)	%	Sulfur-vulcanized PPS-allyl thioglycidyl ether copolymer filled with carbon black	225	(2)
		Sulfur-vulcanized ethylene sulfide propylene sulfide allyloxymethyl thiarne terpolymer filled with carbon black	360	(8)
		PPS filled with carbon black	205	(1)
		PPS cured without carbon black	325	(1)
Hardness	Shore A	Sulfur-vulcanized PPS-allyl thioglycidyl ether copolymer filled with carbon black	80	(2)
		Sulfur-vulcanized ethylene sulfide propylene sulfide allyloxymethyl thiarne terpolymer filled with carbon black	81	(8)
		PPS filled with carbon black	76–70	(1)
		PPS cured without carbon black	38–30	(1)
Modulus	MPa	Modulus at glassy state: viscoelastic method	2200	(31)
		Modulus at rubbery state: viscoelastic method	4.3	(31)
		At 300% elongation, sulfur-vulcanized ethylene sulfide propylene sulfideallyloxymethyl thiarne terpolymer filled with carbon black	13.8	(8)
		At 100% elongation, PPS filled with carbon black	5.8	(1)
		At 100% elongation, PPS cured without carbon black	0.55	(1)
Rebound	%	PPS filled with carbon black	54	(1)
Entanglement molecular weight	g mol^{-1}	M_w: $0.3 - 86 \times 10^4$	2×10^4	(18)
Index of refraction	–	–	1.596–1.597	(30)
		At 23°C	1.594	(18)

Property	Units	Conditions	Value	Reference
Refractive index increment dn/dc	ml g^{-1}	–	8.095×10^{-2}	–
Dipole moment ratio $\langle \mu^2 \rangle / nm^2$	–	In benzene at 25°C, isotactic PPS	0.39	(10)
		In benzene at 25°C, atactic PPS	0.44	
		In carbon tetrachloride at 25°C, isotactic PPS	0.33	
		In carbon tetrachloride at 25°C, atactic PPS	0.37	
Melt viscosity	Pa s	For M_w =		(18)
		7,000 at 30°C	0.006	
		15,000	0.03	
		29,000	0.2	
		60,000	1.6	
		118,000	15.8	
		158,000	54.7	
Weathering test	–	In Toronto, one year's exposure; sulfur-vulcanized ethylene sulfide propylene sulfide allyloxymethyl thiarne terpolymer filled with carbon black	No outward change	(8)
Solvent resistance	% volume swell	7 days at room temperature; sulfur-vulcanized propylene sulfide-allyl thioglycidyl ether copolymer filled with carbon black		(2)
		Ethyl acetate	75	
		Methyl ethyl ketone	101	
		Hexane	10	
		Toluene	173	
Availability		No commercial production		

References

1. Cooper W. *Br. Polym. J.* 1971;3:28–35.
2. Gobran RH. In: Mark HF, et al. ed. *Encyclopedia of Polymer Science and Technology.* Vol 10. New York: Interscience; 1969:324–336.
3. Dumas Ph, Guerin Ph, Sigwalt P. *Nouv. J. Chim.* 1980;4:95–99.
4. Sigwalt P. *Makromol. Chem., Suppl.* 1979;3:69–83.
5. Guerin Ph, Boileau S, Sigwalt P. *Eur. Polym. J.* 1980;16:129–133.
6. Dumas Ph, Sigwalt P, Guerin Ph. *Makromol. Chem.* 1981;182:2225–2231.
7. Dumas Ph, Sigwalt P. *Chirality.* 1991;3:484–491.
8. Adamek S, Wood BBJ, Woodhams RT. *Rubb. Plast. Age* 1965;46 (1):56–62.
9. Guerin P, et al. *E Polym. J.* 1975;11:337–339.
10. Riande E, et al. *Macromolecules.* 1979;12:702–704.
11. Palacios J. *Rev. Soc. Quim. Mex.* 1996;40:147–154.
12. Aida T, Kawaguchi K, Inoue S. *Macromolecules* 1990;23:3887–3892.
13. Sigwalt P. *IUPAC Internal. Symp. Macromol. Chem.* Budapest; 1969:251–280.
14. Marchetti M, et al. *Makromol. Chem.* 1979;180:1305–1312.
15. Boileau S, et al. *J. Polym. Sci., Polym. Lett. Ed.* 1974;12:217–224.
16. Zhi-quan S, et al. *Sci. China B.* 1990;33:553–561.
17. Nevin RS, Pearce EM. *J. Polym. Sci., Part B.* 1965;3:491.

18. Stokes A. *Eur. Polym. Sci.* 1970;6:719–723.

19. Sepulchre M, et al. *J. Polym. Sci., Polym. Chem. Ed.* 1974;12:1683–1693.

20. Rahalkar RR, et al. *J. Polym. Sci., Polym. Phys. Ed.* 1979;17:1623–1625.

21. Sakakihara H. et al. *Macromolecules.* 1969;2:515–520.

22. Chiro A, Raggi E. *Chim. Indust.* 1973;55:512–513.

23. Eskin VE, Nesterov AE. *Vysokomolek. Soedin.* 1966;8(1):141–145.

24. Nash DW, Pepper DC. *Polymer.* 1975;16:105–109.

25. Abe A. *Polym. Prep.* 1979;20(1):460–462.

26. Abe A. *Macromolecules.* 1980;13:541–546.

27. Woodhams RT. *Rep. Prog. Appl. Chem.* 1965;50:480–484.

28. Adamek S, Wood BBJ, Woodhams RT. *Rubber Age.* 1965;96:581–585.

29. Nevin RS, Pearce EM. *J. Polym. Sci. Part B.* 1965;3:487.

30. Lal J, Trick GS. *J. Polym. Sci. Part A-1.* 1970;8:2339–2350.

31. Takahashi M. In: Onogi, ed. *Proc. 5th Int. Congr. Rheol.* Tokyo: University of Tokyo Press; 1970;3:399–407.

Polypropylene, syndiotactic

Charles L. Myers and Anita Dimeska

Acronyms, Trade Names s-PP, sPP, FINAPLAS®

Class Poly(α-olefins)

Structure —[CH$_2$CH(CH$_3$)]—

Major Applications Films and packaging, medical applications, adhesive applications, and as a resin modifier in blends. Metallocene syndiotactic polypropylene grades are now commercially available.

Properties of Special Interest Transparency, flexibility, toughness, heat seal temperature, radiation stability, broad melt processing range, high melt strength, and low level of extractables.[1−5]

Property	Units	Conditions			Value	Reference
Density	g cm^{-3}	Unit cell, 100% crystalline (obsolete cell interpretation)			0.93	(6)
		25°C, experimental sample not defined			0.989–0.91	(6)
		25°C, amorphous, extrapolated from melt temperature			0.856	(6)
		Three s-PP (Fina)				(2, 7)
		% r	% rrrr	% crystallinity XRD		
		91.4	76.5	21	0.87	
		91.9	78.0	22	0.87	
		96.5	91.1	29	0.89	
		Two s-PP (Hoechst)				(3)
		G 1			0.887	
		G 2, 83.6% rrrr, 27% crystalinity			0.885	
Melting temperature	K	rrrr = 72%			392	(8)
		rrrr = 82%			413	
		Three s-PP (Fina)				(2, 7)
		% r	% rrrr	% crystallinity		
		91.4	76.5	21	398	
		91.9	78.0	22	399	
		96.5	91.1	29	421	
		G 2 (Hoechst), 83.6% rrrr			406	(3)
		rrrr = 93%			422	(9)

Property	Units	Conditions			Value	Reference
Melting temperature (equilibrium values, Hoffmann-Weeks)	K	rrrr = 92–95%			433–459	(10)
		r = 94%, rrrr = 86%, M_n > 40,000			433	(11)
		r = 96.8%, rrrr = 92.1%, M_w = 164,000			439	(12)
		r = 91.9–98.0%, rrrr = 81.4–94.5%			408–459	(12)
		–			424–428	(13)
		Extrapolated to 100% syndiotacticity			487	(12)
		–			493	(14)
Glass transition temperature	K	DSC rrrr = 83%, M_w = 60,000 g mol^{-1}			270	(15)
Heat of fusion (equilibrium values, for 100% crystallinity)	kJ mol^{-1}	rrrr = 92–95%			4.4–8.2	(10)
		r = 94%, rrrr = 86%, M_n > 40,000			8.0	(11)
		r = 96.8%, rrrr = 92.1%, M_w = 164,000			6.9	(12)
		–			8.4	(13)
Entropy of fusion	JK^{-1} mol^{-1}	DSC, density			18.8	(16, 17)
Theta temperature	K	M_w = 11,700, cyclohexane			309	(18, 19)
Entanglement molecular weight	g mol^{-1}	463 K, calculated, metallocene			2,170	(20)
Infrared absorption	cm^{-1}	Attributed to				(2, 6)
		Helix			866, 867	
		Helix			977	
		Regularity			962	
Flexural modulus	MPa	Three s-PP (Fina)				(2, 7)
		% r	% rrrr	% crystallinity		
		91.4	76.5	21	380	
		91.9	78.0	22	415	
		96.5	91.1	29	760	
		Homopolymer			359	(1)
		Clear, impact grades			88–250	(1)
		Two s-PP (Hoechst)				(3)
		G 1			790	
		G 2, 83.6% rrrr			600	
Tensile modulus	MPa	Homopolymer			483	(1)
		Clear impact grades			211–244	
Tensile elongation, yield	%	Homopolymer			10.8	(1)
Tensile elongation, break	%	Homopolymer			180	(1)

Unit cell information

Comments	Lattice	Packing	Monomers per Unit Cell	Cell Dimensions (Å)			Space Group	Crystal Density	Reference
				a	b	c			
Form I High order	Orthorhombic	Helical Antichiral	16	14.5	11.2	7.4	Ibca	–	(11, 21–24)
	Monoclinic	Helical Antichiral	16	14.3	11.2	7.5	$P2_1/a$	–	(24, 25)
Disorder	Orthorhombic	Helical Antichiral	8	14.5	5.6	7.4	Pcaa	–	(11, 21–23)
Form II Annealed fiber	Orthorhombic	Helical Isochiral	8	14.5	5.6	7.4	$C222_1$	0.93 0.90	(6, 11, 21–23)
Form III Quenched, cold drawn, metastable	Orthorhombic	Planar Zigzag	4	5.22	11.17	5.06	$P2_1cn$	0.945	(6, 11, 26)
Form IV Metastable	Triclinic	Deformed helix or intermediate	6	5.72	7.64	11.6	–	0.939	(11, 27)
Phase Diagram									(28)

References

1. Shamshoum E, Kim S, Hanyu A, Reddy BR. In: *Metallocenes '96, Proceedings of the 2nd International Conference on Metallocene Polymers* (Düsseldorf, Germany). Skillman, NJ. Schotland Business Research; 1996:259.

2. Shamshoum E, Sun L, Reddy BR, Turner D. *MetCon '94 (Houston, Tex.)*. Spring House Penn: Catalyst Consultants; 1994.

3. Antberg M, et al. *Makromol. Chem., Macromol. Symp.* 1991;48/49:333.

4. *Plastics Technology* 38 (March 1992): 29–31.

5. McLeod MA, Gauthier WJ. In: *International Conference on Polyolefins.* Houston, TX: SPE; February 2004

6. Quirk RP, Alsamarraie MAA. In: Brandrup J, Imergut EH, eds. *Polymer Handbook.* 3rd ed. Vols 27–31. New York: John Wiley and Sons; 1989. (Note: Considerable new information regarding s-PP crystalline polymorphs is available since 1989.)

7. Moore EP Jr, ed. *Polypropylene Handbook.* New York: Hanser Publishers; 1996, chap. 12:406.

8. Ewen JA, et al. *Makromol. Chem., Macromol. Symp.* 1991;48/49:253.

9. De Rosa C, et al. *Macromolecules.* 2004;37:7724.

10. Phillips RA, Wolkowicz MD. In: Moore EP Jr, ed. *Polypropylene Handbook.* New York: Hanser Publishers; 1996, chap. 3.4:144–149.

11. Rodriguez-Arnold J, et al. *Polymer.* 1994;35(9):1884. (Includes review of s-PP polymorphs.)

12. Balbontin G, Dainelli D, Galimberti M, Paganetto G. *Makromol. Chem.* 1992;193:693.

13. Haftka S, Koennecke K. *J. Macromol. Sci., Phys.* 1991;B30(4):319. (Compares s-PP to i-PP of same sequence distribution.)

14. Miller RL, Seely EG. *J. Polymer Sci., Polym. Phys. Ed.* 1982;20:2297.

15. Men Y, Strobl G. *J. Macromol. Sci.-Physics.* 2001;40:775.

16. Galambos A, Wolkowicz M, Zeigler R, Galimberti M. *ACS Preprints, PMSE Div.* April 1991.

17. Mandelkern L, Alamo RG. In: Mark JE, ed. *Physical Properties of Polymers Handbook.* Woodbury, N.Y: AIP Press; 1996, chap. 11:132.

18. Sundararajan PR. In: Mark JE, ed. *Physical Properties of Polymers Handbook.* Woodbury, N.Y: AIP Press; 1996, chap. 15:202.

19. Hirao T, et al. *Polym. J.* 1991;23:925.

20. Eckstein A, et al. *Macromolecules.* 1998;31:1335.

21. Lovinger AJ, Lotz B, Davis DD, Padden FJ Jr. *Macromolecules* 1993;26:3494.

22. DeRosa C, Corradini P. *Macromolecules.* 1993;26:5711.

23. Lovinger AJ, Lotz B, Davis DD, Schumacher M. *Macromolecules.* 1994;27:6603.

24. DeRosa C, Finizia A, Corradini P. *Macromolecules.* 1996;29:7452.

25. DeRosa C, Auriemma F, Vinti V. *Macromolecules.* 1997;30:4137.

26. Chatani Y, et al. *J. Poly. Sci., Part C.* 1990;28:393.

27. Chatani Y, Maruyama H, Asanuma T, Shiomura T. *J. Poly. Sci., Poly. Phys.* 1991;29:1649.

28. Ruiz de Ballesteros O, et al. *Macromolecules.* 2007;40:611.

Poly(pyromellitimide-1,4-diphenyl ether)

Loon-Seng Tan

Acronyms, Trade Names ODA-PMDA, PMDA-ODA, Kapton®, PI2545, Vespel®, Pyralin® (polyamic acid precursor), Pyralux® (flexible laminates); Interra™

Class High performance polymer

Structure

CAS Registry: 25036-53-7

Synthesis Poly(pyromellitimide-1,4-diphenyl ether) is generally prepared from polycondensation of pyromellitic dianhydride and 4,4′-oxydianiline followed by either thermal or chemical (in the presence of acetic anhydride and triethylamine) cyclodehydration of the polyamic acid precursor.

Major Applications Kapton films are used as wire and cable wrap, formed coil wrap, motor-slot liners, substrates for flexible printed circuit boards, insulation for aircraft and spacecraft wiring, automotive applications such as switches and diaphragms, and in seat heaters and airbag seat sensors, magnet-wire insulation, and in transformers and capacitors, pressure-sensitive tapes, bar code labels, and in applications such as the solar array for thermal management.[1] Vespel molded parts are used in automobiles, large on-and-off-road vehicles, farm equipment, business machines, electronic equipment etc.: rotary seal rings, thrust washers and discs, bushings, flanges bearings, printer platen bars, plungers, printer wireguides, stripper fingers, spline couplings, wear strips, locknut inserts, valve seats, check valve balls, thermal and electrical insulators.[2]

Properties of Special Interest Kapton films have excellent thermal stability in air or inert atmosphere, useful mechanical properties over very broad temperature range, outstanding electrical properties and stability of these electrical properties over wide range of relative humidity, insensitive to solvents, excellent radiation resistance; considerable variation in hydrolytic sensitivity, poor hydrolytic resistance in 10% NaOH.[3] Vespel direct-formed parts are resistant to thermally harsh environment, creep, impact and wear and friction at high pressures and velocities.[4]

Poly(pyromellitimide-1,4-diphenyl ether)

Product Names	Product Descriptions	Supplier
Kapton®	Polyimide films available in three types; (a) HN film (b) VN films and (c) FN films. Both HN and VN films are all-polyimide films but FN films are coated on one or both sides with Teflon FEP fluoropolymer resin	DuPont High Performance Films, U.S. Rt.23& DuPont Rd. P. O. Box 89, Circleville, OH 43113.
		Du Pont de Nemours (Luxembourg S.A.) Contern, L-2984 Luxembourg, Grand Duchy of Luxembourg.
		Du Pont Kabushi Kaisha Arco Tower 8-1, Shimomeguro 1-chrome Meguro-ku, Tokyo 153 Japan.
		http://www2.dupont.com/Kapton/en_US/index.html
Vespel®	Available in five compositions: (a) SP-1, unfilled based resin; (b) SP-21, 15% by wt. graphite filler; (c) SP-22, 40% by wt. graphite filler; (d) SP-211, 15% by wt. graphite and 10% by wt. Teflon fluorocarbon resin fillers; (d) SP-3, 15% by wt. Molybdenum disulfide (for lubrication)	Du Pont Engineering Polymers Pencader Site, Newark, DE 19714-6100
		Du Pont de Nemours (Belgium) N. V. Du Pont Engineering Polymers Antoon Spinoystraat 6 B-2800 Mechelen, Belgium
		Du Pont Japan Limited Specialty Polymers VESPEL, Marketing 19-2, Kiyohara, Kogyo Danchi, Utsunomiya, Tochigi, 321–332, Japan.
		http://www2.dupont.com/Vespel/en_US/index.html
Interra™ HK 04	polyimide dielectric laminate with 18–70 μm electro-deposited copper	DuPont Electronic Materials 14 T.W. Alexander Drive Research Triangle Park, NC 27709
		http://www2.dupont.com/Interra/en_US/tech_info/index.html
DuPont™ pyralux®	Flexible, copper cladded Kapton laminates in either adhesive based or adhesiveless form.	DuPont Flexible Circuit Materials 14 T. W. Alexander Drive Research Triangle Park, NC 27709
		http://www2.dupont.com/Pyralux/en_US/index.html

Mechanical properties of Kapton HN film (25 μm)

Property	Units	Conditions	Value	Reference
Ultimate tensile strength	MPa	ASTM D-882, Method A; film size, 25 × 150 mm; 23°C	231	(1, 5)
	MPa	ASTM D-882, 200°C	139	
Yield point at 3%	MPa	ASTM D-882, 23°C	69	(1, 5)
	MPa	ASTM D-882, 200°C	41	
Stress to produce 5% elongation	MPa	ASTM D-882, 23°C	90	(1, 5)
	MPa	ASTM D-882, 200°C	61	
Ultimate elongation	%	ASTM D-882, –195°C	2	(5)
	%	23°C	72	(1, 5)
	%	200°C	83	
Tensile modulus	GPa	ASTM D-882, –195°C	3.5	(5)
	GPa	23°C	2.5	(1, 5)
	GPa	200°C	2.0	(1, 5)
	GPa	Nanoindentation	3.78	(6)

Poly(pyromellitimide-1,4-diphenyl ether)

Property	Units	Conditions	Value	Reference
Folding endurance (MIT)	cycles	ASTM D-2176, 23°C	285,000	(1, 5)
Tear strength-propagating (Elmendorf)	N	ASTM D-1922, 23°C	0.07	(1, 5)
Tear strength-initial (Graves)	N	ASTM D-1004, 23°C	7.2	(1, 5)
Poisson's ratio	–	average of three samples elongated at 5%, 7% and 10%	0.34	(1, 5)
Hardness	GPa	Nanoindentation	0.248	(6)

Thermal properties of Kapton HN film (25 μm)

Property	Units	Conditions	Value	Reference
Melting point	°C	ASTM E-794	None	(1, 5)
Thermal coefficient of linear expansion	10^{-6} °C^{-1}	ASTM D-696; –14–38°C	20	(1, 5)
Thermal coefficient of linear expansion	ppm K^{-1}	Thermomechanical analysis	27.1	(7)
Coefficient of thermal conductivity	W m^{-1} °C^{-1}	ASTM F-433, 296 K	0.12	(1)
Coefficient of thermal conductivity	W m^{-1} °C^{-1}	3ω technique;[8] film thickness, 0.7–12.5 nm	0.22	(9)
Coefficient of thermal expansion	10^{-5} °C^{-1}	ASTM E831 flow X E-5, from –20°C to 150°C	5.95	(10)
Storage modulus	MPa	Dynamic mechanical analysis (DMA)	179.3 (100°C) 116.6 (200°C)	(10)
tan δ_{max}	°C	Dynamic mechanical analysis (DMA)	225.8	(10)
Loss modulus (E″)	°C	Dynamic mechanical analysis (DMA)	223.3	(10)
Specific heat	J g^{-1} °C^{-1}	Differential calorimetry	1.09	(1)
Flammability	–	UL-94	94V-0	(1)
Shrinkage	%	IPC TM 650, Method 2,2.4A; 30 min at 150°C	0.17	(1)
	%	IPC TM 650, Method 2,2.4A; 30 min at 250°C	0.3	(5)
	%	ASTM D-5214; 120 min at 400°C	1.25	(1)
Smoke generation	–	NFPA-258; NBS Smoke Chamber	DM≤ 1	(1)
Glass transition temperature	°C	–	360–410	(1)
Glass transition temperature	°C	TMA	407	(10)
Cut-through temperature	°C	25 μm thickness; at 25°C	435	(1, 5)
Cut-through temperature	°C	50–125 μm thickness; at 25°C	525	(1, 5)

Poly(pyromellitimide-1,4-diphenyl ether)

Optical and spectroscopic properties of Kapton film

Property	Units	Conditions	Value	Reference
refractive index	–	Visible range; film thickness 25 μm)	1.70–1.80	(11, 12)
In-plane refractive index	–	Measured with a prism coupler at 632.8 nm	1.6390	(6)
In-plane refractive index	–	Measured with a prism coupler at 632.8 nm	1.6759	(7)
Out-of-plane refractive index	–	Measured with a prism coupler at 632.8 nm	1.7219	(6)
Out-of-plane refractive index	–	measured using a prism-coupler at 1320 nm	1.6068	(7)
Birefringence (Δn)	–	Measured with a prism coupler at 632.8 nm	0.0829	(6)
Absorption edge (λ_g)	nm	Kapton H-film (75 μm); UV-Visible	536.8	(13)
Band gap energy	eV	spectroscopy (200–800 nm)	2.31	
Characteristic IR peaks	cm^{-1}	FT-IR	$1780_{sym}\nu$(C=O) $1720_{asym}\nu$(C=O) $1370\ \nu$(C—N) 730 (C=O bending)	(14)
Characteristic raman peaks	cm^{-1}	FT-Raman	1121 (imide group) 1389 (imide group) $1788_{sym}\nu$(C=O)	(14)

Film thermo-optic coefficients dn/dT at 1.32 μm

Thickness d (μm)	Refractive Indices		Thermo-Optic coefficient (ppm K^{-1})				Reference
	n_{av}	Δn^a	dn_{TE}/dT	dn_{TM}/dT	dn_{av}/dT	Δ(dn/dT)	
9.1	1.6478	0.0686	−104	−72	−94	−32	(15)

Notes: The subscripts TE, TM, and av. denote in-plane, out-of-plane, and average, in that order.

a. $n_{av}^2 = (2n_{TE}^2 + n_{TM}^2)/3$.

b. in-plane/out-of-plane birefringences, $\Delta n = n_{TE} - n_{TM}$.

c. $\Delta(dn/dT) = dn_{TE}/dT - dn_{TM}/dT$.

Electrical properties of Kapton HN film (25 μm)

Property	Units	Conditions	Value	Reference
Dielectric strength	V μm^{-1}	ASTM D-149; 23°C, 50% RH, 60 Hz, 1/4 in electrodes, 500 V s^{-1} rise	303	(1, 5)
Dielectric constant	–	ASTM D-150; 23°C, 50% RH 10^3 Hz, 23°C, 50% RH	3.4	(1, 5)
	–	ASTM D-150; 23°C, 50% RH 10^3 Hz, 200°C 50% RH	3.0	(1, 5)
Dissipation factor	–	ASTM D-150; 23°C, 50% RH 10^3 Hz, 23°C 50% RH	0.003	(1, 5)
	–	ASTM D-150; 23°C, 50% RH 10^3 Hz, 200°C 50% RH	0.002	(1, 5)
Volume resistivity	ohm cm	ASTM D-257; 23°C, 50% RH	10^{18}	(1, 5)
	–	ASTM D-150; 200°C, 50%RH,	10^{14}	(1, 5)

Poly(pyromellitimide-1,4-diphenyl ether)

Property	Units	Conditions	Value	Reference
Corona start voltage	volts	at 50% RH, 25°C	465	(1, 5)
surface resistivity	ohm	25°C	10^{16}	(1, 5)
Loss tangent: tan δ	–	60 Hz	0.003	(12)
	–	1 KHz	0.0025	(12)
	–	1 MHz	0.011	(12)
Dielectric breakdown	V cm^{-1}	–	1.2×10^5	(12)
voltage (D.C.)	V cm^{-1}	60 Hz	2.76×10^6	(12)

Permeability property of Kapton film (25 μm)

Gas	Units	Conditions	Value	Reference
He	mL m^{-2} day^{-1} MPa^{-1}	ASTM D-1434-82 23°C, 50% RH	63,080	(1, 5)
CO$_2$		ASTM D-1434-82 23°C, 50% RH	6840	(1, 5)
H$_2$		ASTM D-1434-82 23°C, 50% RH	38,000	(1, 5)
N$_2$		ASTM D-1434-82 ASTM D-1434-82 23°C, 50% RH	910	(1, 5)
O$_2$		ASTM D-1434-82 23°C, 50% RH	3800	(1, 5)
Vapor				
H$_2$O	g m^{-2} day^{-1} MPa^{-1}	ASTM E-96-92 23°C	54	(1, 5)

Various treatment method of Kapton films and resulting contact angle values

Treatment	Conditions	Contact Angle (deg.)	Reference
No treatment	measured with a CAM-MICRO contact angle meter (Tantec Inc., Schaumburg, IL)	94 ± 1	(16)
Plasma Asher (100 W)	20 min	< 2	
Untreated	Sessile drop method, ASTM-D2578-67 Ethanol sinse & dry	39 ± 2	(17)
NaOH	Ethanol sinse & dry, 3.1 J L^{-1}, $t = 30$ s	12 ± 2	
	Ethanol sinse & dry, 9.4 g L^{-1}, $t = 30$ s	12 ± 3	
KrF excimer laser	Ethanol sinse & dry, 10 pulses, $f = 10$ pps	16 ± 5	
	Ethanol sinse & dry, 0 pulses, $f = 10^{-1}$pps	23 ± 13	

Property	Units	Conditions	Value	Reference
Limiting oxygen index	%	ASTM D-2863-87	37	(1, 5)
Surface tension	mN m^{-1}	20°C	37.7	(18)
Hygroscopic coefficient of expansion	ppm (%RH)$^{-1}$	23°F, 20–80 % RH	22	(1, 5)
Moisture absorption	%	50% RH, 23°C	1.8	(1, 5)
	%	immersion 24 h at 23°C	2.8	(1,5)

Poly(pyromellitimide-1,4-diphenyl ether)

Property	Units	Conditions	Value	Reference
Density	g cm^{-3}	ASTM D-1505-90	1.42	(1)
Coefficient of friction	–	Kinetic (film-to-film)	0.48	(1)
Coefficient of friction	–	Static (film-to-film)	0.63	(1)
Adhesive strength	Ncm^{-2}	5-μm-thick polyimide film; baked at 250 °C) on top of a silicon wafer and by dry etching in oxygen	$<10^{-3}$	(19)

Mechanical properties of Vespel (SP-1 polyimide resin)

Property	Units	Conditions	Value	Reference
Tensile strength, ultimate	MPa	ASTM-D 1708; 23°C	86.2	(4, 20)
	MPa	260°C	41.4	
Elongation, ultimate	%	ASTM-D 1708; 23°C	7.5	(4, 20)
	%	260°C	6.0	
Flexural strength, ultimate	MPa	ASTM-D 790; 23°C	110.3	(4, 20)
	MPa	260°C	62.1	
Flexural modulus	MPa	ASTM-D 790; 23°C	3102	(4, 20)
	MPa	260°C	1724	(4)
Compressive stress	MPa	ASTM-D 695; 23°C, at 1% strain	24.8	(4)
	MPa	23°C, at 10% strain	133.1	
	MPa	23°C, at 0.1% offset	51.0	
Compressive modulus	MPa	ASTM-D 695; 23°C,	2413	(4)
Axial fatigue endurance limit	MPa	@ 10^3 cycles and 23°C	55.8	(4)
	MPa	@ 10^3 cycles and 260°C	42.1	(4, 20)
	MPa	@ 10^7 cycles and 23°C	55.8	
	MPa	@ 10^7 cycles and 260°C	16.5	
Flexural fatigue endurance limit	MPa	@ 10^3 cycles and 23°C	65.5	(4, 20)
	MPa	@ 10^7 cycles and 23°C	44.8	
Shear strength	MPa	ASTM-D 732; 23°C	89.6	(4, 20)
Impact strength izod, notched	J m^{-1}	ASTM-D 256; 23°C	42.7	(4, 20)
Impact strength izod, unnotched	J m^{-1}	ASTM-D 256; 23°C	747	(4, 20)
Poisson's ratio	–	23°C	0.41	(4, 20)

Wear and friction properties of Vespel (SP-1 polyimide resin)

Property	Units	Conditions	Value	Reference
Friction coefficient	–	steady state, unlubricated in air (PV = 0.875 MPa m s^{-1})	0.29	(4, 20)
Friction coefficient	–	static in air	0.35	(4, 20)
wear rate	cm/1000 h	unlubricated in air	0.64–3.0	(20)

Poly(pyromellitimide-1,4-diphenyl ether)

Thermal properties of Vespel (SP-1 polyimide resin)

Property	Units	Conditions	Value	Reference
Coefficient of linear expansion	10^{-6} °C^{-1}	ASTM-D 696; 23–300°C (m m^{-1})	54	(4, 20)
	10^{-6} °C^{-1}	−62−23°C	45	(4, 20)
Thermal conductivity	W m^{-1} °C^{-1}	40°C	0.35	(4, 20)
Specific heat	J Kg^{-1} °C^{-1}		1130	(4)
Deformation	%	ASTM-D 621; under 2000 psi load	0.14	(4, 20)
Deflection temperature	°C	ASTM-D 648; @ 264 psi	~360	(4, 20)

Electrical properties of Vespel (SP-1 polyimide resin)

Property	Units	Conditions	Value	Reference
Dielectric constant	–	ASTM-D 150;		
		@10^2 Hz, 23°C.	3.62	(4, 20)
	–	@$10^{4.}$ Hz, 23°C	3.64	
	–	@$10^{6.}$ Hz, 23°C.	3.55	
Dissipation factor	–	ASTM-D 150;		
		10^2 Hz, 23°C	0.0018	(4, 20)
	–	10^4 Hz, 23°C	0.0036	
	–	10^6 Hz, 23°C	0.0034	
Dielectric strength	MV m^{-1}	ASTM-D 149; short time, 0.002 m thick	22	(4, 20)
Volume resistivity	ohm m	ASTM-D 257; 23°C	10^{14}–10^{-15}	(4, 20)
surface resistivity	ohm	ASTM-D 257; 23°C	10^{15}–10^{-16}	(4, 20)

Other physical properties of Vespel (SP-1 polyimide resin)

Property	Units	Conditions	Value	Reference
water absorption	%	ASTM-D 570; 24 h at 23°C	0.24	(4, 20)
	%	immersion 48 h at 50°C	0.72	
Equilibqrium-50%RH	–		1.0–1.3	(4, 20)
Specific gravity	–	ASTM-D 792	1.43	(4, 20)
Hardness	Rockwell "E"	ASTM-D 785	45–60	(4, 20)
	Rockwell "M"	ASTM-D 785	92–102	(20)
Limiting oxygen index	%	ASTM-D 2863	53	(4, 20)

Fiber properties of poly(pyromellitimide-1,4-diphenyl ether)*

Property	Units	Conditions	Value	Reference
Tenacity	GPa	Heat treated under tension at 525–575°C	0.45–0.72	(21)
Elongation	%	Heat treated under tension at 525–575°C	9.0–11.7	(21)
Modulus	GPa	Heat treated under tension at 525–575°C	6.4–9.9	(21)

* Fibers were spun from poly(amic acid)/dimethylacetamide solutions and the resultant poly(amic acid) fibers were then thermally converted to polyimide fibers under sufficient tension. The polyimide fiber was finally heat treated at 525–575°C.

Poly(pyromellitimide-1,4-diphenyl ether)

Transition temperatures of poly(pyromellitimide-1,4-diphenyl ether)

Conditions	T_g (°C)	T_m (°C)	Reference
DSC, 20°C min^{-1}; film sample	420	–	(22)
DSC; film thickness ~12.5 μm	410	–	(23)
DSC; film sample	400	–	(24)
Thermomechanical technique	377		(25)
Thermomechanical technique; film thickness ~ 20 μm	–	597	(25, 26)

Secondary-relaxation temperatures of poly(pyromellitimide-1,4-diphenyl ether) adapted from reference[27]

Conditions	T_β (°C)	E_a (kJ mol^{-1})	T_γ (°C)	E_a (kJ mol^{-1})	Reference
Resonance electrostatic method; 15,000 Hz	127	84–105	−23	66	(28)
Resonance electrostatic method; 14,000 Hz	132	–	–	–	(29)
Torsion pendulum; 1 Hz	–	–	−88	44	(30)
Dynamic mechanical analysis, at 5°C min^{-1} & oscillating frequency of 1 Hz	0–200 (broad)	–	–	–	(31)
Thermally simulated depolarization current method; ~125 μm film, poled at 300 K and 16 kV cm^{-1}	124	18	−18	1.5	(32)

Absolute energies and assignments of features in the K-shell carbon of poly(pyromellitimide-1,4-diphenyl ether)

Conditions	Energy (eV)	Assignment	Reference
Secondary electron yield (SEY) measurements with a resolution of 0.2 eV.	285.2	$1s \rightarrow \pi^*$	(33, 34)
	286.6	$1s \rightarrow \pi^*$	
	287.4	$1s \rightarrow \pi^*$	
	291.9	$1s \rightarrow \sigma^*$	
	295.4	$1s \rightarrow \sigma^*$	
	303	$1s \rightarrow \sigma^*$	
Time-resolved experiments using near-edge x-ray absorption fine structure spectroscopy (NEXAFS) at 284 eV (carbon K-edge)	285.3	$1s \rightarrow \pi^*$	(33)
	286.2	$1s \rightarrow \pi^*$	
	288.0	$1s \rightarrow \pi^*$	
	291.7	$1s \rightarrow \sigma^*$	
	295.6	$1s \rightarrow \sigma^*$	
	302	$1s \rightarrow \sigma^*$	

Poly(pyromellitimide-1,4-diphenyl ether)

Unit cell dimensions

Lattice	Monomers Per Unit Cell	Cell Dimensions (Å)			Cell Angles (degrees)			Reference
		a	*b*	*c* (Chain Axis)	α	β	γ	
Orthorhombic	2	6.35	4.05	32.6	90	90	90	(35)
Orthorhombic	2	6.31	3.97	32	90	90	90	(36)
Monoclinic	2	4.66	32.9	5.96	90	100	90	(37)
				15.9				

Structural parameters of Poly(pyromellitimide-1,4-diphenyl ether)

Property	Units	Conditions	Value	Reference
Repeat distance	Å	X-ray diffraction	16	(36)
Mean interchain d-spacing	Å	X-ray diffraction	4.45	(24)
Persistent length	Å	Theoretical calculation	36	(38)
Kuhn segment	Å	Theoretical calculation	72	(38)

References

1. Product Bulletin. "Summary of Properties for Kapton® Polyimide Films ," E. I. du Pont & Co.; http://www2.dupont.com/Kapton/en_US/assets/downloads/pdf/summaryofprop.pdf.
2. Product Bulletin, "Vespel Polyimide Parts and Shapes," E. I. du Pont & Co. http://www2.dupont.com/Vespel/en_US/index.html.1.
3. Sroog CE. *Prog. Polym. Sci.* 1991;16:561–694.
4. Product Bulletin H-15724-1, "Properties of DuPont Vespel Parts," E. I. du Pont & Co.
5. Sroog CE. Chemistry and Properties of Addition Polyimides. In: Wilson D, Stenzenberger HD, Hergenrother PM. *Polyimides.* New York: Chapman & Hall; 1990:254.
6. Lee C. Iver NP, Han H. *J. Polym. Sci. Part B: Polym. Phys.* 2004;42:2202.
7. Terui Y, Matsuda S-I, Ando S. *J. Polym. Sci. Part B. Polym.Phys.* 2005;43:2109.
8. Lee S, Cahill MDG. *J. Appl. Phys.,* 1997;81:2590; (b) Hu C, Morgen M, Ho PS, Jain A, Gill W, Plawsky J, Wayner P, Jr, *Appl. Phys. Lett..* 2000;77:145.
9. Hu C, Kiene M, Ho PS. *Appl. Phys. Lett.* 2001;78:4121.
10. Vora RH, Pallathadka PK, Goh SH, Chung T-S, Lim YX, Bang TK. *Macromol. Mater. Eng.* 2003;288:337.
11. Varma K, Fohlen GM, Parker JA. *U.S. Patent.* 4,276,344(1981).
12. Cassidy PE, Aminabhavi TM. *Polym. News.* 1989;14;362.
13. Virk HS. *Nucl. Instrum. Methods Phys. Res.,Sect. B.* 2002;191:739.
14. Nah C, Han SH, Lee J-H, Lee M-H, Chung KH. *Polym Int.* 2004;53:891.
15. Terui Y, Ando S. *Appl. Phys. Lett.* 2003;83:4755.
16. Prichard HL, Reichert WM, Klitzman B. *Biomaterials.* 2006;28:936.
17. Wehner M, Legewie F, Theisen B, Beyer E. *Appl. Surf. Sci..* 1996;106:406.
18. Sacher E, *J. Appl. Polym. Sci.* 1978;22:2137.

19. Geim K, Dubonos SV, Grigorieva IV, Novoselov KS, Zhukov AA, Shapoval S, Yu. *Nature Mater.* 2003;2:461.

20. Sroog E. Chemistry and Properties of Addition Polyimides(Vespel Property Table). In Wilson D, Stenzenberger HD, Hergenrother PM, eds. *Polyimides*. New York: Chapman & Hall; 1990:262–263.

21. Irwin RS. *U. S. Patent 3,415,782* (du Pont E. I. & Co, 1968).

22. Okamoto K, Tanaka K, Kita H, Nakamura A, Kusuki Y. *J. Polym. Sci.: Part B. Polym. Phys.* 1989;27:2621.

23. Hachisuka H, Tsujita Y, Takizawa A, Kinoshita T. *J. Polym. Sci.: Part B. Polym. Phys.* 1991;29:11.

24. Stern SA, Yamamoto Y, MH, Clai AKS. *J. Polym. Sci. Part B.* 1989;27:1887.

25. Clair TLSt. Structure-Property Relationships in Linear Aromatic Polyimides. In: Wilson D, Stenzenberger HD, Hergenrother PM, eds. *Polyimides*. New York: Chapman & Hall; 1990: Chapter 3, 62–69.

26. Bessonov MI, et. al., *Polyimides: Thermally Stable Polymers* New York: Plenum Publishing; 1987. (b) Bessonov MI, Kuznetsov NP, Koton MM. *Vysokomol. Soedin.* 1978;A20(2):347; (c) *idem., Polym. Sci. U. S. S. R. (Engl. Transl.).* 1978;20:391–400.

27. Fried JR. Sub-Tg Transitions. In: Mark JE, ed. *Physical Properties of Polymers Handbook.* New York Woodbury: American Institute of Physics; 1996:Chapter 13, 166–167.

28. Butta E, de Petris S, Pasquini M. *J. Appl. Polym. Sci.* 1969;13:1073.

29. Baccaredda M, Butta E, Frosini V, De Petris S. *Mater. Sci. Eng.* 1969;3:157.

30. Lim T, Frosini V, Zaleckas V, Morrow D, Sauer JA. *Polym. Eng. Sci..* 1973;13;51.

31. Ragosta G, Musto P, Abbate M, Russo P, Scarinzi G. *Macromolecular Symp.* 2005;228:287.

32. Alagiriswamy, Narayan KS, Raju G. *J. Phys. D: Appl. Phys.* 2002;35:2850.

33. Vogt U, Wilhein T. *Rev. Scientific Instruments.* 2004;75:4606.

34. Jordan-Sweet JL, Kovac CA, Goldberg MJ, Morar JF. *J. Chem. Phys.* 1988;89:2482.

35. Kazaryan LG, Lur'e Ye G, Igonin LA. *Vysokomol. Soedin., Ser, B.* 1969;11:779. (b) Lur'e Ye G, Kazaryan LG, Uchastkina EL, Kovriga VV, Vlasova KN, Dobrokhotova ML, Yemel'yanova LN. *Vysokomol. Soedin., Ser, A.* 1971;13:603; (c) *idem, Polym. Sci. U. S. S. R. (Engl. Transl).* 1971;13:685.

36. Kazaryan LG, Tsvankin D, Ya Ginsburg BM, Tuichiyev Sh, Korzhavin LN, Frenkel S Ya. *Vysokomol. Soedin., Ser, A.* 1972;14:1199; (b) *idem, Polym. Sci. U. S. S. R. (Engl. Transl.).* 1972;14:1344.

37. Ginsburg BM, Volosatov VN, Magdalev Ye T, Tuichiyev Sh. *Vysokomol. Soedin., Ser, A.* 1978;20:900. (b) *idem, Polym. Sci. U. S. S. R. (Engl. Transl.).* 1978;20:1017.

38. Birshtein TM, Goryunov AN. *Vysokomol. Soedin, Ser, A.* 1979;21:1990. (b) *idem, Polym. Sci. U. S. S. R. (Engl. Transl.).* 1979:21:2196.

Polypyrrole

Amit S. Kulkarni, Shrish Rane and Gregory Beaucage

Acronym PPy

Class Polyheterocyclics; conjugated conducting polymers

Structure The pyrrole rings are mainly linked in the α, α' positions giving a planar geometry. There is evidence of other bonding observed through NMR and IR analysis.[1−3]

Major Applications There are potential applications for PPy in display devices, chemical sensors, electrodes in batteries, drug carriers, heating fabrics, deionizers, and as a catalyst.

Properties Of Special Interest Presence of an extended π–bonding system, which imparts electrical properties to the polymer. Doping to either p or n type can enhance these properties. Polypyrrole is stable in air at ambient temperatures, as well as high as 250°C in its doped state. Polypyrrole can also be synthesized in a doped state. It changes color when switched from its conducting to insulating state.

Property	Units	Conditions	Value	Reference
Chemical synthesis		Oxidative polymerization in either solution or vapor phase		(2–6)
		Polymerization on colloidal cerium oxide particles		
		In the presence of ammonium persulfate (oxidant) and dodecylbenzene sulfonic acid (dopant)		
		Metal oxidized oxidative polymerization		(7)
		Polymerization with ionic liquids		(8, 9)
		Interfacial and template polymerization of PPy nano-networks and nano-fibers.		(10)
		PPy Nanowires		(11)
		Emulsion polymerization		(12)
		Redox enzyme initiated polymerization		(13)
		Plasma Polymerization		(14)
		Water Soluble PPy synthesis using horseradish peroxidase enzyme		(15)

Property	Units	Conditions	Value	Reference
Electro-chemical synthesis		Electrodes: anode (platinum, n-type silicon, conducting glass, stainless steel, gold/Glassy Carlson); cathode (copper) Electrolytes: copper sulfate, acetonitrile + p-toluenesulfonic acid (HTSO), lithium perchlorate, sodium perchlorate, sulfuric acid		(1, 16-18)
		In-situ electropolymerization with nanotubes		(19)
		Polymerization using sodium saccharinate		(20)
		Using supercritical TFM		(21)
Conductivity σ	S cm^{-1}	Measured	100	(1, 22, 23)
		Inducing orientation using mechanical methods improves the measured value	~1,000	
Optical Properties		PPy films coated on electrodes undergo a color change when they switch from an oxidized to a reduced state and vice-versa		(1, 24, 25)
Solubility		Neutral PPy is by and large insoluble In its doped state it is soluble in chloroform, DMSO, m-cresol, and NMP		(1, 26)
Thermal stability	K	In air	523	(1)
UV-Vis Spectroscopy	nm	Strong absorption band in the doped state on a platinum electrode	272	(1, 27, 28)
		Neighboring shoulder peaks	368, 381	

Polymer	Peak (ev)
Oxidized PPy	1.0
-do-	3.0
Neutral PPy	1.3
-do-	3.2

^{13}C NMR$^{(29)}$

Polymer	Shifts from TMS (ppm)	Conditions
Neutral PPy	~123	α carbon
	~105	β carbon
	~135	Non-α-α' linkages or chain end groups

Property	Units	Conditions	Values	Reference
IR properties	cm^{-1}	PPy electropolymerized under oxygen free conditions		(1, 28, 30)
		O—H stretch	2,930–2,800	
		C—O stretch	1,750–1,650	
		C—O—O stretch	1,099	
		Electropolymerized PPy-perchlorate first electropolymerized and then reduced to its neutral state		
		Neutral PPy		
		NH band	3,400	
		CH band	3,100	
		C–H stretching	2,870–2,960	
		Pyrrole bands	< 1,800	
		PPy-perchlorate		
		NH band	Absent*	
		CH band	Absent	
		C–H stretching	–	
		Pyrrole bands	<1,800	
		Undoped PPy		
		(NH stretching)	3421	(23)
		(CH stretching)	3100	
		(C=C and C—C stretching)	1535	
		(NH stretching)	1450	
		(CH and NH deformation)	1295	
		(CH deformation)	1050	
		Doped PPy		
		(NH stretching)	No band	
		(CH stretching)	Diminished band	
		(C=C and C—C stretching)	1529	
		(NH stretching)	1445	
		(CH and NH deformation)	1290	
		(CH deformation)	1010	

* Masked by the tail of the 1-eV peak.

Crystallinity[2]*

Lattice	Unit Cell Dimensions (nm)			Cell Angle (°)		
	a	b	c	α	β	Γ
Monoclinic	0.82	0.735	0.682	90	90	117

* PPy shows very low crystallinity

Property	Units	Conditions	Values	Reference
Density	g cm^{-3}	–	1.47	(2)
Molecular weight		It is difficult to measure the molecular weight of PPy because it is insoluble.		(1, 3)
		Indirect methods on substitutes PPs has yielded the number of molecular units as being between 100 and 1,000 contingent upon the polymerization method		
		Soluble PPy doped with di(2-ethylhexyl) sulfosuccinate anion characterized by gel permeation chromatography	303 Pyrrole units	(31, 32)
		Alcohol soluble PPy	158 Pyrrole units	(33)
Morphology		SEM studies on elcetropolymerized PPy show globular particle aggregates. The surface morphology is influenced by the electrolyte used. Its appearance ranges from a rough dendritic structure in CH_3CN to a smoother surface in low amounts of water and the other hydroxylic solvents.		(1, 28, 29, 34)
		STM images of p-toluenesulfonate doped PPy show small islands interconnected by 1.5–2 nm wide fibrils.		

Mechanical properties of PPy-toluenesulfonate films[35,36]*

Electrolyte	Elongation at Break (%)	Young's Modulus (MPa)	Applied Voltage (V)
NaNO$_3$	4–10	2,386–1,930	−0.8 to + 0.4
Mg(NO$_3$)$_2$	3–8	2,014–2,176	−0.8 to + 0.4
KCl	5–14	3,415–1,415	−0.5 to 0
LiCl	7–21	3,666–2,650	−0.5 to 0
NaCl	5–21	3,621–2,193	−0.5 to 0
MgCl$_2$	6–7	2,611–3,609	−0.5 to 0
H$_2$O	9	2,914	−0.5 to 0

* Samples: 30 mm long, 5 mm wide, dumb-bell shaped.

Cosolvent	Ratio	Young's Modulus (Pa)	Elongation at Break (%)
H$_2$O-CAN	1:99	2413.2	4
-do-	1:99	827.4	17
-do-	25:75	482.6	4
EG-CAN	25:75	1,103.2	5
-do-	50:50	896.3	8
H$_2$O-AG-ACN	1:1:98	1,379	8
-do-	2:5:93	1,034.2	8

Cosolvent	Ratio	Young's Modulus (Pa)	Elongation at Break (%)
-do-	5:5:90	896.3	14
-do-	12.5:12.5:75	827.4	7
-do-	25:25:50	1,034.2	6
H$_2$O-EG	50:50	344.7	6
G-CAN	1:99	1,310	8
G-H$_2$O	50:50	1,654.7	6

References (38–42).

References

1. Stockheim TA, ed. *Handbook of Conducting Polymers.* Vol 1. NY: Marcel Dekker; 1986.
2. Geiss H, et al. *IBM J. Res. Dev.* 1983;27(4):321.
3. Street GB, et al. *Mol. Cryst. Liq. Cryst.* 1985;118:137.
4. Galembeck A, Alves OL. *Synth. Met.* 1997;84:151.
5. Sari B, Gok A, Sahin D. *J. App. Polym. Sci.* 2006;101:241.
6. Kim JH, Sharma AK, Lee YS. *Mater. Lett.* 2006;60:1697.
7. Dias HVR, Fianchini M, Rajapakse RMG. *Polymer.* 2006;47:7349.
8. Pringle JM, Ngamna O, Chen J. *Synth. Metals.* 2006;156:979.
9. Pringle JM, Efthimiadis J, Howlett PC. *Polymer.* 2004;45:1447.
10. Acik M, Baristiran C, Sonmez G. *J. Mater. Sci.* 2006;41:4678.
11. Yan HL, Zhang L, Shen JY. *Nanotechnology.* 2006;17:3446.
12. Amaike M, Yamamoto H. *Polymer Journal.* 2006;38:703.
13. Ramanavicius A, Kausaite A, Ramanaviciene A. *Synth. Metals.* 2006;156:409.
14. Zhou J, Fisher ER. *J. Nanoscience and Nanotechnology.* 2004;4:539.
15. Nabid MR, Entezami AA. *J. App. Polym. Sci.* 2004;94:254.
16. Turcu M, et al. *Synth. Met.* 1997;84:825.
17. Pickup G, Osteryoung RA. *J. Am. Chem. Soc.* 1984;106:2294.
18. Wainright S, Zorman CA. *J. Electrochem. Soc.* 1995;142:38
19. Wu TM, Lin SH. *J. Polym. Sci. Part A – Polym. Chem.* 2006;44:6449.
20. Bazzaoui M, Martins JI, Costa SC. *Electrochimica Acta.* 2005;51:4516.
21. Atobe M, Ohsuka H, Fuchigami T. *Chemistry Lett.* 2004;33:618.
22. Funahashi K, Iwata K. *Mol. Cryst. Liq. Cryst.* 1985;118:159.
23. Geetha S, Trivedi DC. *Mater. Chem. Phys.* 2004;88 (2–3):388–397.
24. Yoneyama HK, et al. *J. Electrochem. Soc.* 1985;132(10): 2414.
25. Street B, et al. *Mol. Cryst. Liq. Cryst.* 1982;83:253.
26. Oh J, et al. *Synth. Met.* 1997;84:147.
27. Park S, Shim YB. *J. Electrochem. Soc.* 1993;140(3):609.
28. Diaz F, Hall B. *IBM J. Res. Dev.* 1983;27(4):342.
29. Clark C, et al. *IBM J. Res. Dev.* 1983;27(4):313.
30. Satoh M, et al. *Synth. Met.* 1997;84:167.
31. Oh EJ, Jang KS, MacDiarmid AG. *Synth. Metals.* 2001;125:267.
32. Oh EJ, Jang KS. *Synth. Metals.* 2001;119:109.
33. Jang KS, Han SS, Suh JS. *Synth. Metals.* 2001;119:107.
34. Diaz A. *Chemica Scr.* 1981;17:145.
35. Murray P, et al. *Synth. Met.* 1997;84:847.
36. Diaz A, Hall B. *IBM J. Res. Dev.* 1983;27(4):342
37. Wainright S, Zorman CA. *J. Electrochem. Soc.* 1995;142(2):384.

38. Murao K, Suzuki K. *J. Electrochem. Soc.* 1988;135(6):1415.
39. Yang R, et al. *J. Phys. Chem.* 1989;93:511.
40. Lee JY, et al. *Synth. Met.* 1997;84:137.
41. Kanazawa K, et al. *Synth. Met.* 1979–19801:329.
42. Lei J, Martin CR. *Synth. Met.* 1992;48:331.

Polyquinoline

Amit S. Kulkarni, Shrish Rane and Gregory Beaucage

Acronym PQ

Class Polyheterocyclics; polyaromatics

Structure The structure of PQ can vary from a semi-rigid to a rigid-rod depending on the synthesis conditions.

Major Applications Polyquinoline can be processed into films and fibers and can be spin-coated as an ideal choice for high-performance films, electronic coatings, as a matrix for high-performance composites, and as an interlayer dielectric substrate in multi-chip modules.

Properties of Special Interest The structure of polyquinolines can be altered from semi-rigid to a rigid-rod conformation during synthesis. Although largely an amorphous polymer, some substituted rigid rod members exhibit crystallinity in low amounts. Polyquinolines are also found to poses excellent thermal and oxidative stability, good mechanical properties, low dielectric constants, low values of moisture absorption, and low values of thermal expansion coefficients.

Property	Units	Conditions	Value	Reference
Synthesis		Acid or base catalyzed Friedlander synthesis, Catalytic dehydrogenative polycondensation of non-substituted quinoline oligomers from 1,2,3,4,-tetrahydroquinoline in the presence of transition metal catalysts, Condensation reaction of 3-3′-dibenzoylbenzidine with diacetyl and diphenyl compounds.		(1–4)
		Silicon containing PQ		(5)
		Polypyrrole containing PQ		(6)
		PQ with ether linkage via Friedlander synthesis		(7)

Property	Units	Conditions	Value	Reference
		PQ-hexylfluorene copolymers via Friedlander synthesis		(8)
		PQ with pendant 4′-octyloxybiphenyl group via Friedlander synthesis		(9)
		Liquid crystalline PQ by polycondensation		(10)
		n-type luminescent PQ		(11)
		Synthesis of PQ with a non linear chromophore		(12)
Ionization potential	eV	Calculated on PQ using the Valence Effective Hamiltonian Technique (VEH)		(13–16)
		Gas Phase	7.89	
		Solid Phase	~6.0	
Dielectric constant	–	Range; measured on 28 μm PQ100 (thermoplastic PQ) film	2.5–2.6	(13–16)
Band-gap	eV	Calculated on PQ using the Valence Effective Hamiltonian Technique (VEH)	3.2	(13–16)
Electrooptic coefficient (r_{33}) wavelength	mm	Measured on a 20% wt. RT-9800/PQ film poled at 0.8 MV cm^{-1}	1.3	(15)
Fluorescence, λ_{max}	nm	Absorption		(8)
		POF66	388	
		PSF66	400	
		P1F66	404	
		P2F66	384	
		In solution with tetrahydrofuran		
		POF66	404	
		PSF66	422	
		P1F66	434	
		P2F66	446	
		In films		
		POF66	414	
		PSF66	434	
		P1F66	446	
		P2F66	494, 534	
		PQ with pendant 4′-octyloxybiphenyl group:		(9)
		Blue Fluorescence in dilute solution	449	
		Green Fluorescence in bulk	494, 540	
		Bright yellow light	554	(10)
Conductivity	S cm^{-1}	Vapor phase thermolysed PQ film	400	(1, 2, 17)
		CVD vapor deposition	695–920	
Thermal stability	TGA wt. loss	Air	500–600	(1 ,4, 18, 19)
		N_2	~800	
		Air (less than 5%)	671–809	(6)

Property	Units	Conditions	Value	Reference
Glass transition temperature T_g	K	–	523–663	(1, 4, 18, 19)
			434–612	(5)
			383–527	(20)
			515–612	(6)
			468–516	(8)
Crystallization temperature T_c	K	In case of rigid-rod variety of polyquinilone	688–703	(1, 4, 18, 19)
Melting temperature T_m	K	In case of rigid-rod variety of polyquinilone	721–823	(1, 4, 18, 19)
Thermal Decomposition Temperature T_D	K		673–750	(20)
			>661	(8)
Solubility		The solubility of PQ is dependent on its molecular architecture. Generally it is soluble in a variety of organic solvents (e.g. $CHCl_3$, m-cresol, THF, H_2SO_4, and TCE, DMAC, DCP/m-cresol)		(1, 15, 19, 6)
Tensile strength	MPa	–	97	(1, 4, 19)
	kgf cm^{-2}	–	770–1170	(10)
Tensile modulus	MPa	–	2,680	(1, 4, 19)
Elongation at break	%	–	6.2	(1, 4, 19)
Young's modulus	MPa	Rigid-rod PQ	4,800	(1, 4, 19)
		Semi-rigid PQ	1,900	
Morphology		SEM studies on PQ films show a smooth surface with dense domains without any distinguishing characteristics like fibrils or filaments. TEM on semi-rigid PQ's reveal an amorphous regime without any structure.		(17)
Crystallinity	%	–	20–65	(1, 3, 4)
PQ fiber crystalline d spacings	Å	–	10.23, 10.30, 10.31	(4)

Solution properties*[21,22]

Properties/Parameters	η (ml g^{-1})†	k'†	M_w (g mol^{-1})‡	$R_g^2 \times 10^{10}$ (cm^2)‡	$A_2 \times 10^4$ (ml mol g^{-1})‡	M_e (g mol^{-1})‡
Values	28–61	0.38–0.72	17,000–60,000	0.20–0.19	3–18	23,000–110,000

* Solvents used were m-cresol, chloroform, and THF.

† From intrinsic viscosity measurements.

‡ From light scattering measurements. For a complete description of the samples, see ref. 2.

References

1. Stille JK. *Macromolecules.* 1981;14:870.
2. Chiang LY, et al. *Synth. Met.* 1991;41–43:1425.
3. Agarwal AK, Jenekhe SA. *Macromolecules.* 1993;26:895.
4. Sybert PD, et al. *Macromolecules.* 1981;14:493.
5. Tonzola CJ, Alam MM, Jenekhe SA, *Macromol. Chem. Phys.* 2005; 206:1271.
6. Hou SF, Ding MX, Gao LX. *Macromolecules.* 2003;36:3826.
7. Lee TS, Yang C, Kim, JL, *J. Polym. Sci. Part A Polym. Chem.* 2002;40:1831.
8. Kim JL, Kim JK, Cho HN, Kim DY. Kim CY, Hong SI. *Macromolecules.* 2000;33:5880.
9. Kim JL, Kim JK, Hong SI. *Polym. Bull.* 1999;42:511.
10. Ni CJ, Kim KS, Katsuraya K, Okuyama K, Kato T, Uryu T, *J. Polym. Sci. Part A Polym. Chem.* 1998;36:749.
11. Zhang X. Shetty AS, Jenekhe SA. *Acta Polymerica.* 1998;49:52.
12. Shu CF; Tsai WJ, Chen JY, Jen AKY, Zhang Y, Chen TA. *Chem. Comm.* 1996;19:2279.
13. Hendricks NH, et al. 36th *International SAMPE Symposium*, 1991:42.
14. Thémans B, et al. *Solid State Comm.* 1984;50(12):1047.
15. Kai YM, Jen AKY. *App. Phys. Lett.* 1995;67(3):299.
16. Stille JK. In: *Contemporary Topics in Polymer Science.* Vol 5. NY: Plenum Press; 1984:209.
17. Chiang LY, et al. *Synth. Met.* 1989;29:E483.
18. Wrasidlo W, Stille JK. *Macromolecules.* 1976;9(3):505.
19. Warsidlo W, et al. *Macromolecules.* 1976;9(3):512.
20. Tonzola CJ, Alam MM, Jenekhe SA. *Macromolecules.* 2005;8:9539.
21. Norris SO, Stille JK. *Macromolecules.* 1976;9(3):496.
22. Metzger Cotts P, Berry GC. *J Polym. Sci, Part B, Polym. Phys.* 1986;24(7):1493.

Poly(rotaxane), Example 1

Akira Harada

Acronyms, Alternative Name PR, MN, molecular necklace

Class Cyclic polymers

Structure $(NO_2)_2-C_6H_3-NH-(CH_2CH_2O)_n-CH_2CH_2NH-C_6H_3-(NO_2)_2 + m(C_6H_{10}O_5)_6$

Major Applications Starting materials for tubular polymers. Potential use for curing of PEG.

Properties of Special Interest Stable under ambient conditions. Relatively low cost.

Property	Units	Conditions	Values	Reference
Decomposing point	K	MN-1450	593	(1)
		MN-2000	593	(1)
		MN-3350	603	(2)
Molecular weight	g mol^{-1}	MN-1450	16,500	(1)
		MN-2000	20,000	(1)
		MN-3350	23,500	(2)
		MN-1248	13,244	(3)
		MN-8500	44,000	(4)
Molecular weight	g mol^{-1}	MN-1450	1,375	(1)
(of repeat unit)		MN-2000	1,111	(1)
		MN-3350	1,175	(2)
		MN-1248	1,060	(3)
Number of cyclodextrins	–	MN-1450	15	(1)
		MN-2000	18	(1)
		MN-3350	23	(2)
		MN-1248	12	(3)
		MN-8500	36	(4)
IR (characteristic absorption frequencies)	cm^{-1}	MN-1450	3,386	(1)
			2,923	
			1,153	
			1,077	
			1,029	
UV (characteristic absorption frequencies)	nm $\varepsilon, l\,mol^{-1}\,cm^{-1}$	MN-1450	360 17,950	(1)

Poly(rotaxane), Example 1

Property	Units	Conditions	Values	Reference
^{1}H-NMR	ppm	MN-1450, (DMSO-d6), 270 MHz		(1)
		Phenyl	7.27–8.88	
		Cyclo-dextrin	3.2–5.45	
		PEG	3.52	
^{13}C-NMR	ppm	MN-1450, (DMSO-d6), 125.65 MHz		(1)
		Cyclo-dextrin	71.46–101.90	
		PEG	69.35	
Specific rotation	degrees unit^{-1}	MN-1450, DMSO, 25°C, 589 nm	160	(1)
Solvent	DMSO			(1)
	0.1N NaOH			(5)
Nonsolvent	Organic solvent (benzene, hexane, acetone, chloroform), H_2O			(1)
Water content	w/w%	–	>10	–

References

1. Harada A, et al. *J. Org. Chem.* 1993;58:7524–7528.
2. Harada A, Li J, Kamachi M. *Nature.* 1992;356:325–327.
3. Harada A, Li J, Kamachi M. *J. Am. Chem. Soc.* 1994;116:3192–3196.
4. Harada A. *Supramol. Sci.* 1996;3:19–23.
5. Harada A. In: Semlyen JA, ed. *Large Ring Molecules.* Chichester: John Wiley and Sons; 1996:407.

Poly(rotaxane), Example 2

Akira Harada

Acronym, Alternative Name PR, molecular bracelet

Class Cyclic polymers

Structure

$$(RC_6H_4)_3-CH_2CH_2O(CO-(CH_2)_m-CO-O-(CH_2CH_2O)_nCH_2CH_2O)_x-$$
$$CO-(CH_2)_8-CO-O-(R-C_6H_4)_3 + 30\text{-crown-}10[(CH_2CH_2O)_{10}], \text{ or}$$
$$42\text{-crown-}14[(CH_2CH_2O)_{14}], \text{ or } 60\text{-crown-}20\ [(CH_2CH_2O)_{20}]$$

Major Applications Potential use for curing of polyesters.

Properties of Special Interest Stable under ambient conditions. Soluble in organic solvents.

Property	Units	Conditions*	Value	Reference
Molecular weight	g mol^{-1}	8,10,3-crown-10	4,000	(1)
		8,4,3-crown-10	8,700	
		8,2,3-crown-10	12,200	
		8,2,3-crown-10	24,000	
m/n	–	8,10,3-crown-10	0.15	(1)
		8,4,3-crown-10	0.028	
		8,2,3-crown-10	0.24	
		8,2,3-crown-10	0.31	
Molecular weight (of repeat unit)	g mol^{-1}	8,10,3-crown-10	$440m + 307n$	(2)
		8,4,3-crown-10	$440m + 168n$	
		8,2,3-crown-10	$440m + 268n$	
$[\eta]$	dl g^{-1}	8,2,3-crown-10	0.28	(1)
Mass % cyclic	–	8,10,3-crown-10	16	(1)
		8,4,3-crown-10	5.5	
		8,2,3-crown-10	24	
		8,2,3-crown-10	30	
Melting temperature T_m	K	8,2,3-crown-10	278, 291	(1)
		8,2,3-crown-10	282, 286	
T_c	K	8,2,3-crown-10	264	(1)
		8,2,3-crown-10	265	

Poly(rotaxane), Example 2

Property	Units	Conditions*	Value	Reference
Glass transition temperature T_g	K	8,2,3-crown-10	219	(1)
		8,2,3-crown-10	217.3	
Solvents	–	8,2,3-crown-10	Acetone, THF, CH_2Cl_2	(1)

* m, n,3-crown-10.

References

1. Gibson HW, et al. *J. Am. Chem. Soc.* 1995;117:852.
2. Gibson HW. In: Semlyen JA, ed. *Large Ring Molecules.* Chichester: John Wiley and Sons; 1996:191.
3. Gong C, Gibson HW. *Macromolecules.* 1996;29:7029.
4. Gong C, et al. *Macromolecules.* 1997;30:4807.
5. Gong C, Gibson HW. *J. Am. Chem. Soc.* 1997;119:5862.

Poly(silphenylene-siloxanes)

Petar R. Dvornic and Michael J. Owen

Acronyms, Trade Names Silphenylenes, Silarylene-siloxane polymers, SARSOX Producer: Oxazogen Inc., Midland, MI, USA

Class In-chain modified polysiloxanes

Formula of Repeat Unit

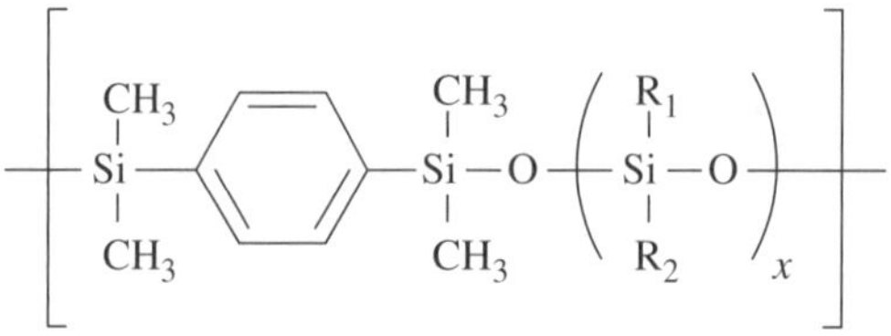

Major Applications Precursors for elastomers having increased thermal and thermo-oxidative stability while retaining low glass transition temperatures (that is gaskets sealants, O-rings). Established method of increasing chain stiffness of polysiloxanes. Gas semi-permeable membranes with increased separation ability at retained high permeability (relative to polysiloxanes). Have been used as laminate impregnating resins in electrical insulation applications where heat resistance is required and as stationary phases for high temperature gas chromatography to improve separation and reduce peak tailing and column bleed.

Properties of Special Interest Partial replacement of siloxane units in polysiloxanes with silphenylene groups increases polymer chain stiffness, glass transition temperature, viscosity, crystallinity, thermal, thermo-oxidative and solvent resistance. Many mechanical properties are also improved although low temperature elasticity is somewhat diminished. The best combination of properties is obtained for $x = 1$ in the above structure (the so-called "exactly alternating silphenylene-siloxane polymers") although derivatives with $x = 0$ through $x = 4$ are also known. Many different homologues with various siloxanylene side groups (R_1 and R_2 in the formula above) have been reported, as well as meta-silphenylenes, diphenyl-ether, diphenyl-dimethylsiloyl, diphenyl, and other more exotic in-chain silarylene derivatives. For these variations, see references 1 and 25.

Preparative techniques

Type of Polymerization	Conditions	Polymer Molecular Weight (g mol^{-1})	Typical Polydispersity (M_w/M_n)	Reference
(a) Self condensation of phenylenedisilanols	In melt or in refluxing solvent. catalysts: NaOH; KOH; LiOH; K$_2$O. Solvents: benzene, toluene.	1,000–100,000	2–6	(1–3)

Poly(silphenylene-siloxanes)

Type of Polymerization	Conditions	Polymer Molecular Weight (g mol^{-1})	Typical Polydispersity (M_w/M_n)	Reference
(b) Solution polycondensation of phenylenedisilanols and dichlorosilanes	In nitrogen. Solvents: THF, toluene or chlorobenzene. Room temperature or below.	30,000–50,000	1,6–2,2	(1–6)
(c) Solution polycondensation of phenylenedisilanes and diaminosilanes	In nitrogen. Solvents: refluxing toluene or benzene.	50,000–500,000	1,8–2,2	(1, 4, 7–9, 22)
(d) Solution polycondensation of phenylenedisilanols and diacetoxysilanes	In nitrogen. Solvents: refluxing toluene. Catalysts: triethylamine; *n*-hexylamine; 2,4,6-trimethylpyridine.	30,000–80,000	1,8–2,2	(1, 4, 10)
(e) Solution polycondensation of phenylenedisilanols and *bis*ureidosilanes	In nitrogen. Solvent: chlorobenzene. Temperature: −20°C to room temperature	100,000–800,000	1,8–2,2	(1, 4, 11)
(f) Catalytic cross-dehydro coupling of 1,4-bis (dimethyl-silyl) benzene	In THF, with water and tris(dibenzylideneacetone) dipalladium(0)-chloroform, palladiumdichloride or hexachloroplatinic acid as catalysts.	15,000–30,000	1.9–2.2	(23, 24)
Typical comonomers	*P*-phenylenedisilanols (a–e) + dichlorosilanes (b); diamino-silanes (c); diacetoxysilanes (d); *bis*ureidosilanes (e)			
(g) Solution self-polycondensation of *m*-phenylenetetra-methyldisilanol	In benzene, with *n*-hexylamine 2-ethylhexoate as catalyst	3,000 Increased to 126,000 on drying at 60°C in vacuum oven for two days	1.4–2.1	(26)

Crystalline state properties

Property	Units	Conditions	Value	Reference
Molecular weight (of repeat unit)	g mol^{-1}	$x = 1$ and $R_1 = R_2 = CH_3$	283	–
Typical molecular weight range of polymer (M_w)	g mol^{-1}	For specific preparative procedures see table above	30,000–150,000	–

Property	Units	Conditions	Value	Reference
Typical polydispersity index (M_w/M_n)	–	–	1.8–2.3	–
Density	g cm^{-3}	$x = 0$; R$_1$ = R$_2$ = CH$_3$; room temperature	1.102 1.103	(19) (3)
Unit cell dimensions	Å	$x = 0$; R$_1$ = R$_2$ = CH$_3$; tetragonal	a = 9.08; b = 9.08; c = 15.38 a = 9.02; b = 9.02; c = 15.43	(19) (21)
Unit cell contents		$x = 0$; R$_1$ = R$_2$ = CH$_3$; tetragonal	4 4	(19) (21)
Heat of fusion (of repeat unit)	kJ mol^{-1}	$x = 0$; R$_1$ = R$_2$ = CH$_3$; depression of Tm in mixtures	18.2	(3)

Solution properties

Property	Units	Conditions	Value	Reference
Solvents		THF; toluene; chlorobenzene		(1)
Non solvents		Methanol		(1)
Mark-Houwink parameters: K and a	K = ml g^{-1} a = None	$x = 0$; R$_1$ = R$_2$ = CH$_3$; toluene/25°C M_w = 70,000–400,000	$K = 1.12 \times 10^{-4}$ $a = 0.75$	(3)
		$x = 1$; R$_1$ = R$_2$ = CH$_3$; THF/30°C M_w = 30,000–109,000	$K = 7.86 \times 10^{-5}$ $a = 0.757$	(9)
		$x = 1$; R$_1$ = CH$_3$; R$_2$ = C$_6$H$_5$; THF/30°C M_w = 38,000–245,000	$K = 5.34 \times 10^{-5}$ $a = 0.749$	(9)
		$x = 1$; R$_1$ = R$_2$ = C$_6$H$_5$; THF/30°C M_w = 76,000–240,000	$K = 3.28 \times 10^{-5}$ $a = 0.821$	(9)

Transition Temperatures

Property	Units	Conditions	Value	Reference
Glass transition temperature	°C	$x = 0$; R$_1$ = R$_2$ = CH$_3$;		
		TBA	−25	(9, 12)
		DSC	−19	(23, 24)
		Dilatometry	−17	(13)
		$x = 1$; R$_1$ = R$_2$ = CH$_3$;		
		TBA	−61	(9)
		DSC (20° min^{-1})	−62	(12, 14)
		DSC (5° min^{-1})	−64	(12, 14)
		$x = 2$; R$_1$ = R$_2$ = CH$_3$;		
		DTA	−63	(7)
		TBA	−88	(9)
		$x = 3$; R$_1$ = R$_2$ = CH$_3$;		
		DTA	−72	(7)
		TBA	−102	(9)

Poly(silphenylene-siloxanes)

Property	Units	Conditions	Value	Reference
		$x = 4$; $R_1 = R_2 = CH_3$;		
		DTA	−80	(7)
		TBA	−109	(9)
		$x = 1$; $R_1 = CH_3$; $R_2 = C_6H_5$;		
		TBA	−25	(9)
		DSC	−32	(9)
		$x = 1$; $R_1 = CH_3$; $R_2 = (CH_2)_2CN$; DSC	−37	(12, 16)
		$x = 1$; $R_1 = CH_3$; $R_2 = (CH_2)_3CN$; DSC	−37	(12, 16)
		$x = 1$; $R_1 = CH_3$; $R_2 = (CH_2)_2CF_3$; DSC	−51	(12, 15)
		$x = 1$; $R_1 = CH_3$; $R_2 = (CH_2)_2(CF_2)_5CF_3$; DSC	−55	(12, 15)
		$x = 1$; $R_1 = CH_3$; $R_2 = (CH_2)_2CH_3$; DSC	−65	(12, 16)
		$x = 1$; $R_1 = CH_3$; $R_2 = (CH_2CH{=}CH_2)$; DSC	−66	(12, 16)
		$x = 1$; $R_1 = CH_3$; $R_2 = (CH{=}CH_2)$;		
		DSC	−69	(12, 14)
		DSC	−75	(12, 14)
		$x = 1$; $R_1 = R_2 = C_6H_5$;		
		TBA	+1	(9)
		DSC	−4	(9, 24)
		$x = 1$; $R_1 = C_6H_5$; $R_2 = (CH{=}CH_2)$; DSC	−31	(12)
		$x = 1$; $R_1 = C_6H_5$; $R_2 = (CH_2CH{=}CH_2)$; DSC	−38	(12)
		$x = 1$; $R_1 = R_2 = (CH_2)_3CN$; DSC	−30	(12, 16)
		$x = 1$; $R_1 = H$; $R_2 = (CH_2)_2CF_3$; DSC	−37	(12, 15)
		m-phenylene polymer $x = 0$; $R_1 = R_2 = CH_3$; DSC	−52	(26)
Melting point	°C	$x = 0$; $R_1 = R_2 = CH_3$	148	(3)
			152	(19)

Stabilities

Property	Units	Conditions	Value*	Reference
Thermal stability in nitrogen	°C	$x = 1$; $R_1 = R_2 = CH_3$; TGA ($10°$ min^{-1})	400	(1, 17)
		$x = 1$; $R_1 = H$; $R_2 = CH_3$; TGA ($10°$ min^{-1})	190	(1)
		$x = 1$; $R_1 = CH_3$; $R_2 = (CH_2)_2(CF_2)_5CF_3$; TGA ($10°$ min^{-1})	150	(1, 18)
		$x = 1$; $R_1 = CH_3$; $R_2 = (CH{=}CH_2)$;		
		TGA ($10°$ min^{-1})	480	(1, 17)
		TGA ($15°$ min^{-1})	547	(22)
		$x = 1$; $R_1 = CH_3$; $R_2 = (CH_2)_2CH_3$; TGA ($15°$ min^{-1})	545	(1, 16)
		$x = 1$; $R_1 = CH_3$; $R_2 = (CH_2CH{=}CH_2)$; TGA ($15°$ min^{-1})	505	(1, 16)

Property	Units	Conditions	Value*	Reference
		$x = 1$; $R_1 = CH_3$; $R_2 = (CH_2)_2CN$; TGA ($15° \, min^{-1}$)	495	(1, 16)
		$x = 1$; $R_1 = CH_3$; $R_2 = (CH_2)_3CN$; TGA ($15° \, min^{-1}$)	525	(1, 16)
		$x = 1$; $R_1 = CH_3$; $R_2 = C_6H_5$; TGA ($15° \, min^{-1}$)	395	(22)
		$x = 1$; $R_1 = R_2 = (CH_2)_3CN$; TGA ($15° \, min^{-1}$)	510	(1, 16)
		$x = 1$; $R_1 = C_6H_5$; $R_2 = (CH_2CH=CH_2)$; TGA ($15° \, min^{-1}$)	534	(1, 16)
		m-phenylene polymer $x = 0$; $R_1 = R_2 = CH_3$; TGA ($10° \, min^{-1}$)	415	(26)
Thermo-oxidative stability in air	°C	$x = 1$; $R_1 = R_2 = CH_3$; TGA ($10° \, min^{-1}$)	345	(1, 17)
		$x = 1$; $R_1 = H$; $R_2 = CH_3$; TGA ($10° \, min^{-1}$)	180	(1)
		$x = 1$; $R_1 = CH_3$; $R_2 = (CH_2)_2(CF_2)_5CF_3$	150	(1, 18)
		$x = 1$; $R_1 = CH_3$; $R_2 = (CH=CH_2)$; TGA ($15° \, min^{-1}$)	431	(22)
		$x = 1$; $R_1 = CH_3$; $R_2 = C_6H_5$; TGA ($15° \, min^{-1}$)	396	(22)
		m-phenylene polymer $x = 0$; $R_1 = R_2 = CH_3$; TGA ($10° \, min^{-1}$)	495	(26)

* Onset of weight loss in dynamic TGA.

Mechanical properties

Property	Units	Conditions	Value*	Reference
Tensile modules	MPa	$x = 1$; $R_1 = R_2 = CH_3$; 20–30% wt. silica filler; <3% cross-linking agent; <3 ppm antioxidant	1.5–3.8	(1, 8, 20)
Tensile strength (ultimate)	MPa	$x = 1$; $R_1 = CH_3$; $R_2 = C_6H_5$; 20–50% wt. silica filler; 10–15% dibutyltin diacetate	28.6–31.4	(1, 8)
		$x = 1$; $R_1 = R_2 = CH_3$; 20–30% wt. silica filler; <3% cross-linking agent; <3 ppm antioxidant	93.8–112.4	(1, 8, 20)
Tensile strength (nominal)	MPa	$x = 1$; $R_1 = CH_3$; $R_2 = C_6H_5$; 20–50% wt. silica filler; 10–15% dibutyltin diacetate	3.9–5.9	(1, 8)
		$x = 1$; $R_1 = R_2 = CH_3$; 20–30% wt. silica filler; <3% cross-linking agent; <3 ppm antioxidant	3.5–13.0	(1, 8, 20)
Elongation at break	%	$x = 1$; $R_1 = R_2 = CH_3$; 20–30% wt. silica filler; <3% cross-linking agent; <3 ppm antioxidant	900–1,150	(1, 8)
		$x = 1$; $R_1 = CH_3$; $R_2 = C_6H_5$; 20–50% wt. silica filler; 10–15% dibutyltin diacetate	530–740	(1, 8)
Surface energy	dyne cm^{-1}	m−phenylene polymer $x = 0$; $R_1 = R_2 = CH_3$	24.4	(26)

References

1. Dvornic PR, Lenz RW. *High Temperature Siloxane Elastomers*. Hüthig and Wepf, New York: Bazel-Heidelberg; 1990.
2. Sweda M, *U. S. Patent.* 2,561,429 and 2,562,000 (1951).
3. Merker RL, Scott MJ. *J. Polym. Sci. Part A.* 1964;2:15.
4. Dvornic PR. *Polym. Bulletin.* 1992;28:339.
5. Lai Y-C, Dvornic PR, Lenz RW. *J. Polym. Sci. Polym. Chem. Ed.* 1982;20:2277.
6. Wu TC. *U. S. Patent.* 1967;3325530.
7. Breed LW, Elliott RL, Whitehead ME., *J. Polym. Sci. Part A-1.* 1967;5:2745.
8. Burks RE Jr, Covington ER, Jackson MV, Curry JE. *J. Polym. Sci. Polym. Chem.Ed.* 1973;11:319.
9. Pittman CU Jr, Patterson WJ, McManus SP, *J. Polym. Sci. Polym. Chem. Ed.* 1976;14:1715.
10. Rosenberg H, Nahlovsky BD. *Polym. Preprints.* 1978;19(2):625.
11. Dvornic PR, Lenz RW. *J. Polym. Sci. Polym. Chem. Ed.* 1982;20:951.
12. Dvornic PR, Lenz RW. *Macromolecules.* 1992;25:3769.
13. Magill JH. *J. Appl. Phys.* 1964;35:3249.
14. Dvornic PR, Lenz RW. *J. Polym. Sci. Polym. Chem. Ed.* 1982;20:593.
15. Dvornic PR, Lenz RW. *Macromolecules.* 1994;27:5833.
16. Hani R, Lenz RW. In: Ziegler JM, Fearon FWG, eds. *Silicon-Based Polymer Science. Advances in Chemistry Series.* Vol 224, Am. Chem. Soc., Washington, D. C., 1990:741.
17. Dvornic PR, Lenz RW. *Polymer.* 1983;24:763.
18. Dvornic PR, Perpall HJ, Uden PC, Lenz RW. *J. Polym. Sci. Part A: Polym. Chem. Ed.* 1989;27:3503.
19. Magil JH. *J. Polym. Sci. Part A-2.* 1967;5:89.
20. Livingston ME, Dvornic PR, Lenz RW. *J. Appl. Polym. Sci.* 1982;27:3239.
21. Gardner KH, Magill JH, Atkins EDT. *Polymer.* 1978;19:370.
22. Zhu HD, Kantor SW, MacKnight WJ. *Macromolecules.* 1998;31:850.
23. Li Y, Kawakami Y, *Macromolecules.* 1999;32:3540.
24. Li Y, Kawakami Y. *Macromolecules.* 1999;32:8768.
25. Guida-Pietrasanta F, Boutevin B. *Adv. Polym. Sci.* 2005;179:1.
26. Zhang R, Pinhas AR, Mark JE. *Polym. Preprints.* 1997;38(2):298.

Poly(silylenemethylenes)

Q. H. Shen and L. V. Interrante

Acronyms, Trade Names, Alternative Names PSM, PSE (**I**), HPCS (**IIa**), AHPCS,
SP-4000, SMP-10 (**II**), poly[silylene(methylene)] (**I**), poly(carbodihydridosilane) (**I**),
poly(silaethylene) (**I**), hydridopolycarbosilane (**IIa**), allylhydridopolycarbosilane (**IIb**),
Starfire Matrix Polymer (**IIb**)

Class Polycarbosilanes

Structure
$[SiH_2CH_2]_n$(**I**); $[SiH_3CH_2-]_w[-SiH_2CH_2-]_x[=SiHCH_2-]_y[\equiv SiCH_2-]_z$, with an
overall SiH_2CH_2 composition (**IIa**);
$[SiR_3CH_2-]_w[-SiR_2CH_2-]_x[=SiRCH_2-]_y[\equiv SiCH_2-]_z$, where R = H, allyl or
vinyl (**IIb**)

Preparation Techniques The linear poly(silylenemethylene) with a "SiH_2CH_2"
$[(CH_4Si)_n]$ formula was synthesized via ring opening polymerization (ROP) of
1,1,3,3-tetrachloro-1,3-disilacyclobutane, followed by reduction of the Si-Cl groups with
$LiAlH_4^{(2)}$. The ROP route yields a high molecular weight linear polymer (labeled **I**). The
branched version of this polymer (**IIa**) and its allyl- or vinyl-substituted derivatives (**IIb**)
(e.g., nominally, $[SiH_xR_yCH_2]$ (where $x/y = 9/1 \rightarrow 1/3$ and $x + y = 2$) were prepared via
the Grignard coupling reaction of Cl_3SiCH_2Cl, followed by a partial substitution with a
Grignard reagent, $CH_2=CHCH_2MgCl$ or $CH_2=CHMgBr$, and finally a reduction by
$LiAlH_4$.[3,8] The Grignard coupling route gives a relatively low molecular weight polymer
with a hyperbranched structure ($[SiR_3CH_2-]_w[-SiR_2CH_2-]_x[=SiRCH_2-]_y$
$[\equiv SiCH_2-]_z$, where R=H, allyl or vinyl).

Major Applications Precursors for SiC ceramics and SiC-matrix composites
(primarily **IIb**).

Properties of Special Interest Relatively high cost. High yield for stoichiometric or
near-stoichiometric SiC ceramic upon pyrolysis to 1000°C under N_2. Viscous liquid,
miscible with hydrocarbons; moderately stable in air at room temperature. Poor resistance to
base and to oxidation by air at elevated temperatures ($>100°C$) or after several weeks at
room temperature. (These properties apply to both **I** and **II**).

Property	Units	Conditions	Value		Reference
			M_n	M_w	
Molecular weight	g mol^{-1}				
		Polymer **I**, gpc, PS standards	24,000	68,000	(1, 2)
		Polymer **I**, NMR	11,400	–	(2)
		Polymer **IIa**, vpo/gpc	740	1,330	(3, 4)

Poly(silylenemethylenes)

Property	Units	Conditions	Value	Reference
IR (characteristic absorption frequencies)	cm^{-1}	For linear PSM (**I**)	2,961; 2,921; 1,881; 2,130; 2,126; 1,406; 1,353; 1,250; 1,036; 946; 925; 856; 840; 756	
		For branched AHPCS (**IIb** (allyl))	2,950; 2,920; 2,870; 2,140; 1,450; 1,350; 1,250; 1,040; 930; 830; 760	
Raman (characteristic absorption frequencies)		For **I**	2,915; 2,875; 2,125; 2,120; 1,528; 1,356; 1,030; 983; 946; 789; 749; 706; 686; 600; 558; 472	(1)
NMR spectra of the linear poly(silylenemethylene)		(**I**)		(1, 2)
^{1}H NMR	ppm	solution(C_6D_6)	−0.15, 4.10	
^{13}C NMR	ppm	solution(C_6D_6)	−9.2	
^{29}Si NMR	ppm	solution(C_6D_6)	−34.4	
NMR spectra of the branched PSM		(**IIa and b**)		(3)
^{1}H NMR	ppm	solution	−0.4 to 1.15, 3.55 to 4.3	
^{13}C NMR	ppm	solution	−12 to 9, 12.5 to 26	
^{29}Si NMR	ppm	solution	−66 to −53, −39 to −26, −14 to −8, 0.0 to 5	
Monoclinic cell dimensions (Å) of the linear PSM		(**I**)	$a = 5.70$; $b = 8.75$; $c = 3.25$; $\gamma = 97.5$	(1)
Heat of fusion	$J\,g^{-1}$	Linear PSM	15.2	(1)
Glass transition	K	Linear PSM	133 to 138	(4)
Melting transition	K	Linear PSM	251 to 298	(4)
Degree of crystallinity		Linear PSM	70	(4)
Degree of branching for PSMs				
Linear PSM				(5)
Branching unit		Methyl	0.5%	
SMP-10 (**IIb**)				(6)
SMP-10	$Si(CH_2{-})_4$	$SiR(CH_2{-})_3$	$SiR_2(CH_2{-})_2$ $SiR_3(CH_2{-})$	
Ratio of branching units	2	8	20 11	
Decomposition Temperatures (K) in N_2	Starting decomposition Temp.	Ending decomposition Temp.		(7)
Linear PSE	473	873		
SMP-10	473	873		

Poly(silylenemethylenes)

Property	Units	Conditions	Value	Reference
	Pyrolysis Temperature(K)			
Pyrolyzability, nature of product from linear and branched PSMs	To 1273 1673–1873	Structure of ceramic formed amorphous SiC ß-SiC		(7)
Pyrolyzability, amount of product under N_2		Pyrolysis product (SiC)		(7)
Linear PSM	1273	87%		
SMP-10	1273	70–90%		
Pyrolyzability, purity of SiC		Pyrolysis product (SiC)		(2, 7)
Linear PSM	1273	1:1 stoichiometric Si to C		
SMP-10	1273	SiC_x where x = ca. 1.1		
Pyrolyzability, gaseous products under N_2		Substances other than desired ceramic		(7)
Linear PSM	473–673	>90% H_2		
Deuteride PSM	473–673	D_2 and HD		
Viscosity	cps	SMP-10, Brookfield Viscometer	50–150	(8)
Important patent		SMP-10	1	(9)
		Linear PSM	1	(10)
Cost	kg	SMP-10	Quoted on request from supplier	
Availability	g to tons	SMP-10 SMP-25 SMP-75		
Supplier	g to tons	SMP-10	Starfire Systems, Inc. 10 Hermes Rd, Suites 100 Malta, NY 12020 www.starfiresystems.com	

References

1. Tsao MW, Pfeifer KH, Rabolt JF, Holt DB, Farmer BL, Interrante LV, Shen Q. *Macromolecules.* 1996;29:7130.
2. Wu HJ, Interrante LV. *Macromolecules.* 1992;25:1840.
3. Whitmarsh CK, Interrante LV. *Organometallics.* 1991;10:1336.
4. Interrante LV, Wu HJ, Apple T, Shen Q, Smith K. *J. Am. Chem Soc.* 1994;116:12086.
5. Shen Q, Interrante LV. *Macromolecules.* 1996;29.
6. Rushkin I, Shen Q, Lehman SE, Interrante LV. *Macromolecules.* 1997;30:3141.
7. Interrante LV, Whitmarsh CK, Sherwood W, Wu HJ. *Better Ceramics Through Chemistry VI*, MRS; 1994;346:593.
8. The viscosity data came from the product specification of Starfire Systems, Inc.
9. Whitmarsh CK, Interrante LV. US patent. Oct. 6, 1992:5153295.
10. Smith TL. *US Patent.* Dec. 23, 1986:4631179.

Polystyrene

Zhengcai Pu

Acronyms, Alternative Name PS, styrofoam

Class Vinyl polymers

Structure

$$\{CH-CH_2\}$$
$$\quad | $$
$$\quad C_6H_5$$

Major Applications One of the most widely used plastics, having applications in industries of packaging, appliances, construction, automobiles, electronics, furniture, toys, house wares, and luggage (1–3).

Properties of Special Interest Crystal clear thermoplastic, hard, rigid, free of odor and taste, ease of heat fabrication, thermal stability, low specific gravity, excellent thermal and electrical properties for insulating purpose, and low cost (4–6).

Producers and/or Suppliers Dow Chemical USA; Huntsman Chemical Corporation; BASF Corporation; Fina Oil and Chemical Company; American Polymers, Inc.; American Polystyrene Corporation; Amoco Chemicals; Arco Chemical Company; Bayer Corporation; Chevron Chemical Company; StyroChem International, Inc. (7, 8)

Property	Units	Conditions	Value	Reference
Abrasion loss factor	mg	DIN 53516	640	(5)
Birefringence dispersion	–	$\dfrac{\Delta n(\lambda)}{\Delta n(546nm)}$	$A + B/\lambda^2 + C/\lambda^4$ $A = 0.8905$ $B = 0.275 \times 10^{-9}$ cm^2 $C = 0.153 \times 10^{-18}$ cm^4	(1, 4)
Ceiling temperature	K	Gas to gas Liquid to amorphous	550 670	(5)
Characteristic ratio $\langle r^2 \rangle_0/nl^2$	–	300 K, various solvents	9.85	(1)
Cohesive energy	kJ mol^{-1}	–	29.6–35.4	(5)
Compressibility coefficient	bar^{-1}	298 K–T_g T_g–593 K	2.7–4.9 ($\times 10^{-5}$) 5.3–11.3 ($\times 10^{-5}$)	(2)
Contact angle θ	°	Water Diiodomethane	85.5 ± 3.8 30.2 ± 1.6	(17)

Property	Units	Conditions	Value	Reference
Critical heat flux, combustion	kW m^2	–	13	(2)
Density ρ	g cm^{-3}	Amorphous	1.04–1.065	(1, 4, 5)
	g cm^{-3}	Crystalline	1.111–1.127	
	g cm^{-3} K^{-1}	$d\rho/dT$		
		$< T_{\mathrm{g}}$	-2.65×10^{-4}	
		$> T_{\mathrm{g}}$	-6.05×10^{-4}	
Dielectric constant	–	At 1 kHz		
		Amorphous	2.49–2.55	(1, 4, 5)
		Crystalline	2.61	(1, 4)
Dielectric loss	–	At 1 kHz		(5)
		Amorphous	15×10^{-4}	
		Crystalline	3×10^{-4}	

Diffusion coefficient $D_0^{(1)}$

Solvent	Temperature (K)	MW (kg mol^{-1})	D_0 (cm^2 s^{-1} ($\times 10^{-7}$))
Acetone	293	1,200–2,450	1.18–0.80
Benzene	298	1.32–3.9	27.9–17.2
Butanone	293	180–5,500	6.4–0.81
Carbon tetrachloride	300	82–1,100	4.43–1.04
Cyclohexane	303	90	4.0
Cyclohexanone	298	200	5.2
Dioxane	303	79.8	3.10
Ethyl acetate	293	117–596	6.23–2.45
Ethyl benzene	300	770	0.96
Tetrahydrofuran	303	198–570	13.41–8.02
Toluene	293	140–2,850	4.30–0.74

Property	Units	Conditions	Value	Reference
Enthalpy of fusion	kJ mol^{-1}	–	8.37–10	(1, 2, 4)
Entropy of fusion	kJ K^{-1} mol^{-1}	–	0.0153–0.0168	(2)
Friction coefficient	–	–	0.38	(5)
G factor	mol J^{-1} ($\times 10^{-8}$)	Cross-linking $G(X)$	7.14–19.2	(10)
		Scission $G(S)$	3.53–7.14	
Glass transition temperature	K	–	373	(2, 4, 5)
Hardness Rockwell hardness	–	R scale	130	(5, 9)
		M scale	75	
Ball indention hardness	MPa	–	110	
Bierbaum scratch hardness	–	–	10.3	
Heat capacity C_{p}	kJ K^{-1} mol^{-1}	$T = 100$ K	0.04737	(2)
		$T = 300$ K	0.12738, 0.13258	(2)
		$T = 400$ K	0.20124	(2)
		$T = 600$ K	0.25430	(2)
dC_{p}/dT	kJ K^{-2} mol^{-1}	$T = 323$ K	4.21×10^{-4}	(1, 4)

Polystyrene

Property	Units	Conditions	Value	Reference
Heat conductivity	$J\,s^{-1}\,m^{-1}\,K^{-1}$	Amorphous	0.13	(5)
Heat of combustion	$kJ\,mol^{-1}$	–	-4.33×10^3	(1, 4)
Ignition temperature	K	–	675	(2)
Impact strength (Izod)	$J\,m^{-1}$	ASTM D256	19.7	(5)

Interaction parameter $\chi^{(1)}$

Solvent	Temperature (K)	Volume Fraction of Polymer	χ
Acetone	298	0.6–1	0.81–1.1
Benzene	298	0.8–0.2	0.26–0.42
Chloroform	298	0.8–0.2	0.17–0.52
Cyclohexane	307	0–0.8	0.50–0.93
Methylcyclohexane	349	0–0.4	0.49–0.67
Methyl ethyl ketone	298	0.4–0.8	0.63–0.77
Propyl acetate	298	0.4–0.8	0.66
Toluene	298	1–0.2	0.16–0.37

Interfacial tension $\gamma_{12}^{(2)}$

Polymer Pair	Temperature (K)	γ_{12} (mN m^{-1})	$-d\gamma_{12}/dT$ (mN m^{-1} K^{-1})
Polystyrene/polychloroprene	413	0.5	–
Polystyrene/poly(methyl methacrylate)	293	3.2	0.013
Polystyrene/poly(vinyl acetate)	293	4.2	0.004
Polystyrene/polyethylene	493	4.4	–
Polystyrene/poly(dimethylsiloxane)	293	6.1	~ 0
Polystyrene/polyethylene, linear	293	8.3	0.020

Mark-Houwink parameters: K and $a^{(1)}$

Solvent	$K \times 10^3$ (ml g^{-1})	a	Temperature (K)	MW Range (kg mol^{-1})
Benzene	11.3	0.73	298	70–1,800
Butyl chloride	15.1	0.659	314	290–1,060
Chlorobenzene	7.4	0.749	299	620–4,240
Chloroform	7.16	0.76	298	120–2,800
Cyclohexane	82	0.50	307	10–700
Dimethylformamide	31.8	0.603	308	4–870
Dioxane	15.0	0.694	307	80–800
Ethylbenzene	17.6	0.68	298	70–1,500
Tetrahydrofuran	11.0	0.725	298	10–1,000
Toluene	12.0	0.71	303	400–3,700

Melt viscosity: molecular weight relationship constant $k^{(4)}$

	Temperature (K)	Molecular Weight Range (kg mol^{-1})	K
Atactic	490	$\geqslant 38$	13.04
Isotactic	554	100–600	14.42

Melt viscosity: temperature relationship constants[4]

*	$T_R = 411$ K	$T_R = 373$ K	Universal Value
C_1^R	6.99	13.35	17.4
C_2^R	81.8	42.00	51.6

* $M_n = 40{,}700$ g mol^{-1}, $M_w/M_n = 2.2$.

Property	Units	Conditions	Value	Reference
Limiting oxygen index (LOI)	%	–	17.8	(11)
Melting point	K	–	513	(1, 4, 5)
Modulus				
Bulk modulus	MPa	–	3,000	(5)
Compressive modulus	MPa	–	3,000	(4)
Flexural modulus	MPa	–	3,100	(5)
Shear modulus G	MPa	–	1,200	(5)
Tensile modulus E	MPa	Unoriented	3,200–3,400	(1, 5, 9)
		Oriented monofilament	4,200	
dE/dT	MPa K^{-1}	Unoriented	−4.48	(1)
Nonsolvents			acetone, acetic acid, alcohols, diethyl ether, diols, ethylene chlorohydrin, glycol ethers, isobutyl phthalate, phenol, saturated hydrocarbons, trichloroethyl phosphate, tricresyl phosphate	(1)
Optical dispersion $\eta_F - \eta_C$	–	$\lambda = 486.1$ nm $\lambda = 656.3$ nm	1.92×10^{-2}	(4, 12)
Permeability coefficient	m^3 (STP) ms^{-1} m^{-2} Pa^{-1}	$T = 298$ K; permeant=		(1)
		H$_2$	17.0×10^{-4}	
		He	14.0×10^{-4}	
		N$_2$	0.59×10^{-4}	
		O$_2$	2.0×10^{-4}	
		H$_2$O	840×10^{-4}	
		CO$_2$	7.9×10^{-4}	
Poisson ratio	–	–	0.325–0.33	(1, 4, 5)
Refractive index n dn/dT	K^{-1}	$\lambda = 589.3$ nm	1.59–1.60 -1.42×10^{-4}	(1, 4)

Property	Units	Conditions	Value	Reference
Refractive index increment dn/dc	ml g^{-1}	Various solvents	0.103–0.225	(1)
Resistivity	ohm cm		10^{20}–10^{22}	(1, 4)
Scattering length density r_n	cm^{-2}	Neutron	1.415×10^{10}	(2)
Solvents			benzene, carbon disulfide, cyclohexane, cyclohexanone, dimethyl phthalate, dioxane, ethyl acetate, ethylbenzene, glycol formal, methyl ethyl ketone, 1-nitropropane, phosphorus trichloride, tetrahydrofuran, tributyl phosphate	(1)

Second virial coefficient A_2[1]

Solvent	Temperature (K)	Molecular Weight (kg mol^{-1})	$A_2 \times 10^4$ (mol cm^3 g^{-2})
Benzene	293	7,100–1,330	3.3–3.6
Bromobenzene	293	1,750–35.5	2.15–6.38
Butanone	293	150	0.0127
Carbontetrachloride	293	150	3.58
Chloroform	–	–	6.56
Dioxane	–	–	2.75
Methyl acetate	303	179.3	−0.235
1-Phenyldecane	295	390	−3.22
Toluene	293	40,200–12,300	1.37–2.32

Shear viscosity, melt[16].

$\eta = a\gamma^{b-1}$ (shear viscosity in unit of Pa s, shear rate γ in the range of 10–1000 s^{-1})

Temperature (K)	Wt % of Blowing Agent	A	b
Hydrochlorofluorocarbon (HCFC) as blowing agent			
403	8	61276	0.2383
423	10	15396	0.3312
443	13	3360	0.4800
70 wt. % Hydrofluorocarbon + 30 wt. % ethanol as blowing agent			
403	7	72769	0.1750
423	8	27085	0.2281
443	9	10857	0.2970

Property	Units	Conditions	Value	Reference
Solubility parameter	(MPa)$^{1/2}$	Various solvents	15.6–21.1	(1, 5, 13)
Sound absorption, longitudinal	dB cm^{-1}	RT, 2 MHz	1.4	(14)

Property	Units	Conditions	Value	Reference
Speed of sound		RT, 1 MHz		(2)
Longitudinal C_L	ms^{-1}		2,400	
$-dC_L/dT$	$ms^{-1} K^{-1}$		1.5	
$dC_L/d \log f$	$ms^{-1} dec^{-1}$		1.4	
$d \ln C_L/dP$	GPa^{-1}		0.9	
Shear C_S	ms^{-1}		1,150	
$-dC_S/dT$	$ms^{-1} K^{-1} GPa^{-1}$		4.4	
$d \ln C_S/dP$			0.5	
Strength				
Compressive strength	MPa	–	95	(5)
Elongation at break	%	–	1–4	(2)
Flexural strength	MPa	–	95	(5)
Tensile strength at break	MPa	–	30–60	(2)
Stress-optical coefficient (brewsters)	–	Monofilament	10.1	(1, 4)
		Extruded sheet	9.5	
		Compression molded	8.3–8.7	

Surface tension γ (mN m^{-1})$^{(1)}$

Molecular Weight (g mol^{-1})	Temperature (K)			$-d\gamma/dT$ (mN m^{-1} K^{-1})
	293	423	473	
$M_v = 44,000$	40.7	31.4	27.8	0.072
$M_n = 9,300$	39.4	31.0	27.7	0.065
$M_n = 1,700$	39.3	29.2	25.4	0.077

Thermal conductivity	W m^{-1} K^{-1}	$T = 273$ K	0.105	(1, 2, 4)
		$T = 323$ K	0.116	(1, 2, 4)
		$T = 373$ K	0.128	(1, 2, 4)
		$T = 473$ K	0.13	(2)
		$T = 573$ K	0.14	(2)
		$T = 673$ K	0.160	(2)
Thermal decomposition	K	Initial temperature	573	(2)
		Half decomposition temperature	637	
Thermal expansion coefficient	K^{-1}	Linear		
		$< T_g$	6–8 ($\times 10^{-5}$)	(1, 4)
		Volume		
		$< T_g$	1.7–2.1 ($\times 10^{-4}$)	
		$> T_g$	5.1–6.0 ($\times 10^{-4}$)	
Theta temperature θ	K	Solvent		(1)
		i-Butyl acetate	227	
		n-Butyl formate	264	
		1-Chlorodecane	279	

Property	Units	Conditions	Value	Reference
		1-Chlorododecane	332	
		1-Chloroundecane	306	
		Cyclohexane	307–308	
		Cyclohexanol	352–361	
		Cyclopentane	293	
		Decalin	285–304	
		Diethyl malonate	304–309	
		Diethyl oxalate	325–333	
		Ethyl acetoacetate	381	
		Ethylcyclohexane	343	
		Methylcyclohexane	333–343	
		3-Methylcyclohexanol	371	
		1-Phenyldecane	301–304	
Unit cell		Isotactic		(1, 4)
Crystallographic system	–		Rhombohedral	
Space group	–		D3D-6	
Cell dimension	Å			
a_0			21.9–22.1	
b_0			21.9–22.1	
c_0	–		6.65–6.63	
Repeat unit per unit cell			18	
Upper use temperature	K	–	333	(2)
Vicat softening point	K	–	373	(4)
Zeta-potential ($\zeta_{plateau}$)	mV		-40.4 ± 5.0	(17)
Zisman critical surface tension	mN m^{-1}	–	32.8	(15)

References

1. Brandrup J, Immergut EH. *Polymer Handbook*. 3rd ed. New York: Wiley-Interscience; 1989.
2. Mark JE, ed. *Physical Properties of Polymers Handbook*. Woodbury, New York: AIP Press; 1996.
3. Windholz M. *The Merck Index: An Encyclopedia of Chemicals, Drugs, and Biologicals*. 10th ed. Rahway, N.J: Merck and Co; 1983.
4. Boyer RF. Mark HF, et al. ed. In *Encyclopedia of Polymer Science and Technology*. Vol 13. New York: John Wiley and Sons; 1970.
5. Van Krevelen DW, Hoftyzer PJ. *Properties of Polymers: Correlations with Chemical Structure*. Amsterdam: Elsevier Publishing Company; 1972.
6. Ulrich H. *Introduction to Industrial Polymers*. 2nd ed. Munich: Hanser Publishers; 1993.
7. *Directory of Chemical Producers*. Menlo Park, Calif: United States of America. SRI International; 1996.
8. *Chem Sources-U.S.A.* Pendleton, S.C: Chemical Sources International; 1997.
9. "Styrene Plastics". In *Technical Data on Plastics*. Washington, D.C: U.S. Manufacturing Chemists' Association; 1957.
10. Parkinson WW, Keyser RM. In: Dole M. *The Radiation Chemistry of Macromolecules*. Vol II. New York: Academic Press; 1973.

11. Cullis CF, Hirschler MM. *The Combustion of Organic Polymers.* Oxford: Clarendon Press; 1981.

12. Boundy RH, Boyer RF, eds. In: *Styrene: Its Polymers, Copolymers and Derivatives.* New York: Reinhold Publishing; 1952.

13. Mangaraj D, Bhatnagar SK, Rath SB. *Makromol. Chem.* 1963;67:75.

14. Wada Y, Yamamoto K. *J. Phys. Soc. Jpn.* 1956;11:887.

15. Ellison AH, Zisman WA. *J. Phys. Chem.* 1954;58:503.

16. Choudhary M, et al. *J. Cell. Plast.* 2005;41:589.

17. Bismarck A, et al. *J. Thermoplastic Composite Mat.* 2005;18:307.

Polystyrene, Head-to-head

Michael. T. Malanga and Ying Li

Acronyms H-H PS, H-H polystyrene

Class Chemical copolymers

Structure

$$[-CH_2-CH(C_6H_5)-CH(C_6H_5)-CH_2-]_n$$

Preparative Technique H—H polystyrene has never been obtained directly from styrene monomer and indirect polymerization techniques have to be used. H–H polystyrene was first synthesized in a four step reaction. It is synthesized by the selective hydrogenation of 1,4-poly(2,3-diphenyl-1,3-butadiene) (PDPB) using potassium/ethanol. PDPB is prepared by the free radical polymerization of 2,3-diphenyl-1,3-butadiene to give a 45% *cis*, 55% *trans* structure. H—H PS is then given in the same ratio of erythro and threo linkages after the chemical reduction of the internal double bond of the PDPB. H—H polystyrene can now be made in larger quantities.[1-4]

Major Applications This polymer is not manufactured commercially by any company in the world at this time. It has only been prepared in laboratory scale quantities. The primary reason for this is that the cost of preparing H—H polystyrene would be very high for the perceived value of its properties. There are no published reports of the mechanical properties of H—H polystyrene at this time.

However, given its measured glass transition temperature and backbone structure it may be anticipated to have similar tensile, modulus, and other mechanical properties to commercial H–T polystyrene.

Properties of Special Interest Blends of H—H PS with H–T PS which have nearly the same T_g (H—H T_g 97°C, H–T T_g 100°C) were miscible. H—H PS is completely miscible with poly(2,6-dimethyl phenylene oxide) in the same way that H–T PS is miscible with that polymer.[5] The T_g of the blends are then intermediate between the two polymers. The thermal stability and glass transition temperature of H—H PS are very similar to those of atactic H–T PS despite the structural differences.[1] Although the H—H linkage has been suggested as a possible "weak link" in the commonly manufactured H–T polystyrene, the thermal stability evidence suggests that this is not the case.[6]

Property	Units	Conditions	Value	Reference
Preparation (see above) hydrogenation of 1,4-poly (2,3-diphenyl-1, 3-butadiene)		Radical polymerization of 2,3-diphenyl-1,3-butadiene followed by chemical reduction of the internal double bond yields the H—H polystyrene structure		(1, 3)
Glass transition temperature	K	DSC with heating rate of $10°C\ min^{-1}$	370	(1)
Thermal decomposition temperature	K	DTG onset of degradation, 10°C heating rate, under nitrogen	620	(1)
Theta temperature	K	Cyclohexane solvent	292	(8)
Mark-Houwink parameters: K and a	$K = ml\ g^{-1}$ a = None	THF solvent at 25°C	$K = 5.3 \times 10^{-2}$ $a = 0.61$	(8)
Second virial coefficient A_2	$mol\ cm^3\ g^{-2}$	Cyclohexane solvent at 35°C, 82,800 weight average molecular weight	2.3×10^{-4}	(8)
Interaction parameter	–	THF solvent at 25°C	0.464	(8)
		Cyclohexane at 35°C	0.471	(8)
Structural analysis: NMR	ppm	^{13}C (obtained in deuterochloroform)	31.0, 51.0, 126.1, and 125.4,127.2 and 127.1, 129.0 and 128.7, and 143.9 and 142.8	(9)
		^{13}C solution in chlorobenzene at 90°C		(1, 7)
		Phenyl 1 carbon	144.3	
		Methine backbone carbon	49.7	
		Methylene backbone carbon	28.9	
		^1H	0.95 (2 H, methylene protons), 2.11 (1 H, methine proton), 6.85(5 H, aromatic protons)	(9)
UV	$l\ (mol\ cm)^{-1}$	dilute solutions in THF	67	(9)
Crystallinity		The polymer shows no crystallinity. It is considered completely amorphous.		(6)

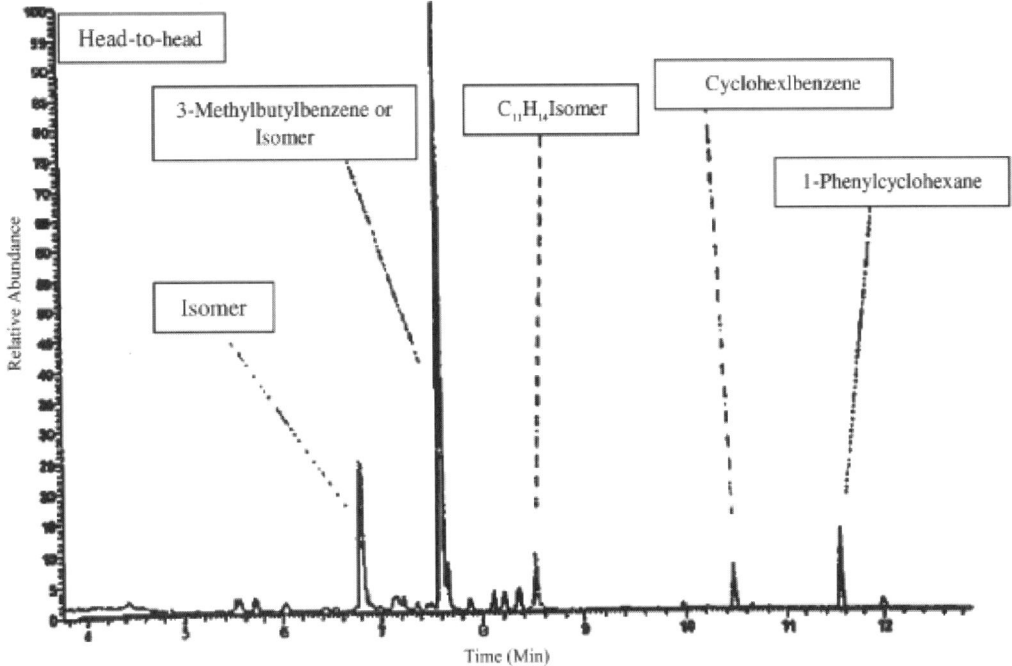

Figure 1. Evolved-gas chromatogram (TG/GC/MS) for initial degradation of head-to-head poly(styrene) at 280°C.[10]

Rate constants for the thermal degradation of head-to-head poly(styrene)[10] ($M_W = 1.4° \times 10^5$, $M_n = 4.9° \times 10^4$ and $M_W/M_n = 2.86$).

Temperature (°C)	$k \times 10^{-5}$ (s^{-1})
280	0.153
320	3.02
350	30.9

References

1. Inoue H, Helbig M, Vogl O. *Macromolecules.* 1977;10(6):1331.
2. Helbig M, Inoue H, Vogl O. *J Polym Sci, Polym Symp Ed.* 1978;63:329.
3. Foldes E, et al. *Eur Polym. J.* 1993;29(2–3):321.
4. Vogl O, Qin MF, Zilkha A. *Progress in Polymer Science.* 1999;24(10):1481.
5. Kryszewski M, et al. *Polymer.* 1982;23:271.
6. Vogl O, Malanga M, Berger W. *Contemporary Topics in Polymer Science.* New York: Plenum Press; 1984;4:35.
7. Bangerter F, Serafini S, Pino P. *Makromol. Chem. Rapid Commun.* 1981;2:109.
8. Strazielle C, Benoit H, Vogl O. *Eur. Polym. J.* 1978;14(5):331.
9. Bob Howell A, Yumin Cui, Duane Priddy B. *Thermochimica Acta.* 2003;396:191.
10. Bob Howell A, Yumin Cui, Duane Priddy B. *Thermochimica Acta* 2003;396:167.

Poly(styrene-*b*-isoprene-*b*-styrene) (unsaturated and hydrogenated)

Juan T. Weaver, Ravi Shankar and Richard J. Spontak

Acronyms SIS: poly[styrene-*b*-(ethylene-*alt*-propylene)-*b*-styrene], SEPS

Class Elastomeric polymers

CAS Numbers 25038-32-8, 25038-32-8 (hydrogenated)

General Chemical Structures

Major Applications Used frequently in synthetic elastomers, pressure sensitive adhesives, and membranes; blended frequently with homopolymers such as polypropylene (PP) and polyethylene (PE) to impart elastomeric properties.

Polymerization Method	Conditions	Reference
Anionic polymerization	Sequential living polymerization initiated by *n*-butyllithium, *sec*-butyllithium, and lithium alkyl initiators in benzene or cyclohexane under high vacuum or high-purity nitrogen. Saturated SIS can be hydrogenated using an octoate/triethylaluminum catalyst in cyclohexane at ~6 atm and 50°C under hydrogen	(1–6)
Radical polymerization	Polymerization initiated by bifunctional alkoxyamine bis-TIPNO.	(7)

Poly(styrene-*b*-isoprene-*b*-styrene) (unsaturated and hydrogenated)

COPOLYMER: SIS

Property	Units	Conditions	Value		Reference
Density	g cm^{-3}	Commercially manufactured (e.g., Kraton and Dexco)	0.92		(8)
Stress relaxation activation energy	kJ mol^{-1}	Measured in a home-built relaxameter	113		(9)
Lamellar interfacial thickness	Nm	Thermodynamic prediction	4.8		(10)
	Nm	Dielectric spectroscopy	~4		(10)
Morphology		Follows the same trend observed for microphase-ordered SI diblock copolymers with shifts in phase boundaries due to added constraints on molecular self-assembly	Increasing S content in the melt: S micelles, S cylinders, gyroid, lamellae, gyroid, I cylinders and I micelles		
	–	In aliphatic oil			(11)
		<20 wt% oil	Lamellar		
		20–54 wt% oil	Cylindrical		
		>54 wt% oil	Micellar		
Self-diffusion coefficient *D* in toluene	10^{-11} cm^2 s^{-1}	Copolymer concentration (wt%)			(12)
		0.13	1300		
		0.23	220		
		0.32	58		
		0.42	26		
		0.51	6.4		
		0.61	0.67		
Radius of gyration R_g	Å	Uniaxial extension (l_f/l_0)	min	max	(13)
		1.0	87 ± 4	109 ± 6	
		1.2	86 ± 4	109 ± 6	
		1.4	84 ± 4	126 ± 7	
		1.6	82 ± 4	149 ± 9	
		1.8	79 ± 4	179 ± 11	
		2.0	77 ± 4	195 ± 13	
		2.2	74 ± 4	204 ± 13	
Ordering half-time upon quenching from 180°C	min	Quench temperature (°C)	Rheology	Scattering	(14)
		120	300	370	
		130	210	230	
		140	720	790	
		150	950	–	
		154	1230	–	

Property	Units	Conditions	Value	Reference
NMR spectroscopy	–	400 MHz solid-state photon spin diffusion	–	(15)
Permeability at 23°C	cm^3-mil $(100\ in^2$-atm-day$)^{-1}$	Penetrant		(16)
		Oxygen	3,170	
		Carbon dioxide	19,300	
Vapor transmission rate at 23 °C	g-mil $(100\ in^2\ h)^{-1}$	Water vapor	22.1	(16)

Grafting monomers

Grafting Monomer	Reaction Conditions	Reference
1-vinylimidazole	Radiation-induced copolymerization	(17)
4-vinylpyridine	Radiation-induced copolymerization	(17)
Acrylic acid (AA)	SIS and AA dissolved in toluene/ethanol mixture. Grafting initiated by benzoyl-peroxide (BPO) at 80 °C under N_2 for 6 h	(18)
1-(4-dimethyl-aminophenyl)-1-phenylethylene (DMADPE)	Styrene polymerized with *sec*-BuLi (initiator) followed by the introduction of DMADPE and sequential addition of isoprene. Diblock molecules are then coupled using $(CH_3)_2SiCl_2$ in benzene	(19, 20)

Crosslinking reactions involving SIS

Crosslinking Molecules	Reaction Conditions	Reference
SIS-*g*-AA and polypyrrole (PPy) doped with dodecylbenzenesulfonic acid (DBSA)	In THF and initiated by tetraethyl orthosilicate (TEOS)	(18)
SIS-*g*-AA and polyaniline doped with DBSA	In THF initiated by TEOS	(21)
SIS and trimethylolpropane mercaptopropionate (TRIS) and/or 1,6-hexanediol diacrylate (HDDA)	After being solvent-casted from toluene, films are exposed to radiation (80 W cm^{-1} medium-pressure Hg lamp in the presence of air)	(22)

Extrusion-blown blends of SIS and polyethylene

Conditions	Morphology	Reference
5 wt% SIS	ellipsoids	(23)
7.5–12.5 wt% SIS	transition between short fibrils and elongated fibrils	(23)
15 wt% SIS	continuous fibrils	(23)
25–45 wt% SIS	nonhomogeneous fibrils	(23)
240°C	co-continuous, elongated domains	(24)
270°C	broken filaments	(24)

Poly(styrene-*b*-isoprene-*b*-styrene) (unsaturated and hydrogenated)

Blends of SIS with various concentrations of poly(vinyl methyl ether) (PVME)[25]

PVME content (wt%)	Miscibility	Opacity
5–15	Miscible	Clear
20–80	Immiscible	Opaque

Dielectric and rheological characteristics of SIS/C$_{14}$ systems in the viscous regime[26]

T (°C)	$10^4 \times \tau_{chain}^a$ (s)	$10^4 \times \tau_G^b$ (s)	$10^5 \times J_c^c$ (Pa^{-1})
60	18	250	6.2
70	9.4	7.8	1.9

a τ_{chain}– relaxation time for SIS chains
b τ_G– viscoelastic relaxation time
c J_c – steady-state compliance

Conductivity and tensile strength of PPy-DBSA/AA-*g*-SIS interpenetrating network[21]

IPN	H$_2$O (g)	TEOS (g)	PPy/AA-*g*-SIS (w/w)	Air Surface	Bottom Surface	Tensile Strength (MPa)	Young's Modulus (MPa)
PPyDBSA/AA-*g*-SIS	0.01	0.1	20/80	0.510	0.200	–	–
PPyDBSA/AA-*g*-SIS	0.01	0.5	20/80	0.210	0.310	8.64	27.2
PPyDBSA/AA-*g*-SIS	0.1	0.1	20/80	1.2	0.310	8.9	24.5
SIS-*g*-AA	0.01	0.1	0/100	Insulating	–	5.73	30.1

Effect of polystyrene (PS) molecular weight and content on the tensile strength of SIS[27]

Styrene Content (wt%)	sis $\overline{M}_w$ (kg mol^{-1})	Tensile Strength (MPa) at 300% Strain	Tensile Strength at Break (MPa)
20	13.7–100.4–13.7	1.8	27.0
20	8.4–63.4–8.4	1.1	16.0
19	7.0–60–7.0	1.3	2.2
11	5.0–80–5.0	∼0	∼0

Physical properties of blends composed of PS, SIS and mineral oil (MO)[28]

Ratio of Components (PS:SIS:MO)	Tensile Stress at Break (MPa)	Elongation at Break (%)	Hardness (shore A)	Haze (%)	Transparency (%)
15:85:0	25.2	910	62.0	84.2	51.2
15:75:10	19.5	990	55.0	88.6	52.0
0:90:10	16.0	1360	47.0	43.1	91.5

Physical properties of blends of high-impact polystyrene (HIPS), high-density polyethylene (HDPE) and SIS[29]

Blend (HIPS:HDPE: SIS by wt%)	Styrene Content in SIS (wt%)	Hardness (Shore A)	Tensile Stress at Break (MPa)	300% Modulus (MPa)	100% Modulus (MPa)	Elongation at Break (%)	Gardner Impact Energy at 23° C (J)	Izod Impact Energy (J m^{-1})	Dtul[1] at 1.83 MPa (°C)
0:0:100	43	88	26	7.9	5.5	985	–	–	–
0:0:100	30	60	30	4.7	3.2	1,100	–	–	–
70:20:10	43	–	22	–	–	–	36.0	107.7[2]	70.7
70:20:10	30	–	20	–	–	–	27.4	119.7[2]	80.8
20:70:10	43	–	17	–	–	–	36.0	48.9[3]	53.7
20:70:10	30	–	17	–	–	–	30.6	33.8[3]	68.7

1 Deflection temperature under load.

Properties of some commercial SIS copolymers (typically possessing 29–33 wt% styrene)[8]

Copolymer	Tensile Strength (MPa)	Elongation at Break (%)	Tensile Strength at 200% Strain (MPa)	Hardness (Shore A)
Shell Kraton 1107	18.6	1,350	1.0	36
Dexco Vector 4113	18.8	1,500	0.4	35
Dexco Vector 4100D	23.3	1,400	0.6	38

Mechanical properties of SIS (Kraton® 1107)/ultralow-density PE (ULDPE) blends given in w/w SIS/ULDPE[8]

Property	100/0	90/10	80/20	76/24	70/30	60/40	50/50	40/60	30/70	20/80	10/90	0/100
Hardness (Shore A)	36	40	45	47	53	64	67	71	79	82	84	86
Tensile strength (MPa)	18.6	16.8	13.5	13.2	9.3	3.2	3.3	3.6	4.1	4.5	4.6	6.6
Tensile strength at 200% strain (MPa)	1.0	0.8	1.2	1.3	1.3	1.9	2.0	2.2	2.9	3.0	3.4	3.7
Elongation at break (%)	1,350	1,450	1,350	1,920	1,375	1,150	900	900	800	850	750	900
Tear resistance (kN m^{-1})	31	27	25	23	26	25	30	30	32	36	35	44

Mechanical properties of SIS (Dexco 4113 or 4100D)/ULDPE blends given in w/w SIS/ULDPE[8]

Property	4113 100/0	4113 90/10	4113 80/20	4100D 100/0	4100D 95/05	4100D 90/10	4100D 80/20
Hardness (Shore A)	35	39	45	38	41	41	46
Tensile strength (MPa)	18.8	18.4	11.0	19.3	23.9	19.2	19.7
Tensile strength at 200% strain (MPa)	0.4	1.1	0.7	0.7	0.7	0.6	1.7
Elongation at break (%)	1,500	1,200	1,200	1,400	1,350	1,300	1,500
Puncture resistance (kN m^{-1})	77	69	68	106	84	53	55
Tear resistance (kN m^{-1})	22	27	27	30	30	29	25

Poly(styrene-*b*-isoprene-*b*-styrene) (unsaturated and hydrogenated)

Mechanical properties of SIS (Dexco 4113)/ULDPE blends given in w/w SIS/ULDPE[8]

Property	100/0	80/20	75/25	70/30
Hardness (Shore A)	35	43	47	45
Tensile strength (MPa)	18.8	18.7	11.2	16.8
Tensile strength at 200% strain (MPa)	0.4	0.6	1.0	0.7
Elongation at break (%)	1,500	1,300	1,100	1,200
Puncture resistance (kN m^{-1})	77	83	83	78
Tear resistance (kN m^{-1})	22	30	25	24

Mechanical properties of SIS (Dexco 4100D)/ULDPE blends given in w/w SIS/ULDPE[8]

Property	100/0	90/10	80/20	70/30	60/40	50/50	40/60	30/70	20/80	10/90	0/100
Hardness (Shore A)	38	39	46	50	55	59	64	67	71	75	78
Tensile strength (MPa)	23.2	27.6	24.8	24.80	18.6	20.4	21.5	23.3	26.2	27.0	28.7
Tensile strength at 200% strain (MPa)	0.6	0.9	1.2	1.4	2.1	2.0	2.3	2.8	2.8	3.1	3.4
Elongation at break (%)	1,400	1,300	1,100	1,100	1,000	900	900	900	700	800	800
Puncture resistance (kN m^{-1})	106	117	122	144	112	132	124	131	126	105	100
Tear resistance (kN m^{-1})	30	38	37	27	28	33	33	38	42	43	46

Physical properties of SIS/polypropylene (PP)/PS blends[30]

Blend (SIS:PP:PS) (phr)	Tensile Strength (MPa)	Tensile Modulus (MPa)	Extension at Break (%)	Flexural Strength (MPa)	Flexural Modulus (MPa)	Impact Strength (J m^{-1})	Linear Shrinkage (%)	E^a (kJ mol^{-1})
2:70:30	33.10	1104.18	108	34.26	1389.24	26.614	1.034	54.91
3:70:30	31.49	988.89	140	34.62	1371.59	31.181	1.106	58.12
5:70:30	30.24	934.89	154	30.33	1323.78	37.953	1.114	57.60
10:70:30	27.81	740.933	183	23.94	792.00	51.570	1.305	50.65

a activation energy for viscous flow

Dependence of electrical resistivity on carbon black (CB) in SIS/CB mixtures[31]

CB Content (phr)	Log (RESISTIVITY) Log (ohm-cm)
0	15.9
3.3	15.8
10.1	14.6
10.2	11.3
15.0	3.7
20.0	3.0

Dependence of resistivity on SIS concentration in HIPS/SIS/CB blends[31]

SIS Content (wt%)	CB Content at 2 phr	CB Content at 4 phr
	Log (RESISTIVITY) Log (ohm-cm)	Log (RESISTIVITY) Log (ohm-cm)
0	16.9	8.4
14	15.7	6.7
21	7.0	–
30	8.2	7.1
30	8.7	7.8
45	15.8	6.8
45	–	5.8
80	–	15.0
100	16	15.2

Physical properties that vary with ferrite content in SIS matrices[32]

Ferrite Content (wt%)	Izod Impact Energy (ft-lb in^{-1})	Resilience	Toughness (kg cm^{-2})	E' (kg cm^{-2})	Shore Hardness (A)
–	3.0	0.75	–	–	–
75	2.9	0.45	15.0	90×10^6	40
80	2.75	0.44	9.6	19.0×10^7	50
85	2.7	0.42	4.5	26.2×10^7	55

Physical properties that vary with ferrite content in SIS matrices[32]

Ferrite Content (vol%)	E' (dyn cm^{-2})	Electrical Conductivity (ohm^{-1} cm^{-1})	Permittivity
0	17.2×10^6	1.17×10^{-14}	0.43×10^3
0.4	63.5×10^6	2.33×10^{-13}	0.49×10^3
0.49	92.1×10^6	4.43×10^{-13}	0.7×10^3
0.58	160.8×10^6	1.01×10^{-12}	1.0×10^3

Dependence of T_g on oil content in SIS (Kraton D1107)/oil (Polyoil 130) mixtures[33]

Oil Content (wt%)	T_g Isoprene Domain (°C)	T_g Styrene Domain (°C)
0	−61.5	65.8
5	−64.4	61.8
10	−65.2	60.9
16	−69.8	61.1

Poly(styrene-*b*-isoprene-*b*-styrene) (unsaturated and hydrogenated)

Frequency dependence of SIS in rheological experiments[34]

Frequency (rad s^{-1})	Viscoelastic Response
$\omega < 10^{-4}$	Terminal region exhibits liquid behavior ($G' << G''$)
$10^{-4} < \omega < 10$	Elastic behavior becomes dominant ($G' > G''$)
$10 < \omega < 10^5$	Rubbery-to-glassy transition region ($G' = 2 \times 10^4$ Pa)
$\omega > 10^5$	Glassy behavior with a modulus value typical of organic glasses ($G_g = 4 \times 10^8$ Pa)

Peel results (sample composed of a rigid Al substrate attached to Al foil by SIS copolymer)[34]

Peel Rate (mm min^{-1})	Result
10^{-2}–1	Cohesive fracture with an adhesive layer remaining on both substrates
1–10^4	Sudden drop of peel force as the adhesive remains on the Al substrate.
10^4–10^{10}	Unstable fracture ('stick-slip') occurs. Cracks propagate by jumps that create periodic oscillations in peel force.
$> 10^{10}$	Adhesive remains on the rigid substrate and the peel force level is small ($F < 10$ N).

Order–Order Transition Temperature[35]

$T_{OOT}(°C)$	$T_{sC}^{a}(°C)$	$T_{sS}^{b}(°C)$
196	202	$\sim$183

a Lower stability limit of cylindrical morphology.
b Upper stability limit of spherical morphology.

Isoprene loop fraction (vs. bridging) in the SIS lamellae morphology in C_{14}[36]

SIS Content (WT%)	Isoprene Loop Fraction
20	0.78
30	0.70
40	0.67
50	0.55

Crosslinked Blends of SIS with TRIS and/or HDDA[22]

SIS + TRIS + HDDA		SIS + TRIS (2wt%)		SIS + HDDA (20wt%)		SIS	
Exposure Time (s)	Gel Fraction (%)	Exposure Time (s)	Gel Fraction (%)	Exposure Time (s)	Gel Fraction (%)	Exposure Time (s)	Gel Fraction (%)
0	0	–	–	–	–	–	–
0.11	18.2	–	–	–	–	–	–
0.20	44.0	0.20	0	–	–	–	–
0.46	68.0	0.43	9.33	0.44	0.44	–	–
0.96	91.6	0.98	76.9	0.99	22.2	0.98	0.89
1.97	97.8	1.96	91.6	1.97	35.1	1.98	1.33
3.00	99.6	3.00	97.8	2.97	48.9	2.98	0.89

Mechanical properties of the bicontinuous gyroid morphology[37]

Stretch Direction	Initial Modulus (MPa)	Necking Behavior	Yield Stress (MPa)	Yield Strain
Isotropic	29 ± 5	Yes	2.6 ± 0.5	0.20 ± 0.02
[111]	48 ± 9	Yes	3.4 ± 0.7	0.25 ± 0.04
Transverse	9.6 ± 3.2	No	0.74 ± 0.09	0.09 ± 0.03

Local and global topological characteristics of the bicontinuous gyroid morphology measured from 3D transmission electron microtomography[38,39]

Property	Units	Experimental	Theoretical[a]
Composition	vol% S	33	32
Area-average mean curvature	nm^{-1}	0.034	0.039
Mean curvature standard deviation	nm^{-1}	0.042	0.023
Channel coordination	%		
3		89.5	100
4		9.0	0
5		1.3	0
Junctions/unit cell		11	15
Interjunction distance/period		0.40 (linear)/0.45 (path)	0.40
Euler-Poincaré characteristic/unit cell		-12.1	$-14.7 \, (-16.0)^{b}$
Genus		7.1	$8.4 \, (9.0)^{b}$

a Relative to a constant thickness model of identical composition.
b Relative to the mathematical definition of the Schoen G surface.

Peel test of SIS and aliphatic oil blends (sample was PET attached to steel with a 50 μm layer of adhesive and measured after 25 mm)[40]

Tackfier Content (wt%)	Peel Adhesion (N)[a]	Peel Adhesion (N)[b]
0	0	13.8
16.7	8.6	17.5
28.7	11.5	17.5
37.3	13.2	18.7
44.3	19.8	22.7
49.7	24.1	27.6
54.8	31.3	33.9

a Pressed five times with a 2-kg roller.
b Pressed five times with a 2-kg roller and then annealed for 2 h at 80°C.

Aging of SIS films at 95°C in air[41]

AGING (min)	Tack (g)	SIS/SI (w/w)
0	70	3.8
35	81	3.3
60	91	2.9
90	104	2.5
120	103	0
150	75	–

Poly(styrene-*b*-isoprene-*b*-styrene) (unsaturated and hydrogenated)

Mechanical alignment of SIS morphologies

Morphology	Conditions	Reference
Cylinders	Reciprocating shear; $\gamma_0 = 0.5$ or 1.0 and $\omega 1 = $ rad s^{-1}	(21)
Cylinders	Ball-milling	(28, 29)
Gyroid	Ball-milling	(30)

Order-disorder transition (ODT) temperature of various SIS copolymers

$\overline{M}_W$ (kg mol^{-1})	Styrene Content (wt%)	T_{ODT} (°C)	Ordered Morphology	Reference
140	14.3	220	Sphere	(42)
113.8	13.0	180	Sphere	(42)
39	30.0	90	Cylinder	(42)
28	52.1	105	Lamellae	(42)
90.6	13	58	Sphere	(43)
106.8	12.2	87	Sphere	(43)
120.1	13.3	102	Sphere	(43)
149	13	158	Sphere	(43)
39.3	51.6	152	Lamellae	(43)
149	13	164	–	(3, 14)

Property development upon sequential step growth from SI diblock to SIS triblock copolymer[44,45]

$\overline{M}_n$ (kg mol^{-1})	T_{ODT} (°C)	Microdomain Period (nm)	G' (kPa)
9.4–46	142	244.0	7.68
9.4–46–2.0	128	240.5	6.06
9.4–46–3.4	107	212.1	5.28
9.4–46–6.5	116	206.0	20.6
9.4–46–8.2	147	205.2	36.4
9.4–46–11.8	191	212.1	94.0
9.4–46–15.9	236	237.6	170.8

COPOLYMER: SEPS

Properties of ethylene–propylene rubber (EPR) and PS blends (55/45) in the presence of SEPS[46]

Property	Units	Conditions	Value
Density	g cm^{-3}	7 wt% S	0.985
		14 wt% S	0.978
Flexural modulus	MPa	7 wt% S	800
		14 wt% S	450
Flexural stress	MPa	7 wt% S	19.7
		14 wt% S	12.3

Properties of ethylene–propylene rubber (EPR) and PS blends (55/45) in the presence of SEPS[46]

Property	Units	Conditions	Value
Tensile modulus	MPa	7 wt% S	874
		14 wt% S	453
Yield stress	MPa	7 wt% S	12.9
		14 wt% S	12.1
Ball indentation	N mm^{-2}	7 wt% S	31
		14 wt% S	23
Vicat softening temperature	°C	7 wt% S	129
		14 wt% S	127
Heat deflection temperature	°C	7 wt% S	54
		14 wt% S	51

Properties of thermoplastic vulcanizates (TPVs) composed of SEPS/poly(styrene-*b*-butadiene-*b*-styrene) (SBS)/ paraffin oil/PP/coupling agent (CA) composites with various levels of SBS/(SBS + SEPS) weight ratios and different concentrations of CA (1.0/1.8 peroxide/acrylic ester weight ratio)[47]

SEPS/SBS/oil/PP/ CA(w/w/w/phr)	Hardness (Shore A)	Tensile Strength (MPa)	100% Modulus (MPa)	Elongation at Break (%)	Compression set (%)	Volume Swell (%)	Gel Content (wt%)
42/0/40/18/2.1	65	13.2	2.2	760	63	120	24
35/5/42/18/2.1	66	9.1	2.5	590	61	90	25
27/10/45/18/2.1	64	6.7	2.5	400	59	67	27
23/13/46/18/2.1	63	6.4	2.6	350	59	52	29
19/15/48/18/2.1	62	5.6	2.6	300	59	48	31
14/18/50/18/2.1	62	4.9	2.6	250	58	39	33
11/21/50/18/2.1	59	3.6	2.5	190	59	34	34
14/18/50/18/0.0	48	3.2	1.7	490	100	88	0
14/18/50/18/1.0	59	4.6	2.2	350	66	57	25
14/18/50/18/3.4	63	4.9	2.9	210	55	32	36

Mechanical and physical properties of TPV composed of TPUs/PP/SEPS composites[48]

Type of TPU	Hardness (Shore A)	Tensile Strength (MPa)	100% Modulus (MPa)	Elongation at Break (%)	Volume Swell (%)	Volume Loss (mm^3)
Pandex T-8180	72	14.2	3.8	550	34	18
Pandex T-1180	69	15.7	3.8	590	18	25
Pandex T-7890N	80	11.6	6.1	490	18	65
Pandex T-R3080	66	10.7	3.1	520	43	47

Poly(styrene-*b*-isoprene-*b*-styrene) (unsaturated and hydrogenated)

Glass transition temperature (in °C) of PP (55.4 kg mol^{-1}) in blends with PS ($127.8 \text{ kg mol}^{-1}$) and SEPS ($23.8 \text{ kg mol}^{-1}$)[49]

SEPS content (phr)	25/75 w/w PP/PS	45/55 w/w PP/PS	50/50 w/w PP/PS	75/25 w/w PP/PS
0.0	−4.90	−3.64	2.13	4.81
1.0	−3.55	−3.72	2.31	5.57
2.0	−1.28	−3.71	2.23	5.58
5.0	3.43	−3.68	3.18	5.61
10.0	5.23	−1.21	3.81	5.65

SEPS properties

Property	Units	Conditions	Value	Reference
Decrease in molecular weight	kg mol^{-1}	Following hydrogenation of SIS	37	(3)
T_{ODT}	°C	SAXS	>300	(50)
Maximum oil uptake	wt% oil	SEPS/oil system	60	(5)
Specific gravity	G cm^{-3}	23°C	0.91	(51)
Melt viscosity	Poise	200°C, 1216 s^{-1}	9100	(51)
Hardness	JIS A	15 s, 23°C	80	(51)
Tensile strength at break	MPa	No. 3 Dumbbell at 23°C	42	(51)
Elongation at break	%	No. 3 Dumbbell at 23°C	480	(51)
Compression set	%	70°C for 22 h	95	(51)

T_{ODT} of various SEPS/oil/clay composites containing different SEPS/oil w/w ratios[52]

Na^+-Clay Content (wt%)	T_{ODT} (°C) (30/70)	T_{ODT} (°C) (40/60)	T_{ODT} (°C) (50/50)
0.0	172	196	226
0.5	170	194	225
1.0	170	193	224
3.0	169	193	223
5.0	169	192	222

Physical properties of 30 wt% SEPS in *i*-PP blend[53]

Styrene Content (wt%)	Yield Strength (MPa)	Craze Strength (MPa)	Tensile Strength (MPa)	Dispersed Phase (vol%)
18.0	23.4	32.1	6.6	9.3
30.2	23.8	33.1	23.7	14.3
65.4	31.0	42.8	27.9	18.0

Properties of mixtures of PP/SEPS and various oils (20:46:34 w/w/w) submerged in oil for 72 h[54]

Oil Type	Hardness (Shore A)	Relative Density	Tensile Strength (MPa)	Modulus at 100 % Elongation (MPa)	Elongation at Break (%)	Volume Swell (%)
No Oil	72	0.948	10.0	2.8	600	0
Paraffinic Oil	53	0.892	2.9	1.6	350	+62
ASTM oil #3	62	0.939	3.3	1.8	290	+37
Automobile grease	73	0.926	6.7	1.9	470	+13
Petrol	82	0.942	7.0	2.5	480	−16

Mechanical properties of SEPS/asphalt blends[55]

Seps Content (wt%)	Tensile Strength (N cm^{-2})	% Elongation
0	30.9	5.0
3	85.5	10.0
5	124.7	15.0
10	263.5	119.4
20	335.9	537.0

Dependence of the dynamic storage moduli (E' and G') on oil concentration in SEPS/aliphatic oil blends[5,56]

Oil Content (wt%)	E' OF Mesogel[a] (MPa)	E' (MPa)	G' (MPa)
0	20.3	20.3	
12	11.1	−	
16	−	6.4	
28	7.3	2.9	
40	−	2.2	
50	−	1.6	
60	3.1	0.6	
70	−	−	2.57
75	−	−	1.00
80	−	−	0.71
85	−	−	0.63
90	−	−	0.23
95	−	−	0.04

a Produced by imbibing oil into microphase-ordered block copolymers.

References

1. Richards RW, Thomason JL. *Polymer.* 1983;24:275.
2. Pixa R, Schirrer R. *Colloid Polym. Sci.* 1981;259:435.
3. Laurer, et al. *Langmuir.* 1999;15:7947.
4. Mather BD, Beyer FL, Long TE. *Polym. Prepr.* 2006;47:456.
5. King MR, White SA, Smith SD, Spontak RJ. *Langmuir.* 1999;15:7886.
6. Holden G, Milkovich R. *U.S. Patent.* 3,265,765. 1964.

7. Gao LC, et al. *Chem. Res. Chinese U.* 2005;21:615.

8. Bigg DM. In: Benedikt GM, Goodall BL. eds. *Metallocene-Catalyzed Polymers – Materials, Properties, Processing and Markets.* Norwich: William Andrew Publishing/Plastics Design Library; 1998:330–336.

9. Wu GW, Hsiue GH, Yang JS. *Mater. Chem. Phys.* 1994;37:191.

10. Alig I, Floudas G, Avgeropoulos A, Hadjichristidis N. *Macromolecules.* 1997;30:5004.

11. Laurer JH, et al. *Langmuir.* 1999;15:7947.

12. Lodge TP, Blazey MA, Liu Z, Hamley IW. *Macromol. Chem. Phys.* 1997;198:983.

13. Richards RW, Welsh G. *Eur. Polym. J.* 1995;31:1197.

14. Adams JL, et al. *Macromolecules.* 1996;29:2929.

15. Marjanski M, Srinivasarao M, Mirau PA. *Solid State Nucl. Mag.* 1998;12:113.

16. In: *Permeability and Other Film Properties of Plastics and Elastomers.* Norwich: William Andrew Publishing/Plastics Design Library; 1995.

17. Pispas S, Hadjichristidis N. *Macromolecules.* 1994;27:1891.

18. Gan LH, Gan YY, Yin WS. *Polymer.* 1999;40:4035.

19. Pispas S, Hadjichristidis N. *J. Polym. Sci. A: Polym. Chem.* 2000;38:3791.

20. Floudas G, Pispas S, Hadjichristidis N, Pakula T. *Macromol. Chem. Phys.* 2001;202:1488.

21. Gan LH, Gan YY, Yin WS. *Polym. Int.* 1999;48:1160.

22. Decker C, Viet TNT. *J. Appl. Polym. Sci.* 2000;77:1902.

23. David C, Getlichermann M, Trojan M, Jacobs R. *Polym. Eng. Sci.* 1992;32:6.

24. Utracki LA. In: Utracki LA, ed. *Polymer Blends Handbook, Volumes 1–2.* Springer, New York: 2002:321.

25. Xie R, Yang BX, Jiang BZ. *J. Polym. Sci. B: Polym. Phys.* 1995;33:25.

26. Watanabe H, et al. *Macromolecules.* 1997;30:5877.

27. In: Legge NR, Holden G, Schroeder HE. eds. *Thermoplastic Elastomers; A Comprehensive Review.* Munich: Hanser Publishers; 1987:75.

28. Uzee AJ. In: *ANTEC 2005 Plastics: Annual Technical Conference, Volume 3: Special Areas.* Society of Plastics Engineers; 2005:3613–3614.

29. Uzee AJ. In: *ANTEC 2000 Plastics: The Magical Solution, Volume 2: Materials.* Society of Plastics Engineers; 2000. Article 669.

30. Raghu P, Nere CK, Jagtap RN. *J. Appl. Polym. Sci.* 2002;88:266.

31. Tchoudakov R, Breuer O, Narkis M, Siegmann A. In: Rupprecht L, ed. *Conductive Polymers and Plastics in Industrial Applications.* Norwich: William Andrew Publishing/Plastics Design Library, 1999:52.

32. Saini DR, Nadkarni VM, Grover PD, Nigam DP. *J. Mater. Sci.* 1986;21:3710.

33. Shea PT, Pietruski RD, Shih C-K, Denelsbeck DA. In: *ANTEC 2002 Plastics: Annual Technical Conference, Volume 3: Special Areas.* Society of Plastics Engineers; 2002. Article 692.

34. Gilbert FX, Allal A, Marin G, Derail CJ. *Adhes. Sci. Technol.* 1999;13:1029.

35. Ryu CY, Lodge TP. *Macromolecules.* 1999;32:7190.

36. Watanabe H, Sato T, Osaki K. *Macromolecules.* 2000;33:2545.

37. Dair BJ, Avgeropoulos A, Hadjichristidis N, Thomas EL. *J. Mater. Sci.* 2000;35:5207.

38. Jinnai H, Nishikawa Y, Spontak RJ, Smith SD, Agard DA, Hashimoto T. *Phys. Rev. Lett.* 2000;84:518.

39. Jinnai H, Kajihara T, Watashiba H, Nishikawa Y, Spontak RJ. *Phys. Rev. E.* 2001;64:10803(R) [Erratum: 69903].

40. Sasaki M, et al. *J. Adhes. Sci. Technol.* 2005;19:1445.

41. Harrison DJ, Johnson JF, Yates WR. *Polym. Eng. Sci.* 1982;22:865.

42. Han CD, Baek DM, Kim JK, Chu SG. *Polymer.* 1992;33:294.

43. Adams JL, Graessley WM, Register RA. *Macromolecules.* 1994;27:6026.

44. Hamersky MW, Smith SD, Gozen AO, Spontak RJ. *Phys. Rev. Lett.* 2005;95:168306.
45. Smith SD, Hamersky MW, Bowman MK, Rasmussen KO, Spontak RJ. *Langmuir.* 2006;22:6465.
46. Schellenberg J. *J. Appl. Polym. Sci.* 1997;64:1835.
47. Tasaka M. Saito S. In *ANTEC 2000 Plastics: The Magical Solution, Volume 3: Special Areas.* Society of Plastics Engineers; 2000:Article 257.
48. Tasaka M, Masubuchi N. In *ANTEC 2000 Plastics: The Magical Solution, Volume 3: Special Areas.* Society of Plastics Engineers; 2000:Article 256.
49. Lee SG, Jae JH, Choi KY, Rhee JM. *Polym. Bull.* 1998;40:765.
50. Vega DA, Sebastian JM, Loo Y-L, Register RA. *J. Polym. Sci., Part B: Polym. Phys.* 2001;39:2183.
51. Marshall D, Ishikawa H, Kawai H, Aoyama T. In: *ANTEC 2004 Plastics: Annual Technical Conference, Volume 3: Special Areas.* Society of Plastics Engineers; 2004:4177.
52. Nakajima T, Akiyama T, Furuya H. *Magn. Reson. Chem.* 2002;40:161.
53. Matsuda Y, et al. *Polym. Eng. Sci.* 2005;45:1630.
54. Tasaka M, Tamura A, Mori R. In: Pourdeyhimi B, ed. *Imaging and Image Analysis Applications for Plastics.* Norwich: William Andrew Publishing/Plastics Design Library; 1999:194–195, 198.
55. Kamiyama S, Tasaka S, Hotta D. *Nippon Kagaku Kaishi.* 2001;3:163.
56. Shankar R Ph.D. *Dissertation.* Raleigh, NC: North Carolina State University; 2007.

Poly(sulfur nitride)

J. F. Rubinson

Acronym, Alternative Name $(SN)_x$, polythiazyl

Class Inorganic and semi-inorganic polymers

Structure [⁻S⁻N⁻]

Major Applications Electrode fabrication in crystalline, film, or paste form. Electrodes are useful in both aqueous and some nonaqueous solvents. Ion-selective electrode. Contact with semiconductors yields high-voltage junction. Photocell and lithium battery fabrication.

Properties of Special Interest A number of the intermediates in its synthesis as well as the dry polymer are explosive under certain conditions. A thorough literature survey of its properties should be undertaken before synthesis or use.[1,2,3] Intrinsic metallic conductor. Undoped polymer is a superconductor at 0.3 K, while doped forms have been made with higher T_c.

Property	Units	Conditions	Value					Reference
Electrochemical breakdown potential ($|i| \geq 0.02$ mA cm^{-2})	V vs.SCE	0.1 M electrolyte: Solvent	Et$_4$NClO$_4$	LiClO$_4$	Me$_4$NClO$_4$	LiCl		(4)
		Ethanol	—	—	−0.51	−0.57		
		Propylene carbonate	−0.48	−0.81	—	—		
		Acetonitrile	−0.40	−0.74	—	—		
		Solvent	Et$_4$NClO$_4$	LiClO$_4$	Me$_4$NClO$_4$	LiCl		
		Ethanol	—	—	+0.80	+0.82		
		Propylene carbonate	+0.87	+0.80	—	—		
		Acetonitrile	+0.95	+0.96	—	—		
Electrochemical breakdown potential ($|i| \geq 1.0$ mA cm^{-2})	V vs.SCE	0.1 M electrolyte: Breakdown type	LiCl	NaCl	KClO$_4$	KPF$_6$	Et$_4$NClO$_4$	(5)
		Cathodic breakdown	−3.6	−3.6	−3.6	−3.6	−3.3	
		Anodic breakdown	2.2	2.2	2.2	3.2	2.1	

Property	Units	Conditions	Value	Reference
Electrical conductivity σ	S cm^{-1}	|| to b-axis, 300 K	2,000	(6)
		$\perp$ to b-axis, 300 K	40	
$\sigma_\|/\sigma_\perp$	—	Room temperature	50	(1)
		20 K	500–1,000	
Electronegativity	—	—	2.9	(6)
Decomposition temperature	K	—	513	(1)
Enthalpy of vaporization	kJ mol^{-1}	—	135.9	(1)
Entropy of vaporization	kJ K^{-1} mol^{-1}	—	0.3388	(1)
Conduction bandwidth	eV	—	2–3	(1)
Young's modulus	MPa	Crystal	3×10^{-16}	(1)
Yield stress	MPa	Crystal	1.450×10^2	(1)
Breaking stress	MPa	Crystal	3.660×10^2	(1)
Specific heat	—	<3.2 K	$C/T = \text{constant} \times T^3$	(1)
		4–20 K	$C/T = \text{constant} \times T^{2.7}$	
		>20 K	$C = \text{constant} \times T$	
Magnetic susceptibility	emu g^{-1}	Crystal	$(0.2 \times 0.1) \times 10^{-6}$	(1)
Paramagnetic susceptibility	emu mol^{-1}	Crystal	$(5.5 - 1.0) \times 10^{-6}$	(1)
Drude edge ($R_{\|}$ vs. $v \times 10^{-3}$)	cm^{-1}	Oriented film or crystal	20,000	(7)
Density	g cm^{-3}	Crystal	2.3	(1)

Preparative techniques

Technique	Conditions	Reference
Plasma	He plasma, S$_4$N$_4$ vapor	(8)
Solution phase	N$_3^-$ + S$_a$N$_b$Cl$_c$ in acetonitrile (-258 K)	(9)
Solution phase	N$_3^-$ + S$_2$NAsF$_6$ in liquid SO$_2$ ($-20°$C)	(9)
Vapor phase	S$_4$N$_4$ sublimation over Ag wool; S$_2$N$_2$ trapped at 77 K, then 273 K; polymerization at temperature	(10)
Solution phase	(NSCl)$_3$ + Me$_3$SiN$_3$ in liquid SO$_3$ ($-18°$C)	(11)
Electrochemical	S$_5$N$_5$Cl in liquid SO$_2$	(12)
Photopolymerization	S$_4$N$_4$ decomposition products, irradiated with γ up to visible range	(13)
Vapor phase	Sputtering of (SN)$_x$ from target in inert gas such as N$_2$ or Ar (RT)	(14)
Solution phase	(SN)$_x$ + dopant in supercritical CO$_2$	(15)

Poly(sulfur nitride)

IR (characteristic absorption frequencies)

	Wavelength (cm^{-1})										Reference
KBr pellet	1,400	1,225		1,010	930	690		600			(8)
Nujol mull	1,400	1,225	1,047	1,015		685	657	600			(16)
Nujol mull					1,000	693	635		500	285	(11)
Film on KBr					1,002	685	625		500, 467	283	(11)

Unit cell dimensisons[6]

Lattice	Monomers per Units Cell	Cell Dimensions (Å)			Cell Angle (Degrees)
		a	b (Chain Axis)	c	β
Monoclinic (P2$_1$/c)	4 (N-S)	4.153	4.439	7.637	109.7

References

1. Labes MM, Love P, Nichols LJ. *Chem. Rev.* 1979;79:1.
2. Rawson JJ, Longridge JJ. *Chem. Soc. Rev.* 1997;26:53.
3. Modelli A, Venuti M, Scagnolari F, Centento M, Jones D. *J. Phys. Chem. A.* 2001;105:219.
4. Nowak RJ, Joyal CL, Weber DC. *J. Electroanal. Chem.* 1983;143:413.
5. Tarby C, Bernard C, Robert G. *Electrochimica Acta.* 1981;26:663.
6. Love P. *Polymer News.* 1981;7:200.
7. Bright AA, et al. *Appl. Phys. Lett.* 1975;26:612.
8. Witt MW, Bailey WI Jr, Lagow RJ. *J. Am. Chem. Soc.* 1983;105:1668.
9. Kennett FA, et al. *J. Chem. Soc. Dalton Trans.* 1982;851.
10. Rubinson JF, Behymer TD, Mark HB Jr. *J. Am. Chem. Soc.* 1982;104:1224.
11. Banister AJ, et al. *J. Chem. Soc. Dalton Trans.* 1982;1986:2371.
12. Banister AJ, Hauptman ZV, Kendrick AG. *J. Chem. Soc., Chem. Commun.* 1983;1016.
13. Love P, Labes MM. *U.S. Patent.* 4,170,477. 1979.
14. Benson MH, Neudecker BJ. *U.S. Pantent.* 6,770,176. 2004.
15. Desimone J, Ni Y. *U.S. Pantent.* 5,855,819. 1998.
16. Banister AJ, Smith NRM. *J. Chem. Soc. Dalton Trans.* 1980;937.

Poly(tetrafluoroethylene)

D. L Kerbow and R. M. Aten

Acronym, Trade Names PTFE, Algoflon, Dyneon, Fluon, Polymist, Polyflon, Teflon, Zonyl

Class Poly(α-olefins)

Structure $[-CF_2-CF_2-]$

Major Applications Granular and fine powder forms are used in electrical wire insulation, seals, and gaskets, and in valve and pipe fittings and linings for harsh chemical applications. Fine powders are also prepared in fiber, filament, and porous fabric forms. Dispersions are used in glass cloth coatings to provide weather protection, mechanical strength, and chemical resistance. Micropowders are used as additives to inks, lubricants, and plastics to provide lubricity, antiburning, and nonstick properties.

Properties of Special Interest Three major forms of PTFE exist: granular, fine powder, and micropowders. Granular is produced by suspension polymerization in the absence of a surfactant. It is a spongy, porous form of irregular particle shape as polymerized, and it is typically ground to a particle size to suit fabrication and end-use needs. Fine powder is coagulated from dispersion which is polymerized in the presence of an emulsifying agent. It can be supplied as the dispersion or in a coagulated form. It is extremely sensitive to mechanical shear. Micropowder can be produced as a low molecular weight form of fine powder or by scission of fine powder products by gamma or electron beam irradiation. It is typically a waxy or friable powder.

Crystalline Repeat Unit The polymer chain in the crystalline matrix exists as a helix, with successive CF_2 units rotated slightly by the steric interference of adjacent fluorine atoms. The repeat distance of the helix is 19.5 Å (15 CF_2 units) at temperatures above 19°C, or 16.9 Å (13 CF_2 units) below 19°C.

Property	Units	Conditions	Value	Reference
Molecular weight (of repeat unit)	g mol^{-1}	–	50.01	–
Tacticity	–	–	None	–
Degree of branching	–	–	None	–
Typical molecular weight range	g mol^{-1}	Polymer form		(1)
		Fine powder	$1-5 \times 10^7$	
		Granular	10^7	
		Micropowder	$2-25 \times 10^4$	

Property	Units	Conditions		Value		Reference
IR (characteristic absorbances	cm^{-1}	Assignment		Strength		
frequencies)		Overtone $(1,152 + 1,213)$: used analytically as a "thickness band"		Very strong	2,367	(2) (Ch. 21)
		–		Very strong	1,242	(2)
		–		Very strong	1,213	(2)
		CF_2 stretch		Very strong	1,152	(2)
		Crystallinity: used analytically to determine % C as noted below		Weak	778	(3)
		C—C—F bend		Strong	638	(2)
		CF_2 bend		Strong	553	(2)
		C—C—F bend		Strong	516	(2)
Coefficient of linear thermal expansion (average)	$K^{-1} \times 10^{-6}$	298–83 K		86		(4)
		298–173 K		112		
		298–273 K		200		
		296–333 K (ASTM D696)		120		
		298–373 K		124		
		298–473 K		151		
		298–573 K		218		
Compressibility	bar^{-1}	Calculated		28.8×10^{-18}		(4)
Solubility parameter	$(MPa)^{1/2}$	Calculated		12.7		(2) (Ch. 16)
Solvents	–	>573 K		Perfluorinated materials		(4)
Crystalline state properties		PTFE exists in multiple forms that are influenced by temperature, pressure, and thermal history. In turn, these forms significantly influence the physical, electrical, and processing properties of the polymer. Particularly, the percent crystallinity and specific gravity have been found to relate to a large number of properties, and since these parameters are influenced by processing history, it is very important to specify precise sample preparation conditions. In equations below, % C = percent crystallinity, and ρ = density.				

Crystal lattice

Crystalline Form	Conditions	Unit Cell Dimensions (nm)			λ (Degrees)
		a	b	c	
Form I	Above 30°C	0.567	0.567	>1.950	–
Form II	Below 19°C	0.559	0.559	1.688	119.3
Form III	High pressure	0.873	0.569	0.262	–
Form IV	19–30°C	0.566	0.566	1.950	–

Crystalline Form	Chain Conformation	Space Group	Crystal Density
Form I	15/7	Trigonal (P3$_1$ or P3$_2$)	2.35
Form II	13/6	Triclinic (pseudohexagonal)	2.30
Form III	2/1	Orthorhombic (Pnam)	2.55
Form IV	15/7	Trigonal (P3$_1$ or P3$_3$)	2.74

Property	Units	Conditions		Value		Reference
Entropy of fusion	kJ K^{-1} mol^{-1}	D4591 (ASTM method)		0.477		(4)
Degree of crystallinity	g cm^{-3}	–		762.5–(1,524.5/ρ)		(4)
Heat of fusion	kJ kg^{-1}	D4591 (ASTM method)		82		(4)
Density	g cm^{-3}	Crystal state				
		Completely amorphous, 298 K		2.0 (calculated)		(4)
		Triclinic, <292 K		2.344		(4)
		Hexagonal, 298 K		2.302		(4)
		As polymerized, 298 K		2.280–2.290		(4)
		Melt, 653 K		1.46		(2)
						(Ch. 24)
Melting point	K		Polymer Form			(4)
		Irreversible	As polymerized	608–618		
		Reversible	Second (and subsequent) melting	600		
		–	Equilibrium	586.9		
		Irreversible	Extended chain	658		
Transition temperature	K		Type of Transition			(4)
		–	Alpha (glass I)	399		
		Crystalline, crystal disordering relaxation	Beta	292		
		Crystal disordering	Beta II	303		
		–	Amorphous 2nd order	243		
		Onset of rotation around C—C bond	Gamma (glass II)	193		
Heat capacity	kJ kg^{-1} K^{-1}			Crystalline	Amorphous	(4)
		DSC, 10 K		1.228		
		100 K		19.37	19.37	
		300 K		45.09	51.42	
		500 K		61.62	66.05	
		605 K (melting point)		67.88	69.54	
		700 K		73.30	72.69	

Poly(tetrafluoroethylene)

Property	Units	Conditions	Value		Reference
Deflection temperature	K	Deflection force (MPa – D648)			(4)
		0.455	405		
		1.82	333		
Thermal conductivity	W m^{-1} K^{-1}	C177	$(4.86 \times 10^{-4})T + 0.253$		(4)
Tensile modulus	MPa	22 K (ASTM D638)	4,100		(4)
		77 K	3,400		
		144 K	2,500		
		200 K	1,800		
		296 K	340		
		373 K	69		
Tensile strength	MPa		Granular	Fine Powder	
		298 K	7–28	17.5–24.5	(1)
		298 K	–	270–0.39	(4)
				(% C) – 99.3ρ	
Yield stress	MPa	22 K (ASTM D638)	131		(4)
		77 K	110		
		144 K	79		
		200 K	53		
		296 K	10		
		413 K	5.5		
		523 K	3.4		
Modulus type Compressive	MPa	ASTM D 695 After 100 h at 6.895 MPa, 23°C	186		(4)
Tensile		–	61		
Flexural		–	2,814 – 158.5 (% C) + 2.919 (% C)^2 – 0.1638 (% C)^3		
Maximum extensibility	%	ASTM D638	Granular	Fine Powder	
		22 K	–	2	(4)
		77 K	–	6	(4)
		144 K	–	90	(4)
		200 K	–	160	(4)
		296 K	100–200	200–600	(1)
Flexural modulus	MPa	ASTM D790	Granular	Fine Powder	
		22 K	–	5,200	(4)
		77 K	–	5,000	(4)
		144 K	–	3,200	(4)
		296 K	350–630	280–630	(1)
		328 K	–	400	(4)
		373 K	–	190	(4)
Flexural strength	MPa	D790	No break		(4)

Property	Units	Conditions	Value		Reference
Impact strength	$J\ m^{-1}$	D256 (notched Izod impact)			(4)
		216 K	107		
		276 K	187		
		350 K	>320		
Hardness	Shore D	D2240	42 + 0.2 (% C)		(4)
Plateau modulus	MPa	653 K	1.7		(2) (Ch. 24)
Entanglement molecular weight	$g\ mol^{-1}$	–	3.7×10^3		(2) (Ch. 24)
Index of refraction	–	η_D^{25}	1.376		(4)
Dielectric constant ε	–	D150	2.1		(4)
Dielectric strength	$V\ mm^{-1}$	D149	2.36×10^4		(1)
Dissipation factor	–	D150 (60 Hz to 2 GHz)	$<3 \times 10^{-4}$		(4)
Resistivity, surface	$ohms\ sq^{-1}$	D257 (100% RH)	3.6×10^6		(4)
Resistivity, volume	ohms cm	D257 (50% RH)	10^{19}		(4)
Surface tension	$mN\ m^{-1}$	293 K	25.6		(2) (Ch. 48)
Thermal conductivity	$W\ m^{-1}\ K^{-1}$	298 K	0.25		(2) (Ch. 10)
Coefficient of sliding friction	–	D 1894	$0.244\ W^{0.163}$ (W = load in grams)		(4)
Static coefficient of friction	–	Against polished steel	0.05–0.08		(1)
Speed of sound	$m\ sec^{-1}$	1 MHz, 298 K			(2) (Ch. 49)
		Longitudinal	1,410		
		Shear	730		
Ignition temperature	K	–	767		(2) (Ch. 42)
Weight loss in air	$\%\ h^{-1}$		Granular	Fine Powder	(4)
		505 K	1–5×10^{-5}	1×10^{-4}	
		533 K	1–2×10^{-4}	6×10^{-4}	
		589 K	5×10^{-4}	5×10^{-3}	
		644 K	4×10^{-3}	3×10^{-2}	
Pyrolysis products	mol %	Vacuum at 783 K			(6)
		CF_4	0.86		
		C_2F_4	93.97		
		C_3F_6	2.55		
		cyclo-C_4F_8	0.73		
Maximum use temperature	K	In air	533		(1)

Poly(tetrafluoroethylene)

Property	Units	Conditions	Value	Reference
Depolymerization rate	g sec^{-1}	Vacuum pyrolysis	3×10^{-19} M$^{(-83,000/RT)}$	(7)
Water absorption	%	D570	0.0	(4)
Flammability	–	UL 94	VE-0	(4)
	%	D2863 (Limiting oxygen index)	>95	(4)
Cost	US$ kg^{-1}	–	11–35	
Availability	106,000 metric tonnes (2007 est.)			
Suppliers	3M, Asahi Glass, Daikin, DuPont, Solvay Solexis			

References

1. Gangal SV. Kroschwitz JI, ed. In *Kirk-Othmer Encyclopedia of Chemical Technology*. 3rd ed. Vol 11. New York: John Wiley and Sons; 1994.
2. Mark JE, ed. In *Physical Properties of Polymers Handbook*. Woodbury, N.Y: AIP Press; 1996.
3. Moynihan RE. *J. Am. Chem. Soc.* 1959;81:1045.
4. Sperati CA. In: Brandrup J, Immergut EH, eds. *Polymer Handbook*. 3rd ed. Vol 35. New York: John Wiley and Sons; 1989.
5. Tadokoro H. *Structure of Crystalline Polymers*. New York: John Wiley and Sons; 1979:354.
6. Siegle JC, Muus LT, Lin T, Larsen HA. *J. Poly. Sci., Part A*. 1964;2:391–404.
7. Settlege PH, Siegle JC. *Phys. Chem. Aerodyn. Space Flight. Conference Proceedings, Philadelphia, 1959*. New York: Pergamon Press; 1961:73–81.
8. Scheirs J, ed. In *Modern Fluoropolymers*. Chichester: John Wiley and Sons Ltd; 1997.
9. Ebnesajjad S, *Fluoroplastics, Volume 1: Non-Melt Processible Fluroplastics*. New York: William Andrew Publishing; 2000.

Poly(tetrahydrofuran)

Qingwen Wendy Yuan-Huffman

Acronyms PTHF

Class Polyethers

Structure $[-CH_2-CH_2-CH_2-CH_2-O-]$

Mol. Wt. of Repeat Unit 72

Property	Unit	Conditions	Values	Reference
Polymerization	None	Cationic ring-opening living polymerization		(1–3)
Solvents	None	Benzene, ethanol, tetrahydrofuran,	Chloroform	(4–6)
Nonsolvents	None	Petroleum ether, hexane, methanol, water		(4–6)
Theta temperature	K	Acetonitrile/benzene (61.5/38.5), CP	298.5	(6, 7)
		Acetonitrile/butanone (38.3/61.7), CP	298.5	(6, 7)
		Acetonitrile/carbon tetrachloride (50.2/49.8), CP	298.5	(6, 7)
		Acetonitrile/chlorobenzene (60.1/39.9), CP	298.5	(6, 7)
		Acetonitrile/tetrahydrofuran (58.7/41/3), CP	298.5	(6, 7)
		Acetonitrile/toluene (61/39), CP	298.5	(6, 7)
		n-Butanol, PE	278.5	(1, 6)
		Butanon, VM	298.5	(6)
		Chlorobenzene, VM	298.5	(6)
		Chlorobenzene/n-octane		
		(25.0/75.0), A, CP	283.5	(6, 8)
		(21.5/78.5), A, CP	299.3	(6, 8)
		(14.5/85.5), A, CP	319.5	(6, 8)
		(13.0/87.0), A, CP	336.5	(6, 8)
		(10.9/89.1), A, CP	353.5	(6, 8)
		Cyclohexane/n-heptane	299.5	(6)
		Diethyl malonate, PE	307.0	(6, 9)
		Ethyl acetate/n-hexane (22.7/77.3)		

Poly(tetrahydrofuran)

Property	Unit	Conditions	Values	Reference
		PE	303.9	(6, 9)
		A	306.5	(6, 10)
		i-Propanol, PE	318.1	(6, 9)
		Toluene, A	301.8	(6, 11)

Note: Abbreviation of method: CP – cloud point titration; PE – phase equilibria; VM – intrinsic viscosity/molar mass; A – virial coefficient

Property	Unit	Conditions	Values		Reference
Second virial coefficient	$\times 10^{-4}$ mol cm^3 g^{-2}	Ethyl Acetate $T = 30°C$ $M_w \times 10^{-3} = 34.6$–1030 g mol^{-1}	6.14–2.47		(6, 10)
Mark-Houwink Parameter and K and α	$K =$ ml g^{-1}	None Mol. Wt. $= 3.5 \times 10^4$ to 1.1×10^6 g mol^{-1}	K	a	
		Benzene, 30°C	131×10^{-3}	0.60	(6, 10)
		Cyclohexane, 30°C	176×10^{-3}	0.54	(6, 10)
		Ethyl acetate, 30°C	422×10^{-4}	0.65	(6, 10)
		Ethyl acetate/n−hexane (22.7/77.3 by wt.), 31.8°C	343×10^{-3}	0.45	(6, 10)
		Toluene, 28°C, $(3–12) \times 10^4$ g mol^{-1} 25.1×10^{-3}	0.78		(6, 11)
Density	g cm^{-3}	Amorphous at 25°C	0.975		(12)
		Amorphous	0.982		(6, 14)
		Crystalline at 25°C	1.07–1.08		(12)
		Crystalline	1.157		(6)
		Crystalline	1.112		(6, 13)
		Crystalline	1.116		(6, 14)
		Crystalline	1.238		(6)
		Crystalline	1.095		(6, 15)
Avrami exponent	None	IR	2.2		(16)
		DSC	2.4		(16)
Glass transition temperature	K	–	187.5		(12)
			189		(6, 17, 18)
			189.5		(19)
Melting temperature	K	–	316.5		(12, 19)
			331.5–333.5		(12)

Property	Unit	Conditions	Values		Reference
Heat capacity	$\times 10^{-3}$ KJ K^{-1} mol^{-1}		Solid	Melt	(6, 20, 21)
		$T = 10$ K	1.47	–	
		20	6.59	–	
		30	12.42	–	
		40	18.32	–	
		50	24.63	–	
		60	29.74	–	
		70	34.61	–	
		80	39.41	–	
		90	43.67	–	
		100	47.27	–	
		110	50.72	–	
		120	54.70	–	
		130	57.99	–	
		140	61.12	–	
		150	64.46	–	
		160	67.96	–	
		170	71.20	–	
		180	74.52	–	
		190	124.20	–	
		200	–	125.92	
		210	–	127.64	
		220	–	129.36	
		230	–	131.09	
		240	–	132.81	
		250	–	134.53	
		260	–	136.26	
		270	–	137.98	
		280	–	139.70	
		290	–	141.42	
		300	–	143.15	
		310	–	144.87	
		320	–	146.59	
		330	–	148.32	
		340	–	150.04	
Tensile strength	MPa	High mol. wt.	29.0		(12)
		Low to high mol. wt.	27.6–41.4		(12)
		Cured	16.8–38.3		(12)
		Cured plasticized high mol. wt.	13.7–19.0		(12)
Elongation	%	High mol. wt.	820		(12)
		Low to high mol. wt.	300–600		(12)
		Cured	400–740		(12)
		Cured plasticized high mol. wt.	450–735		(12)
Modulus of elasticity	MPa	–	97.0		(12)

Poly(tetrahydrofuran)

Property	Unit	Conditions	Values	Reference
Engineering modulus	MPa	Elongation = 300%		
		Low to high mol. wt.	1.6–4.3	(12)
		Cured plasticized high mol. wt.	13.7–19.0	(12)
Hardness	Shore A	–	95	(12)
Thermal expansion coefficient	K^{-1}	$\alpha = (1/V)(\delta V/\delta T)_p$	$4-7 \times 10^{-4}$	(12)
Compressibility	KPa^{-1}	$\beta = (1/V)(\delta V/\delta T)_T$	$4-10 \times 10^{-7}$	(12)
Internal pressure	MPa	–	281	(12)
Coefficient of expansion, dV_s/dT	$cm^3\,g^{-1}\,K^{-1}$	–	7.3×10^{-4}	(12)
Refractive index, n	None	20°C	1.48	(12)
Dielectric constant, ε	None	20°C	5.0	(12)
Specific refractive index increment, dn/dc	$ml\,g^{-1}$	$\lambda_o = 436$ nm $\lambda_o = 546$ nm		(6)
		Chlorobenzene	0.070	
		Ethyl Acetate, 25°C	0.110	(6)
		30°C	0.114	(6)
		Ethyl Acetate/$n-$Hexane (22.7/77.3 wt)		
		31.8°C	0.114	(6)
		Isopropanol, 46°C	0.108	(6)
		Isopropyl Acetate, 22.5 °C	0.098	(6)
		MEK, 30°C	0.102	(6)
		25°C	0.091	(6)
			0.095	(6)
		Methyl Acetate, 25°C	0.101	(6)
		3-Methyl-2-Heptanone, 25°C	0.056	(6)
		2-Pentanone, 25°C	0.084	(6)
		THF, 25°C	0.0625	(6)
		25°C	0.064	(6)

Property	Unit		20°C	150°C	200°C	
Surface tension	$mN\,m^{-1}$	$M = 43000$ g mol^{-1}	31.9	24.0	20.9	(6, 22–23)
		$M = 2500$ g mol^{-1}	38.2	27.9	24.0	(6)

Fractionation[6]

Method	Solvent, or, Solvent/Nonsolvent
Fractional precipitation	Acetone
	Benzene/$n-$Hexane
	Benzene/Methanol
	Toluene, Methanol
Tribidimetric titration	Ethanol/Water
Distribution between immiscible liquids	Cyclohexane-toluene (9:1)/Water-methanol

Method	Solvent, or, Solvent/Nonsolvent
Extraction	Water/Acetone
	Isopropanol/Water
Fractional solution	2-Butanone
	Isopropanol/Water
	Ethyl ether/Petroleum ether
Chromatography	Acetone/Water
	2-Butanone
	Dimethylformamide
	Methanol-water mixture
	Tetrahydrofuran
	Toluene
Sedimentation velocity	Ethyl acetate-n-hexane (22.3/77.7)

Crystalline-state properties

Crystl. Syst. (Lattice)	Space Group	Unit Cell Parameters			Angles (Degree)	Monomer per Unit Cell	Reference
		A (A)	B (A)	C (A)			
Monoclinic	C2H-6	5.48	8.73	12.07	$B = 134.2$	4	(1)
Monoclinic	C2H-6	5.59	8.90	12.07	$B = 134.2$	4	(1, 13)
Monoclinic	–	–	8.89	12.15	–	1, 14	–
Orthohombic	D2-4	12.2	8.75	7.22	–	8	(1)
Monoclinic	C2H-6	5.61	8.92	12.25	$B = 134.5$	4	(1, 15)
Monoclinic	–	5.48–5.61	8.73–8.92	12.97–12.25	$B = 134.2–134.5$	12	–

References

1. Dreyfuss P, Dreyfuss MP. *Adv. Polym. Sci.* 1967;4:526.
2. Furukawa J, Saegusa T. *Polymerization of Aldehydes and Oxides.* New York: Interscience; 1963.
3. Dreyfuss P. *Polytetrahydrofuran.* New York: Gordon and Breach Science; 1982.
4. Weissermel K, Noelken E. *Makromol. Chem.* 1963;68:140.
5. Schelz RC, Wolf R. *Makromol. Chem.* 1966;99:76.
6. Brabdrup J, Immergut EH. *Polymer Handbook.* 3rd Ed. New York: Wiley Interscience; 1989.
7. Elias H-G, Adank G. *Makromol. Chem.* 1963;69:241.
8. Evans JM, Huglin MB. *Europ. Polym. J.* 1970;6:1161.
9. Evans JM, Huglin MB. *Makromol. Chem.* 1969;127:141.
10. Kurata MH, Utiyama K. Kamada. *Makromol. Chem.* 1965;88:281.
11. Ali SM, Huglin MB. *Makromol. Chem.* 1965;84:117.
12. Mark HS, et. al. *Encyclopedia of Polymer Science and Engineering.* Vol 16. Wiley Interscience; 1989.

13. Tadokoro H. *J. Polym. Sci. Part C.* 1966;15:1.
14. Bowman I, Brown DS, Wetten RE. *Polymer.* 1969;10:715.
15. Cesari M, Perego G, Mazzei A. *Makromol. Chem.* 1965;83:196.
16. Shibayama M, Takahashi H, Yamaguchi H, Sakurai S, Nomura S. *Polymer.* 1994;35:No. 14, 2944.
17. Faucher JA, Koleske JV. *Polymer.* 1968;9:44.
18. Miller WG, Saunders JH. *J. Appl. Polym. Sci.* 1969;13:1277.
19. Rodriguez F. *Principles of Polymer Systems.* 4th Ed. Taylor & Francis Publishers; 1996.
20. Guar U, Wunderlich B. *J. Phys. Chem. Ref. Data.* 1981;10(4):1023.
21. Suzuki H, Wunderlich B. *J. Polym. Sci., Polym. Phys. Ed.* 1985;23:1671.
22. Wu S. *J. Polym. Sci.* 1971;C34:19.
23. Roe RJ. *J. Colloid Interface Sci.* 1969;31:228.

Polythiophene

Amit S. Kulkarni, Shrish Rane and Gregory Beaucage

Acronym PT

Class Polyheterocyclics; conjugated conducting polymers

Structure Polythiophene exists in two structures:
(Aromatic)

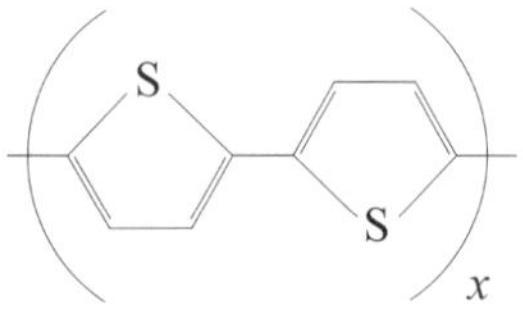

(Quinoid)

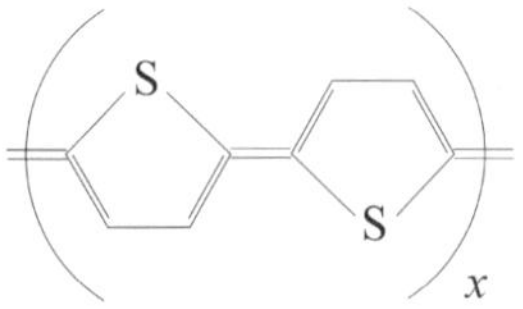

Major Applications Polythiophenes and substituted polythiophenes are utilized in a variety of applications where their conducting properties pose an advantage. They are currently used as anti-static coatings and as films. Research is being done to explore their use in electrochromic and electroluminescent devices. They have also shown promise as materials for biosensors and storage batteries. They are used in making Schottky barrier diodes and field effect transistors.

Properties of Special Interest Presence of an extended π-bonding system, which imparts electrical properties to the polymer. Doping with either p or n type of materials can enhance these properties. First among their class of polymers to be stable to moisture and oxygen in both, their doped and un-doped states. They also exhibit other interesting properties like electrochromism, thermochromism, and pressure induced color change.

Property	Conditions	Reference
Chemical synthesis	Polycondensation of di-functional thiophene in the presence of Ni catalysts. Oxidative coupling reaction of bi-thiophene in the presence of ferric chloride using $AlCl_3$, $CuCl_3$ and organic solvents. Plasma polymerization form 3-methyl thiophene or thiophene	(1–6)

Property	Conditions	Reference
	In-situ oxidative polymerization of PT/nano-tube composites	(7)
	PT copolymers with polyethyleneoxide	(8)
	Chemical synthesis of substituted PT	(9)
	Plasma Polymerization of PT/SiO_2 composites	(10)
	PT with conjugated side chains	(11, 12)
	Atom transfer radical polymerization of PT/polystyrene graft polymers	(13)
	PT with liquid crystal side chains	(14)
	PT synthesis using ionic liquids	(15)
	PT nanowires in mesoporous silica	(16)
	PT nanotubes	(17)
	Plasma Polymerization using different carrier gases	(18)
Electrochemical synthesis	Electrodes (platinum, gold, and Au coated Ni)	(1, 2, 19–23)
	Electrolytes (acetonitrile in tetra-alkylammonium, iodide salts, fluoroborate salts, Bu_4N^+, Et_4N^+, and quaternary ammonium salts)	
	PT copolymers with polyethyleneoxide	(8)
	In the presence of supercritical trifluoromethane	(24)

Property	Units	Conditions	Value	Reference
Conductivity	s cm^{-1}	Iodine doped	6–8	(4, 23)
		$FeCl_3$ doped	0.5	(23)
		$NOSbF_6$ doped	9×10^{-5}	(25)
		$NOPF_6$ doped	2×10^{-5}	(25, 26)
		$SO_3CF_3^-$ doped	50–100	(27, 28)

Property	Conditions	Reference
Chromisms	Two types:	
Thermochromism	An abrupt shift from planar back-bone to a twisted form at high temperatures	(29)
	A continuous modification in the back-bone with increasing temperature	(30, 31)
Electrochromsim	The visible-spectra shows a change during the doping/ de-doping process	(32)
Solvatochromism	Chain conformation changes from non-planar in solid state to coplanar in solution	(33–35)
Ionochromism	Polymer displays an absorption shift with K^+, Na^+ and Li^+	(35–38)
Pressure, light, and electricity induced color changes	PT and its derivatives show different molecular forms under the influence of pressure, light, and electricity	(39, 40)
Optical properties	PT and its derivates display photoluminescence and electroluminescence	(41–43)
Magnetic properties	PT's show variation in their magnetic properties. In the doped state they undergo transition from a paramagnetic state at high temperatures to an ordered phase at low temperatures	(44, 45)
Solubility	PT by itself is insoluble and infusible. Substitution of alkyl units in the 3-position and copolymers of PT increase the solubility and ease of processability, the penalty being some decrease in its conductivity	(1, 2, 27, 46)

Property	Units	Conditions	Value	Reference
Thermal stability	K	In air	523	(1)
		In inert atmosphere or vacuum	1,173	
UV-Vis spectroscopy	nm	Strong absorption band in doped state	480	(1, 27)
		PT with conjugated side chains		(12)
		Side Chain		
		PEHPVT	328	
		PMEHPVT	336	
		PT1	356	
		PT2	330	
		PT3	371	
		PT4	380	
		Main Chain		
		PEHPVT	550	
		PMEHPVT	566	
		PT1	560	
		PT2	552	
		PT3	544	
		PT4	544	
		P3HT		549
IR properties	cm^{-1}	$C{=}C$ stretch		(1, 2, 27)
		Chemical synthesis PT	1,494	
		Electrochemical synthesis PT	1,490	
		$C{-}H$ in plane bend		
		Chemical synthesis PT	1,052	
		Electrochemical synthesis PT	1,058	
		$C{-}H$ out-of-plane bend		
		Chemical synthesis PT	788	
		Electrochemical synthesis PT	755	
		$C_{\gamma}{-}H$		
		Chemical synthesis PT	690	
		Electrochemical synthesis PT	690	
		ν cycle		
		Chemical synthesis PT	1,400	
		Electrochemical synthesis PT	1,408	
		ν cycle		
		Chemical synthesis PT	1,230	
		Electrochemical synthesis PT	1,226	
Crystallinity		PT's appear completely amorphous under XRD scans. Substituted PT's display partial degrees of crystallinity (<5%)		
Hexagonal lattice	Å	–	$a = 9.5\ c = 12.2$	(1)
Density	$g\ cm^{-3}$	–	1.4–1.6	(1)
Morphology		Main determining parameters are monomer structure, dopant, and thickness of the film TEM images of PT films display fibrillar, 'noodle–like' structure. Fibril diameter increases with doping level		(1, 48–51)

Polythiophene

Property	Units	Conditions	Value	Reference
Photoemission Wavelengths	nm	Chemically synthesized substituted PT		(9)
		Poly AZBT	289, 303, 356	
			555, 566, 605	
			782	
		Poly AZEBT	328, 346, 353, 362	
			468	
			565	
			651, 688	
		PBTT	425	

References

1. Stockhein TA. *Handbook of Conducting Polymers*. Vol 1. Marcel Dekker, NY: 1986.
2. Schopf G, Kobmehl G. In *Advances in Polymer Science*. Springer-Verlag, Berlin; 1997.
3. Diaz AF, et al. *J. Phys. Chem.* 1984;88:3333.
4. Yamamoto T, et al. *Polym. J., Tokyo*. 1990;22:187.
5. Ruckenstein E, Park JS. *Synth. Met.* 1991;44:293.
6. Pomerantz E. et al. *Synth. Met.* 1991;41:825.
7. Karim MR, Lee CJ, Lee MS. *J. Polym. Sci. Part A Polym. Chem.* 2006;44:5283.
8. Qi L, Sun M, Dong SJ. *J. App. Polym. Sci.* 2006;102:1803.
9. Radhakrishnan S, Parthasarathi R, Subramanian V. *Comp. Mater. Sci.* 2006;37:318.
10. Nastase F, Stamatin I, Nastase C, Mihaiescu D, Moldovan A. *Prog. Solid State Chem.* 2006;34:191.
11. Zhou EJ, Hou JH, Yang CH. *J. Polym. Sci. Part A Polym. Chem.* 2006;44:2206.
12. Hou JH, Huo LJ, He C, Yang CH, Li YF. *Macromolecules*. 2006;39:594.
13. Shen J, Ogino K. *Chem. Lett.* 2005;34:1616.
14. Zhao XY, Hu X, Zheng PJ. *Thin Solid Films*. 2005;477:88.
15. Shi JH, Yang CH, Gao QY. *Chinese J Chem. Phys.* 2004;17:503.
16. Li GT, Bhosale S, Wang TY. *Angewandte Chemie*. 2003;42:3818.
17. Fu MX, Zhu YF, Tan RQ, Shi GQ. *Adv. Mater.* 2001;13:1874.
18. Silverstein MS, Visoly-Fisher I. *Polymer*. 2002;43:11.
19. Kobmehl G, et al. *Acta Poly.* 1992;43:65.
20. Plieth W, et al. *J. Electroanal. Chem.* 1989;274:213.
21. Rasch B, et al. *Synth. Met.* 1991;43:2963.
22. Bukowska J. *J. Mol. Struct.* 1992;275:151.
23. Yamamoto T, et al. *Synth. Met.* 1991;41:345.
24. Atobe M, Ohsuka H, Fuchigami T. *Chem. Lett.* 2004;33:618.
25. Kobmehl G, Chatzitheodorou G. *Makromol. Chem. Rapid Comm.* 1983;4:639.
26. Heffner GW, Pearson DS. *Synth. Met.* 1991;44:341.
27. Patil AO, et al. *Chem. Rev.* 1988;88:183.
28. Yamamoto T, et al. *Bull. Chem. Soc. Japan*. 1983;56:1497.
29. Robitaille L, Leclerc M. *Macromolecules*. 1994;27:27.
30. Roux C, Leclerc M. *Chem. Mater.* 1994;6:620.
31. Roux C, et al. *Makromol. Chem. Rapid Comm.* 1993;14:461.
32. Collombdunandsauthier MN, et al. *J App. Electrochem.* 1994;24:72.
33. Leclerc M, Daoust G. *J. Chem. Soc. Chem. Commun.* 1990;3:273.
34. Guay J, et al. *J. Electroanal. Chem.* 1993;361:85.

35. Barbarella G, et al. *Adv. Mater.* 1993;5:834.

36. Marsella MJ, Swagger TM. *J. Am. Chem. Soc.* 1993;115:12,214.

37. Miyazaki Y, Yamamoto T. *Chem. Lett.* 1993;1:41.

38. Swagger TM, Marsella M. *J. Adv. Mater.* 1994;6:595.

39. Baeuerle P, Scheib S. *Adv. Mater.* 1993;5:848.

40. Coghlan A, Arthur C. *New Scientist.* 1994;1927:22.

41. Iwasaki K, et al. *Synth. Met.* 1994;63:101.

42. Chosrovian H, et al. *Synth. Met.* 1993;60:23.

43. Berggren M, et al. *Appl. Phys. Lett.* 1994;65:1489.

44. Dyreklev P, et al. *Adv. Mater.* 1995;7:43.

45. Barta P, et al. *Phys. Rev.* 1993;B48:243.

46. Barta P, et al. *Phys. Rev.* 1994;B50:3016.

47. Feldhues M, et al. *Synth. Met.* 1989;28:c487.

48. Toulon G, Garner F. *J. Polym. Sci. Polym. Phys. Ed.* 1984;22:33.

49. Onada M, et al. *Jpn J. Appl. Phys. Part 1.* 1992;31:2265.

50. Miyazakai Y, et al. *Chem. Lett.* 1993;3:415.

51. Fujita W, et al. *Chem. Lett.* 1994;3:511.

Poly(trimethylene oxide)

Qingwen Wendy Yuan-Huffman

Acronyms PTMO

Class Polyethers

Structure $[-CH_2-CH_2-CH_2-O-]$

Molecular Weight of Repeat Unit 58

Property	Unit	Conditions	Values		Reference
Polymerization	None	–	Ring-opening polymerization		(1)
Theta temperature	K	Cyclohexane, phase equilibria method	299.5		(2, 3)
Mark-Houwink parameter: K and α	ml g^{-1} and a None		K	a	
		Acetone, 30°C, $(2.8–20) \times 10^4$ g mol^{-1}	76×10^{-3}	0.59	(2, 4)
		Benzene, 30°C, $(2.8–30) \times 10^4$ g mol^{-1}	21.9×10^{-3}	0.78	(2, 4)
		Carbon tetrachloride, 30°C, $(2.8–25) \times 10^{-4}$ g mol^{-1}	26.7×10^{-3}	0.75	(2, 4)
Solubility parameter	$(\text{Mpa})^{1/2}$	Method: viscosity, 25°C	19.2		(2, 5)
Glass transition temperature	K	–	195		(2, 6–8)
Melting temperature	K	–	308		(2)
Heat capacity	$\times 10^{-3}$ KJ K^{-1} mol^{-1}		Solid	Melt	(2, 9)
		$T = 10$ K	0.86		
		20	4.68	–	
		30	9.31	–	
		40	13.65	–	
		50	17.64	–	
		60	21.37	–	

Property	Unit	Conditions	Values		Reference
Heat capacity	$\times 10^{-3}$ KJ K^{-1}		Solid	Melt	(2, 9)
		$T = 70$ K	25.82	–	
		80	29.59	–	
		90	32.90	–	
		100	35.87	–	
		110	38.58	–	
		120	41.11	–	
		130	43.51	–	
		140	45.83	–	
		150	48.11	–	
		160	50.37	–	
		170	52.65	–	
		180	54.97	–	
		190	57.33	–	
		200	59.01	109.24	
		210	61.08	110.25	
		220	63.15	111.26	
		230	65.23	112.27	
		240	67.30	113.28	
		250	69.37	114.29	
		260	71.44	115.30	
		270	73.51	116.31	
		280	75.59	117.32	
		290	77.66	118.33	
		300	79.73	119.34	
		310		120.35	
		320		121.36	
		330		122.37	
Specific refractive index increment, dn/dc	ml g^{-1}	Solvent: MEK	0.0946		(2, 10)
Fractionation	None	Fractional precipitation	Acetone/water		(2)

Crystalline-state properties

Crystl. Syst. (Lattice)	Space Group	Unit Cell Parameters				Monomer per Unit Cell	Density (g cm^{-3})	Reference
		A (A)	B (A)	C (A)	Angles (°)			
Monoclinic	C2H-3	12.3	7.27	4.80	$B = 91$	4	1.178	(2)
Rhombohedral	C3V-6	14.13	14.13	8.41	–	18	1.941	(2)
Orthohombic	D2-5	9.23	4.82	7.21	–	4	1.203	(2)
				4.79	–	–	–	(2)

Poly(trimethylene oxide)

References

1. Odian G. *Principles of Polymerization.* 3rd Ed. Wiley Interscience; 1991.
2. Brabdrup J, Immergut EH. *Polymer Handbook.* 3rd Ed. New York: Wiley Interscience; 1989.
3. Chiu DS, Takahashi Y, Mark JE. *Polymer.* 1976;17:670.
4. Yamamoto K, Teramoto A, Fujita H. *Polymer.* 1966;7:267.
5. DiPaola-Baranayi G. *Macromolecules.* 1982;15:622.
6. Faucher JA, Koleske JV. *Polymer.* 1968;9:44.
7. Willbourn AH. *Trans Faraday Soc.* 1958;54:717.
8. Saba RG, Sauer JA, Woodward AE. *J. Polym. Sci. A.* 1963;1:1483.
9. Gaur U, Wunderlich B. *J. Phys. Chem. Ref. Data.* 1981:10(4):1015.
10. Yamamoto K, Teramoto A, Fujita H. *Polymer.* 1966;7:267.

Poly(trimethylene terephthalate)

Xiujuan Zhang and Gui Lin

Acronym, Trade Name PTT, PTMT, poly(1,3-propylene terephthalate), PPT, PTET, Corterra, Sorona, 3GT

Class Polyesters; linear and flexible aromatic polyester; thermoplastics

Structure

Major Applications PTT is expected to find industrial applications where PET and PBT are currently dominated such as in the electrical, electronics, automotive, furniture industry. It also has an important application in the textile industry and has been a promising engineering thermoplastic, which that can be spun into both fibers and yarns and has extensive applications in carpeting, textiles and apparel, engineering thermoplastics, nonwovens, films and mono-filament.[1]

Properties of Special Interest PTT contains the property highlights of both PET and PBT. It has the physical properties of PET including strength, stiffness, toughness, and heat resistance, along with the good processing properties of PBT, such as low melt and mold temperatures and rapid crystallization, and good injection molding properties. Resilience, elastic recovery, superior dyeing ability.[1]

Related Polymers Poly(oxybenzoate-p-trimethylene terephthalate) copolymer[2,3]
Poly(trimethylene-m-phenylene)[4]
Poly(trimethylene-co-ethylene terephthalate)[5]
Poly (trimethylene terephthalate)-co-(trimethylene isophthalate)[6]

Preparation PTT is an aromatic polyester made by the polycondensation of propane 1,3-diol with either terephthalic acid or dimethyl terephthalate.[6–9]

Properties Numerous studies on the crystal structure and mechanical properties of PTT have been reported.[2–13] Analysis of the crystalline structure of PTT shows that the aliphatic part of PTT takes a highly coiled structure of gauche– gauche conformation.

Poly(trimethylene terephthalate)

Table 1. Unit cell parameters of PTT (Comparison of WAXD and ED experiments)[6]

	Fiber[a] WAXS	Powder WAXD	ED
a (Å)	4.64	4.53	4.5
b (Å)	6.27	6.20	6.3
c (Å)	18.64	18.7	18.2
α	98.2	97.6	97.5
β	93.0	93.2	91.4
γ	111.1	110.1	111.7
Density (g/cm^3)	1.387	1.408	1.44

^{1}H NMR spectra 8.17 ppm singlet for benzene protons, 2.43 ppm doublet and 4.67 ppm quintuplet for methylene protons.[6]

IR: 811 and 933 cm^{-1} (CH$_2$ rocking), 1037 cm^{-1} gauche C—C glycol residue stretching, 1173 cm^{-1} benzene ring in-plane C—H stretching, 1358 and 1385 cm^{-1} CH$_2$ wagging, and 1505 cm^{-1} benzene ring in plane C—C stretching bands.[10,11]

Table 2. PTT IR band assignments [10,11]

Wavenumber, cm^{-1}	Assignment	Phase
811	CH$_2$ rocking of glycol residue	amorphous
933	CH$_2$ rocking of glycol residue	crystalline
		amorphous
1037	*Ag* C—C stretching	crystalline
1173	ring *Ag* in-plane C—H bending	amorphous
1358	*Bu* CH$_2$ wagging	crystalline
1505	*Ag* CH$_2$ wagging	amorphous
1710	Benzene ring C—C stretch	crystalline
	Bu C=O stretch	

Property	Units	Conditions	Value	References
Molecular conformation			3GT conformation nearly planar	(12)
Molecular weight (of repeat unit)	g mol^{-1}		206	
Number-average molecular weight	g mol^{-1}		18000 $\sim$ 22000 (for use as fibers)	(13)
Specific gravity	g cm^{-3}	X-ray diffraction	1.35	(14, 15)
Density	g cm^{-3}	100% crystalline PTT	1.432–1.44	(12, 16)
		100% amorphous PTT	1.295	(17)
Mark-Houwink parameters: *K* and *a*	K = ml g^{-1} a = None	Solution viscometry, 25°C	$K = 8.2 \times 10^{-2}$ $a = 0.63$	(18)

Poly(trimethylene terephthalate)

Property	Units	Conditions	Value	References
Lattice constant	degrees	X-ray diffraction	a = 4.53 Å/4.5 Å/4.64 b = 6.20 Å/6.3 Å/6.27 c = 18.70 Å/18.2 Å/18.64 α = 97.6°/97.5°/98.4 β = 93.2°/91.4°/93.0 γ = 110.1°/111.7°/111.1	(14, 16, 19–21)
Crystalline density	g cm^{-3}	WAXD	1.387–1.408–1.44	(14, 22)
Unit cell			Triclinic	(14, 23, 24)
Number of chains per unit cell			2	(23–25)
Unit cell volume	nm^3	X-ray diffraction	0.479	(15, 16)
Unit cell density	g cm^{-3}	X-ray diffraction	1.44	(15, 16)
Glass transition temperature T_g	°C	DSC	37–42	(14, 19, 26)
Crystallization temperature	°C	DSC	195.51	(27)
Cold crystalline point	°C	10°C min^{-1}	50	(26)
Melting temperature T_m	°C	DSC	207–232	(14, 26, 27)
Heat of fusion ΔH	kJ mol^{-1}	DSC	15 (semicrystalline) 28~32 (100% crystalline)	(19, 28)
Degradation temperature	°C	DSC	376	(29)
Breaking strength σ_B	MPa	Tensile	55.8–59	(30, 31)
Elongation at break	%	Tensile	19	(30)
Tensile (Young's) modulus E	MPa	ASTM	67.6	(32)
Shear modulus G	GPa		Fibers: 2.38 ± 0.28	(32, 33)
Flexural modulus (rigidity) E	GPa	3-point flexure, ASTM	2.76–2.24	(15, 31)
Flexural strength	MPa	ASTM	65.1	(31)
Ultimate strain, ε_B	%	Tensile	90.2	(18)
Impact strength	J m^{-1}	Notched Izod, ASTM D256–86	48	(15)
Mold shrinkage	m/m	ASTM	0.02	(15)
Heat distortion temperature	°C	At 1.8 MPa	59	(15)
Dielectric strength	V mil^{-1}	ASTM D149	530	(15)
Dielectric constant	1 MHz	ASTD 150	3.0	(15)
Dissipation factor	1 MHz	ASTD150	0.015	(15)

Poly(trimethylene terephthalate)

Property	Units	Conditions	Value	References
Volume resistivity	ohm cm $(\times 10^{16})$	ASTM D257	1.00	(34)
Intrinsic birefringence of PTT crystal			0.029 and 0.206	(34)

References

1. Frisk S, Ikeda M, Chase DB, Kennedy A, Rabolt JF. *Macromolecules.* 2004;37:6027–6036.
2. Cheng YY, Hsuen YL. *Blends Textile Res. J.* 2005;75(2):144–148.
3. Ou CF, Li WC, Chen YH. *J. Appl. Polym. Sci.* 2002;86:1599–1606.
4. Asahara N, Asai Y, Yasue K. *United States Patent.* 4052481, 1977.
5. Lee JW, Lee SW, Lee B, Ree M. *Macromol. Chem. Phys.* 2001;202:3072–3080.
6. Seo YW, Pang K, Kim YH. *Macromol. Mater. Eng.* 2006;291:1327–1337.
7. Nakamura Hideki. Poly(trimethylene terephthalate) compositions with good impact resistance. Jpn. Kokai Tokkyo Koho 2007, JP 2007119594 A 20070517.
8. Stouffer JM, Blanchard EN, Leffew KW. *Production of poly(trimethylene terephthalate)US Patent.* 5,763,104. 1998.
9. Kelsey DR. *Process for preparing poly(trimethylene terephthalate) US Patent.* 6,093,786. 2000.
10. Bulkin BJ, Lewin M, Kim J. *Macromolecules.* 1987;20:830.
11. Ward IM, Wilding MA. *Polymer.* 1977;18:327.
12. Desborough IJ, Hall IH, Neisser JZ. *Polymer.* 1979;20:545.
13. Ben D. *J. Appli. Polym. Sci.* 2003;89(12):3188.
14. Hanzlicek JL. Characterization of poly(trimethylene terephthalate) University of Akron. *College of Polym. Sci. Polym. Eng. Thesis.* 2004.
15. Scheirs J, Long TE. *Modern Polyesters: Chem. and Tech. of Polyesters and Copolyesters.* England: Wiley; 2003.
16. Ho RM, Ke KZ, Chen M. *Macromolecules.* 2000;33:7529–7537.
17. Van Krevelen DW. *Properties of Polymers.* 2nd ed. Amsterdam: Elsevier; 1976:51–62.
18. Abo El Ola M, Kotek R, King M, Kim J, Monticello R, Reeve A. *J. Biomaterials. Sci. Polym. edition.* 2004;15(12):1545.
19. Pyda M, Boller A, Grebowicz J, Chuah H, Lebedev BV, Wunderlich B. *J. Polym. Sci., Part B: Polym. Phys.* 1998; 36:2499.
20. Desborough IJ, Hall IH, Neisser JZ. *Polymer.* 1979;20:545.
21. Chuah HH. *J. Polym. Sci., Part B: Polym. Phys.* 2002;40:1513.
22. Poulin-Dandurand S, Perex S, Revol J-F, Brisse F. *Polymer.* 1979;20:419.
23. Chatani Y, Higashibata N, Takase M, Tadokoro H, Hirahara T. *Ann. Meet. Soc. Polym. Sci. Kyoto, Prepr.* 1977;427.
24. Moss B, Dorset DL J. *Polym. Sci., Polym. Phys. Ed.* 1982;20:1789.
25. Zhang J. *J. Appl. Polym. Sci.* 2004;91(3):1657.
26. Chuah HH. *Macromolecules.* 2001;34:6985–6993.
27. Liu W, Mohanty AK, Drzal LT, et al. *Ind. Eng. Chem. Res.* 2005;44:857–862.
28. Chung WT, Yeh WJ, Hong PD. *J Appl. Polym. Sci.* 2002;83:2426–2433.
29. Shu Y, Hsiao K. *J. Appl. Polym. Sci.* 2007;103(4):2387–2394.

30. Huang J. *J. Appl. Polym. Sci.* 2003;88(9):2247.
31. Lin Z, Mai K. *Polym. Plast. Tech. Eng.* 2007;46:417.
32. Yang J, Jo W. *J. Chem. Phys.* 2001;114(18):8159.
33. Nakamae K, Nishio T, Hata K, Yokoyama F, Matsumoto T. *Zairyo.* 1986;35:1066.
34. Yun JH, Kuboyama K, Ougizawa T. *Polymer.* 2006;47:1715–1721.

Poly[1-(trimethylsilyl)-1-propyne]

Sizhu Wu and Tarek M. Madkour

Acronym PTMSP

Class Conjugated and other unsaturated polymers

Synthesis Polyaddition

Structure

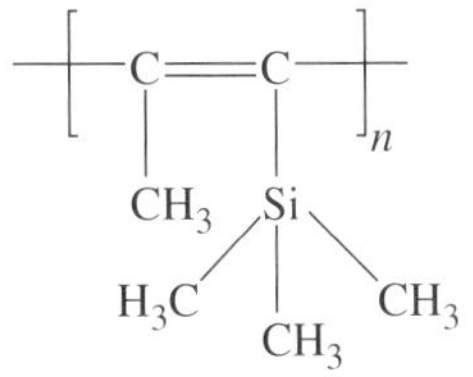

Major Applications Potential applications involve oxygen enrichment applicable to combustion furnaces, car engines, and respiration-aiding apparatuses. Also in the transport of oxygen dissolved in water applied to contact lenses and artificial lungs. In liquid mixture separation associated with ethanol concentration of fermented biomass. Furthermore, in polymer degradation related to resist materials for microlithography.

Properties of Special Interest Glassy ductile polymer with high permeability and low selectivity. A white amorphous silicon containing acetylene stable to air and soluble in nonpolar solvents such as toluene, cyclohexane, and carbon tetrachloride. Thus, it allows for tough film formation by solution casting.

Property	Units	Conditions	Value	References
Molecular weight (of repeat unit)	g mol^{-1}	–	112.25	(1)
Typical molecular weight range	g mol^{-1}	–	1.3–6.1 ($\times 10^5$)	(1)
Typical polydispersity range (M_w/M_n)			1.4–2.4	(1)
Characteristic infrared bands	cm^{-1}	Group assignments SiC—H deformation C—Si stretching	1,240 820,740	(2)
UV absorption maximum (λ_{max})	cm ($\times 10^7$)	–	273	(3)

Property	Units	Conditions	Value	References
UV molar extinction coefficient (ε_{max})	$mol^{-1} cm^{-1}$	–	120	(3, 4)
Density	$g cm^{-3}$	Measured at 21°C	0.964	(5)
Geometric density	$g cm^{-3}$	Geometric density refers to that of thin membranes (usually of lower value than real density)	0.7–0.77	(5)
Mark-Houwink parameter: K and a	$K = ml\ g^{-1}$ $a = None$	–	4.45×10^6 $a = 1.04$	(3)
Glass transition temperature	K	–	503	(6)
Softening point	K	–	613	(3)
Young's modulus	MPa	–	630	(3)
Tensile strength	MPa	–	40	(3)
Elongation at break	%	–	73	(3)
Electrical conductivity	$S\ cm^{-1}$	–	1×10^{-17}	(1)
Permeability coefficients	m^3 (STP) ms^{-1} $m^{-2} Pa^{-1}$	Gas (at 25°C) He H$_2$ O$_2$ N$_2$ CO$_2$ CH$_4$	4.65×10^{-14} 1.24×10^{-13} 6.6×10^{-14} 4.8×10^{-14} 2.64×10^{-13} 1.27×10^{-13}	(7)
Diffusion coefficients	$m^2 s^{-1}$ ($\times 10^{12}$)	Gas (at 25°C) N$_2$ Ar CH$_4$ CO$_2$	3,600 3,900 3,200 3,000	(7)
Kuhn segment value A	cm ($\times 10^8$)		65–93	(8)
Diameter of helix	cm ($\times 10^8$)		8.2	(8)
Optical anisotropy	cm^3 ($\times 10^{25}$)		37.7	(8)

Dual-mode parameters[7]

Gas (at 25°C)	Sorption Parameters			Diffusion Coefficients	
	k_D [m^3 (STP)$m^{-3} atm^{-1}$]	C_H [m^3 (STP)m^{-3}]	b (atm^{-1})	$D_D \times 10^9$ ($m^2 s^{-1}$)	$D_H \times 10^9$ ($m^2 s^{-1}$)
CO$_2$	1.0667	111.7	0.0688	16.2	1.95
CH$_4$	0.6328	58.87	0.0577	–	–
Ar	0.8313	24.23	0.0325	–	–
N$_2$	0.7103	16.10	0.0394	5.23	3.46

Poly[1-(trimethylsilyl)-1-propyne]

Property	Units	Conditions		Value		References
Void volume fraction	–	Gas/Temp. (°C)				(7)
		$N_2/-195$		0.26		
		$CO_2/25$		0.24		
		$SF_6/25$		0.23		
Interchain gap	Å	–		3.3		(7)
Intrinsic viscosity[η]	dl g^{-1}	Polymerized at 80°C and measured in toluene at 30°C				(1)
		Catalyst				
		$NbCl_5$		0.99		
		$NbBr_5$		0.63		
		$TaCl_5$		5.43		
		$TaBr_5$		3.60		
Intrinsic viscosity[η]	cm^3 g^{-1}		Trans-isomer			(9)
		Catalyst	content %	CCl_4	$C_2H_2Cl_4$	
		$NbCl_5$	20	77		
		$NbCl_5$	40	62		
		$TaCl_5$	55	800	600	
		$TaCl_5$	65		420	
Time required for 2% weight loss	min	TGA measurement in air				(10)
		145°C		860		
		161°C		430		
		176°C		225		
		186°C		100		
		198°C		76		
		206°C		51		

References

1. Masuda T, et al. *J. Am. Chem. Soc.* 1983;105:7473.
2. Masuda T, Isobe E, Higashimura T. *Macromolecules.* 1985;18:841.
3. Masuda T, Higashimura T. *Adv. Polym. Sci.* 1987;81:121.
4. Izumikawa H, Masuda T, Higashimura T. *Polym. Bull. Berlin.* 1991;27:193.
5. Plate N, et al. *J. Membr. Sci.* 1991;60:13.
6. Mark JE, ed. *Physical Properties of Polymers Handbook.* Woodbury, N.Y: AIP Press; 1996.
7. Srinivasan R, Auvil S, Burban P. *J. Membr. Sci.* 1994;86:67.
8. Shtennikova IN, et al. *European Polymer Journal.* 1999;35:2037–2078.
9. Langsam M, Robeson L. *Polym. Eng. Sci.* 1989;29:44.
10. Shtennikova IN, et al. *European Polymer Journal.* 2006;42:1325–1329.

Polyurea

L. S. Ramanathan, S. Sivaram
and Munmaya K. Mishra

Acronyms PU, PUR

Class Polyureas

Structure

$$\begin{matrix} & \text{H} & \text{O} & \text{H} & & \text{H} & \text{O} \\ & | & || & | & & | & || \\ \text{-}[\text{R}'\text{-N-C-N-R-N-C}]\text{-} \end{matrix}$$

R = isocyanate unit
R′ = diamine unit

Major Applications The most important practical applications of polyurea elastomers are in the production of automobile parts. High-modulus RIM (reaction-injection molded) and RRIM (reinforced reaction-injection molded) polyureas are suitable for producing high-impact external body panels. It is also useful in the forming microporous films for artificial leather. Ultrathin membranes of polyurea are used in water desalination by reverse osmosis. Polyureas are effective in making lubricant greases, medical equipment and artificial organs. Polyurea is also applied as a wall material for encapsulating drugs, pesticides, catalysts, and other products.

Properties of Special Interest Polyurea fibres have high melting points, low specific gravity, excellent dyeability, and good acid and alkaline resistance. Polyurea coatings have lower solvent and better water resistance compared to polyurethanes. They have good blood compatibility.

Property	Units	Conditions	Value	Reference
Density	g cm^{-3}	1,9-Nonane diamine (NDA)/ethylene bis chloroformate (EBC)	1.175	(1)
		1,10-Decane diamine (DDA)/EBC	1.75	(1)
		Polyisocyanate/polyetheramine/diethyl toluene diamine (DETDA)	1.1	(2)
		Aliphatic-aromatic copolyureas	1.012–1.214	(3)

Unit cell dimensions

Sample	Lattice	Cell Dimensions (Å)			Cell Angles (degrees)			Reference
		a	*b*	*c*	*α*	*β*	*γ*	
4,4′-Dicyclohexyl methane diisocyanate (CHMDI)/1,10 DDA	–	9.30	6.06	45	–	–	–	(4)
4,4′-diphenyl methane diisocyanate (MDI)/1,4-butane diamine (BDA)	Triclinic	4.63	5.83	25.23	90.7	91.58	102.9	(5)

Polyurea

Refractive indices of polyurea before poling[6]

System	Wavelength (μm)	RI	
		^{n}TE	^{n}TM
MDI/4,4′-methylene bis(cyclohexyl amine)	0.532	1.6052	1.5834
	0.6328	1.5962	1.5761
	1.064	1.5762	1.5644
MDI/1,4-diaminocyclohexane	0.532	1.6152	1.6012
	0.6328	1.6089	1.5949
	1.064	1.5838	1.5762
MDI/2,2-dimethyl-1,3-propane diamine	0.532	1.6223	1.6127
	0.6328	1.6091	1.5995
	1.064	1.5919	1.5833
MDI/4,4′-diaminodimethyl sulfone	0.532	1.7088	1.6715
	0.6328	1.6872	1.6564
	1.064	1.6577	1.6340

Contact Angles and Surface Free Energy of Polysulfide-Based Polyureas[7] containing Aminoethylaminopropyl poly(dimethylsiloxane) [AEAPS] segments

Amount of AAEPS (wt %)	θ H$_2$O (°)	θ CH$_2$I$_2$ (°)	γ_s (erg/cm)
0	52.4	41.5	51.9
1	87.5	53.2	33.2
3	98.6	69.2	23.6
6	102.4	72.2	21.7

Property	Units	Conditions	Value	Reference
Piezoelectric 'e' constant	mCm^{-2}	MDI/4,4′-diaminodiphenylmethane(MDA)	15	(8)
Heat capacity	–	MDI/polyether amine/DETDA	0.41	(9)
Glass transition temperature	K	1,9-NDA/EBC	277	(1)
		1,10-DDA/EBC	333.8	(1)
		Octafluorohexamethylene-1,6-diamine (OFHMDA)/1,6-hexamethylene bis(chlorocarbonate) [HMCC]	278	(10)
		Hexamethylene diamine (HMDA)/HMCC	271	(10)
		MDI/polyether amine/DETDA	215	(11)
		Amino terminated polysilanes/MDI/DETDA	186	(12)
		Amino terminated polysilanes/MDI+HMDI/ 1,3-propane sulfonate	180	(12)
		Amino terminated polysilanes/MDI+HMDI/ ethylene diamine (EDA)	176	(12)

Property	Units	Conditions	Value	Reference
Melting temperature	K	4,4′-methylenebis[N-methyl aniline]/2,2-dimethyl-1,3-propanediol bis(chloroformate)	443–463	(13)
		4,4′-methylenebis[N-methyl aniline]/COCl$_2$	523–553	(13)
		ClCON(CH$_3$)C$_6$H$_4$CH$_2$C$_6$H$_4$N(CH$_3$)COCl/ HN(CH$_3$)(CH$_2$)$_6$(CH$_3$)NH	383–453	(13)
		4,4′-diamino-1,3-diphynyl propane/EBC	480	(1)
		4,4′-diamino-1,3-diphynyl butane/EBC	547	(1)
		1,9-NDA/EBC	441	(1)
		1,10-DDA/EBC	447	(1)
		OFHMD A/HMCC	457	(10)
		HMDA/HMCC	443	(10)
Water absorption	%	MDI based polyurea at 25°C, 7 days	2.13	–
Solubility parameter	(MPa)$^{1/2}$	MDI/EDA	24.9	(14)
		MDI/DETDA	23.9	
		MDI/methylene bis(2,6-isopropyl aniline) (MMIPA)	21.6	
		MDI/methylene bis(2-methyl-6-isopropyl aniline) [MDIPA]	20.4	
Tensile strength	MPa	Aminopropyl terminated poly(dimethyl siloxane) [ATPDMS]/MDI	16.6	(15)
		ATPDMS/TDI	10.0	(15)
		ATPDMS/HMDI	9.0	(15)
		MDI/polyether amine/DETDA	4.61	(11)
		MDI/polyether amine/DETDA	15.9	(9)
		Amino terminated polysilanes/MDI/DETDA	9.1	(12)
		Amino terminated polysilanes/MDI+HMDI/ 1,3-propane sulfonate	22.4	(12)
		Amino terminated polysilanes/MDI+HMDI/ED	16.1	(12)
Elongation	%	ATPDMS/MDI	430	(15)
		ATPDMS/TDI	520	(15)
		ATPDMS/HMDI	950	(15)
		MDI/polyether amine/DETDA	276	(9)
		MDI/polyether amine/DETDA	250	(11)
		Amino terminated polysilanes/MDI/DETDA	426	(12)
		Amino terminated polysilanes/MDI+HMDI/ 1,3-propane sulfonate	335	(12)
		Amino terminated polysilanes/MDI+HMDI/EDA	332	(12)
Shore D hardness		Polyisocyanate/polyether amine/DETDA	75	(2)
Tear strength	N m^{-1}	IPDI based polyurea	70×10^3	(16)
		Tetramethyl xylene diisocyanate (TMXDI) based polyurea	45×10^3	
Tensile strength	MPa	AEAPS Modified Polysulfide-Based Polyureas	16.6	(7)

Property	Units	Conditions	Value	Reference
Elongation	%	AEAPS Modified Polysulfide-Based Polyureas	96	(7)
Piezoelectric e' constant	mCm^{-2}	1,5-diisocyanatopentane/pentane-1,5-diamine (thin films by vapor deposition polymerization; poling conditions $E_p = 250$mvm^{-1} $T_{max} = 150^o$c)	10	(17)
		2,4-diisocyanato-1-methylbenzene (TDI)/pentane-1,5-diamine	7	(17)
Glass transition temperature	K	Polyurea based on 3,4-ethylenedioxythiophene		
		(i) 5,7-diisocyanato-2,3-dihydrothieno [3,4-b][1,4]dioxine/ethane-1,2-diamine	499	(18)
		(ii) 5,7-diisocyanato-2,3-dihydrothieno [3,4-b][1,4]dioxine/hexane-1,6-diamine	418	(18)
		(iii) 5,7-diisocyanato-2,3-dihydrothieno [3,4-b][1,4]dioxine/decane-1,10-diamine	366	(18)
Melting temperature	K	Polyurea based on 3,4-ethylenedioxythiophene		
		(i) 5,7-diisocyanato-2,3-dihydrothieno [3,4-b][1,4]dioxine/hexane-1,6-diamine	428	(18)
		(ii) 5,7-diisocyanato-2,3-dihydrothieno [3,4-b][1,4]dioxine/decane-1,10-diamine	403	(18)
		HMDI/4-cyclohexylurazole	369–372	(19)
		IPDI/4-cyclohexylurazole	430–433	(19)
		TDI/4-cyclohexylurazole	506 (dec)	(19)
Ionic conductivity	Scm^{-1}	4,4′-diphenylmethane diisocyanate/Jeffamine-ED2001/3,5-diaminobenzoic acid/0.5mmol of LiClO$_4$(gpolyurea)$^{-1}$		(20)
		(i) at 30°C	2.0e-6	
		(ii) at 60°C	1.3e-5	

References

1. Lyman J, Heller J, Barlow M. *Makromol. Chemie.* 1965;84:64.
2. Harris RF, Anderson RM, Shannon DM. *J. Appl. Polym. Sci.* 1992;46:1547.
3. Ibrahim AM, Mahadevan V, Srinivasan M. *Eur. Polym. J.* 1989;25:427.
4. Barton R Jr. *Bull. Am. Phys. Soc.* 1987;32(3)KU10:701.
5. Born L, Hespe H. *Colloid Polym. Sci.* 1985;263:335 [CA; 103: 7586z].
6. Tao HT, et al. *Macromolecules.* 1995;28:2637.
7. Yi Wu Quan, Wang QJ, Fang JL, Chen QM. *Journal of Applied Polymer Science.* 2003;87(4):584–588.
8. Takahashi Y, et al. *J. Appl. Phys.* 1991;70:6983.
9. Rayan AJ, Stanford JL, Wilkinson N. *Polym. Bull.* 1987;18:517.
10. Malichenko BF, Sheludko YV, Kercha YY. *Polym. Sci. USSR.* 1967;9:2808.
11. Willkokomm WR, Chen ZS, Macosko CW. *Polm. Eng. Sci.* 1988;28:888.
12. Yang CZ, Li C, Cooper SL. *J. Polym. Sc., Polym. Phys. Ed.* 1991;29:75.

13. Foti S, Maravigna P, Mantaudo G. *Macromolecules.* 1982;15:883.

14. Rayan AJ, Stanford JL, Still RH. *Polym. Commun.* 1988;29:196.

15. Tyagi D, et al. *Polymer.* 1984;25:1807.

16. Dominguez RJG, Rice DM, Grigsby RA. *Plas. Eng.* 1987;43(11):41.

17. Takeshi H, Yoshikazu T, Masayuki I, Eiichi F. *Journal of Applied Physics.* 1996;79(3):1713–1721.

18. Uma P, Ojha CRAK. *Journal of Polymer Science Part A: Polymer Chemistry.* 2005;43(23):5823–5830.

19. Shadpour EH, Mallakpour R. *Journal of Applied Polymer Science* 2001;80(8):1335–1341.

20. Shao-Ming Lee, C-YCCCW. *Journal of Polymer Science Part A: Polymer Chemistry.* 2003;41(24):4007–4016.

Polyurethane

L. S. Ramanathan, S. Sivaram
and Munmaya K. Mishra

Acronyms PU, PUR

Class Polyurethanes

Structure

$$-O\!\!\left[\!(R')_{\overline{x}}O-\overset{\overset{\displaystyle O}{\|}}{C}-NH-R-NH-\overset{\overset{\displaystyle O}{\|}}{C}-O\right]_{\!\overline{n}}$$

R = isocyanate unit
R′ = polyol segment

Major Applications Polyurethane flexible foams find applications in protective packaging, gaskets, textile laminates, protective cushioning in automobiles, and two component injection-grouting resins. Rigid polyurethane foams are used as thermal insulating materials in refrigerators, freezers, and water heaters. It is also used as a roof proofing material.

Properties of Special Interest Excellent dampening property, good mechanical and physical properties even at low temperatures, high combustion resistance, and low thermal conductivity.

Property	Units	Conditions	Value	Reference
Density	g cm^{-3}	4,4′-Diphenylmethanediisocyanate (MDI)/1,4-butane diol (BD)	1.297	(1)
Flory-Huggins polymer solvent interaction parameter	–	TDI/1,4-BD (DMF)	0.122	(2)

Unit cell dimensions

Sample	Lattice	Cell Dimensions (Å)			Cell Angles (degrees)			Reference
		a	b	c	α	β	γ	
Hexamethylene diisocyanate (HMDI)/Ethylene glycol (EG)	Triclinic	4.59	5.14	13.9	90	90	119	(3)
HMDI/1,3-propane diol (PD)	Monoclinic	4.70	8.36	33.9	–	–	115	(3)
HMDI/BD	Triclinic	4.98	4.71	19.4	116	105	109	(3)
HMDI/1,5-pentane diol (PtD)	Monoclinic	4.70	8.36	39.0	–	–	115	(3)
HMDI/1,6-hexane diol (HD)	Triclinic	5.05	4.54	21.9	112	108	108	(3)
Trimethylene diisocyanate (TMDI)/BD	Triclinic	5.06	5.04	30.1	112	113	110	(4)
TMDI/HD	Triclinic	5.04	5.04	34.6	111	111	111	(4)
MDI/BD	Triclinic	5.2	4.8	35	115	121	85	(5, 6)
MDI/BD	Triclinic	4.92	5.66	38.4	124	104	86	(7)

Conformational characteristics

Sample	Solvent	Mark-Houwinck Parameters		$[\langle R^2 \rangle / M]^{1/2} \times 10^9$	Reference
		$K \times 10^{-4}$ (ml g^{-1})	a		
Toluene diisocyanate (TDI)/BD	–	5.4	0.74	–	(2)
MDI/EG	100 DMF	3.64	0.71	10.11	(8)
	95/5 DMF/acetone	6.29	0.65	10.19	(8)
	90/10 DMF/acetone	7.19	0.63	10.25	(8)
	85/15 DMF/acetone	10.02	0.59	10.04	(8)
	79/21 DMF/acetone	14.19	0.56	9.97	(8)
	71/29 DMF/acetone	30	0.50	–	(8)
MDI/BD	DMA (at 25°C)	870	1.43	–	(9)

Refractive index gradient

Sample	Condition			dn/dc (ml g^{-1})	Second Virial Coefficient $A_2 \times 10^4$	Reference
	Solvent	λ(nm)	Temp (°C)			
TDI/BD	DMF	546	–	0.14	–	(2)
MDI/1,6-HD	DMF/acetone	–	–	0.159–0.203*	3.0–4.5*	(10)
	DMF/toluene	–	–	0.123–0.154*	2.3–8.0*	(10)

* Variable with respect to DMF volume fraction.

Property	Units	Conditions	Value	Reference
Heat of fusion	kJ mol^{-1}	MDI/BD	5.3	(9)
Heat capacity	cal g^{-1}°C^{-1}	HMDI/BD		(11)
		−50 to 10°C	0.422	
		45–120°C	0.495	
		195–210°C	0.665	
		HMDI/DEG		
		−50 to −5°C	0.422	
		50–100°C	0.512	
		140–160°C	0.623	
Crystallization half time	min	HMDI/BD	6	(12)
		HMDI/diethylene glycol (DEG)	10.5	
Crystallization enthalpy	cal cm^{-3}	HMDI/BD	40	(12)
		HMDI/DEG	45	
Glass transition temperature	K	HMDI/BD	295	(12)
		HMDI/DEG	272	(12)
		HMDI/octafluoro 1,6-hexane diol (OFHD)	271	(13)
		MDI/EG	363	(14)
		Desmodur/1,6-HD	322	(15)
		Desmodur/cyclohexane dimethanol (CHDM)	302	(15)

Polyurethane

Property	Units	Conditions	Value	Reference
Melting temperature	K	HMDI/BD	476	(12)
		HMDI/DEG	396	(12)
		HMDI/OFHD	399	(13)
		TDI/EG	453	(14)
		MDI/EG	498	(14)

Optical properties

System	Temp. (°C)	Solvent	Optical Rotation $[\infty]_D$	Reference
MDI/(1S,2S)-diphenyl propane diol	25	DMSO	−71.6	(16)
HMDI/(1S,2S)-diphenyl propane diol	25	DMSO	−14.7	(16)
HMDI/ (2R,4R)-pentanediol	25	DMSO	−80.6	(16)
MDI/(1S,2S)-(+)-2-acetamido-1-phenyl-1,3-propanediol	25	DMF	−24.6	(17)
TDI/(1S,2S)-(+)-2-acetamido-1-phenyl-1,3-propanediol	25	DMF	−20.6	(17)

NLO properties

Type of Chromophore	λ_{max}	Second Harmonic Generation Coefficient d_{33} pmV^{-1}	Third-order Optical Nonlinearity		Reference
			χ^3 esu (wavelength)	γ, esu	
4-(dicyanomethylene)-2-methyl-6-[p-(dimethylamino)-styryl]-4H-pyran (3 examples studied)	663–750	5–15			(18)
azo chromophores with nitro or sulfonyl moieties as the acceptor (10 examples studied)	437–481	12–82			(19)
C_{60}			9.6×10^{-12} (1830 nm)	4.3×10^{-34}	(20)
C_{60}-PU			9.7×10^{-11} (1550 nm)	9.6×10^{-32}	(20)

Property	Units	Conditions	Value	Reference
Photoconductivity	ohms^{-1} cm^{-1}	Polyurethane with pendant chromophore	1.3×10^{-13}	(21)
Refractive index	–	Polyurethane with pendant chromophore		(21)
		At 532 nm	1.879	
		At 690 nm	1.812	
		At 1,064 nm	1.763	
Water vapor absorption	%	MDI/EG	2.5	(14)
Solubility parameter	(MPa)$^{1/2}$	MDI/BD	27	(1)
		MDI/EG	21	(22)

Property	Units	Conditions	Value	Reference
Elongation	%	MDI/EG	36	(14)
Adhesion strength	psi	Desmodur/1,6-HD	220	(15)
		Desmodur/CHDM	220	
Glass transition temperature	K	Diols derived from chromophore 4-(dicyanomethylene)-2-methyl-6-[p-(dimethylamino)-styryl]-4H-pyran (three examples studied)	456–485	(18)
		Diols containing azo chromophores with nitro or sulfonyl moieties as the acceptor/TDI (10 examples studied)	371–423	(19)
Shore D Hardness	–	Cardanol based diol and triols/MDI/NCO:OH $= 1$, 0.5 wt% dibutyltindilaurate based on hydroxyl monomer weight	65–80	(4)

References

1. Camberlin Y, Pascaut JP. *J. Polym. Sci., Polym. Phys. Ed.* 1984;22:1835.
2. Malichenko BF, et al. *Polym. Sci. USSR.* 1967;9:2975.
3. Sato Y, Nansai S, Kinoshita S. *Polym. J.* 1972;3:113.
4. Sato Y, Hara K, Kinoshita S. *Polym. J.* 1982;14:19.
5. Blackwell J, Gardner KH. *Polymer.* 1979;20:13.
6. Blackwell J, Ross M. *J. Polym. Sci., Polym. Lett. Ed.* 1979;17:447.
7. Born L, et al. *J. Polym. Sci., Polym. Phys. Ed.* 1984;22:163.
8. Beachell HC, Peterson JC. *J. Polym. Sci., Part A-1.* 1969;7:2021.
9. Hwang KKS, et al. *J. Polym. Sci., Polym. Chem. Ed.* 1984;22:1677.
10. Tuzar Z, Beachell HC. *J. Polym. Sci., Polym. Lett. Ed.* 1971;9:37.
11. Godovskii YK, Lipatov YS. *Polym. Sci. USSR.* 1968;10:34.
12. Godovskii YK, Slomimsky GC. *J. Polym. Sci., Polym. Phys. Ed.* 1974;12:1053.
13. Malichenko BF, Sheludks YV, Kercha YY. *Polym. Sci. USSR.* 1967;9:2808.
14. Lyman DJ. *J. Polym. Sci.* 1960;45:49.
15. Chung FH. *J. Appl. Polym. Sci.* 1991;42:1319.
16. Kobayashi T, Kakimoto M, Imai Y. *Polym. J.* 1993;25:969.
17. Chen Y, Tsay J. *Polym. J.* 1992;24:263.
18. Antonio Carella, et al. *Macromolecular Chemistry and Physics.* 2007;208(17):1900–1907.
19. Li Z, et al. *Macromolecules.* 2006;39(20):6951–6961.
20. Kuang L, Chen Q, Sargent EH, Wang ZY. *J. Am. Chem. Soc.* 2003;125(45):13648–13649.
21. Chen M, et al. *Appl. Phys. Lett.* 1994;64:1195.
22. Nishimura H, et al. *Polym. Eng. and Sci.* 1986;26:585.
23. Suresh KI, Kishanprasad VS. *Ind. Eng. Chem. Res.* 2005;44(13):4504–4512.

Polyurethane Elastomers

L. S. Ramanathan, S. Sivaram and Munmaya K. Mishra

Acronyms PU, PUR

Class Polyurethanes

Structure

$$+(R'')_{\overline{x}}O+\overset{\overset{O}{\|}}{C}-NH-R-NH-\overset{\overset{O}{\|}}{C}-O-(R')_{\overline{y}}]_n-O-$$

R = isocyanate unit
R′ = polyol segment
R″ = diol segment

Major Applications Polyurethane elastomers find applications in adhesives, laminates for textiles, covering of conveyor and drive belts, welded bodies, roof underlay sheeting, magnetic tape coatings, water line tubing, and ski boot manufacture. Elastomeric RIM polyurethanes are useful in making automotive parts such as bumpers and fascia. Reinforced RIM polyurethane has been used for car windows door panels and wind shields. Foamed elastomeric polyurethanes are also used in making automotive parts such as arm rests, steering wheels, and rear deck air domes.

Properties of Special Interest Excellent toughness and wear resistance with a broad temperature range for use. Polyurethane has good blood and tissue compatibility.

Property	Units	Conditions	Value	Reference
Density	g cm^{-3}	Oxyester/toluene diisocyanate (TDI)/1,4-butane diol(BD)	1.240	(1)
		Oxyester/4,4′-diphenylmethane diisocyanate (MDI)/BD	1.213	(1)
		Oxyester/hexamethylene diisocyanate (HDI)/BD	1.163	(1)
		Oxyester/Isophorone diisocyanate (IPDI)/BD	1.160	(1)
		Poly(oxyethylene) (PEO) diol/MDI/BD	0.986	(2)
		Poly(tetramethyleneoxide) PTMO/MDI/BD	1.004	(2)
		Poly(propyleneoxide) PPO/MDI/BD	0.976	(2)

Unit Cell Dimensions

Sample	Lattice	Cell Dimensions (Å)			Cell Angles (Degrees)			Reference
		a	b	c	α	β	γ	
PTMO/MDI/BD	Triclinic	5.05	4.67	37.9	116	116	83.5	(3)
PTMO/MDI/Hexane diol (HD)	Triclinic	4.99	–	41.5	114.5	113.8	84.3	(4)
Poly(tetramethyleneadipate) [PTMA]/MDI/HD	Triclinic	5.1	5.1	41.6	116	116	85	(5)

Property	Units	Conditions	Value	Reference
Flory-Huggins interaction parameter	–	PEO/MDI/BD	−0.27	(2)
		PTMO/MDI/BD	−0.33	
		PPO/MDI/BD	−0.08	
Flory-Huggins polymer solvent interaction parameter	–	Chloroform	0.228	(6)
		Benzene	0.333	
		MEK	0.417	
		Dibutyl ether	0.521	
		Acetonitrile	0.606	
		Cyclohexane	0.660	

Conformational characteristics[7]

Sample	Mark-Houwink Parameters		K_0 (g$^{-3/2}$ mol$^{1/2}$ cm^3)	$\bar{R}_0^2$ (Å^2 mol g^{-1})
	K (ml g^{-1})	a		
Poly (caprolactone) diol (PCL)/MDI/BD	0.257	0.54	0.25	1.0
PTMA/MDI/BD	0.043	0.70	0.20	0.84

Refractive index gradient[7,8]

Sample	Conditions			dn/dc (ml g^{-1})
	Solvent	λ (nm)	Temp (°C)	
PCL/MDI/BD	DMF	546	25	0.102
PTMA/MDI/(BD)	DMF	546	25	0.110
TDI/poly (propylene) glycol (PPG)	Benzene	435.8	–	0.031
TDI/PPG	Butanone	435.8	–	0.094
TDI/PPG	Methanol	435.8	–	0.148

Property	Units	Conditions	Value	Reference
Heat of fusion	kJ mol^{-1}	PEO/MDI/BD	197	(9)
		PEO/MDI/BD	155	
Glass transition temperature	K	PPG/MDI/BD	222	(10)
		PCL/MDI/BD	250	(11)
		Poly(ethyleneadipate)[PEA]/MDI/BD	230	(12)
		Hydroxy terminated poly(butadiene) [HTPB]/TDI/BD	246	(13)
		PTMO/2,4-TDI/BD	208	(14)
		PTMO/2,6-TDI/BD	200	(14)
Melting temperature	K	PCL/MDI/BD	358; 426	(11)
		PPG/MDI/BD	462; 468	(10)
		PTMG/poly (dimethylsiloxane) [PDMS]/MDI/EG	505	(15)

Property	Units	Conditions	Value	Reference
Dielectric loss	–	PEA/MDI	0.8	(16)
		PPO/MDI/BD (at 12.5°C)	2.2	(9)
Bulk DC conductivity	ohm^{-1} cm^{-1}	PPO/MDI/BD (at 12.5°C)	22×10^{12}	(9)
Surface resistivity	ohm	Oxyester/MDI/BD	1.8×10^{12}	(1)
Volume resistivity	ohm cm	Oxyester/MDI/BD	6.9×10^{11}	(1)
Contact angle	degrees	Water	89	(17)
		Water/propanol	69	
		α-Br napthalein	25	
Surface free energy	erg cm^{-2}	Estane 5714 FI (BF Goodrich)	21	(17)
Permeation rate	mg cm^{-1} day^{-1}	At 25°C		(18)
		Water	0.33	
		LiCl	0.27	
		NaCl	0.24	
		KCl	0.32	
		CsCl	0.28	
Solubility parameter	(MPa)$^{1/2}$	PPG(1000)/MDI/BD	23	(10)
		PPG(2000)/MDI/BD	23	
		PPG(3000)/MDI/BD	23	
Loss factor tan δ	–	PTMO/MDI/BD	0.072	(19)
Activation energy	kJ mol^{-1}	PEA/MDI/BD	152.5	(16)
		PTMO/MDI/BD	224	(19)
Tensile strength	MPa	PTMO/MDI/BD	45	(18)
		PTMO/MDI/BD (NCO/OH = 2/1)	20.16	(20)
		PTMO/MDI/BD (NCO/OH = 4/1)	37.59	(20)
Elongation	%	PTMO/MDI/BD	850	(21)
		PTMO/MDI/BD (NCO/OH = 2/1)	1,100	(20)
		PTMO/MDI/BD (NCO/OH = 4/1)	649.2	(20)
Shore A hardness	–	Poly(butyleneadipate) [PBA]/MDI/BD	85	(22)
Thermal expansion coefficient	K^{-1}	–	280	(23)
Tensile strength	MPa	Polyurethanes from fully Hydrogenated epoxidized soy oil Polyols	35	(24)
Elongation	%	Polyurethanes from fully Hydrogenated epoxidized soy oil Polyols	8	(24)
		Polycaprolactone diol (PCL)/MDI/BD	550	(25)
		Polycaprolactone diol (PCL)/MDI/BD/7% multiwalled carbonnanotube	100	(25)
Contact angle (stationary)	°	fluoropoly(ether urethane)s	113 ± 1.4	(26)
		fluoropoly(carbonate urethane)s	111 ± 1.1	(26)

Property	Units	Conditions	Value	Reference
Glass transition temperature	K	Polyurethanes from fully Hydrogenated epoxidized soy oil Polyols	324	(24)
		fluoropoly(ether urethane)s	236	(26)
		fluoropoly(carbonate urethane)s	264 and 330	(26)
		Oxazolidine-Functionalized Polyurethanes (Four examples)	301–384	(27)
		1,2-Bis(isocyanate)ethoxyethane/PTMG/BD/TMP	200	(28)
Shore A Hardness		Polyurethanes from fully Hydrogenated epoxidized soy oil Polyols	100	(24)

References

1. Pandya MV, Deshpande DD, Hundiwale DG. *J. Appl. Polym. Sci.* 1986;32:4959.
2. Hwang KKS, Hemker DJ, Cooper SL. *Macromolecules.* 1984;17:307.
3. Blackwell J, Ross M. *J. Polym. Sci., Polym. Let. Ed.* 1979;17:447.
4. Blackwell J, Nagarajan MR, Hoitinic TB. *Polymer.* 1982;23:950.
5. Blackwell J, Lee CD. *J. Polym. Sci., Polym. Phys.* 1984;22:759.
6. Oberth AE. *Rubber Chem. and Technol.* 1990;63:56.
7. Simek L, Tuzar Z, Bondanecky M. *Macromol. Chem. Rapid Commun.* 1980;1:215.
8. Moacanin J. *J. Appl. Polym. Sci.* 1959;1:272.
9. North AM, JC Reid. *Europ. Polym. J.* 1972;8:1129.
10. Petrovic Z, Soda-So, Javani I. *J. Polym. Sci., Polym. Phys.* 1989;27:545.
11. Russo R, Thomas EL. *J. Macromol. Sci. Phys.* 1983;B22:533.
12. VanBogart JWC, Bluemke DA, Cooper SL. *Polymer.* 1981;22:1428.
13. Bengtson B, et al. *Polymer.* 1985;26:895.
14. Schneider NS, Paik Sung CS. *Polym. Eng. and Sci.* 1977;17:73.
15. Shibayama M, et al. *Polymer.* 1990;31:749.
16. Dieldes C, Pethrick RA. *Europ. Polym. J.* 1981;17:675.
17. Busscher HJ, et al. *J. Colloidal and Interface Sci.* 1983;95:23.
18. Wells LA, et al. *Rubber Chem. and Technol.* 1990;63:66.
19. Petrovic ZS, et al. *J. Appl. Polym. Sci.* 1989;38:1929.
20. Bajsic EG, et al. *Polym. Deg. Stab.* 1996;52:223.
21. Petrovic ZS, Simendic JB. *Rubber Chem. and Technol.* 1985;58:701.
22. Nagoshi K. In: Ashida K, Frish KC. eds. *International Progress in Urethanes.* Vol 3. Westport, Conn: Technomic Publishing. 1981:193.
23. Theocaris PS, Varias AG. *J. Appl. Polym. Sci.* 1985;30:2979.
24. Zoran SP, Yang L, Zlatanic A, Zhang W, Javni I. *Journal of Applied Polymer Science.* 2007;105(5):2717–2727.
25. Qinghao Meng, Hu J, Zhu Y. *Journal of Applied Polymer Science.* 2007;106(2):837–848.
26. Tan H, Xie X, Li J, Zhong Y, Fu Q. *Polymer.* 2004;45(5):1495–1502.
27. Robert Andreu G, Lligadas J, Carles R, Marina GV, Cádiz. *Journal of Polymer Science Part A: Polymer Chemistry.* 2007;45(21):4965–4973.
28. Kojio K, Fukumaru T, Furukawa M. *Macromolecules.* 2004;37(9):3287–3291.

Polyurethane Urea

L. S. Ramanathan, S. Sivaram and Munmaya K. Mishra

Acronyms PU, PUU

Class Polyurethanes

Structure

$$-\text{O}\left[\!\left(\text{R}'\right)_{\!\overline{x}}\text{O}-\overset{\displaystyle \text{O}}{\overset{\|}{\text{C}}}-\text{NH}-\text{R}-\text{NH}-\overset{\displaystyle \text{O}}{\overset{\|}{\text{C}}}\right]_{\!\overline{n}}\!\left[\text{NH}-\text{R}''\right]_{\!\overline{m}}$$

x = DP of soft segment; R = isocyanate unit
R′ = soft segment; R″ = amine unit

Major Applications Polyurethane urea is useful in making interior automobile parts like armrests, head rests, gear shifts, knee protection pad, etc. Rigid integral PU foams are used in electronic and construction fields.

Properties of Special Interest High compressive strength, less weight, good weatherability, and excellent properties of electrical insulation.

Property	Units	Conditions	Value	Reference
Density	g cm^{-3}	4,4′-Diphenylmethane diisocyanate (MDI)/polyether polyol/4,4′-diaminodiphenyl methane (MDA)	0.96	(1)
		MDI/polyether polyol/diethyl toluene diamine (DETDA)	0.98	
		MDI/polyetherpolyol/3-chloro-3′methoxy-4,4′diamino diphenylmethane (CMOMDA)	0.94	

Unit cell dimensions[2]

Sample	Lattice	Cell Dimensions (Å)			Cell Angles (Degrees)		
		a	*b*	*c*	α	β	γ
MDI/poly(tetramethylene oxide) (PTMO)/MDA	Monoclinic	4.72	11.33	11.64	–	–	116.5

Property	Units	Conditions	Value	Reference
Electron density (mean square fluctuation)	mol electron cm^{-3}	2,4-Toluenediisocyanate(TDI)/ PTMO/ethylenediamine (EDA)	7.14×10^{-3}	(3)
Bragg spacing	Å	2,4-TDI/PTMO/EDA	140	(3)
Flory-Huggins polymer solvent interaction parameter	–	MDI/polycaprolactone diol (PCL) $M_n = 1,300$/EDA (dimethylacetamide)	0.32	(4)
		MDI/PCL(1,300)/EDA (dimethylformamide)	0.4	
		MDI/PCL(1,300)/EDA (dimethylsulfoxide)	0.45	
		MDI/PCL(2,800)/EDA (dimethylformamide)	0.42	
Partial specific volume	–	At 25°C in DMF MDI/ PCL(1300)/EDA	0.848	(4)
		MDI/PCL(2800)/EDA	0.875	
Heat of fusion	J g^{-1}	MDI/polyether polyol/MDA	26.4	(1)
		MDI/polyether polyol/DETDA	11.9	(1)
		MDI/polyether polyol/CMOMDA	18.6	(1)
		MDI/PTMG/ 1,2-propylenediamine(PDA)	21.49	(5)
Heat capacity	J g^{-1} K	MDI/hydroxy terminated polybutadiene (HTPB)/ 4,4′methylene bis(2-chloroaniline (MOCA)	0.4	(6)
		MDI/HTPB/1,4-butanediamine (BDA)	0.389	(6)
		IPDI/PTMO/methylene bis(2-methyl-6-ethyl aniline) [MBMEA]	0.44	(7)
		IPDI/PTMO/methylene bis(2-methyl-6-isopropyl aniline) [MMIPA]	0.45	(7)
		Trimethyl hexamethylene diisocyanate/PTMO/MBMEA	0.513	(7)
Glass transition temperature	K	MDI/polyether polyol/MDA	215.9	(1)
		MDI/polyether polyol/DETDA	220.9	(1)
		MDI/polyether polyol/CMOMDA	232.8	(1)
		MDI/PTMO/EDA	200	(8)
		MDI/PTMO/MDA	199	(8)
		MDI/PTMO/1,6-hexanediamine (HDA)	200	(8)
		TDI/PTMO/EDA	201	(8)

Property	Units	Conditions	Value	Reference
		2,4-TDI/PTMO(1000)/EDA	220	(9)
		2,4-TDI/PTMO(2000)/EDA	199	(9)
		2,4-TDI/PTMO/EDA	199	(3)
		MDI/PTMO/EDA	225	(10)
		MDI/aminopropyl terminated polycyanoethylmethylsiloxane (ATPCEMS)/1,4-butanediol (BD)	194.1	(11)
		Lysinediisocyanate (LDI)/PCL/ 1,4-BDA	220.9	(12)
		1,4-butanediisocyanate (BDI)/PCL/ 1,4-BDA	216.3	(12)
		1,6-hexanediisocyanate (HDI)/PCL/ 1,4-BDA	222	(12)
		Tetramethyl xylene diisocyanate (TMXDI)/PCL/DETDA	221.7	(13)
		TMXDI/HTPB/DETDA	198.8	(13)
Melting temperature	K	MDI/PTMO/EDA	564	(10)
		MDI/PTMG/1,2-PDA	547	(5)
Melting enthalpy	$J\,g^{-1}$	MDI/PEG(400)/EDA	60	(14)
		MDI/PEG(1500)/EDA	29	
Dielectric loss	–	TMXDI/PCL/DETDA	0.3	(13)
		TMXDI/HTPB/DETDA	0.04	
Dielectric permitivity	–	TMXDI/PCL/DETDA	6.8	(13)
		TMXDI/HTPB/DETDA	3.0	
Activation energy	$kJ\,mol^{-1}$	TDI/POLYESTERDIOL/MOCA	42.7	(15)
		TDI/POLYETHERDIOL/MOCA	66.8	
Tensile strength	MPa	MDI/polyetherdiol/MDA	Brittle	(1)
		MDI/polyetherdiol/DETDA	14.3	(1)
		MDI/polyetherdiol/CMOMDA	10	(1)
		MDI/ATPCEMS/BD	6.65	(11)
		LDI/PCL/1,4-BDA	17	(12)
		BDI/PCL/1,4-BDA	29	(12)
		HDI/PCL/1,4-BDA	38	(12)
		MDI/PPO/DETDA	9.33	(16)
		MDI/PBA/DETDA	14.75	(16)
Elongation	%	MDI/polyetherdiol/MDA	–	(1)
		MDI/ polyetherdiol/DETDA	194	(1)
		MDI/polyetherdiol/CMOMDA	103	(1)
		MDI/PTMG/1,2-PDA	360	(5)
		MDI/ATPCEMS/1,4-BD	256	(11)
		MDI/PPO/DETDA	150	(16)
		MDI/PBA/DETDA	267	(16)

Property	Units	Conditions	Value	Reference
Shore A hardness	–	LDI/PCL/1,4-BDA	800	(12)
		BDI/PCL/1,4-BDA	1042	
		HDI/PCL/1,4-BDA	1168	
Shore D hardness	–	2,4-TDI/PTMG/dimethylthio-2,4-toluenediamine (DM-2,4-TDA)	45	(17)
		2,4-TDI/PTMG/trimethylthio-m-phenylenediamine (TM-m-PDA)	36	
Tearing energy	kg m^2	LDI/PCL/1,4-BDA	36	(12)
		BDI/PCL/1,4-BDA	161	
		HDI/PCL/1,4-BDA	137	
Resilience	–	2,4-TDI/PTMG/DM-2,4-TDA	46	(17)
		2,4-TDI/PTMG/TM-m-PDA	37	
Tensile strength	MPa	Polythio-urethane-urea	5.1	(18)
Elongation	%	Polythio-urethane-urea	411	(18)
Shore A Hardness		Polythio-urethane-urea	67	(18)
Contact angle (Water)	°	Aqueous polyurethane dispersions derived from hydroxyl functional poly(perfluoroalkylethyl acrylate)	109	(19)
Contact angle (Methyl Iodide)	°	Aqueous polyurethane dispersions derived from hydroxyl functional poly(perfluoroalkylethyl acrylate)	83	(19)
Glass transition temperature	K	Polythio-urethane-urea	226	(18)
		Aqueous polyurethane dispersions derived from hydroxyl functional poly(perfluoroalkylethyl acrylate)	236	(19)

References

1. Gao Y, et al. *J. Appl. Polym. Sci.* 1994;53:23.
2. Ishihara H, Kimura I, Yoshihara N. *J. Macromol. Sci. Phys.* 1983–1984;B22:713.
3. Wilkes GL, Abouzahr S. *Macromolecules.* 1981;14:456.
4. Sato H. *Bull. Chem. Soc. Japan.* 1966;39:2335.
5. Shibayama M, et al. *Polym. J.* 1986;18:719.
6. Camberlin Y, Pascault J. *J. Polym. Sci., Polym. Chem. Ed.* 1983;21:415.
7. Knaub P, Camberlin Y. *J. Appl. Polym. Sci.* 1986;32:5627.
8. Hu CB, Ward RS Jr. *J. Appl. Polym. Sci.* 1982;27:2167.
9. Sung CSP, Hu CB. *Macromolecules.* 1981;14:212.
10. Sung CSP, Cooper SL. *Macromolecules.* 1983;16:775.
11. Li C, et al. *J. Polym. Sci., Polym. Phys. Ed.* 1988;26:315.
12. de Groot JH, et al. *Polym. Bull.* 1997;38:211.
13. Capps RN, et al. *J. Appl. Polym. Sci.* 1992;45:1175.
14. Gustafson I, Flodin P. *J. Macromol. Sci. Chem.* 1990;A27:1469.

15. Xiaolie L, Jin L, Dezhu M. *J. Appl. Polym. Sci.* 1995;57:467.

16. Yiu Y, et al. *J. Appl. Polym. Sci.* 1993;48:867.

17. Davis RL, Nalepa CJ. *J. Polym. Sci., Polym. Chem. Ed.* 1990;28:3701.

18. Yiwu Quan, He P, Zhou B, Chen Q. *Journal of Applied Polymer Science.* 2007;106(4):2599–2604.

19. Lim H, Lee Y, Park IJ, Lee S-B. *Journal of Colloid and Interface Science.* 2001;241(1):269–274.

Poly(vinyl acetate)

Jianye Wen

Acronyms, Trade Names PVAC

CAS Number 9003-20-7

Class Vinyl polymers

$$[-CH_2-CH-]$$
Structure
$$|$$
$$OCOCH_3$$

Major Applications Adhesive applications in packaging and wood gluing; chewing-gum bases; PVAC emulsions and resins are used as binder in coatings for paper and as textile finishes.

Properties of Special Interest Tasteless, odorless, and nontoxic

Property	Units	Conditions	Value	Reference
Absorption of water	%	20°C for 24–144 h	3–6	(2)
		23°C	4	(3, 4)
		70°C	6	(3, 4)
^{13}C-NMR spectra	ppm	Selected group assignments		(5)
		Carbonyl C=O	169.7	
		Methane		
		C(mm)H—OCOCH$_3$	67.9	
		C(mm)H—OCOCH$_3$	67.0	
		C(rr)H—OCOCH$_3$	65.9	
		Methylene		
		CH$_2$—CHOCOCH$_3$	38.7	
			38.5	
			38.3	
			38.0	
		Methyl, CH$_3$	20.7	
Coefficient of thermal expansion	10^{-4} K^{-1}			
		At 0°C	2.8	(6)
		At 20°C	2.8	(6)
		At 40°C	7.13	(6)
		At 60°C	7.17	(6)
		At 80°C	7.20	(6)
		At 100°C	7.23	(6)

Poly(vinyl acetate)

Property	Units	Conditions	Value	Reference
Cohesive energy E_{coh}	J mol^{-1}	–	31136	(7)
Cohesive energy density	(MJ m^{-3})$^{1/2}$	–	18.6–19.09	(1(b))
Compressibility	cm^3 (gkPa)$^{-1}$	Glassy state	17.8×10^{-6}	(1(b))
		Glassy state at 0°C	2.9	(6, 8)
		Glassy state at 20°C	3.0	(6, 8)
		At 40°C	5.2	(6, 8)
		At 60°C	5.7	(6, 8)
		At 80°C	6.2	(6, 8)
		At 100°C	6.7	(6, 8)
		At 120°C	7.1	(9)
Decomposition temperature	°C	–	150	(10)
Degradation, thermal ($T^{1/2}$)	°C	T at which the polymer looses 50% of its weight, if heated in vacuum for 30 min.	269	(11)
Degree of crystallinity	%	MW, Annealing temp.		
		2236, 0.623(40°C)	0.701(160°C)	(12)
		4042, 0.514	0.623	(12)
		5246, 0.508	0.562	(12)
		7568, 0.504	0.615	(12)
		16856, 0.487	0.587	(12)
Density	g cm^{-3}	At 0°C	g1.196[a]	(6)
		At 20°C	g1.89	(6)
		At 25°C	1.19	(1(a))
		At 50°C	1.17	(1(a))
		At 120°C	1.11	(1(a))
		At 200°C	1.05	(1(a))
		At T_m	1.28	(1(a))
		35–100°C	$1.2124 - 8.62 \times 10^{-4} T + 0.223 \times 10^{-6} T^2$	
			35–100°C	(6)
Dielectric constant	–	At 2 MHz		
		At 50°C	3.3	(13)
		At 150°C	8.3	(13)
Dielectric loss factor Tan δ	–	At 2MHz		
		At 50°C	150	(13)
		At 120°C	260	(13)
Dielectric strength	V cm^{-1}	At 30°C	3.94×10^5	(14)
		At 60°C	3.07×10^5	(14)
Diffusion coefficients D	10^{-8} cm^2 s^{-1}	Vinyl acetate 25°C	26.8	(15)
		Styrene 25°C	15.4	(15)

Property	Units	Conditions	Value	Reference
Dipole moment	eSU (per monomer unit)	At 20°C	2.3×10^{-18}	(16, 17)
		At 150°C	1.77×10^{-18}	(16, 17)

Emulsion specifications of PVAC[1(b)]

Property	Range
Solids, wt%	48–55
Viscosity, cP	200–4500
pH	4–6
Residual monomer, % max	0.5
Particle size, mm	0.1–3.0
Particle charge	Neutral or negative
Density at 25°C, g cm^{-3}	0.92
Stability to borax	Stable or unstable
Mechanical stability	Good or excellent

Property	Units	Conditions	Value	Reference
FTIR spectra	cm^{-1}	Selected group assignments		(18)
		Isotactic structure	1061	
		Out-of-plane CH_2-CO_2 band	617	
		$C=O$	1734	
		$C-CH3$ bending	1372, 1431	
		Asymmetric $C-O-C$ stretch	1245	
		Aliphatic $C-H$ stretches	2951	
Gas solubility	cm^3(STP) cm^{-3} bar^{-1}	25°C		
		N_2	0.02	(1(b))
		O_2	0.04	(1(b))
		H_2	0.023	(1(b))
Glass-transition temp. (T_g)	°C	–	28–31	(1(b))
		Atactic, $M_n = 3922$	23.6	(1(b))
		Atactic, $M_n = 1.66 \times 10^5$	31.4	(1(b))
		Isotactic, $M_n = 10^5$	25.8	(1(b))
H-NMR spectra	ppm	Selected group assignments		(5)
		Methine: CH-$OCOCH_3$	4.78	
		Methylene CH_2-$CHOCOCH_3$	1.75	
Hardness	Shore units	20°C	80–85	(1(b))
Heat capacity	J (g K)$^{-1}$	–193°C	0.323	(19)
		27°C	1.183	(19)
		47°C	1.841	(19)
		97°C	1.898	(19)
ΔC_p			0.116	(20)
C_v	4.19J (mol K)$^{-1}$		24.409	(21)

Poly(vinyl acetate)

Property	Units	Conditions	Value	Reference
Heat conductivity	$J (s\,m\,K)^{-1}$	–	0.159	(1(b))
Heat distortion point	$°C$	–	50	(1(b))
Heat of polymerization	$kJ\,mol^{-1}$	–	87.5	(1(b))
Huggins coefficients K_H	–	Acetone 25°C	0.37	(22)
		Chlorobenzene 32°C	0.43	
		Chloroform 25°C	0.31	
		Methanol 18°C	0.61	
		Toluene 25°C	0.55	
		Benzene 30°C	0.37	
		Dioxane 25°C	0.29	
Index of refraction, N_D	–	At 20.7°C	1.4669	(1(b))
		At 30.8°C	1.4657	
		At 52.1°C	1.4600	
		At 80°C	1.4480	
		At 142°C	1.4317	
Interfacial tension	$mN\,m^{-1}$	20°C		(1(b))
		With PE	14.5	
		With PDMS	8.4	
		With PIB	9.9	
		With PS	4.2	
Internal pressure	$MJ\,m^{-3}$	At 0°C	255	(1(b))
		At 20°C	284.7	
		At 28°C	397.8	
		At 40°C	431.3	
		At 60°C	418.7	
Interaction parameter χ, with;	–	Acetone 30 to 50°C	0.31 to 0.39	(23)
		100 to 140°C	0.32 to 0.21	(23)
		Benzene 20°C	0.42	(24)
		30 to 50°C	0.30 to 0.26	(23)
		80 to 140°C	0.44 to 0.25	(23)
		N-Butane 100°C	1.97	(23)
		Butanone 25°C	0.44	(24)
		Chloroform 80 to 135°C	−0.17 to −0.09	(23)
		Cyclohexane 100°C	1.18	(23)
		Ethanol 100°C	0.80	(23)
		N-Hexane 100 to 120°C	2.06 to 1.71	(23)
		N-Octane 90–120°C	2.3–1.94	(23)
		1-Propanol 30 to 50°C	1.3 to 1.0	(23)
		Vinyl acetate 30°C	0.41 to 0.22	(23)
		Water 40°C	2.5	(23)

Mark-Houwink constants

Solvent	Temp. (°C)	Mol. Wt. Range ($M \times 10^{-4}$)	$K \times 10^3$ (ml g^{-1})	a	Reference
Acetone	20	−72	15.8	0.69	(25)
	25	−1.3	14.6	0.72	(26)
	–	–	21.4	0.68	(24)
	30	−68	17.4	0.70	(27)
	46	−34	13.8	0.71	(28)
Acetonitrile	25	−215	16.2	0.71	(29)
	30	−153	41.5	0.62	(30)
Benzene	30	−86	56.3	0.62	(31, 32)
	35	−40	21.6	0.675	(33)
Butanone	25	−346	13.4	0.71	(34)
	25	−120	42	0.62	(35)
	30	−120	10.7	0.71	(36)
Chlorobenzene	25	−7	110	0.50	(37)
	53	−34	53.7	0.60	(38)
Chloroform	20	−68	15.8	0.74	(27)
	25	−34	20.3	0.72	(38)
	53	−34	14.7	0.74	(38)
Dioxane	25	−34	11.4	0.74	(39)
Ethanol	56.9 (θ)	−150	90	0.50	(28)
Methanol	6	−150	–	10.1	(40, 41)
	25	−22	38.0	0.59	(39)
	30	−120	31.4	0.60	(42)
MEK	25	–	15.4	0.71	(24)
Tetrahydrofuran	25	−50	16	0.70	(43)
	35	−117	15.6	0.708	(44)
Toluene	25	−15	108	0.53	(39)
	67	−15	156	0.49	(39)

Mechanical properties

Property	Units	Conditions	Value	Reference
Modulus of elasticity	GPa	–	1.275–2.256	(1(b))
			1.6	(45)
Notched impact strength	J m^{-1}		102.4	(1(b))
Elongation at break	%	20°C, %RH	10–20	(1(b))
Poisson's ratio			0.34	(47)
Rubbery shear modulus	N mm^{-2}		13	(1(b))
Tensile strength	MPa		29.4–49.0	(1(b))

Poly(vinyl acetate)

Property	Units	Conditions	Value	Reference
Young's modulus	MPa	–	600	(1(b))
Melting temperature T_m	°C	–	175	1(a)
	–	–	168.4	47
	–	–	167.3	47
	–	–	152.4	47
Molar volume	cm^3 mol^{-1}	25°C	74.25	1(b)
	–	–	72.4	48

Permeability and diffusion coefficients

Permeant	Temp.	$P \times 10^{13c}$	$D \times 10^{6c}$	$S \times 10^{6c}$	Reference
He	10°C	4.95	6.46	0.0784	(49)
	30°C	9.44	9.55	0.101	(49)
H$_2$	10°C	2.99	1.32	0.237	(49)
	30°C	6.84	2.63	0.254	(49)
Ne	10°C	0.838	0.794	0.106	(49)
	30°C	1.97	1.66	0.118	(49)
O$_2$	10°C	0.136	0.0178	0.766	(49)
	30°C	0.367	0.0562	0.637	(49)
	73°C,	0.27	–	–	(1(b))
Ar	10°C	0.0569	0.00479	1.11	(49)
	30°C	0.143	0.0162	0.943	(49)
Kr	10°C	0.0172	0.000602	2.78	(50)
	30°C	0.0582	0.00295	1.96	(50)
CH$_4$	25°C	0.0237	0.0017	1.39	(50)
N$_2$, below T_g	–	0.066	–	–	(1(b))
N$_2$, above T_g	–	0.05	–	–	(1(b))

Property	Units	Conditions	Value	Reference
Refractive index	n_{25}^{D}	–	1.4688	(51)
	–	–	1.467	(52)
Softening temperature	°C	–	35–50	(1(b))
Specific volume	L kg^{-1}	At T = 100–200°C	$0.823+ (6.4 \times 10^{-4})t$	(1(b))
		At T = 28°C (T_g)	0.84	
Solubility parameter	(Mpa)$^{1/2}$	–	18.6–19.9	(1(b))
			19.2	48
		At 25°C	21.07	
		At 50°C	19.4	
		At 125°C	17.9	
		∂_d dispersion forces contribution	19.0	
		∂_p polar forces contribution	10.2	
		∂_h hydrogen bonding contribution	8.2	
		$\partial(\partial_d^2 + \partial_p^2 + \partial_h^2)^{1/2}$	23.1	

Property	Units	Conditions	Value	Reference
Surface resistance	Ω cm^{-1}	–	5×10^{11}	(1(b))
Surface tension	mN m^{-1}	At 20°C	36.5	(53, 54)
		At 140°C	28.6	
		At 150°C	27.9	
		At 180°C	25.9	
		γ_{ds} dispersive	27.4	
		γ_{os} polar	15.4	
		γ_{solid} total	42.85	
	mN (m K)$^{-1}$	$-d\gamma/dT$	0.066	
	χ^P	polarity	0.329	
Thermal conductivity	mW (m K)$^{-1}$	–	159	(1(b))
Second virial coefficient	$A_2 \times 10^4$ (mol. cm^3 g^{-2})	Acetone 25°C $M \times 10^{-4} = 13.71$	6.957	(24)
		30°C $M \times 10^{-3} = 27$ to 845	8.80 to 3.34	(55)
		343 to 722	3.66 to 3.50	(56)
		78 to 660	6.5 to 2.5	(57)
		Methanol 25°C $M \times 10^{-3} = 2360$ to 422900	7.50 to 0.172	(58)
Theta solvent	–	Acetone/isopropanol 23/77	30°C	(59)
		Butanone/isopropanol 73.2/26.8	25°C	(60)
		Cetyl alcohol	123°C	(61)
		Di-i-butyl ketone	136.5°C	(62)
		Ethanol	19°C	(63)
		Ethanol/methanol 80/20	17°C	(64)
		60/40	26.5°C	(64)
		50/50	34°C	(59)
		40/60	36°C	(64)
		Heptanone	29°C	(65)
		Methanol	6°C	(64)

Unperturbed dimension

Conditions	$r_0/M^{1/2} \times 10^4$ (nm)	$r_{of}/M^{1/2} \times 10^4$ (nm)	$\sigma = r_0/r_{of}$	$C_\infty = r_0^2/nl^2$	Reference
3-heptanone, 26.8°C	670	332	2.02	8.15	(66)
methanol, 6°C	720	332	2.17	9.4	(66)
ethanol, 56.9°C	690	332	2.08	8.65	(66)
tetrahydrofuran, 35°C	774 ± 20	–	–	–	(67)

Property	Units	Conditions	Value	Reference
Unperturbed radius of gyration (Q)	Å^2 mol g^{-1}	Linear PVAC	0.107	(1(a))

Poly(vinyl acetate)

Property	Value	Reference
Solvents	Esters, ketones, aromatics, halogenated hydrocarbons, carboxylic acids, alcohols, bzn, toluene, chlorform, carbon tetrachloride/ethanol, chlorobenzene, dichloroethylene/ethanol(20:80), methanol, ethanol/water, allyl alcohol, 2,4-dimethyl-3-pentanol, benzyl alcohol, THF, tetrahydrofurfuryl alcohol, dioxane, glycol ethers, acetone, glycol ether esters, acetic acid, lower aliphatic esters, acetonitrile, nitromethane, DMF, DMSO (chloroform and chlorobenzene for syndiotactic polymers)	(68)
Non-solvents	Saturated hydrocarbons, mesitylene, carbon tetrachloride (sw), ethanol (anhydrous, sw), ethylene glycol, cyclohexanol, diethyl ether (anhydrous, alcohol free), higher esters (C > 5), carbon disulfide, water (sw), dil. acids, dil. alkalies, (benzene and acetone for syndiotactic polymers)	(68)

[a] g = glass.
[b] see Reference (1) for details.
[c] Units are: P in cm^3(273.15 K; 1.013×10^5 Pa) cm cm^{-2} sPa; D in cm^2 s; S in cm^3 (273.15 K; 1.013×10^5 Pa) cm^{-2}Pa^{-1}.

References

1. General references: (a) Brandrup J, Immergut EH. *Polymer Handbook*. Third ed. New York: Wiley-Interscience; 1989; (b) Daniels W. In: Bikales NM, eds. *Encyclopedia of Polymer Science & Technology*. Vol 17. New York: Wiley-Interscience; 1987:402.; (c) Mark JE, Ed. *Physical Properties of Polymers Handbook*. Woodbury, New York: AIP Press; 1996.
2. Schildknecht CE. *Vinyl and Related Polymers*. New York: Wiley; 1952:336.
3. Miyagi Z, Tanaka K. *Colloid Polym. Sci.* 1979;257:259.
4. Johnson GE, Bair HE, Matsuoka S, Scott JE. *ACS Symp. Ser.* 1980;127(Water-Soluble Polym.):451.
5. Huang C, Shieu F, Hsieh W, Chang T. *J. Appl. Polym. Sci.* 2006;100:1457.
6. Yu X, Wang X, Li X, Gao J, Wang H. *J. Polym. Sci. Part B.* 2006;44:409.
7. McKinney JE, Goldstein M. *J. Res. Nat. Bur. Stand.* 1974;78A:331.
8. McKinney JE, Simha R. *Macromolecules*. 1974;7:894.
9. Beret S, Prausnitz JM. *Macromolecules*. 1975;8:536.
10. Mowilith, *Polyvinylacetat*. Frankfurt: Farbwerke Hoechst AG; 1969:214–215.
11. Van Krevelen DW. *Properties of Polymers*. New York: Elsevier Sci. Pub; 1976.
12. Sato T, Okaya T. *Polym. J.* 1992;24:849.
13. Thurn H, Wolf K. *Kolloid Z.* 1956;148:16.
14. Shaw TPG. *Encyclopedia of Chemical Technology*. Vol 14. New York: Interscience Pub; 1955:692.
15. Hornig K, Hemmelmann K, Brandt H, Hergeth WD, Wartewig S. *Acta Polymerica*. 1991;42:601.
16. Meed DJ, Fuoss RM. *J. Am. Chem. Soc.* 1941;63:2839.
17. Broens O, Mueller FH. *Kolloid Z.* 1955;141:20.

18. Mathakiya I, Rakshit AK. *J. Appl. Polym. Sci.* 1998;68:91.

19. Gaur U, Lau SF, Wunderlich BB. *J. Phys. Chem. Ref. Data.* 1983;12:29.

20. Boyer RF. *J. Macromol. Sci., Phys.* 1973;B7:487.

21. Yu X, Wanf X, Gao J, Li X, Wang H, *Polymer.* 2005;46:9443.

22. Stickler M, Sutterlin N. In: Brandrup J, Immergut EH, eds. *Polymer Handbook.* Third ed. New York: Wiley-Interscience; VII191, 1989.

23. Orwoll RA, Arnold P. In Mark JE, Ed. *Physical Properties of Polymers Handbook.* Chap. 14. Woodbury, New York: AIP Press; 1996.

24. Qian JW, Rudin A. *Eur. Polym. J.* 1992;28:725.

25. Tsvetkov VN, Ya S, Kotlyar. *Zh. Fiz. Khim.* 1956;30:1100.

26. Misra GS, Gupta VP. *Makromol. Chem.* 1964;71:110.

27. Fattakhov KZ, Pisarenko ES, Verkotina LN. *Kolloidn. Zh.* 1956;18:101.

28. Ueda M, Kajitani K. *Makromol. Chem.* 1967;108:138.

29. Bevak, Thesis, Mass. Inst. Tech, Cambridge, Mass, USA, 1955.

30. Kalpagam V, Rao R. *J. Polym. Sci.* 1963;A1:233.

31. Nakajima A. *Kobunshi Kagaku.* 1954;11:142.

32. Varadiah VV. *J. Polym. Sci.* 1956;19:477.

33. Berry GC, Hobbs LM, Long VV. *Polymer.* 1964;5:31.

34. Schulz AR. *J. Am. Chem. Soc.* 1954;76:3423.

35. Elias HG, Patat F. *Makromol. Chem.* 1957;25:13.

36. Abe M, Fujita H. *J. Phys. Chem.* 1965;69:3263.

37. Patrone E, Bianchi E. *Makromol. Chem.* 1966;94:52.

38. Ueda M, Kajitani K. *Makromol. Chem.* 1967;108:138.

39. Moore WR, Murphy M. *J. Polym. Sci.* 1962;56:519.

40. Ueda M, Kajitani K. *Makromol. Chem.* 1967;108:138.

41. Naito R, Kagaku K. *Chem. High. Polym.* (Tokyo), 1959;16:7.

42. Matsumoto M, Ohyanagi Y. *J. Polym. Sci.* 1960;46:441.

43. Cane F, Capaccioli T. *Eur. Polym. J.* 1978;14:185.

44. Atkinson CML, Dietz R. *Eur. Polym. J.* 1979;15:21.

45. Konnerth J, Gindl W, Müller U. *J. appl. Polym. Sci.* 2007;103:3936.

46. Bicerano J. *Prediction of polymer properties.* 2nd ed. New York: Marcel Dekker; 1996.

47. Yang J. *J. Physical Chemistry B.* 2006;441(2):137.

48. Yu X, Wang X, Li X, Gao J. *QSAR Comb. Sci.* 2006;25:156.

49. Mears P. *J. Am. Chem. Soc.* 1954;76:3415.

50. Mears P. *Trans. Faraday Soc.* 1957;53:101.

51. magagnini PL. Annali di Chimica (Rome, Italy). 1967;57(7):805.

52. Holder AJ. *QSAR & Combinatorial Science.* 2006;25(10):905.

53. Wu S. *J. Colloid Interface Sci.* 1969;31:153.

54. Roe RJ. *J. Colloid Interface Sci.* 1969;31:228.

55. Matsumoto M, Ohyanagi Y. *J. Polym. Sci.* 1960;46:441.

56. Ohyanagi Y, Matsumoto M. *Chem. High Polym. (Japan).* 1959;16:296.

57. Chinai SN, Scherer PC, Lewi DW. *J. Polym. Sci.* 1955;17:117.

58. Schmidt M, Nerger D, Burchard W. *Polymer.* 1979;20:582.

59. Tsuchiya S, Sakaguchi Y, Sakurada I. *Chem. High Polym. (Japan).* 1961;18:346.

60. Schultz AR. *J. Am. Chem. Soc.* 1954;76:3422.

61. Berry GC, Nakayasu H, Fox TG. *J. Polym. Sci. – Polym. Phys. Ed.* 1979;17:1825.

62. Horii F, Ikada Y, Sakurada I. *J. Polym. Sci. – Polym. Chem. Ed.* 1974;12:323.

63. Candau F, Strazielle C, Benoit H. *Makromol. Chem.* 1973;170:165.

64. Naito R. *Chem. High Polym. (Japan).* 1959;16:7.

65. Matsumoto M, Ohyanagi Y. *J. Polym. Sci.* 1961;50:S1.

66. Ueda M, Kajitani K. *Makromol. Chem.* 1967;108:138.

67. Atkinson CML, Dietz R. *Eur. Polym. J.* 1978;14:867.

68. Fuchs O. In: Brandrup J, Immergut EH, eds. *Polymer Handbook*. Third ed. New York: Wiley-Interscience. 1989:VII379.

Poly(vinyl alcohol)

P. R. Sundararajan and Molla Rafiquel Islam

Acronym, Trade Names PVA, Vinol, Airvol® (Air Products and Chemicals), Elvanol® (du Pont), Gelvatol® (Monsanto), Mowiol® (Hoechst), Poval® (Kuraray, Mowiol (Clariant GmbH, Germany). Poval (Kuraray, Unitika) Japan), Gohsenol® (Nippon Gohsei, Japan), CCP (Chang Chun, Taiwan).

Class Vinyl polymers

Structure $CH_3CHOH(CH_2-CHOH)_n$

Major Applications Paper and textile sizing, oxygen resistant films, adhesives, emulcifiers, colloid stabilizers, base/coatings for photographic films, food wrappings, desalination membranes, electroluminescent devices, and cement coatings, Gels and composites.

General Information Commercial poly(vinyl alcohol) is derived from poly(vinyl acetate). Typical commercial molecular weight ranges for different viscosity grades are: $M_n = 25,000$ (low, 5–7 cP), 40,000 (intermediate, 13–16 cP), 60,000 (medium, 28–32 cP) and 100,000 (high, 55–65 cP). (Viscosities correspond to 4% aqueous solution.)[1]

World-wide production >1000, 000 tons/yr, two-thirds in Japan, China and Taiwan. Price \$2.65 kg^{-1}(1995).[2]

Properties of Special Interest Water soluble; resistant to solvents, oil, and grease; exceptional adhesion to cellulosic and other hydrophilic surfaces. Flexibility and oxygen barrier capacity along with tensile strength.

Synthetic aspects

Stereoregularity	Parent Polymer	Synthetic Conditions	Method of Characterization	Characteristics	Reference
Atactic	PV Ac	Free radical, BEt$_3$/air or AIBN/hv, −78 to 90°C, amyl acetate or MEK solvent	NMR	–	(3)
Syndiotactic	Poly(vinyl trifluoroacetate)	n-Bu$_3$B/air, −78°C, heptane	NMR	m: 39%, r: 61%	(4)
		Benzyl peroxide, 60°C	IR, X-ray diffraction	–	(5)
Syndiotactic	Poly(vinyl pivalate)	Radical polymeriation of VP at −40°C; n-hexane	NMR, DSC	r: 69%	(6)

Dedicated to the memory of my son, Anand.

Stereoregularity	Parent Polymer	Synthetic Conditions	Method of Characterization	Characteristics	Reference
Isotactic	Poly(vinyl t-butyl ether)	BF_3 etherate, $-78°C$, toluene	NMR	m: 67–76%, r: 33–24%	(4)
		BF_3 etherate, $-78°C$, toluene	IR, X-ray diffraction	–	(5)
Isotactic	Poly(vinyl benzyl ether)	Cationic polymerization with BF_3 etherate at $-78°C$			
		In n-heptane/toluene mixture	X-ray diffraction, IR	–	(7)
		In toluene	NMR	m: 93%, r: 7%	(8)
		In nitroethane	NMR	m: 76%, r: 24%	(8)

Spectroscopy	Condition	Chemical Shifts (δ, PPM)*		Reference
NMR				(8, 46[†], 50, 51) (Reviews)
^{1}H (60, 100 and 220 MHz) spectra	PVA from Kuraray Co., in DMSO-d_6, 20–100°C; tacticity analysis; hexamethyldisiloxane as internal standard	OH proton at 50°C: i: 4.52, h: 4.33; s: 4.10 J(H-O-C-H) (Hz): i: 3.1; h: 4.3; s: 5.3		(52)
^{1}H spectra	Gelvatol 2/75 in DMSO-d_6, at 35°C; tacticity analysis; TMS as standard	OH proton: i: 4.63; h: 4.45; s: 4.22		(53)
^{13}C (22.63 MHz) and ^{1}H (220 MHz) spectra	Atactic and isotactic PVA ^{13}C in DMSO-d_6, D_2O and hexafluoroisopropyl alcohol; TMS standard ^{1}H in DMSO-d_6; hexamethyldisiloxane standard	^{13}C: CH_2 peaks: 		(54)

^{13}C: CH_2 peaks:

	DMSO-d_6	D_2O
rrr:	45.8	47.1
$rrm + mrm$:	45.6	46.4
$mmr + rmr$:	45.2	46.1
mmm:	44.8	45.5

CH peaks: 67.8, 66.2, 64.3 (DMSO-d6); 70.4, 69.0, 67.5 (D_2O)

Spectroscopy	Condition	Chemical Shifts		Reference
^{13}C (22.6 and 67.9 MHz) spectra	Pentad tacticity analysis; atactic and isotactic PVA; in DMSO-d_6 at 80°C; TMS standard	$rmmr$: 68.01; $mrrm$: 64.26 (see reference (55) for others)		(55)
^{13}C (100 MHz) spectra	Heptad and hexad sequence analysis; atactic and isotactic PVA; in DMSO-d_6 and D_2O at 50°C; TMS standard			(56)

Atactic: Methine	DMSO-d_6	D_2O
$rrrr$:	64.48	65.53
$mrrm$:	64.18	65.21
Methylene		
$mrrrm$:	45.92	45.07
$rrrrr$:	45.81	44.95

(see reference (56) for others)

Spectroscopy	Condition	Chemical Shifts (δ, PPM)*	Reference
^{13}C CP(MAS) CMX-200	Solid State and Hydrated, Isotactic, syndiotactic		56a
^{1}H (360 MHz), 2D NMR	$M_w = 14{,}000$; 70°C; sodium 3-trimethylsilyl [2,2,3,3] propionate as standard	rr: 4.062; mr: 4.037; mm: 3.985; mmm: 1.769, 1.675; rrr: 1.647	(57)
^{1}H (500 MHz) and ^{13}C(125MHz); 2D NMR	$M_w < 4{,}400$; in D$_2$O at 80°C; ^{13}C assignments to pentad-hexad level	^{1}H spectra: CH group: 3.957 (rr); 3.930 (mr); 3.879 (mm) CH$_2$ group: 1.660 (mmm); 1.539 (rrr) ^{13}C spectra: CH group: 68.18 (rmmr); 65.22 (mrrm) CH$_2$ group: 44.85 (mrrrm); 44.74 (rrrrr) (see reference (58) for others)	(58)
^{1}H (80, 300, and 400 MHz); ^{13}C (100.6 MHz) spectra	$M_w = 50{,}000$; in water at 5–87°C; spin-lattice relaxation times; local chain dynamics; TMS standard	^{13}C spectra at 60°C: CH group: 64.8–65.5 (rr); 66.1–66.9 (mr); 67.7–68.4 (mm) CH$_2$ group: 43.4–43.9 (mmm + mrm); 44.7–45.1 (rrr)	(59)
^{13}C (50 MHz)VT/MAS solid state spectra	DP 1700, 7600 and 15,500 (Kuraray Co.); phase structure of single crystals from triethylene glycol; TMS standard	CH resonance splits into four peaks at 77.5 (two intra H-bonds); 71.5 (one intra h.bond); 65.0 (no intra H-bond); and 62.4 (intermolecular H-bond); fraction of OH groups with intra H-bond is 0.35 for crystalline domains; decreases from 0.66 (DP 1700) to 0.44 (DP 15,500) in noncrystalline regions	(60)
^{13}C (67.8 MHz) CP/MAS solid state spectra	DP 1700 (Kuraray Co.); study of hydrogen bonding in aqueous gels	–	(61)

* m: meso diad; r: racemic diad; i: isotactic triad; h: heterotactic triad; s: syndiotactic triad.

† References (8, 46, 50, 51) are reviews. Reference (46) presents a chronological review of proton and ^{13}C NMR analysis of PVA and spectral assignments.

Unit cell dimensions

Tacticity	Lattice	Monomers (Per Unit Cell)	Cell Dimensions (Å)			Cell Angles (Degrees)			Reference
			a	b*	c	α	β	γ	
Atactic	Monoclinic, P2$_1$/m	2	7.81	2.51	5.51	90	97.7	90	(8, 62)
Atactic	Monoclinic, P2$_1$/m (X-ray and neutron diffraction)	2	7.81	2.52	5.51	90	91.7	90	(63)
Isotactic	–	2	–	2.51	–	–	–	–	(7, 8)

* Chain axis.

Poly(vinyl alcohol)

Property	Units	Conditions/Ethylene Mol %	Value
Density	$g\,cm^{-3}$	EVAL®, 27%	1.20
		47%	1.12
Melting temperature	K	27%	464
		47%	429
Glass transition temperature	K	27%	345
		47%	321
Diffusion coefficient of water	$cm^2\,s^{-1}$	32%, 20°C	6.63×10^{-9}
		32%, 60°C	99.0×10^{-9}
		44%, 20°C	0.74×10^{-9}
		44%, 60°C	34.9×10^{-9}

* See also the entry on *ethylene-vinyl alcohol* in this handbook.

Block copolymers[73, 73a, 73b, 73c, 73d, 73e]

Block Copolymer	Fraction of Other Monomer	Property/Application
PVA–PEO–PVA	25–34 wt%	Low surface tension. Segments crystallize independently
PVA–PPO–PVA	12%	–
PVA–polyacrylic acid	20%	Transparent film with gelatin blends (0–100% blend composition range)
PVA–polyacrylamide-polyacrylic acid	100–95/5	Transparent films with starch (up to 40% (wt) of starch)
Propyl to octadecyl alkanes	–	Prepared by end group modification of PV Ac in the presence of Mercaptan of the alkanes; modifier for surface tension and wetting property; protective colloid
PVA-PS-PVA		Prepared with different syndiotacticity. Nanometer micelle.
PVA-Glycosylate-PVA		Enzymatic synthesis (Glucose, fatty acid)
PVA-polyacrylonitrile-PVA	11-29%	Internal micro domain separation, amphiphilic self associated in water with nanocups.

Compatible polymers in aqueous solutions* [74]

Polymer	Interaction Parameter† α_{23} (ml^{-1})
Carboxy methyl cellulose	0.059
Methyl cellulose	0.128
Hydroxy ethyl cellulose	0.177
Dextrine	0.290
Poly(methyl acrylate) (20% hydrolyzed)	0.006
Poly(ethyl acrylate) (20% hydrolyzed)	0.074

* DP of PVA: 550–1750, concentration of polymers 10–30%; 88% hydrolyzed.
† Smaller value indicates better compatibility.

Blends*

Other Polymer	Conditions	Characterization Method	Morphological Properties	Reference
Poly(N-vinyl-2-pyrrolidone)	PVA $M_w = 25,000$, 98.5% hydrolyzed; PVPy $M_w = 360,000$; films cast from aqueous solutions	^{13}C CP/MAS NMR (100 MHz) and DSC	Miscible over entire composition range; single T_g increasing from 73.1°C (0% PVPy) to 158.9°C (80% PVPy); T_m of PVA depressed from 218.7°C (0% PVPy) to 186.3°C (80% PVPy); chemical shift changes with composition given; intermolecular hydrogen bond between PVA and PVPy	(75, 76)
Polypyrrole	PVA $M_w = 86,000$, 100% hydrolyzed; in situ polymerization of Ppy in PVA matrix	FTIR, X-ray, TGA, DSC, SEM	Miscible over entire composition range; no PVA crystallinity with Ppy > 20%	(77)
Cellulose	PVA: Mowiol 8-88, blend film cast from N-methyl-2-pyrrolidinone/3wt% LiCl	X-ray, dielectric and dynamic mechanical measurements	homogeneous with > 60 wt% of cellulose, no crystallinity	(78)
		^{13}C NMR	–	(79)
Poly (3-hydroxybutyric acid)	P(3HB) $M_w = 380,000$; atactic PVA: DP 2000; syndiotactic PVA: DP 1690; isotactic PVA: DP 7250; films cast from solutions of hexafluoroisopropyl alcohol	FT-IR	Suppression of P(3HB) crystallization is more with syn-PVA than with a-PVA. i-PVA has no influence.	(80)
Starch	Poly(ethylene-vinylalcohol) copolymer, 56% VA; waxy maize, native corn and high-amylose starches; extrusion-blended	X-ray, DSC, SEM, TEM	Phase separated starch domains. Oriented droplets, 0.05–5 μm in length (waxy maize), 0.05–1.2 μm domains (native corn), < 0.25 μm (high amylose)	(81)
Nylon 4,6	Poly(ethylene-vinylalcohol) copolymer, 27 mol% ethylene, 13 mol% vinyl acetate; nylon 4/nylon 6: 69/31 mol%; films cast from formic acid	FT-IR, X-ray, DSC, tensile tests	Miscible when nylon 4,6 < 35 wt%. $C-O \cdots N-H$ hydrogen bond between nylon and EVOH. Increase in tensile strength from 4 for 15/85 wt% nylon/EVOH to 331 kg cm^{-2} for 100/0 blend	(82)

Poly(vinyl alcohol)

Other Polymer	Conditions	Characterization Method	Morphological Properties	Reference
Copolyamide (random 1:1:1 nylon 6/nylon 6,6/nylon 6,10 units)	M_w of PVA: 24,000; solution cast from N,N-dimethyl formamide	FT-IR, DSC	Miscible in the amorphous state; two phases when quenched after DSC scan; blends exhibit LCST behavior; both components show mutual T_m depression	(83)
Soy protein isolate (SPI)	PVA Mw = 9000 SPI = 90% protein films	SEM, XRD, DSC and mechanical measurements	Miscible, plane surface. With time hole evolves. Improved mechanical properties	(75a)
Waste from food/agro industry	–	SEM and mechanical measurements	–	(76a)
Poly(p-dioxanone)	PVA 88% hydrolyzed PPDO, $k = 79 \times 10^{-3}$ cm^3 g^{-1}	SEM, DSC, TGA	Phase separated morphology, Tg and Tm of individual polymers are constant	(77a)
Xylan-MA	PVA Mw = 30,000–70,000	^{13}C NMR, FTIR,TGA	Phase separated morphology	(77b)
Polypropylene	Oriented, PVA = Poval 68.5% – 89.5% hydrolyzed	SEM, FTIR and Mechanical measurements	Long fiber and droplets	(78a)
Nafion	PVA Mw = 86000, 99% hydrolyzed	FTIR, ATR, DSC and conductivity measurements	Morphology of Nafion dominates in the blends	(81a)
PMMA	PVA Mw = 115,000 PMMA = 120,000 and LiBF$_4$	SEM, XRD, DSC, FTIR, CV	Complexation reaction, shows considerably good conductivity	(83a)
Fatty Acids (lauric, myristic, palmitic, and stearic acids)		FTIR, DSC	Dispersed into polymer matrix	(83b , 83c)

* See also gels below for gelation with blends.

Gels

Gelling Agent	Conditions	Features	Reference
Ethylene glycol	Syndiotactic ($r = 57\%$) and atactic ($r = 50\%$)	T_m of wet gel: 131°C for a-PVA and 144°C for s-PVA; T_m of dry gel: a-PVA: 231°C (quick cool), 238.5°C (gradual); s-PVA: 247.5°C (quick cool), 248.5°C (gradual)	(38)
Ethylene glycol/water	DP of PVA: 1700, 99.9% hydrolyzed	Maximum elastic modulus with 35 mol% of EG	(84)

Gelling Agent	Conditions	Features	Reference
Water	PVA blended with poly(styrene sulfonic acid) sodium salt	Dried, drawn blend hydrogels; physical cross-links due to interpolymer complex increasing the Young's modulus with NaPSS content; contraction upon absorbing water, with nonideal rubber elasticity	(85)
Water	PVA blended with poly(styrene sulfonic acid) sodium salt; high water content	Three dimensional honey-comb structure, with bundles or tapes (0.1–0.2 μm); highly transparent; permeability similar to that of commercial soft contact lens	(86)
DMSO/water	DMSO/water: 100/0 to 50/50	–	(87)
Water	Telechelic PVA was used to crosslink with chitosan or PVA	Firm network	(88)
Water	Gel prepared by chemical cross-linking with glutaraldehyde, annealing and then hydrated, or low temperature crystallization from aqueous mixtures of glycerol, ethylene glycol, or DMSO	–	(89)
Borax	0.1% solution wt	Thermally irreversible, bisdiol complex formed	(1)*
Boric acid	Full gelation above pH 6	–	(1)
Congo red	3% (w/w) with fully hydrolyzed PVA	Colored gel	(13)
	DP 1800, 99.96% hydrolyzed	Sol-gel transitions	(22, 93–95)
Resorcinol, 2,4-dihydroxybenzoic acid	–	Colorless, thermoreversible gel	(13)
Water, glycerine, glycol	Moviol (Hoechst) $M_n =$ 48,000; 2% actetate content	Crystalline gels; crystallinity: PVA (initial): 21.5% Water gel: 21.5% Glycerine gel: 34% Glycol gel: 42%	(96)
Ferrous sulfate	Radiation induced	Dose sensitive, depends on temperature	(1a)
Chitosan	Autoclave, freeze-thawing process	Swelling is not dependent on pH	(13a)
Starch/Glycerol	Electron beam irradiation	More flexible, intact network structure	(13a)

Poly(vinyl alcohol)

Gelling Agent	Conditions	Features	Reference
Poly (sodium acrylate)/Potassium persulfate/sodium bisulfite	Interpenetrating Polymer matrix	Water absorbing capacity is high, 700–800 mL/g	(13b)
Methacrylate	Photo cross linked from aqueous solution	Transparent	(13c)
Choindroitin Sulfate/Glutaraldehyde	Cross linked	bioactive	(13d)
Polyacrylic acid/Polysaccharides	PVA/Dextrin, PVA/Chitosan PVA/Starch, PVA/Gellman	Films or gels, the blends actually works as two phase system	(38a)
Succinic acid/ Gluconic acid	Freezing thawing process	Porous structure, open and regular	(96a)
Sulfonate polyester	Freezing thawing process	Crystalline domain	(96b, 96c)

* See references (90–92) for DP and concentration effects.

Comonomers and plasticizers

Additive/Other Monomer	Function	Conditions	Property	Reference
Maleic, fumaric or Itaconic acid	Copolymer with PVA	PVA from solution polymerization of vinyl acetate with comonomers	Increased water solubility in the range of 50–100% hydrolysis, controlling flocculation/dispersion of clay, compatibilization with starch	(97)
Catioinic acrylamide or methacrylamide	Copolymer with PVA	–	Adsorption to pulp surface, protective colloid for emulsion polymerization, affinity to acidic dyestuff (e.g., in ink-jet printing)	(97, 98)
Acrylonitrile, vinylidene chloride, ethylenimine, acrylic esters, vinyl chloride, alkali cellulose	Graft copolymer	In solution, free radical, or ionic catalysts	Reduced water sensitivity, film, and coating applications	(13)
Methyl methacrylate	Graft copolymer	PMMA wt% 23–72	Lamellar phase separated morphology; T_g of PMMA *increased* by 20 K; T_m of PVA decreased with increasing PMMA wt fraction	(99)
N-Succinimido (N) thiocarbonyl acrylamide; acrylamide	Graft copolymer	Grafting using a potassium bromate-thiourea redox system	Grafting efficiency up to 45% with STAA; up to 80% with acrylamide	(100)

Additive/Other Monomer	Function	Conditions	Property	Reference
Styrene	Graft copolymer	Dispersion polymerization of styrene in the presence of PVA-CuCl$_2$ complex	Narrowing of particle size distribution with increased grafting	(101)
		Living polystyrene with vinylsilane end group grafted to PVAc and subsequently saponified	–	(102)
Iodine	Complexing agent	Partially hydrolyzed PVA	Iodine coloring increases with blockiness of acetyl groups (0.05–0.41, arb. units); gold colloid stability increases in parallel	(90)
		Atactic, syndiotactic	Linear polyiodide intercalaction	(103)
		Partially formalized PVA, DP 560	Polyiodide complexation (SAXS); number of iodine atoms per chain increases from 4.2 to 24.9 in the I$_2$ cone, range 4.0×10^{-4} to 3.1×10^{-3} mol^{-1}	(104)
Glycerol	Plasticizer	0 wt% in PVA	T_m of PVA $= 508$ K	(41), 41a, 41b
		12%	497 K	
		20%	488 K	
		60%	468 K	
Ethylene carbonate Propylene carbonate	Plasticizer	Hybrid polymer electrolyte	Ionic conductive, Thermally and mechanically stable	(41a)
Dipropylene glycol, ethylene glycol	Plasticizers	–	–	(41)
Glyoxal, urea-formaldehydes, trimethylolme-lamine	Cross-linking agents	Acid catalysts	–	(1)
Isobutanol, n-butanol, phenol, Ca(SCN)$_2$, NaSCN, NH$_4$SCN	Viscosity stabilizers for aqueous solutions	–	–	(33)

Poly(vinyl alcohol)

Property	Conditions	Value	Reference
Flammability	–	Burns similar to paper	(2)
Thermal stability	–	Gradual discoloration above 100°C; darkens rapidly above 150°C; rapid decomposition above 200°C	(2)
Half decomposition temperature	Temperature at which the polymer loses half its weight, if heated in a vacuum for 30 min	268°C	(105)
Initial decomposition temperature	–	240°C	(105)
Thermal decompositon products	240°C, 4 h		(21, 106)
	Water	33.4%	
	CO	0.12	
	CO_2	0.18	
	Acetaldehyde	1.17	
	Acetone	0.38	
	Ethanol	0.29	
	Others	–	
	98% hydrolyzed, 400–500°C		(2)
	Water	73.88%	
	Methanol	0.56%	
	Acetone	0.85%	
	Ethanol	1.25%	
	Acetic acid	6.98%	
	Others	–	
	As a function of hydrolysis	Detailed data	(107)
Biodegradation	–	Degradation products: water, CO_2	(2, 108), 108a, 108b
		Varieties of microorganism (at least 55 known) degrade PVA (e.g., Acinetobacter, E. coli, Pseudomonas (19 species), Saccharomyces, Lipomyces etc.)	
		Degradable in activated sludge, soil landfills, septic systems	
Biodegradation		Yeast and Fungi	108a, 108b
Photo-degradation	UV- and laser	Effective at acidic and alkaline pH value. Laser degradation proceeds by oxygen	108a, 108b

References

1. Cincera DL. In: Kroschwitz JI, ed. *Kirk-Othmer Encyclopedia of Chemical Technology.* 3rd ed. Vol 23. New York: John Wiley and Sons; 1978:848.

1a. Brendan H, Sven ÅJB, Martin L, John S, Brendan H, Clive B, *Phys. Med. Biol.* 2002;47:4233–4246.

2. Marten FL. In: Kroschwitz JI, ed. *Kirk-Othmer Encyclopedia of Chemical Technology.* 4th ed. Vol 24. New York: John Wiley and Sons; 1997:980.; Marten, FL. In: Mark HF, et al. *Encyclopedia of Polymer Science and Engineering.* Vol. 17. New York: Wiley-Interscience; 1989:167.

3. Friedlander HN, Harris HE, Pritchard JG. *J. Polym. Sci. Part A-l.* 1966;4:649.

4. Harris HE, et al. *J. Polym. Sci. Part A-1.* 1966;4:665.

5. Fujii K, et al. *J. Polym. Sci. Part A.* 1964;2:2327.

6. Fukae R, et al. *Polym. J.* 1997;29:293.

7. Murahashi S, et al. *J. Polym. Sci.* 1962;62:S77.

8. Fujii K. *J. Polym. Sci. Part D, Macromol. Rev.* 1971;5:431.

9. Ohgi H, Sato T. *Macromolecules.* 1993;26:559.

10. Novak BM, Cederstav AK. *Polym. Preprints.* 1995;36(1):548; Cederstav AK, Novak BM. *J. Am. Chem. Soc.* 1994;116:4073.

11. Flory PJ, Leutner FS. *J. Polym. Sci.* 1948;3:880.

12. Wunderlich B, Cheng SZD, Loufakis K. In: Mark HF, et al. *Encyclopedia of Polymer Science and Engineering.* Vol 16. New York: John Wiley and Sons; 1989:799.

13. Leeds M. In: Kroschwitz JI, ed. *Kirk-Othmer Encylopedia of Chemical Technology.* 2nd ed. Vol 21. New York: Wiley-Interscience; 1963:353.

13a. Abd El-Mohdy HL. *J. Appl. Polym. Sci.* 2007;104:504–513.

13b. Yanfeng L, Xianzhen L, Lincheng Z, Xiaoxia Z. Bonian L. *Polym. Adv. Technol.* 2004;15:34–38.

13c. Francesca C, Fausto M, Paolo D, Gaio P. *Biomacromolecules.* 2004;5:2439-2446.

13d. Chih-Ta L, Po-Han K, Yu-Der L. *Carbohydrate Polymers.* 2005;61:348–354.

14. *Gelvatol Polyvinyl Alcohol Resin.* Monsanto Technical Bulletin; 6082F.

15. *Airvol Polyvinyl Alcohol.* Air Products and Chemicals, Inc. Product Bulletin; 152-9312-A.

16. Peyser P. In: Brandrup J, Immergut EH, eds. *Polymer Handbook.* 3rd ed. New York: John Wiley and Sons; 1989:VI-221.

17. Pan R, Cao MY, Wunderlich B. In: Brandrup J, Immergut EH, eds. *Polymer Handbook.* 3rd ed. New York: John Wiley and Sons; 1989:VI-387.

18. Shvarts AG. *Kolloid Z.* 1956;18:755.

19. Grulke EA. In: Brandrup J, Immergut EH, eds. *Polymer Handbook.* 3rd ed. New York: John Wiley and Sons; 1989:VII-554.

20. Orwoll RA, Arnold PA. In: Mark JE, ed. *Physical Properties of Polymers Handbook.* Woodbury, NY: American Institute of Physics Press; 1996:195.

21. Tubbs RK, Wu TK. In: Finch CA, ed. *Polyvinyl Alcohol: Properties and Applications.* Ch. 8. U.K: Wiley-Interscience; 1973.

22. Perrin L, Nguyen QT, Clement R, *J. Neel. Polym. Intl.* 1996;39:251.

23. Lechner MD, Steinmeier DG. In: Brandrup J, Immergut EH, eds. *Polymer Handbook.* 3rd ed. New York: John Wiley and Sons; 1989:VII-61.

24. Mohamed HFM, Ito Y, El-Sayed AMA, Abdel-Hady EE. *Polymer.* 1996;37:1529.

25. Matsuo T, Inagaki H. *Makromol. Chem.* 1962;55:150.

26. Dieu H. *J. Polym. Sci.* 1954;12:417.

27. Sundararajan PR. In: Mark JE, ed. *Physical Properties of Polymers Handbook.* Woodbury, N.Y: American Institute of Physics Press; 1996:205.

28. Wolfram E. *Kolloid Z*. 1968;227:86.

29. Kurata M, Tsunashima Y. In: Brandrup J, Immergut EH, eds. *Polymer Handbook*. 3rd ed. New York: John Wiley and Sons; 1989:VII-12–36.

30. Nakajima A, Yanagawa H. *J. Phys. Chem*. 1963;67:654.

31. Abe H, Prins W. *J. Polym. Sci. Part C*. 1963;2:527.

32. Sakurada I, Nakajima A, Shibatani K. *Makromol. Chem*. 1965;87:103.

33. Toyoshima K. In: Finch CA, ed. *Polyvinyl Alcohol: Properties and Applications*. Ch. 2. U.K: Wiley-Interscience; 1973.

34. Fuchs O. In: Brandrup J, Immergut EH, eds. *Polymer Handbook*. 3rd ed. New York: John Wiley and Sons; 1989:VII–384.

35. Nagashima N, Matsuzawa S, Okazaki M. *J. Appl. Polym. Sci*. 1996;62:1551.

36. Saito S. *J. Polym. Sci. Part A-1*. 1969;7:1789.

37. Frosini V, Butta E, Calamia M. *J. Appl. Polym. Sci*. 1967;11:527.

38. Yamamura K, Kitahara H, Tanigami T. *J. Appl. Polym. Sci*. 1997;64:1283.

38a. Cascone MG, Barbani N, Cristallini C, Giusti P, Ciardelli G. Lazzeri L. *J. Biomater. Sci. Polymer Edn*. 2001;12(3):267–281.

39. Urayama K, Takigawa T, Masuda T. *Macromolecules*. 1993;26:3092.

40. Hammerschmidt JA, et al. *Macromolecules*. 1996;29:8996.

41. Toyoshima K. In: Finch CA, ed. *Polyvinyl Alcohol: Properties and Applications*. Ch. 14. U.K: Wiley-Interscience; 1973.

41a. Chin-An L, Hsiao-Chi T, Te-Hsing K. *Polym-Plas. Tech. and Engi*. 2007;46:689–693.

41b. Jang J, Lee DK. *Polymer*. 2003;44:8139–8146.

42. Yamaura K, Hirata K, Tamura S, Matsuzawa S. *J. Polym. Sci. Polym. Phys. Ed*. 1985;23:1703.

43. Matsumoto M, Imai K. *J. Polym. Sci*. 1957;24:125.

44. Beresniewicz A. *J. Polym. Sci*. 1959;39:63.

45. Tadokoro H. *Structure of Crystalline Polymers*. New York: John Wiley and Sons, 1979.

46. Dunn AS. In: Finch CA, ed. *Polyvinyl Alcohol: Developments*. Ch. 10. U.K: John Wiley and Sons; 1992.

47. Kenney JF, Willcockson GW. *J. Polym. Sci. Part A-1*. 1966;4:679.

48. Nakanishi Y. *J. Polym. Sci. Polym. Chem. Ed*. 1975;13:1223.

49. Peppas NA. *Makromol. Chem*. 1977;178:595.

50. Murahashi S, et al. *J. Polym. Sci. Polym. Letters*. 1966;4:65.

51. Finch CA, ed. *Polyvinyl Alcohol: Properties and Applications*. Ch. 10. U.K: Wiley-Interscience; 1973.

52. Morotani T, Kuruma I, Shibatani K, Fujiwara Y. *Macromolecules*. 1972;5:577.

53. DeMember JR, Haas HC, MacDonald RL. *J. Polym. Sci. Polym. Letters Ed*. 1972;10:385.

54. Wu TK, Overall DW. *Macromolecules*. 1973;6:582.

55. Wu TK, Sheer ML. *Macromolecules*. 1977;10:529.

56. Overall DW. *Macromolecules*. 1984;17:1458.

56a. Hiroyuki O, Hu Y, Toshiaki S, Fumitaka H. *Polymer*. 2007;48:3850–3857.

57. Gippert GP, Brown LR. *Polym. Bulletin*. 1984;11:585.

58. Hikichi K, Yasuda M. *Polym. J*. 1987;19:1003.

59. Petit J-M, Zhu XX. *Macromolecules*. 1996;29:2075.

60. Hu S, Tsuji M, Horii F. *Polymer*. 1994;35:2516.

61. Kobayashi M, Ando I, Ishii T, Amiya S. *Macromolecules*. 1995;28:6677.

62. Wunderlich B. *Macromolecular Physics*. Vol 1. Chs. 2, 4. New York: Academic Press; 1973.

63. Takahashi Y. *J. Polym. Sci. Polym. Phys. Ed*. 1997;35:193.

64. Tubbs RK. *J. Polym. Sci. Part A*. 1965;3:4181.

65. Kenney JF, Holland VF. *J. Polym. Sci. Part A-1*. 1966;4:699.

66. Tsuboi K, Mochizuki T. *J. Polym. Sci. Polym. Letters Ed*. 1963;1:531.

67. Peppas NA, Merrill EW. *J. Polym. Sci. Polym. Chem. Ed*. 1976;14:441.; *J. Appl. Polym. Sci*. 1976;20:1457.; *Eur. Polym. J*. 1976;12:495.

68. Peppas NA, Hansen PJ. *J. Appl. Polym. Sci*. 1982;27:4787.

69. Hong Po-Da, Miyasaka K. *Polymer*. 1994;35:1369.

70. Allegra G, Meille SV, Porzio W. In: Brandrup J, Immergut EH, eds. *Polymer Handbook*. 3rd ed. New York: John Wiley and Sons, 1989:VI–341.

71. Nakamae K, Kameyama M, Matsumoto T. *Polym. Eng. Sci*. 1979;19:572.

72. Okaya T, Ikari K. In: Finch CA, ed. *Polyvinyl Alcohol: Developments*. Ch. 8. U.K: John Wiley and Sons; 1992.

73. Okaya T, T. Sato. In: Finch CA, ed. *Polyvinyl Alcohol: Developments*. Ch. 5. U.K: John Wiley and Sons; 1992.

73a. Rayna B, Nicolas W, Antoine D, Robert J, Christophe D. *J. of Poly. Sci.: Part A: Polymer Chemistry*. 2007;45:81–89.

73a. Xiao L, Xiao W, liu Q, Chen K, Xu X. *J. of Appl. Polym. Sci*. 2001;79:979–988.

73c. Emo C, Andrea C, Salvatore D'A, Roberto S. *Prog. Polym. Sci*. 2003;28:963–1014.

73d. Guang-Hua L, Chang-Gi C. *Colloid. Polym. Sci*. 2005;283:946–953.

74. Toyoshima K. In: Finch CA, ed. *Polyvinyl Alcohol: Properties and Applications*. U.K: Wiley-Interscience; 1973:535.

75. Ping Z-H, Nguyen QT, Neel J. *Makromol. Chem*. 1988;189:437.

75a. Jun-FS Zhen, Kai H, Ling-LF L, Hong-RL. *Polymer Bulletin*. 2007;58:913–921.

76. Feng H, Feng Z, Shen L. *Polymer*. 1993;34:2516.

76a. Emo C, Patrizia C, Federica C, Syed HI. *Macromol. Biosci*. 2004;4:218–231.

77. Wang H-L, Fernandez JE. *Macromolecules*. 1993;26:3336.

77a. Zhi-Xuan Z, Xiu-Li W, Yu-Zhong W, Ke-Ke Y, Si-Chong C, Gang Kl, Jun Li, *Polym. Int*. 2006;55:383–390.

77b. Siriporn T, Somruethai C, Paweena U. *J. Appl. Polym. Sci*. 2006;100:1914–1918.

78. Schartel B, Wendling J, Wendorff JH. *Macromolecules*. 1996;29:1521.

78a. Jang J, Lee DK. *Polymer*. 2004;45:1599–1607.

79. Radloff D, Boeffel C, Spiess HW. *Macromolecules*. 1996;29:1528.

80. Ikejima T, Yoshie N, Inoue Y. *Macromol. Chem. Phys*. 1996;197:869.

81. Simmons S, Thomas EL. *J. Appl. Polym. Sci*. 1995;58:2259.

81a. Nicholas W DeLuca, Yossef AE. *J. Memb. Sci*. 2006;282:217–224.

82. Ha C-S, Ko M-G, Cho W-J. *Polymer*. 1997;38:1243.

83. Zheng S, et al. *Eur. Polym. J*. 1996;32:757.

83a. Sivakumar M, Subadev R, Rajendran S, Wu N-L, Lee J-Y. *Materials Chemistry and Physics*. 2006;97:330–336.

83b. Sari A, Kaygusuz K. *Energy Sources, Part A: Recovery, Utilization, and Environmental Effects*. 29:2007;10:873–883.

83c. Sarı A, KaygusuzK.*Energy Sources, Part A*. 2007;29:873–883.

84. Nishinari K, Watase M. *Polym. J*. 1993;25:463.

85. Nagura M, et al. *Polym J*. 1993;25:833.

86. Nagura M, et al. *Polym. J*. 1994;26:675.

87. Sawatari C, Yamamoto Y, Yanagida N, Matsuo M. *Polymer*. 1993;34:956.

88. Paradossi G, Lisi R, Paci M, Crescenzi V. *J. Polym. Sci. Polym. Chem. Ed*. 1996;34:3417.

89. Cha W-L, Hyon S-H, Ikada Y. *Makromol. Chem*. 1993;194:2433.

90. Okaya T. In: Finch CA, ed. *Polyvinyl Alcohol: Developments*. Ch. 1. U.K: John Wiley and Sons; 1992.

91. Koike A, Nemoto N, Inoue T, Osaki K. *Macromolecules*. 1995;28:2339.

92. Nemoto N, Koike A, Osaki K. *Macromolecules*. 1996;29:1445.

93. Shibayama M, Ikkai F, Moriwaki R, Nomura S. *Macromolecules*. 1994;27:1738.

94. Shibayama M, Ikkai F, Nomura S. *Macromolecules*. 1994;27:6383.

95. Ikkai F, Shibayama M, Nomura S, Han CC. *J. Polym. Sci. Polym. Phys. Ed.* 1996;34:939.

96. Halboth H, Rehage G. *Makromol. Chem.* 1974;38:111.

96a. Miroslawa ElF, Agnieszka P, Wojciech S, Krzysztof JK. *European Polymer Journal*. 2007;43:2035–2040.

96b. Caio MP, Renata NO, Bluma GS, Luiz AP. *Materials Research*. 2007;10(1):43–46.

96c. Caio MP, Bluma GS, Renata NO, Luiz AP, Daniela S. de F, Dario WI, Jose CM. *J. Appl. Polym. Sci.* 2007;105:899–902.

97. Okaya T. In: Finch CA, ed. *Polyvinyl Alcohol: Developments*. Ch. 4. U.K: John Wiley and Sons; 1992.

98. Sato T, Terada K, Yamauchi J, Okaya T. *Makromol. Chem.* 1993;194:175.

99. Yao Y, et al. *Polymer*. 1994;35:3122.

100. Devarajan R, Arunachalam V, Jayakumar E, Selvi P. *J. Appl. Polym. Sci.* 1993;48:921.

101. Nigam S, Bandopadhyay R, Joshi A, Kumar A. *Polymer*. 1993;34:4213.

102. Tezuka Y, Araki A. *Makromol. Chem.* 1993;194:2827.

103. Choi Y-S, Miyasaka K. *J. Appl. Polym. Sci.* 1993;48:313.

104. Hirai M, Hirai T, Ueki T. *Makromol. Chem.* 1993;194:2885.

105. Welsh WJ. In: Mark JE, ed. *Physical Properties of Polymers Handbook*. Woodbury, N.Y: American Institute of Physics Press; 1996:605.

106. Tsuchiya Y, Sumi K. *J. Polym. Sci. Part A-1*. 1969;7:3151.

107. Vasile C, Odochian L, Patachia SF, Popoutanu M. *J. Polym. Sci. Polym. Phys. Ed.* 1985;23:2579.

108. Finch CA, ed. *Polyvinyl Alcohol: Developments*. U.K: John Wiley and Sons; 1992:767.

108a. Jian C, Ying Z, Guo-Cheng D, Zhao-Zhe H, Yang Z. *Enzyme and Microbial Technology*. 2007;40:1686–1691.

108b. Emo C, Andrea C, Salvatore D, Roberto S. *Prog. Polym. Sci.* 2003;28:963–1014.

108c. Vijayalakshmi SP, Madras G. *J. Appl. Polym. Sci.* 2006;102:958–966.

Poly(vinyl butyral)

P. R. Sundararajan and Mostofa Kamal Khan

Acronyms, Trade Names PVB, Butvar® (Monsanto), Butacite® (Du Pont), Vinylite XYHL® (Union carbide), Rhovinal® B (Rhone-Poulenc), Movital® (Hoechst), S'Lec® (Shekisui), Saflex® (Monsanto), Trosofoil (Hüls).

Class Polyvinyl

Structure

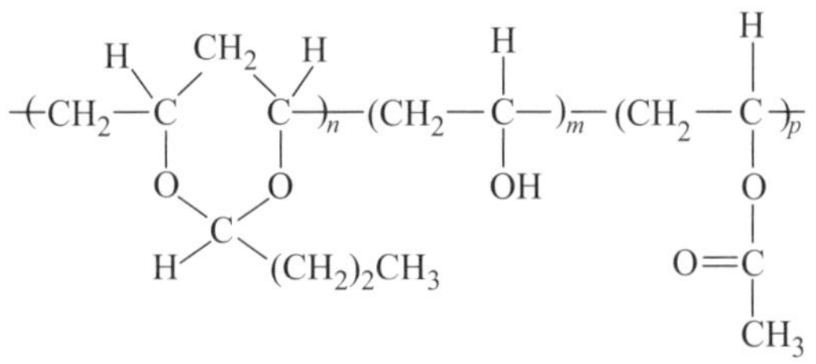

This schematic should not be construed as a block structure.

Major Applications The significant use is in lamination of safety glass (automotive windshields). Others are structural adhesives, binders for rocket propellants, ceramics, in metallized brake linings, lithographic and offset printing plates, magnetic tapes, powder coatings; binder matrix in photoactive, elecrooptic and electronic devices, protective coatings for glass, metal, wood, and ceramics; in wash primers for protecting metal surfaces (e.g., naval vessels); adhesion promoter in inks; dispersions used in textile industry to improve abrasion resistance and reduce color crocking.[1-3]

Properties of Special Interest Resistance to penetration by natural wood oils, film clarity, heat sealability, adhesion to a variety of surfaces, chemical and solvent resistance, physical toughness.

General Information Poly(vinyl butyral) (PVB) is a member of the class of poly(vinyl acetal) resins. It is derived by condensing poly(vinyl alcohol) (PVA) with butyraldehyde in the presence of a strong acid. PVA reacts with the aldehyde, to form six-membered rings primarily between adjacent, intramolecular hydroxyl groups, leading to the structure shown above.

An example of the compositions of a commercial resin (Butvar) is as follows:[1,4]

Poly(vinyl butyral)

Resin Type	Molecular Weight $(M_w) \times 10^{-3}$	Vinyl Alcohol Content (wt%)	Vinyl Acetate Content (wt%)
Butvar B-72	170–250	17–20	0–2.5
Butvar B-76	90–120	11–13	0–1.5

PVB is plasticized for specific applications. Saflex contains 32 phr (parts per hundred resin) of di-*n*-hexyl adipate. Butacite is PVB plasticized with 38.5 phr tetraethylene glycol di-*n*-heptanoate.

Commercial Production   Worldwide unplasticized PVB production was 68,000 tons in 1994. Of this, 66,000 tons were plasticized and extruded for safety glass application. Major interlayer lamination producers are Monsanto (Saflex), Du Pont (Butacite), Shekisui (S'Lec), and Hüls (Trosofoil).[1,5]

Synthetic aspects

Polymer	Synthetic Conditions	Method of Characterization	Characteristics	Reference
Poly(vinyl butyral)	From PVA, 99% hydrolyzed; condensation of butyraldehyde with PVA in ethanol; H_2SO_4 as catalyst; 53–100 g aldehyde to 100 g PVA; 5–7 h at 75–77°C	–	$M = 70,000$; residual PVA decreasing from 25.4–12%, with increasing aldehyde addition	(6)
Poly(vinyl formal)*	From PVA, DP 1000; *i*: 56%, *h*: 32%, *s*: 12%; formalization in 0.1 N HCl aqueous solution at 60°C	^{1}H NMR	Formalization: 84 mol% *Cis*[†] ring: 70% *Trans*[†] ring: 14% Rate constant: 8.8×10^{-2} L mol^{-1} h^{-1}	(7)
	From PVA, DP 1000; *i*: 23%, *h*: 47%, *s*: 30%; formalization in 0.1 N HCl aqueous solution at 60°C	^{1}H NMR	Formalization: 87 mol% *Cis*[†] ring: 59% *Trans*[†] ring: 28% Rate constant: 6.6×10^{-2} L mol^{-1} h^{-1}	
	From PVA, DP 1000; *i*: 17%, *h*: 46%, *s*: 37%; formalization in 0.1 N HCl aqueous solution at 60°C	^{1}H NMR	Formalization: 87 mol% *Cis*[†] ring: 50% *Trans*[†] ring: 37% Rate constant: 6.2×10^{-2} L mol^{-1} h^{-1}	
Poly(vinyl butyral)	Reaction at 10 and 70°C; up to 1,000 h		At 10°C, *cis/trans* ratio is ~5, no significant change with time; at 70°C, ratio decreases from ~5 to ~3 up to 1 h, then increases with time to ~7 after 100 h	(8)
Poly(vinyl butyral)	PVA DP: 1600; 97.5–99.5% hydrolyzed; reaction at 10–60°C, in water, H_2SO_4 or HCl as catalyst (method I); catalyst and aldehyde added to PVA suspension in MEK (method II); aldehyde added in one step or in stages	NMR	Degree of acetalization: 75% (10°C, 3.5 h, method I) 45% (temperature ramp from 18–60°C, 2 h, method I) 85% (30°C, 2 h, method II)	(9)

Polymer	Synthetic Conditions	Method of Characterization	Characteristics	Reference
Poly(vinyl butyral)	From PVA, 98.8 mol % hydrolyzed; condensation of butyraldehyde with PVA in NMP; HCl as catalyst; $2[BA]_o/[OH]_o = 2.6$; $[BA]_o/[HCl]_o = 3$; 3 h at 40°C	FTIR, ^{1}H NMR, ^{13}C NMR	Degree of acetalization 97%	(9a)
Poly(vinyl butyral)	From PVA: M_v 14,000 g/mol; condensation of 7.92 g of butyraldehyde with 4.4 g of PVA; reaction at 90°C, in DMF and benzene (4:1 v/v) mixture, ethyl nitrate dimethyl sulfoxide ($C_2H_5ONO_2$.DMSO) as catalyst, $\sim$24 h;	FTIR, ^{1}H NMR	Degree of acetalization 95 mol% $M_w = 55,000$ (GPC)	(9b)
PVB-g-PVB		SEC, MALS	$M_w = 151,000 \pm 4000$ g/mol $M_n = 348,000 \pm 8000$ g/mol PD $= 3.01 \pm 0.13$	(9c)

Property	Units	Conditions	Value	References
Tensile strength	MPa	Films from Butvar BR aqueous dispersion		(3)
		No plasticizer	41–48	
		40–50 phr plasticizer	14	
		18% hydroxyl content; $M_w \approx 70,000$; plasticized with dibutyl phthalate		(6)[†]
		0 phr DBP	56	
		15 phr DBP	36	
		37.2 phr DBP	38	
Elastic modulus	MPa $\times 10^3$	Butvar		(1, 2)
		B-72	2.28–2.34	
		B-76	1.93–2.0	
Storage modulus	MPa	PVB with 32 phr di-n-hexyladipate (Monsanto Saflex); 1 Hz	3.98×10^2	(15)
		Above, neat resin	1.86×10^3	
Young's Modulus	MPa	Films from PVB dispersed in DMA		(15a)
		Thickness 191 μm, No HCl treatment	52.57	
		Thickness 188 μm, 1.0 M HCl treatment for 1 hr	59.23	
Elongation at break	%	Films from PVB dispersed in DMA		(15a)
		Thickness 191 μm, No HCl treatment	46.56	
		Thickness 188 μm, 1.0 M HCl treatment for 1 hr	59.79	
Elongation at break	%	Butvar		(1)
		B-72	70	
		B-76	110	
		Butacite	>200	(12)

Property	Units	Conditions	Value	References
Elongation at yield	%	Butvar		(1)
		B-72	8	
		B-76	8	
		18% hydroxyl content; $M_w \approx 70,000$; plasticized with dibutyl phthalate		(6)[†]
		0 phr DBP	9	
		15 phr DBP	10	
		37.2 phr DBP	380	
Flexural strength	MPa	Butvar		(1, 2)
		B-72	83–90	
		B-76	72–79	
Impact strength	J m^{-1}	Izod, notched, 1.25 × 1.25 cm, Butvar		(1, 2)
		B-72	58.7	
		B-76	42.7	
Glass transition temperature	K	Butvar		(1, 2)
		B-72	345–351	
		B-76	335–345	
		Triethyleneglycol-di-(2-ethylbutyrate) plasticizer		(14)
		0%	332.3	
		28%	272	
Glass transition temperature	K	Butvar with di-n-hexyl adipate		(16)
		0%	353	
		10 phr	333	
		32 phr (from mechanical loss spectroscopy at 1 Hz)	302	
		18–20% (wt) hydroxyl content	346	(11)
		As above, 5% ionomer	358	
		As above, 15% ionomer	379	
Glass transition temperature	K	Vinyl alcohol content		(11a)
		14%	344	
		45%	353	
		89%	359	
Softening temperature	K	With 25% residual PVA	338	(6)[†]
		As above, with 30 phr dibutyl phthalate	303	
		As above, with 30 phr dibutyl phthalate	291	
Specific heat	J (g K)$^{-1}$	Triethyleneglycol-di-(2-ethylbutyrate) plasticizer		(14)
		0%	1.36	
		28%	1.90	
Thermal conductivity	W (m K)$^{-1}$	Butvar with di-n-hexyl adipate		(16)
		0%	0.236	
		10 phr	0.275	
		32 phr	0.272	
Heat sealing temperature	K	Butvar		(1)
		B-72	493	
		B-76	473	

Property	Units	Conditions	Value	References
Dielectric constant	–	Butvar		(2)
		B-72 50 Hz	3.2	
		B-72 10 MHz	2.7	
		B-76 50 Hz	2.7	
		B-76 10 MHz	2.5	
Dissipation factor	–	Butvar		(2)
		B-72 50 Hz	6.4×10^{-3}	
		B-72 10 MHz	31×10^{-3}	
		B-76 50 Hz	5.0×10^{-3}	
		B-76 10 MHz	15×10^{-3}	
Relaxation temperatures	K	DMA and DSC analysis; PVB with 32 phr di-n-hexyladipate (Monsanto Saflex)	$226 (\beta)$; $285 (\alpha_2)$; $304 (\alpha_1)$	(17)
Surface tension	mN m^{-1}	–	38	(18)
Surface tension	mN m^{-1}	Saflex RB41 Di-2-ethylhexanoate of triethylene glycol as Plasticizer; Pendant drop method		(18a)
		Temperature 240°C	36.4 ± 0.2	
		Temperature 260°C	23.4 ± 0.2	
Critical surface tension	mN m^{-1}	PVB with 12 or 30% hydroxyl content, with polyhydric alcohols	24–25	(5, 19)
Partial specific volume	cm^3 g^{-1}	Amyl alcohol, 20°C; $[\eta] = 122$ cm^3 g^{-1}	0.883	(20)
Second virial coefficient	mol cm^3 g^{-2}	Dioxane, 37°C; $M_w =$		(20)
		57.5–181×10^3	9.45–12.2×10^{-4}	
		73.9–208×10^3	9.15–11.3×10^{-4}	
		89.5–541×10^3	5.53–10.3×10^{-4}	
		$M_w = 68,500$, 25°C		(21, 22)
		Acetic acid	10.4×10^{-4}	
		3:1 MIBK/MeOH	7.9×10^{-4}	
		1:1 MIBK/MeOH	3.6×10^{-4}	
		9:1 MIBK/MeOH	1.6×10^{-4}	
Second virial coefficient	mol cm^3 g^{-2}	PVB, $M_w = 117,000$ g/mol, PD = 1.73; in DMAc/0.5% LiCl	1.396×10^{-3}	(9c)
		PVB-g-PVB, $M_w = 151,000$ g/mol PD = 3.01; in DMAc/0.5% LiCl	1.255×10^{-3}	
Solubility parameter	–	Theoretical estimate of dispersion (δ_d), polar (δ_p) and hydrogen bonding (δ_h) contributions to solubility parameter:		(23)
		δ_d	7.72	
		δ_p	2.90	
		δ_h	3.26	
		δ_{intal}	8.87	

Poly(vinyl butyral)

Property	Units	Conditions	Value	References
		Low hydrogen bonding solvents:		(1)
		Hydroxyl content 9–13%	9.0–9.8	
		Hydroxyl content 17–21%	Insoluble	
		Medium hydrogen bonding solvents:		(1)
		Hydroxyl content 9–13%	8.4–12.9	
		Hydroxyl content 17-21%	9.9–12.9	
		High hydrogen bonding solvents:		(1)
		Hydroxyl content 9–13%	9.7–12.9	
		Hydroxyl content 17–21%	9.7–14.3	

* phr = parts per hundred resin.
† Reference (6) discusses the effect of plasticizers on various acetals.

Solvents and nonsolvents

Conditions	Solvents	Nonsolvents	Partially Soluble In	Reference
Butvar B-72	Acetic acid (glacial), butanol, cyclohexane, dioxane, ethyl Cellosolve, ethylene chloride, methanol, toluene/ethanol (60:40 wt), xylene/butanol (60:40 wt)	Acetone, butyl acetate, carbon tetrachloride, diisobutyl ketone, hexane, methyl ethyl ketone, methyl isobutyl ketone, nitropropane, toluene, xylene	Diacetone alcohol, isophorone, methylene chloride	(1, 3)
Butvar B-76	Acetic acid (glacial), acetone, butanol, butyl acetate, cyclohexane, dioxane, ethyl Cellosolve, ethylene chloride, methyl acetate, methyl ethyl ketone, methyl isobutyl ketone, toluene	Carbon tetrachloride, hexane, methanol, nitropropane, xylene	–	(1, 3)
PVB, 20% hydroxyl content	Acetic acid	Methanol	–	(21, 22)*
70% acetylation	Alcohols, cyclohexane, ethyl lactate, ethyl glycol acetate	Hydrocarbons, methylene chloride, aliphatic ketones	–	(24)
83% acetylation	Methylene chloride, alcohols, ketones, lower esters	Hydrocarbons, methanol, higher esters	–	(24)
PVB				(9a)
30–35% acetylation	NMP, DMSO	THF	–	
40–75% acetylation	NMP, DMSO, THF	–	–	
80% or higher acetylation	NMP, THF, CHCl$_3$	DMSO	–	

* The effect of temperature and solvent on solubility and aggregation is discussed in references (17) and (18).

Mark-Houwink parameters: K and $a^{(25)}$

Solvent	Temp. (°C)	$M \times 10^{-4}$	$K \times 10^4$ (ml g^{-1})	a
Tetrahydrafuran	25	5.8–17 (20% hydroxyl content)	2.89	0.72
Tetrahydrafuran	25	12 (10% hydroxyl content)	2.52	0.72

Spectroscopy

Spectroscopy	Frequency (cm^{-1})	Intensity	Assignment	Observations	Reference
Infrared	1,383, 1,136, 1,111, 1,052, 1,000, 971	–	1,136 and 1,000 cm^{-1} bands to cyclic acetal	–	(5)
	3,448	Weak	Vinyl alcohol	–	(5)
	–	–	–	–	(26)

Other Polymer	Conditions	Techniques Used	Features	Reference
Polyurethane	M_w of PVB: 170,000; 18.5% hydroxyl content. PU: Tecoflex EG-85A (from methylene bis cyclohexyl diurethane and polytetramethylene ether glycol). Extrusion blended	DMA, DSC, TEM	Miscible over entire composition range, due to interaction of PVB with hard segment. Single T_g, decreasing with PU content. Model compounds of the hard segment are also miscible	(23)
Polyaniline	Solution or melt processing	–	Self-assembled network morphology, onset of electrical conductivity with 1% (vol) of PANI	(40)
Ionomeric PVB: poly(vinyl butyral-co-vinyl benzal sodium (or potassium) sulfonate)	Blends up to 50% IPVB; 3% or 5% ion content	NMR, DMA	Storage modulus G' (N m^{-2}) at 25°C increases from 4.2×10^7 to 8.6×10^7 (0–50% IPVB) with 3% ion content	(41)
Polyacetylene	Synthesized in dilute solutions of PVB	–	40–50% *trans* form of polyacetylene; low defec content	(42)
Polysiloxanes	Hydrolysis of tetraethyl orthosilicate in presence of PVB, Ethanol or 1,2-dimethoxyethane as Solvent, Boiling for 3 h, HCl or H$_2$SO4 as catalyst	GLC, FTIR, SEM	–	(42a)

* I_{PTC}: Intensity of positive temperature coefficient, defined as the resistivity ratio ρ_{max}/ρ_{min}; ρ_{max} is the maximum in the temperature-resistivity curve and ρ_{min} is the resistivity at room temperature.

Resins Compatibility with PVB

PVB Type	Compatible	Partially Compatible	Incompatible	Reference
Butvar B-2	Nitrocellulose, epoxy (Epi-Rez 540-C, Araldite 6069), isocyanate, phenolic, shellac	Alkyd, cellulose acetate butyrate, ethyl cellulose, rosin derivatives, silicone, urea formaldehyde	Acrylate, cellulose acetate, chlorinated rubber	(4)*
Butvar B-76	Nitrocellulose, epoxy, phenolic, shellac, silicone	Alkyd, cellulose acetate butyrate, urea formaldehyde, vinyl chloride copolymer	Acrylate, cellulose acetate, chlorinated rubber	(4)*
Shekisui BM-2	Poly (vinylpyrrolidone), poly (vinyl acetate-*co*-N-vinylpyrrolidone), poly (styrene-*co*-maleic acid), poly (styrene-*co*-maleic acid ester) (conditional)	–	Poly(vinylidene chloride), chlorosulfonated polyethylene, polyester, poly(ethylene-*co*-vinyl acetate), poly(butadiene-*co*-styrene), poly-(butadiene-*co*-acrylonitrile), poly(vinyl chloride-*co*-vinyl acetate), polyvinyl chloride-*co*-vinyl propionate)	(43)

* Consult reference (4) for the trade names of the resins applicable to the entries in this table.

PVB as binder polymer in optoelectronic/photoactive devices

Device	Guest Molecule	Application	Reference
Xerographic photoreceptor	Squaraine	Dispersion of squaraine in the charge generating layer	(44)
Electrode/ electrolyte tape	Nickel powder	Molten carbonate fuel cells	(45)
Ceramics	Al_2O_3 powder	Ceramic processing aid	(46)
Ceramics	$BaTiO_3$ powder	Controlling the properties like elongation, tensile strength, porosity, pore size and distribution, and density of the sintered green tapes by proper selection of the molecular weight of the PVB binder.	(46a)
Ceramics	$BaTiO_3$, Li_2O-B_2O_3-BaO-SiO_2	PVB used as binder to these ceramic and glass particles; 2-methyl-2,4-pentanediol used as PVB modifier to protect flocculation of PVB which leads to the formation of gel.	(46b)
Holograms	Cresyl violet	Hologram recording by spectral hole burning	(47)
Holograms	Chlorin (2,3-dihydroporphyrin)	Holographic recording/storage media	(48)

Device	Guest Molecule	Application	Reference
Optical memory	Anthraquinone derivatives	Spectral hole burning	(49)
Optical memory	Perylene	Spectral hole burning	(50)
Optical memory	Chlorin (2,3-dihydroporphyrin)	Holographic recording	(51)
Optical memory	Chlorophyll A	–	(52)
Optical memory	Porphyrin and phthalocyanine derivatives	Spectral hole burning	(53)
Optical memory	–	Holographic recording	(54)
Electrochromic device	LiCl in PVB gel	–	(55)
Fuel cell	–	Interconnect plate for a planar solid oxide fuel cell	(56)
Photochromism	Spirooxazines	–	(57)
Radiation dosimetry/ monitoring	Thymol blue indicator in presence of chloral hydrate	TB/PVB films change their color on exposure to UV-radiation; the dose is dependent on the concentration of chloral hydrate and the irradiation wavelength; sensitivity increases exponentially with the decrease in irradiation wavelength and reaches maximum at 201.2 nm.	(57a), (57b)
Radiation dosimetry/ monitoring	Eosin dye with/without chloral hydrate	Eosin dyed PVB film dosimeters are bleached on exposure to γ-ray photons; the useful dose ranges from 120–250 kGy depending on the eosin and chloral hydrate concentration in the films; dose response is humidity dependent.	(57c)
Organic vapour sensors	NiO-TiO$_2$	Screen-printed thick-films containing NiO-TiO$_2$ dispersed in PVB can function as sensitive alcohol vapor sensors at room temperature.	(57d)
Alcohol vapour sensors	Carbon black	Alcohol vapour sensing in the concentration range of 1000 to 5000 ppm	(57e)
Support film in TEM	Dibutyl phthalate	Support film with high transparency and flexibility and more stability in electron beam.	(57f)
Membrane	Polyethylene glycol with different molecular weight	Membranes with different water permeability, solute rejection and mechanical strength.	(57g)
Membrane	Poly (ether sulfone)	Ultrafiltration membrarnes; performance depends on the PVB concentration, scraper clearance, water temperature and the nature of the casting solvent.	(57h)

Thermal Degradation

Temp. (K)	Method	Machanism	Products/Percentage		Reference
553–583	TGA, FTIR	Oxidation of copolymer	7%		(58)
583–673	–	PVB thermal oxidation	Butanal, C_4 hydrocarbons, CO_2 and water (71%)		(58)
673–733	–	Oxidation of cyclic and cross-linked compounds	9%		(58)
733–823	–	Oxidation of residual carbon	13%		(58)
773	Pyr-GC/mass spectrometry	–	CO, CO_2	10.2%	(59)*
			Acetaldehyde	4.5	
			Acetone	1.0	
			Butanal	60.8	
			H_2O	1.1	
			Benzene	1.2	
			Alkyl aromatics	0.1	
			Butenal	9.3	
			Acetic acid	2.9	

* See reference (59) for the effects of silica, mullite, α-alumina and γ-alumina on the thermal degradation of PVB.

References

1. Knapczyk JW. In: Kruschwitz JI, ed. *Kirk-Othmer Encyclopedia of Chemical Technology.* 4th ed. Vol 24. New York: John Wiley and Sons; 1997:924.

2. Blomstrom TP. In: Mark HF, et al, eds. *Encyclopedia of Polymer Science and Engineering.* Vol. 17, New York: John Wiley and Sons, 1989:136.

3. Lavin E, Snelgrove JA. In: Kruschwitz JI, ed. *Kirk-Othmer Encyclopedia of Chemical Technology.* 3rd ed. Vol 23. New York: John Wiley and Sons; 1983:798.

4. *Butvar Polyvinyl Butyral Resin.* Monsanto Technical Bulletin 8084A, 1991.

5. Lindemann MK. In: Mark HF, et al, eds. *Encyclopedia of Polymer Science and Technology.* Vol 14. New York: John Wiley and Sons; 1971:208.

6. Fitzhugh AF, Crozier RN. *J. Polym. Sci.* 8 1952: 225 (errata in *J. Polym. Sci.* 1952;9:96).

7. Shibatani K, et al. *J. Polym. Sci., Part C.* 1968;23:647.

8. Asahina K. In: Finch CA, ed. *Polyvinyl Alcohol: Developments.* U.K: John Wiley and Sons; 1992, Ch. 19.

9. Toncheva VD, Ivanova SD, Velichkova RS. *Eur. Polym. J.* 1992;28:191.

9a. Ferna'ndez MD, Ferna'ndez MJ, Hoces P. *J. App. Polym. Sci.* 2006;102:5007.

9b. ChetrI P, Dass NN. *J. App. Polym. Sci.* 2001;81:1182.

9c. Striegel AM. *Polym. Int.* 2004;53:1806.

10. Berger PA, Garbow JR, DasGupta AM, Remsen EE. *Macromolecules* 1997;30:5178.

11. Dasgupta AM, David DJ, Misra A. *J. Appl. Polym. Sci.* 1992;44:1213.

11a. Cascone E, David DJ, Di Lorenzo ML, Karasz FE, Macknight WJ, Martuscelli E, David DJ, Di Lorenzo ML, Karasz FE, Macknight WJ, Martuscelli E, and Raimo M. *J. App. Polym. Sci.* 2001;82:2934.

12. *Butacite Polyvinyl Butyral Resin Sheeting.* DuPont Technical Bulletin.

13. Seferis JC. In: Brandrup J, Immergut E, ed. *Polymer Handbook.* 3rd ed. New York: John Wiley and Sons; 1989:VI-451.

14. Wilski H. *Angew. Makromol. Chem.* 1969;6:101.

15. Parker AA, et al. *J. Appl. Polym. Sci.* 1990;40:1717.

15a. Ma X, Sun Q, Su Y, Wang Y, Jiang Z. *Sep. Purif. Technol.* 2007;54:220.

16. Schaefer J, Garbow JR, Stejskal EO, Lefelar JA. *Macromolecules* 1987;20:1271.

17. Parker AA, Hedrick DP, Ritchey WM. *J. Appl. Polym. Sci.* 1992;46:295.

18. Wu S. In: Brandrup J, Immergut E, eds. *Polymer Handbook.* 3rd ed. New York: John Wiley and Sons; 1989:VI-411.

18a. Morais D, Valera TS, Demarquette NR. *Macromol. Symp.* 2006;245–246:208.

19. Newman S. *J. Colloid Interface Sci.* 1967;25:341.

19a. Cho CW, Yeo JG, Jung Y-G, Paik U. *J. Mater. Sci. Lett.* 2003;22:1639.

19b. *Butvar Poly Vinyl Butyral Resin, Properties & Uses.* Solutia Inc. Pub No. 2008284E.

20. Lechner MD, Steinmeier DG. In: Brandrup J, Immergut E, eds. *Polymer Handbook.* 3rd ed. New York: John Wiley and Sons; 1989:VII–61.

21. Paul CW, Cotts PM. *Macromolecules.* 1986;19:692.

22. Paul CW, Cotts PM. *Macromolecules.* 1987;20:1986.

23. Sincock TF, David DJ. *Polymer.* 1992;33:4515.

24. Fuchs O. In: Brandrup J, Immergut E, eds. *Polymer Handbook.* 3rd ed. New York: John Wiley and Sons; 1989:VII-379.

25. Cotts PM, Ouano AC. In: Dubin P, ed. *Microdomains in Polymer Solutions.* New York: Plenum Press, 1985:101.

26. Stolov AA, Kamalova DI, Remizov AB, Zgadzal OE. *Polymer.* 1994;35:2591.

27. Fujii K, et al. *J. Polym. Sci., Polym. Letters Ed.* 1966;4:787.

28. Bruch MD, Jo-Anne K, Bonesteel. *Macromolecules.* 1986;19:1622.

29. Berger PA, Remsen EE, Leo GC, David DJ. *Macromolecules.* 1991;24:2189.

30. Lebek B, et al. *Polymer.* 1991;32:2335.

31. Parker AA, et al. *Polym. Bulletin.* 1989;21:229.

32. Parker AA, et al. *J. Appl. Polym. Sci.* 1993;48:1701.

33. Gopalakrishnan R, Muralikrishna B, Narasimha VV, Rao R, Subba Rao B. *Polimery (Warsaw).* 1992;37:461.

34. David DJ, Rotstein NA, Sincock TF. *Polym. Bulletin.* 1994;33:725.

35. Lee J-C, Nakajima K, Ikehara T, Nishi T. *J. Appl. Polym. Sci.* 1997;64:797.

36. Lee J-C, Nakajima K, Ikehara T, Nishi T. *J. Appl. Polym. Sci.* 1997;65:409.

37. Keith HD, Padden FJ Jr, Russel TP. *Macromolecules.* 1989;22:666.

38. Eguiazabal JI, Iruin JJ, Cortazar M, Guzman GM. *Makromol. Chem.* 1984;185:1761.

39. Järvelä PA, Shucai L, PK. Järvelä. *J. Appl. Polym. Sci.* 1997;65:2003.

40. Heeger AJ. *Trends in Polym. Sci.* 1995;3:39.

41. Dasgupta AM, David DJ, Misra A. *Polym. Bulletin.* 1991;25:657.

42. Kobryanskii VM. *Synth. Met.* 1993;55:797.

42a. Annenkov VV, Danilovtseva EN, Filina EA, Mikhaleva AI, Skotheim TA, Trofimov BA. *Polym. Int.* 2004;53:772.

43. Krause S. In: Paul DR, Newman S, eds. *Polymer Blends.* Vol 1. New York: Academic Press; 1978:Ch. 2.

44. Law KY. *J. Imaging Sci.* 1990;34:38.

45. Niikura J, et al. *J. Appl. Electrochem.* 1990;20:606.

46. Howard KE, Lakeman CDE, Payne DA. *J. Am. Ceram. Soc.* 1990;73:2543.

46a. Cho Y-S, Yeo J-G, Jung Y-G, Choi S-C, Kim J, Paik U. *Mater. Sci. Eng. A.* 2003;362:174.

46b. Cho C-W, Cho Y-S, Yeo J-G, Choi S-C, Kim J, Paik U. *Colloid. Surf. A.* 2003;224:83.

47. Renn A, Meixner AJ, Wild UP. *J. Chem. Phys.* 1990;93:2299.

48. Wild UP, Renn A. *Makromol. Chem., Macromol. Symp.* 1991;50:89.

49. Yoshimura M, Nishimura T, Yagyu E, Tsukada N. *Polymer.* 1992;33:5143.

50. Kanaan Y, Attenberger T, Bogner U, Maier M. *Appl. Phys. B.* 1990;51:336.

51. De Caro C, Renn A, Wild UP. *Appl. Opt.* 1991;30:2890.

52. Altmann RB, Haarer D, Renge I. *Chem. Phys. Lett.* 1993;216:281.

53. Schwoerer H, Erni D, Rebane A, Wild UP. *Adv. Mater.* 1995;7:457.

54. Monroe BM, et al. *J. Imaging Sci.* 1991;35:19.

55. Ozer N, Tepehan F, Bozkurt N. *Thin Solid Films.* 1992;219:193.

56. Sammes NM, Brown MS, Ratnaraj R. *J. Mater. Sci. Lett.* 1994;13:1124; Sammes NM, Ratnaraj R. *J. Mater. Sci. Lett.* 1994;13:678.

57. Kojima K, Hayashi N, Toriumi M. *J. Photopolym. Sci. Technol.* 1995;8:47.

57a. Abdel-Fattah AA, Hegazy El-sayed A, Ezz El-Din H. *J. Photochem. Photobiol. A: Chem.* 2000;137:37.

57b. Abdel-Fattah AA, Hegazy, El-sayed A, Ezz El-Din H. *Int. J. Polym. Mater.* 2002;851:51.

57c. Beshir WB, Abdel-Fattah AA. *Int. J. Polym. Mater.* 2003;52:485.

57d. Arshak KI, Cavanagh LM, Gaidan I, Moore EG, Clifford SA. *IEEE Sensors Journal.* 2007;7(6):925.

57e. Arshak KI, Cavanagh LM, Moore EG. *Mater. Sci. Eng. C.* 2006;26:1032.

57f. Kneissler U, Harendza S, Helmchen U. *J. Elec. Microsc.* 2003;52(3):355.

57g. Fua X, Matsuyamaa H, Teramoto M, Nagai H. *Sep. Purif. Tech.* 2005;45:200.

57h. Shen F, Lu X, Bian X, Shi L. *J. Membr. Sci.* 2005;265 74.

58. Liau LCK, Yang TCK, Viswanath DS. *Polym. Eng. Sci.* 1996;36:2589.

59. Nair A, White RL. *J. Appl. Polym. Sci.* 1996;60:1901.

Poly(N-vinyl carbazole)

John H. Ko

Trade Names PVK (Polysciences, Inc.), Luvican[R] (BASF Corp.)

Class Vinyl Polymers; homopolymers

CAS 25067-59-8

Structure

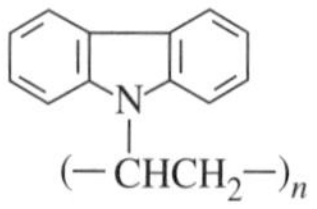

$(-\overset{\displaystyle |}{C}HCH_2-)_n$

Major Application Photoconductor, photoreceptor.

Properties of Special Interest High head distortion temperature and outstanding dielectric properties for electrical uses, such as insulator in continuous high temperature use. High refractive index for optical uses.

Property	Units	Conditions	Value	References
Tensile modulus	MPa	–	$(2.5–4.2 \times 10^3)$	(1–7)
Tensile strength	MPa	–	14	(1–7)
		Oriented	140	
Flexural strength	MPa	–	35–55	(1–7)
Compressive strength	MPa	–	30–35	(1–7)
Shear strength	MPa	–	20–30	(1–7)
Modulus of elasticity	MPa	Tensile test	3,700	(1–7)
Impact strength	$J\,m^{-1}$	DIN 53453	$(5–10) \times 10^5$	(1–7)
Vicat softening temperature	K	–	468	(1–7)
Elongation	%	Amorphous	$\ll 1$	(1–7)
		Oriented	1	
Hardness	MPa	Ball indentation	100	(1–7)
Index of refraction n_D	–	At 20°C	1.69	(1–7)
Density	$g\,cm^{-3}$	Amorphous	1.184	(1–7)
		Oriented	1.191	

Poly(N-vinyl carbazole)

Property	Units	Conditions	Value	References
Melting point			347–351°C	(10)
Glass transition temperature	K	Amorphous	500	(1–7)
		Syndiotactic	549	
		Isotactic	399	
Heat capacity (of repeat units)	kJ K^{-1} mol^{-1}	–	3.47×10^{-2}	(1–7)
Linear coefficient of thermal expansion	K	293–373 K	5×10^{-5}	(1–7)
Thermal conductivity	W cm^{-1} K^{-1}	20°C	1.26×10^{-3}	(1–7)
		170°C	1.68×10^{-3}	
Specific heat	J g^{-1}°C^{-1}	–	1.26	(1–7)
Water absorption	%	–	<0.1	(1–7)
WLF parameters: C_1 and C_2	–	Reference temp. of 220°C	$C_1 = 11.4$ $C_2 = 226.0$	(6)
Color		Transparent, very light yellow		(8)
Solvents		Aromatic hydrocarbons, chloroform, chlorobenzene, methylene chloride, and tetrahydrofuran		(8, 9)
Nonsolvents		Alcohols, esters, ketones, carbon tetrachloride, aliphatic hydrocarbons		(8, 9)
Chemical resistance		Resistant to alkalies, acids, water and salt solutions even at temperature as high as 100°C.		(8)
Dieletric strength	mV cm^{-1}	25–150°C	1.1–0.86	(8, 9)
Resistivity	ohm cm	DIN 57303	10^{16}–10^{17}	(1–5)
		25–150°C	$(0.05–8) \times 10^{15}$	
Loss factor	–	$10^3 - 3 \times 10^8$ Hz	$<10^{-3}$	(1–5)
		10^4 Hz	$(2–6) \times 10^{-4}$	
		10^3 Hz, 200°C	50×10^{-4}	
Permittivity	–	20°C, 50 Hz 1 MHz	3	(1–5)
Dielectric constant	–	104 Hz	3	(1–5)
Breakdown field strength	kV mm^{-1}	–	50	(1–5)
Electric conductivities	S cm^{-1}		$1e^{-10}$–$1e^{-15}$	(11–12)
Optical coefficient	–	Birefrigence/unit Strain at 210–235°C	-5.5×10^{-2}	(6)
Suppliers		Polysciences, Inc., 400 Valley Road, Warrington, PA 18976, USA (PVK) BASF Corp., 36 Riverside Avenue, Rensselaer, NY 12144, USA (Luvican)		

References

1. Mark HF, et al, eds. *Encyclopedia Polymer Science and Engineering.* Vol 17. New York: John Wiley and Sons; 1989:272.
2. Klopffer W. *Kunstoffe.* 1971;61:533.
3. Cornish EH. *Plastics.* 1963;28:61.
4. Pearson JM, Stolka M. *Polymer Monographs.* Vol 6. New York: Gordon and Breach; 1981.
5. Jacobi HJ. *Kunstoffe.* 1959;43:381.
6. Penwell R, Ganugly B, Smith T. *J. Polym. Sci., Macromol Rev.* 1978;13:63.
7. Davidge Jh. *J. Appl. Chem..* 1959;9:553.
8. Data Sheet No. 263, *Poly (N-vinylcarbazole).* Polysciences, Inc., Warrington, Penn. May 1990.
9. BASF Data Sheet, *Polyvinylcarbazole-Luvican* March 1971.
10. Crystal, Richard G. *Macromolecules.* 1971;V4(4):378–384.
11. Yoon, Hyeonseok, *PMSE Preprints.* 2005:V93;677–678.
12. Suzuki, Tetsuyoshi. *JP 62109819 A 1987.*

Poly(vinyl chloride)

Anthony L. Andrady, Taner Z. Sen
and M. Göktuğ Ahunbay

Acronym, Trade Names PVC, Geon (Goodrich), Vinoflex (BASF), Vestolite (Hüls), Airco (Air Products), SCC (Stauffer)

Class Vinyl polymers

Structure [—CH$_2$CHCl—]

Major Applications Poly(vinyl chloride) is used in building applications as rigid formulations in water and sewage pipes, siding, gutters, and downspouts, conduits, and cable coverings. Pipe and conduit application are by far the major use of PVC. It is also used as a plasticized material in membrane roofing, and flooring applications. PVC films are used in packaging of consumer goods.

Property	Units	Conditions	Value	Reference
Suspension polymerization		Diacetyl peroxide, peroxydicarbonates, alkyl peroxyesters and AIBN used as initiator Cellulose derivatives used as protective colloid		(1)
Bulk polymerization		Two-stage reaction process		(2)
Emulsion polymerization				(1)
Typical comonomers		Vinyl acetate (VAM), ~10–15%		–
Molecular weight (of repeat unit)	g mol^{-1}	–	62.5	–
Typical molecular weight range of polymer M_n	g mol^{-1}	Polymerization temperature (°C) 50 57 64 71	67×10^{-3} 54×10^{-3} 44×10^{-3} 33×10^{-3}	(1)
Polydispersity index (M_w/M_n)	–	Determined by GPC for ordinary suspension-polymerized PVC Temp. (°C) $M_n \times 10^{-3}$ 43 58 55 44 75 26	 2.44 2.08 2.01	(3)

Property	Units	Conditions		Value	Reference
Tacticity	Fraction, f, of syndiotactic dyads	Polymerization temperature (°C)			(4)
		55		0.55	
		25		0.57	
		0		0.60	
		−30		0.64	
		−50		0.66	
		−76		0.68	

Property	Units	Conditions	Values				Reference
Degree of branching	%	Polymerization temperature (°C)	$-CH-$ | CH_2Cl	$-CCl-$ | CH_2CH_2Cl	$CCl-$ | $CH_2CHCl(CH_2)_2Cl$	H/Cl | $-C-C-C$ | C	(5)
		45	3.9	<0.1	0.5	<0.1	
		55	4.2	0.2	0.6	0.2	
		65	4.6	0.2	0.8	0.3	
		80	4.9	0.3	1.3	0.3	

Property	Units	Conditions	Value	Reference
Head-to-head and other irregular structures	Per 1,000 repeat units (VC)	Head to head groups	0.2	(6)
		In-chain double bonds	0.1–0.2	
		Chloromethyl branches	4	
		2-Chloroethyl branches	0.5	
		2,4, Dichlorobutyl branches	1	
		Tertiery chlorine	0.5–1.5	
		Long branches	1	
	Per molecule	Total unsaturation	1	
IR (characteristic absorption frequencies)	cm^{-1} band	Planar syndiotactic sequences	603, 638	(7)
		C—Cl units in isotactic sequences	690	(7)
		Discussion of C—Cl region and curve fitting	–	(8)
UV (characteristic absorption frequencies)	Ultraviolet visible absorption bands for polyene sequences with absorption at 306 nm for $n = 4$ sequences. Assignment of peaks.			(9)
^{13}C NMR	–	PVC solution in trichlorobenzene, 380 K	–	(10)

Property	Units	Conditions	Value		Reference
		PVC solution in o-dichlorobenzene, 373 K	–		(11)
		Copolymers with vinyidene chloride	–		(12)
Thermal expansion coefficient	K^{-1}	100°C	4.7×10^{-4}		(1)
		120°C	5.5×10^{-4}		
		140°C	6.2×10^{-4}		
Thermal expansion coefficient	$°C^{-1}$	200°C	1.29		(13)
		220°C	1.50		
		230°C	1.77		
		240°C	1.92		
		250°C	2.14		
		260°C	1.00		
		270°C	0.90		
Thermal conductivity $W\ m^{-1}\ K^{-1}$			0.2		(14)
Compressibility	bar^{-1}	100°C	5.2×10^{-5}		(1)
		140°C	6.4×10^{-5}		
Density	$g\ cm^{-3}$	100°C	1.352		(15)
		120°C	1.338		
		140°C	1.332		
Solvents		Methyl ethyl ketone, cyclohexanone, DMF, toluene, nitrobenzene DMSO, acetone/carbon disulfide			(16, 18)
Nonsolvents		Alcohols, hydrocarbons, acetone, nonoxidizing acids			(16, 18)

Mark-Houwink parameters: K and a	$K = ml\ g^{-1}$ a = None		$(K \times 10^3)^{\dagger}$	a	
		Cyclohexanone, 20°C	13.7	1.0	(19)
		Tetrahydrofurane, 20°C	3.63	0.92	(20)
		Cyclohexanone, 20–60°C	$18.74 - (4.85 \times 10^{-4})T$	0.803	(21)
		Cyclopentanone, 20–60°C	$0.091 - (1.55 \times 10^{-4})T$	0.861	(21)
		Tetrahydrofurane, 20–50°C	$10.87 - (1.67 \times 10^{-4})T$	0.851	(21)
		$M_n = (0.3 - 1.9) \times 10^5$	K	a	
		Chlorobenzene, 30°C	0.0712	0.59	(22)
		Cyclohexane, 25°C	0.0138	0.78	(23)
		Tetrahydrofuran, 25°C	0.0163	0.78	(24)
Second virial coefficient	$mol\ cm^3\ g^{-2}$	Cyclohexanone, 25°C, $M_n = 118,000$	11×10^{-4}		(25)
Interaction parameter χ	–	Toluene, 125–140°C	0.45–0.41		(26)
		2-Propanol, 125–140°C	1.10–0.97		(26)
		Methanol, 125–140°C	1.42–1.24		(26)

Property	Units	Conditions	Value	Reference
		Acetone, 125–140°C	0.77–0.53	(26)
		Benzene, 120°C	0.75	(27)
		Carbon tetrachloride	1.14	(27)
		Chloroform	0.91	(27)
		Dichloromethane	1.63	(27)
Theta temperature Θ	K	Cyclohexanone	324	(28)
		Dimethylformamide	309.5	(29)
		Benzyl alcohol	428.4	(29)
Unit cell dimensions	nm	Orthorhombic unit cell	$a = 1.06$, $b = 0.54$, $c = 0.51$	(30)
Heat of fusion	kJ mol^{-1}	–	11.3	(31)
			3.28	(32)
			3.59	(33)

[†] K values can be canculated from expression given for last three entries.

Degree of crystallinity and density from density measurements[4]

Polymerization Temp. (°C)	Crystallinity (%)	M_n (g mol^{-1})	Density at 20°C (g cm^3)
90	11.3	23,750	1.391
55–60	11.3	75,000	1.391
50	13.2	91,250	1.392
20	15.0	172,250	1.393
−15	57.3	106,300	1.416
−75	84.2	105,300	1.431

From calorimetric measurements[34]

Polymerization Temp. (°C)	Crystallinity (%)	M_n (g mol^{-1})
75	18.4	23,2000
65	15.5	38,700
52	15.3	53,500
52	14.4	66,700
25	11.9	136,000
25	11.8	155.000

Property	Units	Conditions		Value	Reference
Glass transition temperature	K	Effects of tacticity and molecular weight	–		(35)
		Dilatometry		344	(36)
		DSC, 20°C min^{-1}		371	(37)
		By DSC, 32°C min^{-1}			(38)
		Polymerization temp. (°C)	$[\eta]$ (ml g^{-1})*		
		90	–	353	
		50	80	358	
		0	108	370	

Poly(vinyl chloride)

Property	Units	Conditions		Value	Reference
		Polymerization temp. (°C)	$[\eta]$ (ml g^{-1})*		
		−30	125	373	
		−50	−	378	
		−60	90	380	
Melting transition temperature	K	Calorimetry		485–583 473–573 (decomposition)	(33)
Sub-T_g transition temperature	K	Dynamic mechanical		223	(39)

Viscoelastic behaviour in the glass transition range[40]
DSC, 5°C min^{-1}

% additive	T_g (K)		$T_\alpha^\ddagger$ (K)	
	with DOP	with BBP	with DOP	with BBP
0	347	347	347	347
5	330	333	335	336
10	320	321	323	325
15	311	313	312	314
20	299	303	299	306

$\ddagger$Main mechanical relaxation temp.

Property	Units	Conditions	Value	Reference
Heat capacity	kJ K^{-1} mol^{-1}	100°C	0.0268	(41)
		300°C	0.0594	
		360°C	0.0911	
		380°C	0.0981	
Tensile modulus	MPa	As a function of polymerization temp. (°C)		
		−196	7,584	(42)
		−120	5,171	(42)
		−75	3,861	(42)
		20	2,964	(42)
		30	3,000	(43)
		40	2,930	(43)
		50	2,427	(43)
		60	1,551	(43)
		70	276	(43)
		With 3% organic tin, 1% and 0.5% oxidized PE wax	3087	(44)
Elastic modulus	Pa	200°C	1.25	(13)
		220°C	1.20	
		230°C	1.11	
		240°C	1.04	
		250°C	0.96	
		260°C	1.61	
		270°C	1.81	

Property	Units	Conditions	Value			Reference
Tensile strength	MPa	Unplasticized	56.6			(45)
		With 10% dioctylphthalate	55.5			
Elongation	%	Unplasticized	85			(45)
		With 10% dioctylphthalate	104			
		With 3% organic tin, 1%, and 0.5% oxidized PE wax	136			(44)
Yield strength	MPa	With 3% organic tin, 1%, and 0.5% oxidized PE wax	60			(44)
Impact strength	J m^{-1}	With 3% organic tin, 1%, and 0.5% oxidized PE wax	21.2			(44)
Dielectric constant $\varepsilon\prime$	–		60 Hz	1 kHz	10 kHz	(46, 47)
		25°C	3.50	3.39	3.29	
		40°C	3.51	3.40	3.34	
		60°C	3.70	3.61	3.45	
		80°C	4.25	4.09	3.89	
		90°C	6.30	5.05	4.45	
		100°C	10.30	7.77	5.77	
Dielectric loss factor $\varepsilon?$	–		60 Hz	1 kHz	10 kHz	(46, 47)
		25°C	0.110	0.081	0.058	
		40°C	0.116	0.081	0.058	
		60°C	0.125	0.080	0.050	
		80°C	0.172	0.120	0.110	
		90°C	0.410	0.500	0.920	
		100°C	1.20	1.415	1.370	
Permeability coefficient P	m^3(STP)m s^{-1} m^{-2} Pa^{-1} × 10^{-9}	Unplasticized film, 25°C				
		H$_2$	1.3			(48)
		N$_2$	0.0089			(48)
		O$_2$	0.034			(48)
		Ar	0.0086			(48)
		CH$_4$	0.021			(48)
		NH$_3$	3.7			(49)
		H$_2$S	0.14			(49)
		CO$_2$	0.15			(48)
		H$_2$O	0.12			(48)
Permeability coefficient P	m^3(STP)m s^{-1} m^{-2} Pa^{-1} × 10^{-9}	Plasticized with tricresyl triphosphate (TCP), 27°C				(50)
		5% TCP, H$_2$	1.4			
		20% TCP, N$_2$	1.6			
		31% TCP, O$_2$	2.1			
		40% TCP, Ar	2.7			

Poly(vinyl chloride)

Property	Units	Conditions	Value	Reference
Pyrolyzability	–	Dehydrochlorination rate in N_2	–	(51, 52)
		Polyene propagation on degradation	–	(52)
Weathering	–	Change in molecular weight during weathering	–	(53)
Electric conductivity	Siemens cm^{-1}	297 K and 54% relative humidity	10^{-11}	(54)

* In cyclohexanone at 25°C.

References

1. Smallwood PV. In: Mark HF, Bikkales NM, Overberger CG, Menges G, eds. *Encyclopedia of Polymer Science and Engineering.* 2nd ed. Vol 17. New York: John Wiley and Sons; 1987:303.
2. Fitch RM. *Br. Polym. J.* 1973;5:467.
3. Sörvik EM. *J. Appl. Polym. Sci.* 1977;21:2769.
4. Pham QT, Millan J, Madruga EL. *Makromol. Chem.* 1974;175:945.
5. Hjertber T, Sörvik EM. *ACS Symposium Series: Polymer Stabilization.* Vol 280. 1985.
6. Guyot A. *Pure. Appl. Chem.* 1985;57:833.
7. Shimanouchi T. Tasumi M. *Spectrochim. Acta.* 1961;17:755.
8. Baruya A, et al. *J. Polym. Sci., Polym. Lett. Ed.* 1976;14:329.
9. Braun D, Thallmaier M. *Makromol. Chem.* 1966;99:59.
10. Schilling FC. *Macromolecules* 1981;11:1290.
11. Heatley F. In: Ibbett RN, ed. *NMR Spectroscopy of Polymers.* London: Blackie Academic and Professional, Chapman and Hall; 1993:37.
12. Komoroski RA, Shockcor JP. *Macromolecules.* 1983;16:1539.
13. Boztug A, *J. Appl. Polym. Sci.* 2005;96:1635.
14. De Carvalho G, Frollini E, Dos Santos WN. *J. Appl. Polym. Sci.* 1996;62:2281.
15. Rogers PA. *J. Appl. Polym. Sci.* 1993;48:1061.
16. Thinius K. *Analytische Chemie der Plaste.* Berlin: Springer-Verlag; 1963.
17. Nitsche R, Wolf KA. *Struktur und Physikalisches Verhalten der Kunstoffe.* Vol 1. Berlin: Springer-Verlag; 1961.
18. Kurata M, Stockmeyer WH. *Adv. Polymer Sci.* 1963;3:196.
19. Bier C, Kramer H. *Makromol. Chem.* 1955;18–19:151.
20. Batzer H, Nisch A. *Makromol. Chem.* 1957;22:131.
21. Marron SH, Lee MS. *J. Macromol. Sci.* 1973: B7(1):29, 47, 61.
22. Du Y, Xue Y, Frisch HL. In: Mark JE, ed. *Physical Properties of Polymers Handbook.* Woodbury, N. Y: American Institute of Physics Press; 1996:248.
23. Ciampa G, Schwindt H. *Makromol. Chem.* 1954;21:169.
24. Freeman M, Manning PP. *J. Polymer Sci.* 1964;A2:2017.
25. Petrus V. *Collection Czech. Chem. Commun.* 1969;33:119.
26. Merk W, Lichtenthaler RN, Parutsnitz JM. *J. Phys. Chem.* 1980;84:1694.
27. Riedl B, Prudhomme RE. *J. Polym. Sci. Part B: Polym. Phys.* 1986;24:2565.
28. Adamski P. *Polym. Sci. USSR* 1971;13:803.
29. Sato M, Koshiishi Y, Asahina M. *J. Polym. Sci. Polym. Letters.* 1963;1:233.
30. Natta G, Corradini P. *J. Polym. Sci.* 1956;20:215.
31. Kockott D. *Kolloid Z., -Z. Polym.* 1964;198:17.
32. Nakajima A, Hamada H, Hayashi S. *Makromol. Chem.* 1966;95:40.

33. Park HC, Mount EM. In: Mark HF, Bikkales NM, Overberger CG, Menges G, eds. *Encyclopedia of Polymer Science and Engineering.* 2nd ed. Vol 7. New York: John Wiley and Sons; 1987:89.
34. Maron SH, Filisko FE. *J. Macromol. Sci.* 1972;B6(2):413.
35. Daniels CA, Collins EA. *Polym. Eng. Sci.* 1979;19:8.
36. Greiner G, Schwarzl FR. *Rheol. Acta* 1984;23:378.
37. Singh P, Lyngaae-Joergensen J. *J. Macromol. Sci. Phys.* 1981;B19(2):177.
38. Ceccorulli G, Pizzoli M, Pezzin G. *J. Macromol. Sci. Phys.* 1977;B14(4):499.
39. Stephenson RC, Smallwood PV. In: Mark HF, Bikkales NM, Overberger CG, Menges G, eds. *Encyclopedia of Polymer Science and Engineering.* 2nd ed. Vol S. New York: John Wiley and Sons; 1987:858.
40. Dubault A, Bokobza L, Gandin E, Halary JL, *Polym. Int.* 2003;52:1108.
41. Gaur U, Lau SF, Wunderlich BB. *J. Phys. Chem. Ref. Data.* 1983;12:29.
42. Diment J, Ziebland H. *J. Appl. Chem.* 1958;8:203.
43. Orgorkiewics RM. *Engineering Properties of Thermoplastics.* New York: John Wiley and Sons; 1970:251.
44. Tian M, Chen G, Guo S. *Macromol. Mater. Eng.* 2005;290:927.
45. Lutz JT. In: Owen ED, ed. *Degradation and Stabilization of PVC.* New York: Elsevier Applied Science Publishers; 1984:264.
46. Brandup J, Immergut EH, eds. *Polymer Handbook.* 3rd ed. New York: John Wiley and Sons; 1989.
47. Schildknecht CE. *Vinyl and Related Polymers.* New York: John Wiley and Sons; 1952.
48. Tikhomirov BP, Hopfenberg HB, Stannett VT, Williams JL. *Makromol. Chem.* 1968;118:177.
49. Braunisch V, Lenhart H. *Kolloid-Z.* 1961;177:24.
50. Sefcik M, Schafer J, May F, Raucher D. *J. Polym. Sci.* 1983;21:1041.
51. Hjertberg T, Sörvik EM. *Polymer.* 1983;24:673, 685.
52. Hjertberg T, Sörvik EM. In: Owen ED ed. *Degradation and Stabilization of PVC.* New York: Elsevier Applied Science Publishers; 1984:41.
53. Matsumoto S, Oshima H, Hosuda Y. *J. Polym. Sci. Polym. Chem. Ed.* 1984;22:869.
54. Rinaldi AW. *J. Appl. Polym. Sci.* 2005;96:1710.

Poly(vinyl chloride), Head-to-head

Meifang Qin and Ying Li

Acronyms, Trade Names H-H PVC, HH PVC, Cl-*cis*-PBD, Cl-*trans*-PBD, chlorinated PBD rubber

Class Chemical copolymers

Structure $[-CH_2-CHCl-CHCl-CH_2-]$

Major Application H-H PVC is mostly studied in academic field to understand its structure/property relationship, thermal degradation behavior, and mechanism. Its properties are compared to those of commercial head-to-tail PVC. Pure H-H PVC has no significant industrial applications. H-H PVCs containing 40–65 wt% of Cl, also called chlorinated polybutadiene rubber-resins, are used for coating, paint-based applications and the preparation of threads, tires, tubings, and films, etc.

Properties of Special Interest Preparation methods. Toughness and durability. Good compatibility with other polymers and plasticizers. H—H PVC (T_g 70°C) and H—T PVC (T_g 90°C) were immiscible almost over the entire range of compositions and the individual T_gs of the two components of the blends is clearly noticeable. Tacticity and spectrum properties.

Preparative Technique Chlorination of 1,4-polybutadiene solution at room temperature with molecular chlorine, using solvents that favor ionic reaction such as dichloromethane and chloroform. Pure H-H PVC is made by 1,4-PBD with high *cis* content.[1−8] Chlorination of *cis*-l,4-PBD has been studied in detail. H-H PVC can also be prepared by chlorination of 1,4-*trans*-polybutadiene.[9] In the chlorination of *trans*-polybutadiene, chlorine adds directly to the double bond, but that in the case of *cis*-polybutadiene, *cis-trans* isomerization takes place in some double bonds before addition of chlorine. This means that head-to-head poly(vinyl chloride) which has a regular structure is obtained by the chlorination of *trans*-polybutadiene.[8] Improved halogenation techniques for poly(1,4-butadiene) have made H-H PVC accessible in larger quantities and have allowed an extensive characterization.[10] The relative importance of head-to-head versus head-to-tail additions during the propagation of poly(vinyl chloride) is determined by ab initio methods for different chain lengths of the polymer. Their theoretical prediction of the probability of head-to-head addition is 2 per 1000 VC additions.[11]

Property	Units	Conditions	Value					Reference
Tacticity		Chlorinated-*trans*-PBD (Cl-*trans*-PBD)	Diisotactic poly(*erythro*-1,2-dichloro butamer)					(6, 13)
		Chlorinated-*cis*-PBD (Cl-*cis*-PBD)	Disyndiotactic poly(*threo*-1,2-dichloro butamer)					
Infrared absorption at fingerprint region	cm^{-1}		Wavenumbers					
		Cl-*cis*-PB	795	725	680	650	590	(2, 7, 8, 15)
		Cl-*trans*-PB	795		686	650		(8, 15)
NMR	Ppm		CH$_2$		CHCl			
		^{13}C	32.9,33.3		65.7,66.1			(2, 7, 12)
		^{1}H	8.0		5.8			(2, 7)

Transition temperature of partially chlorinated *cis*-1,4-PBD measured by DMA[2]

Degree of Chlorination (%)	β_{low} (K)		α_{low} (K)		β_{high} (K)	
	E''	Tan δ	E''	Tan δ	E''	Tan δ
40	176	178	276	275	321	344
58	173	173	294	294	324	348

Thermal transition temperature of chlorinated *cis*-1,4-PBD measured by DSC

Weight Percent of —CH$_2$—CHCl—CHCl—CH$_2$— units	Temperature (K)				Reference
	T_g (low)	T_{cr} (low)*	T_m (low)†	T_g (high)	
0	165	203	270	–	(2)
0.40	166	201	267	–	(2)
0.61	165	202	266	324	(2)
0.63	164	209	267	322	(2)
0.72	166	201	266	314	(2)
0.81	162	204	266	337	(2)
0.89	–	–	–	324	(2)
0.91	207	–	–	329	(2)
0.93	–	–	–	326	(2)
1.00	–	–	–	347	(2)
1.00‡	–	–	–	336	(3)

* T_{cr} (low) = crystallization temperature of PB domain.

† T_m (low) = melting temperature of PB domain.

‡ Sample made by chlorination of *trans*-1,4-PBD.

Unit cell dimensions[13]

Polymer	Crystal System	Repeat Unit per Unit Cell	Cell Dimensions (Å)			Cal. Density $(g\,cm^{-3})$	Cell Angle (Degrees)	Space Group
			a	b	c			
Cl-*trans*-PBD	Monoclinic	2	7.05	8.05	5.10	1.46	100	P2$_1$/a
Cl-*cis*-PBD	Monoclinic	2	7.37	5.30	10.10	1.46	134	P2/c

Unperturbed molecular dimension, K, and conformational parameter, σ, in different solvents[15]

Sample	Solvent	$K \times 10^3$	σ
Cl-*cis*-PBD	Tetrahydrofuran	1.5	2.3
	Methyl ethyl ketone	2.1	2.5
Cl-*trans*-PBD	Tetrahydrofuran	1.9	2.5
	Dichloroethane	1.6	2.3
PVC	Tetrahydrofuran	3.2	2.9

Conformational population in the chlorinated part of H-H PVC[15]

Sample Form	Solvent	Chlorination of *trans*-PBD		Chlorination of *cis*-PBD	
		Trans	Gauche	Trans	Gauche
Solution	Tetrahydrofuran	64	35	49	51
Solution	Cyclohexanone	45	55	56	44
Unstretched film	None	63	37	62	38
Stretched film	None	–	–	68	32

Mark-Houwink parameters: K and a[16]

Polymer	Solvent	Temperature (°C)	$K \times 10^4$	a
Cl-*cis*-PBD	Tetrahydrofuran	30	2.53	0.71
	Methyl ethyl ketone	30	9.46	0.57
Cl-*trans*-PBD	Tetrahydrofuran	30	6.21	0.61
	Dichloroethane	30	9.46	0.54

Property	Units	Conditions		Value	Reference
Second virial coefficient A$_2$	mol cm^3 g^{-2}	Polymer/Solvent			(16)
		Cl-*cis*-PBD/tetrahydrofuran		1.0×10^{-3}	
		Cl-*cis*-PBD/methyl ethyl ketone		0.5×10^{-3}	
		Cl-*trans*-PBD/tetrahydrofuran		1.1×10^{-3}	
Stabilities					(4, 14, 17)
Initial decomposition temperature	K	–		463	

Property	Units	Conditions	Value	Reference
Degradation product by thermal volatilization analysis	–	–	HCl, ethylene, propylene, benzene, methane	
Activation energy for dehydrochlorination	kcal mol^{-1}	–	23	

Pyrolysis product in helium at 500°C

Pyrolysis Product	Percentage Ratio of Each Peak Height to the Summation of All Peak Heights	
	100% Chlorinated 98% *cis*-1,4-PBD	100% Chlorinated 59% *trans*-1,4, 23%1,2-,18% *cis*-1,4-PBD
Aliphatic hydrocarbons	7.40	24.5
Benzene	32.48	34.5
Toluene	6.31	8.6
Ethylbenzene	1.34	3.3
o-Xylene	1.01	1.15
Monochlorobenzene	31.48	3.2
Styrene	2.33	3.1
Vinyltoluene	2.61	0.9
p-Dichlorobenzene	2.12	0.1
o-Dichlorobenzene	3.7	1.7
1,3,5-Trichlorobenzene	0.84	1.13
1,2,4-Trichlorobenzene	0.65	1.65
Naphthalene	3.35	4.3
α-Methylnaphthalene	0.42	2.35
β-Methylnaphthalene	–	1.5

References

1. Bailey FE Jr, et al. *J Polym. Sci.* Part B, 1964;2:447.
2. Kawaguchi H, et al. *Polymer.* 1982;23:1805.
3. Dall'Asta G, Meneghini P, Gennaro U. *Die Makromolekulare Chemie.* 1972;154:279.
4. Crawley S, McNeill IC. *J. Polym. Sci.* Part A, 1978;16:2593.
5. Uelzmann H, Falls C. *U.S. Patent.* 1968;3392161.
6. Horhold H-H, et al. *Die Makromolekulare Chemie.* 1961;122:145.
7. Qin MF. *Head to Head Vinyl Polymers: Head to Head Poly(vinyl halides) and Chiral Crystallization.* Dissertation, Polytechnic University, 1995.
8. Murayama N, Amagi Y. *J. Polym. Sci. Part B.* 1966;4:119.
9. Murayama N, Amagi Y. *Polymer Letters.* 1966;4:115.
10. Vogl O, Qin MF, Zilkha A. *Progress in Polymer Science.* 1999;24(10):1481.
11. Karen Van Cauter, Veronique Van Speybroeck, and Michel Waroquier. *Chem Phys Chem.* 2007;8:541.
12. Dreyfuss MP, Nevius MR, Manninen PR. *J. Polym. Sci. Part C.* 1987;25:99.

13. Bassi IW, Scordamaglia R. *Die Makromolekulare Chemie.* 1973;166:283.
14. Mitani K, et al. *J. Polym. Sci. Part A.* 1975;13:2813.
15. Kondo S, Takeda M. *Polym. Eng. Sci.* 1985;25(16):1026.
16. Takeda M, Endo R, Matsuura Y. *J. Polym. Sci. Part C.* 1968;23:487.
17. Iida T, Nakanishi M, Goto K. *J. Polym. Sci. Part A.* 1975;13:1381.

Poly(vinylferrocene)

Ian Manners

Class Inorganic and Semi-Inorganic Polymers

Structure $[(C_5H_5)Fe(C_5H_4)CHCH_2)]_n$

Properties of Special Interest Low cost. Interesting redox properties.

Property	Units	Conditions	Value	Reference
UV-Vis Absorption (λ_{max})	Nm	CH_2Cl_2 Solution	440	(1, 2)
UV-Vis Absorption Coefficient (ε)	$M^{-1}\ cm^{-1}$	THF Solution		
Glass Transition Temperature	°C	DMA Experiment		
Glass Transition Temperature	°C	DSC Experiment	222	(3)
Melting Temperature	°C	DSC Experiment		

References

1. Sasaki Y, Walker LL, Hurst EL, Pittmann CU Jr. *J. Polym. Sci. Polym. Chem. Ed.* 1973;11:1213.
2. Pittmann CU Jr, Lai JC, Vanderpool DP, Good M, Prados R. *Macromolecules.* 1970;3:746.
3. George MH, Hayes GF. *J. Polym. Sci. Polym. Chem. Ed.* 1976;14:475.

Poly(vinyl fluoride)

Ronald E. Uschold and Murali K. Kilaru

Acronym, Trade Names PVF, Tedlar® PVF Film, Tedlar® SP Film, PV-116 Resin

Class Vinyl polymers

Repeat Unit $—(CH_2CHF)_n—$

Major Applications As a protective surfacing material for: aircraft interior wall and ceiling panels, architectural fabrics, exterior building panels, wall coverings, reinforced vinyl sheeting for signs and awnings, automotive tubing, thermoformed plastic laminates, truck body panels, solar panels, and green house glazing. As a release sheet for curing: epoxy circuit boards and composite panels.

Properties of Special Interest Weathering resistance, antisoiling properties, chemical resistance, Smooth, glossy finish, UV resistance, and durability.

Property	Units	Conditions	Value	Reference
Preparative techniques	Process			
		Emulsion: 4.0–100 MPa, 46–250°C, water soluble radical initiator, fluorinated surfactant		(1, 2, 3)
		Suspension: 2.5–10MPa, 25–100°C, monomer soluble radical initiator, water soluble suspending agent		(4, 5)
Molecular weight (of repeat unit)	$g\,mol^{-1}$	$CH_2{=}CHF$	46.04	–
Head-to-head sequences	%	$—CH_2CHFCH_2CHF—$	87–89*	(6, 7)
Monomer inversions	%	$—CH_2CHFCHFCH_2—$	11–13*	(6, 7)
Branch points	%	$—CH_2CFCH_2CHF—$ $\mid$ $CH_2CHF—$	0.5–0.7†	(7)
End group	%	$—CH_2CH_2F$	0.2–0.5†	(7)
Tacticity	–	Atactic, Bernoullian distribution	$P_m = 0.43$	(8)
Typical polymer M_w range	$g\,mol^{-1}$	–	$1.43–6.54\,(\times10^5)$	(9)
Typical polydispersity Index (M_w/M_n)	–	–	2.5–5.6	(9)
IR absorption frequencies	cm^{-1}	–	2,940 1,710	(13)

Property	Units	Conditions	Value	Reference
			1,415	
			1,370	
			1,235	
			1,140	
			1,090	
			1,025	
			890	
			830	
			460	
UV/VIS absorption frequencies	cm^{-1}	Transmittance (%)		(13)
		<10	<44,000	
		>80	42,000–25,000	
		>90	25,000–7,000	
Solar energy transmittance (359–2500 nm)	%	ASTM E427-71	90	(15)
NMR signals	ppm	Structure	Chemical Shift	
		$-CH_2CHFCH_2-$	-174 to -184	(6, 7, 10)
		$-CH_2CHFCHFCH_2-$	-188 to -200	(6, 7, 10)
		3°F at branch	-147	(7)
		$-CH_2F$ end group	-220	(7)
Linear expansion coefficient	$°F^{-1}$	–	2.8×10^{-5}	(14)
Thermal aging	Hr	Circulating air oven 150°C (302°F)	300	(15)
Temperature use Continuous Short cycle(1–2hr)	°C	–	-72 to 107°C	(14)
			Up to 175°C	(15)
Shrinkage	%	(Type 2) MD and TD, air oven 30 min	4% at 130°C	(14)
		(Type 3) TD only, air oven 30 min	4% at 170°C	
		(Type 4) TD only, air oven 30 min	2.5% at 170°C	
Density	$g\ cm^{-3}$	Crystallinity (%)		
		Amorphous	1.36	(10)
		20	1.368	(10)
		22	1.370	(10)
		28	1.375	(10)
		32	1.379	(10)
		37	1.383	(10)
		50	1.395	(10)
		61	1.405	(10)
		100	1.44	(11)
	$g\ cc^{-1}$	ASTMD-1505-68	1.37–1.72	(14)
Solvents (above 120°C)		Dimethylacetamide, dimethylformamide, N-methyl pyrrolidone, γ-butyrolactone		(13)

Poly(vinyl fluoride)

Property	Units	Conditions	Value	Reference
Nonsolvents		Alcohols, ketones, esters, ethers, aliphatics, aromatics		(13)
Solubility parameter	$(Pa)^{1/2}$	–	~25	(13)
Mark-Houwink parameters: K and a	$K = $ ml g^{-1} $a = $ None	DMF at 90°C	$K = 6.52 \times 10^{-5}$ $a = 0.8$	(9)
Crystalline state properties				
Lattice	–	–	Hexagonal	(11)
Unit cell dimensions	A°	–	$a = b = 4.93, c = 2.53$	
Unit cell angles	Degrees	–	$\alpha = \beta = 90, \gamma = 120$	
Degree of crystallinity	%	Drawn at 90°C, annealed at 140°C	37	
Glass transition temperature	K	DMA, 1 Hz	337	(13)
Melting point (DSC)	K	Commercial resin	463–466	(13)
		37% crystallinity	470–478	(12)
		45% crystallinity	491–498	(12)
		50% crystallinity	498–508	(12)
Softening point	K	DMA, 1 Hz	398–403	(13)
Tensile modulus	MPa	ASTM D882-80		(14)
		Unoriented, unpigmented film	1,170	
		Oriented, unpigmented film	2,300	
Tensile strength	KJ/m	ASTM D1004-66	129–196	(15)
Ultimate tensile strength	psi	ASTM D-882-80, Method A 100% elong./min—Instron, 22°C	$8–16 \times 10^3$	(14)
Ultimate yield	psi	ASTM D-882-80, Method A 100% elong./min—Instron, 22°C	6000–4900	(14)
Ultimate elongation	%	ASTM D-882-80, Method A 100% elong./min—Instron	90–250	(14)
Loss modulus	MPa	Unoriented, unpigmented film		(13)
		25°C, 1 Hz	16	
		75°C, 1 Hz	60	
		150°C, 1 Hz	100	
Impact strength	lb/mil	Spencer ASTM D-3420-80, 22°C	10–20	(14)
Bursting strength	psi	Mullen, ASTM D-774-67	29–65	(14)
Dielectric strength	kV mil^{-1}	ASTM D-150-81 at 60 Hz		(14)
		TTR20SG4	3.4	
		TWH20BS3	3.5	
Dielectric constant	–	ASTM D-150-81 at 1000 Hz, 22°C		(14)
		TTR20SG4	8.5	
		TWH20BS3	11	

Property	Units	Conditions	Value	Reference
Corona endurance	hr	ASTM -T method, 60 Hz, 1000 V/mil		(14)
		TTR20SG4	2.5	
		TWH20BS3	6.2	
Dissipation factor	%	ASTMD-150-81		(14)
		1000 Hz, 22°C		
		TTR20SG4	1.6	
		TWH20BS3	1.4	
		1000 Hz, 70°C		
		TTR20SG4	2.7	
		TWH20BS3	1.7	
		10,000 Hz, 22°C		
		TTR20SG4	4.2	
		TWH20BS3	3.4	
		10,000 KHz, 70°C		
		TTR20SG4	2.1	
		TWH20BS3	1.6	
Volume resistivity	Ohm cm	ASTM D-257-78, 22°C		(14)
		TTR20SG4	4×10^{13}	
		TWH20BS3	7×10^{14}	
		ASTM D-257-78, 100°C		
		TTR20SG4	2×10^{10}	
		TWH20BS3	1.5×10^{11}	
Surface tension	mN m^{-1}	Contact angle	38	(13)
Coefficient of friction (Film/Metal)	–	ASTM D-1894-78, 22°C	0.18–0.21	(14)
Zero strength	°C	Hot bar	260 to 300	(14)
Index of refraction	–	ASTM D-542-50 Abbé Refractometer	1.46	(14)
Storage modulus	MPa	Unoriented, unpigmented film		(13)
		25°C, 1 Hz	2,000	
		75°C, 1 Hz	400	
		150°C, 1 Hz	50	
Chemical resistance	–	Acids, bases, solvents 1 yr, 25°C	No visible effect	(14)
		Acids, bases, solvents 2 h boiling	No visible effect	
		Soil burial, 5 yr	No visible effect	
Weatherability	–	Florida exposure, facing South at 45° to horizontal	Excellent	(14)
Gas permeability	cc in^{-2} hr^{-1} atm^{-1} mil^{-1}	ASTMD-1434-75, 24°C		(14)
		Carbon Dioxide	0.004625	
		Helium	0.0625	
		Hydrogen	0.0242	
		Nitrogen	1.041×10^{-4}	
		Oxygen, ASTM D-3985-80, 24°C	0.00133	

Poly(vinyl fluoride)

Property	Units	Conditions	Value	Reference
Vapor permeability	$g\ m^{-2}$ $hr^{-1}\ mil^{-1}$	ASTM E-96–80, modified, 24°C		(14)
		Acetic acid	0.45	
		Acetone	100	
		Benzene	0.9	
		Carbon tetrachloride	0.5	
		Ethyl acetate	10	
		Hexane	0.55	
		Ethyl alcohol	0.35	
Moisture absorption	%	Water immersion	<0.5	(14)
Water vapor transmission	$g\ m^{-2}d^{-1}$	ASTM E96-E-80, 39.5°C, 80% RH	9–57	(14)
Availability	–	Standard and custom colors	12–75 μm film up to 3m wide	–
Cost	$US\$\ m^{-2}$	Depends on type and color; $5,000 minimum order	1–5	–
Supplier		DuPont Co., 1007 Market Street, Wilmington, Delaware 19898. (800) 441–7515		–

* Percent monomer units.

† Percent fluorine atoms.

References

1. Frelink JG. *British Patent*. 1969;1161958 to Deutsche Solvay Werke Gesellschaft.
2. Hecht JL. *U.S. Patent*. 1966;3265678 to E. I. DuPont de Nemours and Co.
3. Uschold RE. *U.S. Patent*. 1993;5229480 to E. I. DuPont de Nemours and Co.
4. Johnston FL. *U.S. Patent*. 1950;2510783 to E. I. DuPont de Nemours and Co.
5. James VE. *U.S. Patent*. 1964;3129207 to E. I. DuPont de Nemours and Co.
6. Cais RE, Kometani JM. *Polymer*. 1988;29:168.
7. Ovenall DW, Uschold RE. *Macromolecules*. 1991;24:3235.
8. Tonelli AE, et al. *Macromolecules* 1982;15:849.
9. Wallach ML, Kabayama MA. *J. Polym. Sci., Part A-1*. 1966;4:2667.
10. Goerlitz M, et al. *Angew. Makromol. Chem.* 29/1973;30:137.
11. Golike RC. *J. Polym. Sci.* 1960;42:583.
12. Sianesi D, Caporiccio G. *J. Polym. Sci., Part A*. 1968;6:335.
13. Uschold RE Unpublished data.
14. DuPont Co. Tedlar® PVF Film technical bulletins.
15. DeBergalis M. *J. Fluorine Chem*. 2004;125:1255.

Poly(vinylidene chloride)

Anthony L. Andrady, Andrzej Kloczkowski,
Taner Z. Sen and M. Göktuğ Ahunbay

Acronym, Trade Name PVDC, Saran (copolymer)

Class Vinylidene polymers

Structure $[-CH_2CCl_2-]$

Major Applications Homopolymer and copolymers – usually with vinyl chloride (VC), or methyl acrylate (MA) – used in solvent-based or latex barrier coatings on cellophane, paperboard, plastic film, and rigid food containers. Films of copolymer used as household cling wrap. Also used with other polymers in multilayer barrier films or containers mostly in packaging applications. Also used in fibers and adhesives.

Properties of Special Interest Exceptional barrier properties with very low oxygen and water vapor permeability.

Property	Units	Conditions	Value	Reference
Preparative techniques	Radical polymerization:			
	Photochemical initiation with UV lamp			(1)
	Aqueous emulsion (redox initiators), 32°C			(2)
	Suspension (peroxide initiators)			(3)
Typical comonomers	Vinyl chloride (5–40%)			(4)
Molecular weight (of repeat unit)	$g\,mol^{-1}$	–	96.95	–
Head-to-head content	%	–	>1	(5, 6)
Molecular weight range	$g\,mol^{-1}$	–	DP = 100–10,000	(7)
Polydispersity	–	–	1.5–2.0	(8)
NMR	15% solution in hexamethylphosphoramide, 40°C			(9)
Solvents	THF (hot), tetralin (hot), trichloroethane 1,2 dichlorobenzene, dioxane, DMF, cyclohexanone, butyl acetate, cycloheptanone cyclooctanone, N-acetylpiperidine, N-methyl pyrolidinone, trimethylene sulfide			(10–14)
Nonsolvents	Hydrocarbons, chloroform, alcohols, phenol, THF, carbon disulfide			–

Poly(vinylidene chloride)

Property	Units	Conditions	Value	Reference
Thermal analysis		TGA study of PVDC carbonization temperature (200°C–300°C)		(15)
		TGA study of PVDC carbonization temperature (25°C–800°C)		(16)
		TGA and XPS study of PVDC and its graphite oxide composits		(17)
Mark-Houwink parameters: K and a	$K = \text{ml g}^{-1}$ $a = \text{None}$		$K \times 10^3$ a	
		1-Methyl-2-pyrrolidinone, 25°C	13.1 0.69	(18)
		Tetramethylene sulfoxide, 25°C	13.9 0.69	(18)
		Hexamethylenephosphoramide, 25°C	25.8 0.65	(19)
Unit cell dimensions	Å	Mono	$a = 13.69,$ $b = 4.67,$ $c = 6.296$	(20)
			$a = 22.54,$ $b = 4.68,$ $c = 12.53$	(21)
Heat of fusion	kJ mol^{-1}	At melting point	5.623	(22)
			4.60–7.95	(29)
Entropy of fusion	kJ mol^{-1}	–	0.0120	–
Density (crystalline)	g cm^3	Volumetry during polymerization	1.97	(23)
			1.80–1.97	(19)
			1.948	(20)
			1.958	(21)
Density (amorphous)	g cm^3	–	1.775	(19)
		Molding resin grade	1.65–1.72	
Glass transition temperature	K	Dynamic-mechanical	255	(22, 24)
		Dilatometry	255–258	(2, 24)
		Calorimetry	255	(25)
Melting transition temperature	K	Calorimetry	468	(26)
			471–478	(27)
			473–508	(25)
Sub-T_g transitions	K	β transition	285	(28)
Heat capacity	kJ K^{-1} mol^{-1}	100°C	0.0363	(29)
		200°C	0.0575	
		250°C	0.0690	
Tensile modulus	MPa	Machine direction	483	(30)
		Transverse direction	34.5	
Yield strength	MPa	Machine direction	69	(30)
		VD-VC molding resin grade	19.3–36.2	(19)
Tensile strength	MPa	Machine direction	73	(30)
		VD-VC molding resin grade	24.1–34.5	(29)
		Transverse direction	110	–
		PVDC–PVC copolymer with 43.2% Solid content	3.7	(31)
		PVDC film	37.9	(32)

Property	Units	Conditions		Value	Reference
Elongation	%	Machine direction		55	(30)
		Transverse direction		35	–
		VD-VC molding resin grade		160–240	(19)
		PVDC–PVC copolymer with 43.2% solid content		450	(31)
		PVDC film		8.8	(32)
Young modulus	MPa	–		463.7	(32)
Impact strength, Izod	$J\,m^{-1}$	VD-VC molding resin grade (of notch)		21.35–53.38	(19)
Hardness	Rockwell M	–		50–65	(19)
Permeability coefficient P	$m^3(STP)m\,s^{-1}\,m^{-2}$ $Pa^{-1}(\times 10^{-9})$	Temp (°C)	Gas		
		30	N_2	0.000706	(33)
		30	O_2	0.00383	(33)
		30	CO_2	0.0218	(33)
		25	H_2O	7.0	(34)
Apparent water vapor permeability	$WVPapp$	$g\,ms^{-1}\,Pa^{-1}\,(\times 10^{-12})$		1.1	(32)
Pyrolyzability	–	120–190°C		Only HCL given off (up to 60% available Cl)	(35)
Limiting oxygen index LOI	–	–		60	(36)

References

1. Burnett JD, Melville HW. *Trans. Faraday Soc.* 1950;46:976.
2. Saito S, Nakajima T. *J. Polym. Sci.* 1959;37:229.
3. Heller J, Lyman DJ. *Polym. Lett.* 1963;1:317.
4. Gabbwtt JF, Smith WM. In: Ham G, ed. *Copolymerization.* New York: Interscience; 1964, chap. V.
5. Johnsen U, *Kolloid-Z. Z. Polym.* 1966;210:1.
6. Fisher T, Kinsinger JB, Wilson CW. *Polym. Letters.* 1967;5:285.
7. Matsuo K, Stockmeyer WH. *Macromolecules.* 1975;8:660.
8. Wallach ML. *ACS Polymer Div. Preprints.* 1969;10:1248.
9. Matsuo K, Stockmeyer WH. *Macromolecules.* 1981;14:544.
10. Thinius K. *Analytische Chemie der Plaste.* Berlin: Springer Verlag; 1963.
11. Nitsche R, Wolf KA. *Struktur und Physikalisches Verhalten der Kunststoffe.* Vol 1. Berlin: Springer Verlag; 1961.
12. Roff WJ. *Fibers, Plastics and Rubbers.* New York: Academic Press; 1956.
13. Wessling RA. *J. Appl. Polym. Sci.* 1970;14:1531.
14. Wessling RA. *J. Appl. Polym. Sci.* 1970;14:2263.
15. Eliad L. *Appl. Phys. A: Mater. Sci. Proc.* 2006;82:607.

16. Kim C, Yang KS, Kim YJ, Endo M. *J. Mater. Sci.* 2003;38:2987.
17. Wang J. In: Proceedings of the 9th European Meeting on Fire Retardancy and Protection of Materials, Lille, France, Sept. 17–19, 2003, 161, 2005.
18. Matsuo K, Stockmeyer WH. *Macromolecules.* 1975;8:660.
19. Wessling RA et al. In: Mark HF, et al, eds. *Encyclopedia of Polymer Science and Engineering.* Vol 17. New York: John Wiley and Sons; 1987:492.
20. Reinhardt RC. *Ind. Eng. Chem.* 1943;35:422.
21. Narita S, Okuda K. *J. Polym. Sci.* 1959;38:270.
22. Gaur U, Wunderlich B. *J. Phys. Chem. Ref. Data* 1983;12:29.
23. Arlman EJ, Wagner WM. *Trans. Faraday Soc.* 1953;49:832.
24. Boyer RF, Spencer RS. *J. Appl. Phys.* 1944;15:398.
25. Park HC, Mount EM. In: Mark HF, et al, eds. *Encyclopedia of Polymer Science and Engineering.* 2nd ed. Vol 7. New York: John Wiley and Sons; 1987:89.
26. Okuda K. *J. Polym. Sci., Polym. Chem. Ed.* 1964;2:1749.
27. Wessling RA, Oswald JH, Harrison IR. *J. Polym. Sci., Phys.* 1973;11:875.
28. Schmeider K, Wolf K. *Kolloid-Z. Z. Polym.* 1953;134:149.
29. Gaur U, Lau SF, Wunderlich BB. *J. Phys. Che. Ref. Data.* 1983;12:29.
30. Jack J. *Brit. Plastics.* 1961;34:391.
31. Zhong S, Chen Z. *Cem. Conc. Res.* 2002;32:1515.
32. Herald TJ, Hachmeister KA, Huang S, et al. *J. Food Sci.* 1996;61:415.
33. Waack R, et al. *Ind. Eng. Chem.* 1955;47:2524.
34. Myers AW, et al. *TAPPI* 1961;44:58.
35. Bohme RD, Westling RA. *J. Appl. Polym. Sci.* 1972;16:1761.
36. Johnson DG. *J. Appl. Fire Sci.* 1994–95;4: 185.

Poly(vinylidene fluoride)

Jerry I. Scheinbeim, Qiming Zhang and Shihai Zhang

Acronyms, Trade Names PVDF, PVF2, Kynar, Solef, Hylar, Neoflon, Foraflon, KF, Soltex

Class Vinylidene polymers

Structure $-(CH_2CF_2)_n-$

Major Applications Wire and cable insulation, tubing, piping, sheet and melt-cast films for electrical and electronics, binder for high-quality metal finishes for building components used on exterior wall panels, roofing shingles, and on industrial, commercial and residential buildings, binders for lithium-ion battery electrode, membranes for battery, used in fluid handling systems for solid and lined pipes, fittings, valves, and pumps, in manufacture of microporous and ultrafiltration membranes, chemical-tank lining, piezoelectric sensors for telephone headset, hydrophones, keyboards and, printers, pyroelectric sensor for infrared sensing, high energy density capacitor films.

Properties of Special Interest Excellent mechanical properties and resistance to severe environmental stress, good chemical resistance, good piezoelectric and pyroelectric properties, high dielectric constant and high dielectric breakdown strength.

Preparative Techniques Emulsion polymerization: (a) 300–800 psig, perfluorinated surfactant initiator, 65–85°C, 2–6 h;[1] (b) 200 Ib in^2, 50–110°C, fluorinated surfactant, 17–21 h, iron powder.[2]

Suspension polymerization: suspending agent, reaction accelerator, water soluble initiator, 300–1,000 psig, 35–100°C.[3]

Property	Units	Conditions	Value	Reference
Monomer and molecular weight	g mol^{-1}	$CH_2=CF_2$	64.034	–
Head-to-head sequences	%	$CF_2-CF_2-CH_2-CH_2$	3.5–6	(4–6)
Typical molecular weight range	g mol^{-1}	–	$3.4-40 \times 10^4$	(7)
Typical polydispersity index	–	–	1.62–2.14	(7)
Tacticity	% Isoregic	–	95–97	(8)
Morphology (crystal forms)	–	–	$\alpha, \beta, \gamma, \delta$	(8)

Poly(vinylidene fluoride)

Property	Units	Conditions	Value	Reference
IR (characteristic absorption frequency)	cm^{-1}	α form	530	(9, 10)
			615	(9)
			764	(9)
			796	(9)
		β form	442	(9)
			470	(9)
			484	(9)
			510	(9)
IR (characteristic absorption frequency)	cm^{-1}	γ form	430	(9)
			481	(9)
NMR	–	–	–	(11–15)
Crystal form/density	g cm^{-3}	α form	1.92	(8, 16)
		β form	1.8	
		Amorphous	1.68	(16)
Density (crystalline)	g cm^{-3}	Molded at 170°C (quenched to 0°C)	175–1.78 (47%)	(17)
		Molded at 170°C (quenched to room temperature)	1.779 (60%)	(18)
		Molded at 170°C (quenched to 0°C)	1.768 (65%)	(18)
		Annealed at 120°C for one day	1.769 (69%)	(18)
Thermal coefficient of linear expansion	K^{-1}	–	0.7–1.5($\times 10^{-4}$)	(19)
Thermal conductivity	Wm^{-1} K^{-1}	26–160°C	0.17–0.19	(19)
Compressive strength	MPa	At 25°C	55–90	(19)
Solvents	–	–	Acetone, benzaldehyde, DMF, THF	(20)
		Above 60°C	Acetophenone	(7)
		25°C	DMAc	(7)
		150–190°C	Benzophenone	(7)
Nonsolvents	–	–	Acetic acid, benzyl alcohol, 1,2-dibromoethane, ethanol	(21)
Solubility parameters	MPa	DMA	16.8	(21)
		DMF	17.4	
		DMSO	18.4	
Mark-Houwink parameters: K and a	$K = $ ml g^{-1}	DMAc	$K = 17.8 \times 10^{-6}$, $a = -0.74$	(22)
	$a = $ None	DMF	$K = 31.7 \times 10^{-6}$, $a = -0.70$	(22)
		NMP	$K = 48.8 \times 10^{-6}$, $a = -0.68$	(22)

Property	Units	Conditions	Value	Reference
		Acetophenone, 85° C	$K = 2.13 \times 10^{-4}$, $a = 0.62$	(7)
		Benzophenone, 165°C	$K = 13.6 \times 10^{-4}$, $a = 0.44$	(7)
		Benzophenone, 180° C	$K = 7.54 \times 10^{-4}$, $a = 0.49$	(7)
Second virial coefficient	$cm^3\ g^{-2}\ mol$	Acetophenone, 85°C	$0.3\text{–}7.6\ (\times 10^{-4})$	(7)
Root-mean-square radius of gyration	–	DMAc	$29.5 M_w^{0.55}$	(22)
		DMF	$813 M_w^{0.51}$	
		NMP	$28.2 M_w^{0.45}$	
Unit cell dimensions			Unit cell angles	(16, 23)
Form I (β)	Å	Lattice = orthorhombic; space group = Cm2m-C_{2v}^{14}	$a = 8.58$, $b = 4.91$, $c = 2.56$	
Form II (α)	Å	Lattice = monoclinic; space group = P2$_1$/c-C_{2h}^5	$a = 4.96$, $b = 9.64$, $c = 4.62$	
	Degrees		$\beta = 90$	
Form III (γ)	Å	Lattice orthorhornbic; space group = C2cm	$a = 4.97$, $b = 9.66$, $c = 9.66$	(24, 25)
	Degrees		$\gamma = 91$	
Form IV (δ)	Å	Lattice = orthorhombic; space = P2cn	$a = 4.96$, $b = 9.64$, $c = 4.62$	(24, 25)
Degree of crystallinity	%	–	50	(16, 20)
Heat of fusion	Jg^{-1}	–	30.5	(9)
		Draw ratio of 4	41.4	
Specific heat	$J(g\ k)^{-1}$	23 °C to 100 °C	1.2–1.6	(30)

Avrami exponent [26]

Crystallization Temp. (K)	Avrami Exponent	Kinetic Rate Constant (min$^{-3.94}$)	Half Time of Conversion $t^{1/2}$(min)
407	3.82	0.49	1.09
409	4.62	0.03	2.15
412	3.62	0.02	2.6
414	4.6	87×10^{-5}	5.45
417	3.3	44×10^{-6}	11.60
419	4.35	46×10^{-7}	20.60
420	2.99	10×10^{-7}	30.00
422	4.23	15×10^{-8}	49.00

Property	Units	Conditions	Values	Reference
Melting point	K	Depends on polymorph	443–473	(8)
			451(DSC)	(23)
		Oriented PVF2 Film (β crystal)	439	(23)

Poly(vinylidene fluoride)

Property	Units	Conditions	Values	Reference
Glass transition temperature	K	–	238	(23)
Other transition temperatures (DMA, DSC) (relaxation)	K	(α_2) (α_1) (β) (γ)	323 373 235 203	(27)
Elastic modulus	GNm^{-3}	β phase	1–3	(8)
Acoustic impedance	$ggm^{-2}s^{-1}$	β phase	2–3	(8)
Tensile strength (ultimate elongation)	%	$190\ kg\ mm^{-2}$ fiber, 59% crystallization, crystal melting point 184°C	22	(28)
		42–58 MPa (25°C)	50–300	(19)
		34.5 MPa (100°C)	200–500	(19)
Tensile yield strength	MPa	Commercial grade	42.8	(29)
Ultimate tensile strength	MPa	Commercial grade	43.8	(29)
Tensile modulus	MPa	Commercial grade	1,194.4	(29)
Elongation at break	%	Commercial grade	43	(29)
		Commercial high molecular weight grade, ASDM D638 23°C, 50 mm/min	100	(30)
Yield point	MPa	At 25°C At 100°C	38–52 17	(19)

Elastic modulus	GPa	At 25°C
		    Tensile modulus
		    Compression modulus
			
			 1.0–2.3
			 1.0–2.3
				(19)

Flexure modulus	GPa	ASTM D790, 23°C, 50 mm/min^{-1}	1.3–2.3	(30, 31)
Flexure modulus	MPa	ASTM D790, 23°C, 50 mm/min^{-1}	55–76	(30, 31)
Notched Izod Impact Strength	J/m^{-1}	ASTM D256, 23°C	107	(30, 31)
Unnotched Izod Impact Strength	J/m^{-1}	ASTM D256, 23°C	1070	(30, 31)
Abrasion resistance	–	Tabor CS-17, 0.5 kg load, mg $(1,000\ cycles)^{-1}$	17.6	(19)
Index of refraction	–	–	1.42	(19)
Hardness, Shore D		ASTM D2240	75	(30)

Dielectric constant	–	At 25°C
		    60 Hz
		    10^3 Hz
		    10^6 Hz
		    10^9 Hz
			
			 9–10
			 8–9
			 8–9
			 3–4
				(19)

Property	Units	Conditions	Values	Reference
Dissipation factor	%	–	3–5	(19)
			5–2	
			3–5	
			9–11	
Volume resistivity	Ohm-cm	–	5×10^{14}	(ASTM D 257)
Dielectric strength	V/mil^{-1}	0.003175m thickness	260	(19)
		0.000203 m thickness	1,300	
Piezoelectric coefficient	cgs esu	α phase, 38% crystallinity poled at 140°C	0.32×10^{-7}	(32)
	pC/N	β phase	20–30	(8)
	pC/N	δ phase	2–3	(8)
Pyroelectric coefficient	μ Ckm^{-2}	β phase	30–40	(8)
Optical transmittance	–	Visible/UV	80%	(17)
Specular transmittance	%	At 0.5° cone	85–90	(17)
Coefficient of friction	–	PVF2 to steel	0.14–0.17	(19)
Contact angle	Degrees	Water	82	(33)
		Methylene oxide	63	
		Formaldehyde	59	
		α–Bromonaphthalene	42	
		Glycerol	75	
		Tricresyl phosphate	28	
Solid surface tension	–	Harmonic means	37.4	(33)
		Geometric means	36.2	
		Critical surface tension	36.5	
		Equation of state	25	
Thermal decomposition	°C	–	390	(19)
		Charring	480	
Chemical resistance	–	Inorganic acids	No effect	(19)
		Halogens	No effect	(19)
		Oxidants	No effect	(19)
		Weak bases	No effect	(19)
		Aliphatic, aromatic and chlorinated solvents	No effect	(19)
		Strong bases	Softening	(19)
		Amines, esters, and ketones	Swelling and dissolution	(19)
		Acetone (30 min at room temperature)	Etching	(32)
Water absorption	%	–	0.01–0.04	(19)
Flammability	–	–	Low to none	–

Poly(vinylidene fluoride)

Property	Units	Conditions	Values	Reference
Flame propagation rate	ft $(20\ \text{min})^{-1}$	Maximum flame spread	2.0	(34)
Intrinsic viscosity	–	Commercial grade, 35% crystallinity, melting point 160°C	1.40–1.43	(29)
Viscosity	dlg^{-1}	DMAc	1.29	(21)
		NMP	1.28	
		DMF	1.17	
		DMSO	1.05	
Melt viscosity	poise	Commercial grade, 250°C, shear rate $= 10^3 \text{s}^{-1}$	62×10^2	(29)
	poise	Commercial grade, 450°F, shear rate $= 100 \text{s}^{-1}$	$4\text{–}30 \times 10^3$	(31)
Moisture vapor permeability	$\text{g day}^{-1}\ \text{m}^{-2}$	1 mm thickness	2.5×10^{-2}	(19)
Gas permeability	$\text{cm}^3(\text{STP})/$ (cms mmHg)	Argon, 25°C, 5.21×10^{-3} cm thickness	$2 \times 10^{-1}2$	(35)
Diffusivity	cm^2s^{-1}	Argon, 25°C, 5.21×10^{-3} cm, thickness	4×10^{-9}	(35)
Cost	kg^{-1}	–	15–17	
	lb^{-1}	–	8~10	
Suppliers and trademarks		Arkema, France	Kynar	(19)
		Daikin Kogyo Co., Japan	Neoflon	
		Kureha Chemical Co., Japan	KF	
		3M, USA	Hylar	
		Solvay and Cie, Belgium	Solef/Solexis	

References

1. McCain GC, Semancik JR, Dietrich JJ. *U.S. Patent.* 1969;3475396, to Diamond Shamrock Corporation.
2. Iserson H. *U.S. Patent.* 1966;3245971, to Pennwalt Chemical Corporation.
3. Dohany J. *U.S. Patent.* 1973;37781265, to Pennwalt Corporation.
4. Bachmann MA, et al. *J. Appl. Phys.* 1979;50:6106.
5. Mattern DE, Fu-Tyan L, Hercules DM. *Anal. Chem.* 1984;56:2762–2769.
6. Lovinger AJ, et al. *Polymer* 1987;28:617–626.
7. Welch GJ. *Polymer* 1974;15:429.
8. Lovinger A. *Science* 1983;220:1115.
9. Mead WT, et al. *Macromolecules.* 1979;12(3): 473.
10. Liepins R, et al. *1. Polym. Sci, Polym. Chem. Ed.* 1978;16.
11. Katoh E, Ogura K, Ando I. *Polym. J.* 1994;26(12):1352.
12. Cais RE, Kometani JM. *Macromolecules.* 1985;18:1357.
13. McBrierty VJ, Douglass DC Weber TA. *J. Polym. Sci., Polym. Phys. Ed.* 1976;14:1271.
14. Clements J, Davies GR, Ward IM. *Polymer.* 1985;26(2);208.

15. Lin F. *J. Macromol. Sci.* 1989;A26(1):1–16.
16. Nagakawa K, Ishida Y. *Kolloid Z.Z. Poly.* 1973;251:1003.
17. *Plastic Film Performance Improvement for Heliostats.* Report SAND 79–8185, Sandia National Laboratories, Albuquerque, July 1980.
18. Enns JB, Simha R. *J. Macromol. Sci.-Pliys.* 1977;B13(1):11–24.
19. Dohany JE, Humphrey JS. In: Mark HF, et al, *Encyclopedia of Polymer Science and Engineering.* Vol 17. New York: John Wiley and Sons; 1989:532.
20. Pae KD, Bhateja SK, Gilbert JR. *J. Polym. Sci., Part B: Polym. Phys.* 1987;25:717.
21. Botlino A, et al. *J. Polym. Sci., Part B: Polym. Phys.* 1988;26:785.
22. Ali S, Raina AK. *Makromol. Chem.* 1978;179:2925.
23. Kobayashi M, Tashiro K, Tadokoro H. *Macromolecules.* 1975;8(2):158.
24. Weinhold S, Litt MH, Lando JB. *Macromolecules.* 1980;13(5):1178.
25. Bachmann MA, et al. *J. Appl. Phys.* 1980;51(10):5095.
26. Mancarella C, Martuscelli E. *Polymer* 1977;18:1240–1242.
27. Lovinger A, Wang TT. *Polymer* 1979;20:725.
28. Mizuno T, Murayama N. *U.S. Patent* 1985;4546158, to Kureha Kagaku Kogyo Kabushiki Kaisha Chemical Company, Tokyo.
29. Stallings JP. *U.S. Patent.* 1973;3780007, to Diamond Shamrock Corporation.
30. Solvay Solef (1010) PVDF homopolymer datasheet.
31. Kynar PVDF homopolymer datasheet.
32. Murayama N, et al. *J. Polym. Sci., Polym. Phys Ed.* 1975;13:1033.
33. Dalal EN. *Langmuir* 1987;3:1009.
34. Bretz PE, Hertzberg RW, Manson JA. *Polymer* 1981;22:1272.
35. Odhner OR, Michaud JW. *U.S. Patent,* 1983;4401845, to Pennwalt Corporation.
36. Fujii M, Stannett V, Hopfenberg HB. *J. Macromol. Sci.-Phys.* 1978;B15(3):421.

Poly(vinyl methyl ether)

Jianye Wen

Acronyms, Trade Names PVME, PVM, Lutonal M, Gantrez M

CAS Number 9003-09-2

Class Vinyl polymers

Structure

$$[- CH_2 - CH -]$$
$$|$$
$$OCH_3$$

Major Applications Plasticizer for coatings; aqueous tackifier; adhesion promoter of nonadhering materials to glass, metal, and plastics; copolymers used in pharmaceuticals, lens arrays for optical device (as thermographic copying material).

Properties of Special Interest Viscous and balsamlike; high adhesion to high and low surface-energy free substrates.

Property	Units	Conditions	Value	Reference
Bulk density	g cm^{-3}	Gantrez M-154^{a}	1.03	(2)
		Gantrez M-574	0.96	(2)
		Gantrez M-555	0.94	(2)
		Gantrez M-550	0.94	(2)
		At 20°C	1.0580	(3)
		At 40°C	1.0436	(3)
		At 60°C	1.0294	(3)
		At 80°C	1.0152	(3)
		At 100°C	1.0011	(3)
		At 120°C	0.9871	(3)
		25–120°C	$1.725 - 7.259 \times 10^{-4} T$ $+ 0.116 \times 10^{-6} T^2$	(3)
Coefficient of thermal expansion	10^{-4} K^{-1}			
		At 40°C	6.87	(3)
		At 60°C	6.92	(3)
		At 80°C	6.96	(3)
		At 100°C	7.01	(3)
		At 120°C	7.06	(3)
Cohesive energy E_{coh}	J mol^{-1}		17299	(4)

Poly(vinyl methyl ether)

Crystallographic data[5,6]

System	Crystal Space Group	Unit Cell Parameter			Density (g cm^{-3})	Chain. Conf. $N*P/Q$
		A	B	C		
RHO	D3D-6	16.20	16.20	6.50	1.175	2*3/1
RHO	D3D-6	16.25	16.25	6.50	1.168	2*3/1

Property	Units	Conditions	Value	Reference
Flash point	°C	Gantrez M-154	None	(2)
		Gantrez M-574	22	(2)
		Gantrez M-555	22	(2)
		Gantrez M-550	22	(2)
FTIR spectra	cm^{-1}	Selected group assignments		(7)
		C—CH2 bending	1381, 1457	
		C—O—C stretch	1189	
		Aliphatic C—H stretches	2934	
Glass transition temperature T_g	°C		−34	(8)
			−31	(9)
			−33	(10)
			−26	(11)
			−25	(12)
			−24	(13)
Isothermal compressibilities	$\times 10^{-5}$ bar^{-1}	At 20°C	5.3	(14)
		At 40°C	5.8	(14)
		At 60°C	6.4	(14)
		At 80°C	7.2	(14)
		At 100°C	8.1	(14)
		At 120°C	9.2	(14)

Mark-Houwink coefficients[15]

Solvent	Temp. (°C)	Mol. Wt. Range ($M \times 10^4$)	$K \times 10^3$ (ml g^{-1})	a
Benzene	30	−45	76	0.60
Butanone	30	−45	137	0.56

Property	Units	Conditions	Value	Reference
Melting temperature T_m	°C		144	(8)
			144/114	(16)
Radius of gyration	Nm	$M_w = 57800$		
		$M_n = 8300$	18	(17)
Refractive index, n_D		30°C, isotactic	1.4700	(18)
		25°C	1.469–1.478	(19)

Poly(vinyl methyl ether)

Property	Units	Conditions	Value	Reference
Solvent		Water, toluene		(2)
		Halogenated hydrocarbons, benzene, n-butanol, methyl ethyl		(20)
		Ketone, ketone, ethanol, acetone, ethylacetate, water (cold)		(20)
Non-solvents		Heptane, ethylene glycol, ethyl ether, water (hot)		(20)
		(methanol, acetone, and water for crystalline polymer)		(20)
Specific viscosity η_{sp}	1 g in 100 ml	Gantrez M-154	0.47	(2)
		Gantrez M-555	0.77	(2)
		Lutonal M[c]	0.68	(2)
Solubility parameter	$(J\ cm^{-3})^{1/2}$	19.66		(21)
Surface tension	mN m^{-1}	$M_n = 46,500\ M_w = 99,000$		(22)
		at 20°C	31.8	
		at 150°C	22.1	
		at 200°C	18.3	
	mN (m K)$^{-1}$	$-d\gamma/dT$	0.075	
Viscosity	P	Gantrez M-555	Ca 15	(2)
		Gantrez M-574	Ca 30	(2)
		Gantrez M-154	Ca 40	(2)

[a] Products of GAF Corp.

[b] See ref. 14 for details.

[c] Products of BASF Corp.

Unperturbed dimension[b(23,24)]

Conditions	$r_0/M^{1/2} \times 10^4$ (nm)	$r_{of}/M^{1/2} \times 10^4$ (nm)	$\sigma = r_0/r_{of}$	$C_\infty = r_0^2/nl^2$
Benzene; butanone 30°C	900 ± 30	404	2.23 ± 0.13	9.95

References

1. General references: (a)In: Brandrup J, Immergut EH, eds. *Polymer Handbook*. Third ed. New York: Wiley-Interscience; 1989; (b) Biswas M, Mazumdar A, Mitra P. In: Bikales NM, ed. *Encyclopedia of Polymer Science & Technology*. Vol 17. New York: Wiley-Interscience; 1987:447; (c) Mark JE, ed. *Physical Properties of Polymers Handbook*. Woodbury, New York: AIP Press; 1996.
2. Gantrez M. *Tech. Bull.*, 8740, GAF Corp., 1970.
3. Orwoll RA, In: Mark JE, ed. *Physical Properties of Polymers Handbook*. Chap. 7, Woodbury, New York: AIP Press; 1996.
4. Yu X, Wang X, Li X, Gao J, Wang H. *J. Polym. Sci. Part B*. 2006;44:409.
5. Bassi IW, Atti. *Accad. Nazl. Lincei, Cl. Sci. Fis., Mat. Nat., Rend.* 1960;29:193.

6. Corradini P, Bassi IW. *J. Polym. Sci. part C.* 1968;16:3233.

7. The Aldrich library of FT-IR spectra , C. J. Charles, Ed. 1 Milwaukee, Wis. Aldrich Chemical Co, V2, P1186,1985.

8. ND Field, Lorenz DH, In: Leonard EC, ed. *Vinyl and Diene Monomers Part 1.* New York: Wiley-Interscience; 1970:365.

9. Nielson LE. *Mechanical Properties of Polymers.* New York: Reinhold; 1962.

10. Arzondo L. *Macromolecular Rapid Communications.* 2005;26:632.

11. Schwartz GA. *Macromolecules.* 2006;39:3931.

12. Pyda M. *J. Polym. Sci., Part B: Polymer Physics.* 2005;43:2141.

13. Schwartz GA. *J. Physical Chemistry.* 2006;39:3581.

14. Shiomi T, Hamada F, Nasako T, et al. *Macromolecules.* 1990;23:229.

15. Manson JA, Arquette GJ. *Makromol. Chem.* 1960;37:187.

16. Bassi IW. *Atti Accad. Nazl. Lincei, Cl. Sci. Fis., Mat. Nat., Rend.* 1960;29:193.

17. Querner C, Schmidt T, Arndt K-F. *Lamgmuir.* 2004;20:2883.

18. Seferis JC. In: Brandrup J, Immergut EH, eds. *Polymer Handbook.* Third ed. New York: Wiley-Interscience; 1989:VI461.

19. Yu X, Wang X, Li X, Gao J. *QSAR Comb. Sci.* 2006;25:156.

20. Fuchs O. In: Brandrup J, Immergut EH, ed. *Polymer Handbook.* Third ed. New York: Wiley-Interscience; 1989:VII379.

21. Hegewald J. *Langmuir.* 2006;22(11):5152.

22. Koberstein JT. *Chem. Eng. Dept.* The Princeton Univ., Princeton, New Jersey, private communication, 1986.

23. Kurata M, Tsunashima Y. In: Brandrup J, Immergut EH, ed. *Polymer Handbook.* Third ed. New York: Wiley-Interscience; 1989:VII1.

24. Kurata M, Stockmayer WH. *Fortschr. Hochpolymer. Forsch.* 1963;3:196.

Poly(4-vinyl pyridine)

Osei A. Owus and John H. Ko

Acronyms, Alternate Names, Trade Names Pyridine, 4-ethenyl homopolymer, Reilline, Poly (4-pyridylethylene), Poly (p-vinylpyridine) P4VP

Class Vinyl Polymers; homopolymers

CAS 25232-41-1, 9017-40-7

Structure

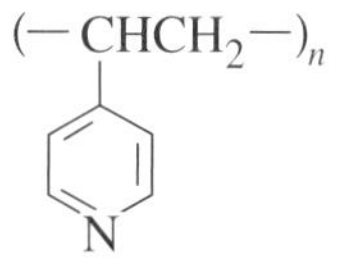

$$(-CHCH_2-)_n$$

Major Application Poly (4-vinylpyridine) P4VP with its nucleophilic and weakly basic ring nitrogen has found uses in the areas of metal recovery (complex), and pollution control for removal of acidic and neutral materials. It is also used as an acid scavenger and catalyst and catalyst support. Commercial resin beads are mostly prepared by suspension polymerization with crosslinker such as divinylbenzene.[1,2]

Property	Units	Conditions	Value	Reference
Solubility	%	Soluble in carboxylated polysulfone with degrees of carboxylation of 0.43, 0.93, 1.38 and 1.93		(3)
		Soluble in water when pH < 4.7		(4)
Basicity, pKa		45 % ethanol/55 % water	3.25	(5)
Density	$g\ cm^{-3}$	At 20°C	1.114	(5)
Flash point	°C		93.3	(5)
Melting point	K	DSC, isotactic P2VP crystallized at 130°C	450, 472.5 (two peaks)	(6)
Heat of fusion	$kJ\ mol^{-1}$	Isotactic P2VP	2.07	(6)
Glass transition temperature	°C	P4VP at 200,000 g mol^{-1}	160	(7)
		P2VP at 150,000 g mol^{-1}	106	(8)

Poly(4-vinyl pyridine)

Property	Units	Conditions	Value	Reference
Permeability		Barriers at 35°C and 619 cm Hg		(8)
		He	12.36	
		H_2	12.64	
		CO_2	3.31	
		O_2	0.84	
		N_2	0.13	
		CH_4	0.14	
Dielectric constant		10 kHz, 50 K	2.88	(9)
DC Conductivity	$S\ cm^{-2}$	25°C	10^{-14}	(10)

Suppliers Aldrich Chemical Co., 1001 West St Paul Ave., Milwaukee, WI 53233, USA.
Reilly Chemicals, SA, Rue Defacqz 115, Boite 19, B-1050 Brussels, Belgium.
Carbomer, Inc., P. O. Boxx 721, Westborough, MA 01582, USA.

References

1. Frechet J, Vivas de Meftahi M. *Br. Polym. J.* 1984;16:193.
2. Sugii A, Ogawa N, Iinuma Y, Yamamure H. *Talanta.* 1981;28:551.
3. Goh SH, Lau WY, Lee CS. *Polym. Bulletin.* 1992;29:521.
4. Sidorov SN, Bronstein LM, Kabachii YA, Valetsky PM, Soo PL, Maysinger D, et al. *Langmuir.* 2004;20: 3543.
5. Frosini V, Petris S. *Chim. Ind.* 1967;49: 1178.
6. Aberda van Ekenstein C, Tan Y, Challa G. *Polymer.* 1985;28:283.
7. Rodrigues JR, Goncalves D, Mangrich S. *Adv. Polym. Tech.* 2000;19:113.
8. Jyh-Jeng S. *NAMS Gas Separation Symp.* 2001.
9. Shimizu K, Yano O, Wada Y. *J. Polym. Sci.* 1975;13:2357.
10. Vijayalakshmi R, Ashokan PV, ShridharMH. *Matl. Sci. Eng.* 2000;276:266.

Poly(N-vinyl pyrrolidone)

John H. Ko

Acronyms, Alternate Names, Trade Names PVP, Povidone, Crospovidone, Luviskol®, Kollidon®, Divergan®, Plasdone®, Biodone®, Polyclar®, Albigen®, Peregal®

Class Vinyl polymers; homopolymers

CAS 9003-39-8

CAS No For cross-linked polyvinylpyrrolidone 25249-54-1

Structure

$$(-CHCH_2-)_n$$

Major Application Additives (clarifying agent, stabilizing agent, viscosity modifier), adhesives, agriculture products, coatings (paints and surface coatings, inks, paper printing); personal care products (detergent, shampoos, toothpastes, hair spraying agent, dye); medical devices (contact lenses, lubricious coating, biocompatibility coating, complex, blood plasma expander); pharmaceuticals (control release as a binder in many pharmaceutical tablets, stabilizer for polymerization, thickener), Active Ingredients and Advanced Intermediates. The main products are caffeine, povidone iodine; ink-jet papers and transparencies, inks for inkjet printers, photography.

Properties of Special Interest PVP power is white, stable, hygroscopic and water soluble. Forms complexes with many substances.[1,2,3] Coated PVP solution forms brittle, clear and glassy films.

Property	Units	Conditions	Value	Reference
Solubility	%	Soluble	>10	(4)
		Insoluble	<1	
Solvents		Water, alcohol (methanol, ethanol, propanol, butanol, glycol); ester alcohol (ethylene glycol monoethylether, diethylene glycol, polyethylene glycol, 1, 4-butanediol); chlorinated hydrocarbons (dichloromethane, chloroform); Amine (butylamine, ethylenediamine); acid (formic, acetic, propionic); dilute acid, base, low salt solutions.		
Nonsolvents		Hydrocarbons (benzene, hexane, pentane, cyclohexane, toluene, xylene, mineral oil); ethers (dioxane, diethyl ether, ethyl vinylether); ketones (acetone, cyclohexanone); esters (ethyl acetate, methyl acetate); chlorinated hydrocarbons (carbon tetrachloride, chlorobenzene)		

Property	Units	Conditions	Value			Reference
K-value*			$M_w \times 10^{-4}$	$n \times 10^{-4}$	$M_v \times 10^{-4}$	
		17	0.9	0.25	0.92	
		30	0.45	1	4	
		60	35	9	30	
		80	90	28.5	60	
		90	120	36	110	
Coil dimension (end-to-end distance)	nm	In 0.9% NaCl				
		$K = 12$	2.2			
		$K = 17$	5			
		$K = 20$	7			
		$K = 90$	100			
pH	–	5% in water	3–7			(3, 4)
		10% in water ($K = 27$–103)	5–9			(12)
Theta temperature	K	$M_n = 3.26 \times 10^5$				(6)
		0.55 M Na_2SO_4/water	301			
		Water	297			
		2-Propanol	297			
Second virial coefficient	mol cm^3 g^{-2}	Osmotic pressure in 2-propanol				(6)
		$M_n = 3.26 \times 10^5$	0.58			
		$M_n = 1.68 \times 10^5$	0.63			
		$M_n = 0.99 \times 10^5$	0.78			
Mark-Houwink parameter	–	Water	0.82			(6)
Heat of solution	kJ mol^{-1}	Water	~16.6			(6)
		0.2 molal Na_2SO_4/water	~11.6			
Melting point		MW ~29,000	~300°C			
Glass transition temperature	K	DSC, M_w, light scattering				(4)
		$M_w = 9 \times 10^3$	382			
		$M_w = 4.5 \times 10^4$	448			
		$M_w = 3.5 \times 10^5$	449			
		$M_w = 9 \times 10^5$	452			
		$M_w = 1.2 \times 10^6$	452			
		$M_v = 7.5 \times 10^5$, viscometry				(7, 8)
		0%	448			
		2%	427			
		8%	368			
		16%	318			
		Cross-linked PVP (Crospovidone)	463–468			(5)
Density	g cm^{-3}	25°C	1.25			(9)
Diffusion coefficient	cm^2 s^{-1}	Electrophoresis	4.81×10^{-7}			(10)

Poly(N-vinyl pyrrolidone)

Property	Units	Conditions	Value	Reference
Index of refraction n_D	–	At 25°C	1.53	(9)
Index of refraction n	–	Ultrasonic at 30°C Concentration (in water) 0.000875 mol l^{-1} 0.001500 mol l^{-1} 0.004000 mol l^{-1}	 1.339 1.343 1.357	(11)
Diffusion coefficients	cm^2 min^{-1}	Dose (KGy), $M_w = 700,000$ 26 64 96 124	$D \times 10^7$ 2.51 3.46 4.60 6.56	(12)

* K-value represents a viscosity index relating to molecular weight and is calculated by the following Fikentscher's formula with relative viscosity which is measured by capillary viscometer at 25°C.

Suppliers BASF Corp., 100 Cherry Hill Road, Parippany, NJ 07054, USA.
ISP, 1361 Alps Road, Wayne, NJ 07470, USA.
Nippoh Chemicals Co., Ltd. CM Bldg., 3-3-3.[13]
Nihonbashi-Muromachi,Chuo-ku, Tokyo.

References

1. Blecher L, et al. In: *Handbook of Water-Soluble Gums and Resins*. New York: McGraw-Hill; 1980.
2. Gargallo L, Radic D. In: *Polymeric Materials Encyclopedia*. Vol 9. Boca Raton, Fla: CRC Press; 1996.
3. Hort E, Waxman B. In: Kroschwitz JI, ed. *Kirk-Othmer Encyclopedia of Chemical Technology*. 3rd ed. Vol 23. New York: John Wiley and Sons; 1983:960.
4. BASF product literature: Luviskol® PVP Polymers and Kollidon® PVP Polymers, 1993.
5. Haaf F, Sanner A, Straub F. *Polym. J.* 1985;17(1):143.
6. Meza R, Gargallo L. *Eur. Polym. J.* 1977;13:235.
7. Tan Y, Challa G. *Polymer*. 1976;17:739.
8. del Pilar Buera M, Levi G, Karel M. *Glass Transition in PVP: Effect of Molecular Weight and Dilunets*. CS amd AICE, 1982:144138.
9. Schidknecht C. *Vinyl and Related Polymers*. New York: John Wiley and Sons; 1952.
10. Miller L, Hamm F. *J. Phys Chem*. 1953;57:110.
11. Rajulu A, et al. *Acustica*. 1991;75:213.
12. Nippon Shokubai website: http://www.shokubai.co.jp/eng/products/pvp.html
13. Hatice Kaplan, Ali Güner. *J. Appl. Polym. Sci.*, Vol. 78, Issue 5, 2000. pp. 994–1000.

Poly(*p*-xylylene)

Amit S. Kulkarni, Shrish Rane and Gregory Beaucage

Acronym, Trade Name PPX, Parylene N (Union Carbide)

Class Polyaromatics

Structure

Major Applications Films and coatings on electronic components; insulating applications.

Special Properties High thermal stability; excellent dielectric and barrier properties; high resistance to electronic irradiation, solvent resistance, high degree of crystallinity.[1,2]

Property	Conditions	Reference
Synthesis	Pyrolitic decomposition polymerization of cyclic di-*p*-xylylene.	(3–7)
	Plasma decomposition of cyclic di-*p*-xylylene.	
	Vapor phase pyrolysis of di-*p*-xylylene or diesters of α,α'-dihydroxy-1,4-xylylenes or α,α'-dibromo-1,4-xylylenes in the presence of Zn/Cu.	
	Electrochemical polymerization of α,α'-dibromo-1,4-xylylenes in DMF and tetra-ethyl-ammonium-bromide (TEAB) as the electrolyte.	
	By the "Wessling Process": heating high molecular weight water soluble precursor electrolyte (α,α'-bis-tetrahydrothiophenium chloride)-p-xylylene with NaOH.	(8, 9)
	Chemical vapor deposition synthesis of PPX and PPX copolymers and derivatives	(10)
	PPX by Gorham process via vapor deposition	
	Polymerization by metal ion mediated self assembly	(11)
	PPX derivatives by Gilch-polymerization (graft copolymers)	(12)
	Controlled molecular weight synthesis using Gilch process with dioxane and dioxane/co-solvent mixtures	(13)
	Photopolymerization of metal-PPX mixtures	(14)
	Photopolymerization of PPX with Mn	(15)
	Reductive coupling polymerization of hydroxyl functionalized PPX	(16)
	PPX by base induced dehydrohalogenation	(17)
	Review of different synthesis techniques	(18)

Poly(p-xylylene)

IR properties[7,19]

Values of ν (cm^{-1})	Types of Vibrations
3,150; 3,110; 3,060; 3,030; 2,995	C—H stretch if aromatic ring
2,950; 2,935; 2,870	Asymmetrical and symmetrical C—H stretch of —CH$_2$—
1,900; 1,795	Characteristic bands for a 1,4 substituted aromatic ring
1,497, 1,350	Deformation of C—H from —CH$_2$—
1,210; 1,142; 1,080; 1,021	Planar vibration of C—H from aromatic ring
820	Extra planar vibration of C—H from aromatic ring
540	Extra planar vibration of C—C from aromatic ring

Morphology[3,5,20,21]

PPX	Morphology	Lamellar width (nm)	Lamellar thickness (nm)	Comments
Fiber	Main fiber is made of secondary fibrils	~500 (fibril width)	~10 (fibril thickness)	Low magnification electron microscopy
Fiber	Shish-Kebab	~100	~25	High magnification electron microscopy
Melt crystallized films	Spherulitic	–	~8	–
Solution grown films	Lamellae	–	~12	–

Crystal structure

Unit Cell Type	Unit Cell Dimensions (nm)			
	a	b	c (Fiber Axis)	
α-Monoclinic	0.592	1.064	0.655	$\beta = 134.7°$
β-Trigonal	2.052	2.052	0.655	$\gamma = 120°$

$\alpha \xrightarrow[231°C]{} \beta_1$ condis crystal $\xrightarrow[287°C]{} \beta_2$ condis crystal $\xrightarrow[427°C]{}$ Melt

Property	Units	Conditions	Values	Reference
Number average/Weight average molecular weight, M_n/M_w	g mol^{-1}	By Gilch process using different solvent mixtures		(13)
		dioxane	232,000/601,000	
		dioxane/n-hexane	105,000/220,000	
		dioxane/cyclohexane	149,000/411,000	
		dioxane/triethylamine	77,000/152,000	
		dioxane/triethylamine	19,000/53,000	
		dioxane/triethylamine	31,000/48,000	
		dioxane/triethylamine	12,000/21,000	
		dioxane/triethylamine	5,600/9,600	
		dioxane/triethylamine	2,900/3,500	

Property	Units	Conditions	Values	Reference
Degree of crystallinity	%	–	35–66	–
Chains per unit cell	–	–	2	–
Packing density, ρ	g cm^{-3}	–	0.705	–
Solubility	–	Generally PPX is insoluble in most organic solvents. Phenyl derivates such as phenyl substituted PPX is found to be soluble in common solvents like THF and CHCl$_3$.		(7)
Dielectric loss tan δ	–	At 1 KHz	1.5×10^{-4}	(22)
Glass transition temperature, T_g	K	–	286	(21, 23)
Crystalline melting temperature, T_m	K	–	700	(21, 23)
		PPXN (PPX)	693	(24)
		PPXC (chlorinated PPX)	566	
		VT-4 (fluorinated PPX)	675	
		PPXD	653	
Crystallographic Data	Å	*b* axis lattice parameter (monoclinic) $T < 150°$C		(24)
		α−PPXN	10.64	
		PPXC	12.77	
		PPXD	13.82	
		a axis lattice parameter (hexagonal) $T > 380°$C		
		β−PPXN	20.52	
		tetrafluoro-PPX	21.09	
Young's modulus E	MPa	Isotropic film	600–1,400	(1, 3, 6,
		Oriented film	90,000–100,000	21, 23)
		Fibers	102,000	
			24,000	
Tensile stress	MPa	Isotropic film	25–62	(1, 3, 6,
		Oriented film	1,800	21, 23)
		Fibers	3,000	
			47	
Elongation at break	%	Isotropic film	18–330	(1, 3, 6,
		Oriented film	–	21, 23)
		Fibers	–	

References

1. Beach WF, Xylylene Polymers. In: Kroschwitz J, ed. *Encyclopedia of Polymer Science and Technology*. 3rd ed. Vol 12. New York: John Wiley & Sons; 2004:587.
2. Greiner A, Mang S, Scha¨fer O, Simon P. *Acta Polym.* 1997;48: 1.
3. Mailyan KA, et al. *Polym. Sci.* 1991;33:1420.
4. Krasovsky AM, et al. *J. App. Polym. Sci.* 1995;57:117.
5. Li H, et al. In: *ANTEC Conference Proc.* 1991;37: 2203.
6. Liu D, et al. *App. Polym. Sci.* 1990;40:1795.

7. Schäfer O, Greiner A. *Macromolecules*. 1996;29:6047.

8. Pu HT, Sun XR. *Acta Polymerica Sinica*. 2006;4:603.

9. Simon P, Mang S, Hasenhindl A, Gronski W, Greiner A. *Macromolecules*. 1998;31:8775.

10. Hanefeld P, Sittner F, Ensinger W, Greiner A. *E-Polymers*. 2006;026.

11. Knapton D, Iyer PK, Rowan SJ, Weder C. *Macromolecules*. 2006;39:4069.

12. Madan R, Greiner A. *Designed Monomers and Polymers*. 2006;9:81.

13. Brink-Spalink F, Greiner A. *Macromolecules*. 2002;35:3315.

14. Alexandrova L, Sansores E, Martinez E, Rodriguez EE, Gerasimov G, *Polymer*. 2001;42:273.

15. Alexandrova L, Likhatchev D, Muhl S, Salcedo R, Gerasimov G, Kardash I. *J. Inorg. Organometallic Polym*. 1998;8:157.

16. Brandukova-Szmikowski NE, Greiner A. *Acta Polymerica*. 1999;50: 141.

17. Schafer O, Brink-Spalink F, Greiner A. *Macromol. Rapid Comm*.1999;20:190.

18. Greiner A, Mang S, Schafer O, Simon P. *Acta Polymerica*. 1997;48:1.

19. Sochilin VA, et al. *Polym. Sci.*. 1991;33:1426.

20. Van Der Werf H, et al. *J. Mater. Sci. Lett.* 1989;8:1231.

21. Van Der Werf H, Pennings AJ. *Polym. Bull.* 1988;19:587.

22. Mori T, et al. *J. Phys. D. Appl. Phys.* 1989;22:1518.

23. Mailyan KA, et al. *Polym. Sci. Ser.* 1997;A 39(5):538.

24. Senkevich JJ, Desu SB, Simkovic V. *Polymer* 2000;41:2379.

Silicon (germanium) Oxo Hemiporphyrazine Polymers

Martel Zeldin and Yuli Zhang

Class Cofacial polymers

Structure $[M(hp)O]_n$:

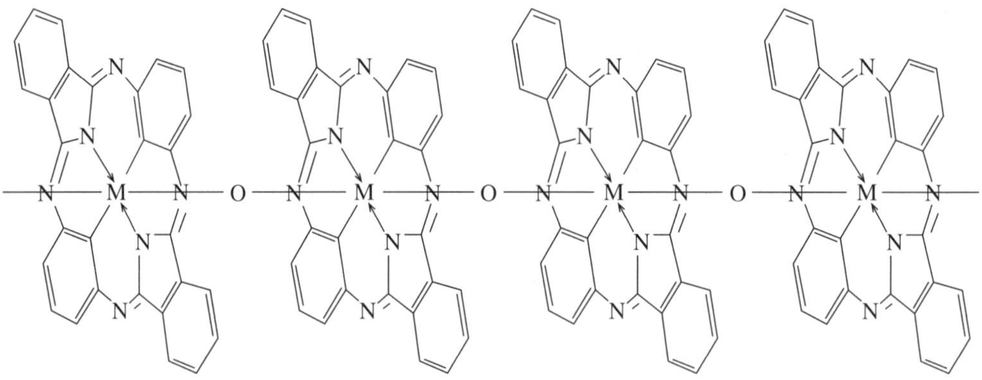

Synthesis Preparation of $[Si(hp)O]_n$ and $[Ge(hp)O]_n$.[1]

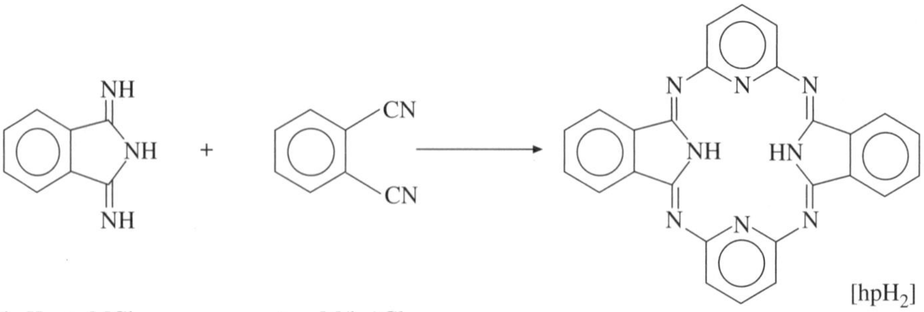

[hpH$_2$]

$$hpH_2 \ + \ MCl_4 \ \longrightarrow \ M(hp)Cl_2$$

$$M(hp)Cl_2 \ \xrightarrow{\text{Hydrolysis}} \ M(hp)(OH)_2$$

$$M(hp)(OH)_2 \ \xrightarrow[\text{Vacuum}]{\text{Heat}} \ M(hp)(O)_n$$

Infrared spectroscopy[1]

Compound	IR Spectral Data (cm^{-1})*
[Si(hp)O]$_n$	3092(vw), 3042(w), 1684(m), 1640(s), 1589(vs), 1553(m/s), 1437(s), 1323(m), 1293(m), 1264(m), 1209(m), 1170(w), 1160, 1152(w), 1105(s), 990(m-s), 909(w), 811(vs), 771(m), 744(w), 710(vs), 703(vs), 681(m), 503(m), 485(m), 406(w/m)
[Ge(hp)O]$_n$	3080(vw), 3040(vw), 1672(m), 1629(s), 1603(vs), 1582(vs), 1549(s), 1433(vs), 1318(s), 1290(w), 1256(m), 1206(m), 1190(vw), 1153(w), 1110(vs), 1004(vw), 900(m), 808(vs), 768(w), 734(vw), 700(vs), 678(w), 485(vw), 472(vw)

* Peaks not readily assigned to M(hp) moiety; s = strong, m = medium, w = weak, v = very.

Densities[1]

Polymer	y	Density (g cm^{-3})
[Si(hp)O]$_n$	Nondoped	1.63
[Ge(hp)O]$_n$	Nondoped	1.62

Reference

1. Dirk CW, Marks TJ. *Inorg. Chem.* 1984;23(25):4325–4332.

Silk Protein

Stephen A. Fossey, Peggy Cebe and David L. Kaplan

Alternative Names Silk, fibroin, spidroin

Class Polypeptides and protein

Structure

$$\begin{array}{c} R \\ | \\ [\ -NH-CH-COO-\ ]_n \end{array}$$

(R = H, CH_3, or CH_2OH in crystalline domains)

Major Applications Clothing, sutures, biomaterial

Properties of Special Interest Natural fibers with high strength and compliance, high energy absorption before failure, durable fibers with high luster, resistant to proteolysis.

Preparative Techniques Type of polymerization: biosynthesis (enzymatic), ambient conditions of temperature and pressure.

Property	Units	Conditions	Values	Reference
Molecular weight (of repeat unit)	g mol^{-1}	–	~64	–
Tacticity	–	Enzymatic polymerization (all L-amino acids)	100% isotactic	–
Degree of branching	%	Linear protein	None	–
Molecular weight	g mol^{-1}	–	390,000	(1)
Polydispersity index M_w/M_n	–	Monodisperse due to genetic controls	1.0	–
Morphology in multiphase systems	–	Crystalline blocks with amorphous blocks	Block copolymers	(2)
IR (characteristic absorption frequencies)	cm^{-1}	Amide I Amide II Amide III	1,624 1,522 1,258	(3)
UV (characteristic absorption frequencies)	nm	Tyrosine	280	–
NMR	–	^{13}C NMR ^{2}H NMR ^{15}N NMR		(4, 5) (6) (7)

Property	Units	Conditions	Values	Reference
Coefficient of linear thermal expansion	K^{-1}	For dry film range $50 \pm 150°C$	0.461×10^{-4}	(8)
Solvents	0.06 g ml^{-1}	silkworm silk in 9.3 M LiBr in H2O		(9)
		0.28 g mol^{-1} silkworm silk in 75% wt/wt $Ca(NO_3)_2$/MeOH		(10)
		0.1 g ml^{-1} *N. clavipes* dragline silk in hexafluoroisopropanol (highest reported solubility)		(11)
Nonsolvents		Methanol, ethanol, nonpolar hydrocarbons		(12)
Second viral coeffcient	$\text{mol cm}^3 \text{ g}^{-2}$	*Nephila clavipes* spider dragline silk in hexafluoroisopropanol with 10 mM trifluoroacetic acid	3.0×10^{-3}	(13)
Mark-Houwink parameters: K and a	$K = \text{ml g}^{-1}$	–	$K = 1.8 \times 10^{-4}$	(13)
	$a = \text{None}$		$a = 0.81$	

Lattice and unit cell dimensions

Lattice	Unit Cell Dimensions (nm)				Reference
	a	b	c (chain axis)		
Silk I Orthorhombic	0.465	1.424	0.888	4 residues	(14)
Silk II Orthorhombic	0.944	0.920	0.695	4 residues	(15, 16)
Silk III Hexagonal	0.456	0.456	0.867	3 residues	(17)

Degree of crystallinity	%	Silkworm silk	$\sim38 - 66$	(18)
		Spider silk	$\sim20 - 45$	(19)
Density (crystalline)	g cm^{-3}	Fiber in benzene	1.351	(20)
		In water	1.421	
Crystallite size (typical)	Nm	Silkworm silk	1.0–2.5	(18)
		N. clavipes dragline	$2 \times 5 \times 7$	(21)
Glass transition temperature T_g	K	0% RH at 23–26°C, absorbed moisture = 0 g/100 g silk	451	(22)
		75% RH at 23–26°C, absorbed Moisture = 21 g/100 g silk	312	
Melting point	–	Degrades prior to melting		
Mesomeric transition	–	Room temperature	Lyotropic	(23)
Heat capacity	$\text{J g}^{-1} \text{ K}^{-1}$	Specific heat is temp. dependent For *B. mori* non-crystalline fibroin: $C_p(T)^{\text{solid}} = 0.130 + 3.70 \times 10^{-3} \text{ T}$ $[\text{J (g K)}^{-1}]$ Increment in heat capacity at glass transition: $\Delta C_{p0}(T_g) = 0.478 \pm 0.005$ $[\text{J (g K)}^{-1}]$		(24)

Property	Units	Conditions	Value	Reference
Polymers with which compatible	–	–	Nylon	–
Thermal stability	–	Spider dragline silk (*N. clavipes*)	5% weight loss to 234°C	(25)
		Silkworm silk (*B. mori*)	5% weight loss to 250°C	(26)
Tensile strength	MPa	Silkworm (*Bombyx mori*) silk	513	(27)
Maximum extensibility	%	Silkworm (*Bombyx mori*) silk	23.4	(27)
Work to rupture	MPa	Silkworm (*Bombyx mori*) silk	80.6	(27)
Tensile modulus	MPa	Silkworm (*Bombyx mori*) silk	9,860	(27)
Yield stress	MPa	Silkworm (*Bombyx mori*) silk	211	(27)
Yield strain	%	Silkworm (*Bombyx mori*) silk	3.3	(27)
Storage modulus	MPa	Silkworm (*Bombyx mori*) silk, for $80°C < T < 160°C$	70,000	(28)
Loss modulus	MPa	Silkworm (*Bombyx mori*) silk, for $80°C < T < 160°C$	1,600	(28)

Spider dragline silks

	Tensile Strength (GPa)	Initial Tensile Modulus (GPa)	Ultimate Elongation (%)	Shear Modulus (GPa)	Transverse Comp. Modulus (GPa)	Reference
Nephila clavipes	1.1	22	9	–	–	(25)
Nephila clavipes	0.85	12.7	20	3.58	0.58	(29)
Argiope aurantia	0.5–1.3	6–24	18.3–21.5	–	–	(30)
Araneus sericatus	1.0	10	30	–	–	(30)

Index of refraction n	–	Parallel to fiber	1.591	(24)
		Perpendicular to fiber	1.538	(24)
Piezoelectric coefficient	pN/C	$1 = d_{14}^0$	3.3	(31)
Speed of sound	M s^{-1}	–	540	(32)
Biodegradability		Ubiquitous microorganisms, proteases, soil, water, human body		
Maximum use temperature	K	*N. clavipes* dragline silk	443	(25)
		B. mori silkworm silk	443	(26)
Decomposition temperature	K	*N. clavipes* dragline silk	507	(25)
		B. mori silkworm silk	523	(26)
Water absorption		Some spider silks supercontract	~50% in water	(33)
Scission	–	UV light		(34)
Important patents				(35)

References

1. Zhou CZ, et al. *Nucleic Acids Res.* 2000;28:2413–2419.
2. Bini E, Knight DP, Kaplan DL. *J. Mol. Biol.* 2004;335:27–40.
3. Yoshimizu H, Asakura T. *J. Appl. Polym. Sci.* 1990;40:127–134.
4. Asakura T, et al. *Macromolecules.* 1985;18:1841–1845.
5. Ishida M, et al. *Macromolecules.* 1990;23:88–94.
6. Simmons AH, Michal CA, Jelinski LW. *Science* 1996;271:84–87.
7. Nicholson LK, et al. *Biopolymers.* 1993;33:847–861.
8. Mogoshi J, et al. *J. Polymer Sci., Polym. Phys. Ed.* 1977;15:1675–1683.
9. Yamura K, et al. *J. Appl. Polym. Sci.: Appl. Polym. Symp.* 1985;41:205.
10. Bagga A. *Masters Thesis.* Department of Textile Chemistry, North Carolina State University, 1995.
11. Jackson C, O'Brien JP. *Macromolecules.* 1995;28:5975–5980.
12. Kaplan DL, et al, eds. In: *Silk Polymers: Materials Science and Biotechnology.* Vol 544. American Chemical Society Symposium Series, 1994:1–358.
13. Jackson C, O'Brien JP. *Macromolecules.* 1995;29:5975–5977.
14. Asakura T, et al. *Macromolecules.* 2005;38:7397–7403.
15. Takahashi Y, Gehoh M, Yuzuriha K. *J. Polym. Sci., Polym. Phys. Ed.* 1991;29:889–891.
16. Marsh RE, Corey RB, Pauling L. *Biochim. Biophys. Acta.* 1955;16:1–34.
17. Valluzzi R, et al. *Macromolecules.* 1996;29:8606–8614.
18. Bhat NV, Nadiger GS. *J. Appl. Polym. Sci.* 1980;25:921–932.
19. Warwicker JO. *J. Mol. Biol.* 1960;2:350–362.
20. Lucas F, Shaw JT, Smith SG. *J. Textile Institute.* 1955;46:T440–T452.
21. Grubb DT, Jelinksi LW. *Macromolecules.* 1997;30:2860–2867.
22. Agarwal N, Hoagland DA, Farris RJ. *J. Appl. Polym. Sci.* 1997;63:401–407.
23. Willcox PJ, et al. *Macromolecules.* 1996;29:5106–5110.
24. Xu X, Kaplan DL, Cebe P. *Macromolecules.* 2006: 39: 6161–6170.
25. Cunniff PM et al. *Polym. Adv. Technol.* 1994;5:401–410.
26. Nakamura S, Magoshi J, Magoshi Y. In: *Silk Polymers: Materials Science and Biotechnology.* Vol 544. American Chemical Society Symposium Series, 1994:292–310.
27. Robson RM. In: Lewin M, Pearce E, eds. *Fiber Chemistry Handbook of Science and Technology.* Vol IV. New York: Marcel Dekker; 1970:647–700.
28. Tsukada M, Freddi G, Kasai N. *J. Polym. Sci.: Part B: Polym. Phys.* 1994;32:1175–1182.
29. Kawabata S, et al. In: Proc. 11th Internatl. Conf. Composite Matls., Australia, July 1997.
30. Kaplan DL, et al. In: Byrom D, ed. *Biomaterials.* New York; Stockton Press; 1991:3–52.
31. Ando Y, et al. *Reports Prog. Polym. Phys. Japan.* 1980;23:775–778.
32. Laible RC, In: *Ballistic Materials and Penetration Mechanics.* Amsterdam: Laible. Elsevier; 1980:73–115.
33. Work R. *Tex. Res. J.* 1985;47:650–662.
34. Becker MA, Tuross N. In: Kaplan DL, et al, eds. *Silk Polymers: Materials Science and Biotechnology.* Vol 544. American Chemical Society Symposium Series; 1994:270–282.
35. Ferrari FA, et al. *U.S. Patent.* 1992;5243038.

Siloxane Dendrimers

Aziz M. Muzafarov

Acronym PMSQD

Class Dendritic polymers

Structure

$$CH_3SiO_{1,5}\left\{\left[CH_3SiO_{1,5}\right]_{2^{(G-1)}-1}\!\!\!\!-\!\!\left[O_{0,5}Si(OC_2H_5)_2\right]_{2^{(G-1)}}\right\}_3$$

G – generation number

Introduction All three known siloxane dendrimer structures are based on methylsilsesquioxane core and branching units. The first siloxane dendrimer was synthesized in 1989[1] as pure PMSQ structure. Two other approaches appeared in 1990[2] and 1991[3]. They used D-units to decrease the density of the molecular structure. Since then, hybrid-type dendrimers combining siloxane units with links of other chemical nature were the focus of interest due to simplification of chemical technique in that case. At the same time, the benefits of pure siloxane structure leave no doubt in further progress of this area.

Related Polymers Siloxane-containing dendrimers were synthesized using both divergent and convergent approaches or mixed synthetic method. Most of them related to hybrid structure as far as molecular skeleton of those systems were presented by siloxane links alternating with carbosilane units[4−9]. Other kinds of siloxane dendrimers are composed as core-shell structures with carbosilane[10,11] or organic[12−14] core and methylsilsesquioxane outer shell.

Preparative Techniques Using unique properties of 'Rebrov salt'[15,16]— sodiumoxymethyldiethoxysilane — the synthesis of siloxane dendrimers was performed as a two-stage repetitive cycle of classical 'protection–deprotection' type. The proton NMR and Cl-ion titration analysis were used as effective controls of each stage conversion. Due to very compact structure, the first three generations of ethoxy-terminated dendrimers were isolated by vacuum distillation in high yield[1,17]. The key stages of two other kinds of siloxane dendrimer synthesis were performed using Si–C$_6$H$_5$ and Si–H groups as 'protected' functions with further transformation to Si–Br and Si–Cl correspondingly[2,3].

Major Applications A number of practical applications were highlighted for dendrimers considering flexibility, thermostability, and bioinertness of methylsiloxane structure. As highly functional matrixes soluble in most organic solvents they were proposed for immobilization of catalytic, LC, and surface active ensembles. Si, O, C-ceramics, membrane materials, and drug carriers were also highlighted for siloxane dendrimers [3,17].

Siloxane Dendrimers

Properties of Special Interest The properties of siloxane dendrimers are studied rather poorly as yet. The most peculiar system of this type — PMSQD is fully acyclic alternative to highly cyclic or even polyhedral well-known PMSQ polymers. Good solubility in most of the organic solvents reflected in ref. 1 and some results of characteristic constants presented in the table are the only available data till now.

PMSQD physicochemical parameters

PMSQD Generation Number	Refractive Index, n_p^{25}	Density, g/cm^3	B.p., °C (mm Hg)
G1-Si$_4$^{6}(OC$_2$H$_5$)	1.3956	1.0007	98–99(1)
G2-Si$_{10}$^{12}(OC$_2$H$_5$)	1.4031	1.0409	186–187.5(1)
G3-Si$_{22}$^{24}(OC$_2$H$_5$)	1.4100	1.0717	299–305(1)
G4-Si$_{46}$^{48}(OC$_2$H$_5$)	1.4110	1.0815	–

References

1. Rebrov EA, et al. *Dokl. Akad. Nauk. SSSR.* 1989;309(2):376.
2. Uchida H, et al. *J. Am. Chem. Soc.* 1990;112(19):7077.
3. Morikawa A, Kakimoto M, Imai Y. *Macromolecules.* 1991;24(12):3469.
4. Morikawa A, Kakimoto M, Imai Y. *Macromolecules.* 1992;25(12):3247.
5. Ignat'eva GM, et al. *Polym. Sci.* 1997;A39:843.
6. Sinevich EA, et al. *Polym. Sci.* 1996;A38(11):1256.
7. Sheiko SS, et.al. *Macromol. Rapid Commun.* 1996;17:283.
8. Kim C, An K. *J. Organomet. Chem.* 1997;547:55.
9. Brüning K, Lang H. *J. Organomet. Chem.* 1998;571:145.
10. Bystrova AV, et al. *Polymer Sci.* 2005;A47(8):820.
11. Bystrova AV, et al. *Rus. Nanotechnologies.* 2007;2(1):83.
12. de Leuze-Jallouli, et al. *Polym. Mater. Sci. Eng.* 1997;77:67.
13. Dvornic PR, et al. *Polym. Mater. Sci. Eng.* 1999;81:187.
14. Dvornic PR, et al. *Macromolecules* 2000;33:5366.
15. Rebrov EA, et al. *Dokl Akad. Nauk SSSR.* 1998;302(2):346.
16. Rebrov EA, Muzafarov AM. *Heteroat. Chem.* 2006;17(6):514.
17. Muzafarov AM, et al. *Russ. Chem. Rev.* 1991;60(7):807.

Starch

W. Brooke Zhao and Zongming Gao

Starch is the cheapest and most abundant food biopolymer worldwide. It is a complex carbohydrate which is insoluble in water; it is used by plants as a way to store excess glucose. Starch is often found in the fruit, seeds, rhizomes or tubers of plants. The four major resources for starch production and consumption in the USA are corn, potatoes, rice, and wheat. Pasta is an important dietary source of starch which is commonly prepared from wheat, rice or beans. Bread is another important source of starch and is commonly prepared from wheat.

Properties	Units	Conditions	Values	Reference
Size of granules	μm	Source and Shape		(1)
		Corn, polygonal or round	5–25 (avg. 15)	
		Maize, polygonal or round	5–25 (avg. 15)	
		Potato, oval or egg-shaped	15–100	
		Rice, polygonal	3–8	
		Tapioca, rounded, truncated at one end	20, 15–25	
		Wheat, flat, round, or elliptical	2–10, 20–35	
Gelatinization temperature	K	Source		(1)
		Corn	335–345	
		Sorghum	341.5–348	
		Wheat	325–336	
		Rice	334–350.5	
		Waxy maize	336–345	
		Tapioca	331.5–343	
		Potato	329–339	
Enthalpy of gelatinization $(-\Delta H_G)$	kJ mol^{-1}	Corn, A (X-ray pattern)	2.8–3.3	
		Wheat, A (X-ray pattern)	2.0	
		Rice, A (X-ray pattern)	2.3–2.6	
		Dasheen, A (X-ray pattern)	2.9	
		Waxy maize, A (X-ray pattern)	3.2	
		Compacted com, A (X-ray pattern)	1.5	
		High amylose corn, B (X-ray pattern)	4.5	
		Potato, B (X-ray pattern)	3.0	
		Arrowroot, C (X-ray pattern)	3.1	
		Tapioca, C (X-ray pattern)	2.7	

Properties	Units	Conditions	Values		Reference
Density	g cm^{-3}	Maize			
		Pycnometric	1.637		(3)
		Buoyant	1.50		(3)
		Potato			
		Pycnometric	1.617		(3)
		Perfect evacuation	1.594		(3)
		Wheat			
		Pycnometric	1.650		(4)
		Corn	1.5		(5)
		Sorghum	1.5		(5)
		Rice, nonwaxy			
		Xylene displacement	1.49–1.51		(6)
		Rice, waxy			
		Xylene displacement	1.48–1.50		
		Perfect evacuation	1646		(4)
Molecular weight (of repeat unit)	g mol^{-1}	Exclusion chromatography	M_n	M_w	(7)
		Regular dent corn starch	2.14×10^5	1.45×10^7	
		Waxy maize	1.48×10^5	2.18×10^7	
		Amylomaize (70–75% amylose)	4.8×10^4	3.96×10^6	
		Amylomaize (52% amylose)	5.4×10^4	5.75×10^6	
Polydispersity index (M_w/M_n)	–	Regular dent corn starch	68		(7)
		Waxy maize	147		
		Amylomaize (70–75% amylose)	82		(7)
		Amylomaize (52% amylose) in DMSO solution, GPC	106		
		$M_w = 7.11 \times 10^6$; $M_n = 1.35 \times 10^6$	5.27		(8)
Polymorphs	–	Cereal grain starches, such as from maize, wheat, and rice	A		(3)
		Tuber, fruit, and stem starches, such as from potato, sago, and banana	B		
		Mixture of A- and B-type crystallites	C		
NMR (^{13}C chemical shift)	ppm	Solid state CP/MAS			(9)
		A polymorph	102.3 (0.3), 101.5(0.4), 100.3(0.4) (t) (C-1) 62.8(02) (C-6) 101.4(0.4), (C-1), 100.4(0.4) (d) (C-1)		
		B polymorph	62.1(02) (C-6)		
Glass transition temperature T_g	K	Corn starch	496		(10)

Starch

Properties	Units	Conditions	Values	Reference
Heat capacity increment at T_g (ΔC_P)	kJ K^{-1} mol^{-1}	Corn starch	0.47	(10)
Melting temperature T_m	K	A + B polymorphs	530	(10)
		Maize, A	460	(3)
		Wheat, A	454	(3)
		Waxy maize, A	470	(3)
		Potato, B	441	(3)
Enthalpy of melting ΔH_m	kJ mol^{-1}	Maize, A	57.7	(3)
		Wheat, A	52.7	
		Waxy maize, A	61.1	
		Potato, B	59.8	
Heat of hydration	Jg^{-1}	Potato	116.7	(3)
		Wheat	105.4	
		Maize	103.3	
		Rice	101.7	
Activation energy for hydration	kJ g^{-1}	Wheat starch dielectric absorption	42.3	(3)
Flory-Huggins interaction parameter χ	–	–	0.5	(10)
Birefringence	–	In water and alcohol	0.0131–0.0139	(3)
		In aldehydes	0.0135–0.0143	
		In hydrophobic liquids	0.0134–0.0135	
Refractive indexes	–	Potato starch, 25°C, $\lambda = 589$ nm	1.523, 1.535	(3)
Surface area	m^2 g^{-1}	Dasheen		(3)
		N$_2$ adsorption	2.62	
		Photomicrographic	2.64	
		Corn		
		N$_2$ adsorption	0.70	
		Photomicrographic	0.48	
		Tapioca		
		N$_2$ adsorption	0.28	
		Photormicrographic	0.25	
		Potato		
		N$_2$ adsorption	0.11	
		Photomicrographic	0.15	
Surface tension	dynes cm^{-1}	Corn starch	39	(11)
Tensile strength	MPa ($\times 10^{-3}$)	Waxy maize	34.9	(12)
		Tapioca	44.0	
		Potato	44.2	
		Wheat	46.3	

Properties	Units	Conditions	Values	Reference
		Corn (A)	46.1	
		Com (B)	46.7	
		High-amylose com	50.3	
		15-F acid-modified	44.7	
		34-F acid modified	44.5	
		50-F acid modified	49.4	
		71-F acid modified	45.7	
		89-F- acid modified	45.8	
		Hypochlorite-oxidized	48.7	
		Hypocholoritc-oxidizcd	45.0	
		0.05-D.S. hydroxyethyl corn	47.4	
		0.05-D.S. hydroxyethyl, acid-modified corn	41.8	
Elongation at break	%	Waxy maize	1.7	(12)
		Tapioca	3.4	
		Potato	3.1	
		Wheat	2.9	
		Corn (A)	2.5	
		Corn (B)	3.2	
		High-amylose corn	2.5	
		15-F acid-modified	2.7	
		34-F acid modified	2.6	
		50-F acid modified	2.7	
		71-F acid modified	2.9	
		89-F acid modified	2.2	
		Hypochlorite-oxidized	3.0	
		Hypocholorite-oxidized	2.3	
		0.05-D.S. hvdroxyethyl corn	2.5	
Elongation at break	%	0.05-D.S. hydroxyethyl, acid-modified com	2.6	
Biodegradation	–	Films obtained after extrusion of native potato starch and glycerol		(13)
		Enzymatic test	100% weight loss after 24 h	
		Head-space test	100% CO_2 evolution after 50 days	
		Compost test	100% weight loss after 49 days	

Class [14] (CAS# 9005-25-8) Starch contains a mixture of two molecules: amylose and amylopectin. Usually these are found in a ratio of 30:70 or 20:80, with amylopectin found in larger amounts than amylose.

Structure [15] Native starches from different botanical sources vary widely in structure and composition, but all granules consist of two major molecular components, amylose (20–30%) and amylopectin (70–80%), both of which are polymers of α-D-glucose units in the $^4C_1$ conformation. In amylose (*Figure 1*), these are linked –(1 → 4)–, with the ring oxygen atoms all on the same side, whereas in amylopectin about one residue in every twenty is also linked –(1 → 6)– forming branch-points as shown in *Figure 2.*

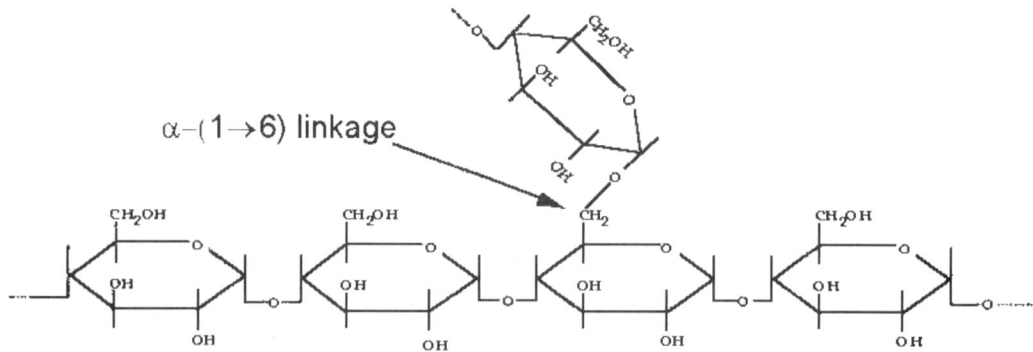

Figure 1. Amylose molecule

Figure 2. Amylopectin molecule

Physical Properties [15] Of the two components of starch, amylose has the most useful functions as a hydrocolloid. Its extended conformation causes the high viscosity of water-soluble starch and varies relatively little with temperature. The extended loosely helical chains possess a relatively hydrophobic inner surface that is not able to hold water well and more hydrophobic molecules such as lipids and aroma compounds can easily replace this. Amylose forms useful gels and films. Its association and crystallization (retrogradation) on cooling and storage decreases storage stability causing shrinkage and the release of water (syneresis). Increasing amylose concentration decreases gel stickiness but increases gel firmness. Amylopectin interferes with the interaction between amylose chains (and retrogradation) and its solution can lead to an initial loss in viscosity and followed by a more slimy consistency. Mixing with κ-carrageenan, alginate, xanthan gum and low molecular weight sugars can also reduce retrogradation. At high concentrations, starch gels are both pseudoplastic and thixotropic with greater storage stability. Their water binding ability (high but relatively weak) can provide body and texture to foodstuffs and is encouraging its use as a fat replacement.

Starch is often used as an inherent natural ingredient but it is also added for its functionality. It is naturally found tightly and radially packed into dehydrated granules (about one water per glucose) with origin-specific shape and size (maize, 2–30 μm; wheat, 1–45 μm; potato, 5–100 μm).[16] The size distribution determines its swelling functionality with granules being generally either larger and lenticular (lens-like, **A**-starch) or smaller and spherical (**B**-starch)[17] with less swelling power. Granules contain "blocklets" of amylopectin containing both crystalline (~30%) and amorphous areas. As they absorb water, they swell, lose crystallinity and leach amylose. The higher the amylose content, the lower is the swelling power and the smaller is the gel strength for the same starch concentration.

To a certain extent, however, a smaller swelling power due to high amylose content can be counteracted by a larger granule size.[18]

A significant proportion of starch in the normal diet escapes degradation in the stomach and small intestine and is labeled "resistant starch" (for a recent review see),[19] but this portion is difficult to measure and depends on a number of factors including the form of starch and the method of cooking prior to consumption. Nevertheless resistant starch serves as a primary source of substrate for colonic microflora, and may have several important physiological roles. Resistant starch has been categorized as physically inaccessible (RS_1), (raw) ungelatinized starch (*e.g.* in banana; RS_2), thermally stable retrograded starch (*e.g.* as found in bread, especially stale bread, mainly amylose; RS_3) and chemically modified starch (RS_4). Resistant starch should be considered a dietary fiber. Although not exactly quantifiable due to its heterogeneous nature, some is determined by the official Association of Official Agricultural Chemists (AOAC) method.

Starch can be hydrolyzed into simpler carbohydrates by acids, various enzymes, or a combination of the two. The extent of conversion is typically quantified by dextrose equivalent (DE), which is roughly the fraction of the glycoside bonds in starch that have been broken. Food products made in this way include

Maltodextrin, a lightly hydrolyzed (DE 10–20) starch product used as a bland-tasting filler and thickener. Various corn syrups (DE 30–70), viscous solutions used as sweeteners and thickeners in many kinds of processed foods. Dextrose (DE 100), commercial glucose, prepared by the complete hydrolysis of starch. High fructose syrup, made by treating dextrose solutions to the enzyme glucose isomerase, until a substantial fraction of the glucose has been converted to fructose.

Testing Methods [14] Starch solution is used to test for iodide ions. A blue/black color indicates the presence of iodide ions in starch solution. The details of this reaction are not yet fully known, but it is thought that the iodine fits inside the coils of amylose. A 0.3% w/w solution is the standard concentration for a dilute starch indicator solution. It is made by adding 4 g of soluble starch to 1 l of heated water; the solution is cooled before use (starch-iodine complex becomes unstable at temperatures above 35°C). This complex is often used in redox titrations: in presence of an oxidizing agent the solution turns blue, in presence of reducing agent blue color disappears because I_3- ions break up into iodine and iodide.

Under the microscope, starch grains show a distinctive Maltese cross effect (also known as "extinction cross" and birefringence) under polarized light.

Swelling power is determined after heating the starch in excess water and is the ratio of the wet weight of the (sedimented) gel formed to its dry weight. It will depend on the processing conditions (temperature, time, stirring, centrifugation) and may be thought of as its water binding capacity.

Applications Starch is a versatile and cheap, and has many uses as thickener, water binder, emulsion stabilizer and gelling agent. Many functional derivatives of starch are marketed including cross-linked, oxidized, acetylated, hydroxypropylated and partially hydrolyzed material. For example, partially hydrolyzed (*i.e.* about two bonds hydrolyzed out of eleven) starch (dextrin) is used in sauces to control viscosity.

Food Applications [14] Starch (in particular cornstarch) is used in cooking for thickening foods such as sauce. In industry, it is used in the manufacturing of adhesives, paper, textiles and as a mold in the manufacture of sweets such as wine gums and jelly beans. It is a white powder, and depending on the source, may be tasteless and odourless. As an additive for food processing, arrowroot and tapioca are commonly used as well. Commonly used starches around the world are: arracacha, buckwheat, banana, barley, cassava, kudzu, oca, sago, sorghum, sweet potato, taro and yams. Edible beans, such as favas, lentils and peas, are also rich in starch. When a starch is pre-cooked, it can then be used to thicken cold foods. This is referred to as a pregelatinized starch. Otherwise starch requires heat to thicken, or "gelatinize." The actual temperature depends on the type of starch.

A modified food starch undergoes one or more chemical modifications that allow it to function properly under high heat and/or shear frequently encountered during food processing. Food starches are typically used as thickeners and stabilizers in foods such as puddings, custards, soups, sauces, gravies, pie fillings, and salad dressings, but have many other uses.

Non-Food Applications   Clothing starch or laundry starch is a liquid that is prepared by mixing a vegetable starch in water (earlier preparations also had to be boiled), and is used in the laundering of clothes. Starch was widely used in Europe in the 16th and 17th centuries to stiffen the wide collars and ruffs of fine linen which surrounded the necks of the well-to-do. During the 19th century and early 20th century, it was stylish to stiffen the collars and sleeves of men's shirts and the ruffles of girls' petticoats by applying starch to them as the clean clothes were being ironed.

Aside from the smooth, crisp edges it gave to clothing, it served practical purposes as well. Dirt and sweat from a person's neck and wrists would stick to the starch rather than fibers of the clothing, and would easily wash away along with the starch. After each laundering, the starch would be reapplied.

Starch glues are widely used in the bonding of paper, wood and cotton.[20] Starch is also used to make some packing peanuts, and some ceiling tiles.

References

1. Wurzburg OB, ed. *Modi®ed Starches: Property and Uses.* Boca Raton, Fla: CRC Press; 1986:4.
2. Zobel HF. In: Whistler RL, Bemiller JN, Paschall EF, eds. *Starch: Chemistry and Technology.* 2nd ed. Orlando, Fla: Academic Press; 1984.
3. French, D. In: Whistler RL, Bemiller JN, Paschall EF. *Starch: Chemistry and Technology.* 2nd ed. Orlando, Fla: Academic Press; 1984.
4. Takei B. *Mem. Coll. Sci. Kyoto Imp. Univ.* 1935;A18:169.
5. Watson SA. In: Whistler RL, Bemiller JN, Paschall EF. *Starch: Chemistry and Technology.* 2nd ed. Orlando, Fla: Academic Press; 1984.
6. Juliano BO. In: Whistler RL, Bemiller JN, Paschall EF. *Starch: Chemistry and Technology.* 2nd ed. Orlando, Fla: Academic Press; 1984.
7. Young AH. In: Whistler RL, Bemiller JN, Paschall EF. *Starch: Chemistry and Technology.* 2nd ed. Orlando, Fla: Academic Press; 1984 (and references therein).
8. Salemis P, Rinaudo M. *Polym. Bull.* 1984;11:397.
9. Veregin RP, Fyfe CA, Marchessault RH, Taylor MG. *Macromolecules.* 1986;19:1030.
10. Whittman MA, Neol TR, Ring SG. In: Dickson E, ed. *Food Polymers, Gels and Colloids* (Spec. Publ. No. 82). Cambridge, U.K: The Royal Chemical Society; 1991.
11. Ray, BR, Anderson JR, Scholz JJ. *J. Phys. Chem.* 1958;62:1220.
12. Lloyd N, Kirst LC. *Cereal Chem.* 1963;40:155.

13. Vikman M, Itavaara M, Poutanen K. In: Albertsson A, Huang SJ, eds. *Degradable Polymers, Recycling and Plastics Waste Management.* New York: Marcel Dekker; 1995.

14. From Wikipedia, the free encyclopedia, http://en.wikipedia.org/wiki/Gelatin, Accessed March 2, 2007.

15. Starch structure, http://www.cheng.cam.ac.uk/research/groups/polymer/ RMP/nitin/Starchstructure.html, Accessed June 2, 2007.

16. Jobling S, Improved starch for food and industrial applications. *Curr. Opinion Plant Biol.* 2004;7:210–218.

17. Heyrovská R, Volumes of ions, ion pairs, and electrostriction of alkali halides in aqueous solutions at 25°C. *Marine Chem.* 2000;70:49–59.

18. (a) Li J-Y, Yeh A-I, Relationships between thermal, rheological characteristics and swelling power for various starches. *J. Food Engineering.* 2001;50:141–148.
    (b) Singh N, Singh J, Kaur L, Singh Sodhi N, Singh Gill B. Morphological, thermal and rheological properties of starches from different botanical sources. *Food Chem.* 2003;81:219–231.

19. Sajilata MG, Singhal RS, Kulkarni PR. Resistant starch – A review. *Comp. Rev. Food Sci. Food Safety.* 2006;5:1–17.

20. Jones, Orlando, "US2000 Improvement in the manufacture of starch". (Class: 127/68; 48/119; 127/69). Middlesex, England,

Styrene-acrylonitrile Polymers

Shuhong Wang

Acronym SAN

Class Chemical copolymers

Trade Names Lustran® (Lanxess), Tyril® (Dow Chemical Co.)

Structure $-[CH-CH_2]_m-[CH-CH_2]_n-$
$\qquad\qquad\;\; |\qquad\qquad |$
$\qquad\qquad C_6H_5\qquad\; CN$

Typical Comonomers Styrene, acrylonitrile

Polymerizations Emulsion, suspension, and continuous processes

Major Applications Incorporated in ABS (≥80%). Appliances, housewares, packing materials, automotive features, industrial applications, medical devices, and custom molding products.

Properties of Special Interest Rigidity, resistance to heat and chemicals (acids, alkalis, fat, grease, oil, gasoline, alcohol, and some solvents), and high optical clarity.

Property	Units	Conditions	Value	Reference
Density	g cm^{-3}		1.07–1.09	(1)
Glass transition temperature, T_g	°C	20 mol% AN	~103	(2)
		40 mol% AN	~108	(2)
		50 mol% AN	~110	(2)
		75 mol% AN	~109	(2)
Rockwell Hardness		Lustran®-35, ASTM D785	83	(3)
		Tyril®-880, ASTM D785	80	(4)
Tensile strength	MPa	5.5% AN	42.27	(2)
		9.8% AN	54.61	(2)
		14.0% AN	57.37	(2)
		21.0% AN	63.68	(2)
		27.0% AN	72.47	(2)
		Lustran®-35, ASTM D638	79.4	(3)
		Tyril®-880, ASTM D638	82.1	(4)

Property	Units	Conditions	Value	Reference
Elongation	%	5.5% AN	1.6	(2)
		9.8% AN	2.1	(2)
		14.0% AN	2.2	(2)
		21.0% AN	2.5	(2)
		27.0% AN	3.2	(2)
		Lustran®-35, ASTM D638	3.0	(3)
		Tyril®-880, ASTM D638	3.0	(4)
Impact strength	$J\,m^{-1}$	5.5% AN, notch	26.6	(2)
		9.8% AN, notch	26.0	(2)
		14.0% AN, notch	27.1	(2)
		21.0% AN, notch	27.1	(2)
		27.0% AN, notch	27.1	(2)
Izod impact strength	$J\,m^{-1}$	Lustran®-35, ASTM D256	24.0	(3)
		Tyril®-880, ASTM D256	26.7	(4)
Heat distortion	°C	5.5% AN	72	(2)
		9.8% AN	82	(2)
		14.0% AN	84	(2)
		21.0% AN	88	(2)
		27.0% AN	88	(2)
Deflection temperature	°C	Lustran®-35, ASTM D648	104.4	(3)
		Tyril®-880, ASTM D648	103.3	(4)
Vicat softening point	°C	Lustran®-35, ASTM D1525	111	(3)
		Tyril®-880, ASTM D1525	111	(4)
Melt-flow rate	g per 10 min	Lustran®-35, ASTM D1238, cond. 1	7.0	(3)
		Tyril®-880, ASTM D1238, cond. 1	3.0	(4)
Coefficient of linear thermal expansion	$cm\,(cm\,°C)^{-1}$	Lustran®-35, ASTM D696	6.8×10^{-5}	(3)
		Tyril®-880, ASTM D696	6.6×10^{-5}	(4)
Flammability	$cm\,min^{-1}$	Tyril®-880, ASTM D635	2.0	(4)
Specific heat	$J\,(g\,°C)^{-1}$	Tyril®-880, Dow Test	1.30	(4)
Dielectric constant	kHz	Tyril®-880, ASTM D150	3.18	(4)
Dissipation factor	kHz	Tyril®-880, ASTM D150	0.007	(4)
Refractive index, n_D		Lustran®-35, ASTM D542	1.57	(3)
		Tyril®-880, ASTM D542	1.57	(4)
Water absorption	%	Lustran® −35, ASTM D570(24 h)	0.25	(3)
		Tyril®-880, ASTM D570(24 h)	0.35	(4)
Solution viscosity	mPa s	5.5% AN, 10% in MEK	11.1	(2)
		9.8% AN, 10% in MEK	10.7	(2)
		14.0% AN, 10% in MEK	13.0	(2)
		21.0% AN, 10% in MEK	16.5	(2)
		27.0% AN, 10% in MEK	25.7	(2)

Styrene-acrylonitrile Polymers

Property	Units	Conditions	Value	Reference
Specific gravity		Lustran®-35, ASTM D792	1.07	(3)
		Tyril®-880, ASTM D792	1.08	(4)
Mold shrinkage	cm cm^{-1}	Lustran®-35	0.003–0.004	(3)
		Tyril®-880	0.003–0.007	(4)
Thermal conductivity	W (m K)$^{-1}$	33% glass fiber	0.28	(5)
Theta temperature, θ	°C	51% AN ethyl acetate	43.0	(6)

References

1. Mark JE, ed. *Physical Properties of Polymers Handbook*. 2nd ed. New York, NY: Springer Science Business Media, LLC; 2007.
2. Johnston NW. *Am, Chem. Soc. Div. Polym. Chem. Prepr.* 1973;14:46.
3. Lanexss product data sheets
4. Dow product data sheets
5. Harper CA, ed. *Handbook of Plastics, Elastomers, and Composites*. New York: McCraw-Hill; 1992.
6. Mangalam PV, Kalpagam V. *J. Polym. Sci, Polym. Phys. ed.* 1982;20:773.

Styrene-butadiene Elastomers

Shuhong Wang

Acronyms SBR, SB

Class Chemical copolymers

Trade Names Ameripol®, Synpol® (Ameripol Synpol); Copo®, Carbomix® (Lion Copolymer, LLC); Darex® (W.R. Grace); Duradene®, Srereon® (Firestone); Humex® (Hules Mexicanos, S.A.); JSR (Japan Synthetic Rubber Co.); NIPOL (Nippon Zeon Co.); Plioflex®, Pliolite® (Goodyear Tire & Rubber); Polysar SS®, Polysar S® (Bayer AG) Solprene® (Negromex, S.A.); Tylac® (Reichhold Chemicals)

Structure

$$-[CH_2-CH=CH-CH_2]_m-[CH_2-CH]_n- $$
$$\underset{C_6H_5}{\overset{|}{}}$$

Typical Comonomers Styrene, butadiene

Major Applications Tires (75%), shoes and other footwear, mechanical goods, sponge and foamed products, waterproofed materials, hose, belting, adhesives, etc.

Properties of Special Interest Standard emulsion SBR is a general-purpose rubber. Most widely used synthetic rubber in the world. Better tire tread-wear and aging properties than NR. Good abrasion resistance and crack initiation resistance. Poor in tack and heat build-up. Physical properties are poor without reinforcing fillers. Solution SBR is a specialty rubber and more expensive than emulsion SBR. Solution SBR with high vinyl and styrene level is used in high performance tire treads to improve wet traction. Also used as impact modifier in plastics and as thermoplastic elastomers.

Polymerization

Emulsion Polymerization Used for standard SBR. Monomer is emulsified in water with emulsifying agents. Polymerization is initiated by either decomposition of a peroxide or a peroxydisulfate. Hot SBR is initiated by free radicals generated by thermal decomposition of initiators at 50°C or higher, while cold SBR by oxidation-reduction reactions (redox) at temperatures as low as –40°C. Styrene content normally is 23%. Copolymer is randomly distributed. Structure of butadiene contents is about 18% cis-1,4, 65% trans-1,4 and 15 to 20% vinyl.

Typical polymerization condition

Type	HOT	COLD
	1000	1500
Monomer ratio S/B	71/29	71/29
Water/monomer	2/1	2/1
Emulsifier	Fatty acid	Rosin acid
Coagulation	Acid/amine	Acid/amine
Temperature, °C	50	5
Conversion, %	72	60–65
Styrene content	24	24
Mooney at 100 °C	48	46–58

Commercial grades (Institute numbering system)

1000	Hot SBR
1500	Cold SBR
1600	Cold SBR black masterbatch with 14 or less phr oil
1700	Cold SBR oil-masterbatch
1800	Cold oil-black masterbatch with more than 14 phr oil
1900	Miscellaneous resin rubber masterbatches
2000	Hot latexes
2100	Cold latexes

Solution Polymerization Solution SBR typically made in hydrocarbon solution with alkyl lithium-based initiator. In this stereo-specific catalyst system, in principle, every polymer molecule remains live until a deactivator or some other agent capable of reacting with the anion intervenes. Able to control molecular weight, molecular weight distribution, branching. Able to make random and block copolymers with designed chain sequence. Able to make copolymer with controlled styrene content. Able to control the butadiene structure of vinyl/cis/trans. Higher purity due to no soap added.

Property	Units	Conditions	Value	Reference
Density	$g\,cm^{-3}$	Emulsion SBR, 23–25% styrene solution SBR, vinly % 8 to 77	0.93	(1)
		Styrene 13 to 27%	0.92–0.95	(2)
Mark-Houwink parameter, K	$ml\,g^{-1}$	Emulsion hot SBR measured in toluene at 30°C	5.4×10^{-4}	(3)
Mark-Houwink parameter, α		Emulsion hot SBR measured in toluene at 30°C	0.66	(3)
Refractive index		Solution SBR, block copolymer, 30% styrene	1.53	(4)
Service temperature (max)	°C	Emulsion SBR, 23% styrene	70	(5)
		Solution SBR, block copolymer, S/B = 1:100	65	(5)

Property	Units	Conditions	Value	Reference
Solubility parameter	MPa$^{1/2}$	Emulsion SBR, 15% styrene	17.39	(1)
Thermal conductivity	W (m K)$^{-1}$	33% carbon black loaded	0.300	(1)
Theta temperature, θ	°C	Emulsion SBR, 23.9% styrene in methyl n-propyl ketone	21.0	(1)
		Emulsion SBR, 23.9% styrene in methyl isobutyl ketone	46.0	(1)
		Emulsion SBR, 25% styrene in n-octane	21.0	(1)
Glass transition temperature, T_g	°C	Emulsion SBR, 23% styrene,	−52	(3)
		Emulsion SBR, 36% styrene	−38	(3)
		Emulsion SBR, 53% styrene	−14	(3)
		Emulsion SBR, 75% styrene	13	(3)
		Solution SBR, block copolymer, S/B = 1:100	−70, 90	(3)
		Solution SBR, vinly % 8 to 77 styrene 13 to 27%	−35 to −7	(2)
		Emulsion SBR at 50°C	$T_g = (-85 + 135\,S)/(1 - 0.5\,S)$ S = wt fraction of styrene	(6)
		Emulsion SBR at 5°C	$T_g = (-78 + 128\,S)/(1 - 0.5\,S)$ S = wt fraction of styrene	(6)
		Solution SBR, assume T_g of styrene as 100°C T_g of polybutadiene as −100°C T_g of all-vinyl polybutadiene as 0°C S = wt fraction of styrene in the polymer	$T_g = [1/(0.00578 - 0.0031\,S - 0.00212\,V + 0.00212\,VS)] + 273$	
		V = (% vinyl in total polymer × 100)/(% butadiene in polymer)		(6)
Tensile	MPa	Unfilled vulc. emulsion hot SBR, 23–25% styrene	1.4–2.8	(1)
		1006 in ASTM 3185 1A, see below	21.4	(7)
		1500 in ASTM 3185 1A, see below	23.5	(7)
		1605 in ASTM 3186, see below	19.3	(7)
		1712 in ASTM 3185 2B, see below	19.0	(7)
		1805 in ASTM 3186, see below	18.6	(7)
Elongation	%	Unfilled vulc. emulsion hot SBR, 23–25% styrene	450–600	(1)
		1006 in ASTM 3185 1A, see below	325	(7)
		1500 in ASTM 3185 1A, see below	450	(7)

Property	Units	Conditions	Value	Reference
		1605 in ASTM 3186, see below	350	(7)
		1712 in ASTM 3185 2B, see below	525	(7)
		1805 in ASTM 3186, see below	350	(7)
Modulus	MPa	Unfilled vulc. unfilled vulc. emulsion hot SBR, 23–25% styrene	1–2	(1)
		1006 in ASTM 3185 1A, see below	13.8–17.9	(7)
		1500 in ASTM 3185 1A, see below	10.4–14.5	(7)
		1605 in ASTM 3186, see below	13.5–17.6	(7)
		1712 in ASTM 3185 2B, see below	6.2–10.4	(7)
		1805 in ASTM 3186, see below	9.0–13.1	(7)

Conditions

SBR Type	Styrene	Mooney	Carbon Black	phr	Oil	phr
1006A	23.5	49				
1500	23.5	52				
1605	23.5	62	N550	50	–	–
1721	23.5	55	–	–	Aromatic	37.5
1805	23.5	58	N330	75	Naphthenic	37.5

SBR Test Compounds

ASTM 3185 1A	
Polymer	100 phr
Furnace black	50
Stearic acid	1
Zinc oxide	3
Sulfur	1.75
TBBS	1
ASTM 3185 2B	
Polymer	137.5 phr
Furnace black	68.75
Stearic acid	1
Zinc oxide	3
Sulfur	1.75
TBBS	1.38
ASTM 3186	
Polymer	162 phr
Stearic acid	1.5
Zinc oxide	3
Sulfur	1.75
TBBS	1.25

References

1. Mark JE, ed. *Physical Properties of Polymers Handbook*. 2nd ed. New York, NY: Springer Science Business Media, LLC; 2007.
2. *Manual for the Rubber Industry*. 2nd ed. Bayer AG; 1993.
3. Hibbs J. "Styrene-Butadiene Rubbers" in *The Vanderbilt Rubber Handbook*. 13th ed. 1990.
4. Product Bulletin #301-665-687, The Dow Chemical Company, Midland, Mich.
5. Ohm RF. "Introduction to the Structure and Properties of Rubber" in *The Vanderbilt Rubber Handbook*. 13th ed. 1990.
6. Henderson JN. "Styrene-Butadiene Rubber" In: Morton M, ed. *Rubber Technology*. 3rd ed. New York: Van Nostrand Reinhold; 1987.
7. Ameripol Synpol Corporation products data sheets.

Styrene-methylmethacrylate Copolymer

Shuhong Wang

Acronym SMMA

Class Chemical copolymers

Trade Names ACRYSTEX® (Chi Mei Co.)

Structure

$$-[CH_2-CH]_m-[CH_2-\underset{\underset{COOCH_3}{|}}{\overset{\overset{CH_3}{|}}{C}}]_n-$$
$$\underset{C_6H_5}{|}$$

Typical Comonomers Styrene, methylmethacrylate

Major Applications Blend with other polymers to produce a variety of products. Blends normally have both transparency and impact resistance. Typical uses include household goods, building materials, optical, toy and food-packing applications.

Properties of Special Interest Transparency, weather resistance, low water absorption, low residuals. Better weatherability and solvent resistance comparing to polystyrene homopolymer. Properties fall between those of the individual homopolymers.

Property	Units	Conditions	Value	Reference
Tensile strength	MPa	ASTM D638		
		P-205 UVA extrusion grade	68.2	(1)
		NAS injection-molding grade	57.2	(1)
			68.7	(3)
Tensile elongation	%	ASTM D638		
		P-205 UVA extrusion grade	5.0	(1)
		NAS injection-molding grade	2.0	(1)
			8.0	(3)
Tensile modulus	MPa	ASTM D638		
		P-205 UVA extrusion grade	3300	(1)
		NAS injection-molding grade	3500	(3)

Styrene-methylmethacrylate Copolymer

Property	Units	Conditions	Value	Reference
Flexural strength	MPa	ASTM D790		
		P-205 UVA extrusion grade	116	(1)
		NAS injection-molding grade	103	(1)
			103	(3)
Flexural modulus	MPa	ASTM D790		
		P-205 UVA extrusion grade	3300	(1)
		NAS injection-molding grade	3500	(1)
			3200	(3)
Izod Impact	$J\,m^{-1}$	ASTM D256		
		P-205 UVA extrusion grade	20	(1)
		NAS injection-molding grade	20	(1)
			16.7	(3)
Specific gravity		ASTM D792		
		P-205 UVA extrusion grade	1.13	(1)
		NAS injection-molding grade	1.09	(1)
			1.12	(3)
Rockwell Hardness		ASTM D785		
		P-205 UVA extrusion grade	80	(1)
		NAS injection-molding grade	64	(1)
Deflection temperature	°C	ASTM D648		
		P-205 UVA extrusion grade	99	(1)
		NAS injection-molding grade	98	(1)
		Unannealed	88.9	(3)
Vicat Softening Point	°C	ASTM D1525	105	(3)
Water absorption	%	ASTM D570, 24 h		
		P-205 UVA extrusion grade	0.17	(1)
		NAS injection-molding grade	0.15	(1)
Light transmission	%	P-205 UVA extrusion grade	90	(1)
		NAS injection-molding grade	90	(1)
Haze	%	ASTM D1003, 125 mil	0.4	(3)
Transmission, Visible	%	ASTM D1003, 125 mil	91	(3)
Refractive index, n_D		ASTM D542		
		P-205 UVA extrusion grade	1.53	(1)
		NAS injection-molding grade	1.56	(1)
Melt flow rate	g per 10 min	ASTM D1238		
		P-205 UVA extrusion grade, 190°C/10 kg	0.2	(1)
		NAS injection-molding grade, 190°C/10 kg	4.3	(1)
		P-205 UVA extrusion grade, 230°C/3.8 kg	0.7	(1)
		P-205 UVA extrusion grade, 230°C/1.2 kg	0.13	(1)
		200C	1.5	(3)

Property	Units	Conditions	Value	Reference
Theta temperature, θ	°C	29.3% Styrene, $M_n = 4.7–59.2$		
		2-ethoxy ethanol	40.0	(2)
		Cyclohexanol	68.0	(2)
		56.2% Styrene, $M_n = 3.4–50$		
		2-ethoxy ethanol	58.4	(2)
		Cyclohexanol	61.3	(2)
		70.2% Styrene, $M_n = 4.0–43$		
		2-ethoxy ethanol	72.8	(2)
		Cyclohexanol	63.0	(2)
		76.3% Styrene		
		benzene/n-hexane 44/56	20.0	(2)
		benzene/isopropanol 57/43	20.0	(2)
		n-hexane/3-methyl butanone 40/60	20.0	(2)
		58.1% Styrene		
		benzene/n-hexane 51/49	20.0	(2)
		benzene/isopropanol 51/49	20.0	(2)
		n-hexane/3-methyl butanone 34/66	20.0	(2)
		42.3% Styrene		
		benzene/n-hexane 59/41	20.0	(2)
Theta temperature, θ	°C	benzene/isopropanol 48/52	20.0	(2)
		n-hexane/3-methyl butanone 29/71	20.0	(2)
		26.1% Styrene		
		benzene/n-hexane 62/38	20.0	(2)
		benzene/isopropanol 41/59	20.0	(2)
		n-hexane/3-methyl butanone 24/76	20.0	(2)

References

1. Taught TAD. "Styrene Polymers". In: Mark HF, et al. ed. *Encyclopedia of Polymer Science and Engineering*. New York: Wiley-Interscience Publication; 1989.
2. Sundararajan PR. "Theta Temperatures". In: Mark JE, ed. *Physical Properties of Polymers Handbook*. 2nd ed. New York, NY: Springer Science Business Media, LLC; 2007.
3. Chi Mei Corporation Product data sheet.

Sulfo-ethylene-propylene-diene Monomer Ionomers

Ruskin Longworth

Alternative Name, Trade Name Sulfo-EPDM ionomers, Vistalon® derivative (Exxon Chemical Co.)

Class Chemical copolymers; EPDM rubber derivatives

Structure[1]

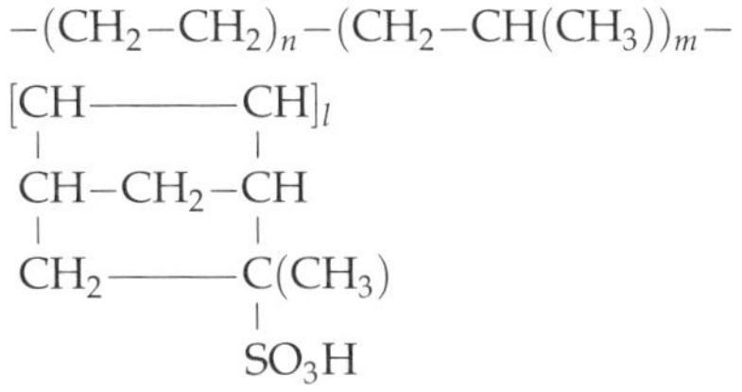

$$-(CH_2-CH_2)_n-(CH_2-CH(CH_3))_m-$$

$$[CH\text{———}CH]_l$$
$$\overset{|}{CH}-CH_2-\overset{|}{CH}$$
$$\overset{|}{CH_2}\text{———}\overset{|}{C(CH_3)}$$
$$\overset{|}{SO_3H}$$

$n + m + l = 100$; $n = 52$; $m = 43$; $l = 5$.

General Information These polymers consist of sulfonated derivatives of ethylene-propylene-diene terpolymers. The ionic associations induced by the sulfonate groups are significantly stronger than is the case with the carboxylated products. For a detailed comparison, see reference (2). Even after several years of active development, sustainable commercial uses have not emerged. Thus, these products are no longer being produced even though they are of considerable technical interest.

Major Applications Drilling mud additives.

Preparative techniques[1]

Sulfonation of rubber	Acetyl sulfate added to cold solution of rubber in hydrocarbon solvent
Neutralization of sulfo-EPDM	Addition of excess solution of metal acetate in water/methanol solvent
Comonomer	Ethylidene norbornene

Property	Units	Conditions	Value	Reference
Molecular weight (of repeat unit)	g mol^{-1}	–	44	(1)
Tacticity	–	–	Slightly nonrandom	(1)
Molecular weight (of ionomer)	g mol^{-1}	Mooney viscosity, ML, H8, 373 K	20	(1)
Solvents	–	Sulfo-EPDM, 273 K EPDM, 273 K	Toluene/methanol (95/5) Hexane	(1)

Property	Units	Conditions	Value		Reference
Crystalline state properties	EPDM and sulfo-derivatives are amorphous				
Tensile strength	MPa $(\times 10^{-3})$	(a) Effect of neutralizing ion (base polymer: sulfo-EPDM, 2.7% sulfonic acid; 100% neutralized)			(1)
		Ion: Mg	2.21		
		Co	8.13		
		Pb	11.03		
		Zn	10.20		
		(b) Effect of plasticizer (base polymer: sulfo-EPDM, 3.8% zinc sulfonate)	298 K	343 K	(3)
		Zinc stearate (%): 0	6.76	1.72	
		16	21.0	4.48	
		27	25.2	6.41	
		36	22.4	7.93	
Elongation	%	Effect of neutralizing ion (base polymer: sulfo-EPDM, 2.7% sulfonic acid; 100% neutralized)			(1)
		Ion: Mg	70		
		Co	290		
		Pb	480		
		Zn	400		
Melt viscosity	Pas $(\times 10^{-3})$	(a) Effect of cation on melt viscosity (base polymer: sulfo-EPDM, 2.7% sulfonic acid)			(1)
		Ion: Ca	5.32		
		Li	5.15		
		Na	5.06		
		Pb	3.28		
		Zn	1.20		
	gs^{-1} $(\times 10^{-3})$	(b) Effect of stearic acid on melt viscosity (base polymer: sulfo-EPDM, 3.8% zinc sulfonate)	Viscosity as melt index at 463 K, 1.72 kPa		(3)
		Stearic acid (%): 0	<0.1		
		16	0.50		
		27	1.60		
		36	5.00		
	gs^{-1} $(\times 10^{-3})$	(c) Effect of zinc stearate on melt viscosity (base polymer: sulfo-EPDM, 3.8% zinc sulfonate)	Viscosity as melt index at 463 K, 1.72 kPa		(3)
		Zinc stearate (%): 0	<0.1		
		16	0.50		
		27	2.3		
		36	6.3		

Sulfo-ethylene-propylene-diene Monomer Ionomers

Property	Units	Conditions	Value	Reference
Water absorption (gain)	%	(a) Base polymer: sulfo-EPDM, 3.8% zinc sulfonate; at 323 K for:		(2)
		24 h	5	
		72 h	7	
		144 h	8	
		310 h	11	
		(b) Plasticized composition: sulfo-EPDM, 3.8% zinc sulfonate plasticized with 36% zinc stearate; at 323 K for:		(3)
		24 h	3.3	
		72 h	4.1	
		144 h	4.9	
		310 h	5.4	
Important patents		O'Farrell, C. P., and G. E. Serniuk. *U.S. Patent* 3,836,511 (1972), assigned to Esso Research and Engineering Co. Canter, N. H., and D. J. Buckley, Sr. *U.S. Patent* 3,847,854 (1974), assigned to Esso Research and Engineering Co. Canter, N. H. *U.S. Patent* 3,642,728 (1974), assigned to Esso Research and Engineering Co.		
Cost and availability	–	–	Unavailable	–

References

1. Makowski HS, et al. In: Eisenberg A, ed. *Advances in Chemistry, No. 187. Ions in Polymers.* Washington, D.C: American Chemical Society; 1980.
2. Lundberg RD, Makowski HS. In: Eisenberg A. ed. *Advances in Chemistry, No. 187. Ions in Polymers.* Washington, D.C: American Chemical Society; 1980.
3. Makowski HS, Lundberg RD. In: Eisenberg A, ed. *Advances in Chemistry, No. 187. Ions in Polymers.* Washington, D.C: American Chemical Society; 1980.
   The author gratefully acknowledges Dr. R. D. Lundberg for his assistance in preparing this entry.

Syndiotactic Polystyrene

Junzo Masamoto and Takashi Iwamoto

Acronym, Trade Name SPS, XAREC (Idemitsu Kosan)

Class Vinyl polymers

Structure $[-CH_2CH(C_6H_5)]$

Major Applications Thermoplastics as an engineering plastic, usually reinforced with glass fiber, and used for automobile parts, electrical and electronic parts. SPS neat resin film is available for sheet and tape.

Properties of Special Interest Crystalline engineering plastics starting from a commodity monomeric material of styrene. Quite different properties compared to conventional amorphous polystyrene. High melting point (463 K) and good solvent resistance. Excellent electrical properties with low dielectric loss. High heat deflection temperature, low water absorption and hydrolytic resistance. Excellent dimensional precision during injection molding because of equal density of amorphous and crystalline parts.

Preparative Technique Metallocene polymerization: Combination of Cp*Ti(OiPr)$_3$ [pentamethyl cyclopentadienyl titanium triisopropoxide] and MAO [methyl almoxane], or combination of Cp*Ti(Me)$_3$ [pentamethyl cyclopentadienyl titanium trimethyl] and B(C$_6$F$_5$)$_3$ [tris-(pentafluorophenyl) borane] usually polymerized around 70–90°C, under bulk polymerization conditions.[1–3]

Property	Units	Conditions	Value	Reference
Molecular weight (of repeat unit)	g mol^{-1}	–	104	–
Tacticity (stereo regularity)	%, pentad, syndiotacticity	Metallocene polymerization	>98	(4)
Typical molecular weight range of polymer M_w	g mol^{-1}	Metallocene polymerization: combination of Cp*Ti(OMe)$_3$ [pentamethyl cyclopentadienyl titanium trimethoxide] and MAO [methyl almoxane], or combination of Cp*Ti(Me)$_3$ [pentamethyl cyclopentadienyl titanium trimethyl] and B(C$_6$F$_5$)$_3$ [tris-(pentafluorophenyl)borane] usually polymerized around	2–5 ($\times 10^5$)	(1–3)

Property	Units	Conditions	Value	Reference
		70–90°C, under bulk polymerization conditions		
Typical polydipersity index (M_w/M_n)	–	–	2	(1–3)
IR (characteristic absorption frequencies)	cm^{-1}	Planar zigzag conformation (T4)	1,224	(5)
		Helex conformation (TTGG)	935	
NMR		The 67.8 MHz ^{13}C-NMR: 1,2,4-trichrolobenzene at 130°C with JNMGX-270 spectrometer		(4)
		The 270 MHz ^{1}H-NMR: 1,2,4-trichlorobenzene at 130°C with JNMGX-270 spectrometer		
Thermal expansion coefficients	K^{-1}	Neat SPS	9.2×10^{-5}	(6)
		30% glass fiber filled SPS	2.5×10^{-5}	
Solvent	–	130°C	Trichlorobenzene	(4)
Nonsolvents	–	At its boiling point	Methanol, methyl ethyl ketone	(4)

Unit cell dimensions

	Lattice	Polymer Chain Per unit Cell	Cell Dimension (Å)			Cell Angles			Reference
			a	b	c (Chain Axis)	α	β	γ	
α	Hexagonal	18	26.25	–	5.045	–	–	–	(7)
β	Orthorhombic	4	8.81	28.82	5.06	–	–	–	(8)
γ	Monoclinic	2	–	–	–	–	–	–	(9)
δ	Monoclinic	2	17.58	13.26	7.71	–	–	121.2	(9)

Property	Units	Conditions	Value	Reference
Space group	–	–	α P62C	(7)
			β Pbnm	(8)
			δ P2$_1$/a	(9)
Chain conformation	–	α	(T$_4$)	(7)
		β	(T$_4$)	(8)
		γ	(TTGG)$_2$	(9)
		δ	(TTGG)$_2$	(9)
Degree of crysatallinity	%	Quenched from 320°C in ice water	∼0	(10)
		Injection molded sample at the mold temperature of 140°C	50	
Heat of fusion	kJ mol^{-1}	–	5.8	(11)
	mJ mg^{-1}		53	(10)

Property	Units	Conditions	Value	Reference
Density	g cm^{-3}	Neat SPS	1.05	(12)
		α crystal	1.033	(7)
		β crystal	1.08	(8)
		δ crystal (molecular compound with toluene)	1.11	(9)
Polymorphs	–	–	α crystal (T$_4$)	(7)
			β crystal (T$_4$)	(8)
			γ crystal (TTGG)$_2$	(9)
			δ crystal (TTGG)$_2$	(9)
Glass transition temperature	K	–	373	(1)
Melting point	K	DSC, 20°C min^{-1}	543	(2)
Equilibrium melting point	K	Crystallization temperature vs. polymer melting point	548	(13)
			558	(14)
		Lammela thickness vs. polymer melting point	583	(15)
Mesomeric transition temperature	K	From helix (TTGG)$_2$ to planar zigzag (T$_4$)	463	(16)
Heat capacity	kJ K^{-1} mol^{-1}	–	0.140	(17)
Deflection temperature	K	Neat SPS, 18.3 kg cm^{-2}	372	(12)
		30% glass fiber filled SPS, 18.3 kg cm^{-2}	522	
Polymer with which compatible	–	SPS, $M_w = 680,000$	Poly(2,6-dimethyl-1,4-phenyleneoxide)	(18)
Tensile modulus	MPa	Neat SPS	3,440	(12)
		30% glass fiber filled SPS	10,000	
Tensile strength	MPa	Neat SPS	41	(12)
		30% glass fiber filled SPS	121	
Yield stress	MPa	Neat SPS	41	(12)
		30% glass fiber filled SPS	121	
Yield strain	%	Neat SPS	1.0	(12)
		30% glass fiber filled SPS	1.5	
Flexural modulus	MPa	Neat SP	39	(12)
		S30% glass fiber filled SPS	97	
Flexural strength	MPa	Neat SP	71	(12)
		S30% glass fiber filled SPS	166	
Izod impact	kJ m^{-1}	Neat SPS	11	(12)
		30% glass fiber filled SPS	96	(19)

Property	Units	Conditions	Value	Reference
Dielectric constant	–	Neat SPS [1 MHz]	2.6	(20)
		30% glass fiber filled SPS	2.9	
Dielectric loss	–	Neat SPS	<0.001	(20)
		30% glass fiber filled SPS	<0.001	
Breakdown strength	kV mm^{-1}	Neat SPS	66	(21)
		30% glass fiber filled SPS	48	
Resitivity	ohm cm	Neat SPS, ASTM D 257	>10^{16}	(6)
		30% glass fiber filled SPS, ASTM D 257	>10^{16}	
Maximum use temperature (long term)	K	–	400	(6)
Water absorption	%	Neat SPS, 24 h equilibrium, ASTM D 570	0.04	(6)
		30% glass fiber filled SPS, 24 h equilibrium ASTM D 570	0.05	
Important patent	U.S. Patent 5,502,133			
	U.S. Patent 4,680,353			
Availability	kg yr^{-1}	–	5,000,000	(22)
Suppliers	Idemitsu Kosan, 3-1-1, Marunouchi, Chiyoda-ku, Tokyo 100-8321, Japan (5,000,000 kg yr^{-1})			

References

1. Ishihara N, Kuramoto M, Uoi M. *Macromolecules*. 1988;21:3356.
2. Campbell RE, Newman TH, Malanga MT. *Macromol. Sym.* 1995;97:151.
3. Pellecchia C, Longo P, Proto A, Zambelli A. *Makromol. Chem. Rapid Commun.* 1992;13:265.
4. Ishihara N, Seimiya T, Kuramoto M, Uoi M. *Macromolecules*. 1986;19:2465.
5. Kobayashi M, Nakaoki T, Ishihara N. *Macromolecules*. 1989;22:4377.
6. Uoi M. *Seikei Kako (Polymer Processing, Jpn)*. 1996;8:167.
7. Greis O, Xu Y, Asano T, Petermann J. *Polymer*. 1989;30:590.
8. Chatani Y, Shimane Y, Ijitsu T, Yukinari T. *Polymer*. 1993;34:1625.
9. Chatani Y, et al. *Polymer*. 1993;34:1620.
10. Krzystowczyk DH, Niu X, Wesson RD, Collier JR. *Polym. Bull.* 1994;33:109.
11. de Candia F, Filho AR, Vittoria V. *Colloid Polym. Sci.* 1991;269:650.
12. Newmann TH, Campbell RE, Malanga MT. In *Metcon '93*. Houston, Tex: 26–28 May 1993.
13. Cimmino S, Pace ED, Martuscelli E, Silvestre C. *Polymer*. 1991;32:1080.
14. Arnaunts J, Berghmans H. *Polym. Commun.* 1990;31:343.
15. Abe T. *Polym. Prep. Jpn*. 1993;42:4309.
16. Doherty DC, Hopfinger AJ. *Comput. Polym. Sci.* 1991;1:107.
17. Liu TM, et al. *J. Appl. Polym. Sci.* 1996;62:1807.

18. Guerra G, Rosa CR, Petracone V. *Networks Blends.* 1992;2:1145.
19. Bank D, Brentin R. *SPS Crystalline Polymer: A New Material for Automotive Interconnect Systems.* SAE 970305, 1997:71.
20. Ishihara N, Kuramoto M. *Stud. Surf. Sci. Catal.* 1994;89:339.
21. Yamato H. In *Styrenics '93.* Session 1, 6–8 December 1993.
22. News Release of Idemitsu Kosan: http://www.idemitsu.co.jp/company/information/news/2006/061101.html.

Vinylidene Fluoride-hexafluoropropylene Elastomers

Rahul D. Patil and Andrew D. Vogt

Acronyms, Alternative Names, Trade Names Poly(vinylidene fluoride-co-hexafluoropropylene), poly(vdf-hfp), Dai-el, Fluorel, Tecnoflon, Viton[1,2]

Class Chemical copolymers; fluoroelastomers

CAS Registry Number [9011-17-0]

Structure

$$-(CF_2-CH_2)_n-(\overset{\overset{\textstyle CF_3}{|}}{CF}-CF_2)_m-$$

Major Applications Poly(vdf-hfp) is a synthetic, noncrystalline polymer that exhibits elastomeric properties when cross-linked. Known to be chemically inert, it is designed for demanding service applications in hostile environments and commonly used as a sealant in hot and corrosive environments,[1,3] such as used in aircraft, aerospace, chemical and petroleum industries. Also used for biomedical applications as coatings.[4]

Membranes of Poly(vdf-hfp) are used in biomedical applications, catalytic reactors, and filtration applications.[5]

Commercial Use Poly(vdf-hfp) has found its niche in industry. Once considered exotic and too expensive, it has proven to be the most cost-effective answer to modern sealing needs. Commonly used as a sealant in hot and corrosive environments.[3]

Properties of Special Interest Poly(vdf-hfp) contains approximately 30 ± 40 mol% hexafluoropropylene. When the copolymer contains less than 30 mol% hexafluoropropylene it tends to become nonelastic; at less than 15 mol% the copolymer has thermoplastic properties.[2,5–9]

Preparative technique.[8]

Polymerization type	Emulsion, aqueous
Process	Batch or continuous
Temperature	80–120°C
Pressure	1.72–10.34 MPa
Comonomers	Vinylidene fluoride and hexafluoropropylene
Initiator	Ammonium persulphate, organic and inorganic peroxides
Catalyst	Sodium bisulphate

^{19} F NMR* $^{(10)}$

Group	Microstructure Sequence†	Peak (ppm)
CF$_3$	HTXTX	71.4
	TTXTH	75.9
CF$_2$	THTHT	91.9
	HHTHT	95.7
	THTXT	103.7
	THTTH	114.0
	HTTHH	116.3
	HTTXH	118.9
CF	HTXTH	182.3
	TTXHT	184.9

* In 50% acetone at 30°C.

† H $= -CH_2-$; T $= -CF_2-$; X $= -CF(CF_3)-$.

Property	UNITS	Conditions	Value	Reference
Average molecular weight M_w	g mol^{-1}	–	85,000	(11)
Mooney viscosity	ML 1 + 10	At 121°C	22	(9)
Molar ratio of comonomers	–	Vinylidene fluoride/hexafluoro-propylene	3.5	(9)
Total fluorine content	%	Viton A	66	(9)
Specific gravity	g cm^{-1}	Gumstock	1.54–1.88	(1)
		Viton A	1.82	(9)
Solubility parameters	(MPa)$^{1/2}$	Total	17.8	(12)
		Nonpolar	15.3	
		Polar	7.2	
		Hydrogen-bonding	5.3	
Solubility	–	In carbon dioxide, at 100°C and 1,000 bar	Soluble	(11)
		In C$_3$F$_3$, at 163°C and 2,750 bar	Soluble	(13)
		In CClF$_3$, at 230°C and 1,500 bar	Soluble	(13)
		In CHF$_3$, at 30°C and 1,500 bar	Soluble	(13)
Intrinsic viscosity [η]	dl g^{-1}	For various compositions at 30°C in methyl ethyl ketone		(14)
		11 mol% of HFP	1.5	
		16.3 mol% of HFP	1.7	
		22.6 mol% of HFP	1.3	
		27.8 mol% of HFP	1.0	

Vinylidene Fluoride-hexafluoropropylene Elastomers

Property	Units	Conditions	Value		Reference
Parameters for Flory-Rehner equation Describing sorption isotherms	–	At 25°C in acetone			(3)
		χ_0	1.596		
		χ_1	−3.319		
		χ_2	1.514		
		K	−0.033		
Glass transition temperatures T_g	K	Mol% of HFP			(14)
		11	240.5		
		16.3	244.0		
		22.6	250.0		
		27.8	255.5		
		33.8	261.3		
		40.7	267.8		
Relaxation processes	K	Mol% of HFP	β process (T_β)	α_L process ($T_{\alpha,L}$)	(15)
		19.2	185.35	256.15	
		−87.8	185.35	257.15	
		24.2	183.65	258.15	
		26.0	183.65	260.15	
		30.3	189.15	266.15	
		34.7	189.15	268.15	
		39.0	192.15	271.15	
		39.2	189.15	271.15	
Change in volume	%	At room temperature after 72 hours of immersion in water and various alcohols*			(16)
		Solvent			
		Water	−1		
		Methanol	89		
		Ethanol	0.5		
		n-Propanol	0.7		
		n-Butanol	0.4		
		n-Pentanol	0.1		
		n-Hexanol	0.1		
		n-Heptanol	0.2		
		n-Octanol	0.3		

Vinylidene Fluoride-hexafluoropropylene Elastomers

Property	Units	Conditions	Value	Reference
Change in volume	%	At exposure to methanol at various temperatures* Temperature (K)		(16)
		233.15	172	
		253.15	149	
		273.15	121	
		294.15	84	
		303.15	76	
		308.15	68	
		313.15	58	
		323.15	54	
		328.15	39	
		333.15	16	

* Sample used in reference (16) is Viton A, cross-linked and reinforced with carbon black filler. The description of the cross-linking method is given in detail in the same reference.

Effect on mechanical properties* [16]†

Group	% Alcohol	Tensile Strength (MPa)	Elongation (%)	Modulus at 100% Elongation (MPa)
None	Original properties	16.8	200	5.7
Methanol	0	15.8	232	5.3
	2	12.1	199	4.5
	5	11.7	219	4.0
	10	8.7	178	3.7
	25	5.3	127	3.9
	50	5.6	116	4.6
	75	4.8	96	–
	100	4.3	87	–
Ethanol	0	15.2	153	9.5
	5	12.7	153	6.3
	10	12.3	153	5.8
	15	12.7	150	7.0
	25	12.3	150	6.2
	50	13.0	158	6.3
	75	12.9	152	6.7
	100	14.6	160	7.1

* After 72 h of immersion in mixtures of methanol/indolene and ethanol/indolene at room temperature.
† Sample used in reference (16) is Viton A, cross-linked and reinforced with carbon black filler. The description of the cross-linking method is given in detail in the same reference.

Vinylidene Fluoride-hexafluoropropylene Elastomers

Mechanical properties[8]

Property	Units	Conditions	Temperature (K)	Value
Tensile strength at break	MPa	Dry*	350	11.79
		Dry*, after 72 h	475	10.27
		Wet†	298	16.20
		–	343	12.76
		–	373	7.23
Elongation at break	%	Wet†	298	390
		Dry*	350	625
		Dry*, after 72 h	478	615
		–	343	490
		–	373	500
Modulus-300%	Mpa	Dry*	350	3.44
		Dry*, after 72 h	475	2.90
		Wet†	298	10.00
		–	343	5.52
		–	373	3.27

* The sample used in the dry tests was compounded with dibasic lead phosphite and cured for 1 h at $120 \pm 150°C$. See reference (8) for details.
† The sample used in the wet tests was compounded with silica and cured for one hour at 120°C. See reference (8) for details.

Thermal degradation[17]

Type of Elastomer	Atmosphere	Temperature (K)				Final Total
		Initial Weight Loss	1% Weight Loss	Initial F Yield	1% F Yield	Yield of F (%)
Viton A	Air	613	668	416	704	54.2
	Nitrogen	543	623	409	730	12.9
Viton A-HV	Air	653	693	468	690	54.7
	Nitrogen	623	683	414	703	13.2
Viton A cross-linked	Air	473	633	493	656	56.2
	Nitrogen	443	593	461	634	28.5

Suppliers

Trade Name	Supplier
Viton	DuPont Dow Elastomers, 300 Bellevue Parkway, Suite 300, Wilmington, Delaware 19809, USA
Tecnoflon	Solvay Solexis, 10 Leonard Lane, Thorofare, New Jersey 08086, USA
Dai-el	Daikin America, 20 Olympic Drive, Orangeburg, New York 10962, USA
Fluorel	Dyneon, 3M-Hoechst Enterprise, 6744 33rd Sreet North, Oakdale, Minnesota 55128, USA

References

1. Grootaert WM, Millet GH, Worm AT. In: Kroschwitz JI, ed. *Kirk-Othmer Encyclopedia of Chemical Technology.* 4th ed. Vol 8. New York: John Wiley and Sons; 1989.
2. Dohany JE, Humphrey JS. In: Mark HF, et al. ed. *Encyclopedia of Polymer Science and Engineering.* 2nd ed. Vol 17. New York: John Wiley and Sons; 1989.
3. Wang P, Sung N. *Polym. Mater. Sci. Eng.* 1993;69:372.
4. Llanos GH, Narayanan P, Roller MB, Scopelianos A. *U.S. Patent.* 6,746,773.
5. Reverchon E, Cardea S. *Ind. Eng. Chem. Res.* 2006;45:8939.
6. Worm AT. *Machine Design.* 1990;62:46.
7. Elleithy R, Aglan H, Letton A. *J. Elasto. Plast.* 1996;28:199.
8. Rexford DR. *U.S. Patent.* 3,051,677. 1962.
9. Theodore AN, Zinbo M, Carter RO III. *J. Appl. Polym. Sci.* 1996;61:2065.
10. Ferguson R, Kaut C. *Gummi.* 1965;18:723.
11. Rindfleisch F, DiNoia TP, McHugh MA. *J. Phys. Chem.* 1996;100:15581.
12. Beerbower A, Dickey JR. *ASLE Trans.* 1969;12:1.
13. Mertdogan CA, DiNoia TP, McHugh MA. *Macromolecules.* 1997;30:7511.
14. Bonardelli P, Moggi G. *Polymer.* 1986;27:906.
15. Ajroldi G, et al. *Polymer.* 1989;30:2180.
16. Myers ME, Abu-Isa IA. *J. Appl. Polym. Sci.* 1986;32:3515.
17. Knight GJ, Wright WW. *J. Appl. Polym. Sci.* 1972;16:683.

Contributors and Polymers

Reda Abouhussein
Department of Chemistry, Kent State University, OH
Nylon 6,12

Zahoor Ahmad
Department of Chemistry, Faculty of Science, Kuwait University, Safat, State of Kuwait
Nylon 6,10

M. Göktuğ Ahunbay
Department of Chemical Engineering, Istanbul Technical University, Turkey
Poly(acrylonitrile)
Poly(chlorotrifluoroethylene)
Poly(vinyl chloride)
Poly(vinylidene chloride)

Rufina G. Alamo
Professor of Chemical Engineering, Florida Agricultural and Mechanical University/Florida State University College of Engineering, Tallahassee, FL.
Polyethylene, linear high-density

Harry Allcock
Department of Chemistry, The Pennsylvania State University, University Park, PA
Polyphosphazenes

Anthony L. Andrady
Senior Research Scientist, Research Triangle Institute, Durham, NC
Poly(acrylonitrile)
Poly(chlorotrifluoroethylene)
Poly(vinyl chloride)
Poly(vinylidene chloride)

George Apgar
Technical Manager, Technical Polymers Research and Development, Elf Atochem

North America, Inc., King of Prussia, PA
Nylon 11

Ralph M. Aten
DuPont Experimental Station, Wilmington, DE
Poly(tetrafluoroethylene)

Zongwu Bai
University of Dayton Research Institute, Dayton, OH
Amino resins

Ronald H. Baney
Courtesy Visiting Scientist, Department of Materials Science and Engineering, University of Florida, Gainesville, FL
Poly(hydridosilsesquioxane)
Poly(methylsilsesquioxane)
Poly(phenylsilsesquioxane)

Greg Beaucage
Department of Chemical and Materials Engineering, The University of Cincinnati, OH
Polypyrrole
Polyquinoline
Polythiophene
Poly(p-xylylene)

Jennifer L. Braun
Department of Wood Science, University of British Columbia, Vancouver, BC
Amylopectin
Amylose
Alkyd resins
Cellulose
Chitin
Epoxy resins

Glycogen
Polyesters, unsaturated

Witold Brostow
*Departments of Materials Science and Physics,
    University of North Texas, Denton, TX*
Ethylene-propylene-diene monomer
    elastomers

Peggy Cebe
*Department of Physics, Tufts University,
    Medford, MA*
Silk protein

Yong S. Chong
*United States Patent and Trademark Office,
    Alexandria, VA*
Polyacrylamide
Poly(acrylic acid)

Stephen J. Clarson
*Department of Chemical and Materials
    Engineering, The University of
    Cincinnati, OH*
Poly(dimethylsiloxanes), cyclic
Polymeric selenium
Polymeric sulfur
Poly(phenylmethylsiloxanes), cyclic
Poly(vinylmethylsiloxanes), cyclic

Thuy D. Dang
*U.S. Air Force Research Laboratories, Wright
    Patterson Air Force Base, OH*
Benzimidazobenzophenanthroline-type
    Ladder Polymer (BBL)
Poly(benzobisoxazole), naphthalene
    derivative
Poly(benzobisthiazole), naphthalene
    derivative

Tea Datashvili
*Department of Materials Science, University of
    North Texas, Denton, TX*
Ethylene-propylene-diene monomer
    elastomers

James L. DeRudder
SABIC Innovative Plastics, Mt. Vernon, IN
Polycarbonate

Anita Dimeska
*Lummus Novolen Technology,
    Cincinnati, OH*

Polypropylene, atactic
Polypropylene, elastomeric (stereoblock)
Polypropylene, isotactic
Polypropylene, syndiotactic

Leiyuchuan Ding
*Department of Chemistry, The University of
    Cincinnati, OH*
Polyphosphates
Poly(phosphonates)

Yong Ding
*C. S. Marvel Laboratories, Department
    of Chemistry, University of Arizona,
    Tucson, AZ*
Poly(2,6-dimethyl-1,4-phenylene oxide)
Poly(methylene oxide)
Poly(p-phenylene oxide)

Abraham J. Domb
*Professor, Hebrew University of Jerusalem,
    School of Pharmacy, Faculty of Medicine,
    Israel*
Poly(1,3-bis-p-carboxyphenoxypropane
    anhydride)
Poly(erucic acid dimer anhydride)
Polyesters of ricinoleic acid and lactic acid
Poly(ester-co-anhydride) of ricinoleic acid
    and sebacic acid

Petar R. Dvornic
Michigan Molecular Institute, Midland, MI
Poly(silphenylene-siloxanes)

Richard E. Fernandez
Du Pont Fluoroproducts, Wilmington, DE
Perfluorinated ionomers

Warren T. Ford
*Department of Chemistry, Oklahoma State
    University, Stillwater, OK*
Carbon nanotube-containing polymers
Fullerene-containing polymers

Stephen A. Fossey
*U.S. Army Natick Research, Development and
    Engineering Center, Massachusetts, MA*
Silk protein

Bruce M. Foxman
*Professor of Chemistry, Brandeis University,
    Waltham, Massachusetts, MA*

Polyphosphates
Poly(phosphonate)

J. R. Fried

*Chemical and Materials Engineering, and the
Ohio Molecular Computation and
Simulation Network and Center for
Computer-Aided Molecular Design,
University of Cincinnati, OH*

Poly(ether ether ketone)
Poly(ether ketone)
Poly(methylacrylonitrile)

Michelle K. Gaines

*Department of Materials Science &
Engineering, North Carolina State
University, Raleigh, NC*

Poly(α-methylstyrene)

Paul G. Galanty

Plastics Industry Consultant, West Orange, NJ

Nylon 6

Vassilios Galiatsatos

Equistar Chemicals, Cincinnati, OH

Polychloroprene
Poly(norbornene)
Polyoctenamer
Polypentenamer

Robert R. Gallucci

SABIC Innovative Plastics, Mt. Vernon, IN

Polycarbonate

Zongming Gao

Food and Drug Administration, St. Louis, MO

Gelatin
Starch

Yuli K. Godovsky

*Professor of Polymer Science, Karpov Institute
of Physical Chemistry, Moscow, Russia*

Poly(di-*n*-butylsiloxane)
Poly(diethylsiloxane)
Poly(di-*n*-hexylsiloxane)
Poly(di-*n*-pentylsiloxane)
Poly(di-*n*-propylsiloxane)

Douglas G. Gold

*Senior Research Chemist, 3M Company,
St. Paul, MN*

Poly(L-alanine)
Poly(γ-benzyl-L-glutamate)
Polyglycine

Richard V. Gregory

*School of Textiles, Fiber and Polymer Science,
and Center for Advanced Engineering
Fibers and Films, Clemson University, SC*

Polyaniline

Anna D. Gudmundsdottir

*Department of Chemistry, The University of
Cincinnati, OH*

Polyaniline

Julio Guzmán

*Professor, Institute de Ciencia y Technologid
de Polimeros, Madrid, Spain*

Poly(1,3-dioxepane)
Poly(1,3-dioxolane)

Akira Harada

*Professor of Polymer Science, Osaka
University, Japan*

Poly(rotaxane), example 1
Poly(rotaxane), example 2

Stephen S. Hardaker

*School of Textiles, Fiber and Polymer Science,
and Center for Advanced Engineering
Fibers and Films, Clemson University, SC*

Polyaniline

Mohamad Hassan

*Department of Chemistry, Bani Suef
University, Bani Suef, Egypt*

Nylon 11

Allan S. Hay

*Tomlinson Chair in Chemistry, McGill
University, Montreal, Quebec, Canada*

Poly(2,6-dimethyl-1,4-phenylene oxide)
Poly(methylene oxide)
Poly(*p*-phenylene oxide)

David V. Howe

Amoco Polymers, Inc., Alpharetta, GA

Polypropylene, isotactic

Shaw Ling Hsu

*Polymer Science and Engineering, University
of Massachusetts, Amherst, MA*

Aromatic Polyamides
Poly(methyl methacrylate)

L. V. Interrante

*Professor of Chemistry, Rensselaer Polytechnic
Institute, Troy, NY*

Poly(methylsilmethylene)

Poly(silylenemethylene)

Jude O. Iroh

Professor of Materials Science (Polymers), University of Cincinnati, OH

Poly(butylene terephthalate)

Poly(ε-caprolactone)

Poly(ethylene-2,6-naphthalate)

Poly(ethylene terephthalate)

M. R. Islam

Department of Chemistry, Carleton University, Ottawa, Ontario, Canada

Poly(vinyl alcohol)

Takashi Iwamoto

Performance Plastics Development Department, Asahikasei Chemicals Corporation, Sodegaura-shi, Japan

Syndiotactic polystyrene

Karl I. Jacob

Department of Polymer, Textile, and Fiber Engineering, The Georgia Institute of Technology, Atlanta, GA

Nylon 6

Abhijit V. Jadhav

Department of Chemistry, The University of Cincinnati, OH

Polyaniline

Chandima Kumudinie Jayasuriya

Department of Chemistry, University of Kelaniya, Sri Lanka

Collagen

Poly(n-butyl isocyanate)

Polychloral

Poly(n-hexyl isocyanate)

Poly(α-phenylethyl isocyanide)

John F. Kadla

Department of Wood Science, University of British Columbia, Vancouver, BC

Amylopectin

Amylose

Cellulose

Chitin

Glycogen

Lisaleigh Kane

Department of Materials Science and Engineering, North Carolina

State University, Raleigh, NC

Poly(α-methylstyrene)

David L. Kaplan

Biotechnology Center, Department of Bioengineering, Tufts University, Medford, MA

Silk protein

D. L. Kerbow

Technology Fellow, Du Pont de Nemours and Company, Wilmington, DE

Poly(tetrafluoroethylene)

M. K. Khan

Department of Chemistry, Carleton University, Ottawa, Ontario, Canada

Poly(vinyl butyral)

Vladyslav Kholodovych

Department of Pharmacology, Robert Wood Johnson Medical School, Piscataway, NJ

Benzimidazobenzophenanthroline-type Ladder Polymer (BBL)

Poly(benzimidazole)

Poly(benzobisoxazole)

Poly(benzobisoxazole), naphthalene derivative

Poly(benzobisthiazole)

Poly(benzobisthiazole), naphthalene derivative

Murali Kilaru

Department of Electrical and Computer Engineering, The University of Cincinnati, OH

Poly(vinyl fluoride)

Andrzej Kloczkowski

L.H. Baker Center for Bioinformatics and Biological Statistics, Iowa State University, Ames, IA

Poly(L-alanine)

Poly(γ-benzyl-L-glutamate)

Polyglycine

Poly(vinylidene chloride)

John H. Ko

3M Security Systems Division, St. Paul, MN

Poly (N-vinyl carbazole)

Poly(4-vinyl pyridine)

Poly(N-vinyl pyrrolidone)

Melvin I. Kohan

Engineering Thermoplastics Consultant, MIK
*   Associates, Wilmington, DE*

Nylon 6,10

Amit S. Kulkarni

Department of Chemical and Materials
*   Engineering, The University of Cincinnati,*
*   OH*

Polypyrrole

Polyquinoline

Polythiophene

Poly(*p*-xylylene)

Alex C.M. Kuo

Dow Corning Taiwan, Taipei, Taiwan, ROC

Poly(dimethylsiloxane)

Poly(methylphenylsiloxane)

Robert Langer

Department of Chemical and Biomedical
*   Engineering, Massachusetts Institute of*
*   Technology, Cambridge, MA*

Poly(1,3-bis-*p*-carboxyphenoxypropane
   anhydride)

Poly(erucic acid dimer anhydride)

Polyesters of ricinoleic acid and lactic acid

Poly(ester-co-anhydride) of ricinoleic acid
   and sebacic acid

Jonathan H. Laurer

Department of Materials Science and
*   Engineering, North Carolina State*
*   University, Raleigh, NC*

Poly(*p*-chlorostyrene)

Sung Woo Lee

Department of Chemical Engineering,
*   Yeungnam University, Gyongsan, Korea*

Nylon MXD6

Ying Li

School of Chemical Engineering, Shijiazhuang
*   University, Shijiazhuang, China*

Polystyrene, head-to-head

Poly(vinyl chloride), head-to-head

Poly(vinyl chloride)

Gui Lin

Department of Chemistry, The University of
*   Cincinnati, OH*

cis-1,4-Polybutadiene

Poly(hydridosilsesquioxane)

Poly(methylene terephthalate)

Poly(methylsilsesquioxane)

Poly(phenylsilsesquioxane)

Poly(trimethylene terephthalate)

Ralph Lloyd

DuPont Nafion(R), Fayetteville, NC

Perfluorinated ionomers

David J. Lohse

Distinguished Research Associate, ExxonMobil
*   Research and Engineering Company,*
*   Annandale, NJ*

Ethylene-propylene-diene monomer
   elastomers

Polyisobutylene, butyl rubber, and halobutyl
   rubber

Ruskin Longworth

DuPont Company (retired), and
*   Teixido-Longworth Enterprises, Greenville,*
*   DE*

Sulfo-ethylene-propylene-diene monomer
   ionomers

Robert D. Lousenberg

DuPont Experimental Station,
*   Wilmington, DE*

Perfluorinated ionomers

Lichun Lu

Orthopedics and Biomedical Engineering,
*   Mayo Clinic College of Medicine,*
*   Rochester, MN*

Poly(glycolic acid)

Poly(lactic acid)

Chi-Hao Luan

Department of Biochemistry, Molecular
*   Biology and Cell Biology, Northwestern*
*   University, Evanston, IL*

Elastic, plastic, and hydrogel-forming
   protein-based polymers

Tarek M. Madkour

Department of Chemistry, Helwan University,
*   Cairo, Egypt*

Bisphenol-A polysulfone

Poly(ether sulfone)

Poly(ethylene imine)

Poly(4-hydroxy benzoic acid)

Poly[l-(trimethylsilyl)-l-propyne]

Joseph H. Magill
*Professor Emeritus, Material Science and
Engineering Department, University of
Pittsburgh, PA*
Poly(aryloxy)thionylphosphazenes
Poly(phosphazene), bioerodible
Poly(phosphazene), elastomers
Poly(phosphazene), semicrystalline

Michael T. Malanga
*Scientist, Dow Chemical Company,
Midland, MI*
Polystyrene, head-to-head

Leo Mandelkern
*Department of Chemistry and Institute of
Molecular Biophysics, Florida State
University, Tallahassee, FL*
Polyethylene, linear high-density

Ian Manners
*Department of Chemistry, University of
Bristol, UK*
Poly[(n-butylamino)thionylphosphazene]
Poly(dimethylferrocenylethylene)
Poly(ferrocenyldimethylsilane)
Poly(vinylferrocene)

Rachel Mansencal
*Givaudan Flavors Corporation,
Cincinnati, OH*
Cellulose
Chitin
Glycogen

Robert H. Marchessault
*Department of Chemistry, McGill University,
Montreal, Quebec, Canada*
Poly(hydroxybutyrate)

James E. Mark
*Department of Chemistry, University of
Cincinnati, OH*
Cis-1,4 Polyisoprene

Junzo Masamoto
*Professor, Fukui University of Technology,
Fukui-shi, Japan*
Nylon 3
Polychloral
Poly(ethylene sulfide)
Poly(n-hexyl isocyanate)
Poly(p-phenylene sulfide)

Poly(propylene sulfide)
Syndiotactic polystyrene

Steven Mazur
*DuPont Experimental Station,
Wilmington, DE*
Perfluorinated ionomers

Dale J. Meier
*Professor, Michigan Molecular Institute,
Midland, MI*
Poly(diphenylsiloxane)
Poly(phenyl/tolylsiloxane)

Antonios G. Mikos
*Department of Chemical Engineering, Rice
University, Houston, TX*
Poly(glycolic acid)
Poly(lactic acid)

Wilmer G. Miller
*Professor, Department of Chemistry, University
of Minnesota, Minneapolis, MN*
Poly(L-alanine)
Poly(γ-benzyl-L-glutamate)
Polyglycine

Munmaya K. Mishra
*Ethyl Corporation Research and Development,
Richmond, VA*
Polyurea
Polyurethane
Polyurethane elastomers
Polyurethane urea

Alexander B. Morgan
*University of Dayton Research Institute,
Dayton, OH*
Phenolic resins

Barry Morris
*DuPont Packaging & Industrial Polymers,
Wilmington, DE*
Ethylene acid copolymer metal salts
(ionomers)

Suresh Murugesan
Collinsink Corporation, Cincinnati, OH
Polyisobutylene, butyl rubber, and halobutyl
rubber

Aziz Muzafarov
ISPM RAS, Moscow, Russia
Carbosilane dendrimers
Siloxane dendrimers

Charles L. Myers
Research Associate, Amoco Chemical Company, Naperville, IL
Polypropylene, atactic
Polypropylene, elastomeric (stereoblock)
Polypropylene, syndiotactic

Donna M. Narsavage-Heald
Assistant Professor of Chemistry, University of Scranton, PA
Hydridopolysilazane
Poly(N-methylcyclodisilazane)

Linda S. Nixon
Central Michigan University, Mt. Pleasant, MI
Poly(amidoamine) dendrimers
Poly(ester-acrylate/amine)

Isao Noda
Beckett Ridge Technical Center, The Procter & Gamble Company, West Chester, PA
Poly(hydroxybutyrate)

Robert A. Orwoll
Department of Chemistry, College of William and Mary, Williamsburg, VA
Polyacrylamide
Poly(acrylic acid)

Michael J. Owen
Michigan Molecular Institute, Midland, MI
Fluorosiloxane polymers
Poly(silphenylene-siloxanes)

Osei A. Owus
PC&RP Division, 3M Company, St. Paul, MN
Poly(4-vinylpyridine)

Ali E. Ozcam
Department of Chemical & Biomolecular Engineering, North Carolina State University, Raleigh, NC
Poly(*p*-methylstyrene)

D. R. Panse
Department of Materials Science and Engineering, University of Tennessee, Knoxville, TN

Poly(butene-l)
Poly(hexene-l)
Poly(4-methyl pentene-1)

Vladimir S. Papkov
Institute of Organo-Element Compounds, Russian Academy of Sciences, Moscow
Poly(di-*n*-butylsiloxane)
Poly(diethylsiloxane)
Poly(di-*n*-hexylsiloxane)
Poly(di-*n*-pentylsiloxane)
Poly(di-*n*-propylsiloxane)

Soo-Young Park
Kyungpook National University, Daegu South Korea
Poly(benzobisoxazole), naphthalene derivative
Poly(benzobisthiazole), naphthalene derivative

Anand K. Patel
Department of Chemical & Biomolecular Engineering, North Carolina State University, Raleigh, NC
Poly(*p*-chlorostyrene)

Rahul D. Patil
Abbott Laborataories, Abbott Park, IL
1,2-Polybutadiene
Vinylidene fluoride hexafluoropropylene elastomers

Rahul P. Patki
Department of Chemical and Materials Engineering, The University of Cincinnati, OH
Poly(butene-l)
Poly(hexene-l)
Poly(4-methyl pentene-1)

Dinesh V. Patwardhan
Division of Chemistry and Materials Science, The US Food and Drug Administration
Nylon 4,6

John Pennias
DuPont Packaging & Industrial Polymers, Wilmington, DE
Ethylene acid copolymer metal salts (ionomers)

Nicholas A. Peppas
Department of Chemical Engineering, University of Texas at Austin, TX
Poly(2-hydroxyethyl methacrylate)
Poly(N-isopropyl acrylamide)

Edward N. Peters
Principal Scientist, SABIC Innovative Plastics, Selkirk, NY
Poly(2,6-dimethyl-1,4-phenylene oxide)
Carborane-containing polymers

Gus G. Peterson
Advisory Scientist, IBM Corporation, San Jose, CA
Nylon 6,12

Paul J. Phillips
Department of Chemical and Materials Engineering, The University of Cincinnati, OH
Poly(butene-1)
Poly(hexene-1)
Poly(4-methyl pentene-1)

Abaneshwar Prasad
Cabot Microelectronics, Aurora, IL
Polyethylene, elastomeric (very highly branched)
Polyethylene, linear low-density
Polyethylene, low-density
Polyethylene, metallocene linear low-density

Paras N. Prasad
Department of Chemistry, and Photonics Research Laboratory, State University of New York, Buffalo, NY
Poly(1,4-phenylene)
Poly(1,4-phenylene vinylene)

Jagath K. Premachandra
Department of Chemical and Process Engineering, University of Moratuwa, Sri Lanka
Collagen
Poly(n-butyl isocyanate)
Polychloral
Poly(n-hexyl isocyanate)
Poly(α-phenylethyl isocyanide)

Zhengcai Pu
Life Technologies Corp., Carlsbad, CA
Barex
trans-1,4-Polybutadiene
Poly(m-phenylene isophthalamide)
Polystyrene

Meifang Qin
Senior Research Scientist, AlliedSignal, Inc., Morristown, NJ
Poly(vinyl chloride), head-to-head

Guru Sankar Rajan
Tata Autocomp Systems, Ltd., Tal. Mulshi, Pune, India
Poly(p-benzamide)
trans-1,4-Polyisoprene

L. S. Ramanathan
National Chemical Laboratory, Division of Polymer Chemistry, Pune, India
Polyurea
Polyurethane
Polyurethane elastomers
Polyurethane urea

Shrish Rane
LNP-GE Plastics, Columbus, IN
Polypyrrole
Polyquinoline
Polythiophene
Poly(p-xylylene)

Evaristo Riande
Institute de Ciencia y Technologid de Polimeros, Madrid, Spain
Poly(1,3-dioxepane)
Poly(1,3-dioxolane)

H. Ulf W. Rohde-Liebenau
Hills AC. (retired), Marl, Germany
Nylon 12

C. M. Roland
Polymer Physics Section, Naval Research Laboratory, Washington, DC
Kraton D1100 SBS block copolymers
Kraton G1600 SEBS block copolymers

Margaret Rookmaker
DSM, Heerlen, The Netherlands
Poly(1,3-trimethyleneimine) dendrimers

Judith F. Rubinson

Department of Chemistry, Georgetown University, Washington, DC

Poly(sulfur nitride)

Jerry I. Scheinbeim

Professor and Director, Polymer Electroprocessing Laboratory, Department of Chemical and Biochemical Engineering, Rutgers University, Piscataway, NJ

Poly(vinylidene fluoride)

Taner Z. Sen

Department of Genetics, Development and Cell Biology, Iowa State University, Ames, IA

Poly(acrylonitrile)

Poly(L-alanine)

Poly(γ-benzyl-L-glutamate)

Poly(chlorotrifluoroethylene)

Polyglycine

Poly(vinyl chloride)

Poly(vinylidene chloride)

Ravi Shankar

Fiber & Polymer Science Program and Department of Materials Science & Engineering, North Carolina State University, Raleigh, NC

Poly(styrene-b-isoprene-b-styrene) (unsaturated and hydrogenated)

M. A. Sharaf

Department of Chemistry, Helwan University, Ain Helwan, Cairo, Egypt

cis-1,4-Polybutadiene

Mee Y. Shelley

Visiting Assistant Professor, Department of Chemistry and Biochemistry, Miami University, Oxford, OH

Alkyd resins

Epoxy resins

Polyesters, unsaturated

Q. H. Shen

Starfire Systems, Inc., Malta, NY

Poly(methylsilmethylene)

Poly(silylenemethylene)

Ariella Shikanov

Hebrew University of Jerusalem, School of Pharmacy, Faculty of Medicine, Israel

Polyesters of ricinoleic acid and lactic acid

Poly(ester-co-anhydride) of ricinoleic acid and sebacic acid

S. Sivaram

Division of Polymer Chemistry, National Chemical Laboratory, Pune, India

Polyurea

Polyurethane

Polyurethane elastomers

Polyurethane urea

Archie P. Smith

Department of Materials Science and Engineering, North Carolina State University, Raleigh, NC

Poly(p-methylstyrene)

Milind Sohoni

Business Analyst, Nutraceutical Department, Cargill, Minneapolis, MN

Amino resins

Phenolic resins

Richard J. Spontak

Chemical & Biomolecular Engineering and Materials Science & Engineering, North Carolina State University, Raleigh, NC

Poly(styrene-b-isoprene-b-styrene) (unsaturated and hydrogenated)

Poly(p-chlorostyrene)

Poly(α-methylstyrene)

Poly(p-methylstyrene)

P. R. Sundararajan

Department of Chemistry, Carleton University, Ottawa, Ontario, Canada

Poly(vinyl alcohol)

Poly(vinyl butyral)

Gil Sur

Department of Chemical Engineering, Yeungnam University, Gyongsan, Korea

Nylon MXD6

Jacek Swiatkiewicz

Senior Research Scientist, Photonics Research Laboratory, State University of New York, Buffalo, NY

Poly(1,4-phenylene)
Poly(1,4-phenylene vinylene)

Loon-Seng Tan
Polymer Research Group Leader, U.S. Air
*   Force Wright Laboratory,*
*   Wright-Patterson Air Force Base, OH*
Poly(amide imide)
Poly(bis maleimide)
Poly(ether imide)
Poly(pyromellitimide-1,4-diphenyl ether)

Maxim N. Tchoul
Department of Chemistry, Oklahoma State
*   University Stillwater, OK*
Carbon nanotube-containing
polymers

Mikio Terada
Research Chemist, Rengo Company Ltd.,
*   Osaka, Japan*
Poly(hydroxybutyrate)

Donald A. Tomalia
Central Michigan University, Mt. Pleasant,
*   MI*
Poly(amidoamine) dendrimers
Poly(ester-acrylate/amine)
Poly(propylene imine) dendrimers

Dan W. Urry
BioTechnology Institute, College of Biological
*   Sciences, University of Minnesota, St. Paul,*
*   MN*
Elastic, plastic, and hydrogel-forming
    protein-based polymers

Ronald E. Uschold
Research Fellow, Du Pont Fluoroproducts,
*   Wilmington, DE*
Poly(vinyl fluoride)

Boris Vaisman
Hebrew University of Jerusalem, School of
*   Pharmacy, Faculty of Medicine, Israel*
Poly(ester-co-anhydride) of ricinoleic acid
    and sebacic acid
Polyesters of ricinoleic acid and lactic
    acid

Narayanan Venkatasubramanian
University of Dayton Research Institute,
*   Dayton, OH*

Amino resins
Benzimidazobenzophenanthroline-type
    Ladder Polymer (BBL)
Poly(benzobisoxazole), naphthalene
    derivative
Poly(benzobisthiazole), naphthalene
    derivative

Gary W. Ver Strate
Exxon Chemical Company, Linden, NJ
Ethylene-propylene-diene monomer
    elastomers
Poly(isobutylene), butyl rubber, and
    halobutyl rubber

Brent D. Viers
Department of Chemistry and Physics,
*   Radford University, Radford, VA*
Kevlar
Nylon 6,6

Andrew D. Vogt
Abbott Laborataories, Abbott Park, IL
Vinylidene fluoride hexafluoropropylene
    elastomers

David Walsh
DuPont Packaging and Industrial Polymers,
*   Experimental Station, Wilmington,*
*   DE*
Ethylene acid copolymer metal salts
    (ionomers)

Shuhong Wang
DuPont Performance Elastomers,
*   Wilmington, DE*
Acrylonitrile-butadiene elastomers
Polyacetylene
Styrene-acrylonitrile polymers
Styrene-butadiene elastomers
Styrene-methylmethacrylate copolymer

Juan T. Weaver
Department of Chemical & Biomolecular
*   Engineering, North Carolina State*
*   University, Raleigh, NC*
Poly(styrene-b-isoprene-b-styrene)
    (unsaturated and hydrogenated)

William J. Welsh

Department of Pharmacology,
Robert Wood Johnson Medical School,
Piscataway, NJ

Benzimidazobenzophenanthroline-type
Ladder Polymer (BBL)
Poly(benzimidazole)
Poly(benzobisoxazole)
Poly(benzobisoxazole), naphthalene
derivative
Poly(benzobisthiazole)
Poly(benzobisthiazole), naphthalene
derivative

Jianye Wen

Alza Corporation, Mountain View, CA
Poly(cyclohexyl methacrylate)
Poly(ethyl acrylate)
Poly(methacrylic acid)
Poly(methyl acrylate)
Poly(vinyl acetate)
Poly(vinyl methyl ether)

Robert West

Department of Chemistry, University of
Wisconsin, Madison, WI
Poly(di-*n*-hexylsilylene)
Poly(dimethylsilylene)
Poly(dimethylsilylene-co-
phenylmethylsilylene)
Polygermanes
Poly(methylphenylsilylene)

Sizhu Wu

College of Materials Science and
Technology, Beijing University of
Chemical Technology, China
Bisphenol-A polysulfone
Poly(ether sulfone)
Poly(ethylene imine)
Poly(4-hydroxy benzoic acid)
Poly[l-(trimethylsilyl)-l-propyne]

Ping Xu

W. L. Gore and Associates, Inc., Elkton, MD
Ethylene-vinyl acetate copolymer
Ethylene-vinyl alcohol copolymer
Polyacetylene

Yong Yang

Benjamin Moore and Company,
Flanders, NJ
Cellulose acetate
Cellulose butyrate
Cellulose nitrate
Ethylcellulose
Hydroxypropylcellulose

Qingwen Wendy Yuan-Huffman

Akzo Noble Surface Chemistry
Company, Bridgewater,
NJ
Poly(epichlorohydrin)
Poly(ethylene oxide)
Poly(propylene oxide)
Poly(tetrahydrofuran)
Poly(trimethylene oxide)

Martel Zeldin

Department of Chemistry, University of
Richmond, VA
Metallophthalocyanine polymers
Silicon (germanium) oxo hemiporphyrazine
polymers

Qiming Zhang

Department of Electrical
Engineering, Pennsylvania State
University, PA
Poly(vinylidene fluoride)

Ruzhi Zhang

AZ Electronic Materials USA Corp,
Somerville, NJ
cis-1,4-Polyisoprene
Methacrylate polymers containing
adamantane
Poly(4-hydroxystyrene)

Shihai Zhang

Department of Electrical
Engineering, Pennsylvania State
University, PA
Poly(vinylidene fluoride)

Xiujuan Zhang

Department of Chemistry, The University of
Cincinnati, OH
Poly(methylene terephthalate)
Poly(trimethylene terephthalate)

Yuli Zhang

*Research Assistant, Department
of Chemistry, College of
Staten Island, City University
of New York*

Metallophthalocyanine polymers

Silicon (germanium) oxo
hemiporphyrazine polymers

W. Brooke Zhao

*Research Scientist, HMT Technology
Corporation, Fremont, CA*

Amylopectin

Amylose

Gelatin

Nylon 6,12

Starch

Index